T0401023

CHEMICAL ENGINEERING METHODS AND TECHNOLOGY

THE SOL-GEL PROCESS: UNIFORMITY, POLYMERS AND APPLICATIONS

CHEMICAL ENGINEERING METHODS AND TECHNOLOGY

Additional books in this series can be found on Nova's website under the Series tab.

Additional E-books in this series can be found on Nova's website under the E-books tab.

MATERIALS SCIENCE AND TECHNOLOGIES

Additional books in this series can be found on Nova's website under the Series tab.

Additional E-books in this series can be found on Nova's website under the E-books tab.

CHEMICAL ENGINEERING METHODS AND TECHNOLOGY

THE SOL-GEL PROCESS: UNIFORMITY, POLYMERS AND APPLICATIONS

RACHEL E. MORRIS
EDITOR

Nova Science Publishers, Inc.

New York

For permission to use material from this book please contact us:
Telephone 631-231-7269; Fax 631-231-8175
Web Site: http://www.novapublishers.com

Additional color graphics may be available in the e-book version of this book.

LIBRARY OF CONGRESS CATALOGING-IN-PUBLICATION DATA

The sol-gel process : uniformity, polymers, and applications / editor, Rachel E. Morris.
p. cm.
Includes index.
ISBN 978-1-61761-321-0 (hardcover)
1. Colloids. I. Morris, Rachel E.
QD549.S546 2010
660'.294513--dc22
2010031450

Published by Nova Science Publishers, Inc. † New York

CONTENTS

PREFACE

The sol-gel process, also known as chemical solution deposition, is a wet-chemical technique widely used in the fields of materials science and ceramic engineering. Such methods are used primarily for the fabrication of materials (typically a metal oxide) starting from a chemical solution which acts as the precursor for an integrated network (or gel) of either discrete particles or network polymers. This book presents current research from around the globe in the study of the sol-gel process, including sol-gel based materials for biomedical applications; methods for prevention, diagnosis and treatment achieved with the aid of sol-gel chemistry; protein sol-gel encapsulation with polymer additives; the application of a sol-gel based nanostructured ceramic membrane for hydrogen separation for CO2 capture purposes; and sol-gel titania.

Chapter 1 – The sol-gel method is the most used and studied among the nonconventional processes for the preparation of oxide materials. It allows the control of the final properties of the product by varying the experimental conditions in the stage of the sol and ensures the synthesis of tailor made materials (monolithic gels, fibers, films, powders) with high homogeneity and purity. Titania, one of the most often sol-gel prepared nanomaterials, has attracted significant interest due to its photocatalytic activity, excellent functionality, thermal stability and non-toxicity. These properties make it useful in different fields, ranging from optics to gas sensors via solar energy. The photocatalytic degradation of pollutants using TiO2 (nanopowders or thin films) is very attractive for applications to environmental protection, as a possible solution for water depollution. Impurity doping is one of the typical approaches to extend the spectral response of a wide band gap semiconductor to visible light. The sol-gel method easily allows the preparation of nanocomposite materials where a metal phase could be highly dispersed in the inorganic matrices. On the other hand, titania-based perovskite systems, as BaTiO3 and its iso- or aliovalent solid solutions obtained as ceramics or thin films by the sol-gel process are of great interest. Some electronic and optoelectronic applications are the following: multilayer ceramic capacitors (MLCC), piezoelectric actuators, electroluminescent elements, pyroelectric detectors, temperature sensors and controllers based on the positive temperature coefficient of the electrical resistance (PTCR), dynamic, non-volatile or ferroelectric random access memories (DRAM, NVRAM, FeRAM).

The present work describes a comparative structural and chemical study of un-doped TiO2 and of corresponding S-, Ag-, and Pd-doped materials. Sol-gel Pd- and Ag-doped TiO2 coatings obtained from nanopowders were also compared. The relationship between the synthesis conditions and the properties of titania nanosized materials, such as thermal stability, phase composition, crystallinity, morphology and size of particles and the influence of dopant were investigated. Their structural evolution and crystallization behavior (lattice parameters, crystallite sizes, internal strains) with thermal treatment were performed. The complex TiO_2-based oxide systems, e.g., $BaO \cdot TiO_2$ (barium titanate, BT), lanthanum-doped barium titanate (BLT) and $BaO \cdot ZrO_2 \cdot TiO_2$ (BZT) were also studied. The dielectric behavior of some BT and La-doped BT nanopowders and related ceramics, as well as the topography and optical characteristics of the BZT thin films prepared via sol-gel route in relationship with their structural and microstructural properties are reported.

Chapter 2 – In practice, the development of high-performance materials for electronics, optics and ceramics fabrication is restrained by the traditional mode of their production. Many of the basic technologies for materials include high-temperature physical and chemical processes and require special conditions to attain the desired properties of the final products. Advanced preparation methods for materials with new features are feasible on the basis of colloid-chemical processes and nanochemistry. In this respect, the sol-to-gel transformation at molecular level and low temperature, followed by solidification and chemical modification is of great interest to attain new materials with non-traditional physical, chemical and functional properties of special interest, highly homogeneous and pure.

Chapter 3 – In this chapter, significant results obtained (past 5 years) in their laboratory related to the development of sol-gel processed nanocrystalline anatase-titania and its use as a photocatalyst for the industrial dye-removal application has been reviewed in detail. The pure nanocrystalline sol-gel anatase-titania has been processed using an alkoxide-precursor and utilized for the removal of methylene blue, a model catalytic dye-agent, from an aqueous solution via photocatalytic degradation mechanism under the ultraviolet- and solar-radiation exposure. The effect of various processing (conventional and modified sol-gel) and material parameters, such as 'R' (the ratio of molar concentration of water to that of the alkoxide precursor), calcination temperature, polymer concentration, specific surface-area, average nanocrystallite size, surface-purity, and volume fraction of the rutile phase on the photocatalytic activity of the nanocrystalline sol-gel anatase-titania has been taken into account. The role of surface-functionalization of the nanocrystalline sol-gel anatase-titania in enhancing its photocatalytic activity has been discussed by considering the deposition of silver via ultraviolet- and chemical-reduction techniques. The use of sol-gel titania photocatalyst as a surface-sensitizer for depositing the conducting-metals, such as copper and silver, on the non-conducting flyash particle surface has been demonstrated. In order to ease the separation of nanocrystalline sol-gel anatase-titania, from the treated effluent by applying an external magnetic field, the processing and the use of a conventional magnetic photocatalyst have been systematically described, in which a ceramic magnet has been introduced in between the nanocrystalline particles of the sol-gel anatase-titania. It has been shown that, the major limitations associated with the conventional magnetic photocatalyst can be overcome using a newly developed magnetic dye-adsorbent catalyst. It has been revealed that, the new magnetic dye-adsorbent catalyst removes an organic dye from an aqueous solution much quickly via surface-adsorption mechanism under the dark-condition, which is in contrast to the photocatalytic degradation mechanism, under the ultraviolet-radiation exposure, associated with the conventional magnetic photocatalyst. It appears that, the morphology of the nanocrystalline sol-gel titania plays a significant role in governing the predominant dye-removal mechanism, and hence, the overall efficiency and the cost of the dye-removal process.

Chapter 4 – Ordered nanoporous alumina and titania represent very interesting molecular sieves exhibiting a narrow pore size distribution with higher surface areas compared to conventional alumina and titania used as a support for catalytically active species and photocatalyst in numerous large-scale industrial processes. This contribution encompasses various synthesis approaches to ordered nanoporous alumina and titania, description of their structures and properties, and characterization by various experimental techniques. Self-assembly mechanism, micelle formation in solution is described as well. Unlike siliceous materials with ordered nanoporosity that are commonly prepared through the sol-gel process with surfactants as structure-directing agents or by utilizing the nanocasting method with silica or carbon materials as hard templates, synthesis of ordered and thermal stable alumina and titania represents a much more complex problem due to its susceptibility for hydrolysis as well as to the phase transitions accompanying the thermal breakdown of the ordered structure. Effects of synthesis conditions, calcination temperature, removal of surfactants and drying is discussed. The potential of the ordered alumina and titania with respect to use in catalysis and photocatalysis is considered.

Chapter 5 – The preparation of nanoparticles and nanocrystallites by conventional synthesis methods is difficult to achieve due to their strong tendency to agglomerate into bigger aggregates. This fact leads to the loss of unique properties generated by the large number of atoms located at the surface of the nanomaterial compared to the bulk material. The embedding of the nanocrystallites in amorphous matrices represents one of the possibilities to limitate the nanocrystallite's growth and agglomeration and to achieve a narrow size distribution. The sol-gel method is one of the few synthesis methods which lead to nanocomposites at low temperatures, by embedding or formation of oxidic particles within amorphous matrices. Using inorganic matrices (silica dioxide, titanium dioxide) as host materials, represents an efficient preparation way of oxidic nanoparticles with controlled dimensions and shapes, i.e. of nanocomposites with pre-defined properties. Since the properties of the nanocomposites significantly depend on the interface interaction between the oxidic particle and matrix, an important role is played by the nature and morphology of the matrix. This chapter will focus on a study regarding the obtaining of nanocomposites metal oxides/silica with controlled magnetic properties, using an original modified sol-gel method. The synthesis method used consists in the obtaining of some gels in the system tetraethyl orthosilicate-diol-metal nitrates, using a controlled thermal treatment. The diol present in the system interacts with both tetraethyl orthosilicate (TEOS) and metal nitrates. The interaction of the diol with TEOS during the gelation process influences the formation of the gels and finally the morphology of the resulted silica matrix. During the thermal treatment of the gels, the diol interacts with the metal nitrates by a redox reaction, leading to the formation of some carboxylate type compounds embedded in the silica matrices pores (precursors for nanocomposites). The controlled thermal treatment applied to these precursors leads to the formation of nanocrystalline oxidic phases embedded within the silica matrix. The first part of this chapter will present a systematic study on the interaction of the diols with the hydrolysis products of TEOS and the effect of this interaction on the silica matrices morphology. The second part will focus on the preparation of different nanocomposites with magnetic properties. The synthesis method involves particular conditions (the diols nature, molar ratio of reactants, concentration of metal oxide within the matrix, thermal treatment) for every studied oxidic system, especially in the cases of those that contain ions with variable valences.

Chapter 6 – The solgel process has several well-known advantages such as high purity precursors, homogeneity of the obtained material and especially the possibility of making hybrids and composite materials with new chemical and mechanical properties. This emerging field of material science has generated considerable and increasing interest from researchers of various disciplines. Recent advances due to the generation of multidisciplinary investigations have allowed real interesting advances in biology and medicine applications. The relation with the biomedical applications involves principally the development of methods for treatment, diagnosis, monitoring and control of biological systems. The first part of this chapter is dedicated to illustrate latest development in the immobilization of bactericide ions such as silver, copper or zinc and organic compounds with proven antimicrobial properties to obtain antimicrobial coatings, since microbial surface contamination control is an important tool for the prevention of infectious diseases. Secondly, due to the fact that early detection of infectious diseases is of great importance for the implementation of efficient treatments the entrapment of proteins, bacteria and parasites for the detection of infectious diseases is presented. The final section is an overview of the state of the art in materials for the treatment of human diseases including drug delivery systems and materials for tissue engineering achieved with the aid of sol-gel chemistry.

Chapter 7 – The solvent diffusion process in silica based sol-gel nano-thin film was examined as the film was immersed in water, methanol, ethanol, 2-propanol, and glycerol by observing the luminescence decay time constant of the photo-excited 3MLCT (metal-ligand charge-transfer) state of the [Ru(bpy)3]2+ ion dopant in the film. The luminescence decays of the films in the methanol, 2-propanol, and glycerol were better explained by the KWW model, while the luminescence decay of film immersed in water and ethanol were both well explained by a single, exponential decay. The water and ethanol were the only two solvents that exhibited solvent dependence in the aging effect of the film. The resulting photodynamics were influenced by the interaction between dopant and residual water/ethanol solvents co-encapsulated inside the cavity. The authors further investigated the effect of the diffusion speed of aqueous acid solution as a function of sizes of encapsulated gold nanocolloidal particles ranging from 5 nm to 100 nm by observing the color change of gold colloids as a function of time. While the maximum rate was found for silica sol-gel matrix encapsulating 15 nm gold colloid, 60 nm gold colloid doped silica sol-gel showed the smallest acid interaction. It was speculated that the surface of these gold colloids was homogeneously covered by the silica gel layer, thus avoiding direct contact of the acid with the surface of the gold colloid. The close match of the average size of the cavity with the size of the gold colloids was considered to prevent acid penetration. This study revealed that the nanoscale dopant size affects the rate of solvent penetration into a sol-gel cavity.

Chapter 8 – The precise organization of functional matters on the chemically inert surfaces of varied substrates for the fabrication of uniform hybrid architectures in the meso- and nanoscale is one of the new challenging frontiers in materials chemistry. Sol-gel process provides a promising synthesis route towards such fascinating materials because it not only endows the resulting hybrids with cooperatively distinct and superior properties, but also offers the refreshed surfaces to facilitate linking diverse functionalities targeted to specific applications. In this chapter, the authors mainly focused on the sol-gel processes on the chemically inert surfaces of carbon nanotubes and polystyrene colloids. The newly developed preparation protocols and tentative application explorations of such novel hybrid materials were reviewed. Three sections are involved in the context of this chapter: 1) an overview of the relative research background; 2) recent advances in the typical fabrication protocols, the correspondingly fascinating properties and representative applications for these novel hybrid materials (carbon nanotube- and PS-inorganic oxides) with well-defined architectures; 3) an account on the correspondingly recent researches, the perspectives and outlook for the future research directions.

Chapter 9 – This review article describes how the sol-gel process can be modified to encapsulate proteins in a silica sol-gel matrix. The entrapment of intact and functional proteins leads to intriguing applications in biocatalysis and biosensing. The main challenges are protein deactivation and limited material transport. Proteins are prone to deactivation by the release of alcohol during the sol-gel process or by steric or chemical interactions with the sol-gel cage. The fine pores of the sol-gel matrix result in a very efficient encapsulation of the protein molecules but also restrict the transport of substrate or analyte molecules to the proteins' active sites. A major goal in the further development of sol-gel based biocatalysts and biosensors is to achieve better pore size control in more uniform sol-gel matrices. This review article focuses on recent efforts to improve the sol-gel process with the help of polymer additives for applications in biocatalysis and biosensing. The effect of the polymer additive on the encapsulated protein and the sol-gel host will be discussed. As more sophisticated spectroscopic, physisorption, and imaging tools are employed to characterize these novel hybrid materials, their understanding about the effect of polymer additives increases. Together with the systematic exploration of polymers with different chemical and physical properties, these technological advancements will result in the emergence of improved sol-gel based biocatalysts and biosensors.

Chapter 10 – Combining ferromagnetism and ferroelectricity in one material and coupling between ferromagnetism and ferroelectricity are two big challenges in Multiferroics. Hybrid sol-gel processing inherently has advantages in making biphasic nanocomposites which exhibit coexisting ferroelectric and ferromagnetic properties. In this chapter, multiferroic CoFe2O4-Pb(Zr0.53Ti0.47)O3 (CFO-PZT) composite thick films have been fabricated onto Pt/Ti/SiO2/Si by the hybrid sol-gel processing for their multifunctional applications. This chapter includes the development of deposition technique of thick films on silicon substrate as well as the continuous optimization for multiferroic properties. In the first part, a suitable route for sol-gel deposition of CFO-PZT composite thick films on silicon substrate has been developed, and then deposition of dense, crack-free CFO-PZT thick films with desired thickness range (3~5 μm) along with a coexistence of ferromagnetic and ferroelectric properties have been achieved. In the second part, the authors study the effect of polyvinylpyrrolidone (PVP) on microstructures and ferromagnetic, ferroelectric and dielectric properties of CFO-PZT composite thick films. It is observed that PVP is beneficial for enhancing the film thickness remarkably and avoids the cracks probably induced by different thermal expansion coefficients or lattice mismatch between different materials. However, the disadvantage of PVP exists in diluting the ferroelectric and dielectric properties of composite thick films. To deal with this issue, the third part of this chapter introduces the sol infiltration to enhance the dielectric and ferroelectric properties of the composite thick films. It is found that enough sol infiltration between slurry layers, the maximum layers investigated here are four layers, dramatically improve the multiferroic properties as well as the density of the microstructure. Based on the obtained microstructures and multiferroic properties, the last part optimizes the deposition sequence and ferromagnetic phase content. Experimental results indicate that 8 wt% of CFO content along with an optimized deposition sequence gives satisfied coexisting ferromagnetic and ferroelectric properties, which is a precondition for exploring multiferroic devices.

Chapter 11 – The two most significant industrial processes for hydrogen production from fossil fuels are coal gasification and steam reforming of natural gas. A membrane process for CO2-H2 separation is an attractive alternative to conventional technologies such as solvent scrubbing, pressure swing adsorption, and cryogenic distillation for capturing the CO2. The advantages of using membrane reactors, i.e., the combination of a chemical reactor and membrane separation as one unit operation, are highly evident, e.g., the possibility to shift the chemical equilibrium towards the product side, the improvement of hydrogen production, along with high fossil fuel conversion at milder operating conditions and thus a lower efficiency penalty and lower cost. In principle, two different types of membranes are under development, dense Pd-alloy based and porous inorganic ceramic ones. The non-metallic inorganic membranes are in principle more robust, thermally and mechanically, but suffer from the drawback that the pore size has to be tuned to the kinetic diameter of hydrogen in order to obtain sufficient selectivity, for instance by Atomic Layer Deposition (ALD). The present contribution provides a literature overview of ceramic membranes and their performance which are currently under investigation for hydrogen separation at high temperatures, as well as recent research results by the authors. The scope of materials presented here is limited to those that are primarily based on sol-gel synthesis.

Chapter 12 – This work is devoted to the development of a new generation of sol-gel hybrid materials, their driving forces on the material science and technology and especially on their impact on possible advanced applications. The motivation of the carried work was to point out that the choice of the appropriate precursors can facilitate the connection between application and science, the future of the sol-gel technology. In this context, their group decided to combine the properties of new generation polymeric products commercialized in the last few years: the isocyanate modified silanes with traditional sol-gel products such as tetramethoxysilane in order to synthesize several attractive inorganic/organic networks. Among the group of isocyanate modified silanes, a trimethylsilyl isocyanate (TMSI) has not been used before in the sol-gel technology. The obtained nanostructured gel siliceous backbone with urea (-NHC(-O)NH-) or urethane (-NH(C-O)O-) linkages could be classified as ureasiles and urethanesiles, respectively. This kind of hybrid products has application in the domains of optics, magnetism and electrochemistry. Here the authors show that they possess unique behavior when modified by Ti, Zr, B and Al. The coordination ability of the designed hybrid Si-O-C-N host matrix extends its well-known applications to the optic field and makes urea and urethanesile sol-gel materials applicable in the area of thin films, adsorbents and hybrid membranes. When the obtained nanostructured gel products were subjected to heat treatment from 60 up to 1100oC in an inert atmosphere, the structural changes such as glass formation and crystal nucleation proved that they can be suitable for high temperature use. These features were established with the help of different traditional and advanced methods such as XRD, FTIR, 29Si, 27Al and 11B MAS NMR, SEM, AFM, TG and BET. In general, the obtained results provide conclusive evidence that the nature of the organic group incorporated in the siloxane structure plays a major role in the understanding of the relationships between design of new sol-gel materials and their application. This study is very interesting since it refers to the application of sol-gel process as synthesis route to advanced materials development and engineering.

Chapter 13 – Advanced oxidation processes are used for treatment of recalcitrant and non-biocompatible pollutants in soil, air, and water. Two of the key AOPs are heterogeneous photocatalysis and electrochemical oxidation where interactions between the polluted media and a surface play the main role. In this chapter, research are presented where XPS has been applied for characterization of differently prepared photocatalytic TiO2 films before and after UV irradiation for specific determination of the amount of OH groups on the surface. When investigated under different humid conditions, the spectral analyses have been applied for studies concerning the mechanisms for photocatalysis and photoinduced super hydrophilicity. Platinum based metal alloys are a widely used electrode material also considered for water treatment by electrochemical oxidation due to its inert and catalytic properties. XPS has in this context been applied for studying the extent of oxidation of a Ti/Pt90-Ir10 anode surface before and after electrolysis under oxygen and chlorine evolution conditions. The results provide insight in both the risk of corrosion of the surface in these oxidative environments and support a suggested mechanism for the oxidation of organics in electrolytic water treatment.

Chapter 14 – The chemistry of the sol-gel method in aqueous medium in systems containing transition metals can be very complicated as several molecular species depending on the oxidation state of the metal, the pH or the concentration could be formed. In the aqueous medium during the reaction of hydrolysis of inorganic salts, new ionic species or precipitates can occur. To prevent this precipitation along the metal salts dissolved in water, organic ligands are used. In this synthesis pathway, the metal ions are coordinated by organic ligands with a strong electron donor character which prevents their precipitation in the solution. By water evaporation, an amorphous inorganic-organic network is formed. In this network, the metal ions are homogeneously dispersed and linked by coordinative bonds. The sol-gel method in aqueous medium is a low cost procedure and is an environmental friendly alternative compared to the classical sol-gel technique that uses organic solvents and metal alkoxides.

In the present work, the sol-gel chemistry in aqueous medium of cobalt-based system (Li-Co and La-Co) with citric acid as chelating agent is presented. Information concerning the influence of the precursors on the sol-gel method and the properties of the resulted gels and powders are discussed.

Chapter 15 – Nanotechnology, which employs physical or chemical techniques to construct materials, devices or systems on a nanometre scale, is recently being introduced in preparation of functional textiles. It is a technology that can significantly improve properties of materials compared with those of conventional ones. Various functionalities can be enhanced or imparted to a textile material by applying different coatings of nano particles; e.g. self-cleaning properties and antimicrobial activity. Numerous techniques for the production of nanocoatings exist. Sol-gel process is one of the most applicable techniques to manufacture thin films. One of the most advantageous and attractive characteristics of the sol-gel process is the ability to control and manipulate composition and microstructure of the material at the molecular level and at room temperature.

Two basic coating processes will be explained; (i) process of direct (in situ) formation of nanocoatings on fibres' surfaces and (ii) process of nanocoatings' preparation starting from TiO2 P25 powder. Conditions, under which the processes for nanocoating can give rise to self-cleaning cellulose materials, were studied. In the first phase, the preparation technique and properties of nano titanium dioxide (TiO2) coatings generated directly on the surface of materials via sol-gel process have been studied extensively. Process was optimized in order to obtain coatings with desired structure and properties. The influence of sol-gel process conditions on the particle size and effectiveness was examined. When applying the procedure to the fibres some problems could occur, i.e. high temperature treatment conditions, problems connected with the formation of required polymorph TiO2 form, fibre damage risks, problems connected with durability of the modification, etc. For that reason, in the second part of research, already formed TiO2 P25 nanoparticles were used for obtaining self-cleaning modified surfaces. In addition to that, the process for attaching composite TiO2-SiO2 nanoparticles, where SiO2 acts as a binding agent was applied; furthermore SiO2 protects the fibres against possible photocatalytic influence of photocatalyst, since TiO2 P25 is known as a photocatalyst with high photocatalitic activity. In spite of the fact, that investigations indicated self-cleaning effectiveness of all TiO2 coated samples, the self-cleaning effectiveness was the highest in the case of samples, treated using stable dispersions of TiO2 P25 nanoparticles. According to functional effectiveness and yielded technologically-applicable properties, the process of preparing composite TiO2-SiO2 nanocoatings was the most appropriate.

Chapter 16 – The development of inorganic polymers is a new promising technology that may be used for many applications including inorganic coatings with protective or self-cleaning properties. In this chapter the development of an inorganic polymer based on amorphous silica and potassium hydroxide is described. The inorganic polymers are synthesized by a sol–gel process in which the silicate precursors are formed in-situ by dissolution of amorphous silica particles (microsilica, Elkem 983 U) by an alkaline solution. In this work potassium hydroxide with a molar concentration between 0.75 M and 4 M was used. In standard experiments equal amounts (w/w %) of silica and potassium hydroxide solution was used.

Using experimental data obtained from ESI-MS, FT-IR, XPS, NMR, SEM, and XRD a model for the gelation is suggested based on hydrolysis and condensation reactions occurring during synthesis. In addition the optimal composition of the binder system was determined from compressive strength tests and solubility experiments.

In the developed physico-chemical model the silica species polymerize through a condensation reaction resulting in formation of oligomers. The size and number of oligomers in the system increase until they extent throughout the solution and results in the formation of a gel. It was found that the dissolution proceeds for a longer time period when higher hydroxide concentrations are used resulting in higher dissolution and higher concentration of monomers. The gelation of the inorganic polymers synthesized from different concentrations of KOH was studied by viscosity measurements. It was found that two mechanisms contributed to the time required for gelation. At high hydroxide concentration the increased concentration of monomers in the system increase the rate of polymerization and therefore reduce the gelation time. At low hydroxide concentrations the surface charge of the silica particles is limited and the particles coagulate due to destabilization by the presence of potassium ions. A turn over point for these two mechanisms seems to be found at hydroxide concentrations between 0.75 and 1.5 M in the investigated systems.

Chapter 17 – One of the potential areas of research in the development of biosensors is the production of analytical devices based on the use of immobilized multienzymatic systems. Recently, the recovery of sol-gel chemistry has provided a new versatile method for immobilizing and stabilizing a wide range of enzymes and other biological molecules in transparent inorganic matrices. Compared to other immobilization methods, such as the adsorption to solid supports, the covalent attachment and the polymer entrapment, the sol-gel glasses show numerous advantages, including entrapment of a large amount of enzymes, thermal and chemical stability of the matrix, enhanced stability of the encapsulated biomolecules, excellent optical transparency and flexibility in controlling the pore size and the geometry. Furthermore, thanks to the porous nature of the matrix, the immobilized proteins remain accessible to interact with external specific analytes with negligible protein leaching, making possible the conduction of multienzymatic reactions. Indeed, the application of multienzymatic systems is an area of large interest in the development of biosensors, due to the fact that the product of one reaction can become the substrate of another one; regarding to this, the sol-gel process offers an excellent alternative for multienzymatic immobilization. In addition, thanks to the exceptional optical transparency that sol-gel matrices present, the use of fluorescent indicators coupled to enzymatic reactions is possible, which represents a successful advantage to solve some problems related with the specificity and sensitivity of this type of biosensors. In the present work, the authors revised the development of fluorescent multienzymatic biosensors based on sol-gel technology.

Chapter 18 – Bromocresol purple (CPR) and chlorophenol red (BCP) encapsulated in the sol-gel matrix were successfully applied to optical response towards aliphatic amines. The sol-gel matrix obtained by acidic hydrolysis of tetraethoxysilane and phenyltriethoxysilane in the presence of selected dyes was further spin coated onto glass slides. The coating process was repeated as needed. The optical sensors with a single coating and double coatings showed similar long-term stability, but they differed in response times and sensitivities. These sensors' selectivity sequence towards amines is: methylamine > ethylamine > propylamine > butylamine > triethylamine. The response time for methylamine (t95%) is 75 s, and its detection limit is ~5 ×10-5 M in aqueous solution. The CPR and BCP doped sol-gel sensors showed good reproducibility, with a useful life-time of four weeks.

Chapter 19 – The first part of the chapter is associated with the necessary and sufficient back ground which is quite often needed to understand the rest of the matter. It starts from solution based dye laser system with detail discussion and ends on the concept of solid state dye laser.

The second part has been devoted only on the development of sol-gel based solid state dye laser. It begins with an important introductory section on sol-gel and then dye molecules impregnation process; spectroscopic properties, lasing performance, and photostability are discussed into detail in different sections. Ultimately, the chapter ends with open questions and future research direction.

Chapter 20 – The use of sol-gel deposition for the fabrication of ZnO nanoparticles and thin films is reviewed. Low-agglomeration ZnO nanoparticles are obtained, and the effects of citric acid concentration, pH value, and various surfactants on grain size and distribution are discussed. ZnO thin films are then deposited and annealed in various environments, and the relationships between the annealing environments and the properties of the ZnO thin films are investigated. The roles that kinetic factors play in the film transformation process are also reviewed. Finally, the composition-related structural, microstructure, electrical, and optical properties of MgxZn1-xO thin films are discussed in detail.

Chapter 21 – Sol-gel silica in the wet form formulated in the shape of microsphere was investigated for the controlled release of protein drugs and in particular human growth hormone (r-hGH). The influence of the gel SiO2 content on protein conformation, load and release rate was investigated. Protein fold upon gel embedment was measured by circular dichroism analysis and was found to be unaffected up to a concentration of SiO2 equal to 12% w/v, while minor loss of r-hGH a-helix occurred at higher silica concentration. Several r-hGH loaded microspheres containing 5, 8, 10, 12 and 15% SiO2 (w/v) were synthesized using a surfactant-free W/O pseudo-emulsion method, purposely selected to minimize the risk of protein unfolding. The amount of protein that can be incorporated in the polymer and is released in a controlled way was found to depend on the matrix silica content, varying form 0.6 mg to more than 5.3 mg/ml of wet gel for the 5% and 12-15% SiO2 formulations respectively. At low silica concentration, total release of r-hGH from microspheres occurs within 12 hours and it is mostly driven by diffusion through the gel pores. At higher SiO2 content, release is significantly slower and it is mostly dominated by polymer erosion with a time scale that varies between 100 and 150 hours depending on the formulation under investigation.

Chapter 22 – In the present time most researchers believe that progress in medicine is connected with the development of drug delivery systems using nanoparticles. Very important drugs are immune-modulators which are natural or synthetic compounds. They often have protein nature. The efficiency of immune-modulator delivery into cells using nanocarriers will depend on both physical-chemical and biological properties of the drug, and the properties of nanocarriers themselves. Silica materials with different surface functionalities have been synthesized by sol-gel method. Physical properties of the silica materials in general have been studied to date to enlighten the molecular interactions and biocompatibility between host matrices and biomolecules (enzymes, proteins). These properties will be reported in this Chapter. The authors will also mainly report the work that is a part of original study concerning with the potential application of silica particles as nanocarriers of immune-modulator. For instance, for drug delivery applications, the adsorption ability of the silica materials has been studied and will be also described specifically using human serum albumin as a model compound of the drug. The studies of physical and adsorption properties of the silica materials as well as their interactions with immune cells in vitro is in the forefront of the research in the field as it will allow designing and selecting the most promising, and efficient silica carrier of immune-modulator proteins for drug delivery.

Chapter 23 – Calcium phosphate ceramics have been intensively studied during the last decades because of their great potential use for human dental and bone implants. The most used synthesis methods in the field are the wet chemistry methods, namely sol-gel and wet precipitation, due to a series of advantages comparing with other methods. The sol-gel synthesis method involves the formation of a colloidal sol which will then turn into a gel, this method requiring no special energy conditions for the formation of the desired compound; meanwhile, the chemical precipitation method means the system needs to be offered special conditions in order for the precipitation to take place, like certain values for the pH or for temperature. Based on previous working experience and observations, the authors would like to try a different, less-conventional approach, which consists in the study of the line between the two chemical synthesis routes. The authors have started the synthesis like a classic wet precipitation, with calcium chloride and phosphoric acid as calcium and phosphorus precursors, but the authors did not precipitate the desired calcium phosphate phase by pH control, but allowed the reaction to take place in time, in aqueous media at room temperature and pressure, approach which is specific to the sol-gel way. This new approach is based on the notice that though at the first sight they seem very similar, the experience proves us that actually a very fine border exists between them, and that is what the authors wanted to see and prove. The question that arises is: what is the actual border between sol-gel and wet precipitation? The idea of this question actually came from the study of the literature, which presented inconsistencies and differences in the opinions of different researchers in the field and inappropriate use of terms when describing synthesis work, probably due to insufficient knowledge and differentiation between one method or another.

Chapter 24 – Biphasic calcium phosphate (BCP) is a mixture of non-resorbable hydroxyapatite (HA) and the resorbable tricalcium phosphate (TCP) is an interesting material for bone implant as it shows biocompatibility and bioactivity to tissue bone. More efficient bone repair was widely been known in BCP than HA alone. Good implant materials should be biodegradable as it can degrade inside the bone and defect simultaneously with the formation of a new bone.

In this study BCP has been doped with magnesium through sol-gel method. Doping of magnesium ions into BCP will results in biological improvement as the ion will cause the acceleration of nucleation kinetics of bone minerals. Magnesium depletion adversely affects all stages of skeletal metabolism, leading to decrease in osteoblastic activities and bone fragility.

Magnesium–doped biphasic calcium phosphate (Mg-BCP) powders were successfully prepared using Ca(NO3)2.4H2O and (NH4)2HPO4 as the precursors and Mg(NO3)2.6H2O as the source of the dopant. Morphological evaluation by FESEM measurement showed that the particles of Mg-BCP were tightly agglomerated, with primary particulates of 50-150 nm diameters. FESEM result also showed that doping of magnesium into BCP particles caused fusion of particles leading to more progressive densification of particles as shown by higher concentration of magnesium doping. Successful incorporation of Mg into BCP lattice structure was confirmed by higher crystallinity of Mg-BCP and by shifting of tricalcium phosphate (TCP) peaks in XRD patterns to higher 2θ angles as the Mg content increased. XRD and FTIR measurement showed that the increment of crystallinity was directly proportional to the amount of the dopant. Both analyses also revealed that TCP appeared only after calcination of 700°C and above. All the powder exhibited highly crystalline BCP characteristics after calcination at 900 °C. XRD analysis revealed that β-TCP peak increased in intensity with the increased level of doped Mg, meanwhile HA peak was almost no change in intensity. With the increasing Mg concentration into the BCP, the solubility limit of the Mg in the β-TCP decreases and Mg starts segregated as free MgO or incorporated into the HA. A significant contraction has been observed in the calculated lattice parameter which may reflect the addition of Mg into the β-TCP phase and the differences of the lattice parameters and c/a ratio of β-TCP were much bigger than that of HA. FT-IR analysis confirms the formation of biphasic mixtures of HA and Mg stabilized β-TCP when calcined at high temperatures as bands of HPO4-2 and P2O7-4 decreased. Thermal analysis showed that the particles crystallize faster with more magnesium added. This study showed that magnesium doping into BCP through sol-gel method has improved crystal growth and fusion of BCP particulates.

Chapter 25 – Improved sol-gel method has been applied to prepare highly crystalline TiO2 colloidal nanoparticles with high photocatalytic reactivity. The precursor solution contained titanium (IV) isopropoxide (TTIP), 2-propanol. The sol was obtained through the hydrolysis of TTIP under the optimized conditions. FTIR, TEM and XRD were used to study the morphology, size, shape and crystallinity of prepared TiO2 sol. Experimental results have shown that the prepared sample has an anatase structure. It has a narrow size distribution between 2–5 nm which has been confirmed by X-ray diffraction pattern. To demonstrate the photocatalytic properties of TiO_2 colloidal nanoparticles, the Methylene Blue (MB) photodecomposition test has been used. In this approach, the obtained photodecomposition reaction rate of Methylene blue under UV light is high and shows the prepared colloidal TiO_2 sample has enough potential for photocatalytic applications. The photocatalytic property of the dried sol has been evaluated by inactivation test of Escherichia coli.

Chapter 26 – The sol-gel process as a chemical route is widely used in the fields of materials science and engineering. Historically, the sol-gel techniques have been developed during the past 40 years as an alternative process for glasses and ceramics production [1]. It is reported that the sol-gel process had been introduced by Ebelman [2], who had prepared a transparent SiO_2 thin film from slow hydrolysis of an ester of silicic acid in 1845. In comparison with other techniques such as CVD or sintering, the sol-gel technique is a very simple process with considerable advantages. This process is low cost, has high controllability and potential application for preparation of different materials and devices. These specifications of the sol-gel process caused it to be employed as a unique method for preparation of metallic, ceramic, hybrid, composite, fiber, and glass substances with different size and morphology with a wide range of applications. During several decades, different forms of the sol-gel process have been introduced by researchers. These methods are different in some aspects of sol-gel principles. But, the preparation of sol or gel as an intermediate stage of the process is common in all of them. In figure (1), the different kind of sol-gel processes are shown. The hydrolysis and condensation of metal alkoxide can be completely considered as a fully sol-gel process. In such a method, which is used primarily for the fabrication of materials (typically a metal oxide), the process is started by preparation of a chemical solution which acts as the precursor for an integrated network (or gel) of either discrete particles or network polymers. Typical precursors are metal alkoxides and metal chlorides, which undergo various forms of hydrolysis and polycondensation reactions.

Chapter 27 – Chemical industry is facing major challenges to dispose environmentally hazardous waste materials. Heterogeneous catalysis is a fascinating field which offers eco-friendly treatment steps and minimum waste; heterogeneous photocatalytic degradation of pollutants being the most desired method in this direction. Titania is well known as a cheap, stable, nontoxic, and efficient photocatalyst without secondary pollution. However, pure TiO_2 materials usually have very low quantum efficiency and poor performance in the visible region of light, thus restricting its extended applications in photocatalysis. Use of nonmetals or other metals as dopant systems is the most practicable approach for shifting the absorbance of TiO_2 to the visible region. These dopant materials bring forth new properties and enhanced activity due to structural and electronic modifications on TiO_2. Sol-gel synthesis of metal oxides is a widely accepted method for the preparation of such materials, particularly to incorporate ions in the gels. This short communication discusses the sol - gel preparation of TiO_2, its properties, and photocatalytic applications in waste remediation. A brief review on the modifications adopted for developing TiO_2 photocatalysts of improved performance is presented with examples.

Chapter 28 – Tetraethoxysilane (TEOS) is a commonly used precursor for preparation of sol-gel derived silica membranes. In this chapter, the authors propose the use of a new type of alkoxide, a bridged alkoxide which contains siloxane bonding or organic group between 2 silicon atoms, such as bis (triethoxysilyl) ethane (BTESE) and 1,1,3,3-tetraethoxy-1,3-

dimethyldisiloxane (TEDMDS), for the development of a highly permeable hydrogen separation membrane. The concept for improvement of hydrogen permeability of a silica membrane is to design a loose silica network using bridged alkoxides, i.e., to shift the silica networks to a larger pore size for an increase in H2 permeability. BTESE silica membranes showed approximately one order magnitude high H2 permeance (0.2-1.0x10-5 mol/(m2·s·Pa)) compared with previously reported TEOS-derived silica membranes, and high permeance ratios of H2 to SF6 (α(H2/SF6)=1,350-36,300) with low permeance ratio of H2 to N2 (α(He/N2) ~10). TEDMDS membranes also showed loose silica networks. The present result confirms the new concept of controlling silica networks for designing silica membranes.

Chapter 29 – A short but detailed introduction to the statistical theory of polymer network formation is given, including gel formation, gel structure, and sol fraction. Focus is put on the use of probability generating functions, and results that are of interest for polymer network elasticity are emphasized. Detailed derivations are supplied, and a simple 6-step procedure is provided, so that the reader is able to adapt and apply the theory to his own chemical systems, even if examples are given on polyurethanes essentially.

In: The Sol-Gel Process
Editor: Rachel E. Morris

ISBN 978-1-61761-321-0

Chapter 1

SOL-GEL TiO_2 - BASED OXIDE SYSTEMS

Maria Crişan[1*], Mălina Răileanu[1], Adelina Ianculescu[2], Dorel Crişan[1] and Nicolae Drăgan[1]

[1]Romanian Academy, Institute of Physical Chemistry Ilie Murgulescu, 202 Splaiul Independenței, 060021, Bucharest, Romania

[2]Department of Oxide Materials Science and Engineering, "Politehnica" University of Bucharest, 1-7 Gh. Polizu, P.O. Box 12-134, 011061 Bucharest, Romania

ABSTRACT

The sol-gel method is the most used and studied among the nonconventional processes for the preparation of oxide materials. It allows the control of the final properties of the product by varying the experimental conditions in the stage of the sol and ensures the synthesis of tailor made materials (monolithic gels, fibers, films, powders) with high homogeneity and purity. Titania, one of the most often sol-gel prepared nanomaterials, has attracted significant interest due to its photocatalytic activity, excellent functionality, thermal stability and non-toxicity. These properties make it useful in different fields, ranging from optics to gas sensors via solar energy. The photocatalytic degradation of pollutants using TiO_2 (nanopowders or thin films) is very attractive for applications to environmental protection, as a possible solution for water depollution. Impurity doping is one of the typical approaches to extend the spectral response of a wide band gap semiconductor to visible light. The sol-gel method easily allows the preparation of nanocomposite materials where a metal phase could be highly dispersed in the inorganic matrices. On the other hand, titania-based perovskite systems, as $BaTiO_3$ and its iso- or aliovalent solid solutions obtained as ceramics or thin films by the sol-gel process are of great interest. Some electronic and optoelectronic applications are the following: multilayer ceramic capacitors (MLCC), piezoelectric actuators, electroluminescent elements, pyroelectric detectors, temperature sensors and controllers based on the positive temperature coefficient of the electrical resistance (PTCR), dynamic, non-volatile or ferroelectric random access memories (DRAM, NVRAM, FeRAM).

[*] E-mail: mcrisan@icf.ro

The present work describes a comparative structural and chemical study of un-doped TiO_2 and of corresponding S-, Ag-, and Pd-doped materials. Sol-gel Pd- and Ag-doped TiO_2 coatings obtained from nanopowders were also compared. The relationship between the synthesis conditions and the properties of titania nanosized materials, such as thermal stability, phase composition, crystallinity, morphology and size of particles and the influence of dopant were investigated. Their structural evolution and crystallization behavior (lattice parameters, crystallite sizes, internal strains) with thermal treatment were performed. The complex TiO_2-based oxide systems, e.g., $BaO \cdot TiO_2$ (barium titanate, BT), lanthanum-doped barium titanate (BLT) and $BaO \cdot ZrO_2 \cdot TiO_2$ (BZT) were also studied. The dielectric behavior of some BT and La-doped BT nanopowders and related ceramics, as well as the topography and optical characteristics of the BZT thin films prepared via sol-gel route in relationship with their structural and microstructural properties are reported.

1. INTRODUCTION

1.1. State of the Art Regarding the Sol-Gel Process

Remarkable progress has been made in the last three decades in the science of processing ceramic materials as a result of the increasing use of "wet chemistry". Considered as unique and fascinating both from a scientific and practical point of view, the sol-gel process has gained, in the last years, more and more importance in the materials science field, being unanimously recognized for the uniqueness of its advantages in preparing of some special materials and biomaterials with remarkable properties (electric, magnetic, optic or of sensing, etc.).

The words "sol-gel methods" have been widely used to refer to methods of glass preparation not involving melting. For the present discussions, it is arbitrarily considered that the sol method involves a suspension of fine solid particles such as colloids in solution, whereas the gel methods involve only liquids without solid suspensions. This report is primarily concerned with the "gel" method [1]. Interest in the sol-gel processing of inorganic ceramic and glass materials began as early as the mid-1800s with Ebelmen and Graham's studies on silica gels [2-4]. These early investigators observed that the hydrolysis of tetraethyl orthosilicate (TEOS), $Si(OC_2H_5)_4$, under acidic conditions yielded SiO_2 in the form of a "glass-like material". However, extremely long drying times of one year or more were necessary to avoid the silica gels fracturing into a fine powder, and consequently there was little technological interest [5]. The average composition of the polyethoxysiloxanes formed was obtained by evaluating the degree of conversion of the monomer by Aelion et al. [6]. The first practical application of the sol-gel method dated from 1939 when Geffcken and Berger [7] have obtained the first antireflection TiO_2 based coatings. Roy et al. [5, 8, 9] recognized the potential for achieving very high levels of chemical homogeneity in colloidal gels and used the sol-gel method in the 1950s and 1960s to synthesize a large number of novel ceramic oxide compositions, involving Al, Si, Ti, Zr, etc., that could not be made using traditional ceramic powder methods.

Filho and Aegerter [10] defined the gelation process in 1988 as that process which supposes the transformation of a sol into a wet gel. Four years later, Pierre [11] declared that practically there are as many definitions of the gelation process, as authors. He adopts, however, an alternative in which the idea of Filho can be recovered. Thus, he defines the gelation as a phenomenon through which a sol or a solution changes into a gel. The transformation supposes the establishment of some bonds either between the particles of the sol, or between the molecules of the solution, in order to form a final solid tri-dimensional network. But the situation is different from the classical solidification of a liquid, because in the case of the gelation, the solid structure remains opened and impregnated with the sol liquid or with the initial solution. This mixed composition "liquid-solid" confers to the gels some particular properties.

One of the most accepted definitions of the sol-gel process belongs to Schmidt [12]. He considers that the essence of the process, at least in a first stage, consists from the synthesis of an inorganic amorphous network by a number of chemical reactions in solutions, at low temperatures. In a second stage, the inorganic network can be converted to a glass at temperatures much lower than the melting ones corresponding to the component oxides, or to a crystallized material at a temperature inferior to that used in the conventional methods. The essential stage of the process, respectively the transition from a liquid (solution or colloidal sol) to a solid (the di- or multi-phasic gel) led to the expression "sol-gel process", giving thus practically its name.

The definition given by Lopez and co-workers in 1990 [13] is assumed by Ward and co-workers [14]: "sol-gel" is the name given to a great number of processes which suppose the existence of a solution or sol that turns into a gel.

No matter what definition is adopted, it is evident that the gelation process means "sols" on the one hand and "gels" on the other hand. In such conditions an explanation of these terms is imposed. So, what are in fact, sols and gels? These constitute forms of manifestation of the matter which exist in the natural state and which are known since the oldest times but only in the last period, beginning with the appearance and the development of the material sciences, have started to present scientific interest.

The definition of a sol, from 1979, which belongs to Iler [15], is assumed by Pierre [11] in 1992. In accordance with this, the sol represents a stable dispersion of (discrete, colloidal) particles in a liquid. Recently, Pachulski [16] extended the definition by specifying the meaning of the word "stable" in the context: stable means hear that there is no settling or agglomeration of the particles. Zarzycki [17] considered in 1990 the wet gel as a material consisting of a solid skeleton impregnated with an interstitial liquid. In the same year, Hench and West [5] defined the gel as a rigid, interconnected network, presenting sub-micrometric pores and polymeric chains with medium length bigger than 1μm. In 1992 Pierre [11] supplements this definition, specifying that the solid, tri-dimensional, interconnected network is developed in a stable manner inside of a liquid medium. The liquid which is present in the voids of the solid network is in thermo-dynamic equilibrium with the solid. According to the Science and Engineering Polymers Encyclopaedia, Almdal and co-workers [18] define the gel as a branched network of polymers which develops in a liquid medium, their properties being strongly dependent of interactions between the two components. The same authors quote a definition from the Webster dictionary, in accordance with which a gel is a gelatinous substance, formed from a colloidal solution that becomes a solid phase. It is the opposite of a sol. This last wording evidences the fact that the gel is a solid or semisolid material consisting of minimum two components, from which one is soft and elastic. In 2007 Pachulski [16] defined the gel as a continuous solid skeleton made of colloidal particles/polymers enclosing a continuous liquid phase.

The condition under which a sol becomes a gel (as a result of the gelation process) is that the sol passes through the so-called "gelation point". In 1988 Filho and Aegerter [10] pointed out that the definition of the gelation point in the literature is arbitrary and qualitative, without specifying some dynamic flow properties. The "gelation time" corresponding to the gelation point was qualitatively determined, at the room temperature, by visual inspection. It was considered to be the time after which the sol does not cool any more under the gravity influence. Sakka and Kamiya [19] define the gelation point as the moment when the sol loses its fluidity, and Yu and co-workers [20] consider it as the recorded time after which the surface of the sol remains unmodified at the inclination of the container during two minutes. Similar observations belong to Hench and West [5]. They specify that the gelation point, respectively the gelation times are easy to be observed from a qualitative point of view and easy to be defined in abstract terms but very hard to be analytically measured. As the particles of the sol grow and collide together, the condensation and the formation of macro-particles take place. The sol becomes a gel when it can bear an elastic stress. This is the moment which literature defines typically as the gelation point (or time) of the sol. There is not an activation energy which could be measured and it could not be defined with precision the moment in which the sol turns from a viscous fluid into an elastic gel. The change takes place gradually as more and more particles become interconnected. Some authors consider that the gelation is produced in the moment when the meniscus of the reaction mixture remains un-deformed at the inclination of the recipient [21].

The structure of the gel is established at the time of gelation. Subsequent processes such as aging, drying, stabilization and densification depend on the gel structure. The gel-to-glass conversion is dependent on the physicochemical properties of the dried gel and is, therefore, largely determined by the gelation process itself [22].

1.2. The Advantages and Drawbacks of the Sol-Gel Process

Solution chemistry offers many possible routes for "chemical manipulation" and allows various combinations in the synthesis of solids of diverse structures, compositions and morphologies [23]. The motivation for sol-gel processing is primarily the potentially higher purity and homogeneity and the lower processing temperature associated with sol-gels compared with traditional glass melting or ceramic powder methods [5].

A number of potential advantages and disadvantages of sol-gel methods are summarized in the literature [1, 11, 12, 24-29]. Essentially, the advantages are the following:

- it ensures final reaction products very pure by using purified precursors (by distillation, crystallization or electrolytic);
- homogeneous multi-component systems can be easily obtained by mixing the molecular precursor solutions;
- temperatures required for material processing can be noticeably lowered leading to unusual glasses or ceramics; this minimizes the chemical interactions with the walls of the recipient which is different from the case of the synthesis at high temperatures;

- the kinetic of the different chemical reactions can easy be controlled because of the low temperatures and of the frequent dilution conditions which is considered one of the major advantage of the sol-gel method;
- controlling the kinetic of the hydrolysis-condensation reactions, the gel structure could be modified for the same chemical composition; the effect of the molecular structure variation in the gel is maintained in the derived oxide materials allowing it to obtain different properties without modifying the chemical composition;
- it allows the formation of a “pre”-inorganic network in solution, which allows the densification to inorganic solids at low temperatures;
- it allows the control of the nucleation and growth of primary colloidal particles in order to obtain certain shapes, sizes and size distributions in a sub-microscopic domain; the sol-gel process guarantees the possibility to obtain nanometric reaction products which represents one of the most special achievements of the sol-gel method;
- the ability to go all the way from the molecular precursors to the product, allowing a better control of the whole process and the synthesis of “tailor-made” materials; depending on the experimental conditions of the process different forms of the final reaction products could be obtained: monoliths, powders, fibers, wires, coatings;
- it allows us to obtain not only any oxide-base composition but very special reaction products which also are of great and present-day interest, as: complex systems of mixed oxides whose homogeneity can be controlled until atomic level, doped-oxide systems, and inorganic-organic hybrid materials obtained by introducing some organic permanent groups.

Some drawbacks of the sol-gel method are:

- the high cost of raw materials;
- the large shrinkage during processing;
- the presence of residual fine pores, hydroxyl groups and carbon in the final product;
- the long processing times;
- the difficulties in the synthesis of monoliths;
- the health hazards of organic solutions.

Despite the mentioned drawbacks, based on the above advantages which are prevalent, the sol-gel process has gained very much scientific and technological attention during the last decades.

1.3. The Chemistry of the Sol-Gel Process

1.3.1. Precursors

According to Schmidt’s definition [12] the obtaining of the oxide materials through sol-gel method involves two distinct stages:

- the transition from the solution to gel;

- the transition from the gel to vitreous or crystalline solid.

There are two important sol-gel processes depending on the nature of the precursor, namely: the alkoxide that uses the molecular solutions (of the organometallic compounds) [26, 30] and the colloidal one (that uses colloidal solutions-the aqueous solutions of the inorganic salts) [26, 31]. A third one, intermediate between the mentioned ones, but suitable only for silicate systems and based on the chemistry of amine-silicate solutions, has been re-proposed [31, 32]. The sol-gel process which uses exclusively acetates also begins to present interest from both theoretical and practical point of view [33]. Indeed, alkoxides are undoubtedly the most common and well established presursors used in sol-gel, but not strictly the only ones. Other compounds have been and are currently used, together with alkoxides, for introducing some elements in multicomponent oxide systems. The most used non-alkoxide precursors are inorganic salts-nitrates, chlorides-or organic salts like acetyla-cetonates and acetates. Use of these products instead of alkoxides involved a different chemistry, introducing various problems such as the anion removal from the system. In general, different chemical and physical properties, structure and the microstructure of the final products may be expected [31].

The alkoxide route represents, in fact, the most common way of synthesis (the so-called "classical") of the sol-gel process. As its name suggests, the precursors that are used in the synthesis consist of alkoxides (alcoolates). The precursors must be liquids able to form reactive monomers or oligomers. Their simplified chemical formula is $M(OR)_n$, which indicates the fact that they represent the result of a direct chemical reaction between a metal and an alcohol [11]. Their developed chemical formulas are not always known. Anyway, in all cases, they suppose "metal-oxygen" (M-O) bonds. In the indicated general formula "M" represents a metal (Si, Ti, Al, etc.) and "R" indicates an organic radical as methyl, ethyl, propyl, buthyl or another alchylic group. These organic groups are very important because they are the only ones which could introduce a certain degree of contamination. They must confer enough stability and volatility to the alkoxide in order to allow its manipulation. Furthermore, they must yield to the net breaking of the M-OR or MO-R bonds early enough during the chemical process in order to form, finally, an inorganic, pure polymer based on M-O-M type bonds. In the case of alkoxides, the degree of ionicity of the M-O bond, which depends on the size and on the electro-negativity of the metallic atom, is also very important.

The chemical elements which can realize the final product by means of the sol-gel process must be hydrolizable, that is to allow the construction of a complex scale of molecules containing O or OH groups as ligands, a condition that is fulfilled by the alkoxides. Normally they dissolve into an alcohol and hydrolyse when water is added, in acid or alkaline conditions. The role of hydrolysis consists just in the transformation of the alkoxide-type ligand into a hydroxyl-type one. After that, further condensation reactions will lead to the obtaining of M-O-M, respectively M-($_{\mu}$OH)-M (where $_{\mu}$OH refers to hydroxy bound groups) based polymers [34].

1.3.2. The Transition from the Solution to Gel

The chemistry of alkoxides hydrolysis and polymerization is very complex. The chemical reactions which take place are [11]:

- hydrolysis reactions:

$$M(OR)_n + x\ H_2O \rightarrow M(OH)_n(OR)_{n-x} + x\ ROH \tag{1}$$

polymerization-condensation reactions through dehydration:

$$-M\text{-}OH + HO\text{-}M- \rightarrow -M\text{-}O\text{-}M- + H_2O \tag{2}$$

- polymerization-condensation reactions through alcohol elimination:

$$-M\text{-}OH + RO\text{-}M- \rightarrow -M\text{-}O\text{-}M- + ROH \tag{3}$$

- polymerization-condensation reactions through ether elimination:

$$-M\text{-}OR + RO\text{-}M- \rightarrow -M\text{-}O\text{-}M- + ROR \tag{4}$$

All these reactions take place in a solvent which is generally the corresponding alcohol. Moreover, they are often reversible. One can see that, since alcohol is involved as a reacting component, the concentration of alcohol is able to be involved in the hydrolysis equilibrium, according to eqs. (1), (3). This demonstrates that the reaction from a metal alkoxide to a solid material is not a simple one, but many different intermediates are possible. That means that it is very hard to give an exact thermodynamical description of what might be possible or not in this complicated reaction path. As a consequence, it is not possible to give detailed thermodynamically based prognoses, and the effect of the different reaction parameters (organic radical of the OR group, solvent, catalyst, temperature, concentration) has to be investigated experimentally, and the conclusions have to be drawn from these results [35]. Depending on the relative kinetics of the condensation and hydrolysis reactions, it is possible to result either linear polymers or dense colloidal particles or intermediate colloidal particles consisting from "balls" of weak reticulate polymers. These structures could be confirmed by the SAXS method (Small Angle X-ray Diffraction) or by viscosity measurements [11].

It is considered that a slow hydrolysis reported to condensation in the case when the polymerization products do not re-dissolve favors the formation of linear polymer structures and leads to the obtaining of polymeric gels. This polymerization can be catalyzed by acids or alkalis. The weak hydrolysis can be due to some chemical characteristics of the alkoxide or can be artificially produced for example by choosing a hydrolysis water quantity enough for the imposed stoichiometry of the alkoxide. It is also possible a second alternative in which the hydrolysis is rapid, the polymerization-condensation reactions being weaker or being possible to be revoked by re-dissolution. In this case either hydrate colloidal oxides (when the condensation products re-dissolve) or aggregates of hydrolyzed monomers as a result of reactions (5) and (6) could be obtained [11, 30].

$$M(OR)_n + n\ H_2O \rightarrow M(OH)_n + n\ ROH$$

$$M(OH)_n \rightarrow MO_{n/2} + n/2\ H_2O \tag{5, 6}$$

The encouragement of hydrolysis in a first step by using an excess of water facilitates the obtaining of relative massive monoliths, the polymers network being reticulate in this case.

Thus, one can see that the sol-gel process is based on the hydrolysis and condensation of molecular precursors. The chemical design of these precursors provides an interesting tool to control condensation reactions and tailor the nanostructure of the oxide materials [36]. A recent review regarding the evolution from the classical sol-gel route to advanced chemical nanotechnologies belongs to Schmidt [37].

1.3.3. The Transition from the Gel to Vitreous or Crystalline Solid

The gel to glass conversion is dependent on the physicochemical properties of the dried gel and is, therefore, largely determined by the gelation process itself. The dehydrated gels are essentially porous materials and a densification step is necessary to convert them into solid glasses devoid of residual porosity. Gels which are originally non-crystalline may crystallize during the drying process leading to the production of microcrystalline material suitable for some ceramic applications.

The dried amorphous gel differs from a glass by its texture. The gel is essentially an agglomerate of elementary particles, the size of which may be of the order of 100 Å arranged more or less compactly. The porosity may vary considerably according to the method of preparation. The residual space represents the pores which may be closed in the case of dense stacking of the particles or open when the texture consists of more or less regular "lattices" of particles leaving large interstices [38]. The constituent particles are coated with residual OH groups which are partly eliminated during the transition from a particulate texture towards a continuous solid; they may be detected by conventional infrared spectroscopic techniques. The transformation of gel into glass is basically a sintering process: at a suitable temperature the elementary particles weld together and the pores between them become nearly spherical. The driving force for this process arises from the excess surface free energy of the gel; an application of external pressure can be made to speed up the densification.

During the gel to glass conversion, both chemical and structural transformations take place which can be summarized as follows [22, 27]: (1) physical desorption of water and solvents from micropore walls (between 100-150°C); (2) carbonization (between 200-300°C) and combustion of residual organic groups (between 275-400°C); (3) condensation polymerization; (4) volume relaxation; and (5) viscous sintering. From these transformations (3), (4) and (5) result in densification; therefore, it is expected that gels of different chemical compositions and morphologies should show markedly different densification behavior. Results that the chemical processes are predominant at low temperature, while the physical processes are predominant at high temperatures. There is, also a temperature range, between 400-700°C, in which the two types of processes are overlapped. Both relaxation processes and micropore collapse combined with additional cross-linking are expected to produce a dense gel at low temperatures. Therefore, during the so-called "gel-to-glass" conversion, the desiccated gel will change to become more highly crosslinked while reducing its free volume (structural relaxation) and surface area (viscous sintering). Thus not only microstructure but also (and possibly more importantly) local chemical structure must be considered in modeling gel densification [39, 40]. Taking into account the mentioned above ranges of temperature and the structure of the prepared gels, specific thermal treatment curves (schedules) were established for each case. Generally, very low heating rates (0.5-1°C/min) with prolonged plateau in the ranges of temperature in which the physical and chemical processes are placed were used. The kinetics of the gel-to-glass conversion depends strongly on thermal history [41].

1.4. Specific Features of the Sol-Gel Chemistry of Transition Metal Oxides

The chemistry of alkoxides was extensively studied for the silica, in which case, depending on the adopted chemical protocol, different types of silica gels could be obtained. The parameters which influence the chemistry of the sol-gel process are the same for the transition metal based systems. Their increased reactivity due to the high electropositive character of metal atoms relative to that of silicon makes difficult to establish the experimental parameters by means of direct investigation methods. A lot of papers [26, 42-51] studied the behavior of transition metal alkoxide in the sol-gel process.

1.4.1. Molecular Metallorganic Precursors

The nature of the starting metallorganic precursor undoubtedly constitutes an important parameter which influences all the steps of the sol-gel process. As mentioned before, the most frequently used and studied precursors are metal alkoxides $M(OR)_n$ which are very reactive toward nucleophilic reagents such as water. They are known for almost all transition metal elements, including the lanthanides. The number and stability of transition metal alkoxides decrease from left to right with the increase of the number of the group in the periodic table. The alkoxy group –OR (R = saturated or unsaturated organic group) is a strong π donor and stabilizes the highest oxidation state of the metal. Consequently, the alkoxides of the elements from the principal groups and from the d^o transition metals (Ti, Zr) are rather well known, while those corresponding to the d^n transition metals have been much less studied [26, 44]. The transition metal alkoxides, mostly $Ti(OR)_4$ appear to be much more reactive than silicon ones. The main differences arise from the following reasons:

- the lower electronegativity of transition elements which leads to a much higher electrophilic character of the metal;
- the possibility to exhibit several coordinations (typical for the most transition metals) so that full coordination is usually not satisfied in the molecular precursor, which allows coordination expansion. As a result of the latter property, coordination expansion spontaneously occurs when the metal alkoxide reacts with water [26, 46-48]. Consequently, transition metal alkoxides are much more reactive. They must be handled with care, in the absence of moisture. They readily form precipitates rather than gels when the water is added.

1.4.2. Hydrolysis and Condensation of Transition Metal Alkoxides

Two chemical processes, namely hydrolysis and condensation, are involved in the formation of an oxide network from metal alkoxides. Hydrolysis of the alkoxide occurs upon adding water or a water/alcohol solution, and a reactive M-OH hydroxo group is generated. A three steps mechanism is usually proposed in literature [26, 45].

$$\underset{(a)}{H_2O + M\text{-}OR} \longrightarrow \underset{(b)}{H_2O{:}\rightarrow M\text{-}OR} \longrightarrow \underset{(c)}{HO\text{-}M \leftarrow O(H)R} \longrightarrow \underset{(d)}{M\text{-}OH + ROH} \qquad (7)$$

The first step (a) is a nucleophilic addition of a water molecule to the positively charged metal atom M. This leads to a transition state (b) where the coordination number of M has increased by one. The second step involves a proton transfer within (b) leading to the intermediate (c). A proton from the entering water molecule is transferred to the negatively charged oxygen of an adjacent OR group. The third step is the departure of the better leaving group which should be the most positively charged species within the transition state (c). The whole process, (a) to (d), follows a nucleophilic substitution mechanism. Charge distribution governs the thermodynamics of this reaction which will be highly favored when:

- the nucleophilic character of the entering molecule and the electrophilic character of the metal atom are strong: δ (0) << 0 and δ (M) >> 0;
- the nucleofugal character of the leaving molecule is high: δ (ROH) >> 0.

On the other hand, the rate of the nucleophilic substitution depends on:

- the coordination unsaturation of the metal atom in the alkoxide given by the difference between the maximum coordination number N of the metal atom in the oxide and its oxidation state Z. A larger (N-Z) value leads to a lower activation energy associated to the nucleophilic addition of step (a);
- the ability of the proton to be transferred within the intermediate (b). The activation energy associated with this transfer will be lower as the proton will be more acidic.

Condensation is also a complex process and can occur as soon as hydroxo groups are generated. Depending on the experimental conditions, three competitive mechanisms can be considered: alcoxolation, oxolation and olation.

Alcoxolation is a reaction by which a bridging oxo group is formed through the elimination of an alcohol molecule. The mechanism is basically the same as for hydrolysis with M replacing H in the entering group:

```
                                                   R
                                                   /
M—O + M-OR ——→ M-O:→M-OR ——→ M-O-M←O ——→ M-O-M + ROH
  |                 \                \
  H (a)             H (b)        (c)  H          (d)
```

(8)

Consequently, the thermodynamics and kinetics of this reaction are governed by the same parameters as for hydrolysis.

Oxolation follows the same mechanism as alcoxolation, but the R group of the leaving species is a proton:

```
                                                    H
                                                    /
M—O + M-OH ——→ M-O:→M-OH ——→ M-O-M←:O ——→ M-O-M + H2O
  |                 \                 \
  H  (a)            H (b)        (c)   H          (d)
```

(9)

The leaving group is thus a water molecule.

Olation can occur when the full coordination of the metal atom is not satisfied in the alkoxide (N-Z $\neq$ 0). In this case bridging hydroxo groups can be formed through the elimination of a solvent molecule. This latter can be either H_2O or ROH depending on the water concentration in the medium:

$$\begin{cases} M\text{-}OH + M \leftarrow \overset{H}{\underset{R}{O}} \longrightarrow M\text{-}\overset{H}{O}\text{-}M + ROH \\ M\text{-}OH + M \leftarrow \overset{H}{\underset{H}{O}} \longrightarrow M\text{-}\overset{H}{O}\text{-}M + H_2O \end{cases} \qquad (10, 11)$$

The reaction is strongly favored when the nucleophilic character of the entering group and the electrophilic strength of the metal are high: $\delta(0) << 0$ and $\delta(M) >> 0$. Moreover, since no proton transfer is involved within the transition state and since the metal coordination is not saturated, the reaction rate is usually very fast.

These four reactions (hydrolysis, alcoxolation, oxolation and olation) may be involved in the transformation of the molecular precursor into an oxide network. The structure and morphology of the resulting oxide strongly depend on the relative contribution of each reaction. These contributions can be optimized by carefully adjusting the experimental conditions which are related to both internal (nature of metal atom and alkyl groups, structure of the molecular precursor) and external (water/alkoxide ratio, catalyst, concentration, solvent, temperature) parameters.

1.4.3. Nature of the Metal Atom

Since transition elements are more electropositive than silicon, hydrolysis of transition metal alkoxides is much easier. The reaction is strongly exotermic and is observed as soon as the alkoxide is brought into contact with water. The electrophilic character of some alkoxides $M(OR)_n$ (expressed as the partial positive charge $\delta(M)$ on the metal atom) together with their degree of insaturation (expressed by the difference N-Z where N is the coordination number of the metal in oxide and Z is its oxidation state) are presented in Table 1 [26, 46]. It can be observed that the partial positive charge $\delta(M)$ on the central atom in a series of metal alkoxides is much higher for transition metals then for silicon. This explains why transition metal alkoxides are very unstable towards hydrolysis [26, 42, 45, 46]. They must be handled very carefully, in a dry environment and stabilizing agents are often added in the sol-gel processing of transition metal oxides.

Table 1. Positive partial charge on M for some metal alkoxides.

Alcoxide	$Ce(OPr^i)_4$	$Zr(OEt)_4$	$Zr(OPr^i)_4$	$Ti(OEt)_4$	$Ti(OPr^i)_4$	$Nb(OEt)_5$
$\delta(M)$	+0.75	+0.65	+0.64	+0.63	+0.60	+0.53
N-Z	4		3; 4	2	2	

Alcoxide	$Ta(OEt)_5$	$VO(OEt)_3$	$W(OEt)_6$	$Si(OEt)_4$	$Si(OPr^i)_4$	$PO(OEt)_3$
$\delta(M)$	+0.49	+0.46	+0.43	+0.32	+0.32	+0.13
N-Z		1; 2		0	0	0

The chemical reactivity of alkoxides towards nucleophilic reactions increases when both N-Z and δ(M) increase.

1.4.4. Nature of the Organic Ligand

The R group influences the hydrolysis and polycondensation kinetically as well as thermodynamically. Thus, the hydrolysis rate will decrease with increasing chain length (in the case of a linear R group) and also with increasing branching due to steric reactions [42]. Table 2 shows that the partial charge distribution in the alkoxide depends on the alkyl group, giving rise to more or less polar M-R bonds [26].

Table 2. Charge distribution in $Ti(OR)_4$ and $Si(OR)_4$ n-alcoxides.

R	δ(Ti)	δ(OR)	δ(H)	δ(Si)	δ(OR)	δ(H)	$k_h 10^2 M^{-1} s^{-1} [H^+]^{-1}$
CH_3	+0.66	-0.16	+0.12	+0.36	-0.09	+0.14	-
C_2H_5	+0.63	-0.16	+0.10	+0.32	-0.08	+0.11	5.1
n-C_4H_9	+0.61	-0.15	+0.09	+0.30	-0.08	+0.09	1.9
n-C_6H_{13}	+0.60	-0.15	+0.08	+0.29	-0.07	+0.08	0.83
n-C_9H_{19}	+0.59	-0.15	+0.07	+0.28	-0.07	+0.08	0.3
$Ti(OR)_4$				$Si(OR)_4$			

The positive partial charge of the metal atom (M = Si, Ti) decreases with the length of the alkyl chain. The sensitivity of the alkoxide towards hydrolysis should then decrease, in agreement with experiments [35, 52-54]. The positive partial charge of the hydrogen atom decreases in the same way. Proton transfer should then become more difficult, which is an effect that could be related to the decrease of the kinetic constant [35]. Condensation is also strongly affected by the nature of the alkyl chain. For transition metal alkoxides, under neutral or basic conditions, and without any chemical modification, gelation is never possible. Depending of the chain length precipitates or polymer colloids are formed. Precipitation of TiO_2 from $Ti(OR)_4$ is observed when R = Et [55, 56] or i-Pr while linear polymers seem to be formed when R = n-Bu or R = n-Am [26].

The characteristics of sol-gel oxide powders such as particle size, surface area, morphology and crystalline phases, obtained by hydrolysis and condensation of metal alkoxides, strongly depend on the type of the alkyl group. Thus, the ratio anatase/rutile of the titania powders resulted after thermal treatment of the gel can be varied by changing the molecular weight of the titanium alkoxide precursor [57]. Dense nanostructured titania with an average grain size of less than 60 nm, can be prepared by sintering a sol-gel titanium oxide near the anatase-rutile phase transformation temperature, about 600°C, suggesting that this method could be used more generally to produce nanophase materials with near theoretical densities.

1.4.5. Molecular Structure of the Alkoxide

The molecular structure of the precursor has to be taken into account in order to describe its reactivity. The full coordination of many metals can often not be satisfied in metal alkoxides $M(OR)_n$. This is due to the fact that the oxidation state Z of the metal is smaller than its usual coordination number N. In such cases coordination expansion of the metal occurs via oligomerization or change transfer complex formation [42]. The greater ability of metals to vary their coordination number and geometry depends on the size and charge of the ion, the number of d-electrons, the crystal field stabilization energy and the nature of the surrounding ligands. This provides great versatility of reaction mechanisms via a wide range of permissible transition state coordination geometry [51]. Titanium alkoxides $Ti(OR)_4$ provide a good example of a such behaviour. The oxidation state of titanium is Z = 4 while its usual coordination number is N = 6. Therefore the primary $Ti(OR)_4$ alkoxides exhibit oligomeric structures [42, 45, 58]. The degree of association depends on the nature of the metal atom. Within a given group, the molecular complexity increases with the atomic size of the metal. Titanium alkoxides with primary alkoxy groups are trimeric species with a pentacoordinated metal atom while those with secondary or tertiary alkoxy groups are monomers with titanium in a tetrahedral environment [46]. The fourfold coordinated monomeric species $Ti(O\text{-}i\text{-}Pr)_4$ is much more reactive. Consequently, as soon as one O-i-Pr group is hydrolized, condensation occurs. Hydrolysis and condensation proceed simultaneously so that precipitation leads to small polydispersed particles. Monodispersed oxide powders are usually obtained from oligomeric alkoxides with small OR groups rather than monomeric species with bulky OR groups.

Alkoxy-bridging is not the only way for coordination expansion. Metal alkoxides are often dissolved in organic solvents before hydrolysis is performed. These solvents usually correspond to the parent alcohol and in general, dilution should lead to lower association. The nature of the solvent has to be taken into account. Nucleophilic properties of the alcohol contribute to dissociation and solvation of the oligomer:

$$2[Ti_3(O\text{-}C_2H_5)_{12}] + 6\ C_2H_5OH \rightarrow 3[Ti_2(O\text{-}C_2H_5)_8 \cdot 2C_2H_5OH] \qquad (12)$$

The stability of such solvates increases with the positive charge of the metal atom and its tendency to reach a higher coordination number.

No solvate has ever been characterized for $Si(OEt)_4$ in ethanol. The geometry of silicon is generally that of a tetrahedral covalent bonded species modified by the availability of vacant d orbitals and the reaction mechanisms are more limited.

Chemical additives are almost always used in order to control and improve the sol-gel process. In most cases, these are nucleophilic hydroxylated ligands such as organic acids [46, 59, 60], polyols [61], β-diketones [45, 46, 59], hydrogen peroxide [45], and allied derivatives [46]. Upon hydrolysis a metal oxide is obtained when the new ligand is hydrolyzable while a mixed organic-inorganic network is formed when the metal-ligand bond cannot be broken upon hydrolysis. Mixed organic-inorganic materials open up wide opportunities for new materials [46].

1.4.6. Hydrolysis Ratio

Condensation could be adjusted by careful control of the hydrolysis ratio: $h = [H_2O]/[M(OR)_z]$. Three main domains could be considered in a rough qualitative analysis [26, 46]:

- $h<1$; in this range condensation is mainly governed by alkoxy bridging oxolation, with preferential elimination of ROH and in some cases olation reactions. Gelation or precipitation cannot occur as long as hydrolysis remains carefully controlled. Both processes alkoxy bridging and oxolation, lead to molecular oxo-alkoxides.

- $1<h<Z$; under such conditions polymeric gels can be obtained according to the following simplified scheme:

$$nM(OH)(OR)_3 \longrightarrow \cdots\text{-O-}\underset{OR}{\overset{OR}{M}}\text{-O-}\underset{OR}{\overset{OR}{M}}\text{-O-}\underset{OR}{\overset{OR}{M}}\text{-O-}\underset{OR}{\overset{OR}{M}}\text{-O-}\underset{OR}{\overset{OR}{M}}\text{-O-}\cdots + nROH \qquad (13)$$

  Under similar conditions, spinnable sols were synthetized by Sakka and Kamiya [19, 30], from which SiO_2 or TiO_2 fibers could be drawn. The formation of olated polymers in this domain is strongly supported by the fact that upon ageing, solvent is released by syneresis.
- $h>Z$; cross-linked polymers, particulate gels or precipitates can be obtained when an excess of water is added to the alkoxide. Olation and not oxolation is the predominant way for condensation.

1.4.7. The Influence of Catalyst

Another way to control hydrolysis and condensation process is to adjust the pH of the water used to perform hydrolysis. This can be done with an acid such HCl or HNO_3, or a base such as NH_3 or NaOH.

Similarly the effect of acid and base catalysts on the progress of condensation reactions can be predicted by calculating the charges on the species at various stages in the reaction. Thus for titanium alkoxide species the following charge distribution (δ) have been calculated [51]:

Table 3. Charge distribution for titanium oxo-polymer.

Species	δ(OR)	δ(Ti)
$Ti(OR)_3O$-	-0.08	+0.68
$Ti(OR)_2(OH)O$-	-0.01	+0.70
$Ti(OR)_2(O\text{-})_2$	+0.04	+0.71
$Ti(OR)(O\text{-})_3$	+0.22	+0.76

In acid catalysis, the least positively charged species will react fastest, i.e. the order of reactivity will be descending order in Table 3, so chain end sites will be more reactive than chain centre sites, and long chains with little branching will be produced. Conversely for base catalyzed reactions the most positively charged species will react fastest, i.e. in ascending order in Table 3, so chain centre sites will be most reactive, leading to highly branched chains. This clearly provides control of the evolving oxide structure.

1.4.8. Other Parameters

The hydrolysis ratio and the nature of the catalysts are the most important external parameters in sol-gel processing. However, other parameters such as concentration, nature of the solvent, and temperature can also play a decisive role in the progress of the reactions [26, 62].

The molecular separation of species during the hydrolytic polycondensation by dilution was found to affect the densification rate as well as the crystallization of TiO_2 samples [43]. Another effect of dilution is to prevent growth through aggregation. According to Yoldas [43, 62-64], the mean polymer size decreases as the precursor concentration increases for $Ti(OR)_4$ and $Zr(OR)_4$ systems. This is connected to the occurrence of sol-gel transition which is strongly affected by aggregation processes.

Generally, the increase of temperature activates both hydrolysis and condensation processes. For strongly reactive precursors, such as transition metal alkoxides, the temperature of reaction must be lowered in order to slow down hydrolysis and condensation process. So, sol-gel methods are currently applied for binary systems such as $BaO\text{-}TiO_2$ using the alkoxide precursors. The control of hydrolysis kinetics using a weak concentration and a low temperature leads to monolithic gel [65]. Temperature lowering of the reaction medium reduces reactions rates favoring the extension of polycondensation reactions. A monolithic non-crystalline solid was obtained after the drying process and by thermal treatment at 725°C the crystallization of $BaTiO_3$ with a particular shape was noted.

1.5. TiO_2-Based Oxide Materials Obtained by Sol-Gel Method

TiO_2 gels have been known for a long time. They can be obtained either by dissolving sodium titanate in concentrated HCl followed by adding of a weak base such as K_2CO_3, $(NH_4)_2CO_3$ or Na_2CO_3 in order to avoid high pH variations, or through thermohydrolysis of $TiCl_4$ or $TiO(NO_3)_2$ under acidic conditions. The colloidal particles are crystalline and have anatase or rutile structure depending on the pH and the nature of the counter-ions [26].

Most recent studies are based on sol-gel procedure using alkoxide precursors $Ti(OR)_4$. Monolithic TiO_2 gels can be synthesized from $Ti(OR)_4$ (R = Et, n-Pr, i-Pr, n-Bu, s-Bu) using substoichiometric hydrolysis ratios ($1<h<4$) and anorganic acid catalysts (HCl, HNO_3). The preparation routes and the effect of the different parameters during the hydrolysis-condensations reactions have been detailed presented in chapter 1.4.

Mackenzie [66] classified gel-derived amorphous solids in two groups: those which are likely to vitrify and those which are likely to crystallize. This is an important distinction since the application of any sol-gel process may produce either a glass-like solid, a polycrystalline ceramic or a "glass-ceramic". TiO_2 belongs to non-glass forming oxides.

Among the transition metals, titanium, as Ti oxides has found wide applications in sol-gel technology. This is due to high refractive index, good optical transmission in VIS and NIR regions, good physical and chemical stability, high electrical resistance and dielectric constant, the catalytic and photocatalytic aptitude and semiconductive properties. In their applications, TiO_2 resulted from sol-gel process is present both as amorphous gel and crystalline material as anatase, brookite and rutile. The temperature at which anatase completely transforms in rutile depends on many factors such as the method of preparation, the impurities present in anatase, the oxygen-metal coordination in the precursor, the oxygen-metal bond length in the precursor gel, and the texture and primary particle size of anatase [67]. Both these phases have tetragonal symmetry (anatase has a body centered tetragonal structure whereas rutile has simple tetragonal). The anatase phase of TiO_2 is known for its applications as photocatalysts, gas sensors, solar cells and electrochemical devices. The rutile phase of TiO_2 has found applications in capacitors, filters, power circuits and condensers because of its high dielectric constant.

The diversity of the sol-gel TiO_2-based materials is very high due to the possibility to control the hydrolysis-polycondensation parameters and the densification process (see Figure 1).

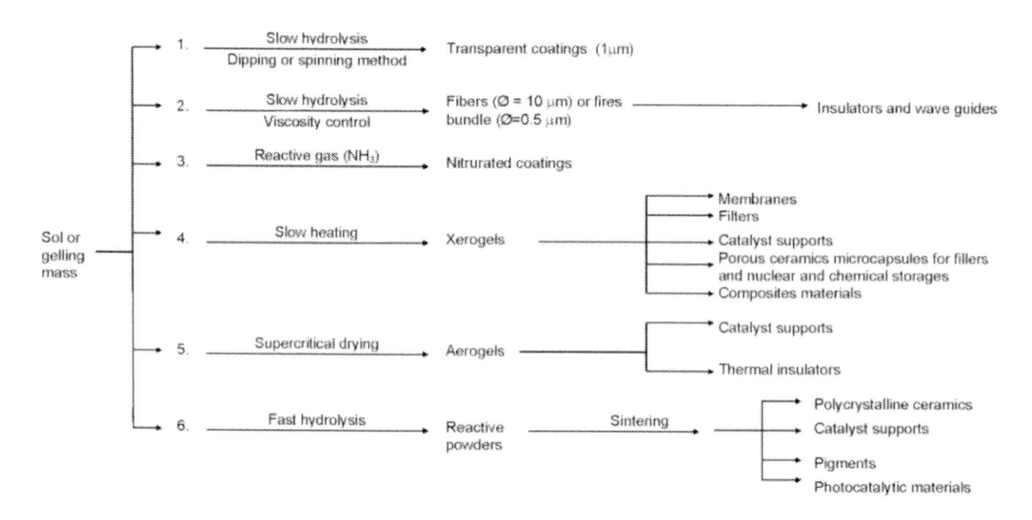

Figure 1. Diversity of the sol-gel process.

This diversity can also be observed in Table 4 in which the correlation between the preparation conditions and applications is obvious. Some TiO_2 compositions from literature data (also including proper experimental results) have been recalculated as molar ratios (water/alkoxide, solvent/alkoxide and catalyst/alkoxide) for an unitary presentation. From the presented table it can be clearly seen the advantage of the sol-gel process in preparation of predetermined structure materials.

The higher reactivity of the transition metals due to the high electropositive character relative to silicon makes it difficult to establish the experimental parameters by means of direct investigation methods as those used for the study of the SiO_2-based systems. Molecular-structural variations can be introduced into metal-organic derived polymer gels by variations of parameters. The effect of these variations extends to the glass and ceramic materials obtained from these gels and can even modify their high temperature properties, e.g. sintering, crystallization and viscosity [62]. Therefore the effect of molecular structural variation in the gel, by modification of the reaction parameters, can be evaluated in principle by establishment of the properties of the resulted oxide materials.

The present work describes a comparative structural and chemical study of un-doped TiO_2 and the corresponding S-, Ag-, and Pd-doped materials. Sol-gel Pd-, and Ag- doped TiO_2 coatings obtained from nanopowders were also compared. The photocatalytic degradation of pollutants using TiO_2 (nanopowders or thin films) for applications to environmental protection was established by testing the degradation of organic chloride compounds from aqueous solutions.

Table 4. Experimental conditions and some applications of TiO_2 obtained by sol-gel method.

No.	Experimental conditions for gel preparation					Applications	References
	Molar ratios			Conditions of reaction			
	H_2O / $Ti(OR)_4$	R'OH / $Ti(OR)_4$	Catalyst / $Ti(OR)_4$	t (min)	T (°C)		
1	1.82	55.45	0.113	30	40	Relative uniform coatings	[68]
		65.68				Uniform coatings	
		84.32				Coatings with more interference fringes	
	1.55	69.78	0.055	30	45	Uniform coatings	
		79.78					
		89.78					
2	2.00	61.54	0.05		20	Reflective coatings in IR and VIS regions	[69]
3	1.00	8.00	0.08		20	Coloured coatings: Coatings with electrochemical properties	[70]
4	1.004	36.07	0.473	30	40	Antireflective coatings with chemical durability	[71]
5	2.00	43.41	0.022			Antireflective coatings	[72]
6	2.00	44.40	0.1			Optical coatings	[73]
7	2.0	5.00	0.3	30	20	Optical coatings	[74]
	3.00		0.5				
	2.00		1.0				
	3.00		1.0				
8	1.42	44.35	0.113	30	40	Coating and un-supported gel for membranes	[75]
9	1.50	55.00	0.173[a]	30	50	Optical coatings. pH = 4.1 a – C_2H_5OH b - i-C_3H_7OH c – C_4H_9OH	[53]
			0.192[b]	30	50		
			0.177[c]	30	50		
10	1.50	35.00	0.05			Coatings for electronic devices, photocatalysts, photoanods, sensors	[76]
11	1.82	65.50	0.113 pH = 4÷4.5	30	40	Optical coatings	[77]
12	0.33 – 33.33	5.00 – 41.67	0.01		20	Coatings	[78-80]
13	0.38 – 3.76	0.00 – 2.00	0.10 – 1.00	30		The transparency of the fibers decreased with the appearance of the rutile phase. Higher amounts of HCl favour the formation of rutile phase.	[81]

Table 4. (Continued)

No.	Experimental conditions for gel preparation					Applications	References
	Molar ratios			Conditions of reaction			
	$\frac{H_2O}{Ti(OR)_4}$	$\frac{R'OH}{Ti(OR)_4}$	$\frac{H_2O}{Ti(OR)_4}$	$\frac{R'OH}{Ti(OR)_4}$	$\frac{H_2O}{Ti(OR)_4}$		
14	0.50	0.50 – 2.00	0.05 – 0.07	30÷60	30 and 60 in air	The fibres heated to 700°C are transparent; they become opaque at 750°C.	[82]
	1.00	0.10-5.00	0.01-0.21				
	2.00	0.10 – 2.00	0.30 – 0.42				
	3.00	0.10-10.00	0.30 – 0.64				
	4.00	0.50-5.00	0.55-0.80				
	7.00	0.50-10.00	0.70 – 1.10				
	10.00	1.00-2.00	1.00 -1.20				
15	3.00	2.00	0.55	60	40	Titanium nitride fibres	[83]
16	16.00	3.25	pH = 3.0[a]		70	TiO_2gel used as adsorbant, pigment, catalyst support a – HCl b – CH_3COOH c – $H_2C_2O_4$ d - H_2O e - NH_4OH	[84]
			pH = 5.0[b]				
			pH = 5.0[c]				
			pH = 7.0[d]				
			pH = 9.0[e]				
17	20.00	20.00-85.26	0.2	120	25 -75	Catalyst support	[85]
18	9.00	17.00	0.2		75	Catalyst support	[86]
	20.00	17.00			25		
	20.00	42.00					
19	200.00		0.2 – 0.7	120	80	Catalyst support	[87]
	400.00						
20	410	31.89	0.08	10 in N_2	265	Catalyst support	[88]
	4.20	31.89	0.12				
21	40.00		0.35 ÷0.60	720		Ceramic membranes	[89]
	100.00		0.35÷ 0.40				
	20.00						
22	22..22	82..22	0.5	20	20 N_2	Non-supported and supported α-alumina ceramic membranes obtained from colloidal gels.	[90]
23	1.50	55.00	0.173 - 0.192	30	50	Polymeric gels. Non-supported ceramic membranes for gas separation.	[52]
	3.00		0.171-0.256				
	5.00		0.171–0.336				

Table 4. (Continued)

No.	Experimental conditions for gel preparation					Applications	References
	Molar ratios			Conditions of reaction			
	$\frac{H_2O}{Ti(OR)_4}$	$\frac{R'OH}{Ti(OR)_4}$	$\frac{H_2O}{Ti(OR)_4}$	$\frac{R'OH}{Ti(OR)_4}$	$\frac{H_2O}{Ti(OR)_4}$		
24	1.00-16.00	50.0	0.08	30	25	Ceramic membranes obtained from polymeric gels	[91]
	1.00	8.0	0.08				
	1.00 – 4.00	28.0	0.025				
25	1.5	30 - 70	0.113 – 0.278	30	50	Polymeric gels. Non-supported and supported (alumina support) ceramic membranes for gas separation	[92] [93]
26	1.5	30	0.155	30	50	Porous coatings deposited on silicon wafer	[54]
	1.5	70	0.205				
	5.0	55	0.171				
27	1.5	55	0.148	30	50	Porous coatings deposited on different types of alumina supports. Polymeric membranes for separation processes	[94]
28	1.5	34..36	0.05		0	Coatings with photocatalytic properties	[95]
29	1.5	55	0.034 – 0.173	30	50	Coatings with electrochemical properties. TiO_2-V^{3+}/Ti electrode has been tested as sensor of redox potential for a large range of potential values (-531 and 293 mV). A very good indication of the redox potential was obtained in the pH range 2.34-11.5	[96]
30	1.0	21.0	0.11		10	Coatings and gels for humidity sensors in electronic devices. Humidity range = 5 ÷ 85%.	[97]
31	16.00	6.87				TiO_2 with semiconductive properties	[98]
32	2.00	30.93	0.066	0.5		Powder	[99]
33	5.00	80.77			20	Powder	[100]
34	5	85		0.5 – 2.0	25	TiO_2 powders and polyethylene-coated titania as reversed-phase support in chromatography	[101] [102]
35	20 - 100		pH = 11	120	20	TiO_2 powders for photosensitized oxidation of ethanol	[103] [104]
36	1	34.36	2			Coatings used for photocatalytic decomposition of acetic acid solutions	[105] [106]
37	166.7	2.27	2.00	720	80	Photocatalytic coatings for water depollution	[107]
38	1 - 4	15	0.3			TiO_2 nanocrystalline powders used as gas sensors, with excellent dielectric properties and catalysis applications	[108]
39	1	15	0.3	0.12		TiO_2 nanopowders used as humidity sensors, gas sensors (H_2 and O_2) and some catalytic applications	[109] [110]

The relationship between the synthesis conditions and the properties of titania nanosized materials, such as thermal stability, phase composition, crystallinity, morphology and size of particles and the influence of dopant were investigated. Their structural evolution and crystallization behavior (lattice parameters, crystallite sizes, internal strains) with thermal treatment were performed. The complex TiO_2-based oxide systems, e.g., $BaO{\cdot}TiO_2$ (barium titanate, BT), lanthanum-doped barium titanate (BLT) and $BaO{\cdot}ZrO_2{\cdot}TiO_2$ (BZT) were also studied. The dielectric and ferroelectric behavior of some BT and La-doped BT nanopowders and related ceramics, as well as the topography and optical characteristics of the BZT thin films prepared via sol-gel route in relationship with their structural and microstructural properties are reported.

2. Study of Sol-Gel Un-Doped and Doped TiO_2 System

Titania represents one of the most studied inorganic compounds in chemistry. Its preparation, physical and chemical properties together with its applications for all crystallographic forms (anatase, rutile and brookite) have been the object of numerous studies. The literature supplies thousands of articles dedicated to TiO_2. Let us mention at least some reviews [111-121] and books [122-126]. The explanation of such a huge interest is simple. It is well known that among the various photocatalysts, titania occupies a very important place, due to its high photocatalytic activity, excellent functionality, high chemical stability, thermal stability and non-toxicity. In 1972, Fujishima and Honda discovered the photocatalytic splitting of water on TiO_2 electrodes under UV light [127, 128], marking the beginning of a new period in photocatalysis field. Since then, enormous studies have been focused to the research of TiO_2 material, which led to many promising applications in different fields, ranging from optics to gas sensors via solar energy. These applications can be roughly divided into "energy" and "environmental" categories, many of which depending not only on the properties of the TiO_2 material itself but also on the modifications of the TiO_2 material host and of the interactions of TiO_2 materials with the environment [119]. All the mentioned properties are improved in the case of nanostructured TiO_2, which can be easily obtained by the sol-gel method. An advantage of nanometer sized materials stems from the changes in electrochemical potentials of the photogenerated charge carriers that accompany decreasing particle size. The sol-gel method facilitates the preparation of the photocatalyst in both forms (powders and coatings). Moreover, it is very convenient for producing nanocomposites materials in which different phases could be highly dispersed in an inorganic matrix, that is for the doping of titania with different metal or non-metal ions.

Recently, titanium dioxide has been extensively used for the decomposition of environmental pollutants as a possible alternative to conventional water treatment technologies. In order to clean the water from chemically stable synthetic organic compounds, so-called Advanced Oxidation Methods (AOMs) have been developed. TiO_2 photocatalysis belongs to AOMs processes that use energy to produce highly reactive intermediates of high oxidizing or reducing potential, which destroy the target compounds [129, 130]. The reason for the increased interest in this method is that the process can be carried out under ambient conditions and may lead to total mineralization of organic carbon to CO_2 [131].

In order to undertake comparative studies, in our paper, sol-gel doped TiO_2 materials have been prepared in both forms (coatings and powders) and two kinds of dopants (non-metals: S and metals: Ag and Pd) have been used. To a better assess of the effect of dopant on the final properties of the obtained products, all the doped materials have also been compared with the corresponding materials in the absence of the dopant.

2.1. Photocatalytic Effect

When TiO_2 is irradiated by UV rays, pairs of electrical charges-holes in the valency band and electrons in the conductivity band are created. The holes react with water molecules or with the hydroxyl ions and hydroxyl radicals are formed, which are strong oxidants of the organic molecules [132].

In principle, a photocatalytic reaction may proceed on the surface of TiO_2 powders via several steps, namely:

a) production of electron-hole pairs, photogenerated by exciting semiconductor with light energy;
b) separation of electrons and holes by trap available on the TiO_2 surface;
c) a redox process induced by the separated electrons and holes with the adsorbants present on the surface;
d) desorption of the products and reconstruction of the surface.

It has been shown that the photocatalytic activity of TiO_2 is influenced by the crystal structure (anatase, rutile), surface area, size distribution, porosity, surface hydroxyl group density, etc [133]. These characteristics influence the production of electron-hole pairs, the surface adsorption-desorption and the redox process. Electron-hole recombination is in direct competition with the trapping process (step b). The rate of trapping and the photocatalytic activity of TiO_2 will be enhanced by retarding the electron-hole recombination. The principal method of slowing electron-hole recombination consists in the loading metals onto the surface of the TiO_2 particles. The mechanisms of photocatalysis are discussed in recent reviews [111, 116, 134-137].

A simplified mechanism for the photo-activation of a semiconductor catalyst is presented in Figure 2 [116].

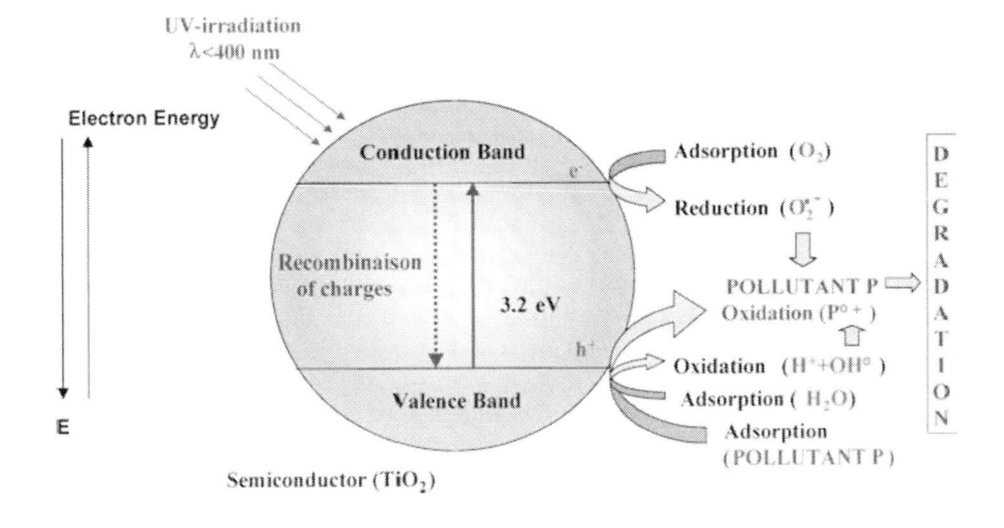

Figure 2. Simplified mechanism for the photo-activation of a semiconductor catalyst.

A model for photocatalytic mechanism is presented below.

$TiO_2 + h\nu \rightarrow e^- + h^+$	1. Photo-generation electron/hole pairs
$e^- + h^+ \rightarrow TiO_2$ + (energy)	2. Electron/hole recombination.
$M + e^- \rightarrow M(e^-)$	Metal attracts free electron from TiO_2 conduction band and slows recombination and promotes radical formation
$h^+ + H_2O \rightarrow OH\bullet + H^+$	
$e^- + O_2 \rightarrow O_2^-\bullet$	3. Formation of radicals (symbol O_x)
$O_2^-\bullet + H^+ \rightarrow HO_2\bullet$	
TOC + $O_x \rightarrow$ TOC (partially oxidized species) + $CO_2 + H_2O$	4. Radical oxidation of organic compounds

Titanium dioxide is a photoactive semiconductor. When it is illuminated with UV light, an electron from its valence band is promoted to its conduction band, generating an electron deficiency or a hole in its valence band, thus overloading its conduction band. The conduction band in crystalline TiO_2 is composed from unoccupied titanium 3d orbitals while the valence band is formed by filled oxygen 2p orbitals. It is confirmed that the oxygen vacancies can be treated as electron donors, which determine the n-type conductivity. In the nanocrystalline porous TiO_2 the oxygen adsorption has a great influence on photoconductivity [140]. The generated electron-hole pairs react with water or oxygen to produce radicals (Ox) in the surface of the semiconductor. As the organic species are adsorbed and desorbed on the surface of the catalyst, the formed radicals readily react with the adsorbed species, oxidizing them and creating CO_2 and H_2O and partially oxidized species.

However, conventional TiO_2 powder catalysts present the disadvantages of agglomeration and of a difficult separation of the final particle-fluid for the catalyst recycling. Thus, the application of TiO_2 thin films has attracted much attention in the last years.

2.2. Sol-Gel Coatings

Titania thin films have attracted considerable attention, because of their application to many types of devices, such as photovoltaic cells, electrochemical photolysis of water and organic substances, gas sensors and electrochromic displays. The most common application is a surface treatment of the glass with TiO_2 to give self-cleaning windows. The grime deposited on the surface can be removed by oxidation under sunlight. Because increasing the surface area of the film is favorable to enhance the efficiency of such devices, several attempts to fabricate porous TiO_2 films have been performed via powder-sintering and sol-gel methods. At the same time, porous films can be applied for new devices such as thin-film chromatography and overhead display, when the size and distribution of pores are fully controlled [141]. An increasing interest toward sol-gel synthesis of nanostructured oxide thin films is due to method advantages: simple and cheap technological equipment, low processing temperature for film densification and wide scope for varying film properties by changing the composition of the precursor solution and deposition conditions.

The method involves the hydrolysis of alkoxides in alcohol with a small amount of water, in order to form soluble intermediate species which condense, generating inorganic polymers.

The sol to gel transition occurs during deposition due to solvent evaporation which accelerates the reaction rate among the precursor oligomers. TiO_2 based vitreous films were among the first prepared ones [142-144]. We have previously published results on vitreous reflecting TiO_2 films on aluminized commercial glass [68], TiO_2 films containing transition metals [77, 96, 132, 145], TiO_2 films obtained from different Ti-alkoxides [53, 54] and their atomic force microscopy (AFM) study [146].

In a review with the same subject, Dislich [147] one of the initiator of the sol-gel method mentioned: "The fascinating feature of coatings on glass is their manifold chemical and operational possibilities corresponding to a multitude of demands. There is usually a solution to each problem". Thus, there are optical coatings, anticorrosion coatings, with chemical resistance, for coloration and decoration, and porous. Figure 3 presents the scheme of TiO_2 coatings classification.

Compared with conventional thin film forming processes such as chemical vapor deposition, evaporation, or sputtering, sol-gel dip coating requires considerably less equipment and it is potentially less expensive. However, the most important advantage of sol-gel over conventional coating methods is the ability to tailor the microstructure of the deposited film [148].

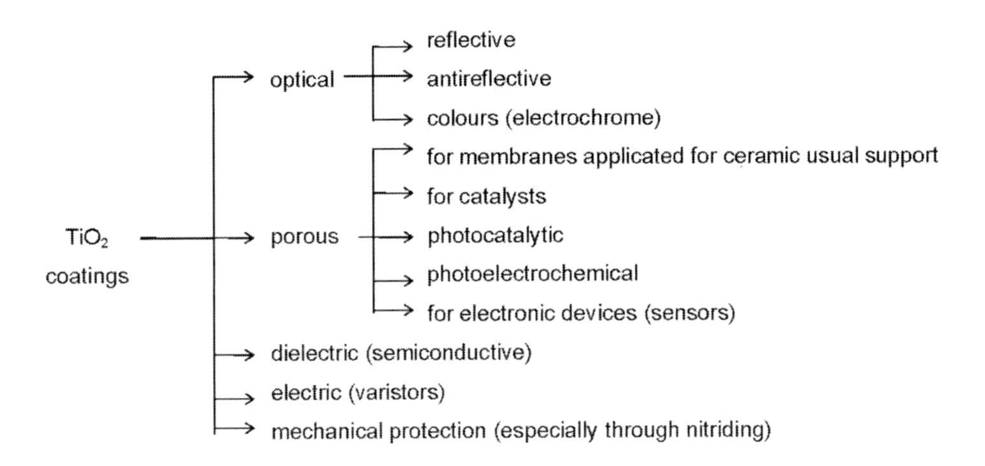

Figure 3. Scheme of TiO_2 coatings classification.

The stages of coating layer deposition are: formation of a liquid film on the substrate; the gelling of liquid and conversion of the gel to oxide compound. The hydrolysis and polycondensation reactions induce continuous gelling during the heating treatment required for film densification. Removal of the solvent leaves an interconnected porosity in the film, and final pore size depends upon the polymer size and topology prior to film formation [149]. If the heat treatment temperature is maintained below the densification temperature, a porous glass film remains which will have a lower effective index of refraction than that of the un-supported glass material. The refractive index and thickness of a film can be controlled by varying the composition, the annealing temperature, and the coating solution chemistry.

Spin and dip coating procedures are used to deposite the sol onto substrate. There are many aspects to the processing of films which are common to all deposition techniques. Schroeder [150-152] has outlined the necessary conditions for thin film formation:

- the solution must wet the substrate (hydrophilic solution);
- it must remain stable with ageing;
- it must solidify as an homogeneous transparent film;
- it should have some tendency towards crystallization into a stable high-temperature phase;
- for multiple layers the previous layers must be either insoluble or heat treated to make them insoluble before subsequent depositions.

The alkoxide solutions fulfill these conditions.

The basic characteristic of the solution which substantially influences coating thickness and its uniformity during deposition is viscosity. The viscosity of the solution is not affected only by the concentration but also by the degree of hydolysis, type of polymerization, time of aging, and type of liquid medium [144]. These factors and other ones related to the preparation of the solution are presented in Figure 4.

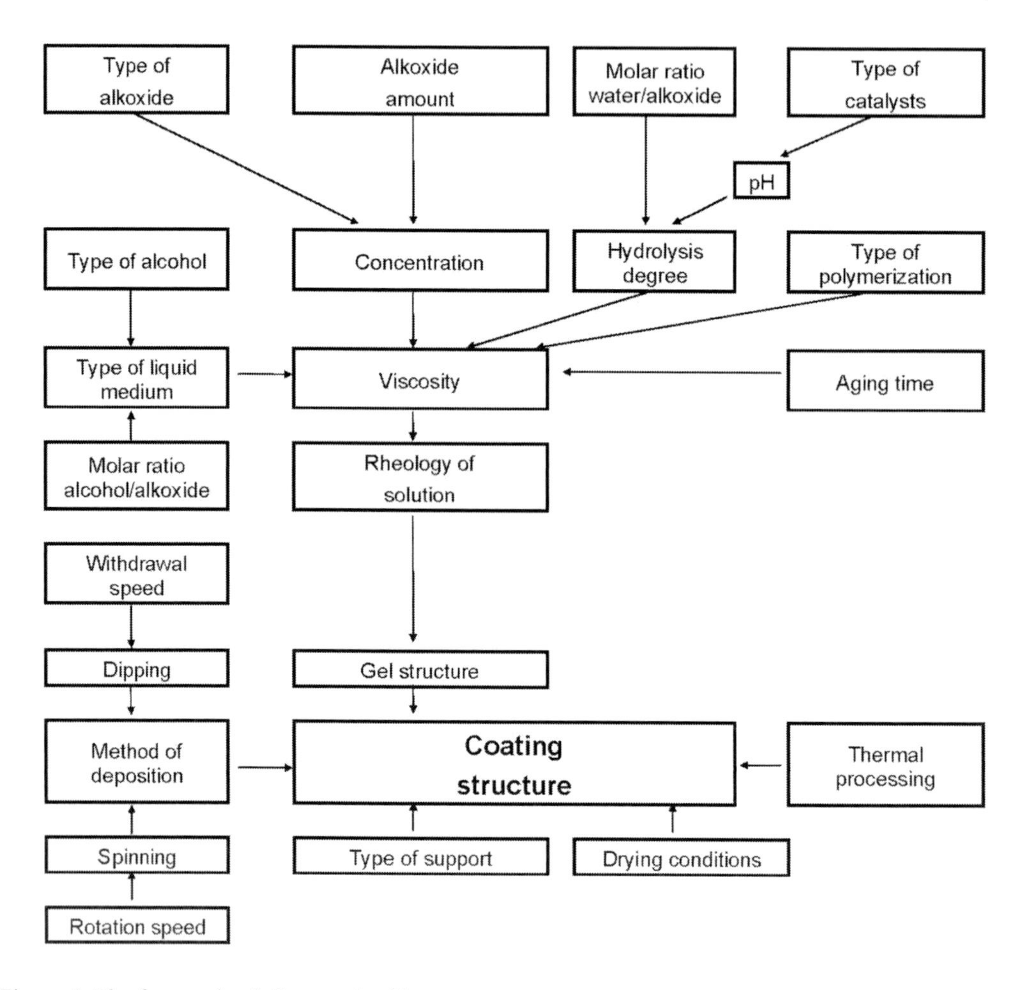

Figure 4. The factors that influence the film structure.

Gel-formation is not a desirable process in thin-layer preparation. There are various steps which are undertaken to stabilize the sol solution: the addition of stabilizers and of organic additives or maintaining low concentrations of the initial constituents. Low concentrations of the film-forming components in the sol solution may be guaranteed by using compositions rich in ethyl alcohol and lacking in water, which enables a suitable viscosity of the solution (below 10 mPa·s). It was established that solutions with a viscosity over 10 mPa·s produce microcracks in the coatings during thermal treatment. A high stability of the sol solutions with time is achieved at a pH value between 1 and 4 [52-54, 78].

To obtain oxide layers whith good optical and mechanical properties, hydrolysis and polycondensation must proceed slowly and simultaneously. In time, the hydrolysis is suppressed, being stimulated by the next thermal treatment. Good adhesion of the layer to the glass substrate results from the surface reaction of some building groups with the glass surface. A chemical anchorage of the coating on the glass is realized because Si-OH groups from the surface of the support react in accordance with reaction:

$$\text{Glass-Si-OH} + \text{R-O-Ti-} \rightarrow \text{glass-Si-O-Ti-} + \text{ROH}\uparrow \quad (14)$$

Because the adhesion of the film depends on the reactions between alkoxide solution and silanol groups localized at the surface of the glass support, the cleaning of the support is very important. After deposition, heat treatment is used to pyrolyze residual organics and densify and crystallize the coating.

It was found that, unlike other oxide coatings similarly deposited, the TiO_2 coating is either denser at low temperatures corresponding to the deposition, or its porosity is such that no penetration of external corrosive agents take place. For this reason TiO_2 coatings are preferred. This property of the TiO_2 film is in addition to the excellent natural resilience of titanium oxide material [73].

Another advantage of the sol-gel process for the film preparation is the high homogeneity and the uniformity of the layers and the fact that coatings with variable thickness can be obtained. For the films with optical properties, the coatings with a thickness between 200 and 5000 Å are prefered.

2.3. Dopants

As it was mentioned at the beginning of Chapter 2 the preferential use of TiO_2 in the photocatalytic degradation of organic pollutants is based on its high oxidative power, photostability and nontoxicity. However, a drawback of this semiconductor material is the large band gap (3.2 eV for anatase and 3.05 eV for rutile) which does not fit the solar energy spectrum and limits the photosensitivity to the UV region. Thus the yield for solar conversion of these oxides is quite low (1%). Hence, in order to harvest maximum solar energy, it is necessary to shift the absorption threshold towards visible region. Dye sensitization is an alternative method in which dye (sensitier) adsorbed on TiO_2 surface gets excited by adsorbing visible light and effects charge transition at sub-bandgap excitation to permit photocatalytic process [153]. In presence of visible light, surface adsorbed 8-hydroxyquinoline (HOQ) acts as a sensitizer in the photocatalytic electron transfer process [154]. Dye sensitization in photoelectrochemical systems has been extensively explored, but cannot be used for detoxification of waste waters since the dye molecules are also degradeable [155].

Another disadvantage of titanium dioxide is that the electron-hole (charge carrier) recombination takes place in a time span of several nanosecundes and in the absence of the promotors (e.g. Pt or RuO_2) the catalytic activity decreases. Thus, the dopping is recommended.

Moreover, it is very convenient to produce nanocomposites materials in which different phases could be highly dispersed in an inorganic matrix, that is to dope titania with different metal or non-metal ions. Recent studies [116, 117, 156] pointed out the doping procedure as a possibility to enhance the efficiency of the photocatalytic process. The dopant ions can behave as both hole and electron traps or they can mediate interfacial charge transfer [156]. Also known as "impurity doping", the procedure ensures the extension of the spectral response of a wide band gap semiconductor to visible light. An advantage of using the sol-gel method is the ability to control in a simple way the concentration of the dopant in the nanostructure of titanium dioxide.

An enhancement of photocatalytic activity of the noble metal (Pt, Pd, Au, Ag, etc) modified TiO_2 has been explained in terms of a photoelectrochemical mechanism in which the electrons generated by UV irradiation on the TiO_2 semiconductor transfer to the loaded metal particles, while the holes remain in the semiconductor, resulting in a decrease in the electron-hole recombination [157]. Choi et al. [156] have completely studied the doping effect of twenty-one kinds of transitional metal ions on nanocrystalline TiO_2. The dependence of photocatalytic activity versus the concentration of metal ions doping has been established. It can be found that there is an optimum dopant concentration for the best photocatalytic performance of doped-TiO_2 material. Above this value the metal dopants can act as electron-hole recombination centers which are detrimental to the photocatalytic activity.

Highly dispersed nanoparticles of noble metals, such as Pt, Pd, Rh, Ru and Au in mesoporous supports as titania, alumina, silica are widely used as catalysts in organic synthesis, petrochemistry, etc. In a study concerning the photocatalytic activity of noble-metal-loaded TiO_2, Ranjit et al. [158] suggest that an ohmic contact is formed between the metal and semiconductor. Hence electrons can easily flow to the metal sites on TiO_2 under irradiation and the role of the metal is to act as an electron sink and thus to enhance the activity. However, it is difficult to introduce metal nanoparticles into mesopores by traditional impregnation methods, because they tend to richly deposit on outer surface of mesoporous materials and moreover it is difficult to control the loading amount by impregnation [159]. The sol-gel method easily allows the preparation of nanocomposite materials such as inorganic matrices in which a metal phase could be highly dispersed.

Different dopants may not have the same effect on trapping electrons and/or holes on the surface or during interface charge transfer because of the different positions of the dopant in the host lattice. Consequently, the photocatalytic efficiency would be different for different types of dopants [160]. Selection of the dopants depends on the reaction of interest. Not all dopants work efficiently for all reactions. For example, Fe^{3+} works well for the catalysis of $CHCl_3$, but not for 2-chlorophenol [161].

Moreover, the role of the dopants in general and in particular in the case of the films is to reduce the internal strains in the thin coatings which can modify optical, mechanical or electrical properties. The dopants hinder the crystalline structure formation, leading to densification and non-crystalline structures as the dopant occupies and completes unsatisfied chemical and mechanical bonds.

The sol-gel method was extensively used in order to improve the photocatalytic activity of TiO_2 by doping with:

a) different metals: Pt [160, 162-164]; Pd [160, 165-168]; Au [169-173]; Ag [174-178]; V [96, 179]; Cr [180, 181]; Mn [182]; Fe [132, 160, 183-190]; Ni [191, 192]; Co [193-195]; Cu and Co [77]; Nb [196-198]; Mo [199]; W and Al [200-202]; lanthanides [160, 161, 203-208]; Sb [209]; Sn [210];
b) different non-metals: C [211-218]; N [218-224]; S [177, 216-218, 222, 224-239]; B [239, 240];
c) co-doping: Zr-S [241]; N-S [242]; B-N [243]; Br-N [244]; B-C and S-N-C [245].

Besides photocatalytic enhancement, the modification through dopants (metals, or non-metals) enables TiO_2 coatings or nanoparticles to be active in the energy range of visible light.

2.4. S-doped TiO_2 Coatings

Concerning the S-dopant, the literature data report that it can act both as anion, replacing the lattice oxygen in TiO_2 [228] and cation, replacing Ti ions [227, 229, 230]. Very few data could be found regarding the S-doping of TiO_2 that refer to the preparation of thin coatings [236]. Taking into account that prevention and control of the effects due to different environmental pollutants represent one of the most acute problems of the humanity in our days and that the pollution of the waters can be considered as one of the most important aspects, researches in the domain of materials with catalytic properties for water pollutants degradation represent a very important topic. Nanostructured coatings are preferred for environmental protection due to their high surface/volume ratio, taking into account that photocatalytic process is based on the chemical reactions at the surface of the material. In addition, sol-gel processing allows control of the microstructure of the film.

2.4.1. Samples Preparation

Pure TiO_2 sol-gel coatings (C-TiO_2) have been prepared by controlled hydrolysis-condensation of tetraethylorthotitanate, $Ti(OC_2H_5)_4$. Absolute ethanol has been used as solvent. The hydrolysis of the Ti alkoxide was provided by water addition in substoichiometric quantity and nitric acid as catalyst. The molar ratios used in order to obtain coatings were: water/alkoxide = 1.5, alcohol/alkoxide = 55 and catalyst/alkoxide = 0.258, at pH=3.5. The reaction temperature was 323 K. The hydrolysis reaction was carried out in a closed system in nitrogen atmosphere under vigorous stirring. The reaction time was 1 hour.

The corresponding sol-gel S-doped TiO_2 coatings (C-S-TiO_2) have been obtained in similar conditions with C-TiO_2 sample. They were doped with 2 and 5 wt. % S related to the TiO_2 content (C-S2-TiO_2 and C-S5-TiO_2 samples). The sulfur source was thiourea (H_2NCSNH_2) which was introduced in the reaction mixtures dissolved in the corresponding alcohol.

In all cases single and double-layers coatings have been deposited on glass substrates using dip-coating procedure, with a withdrawal rate of 5 cm/min. Continuous and homogeneous coatings with good adherence to the support were obtained. The conditions used for thermal processing of the films were determined taking into account the thermal behaviour (DTA/TG/DTG) of the bulk gels obtained from the sol used for coatings preparation [177, 236]. Thus, thermal treatments at 573, 673 and 773 K have been applied with heating rate of 1 K $min.^{-1}$ and 1 h plateau in all cases. The second layer was deposited after the densification of the first one at 573 K.

2.4.2. Results

The presence of sulfur at 773 K was confirmed by X-ray fluorescence spectra (XRF) presented in Figure 5 [236]. The XRF measurements were performed on gels because in the case of the films the characteristics lines of S overlap with the Si ones from the glass substrate.

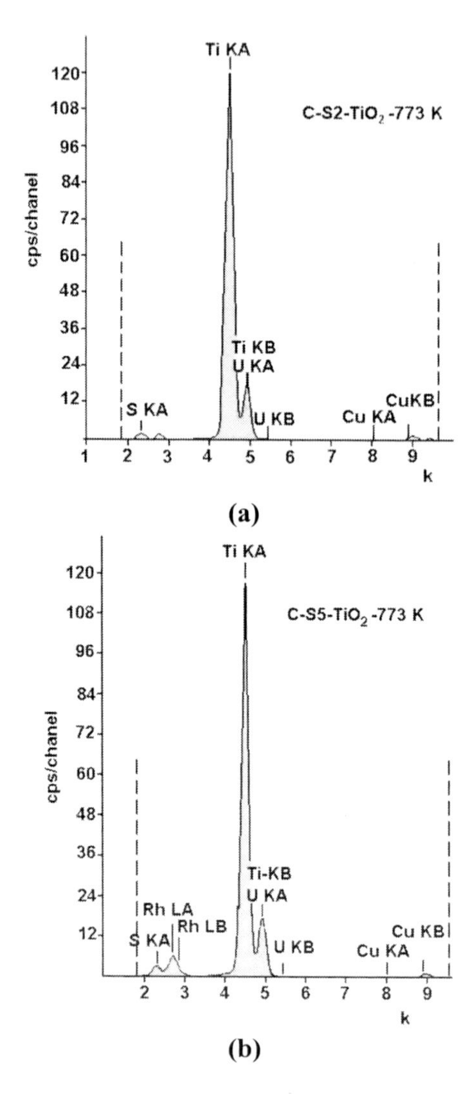

Figure 5. XRF spectra of S-doped TiO_2 samples thermally treated at 773 K.

FT-IR spectra of un-doped and S-doped TiO_2 films thermally treated at 573 and 773K point out four-fold Ti coordination in vitreous matrix (Ti-O bond vibration at 920 cm^{-1} together with Si-O-Si vibration bands characteristic to glass substrate (760 and 570 cm^{-1})) [68, 236]. A similar profile is obtained for mono- and bi-layers films.

IR spectra of the thermally treated (573-773 K) un-doped and S-doped gels show the following vibration bands: 3440 cm^{-1} (ν_{OH}), 1640 cm^{-1} (δ_{HOH}), 670 cm^{-1} (ν_{Ti-O} of isolated tetrahedral TiO_4), 540 cm^{-1} (ν_{Ti-O} of condensed octahedral TiO_6) and 360 cm^{-1} (ν_{Ti-O} of isolated octahedra TiO_6) in agreement with literature data [246]. IR spectroscopy has evidenced the Ti-O bond formation. With the increase of the temperature, the intensity of the vibration bands due to Ti-O bonds increases and they are better defined. In the same time the vibration bands due to the presence of molecular water and structural OH^- diminish.

Because the coatings are very thin (thickness obtained from spectroellipsometry measurements: 361-509 Å for one layer and 632-775 Å for two layers [236]) there is an overlap between their diffraction spectra and the diffraction spectra of the support, which is amorphous, irrespective of the thermal treatment. For this reason, the phase composition was established on un-supported gels obtained from the gelation of the sols used for coatings preparation. Structural evolution (XRD results) with temperature of the S-doped TiO_2 gels compared with the pure TiO_2 gel is presented in Figure 6 and in Table 5.

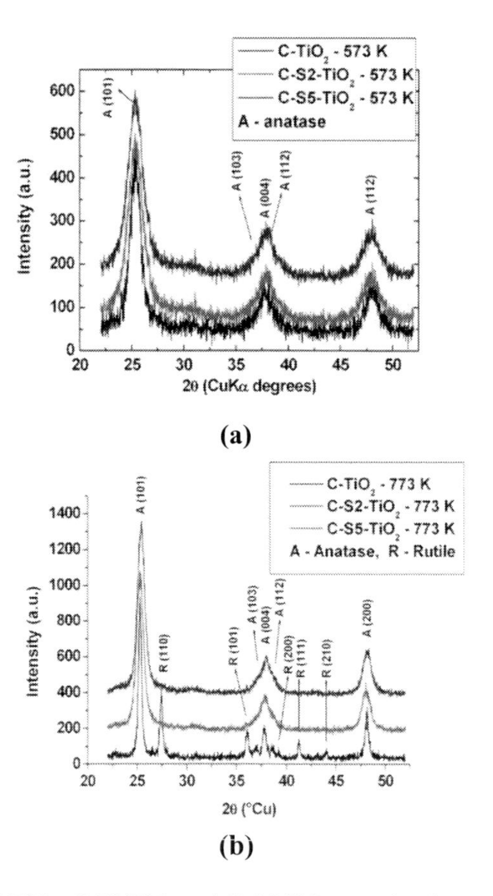

Figure 6. XRD patterns of C-TiO_2, C-S2-TiO_2 and C-S5-TiO_2 samples thermally treated at 573 (a) and 773 K (b).

Table 5. Calculated values of microstructural factors obtained from computerized analysis of XRD spectra of un-doped and S-doped TiO_2 samples.

Sample	T[K]	Identified phases [%]	Lattice constants			Microstructural factors	
			a [Å]	c [Å]	u.c.v. [$Å^3$]	<D>	10^{+3}x<S>
C-TiO_2	573	A	3.7954(68)	9.5659(271)	137.79(88)	173(80)	2.1(2.8)
C-S2-TiO_2		A	3.7849(34)	9.4671(133)	135.62(44)	100(36)	3.0(3.1)
C-S5-TiO_2		A	3.7915(18)	9.5050(69)	136.6(22)	112(54)	4.2(3.8)
C-TiO_2	673	A > 95	3.7721(27)	9.4821(97)	134.91(33)	153(11)	2.27(32)
		R < 5	-	-	-	-	-
C-S2-TiO_2		A	3.7775(26)	9.4850(91)	135.35(31)	107(20)	3.0(1.2)
C-S5-TiO_2		A	3.7700(34)	9.5147(121)	135.23(41)	109(7)	0.89(18)
C-TiO_2	773	A – 73.96	3.7801(3)	9.5111(12)	135.90(4)	442(82)	1.1(4)
		R – 26.04	4.5914(12)	2.9564(16)	62.32(7)	405(246)	1.2(1.4)
C-S2-TiO_2		A	3.7856(62)	9.4771(242)	135.81(79)	161(78)	2.7(2.6)
C-S5-TiO_2		A	3.7779(20)	9.4709(80)	135.17(26)	199(37)	1.2(9)

A: Anatase; R: Rutile; a, c: lattice parameters; u.c.v.: unit cell volume; D: crystallite size; S: internal strain

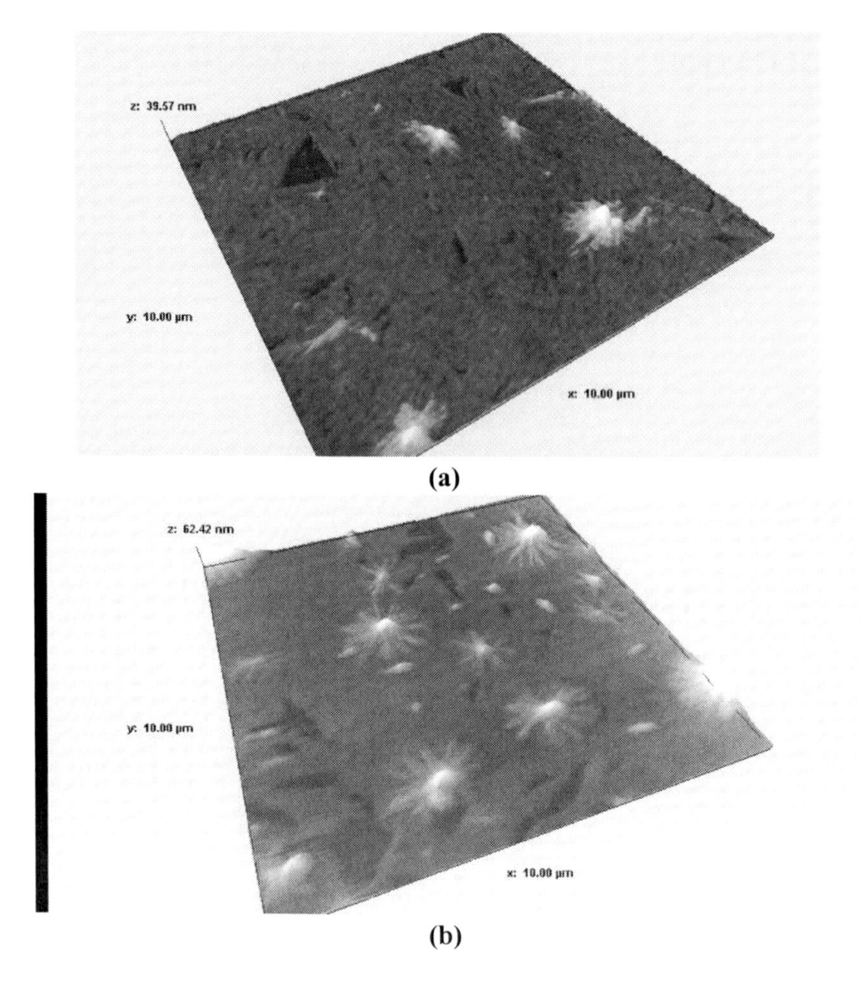

Figure 7. AFM images of S-doped TiO_2 bi-layers films thermally treated at 573 K; C-S2-TiO_2 (a), C-S5-TiO_2 (b), measured over an 10x10 µm area; Root-Mean-Square (RMS) Deviation: 1.81 and 4.36 nm, respectively.

The phase composition established from XRD results (Table 5) point out the presence of anatase as single crystalline phase in all un-doped and S-doped samples thermally treated at 573 K. At 773 K rutile phase could be observed only in the un-doped sample. C-S2-TiO_2 samples have the smallest crystallite sizes, no matter of the temperature of the thermal treatment. It can also be remarked the important effect of the 2 % S dopant on the distortion of anatase lattice even at 773 K, when its lattice is well-formed (D=442 Å). The lattice strain at 773 K (S=$2.7x10^{-3}$) is comparable with that at 573 K (S=$3.0x10^{-3}$), when the anatase phase is not well formed.

Figure 7 presents AFM images of S-doped TiO_2 bi-layers films thermally treated at 573 K. Small values of the surface roughness (<10 nm) are observed; these ensure good hydrophilicity properties of the films. An increase of structural order in the matrix of the samples C-S5-TiO_2 is well correlated with XRD data.

The tests of photocatalytic activity of the prepared films for chlorobenzene removal from water at neutral pH have shown that for the same irradiation time, the better chlorobenzene removal yield was obtained with a greater number of layers and with an increase of the annealing temperature of the films. The following experimental conditions have been used: sample volume = 1300 mL; surface area of photocatalyst (film) = 175 cm^2; irradiation time = 60 min.; type of the lamp = TQ1(P=150W)-Heraeus; λ = 200÷280, 400÷450 nm; O_2 = 7 mg/L; irradiation pathlength l = 2 cm; the initial concentration of chlorobenzen, $[CB]_0$ = 10.8 mg/L = $0.96 \cdot 10^{-4}$ M; pH = 7. The results are presented in Table 6.

Table 6. The effect of the annealing temperature of the photocatalyst on the efficiency of photocatalytic depollution of contaminated water with chlorobenzene at pH = 7, using un-doped and S-doped TiO_2 coatings.

Photocatalyst type	Number of layers	[CB] after irradiation		Film thermally treated at 573 K η_{CB} (%)	Film thermally treated at 673 K η_{CB} (%)
		mg/L	$x10^3$, M		
C-TiO_2	1	1.13	0.010	89	89.6
	2	0.68	0.006	93	93.8
C-S2-TiO_2	1	0.56	0.005	94.3	94.8
	2	0.11	0.001	97	98.9
C-S5-TiO_2	1	0.79	0.007	93	93.2
	2	0.34	0.003	95.5	96.7

An improvement of the photocatalytic activity for C-S2-TiO_2 coatings can be observed. The chlorobenzene removal yield is better for the S-doped TiO_2 films than for the un-doped ones, in all cases, no matter the S concentration, the number of layers or annealing temperature. According to XRD data, this could be explained by the presence of anatase as single phase in S-doped samples. The smallest crystallite size of C-S2-TiO_2 sample is responsible for its better photocatalytic activity. The important effect of the 2 % S dopant on the distorsion of anatase lattice even at 773 K, when its lattice is well-formed could also be taken into consideration. In all cases, the increase of annealing temperature of the films led to better photocatalytic activity.

2.5. Ag-doped TiO_2 Coatings

Knowing that TiO_2 is an excellent photocatalyst that allows the degradation and finally the mineralization of stable organic pollutants (detergents, dyes, pesticides) in water, the present work has extended the field of investigation from S to Ag-doped TiO_2, aiming at accomplishing a comparison of their efficiency for the removal of chloride organic compounds from water. The selection of silver as metal ion dopant for TiO_2 was based on two well known properties: it enhances the photoactivity in the visible domain [174] and it produces the highest Schottky barrier among the metals, which facilitates the electron capture [176].

2.5.1. Samples Preparation

The corresponding sol-gel Ag-doped TiO_2 coatings, C-Ag-TiO_2, have been obtained in situ by cogelation of the precursors, under similar conditions to un-doped and S-doped TiO_2 samples (see paragraph 2.4.1.). The concentration of the dopant was 2 wt. % Ag related to the TiO_2 content. The silver source was $AgNO_3$, which was introduced in the reaction mixture dissolved in the quantity of water required by the hydrolysis process.

The conditions used for thermal processing and the characterization of the films were similar with those for S-doped coatings.

2.5.2. Results

Structural evolution (XRD results) with temperature of the S- and Ag-doped TiO_2 gels compared with the pure TiO_2 gel is presented in Figure 8 and Table 7.

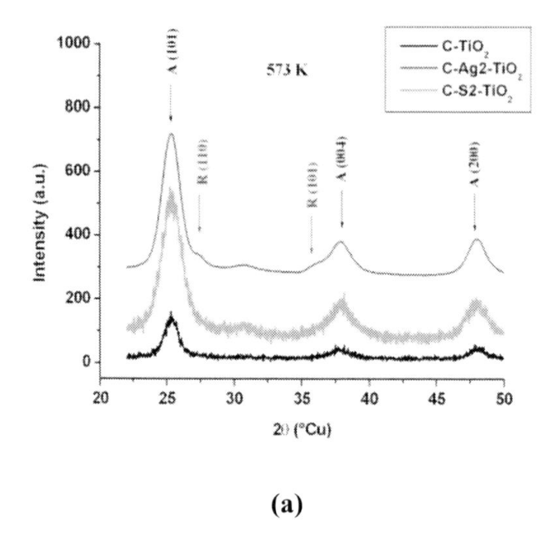

(a)

Figure 8. Continued.

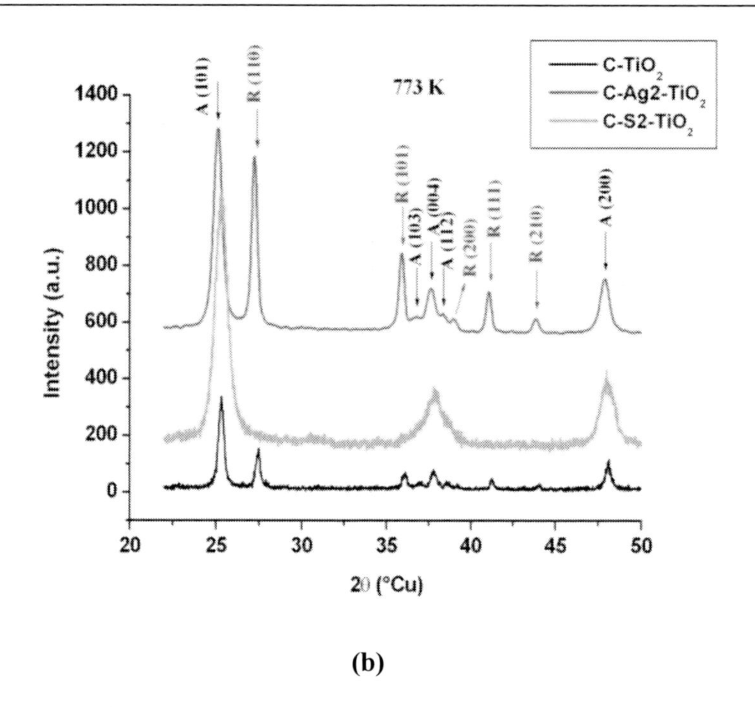

(b)

Figure 8. XRD patterns of C-TiO_2, C-S2-TiO_2 and C-Ag2-TiO_2 samples thermally treated at 573 K (a) and 773 K (b).

Table 7. Calculated values of microstructural factors obtained from computerized analysis of XRD spectra of un-doped and S- and Ag-doped TiO_2 samples.

Sample	T [K]	Identified phases [%]	Lattice constants			Microstructural factors	
			a [Å]	c [Å]	u.c.v. [Å^3]	<D>[Å]	10^{+3}x<S>
C-TiO_2	573	A	3.7954(68)	9.5659(271)	137.79(88)	173(80)	2.1(2.8)
C-S2-TiO_2		A	3.7849(34)	9.4671(133)	135.62(44)	100(36)	3.0(3.1)
C-Ag2-TiO_2		A > 97	3.7838(52)	9.5215(184)	136.32(64)	93(11)	4.0(7)
		R < 3	-	-	-	-	
C-TiO_2	673	A > 95	3.7721(27)	9.4821(97)	134.91(33)	153(11)	2.27(32)
		R < 5	-	-	-	-	-
C-S2-TiO_2		A	3.7775(26)	9.4850(91)	135.35(31)	107(20)	3.0(1.2)
C-Ag2-TiO_2		A – 89	3.7788(32)	9.4936(112)	135.56(39)	111(24)	2.7(1.3)
		R – 11	4.6020(106)	2.9528(69)	62.54(43)	389(128)	8.2(6.8)
C-TiO_2	773	A – 73.96	3.7801(3)	9.5111(12)	135.90(4)	442(82)	1.1(4)
		R – 26.04	4.5914(12)	2.9564(16)	62.32(7)	405(246)	1.2(1.4)
C-S2-TiO_2		A	3.7856(62)	9.4771(242)	135.81(79)	161(78)	2.7(2.6)
C-Ag2-TiO_2		A – 54	3,7859(22)	9.4875(78)	135.99(27)	267(112)	0.6(5)
		R – 46	4.5978(47)	2.9602(38)	62.58(21)	619(223)	1.4(6)

A: Anatase; R: Rutile; a, c: lattice parameters; u.c.v.: unit cell volume; D: crystallite size; S: internal strain

Table 8. The influence of the number of layers and the irradiation time of C-S2-TiO_2 and C-Ag2-TiO_2 photocatalysts, thermally treated at 573 K in the photodegradation of chlorobenzene in water depollution process.

Photocatalyst	Number of layers	Irradiation time (h)	Remanent chlorobenzene (mg/L)	Chlorobenzene removal yield (%)
C1-TiO_2	1	0.5	2.70	75.0
C1-Ag2-TiO_2			2.75	75.9
C1-S2-TiO_2			2.37	78.0
C1-TiO_2		1	1.19	89.0
C1-Ag2-TiO_2			1.08	90.5
C1-S2-TiO_2			0.62	94.3
C2-TiO_2	2	0.5	2.59	76.0
C2-Ag2-TiO_2			2.46	78.4
C2-S2-TiO_2			1.95	82.0
C2-TiO_2		1	0.76	93.0
C2-Ag2-TiO_2			0.58	94.9
C2-S2-TiO_2			0.32	97.0

The Ag-doping favors TiO_2 crystallization in anatase simultaneously with the transformation of anatase to rutile, even at 573 K. This fact explains the decrease of the catalytic activity in comparison with S-doped sample (see Table 8). In the Ag-doped TiO_2 samples, annealed at 673 and 773 K, the anatase phase exhibits a larger mean crystallite size and smaller internal strains than in the case of similar S-doped samples. Moreover, from XPS measurements, the surface quantities of the Ag dopant are significantly lower than the S surface concentrations, which explain the better photocatalytic properties of sulfur as dopant [177].

The results of the photocatalytic tests are presented in Table 8 [177]. The following experimental conditions have been used: sample volume = 1300 mL; surface area of photocatalyst (film) = 175 cm^2; medium pressure mercury lamp TQ-Z1 (λ = 200÷280; 400÷450 nm); O_2 = 7 mg/L; irradiation pathlength l = 2 cm; the initial concentration of chlorobenzen, $[CB]_0$ = 11.4 mg/L; pH = 7.

The chlorobenzene removal yield increases with the increase of irradiation time and the number of layers and is slightly better for C-S2-TiO_2 coating.

2.6. Pd-doped TiO_2 Coatings

As it was mentioned in chapter 2.3. the sol-gel method easily allows the preparation of nanocomposite materials such as inorganic matrices in which a metal phase could be highly dispersed. The position of dopants is determined by the size differences between the host Ti^{4+} ionic radius and the ionic radii of the dopants. Pd^{2+} acts as an interstitial dopant. Large disturbance of the potential energy results in the creation of localized positive charge around Ti and/or formation of oxygen vacancy which enhance the electron trapping efficiency [160]. Concerning Pd doping, generally, its dispersion by sol-gel method was made in SiO_2 [247,

248], Al_2O_3 [249, 250] and in vitreous matrices [251]. There are few references so far regarding the use of Pd^{2+} as a dopant in the case of a sol-gel prepared TiO_2 matrix [165-168].

2.6.1. Samples Preparation

The corresponding sol-gel Pd-doped TiO_2 coatings, C-Pd-TiO_2, have been obtained in situ by cogelation of the precursors, under similar conditions to un-doped and S- and Ag-doped TiO_2 samples (see. paragraphs 2.4.1. and 2.5.1.). The source of Pd was Pd acetylacetonate, $Pd(acac)_2$, $C_{10}H_{14}O_4Pd$, which was introduced in the reaction mixtures dissolved in the corresponding alcohol. Two Pd-TiO_2 samples containing 0.5 and 1 wt. % Pd related to TiO_2 were prepared. The conditions used for thermal processing and the characterization of the films were similar with those for S- and Ag-doped coatings.

2.6.2. Results

Structural evolution (XRD results) with temperature of the Pd-doped TiO_2 gels compared with the pure TiO_2 gel is presented in Figure 9 and Table 9.

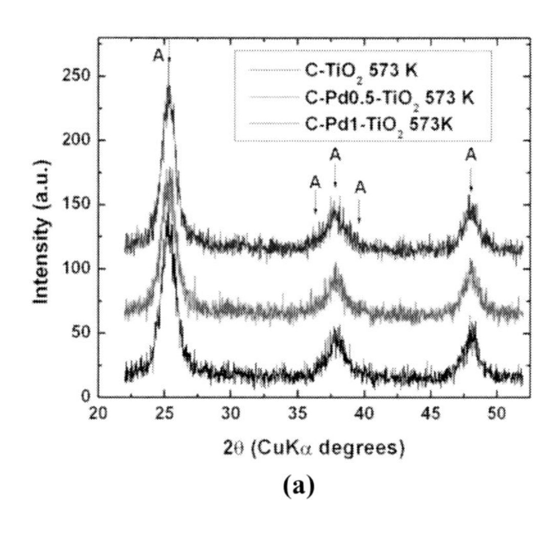

(a)

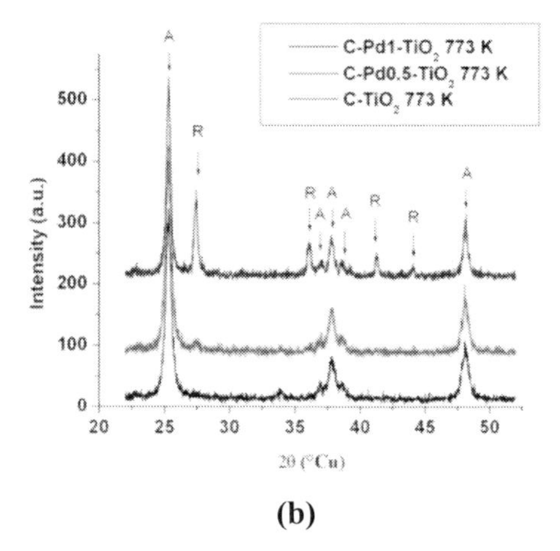

(b)

Figure 9. XRD patterns of C-TiO_2, C-Pd0.5-TiO_2 and C-Pd1-TiO_2 samples thermally treated at 573 (a) and 773 K (b).

Table 9. Calculated values of microstructural factors obtained from computerized analysis of XRD spectra of un-doped and Pd-doped TiO_2 samples

Sample	T [K]	Identified phases [%]	Lattice constants			Microstructural factors	
			a [Å]	c [Å]	u.c.v. [$Å^3$]	<D> [Å]	10^{+3}x<S> [ad]
C-TiO_2	573	A	3.7954(68)	9.5659(271)	137.79(88)	173(80)	2.1(2.8)
C-Pd0.5- TiO_2		A	3.7940(72)	9.5046(284)	136.81(93)	185(10)	6.6(4)
C-Pd1-TiO_2		A	3.7910(83)	9.5091(326)	136.7(1.1)	169(124)	4.5(4.7)
C-TiO_2	673	A > 95	3.7721(27)	9.4821(97)	134.91(33)	153(11)	2.27(32)
		R < 5	-	-	-	-	-
C-Pd0.5-TiO_2		A	3.7794(22)	9.4614(78)	135.15(27)	126(14)	5.62(69)
C-Pd1-TiO_2		A	3.7700(45)	9.4603(159)	134.46(55)	161(17)	3.39(54)
C-TiO_2	773	A – 73.96	3.7801(3)	9.5111(12)	135.90(4)	442(82)	1.1(4)
		R – 26.04	4.5914(12)	2.9564(16)	62.32(7)	405(246)	1.2(1.4)
C-Pd0.5-TiO_2		A – 93.98	3.7783(32)	9.5030(127)	135.66(41)	307(68)	1.9(5)
		R – 6.02	4.7309(504)	2.9143(533)	65.2(2.6)	295(48)	1.6(5)
C-Pd1-TiO_2		A	3.7819(10)	9.5279(38)	136.28(12)	348(144)	1.8(9)

A: Anatase; R: Rutile; a, c: lattice parameters; u.c.v.: unit cell volume; D: crystallite size; S: internal strain

XRD results evidence the presence of anatase as single crystalline phase in all samples thermally treated at 573 K, respectively un-doped and Pd-doped TiO_2, no matter of the dopant concentration. At 773 K beside anatase, rutile phase could be observed, its presence depending strongly on the dopant concentration. With the Pd concentration increase, the rutile crystallinity decreases, vanishing at 1 % Pd.

The profile analysis of the XRD peaks shows that, for 0.5 % Pd concentration in the material annealed at 773 K, the anatase crystallite size has the smallest value (307 Å), and the local strain is the highest (1.9×10^{-3}). Our calculations show a slight decrease (~5 %) of the interplanar spacing along the [200] and [004] crystallographic axes. This involves both a local densification of the anatase lattice and a decrease of the rutile cristallinity compared to the un-doped sample. The further increase of Pd concentration to 1 % leads to a slight relaxation of the lattice and, consequently to the increase of the crystallites size.

Figure 10 (a-c) presents AFM images of pure and Pd-doped TiO_2 coatings thermally treated at 773 K [167].

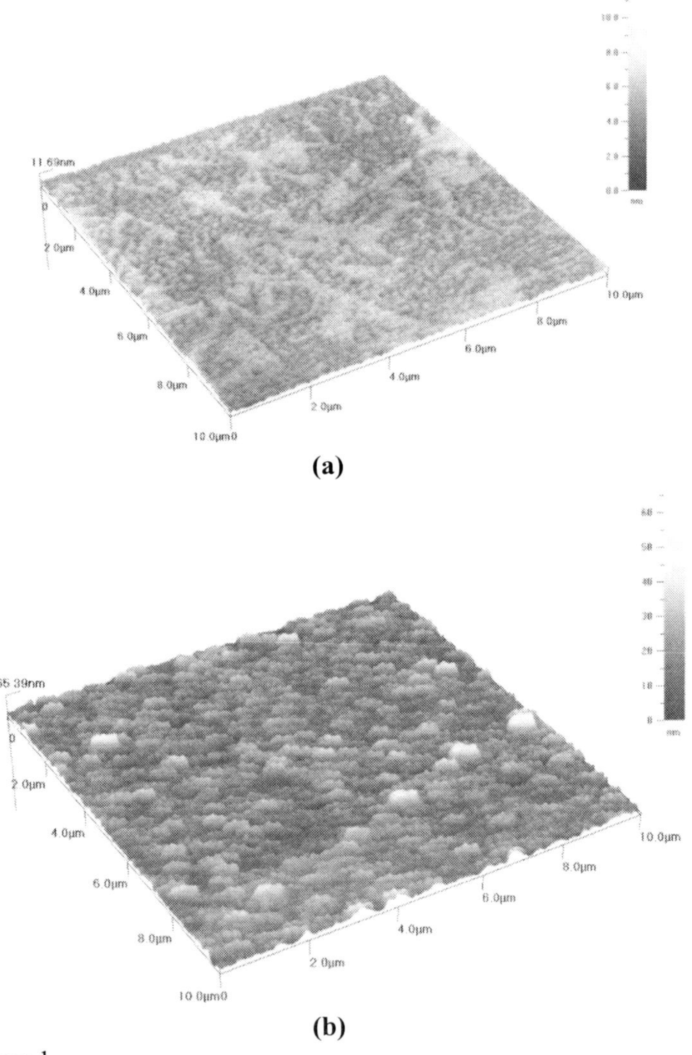

Figure 10. Continued.

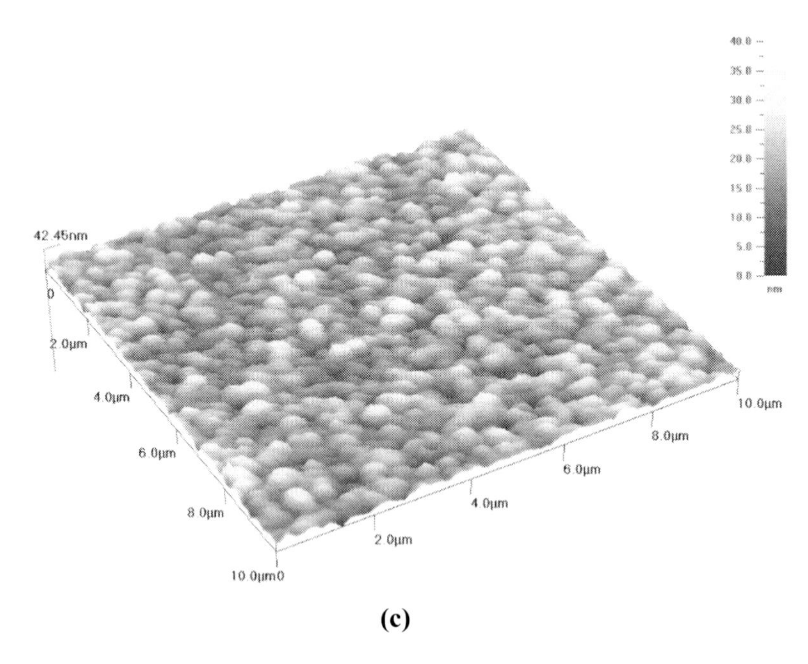

(c)

Figure 10. AFM images of C-TiO_2 (a), C-Pd0.5-TiO_2 (b) and C-Pd1-TiO_2 (c) coatings thermally treated at 773 K, measured over an 10x10 µm area; Root-Mean-Square (RMS) Deviation: 1.18, 6.40 and 5.29 nm, respectively.

In spite of different mechanisms, the induced photo-catalytic activity and hydrophilicity of the surface are two phenomena strongly inter-related [113]. Therefore, CA measurements (contact angle between the deionized water and the film surface) are often used as an (indirect) indication of the photocatalytic activity of the thin films. The hydrophylicity results are presented in Figure 11 [167].

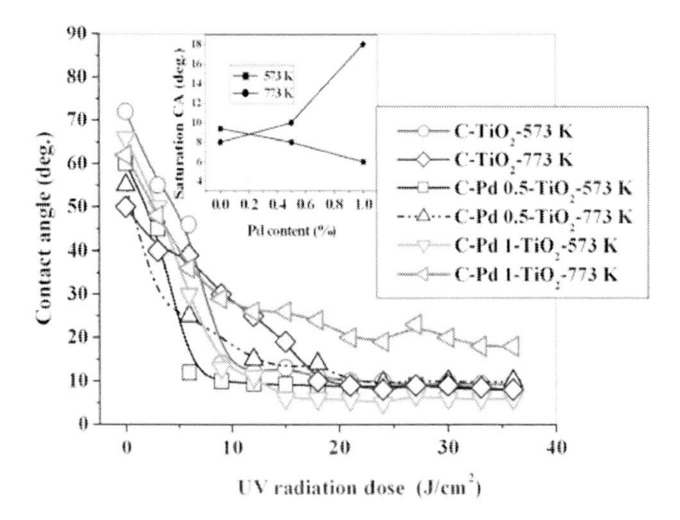

Figure 11. Contact angle versus UV irradiation dose dependences of the un-doped and Pd-doped TiO_2 bi-layers films, thermally treated at 573 and 773 K. An inset was added, where the saturation values of the CA were plotted with respect to Pd content (0-1%), for the thermally treated films.

All the films present super-hydrophilicity properties which were connected with their structure, composition, and surface morphology. The thermal treatment of the samples at 573 K leads to an increased super-hydrophilic performance, especially for the sample with the highest Pd concentration. At 775 K, the annealing of the films results in a marked degradation of the hydrophylicity of the surface. The slower decrease of the CA values versus UV irradiation dose and its high final value (18 deg at 36 J/cm^2) for 1 % Pd concentration can be explained by a slight relaxation of the anatase lattice and consequently by the increase of the crystallites size [167].

2.7. Doped TiO_2 Coatings Obtained from Nanopowders

In order to increase the thickness of the coatings many depositions are necessary, but these induce tensile stress by shrinkage of the layers during drying and thermal processing. This problem can be resolved either by increasing the viscosity of the starting solution, or by embedding in a matrix a secondary phase (e.g. precalcinated powders, fibers, etc) as a filler, to reinforce the resulting composite structure [252]. The physical properties of the matrix can also be used to amend the catalytic process [253]. The processing of sol-gel derived titanium dioxide composites coatings is pointed out in the literature [168, 254-256].

2.7.1. Ag-doped TiO_2 Coatings Obtained from Nanopowders

2.7.1.1. Samples Preparation

Ag-doped TiO_2 coatings from nanopowders (CN-Ag-TiO_2) were obtained by embedding the 2 wt.% Ag-doped TiO_2 nanopowder in TiO_2 matrix, after the aging of the sol. The powder was prepared by ultrasonically assisted hydrothermal synthesis. The silver source was $AgNO_3$.

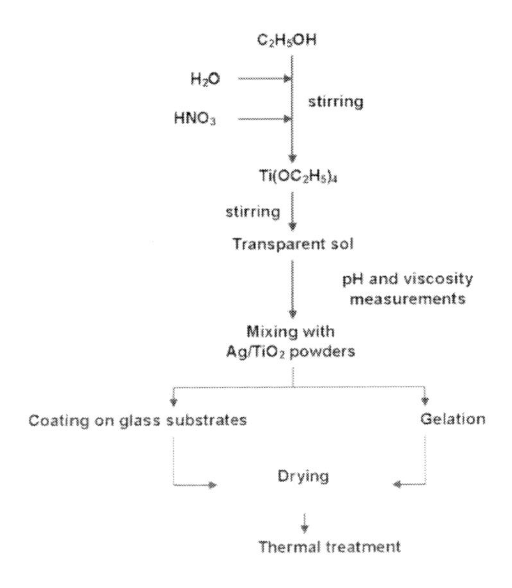

Figure 12. The scheme of preparation of the Ag-doped TiO_2 coatings obtained from nanopowders, CN-Ag-TiO_2.

This matrix was directly obtained from tetraethylorthotitanate $Ti(OC_2H_5)_4$, using absolute ethanol as solvent, water for hydrolysis and nitric acid as catalyst. A very rigorous pH and viscosity control of the TiO_2 sol was necessary in order to establish the most propitious moment for including the nanopowder. The scheme of preparation for the Ag-doped TiO_2 coatings obtained from nanopowders is presented in Figure 12. Single layer coating has been deposited on glass substrates using the dip-coating procedure with a withdrawal rate of 5 cm/min. Continuous and homogeneous coatings, labeled CN-Ag-TiO_2, with good adherence to the support were obtained. The thermal processing of the films was determined from the thermal behavior of the gels obtained from the gelation of the sol used for coatings preparations [177]. Thus, the obtained films have been dried at room temperature and then densified by thermal treatments at 573, 673 and 773 K for 1h, using in all cases the same heating rate of 1K min^{-1}. In order to compare their structural behavior with that of the TiO_2 matrix the same thermal schedules have been applied.

2.7.1.2. Results

The structural evolution (XRD results) of the CN-Ag-TiO_2 gel with temperature is presented in Figure 13 (a-c) and Table 10 and compared to that of the pure TiO_2 matrix (TiO_2-M) and of C-Ag-TiO_2 gel obtained from co-gelation.

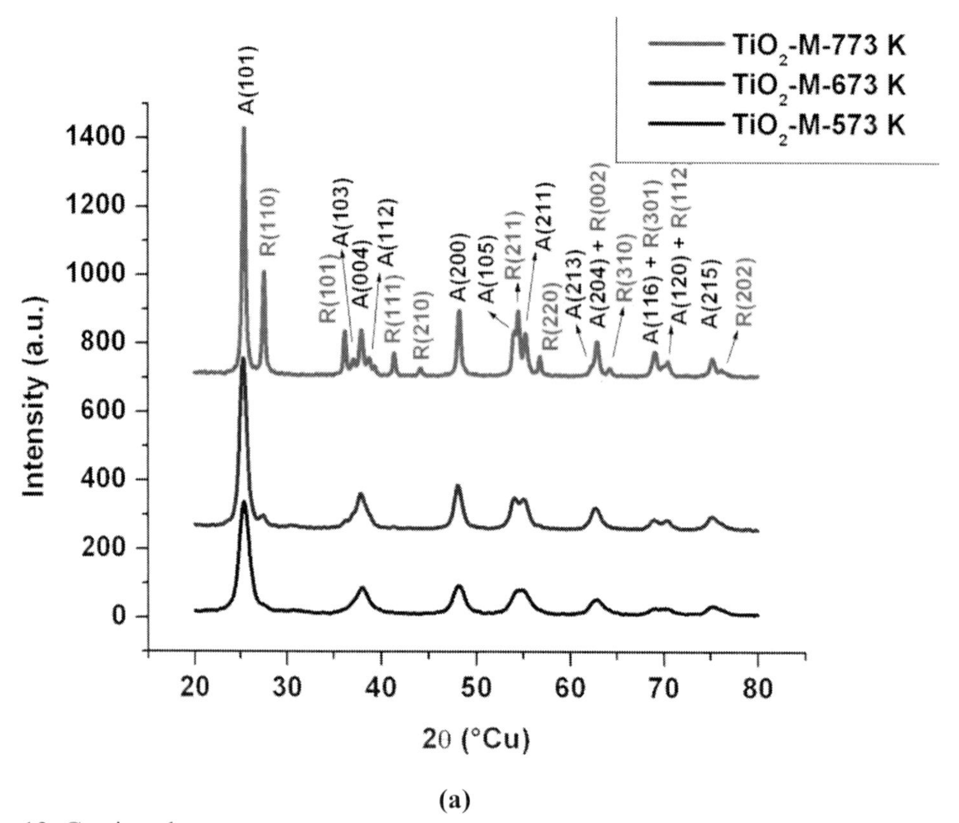

(a)

Figure 13. Continued.

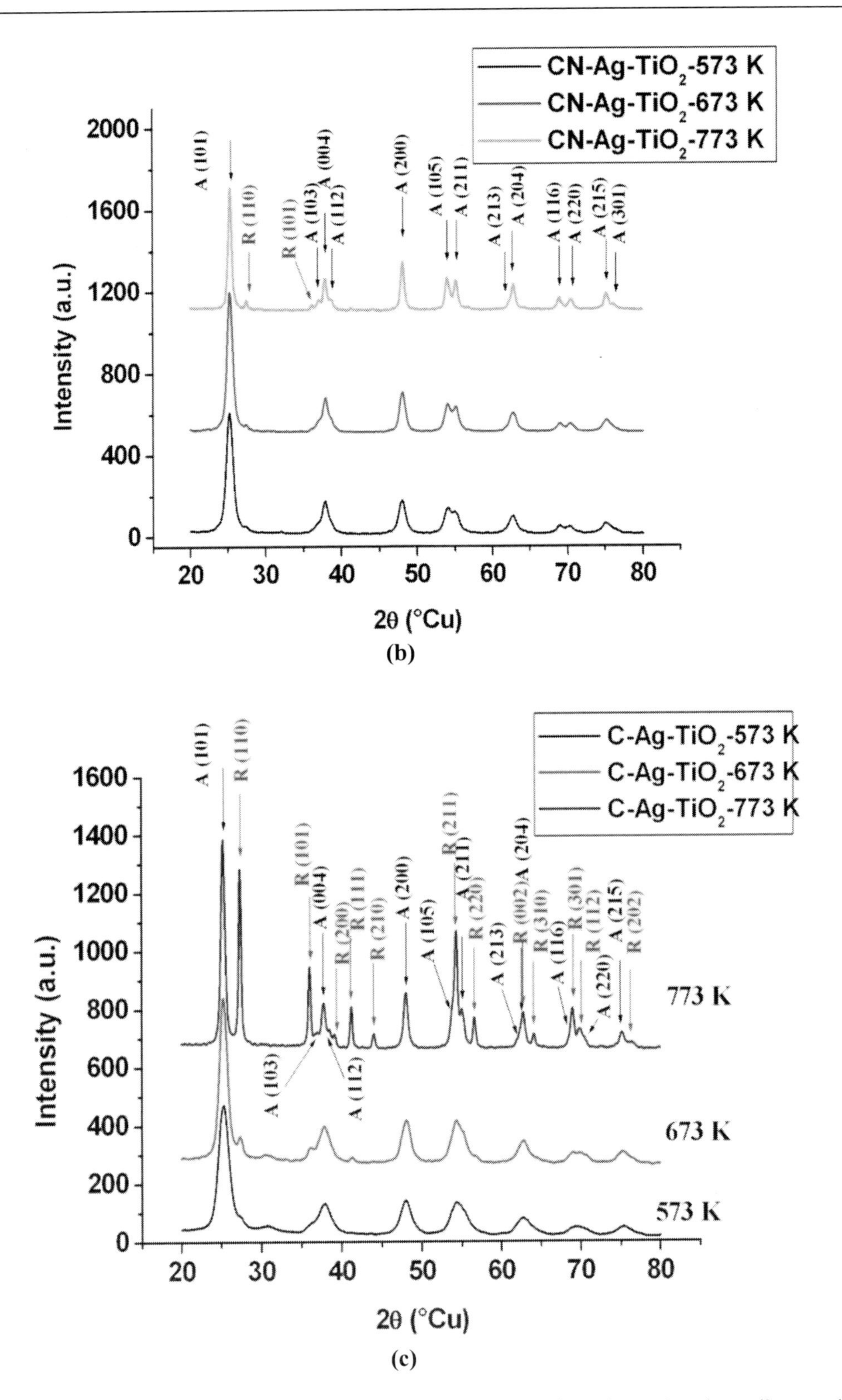

Figure 13. XRD patterns of TiO_2-M (a), CN-Ag-TiO_2 (b) and C-Ag-TiO_2 (c) samples, thermally treated at 573, 673 and 773 K, respectively.

Table 10. Microstructural factors calculated from the computerized profiles analysis of the XRD spectra of TiO_2 matrix and of Ag-doped TiO_2 gels obtained from nanopowders and co-gelation [177].

Sample	Identified phases	Microstructural factors				
		a [Å]	c [Å]	u.c.v. [$Å^3$]	<D> [Å]	10^3 x <S>
TiO_2-M-573 K	A	3.7770(12)	9.4651(42)	134.90(14)	127(25)	3.7(1.1)
TiO_2-M-673 K	A	3.7835(19)	9.4883(67)	135.83(23)	161(26)	1.3(7)
TiO_2-M-773 K	A – 70%	3.7756(16)	9.4740(57)	135.05(19)	443(92)	0.5(4)
	R – 30%	4.5886(38)	2.9561(30)	62.24(17)	685(193)	1.1(4)
CN-Ag-TiO_2-573 K	A	3.7836(17)	9.5069(61)	136.10(21)	178(31)	3.4(1.0)
CN-Ag-TiO_2-673 K	A	3.7717(30)	9.4896(106)	135.00(36)	165(27)	1.9(6)
	R < 3%	-	-	-	-	-
CN-Ag-TiO_2-773 K	A – 90%	3.7731(18)	9.4449(61)	134.46(21)	198(42)	1.8(7)
	R – 10%	4.6048(134)	2.9487(64)	62.52(50)	610(31)	1.5(3)
C-Ag-TiO_2 -573 K	A	3.7838(52)	9.5215(184)	136.32(64)	93(11)	4.0(7)
C-Ag-TiO_2-673 K	A – 89%	3.7788(32)	9.4936(112)	135.56(39)	111(24)	2.7(1.3)
	R – 11%	4.6020(106)	2.9528(69)	62.54(43)	389(128)	8.2(6.8)
C-Ag-TiO_2-773 K	A – 54%	3.7859(22)	9.4875(78)	135.99(27)	267(112)	0.6(5)
	R – 46%	4.5978(47)	2.9602(38)	62.58(21)	619(223)	1.4(6)

A: Anatase; R: Rutile a, c: lattice parameters; u.c.v.: unit cell volume; D: crystallite size; S: internal strain

The Ag-doping favors the transformation of TiO_2 from anatase to rutile, the presence of rutile being detected even at 673 K in both samples (C-Ag-TiO_2, CN-Ag-TiO_2), which does not happen for the S-doped samples. This process is more evident for the sample prepared by co-gelation. In this case the rate of growth of the anatase crystallites dimension is much greater, and is accompanied by a decrease of the tensile strains of the anatase phase with increasing temperature.

The morphological characterization by TEM of the CN-Ag-TiO_2 gel, CN-Ag-TiO_2 coating and TiO_2-M coating are presented in Figures 14-16 [177]. In order to illustrate the transition zone between quasi-amorphous and crystalline structure for CN-Ag-TiO_2 coating treated at 773 K, the SAED (selected area electron diffraction) image is also included between the TEM ones (Figure 15 (c)).

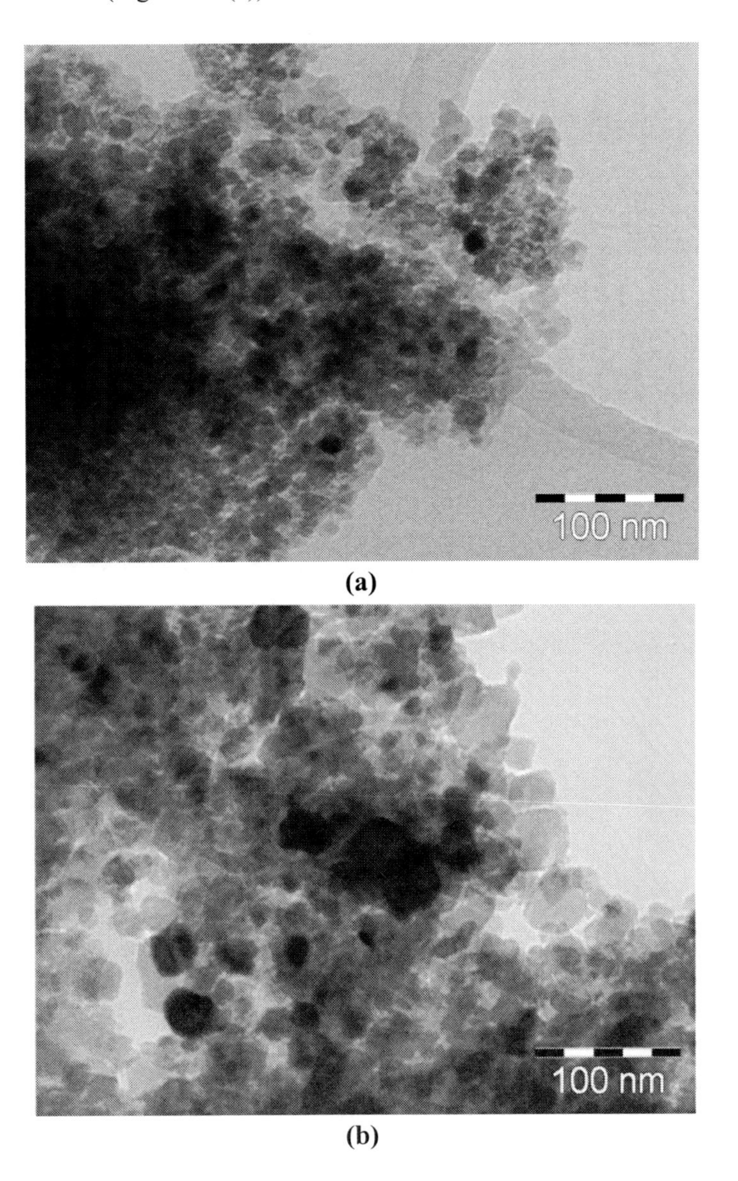

Figure 14. TEM images of CN-Ag-TiO_2 gel thermally treated at 573 K (a) and 773 K (b).

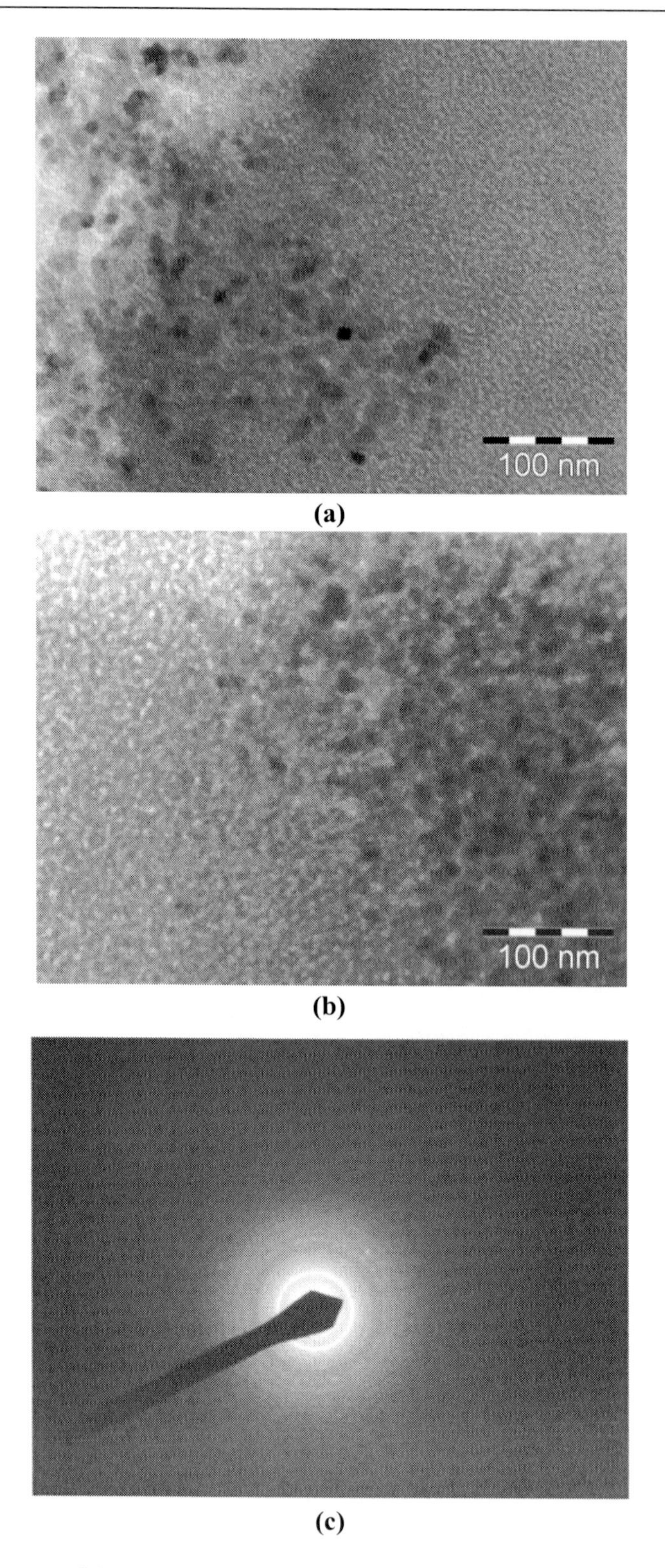

Figure 15. TEM images of CN-Ag-TiO_2 coating thermally treated at 573 K (a) and 773 K (b); SAED image of CN-Ag-TiO_2 coating thermally treated at 773 K (c).

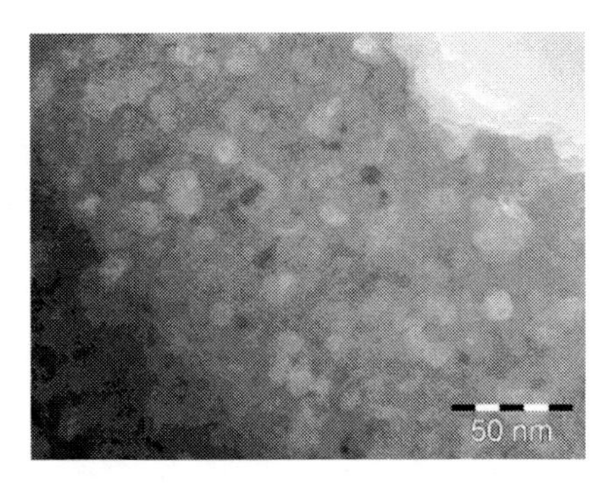

Figure 16. TEM image of C-TiO_2-M coating thermally treated at 573 K.

The un-supported gels CN-Ag-TiO_2 exhibits aggregates of nanocrystallites with the nanocrystallite size in the range 10-30 nm at 573 K (Figure 14 (a)) and 20-50 nm at 773 K (Figure 14 (b)). The CN-Ag-TiO_2 coatings have a mixed structure. There are quasi-amorphous zones with granular structure of a few nanometers, together with crystalline zones with the crystallite size of 5-10 nm at 573 K (Figure 15 (a)) or 10-20 nm at 773 K (Figure 15 (b)). The SAED pattern (Figure 15 (c)) is taken from the area of the transition zone between quasi-amorphous to crystalline structure shown in Figure 15 (b). For comparison, Figure 16 presents the TEM image of the pure TiO_2 film at 573 K, obtained from the sol matrix (M). TiO_2 films are amorphous and their morphology evidences a nanometric porosity. The increase of temperature at 773 K determined the beginning of the crystallization and led to almost a doubling in the sizes of pores.

AFM images of Ag-doped TiO_2 coatings obtained by co-gelation and from nanopowders, measured over a 9.15 x 9.15 μm area, are presented in Figure 17 [177]. The roughness values Root-Mean-Square (RMS) deviation for CN-Ag-TiO_2 coatings were greater (20 nm) than for C-Ag-TiO_2 ones (14 nm). Regardless of method used to prepare the coating (from nanopowder or by co-gelation) the roughness values of the Ag-doped films are greater than those for the S-doped ones.

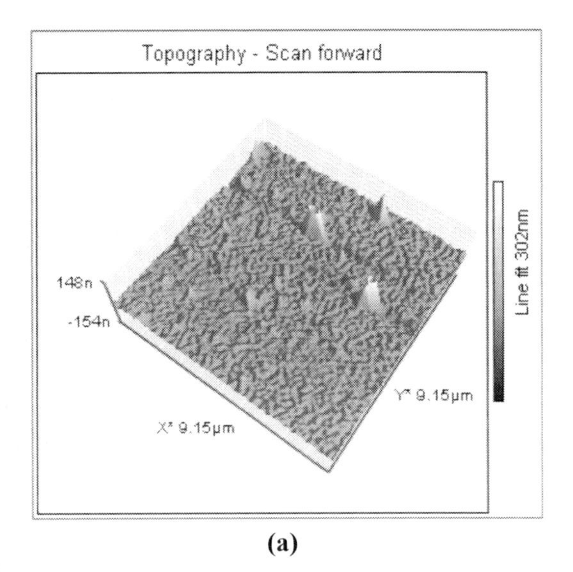

(a)

Figure 17. Continued.

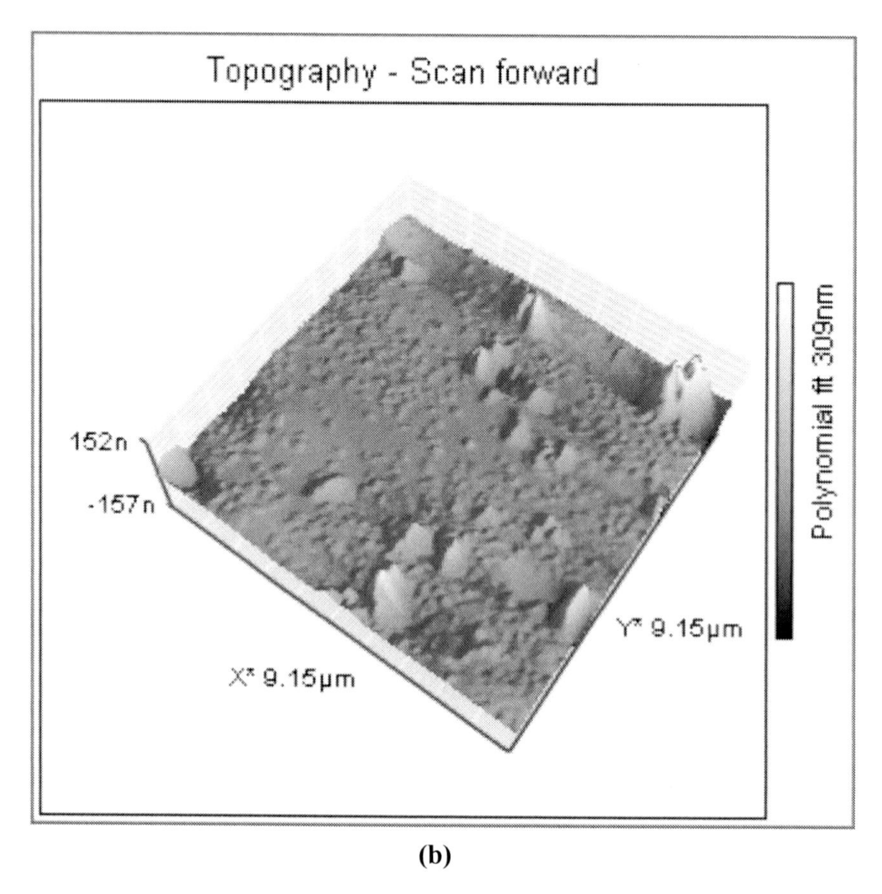

(b)

Figure 17. AFM images of C-Ag-TiO_2 (a) and CN-Ag-TiO_2 (b) coatings, thermally treated at 573 K, measured over an 9.15 x 9.15 µm area; Root-Mean-Square (RMS) deviation: 14.32 and 19.79 nm respectively.

2.7.2. Pd-doped TiO_2 Coatings Obtained From Nanopowders

2.7.2.1. Samples Preparation

Pd-doped TiO_2 coatings from nanopowders (CN-Pd-TiO_2) were obtained by embedding the 1 wt. % Pd-doped TiO_2 nanopowder in TiO_2 matrix, after the aging of the sol. The powder was prepared by sol-gel method, the Pd source was Pd acetylacetonate, $Pd(acac)_2$, $C_{10}H_{14}O_4Pd$. The preparation procedure is the same with the one described for CN-Ag-TiO_2 (paragraph 2.7.1.1.) and the scheme of preparation has been presented in our previous work [168].

2.7.2.2. Results

The structural evolution (XRD results) of the CN-Pd-TiO_2 gel with temperature is presented in Table 11 and compared to that of the pure TiO_2 matrix (TiO_2-M).

The presence of Pd hinders anatase crystallization: at 573 K the sample is amorphous and at 773 K palladium decreases the crystallite size of anatase and increases the internal strains of the material, confirming the fact that Pd is inside of the TiO_2 network.

Table 11. Microstructural factors calculated from the computerized profile analysis of the XRD spectra of TiO_2 matrix and of Pd-doped TiO_2 gels obtained from nanopowders.

Sample	Identified phases	Microstructural factors				
		a [Å]	c [Å]	V [Å^3]	D [Å]	10^{+3} x S
TiO_2-M-573 K	Anatase	3.7770(12)	9.4561(42)	134.90(14)	135(11)	3.61(45)
TiO_2-M-773 K	Anatase	3.7756(16)	9.4740(57)	135.05(19)	453(48)	0.57(19)
	Rutile	4.5886(38)	2.9561(30)	62.24(17)	685(148)	1.10(32)
CN-Pd-TiO_2-573 K	Amorphous	-	-	-	-	-
CN-Pd-TiO_2-773 K	Anatase	3.7725(17)	9.4533(62)	134.54(21)	351(8)	0.84(4)

a, c - lattice parameters; V - unit cell volume; D - crystallite size, S - internal strain

The TEM micrographs corresponding to the CN-Pd-TiO_2 coatings thermally treated at 673 and 773 K are presented in Figure 18 (a, b) [168].

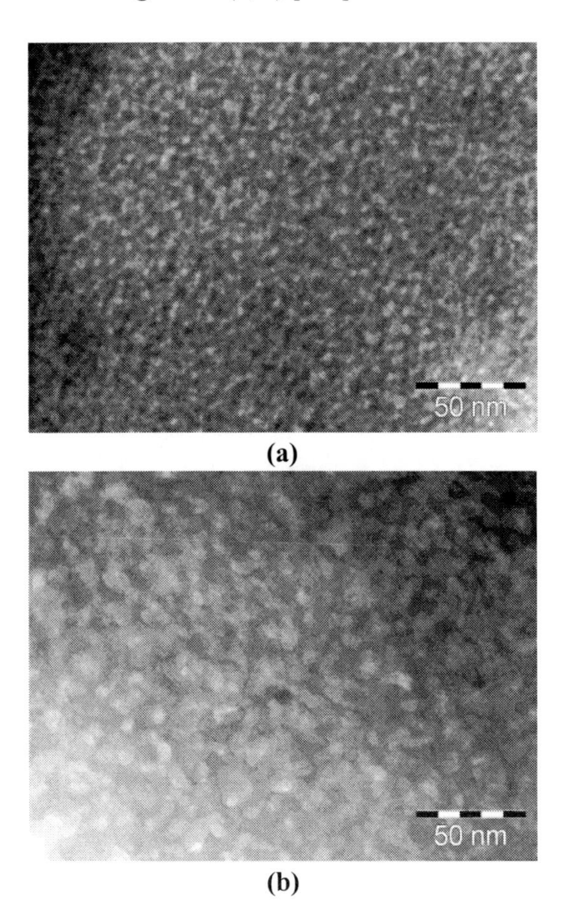

Figure 18. TEM image of CN-Pd-TiO_2 coating thermally treated at: 673 K (a) and 773 K (b).

A beginning of crystallization can be observed at 773 K, compared to 673 K, when the film is amorphous.

AFM images of the C-TiO_2-M and CN-Pd-TiO_2 coatings, thermally treated at 773 K are presented in Figure 19 (a, b).

The un-doped TiO_2 films present the greatest value of RMS.

The hidrofilicity test is presented in Figure 20 [168].

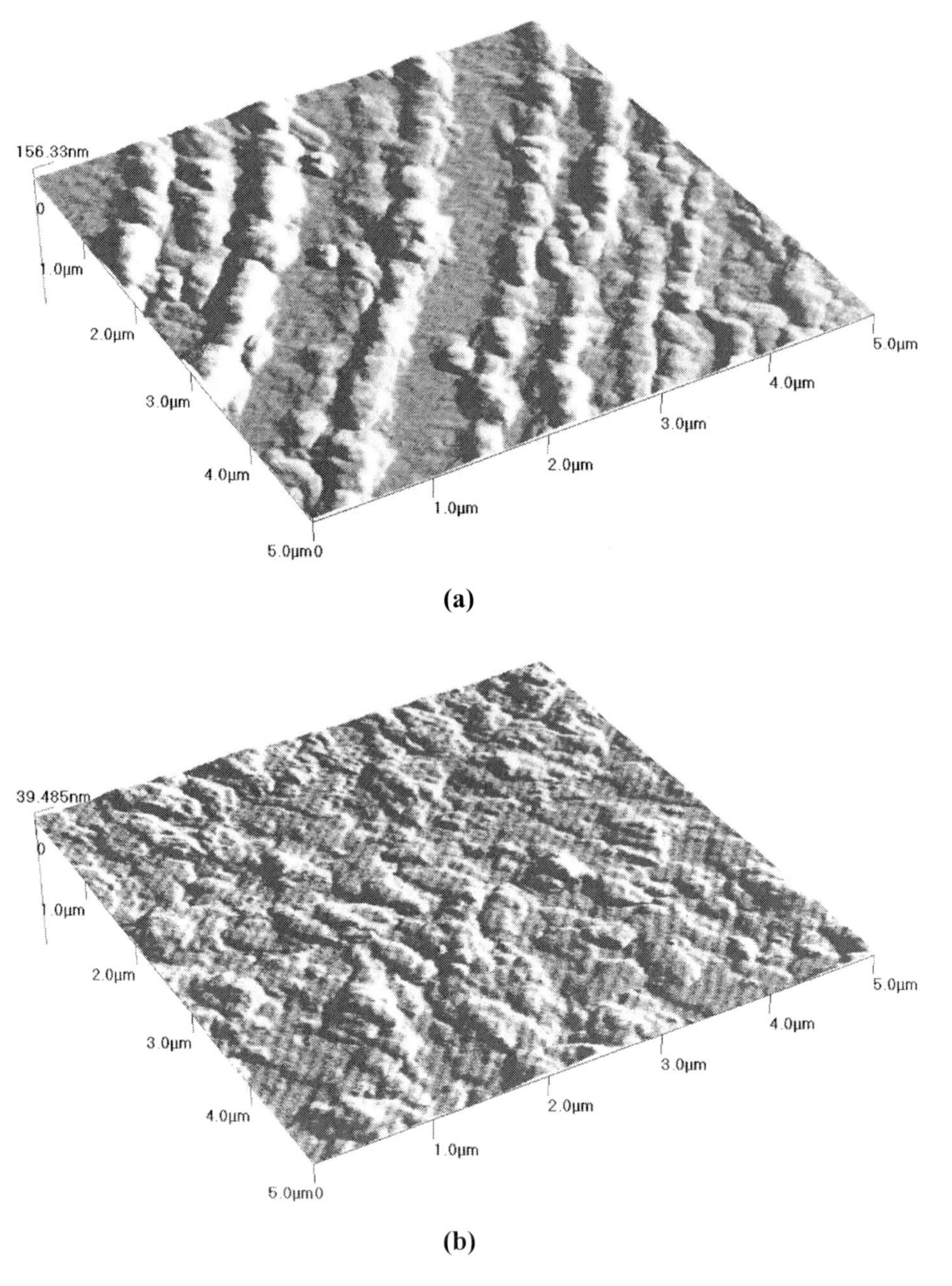

Figure 19. AFM images of C-TiO_2-M and CN-Pd-TiO_2 coatings thermally treated at 773 K, measured over an 5x5 µm area; Root-Mean-Square (RMS) deviation (roughness): 19.87 and 4.97 nm, respectively.

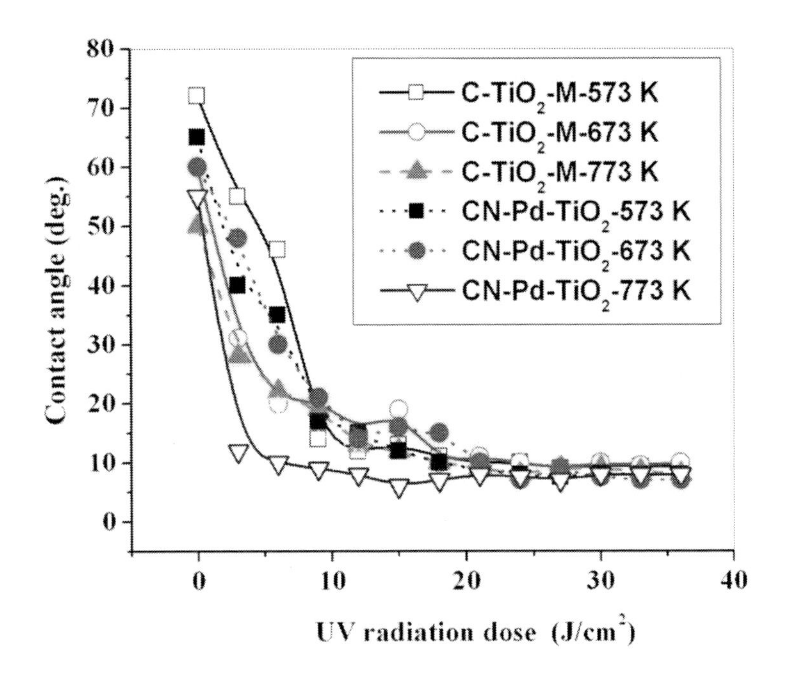

Figure 20. Contact angle as a function of the UV irradiation dose for $C\text{-}TiO_2\text{-}M$ and $CN\text{-}Pd\text{-}TiO_2$ coatings.

All the studied films become super-hydrophilic by UV irradiation. It can be observed a sharp decrease of the contact angle towards values lower than 10 degrees, for UV doses ranging between 5 and 15 Jcm^{-2} [168].

2.8. Sol-Gel Oxide Powders

The oxide powders could be obtained by numerous synthesis methods, as the classical ceramic method, the ceramic with additives and the co-precipitation. Between them, the sol-gel process occupies a very important place.

The potential advantages of the sol-gel powders, compared to the conventional ones, are: the controlled shape and size (usually, mono-disperse), the homogeneity at the molecular scale, the increased purity and reactivity (which leads to lower processing temperatures) and the nanometer size of the particles. All these advantages are very significant from point of view of applications and must surpass the drawbacks of the high costs, long processing time and low yield characteristic to the sol-gel method.

The importance of the sol-gel powders results from their numerous practical applications. They serve as components for the high technique polycrystalline ceramics, pigments, catalysts, abrasive materials, chromatographic supports and constituents of electro-optic and magnetic devices. The dependence relationship between the properties of the molecular precursors used in the sol-gel synthesis and the characteristics of the final macromolecular obtained materials allow the extension of new expectations regarding the applications.

In the case of the sol-gel powders, the sol-gel process refers to the processing into solution of some solid materials which do not deposit, so they are not precipitates. They can be sols (composed from discrete units which remain dispersed in the liquid) or gels (make up from solid tridimensional networks spread into the whole liquid matrix). The result of the succession of reactions starting from an initial chemical compound containing a cation ("M") which is called "precursor" has already been presented in Sections 1.3.1. and 1.4.5. It supposes the following transformations of the precursors in an aqueous medium: hydrolysis → polymerization → nucleation → growth. In the case of un-ionized precursors, as alkoxides, the hydrolysis and polymerization-condensation reactions lead to some linear polymeric structures containing M-O-M bonds, favored by a slow hydrolysis compared to the polymerization.

No matter of the type of the precursor, big and complex poly-nuclear species are progressively formed in the solution, which contain the cation "M" and which lead to the building of some discrete, solid units, named primary particles. The simplest theories regarding their formation are based on their size, no matter of the internal structure. One can distinguish two steps: (1) the nucleation, which consists in the statistical and unstable association between the poly-nuclear species present in the solution in order to form nuclei and (2) the growth, which corresponds to a certain stable dimension for the growth of the nuclei [11].

The nucleation can be heterogeneous, if it develops in the presence of some foreign inclusions such as the dust or some un-wanted previous hydrolysis products of precursors which make difficult to obtain some well defined dimensions of the particles. The ideal case corresponds to a homogeneous nucleation in the solution, which can be obtained by monitoring the temperature and depending on the nature and concentration of the ligands. The type of ions also very much influence the obtained results.

The lack of a very careful pursuit of the process allow to the particles which initially nucleates to have enough time to grow in the same time with the formation of new nuclei. Thus, particles in a large domain of dimensions can be synthesized. The control of process parameters can lead to mono-disperse sol-gel powders that is with a very narrow distribution of the mean size of particles. The typical standard deviation of the size distribution is of about 15% [257].

The easiest method to prepare mono-disperse powders is the forced hydrolysis. It offers enough time to nucleation to proceed at the initial cation concentration. So, the total number of the corresponding complexes which can be formed is finite. Their growth begins only when their concentration decreases below the necessary minimum for the nucleation.

The precursors are often metallic salts containing anions which favor the polymerization. In the case of alkoxides, the concentration of precursors and the quantity of necessary water for hydrolysis can be controlled by dissolution in a solvent, which usually is the parental alcohol of the precursor. The preparation of mono-disperse particles can be realized for a great number of oxide materials.

The possibility to control both nucleation and growth of the particles during the preparation process allows the preparation of special powders with a much more complex distribution of dimensions. For example, bi-disperse powders could be obtained, by adding an excess of anions at different times, which lead to the accomplishment of two nucleation steps. They are composed from mixtures of two monodisperse powders, having different mean radii. In the case of powders containing more than two metals, the process is more complex, so much difficult. The ideal solution for such situations supposes the use of a double alkoxide, which contains both metals in an only one molecule. In the lack of mixed alkoxides it can be used a mixture of alkoxides, in the circumstances of a slow hydrolysis which can favor the formation of metaloxan bonds M-O-M' between the two metals (M and M'). This technique was successfully applied for many binary and ternary compositions.

The most studied sol-gel oxide powders belong to the silica and titania systems, respectively. Due to their scientific and technological importance, their study was extended from mono- to bi- and poly-component systems, obtained by doping.

2.8.1. Sol-gel TiO_2 Powders

The sol-gel prepared titania can be obtained both as amorphous gel and crystalline material (anatase, rutile or brookite). The temperature corresponding to the anatase transformation into rutile is dependent on many factors: the coordination of the metal in the precursor, the length of the metal-oxygen (M-O) bond in the precursor gel, the method of preparation, the presence or the absence of the impurities as well as the texture and the size of the primary particles of anatase [67].

The sol-gel process can provide submicron, monodisperse TiO_2 powders, no matter of the used precursor (inorganic or organic). The inorganic method supposes the thermohydrolysis of titanile sulfate ($TiOSO_4$) or of titanium chloride ($TiCl_4$) in the presence of sodium sulfate (Na_2SO_4) [258, 259]. In both cases, monodisperse spheres with diameters around 0.4 nm are obtained. Starting from $TiOSO_4$ in sulfuric acid medium, titania powders have been obtained deposed on monodisperse silica particles, obtained by the Stőber procedure (the hydrolysis of tetraethylorthotitanate in alcoholic medium in the presence of water and ammonia). They have been applied as white pigments in the bleaching of the paper [260]. In order to avoid the particle agglomeration, 0.05 wt. % of polyvinylpyrrolidone and hydroxypropylcellulose must be added.

The hydrolysis of alkoxides (the organic method) represents the most common way for the preparation of TiO_2 particles. The chemical reactions can be schematically presented as follows:

$$Ti(OR)_4 + xH_2O \rightarrow Ti(OH)_{4-x}OH_x + xROH \quad \text{(hydrolysis)} \tag{15}$$

$$Ti(OH)_{4-x}OH_x + Ti(OR)_4 \rightarrow (OR)_{4-x}TiO_xTi(OR)_{4-x} + xROH \quad \text{(condensation)} \tag{16}$$

where R is an organic radical, such as: ethyl, propyl, i-propyl, n-butyl, etc.

As the hydrolysis and the polycondensation take place in the same time, being in competition, there is possible to change in some manner their relative rate. There are many factors which influence the process. The most studied are: the alkoxy groups belonging to the alkoxides, the reactants concentration, the solvent, the quantity of water used for hydrolysis, the pH of the solution, the temperature of hydrolysis, the nature of catalyst, and the presence of additives.

The methods which start from alkoxides comprise two routes:

1) the controlled precipitation of titanium alkoxides in the parent alcohols in excess of water [26, 63, 99, 100, 102, 103, 261-263]. The hydrolysis-polycondensation of Ti alkoxides in alcoholic medium have been studied, both for the classical sol-gel

process, and for the modified one, in which hydroxypropylcellulose [261], chelating agents [264], or reverse micellia [262, 265] have been included.

2) the hydrolysis of $Ti(O\text{-}C_2H_5)_4$ or $Ti(O\text{-}iC_3H_7)_4$ aerosols, which lead to monodisperse spheres with diameters ranging from 0.06 to 0.6 μm.

All powders can be sintered in order to obtain dense ceramic masses, the corresponding temperature being dependent on the physical properties of the powder.

Table 12 summarizes the characteristics of some TiO_2 powders prepared by hydrolysis of different alkoxides, together with the sintering temperatures and the corresponding densifications. The last row from the table refers to a commercial titania powder sintered through the classical method. It can be observed that the sol-gel obtained powders can reach comparable densifications as the commercial one but at much lower temperatures, which represents the potential and advantages of wet chemical routes in the preparation of the oxides.

Table 12. TiO_2 powders characterization.

Organic radical*	Added element [%]	Specific surface area [m^2/g]	Diameter of particle [μm]	Crystalline phase [%]		Sintering temperature/ Densification [°C/%]
				Anatase	Rutile	
Et	-	10	0.150	43	57	-
Et	Ta (0.25 %)	82	0.019	-	100	940/-
i-Pr	-	52	0.030	3	97	-
Bu	-	24	0.064	7	93	900/99
Bu	Al (0.5 %)	44	0.360	-	100	850/97
Fischer		7.8	0.200	-	100	1230/96

*Et – ethoxide; i-Pr – isopropoxide; Bu – butoxide.

The morphology of the TiO_2 powders strongly depends on the nature of the alkoxy group. Titanium alkoxides with primary alkoxy groups are trimeric species with a pentacoordinated metal atom while those with secondary or tertiary alkoxy groups are monomers with titanium in a tetrahedral environment. Spherical, monodisperse titania particles can be obtained by controlled hydrolysis in diluted ethylic alcohol of tetraethylorthotitanate $Ti(O\text{-}C_2H_5)_4$ [55]. The use of tetraisopropylorthotitanate $Ti(O\text{-}iC_3H_7)_4$ leads to irregular, polydisperse particles [266]. The differences between the molecular structures of these alkoxides lead to different hydrolysis and condensation behaviors which explain the phenomena. The monomer, tetracoordinated $Ti(O\text{-}iC_3H_7)_4$ species is very reactive. The hydrolysis and condensation take place simultaneously, thus the precipitation leads to small, polydisperse particles (~ 0.4 μm). The hydrolysis of $Ti(O\text{-}C_2H_5)_4$ accomplished in similar conditions leads to bigger monodisperse particles (~ 0.7 μm).

Generally, the formation of the monodisperse particles is favored when the nucleation and growth stages can be separated. Olygomerization has proved to be a good solution for this separation. So, in order to obtain monodisperse oxide powders, the use of olygomeric alkoxides with small OR groups is preferable to monomer species with branched OR groups [267].

TiO_2 powders present, as well as the films, photocatalytic properties [103, 268, 269]. They were examined by the quantitative analysis (gas chromatography) of the chemical species resulted from the photosensitive oxidation of ethylic alcohol to acetaldehyde and acetic acid. The photocatalytic activity depends on the crystallite size and the microstructure of the powder. Crystallite dimensions belonging to the 20-30 nm range ensure good photocatalytic properties. However, in the case of rutile, even relatively big crystallites do not grant a high photocatalytic activity [104]. In order to regenerate the photocatalyst after photoreaction, fine TiO_2 particles have been anchored on ceramic surfaces consisting from porous alumina [268]. TiO_2 particles precipitated from the solution at pH=8 (in the domain between the isoelectric points of TiO_2 and Al_2O_3) have been homogeneously dispersed on the whole porous alumina ceramic support. This ceramic material can be applied for the purification of a system comprising chemical species harmful for the environment. Many researchers [108, 110, 270-274] recommend the use of titania powders as nanocrystallites having sizes of nanometers or tenths of nanometers. Moreover, regarding the photocatalytic reactions of titania, a difference has been observed between the crystalline phases of TiO_2, anatase proving to be more active than rutile [132, 275]. That is why, the process must be managed so as to delay the anatase to rutile transformation.

Although there are well known the factors which can improve the photocatalytic activity of the titania powders (such as the crystalline structure, the specific surface area, the distribution of the sizes of particles, the density of the hydroxyl groups on the surface), TiO_2 still presents the disadvantage of absorbing a too small portion from the solar spectrum in the UV region. As it was already mentioned in Section 2.3., the doping represents a solution in order to enhance the photocatalytic properties of titania powders.

The gelation process takes place depending on a multitude of reaction parameters (previously mentioned) which influence the final structural properties of the prepared powders, their specific surface areas and morphologies. The next section presents some original results regarding the preparation and characterization of monodisperse TiO_2 powders, using the classical, alkoxide route of the sol-gel process.

2.8.1.1. Un-doped TiO_2 Powders

Three different sol-gel titania powders have been prepared using identical experimental conditions but varying only the type of the alkoxide precursor, considering that its influence on the final physico-chemical characteristics of the powders is the most important. The alkoxides were: tetraethylorthotitanate $Ti(OC_2H_5)_4$ for sample Et-TiO_2, tetraisopropy-lorthotitanate $Ti(O\text{-}iC_3H_7)_4$ for sample iPr-TiO_2, and tetrabutylorthotitanate $Ti(O\text{-}C_4H_9)_4$, respectively, for sample Bu-TiO_2. As solvents and reaction media the corresponding parent alcohols have been used: absolute ethylic alcohol for sample Et-TiO_2, 2-propylic alcohol for sample iPr-TiO_2, and normal butylic alcohol, respectively, for sample Bu-TiO_2. The hydrolysis of Ti alkoxides took place in excess of water and in un-catalyzed conditions. In order to point out exclusively the role of the alkoxide type on the structural, textural, and morphological properties of the titania powders, all the other reaction parameters have been unchanged, such as the following molar ratios: water/alkoxide = 5, solvent/alkoxide = 85, and pH of 4. The reactions took place at the room temperature. Depending on the type of the alkoxide precursor, the reaction times varied between 1 and 6 hours. All samples have been dried at 353 K.

The differences between the molecular structures of the used alkoxides lead to different behaviors both in the case of hydrolysis and condensation reactions, which explain the different structures and morphologies observed for the prepared powders.

No matter of the alkoxide type, the synthesis led to dry and amorphous samples. They have been annealed for 1 h at 573 K with a heating rate of $1 Kmin^{-1}$ in order to remove the organic matter and to allow the investigation of the structure and morphology of the resulted powders. Supplementary thermal treatments (at 673 K and 1273 K with the same heating rate and plateau) have been accomplished in order to study the influence of the crystallization process on the shape and size of particles. The temperatures of the thermal treatments have been chosen based on the thermal analysis behavior study. The phase compositions established by XRD measurements are presented together with the BET surface area results, in Table 13 [234].

Table 13. Structural and textural properties of TiO_2 powders.

Sample	Temperature [K]	Phase composition	Degree of crystallinity* [%]	D_{101} [nm]	BET surface area [m^2g^{-1}]
$Et-TiO_2$	353	amorphous			251
	573	anatase weak crystallized			130
	673	anatase	100	16.8	94
	1273	rutile			1
$iPr-TiO_2$	353	amorphous			344
	573	anatase weak crystallized			141
	673	anatase	97	13.3	115
	1273	rutile			1
$Bu-TiO_2$	353	amorphous			188
	573	amorphous			111
	673	anatase	65	17.2	80
	1273	rutile			1

*The degree of crystallinity was estimated using sample $Et-TiO_2$ as internal reference sample from the sum of integrated intensities (areas under deconvoluted peaks) of the (101), (103), (004), (112), (200), (105) and (211) reflections

As it can be seen, TiO_2 powders present a pronounced crystallization tendency with the temperature. As a consequence, the BET surface area decreases with the increase of temperature. Also, the sizes of particles grow due to the applied thermal treatments, depending on the nature of the used precursor, as follows: isopropylate < ethylate < butylate. Figures 21-23 present the HRTEM and TEM images of the TiO_2 powders thermally treated at 673 K for 1h as they were presented in a previous work [234].

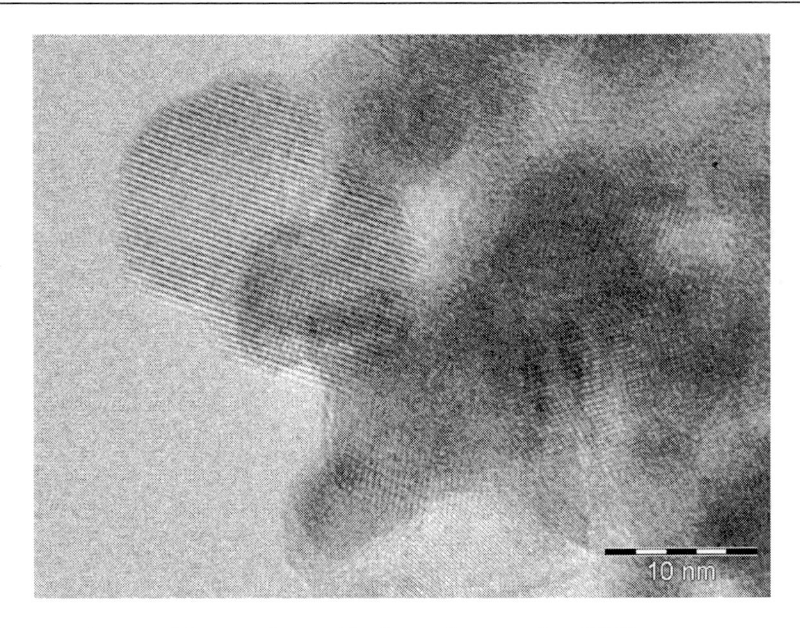

Figure 21. HRTEM image of the sample Et-TiO_2 thermally treated at 673 K.

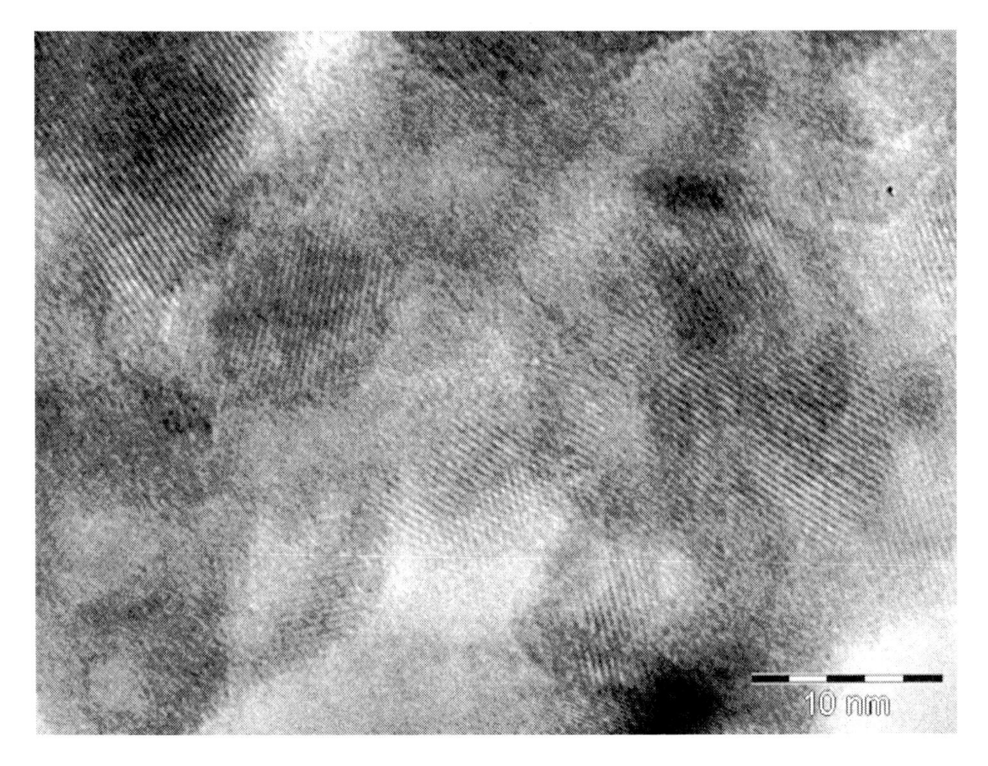

Figure 22. HRTEM image of the sample iPr-TiO_2 thermally treated at 673 K.

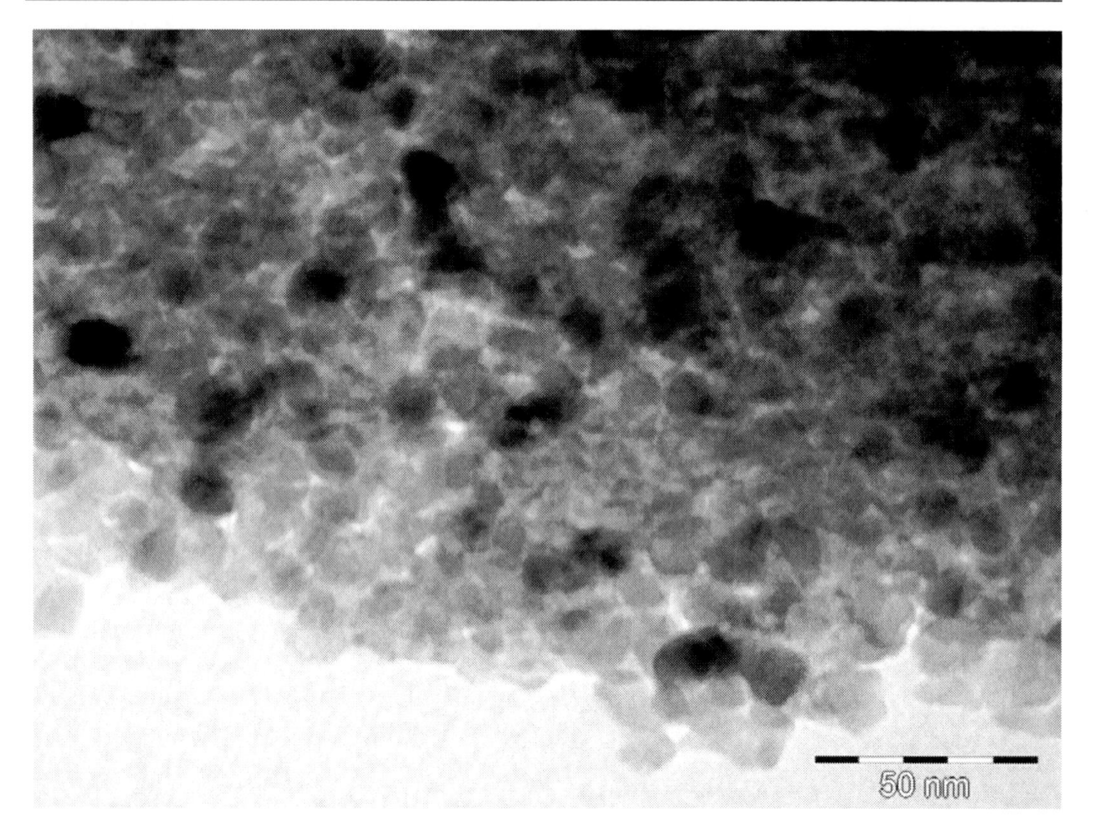

Figure 23. TEM image of the sample Bu-TiO_2 thermally treated at 673 K.

It can be seen that the sizes of the crystallized particles obtained starting from tetraisopropylorthotitanate are in the range of 6-12 nm, those obtained from tetraethylo-rthotitanate have approximately 10-15 nm, and those resulting from tetrabutylorthotitanate corresponds to the 10-20 nm domain. These results were in good agreement both with the XRD data and the specific area measurements, as it can be observed from Table 13. Sample iPr-TiO_2, with the smallest particles, presents the highest specific surface area.

2.8.1.2. S-doped TiO_2 Powders

The benefic effect of dopants addition in the titania powders has already been to the full described and explained in Section 2.3. In order to establish the manner in which S influences the structural and photocatalytic properties of the TiO_2 powders, as well as their morphology, the corresponding sol-gel S-doped TiO_2 powders have been prepared, under similar conditions as samples previously described in Section 2.8.1.1. They were doped with 5 wt. % S related to the TiO_2 content. The sulfur source was thiourea (H_2NCSNH_2), which was introduced in the reaction mixtures dissolved in the corresponding alcohol. The resulted samples have been noticed as S-Et-TiO_2, S- iPr-TiO_2, and S-Bu-TiO_2, respectively.

The phase compositions established by XRD measurements are presented in Table 14 for the S-doped titania powders thermally treated at 673 K for 1h [234].

Table 14. Structural properties of the thermally treated S-doped TiO_2 samples

Sample	Temperature [K]	Phase composition	Degree of crystallinity [%]
S-Et-TiO_2	673	anatase	98
S- iPr-TiO_2	673	anatase	73
S-Bu-TiO_2	673	anatase	56

As it can be seen, the presence of S leads to the formation of one single crystalline phase. Thus, in all cases, only anatase has been detected. The dopant also decreases the degree of crystallinity and inhibits the rate of growing of TiO_2 nanocrystals.

The size distributions and the mean size of particles have been obtained from the statistical processing of the TEM images. The results are presented in Figures 24-26 and in Table 15.

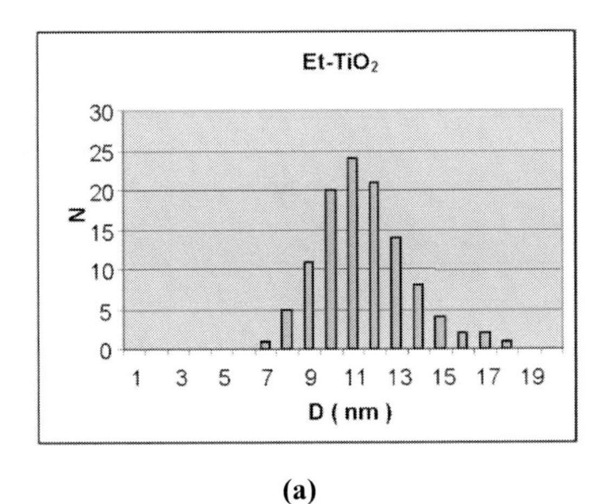

(a)

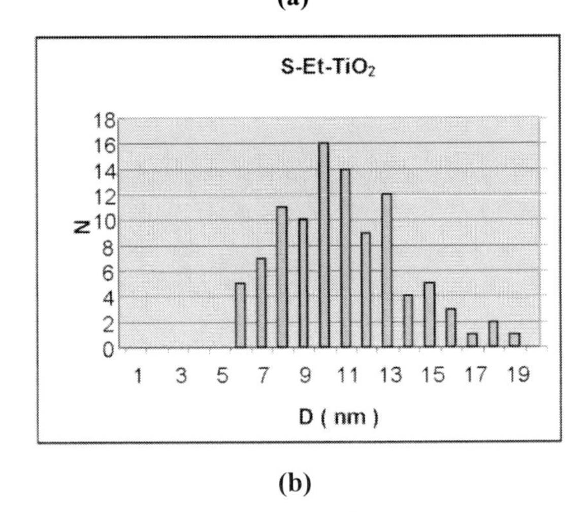

(b)

Figure 24. Size distributions from the statistical processing of the TEM images for samples Et-TiO_2 (a) and S-Et-TiO_2 (b).

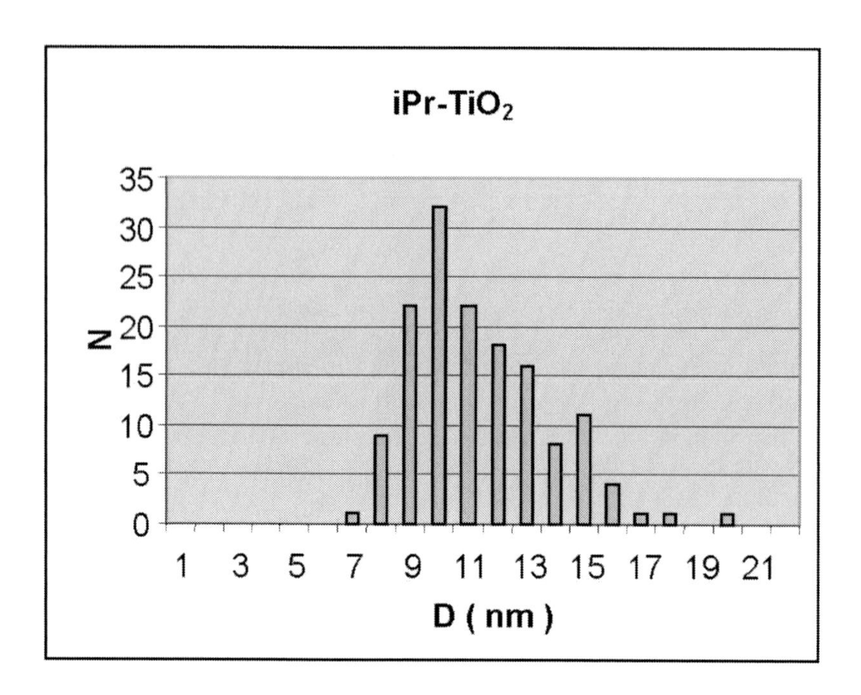

(a)

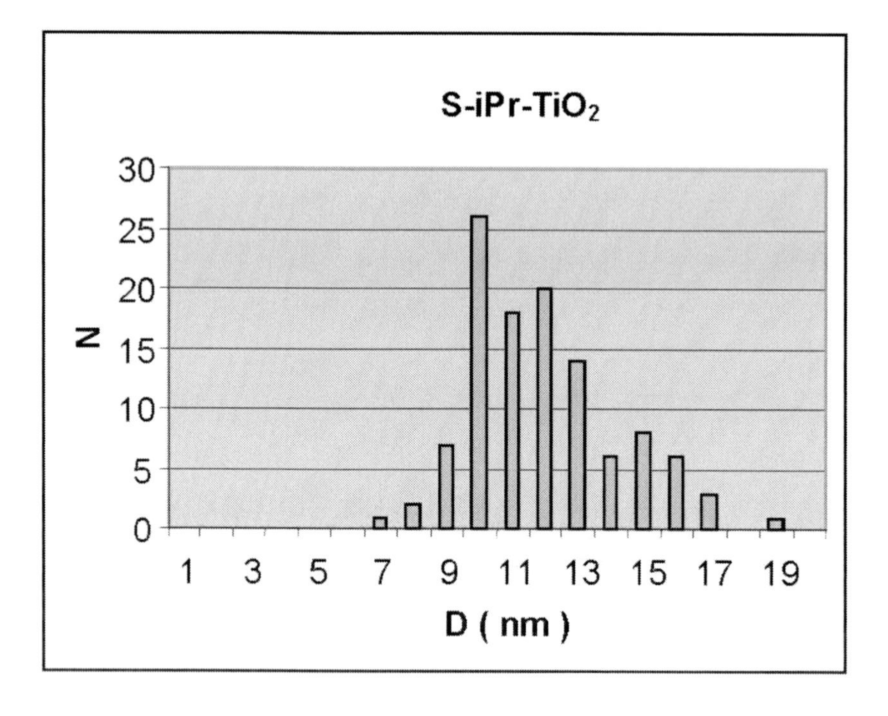

(b)

Figure 25. Size distributions from the statistical processing of the TEM images for sample iPr-TiO_2 (a) and S-iPr-TiO_2 (b).

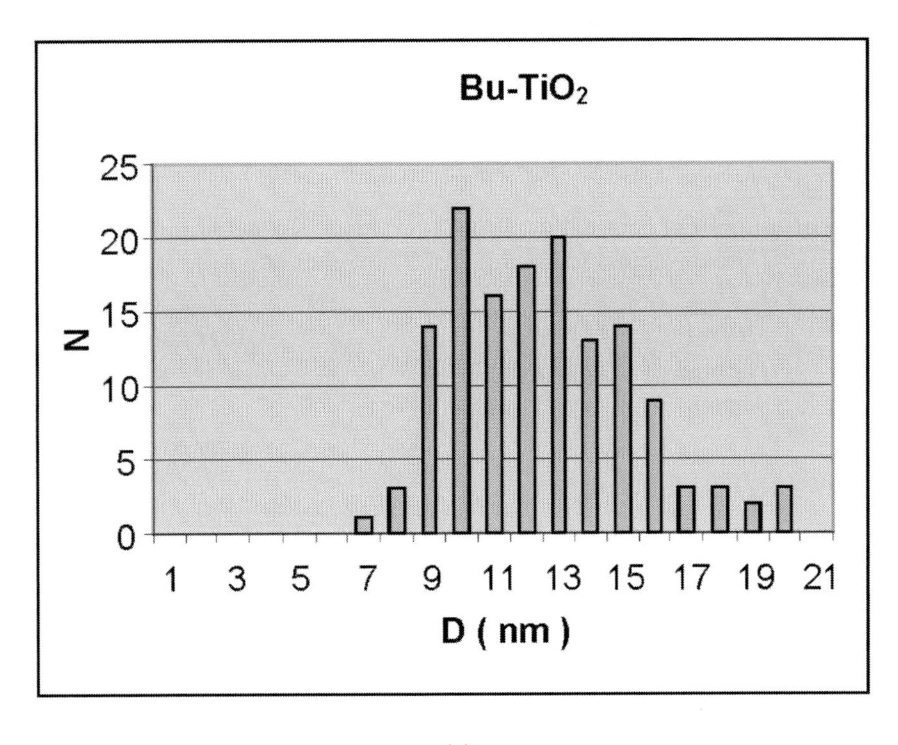

(a)

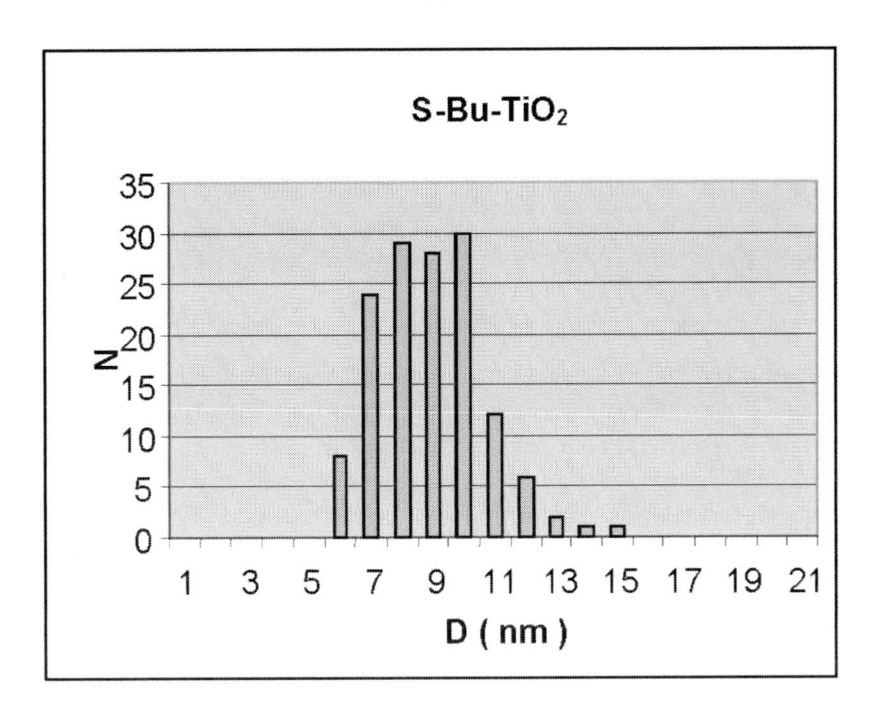

(b)

Figure 26. Size distributions from the statistical processing of the TEM images for sample Bu-TiO_2 (a) and S-Bu-TiO_2 (b).

The differences in size distributions for the three types of Ti-alkoxides and the effect of the presence of sulfur as dopant have been pointed out.

Table 15 contains some details in what the experimental conditions are concerned (as the number of the counted particles) and the obtained mean values of particles, together with some observations for each of the prepared samples.

Table 15. The number of counted particles, their mean diameter and observations for the un-doped and S-doped TiO_2 prepared powders.

Sample	No. of counted particles	Mean diameter of particle, [nm]	Observations
Et-TiO_2	113	11.7	Spherical aggregates of nanocrystallites with sizes in the 300-800 nm range. Seldom lamellar aggregates.
iPr-TiO_2	146	11.4	Un-shaped aggregates. Seldom some spherical aggregates. There is amorphous phase.
Bu-TiO_2	139	12.7	There are both amorphous and crystalline components. Un-shaped nanocrystalline aggregates.
S-Et-TiO_2	100	10.7	Spherical aggregates of nanocrystallites which are joined together. Compact sample.
S-iPr-TiO_2	112	11.3	Un-shaped aggregates but more compact than in the S absence. The presence of sulfur seems to grow the nanocrystallites.
S-Bu-TiO_2	141	8.9	The amorphous phase disappears and the aggregates are more compact than in the S absence.

It can be observed that the mean diameter of particles decreases in the presence of sulfur, no matter what Ti-alcoxide has been used as TiO_2 precursor.

Taking into account that both powders obtained from titanium ethoxide, the un-doped (Et-TiO_2) and the S-doped one (S-Et-TiO_2) basically consist from spherical aggregates this is the pair of samples which was chosen for detailed experiments [177]. Their TEM images are presented in Figure 27.

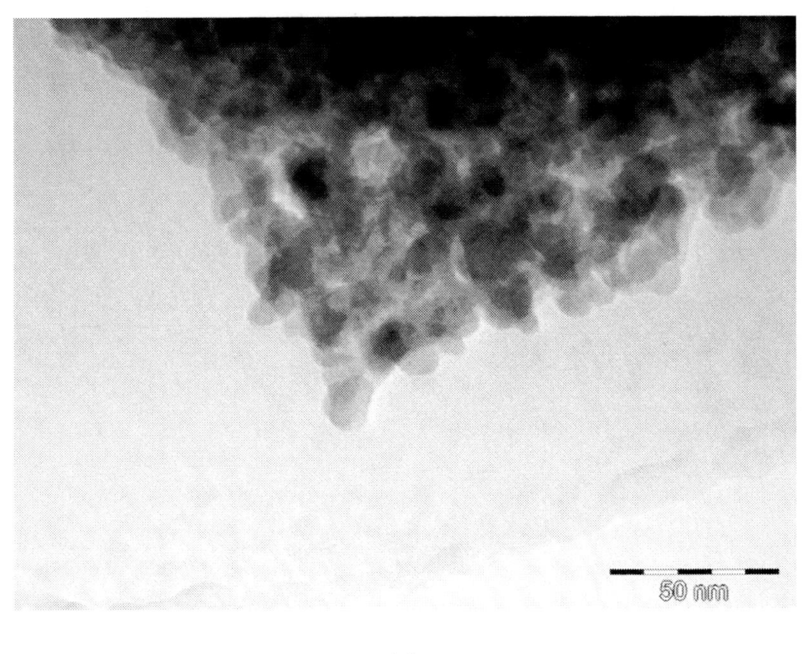

(a)

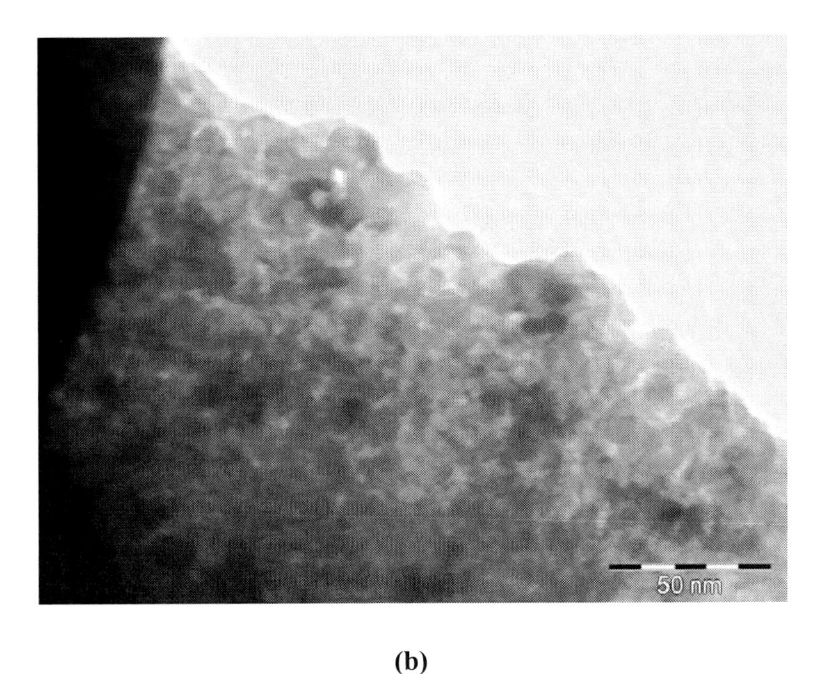

(b)

Figure 27. TEM images of un-doped (Et-TiO_2) powder (a) and of S-doped (S-Et-TiO_2) powder (b).

The structural evolution (XRD) of samples Et-TiO_2 and S-Et-TiO_2 as a function of temperature is presented in Table 16 [177] and in Figure 28. The XRD analysis supplied valuable information regarding the influence of S on TiO_2 crystallization by means of lattice constants and microstructural factors.

Table 16. Calculated values of microstructural factors obtained from computerized analysis from XRD spectra.

Sample/ T [K]	Identified phases	Lattice constants*			Microstructural factors*	
		a [Å]	c [Å]	u.c.v. [$Å^3$]	<D>[Å]	10^{+3}x<S>
Et-TiO_2 - 573	anatase	3.7752(45)	9.4973(164)	135.36(56)	74(7)	10.0(8)
Et-TiO_2 - 673	anatase	3.7815(13)	9.4709(47)	135.43(16)	178(34)	1.4(6)
Et-TiO_2 - 773	anatase	3.7750(19)	9.4828(68)	135.14(23)	346(51)	0.3(3)
S-Et-TiO_2 - 573	anatase	3.7812(18)	9.4778(64)	135.51(22)	172(25)	1.6(5)
S-Et-TiO_2 - 673	anatase	3.7783(16)	9.4735(58)	135.24(20)	247(37)	1.2(4)
S-Et-TiO_2 - 773	anatase	3.7676(55)	9.4561(193)	134.23(66)	302(60)	1.5(4)

*a, c - lattice parameters; u.c.v. - unit cell volume; D - crystallite size; S - internal strain

The phase composition points out the presence of anatase as the single crystalline phase for both samples thermally treated in the range of temperatures between 573 and 773 K.

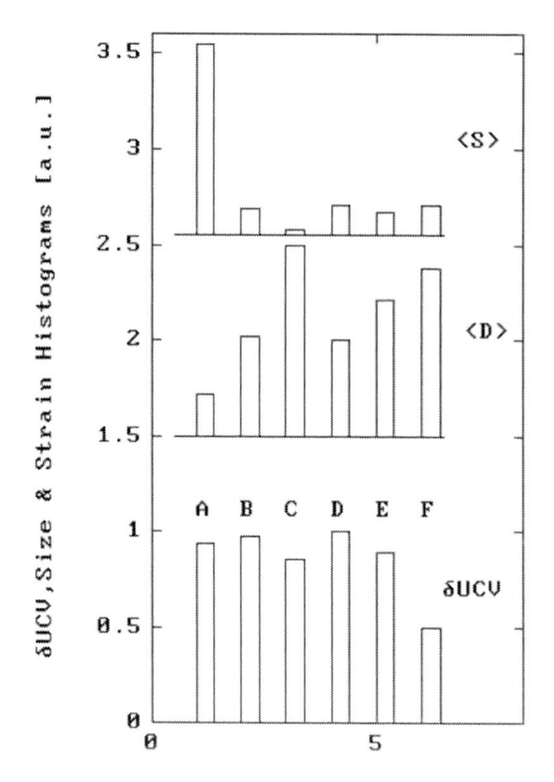

Figure 28. The histograms of lattice strain <S>, crystallite size <D> and unit cell volume (u.c.v.) variations for samples Et-TiO_2 and S-Et-TiO_2, thermally treated; S_{max} = 10.0 x 10^{-3}; S_{min} = 0.3 x 10^{-3}; D_{max} = 346Å; D_{min} = 74 Å; u.c.v. expressed as δ gradient (V_{max} = 135.5 $Å^3$; V_{min}= 134.2$Å^3$); A-C: sample Et-TiO_2 thermally treated at 573, 673, 773 K; D-F: sample S-Et-TiO_2 thermally treated at 573, 673, 773 K.

From the presented data it can be established that:

a) in the case of un-doped TiO_2 (sample Et-TiO_2) the evolution of anatase lattice with temperature is the expected one: the lattice becomes more and more structurated as the temperature increases; so the lattice strains quickly decrease;
b) in the same time and in agreement with <D> and <S> evolution, the geometry of the elementary cell (EC) tends to evaluate towards the most stable state; in this case the tendency of EC is to slightly contract with temperature;
c) for the S-doped TiO_2 (sample S-Et-TiO_2), the effect of sulfur dopant interferes with the effect of temperature which arranges the lattice; indeed the rate of crystallite growth is bigger for sample Et-TiO_2 compared to sample S-Et-TiO_2;
d) concerning δ u.c.v. the decrease tendency with temperature is lower for sample Et-TiO_2 compared to sample S-Et-TiO_2.

A very good correlation between the XRD and XPS results (presented in Figure 29) was put in evidence [177].

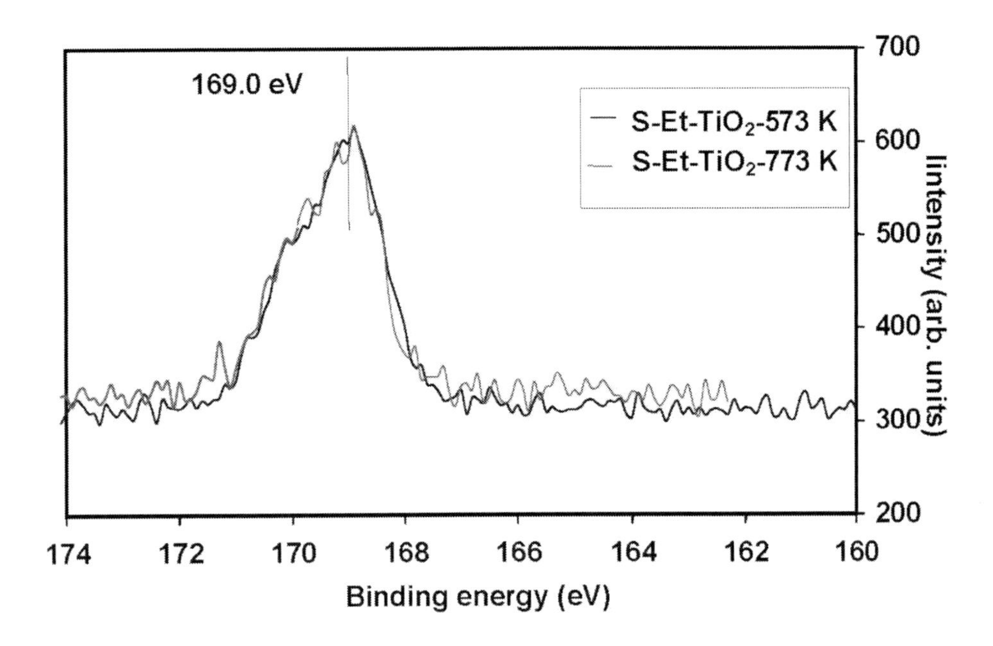

Figure 29. The XPS spectra of sample S-Et-TiO_2 thermally treated at 573 and 773 K.

XPS measurements of the S 2p core level indicate the presence of SO_4^{2-} species on the sample surface (S 2p core level binding energy of about 169 eV), and the absence of sulfide species S^{2-} (S 2p core level binding energy of about 162 eV), both at 573, and 773 K. This indicates that the sulfur is present as S^{6+} ion. These data may be correlated with the u.c.v. variations which suggest that the substitution of some Ti^{4+} ions (ionic radius = 0.68 Å) by S^{6+} ions (ionic radius = 0.29Å) is the most probably occurring. However, the accommodation of S^{6+} ions in the anatase lattice results in an increase in the lattice strains. The difference between the ionic radii and change of Ti^{4+} and S^{6+} leads to a more distorted crystalline lattice. This supplementary positive charge must be compensated to maintain the electro-neutrality of the lattice, presumably by O^{2-} ions from the lattice or from the atmosphere. The XPS S/Ti atomic ratio is 0.15 at 573 K and 0.08 at 773 K, characteristic of S diffusion in the material after annealing at higher temperatures.

The un-doped and S-doped TiO_2 powders have been studied comparatively in order to establish the influence of dopant on the photocatalytic properties. Table 17 presents the influence of the temperature of the thermal treatment of samples Et-TiO_2 and S-Et-TiO_2 on the chlorobenzene (CB) photodegradation and on the mineralization of organic chlorine during water purification [177].

Table 17. The influence of the temperature of the thermal treatment of the un-doped (Et-TiO_2) and S-doped (S-Et-TiO_2) powders on the chlorobenzene (CB) photodegradation and on the mineralization of organic chloride during water purification.

Photo-catalyst	Temperature [K]	CB and Cl content in contaminated water				η_{CB} [%]	η_{Cl^-} [%]
		[CB]		$[Cl^-]$			
		mg/L	Mx10^3	mg/L	Mx10^3		
Et-TiO_2	573	0.34	0.0030	2.56	0.072	97	70
S-Et-TiO_2	573	0.23	0.0020	2.27	0.064	98	70
S-Et-TiO_2	673	0.18	0.0016	2.49	0.070	98	77
S-Et-TiO_2	773	0.11	0.0010	2.66	0.075	99	82

The results presented in Table 17 have been obtained in the following experimental conditions: pH = 7; $[CB]_o = 0.091 \times 10^{-3}$ M; photocatalyst (powder) = 50 mg/L; O_2 = 7 mg/L; λ = 200-280; 400-450 nm; irradiation time = 3600 s.

The presence of S dopant in the TiO_2 powders slightly improved their photocatalytic activities and led to better chlorobenzene removal yields ([CB] and η_{CB}) than for the un-doped samples. The temperature of the thermal treatment also slightly enhances the photocatalytic activity of the sol-gel powders. The mineralization yield of the organic chlorine (η_{Cl^-}) increases too with the increasing temperature.

Previous studies [177] have highlighted the catalytic properties of the S-doped TiO_2 powders which were also tested for the degradation of different chloride organic compounds from aqueous solutions. The results are presented in Table 18, together with the influence of the irradiation time, for sample S-Et-TiO_2. The experiments were performed with the photocatalyst thermally treated at 673 K, in the following experimental conditions: pH = 7; $[pollutant]_0 = 0.1 \times 10^{-3}$ M; O_2 = 7 mg/L; photocatalyst (powder) = 50 mg/L; λ = 200-280; 400-450 nm; irradiation pathlenght = 1 cm.

The performance of the catalyst for the different types of chloride organic pollutant is noted.

Figure 30 (a-c) presents the TEM images of the S-Et-TiO_2 powder before (a) and after 4 cycles testing in the photocatalytic process (b, c).

Table 18. The influence of the irradiation time of S-Et-TiO_2 photocatalyst in the degradation of different chloride organic compounds during water purification.

Pollutant-chloride organic compound	Time, [min]	CB and Cl content in contaminated water				$\eta_{pollutant}$ [%]	η_{Cl^-} [%]
		Pollutant		Cl^-			
		mg/L	$Mx10^{-3}$	mg/L	$Mx10^{-3}$		
MCB	0	10.23	0.091	-	-	-	-
	60	0.18	0.0016	2.49	0.070	98	77
	90	0.04	0.0004	2.53	0.071	100	78
1,2 DCB	0	18.59	0.126	-	-	-	-
	60	1.65	0.011	7.39	0.210	91	82
	90	0.80	0.0064	7.80	0.220	96	87
	120	0.17	0.001	8.05	0.227	99	90
1,3 DCB	0	14.40	0.098	-	-	-	-
	60	0.77	0.005	5.94	0.171	97	85
	90	0.22	0.002	6.15	0.173	99	88
	120	0.04	0.0003	6.29	0.177	100	90
1,2,4 TCB	0	17.24	0.095	-	-	-	-
	60	0.87	0.005	8.17	0.230	94	81
	90	0.33	0.002	8.49	0.239	98	84
	120	0.10	0.0006	8.62	0.243	99	85

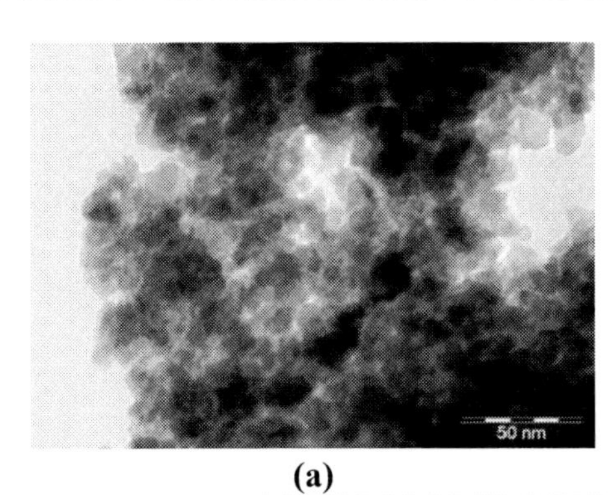

(a)

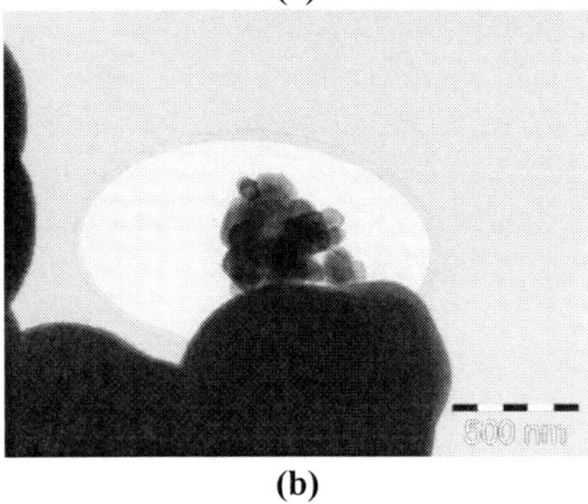

(b)

Figure 30. Continued.

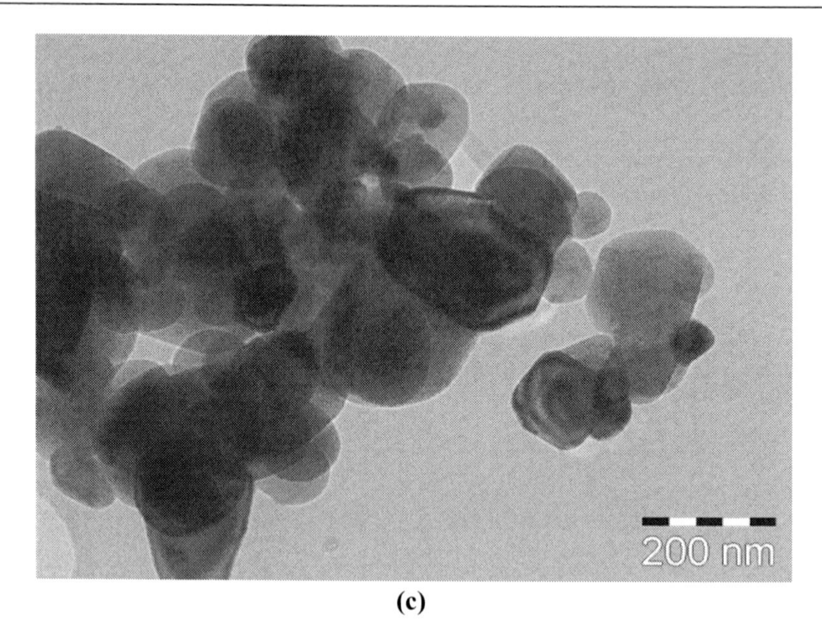

(c)

Figure 30. The TEM images of S-Et-TiO_2 powder before (a) and after 4 cycles testing in the photocatalytic process (b, c).

The changes in the crystallite sizes could be observed. Before of its use in the catalytic test, the sample presents spheroidal TiO_2 nanocrystallites, joined in aggregates with spherical shape, with sizes between 300 and 800 nm (Figure 30a). After testing, the sample presents the aggregates of nanoparticles with sizes between 800 and 1000 nm (Figure 30b). The crystallites increased very much and have the sizes between 50 and 200 nm (Figure 30c).

From the presented experimental results, it can be concluded that the presence of S as dopant in the TiO_2 powders (5 wt. % S related to the TiO_2 content), no matter what Ti alkoxide has been used as precursor leads to better photocatalytic properties (expressed as higher chlorobenzene removal yields), compared to the un-doped samples. This finding allows us to propose our sol-gel S-doped TiO_2 powders as materials with potential for water purification.

2.8.1.3. Pd-doped TiO_2 Powders

Un-doped and Pd-doped TiO_2 powders have been prepared in order to establish the influence of Pd as dopant on the properties of the final product. Both powders have been synthesized using the alkoxide route of the sol-gel method. The precursor of TiO_2 was tetraethylorthotitanate $Ti(OC_2H_5)_4$ and the source of Pd was Pd acethylacetonate $Pd(acac)_2$, $C_{10}H_{14}O_4Pd$. The concentration of Pd was of 1 wt. % related to the TiO_2 content. The ethylic alcohol has been used as solvent and the hydrolysis of the Ti alkoxide took place with water excess. Details concerning the sol-gel synthesis of the TiO_2 nanopowders are available in one of our previous work [234]. Both powders have been dried at 353 K and then thermally treated at 573, 673, and 773 K for 1 h, according to the thermal analysis results.

The structural evolution with temperature of the Pd-doped sample is presented as XRD patterns in Figure 31.

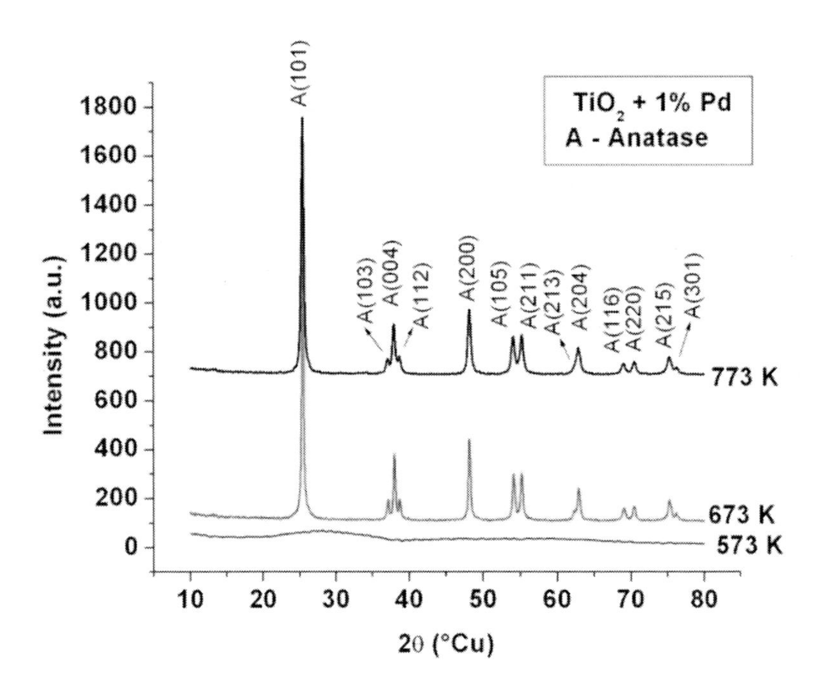

Figure 31. XRD patterns of Pd-doped TiO_2 powder thermally treated at 573, 673, and 773 K, respectively.

The un-doped and Pd-doped TiO_2 powders have presented different structures and morphologies, as it can be observed from Table 19 and from Figures 32 and 33. The XRD data from Table 19 refers to the Pd-doped powders thermally treated at 573 and 773 K.

Table 19. The profile analysis results from the XRD spectra of un-doped and Pd-doped TiO_2 powders.

Sample	T = 573 K			T = 773 K		
	10^{+3} x <S>	<D> [Å]	u.c.v. [Å^3]	10^{+3} x <S>	<D> [Å]	u.c.v. [Å^3]
Un-doped TiO_2	9.9 (1.5)	79 (14)	135.36 (56)	0.24 (4)	346 (8)	135.14 (23)
Pd-doped TiO_2	amorphous			0.84 (4)	351 (8)	134.54 (21)

Studying the presented data, it comes out that:

1) In the absence of the dopant, the increase of temperature leads to the rise of crystallinity degree as expected; the crystallite size grows from 79 Å to 346 Å in the same time with a strong decrease of the tensile strain (~ 40 times smaller);
2) In the presence of Pd as dopant at 573 K the crystalline lattice of anatase strongly deteriorates, becoming amorphous; increasing the temperature to 773 K, the size of crystallite becomes measurable (351 Å), being a little higher than the corresponding dimension in the absence of the dopant (346 Å);

3) Although the increase of temperature leads to a higher crystallinity, in the case of the Pd-doped powder the presence of dopant in the anatase lattice induces tensile strains bigger than in its absence (~ 4 times bigger);
4) It can be concluded that Pd as dopant promotes a sudden and more rapid grow of the crystallinity than in the case of the un-doped TiO_2 sample where the effect of growth is probably continuous and more slow; as a consequence of this behavior the microstrain tensile from the anatase lattice will grow. It is possible that both the ionic radius of the Pd^{2+} (0.80 Å) which is bigger than the ionic radius of the Ti^{4+} (0.68 Å) and the local electrostatic unbalance due to the substitution (deficit of positive charge due to Pd^{2+}) be responsible of the local deformation of the crystalline plans which leads to the increase of the microtensile strain.

TEM and SAED images of the un-doped and Pd-doped TiO_2 powders thermally treated at 673 K for 1 h are presented in Figures 32 and 33.

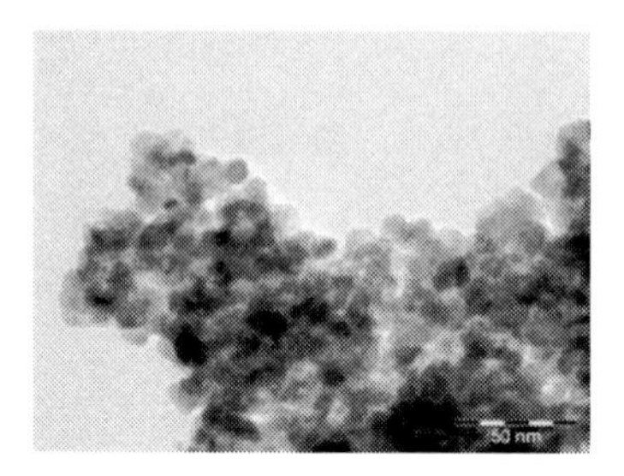

Figure 32. TEM image of un-doped TiO_2 powder thermally treated at 673 K.

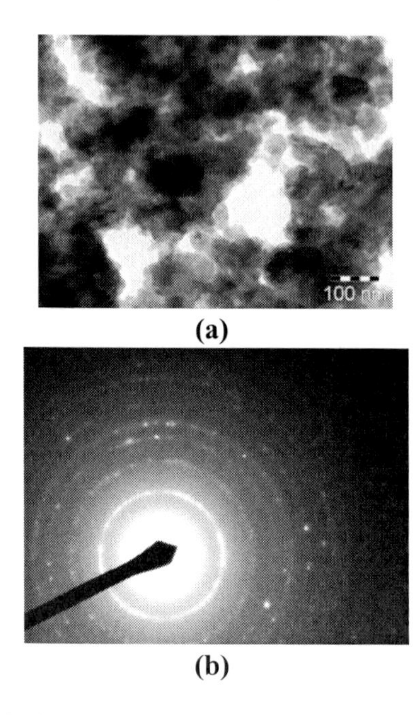

Figure 33. TEM (a) and SAED (b) images of Pd-doped TiO_2 powder thermally treated at 673 K.

The TEM images of un-doped TiO_2 nanopowder evidence the presence of spheroidal TiO_2 nanocrystallites jointed in aggregates of 300-800 nm. The Pd-doped TiO_2 nanopowder presents anatase crystallites of 20-50 nm with a lot of internal defects.

The presented studies have confirmed that Pd is inside the TiO_2 network and pointed out its influence as dopant on the final properties of the TiO_2 powder.

CONCLUSION

The sol-gel procedure is one of the most promising methods in the case of coatings preparation, the morphology of the film being varied by changing the composition of the solution and the conditions of deposition.

For both types of synthetized-materials, doped TiO_2 coatings with non-metals (S) and metals (Ag, Pd) the applicative photocatalytic properties (in depollution of water or hydrofilicity) have been correlated with structure, composition and morphology of surfaces.

The sol-gel process has proved to be one of the most suited methods for preparing un-doped and doped TiO_2 powders with scientific and technological importance. The photocatalytic activity was influenced by the lattice defects induced by the dopants, both in the case of S and Pd. The S-doped TiO_2 powders had good properties in water purification from different organic chloride compounds.

3. STUDY OF SOL-GEL $BaTiO_3$-BASED SYSTEMS

Barium titanate-based compositions represent one of the most widely studied classes of oxide materials, due to their unique and multiple useful properties as ferroelectricity, high dielectric permittivity, positive temperature coefficient of resistivity (PTCR effect), high-voltage tunability, piezoelectricity and pyroelectricity [276-284]. These characteristics make the $BaTiO_3$-derived bulk ceramics and thick films attractive for various applications including multilayer ceramic capacitors (MLCC), posistors, piezoelectric and ultrasonic actuators, pyroelectric detectors, temperature sensors and controllers, tunable elements in the microwave circuits for telecommunication [285-292]. Besides, pure and doped $BaTiO_3$-based thin films have received also much attention in recent years for their potential applications in microelectronics and integrated optics technologies, ranging from buffer layers for the integration of high-temperature superconductors and voltage tunable microwave filters, to pyroelectric imaging arrays and multifunctional highly integrated CMOS-based devices, such as nonvolatile random access memories (NVRAM) and dynamic random access memories (DRAM) [293-299].

Undoped $BaTiO_3$ ceramics are ferroelectric at room temperature, exhibiting high dielectric and piezoelectric constants, which strongly depend on the microstructure. Polycrystalline $BaTiO_3$ has a Curie temperature of ~130°C, where these constants present important thermal anomalies. The remanent polarization, the dielectric and piezoelectric constants are not as high as those reported for the ferroelectric lead-base perovskites as PZT or Pb-based relaxors. Still these properties can be significantly improved by doping with various elements or by suitable modifications of the microstructural features.

The temperature of the ferroelectric – paraelectric phase transition (Curie temperature T_C) can be shifted to lower temperature values via partial substitution of Ba ions (*A*-site doping), Ti ions (*B*-site doping) or both Ba and Ti ions (simultaneous *A* and *B*-site doping) with suitable homovalent or aliovalent dopants. Thus, according to the Goldschmidt formula, cation species of lower ionic radius than that one corresponding to Ba^{2+} or of higher ionic radius than that one of Ti^{4+}, acting as *A*-site or *B*-site dopants in the perovskite lattice are required, in order to obtain ceramics with lower Curie temperatures than that one specific to the pure $BaTiO_3$.

Because of the lower solubility of the aliovalent dopants induced by the valence dissimilarity between the solutes and the host cations, smaller concentrations of aliovalent dopants may have the same efficiency in the drop of the Curie temperature as the higher concentrations of homovalent admixtures. That is the reason why in the case of the substituted $BaTiO_3$ with homovalent species one can speak in terms of solid solutions, whereas in the case of the aliovalent solutes one can speak rather in terms of the doping level.

Due to the high flexibility of the perovskite structure, many aliovalent dopants can be easily accommodated in the $BaTiO_3$ crystalline lattice. Cations with lower oxidation state replacing Ba^{2+} or Ti^{4+} and acting as donor dopants are well known for inducing PTCR or IBBL (internal barrier boundary layer) effects in the related ceramics [300].

The donor dopants strongly influence the defect chemistry in the perovskite lattice by generating, as a function of their concentration, electrons or complementary compensating ionic defects (cation vacancies), in order to preserve the charge balance. One of the most efficient donor dopant for inducing PTCR behavior in the $BaTiO_3$ ceramic is lanthanum, partially replacing the host Ba^{2+} ions. As the lanthanum content is lower than a critical concentration ($x_c < 0.3$ at. %), then the additional charge of the donor dopant is compensated by electrons and the ceramic gains semiconducting character. A higher donor doping level (concentrations exceeding the critical value, but still in the compositional range of the dopant solubility) determines the change of the compensating mechanism. In this case, ionized cation vacancies are the most likely defects which compensate the lanthanum solute, inducing an insulating behavior for the ceramic material.

3.1. Pure and Lanthanum Doped Barium Titanate Sol-Gel Products

3.1.1. Choice of Compositions

The conventional method for preparation of pure, doped or highly-substituted $BaTiO_3$ powders by solid-state reaction between $BaCO_3$, TiO_2 and dopant metal oxides, generally leads to coarse and agglomerated particles with broad particle size distributions. These starting powders are not suitable for processing fine and dense ceramics with controlled microstructure, required for advanced applications. Particularly, in the case of donor-doped $BaTiO_3$, the solid state route offers little potential for control over the grain size and structure of the resulting material, because of abnormal grain growth taking place during the sintering process [301-305].

Pure and fine $BaTiO_3$-derived powders with controlled stoichiometry and narrow particle size distribution can be obtained by using several innovative methods including hydrothermal process [306-309], coprecipitation [310-313], polymeric precursor methods (PPM) [314-316], sol-gel and sol-precipitation techniques [317-320], microemulsion route [321], auto-combustion [322, 323], molten salt [324], as well as syntheses performed by mechanically activated processes [325, 326].

Beside the decrease of the sintering temperature of the ceramics, an important advantage of the sol-gel technique consists in the preparation of very complex, stoichiometric, multicomponent products. Especially the alkoxide route ensures the facility of the doping process, which leads to uniform dopant distributions, due to the atomic level of the mixing. This aspect is crucial mainly for materials with electronic applications such as $BaTiO_3$-based ceramics. Despite the mentioned advantages, only few papers reported the synthesis and functional properties of the sol-gel La-doped $BaTiO_3$ [327-330] and from these, only one pointed out the role of the size effects on the PTCR behavior exhibited by the $Ba_{1-x}La_xTiO_3$ (x = 0.17 at. % La^{3+}) composition prepared by an innovative sol-gel method [330]. Taking into account the already mentioned aspects, in this study the formation mechanism and the characteristics of the pure and lanthanum doped barium titanate powders prepared by the alkoxide variant of the sol-gel method were analyzed.

From the donor concentration point of view, the investigated compositions were chosen in such a way to result oxide powders and subsequent ceramic products exhibiting semiconducting, as well as insulating properties. The selected compositions are listed in Table 20. Unlike the compositions with lower dopant concentrations (no. 2 and 3) described by the formula $Ba_{1-x}La_xTiO_3$ and characterized by conductive properties of the grain cores (even after sintering in air) and leading to the PTCR effect, the highly-La doped composition (no. 4), exhibiting dielectric behavior corresponds to the formula $Ba_{1-x}La_xTi_{1-x/4}O_3$, in order to ensure the charge balance by cation vacancy compensation and thus avoiding the segregation of some undesirable secondary phases.

In the first case, the electronic compensation mechanism which occurs can be expressed by the following equation, using the Kröger-Vink notation:

$$La_2O_3 + 2TiO_2 \rightleftarrows 2La^{\bullet}_{Ba} + 2Ti^{\times}_{Ti} + 6O^{\times}_{O} + \frac{1}{2}O_2(g) + 2e' \quad (17)$$

For highly donor doped $BaTiO_3$ ceramics, several compensating mechanisms, involving the formation of complementary ionic defects, are possible, according to equations (18)-(21):

formation of interstitial anions:

$$La_2O_3 + 2TiO_2 \rightleftarrows 2La^{\bullet}_{Ba} + 2Ti^{\times}_{Ti} + 6O^{\times}_{O} + O''_{i} \quad (18)$$

formation of cation vacancies:

- barium vacancies:

$$La_2O_3 + 2TiO_2 \rightleftarrows 2La^{\bullet}_{Ba} + 2Ti^{\times}_{Ti} + V''_{Ba} + 7O^{\times}_{O} \quad (19)$$

Table 20. Selected compositions and processing parameters of the sol-gel syntheses

Sample no.	Composition	Metallic precurors	Molar ratio			pH of mixture	Reaction conditions	
			$\frac{i\text{-}C_3H_7OH}{\sum \text{metallic precursors}}$	$\frac{H_2O}{\sum \text{metallic precursors}}$	$\frac{HNO_3}{\sum \text{metallic precursors}}$		T (°C)	t (h)
1.	$BaTiO_3$	$Ba(O\text{-}iC_3H_7)_2$ + $Ti(O\text{-}iC_3H_7)_4$	10.59	27.78	0.27	9 - 10	20	2
2.	$Ba_{0.9975}La_{0.0025}TiO_3$	$Ba(O\text{-}iC_3H_7)_2$ + $Ti(O\text{-}iC_3H_7)_4$ + $La(NO_3)_3{\cdot}6H_2O$	16.40			9 - 10		
3.	$Ba_{0.995}La_{0.005}TiO_3$		15.17			8 - 9		
4.	$Ba_{0.975}La_{0.025}Ti_{0.99375}O_3$		15.09			8		

$Ba(O\text{-}iC_3H_7)_2$ = Barium isopropoxide with 51.3 % Ba content (EDTAH titration 51.3% Ba complexometric);
$Ti(O\text{-}iC_3H_7)_4$ = Tetraisopropylorthotitanat;
$La(NO_3)_3{\cdot}6H_2O$ = Lanthannitrat hexahydrat;
HNO_3 = Nitric acid 65 %;
$i\text{-}C_3H_7OH$= 2 Propanol.

- equal number of barium and titanium vacancies:

$$3La_2O_3 + 6TiO_2 \rightleftarrows 6La^{\bullet}_{Ba} + 6Ti_{Ti} + V''_{Ba} + V''''_{Ti} + 21O^{\times}_{O} \tag{20}$$

- titanium vacancies:

$$2La_2O_3 + 4TiO_2 \rightleftarrows 4La^{\bullet}_{Ba} + 4Ti^{\times}_{Ti} + V''''_{Ti} + 14O^{\times}_{O} \tag{21}$$

For energetic reasons, the formation of anti-Schottky defects (described by eq. 18) consisting of oxygen ions placed on interstitial sites is unlikely in the close-packed perovskite lattice of $BaTiO_3$ [331, 332]. Although the formation of cation vacancies was considered the valid compensating mechanism for the supplementary charge induced by the donor dopant in highly-doped barium titanate [333, 334], however the type of these cation vacancies was a controversial topic in the literature. Some authors sustained that the charge balance is achieved by means of double-ionized barium vacancies (eq. 19), or equal numbers of barium and titanium vacancies (eq. 20), claiming that the exclusive formation of titanium vacancies is unlikely because of the high effective charge [335, 336], while other researchers indicated the predominance of the titanium vacancies as compensating defects in lanthanum-doped $BaTiO_3$ (eg. 21) [337]. Later calculations of defect energetics in barium titanate using computer simulation techniques reported by Lewis and Catlow [331] confirmed the hypothesis of the preponderant compensation of the donor dopant by tetra-ionized titanium vacancies.

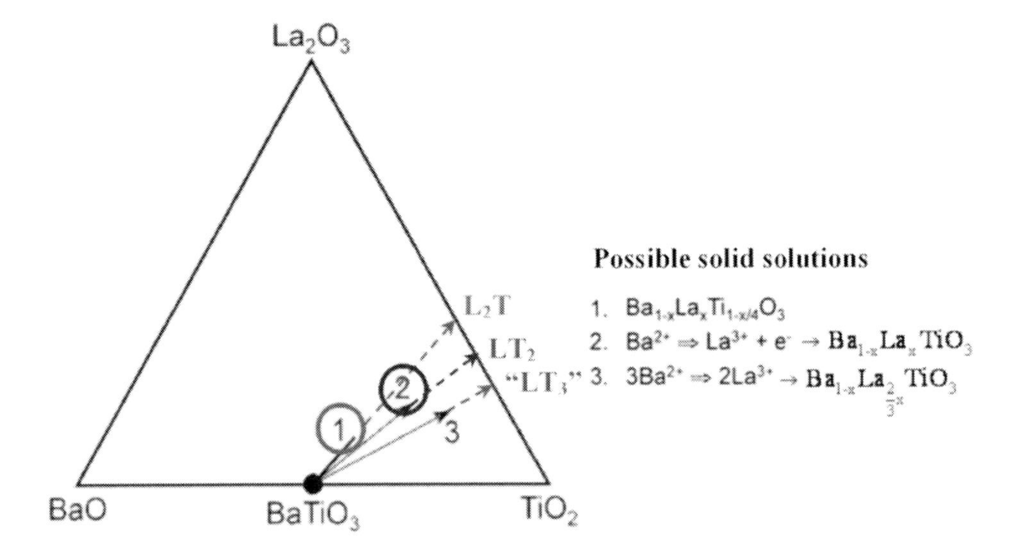

Figure 34. The place in the ternary BaO-La_2O_3-TiO_2 system of the potential solid solutions formed by lanthanum incorporation in $BaTiO_3$ lattice.

Figure 34 shows the place in the ternary BaO-La_2O_3-TiO_2 system of the potential solid solutions formed by lanthanum incorporation in $BaTiO_3$ lattice, as a function of the compensating mechanism [338]. The most probable compensating mechanism by means tetra-ionized titanium vacancies acting in the highly-doped $BaTiO_3$ (denoted (1)) involves the formation of $Ba_{1-x}La_xTi_{1-x/4}(V_{Ti}^{''''})_{x/4}O_3$ solid solutions placed on the $BaTiO_3$-$La_4Ti_3O_{12}$ tie line (composition no. 4 of Table 20), while the electronic compensating mechanism acting in slightly-doped $BaTiO_3$ (denoted (2)) corresponds to the solid solutions with the formula $Ba_{1-x}La_xTiO_3$, placed on the potential $BaTiO_3$-$La_2Ti_2O_7$ tie line (composition no. 2 and 3, respectively – Table 20). The unlikely compensating mechanism implying the formation of barium double-ionized vacancies (denoted (3)) corresponds to the formation of $Ba_{1-x}La_{2x/3}(V_{Ba}^{''})_{x/3}TiO_3$ solid solutions placed on the uncertain tie line between $BaTiO_3$ and the hypothetical $La_2Ti_3O_9$ compound.

More recent phase equilibria studies [339] shown that the uncertain "LT_3" compound and even the $BaTiO_3$-$La_2Ti_2O_7$ tie line does not exist and only small domains of adjacent ternary solid solutions were formed (Figure 34).

3.1.2. Preparation and Characterization of Powders and Related Ceramics

The preparation flow chart of the lanthanum doped barium titanate nanopowders and related ceramics is presented in Figure 35.

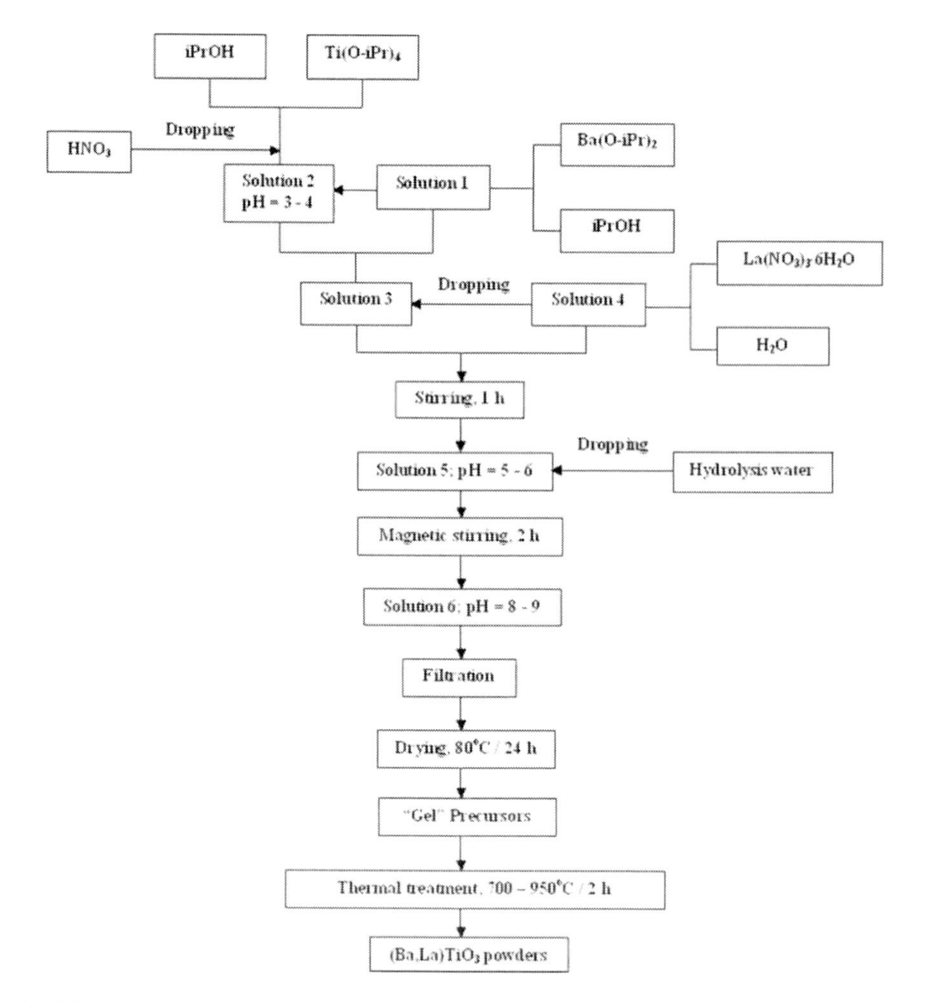

Figure 35. Flowchart for the preparation of the lanthanum doped barium titanate sol-gel powders.

The raw materials, as well as the synthesis conditions are summarized in Table 20. The precursor powders were thermally treated in air, at temperatures ranged between 700–950°C with a plateau of 2 hours and using a heating rate of 5°C/min.

The nanopowders resulted after annealing at 900°C for 2 hours, were shaped by uniaxial pressing as pellets with a diameter of ~ 13 mm and a thickness of ~ 2 mm. The green bodies were sintered in air at 1300°C with 3 hours plateau, using a heating rate of 5°C/min and then were slowly cooled (with the normal cooling rate of the furnace) at room temperature, in order to obtain ceramic samples. The flowchart for the preparation of the lanthanum doped barium titanate ceramics is presented in Figure 36.

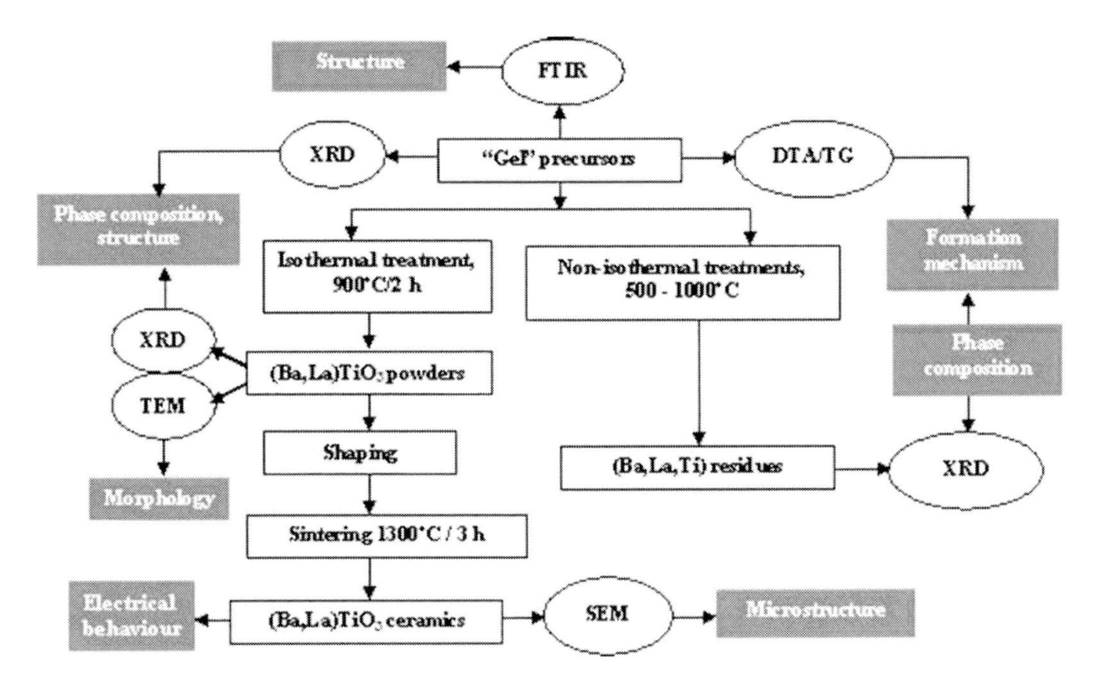

Figure 36. Flow chart for preparation and characterization of precursors, powders and related ceramics.

IR spectroscopy, X-ray diffraction and thermal analysis methods were used to investigate the structure, the phase composition and the thermal behavior of the precursors.

To study the evolution of the precursors decomposition and to identify the intermediate phases which were formed during this process, XRD analyses were performed on the powdered residues resulted after non-isothermal treatments at various temperatures ranged between 500–1000°C followed by subsequent quenching, in order to preserve the high temperature phase composition. The final oxide powders thermally treated at temperatures ranged between 700-950°C for 2 hours and slowly cooled at room temperature were analyzed from structural, phase composition and morphological point of view using XRD and transmission electron microscopy (TEM). The microstructure of the ceramic samples obtained after sintering was investigated by scanning electron microscopy. The conducting or insulating behavior was estimated by electrical measurements performed at different temperatures at 1kHz frequency.

For all the products (precursors, powders and ceramics), the investigation techniques used and the characteristics studied were presented in Figure 36.

3.1.3. Characteristics of Precursor Powders

3.1.3.1. Phase Composition

The X-ray diffraction patterns corresponding to the undoped precursor show that even at room temperature, the barium titanate represents the major crystalline phase. Orthorhombic barium carbonate (witherite) and small amounts (at the detection limit) of nanocrystalline TiO_2 were also identified as secondary phases. The high background of the X-ray diffraction pattern clearly indicates the presence of a significant amount of amorphous phase, probably rich in titanium, with respect to the initial stoichiometric Ba/Ti ratio (Figure 37).

Unlike this case, in the lanthanum doped precursors, the perovskite phase is not yet formed, which means that the addition of the lanthanum nitrate inhibited the barium titanate crystallization. These precursors are predominantly amorphous and only slightly crystallized amounts of $BaCO_3$ were identified in the related XRD patterns (Figure 37).

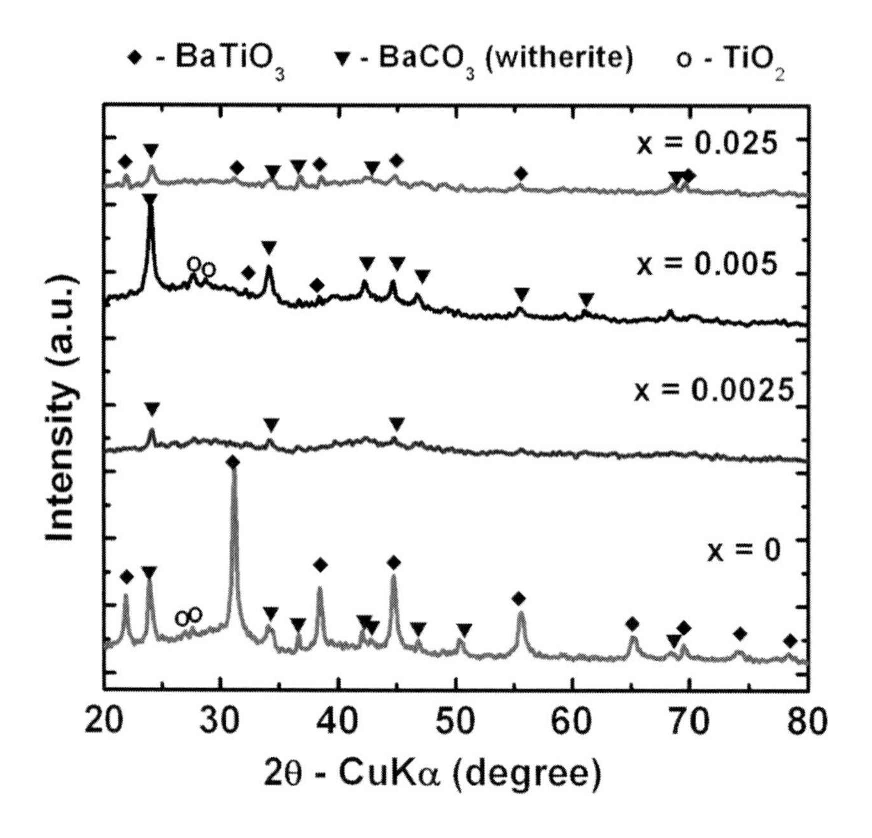

Figure 37. Room temperature X-ray diffraction pattern for (Ba, Ti) and (Ba, Ti, La) precursors.

3.1.3.2. Structure

The FTIR spectrum of the undoped precursor powder presented in Figure 38(a) shows the presence of the band located at ~ 588 cm^{-1} specific to the vibration mode of the Ti-O bond in the TiO_6 octahedra. The main bands specific to the ionic CO_3^{2-} (693, 854, 1059 and 1420 cm^{-1}), as well as those corresponding to the NO_3^- ion (1385 cm^{-1} and 860 cm^{-1}) introduced in the system by the catalyst and by the lanthanum source, $La(NO_3)_3 \cdot 6H_2O$ in the case of the doped precursors, were also identified (Figure 38(a)-(d)). Bending vibrations of the CH_3 group are present at 1457 cm^{-1} for all samples and nitro compounds signal at 1545 cm^{-1} only for the doped ones. Bending mode of the coordinated water at 1623-1652 cm^{-1} have also been noticed. Impurities as CH_2 and CH radicals were pointed out by the weak bands located at 2823 and 2913 cm^{-1}, respectively. The broad band located at 3420 cm^{-1} and the sharper bands at ~3750 and ~3900 cm^{-1} were assigned to the vibration modes of the surface and internal hydroxyl ions.

The assignments of the main FTIR bands of the spectra presented in Figure 38 are summarized in Table 21.

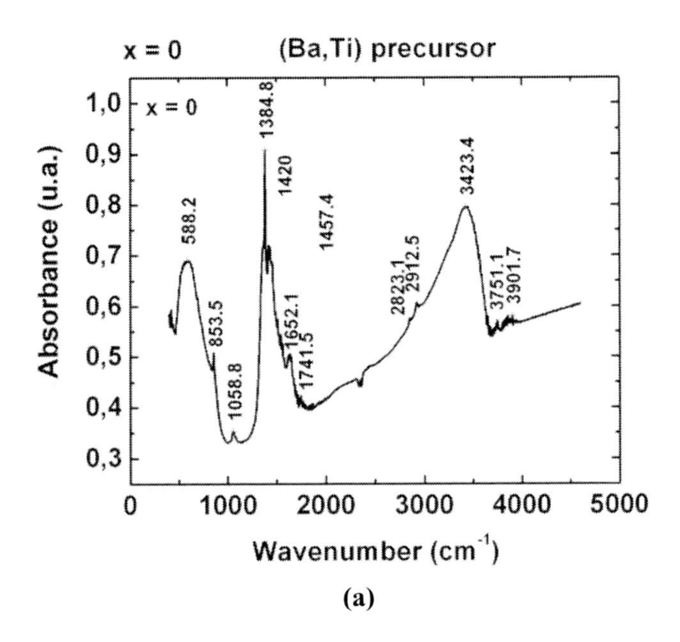

(a)

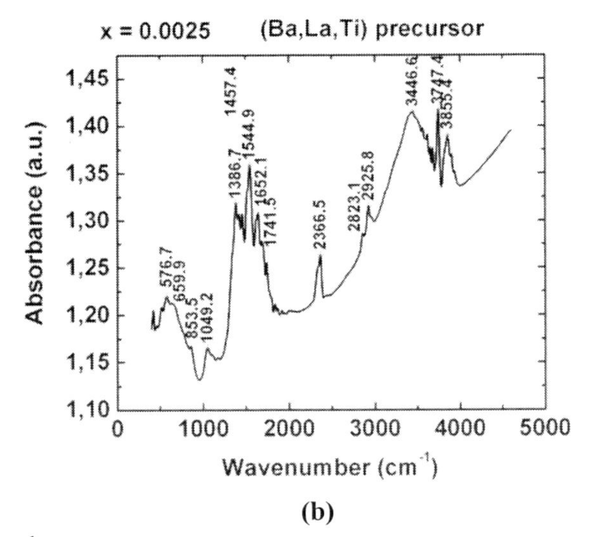

(b)

Figure 38. Continued.

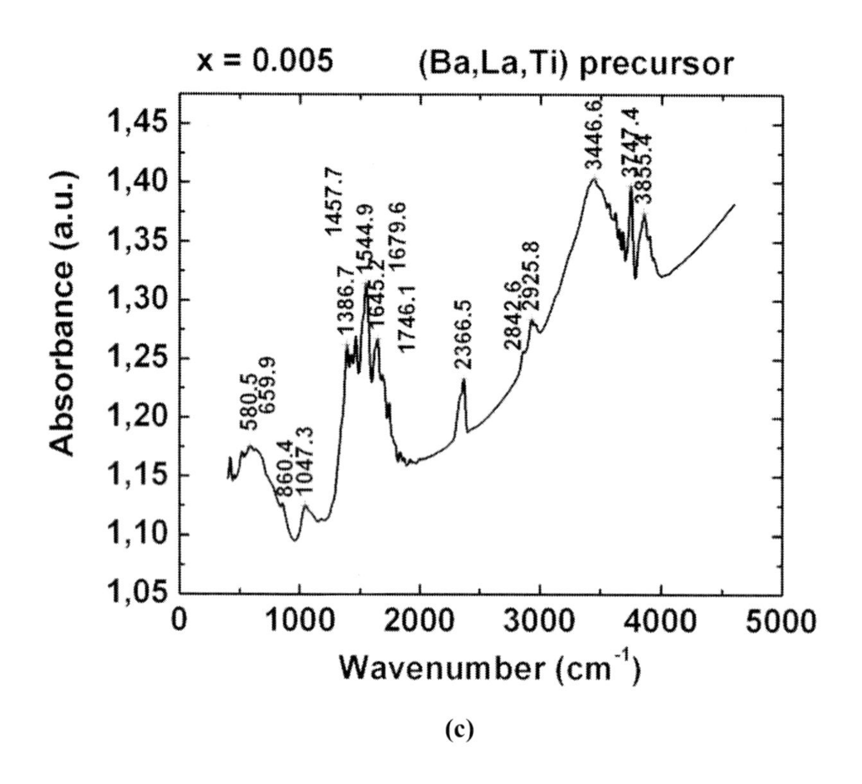

(c)

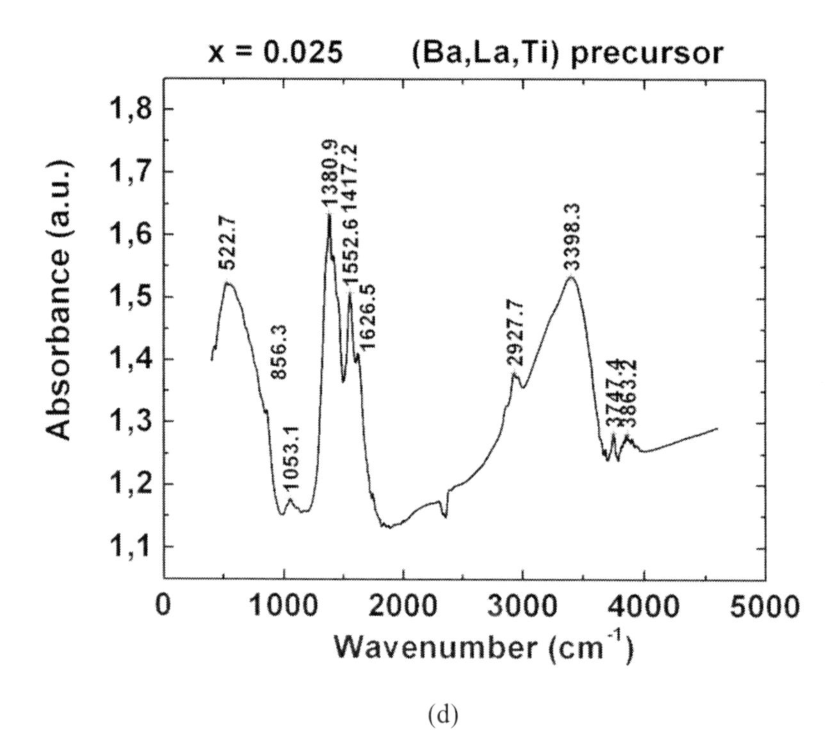

(d)

Figure 38. FTIR spectra of the undoped precursor (a) and lanthanum doped precursors: x = 0.0025 (b); x = 0.005 (c); x = 0.025 (d).

Table 21. Assignment of the main FTIR bands.

Wavenumber (cm^{-1}) / composition				Assignment
x = 0	x = 0.0025	x = 0.005	x = 0.025	
588	577	581	523	ν(Ti – O) – vibration mode in TiO_6 octahedra
693	656	660	689	δ_{ip} (O – C = O) – "in plane" deformation mode for CO_3^{2-} ion
854	854	860	856	δ_{op} (O – C = O) – "out of plane" deformation mode for CO_3^{2-} ion ν - NO^{3-}
1059	1049	1047	1053	ν_s (C = O) – symmetric vibration mode for CO_3^{2-} ion
1385	1387	1387	1381	ν - NO^{3-}
1420	1421	1421	1417	ν_{as} – asymmetric vibration mode for CO_3^{2-} ion
1457	1457	1458	1453	δ_a (CH_3) – bending vibrations of the CH_3 group
-	1545	1545	1553	Nitro compounds –NO_2 (1515-1560)
1652	1652	1645	1623	δ H-O-H – bending mode of the coordinated water
1741	1742	1746	1748	$\nu_s + \delta_{ip}$ – symmetric stretching mode and "in plane" deformation mode of CO_3^{2-}
1741	1742	1746	1748	$\nu_s + \delta_{ip}$ – symmetric stretching mode and "in plane" deformation mode of CO_3^{2-}
-	2367	-	-	ν - vibration mode of adsorbed CO_2 on metallic cations
2851	2855	2857	2857	ν -CH_2– stretching mode of – CH_2 groups
2924	2926	2926	2928	ν -CH – stretching mode of – CH groups
3423	3447	3447	3398	ν - vibration mode of adsorbed H_2O
3751	3747	3747	3747	ν_s -OH^- – symmetric stretching mode of internal -OH groups
3902	3855	3855	3863	ν_{as} -OH^- – asymmetric stretching mode of internal -OH groups

3.1.4. Thermal Decomposition of the Precursors and Formation Mechanism of Pure and Lanthaunm Doped Barium Titanate

3.1.4.1. Thermal Behavior of Precursors

In the low temperature range, the thermal analysis performed on the undoped precursor pointed out an endothermic effect with a maximum at ~85°C accompanied by a slight mass loss on the TG curve, which was attributed to the release of the adsorbed water (Figure 39(a)). At temperatures ranged between 200–400°C the combustion of the organic matter takes place in two different stages indicated by two exothermic effects, with their maxima at 275 and

351°C, respectively. At higher temperatures (~543, 581, 804, 951°C) several flattened endothermic maxima were recorded on the DTA curve. This shows that the decomposition process which leads to the $BaTiO_3$ formation occurs also in several steps in a large temperature range and is completely ended at ~940°C. The total mass loss is of 18.42 %, as indicated the TG curve (Figure 39(a)).

The thermal behavior of the slightly La-doped precursor (x = 0.0025) seems to be more complicated. The combustion process takes place in three distinct steps (297, 362 and 428°C). The additional exothermic effect with its maximum recorded at ~549°C and accompanied by a very slight mass gain, closely followed by a more visible mass loss on the TG curve represents the resultant of several processes, as the slight oxidation of some Ti^{3+} ions (probably appeared due to the reducing atmosphere induced by the combustion of the organic matter), a potential crystallization and a decomposition, which occur almost simultaneously in the temperature range of 500–600°C (Figure 39(b)). The main decomposition process of the intermediates takes place in several different steps (735, 780, 813 and 896°C) and is emphasized by the slope changes recorded on the TG curve and especially by the more complex profile of the DTG curve in the temperature range of 700–940°C. Unlike the undoped precursor, in this case a higher mass loss (of 30.71 %) was recorded (Figure 39(b)). This is in good agreement with the XRD data which clearly indicated the amorphous nature of the La-doped precursor generating a higher mass loss, in comparison with the predominantly crystalline undoped precursor, in which the decomposition of a smaller amount of amorphous phase resulted in a significantly smaller mass loss.

The thermal behavior of the highly La-doped precursor (x = 0.025) is almost similar (Figure 39(c)). The only difference consists in the more attenuated high temperature endothermic effects due to the formation of a highly-substituted solid solution.

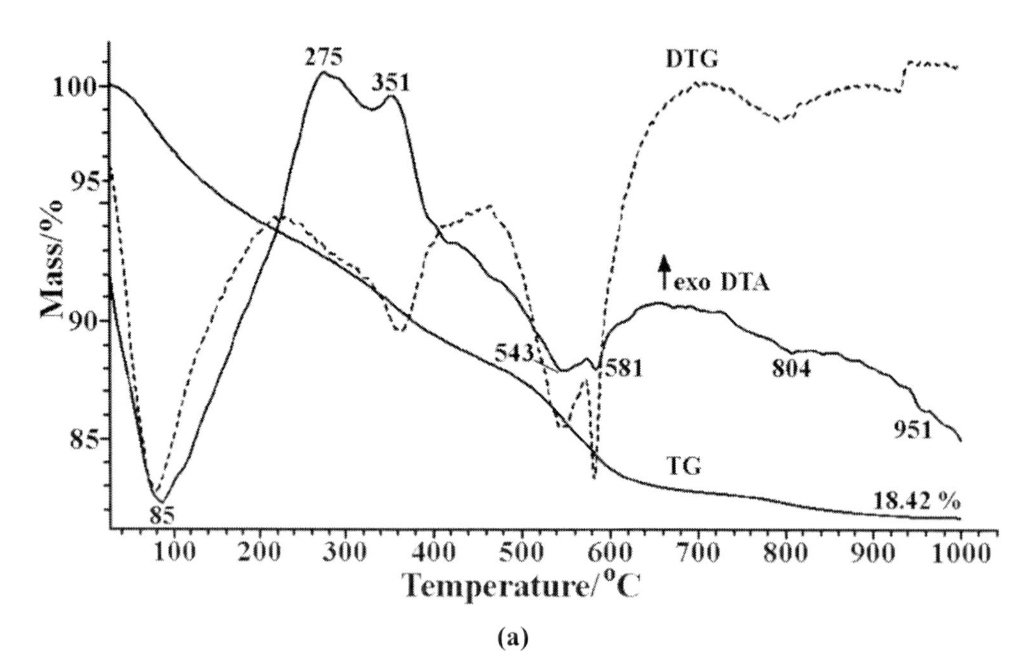

(a)

Figure 39. Continued.

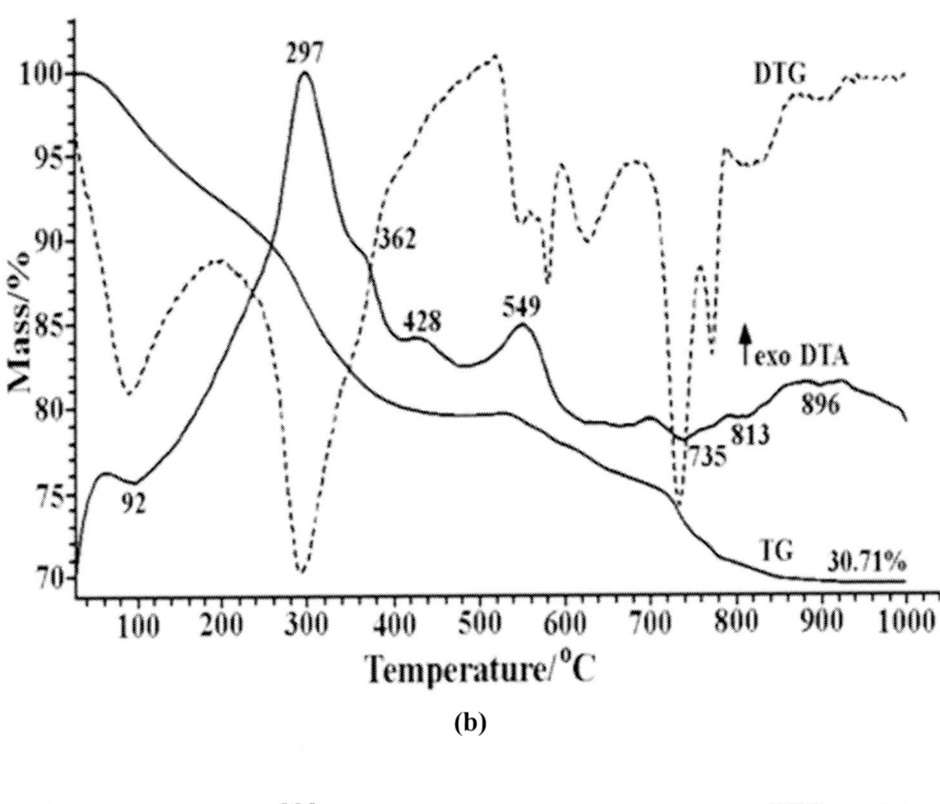

(b)

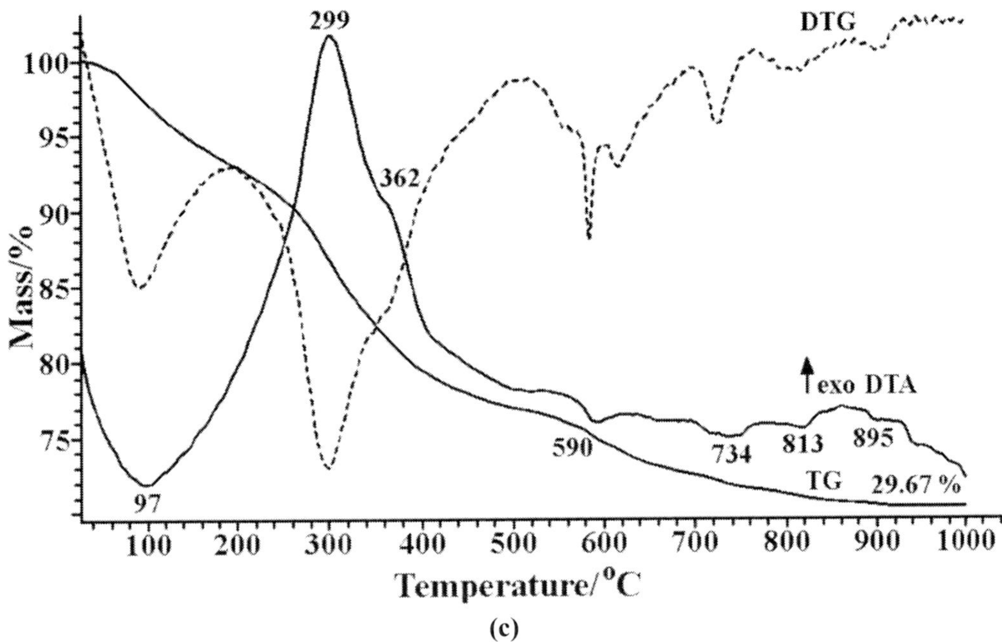

(c)

Figure 39. Thermal analysis curves for the undoped precursor (a) and lanthanum doped precursors: x = 0.0025 (b); x = 0.025 (c).

3.1.4.2. Phase Composition of the Residues

The X-ray diffraction analyses performed on the residues resulted after non-isothermal treatments of the undoped precursor at various temperatures indicated the presence of the well-crystallized perovskite beside small amounts of $BaCO_3$ (witherite) identified as secondary phase (Figure 40). The intensity of the diffraction peaks corresponding to the witherite decreases as the thermal treatment temperature increases due to the progress of the solid state reaction between $BaCO_3$ and the amorphous nano-sized TiO_2 particles, according to the reaction:

$$BaCO_3 + TiO_2 \rightarrow BaTiO_3 + CO_2\uparrow \qquad (22)$$

Consequently, all the high temperature endothermic effects recorded on the DTA curve of Figure 39(a) and accompanied by mass loss can be assigned to the decarbonation process which takes place simultaneously with the quantitative formation and crystallinity increase of the $BaTiO_3$ phase.

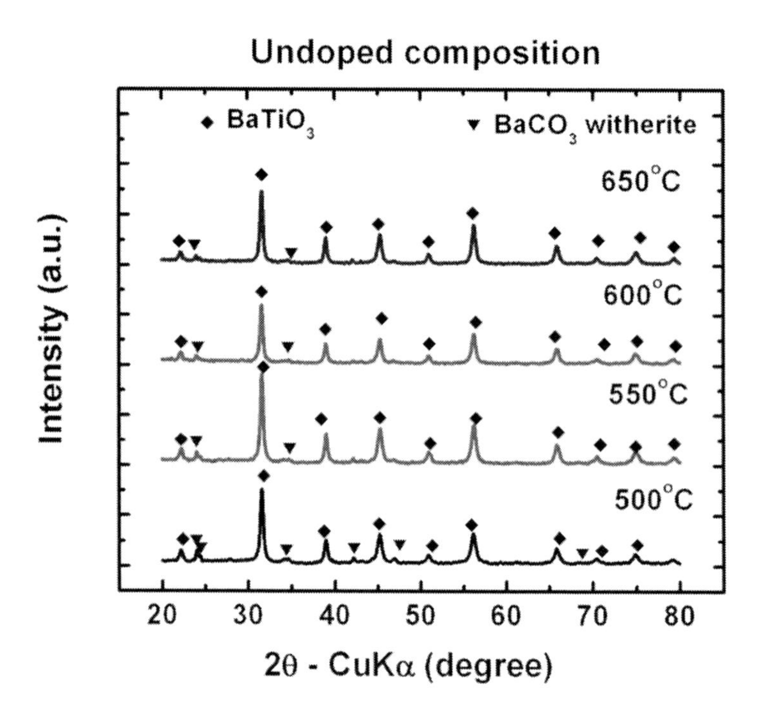

Figure 40. X-ray diffraction patterns of the residues resulted after non-isothermal treatments of the undoped precursor (x = 0) at different temperatures.

The XRD data obtained for the residues of the slightly La-doped amorphous precursor (x = 0.0025) sustain the thermal analysis results, showing a more complex decomposition process. Thus, after non-isothermal treatment at 500°C, the composition of the corresponding residue consists of a mixture of carbonated phases, i.e. the ionic $BaCO_3$ and a mixed (Ba, La, Ti) oxycarbonate phase probably described by the $Ba_{2-2x}La_{2x}Ti_2O_5{\cdot}CO_3$ formula and identified by its main peak located at $2\theta = 26.4^{\circ}$ on the related X-ray diffraction pattern, and of a significant amount of Ti-rich amorphous phase (Figure 41(a)). The existence of this kind of metastable, oxycarbonate phase reported by several authors as intermediate in the decomposition process of some molecular or polymeric nano-sized (Ba, Ti) precursor powders remains a controversial topic still under debate in the literature concerning the $BaTiO_3$ preparation by soft chemistry routes, as the oxalate/citrate procedures [340-342].

From this point of view, two different mechanisms have been proposed for the decomposition of the organic precursors: (*i*) the first one takes into account the existence of the mentioned oxycarbonate-like intermediate phase, $Ba_2Ti_2O_5CO_3$ [343-345] (which decomposes directly in $BaTiO_3$), as an important step during the precursor thermolysis, and (*ii*) contrarily, the second one concludes that $BaTiO_3$ formation occurs by solid state diffusion between the very reactive and fine $BaCO_3$ and TiO_2 particles [346-349]. Despite of the additional evidences for the existence of an intermediate $Ba_2Ti_2O_5CO_3$ phase, provided by different techniques like nuclear magnetic resonance [350, 351] or electron energy loss spectroscopy (EELS) [352] and quantum mechanical calculation by density functional theory (DFT) [352] used in the most recent works, the nature and the stability of this phase still remains ambiguous [353, 354]. For example Dutta et al. [353] suggested by mean of Raman analyses that an appropriate description of the Ba-Ti intermediate is that the Ba^{2+} ions are inserted into a Ti – O – Ti polymer network, while Ischenko et al. [354] demonstrated in their study based on highly local analytical methods (EELS in the STEM mode) that the crystalline phase commonly attributed to the mixed (Ba, Ti)-oxycarbonate is actually a barium oxycarbonate, containing no titanium.

In the case of the slightly La-doped (Ba, Ti) precursor obtained by the alkoxide route, the XRD data indicated that the perovskite phase starts to form in the temperature range of 500 – 550°C and its crystallization progresses with the temperature increase (Figure 41(a)). Taking into account that the intensities of the main diffraction peaks corresponding to both $BaCO_3$-witherite (2θ = 23.9°) and oxycarbonate intermediate (2θ = 26.4°) remain practically unchanged for thermal treatment temperatures between 500–550°C, one can assume that the formation of the perovskite via crystallization from the amorphous phase was the prevalent mechanism in this temperature range. These results are in agreement with the thermal analyses data and support the supposition that the exothermic effect recorded at 549°C on the DTA curve presented in Figure 39(b) was mainly determined by a crystallization process.

In the case of the residue obtained after thermal treatment at 600°C, only a defective $BaTiO_3$ and witherite were identified as crystalline phases. Besides, the high background of the XRD pattern still indicates the presence of a certain amount of Ti-rich amorphous phase. This evolution of the phase composition suggests that in the temperature range of 550–600°C the increase of the amount of the perovskite phase (pointed out by the significant enhancement of the characteristic diffraction peaks) is not determined by a solid state reaction between $BaCO_3$ and the nano-sized amorphous titania particles as in the case of the undoped precursor, but it occurred concurrently at the expense of consuming the mixed oxycarbonate intermediate, as described in the following equation :

$$Ba_{2-2x}La_{2x}Ti_2O_5{\cdot}CO_3 \rightarrow 2\ Ba_{1-x}La_xTiO_3 + CO_2\uparrow \tag{23}$$

Consequently, the mass loss recorded on the TG curve of Figure 39(b) is due to the gradual release of carbon dioxide. This process starts at the surface of the particles, leading to the formation of a perovskite layer surrounding the unreacted particle core. As the carbon dioxide is given off, the surface layer becomes thicker, moving more and more towards the particle center and leading initially to the formation of a perovskite (Ba, La)$TiO_{3-\delta}$ phase with hexagonal symmetry proved by the presence of the peak placed at $2\theta = 26.55^o$ in the related XRD pattern. Usually, the hexagonal form of barium titanate is stable at high temperature (≥ 1462°C) and contains Ti^{3+} ions, which are electrically compensated by double-ionised oxygen vacancies. Eror et al. reported for the first time the stabilization of hexagonal polymorph below 700°C in the low temperature chemically-prepared $BaTiO_3$ [355]. It seems that during the decomposition of the polymeric precursors this hexagonal form is stabilized by a significant increase of the surface energy to the detriment of the volume free energy because of the small particle size. Thus, one can conclude that in such nano-sized systems the hexagonal barium titanate formation precedes the crystallization of the normal $BaTiO_3$ with pseudocubic structure. Some studies showed that the so-called "hexagonal" phase represents actually a defective $BaTiO_3$ containing {111} twins as stacking faults determined by the double-ionized oxygen vacancies induced by the reducing environment [356, 357]. The corresponding satellite peak ($2\theta = 26.55^o$) already mentioned was clearly identified in the XRD pattern of the residue obtained after treatment in nonisothermal conditions at 600°C and it was also found at the detection limit even in the XRD pattern of the residue resulted after the thermal treatment at 730°C (Figure 41(b)). Therefore, the $BaTiO_3$ formation governed by gas release as described by eq. (23) takes place continuously in a large temperature range (550-730°C). As long as the stacking faults still exist, some of CO_2 resulted after the main decomposition step is chemically adsorbed and reacts with the surface barium, leading to the formation of a residual $BaCO_3$. This reaction is possible only in the presence of the hexagonal barium titanate, due to the structural similarity among the spatial groups of these two compounds (D_{6h}^4 for hexagonal $BaTiO_3$ and D_{2h}^{16} for $BaCO_3$). As soon as the hexagonal barium titanate changes toward the cubic form (the <111> stacking faults completely disappeared) the $BaCO_3$ formation is prevented. Eror et al. [355] showed that the adsorption-desorption of the surface $BaCO_3$ is a reversible process and they even postulated an epitaxy relation between the defective barium titanate and the CO_2 retained at the surface. As a result of the stoichiometry deviation created by the surface $BaCO_3$ formation, a small amount of a titanium rich secondary phase, i.e. $BaTi_2O_5$, was segregated at the surface and was identified at the detection limit in the XRD patterns of the higher temperature residues. As the TG curve of Figure 39(b) indicated, at higher temperatures the solid state reaction between the residual $BaCO_3$ and $BaTi_2O_5$ occurs in several steps, in the temperature range of 730–920°C, until the complete conversion into the $Ba_{1-x}La_xTiO_3$ perovskite phase occurs, according to the equation:

$$BaCO_3 + (Ba_{1-2x}La_{2x})Ti_2O_5 \rightarrow 2\ Ba_{1-x}La_xTiO_3 + CO_2\uparrow \qquad (24)$$

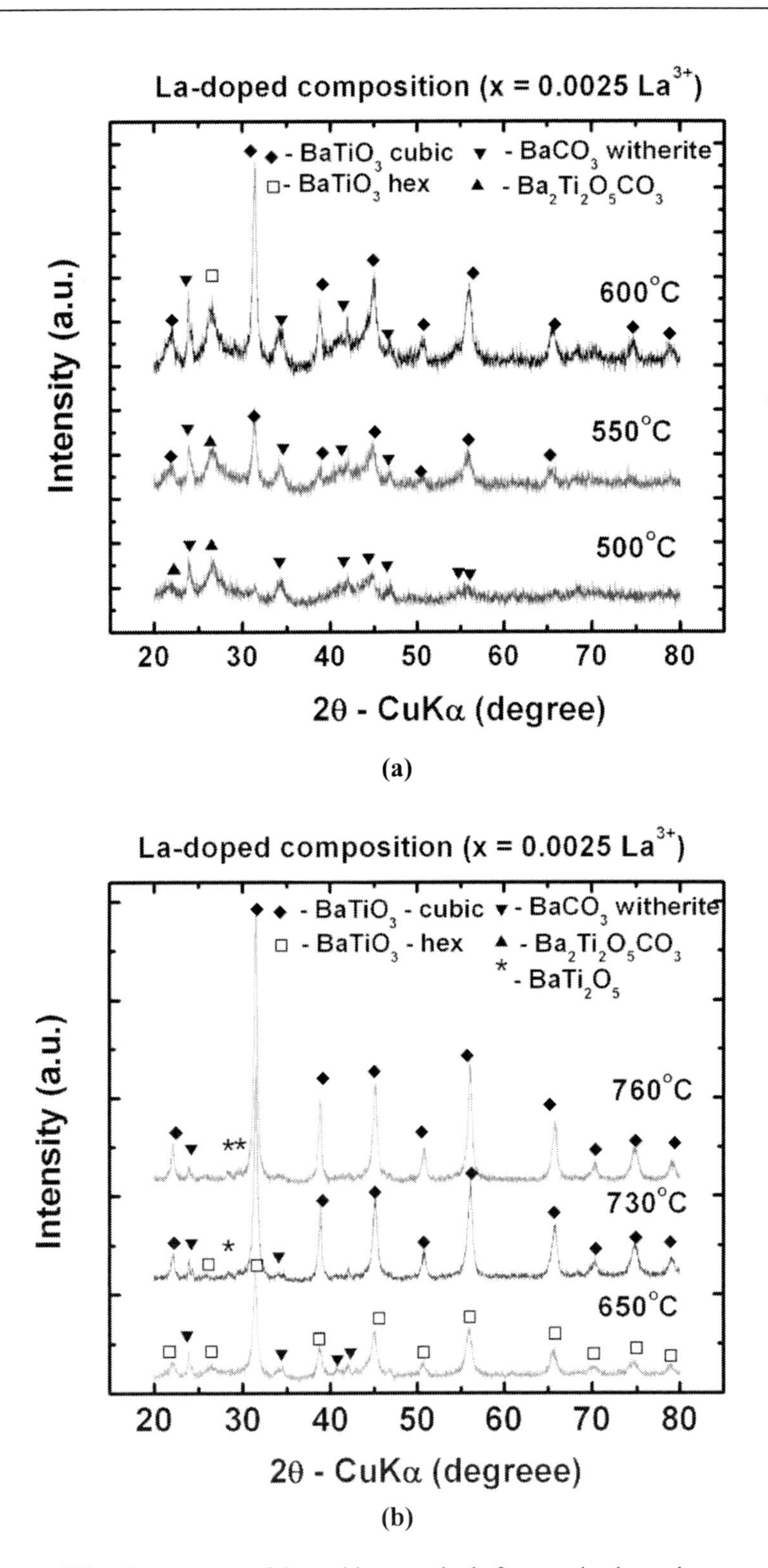

Figure 41. X-ray diffraction patterns of the residues resulted after non-isothermal treatments of the slightly doped precursor (x = 0.0025) at different temperatures: temperature range of 500 - 600°C (a) and temperature range of 650 - 760°C (b).

Taking into account the thermal behavior pointed out by the thermal analysis data, a similar reaction sequence for the decomposition process is expected in the case of the highly La-doped precursor.

3.1.5. Characteristics of Oxide Powders

3.1.5.1. Phase Composition

In order to obtain single phase oxide powders isothermal treatments were carried out at various temperatures in the temperature range of 750–900°C, using a plateau of 2 hours. For the undoped oxide powder the XRD analysis showed that a single phase composition was obtained after thermal treatments at temperatures above 800°C (Figure 42(a)). At lower temperatures, small amounts of $BaCO_3$ were still detected. The same evolution toward the unique phase was noticed for the lanthanum doped compositions (Figure 42(b), (c)).

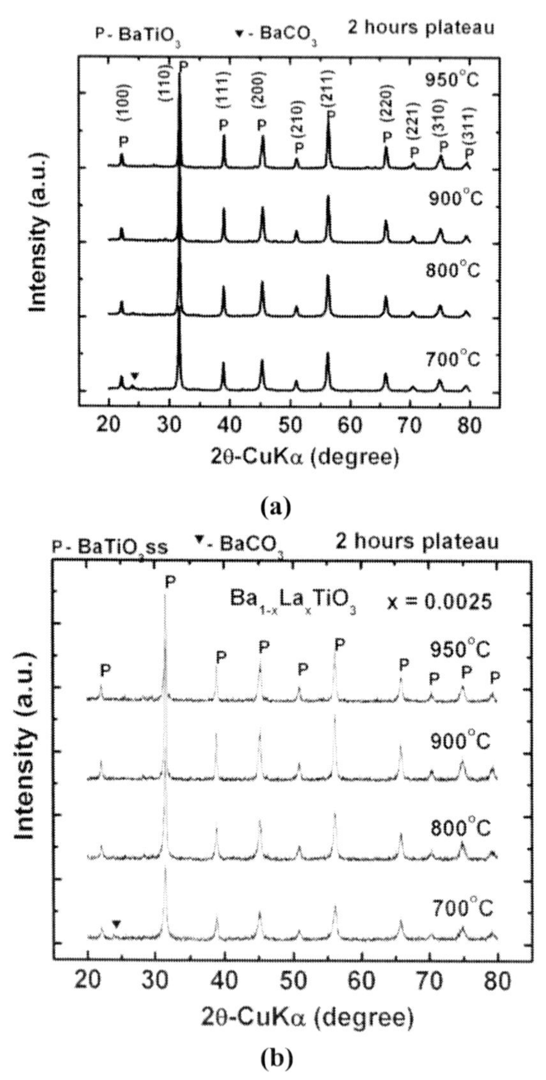

Figure 42. Continued.

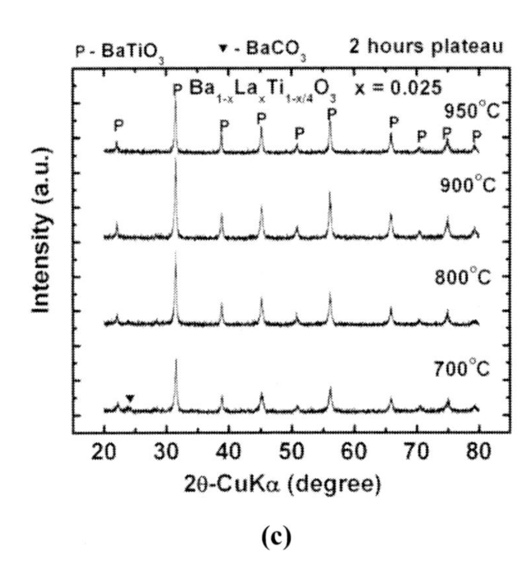

(c)

Figure 42. Room temperature X-ray diffraction patterns for the powders resulted after isothermal treatment performed at various temperatures with 2 hours plateau: undoped $BaTiO_3$ (a); and lanthanum doped compositions: x = 0.0025 (b); x = 0.025 (c).

3.1.5.2. Structure

From the structural point of view, a concurrent evolution of the unit cell parameters against the lanthanum content was noticed for the powders thermally treated at 900°C for 2 hours (Figure 43). This kind of variation leads to a decrease of the tetragonality expressed by the c/a ratio, which shows that a higher lanthanum concentration tends to change the unit cell symmetry from a tetragonal toward a cubic one (Figure 44). Regarding the average crystallite size, an obvious decrease of this structural feature against the dopant concentration was noticed. This evolution is significant in the case of the powder with the highest La content, for which the crystallite size is reduced to almost half the corresponding value for the undoped sample (350 Å compared to 669 Å) – Figure 44.

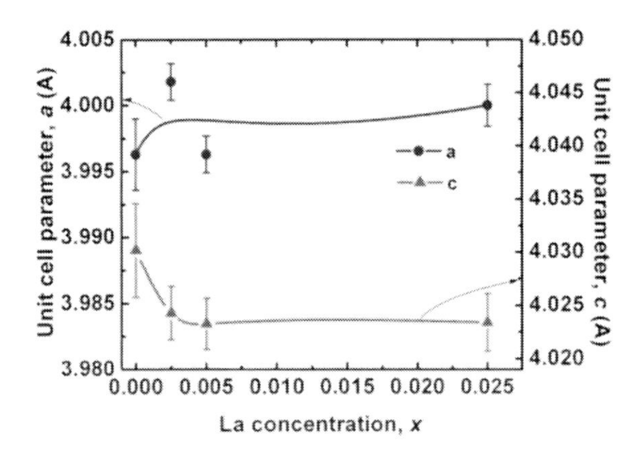

Figure 43. Evolution of the unit cell parameters against the lanthanum content in the oxide powder resulted after thermal treatment at 900°C/2 hours.

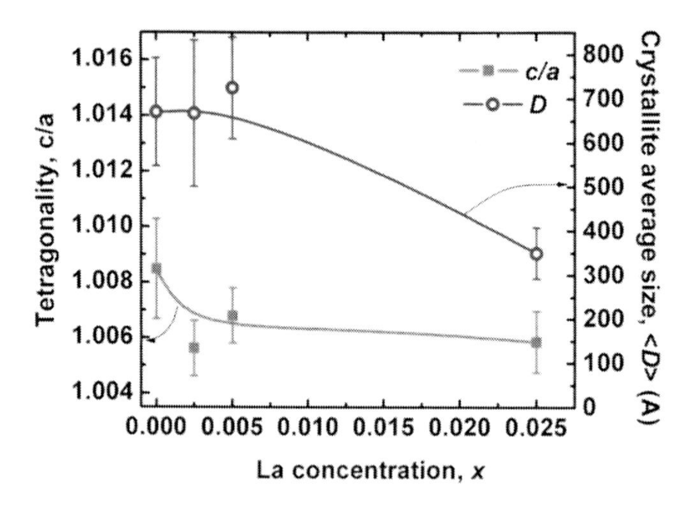

Figure 44. Evolution of the tetragonality (c/a) and of the average crystallite size, D against the lanthanum content in the oxide powder resulted after thermal treatment at 900°C/2 hours.

3.1.5.3. Morphology

Transmission electron microscopy (TEM) investigations performed on the undoped oxide powder resulted after thermal treatment at 900°C/2 hours pointed out the presence of isolated particles with well defined boundaries. An average particle size of ~ 65 nm was estimated from the TEM image of Figure 45. This value is in good agreement with the average crystallite size calculated from XRD data (~ 67 nm) proving the single crystal nature of these particles.

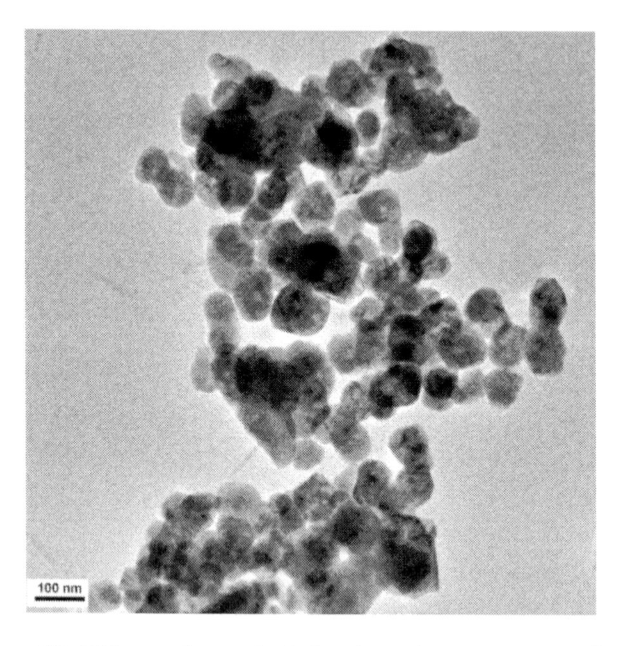

Figure 45. TEM image of $BaTiO_3$ powder resulted after thermal treatment at 900°C for 2 hours (bar = 100 nm).

The high crystallization degree of the particles is clearly indicated by the well marked spots in the selected area electron diffraction (SAED) pattern of Figure 46(a)), as well as by the highly ordered fringes corresponding to the crystalline planes of the perovskite structure, as resulted from HRTEM image presented in Figure 46(b).

For powder with low lanthanum content (x = 0.0025), the TEM image of Figure 47 shows particle size and morphology similar to that of undoped powder. However, unlike the pure $BaTiO_3$ powder, with rather isolated particles, in this case there is a clear trend of formation of polycrystalline hard dispersible aggregates.

Agglomeration and aggregation tendency seems to increase with the rise of the dopant concentration at x = 0.005 (Figure 48). A higher lanthanum content, also causes a clear decrease of the average particle size, which for the powder with x = 0.025 reached a value of ~ 37 nm (Figure 49). As for the undoped powder, this value fits well with the mean size of crystallites (~ 35 nm) determined from X-ray diffraction data.

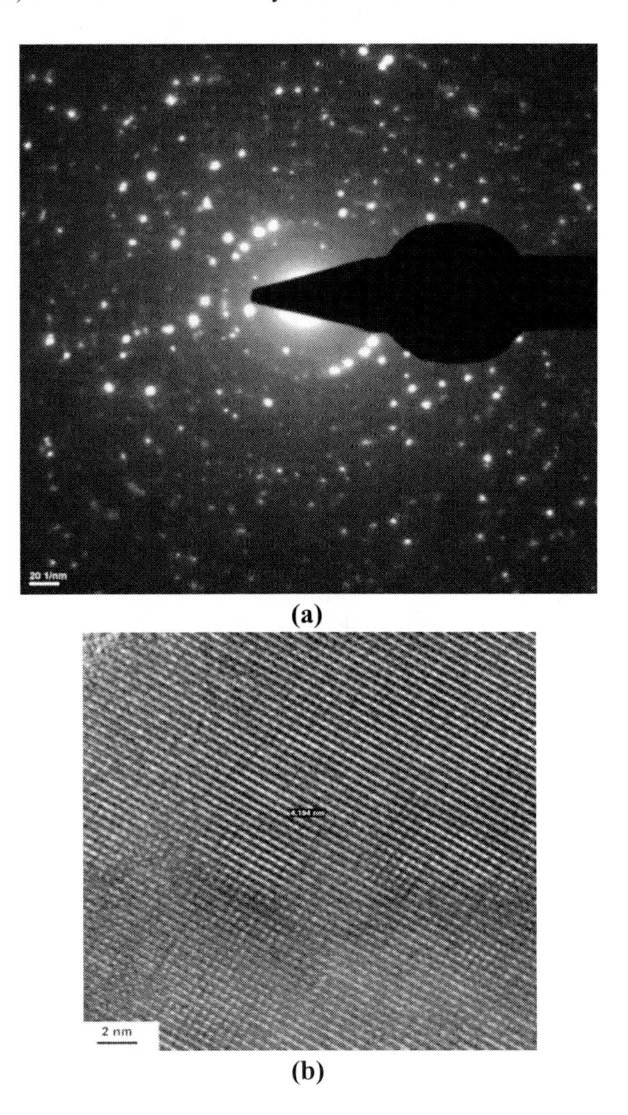

Figure 46. SAED pattern (a) and HRTEM image (bar = 2 nm) (b) of the undoped $BaTiO_3$ powder.

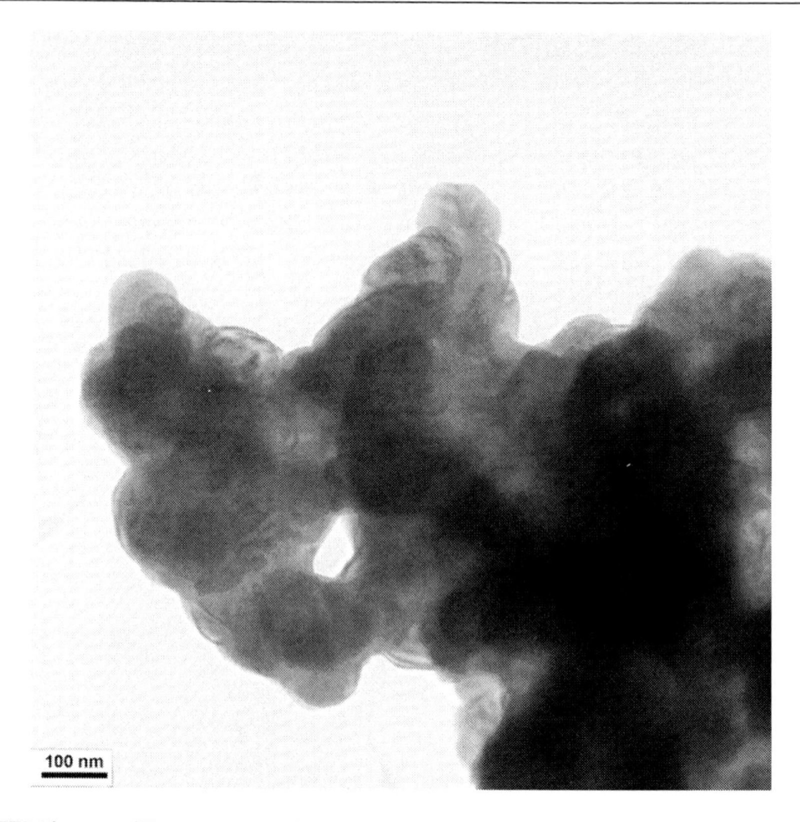

Figure 47. TEM image of $Ba_{0.9975}La_{0.0025}TiO_3$ powder (bar = 100 nm).

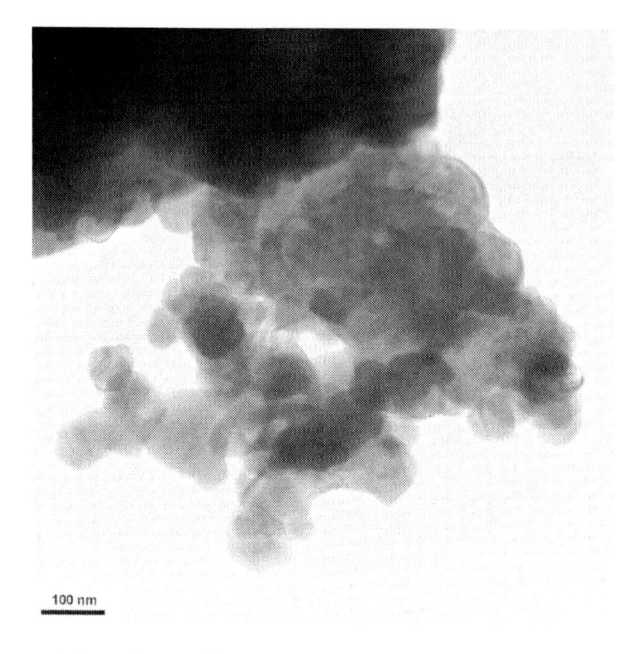

Figure 48. TEM image of $Ba_{0.995}La_{0.005}TiO_3$ (bar = 100 nm).

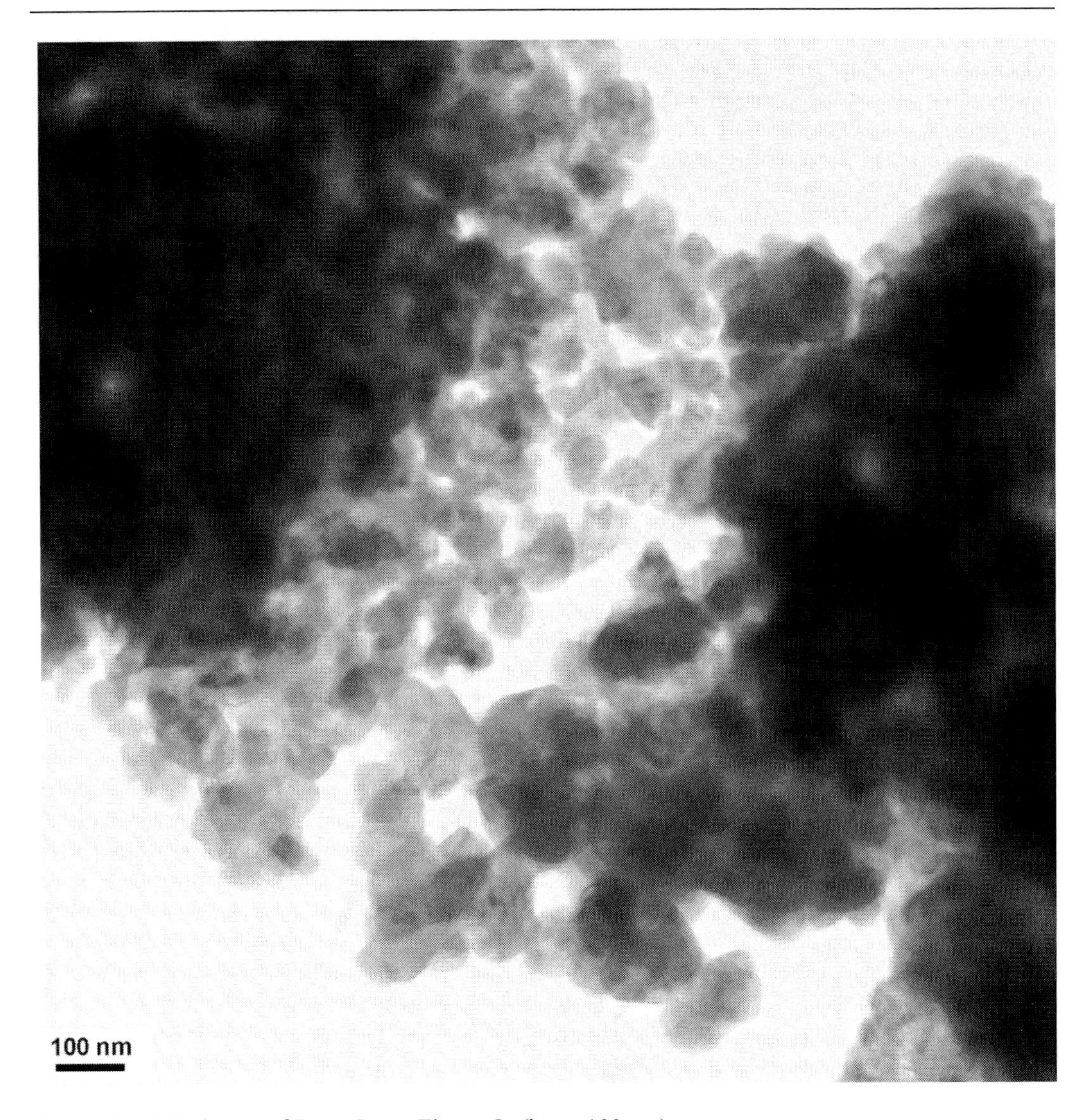

Figure 49. TEM image of $Ba_{0.975}La_{0.025}Ti_{0.99375}O_3$ (bar = 100 nm).

3.1.6. Characteristics of ceramics

3.1.6.1. Microstructure and density

Ceramic materials obtained after sintering at 1300°C for 3 hours were analyzed from microstructural point of view. The undoped ceramic exhibits a dense microstructure, consisting of polyhedral grains, nonuniform as shape and size, but with well defined grain boundaries and almost perfect triple junctions (Figure 50). The estimated average grain size was of ~ 25 μm.

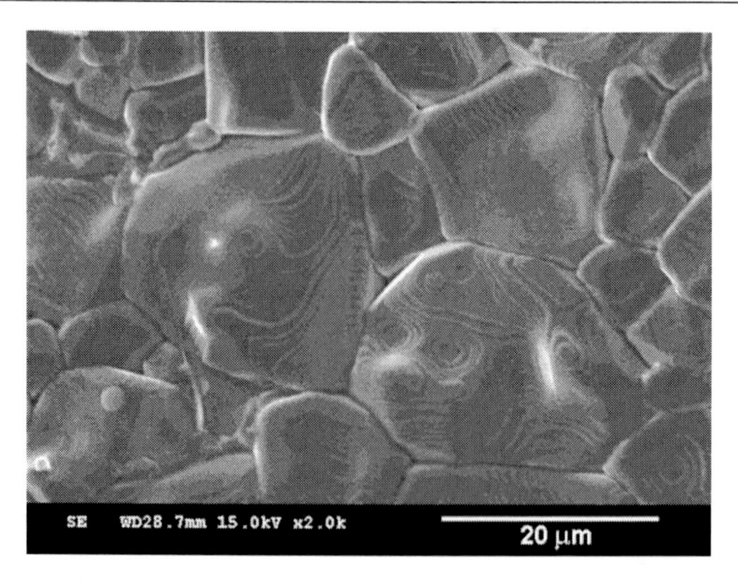

Figure 50. SEM image of $BaTiO_3$ ceramic resulted after sintering at 1300°C for 3 hours.

For sample with low lanthanum concentration a drastic change of the microstructural features was noticed. The presence of the dopant seems to inhibit the grain growth process, so that a bimodal grain size distribution was obtained. This is probably a result of a discontinuous grain growth, which usually occurs in the titanates systems. Therefore, the ceramic consists of well-faceted, polyhedral larger grains with sizes ranged between 10–30 μm, which coexist with more uniform, smaller grains, with an average size of only 2 μm (Figure 51).

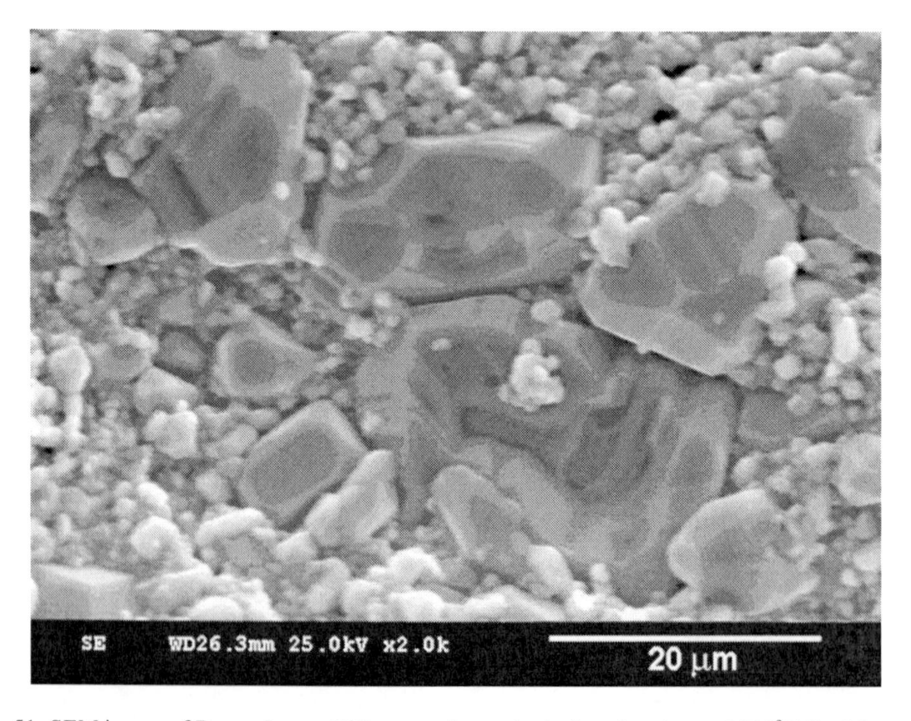

Figure 51. SEM image of $Ba_{0.9975}La_{0.0025}TiO_3$ ceramic resulted after sintering at 1300°C for 3 hours.

The increase of the lanthanum content leads to further microstructural disturbance by preventing grain growth and generating intergranular pores (Figure 52). The densification is affected by the porosity increase, especially in the case of the highly-doped sample. Thus, for the ceramic with x = 0.025 a more uniform microstructure consisting of small grains (of ~ 1 μm) and a rather mono-modal grain size distribution was observed (Figure 53).

Figure 52. SEM image of $Ba_{0.995}La_{0.005}TiO_3$ ceramic resulted after sintering at 1300°C for 3 hours.

Figure 53. SEM image of $Ba_{0.975}La_{0.025}Ti_{0.99375}O_3$ ceramic resulted after sintering at 1300°C for 3 hours.

The influence of the lanthanum concentration on the average grain size and relative density (determined by the hydrostatic method) is presented in Figure 54.

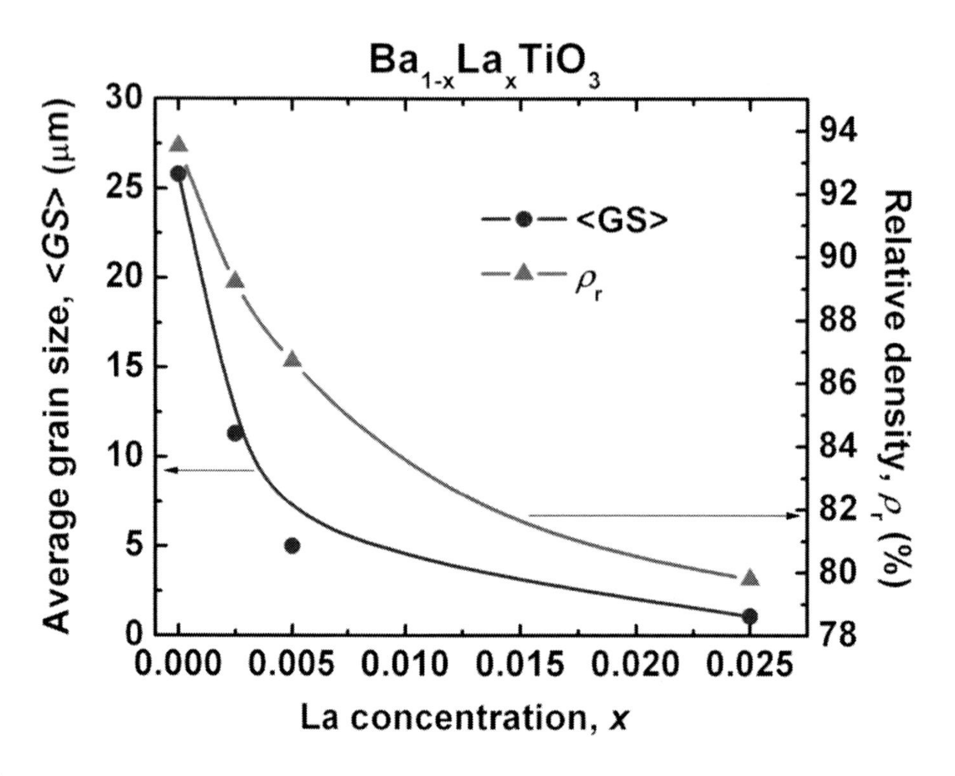

Figure 54. Evolution of grain size and relative density against the lanthanum content for the ceramics sintered at 1300°C for 3 hours.

3.1.6.2. Electrical Behavior

The dielectric properties (relative permittivity and dielectric losses expressed by tan δ) show higher values for samples with lower dopant concentration (Figure 55(a), (b)). The highest effective permittivity for the ceramic with x = 0.005 is probably caused by an increase of the interfacial polarization induced by the heterogeneous distribution of the defects in the crystalline grains (Figure 55(a)). In this case, the grains consist of semiconducting cores because of the electronic compensation of the additional charge of the donor dopant and highly resistive grain boundaries called also „depletion layers", rich in acceptor centers as cation vacancies, which are ionized by trapping electrons and thus prevent conduction [335]. Consequently, in this case of ceramics prepared by the sol-gel alkoxide route, the change of the charge compensation from an electronic toward an ionic mechanism seems to take place at a higher lanthanum concentration than that one (x = 0.3 at.%) reported for the ceramics resulted after sintering in similar conditions, but processed by the conventional solid state reaction method [358]. As the lanthanum content increases to x = 0.025, the permittivity decreases in all the temperature range analyzed here. In this case, the evolution of the dielectric constant must be correlated with the microstructure. Thus, for highly donor doped $BaTiO_3$, because of the drastic decrease of the grain size, the resistive depletion layer is extended toward the grain centre inducing a homogeneous distribution of the cation vacancies into the grains, so that all the dopant donors are compensated by ionized cation vacancies and the material behaves predominantly as a classical dielectric. The disappearance of the two adjacent regions with different conducting properties in the ceramic grains leads not only to the conduction suppressing, but also to the cancellation of the interfacial polarization, this resulting in a decrease of the overall dielectric permittivity to normal values for the substituted $BaTiO_3$ ceramics. Thus, for the ceramic with x = 0.025, the dielectric constant at room temperature presents a value of ~ 6950, while the value of this property at the ferroelectric - paraelectric phase transition temperature (T_C) is of ~ 13800.

The highest dielectric losses (tan $\delta \sim 6.5 \times 10^{-1}$ at room temperature and at 1 kHz frequency) were recorded for the composition with x = 0.005 because of the already mentioned conducting character of the grain cores. For the ceramic with a higher La concentration (x = 0.025), the lower room temperature dielectric losses (tan $\delta \sim 5.4 \times 10^{-2}$), closer to the value specific to the undoped sample (tan $\delta \sim 1.2 \times 10^{-2}$), originate in the decrease of the conducting properties of the material (Figure 55(b)).

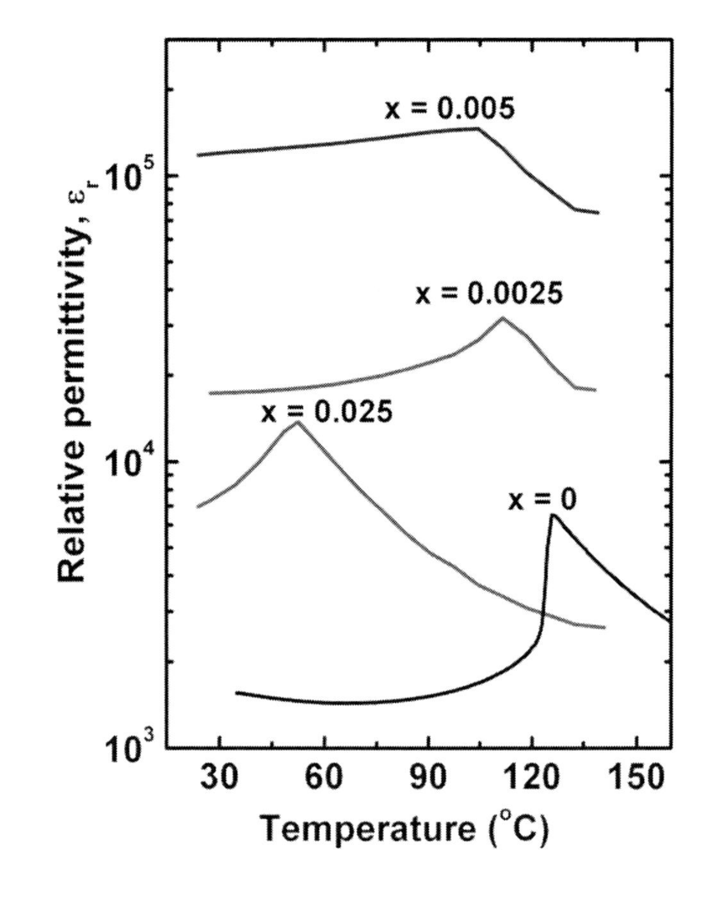

(a)

Figure 55. Continued.

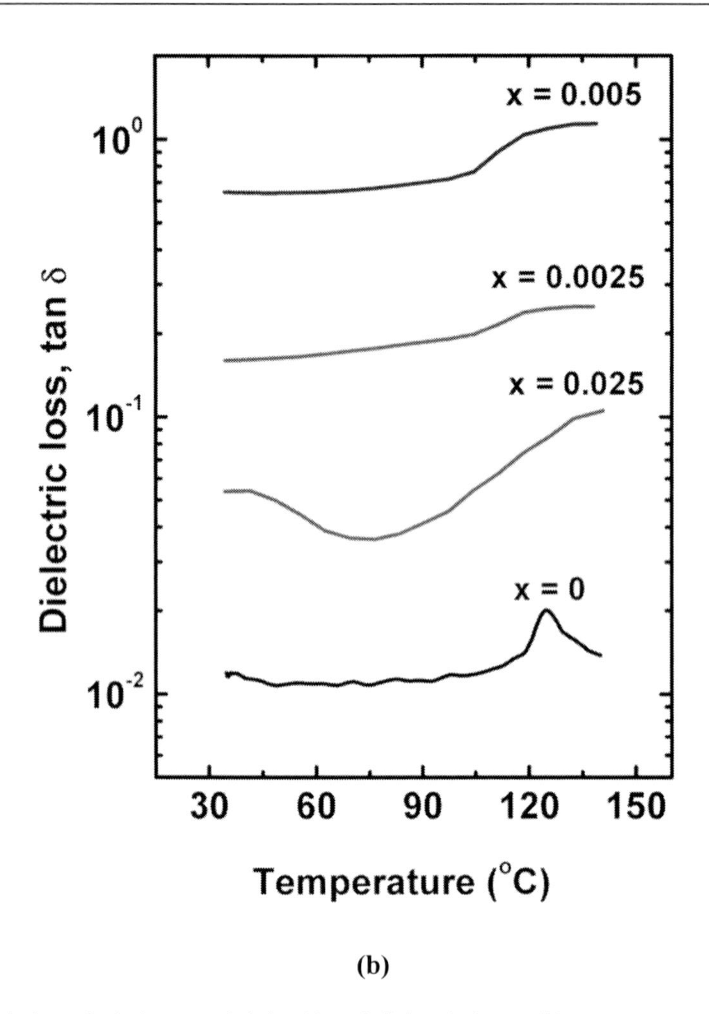

(b)

Figure 55. Evolution of relative permittivity (a) and dielectric losses (b) versus temperature.

Regarding the influence of the temperature on the dielectric properties, a shift of the paraelectric-ferroelectric phase transition temperature with the rise of the lanthanum content was noticed (Figure 56). It has been postulated that the decrease in T_C with increasing La^{3+} doping is a result of the conjugated effect of larger Ba^{2+} ions ($r(Ba^{2+})$ = 1.61 Å) partial replacing and Ti^{4+} vacancies formation caused by substitution with trivalent ions of a smaller ionic radii ($r(La^{3+})$ = 1.36 Å).

For the ceramic with the highest lanthanum concentration (x = 0.025) investigated here, the Curie temperature of 52.5°C is frequency independent and significantly smaller to those one of ~ 75°C (at 100 kHz) found by West et al. [300] for the ceramic with the same composition processed by the solid state reaction method. In our case, this variation corresponds to a dramatic -29 °C decrease in T_C per at. % La^{3+}, higher than that one of -24°C per at. % La^{3+} reported by Morisson et al. [359].

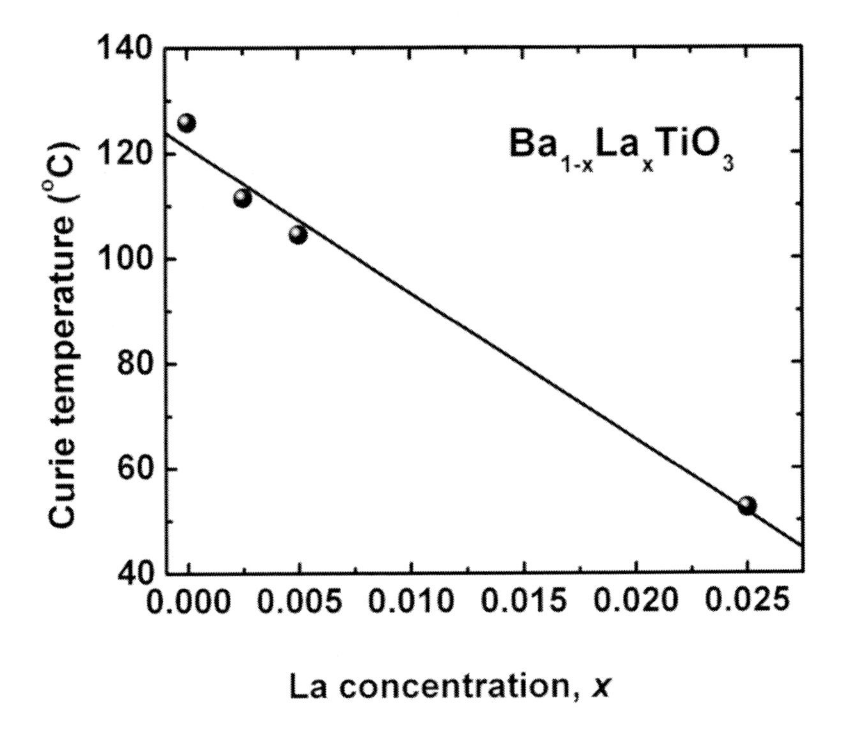

Figure 56. Curie temperature versus dopant concentration in sol-gel La-doped $BaTiO_3$ sintered at 1300°C for 3 hours.

Further electrical measurements for $BaTiO_3$ ceramics doped with very small amounts ($x < 0.003$) of lanthanum are required in order to determine the frequency and temperature dependence of the conductivity and the magnitude of the related PTCR jump.

3.2. Sol-gel Ba(Ti, Zr)O_3 Thin Films

3.2.1. Choice of Compositions

Unlike the aliovalent dopants, the solubility of the homovalent cation species in the perovskite ABO_3 lattice of barium titanate is significantly higher, so that, in this case, is more correct to discuss in terms of solid solutions than in terms of doping.

With suitable *B*-site substitutions in $BaTiO_3$ ceramics, not only the Curie temperature is shifted toward lower values, but also the cooperative dipolar long-range order formed by the off-center displacement of Ti^{4+} ions in their TiO_6 octahedra is disrupted. This often leads to a significant broadening of the permittivity – temperature dependence in the range of Curie temperature, resulting in the so-called *diffuse phase transition*. From this point of view, zirconium is well known as one of the most effective solutes in $BaTiO_3$ [360]. Zr^{4+} ion has a larger ionic radius (0.87 Å) than that of Ti^{4+} (0.68 Å), inducing the expanding of the perovskite lattice. In $Ba(Ti_{1-x}Zr_x)O_3$ (BTZ) system, Zr/Ti ratio is a very important parameter which tailors the type of the ferroelectric – paraelectric phase transition and its characteristic Curie temperature [276]. Thus, it was found that BTZ bulk ceramic exhibits a pinched phase transition at $x = 0.15$, where all the three dielectric constant peaks coalesce into a single broad maximum [361, 362]. Moreover, Zr^{4+} ion is chemically more stable than Ti^{4+}. Consequently, replacing Ti by Zr would depress the conduction by small polarons hopping between Ti^{4+} and Ti^{3+} and it would also decrease the leakage current. This makes the barium zirconate titanate $Ba(Ti, Zr)O_3$ (BTZ) a promising candidate as an alternative material to barium strontium titanate $(Ba, Sr)TiO_3$ in DRAM, NVRAM and other microwave applications.

For this reason, the properties of BTZ thin films, deposited by various methods as magnetron sputtering [363-366], pulsed laser deposition (PLD) [367] and chemical solution deposition (CSD) [368] were also intensively studied in the last decade. Several authors reported data regarding the structural, microstructural, morphological, surface characteristics, as well as dielectric and non-linear properties of BTZ thin films prepared by the sol-gel method [369-379]. However, the only few papers describing the microstructure and optical properties, refers either to BTZ thin films with low Zr content ($x = 0.05$) [379, 380], or to those of higher Zr concentration ($0.2 \leq x \leq 0.5$) [381-386]. No papers reporting optical properties of the sol-gel $BaTi_{0.85}Zr_{0.15}O_3$ films were found in the literature.

3.2.2. Preparation of BTZ Thin Films

BTZ films with the nominal $BaTi_{0.85}Zr_{0.15}O_3$ composition have been prepared by the alkoxide route of the sol-gel method, using the following precursors: barium acetate $C_4H_6BaO_4$, tetraethyl orthotitanate (TEOT) $Ti(OC_2H_5)_4$, and zirconium propoxide solution (70% in propanol) $Zr(OC_3H_7)_4$. Each of the mentioned solid precursors has required its own solvent, *i.e.* acetic acid ($C_2H_4O_2$) for the barium acetate and ethanol C_2H_5OH for TEOT, respectively. The hydrolysis and condensation reactions have been acid catalyzed by the nitric acid solution 65%. The same acid was used in order to establish the desired value of the pH.

The synthesis has been performed in a three-necked flask placed in a water bath (in order to ensure and control the temperature) and in conditions of continuous stirring. The mechanical stirrer was provided with the possibility of heating. In the central neck of the synthesis vessel, a condenser has been placed, in order to ensure the back-flow. A dropping funnel has been attached to one of the lateral necks, to ensure the dropwise addition of the reacting substances. The third neck is necessary for allowing the simultaneously addition of the reagents in the reaction mixture.

The synthesis of the sol from which the films have been obtained was accomplished under continuous stirring, in more steps. First of all, TEOT has been added drop by drop in a slightly acidulated quantity of ethanol. Then, the zirconium alkoxide has been included in the reaction mixture, followed by the simultaneous addition of the water and of the nitric acid until the total disappearance of the opalescence. From this moment the heating and the back flow were started. Separately the barium acetate was solved in acetic acid and the resulting solution was the last introduced in the three-necked flask. The reaction mixture has been maintained under stirring and back flow for three hours, at 50°C. The final pH was of 5.5.

The resulting sol has been used in order to obtain the BTZ films. They have been deposed on Si/SiO_2 wafers, using the dipping procedure, with a withdraw rate of $5\ cm \cdot min^{-1}$. Multi-layers coatings have been prepared (from 1 to 6 layers), each of one being thermally treated at 300°C for 1 min, with a heating rate of $1°C \cdot min^{-1}$. A final thermal treatment of 2 hours at 800°C, with a heating rate of $5°C\ min^{-1}$ has been accomplished for all the prepared coatings.

3.2.3. Characterization of BTZ Thin Films

3.2.3.1. Phase Composition and Structure

The X-ray diffraction patterns corresponding to $BaTi_{0.85}Zr_{0.15}O_3$ thin films of different thickness induced by a different number of deposits were presented in Figure 57. A predominant amorphous nature was noticed for thinner films, consisting of one and two deposits, respectively. As the thickness increases due to an increased number of deposits, the crystallization process progresses, so that in the thin film with three deposited layers, the appearance of the perovskite phase was noticed (Figure 57(a)). Beside this major phase, some non-equilibrium compounds as TiO_2 (rutile) and Ba_2TiO_4, were also identified. The small angle higher background of the X-ray diffraction pattern shows that a certain amount of amporhous phase is still present in the three layered-films.

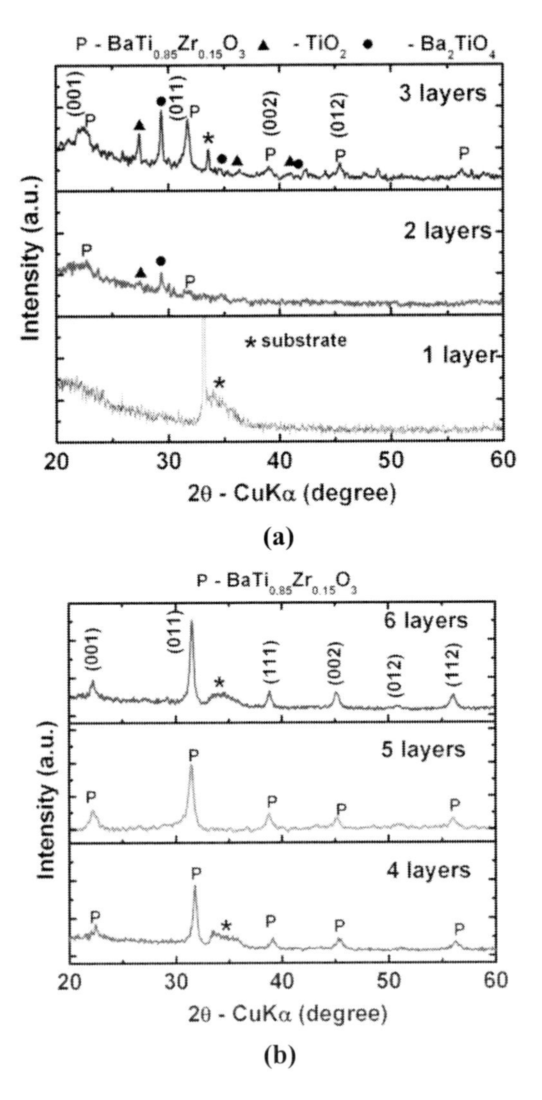

Figure 57. X-ray diffraction patterns of $BaTi_{0.85}Zr_{0.15}O_3$ thin films with different number of deposits: 1 - 3 layers (a); 5 - 6 layers (b).

As more layers were deposited, leading to the overall thickness increase, an enhancement of the main diffraction peaks of the perovskite, beside the evolution toward a single phase composition was pointed. Thus, only the phase corresponding to the $BaTi_{0.85}Zr_{0.15}O_3$ solid solution was identified for the well-crystallized multilayer thin films consisting of 4–6 deposits (Figure 57(b)). This occurred most likely due tot the successive thermal treatments carried out after the deposition of each layer, leading to the quantitative reaction between TiO_2 and the rich-barium compound ($BaTi_2O_4$) and thereby resulting in the increase of the crystallinity of the multi-deposit BTZ films.

Parameters to define the position, magnitude, shape and integral breadth or full width at half maximum of profile (FWHM) of the individual peaks are obtained using the pattern fitting and profile analysis of the original X-ray 5.0 program. The lattice constants calculation is based on the Least Squares Procedure (LSP) using the linear multiple regressions for several XRD lines, depending on the unit cell symmetry. To deconvolute size-D and strain-S broadening from the XRD spectra, the multiple line analysis and integral breadth methods applied to the observed profiles (properly approximated with the Pearson's VII analytic functions) were used.

A pseudocubic symmetry of the unit cell was found for all the thin films analyzed here. However, the profiles of the diffraction peaks are quite complicated and for this reason, it is more likely that this so-called pseudocubic phase refers not to a normal cubic symmetry, but rather to a distorted structure, or to a mixture of several BTZ phases of different symmetries, as it was also reported for undoped $BaTiO_3$ multilayer thin films [387]. It is well known that the calculation of the lattice parameters based on the XRD data involves an averaging over more than ~ 10000 unit cells. Therefore, this method is not very sensitive to small structural deviations in local ordering of cations. In order to provide more detailed information concerning the structural features, further Raman investigation are required.

It has been noticed that, as the number of deposits increased, an obvious decreasing trend of the lattice parameter *a*, inducing the consequent shrinkage of the unit cell volume occurred (Figure 58(a)). This variation is normal, taking into account that the multi-deposit films exhibit higher crystallization degree, which causes the reducing of the inter-atomic spaces. The decrease of the lattice constant is not linear, being more pronounced between the sample containing 3 and 4 layers, respectively. This corresponds to a "critical" number of layers, for which the composition of the system changes, from a high amount of amorphous phase toward a well-crystallized perovskite. Thus, no difference between the unit cell volume of the samples BTZ5 and BTZ6 was noticed. Therefore, we can conclude that, for the five-layer thin film, the crystalline structure is already completed and further deposition processes can contribute only to some microstructural changes, as porosity, grain size and surface topography.

Figure 58(b) shows that the evolution of the average crystallite size, $<D>$ versus the deposits number shows a maximum for the four-deposit BTZ4 sample, whereas the internal microstrains are almost independent on the number of the constituent layers in BTZ thin films.

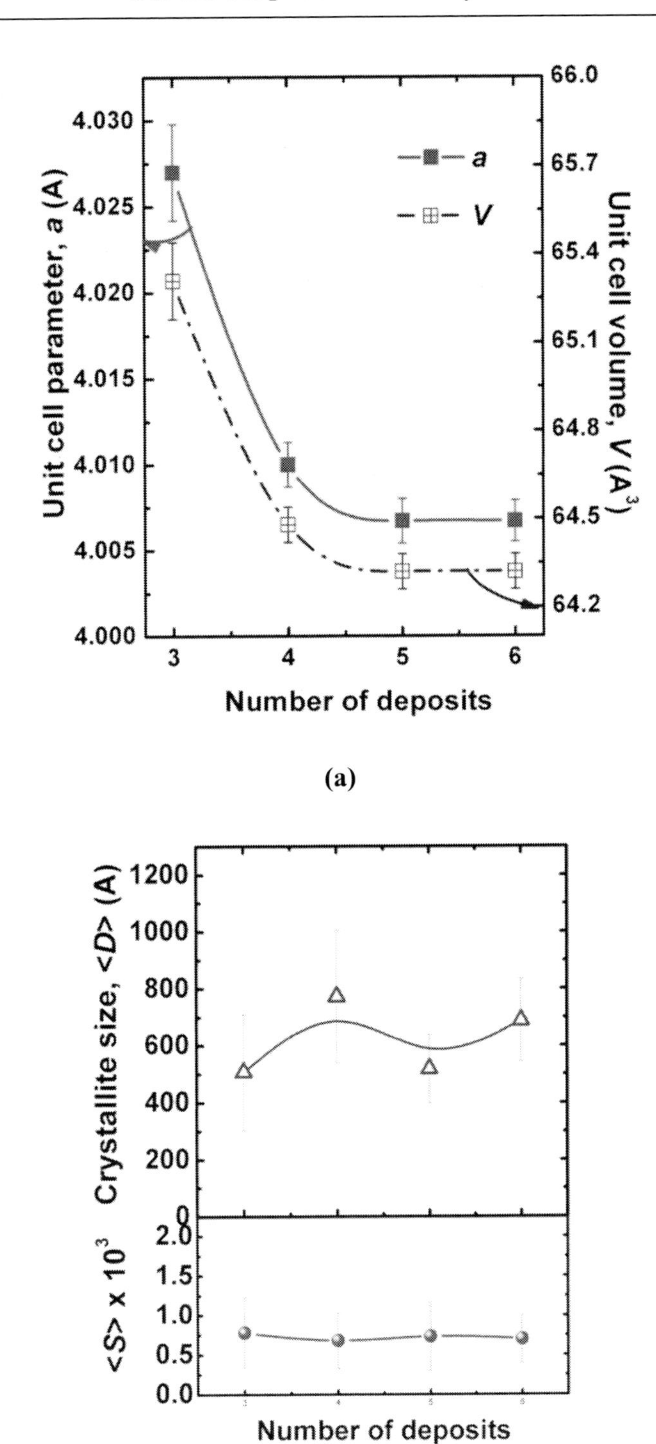

(a)

(b)

Figure 58. Evolution of the structural parameters against the film thickness: lattice constant, a, unit cell volume, V – (a) and average crystallite size, <D>, internal strains, <S> – (b).

3.2.3.2. Surface topography and microstructure

Figure 59(a), (b) depicts the 2D and 3D topographic AFM images of the four layered (BTZ4) and six-layered (BTZ6) thin films. As can be seen from Figure 59 the surfaces of the BTZ thin films are cracks-free, without exfoliations (meaning a good adherence of the deposited layers) and covered with grains whose morphology depends on the preparation conditions. The four-layered BTZ4 film shows larger and smaller grains (Figure 59(a)), while the six-deposit BTZ6 film exhibits a more uniform surface, with grains of similar shape and size (Figure 59(b)).

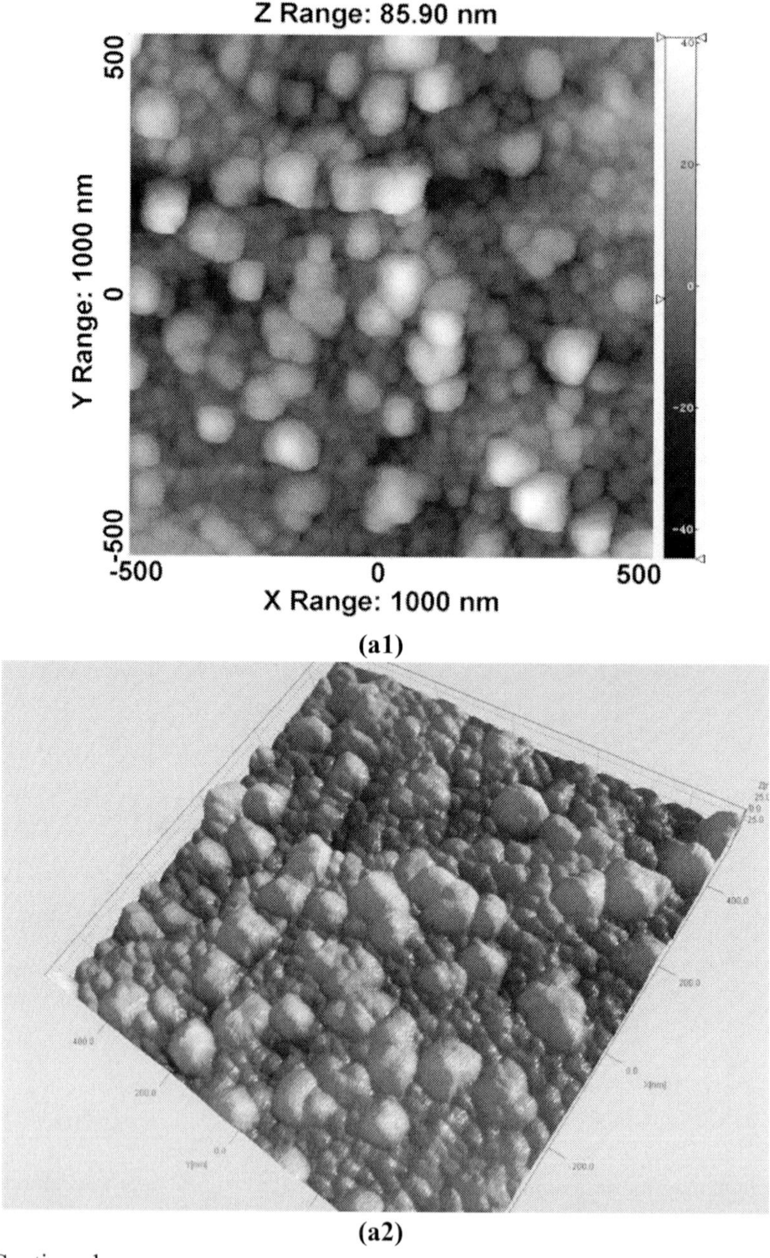

Figure 59. Continued.

(b1)

(b2)

Figure 59. 2D and 3D topographic AFM images of BTZ4 (a) and BTZ6 (b) thin films.

Further, a statistic grain size distribution analysis was performed, in order to obtain quantitative data. Images from Figure 59 were analyzed based on the so-called "watershed method without gradient". The 2D and 3D topographic AFM images from Figure 59(a) indicate that the BTZ4 sample exhibits a complex, bimodal grain size distribution, as a result of a discontinuous grain growth process. Consequently, the microstructure consists of smaller grains with a mean diameter of 36 nm (Figure 60 (a)), which coexist with larger grains (with a mean size of 95 nm), formed by the coalescence of the smaller ones (Figure 60(b)). Anyway, in the BTZ4 thin film, the larger grains seem to predominate. If we take into account the average crystallite size of 77.3 nm determined by XRD, and the mean diameter values of larger and smaller grains estimated from the histograms of Figures 60(a), (b), then an approximate calculation indicates that the BTZ4 sample consists of ~70 % larger grains and only 30 % smaller grains. This could explain also the maximum value of the average crystallite size calculated from XRD data for the BTZ4 sample.

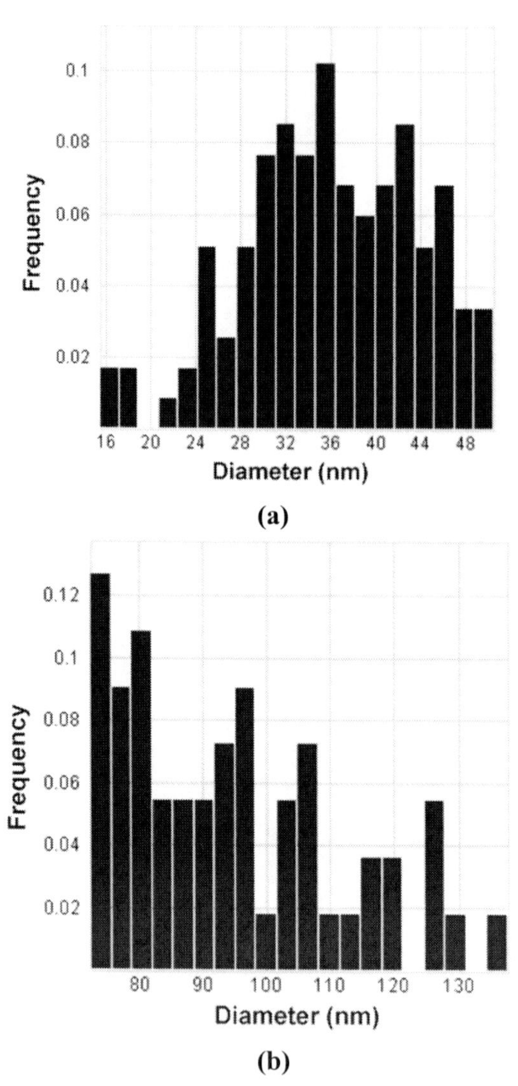

Figure 60. Continued.

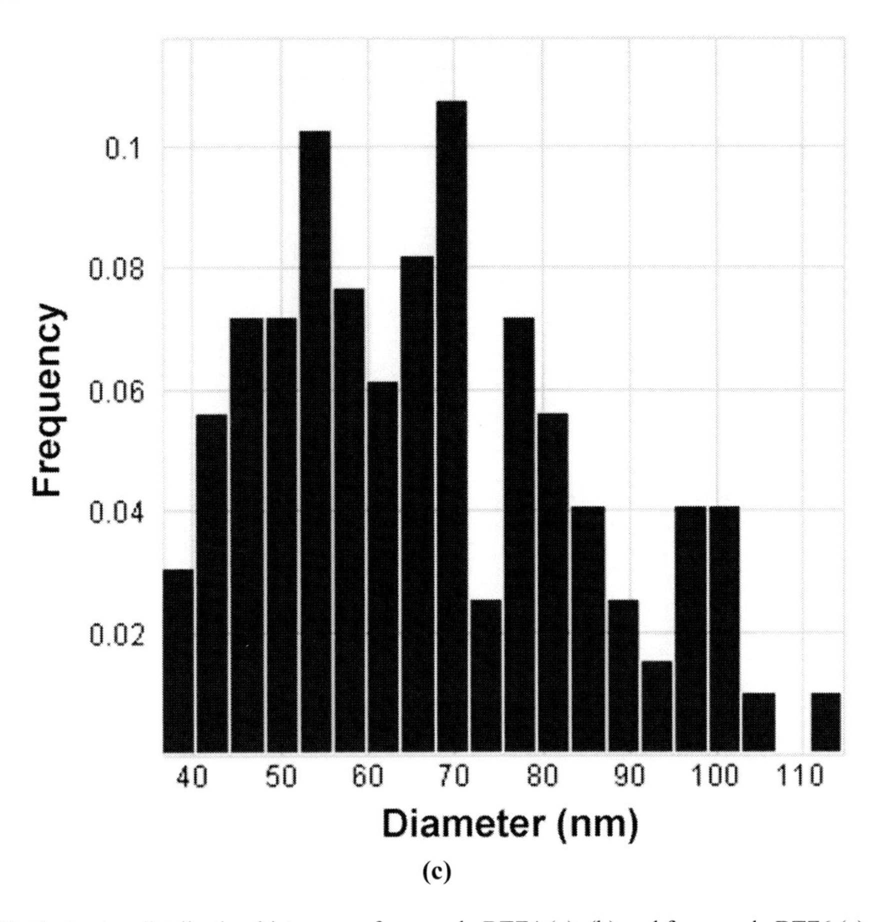

Figure 60. Grain size distribution histograms for sample BTZ4 (a), (b) and for sample BTZ6 (c).

For the six-deposit thin film (BTZ6), the surface microstructure seems to be more uniform (Figure 59(b)), tending toward a large but monomodal grain size distribution. In this case, a mean diameter of 66 nm was found, with the observation that most of the grains are around 70 nm, as can be noticed in the corresponding histogram of Figure 60(c). This microstructural feature of the film surfaces indicates the change toward a continuous grain growth, probably activated by surface diffusion mechanism and induced by the successive thermal treatments subjected by the bottom layers in the thin multi-deposit BTZ6 thin film. The mean grain size value estimated from the histogram of Figure 60(c) is consistent with the average crystallite size of 68.8 nm calculated from the XRD data, proving the single crystal nature of the grains of BTZ6 film.

Amplitude parameters giving information about the statistical average properties have been calculated with the SPIP program, based on the images shown in Figure 59, at the scale of 1×1 μm^2, as follows:

- root mean square (RMS) roughness, $\mathbf{S_q}$, defined as:

$$S_q = \sqrt{\frac{1}{MN}\sum_{k=0}^{M-1}\sum_{l=0}^{N-1}\left[z(x_k, y_l) - \mu\right]^2}\,, \tag{25}$$

where μ is the mean height, defined as:

$$\mu = \frac{1}{MN}\sum_{k=0}^{M-1}\sum_{l=0}^{N-1} z(x_k, y_l); \tag{26}$$

- surface skewness, $\mathbf{S_{sk}}$, defined by the formula:

$$S_{sk} = \frac{1}{MNS_q^3}\sum_{k=0}^{M-1}\sum_{l=0}^{N-1}\left[z(x_k, y_l) - \mu\right]^3; \tag{27}$$

- surface kurtosis, $\mathbf{S_{ku}}$, defined as:

$$S_{ku} = \frac{1}{MNS_q^4}\sum_{k=0}^{M-1}\sum_{l=0}^{N-1}\left[z(x_k, y_l) - \mu\right]^4; \tag{28}$$

- peak-peak height, $\mathbf{S_y}$, defined by:

$$S_y = z_{max} - z_{min} \tag{29}$$

(z is the normal axis to the (x, y) plane of the surface);

Due to the fact that the parameters depend on the definition of a local minimum and a local maximum, it should be mentioned that a local minimum is defined as a pixel where all eight neighboring pixels are higher, while a local maximum as a pixel where all eight neighboring pixels are lower. The values of these parameters are shown in Table 22.

Table 22. Amplitude parameters giving information about the statistical average properties on the surfaces of BTZ4 and BTZ6 thin film.

Sample	S_q (RMS) [nm]	S_{sk}	S_{ku}	S_y [nm]
BTZ4	12.71	1.40	5.62	85
BTZ6	5.76	0.77	4.76	44.41

For BTZ films, the increase of the deposits number (from 4 to 6) induces a smoothing of the surface, so that a smaller value of the roughness (of only 5.76 nm) was obtained for BTZ6 compared to more than double the value corresponding to BTZ4 (S_q = 12.71 nm). This difference also originates in the dissimilar grain size distribution of the mentioned BTZ samples. Thus, in the case of the film with four deposits (BTZ4), the bimodal grain distribution generates a more corrugated surface, than that one specific of the six-deposit thicker film (BTZ6).

Since the surface skewness, S_{sk}, describing the asymmetry of the height distribution histogram is positive, for both BTZ4 and BTZ6 samples one can speak in terms of a flat surface with peaks originating in the grains (crystallites) formation. For lower S_{sk} values (ideally close to zero) it can be assumed a (near perfect) Gaussian distribution of the grains height.

The surface kurtosis, S_{ku}, related to the "peakedness" of the surface topography, was found to have values larger than three (the typical value for a Gaussian height distribution), especially for sample BTZ4. This shows sharper height distributions (maybe originate in the bimodal grain size distribution consisting of larger and smaller grains).

Regarding the peak-peak height, S_y, defined as the height difference between the highest and the lowest pixels from the analyzed images, it can be observed a tendency which is in agreement with the roughness and grains size curves. Here a much pronounced value is found also for sample BTZ4 that could be also due to the bimodal grain size distribution.

In order to investigate the self-similarity and the isotropy/anisotropy of the BTZ films, an investigation with the help of the SPIP™ software package has been also performed. This is a useful approach for evaluating the morphology of the whole surface of a film based on AFM images. The mean fractal dimension (MFD), calculated at different angles by analyzing the Fourier amplitude spectrum, is shown in Table 23 for both BTZ films and exemplified for sample BTZ6 in Figure 61(a).

Table 23. Spatial parameters of the analyzed BTZ thin films: MFD means mean fractal dimension, S_{tdi} – texture direction index and Str_{37} – texture aspect ratio

Sample	MFD	S_{tdi}	Str_{37}
BTZ4	2.92	0.75	0.83
BTZ6	2.84	0.66	0.68

As it can be observed, all surfaces exhibit a self-similar behavior, reflecting the property that a part of the surface is similar to the whole surface.

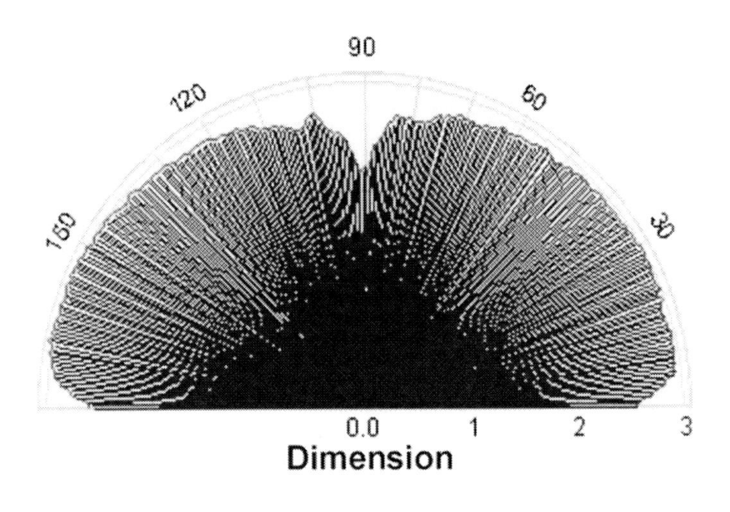

(a)

Figure 61. Continued.

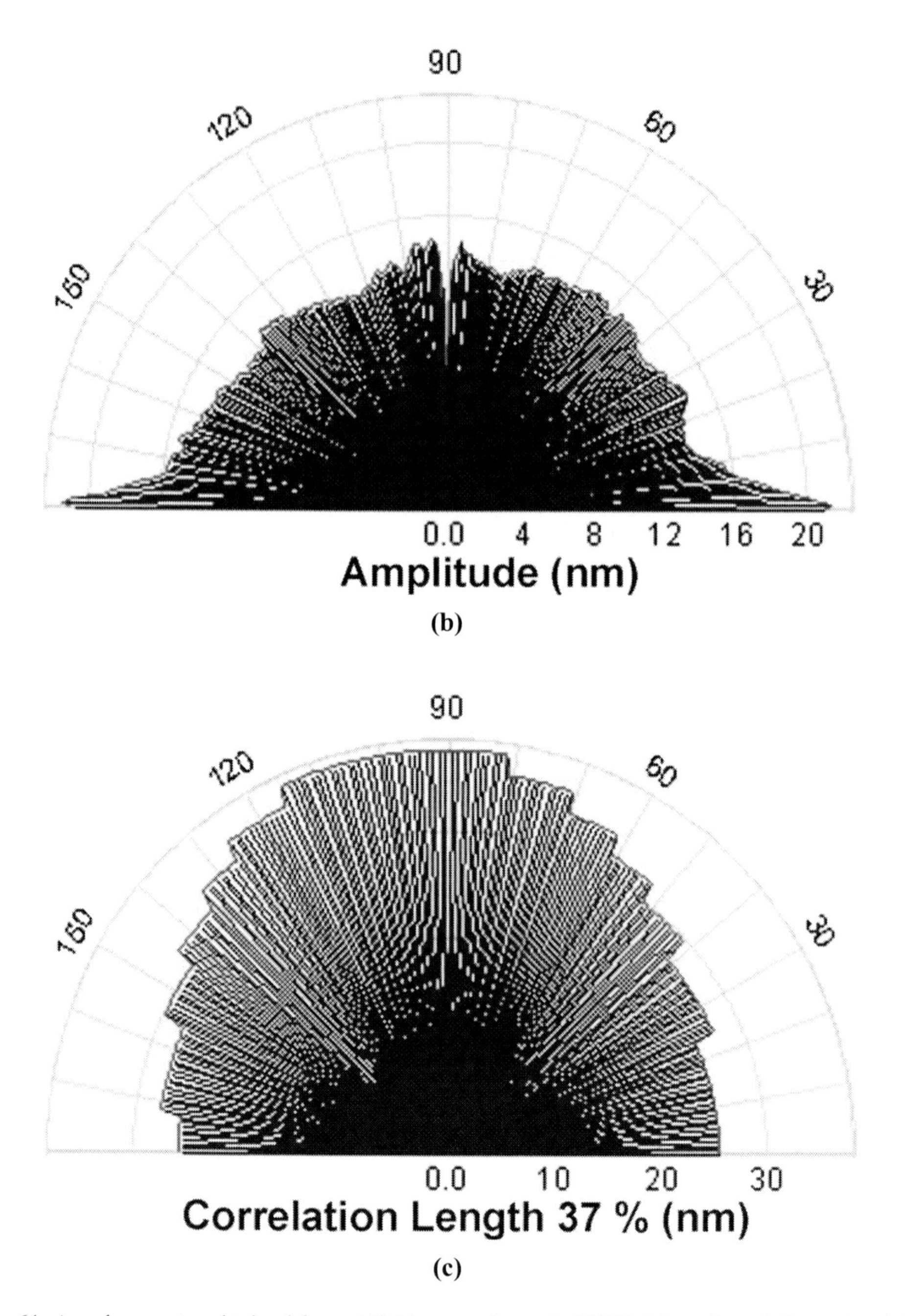

(b)

(c)

Figure 61. Angular spectra obtained from AFM image of sample BTZ6: Mean fractal dimension (a), amplitude (b), and correlation length plots (c).

The upper limit of MFD, found for sample BTZ4, can be explained as a result of a rough surface (due to the spatial distribution of the grains and pores). From the angular spectrum exemplified in Figure 61(b) for sample BTZ6, it was calculated the texture direction index, S_{tdi}, which is a measure of the weight of the dominant direction (mathematically, its expression is defined as the average amplitude sum divided by the amplitude sum of the dominating direction). Since all S_{tdi} indices were found to be higher than 0.6, it can be assumed that the amplitude sum of all directions are similar, or, in other words, the surface texture is isotropic.

The texture ratio aspect parameters, Str_{37}, obtained from the correlation length plot shown in Figure 61(c) also for sample BTZ6 could be used to identify the texture strength (uniformity of the texture aspect). It is defined as the ratio of the fastest to slowest decay to correlation 37% (equivalent of 1/e) of the autocorrelation function respectively. Since all of the Str_{37} indices were found also to be higher than 0.6, it can be again assumed a spatially uniform texture (this parameter equals the unity for a perfect spatially isotropic texture).

Cross-section high resolution scanning electron microscopy analysis with field emission gun (SEM-FEG) performed for the five-deposit thin film (BTZ5) emphasized an almost uniform thickness and a rather polycrystalline "layer by layer" growth, than a columnar growth mechanism (Figure 62).

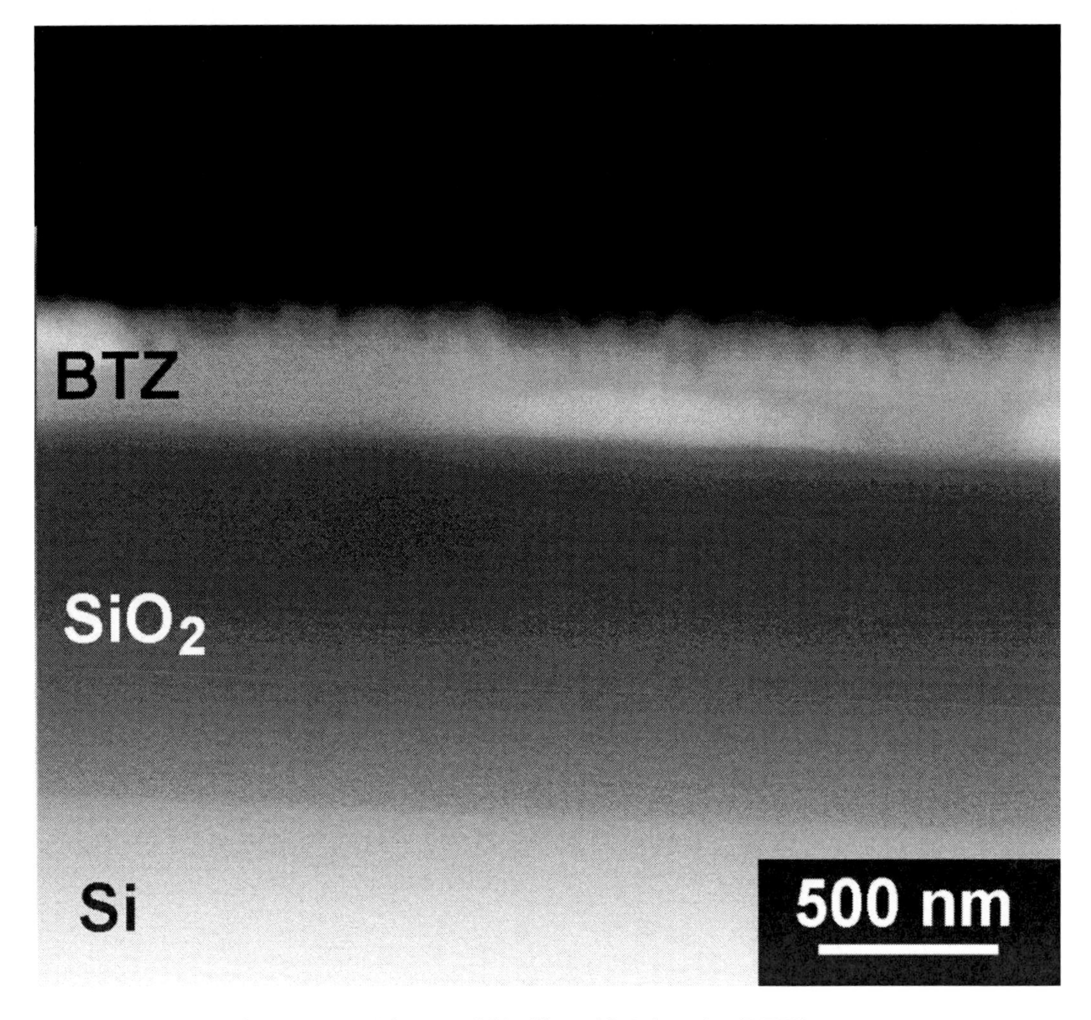

Figure 62. Cross-section SEM-FEG image of thin film with 5 deposits (BTZ5).

3.2.3.3. Optical properties

The optical proprieties of the films were studied by spectroscopic ellipsometry (SE) in visible range. The measurements were carried out in air, at an angle of incidence of 70°.

Experimental spectra were simulated with Bruggemann Effective Medium Approximation (B-EMA) model for two layers: BTZ film / interfacial SiO_2 / Si substrate [388]. Optical constants (*n*, *k*), thickness (*d*), void content and optical gap (*Eg*), obtained from the best fit (between experimental and simulated data) are presented in Figures 63 and 64, respectively. It can be observed that optical constants and thickness of BTZ films increase with the number of the deposits, while the void content exhibits a reverse evolution. The presence of a high percentage of voids in the first deposited layer is a common feature for sol–gel-derived oxide thin films.

The decrease of the void content and implicitly the increase of the refractive index (*n*) could be explained by the increase of thermal treatments number with each new layer and also with the fact that the new layer is growing on a more "ordered" old layer. These evolutions are in agreement with the XRD data and microstructural observations, revealing higher densification and crystallization degree for the multilayer BTZ thin films.

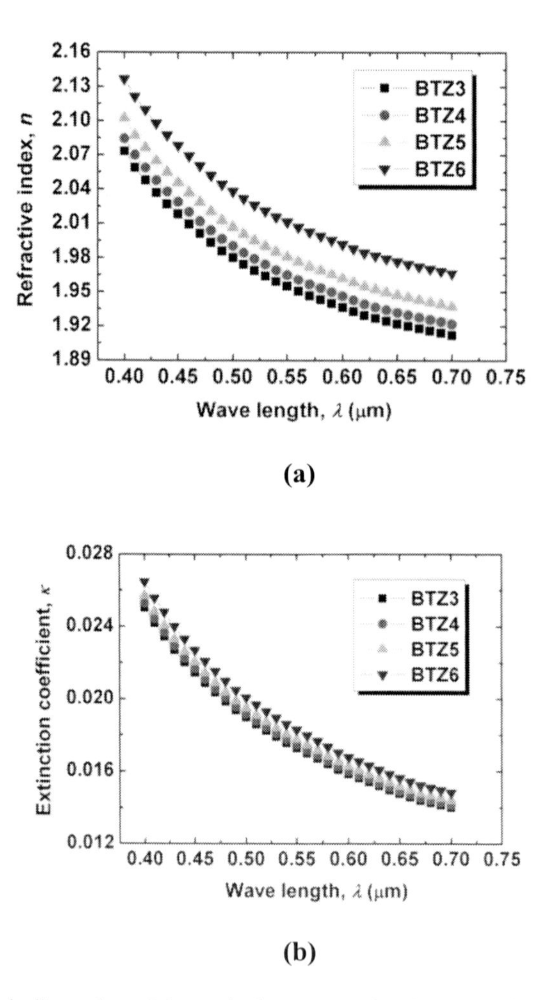

Figure 63. Wave length dispersion of the optical constants for the BTZ films containing three to six layers (BTZ3 - BTZ6): refractive index, *n* (A) and extinction coefficient, *k* (B).

The values of the refractive index rise obviously towards shorter wavelength, showing the typical shape of dispersion curve near an electronic interband transition [389]. This evolution is close tot that reported for the $BaTiO_3$ undoped films, but the refractive index values are lower [390]. On the other hand, these values are slightly higher than those reported by Liu et al. [384] for their sol-gel $BaTi_{0.80}Zr_{0.20}O_3$ $BaTi_{0.70}Zr_{0.30}O_3$ thin films, but lower than those obtained by Tang et al. [380] for the sol-gel films with the nominal composition $BaTi_{0.95}Zr_{0.05}O_3$. This suggests that higher crystallization degrees are obtained for films with smaller zirconium content.

In the case of thickness the increase is "almost" linear, that means the layers are stable and not react between them. The increase is not "perfect" linear because the voids content is not diminished "linearly" (Figure 64). Anyway, the thickness values determined by spectroscopic ellipsometry measurements are in good agreement with those estimated from the cross-section SEM-FEG images.

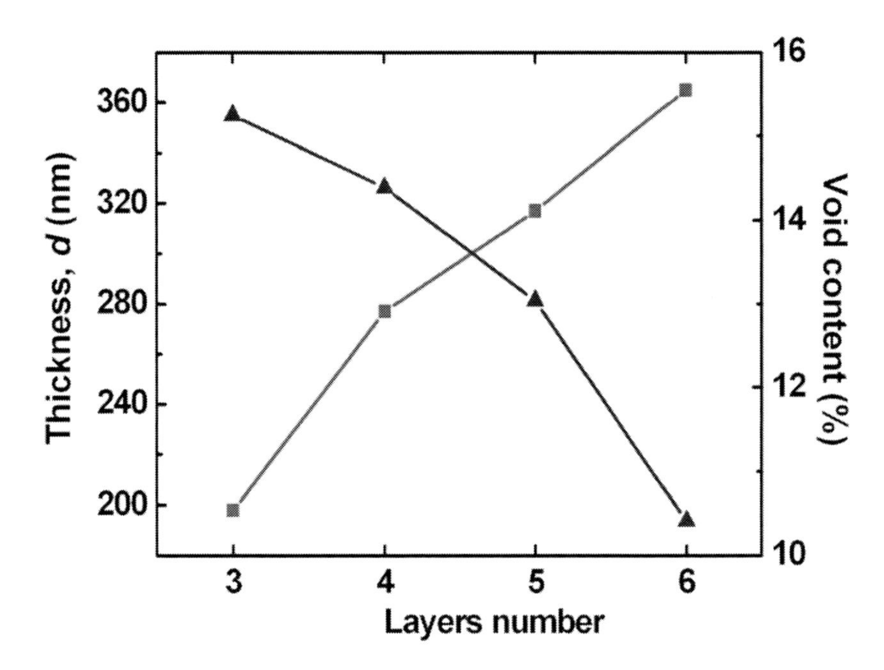

Figure 64. Evolution of thickness and voids content as a function of the layers number in the sol-gel $BaTi_{0.85}Zr_{0.15}O_3$ thin films.

The optical energy band gap is related to the absorbance and the photon energy by the equation proposed by Wood and Tauc [391]:

$$(\alpha h\nu)^{i} = (h\nu - Eg) \tag{30}$$

where α is the absorbance, h is the Planck constant, ν is the frequency and Eg is the optical band gap. For i = 2 the direct optical band gap (Eg_d) is obtained, while the indirect optical band gap (Eg_i) resulted for i = 1/2 indirect optical band gap (Eg_i).

In the case of our films, values around 2.51 eV and 1.1 eV were obtained for the Eg_d and Eg_i, respectively. The smaller Eg_d values compared with those reported by other authors [379,

385] can be caused either by a lower crystallinity or, more likely, by the presence of a large concentration of defects, especially deep holes situated near from the valence band, as Cavalcante et al. [381, 383] pointed out by using photoluminescence spectroscopy, for their sol-gel $BaTi_{0.75}Zr_{0.25}O_3$ thin films. Therefore, in order to improve the quality and to obtain denser and defect-free multilayer BTZ films, a further increase of the deposits number will be considered.

Conclusion

Concerning the undoped and lanthanum doped sol-gel ceramics prepared via alkoxide route, the following aspects have to be pointed out:

- For the undoped composition, the described preparation procedure induces crystallization of $BaTiO_3$ as major phase at room temperature.
- In the doped compositions, the presence of lanthanum prevents the crystallization of the perovskite phase, so that the corresponding precursors are predominantly amorphous.
- For the lanthanum-doped composition the formation mechanism of the perovskite during the thermal decomposition of the precursors is more complex, involving beside the higher temperature of the solid state reaction, a lower temperature process which leads to the formation of metastable intermediates, as the mixed (Ba, Ti, Zr) oxycarbonate phase, by continuous CO_2 release in a large temperature range.
- Single phase powders were obtained after annealing in air, at temperatures above 800°C, with two hours plateau.
- The increase of the dopant content causes significant structural and morphological changes in the oxide powders (evolution of the unit cell from tetragonal toward a cubic symmetry, particle size decrease, increase of the agglomeration tendency) and dramatic microstructural modifications (decrease of the average grain size, reducing densification by increasing the intergranular porosity) of the related ceramics.
- The rise of the lanthanum concentration in the sol-gel $Ba_{1-x}La_xTiO_3$ system induces a more significant shift of the Curie temperature toward lower values, than that one reported in literature for ceramics of similar compositions, but processed by the conventional solid state method.

Continuous and homogeneous multilayered BTZ films were successfully deposited on SiO_2/Si substrates, as proved by XRD, AFM and SE analysis. The surfaces of all $BaTiO_3$-based samples are cracks-free, without exfoliations and covered with nanometric grains with dimension and distribution depending on the preparation conditions.

- X-ray diffraction analysis pointed out the intensity enhancement of the main XRD peaks of the perovskite structure with the increase of the number of the deposits, proving the rise of the crystallinity degree.

- Beside the shrinkage of the unit cell, the increase of the number of the deposits in the BTZ films induces also a smoothing of the surface, so that a smaller roughness value was obtained for the six-deposit BTZ film.
- The multilayer BTZ films present a more uniform distribution of the grain size; thereby, while the four-deposit BTZ film shows a bimodal grain size distribution consisting of larger grains of 93 nm and of smaller grains of ~ 36 nm, the six-deposits BTZ film clearly shows a uniform surface from the grain size point of view.
- All surfaces are characterized by self-similarity with normal-scaling behavior, with mean fractal dimension in the range of 2.84-2.93. No preferential dependency of the surface texture of the BTZ films was emphasized.
- The increase of the thickness induced by the number of the deposits leads to a higher compactness for the BTZ films. The refractive index shows normal values indicating a higher crystallinity for the films. However, the lower values of the direct optical gap indicated a high concentration of point defects in the sol-gel films with the nominal composition $BaTi_{0.85}Zr_{0.15}O_3$.

ACKNOWLEDGMENTS

This work was supported by the Romanian Excellence grants CCEX-FOTOX, CEEX-NANOCRIS and CEEX-FEROCER.

REFERENCES

[1] Mackenzie, J.D. Glasses from melts and glasses from gels. *J. Non-Crystalline Solids*, 1982, 48, 1-10.

[2] Ebelmen, M. Recherches sur les combinaisons des acides borique et silicique avec les éthers. *Ann. Chimie Phys.*, 1846, 16, 129-166.

[3] Ebelmen, M. Sur l'hyalite artficielle et l'hydrophane. *C.R. Acad. Sci.*, 1847, 25, 854-856.

[4] Graham, T. On the properties of silicic acid and other analogous colloidal substance. *J. Chem. Soc.*, 1864, 17, 318-326.

[5] Hench, L.L; West, J.K. The sol-gel process. *Chem. Rev.*, 1990, 90, 33-72.

[6] Aelion, R; Loebel, A; Eirich, F. Hydrolysis of ethyl silicate. *J. Amer. Chem. Soc.*, 1950, 72, 12, 5705-5711.

[7] Geffcken, W; Berger, E. Single oxide coatings by sol-gel process. Dtsch.Reichpatent 736 411 (1939), Jenaer Glaswerk Schott & Gen., Jena.

[8] Roy, R. Aids in hydrothermal experimentation: II. Methods of making mixtures for both "dry" and "wet" phase equilibrium studies. *J. Am. Ceram. Soc.*, 1956, 39, 145-146.

[9] Roy, R. Gel route to homogeneous glass preparation. *J. Am. Ceram. Soc.*, 1969, 52, 344-344.

[10] Filho, O.K; Aegerter, M.A. Rheology of the gelation process of silica gel. *J. Non-Cryst. Solids*, 1988, 105, 191-197.

[11] Pierre, A.C. *Introduction aux procedes sol-gel*. Editions SEPTIMA. Paris, 1992.

[12] Schmidt, H. Chemistry of material preparation by the sol-gel process. *J. Non-Cryst. Solids*, 1988, 100, 51-64.
[13] Lopez, T; Mendez, J; Zamudio, T; Villa, M. Spectroscopic study of sol-gel silica doped with iron ions. *Mater. Chem. Phys.*, 1990, 30, 161-167.
[14] Ward, D.A; Ko, E.I. Preparing catalytic materials by the sol-gel method. *Ind. Eng. Chem. Res.*, 1995, 34, 421-433.
[15] Iler, R.K. *The chemistry of silica*, Ed. John Wiley & Sons. New York, 1979.
[16] Pachulski, N; Ulrich, J. Production of tablet-like solid bodies without pressure by sol-gel processes. *Lett. Drug Des. Discov.*, 2007, 4, 78-81.
[17] Zarzycki, J. Structural aspects of sol-gel synthesis. *J. Non-Cryst. Solids*, 1990, 121, 110 -118.
[18] Almdal, K; Dyre, J; Hvdit, S; Kramer, O. Towards a phenomenological definition of the term “gel”. *Polymer Gels and Networks*, 1993, 1, 5-17.
[19] Sakka, S; Kamiya, K. The sol-gel transition in the hydrolysis of metal alkoxides in relation to the formation of glass fibers and films. *J. Non-Cryst. Solids*, 1982, 48, 31-46.
[20] Yu, P; Liu, H; Wang, Y. Hydrolysis and gelling process of silicon ethoxide. *J. Non-Cryst. Solids*, 1982, 52, 511-520.
[21] Dokter, W.H; Beelen, T.P.M; van Garderen H.F; van Santen R.A. Structural development in silica systems. *Mat. Res. Soc. Symp. Proc.*, 1994, 346, 217-222.
[22] Brinker, C.J; Keefer, K.D; Schaefer, D.W; Ashley, C.S. Sol-gel transition in simple silicates. *J. Non-Cryst. Solids*, 1982, 48, 47-64.
[23] Jolivet, J.P; Henry, M; Livage, J. *Metal oxide chemistry and synthesis-From solution to solid state.* John Wiley &Sons Ltd. England: Rowe, A; 2000.
[24] Mehrotra, R.C. Synthesis and reactions of metal-alkoxides. *J. Non-Cryst. Solids*, 1988, 100, 1-15.
[25] Mukherjee, S.P. Sol-gel processes in glass science and technology. *J. Non-Cryst. Solids*, 1980, 42, 477-488.
[26] Livage, J; Henry, M; Sanchez, C. Sol-gel chemistry of transition metal oxides. *Prog. Solid State Chem.*, 1988, 18, 259-341.
[27] Dislich, H. New routes to multicomponent oxide glasses. *Angew. Chem. Internat. Edit.*, 1971, 10, 363-370.
[28] Brinker, C.J; Scherrer, G. *Sol-gel science, the physics and chemistry of sol-gel processing.* Academic Press. New York, 1990.
[29] Yoldas, B.E. Preparation of glasses and ceramics from metal-organic compounds. *J. Mater. Sci.*, 1977, 12, 1203-1208.
[30] Sakka, S; Kamiya, K. Glasses from metal alcoholates. *J. Non-Cryst. Solids*, 1980, 42, 403-422.
[31] Guglielmi, G; Carturan, G. Precursors for sol-gel preparations. *J. Non-Cryst. Solids*, 1988, 100, 16-30.
[32] Guglielmi, G; Scarinci, G; Maliavski, N; Bertoluzza, A; Fagnano, C; Morelli, M.A. SiO_2 gel-glasses prepared from alkoxide, colloidal and amine-silicate solutions, a comparison. *J. Non-Cryst. Solids*, 1988, 100, 292-297.
[33] Marta, L; Sârbu, C; Zador, L; Zaharescu, M; Crişan, D. Contributions to the gelling process in the Bi-Sr-Ca-Cu-O system, via acetate route. *J. Sol-Gel. Sci.Technol.*, 1997, 8, 681-684.

[34] Brinker, C.J; Smith, D.M; Deshpande, R; Davis, P.M; Hietala, S; Frye, G.C; Ashley, C.S; Assink, R.A. Sol-gel processing of controlled pore oxides. *Catal. Today*, 1992, 14, 155-163.
[35] Schmidt, H; Scholze, H; Kaiser, A. Principles of hydrolysis and condensation reaction of alkoxysilanes. *J. Non-Cryst. Solids*, 1984, 63, 1-11.
[36] Livage, J; Beteille, F; Roux, C; Chatry, M; Davidson, P. Sol-gel synthesis of oxide materials. *Acta Mater.*, 1998, 46, 743-750.
[37] Schmidt, H. Considerations about the sol-gel process: From the classical sol-gel route to advanced chemical nanotechnologies. *J. Sol-Gel Sci. Techn.*, 2006, 40, 115-130.
[38] Zarzycki, J. Gel-glass transformation. *J. Non-Cryst. Solids*, 1982, 48, 105-116.
[39] Brinker, C.J; Scherer, G.W. Sol → Gel → Glass: I. Gelation and gel structure. *J. Non-Cryst. Solids*, 1985, 70, 301-322.
[40] James, P.F. The gel to glass transition: chemical and microstructural evolution. *J. Non-Cryst. Solids*, 1988, 100, 93-114.
[41] Klein, L.C; Gallo, T.A. Densification of sol-gel silica: constant rate heating, isothermal and step heat treatments. *J. Non-Cryst. Solids*, 1990, 121, 119-123.
[42] Bradley, D.C; Mehrota R.C, Gaur, D.P. *Metal Alkoxides*, Ed. Academic Press. New York, 1978.
[43] Yoldas, B.E. Hydrolysis of titanium alkoxide and effect of hydrolytic polycondensation parameters. *J. Mater. Sci.*, 1986, 21, 1087-1092.
[44] Hubert-Pfalzgraf, L.G. Alkoxides as molecular precursors of oxide-based inorganic materials: opportunities for new materials. *New. J. Chem.*, 1987, 11, 663-675.
[45] Sanchez, C; Livage, J; Henry, M; Babonneau, F. Chemical modification of alkoxide precursors. *J. Non-Cryst. Solids*, 1988, 100, 65-76.
[46] Sanchez, C; Livage, J. Sol-gel chemistry from metal alkoxide precursors. *New. J. Chem.*, 1990, 14, 513-521.
[47] Livage J; Sanchez, C. Sol-gel chemistry. *J. Non-Cryst. Solids*, 1992, 145, 11-19.
[48] Henry, M; Jolivet, J.P; Livage, J. Aqueous chemistry of metal cations: hydrolysis, condensation and complexation. In: Reisfeld, R. editor. *Chemistry, spectroscopy and applications of sol-gel glasses. Structure and Bonding*, vol.77. Berlin: Springer-Verlag; 1992, 153-206.
[49] Livage, J. Sol-gel synthesis of multifunctional mesoporous materials.-I. Metal-organic precursors. In Sequeira, C.A.C; Hudson, M.J, editors. *Multifunctional Mesoporous Inorganic Solids*. Netherland: Kulwer Academic Publishers; 1993, 305-319.
[50] Livage, J; Henry, M; Jolivet, J.P. Sol-gel synthesis of multifunctional mesoporous materials.-II Inorganic precursors. In: Sequeira, C.A.C; Hudson, M.J, editors. *Multifunctional Mesoporous Inorganic Solids*. Netherland: Kulwer Academic Publishers; 1993, 321-338.
[51] Wright, J.D; Sommerdijk, N.A.J.M. *Sol-gel materials – Chemistry and applications*. Ed. Gordon and Breach Science Publishers, Amsterdam; 2001.
[52] Crişan, M; Zaharescu, M; Crişan, D; Simionescu, L. TiO_2 porous materials obtained from different Ti-alkoxides. In: Editor Vincenzini, P. *Adv. Sci. Technol., B_3 (Ceramics Charting the Future)*. Faenza (Italy): 1995, 2805-2812.
[53] Zaharescu, M; Crişan, M; Crişan, D; Gartner, M; Moise, F. TiO_2 films obtained from different Ti-alkoxides. *Rev. Roum. Chim.*, 1995, 40, 993-1001.

[54] Zaharescu, M; Crişan, M; Simionescu, L; Crişan, D; Gartner, M. TiO_2-based porous materials obtained from gels, in different experimental conditions. *J. Sol-Gel Sci. Technol.*, 1997, 8, 249-253.

[55] Barringer, E.A; Bowen, H.K. High-purity, monodisperse TiO_2 powders by hydrolysis of titanium tetraethoxide. 1. Synthesis and physical properties. *Langmuir*, 1985, 1, 414-420.

[56] Barringer, E.A; Bowen, H.K. High-purity, monodisperse TiO_2 powders by hydrolysis of titanium tetraethoxide. 2. Aqueous interfacial electrochemistry and dispersion stability. *Langmuir*, 1985, 1, 420-428.

[57] Springer, L; Yan, M.F. Sintering of TiO_2 from organometallic precursors. In: Hench, L.L; Ulrich, R.editors. *Ultrastructure Processing of Ceramics, Glasses, and Composites*. Wiley, New York, 1984, 464-475.

[58] Babonneau, F; Doeuff, S; Leaustic, A; Sanchez, C; Cartier, C; Verdaguer, M. XANES and EXAFS study of titanium alkoxides. *Inorg. Chem.*, 1988, 27, 3166-3172.

[59] Pouskouleli, G. Metallorganic compounds as preceramic materials II. Oxide ceramics. *Ceram. Int.*, 1989, 15, 255-270.

[60] Doeuff, S; Henry, M; Sanchez, C; Livage, J. Hydrolysis of titanium alkoxides: modification of the molecular precursor by acetic acid. *J. Non-Cryst. Solids*, 1987, 89, 206-216.

[61] Jiang, X; Herricks, T; Xia, Y. Monodispersed spherical colloids of titania: synthesis, characterization, and crystallization, *Adv. Mater.* 2003, 15, 1205-1209.

[62] Yoldas, B.E. Modification of polymer-gel structures. *J. Non-Cryst. Solids*, 1984, 63, 145-154.

[63] Yoldas, B.E. Effect of variations in polymerized oxides on sintering and crystalline transformations. *J. Amer. Ceram. Soc.*, 1982, 65, 387-393.

[64] Yoldas, B.E. Zirconium oxides formed by hydrolytic condensation of alkoxides and parameter that affect their morphology. *J. Mater. Sci.*, 1986, 21, 1080-1086.

[65] Rehspringer, J.L; Poix, P; Bernier, J.C. Synthesis of glass 'precursor' BaO, TiO_2, nH_2O by gel processing. *J. Non-Cryst. Solids*, 1986, 82, 286-292.

[66] Mackenzie, J.D. Applications of sol-gel process. *J.Non-Cryst.Solids*, 1988, 100, 162-168.

[67] Kumar, K.N.P; Keizer, K; Burggraaf, J.A. Textural evolution and phase transformation in titania membranes: Part 1. Unsupported mambranes. *J. Mater. Chem.*, 1993, 3, 1141-1149.

[68] Zaharescu, M; Pârlog, C; Crişan, M; Sahini, M; Moraru, D. Hertstellung von Glasschichten durch Sol-Gel Technik. *Silikattechnik*, 1986, 37, 165-166.

[69] Hiroshi, H; Kusaka, T. Structure of titania sol-gel coatings. *SPIE Proceedings Series - Sol-Gel Optics II*, 1992, 1758, 67-76.

[70] Sakka, S; Kamiya, K; Yogo, Y. Sol-gel preparation and properties of fibers and coating films. In: Zeldin, M; Wynne, K.J; Alco, H.R. editors. *Inorganic and Organometallic Polymers*. ACS Symposium Series, 1988, 360, 345-353.

[71] Brinker, C.J; Harrington, M.S. Sol-gel derived antireflective coatings for silicon. *Solar Energy Materials*, 1985, 5, 159-172.

[72] Yoldas, B.E. Polymerized solutions for depositing optical oxide coatings. Patent U.S. 4.346.131, 1980.

[73] Yoldas, B.E; Filippi, A.M; Buckman, R.W Jr. Oxide Protected Mirror. Patent U.S. 4.272.588, 1981.

[74] Kamiya, K; Nishijima, T; Tanaka, K. Nitridation of the sol-gel derived titanium oxide films by heating in ammonia gas. *J. Am. Ceram. Soc.*, 1990, 73, 2750-2752.

[75] Zaharescu, M; Crişan, M; Pârlog, C; Crişan, D; Drăgan, N; Simionescu, L. TiO_2 porous materials obtained by sol-gel method. *Rev. Roum. Chim.*, 1996, 41, 63-70.

[76] Masuda, H; Nishio, K; Baba, N. Fabrication of porous TiO_2 films using two-step replication of microstructure of anodic alumina. *Jpn J. Appl. Phys.-Part.2*, 1992, 31, L1775-L1777.

[77] Zaharescu, M; Pîrlog, C; Crişan, M; Gartner, M; Vasilescu, A. TiO_2-based vitreous coatings obtained by sol-gel method. J. Non-Cryst. Solids, 1993, 160, 162-166.

[78] Samuneva, B; Kozhukharov, V; Trapalis, C; Kranold, R. Sol-gel processing of titanium-containing thin coatings. 1. Preparation and structure. *J. Mater. Sci.*, 1993, 28, 2353-2360.

[79] Trapalis, C.C; Kozhukharov, V; Samuneva, B; Stefanov. P. Sol-gel processing of titanium-containing thin coatings. 2. XPS studies. *J. Mater. Sci.*, 1993, 28, 1276-1282.

[80] Kozhukharov, V; Trapalis, C; Samuneva, B. Sol-gel processing of titanium-containing thin coatings. 3. Properties.; *J. Mater. Sci.*, 1993, 28, 1283-1288.

[81] Aizawa, M; Nakagawa, Y; Nosaka, Y; Fujii, N; Miyama, H. Preparation of hollow TiO_2 fibers. *J. Non-Cryst. Solids*, 1990, 124, 112-115.

[82] Kamiya, K; Tanimoto, K; Yoko, T. Preparation of TiO_2 fibers by hydrolysis and polycondesation of $Ti(O\text{-}i\text{-}C_3H_7)_4$. *J. Mater. Sci. Lett.*, 1986, 5, 402-404.

[83] Kamiya, K; Yoko, T; Bessho. M. Nitridation of TiO_2 fibers prepared by the sol-gel method. *J. Mater. Sci.*, 1987, 22, 937-941.

[84] Lopez, T; Sanchez, E; Bosch, P; Means, Y; Gomez, R. FTIR and UV-Vis (diffuse reflectance) spectroscopic characterization of TiO_2 sol-gel. *Mater. Chem. Phys.*, 1992, 32, 141-152.

[85] Montoya, J.A; Viveros, T; Dominguez, J.M; Canales, L.A; Schifter, J. On the effect of the sol-gel synthesis parameters on textural and structural characteristics of TiO_2. *Catal. Lett.*, 1992, 15, 207-215.

[86] Montoya, J.A; Viveros, T; Canales, A; Dominguez, J.M; I.Schifter, I. Estructura y morfologia del TiO_2 sintetizado o partir de un alcoxido de titano. Actas XIII Simposio Iberoamericano de Catalysis - Segovia, Spain, July 1992, 1992, 145-150.

[87] Rodriguez, O; Gonzales, F; Bosch, P; Portilla, M; Viveros, T. Physical characterization of TiO_2 and Al_2O_3 prepared by precipitation and sol-gel methods. *Catalysis Today*, 1992, 14, 243-252.

[88] Schneider, M; Baiker. A. High-surface-area titania aerogels. Preparation and structural properties. *J. Mater. Chem.*, 1992, 2, 587-589.

[89] Lijzenga, C; Zaspalis, V.T; Ransijin, C.D; Kumar, K.P; Keizer, K; Burggraaf, A.J. Nanostructure characterization of titania membranes. *Key Eng. Mater.*, 1991, 61 & 62, 379-382.

[90] Zaspalis, V.T; Van Praag, W; Keiser, K; Ross, J.H.R; Burggraaf, A.J. Synthesis and characterization of primary alumina, titania and binary membranes. *J. Mater. Sci.*, 1992, 27, 1023-1035.

[91] Anderson, M.A; Gieselmann, M.J; Xu, Q.Y. Titania and alumina ceramic membranes. *J. Membrane Sci.*, 1988, 39, 243-258.

[92] Crişan, M; Zaharescu, M; Simionescu, L; Crişan, D. Sol-gel TiO_2 ceramic membranes obtained by polymeric method. In: Bowen, W.R; Field, R.W; Howell, J.A. editors. *Proceedings of EUROMEMBRANE'95*, Bath, England, September 1995, University of Bath, 1995; 1, I-341- I-344.

[93] Zaharescu, M; Crişan, M; Crişan, D; Simionescu, L. Membrane ceramice de TiO_2 obținute prin procedeul sol-gel ruta polimerică- Zilele Academice Timişene, Ediția a IV-a, Timişoara, 25-27 Mai 1995, Ed. Mirton, Timişoara, 1995, 1, 228-231.

[94] Crişan, M; Zaharescu, M; Simionescu, L; Crişan, D; Toth, A. TiO_2 membrane obtained by sol-gel method deposited on Al_2O_3 supports. *Key. Eng. Mater.*, 1997, 132-136, 1766-1769.

[95] Hamasaki, Y; Ohkubo, S; Murakami, K; Sei, H; Nogami, G. Photoelectrochemical properties of anatase and rutile films prepared by the sol-gel method. *J. Electrochem. Soc.*, 1994, 141, 660-663.

[96] Crişan, M; Zaharescu, M; Crişan, D; Ion, R; Manolache, M. Vanadium doped sol-gel TiO_2 coatings. *J Sol-Gel Sci. Technol.*, 1998, 13, 775-778.

[97] Gusmano, G; Montesperelli, G; Nunziante, P; Traversa, E; Montenero, A; Braghini, M; Mattogno, G; Bearzotti, A. Humidity-sensitive properties of titania films prepared using the sol-gel process. *J. Ceram. Soc. Japan*, 1993, 101, 1095-1100.

[98] Lopez, T; Sanche, E; Effect of hydrolysis catalyst and thermal treatment on titania synthesis via sol-gel. *React. Kinet. Catal. Lett.*, 1992, 48 , 295-300.

[99] Komarneni, S; Breval, E; Roy, R. Structure of solid phases in titania and zirconia gels. *J. Non-Cryst. Solids*, 1986, 79, 195-203.

[100] Chappel, J.S; Procopio, L.J; Birchall, J.D. Observation on modifying particle formation in the hydrolysis of titanium (IV) tetraethoxide. *J. Mater. Sci. Lett.*, 1990, 9, 1329-1331.

[101] Zaharescu, M; Cserhati, T; Forgacs, E. Retention characteristics of titanium dioxide and polyethylene-coated titanium dioxide, as reversed-phase supports. *J. Liq. Chrom. Rel. Technol.*, 1997, 20, 2997-3007.

[102] Crişan, M; Jitianu, A; Crişan, D; Bălăşoiu, M; Drăgan, N; Zaharescu, M. Sol-gel monocomponent nano-sized oxide powders. *J Optoelectron. Adv. M.*, 2000, 2, 339-344.

[103] Kato, K. Morphology and photocatalytic property of alkoxy-derived TiO_2 powders. *Ceramic Transactions*, 1991, 22, 63-68.

[104] Kato, K. Photosensitized oxidation of ethanol on alkoxy-derived TiO_2 powders. *Bull. Chem. Soc. Japan*, 1992, 65, 34-38.

[105] Kato, K; Tsuzuki, A; Taoda, H; Torii, Y; Kato, T; Butsugan, Y. Crystal structures of TiO_2 thin coatings prepared from the alkoxide solution via the dip-coating technique affecting the photocatalytic decomposition of aqueous acetic acid. *J. Mater. Sci.*, 1994, 29, 5911-5915.

[106] Kato, K; Tsuzuki, A; Torii, Y; Taoda, H; Kato, T; Butsugan, Y. Morphology of thin anatase coatings prepared from alkoxide solutions containing organic polymer, affecting the photocatalytic decomposition of aqueous acetic acid. *J. Mater. Sci.*, 1995, 30, 837-841.

[107] Cui, H; Shen, H.S; Gao, Y.M; Dwight, K; A.Wold, A. Photocatalytic properties of titanium oxide thin films prepared by spin coating and spray pyrolysis. *Mat. Res. Bull.*, 1993, 28, 195-201.

[108] Ding, X.Z; Qi, Z.Z; He, Y.Z. Effect of hydrolysis water on the preparation of nano-crystalline titania powders via a sol-gel process. *J. Mater. Sci. Lett.*, 1995, 14, 21-22.

[109] Ding, X.Z; Liu, L; Ma, X.M; Qi, Z.Z; He, Y.Z. The influence of alumina dopant on the structural transformation of gel-derived nanometer titania powders. *J. Mater. Sci. Lett.*, 1994, 13, 462-464.

[110] Ding, X.Z; Liu, X.L; He, Y.Z. Grain size dependence of anatase-to-rutile structural transformation in gel-derived nanocrystalline titania powders. *J. Mater. Sci. Lett.*, 1996, 15, 1789-1791.

[111] Hoffmann, M.R; Martin, S.T; Choi, W; Bahnemann, D.W. Environmental applications of semiconductor photocatalysis. *Chem. Rev.*, 1995, 95, 69-96.

[112] Linsebigler, A.L; Lu, G; Yates, J.T. Jr. Photocatalysis on TiO_2 surfaces: Principles, mechanisms, and selected results. *Chem. Rev.*, 1995, 95, 735-758.

[113] Fujishima, A; Rao, T.N; Tryk, D.A. Titanium dioxide photocatalysis. *Photochem. Photobiol C: Photochem. Rev.*, 2000, 1, 1-21.

[114] Yoshida, N; Watanabe, T. Sol-gel processed photocatalytic titania films. In Sakka, S; editor. *Handbook of sol-gel science and technology – Processing characterization and application.* Vol 3. Applications of sol-gel technology, 2007, 355-383

[115] Diebold, U. The surface science of titanium dioxide. *Surf. Sci. Rep.*, 2003, 48, 53-229.

[116] Herrmann, J.M. Heterogeneous photocatalysis: state of the art and present applications. *Top. Catal*, 2005, 34, 49-65.

[117] Hashimoto, K; Irie, H; Fujishima, A. TiO_2 Photocatalysis: A historical overview and future prospects. *Jpn. J. Appl. Phys.*, 2005, 44, 8269-8285.

[118] Litter, M.I. Introduction to photochemical advanced oxidation processes for water treatment. In: Boule, P; Bahnemann, D; Robertson, P.J.K. editors. *Environmental Chemistry part II* (The Hanbook of Environmental Chemistry) vol. 2M. Berlin: Springer; 2005; 325-366.

[119] Chen, X; Mao, S.S. Titanium dioxide nanomaterials: Synthesis, properties, modifications and applications. *Chem. Rev.*, 2007, 107, 2891-2959.

[120] Han, F; Kambala, V.S.R; Srinivasan, M; Rajarathnam, D; Naidu, R. Tailored titanium dioxide photocatalysts for the degradation of organic dyes in wastewater treatment: A review. *Appl. Catal. A: General*, 2009, 359, 25-40.

[121] Akpan, U.G; Hameed, B.H. The advancements in sol-gel method of doped-TiO_2 photocatalysts. *Appl. Catal. A: General*, 2010, 375, 1-11.

[122] Ollis, D.F; Pellizetti, E; Serpone, N. Heterogeneous photocatalysis in the environment: Application to water purification. In: Serpone, N; Pelizzeti, E.(editors). *Photocatalysis fundamentals and applications.*Wiley-Inter-Science, Amsterdam, 1989.

[123] Watanabe, T; Kitamura, A; Kojima, E; Nakayama, C; Hashimoto, K; Fujishima, A. Photocatalytic activity of TiO2 thin film under room light. In: Ollis, D.F; Al-Ekabi, H.(editors). *Photocatalytic purification and treatment of water and air.* Elsevier, Amsterdam, 1993, 747-751.

[124] Fujishima, A; Hashimoto, K; Watanabe, T. *TiO_2 photocatalysis: fundamental and applications.* BKC Inc., Chiyoda-Ku, Tokyo, 1999.

[125] Herrmann, J.M. *Catalytic science series*, vol.1: Environmental Catalysis. Imperial College Press, London, 1999.

[126] Kaneko, M; Okura, I. *Photocatalysis - science and technology.* Springer - Verlag, Berlin, Heidelberg, New York, 2003. ISBN 3-540-43473-9.

[127] Fujishima, A; Honda, K. Electrochemical photolysis of water at a semiconductor electrode. *Nature*, 1972, 238, 37-38.

[128] Tryk, D.A; Fujishima, A; Honda, K. Recent topics in photoelectrochemistry: achievements and future prospects. *Electrochim. Acta*, 200, 45, 2363-2376.
[129] Černigoj, U; Štangar, U.L; Trebše, P; Ribič, R. Comparison of different characteristics of TiO_2 films and their photocatalytic properties. *Acta Chim. Slov.*, 2006, 53, 29-35.
[130] Sobczyński, A; Dobosz, A. Water purification by photocatalysis on semiconductors. *Polish J. Environmental Studies*, 2001, 10, 195-205.
[131] Kiriakidou, F; Kondarides, D.I; Verykios, X.E. The effect of operational parameters and TiO_2 – doping on the photocatalytic degradation of azo-dyes. *Catal. Today*, 1999, 54, 119-130.
[132] Trapalis, C.C; Keivanidis, P; Kordas, G; Zaharescu, M; Crişan, M; Szatvanyi, A; Gartner, M. TiO_2 (Fe^{3+}) nanostructured thin film with antibacterial properties. *Thin Solid Films* 2003, 433, 186-190.
[133] Lee, W; Shen, H.S; Dwight, K; Wold, A. Effect of silver on the photocatalytic activity of TiO_2. *J. Solid State Chem.*, 1993, 106, 288-294.
[134] Mills, A; Davies, R.H; Worsley, D. Water purification by semiconductor photocatalysis. *Chem. Soc. Rev.*, 1993, 417-425.
[135] Fox, M.A; Dulay, M.T. Heterogeneous photocatalysis. *Chem. Rev.*, 1993, 93, 341-357.
[136] Mills, A; Le Hunte, S. An overview of semiconductor photocatalysis. *J. Photochem. Photobiol. A: Chemistry*, 1997, 108, 1-35.
[137] Litter, M.I. Heterogeneous photocatalysis. Transition metal ions in photocatalytic systems. *Appl. Catal. B: Environmental*, 1999, 23, 89-114.
[138] Bhatkhande, D.S; Pangarkar, V.G; Beenackers, A.A.C.M. Photocatalytic degradation for environmental applications - A review. *J. Chem. Technol. Biotechnol.*, 2001, 77, 102-116.
[139] Stasinakis, A.S. Use of selected advanced oxidation processes (AOPs) for wastewater treatment – A mini review. *Global NEST J.*, 2008, 10, 376-385.
[140] Brajsa, A; Szaniawska, K; Barczyński, R.J; Murawski, L; Kościelska, B; Vomvas, A; Pomoni, K. The photoconductivity of sol-gel derived TiO_2 films. *Optical Materials*, 2004, 26, 151-153.
[141] Kajihara, K; Nakanishi, K; Tanaka, K; Hirao, K; Soga, N. Preparation of macroporous titania films by a sol-gel dip-coating method from the system containing poly(ethylene glycol). *J. Am. Ceram. Soc.*, 1998, 81, 2670-2676.
[142] Dislich, H; Hussmann, E. Amorphous and crystalline dip-coatings obtained from organometallic solution: procedures, chemical processes and products. *Thin Solid Films*, 1981, 77, 129-139.
[143] Dislich, H. Glassy and crystalline systems from gels, chemical basis and technical application. *J. Non-Cryst. Solids*, 1984, 63, 237-241.
[144] Yoldas, B.E; O'Keeffe, T.W. Antireflective coatings applied from metal-organic derived liquid precursors. *Appl. Opt.*, 1979, 18, 3133-3138.
[145] Crişan, M; Gartner, M; Szatvanyi, A; Zaharescu, M. Structural and optical study of the Fe^{3+} doped TiO_2 sol-gel coatings. *Rev. Roum. Chim.*, 2002, 47, 123-130.
[146] Zaharescu, M; Crişan, M; Muševič, I. Atomic force microscopy study of TiO_2 films obtained by sol-gel method. *J. Sol-Gel Sci. Technol.*, 1998, 13, 769-773.
[147] Dislich, H. Coatings on glass. In: Kriedl, N.J; Uhlman, D.R. editors. *Glass: Science and Technology*, Ed. Academic Press, Boston, 1984, vol.2, 251-283.

[148] Brinker, C.J; Frye, G.C; Hurd, A.J; Ashley, C.S. Fundamentals of sol-gel dip coating. *Thin Solid Films*, 1991, 201, 97-108.

[149] Pettit, R.B; Brinker, C.J. Use of sol-gel films in solar energy applications. *SPIE Proceedings\ Series - Optical Material Technology for Energy Efficiency and Solar Energy Conversion IV*, 1985, 562, 256-268.

[150] Bach, H; Schroeder, H. Kristallstruktur und optische eigenschaften von dünnen organogenen titanoxyd-schichten auf glasunterlagen. *Thin Solid Films*, 1968, 1, 255-276.

[151] Schroeder, H. Oxide layers deposited from organic solutions. In: Hass, G; Thun, R.T; editors. *Physiscs of Thin Films: Advances in Research and Development*. New York: Academic Press; 1969, vol.5, 87-140.

[152] Gallagher, D; Ring, T.A. Sol-gel processing of ceramic films. *Chimia*, 1989, 43, 298-304.

[153] Chatterjee, D; Mahata, A. Photoassisted detoxification of organic pollutants on the surface modified TiO_2 semiconductor particulate system. *Catal. Commun.*, 2001, 2, 1-3.

[154] Chatterjee, D; Bhattacharya, C. Photocatalytic destruction of organic pollutants in a Pt/TiO_2 semiconductor particulate system. *Ind. J. Chem.*, 1999, 38A, 1256-1258.

[155] Zang, L; Macyk, W; Lange, C; Maier, W.F; Antonius, C; Meissner, D; Kisch, H. Visible-light detoxification and charge generation by transition metal chloride modified titania. *Chem. Eur. J.*, 2000, 6, 379-384.

[156] Choi, W.Y; Termin, A; Hoffmann, M.R. The role of metal ion dopants in quantum-sized TiO_2: Correlation between photoreactivity and charge carrier recombination dynamics. *J. Phys. Chem.*, 1994, 98, 13669-13679.

[157] Zhang, Y; Xiong, G; Yao, N; Yang, W; Fu, X. Preparation of titania-based catalysts for formaldehyde photocatalytic oxidation from $TiCl_4$ by the sol-gel method. *Catal Today*, 2001, 68, 89-95.

[158] Ranjit, K.T; Varadarajan, T.K; Viswanathan, B. Photocatalytic reduction of dinitrogen to ammonia over noble-metal-loaded TiO_2. *J. Photochem. Photobiol. A - Chemistry*, 1996, 96A, 181-185.

[159] Yuan, S; Sheng, Q; Zhang, J; Chen, F; Anpo, M; Dai, W. Synthesis of Pd nanoparticles in La-doped mesoporous titania with polycrystalline framework. *Catal. Lett.*, 2006, 107, 19-24.

[160] Shah, S.I; Li, W; Huang, C.P; Jung, O; Ni, C. Study of Nd^{3+}, Pd^{2+}, Pt^{4+}, and Fe^{3+} dopant effect on photoreactivity of TiO_2 nanoparticles. *PNAS*, 2002, 99, 6482-6486.

[161] Burns A; Li, W; Baker, C; Shah, S.I. Sol-gel synthesis and characterization of neodymium-ion doped nanostructured titania thin films. *Mat. Res. Soc. Symp. Proc.*, 2002, 703, V.5.2.1-V.5.2.6.

[162] Facchin, G; Carturan, G; Campostrini, R; Gialanella, S; Lutterotti, L; Armelao, L; Marci, G; Palmisano, L; Sclafani, A. Sol-gel synthesis and characterisation of TiO_2-anatase powders containing nanometric platinum particles employed as catalysts for 4-nitrophenol photodegradation. *J. Sol-Gel Sci. Technol.*, 2000, 18, 29-59.

[163] López, T; Gómez, R; Pecci, G; Reyes, P; Bokhimi, X; Novaro, O; Effect of pH on the incorporation of platinum into the lattice of sol–gel titania phases. *Mater. Lett.,* 1999, 40, 59–65.

[164] López, T; Gómez, R; Romero, E; Schifter, I. Phenylacetylene hydrogenation on Pt/TiO_2 sol-gel catalysts. *React. Kinet. Catal. Lett.*, 1993, 49, 95-101.

[165] Jin, S; Shiraishi, F. Photocatalytic activities enhanced for decompositions of organic compounds over metal-photodepositing titanium dioxide. *Chem. Eng. J.*, 2004, 97, 203-211.

[166] Erkan, A; Bakir, U; Karakas, G. Photocatalytic microbial inactivation over Pd doped SnO_2 and TiO_2 thin films. *J. Photochem. Photobiol. A: Chemistry*, 2006, 184, 313-321.

[167] Crişan, D; Drăgan, N; Crişan, M; Răileanu, M; Brăileanu, A; Anastasescu, M; Ianculescu, A; Mardare, D; Luca, D; Marinescu, V; Moldovan, A. Crystallization study of sol– gel un-doped and Pd-doped TiO_2 materials. *J. Phys. Chem. Solids,* 2008, 69, 2548– 2554.

[168] Crişan, M; Brăileanu, A; Crişan, D; Răileanu, M; Drăgan, N; Mardare, D; Teodorescu, V; Ianculescu, A; Bîrjega, R; Dumitru, M. Thermal behaviour study of some sol–gel TiO_2 based materials. *J. Therm. Anal. Cal.*, 2008, 92, 7-13.

[169] Li, F.B; Li, X.Z. Photocatalytic properties of gold/gold ion-modified titanium dioxide for wastewater treatment. *Appl. Catal. A: General*, 2002, 228, 15-27.

[170] Manera, M.G; Spadavecchia, J; Buso, D; de Julián Fernández , C; Mattei, G; Martucci, A; Mulvaney, P; Pérez-Juste, J; Rella, R; Vasanelli, L; Mazzoldi, P. Optical gas sensing of TiO_2 and TiO_2/Au nanocomposite thin films. *Sensors and Acuators B: Chemical,* 2008, 132, 107-115.

[171] Du, Z; Feng, C; Li, Q; Zhao, Y; Tai, X. Photodegradation of NPE-10 surfactant by Au-doped nano-TiO_2. *Colloids and Surfaces A: Physicochem Eng. Aspects*, 2008, 315, 254-258.

[172] Hernandez-Fernandez, J; Aguilar-Elguezabal, A; Castillo, S; Ceron-Ceron, B. Oxidation of NO in gase phase by Au-TiO_2 photocatalysts prepared by the sol-gel method, *Catal. Today*, 2009, 148, 115-118.

[173] Kafizas, A; Kellici, S; Darr, J.A; Parkin, I.P. Titanium dioxide and composite metal/metal oxide titania thin films on glass: A comparative study of photocatalytic activity, *J. Photochem. Photobiol. A: Chemistry*, 2009, 204, 183-190.

[174] Traversa, E; Di Vona, M.L; Nunziante, P; Licoccia, S; Sasaki, T; Koshizaki, N. Sol-gel preparation and characterization of Ag-TiO_2 nanocomposite thin films. *J. Sol-Gel Sci. Techn.*, 2000, 19, 733-736.

[175] Falaras, P; Arabatzis, I.M; Stergiopoulos, T; Bernard, M.C. Enhanced activity of silver modified thin-film TiO_2 photocatalysts. *Int. J. Photoenergy*, 2003, 5, 123-130.

[176] Esfahani, M.N; Habibi, M.H. Silver doped TiO_2 nanostructure composite photocatalyst film synthesized by sol-gel spin and dip coating technique on glass. *Int. J. Photoenergy*, 2008, Article ID 628713 Doi:10.1155/2008/628713

[177] Răileanu, M; Crişan, M; Drăgan, N; Crişan, D; Galtayres, A; Brăileanu, A; Ianculescu, A; Teodorescu, V.S; Niţoi, I; Anastasescu, M. Sol-gel doped TiO_2 nanomaterials: a comparative study. *J. Sol-Gel Sci. Technol.*, 2009, 51, 315-329.

[178] Pan, X; Medina-Ramirez, I; Mernaugh, R; Liu, J. Nanocharacterization and bactericidal performance of silver modified titania photocatalyst. *Colloid and Surfaces B: Biointerfaces*, 2010, 77, 82-89.

[179] An, J.H; Kim B.H, Jeong, J.H; Kim, D.M; Jeon, Y.S; Jeon, K.O; Hwang K.S. Preparation of vanadium-doped TiO_2 thin films on glass substrates. *J. Ceram. Proc. Res.*, 2005, 6, 163-166.

[180] Droubay, T; Heald, S.M; Shutthanandan, V; Thevuthasan, S; Chambers, S.A; Osterwalder, J. Cr-doped TiO_2 anatase: A ferrromagnetic insulator. *J. Appl. Phys.*, 2005, 97, 046103 Doi:10.1063/1.1846158.
[181] Zhu, J; Deng, Z; Chen, F; Zang, J; Chen, H; Anpo, M; Huang, J; Zang, L. Hydrothermal doping method for preparation of Cr^{3+}-TiO_2 photocatalysts with concentration gradient distribution of Cr^{3+}. *Appl. Catal. B: Environmental*, 2006, 62, 329-335.
[182] Zhang, K; Xu, W; Li, X; Zheng, S; Xu, G. Effect of dopant concentration on photocatalytic activity of TiO_2 film doped by Mn non-uniformly. *Central Eur. J. Chem.*, 2006, 4, 234-254.
[183] Marugan, J; Christensen, P; Egerton, T; Purnama, H. Synthesis, characterization and activity of photocatalytic sol-gel TiO_2 powders and electrodes. *Appl. Catal. B: Environmental*, 2009, 89, 273-283.
[184] Sugimoto, K; Kim, H; Akao, N; Hara, N. Corrosion resistance of Fe_2O_3-TiO_2 thin films prepared by low pressure MOCVD. In: Bardwell, J. editor. Surface Oxide Films, PV 96-18, *The Electrochemical Society Proceedings Series*, Pennington, NJ; 1994; 194-205.
[185] Kim, H; Akao, N; Hara, N; Sugimoto, K. Comparison of Corrosion Resistances Between Fe_2O_3-TiO_2 Artificial passivation film and passivation film on Fe-Ti Alloy. *J. Electrochem. Soc.*, 1998, 145, 2818-2826.
[186] Zaharescu, M; Crişan, M; Szatvanyi, A; Gartner, M. TiO_2-based nanostructured sol-gel coatings for water treatment. *J. Optoelectron. Adv. M.*, 2000, 2, 618-622.
[187] Crişan, M; Gartner, M; Szatvanyi, A; Zaharescu, M. Structural and optical study of the Fe^{3+} doped TiO_2 sol-gel coatings. *Rev. Roum. Chim.*, 2002, 47, 123-130.
[188] Barău (Szatvanyi), A; Crişan, M; Gartner, M; Crişan, D; Zaharescu, M. TiO_2 /ITO photoanod materials prepared by sol-gel method. *Nanoscience and Nanotechnology*, 2004, 4, 194-197.
[189] Sonawane, R.S; Kale, B.B; Dongare, M.K. Preparation and photo-catalytic activity of Fe–TiO_2 thin films prepared by sol–gel dip coating. *Mat. Chem. Phys.*, 2004, 85, 52-57.
[190] Gauthier, V; Bourgeois, S; Sibillot, P; Maglione, M; Sacilotti, M. Growth and characterization of AP-MOCVD iron doped titanium dioxide thin films. *Thin Solid Films*, 1999, 340, 175-182.
[191] Matsuo, S; Sakaguchi, N; Obuchi, E; Nakano, K; Perera, R.C; Watanabe, T; Matsuo, T; Wakita, H. X-ray absorption spectral analyses by theoretical calculations for TiO_2 and Ni-doped TiO_2 thin films on glass plates. *Anal. Sci.*, 2001, 17, 149-153.
[192] Ramírez-Meneses, E; García-Murillo, A; Carrillo-Romo, F. de J; García-Alamilla, R; Del Angel-Vicente, P; Ramírez-Salgado, J; Pérez, P.B. Preparation and photocatalytic activity of TiO_2 films with Ni nanoparticles. *J Sol-Gel Sci. Technol.*, 2009, 52, 267-275.
[193] Matsumoto, Y; Murakami, M; Shono, T; Hasegawa, T; Fukumura, T; Kawasaki, M; Ahmet, P; Chikyow, T; Koshihara, S; Koinuma, H; Room-temperature ferromagnetism in transparent transition metal-doped titanium dioxide. *Science*, 2001, 291, 854-856.
[194] Kim, D.H; Yang, J.S; Kim, Y.S; Chang, Y.J; Noh, T.W; Bu, S.D; Kim, Y.W; Park, Y.D; Pearton, S.J; Park, J.H. Superparamagnetism in Co ion-implanted epitaxial anatase TiO_2 thin films. *Annalen der Physik*, 2004, 13, 70-71.

[195] Subramanian, M; Vijayalakshmi, S; Venkatraj, S; Jayavel, R. Effect of cobalt doping on the structural and optical properties of TiO_2 films prepared by sol–gel process. *Thin Solid Films*, 2008, 516, 3776-3782.

[196] Carrota, M.C; Ferroni, M; Gnani, D; Guidi, V; Merli, M; Martinelli, G; Casale, M.C; Notaro, M. Nanostructured pure and Nb-doped TiO_2 as thick film gas sensors for environmental monitoring. *Sensors and Actuators B: Chemical*, 1999, 58, 310-317.

[197] Devi, G.S; Hyodo, T; Shimizu, Y; Egashira, M. Synthesis of mesoporous TiO_2-based powders and their gas-sensing properties. *Sensors and Actuators B: Chemical*, 2002, 87, 122-129.

[198] Zhao, L; Zhao, X; Liu, J; Zhang, A; Wang, D; Wei, B. Fabrications of Nb-doped TiO_2 (TNO) transparent conductive oxide polycristalline films on glass substrates by sol-gel method. *J. Sol-Gel Sci. Technol.*, 2010, 53, 475-479.

[199] Chary, K.V.R; Reddy, K.R; Kumar, C.P. Dispersion and reactivity of molybdenum oxide catalysts supported on titania. *Catal. Commmun.*, 2001, 2, 277-284.

[200] Garzella, C; Comini, E; Bontempi, E; Depero, L.E; Frigeri, C; Sberveglieri, G. Sol-gel TiO_2 and W/TiO_2 nanostructured thin films for control of drunken driving. *Sensors and Actuators B: Chemical*, 2002, 83, 230-237.

[201] Hwang, Y.K; Patil, K.R; Kim, H.K; Sathaye, S.D; Hwang, J.S; Park S.E; Chang, J.S. Photoinduced superhydrophilicity in TiO_2 thin films modified with WO_3. *Bull Korean Chem. Soc.*, 2005, 26, 1515-1519.

[202] Lee, Y.C; Hong, Y.P; Lee, H.Y; Kim, H; Jung, Y.J; Ko, K.H; Jung, H.S; Hong, K.S. Photocatalysis and hydrophilicity of doped TiO_2 thin films. *J. Colloid Interface Sci.*, 2003, 267, 127-131.

[203] Palomino-Merino, R; Conde-Gallardo, A; Garcia-Rocha, M; Hernández-Calderón, I; Castaño, V; Rodriguez, R. Photoluminescence of TiO_2 : Eu^{3+} thin films obtained by sol–gel on Si and Corning glass substrates. *Thin Solid Films*, 2001, 401, 118-123.

[204] Bamwenda, G.R; Uesigi, T; Abe, Y; Sayama, K; Arakawa, H. The photocatalytic oxidation of water to O_2 over pure CeO_2, WO_3, and TiO_2 using Fe^{3+} and Ce^{4+} as electron acceptors. *Appl. Catal. A: General*, 2001, 205, 117-128.

[205] Xu, A.N; Gao, Y; Liu, H.Q. The preparation, characterization and their properties of rare-earth-doped TiO_2 nanoparticles. *J. Catal.,* 2002, 207, 151-157.

[206] Li, F.B; Li, X.Z; Hou, M.F; Cheah, K.W; Choy, W.C.H. Enhanced photocatalytic activity of Ce^{3+}–TiO_2 for 2-mercaptobenzothiazole degradation in aqueous suspension for odour control. *Appl. Catal. A: General*, 2005, 285, 181-189.

[207] Manjumol, K.A; Smitha, V.S; Shajesh, P; Baiju, K.V; Warrier, K.G.K. Synthesis of lanthanum oxide doped photocatalytic nano titanium oxide through aqueous sol-gel method for titania multifunctional ultrafiltration membrane. *J. Sol-Gel Sci. Technol.*, 2010, 53, 353-358.

[208] Cao, B.S; Feng, Z.Q; He, Y.Y; Li, H; Dong, B. Opposite effect of Li^+ codoping on the conversion emissions of Er^{3+}-doped TiO_2 powders. *J. Sol-Gel Sci. Technol.*, 2010, 54, 101-104.

[209] Bei, Z; Ren, D; Cui, X; Shen, J; Yang, X.; Zhang, Z. Photoelectrochemical properties and crystalline structure change of Sb-doped TiO_2 thin films prepared by sol-gel method. *J. Mater. Res.*, 2004, 19, 3189-3195.

[210] Sayilkan, F; Asiltürk, M; Kiraz, N; Burunkaya, E; Arpaç, E; Sayilkan, H. Photocatalytic antibacterial performance of Sn^{4+}-doped TiO_2 thin films on glass substrate. *J. Hazard. Mater.*, 2009, 162, 1309-1316.

[211] Matos, J; Laine, J; Herrmann, J.M. Synergy effect in the photocatalytic degradation of phenol on a suspended mixture of titania and activated carbon. *Appl. Catal. B: Environmental,* 1998, 18, 281-291.

[212] Colón, G; Hidalgo, M.C; Navío, J.A. A novel preparation of high surface area TiO_2 nanoparticles from alkoxide precursor and using active carbon as additive. *Catal. Today*, 2002, 76, 91-101.

[213] Tryba, B; Morawski, A.W; Inagaki, M. Application of TiO_2-mounted activated carbon to the removal of phenol from water. *Appl. Catal. B: Environmental*, 2003, 41, 427-433.

[214] Toyoda, M; Nanbu, Y; Kito, T; Hirano, M; Inagaki, M. Preparation and performance of anatase-loaded porous carbons for water purification. *Desalination*, 2003, 159, 273-282.

[215] Irie, H; Watanabe, Y; Hashimoto, K. Carbon-doped anatase TiO_2 powders as a visible-light sensitive photocatalyst. *Chem. Lett.*, 2003, 32, 772-773.

[216] Ohno, T; Tsubota, T; Toyofuku, M; Inaba, R. Photocatalytic activity of a TiO_2 photocatalyst doped with C^{4+} and S^{4+} ions having a rutile phase under visible light. *Catal. Lett.*, 2004, 98, 255-258.

[217] Tachikawa, T; Tojo, S; Kawai, K; Endo, M; Fujitsuka, M; Ohno, T; Nishijima, K; Miyamoto, Z; Majima, T. Photocatalytic oxidation reactivity of holes in the sulfur- and carbon-doped TiO_2 powders studied by time-resolved diffuse reflectance spectroscopy. *J. Phys. Chem. B.*, 2004, 108, 19299-19306.

[218] Wang, H; Lewis, J.P. Second-generation photocatalytic materials: anion-doped TiO_2. *J. Phys: Condens. Matter.*, 2006, 18, 421-434.

[219] Asahi, R; Morikawa, T; Ohwaki, T; Aoki, K; Taga, Y. Visible-light photocatalysis in nitrogen-doped titanium oxides. *Science*, 2001, 293, 269-271.

[220] Irie, H; Watanabe, Y; Hashimoto, K. Nitrogen-concentration dependence on photocatalytic activity of $TiO_{2-x}N_x$ powders. *J. Phys. Chem. B*, 2003, 107, 5483-5486.

[221] Nakamura, R; Tanaka, T; Nakato, Y. Mechanism for visible light responses in anodic photocurrents at N-doped TiO_2 film electrodes. *J. Phys. Chem. B*, 2004, 108, 10617-10620.

[222] Bacsa, R; Kiwi, J; Ohno, T; Albers, P; Nadtocchenko, V. Preparation, testing and characterization of doped TiO_2 active in the peroxidation of biomolecules under visible light. *J. Phys. Chem. B*, 2005, 109, 5994-6003.

[223] Xu, M; Lin, S; Chen, X; Peng, Y. Studies on characteristics of nanostructure of N-TiO_2 thin films and photo-bactericidal action. *J. Zhejiang Univ. Sci. B*, 2006, 7, 586-590.

[224] Nishijima, K; Kamai, T; Murakami, N; Tsubota, T; Ohno, T. Photocatalytic hydrogen or oxygen evolution from water over S- or N-doped TiO_2 under visible light. Int J Photoenergy, 2008, Article ID 173943 Doi:10.1155/2008/173943.

[225] Hebenstreit, E.L.D; Hebenstreit, W; Diebold, U. Adsorption of sulfur on TiO_2 (110) studied with STM, LEED and XPS: temperature-dependent change of adsorption site combined with O–S exchange. *Surf. Sci.*, 2000, 461, 87-97.

[226] Hebenstreit, E.L.D; Hebenstreit, W; Diebold, U. Structures of sulfur on TiO_2 (110) determined by scanning tunneling microscopy, X-ray photoelectron spectroscopy and low-energy electron diffraction. *Surf. Sci.*, 2001, 470, 347-360.

[227] Ohno, T; Mitsui, T; Matsumura, M. Photocatalytic activity of S-doped TiO_2 photocatalyst under visible light. *Chem. Lett.*, 2003, 32, 364-365.
[228] Umebayashi, T; Yamaki, T; Tanala, S; Asai K. Visible light-induced degradation of methylene blue on S-doped TiO_2. *Chem. Lett.*, 2003, 32, 330-331.
[229] Ohno, T; Akiyoshi, M; Umebayashi, T; Asai, K; Mitsui, M; Matsumura M. Preparation of S-doped TiO_2 photocatalysts and their photocatalytic activities under visible light. *Appl. Catal. A: General*, 2004, 265, 115-121.
[230] Ohno, T. Preparation of visible light active S-doped TiO_2 photocatalysts and their photocatalytic activities. *Water Sci. Technol.*, 2004, 49, 159-163.
[231] Sun, H; Liu, H; Ma, J; Wang, X; Wang, B; Han L. Preparation and characterization of sulfur-doped TiO_2/Ti photoelectrodes and their photoelectrocatalytic performance. *J. Hazard. Mater.*, 2008, 156, 552-559.
[232] Yu, J.C; Ho,W; Yu, J; Yip, H; Wong, P.K; Zhao J. Efficient visible-light-induced photocatalytic disinfection on sulfur-doped nanocrystalline titania. *Environ. Sci. Technol.*, 2005, 39, 1175-1179.
[233] Ho, W; Yu, C.J; Lee S. Low-temperature hydrothermal synthesis of S-doped TiO_2 with visible light photocatalytic activity. *J. Solid State Chem.*, 2006, 179, 1171-1176.
[234] Crişan, M; Brăileanu, A; Răileanu, M; Crişan, D; Teodorescu, V.S; Bîrjega, R; Marinescu, V.E; Madarász, J; Pokol, G. TiO_2-based nanopowders obtained from different Ti-alkoxides. *J. Therm. Anal. Cal.*, 2007, 88, 171-176.
[235] Liu, S; Chen, X. A visible light response TiO_2 photocatalyst realized by cationic S-doping and its application for phenol degradation. *J. Hazard Mater.*, 2008, 152, 48-55.
[236] Crişan, M; Brăileanu, A; Răileanu, M; Zaharescu, M; Crişan, D; Drăgan, N; Anastasescu, M; Ianculescu, A; Niţoi, I; Marinescu, V.E; Hodorogea, S.M. Sol–gel S-doped TiO_2 materials for environmental protection. *J. Non-Cryst. Solids*, 2008, 354, 705-711.
[237] Randenija, L.K; Murphy, A.B; Plumb, I.C. A study of S-doped TiO_2 for photoelectrochemical hydrogen generation from water. *J. Mater. Sci.*, 2008, 43, 1389-1399.
[238] Tajammul Hussain, S; Khan, K; Hussain, R. Size control synthesis of sulphur doped titanium dioxide (anatase) nanoparticles, its optical property and its photo catalytic reactivity for CO_2+H_2O conversion and phenol degradation. *J. Natural Gas Chemistry*, 2009, 18, 383-391.
[239] Klauson, D; Portjanskaya, E; Budarnaja, O; Krichevskaya, M; Preis, S. The synthesis of sulphur and boron-containing titania photocatalysts and the evaluation of their photocatalytic activity. *Catal. Commun.*, 2010, 11, 715-720.
[240] Deng, L; Chen, Y; Yao, M; Wang, S; Zhu, B; Huang, W; Zhang, S. Synthesis, characterization of B-doped TiO_2 nanotubes with high photocatalytic activity. *J. Sol-Gel Sci. Technol.*, 2010, 53, 535-541.
[241] Kim, S.W; Khan, R; Kim, T.J; Kim, W.J. Synthesis, characterization, and application of Zr, S co-doped TiO_2 as visible-light active photocatalyst. *Bull. Korean Chem. Soc.*, 2008, 29, 1217-1223.
[242] Wei, F; Ni, L; Cui, P. Preparation and characterization of N–S-codoped TiO_2 photocatalyst and its photocatalytic activity. *J. Hazard Mater.*, 2008, 156, 135-140.

[243] Liu, G; Zhao, Y; Sun, C; Li, F; Lu, G.Q; Cheng, H.M. Synergistic effects of B/N doping on the visible-light photocatalytic activity of mesoporous TiO_2. *Angew. Chem.*, 2008, 120, 4592-4596.

[244] Shen, Y; Xiong, T; Du, H; Jin, H; Shang, J; Yang, K. Investigation of Br-N co-doped TiO_2 photocatalysts: preparation and photcatalytic activities under visible light. *J. Sol-Gel Sci. Technol.*, 2009, 52, 41-48.

[245] Zaleska, A. Characteristics of doped-TiO_2 photocatalysts. *Physicochemical Problems of Mineral Processing*, 2008, 42, 211-222.

[246] Crişan, M; Jitianu, A; Zaharescu, M; Mizukami, F; Niwa, Sol-gel mono- and poly-component nanosized powders in the Al_2O_3-TiO_2-SiO_2-MgO system. *J. Disper. Sci. Technol.*, 2003, 24, 129-144.

[247] López, T; Morán, M; Navarrete, J; Herrera, L; Gómez, R. Synthesis and spectroscopic characterization of Pt and Pd silica supported catalysts. *J. Non-Cryst. Solids,* 1992, 147&148, 753-757.

[248] López, T; Bosch, P; Navarrete, J; Asomoza, M; Gómez, R. Structure of Pd/SiO_2 sol-gel and impregnated catalysts. *J. Sol-Gel Sci. Technol.*, 1994, 1, 193-203.

[249] Othman, M.R; Sahadan, I.S. On the characteristics and hydrogen adsorption properties of a Pd/γ-Al_2O_3 prepared by sol-gel method. *Microporous and Mesoporous Materials,* 2006, 91, 145-150.

[250] Noh, J; Yang, O.B; Kim, D.H; Woo, S.I. Characteristics of the Pd-only three-way catalysts prepared by sol-gel method. *Catal. Today*, 1999, 53, 575-582.

[251] Carturan, G; Facchin, G; Gottardi, V; Guglielmi, M; Navazio, G. Phenylacetylene half-hydrogenation with Pd supported on vitreous materials having different chemical composition. *J. Non-Cryst. Solids*, 1982, 48, 219-226.

[252] Habibi, M.H; Esfahani, M.N; Egerton, T.A. Photochemical characterization and photocatalytic properties of a nanostructure composite TiO_2 film. *Int. J. Photoenergy*, 2007, Article ID 13653, Doi: 10.1155/2007/13653.

[253] Corriu, R; Nguyên, T.A. *Molecular chemistry of sol-gel derived nanomaterials*. John Wiley & Sons, Ltd. 2009, ISBN: 978-0-470-72117-9.

[254] Keshmiri, M; Mohseni, M; Troczynski, T. Development of novel TiO_2 sol-gel-derived composite and its photocatalytic activities for trichloroethylene oxidation. *Appl. Catal. B: Environmental*, 2004, 53, 209-219.

[255] Habibi, M.H; Nasr-Esfahani, M; Egerton, T.A Preparation, characterization and photocatalytic activity of TiO_2 / Methylcellulose nanocomposite films derived from nanopowder TiO_2 and modified sol–gel titania. *J. Mater. Sci.*, 2008, 42, 6027-6035.

[256] Traversa, E; Di Vona, M.L; Nunziante, P; Licoccia, S; Yoon, J.W; Sasaki, T; Koshizaki, N. Photoelectrochemical properties of sol-gel processed Ag-TiO_2 nanocomposite thin films. *J. Sol-Gel Sci. Techn.*, 2001, 22, 115-123.

[257] Matijevic, E. Monodispersed Colloidal Metal Oxides, Sulfides, and Phosphates . In: Hench, L.L; Ulrich, R.editors. *Ultrastructure Processing of Ceramics, Glasses, and Composites.* Wiley, New York, 1984, 334-352

[258] Duncan, J.F; Richards, R.G. Hydrolysis of titanium (IV) sulphate solutions. 2. Solution equilibria, kinetics and mechanism. *New Zealand J. Sci.*, 1976, 19, 179-183.

[259] Santacesaria, E; Tonello, M; Storti, G; Pace, R.C; Carra, S. Kinetics of titanium dioxide precipitation by thermal hydrolysis. *J. Colloid Interface Sci.*, 1986, 111, 44-53.

[260] Hsu, W.P; Yu, R; Matijevic, E. Paper whiteners: I. Titania coated silica. *J. Colloid Interface Sci.*, 1993, 156, 56-65.
[261] Nagpal, V.J; Davies, R.M; Desu, S.B. Novel thin films of titanium dioxide particles synthesized by a sol-gel process. *J. Mat. Res.*, 1995, 10, 3068-3078.
[262] Papousti, D; Lianos, P. Pyrene in mixed titania-surfactant films made by hydrolysis of titanium isopropoxide in the presence of reversed micelles. *Langmuir*, 1995, 11, 1-4.
[263] Zaharescu, M; Crişan, M; Crişan, D; Drăgan, N; Jitianu, A; Preda, M. Al_2TiO_5 preparation starting with reactive powders obtained by sol-gel method. *J. Eur. Ceram. Soc.*, 1998, 18, 1257-1264.
[264] Antonelli, D.M; Ying, J.Y. Synthesis of hexagonally packed mesoporous TiO_2 by a modified sol-gel method. *Angew. Chem. Int. Ed. Engl.*, 1995, 34, 2014-2017.
[265] Baek, S.Y; Chai, S.Y; Hur, K.S; Lee, W.I. Synthesis of highly soluble TiO_2 nanoparticles with narrow size distribution. *Bull. Korean, Chem. Soc.*, 2005, 26, 1333-1334.
[266] Barringer, E.A; Bowen, H.K. Formation, packing, and sintering of monodisperse TiO_2 powders. *J. Amer. Ceram. Soc.*, 1982, 65, C 199-C 201.
[267] Ogihara, T; Ilkemoto, T; Mitzutani, N. Kato, M; Mitarai, Y. Formation of monodispersed Ta_2O_5 powders. *J. Mater. Sci.*, 1986, 21, 2771-2774.
[268] Kato, K. Photocatalytic property of titania anchored on porous alumina ceramic support by the alkoxide method. *J. Ceram. Soc. Jpn.*, 1993, 101, 245-249.
[269] Lee, W; Gao, Y.M; Dwight, K; Wold, A. Preparation and characterization of titanium (IV) oxide photocatalysts. *Mat. Res. Bull.*, 1992, 27, 685-692.
[270] Poniatovski, E.H; Talavera, R.R; De la Cruz Heredia, M; Cano-Corona, O; Arroyo-Murillo, R. Crystallization of nanosized titania particles prepared by the sol-gel process. *J. Mater. Res.*, 1994, 9, 2102-2108.
[271] Sberveglieri, G; Depero, L.E; Ferroni, M; Guidi, V; Martinelli, G; Nelli, P; Perego, C; Sangaletti, L. A novel method for the preparation of nanosized TiO_2 thin films. *Adv. Mater.*, 1996, 8, 334-337.
[272] Kim, C.S; Moon, B.K; Park, J.H; Chung, S.T; Son, S.M Synthesis of nanocrystalline TiO_2 in toluene by a solvothermal route. *J. Crystal Growth*, 2003, 254, 405-410.
[273] Li, Y; Lee, N.H; Lee, E.G; Song, J.S; Kim, S.J. The characterization and photocatalytic properties of mesoporous rutile TiO_2 powder synthesized through self-assembly of nano crystals. *Chem. Phys. Lett.*, 2004, 389, 124-128.
[274] Baolong, Z; Baishun, C; Keyu, S; Shangjin, H; Xiaodong, L; Zongjie, D; Kelian, Y. Preparation and characterization of nanocrystal grain TiO_2 porous microspheres. *Appl. Catal. B-Environmental*, 2003, 40, 253-258.
[275] Fang Chen, Y; Young Lee, C; Yu Yeng, M; Tien Chiu, H. The effect of calcination temperature on the crystallinity of TiO_2 nanopowders. *J. Crystal Growth*, 2003, 247, 363-370.
[276] Ianculescu, A.; Mitoşeriu, L. $Ba(Ti,Zr)O_3$ - functional materials: from nanopowders to bulk ceramics (Chap. 2) in *Advances in Nanotechnology*, vol. 3, Ed. Z. Bartul & J. Trenor, NovaScience Publisher's, Inc., Hauppauge New York, USA, 2010, in press.
[277] Mitoşeriu, L.; Buscaglia, V.; Buscaglia, M.T.; Viviani, M.; Nanni, P. $BaTiO_3$ nanopowders and nanocrystalline ceramics: II. Nanoceramics (Chap. V) in *New Developments in Advanced Functional Ceramics*, Ed. L. Mitoşeriu, Transworld Research Network India, 2007, pp. 119-144.

[278] Ianculescu, A.; Mitoşeriu, L.; Berger, D.; Ciomaga, C. E.; Piazza, D.; Galassi, C. Composition dependent ferroelectric properties of $Ba_{1-x}Sr_xTiO_3$ ceramics. *Phase Transit.*, 2006, 79, 375-388.
[279] Ianculescu, A.; Berger, D.; Mitoşeriu, L.; Curecheriu, L. P.; Drăgan, N.; Crişan, D.; Vasile, E. Properties of $Ba_{1-x}Sr_xTiO_3$ ceramics prepared by the modified-Pechini method. *Ferroelectrics*, 2008, 369, 22-34.
[280] Tufescu, F. M.; Curecheriu, L.; Ianculescu, A.; Ciomaga, C.E.; Mitoşeriu L. High-voltage tunability measurements of the $BaZr_xTi_{1-x}O_3$ ferroelectric ceramics. *J. Optoelectron. Adv. Mater.*, 2008, 10, 1894-1897.
[281] Curecheriu, L.P.; Mitoşeriu, L.; Ianculescu A. Non-linear dielectric properties of $Ba_{1-x}Sr_xTiO_3$ ceramics. *J. Alloy Compd.*, 2009, 482, 1-4.
[282] Curecheriu, L. P.; Tufescu, F. M.; Ianculescu, A.; Ciomaga, C. E.; Mitoşeriu, L.; Stancu, A. Tunability characteristics of $BaTiO_3$-based ceramics: Modeling and experimental study. *J. Optoelectron. Adv. Mater.*, 2008, 10, 1792-1795.
[283] Curecheriu, L. P.; Mitoşeriu, L.; Ianculescu, A. Temperature dependent tunability data and modeling in the paraelectric $Ba_{0.70}Sr_{0.30}TiO_3$ solid solutions. *Proc. Applic. Ceram.*, 2009, 3, 43-46.
[284] Curecheriu, L. P.; Frunză, R.; Ianculescu, A. Dielectric properties of the $BaTi_{0.85}Zr_{0.15}O_3$ ceramics prepared by different techniques. *Proc. Applic. Ceram.*, 2008, 2, 81–88.
[285] Tian, Z.; Wang, X.; Shu, L.; Wang, T.; Song, T.-H.; Gui, Z.; Li, L. Preparation of Nano $BaTiO_3$-Based Ceramics for Multilayer Ceramic Capacitor Application by Chemical Coating Method. *J. Am. Ceram. Soc.*, 2009, 92, 830-833.
[286] Park, S.-S.; Ha, J.-H.; Wadley, H. N. Preparation of $BaTiO_3$ Films for MLCCs by Direct Vapor Deposition. *Integr. Ferroelectr.*, 2007, 95, 251–259.
[287] Sauer, H. A.; Fisher, J. R. Processing of PTC Thermistors. *J. Am. Ceram. Soc.*, 1960, 43, 297-301.
[288] Heywang, W. Semiconducting barium titanate. *J. Mater. Sci.*, 1971, 6, 1214-1224.
[289] Jonker, G, H. Some aspects of semiconducting barium titanate. *Solid State Electron.*, 7, 1964, 895-903 .
[290] Brahmecha, B. G.; Sinha, K. P. Resistivity Anomaly in Semiconducting $BaTiO_3$. *Jpn. J. Appl. Phys.*, 1971, 10, 496-504.
[291] Ianculescu, A.; Guillemet-Fritsch, S.; Durand, B. $BaTiO_3$ thick films obtained by tape casting from powders prepared by the oxalate route. *Proc. Applic. Ceram.*, 2009, 3, 65-70.
[292] Guillemet, S.; Ianculescu, A.; Calmet, C.; Sarrias, J.; Durand B.; Lebey, Th. Influence of Powder Quality on the Dielectric Properties of $BaTiO_3$ Ceramics and Thick Films for Powder Integration, *Advances in Science and Technology 33*, Vol. 4, 10th International Ceramics Congress - Part D, Ed. P. Vincenzini, Techna Srl, 2003, pp. 475-482.
[293] Scott, J. F. Device physics of ferroelectric thin-film memories. *Jpn. J. Appl. Phys.*, 1999, 38, 2272–2274.
[294] Ramesh, R.; Aggarwal, S.; Auciello, O. Science and technology of ferroelectric films and heterostructures for non-volatile ferroelectric memories. *Mater. Sci. Eng.*, 2001, 32, 191-236.

[295] Rose, T. L.; Kelliher, E. M.; Scoville, A. N.; Stone, S. E.. Characterization of rf - sputtered $BaTiO_3$ thin films using a liquid electrolyte. *J. Appl. Phys.*, 1984, 55, 3706-3714.

[296] Kuroiwa, T.; Tsunemine, Y.; Horikawa, T.; Makita, T.; Tanimura, J.; Mikami, N.; Sato, K. Dielectric properties of $(Ba_xSr_{1-x})TiO_3$ thin films prepared by rf sputtering for dynamic random access memory application. *Jpn. J. Appl. Phys.*, 1994, 33, 5187-5191.

[297] Ianculescu, A.; Despax, B.; Bley, V.; Lebey, Th.; Gavrilă, R.; Drăgan, N. Structure - properties correlations for barium titanate thin films obtained by rf-sputtering. *J. Eur. Ceram. Soc.*, 2007, 27, 1129 – 1135.

[298] Ianculescu, A.; Gartner, M.; Despax, B.; Bley, V.; Lebey, Th.; Gavrilă, R.; Modreanu, M. Optical Characterization and Microstructure of $BaTiO_3$ Thin Films Obtained by Rf-Magnetron Sputtering. *Appl. Surf. Sci.*, 2006, 253, 344-348.

[299] Preda, L.; Despax, B.; Courselle, L.; Bandet, J.; Ianculescu, A. Structural Characteristics of Rf -Sputtered $BaTiO_3$ Thin Films. *Thin Solid Films*, 2001, 389, 43-50.

[300] West, A. R.; Adams, T. B.; Morrison, F. D.; Sinclair, D. C. Novel high capacitance materials:- $BaTiO_3$:La and $CaCu_3Ti_4O_{12}$. *J. Eur. Ceram. Soc.*, 2004, 24 1439-1448.

[301] Huybrechts, B.; Ishizaki, K.; Takata, M.; The Positive Temperature Coefficient of Resistivity in Barium Titanate. *J. Mater. Sci.*, 1995, 30, 2463-2474.

[302] Glinchuk, M. D.; Bykov, I. P.; Kornienko, S. M.; Laguta, V. V.; Slipenyuk, A. M.; Bilous, A. G.; V'yunov, O. I.; Yanchevskii, O. Z. Influence of impurities on the properties of rare-earth-doped barium-titanate ceramics. *J. Mater. Chem.*, 2000, 10, 941-947.

[303] Urek, S., Drofenik, M.; Makovec, D. Sintering and Properties of Highly. Donor-Doped Barium Titanate Ceramics, *J. Mater. Sci.*, 2000, 35, 895-901.

[304] Peng, C.-J.; Lu, H.-Y. J. Compensation effect in semiconducting Barium titanate. *Am. Ceram. Soc.*, 1988, 71, 44-46.

[305] Drofenik, M. Oxygen Partial Pressure and Grain Growth in Donor-Doped $BaTiO_3$. *J. Am. Ceram. Soc.*, 1987, 70, 311-314.

[306] Hennings, D. Review of chemical preparation routes for barium titanate. *Br. Ceram. Proc.*, 1989, 41, 1-10.

[307] Ianculescu, A.; Guillemet-Fritsch, S.; Durand, B.; Brăileanu, A.; Crişan, M.; Berger, D. $BaTiO_3$ nanopowders and nanocrystalline ceramics: I. Nanopowders (Chap. IV) in *New Developments in Advanced Functional Ceramics*, Ed. L. Mitoşeriu, Transworld Research Network India, 2007, pp. 89-118.

[308] Roeder, R. K.; Slamovich, E. B. Stoichiometry control and phase selection in hydrothermally derived $Ba_xSr_{1-x}TiO_3$ powders. *J. Am. Ceram. Soc.*, 1999, 82, 1665–1675.

[309] Ianculescu, A.; Guillemet-Fritsch, S.; Durand, B.; Calmet, C. Influence of processing parameters on the characteristics o some $BaTiO_3$ powders obtained by the hydrothermal method. *Rom. J. Mater.*, 2006, 36, 251-260.

[310] Ianculescu, A.; Berger, D.; Matei, C.; Budrugeac, P.; Mitoşeriu, L.; Vasile, E. Synthesis of $BaTiO_3$ by soft chemistry routes. *J. Electroceram.*, 2010, 24, 46-50.

[311] Suasmoro, S.; Pratapa, S.; Hartanto, D.; Setyoko, D.; Dani, U. M. The characterization of mixed titanate $Ba_{1-x}Sr_xTiO_3$ phase formation from oxalate coprecipitated precursor. *J. Eur. Ceram. Soc.*, 2000, 20, 309-314.

[312] Nanni, P.; Viviani, M.; Buscaglia, V. Synthesis of Dielectric Ceramic Materials in *Handbook of Low and High Dielectric Constant Materials and Their Applications*, Nalwa, H.S. (Ed.), Acad. Press, San Diego, CA, 1999, pp. 429-455.

[313] Valdez-Nava, Z.; Guillemet-Fritsch, S.; Tenailleau, Ch., Lebey, T.; Durand, B.; Chane-Ching, J. Y. Colossal dielectric permittivity of $BaTiO_3$-based nanocrystalline ceramics sintered by spark plasma sintering. *J. Electroceram.*, 2009, 22, 238–244.

[314] Durán, P.; Capel, F.; Tartaj, J.; Moure, C. $BaTiO_3$ formation by thermal decomposition of (BaTi)-citrate polyester resin in air. *J. Mater. Res.*, 2001, 16, 197-209.

[315] Kao, C.-F.; Yang,W.-D. Preparation of barium strontium titanate powder from citrate precursor. *Appl. Organometal. Chem.*, 1999, 13, 383–397.

[316] Ianculescu, A.; Berger, D.; Viviani, M.; Ciomaga, C.; Mitoşeriu, L.; Vasile, E.; Drăgan, N.; Crişan, D. Investigation of $Ba_{1-x}Sr_xTiO_3$ ceramics prepared from powders synthesized by the modified Pechini route. *J. Eur. Ceram. Soc.*, 27, (2007), 3655-3658.

[317] Beck, H. P.; Eiser, W.; Haberkorn, R. Pitfalls in the synthesis of nanoscaled perovskite type compounds. Part I: Influence of different sol-gel preparation methods and characterization of nanoscaled $BaTiO_3$. *J. Eur. Ceram. Soc.*, 2001, 21, 687-693.

[318] Shiibashi, H.; Matsuda, H.; Kuwabara, M., Low-temperature preparation of $(Ba,Sr)TiO_3$ perovskite phase by sol-gel method. *J. Sol-Gel Sci. Technol.*, 1999, 16, 129-134.

[319] M. García-Hernández, A. García-Murillo, F. de J. Carrillo-Romo, D. Jaramillo-Vigueras, G. Chadeyron, E. De la Rosa, D. Boyer. Eu-Doped BaTiO3 Powder and Film from Sol-Gel Process with Polyvinylpyrrolidone Additive. *Int. J. Mol. Sci.*, 2009, 10, 4088-4101

[320] Ianculescu, A.; Brăileanu, A.; Crişan, M.; Budrugeac, P.; Drăgan, N.; Voicu, G.; Crişan, D.; Marinescu, V. Influence of Barium Source on the Characteristics of Sol-Precipitated $BaTiO_3$ Powders and Related Ceramics. *J. Therm. Anal. Calorim.*, 88, (2007), 251-260.

[321] Wang, J.; Fang, J.; Ng, S.; Gan, L.; Chew, C.; Wang, X.; Shen, Z. J. Ultrafine barium titanate powders via microemulsion processing routes. *J. Am. Ceram. Soc.*, 1999, 82, 873-881.

[322] Komarov, A. V.; Parkin, I. P.; Odlyha, M., Self-propagating high temperature synthesis of $SrTiO_3$ and $Sr_xBa_yTiO_3$ $(x + y = 1)$. *J. Mater. Sci.*, 1996, 31, 5033–5037.

[323] Grohe, B.; Miehe, G.; Wegner, G. Additive controlled crystallization of barium titanate powders and their application for thin film ceramic production. I. Powder Synthesis. *J. Mater. Res.*, 2001, 16, 1901-1910.

[324] Ito, Y.; Shimada, S.; Takahashi, J.; Inagaki, M. Phase transition of $BaTiO_3$-$Ba_{1-x}Pb_xTiO_3$ composite particles prepared by the molten salt method. *J. Mater. Chem.*, 1997, 7, 781-785.

[325] Kong, L.B.; Ma, J.; Huang, H.; Zhang, R.F.; Que, W.X. Barium titanate derived from mechanochemically activated powders. *J. Alloy Compd.*, 2002, 337, 226-230.

[326] Xue, J.; Wang, J.; Wan, D. M. Nanosized barium titanate powder by mechanical activation. *Am. Ceram. Soc.*, 2000, 83, 232-234.

[327] Liu, Y.; Feng, Y.; Wu, X.; Han, X. Microwave absorption properties of La doped barium titanat. e in X-band. *J. Alloy Compd.*, 2009, 472, 441-445.

[328] Liu J.-B.; Li W.-C.; Wang Z.-M.; Zheng C.-P. Preparation and characterisation of lanthanum doped $BaTiO_3$ nanosize polycrystals by sol-gel processing. *Mater. Sci. Technol.*, 2001, 17, 606-608.

[329] Zhao, X.; Ma Z.; Xiao, Z.; Chen, G. Preparation and Characterization on Nano-Sized Barium Titanate Powder Doped with Lanthanum by Sol-Gel Process. *J. Rare Earth,* 2006, 24, 82-85.

[330] Brutchey, L. R.; Guosheng, C.; Gu, Q.; Morse D. E. Positive Temperature Coefficient of Resistivity in Donor-Doped $BaTiO_3$ Ceramics Derived from Nanocrystals Synthesized at Low Temperature. *Adv. Mater.*, 2008, 20, 1029–1033.

[331] Lewis, C. V.; Catlow, C.R.A. Defect studies of doped and undoped *barium titanate using* computer simulation techniques. *J. Phys. Chem. Solids*, 1986, 47, 89-97.

[332] Lewis, C. V.; Catlow, C.R.A.; Casselton, R. E. W. PTCR Effect in $BaTiO_3$. *J. Am. Ceram. Soc.*, 1985, 68, 555-558.

[333] Chan, H.-N.; Smyth, D.M. Defect Chemistry of Donor-Doped $BaTiO_3$. J. Am. Ceram. Soc., 1984, 67, 285-288.

[334] Chan, H.-N.; Harmer, M.P.; Smyth, D.M. Compensating Defects in Highly Donor-Doped $BaTiO_3$. *J. Am. Ceram. Soc.*, 1986, 69, 507-510.

[335] Daniels, *J.;* Härdtl, *K*.-H.; Hennings, D.; Wernicke, R. *Defect Chemistry* and Electrical Conductivity of *Doped Barium Titanate Ceramics. Phys. Res. Rep.*, 1976, 31, 487-559.

[336] Nasrallah, M. M.; Anderson, H. U.; Agarwal, A. K.; Flandermeyer, B. F. Oxygen activity dependence of the defect structure of La-doped $BaTiO_3$. *J. Mat. Sci.*, 1984, 19, 3159-3165.

[337] Jonker, G.H.; Havinga, E. E. Mater. Res. Bull., The influence of foreign ions on the crystal lattice of barium titanate. *Mater. Res. Bull.*, 1982, 17, 345-350.

[338] Makovec, D.; Samardžija, Z.; Delalut, U.; Kolar, D. Defect structure and phase relations of highly lanthanum-doped barium titanate. *J. Am. Ceram. Soc.*, 1995, 78 2193-2197.

[339] Skapin, S. D.; Kolar, D.; Suvorov, D.; Samardzija, Z. Phase equilibria in the $BaTiO_3$-La_2TiO_5-TiO_2 system. *J. Mater. Res.*, 1998, 13, 1327-1334.

[340] Zhong, Z.; Gallagher, P.K. Combustion synthesis and characterization of $BaTiO_3$. *J. Mater. Res.*, 1995, 10, 945-952.

[341] Gopalakrishnamurthy, H. S.; Rao, M. S.; Kutty, T.R.N. Thermal decomposition of titanyl oxalates – I. Barium titanyl oxalate. *J. Inorg. Nucl. Chem.*, 1975, 37, 891-898.

[342] Gallagher P.K.; Thomson, Jr., J. Thermal Analysis of some barium and strontium titanyl oxalates. *J. Am. Ceram. Soc.*, 1965, 48, 644-647.

[343] Kumar, S.; Messing, G. L.; White, W. B. Metal organic resin derived barium titanate: I. Formation of barium titanium oxycarbonate intermediate. *J. Am. Ceram. Soc.*, 1993, 76, 617-624.

[344] Otta, S.; Bhattamistra, S. D. Kinetics and mechanism of the thermal decomposition of barium titanyl oxalate. *J. Thermal Analysis*, 1994, 41, 419-433.

[345] Arima, M.; Kakihana, M.; Nakamura, Y.; Yashima, M.; Yoshimura, M. Polymerized complex route of barium titanate powders using barium-titanium mixed-metal citric acid complex. *J. Am. Ceram. Soc*, 1996, 79, 2847-2856.

[346] Vasyl'kiv, O. O.; Ragulya, A. V.; Skorokhod, V. V. Synthesis and sintering of barium titanate powder under nonisothermal conditions. II. Phase analysis of the decomposition products of barium titanyl-oxalate and the synthesis of barium titanate. *Powder Metall. Met. C+*, 1997, 36, 277-282.

[347] Stockenhuber, M.; Mayer, H.; Lercher, J.A. Preparation of barium titanates from oxalates. *J. Am. Ceram. Soc.*, 1993, 76, 1185-1190.

[348] Cho, W.-S. Structural evolution and characterization of $BaTiO_3$ nanoparticles synthesized from polymeric precursor. *J. Phys. Chem. Solids*, 1998, 59, 659-666.

[349] Hennings, D.; Mayr, W. Thermal decomposition of (BaTi) citrates into barium titanate. *J. Solid State Chem.*, 1978, 26, 329-338.

[350] Durán, P.; Gutierrez, D.; Tartaj, J.; Bañares, M. A.; Moure, C. On the formation of an oxycarbonate intermediate phase in the synthesis of $BaTiO_3$ from (Ba,Ti)-polymeric organic precursors. *J. Eur. Ceram. Soc.*, 2002, 22, 797-807.

[351] Durán, P.; Capel, F.: Gutierrez, D.; Tartaj, J.; Bañares, M. A.; Moure, C. Metal citrate polymerized complex thermal decomposition leading to the synthesis of $BaTiO_3$: effects of the precursor structure on the $BaTiO_3$ formation mechanism. *J. Mater. Chem.*, 2001, 11, 1828-1836.

[352] Gablenz, S.; Abicht, H.-P.; Pippel, E.; Lichtenberger, O.; Woltersdorf, J. New evidence of an oxycarbonate phase as an intermediate step in $BaTiO_3$ preparation. *J. Eur. Ceram. Soc.*, 2000, 20, 1053-1060.

[353] Dutta, P. K.; Gallagher, P. K.: Twu, J. Raman spectroscopic study of the formation of barium titanate from an oxalate precursor. *Chem. Mater.*, 1993, 5, 1739-1743.

[354] Ischenko, V.; Pippel, E.; Köferstein, R.; Abicht, H.-P.; Woltersdorf, J. Barium titanate via thermal decomposition of Ba,Ti-precursor complexes: The nature of the intermediate phases. *Solid State Sci.*, 2007, 9, 21-26.

[355] Eror, N.G.; Loehr, T.M. Cornilsen, B.C. Low temperature hexagonal $BaTiO_3$ polymorph and carbonate adsorption. *Ferroelectrics*, 1980, 28, 321-324.

[356] Cho, W.-S.; Hamada, E. Planar defects and luminescence of $BaTiO_3$ particles synthesized by a polymerized complex method. *J. Alloy Compd.*, 1998, 268, 78-82.

[357] Rečnik, A. Twins in barium titanate. *Acta Chim. Slov.*, 2001, 48, 1-50.

[358] Desu, S. B.; Payne, D.., Segregation in Perovskites: III, Microstructure and Electrical Properties, *J. Am. Ceram. Soc*, 1990, 73, 3407-3415.

[359] Morrison, F. D.; Sinclair, D. C.; West, A. R. *Electrical and structural characteristics* of lanthanum-doped *barium titanate ceramics. J. Appl. Phys.* 1999, 86, 6355-6366.

[360] *Kell, R. C.; Hellicar, N. J.* Structural Transitions in Barium Titanate-.Zirconate Transducer Materials. *Acustica, 1956, 6, 235*-238.

[361] Hennings, D.; Schnell, A.; Simon, G. Diffuse Ferroelectric Phase Transitions in $Ba(Ti_{1-y}Zr_y)O_3$ Ceramics. *J. Am. Ceram. Soc.*, 1982, 65, 539-544.

[362] Yu, Z.; Ang, C.; Guo, R.; Bhalla, A. S. Piezoelectric and strain properties of $Ba(Ti_{1-x}Zr_x)O_3$ ceramics. *J. Appl. Phys.*, 2002, 92, 1489-1493.

[363] Wu, D.; Sciau, Ph.; Schamm, S.; Gloux, F.; Fernandez Varela, M. Preparation and microstructures of $BaTi_{1-x}Zr_xO_3$ hetero-epitaxial thin films on $SrTiO_3$ substrates. *J. Phys. D: Appl. Phys.*, 2007, 40, 4701-4706.

[364] Reymond, V. Structural and electrical properties of $BaTi_{1-x}Zr_xO_3$ sputtered thin films: effect of the sputtering conditions. *Thin Solid Films*, 2004, 467, 54-58.

[365] Reymond, V.; Payan, S.; Michau, D.; Manaud, J. P.; Maglione, M. Structural and electrical properties of $BaTi_{1-x}Zr_xO_3$ sputtered thin films: effect of the sputtering conditions. *Thin Solid Films*, 2004, 467, 54-58.

[366] Wang, B.; Yang, C.; Chen, H.; Zhang, J.; Yu, A.; Zhang, R. Effects of oxygen to argon ratio on $Ba(Zr_{0.2}Ti_{0.8})O_3$ thin films prepared by RF magnetron sputtering. *J. Mater. Sci.: Mater. Electron.*, 2009, 20, 614-618.

[367] Zhang, W.; Tang, X. G.; Wong, K. H.; Chan, H. L. W. Dielectric properties and high tunability of (100)-oriented $Ba(Zr_{0.2}Ti_{0.8})O_3$ thin films prepared by pulsed laser deposition. *Scripta Mater.*, 2006, 54, 197-200.

[368] Sakamoto, W.; Mimura, K.; Naka, T.; Shimura, T.; Yogo, T. Chemical solution processing and characterization of $Ba(Zr,Ti)O_3/LaNiO_3$ layered thin films. *J. Sol-Gel Sci. Techn.*, 2007, 42, 213-220.

[369] Dixit, A.; Majumder, S. B.; Dobal, P. S.; Katiyard, R. S.; Bhalla, A. S. Phase transition studies of sol-gel deposited barium zirconate titanate thin films. *Thin Solid Films*, 2004, 447-448, 284-288.

[370] Dixit, A.; Majumder, S. B.; Savvinov, A.; Katiyar, R. S.; Guo, R.; Bhalla. A. S. Investigations on the sol-gel-derived barium zirconium titanate thin films. *Mater. Lett.*, 2002, 56, 933-940.

[371] Cavalcante, L. S.; Anicete-Santos, M.; Pontes, F. M.; Souza, I. A.; Santos, L. P. S.; Rosa, I. L. V.; Santos, M. R. M. C.; Santos-Júnior, L. S.; Leite, E. R.; Longo, E. Effect of annealing time on morphological characteristics of $Ba(Zr,Ti)O_3$ thin films. *J. Alloy Compd.*, 2007, 437, 269-273.

[372] Jiang, L. L.; Tang, X. G.; Kuang, S. J.; Xiong, H. F. Surface chemical states of barium zirconate titanate thin films prepared by chemical solution deposition. *Appl. Surf. Sci.*, 2009, 255, 8913-8916.

[373] Marques, L. G. A.; Cavalcante, L. S.; Simões, A. Z.; Pontes, F. M.; Santos-Junior, L. S.; Santos, M. R. M. C.; Rosa, I. L. V.; Varela, J. A.; Longo, E. Temperature dependence of dielectric properties for $Ba(Zr_{0.25}Ti_{0.75})O_3$ thin films obtained from the soft chemical method. *Mater. Chem. Phys.*, 2007, 105, 293-297.

[374] Pontes, F. M.; Escote, M. T.; Escudeiro, C. C.; Leite, E. R.; Longo, E.; Chiquito, A. J.; Pizani, P. S.; Varela, A. J. Characterization of $BaTi_{1-x}Zr_xO_3$ thin films obtained by a soft chemical spin-coating technique. *J. Appl. Phys.*, 2004, 96, 4386-4391.

[375] Dixit, A.; Agrawal, D. C.; Mohapatra, Y. N.; Majumder, S. B.; Katiyar, R. S. Studies on the dielectric and relaxor behavior of sol-gel derived barium strontium zirconate titanate thin films, *Mater. Lett.*, 2007, 61, 3685-3688.

[376] Zhai, J.; Hu, D.; Yao, X.; Xu, Z.; Chen, H. Preparation and tunability properties of $Ba(Zr_xTi_{1-x})O_3$ thin films grown by a sol-gel process. *J. Eur. Ceram. Soc.*, 2006, 26, 1917-1920.

[377] Zhai, J.; Gao, C.; Yao, X.; Xu, Z.; Chen, H. Enhanced dielectric tunability properties of $Ba(Zr_xTi_{1-x})O_3$ thin films using seed layers on $Pt/Ti/SiO_2/Si$ substrates. *Ceram. Int.*, 2008, 34, 905-910.

[378] Gao, L. N.; Song, S. N.; Zhai, J. W.; Yao, X.; Xu, Z. K. Effects of buffer layers on the orientation and dielectric properties of $Ba(Zr_{0.20}Ti_{0.80})O_3$ thin films prepared by sol-gel method. *J. Cryst. Growth*, 2008, 310, 1245-1249.

[379] Cheng, W. X.; Ding, A. L.; He, X. Y.; Zheng, X. S.; Qiu, P. S. Characterization of $Ba(Zr_{0.05}Ti_{0.95})O_3$ thin film prepared by sol-gel process, *J. Electroceram.*, 2006, 16, 523-526.

[380] Tang, X. G.; Chan, H. L. W.; Ding, A. L. Structural, dielectric and optical properties of $Ba(Ti, Zr)O_3$ thin films prepared by chemical solution deposition. *Thin Solid Films*, 2004, 460, 227-231.

[381] Cavalcante, L. S.; Sczancoski, J. C.; De Vicente, F. S.; Frabbro, M. T.; Siu Li, M.; Varela, J. A.; Longo, E. Microstructure, dielectric properties and optical band gap

controlon the photoluminescence behavior of $Ba[Zr_{0.25}Ti_{0.75}]O_3$ thin films. *J Sol-Gel Sci. Technol.*, 2009, 49, 35-46.

[382] Cavalcante, L. S.; Gurgel, M. F. C.; Paris, E. C.; Simões, A. Z.; Joya, M. R.; Varela, J. A.; Pizani, P. S.; Longo, E. Combined experimental and theoretical investigations of the photoluminescent behavior of $Ba(Ti,Zr)O_3$ thin films. *Acta Mater.*, 2007, 55, 6416-6426.

[383] Cavalcante, L. S.; Gurgel, M. F. C.; Simões, A. Z.; Longo, E.; Varela, J. A.; Joya, M. R.; Pizani, P. S. Intense visible photoluminescence in $Ba(Zr_{0.25}Ti_{0.75})O_3$ thin films. *Appl. Phys. Lett.*, 2007, 90, 011901.

[384] Liu, A.; Xue, J.; Meng, X.; Sun, J.; Huang, Z.; Chu, J. Infrared optical properties of $Ba(Zr_{0.20}Ti_{0.80})O_3$ and $Ba(Zr_{0.30}Ti_{0.70})O_3$ thin films prepared by sol-gel method. *Appl. Surf. Sci.*, 2008, 254, 5660-5663.

[385] Anicete-Santos, M.; Cavalcante, L. S.; Orhan, E.; Paris, E. C.; Simões, L. G. P.; Joya, M. R.; Rosa, I. L. V.; De Lucena, P. R.; Santos, M. R. M. C.; Santos-Júnior, L. S.; Pizani, P. S.; Leite, E. R.; Varela, J. A.; Longo, E. The role of structural order-disorder for visible intense photoluminescence in the $BaZr_{0.5}Ti_{0.5}O_3$ thin films. *Chem. Phys.*, 2005, 316, 260-266.

[386] Xu, J. B.; Gao, C.; Zhai, J. W.; Yao, X.; Xue, J. Q.; Huang, Z. M. Structure-related infrared optical properties of $Ba(Zr_xTi_{1-x})O_3$ thin films grown on $Pt/Ti/SiO_2/Si$ substrates by low-temperature processing. *J. Cryst. Growth*, 2006, 291, 130-134.

[387] Ianculescu, A.; Despax, B.; Bley, V.; Lebey, T.; Gavrilă, R.; Drăgan, N. Structure – properties correlations for barium titanate thin films obtained by rf-sputtering. *J. Eur. Ceram. Soc.*, 2007, 27, 1129 – 1135.

[388] Bruggeman, D.A.G. Berechnung verschiedener physikalischer Konstanten von heterogenen Substanzen. *Ann. Phys.*, 1935, 24, 636-679.

[389] Wohlecke, M.; Marello, V.; Onton, A. Refractive-Index of $BaTiO_3$ and $SrTiO_3$ Films. *J. Appl. Phys.*, 1977, 48, 1748-1750.

[390] Ianculescu, A.; Gartner, M.; Despax, B.; Bley, V.; Lebey, T.; Gavrilă, R.; Modreanu, M. Optical Characterization and Microstructure of $BaTiO_3$ Thin Films Obtained by Rf-Magnetron Sputtering, *Applied Surface Science*, 2006, 253, 344-348.

[391] Wood, D. L., Tauc, J. Weak Absorption Tails in Amorphous Semiconductors. *Phys. Rev. B*, 1972, 5, 3144-3151.

In: The Sol-Gel Process
Editor: Rachel E. Morris
ISBN 978-1-61761-321-0

Chapter 2

SOL-GEL-BASED MATERIALS FOR BIOMEDICAL APPLICATIONS

Aurica P. Chiriac, Iordana Neamtu, Loredana E. Nita and Manuela T. Nistor
Petru Poni Institute of Macromolecular Chemistry, Iasi, Romania

INTRODUCTION

In practice, the development of high-performance materials for electronics, optics and ceramics fabrication is restrained by the traditional mode of their production. Many of the basic technologies for materials include high-temperature physical and chemical processes and require special conditions to attain the desired properties of the final products. Advanced preparation methods for materials with new features are feasible on the basis of colloid-chemical processes and nanochemistry. In this respect, the sol-to-gel transformation at molecular level and low temperature, followed by solidification and chemical modification is of great interest to attain new materials with non-traditional physical, chemical and functional properties of special interest, highly homogeneous and pure.

BASIC DATA. ADVANTAGES AND APPLICABILITY

The sol-gel chemical process is self-described in the definition of a sol, a gel, and a summary of the process in which a sol evolves into a gel [1]. A *sol* is a colloidal dispersion in a liquid of small particles with the diameter of some nanometers, consisting of precursors. By chemical reactions (hydrolysis and condensation) most often at room temperature, the sol transforms into a continuous network with infinite viscosity, encompassing a continuous liquid phase, named *gel* [2, 3].

The method permits the formation of materials with different configurations (monoliths, thin films, fibers, powders). The great diversity of materials makes the process very attractive in many domains of applications: optical, electrical and electronic, biomaterials, sensors, separation (chromatographic).

The distinction between the sol-gel process and the traditional methods of materials' forming is:

- High chemical homogeneity and purity component, containing in a final material on a molecular scale;
- Flexibility and controllability of the process;
- Forming of silicate and ceramic matrices at lower temperatures.

The sol-gel technology is versatile, simply, cheaply and ecologically. There is believed as energy and resource saving process. One more advantage is the simplicity of the necessary equipment, too [1].

Gupta [4,5] has also underlined, the sol-gel technology constitutes an innovative way in science that requires a multidisciplinary approach for its various applications.

The motivations for sol-gel processing are primarily the potentially higher purity and homogeneity and the lower processing temperatures associated with sol-gels compared with traditional methods. The final homogeneity is directly obtained in solution on a molecular scale. In the sol-gel route the wet gel may in principle be prepared in stoichiometric conditions and with a degree of purity which depends only on the starting ingredients. At the same time, the lower temperature reduces the risks of contamination and loss of more volatile components.

The reaction is easy to perform, does not require special conditions (can be done on the bench top in a beaker), and offers the possibility of various forming processes.

At the same time there are recognized the improvements in the ability to control/tailor the process and in the production rates, the reproducibility and the control of porosity, as well as the greater automation, the homogeneity and ability to control phase development in films.

Another unique feature of the process is the monitoring of pore–solid architecture. There is extraordinary control not only of the size (mesopores of 2–50 nm) but also the arrangement of pores within the inorganic (or organic/inorganic) framework. The design of materials with specific architectures is enabling researchers to obtain unique properties in such diverse areas as drug delivery and electrochemistry.

Sol-gel science and technology has the potential to make a significant impact in modification of the properties of materials. One of the most significant benefits of sol-gel science is its use of room temperature conditions and insensitivity to the atmosphere. These features will allow its use with various materials which cannot tolerate high temperatures. The sol-gel materials can increase thermal and mechanical properties of composites if the two systems are mixed homogeneously on the molecular level. Sakka [6] as an important attorney in the new trend of the technological development of the sol gel process, by underlining its close link with nanotechnology and with life sciences, shows all the porous gels with pores of nano-size diameter may consist of nano-particles or nano-composites, with nano-size organic pigments, noble metal colloids or semi-conductor nano-particles, or organic-inorganic hybrids, with nano-size organic networks, connected with nano-size inorganic networks.

At the same time, there are current results in the polymers' area that utilize sol-gel chemistry. These include silane epoxies for adhesives, silane urethanes for controlled porosity materials for water filtration, silane zeolites for modifying polymer membranes for fuel cells or chemical/biological resistant materials, and silsesquioxane materials for compatibilizing organic-inorganic composites.

By applying the sol-gel technique, a significant impact in modification of the materials properties can be achieved (Figure 1): mechanical (modulus, strength, toughness, impact, fracture, fatigue, abrasion, scratch), thermal (degradation, glass transition Tg), chemical (degradation, solvent resistance), permselectivity, optical, electrical.

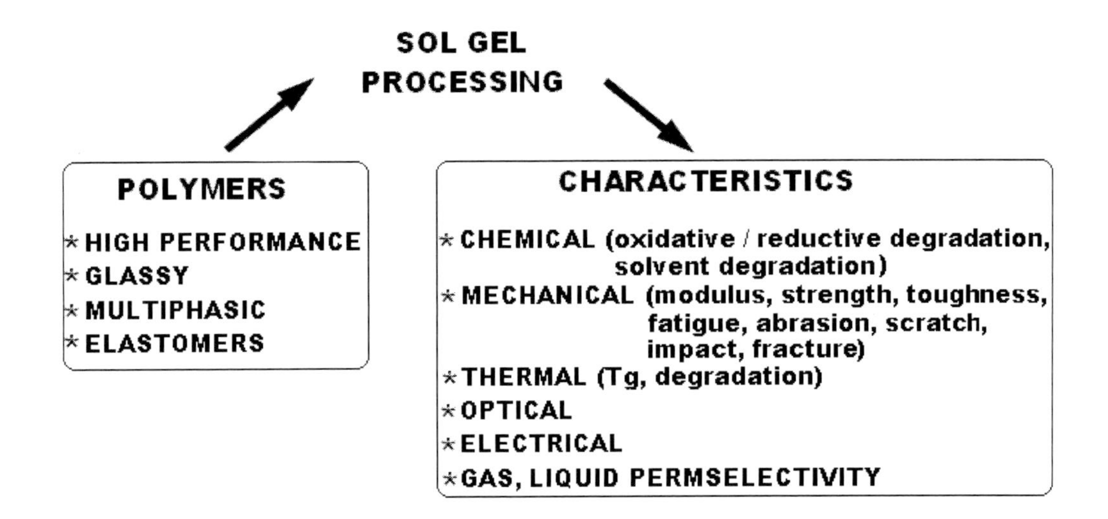

Figure 1. Sol-gel process applications in the material modification.

According to Uhlmann and Teowee [7], the sol-gel processing has a great potential for various application. Products, such as molded gels [8, 9], spun fibers [10, 11], thin films [12 - 14], molecular cages [15, 16], and xerogels [17] have been developed for utility in different areas as gas separations [13, 18], elastomers [19, 20], coatings [21, 22], and laminates [18, 23].

It has been used for the fabrication of optical fibers, optical coatings, electrooptic materials, nanocrystalline semiconductor-doped xerogels, colloidal silica powders for chromatographic stationary phase and as catalytic support, nanoporous carbon xerogels and aerogels as hydrogen storage materials, luminescence concentrators, tunable lasers, active wave guides, semiconducting devices, sunscreen formulations (sol–gel pearls) and chemical sensors for detecting gases, heavy metals and pH, and for many biosensor applications. Potential applications of sol–gel technology in the areas of defense, nanotechnology, environmental monitoring and biomedical devices are now continuously emerging [24-27].

Table 1 summarizes the correlation between: modified material function – domain of application and benefit/difficulty intervened in the sol – gel process.

Table 1. Modified material function – domain of application and benefit/difficulty correlation in the sol – gel process.

Material function	Domain of application	Benefit / difficulty offered by the process
Optical and photonic	Fluorescence solar collector, solar cell Laser element, light guide Optical switching, light amplification, antireflecting coatings, non-linear optical effect (second generation)	- The ease of films deposition, - The ability to provide coatings over large areas with high purity and homogeneity, - The ability to tailor refractive index, - Low cost process, - The ability to provide novel functionalities, - Small capital cost
Electrical and electronic (ferroelectricity electronic and ionic conduction)	Capacitor, piezoelectric transfer Non-volatile memory, transparent semiconductors Solid electrolyte (battery, fuel cell) Electrochromic films, antistatic coatings on plastics Dielectric films, multicomponent films, high dielectric constant films, porous silica low dielectric constant films, and electrode materials with high surface area for secondary batteries.	- The ease of preparation of sol-gel coatings, - The capability (with appropriate apparatus) of making multi-layered coatings, - The difficulty of processing, especially multilayered structures, - Problems with the densification of individual layers, - The difficulty of making competitive transparent conductors
Mechanical/Wear	Protection with hard coat, strong ceramics abrasive Protective/scratch resistant/wear-resistant coatings on plastics High temperature rubbers, protective coatings on glass, nanocomposites, coatings on monumental stones, and ZrO2.	- The achieving of dense films at low temperatures
Thermal	Monolithic aerogels as thermal insulation Thin film aerogels, IR reflective glazing Refractory ceramics, fibers wood Low expansion ceramics Thermally resistant paints Polymers with increased thermal durability	- Mass production of aerogels without supercritical drying, - Hybridization with inorganic Substances, - Thermal durability obtained by infiltrating sols, - Too expensive
Passivation.	Protective coatings on polymers and metals	- By pre-coating with sol-gel SiO_2 or similar films (presumably

Table 1. (Continued)

Material function	Domain of application	Benefit / difficulty offered by the process
Passivation.	Flaw-heating coatings on glass Protection of functional coatings on glass Passivation of semiconductors	referring to uses such as the protection of ITO when deposited on soda-lime-silicate substrates).
Release/Wetting	Biomedical applications Slow release of materials/drugs Antistick coatings, easy-to-clean coatings Antigraffiti coatings Antifouling coatings Repellent films	- The thermal, mechanical and chemical stability of release /wetting coatings can be improved, - Scratch resistance of hydro-phobic coatings can be improved as good as that of other hard coatings
Sensors	Chemical sensors, biosensors, thin film sensors, pH sensors, temperature sensors, gas sensors, fiber sensors, photochromic sensors Lead Zirconate Titanate (PZT)-related materials IR detection using piezoelectric properties Selective adsorbents incorpo-rating optical indicators	- Application in niche market, - Easily-formed microsensor elements, - Highly porous films synthesized without supercritical drying, - Nanocomposite and hybrid films containing functional molecules
Chemical	Aerogel catalyst, catalyst supports, membrane, gas barrier, repellent film Liquid chromatography elements microfilters, and controlled pore materials	- Custom tailored porosity and interfacial chemistry, - Size and arrangement control of pores
Biomedical	Biomaterials, biocatalysts, biospacers controlled drug release capsules, Selective bioadsorbents Coated implants Medical tests	- Entrapment of enzymes, cells, antibodies and bioactive substances

SHORT HISTORY

The transition mechanism from a liquid 'sol' into a solid 'gel' phase describing sol–gel processes has been the subject of many scientific works.

In a comprehensive review, Hench [28] cited the revolutionary discovery of Ebelman that firstly reported in 1845 the formation of a transparent solid "glass-like material" of SiO_2, by slow hydrolysis of tetraethyl orthosilicate, and wrote that *"it is permitted to hope that it could be used in the construction of optical instruments"*. From that moment, the scientific way of the sol-gel science was very progressive and, the initial impulse was of a practical, if not technical, nature.

Thus, the synthesis of SiO_2 from liquid silicon metal–organic precursors is probably the oldest and most investigated sol–gel process. Typically, it involves hydrolysis and condensation of a metal–organic precursor, such as tetraethoxysilane, in an appropriate solvent, such as ethanol, with or without the use of a catalyst. As these reactions proceed and the viscosity of the solution increases, a gel is formed. It is typically made of Si–O–Si bonds, forming a network within which reaction products are trapped. When such products are removed via evaporation, a nanoporous structure results. Pore size, distribution, and interconnectivity are affected by processing parameters, such as the type and amount of solvent and/or catalyst, temperature, and in some cases, the presence and arrangement of templating molecules. By themselves, these porous gels constitute a very interesting class of nanomaterials. The nanoporous architecture may be tailored for connectivity, orientation, arrangement, or size and exploited for various applications, such a chromatography columns, sensors, catalyst supports, low dielectric constant materials, or controlled release of reactants.

Over the past several decades the sol-gel method has a remarkable growth. The experience accumulated by many scientists with the participation of world-known names in the field of sol-gel process is summarized in many books, monographs and reviews, with the main types of precursors, methods of synthesis, structure and characterization and the most important practical applications, the current state of the problem and the perspectives in the domain. The literature concerning sol-gel chemistry also includes the great potential of technical applications that can be realized with sol-gel derived products. The scientists have the challenge to make competitive products by doping sol-gel derived materials with functional moieties, using the aerogel or xerogel as the host matrix. Due to the fact that aerogels and xerogels have physical properties such as high porosities, high surface areas, etc., they lend themselves to this role quite well.

In the recent years a number of reviews on sol–gel technology have appeared with specific applied areas [29 – 34]. Thus, Mackenzie [35] summarized a number of potential advantages and disadvantages and the relative economics of sol-gel methods in general considered the main achievements associated with the sol-gel process, showing the research in the past two decades is divided into three generations of solid gels: oxides, organic-inorganic hybrids and composites made from the suspension of a variety of solids in the hybrids. Starting from a chemical and physical variety of precursors 'structure – above all alkoxides, there were obtained oxide gels in the First Generation Process (represented in Figure 2).

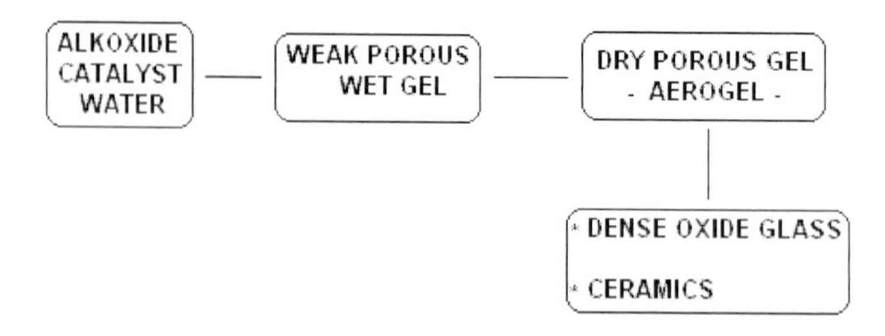

Figure 2. Simplified overview of the First Generation Sol-Gel Process.

The obtaining of organic-inorganic hybrids belongs to the Second Generation Process (represented in Figure 3).

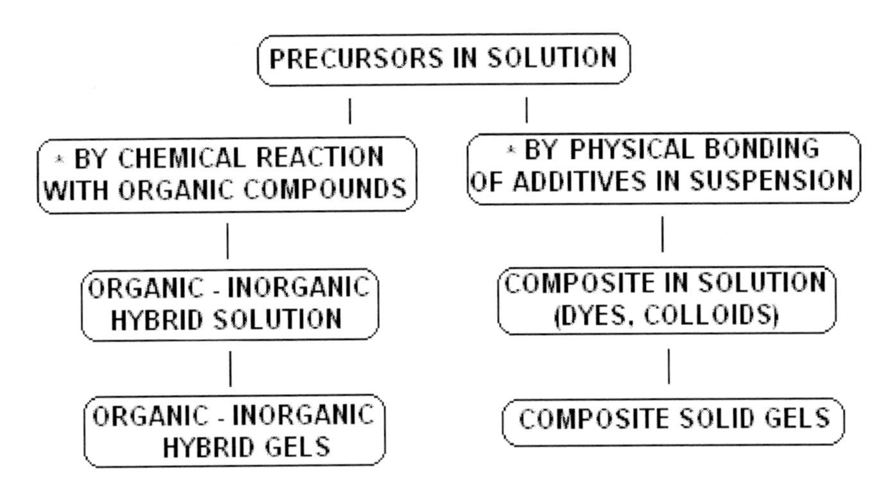

Figure 3. Simplified overview of the Second Generation Sol-Gel Process.

In the pioneer studies of Schmidt [36-38] on the preparation and characterization of the organic – inorganic hybrid structures, a new type of noncrystalline materials (named ORMOSIL and ORMOCER- Organically Modified Ceramics) with strong chemical bonds between organic and inorganic components, were obtained.

Also an innovative solution was the investigation of Avnir et al. [39] on another kind of hybrid materials obtained according to the sol – gel method and by incorporating organic dyes in an oxide gel matrix.

Because the chemical bonds between the inorganic matrix and the organic molecules of dyes are weakly, the products are named nanocomposites. On this basis, there are developed compositions with luminescent dyes and SiO_2, for self-tuning lasers [40].

Irrespectively of the initial precursors and the sol – gel route, the system follows the following steps:

$$GelFormation \Rightarrow GelAging \Rightarrow Solvent\mathrm{Re}moval \Rightarrow HeatTreatment(eventually)$$

The method schematic represented in Figure 4 enables the formation of gels with different configurations (porous solid monoliths, thin films, glass, fibres, ceramic or powders), as a function of the used drying:

- drying in mild conditions: the gel hardens and becomes compacted: xerogel (glasses and dense ceramics);
- solvent evaporation in supercritical conditions: the formation of a less compact gel named aerogel;
- spreading the sol on a surface to obtain thin films of xerogel (by different coating techniques).

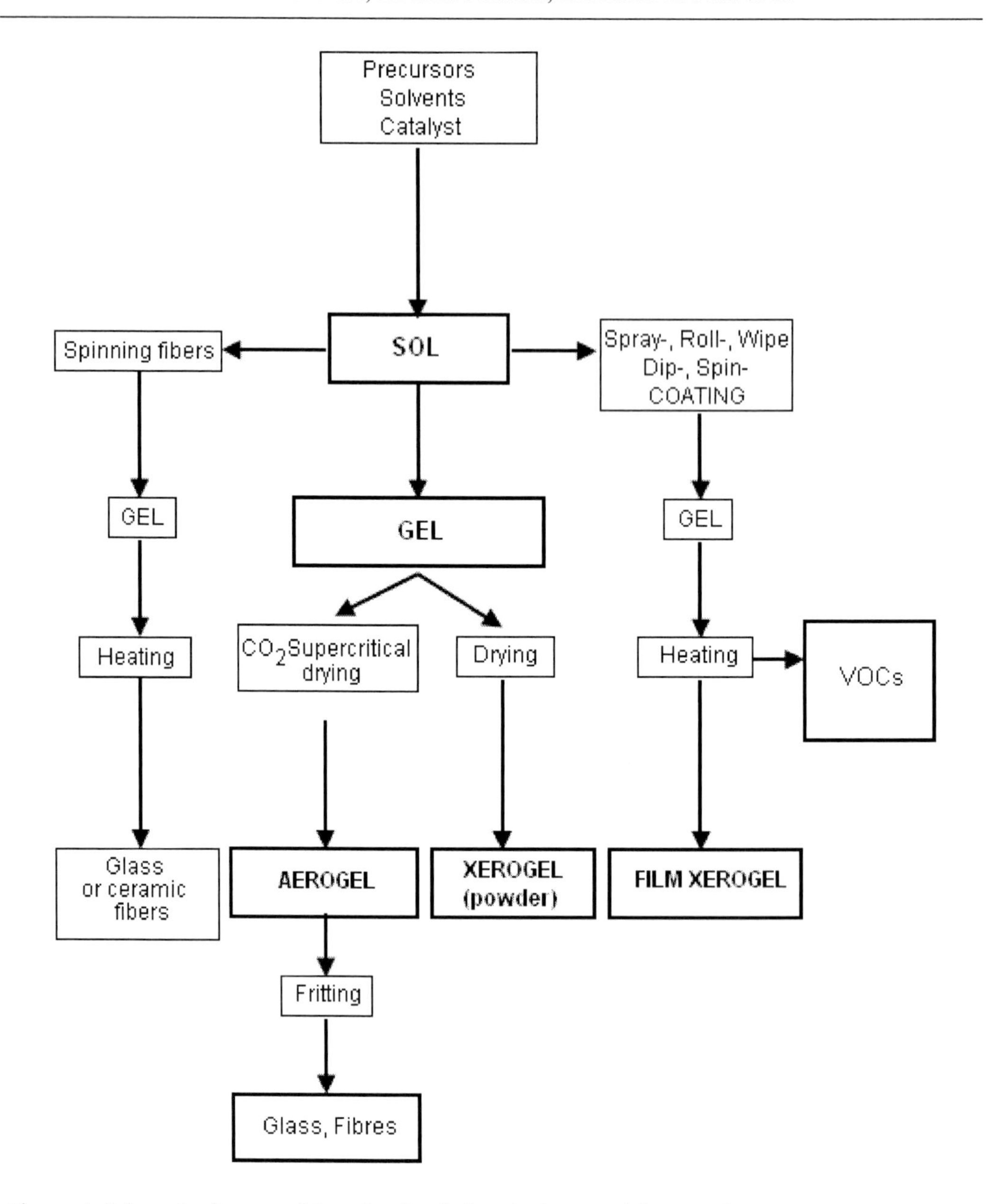

Figure. 4. Schematic diagram of the sol-gel technique in the material processing.

Basic Concepts in Sol – Gel Synthesis

Initially, the sol gel technologies were developed during many years as an alternative for the preparation of glasses and ceramics at considerably lower temperatures. Sol-gel processing is sufficiently flexible, that it is particularly appealing to material scientists in the production of a wide range of highly porous inorganic oxide and carbon based networks. The initial system represents a solution where different polymerization and polycondensation processes lead to the gradual formation of the solid phase network.

The solid is first subjected to a series of operations: gelling, drying, pressing, drawing and casting, which results in various structural and phase transformations. This permits formation of powders, fibers, coatings, bulk monolithic products, etc., from the same initial composition.

Based on the scientific investigations, many routes of synthesis have been investigated. The schemes of synthesis are dependent of the initial precursors: aqueous solutions of metal salts; metal alkoxide solutions; mixed organic and inorganic precursors. The process may be realized by alkoxide route and by colloidal route.

The most investigated is the *alkoxide route*, especially for inorganic systems. The organically modified alkoxides or organo alkoxy silanes are characterized by having a Si-C bond that survives, in general, hydrolysis and condensation processes.

The addition of organic monomers or oligomers may lead to a so-called hybrid network either by being linked to the inorganic backbone or independently of it (interpenetrating network). The introduction of any type of functional groups can be achieved by linking complex organic structures to the alkoxides.

The inorganic polymerization proceeds by hydrolysis of the alkoxide precursors to introduce a reactive hydroxyl group on the metal. This step is followed by the formation of metal oxo-bridges or metal hydroxo-bridges by condensation or addition reaction, respectively. The formation of hydroxo-bridges occurs when the coordination number of the metal is higher than the valence state.

- Hydrolysis :

$$M-OR+H_2O\rightarrow M-OR+R-OH$$

- Condensation :

$$M-OH+XO-M\rightarrow M-O-M+X-OH,\ X=HorR$$: *Formation of oxo-bridges by oxolation reactions*

$$M-OH+XO-M\rightarrow M-OH-M-OX,\ X=HorR$$: *Formation of hydroxo-bridges by olation reactions*

Alkoxides are not miscible with water; a common solvent usually an alcohol is used. The oxo metallic network progressively grows from the solution, leading to the formation of oligomers, oxopolymers, colloids (sols or gels) and a solid phase. These reactions can be described as S_N2 nucleophilic substitutions and the chemical reactivity of metal alkoxides toward hydrolysis and condensation depends mainly on the electronegativity of the metal ion and the ability to increase the coordination number [1, 26].

Sol Gel Route and Alkoxysilane Precursor

Sol-gel reactions promote the growth of colloidal particles (sol) and their subsequent network formation (gel) through the hydrolysis and condensation reactions of inorganic alkoxide monomers [41]. The precursors for synthesizing these colloids consist of a metal or metalloid element surrounded by various reactive ligands. Metal alkoxides are most popular because they react readily with water. The most widely used metal alkoxides are the alkoxysilanes, such as tetramethoxysilane (TMOS) and tetraethoxysilane (TEOS). However, other alkoxides such as aluminates, titanates, zirconates, and borates are also commonly used in the sol-gel process, either alone or in combination with other alkoxides, such as TEOS.

Sol-gel reactions are a series of hydrolysis and condensation reactions of an alkoxysilane, which proceed according to the reaction scheme shown in Figure 5 [41]. Hydrolysis is initiated by the addition of water to the TEOS solution under acidic, neutral, or basic conditions.

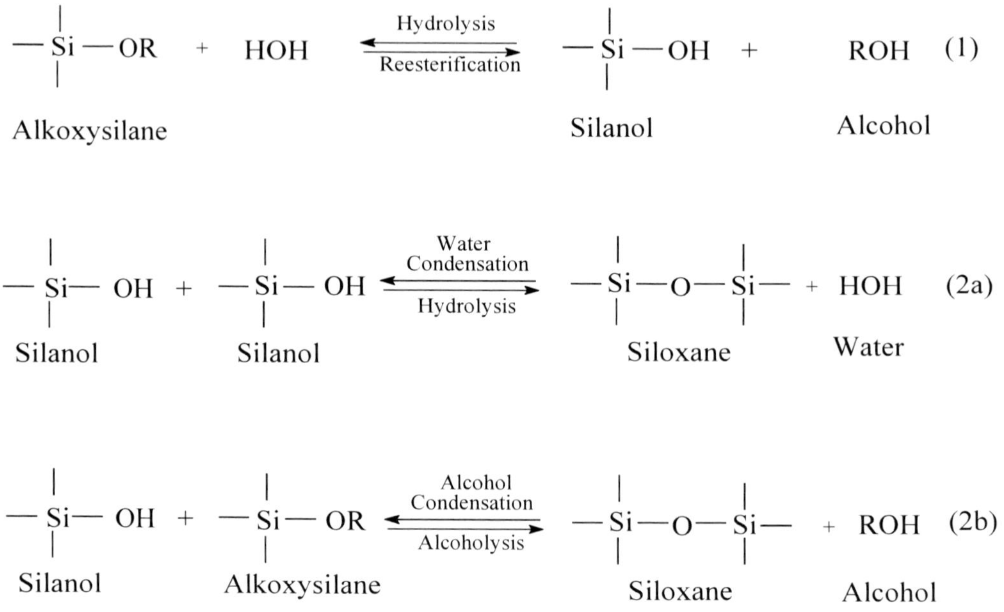

$$\underset{\text{Alkoxysilane}}{\equiv\mathrm{Si-OR}} + \mathrm{HOH} \underset{\text{Reesterification}}{\overset{\text{Hydrolysis}}{\rightleftharpoons}} \underset{\text{Silanol}}{\equiv\mathrm{Si-OH}} + \underset{\text{Alcohol}}{\mathrm{ROH}} \qquad (1)$$

$$\underset{\text{Silanol}}{\equiv\mathrm{Si-OH}} + \underset{\text{Silanol}}{\mathrm{HO-Si}\equiv} \underset{\text{Hydrolysis}}{\overset{\text{Water Condensation}}{\rightleftharpoons}} \underset{\text{Siloxane}}{\equiv\mathrm{Si-O-Si}\equiv} + \underset{\text{Water}}{\mathrm{HOH}} \qquad (2a)$$

$$\underset{\text{Silanol}}{\equiv\mathrm{Si-OH}} + \underset{\text{Alkoxysilane}}{\equiv\mathrm{Si-OR}} \underset{\text{Alcoholysis}}{\overset{\text{Alcohol Condensation}}{\rightleftharpoons}} \underset{\text{Siloxane}}{\equiv\mathrm{Si-O-Si}\equiv} + \underset{\text{Alcohol}}{\mathrm{ROH}} \qquad (2b)$$

Figure 5. Sol – gel process: the general reaction scheme [41].

In the first step – hydrolysis - a silanol group is generated [(Figure 5, reaction (1)]. As the acid catalysis is used in this example, the consideration of the sol-gel reactions will be made to acid catalysis only [3]. A positive charge develops on the alkoxysilane through the attack of an acid catalyst. The alkoxysilane then is hydrolyzed in a nucleophilic substitution S_N2-type reaction, forming a silanol moiety.

The mechanism for reaction (1) is depicted in Figure 6. TEOS is not soluble in water. Therefore, hydrolysis is promoted by the addition of organic cosolvents, such as alcohols [3] that aid in the mixing of the alkoxide molecules with water molecules in solution.

In the second step, the silanol group can undergo condensation with either an alkoxide or another silanol group [reaction (2a) or (2b) in Figure 5], which results in the formation of strong siloxane linkages and produces either alcohol (ROH) or water *in situ*. As the number of

Si-0-Si bridges increases, the siloxane particles can aggregate into a sol. This consists in the formation of small silicate clusters dispersed in solution. Gel formation occurs when the sol particles undergo sufficient condensation reactions such that a network (a gel) is formed, trapping the aqueous and alcohol by-products. After gel-network formation is complete, the by-products are removed by heat and vacuum, yielding a vitrified and densified network.

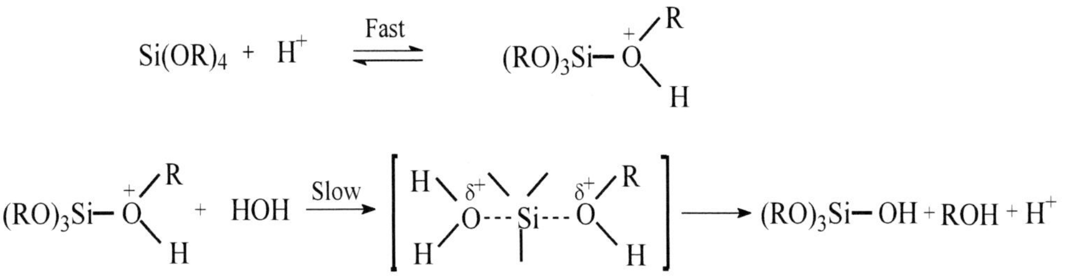

$$Si(OR)_4 + H^+ \underset{}{\overset{\text{Fast}}{\rightleftharpoons}} (RO)_3Si-\overset{+}{O}(R)(H) \quad (3)$$

$$(RO)_3Si-\overset{+}{O}(R)(H) + HOH \xrightarrow{\text{Slow}} \left[\begin{matrix} H \\ H \end{matrix} \overset{\delta+}{O}\cdots Si \cdots \overset{\delta+}{O}\begin{matrix} R \\ H \end{matrix} \right] \longrightarrow (RO)_3Si-OH + ROH + H^+ \quad (4)$$

Figure 6. Mechanism of tetraalkoxysilane hydrolysis reaction [41].

Factors Influencing the Sol-Gel Reaction

A number of conditions can influence the hydrolysis and condensation reactions. Of these, the most relevant include water-to-alkoxide ratio, type and amount of catalyst, type of network modifier, and solvent effects [3, 41, 42]. Their influence is briefly presented in the sequential sections.

Water / Alkoxide Ratio

The effect of the water/alkoxide ratio for the sol-gel process is such, that as the ratio increases, so does the SiO_2 content of the gel [41]. Therefore, for complete hydrolysis, there must be at least one mole of water for every alkoxide group. Some researchers have gone further to state that if more than one mole of water per alkoxide group is used, the reverse reaction, reesterification, will occur faster than the forward reaction [3]. However, in a recent article by McCormick et al. [43], whose experiments were conducted over a wide range of water/TEOS ratios, there was no correlation between the water/alkoxide ratio and the achievement of complete hydrolysis. The validity of these experiments on the effect of water/alkoxide ratio is correctly because water is *in situ* generated in the reaction and therefore the reaction, once catalyzed, self-propagates the hydrolysis.

Type and Amount of Catalyst

One important subject for consideration when deciding the concentration of catalyst is whether a precise concentration is needed [41]. In the sol-gel process, water is *in situ* generated through condensation reactions. This makes difficult the addition of a precise amount of catalyst. In addition to water / TEOS ratio experiments, McCormick et al. [43] synthesized sol-gel films, adding a wide range of acid concentrations. Their results indicated no correlation between the acid concentration and acid initiation of the sol-gel reaction. In the experiments, different acid concentrations were used with each sample and, as previously discussed, the condensation reaction caused the *in situ* generation of water that will dilute the initial concentration of acid. The nuclear magnetic resonance (NMR) spectroscopy for the different overall sol-gel reactions displayed no evident difference between the sol-gel structures.

The results show that a minimum catalytic amount of acid is necessary in all the experiments for the self propagation of the reaction. The kinetic of the reaction could be modified, but not the basic structure of the overall network.

Hydrolysis and condensation reactions of most inorganic alkoxides can be carried out without catalyst because of the extremely fast rates of reaction. However, alkoxysilanes hydrolyze more slowly, requiring the addition of either an acid or base catalyst [41] (Figure 7). Acid-catalyzed reactions, having a particle nucleation rate-determining process, tend to yield more linear-like networks due to the fast hydrolysis. Therefore, acid catalyzed systems have a less completely formed network of siloxane bonds with a higher concentration of unreacted silanols. Base-catalyzed reactions, on the other hand, yield highly dense materials due to the longer time that the sol particles have to aggregate and arrange themselves in the most thermodynamically stable arrangement.

This process leaves fewer unreacted silanol groups in the overall network and a more highly densified network. Acid catalysis, in this example, increases the rate of the hydrolysis reaction, as shown in Figure 6. In the first step [Figure 6, equation (4)], hydrolysis occurs through the protonation of the oxygen atom of the alkoxy group due to the presence of extra electrons. This causes a shift in the electron cloud of the Si-0 bond toward the oxygen, resulting in a positive charge on the silicon atom [28, 41].

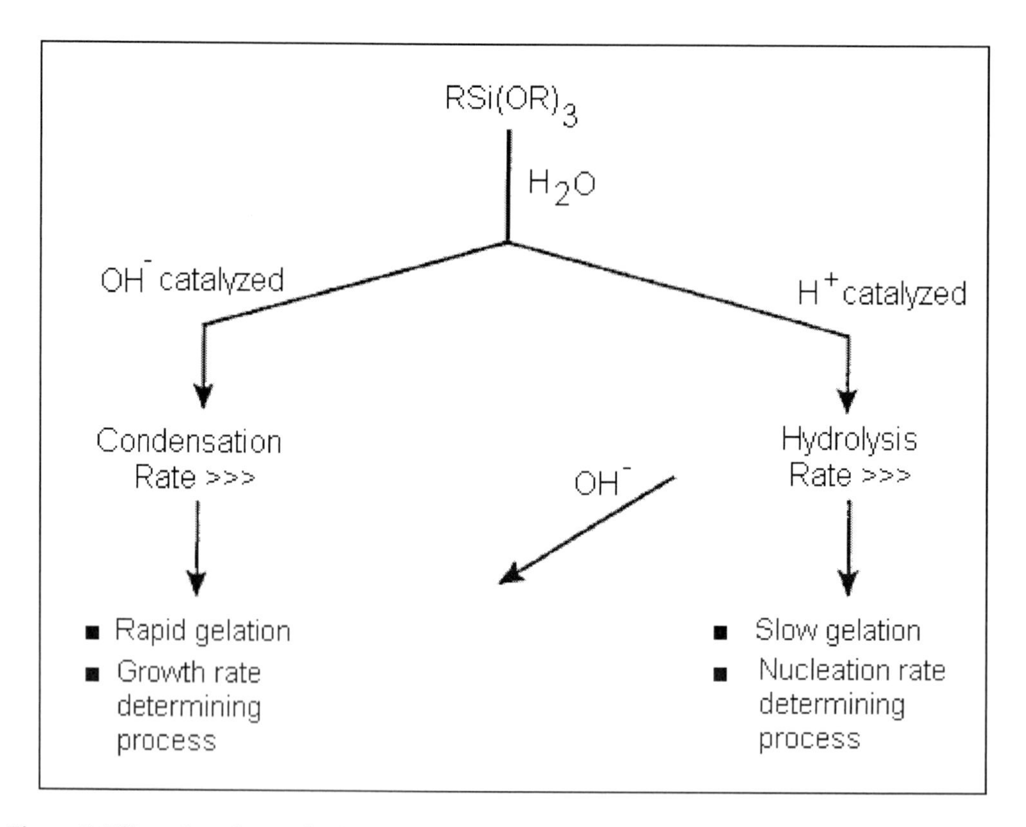

Figure 7. Effect of catalyst on hydrolysis and condensation [41].

In the rate-determining step [Figure 6, equation (4)], oxygen from the water attacks the silicon atom, which has low electron density. This results in the formation of a pentacoordinate transition state in which partial positive charges are developed. Steric effects, such as the presence of bulky and/or long alkyl substituents, will affect the rate of the hydrolysis reaction by hindering the inversion of the S_N2 transition state [28, 41].

In a presentation [41] - not to scale - in Figure 8, there are some pH-dependent rate profiles for the hydrolysis and condensation reactions; thus, there is observed [3, 41, 42, 44,45] the reaction rates are in essence dependent on pH: at pH = 7, hydrolysis occurs at a slow rate, while condensation occurs at a fast rate. It is this inverse correlation between the rates of the hydrolysis and condensation reactions, that controls the kinetics of the reactions and as a result, the final network structure.

Type of Network Modifier

When TEOS is commixed with a silicon alkoxide comonomer containing organic groups directly bonded to the silicon atom, organically modified silicates (ORMOSIL®s) are created. The result of synthesizing ORMOSIL®s is to modify the network connectivity. Organic groups cause coordination centers with functionality less than four and influence the reactivity of the alkoxy groups and therefore the connectivity of the sol-gel network in two ways [41]:

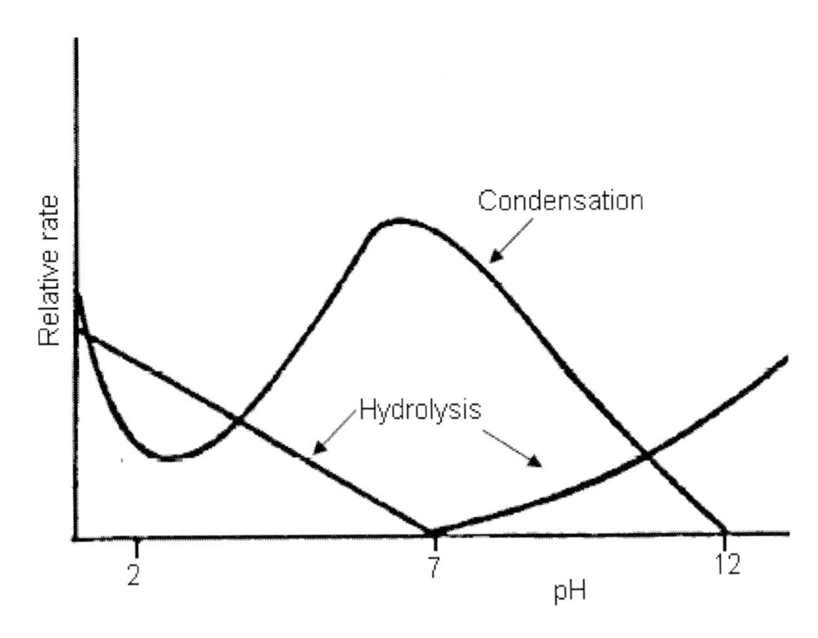

Figure 8. Effect of pH on hydrolysis and condensation rates.

(1) Formation of siloxane bonds requires the diffusion of partially hydrolyzed molecules. The larger the alkyl group attached to the silicone, the slower the rate of diffusion and therefore the less interconnection within the network.

(2) It is also reported that larger alkyl groups produce polymers with higher surface area [3]. Larger surface area allows for higher unreacted oxide concentration in the gels. This produces a branching effect in the sol-gel network. In addition to these two points, it was previously discussed that steric effects, such as the presence of bulky

and/or long alkyl substituents, will affect the rate of the hydrolysis reaction by hindering the inversion of the S_N2 transition state [28, 41].

Solvent Effects

The effect of solvents on hydrolysis and condensation reactions is not usually discussed primarily because solvents other than water and simple alcohols, which are produced internally, are not often used. However, addition of external solvents can have an important effect on controlling hydrolysis and condensation rates [41]. These solvents that operate like additives in the process, include tetrahydrofuran, formamide, dimethylformamide, and oxalic acid. They may be used in addition to alcohol in order to promote slow drying of the silica monolithic gel [3, 28]. In bulk films, slow drying is necessary to prevent cracking of the film.

Condensation rates may be radically affected by solvent type [41] (Figure 9). Condensation, via acid catalysis devolves through a two-step S_N2 mechanism.

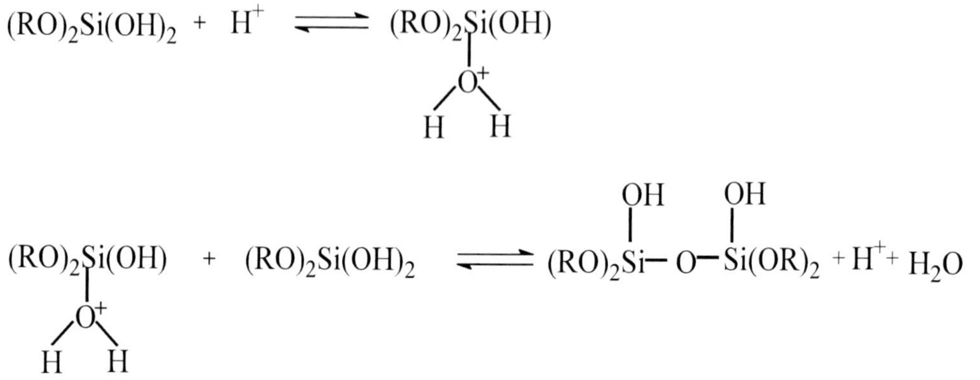

$$(RO)_2Si(OH)_2 + H^+ \rightleftharpoons (RO)_2Si(OH)(OH_2^+) \quad (5)$$

$$(RO)_2Si(OH)(OH_2^+) + (RO)_2Si(OH)_2 \rightleftharpoons (RO)_2Si(OH){-}O{-}Si(OH)(OR)_2 + H^+ + H_2O \quad (6)$$

Figure 9. Tetrafunctional silicate condensation mechanism

Protic solvents have little or no effect on condensation rates. Polar aprotic solvents, such as those previously mentioned, solvate the protonated silanol ($Si\text{-}OH^+$). The cation is solvated by orienting the negative end of the polar aprotic around the cation and by donating unshared electron pairs to the vacant orbitals of the cation. This process stabilizes the transition state and increases the rate of condensation.

Sol gel route and hybrid precursors ORMOCER®

The strategy to construct hybrid materials consists of making intentionally strong bonds (covalent or iono-covalent) between the organic and inorganic components. Organically modified metal alkoxides are hybrid molecular precursors that can be used in this view. The chemistry of hybrid organic-inorganic networks is mainly developed around silicon containing materials. Currently, the most common way to introduce an organic group into an inorganic silica network is to use organo-alkoxysilane molecular precursors (low molecular weight alkoxysilanes, such as tetraethoxysilane –TEOS- or tetramethoxysilane –TMOS- as silica precursors) or oligomers of general formula:

$$R'_nSi(OR)_{4-n} \; or \; (OR)_{4-n}Si - R'' - Si(OR)_{4-n} \; with \; n = 1,2,3$$

The reactions of alkoxysilanes can be summarized in terms of three steps: hydrolysis of the alkoxide, silanol-silanol condensation, and silanol-ester condensation. Hydrolysis occurs by the nucleophilic attack of the oxygen contained in water in the silicon atom. Subsequent condensation reactions take place producing siloxane bonds. The polymerization stages may be described as, (1) polymerization of monomers to polymers, (2) condensation of polymers to primary crystals, (3) growth or agglomeration of primary crystals to particles and (4) linking of particles into chains and to three dimensional network [1].

Networks of the chains extend throughout the liquid medium, thickening the network into a gel. In the last stage water and alcohol are evacuated from the network structure causing gradual shrinkage and even cracking of the monolithic gel.

Like it is already underlined, in the most sol-gel conditions the Si-C bond remains stable towards hydrolysis and the R' group introduces focused new properties to the inorganic network (flexibility, hydrophobicity, refractive index modification, optical response, etc).

Organic groups R' can be introduced into an inorganic network in two different ways: as network modifiers or network formers. Both functions have been achieved in the so-called ORMOCER®s - organically modified ceramics - (registered trademark of Fraunhöfer-Gesellschaft zur Förderung der angewandten Forschung e.V. in Germany). Since the eighties, these products have been extensively studied and developed by the Fraunhöfer Institut für Silicatforschung, Würzburg [36, 46-48]. The organic group R' can be any organofunctional group. If R' is a simple nonhydrolizable organic group (Si-CH_3, Si-phenyl, etc) it will have a modifying effect. Moreover if R' can react with itself (R' contains a vinyl, a methacryl or an epoxy group) or additional polymerisable monomers, it acts as a network former. Sanchez et al. [49] were gathered examples of network formers and network modifiers. Polymeric components can also be introduced in the hybrid nanocomposites by using functionalized macromonomers (R'') of general formula:

$$(OR)_{4-n}Si\text{-}R''\text{-}Si(OR)_{4-n} \text{ with } n=1,2,3.$$

Whereas most of the investigations were carried out in organic solution due to the solubility of alkoxides in these solvents, aqueous route or *colloidal route* also was investigated, especially if one moves to the sol phase, because inorganic sols (stabilized colloids) were known since a long time, especially in the case of silica [1].

Like it was already underlined, sol-gel technology shows many promises and offers many important advantages in materials processing. The nanometer structure of the gels permits low temperature processing of materials so that they can be combined in hybrid materials. The introduction of metal alkoxides precursors for sol-gels made possible the production of high purity materials with improved qualities. The pore structure and large surface areas associated with sol-gel materials has been essential to the development of catalyst and adsorbent materials applied in many other technologies. Therefore, the scientific community has now an important tool of chemical and processing methods to tailor sol-gels and to tackle new materials technologies.

Though it is very difficult to include all the aspects of sol–gel science and technology and cover every leading researcher's report, however, the authors of the chapter tried to highlight the most important aspects with respect to the application, recent development and future perspectives in the biomedical domain of the sol-gel synthesized materials.

Bioencapsulation of Biomolecules by Sol Gel Technique

Introductive Notions

The encapsulation or generation of new surfaces that can fix biomolecules firmly without altering their original conformations and activities is still challenging. The sol–gel chemistry offers new and interesting possibilities for the promising encapsulation of heat-sensitive and fragile biomolecules (enzyme, protein, antibody and whole cells of plant, animal and microbes) because it is an inherent low temperature and biocompatible process [50]. In aqueous solutions, biomolecules such as enzymes, antibodies, protein, lose their functionality. These problems can be minimized considerably by biomolecules immobilization. Taking into account these advantages, since the 1960s, an extensive variety of techniques have been developed to immobilize biomolecules, including adsorption, covalent attachment and entrapment in various polymers [50]. Generally, adsorption techniques are easy to perform, but the bonding of the biomolecules is often weakly and such biocatalysts lack the degree of stabilization and easy leakage from the matrix. The covalent linkage method increases the stability but often requires several chemical steps, and sometimes the involved compounds inactivate or reduce the activity of biomolecules.

Direct immobilization of active biological substances in porous metal oxide carrier by physical entrapment via the sol–gel processes has drawn a great interest in recent years. This is due to its simplicity of preparation, low-temperature encapsulation, easy for immobilization, chemical inertness, tunable porosity, optical transparency, mechanical stability and negligible swelling behavior [51 - 54].

The two major advantages with a sol–gel system is that it can retain a large content of water; this feature makes the encapsulated bio-recognition agents or enzyme catalytic centers long-term stable [55] and the process can be performed at room temperature. Other advent-ages of silica supports include biocompatibility and resistance to microbial attack. Moreover, the preparation conditions of a sol–gel have a remarkable effect on the activity of the entrapped active biomolecules [50].

The ability to form doped inorganic glasses under aqueous, room-temperature conditions (at which proteins and cells are active) opened up the possibility of extending sol–gel processing to the encapsulation of biomolecules. An array of substances, including catalytic antibodies, DNA, RNA, antigens, live bacterial, fungal, plant and animal cells, and whole protozoa, have been encapsulated in silica, metal-oxide, organosiloxane and hybrid sol–gel polymers [56]. The sol–gel process as a route to form inorganic glasses has been known for over a century, however, the first report demonstrating the use of silicate materials for the entrapment of a biological moiety did not appear until the mid-1950s, when Dickey showed that several enzymes could be entrapped into silicic acid-derived glasses with partial retention of biological activity [57]. Unfortunately, the importance of this finding was not realized at the time, and the development of sol–gel derived biomaterials was not revisited for over three decades. Venton et al. showed in 1984 that antiprogesterone antibodies could be trapped within monolithic silica–poly(3-aminopropylsiloxane) sol–gel polymers and fully retain their native recognition and binding functions [58].

An year after, in 1985, Glad et al. showed that in monolithic and thick-film organic–inorganic sol–gel matrices comprising silica–poly[*N*,*N*-bis(29-hydroxyethyl)-3-aminopropylsiloxane] could be entrapped functional glucose oxidase, horseradish peroxidase, trypsin and alkaline phosphatase [59]. These studies allow Glad and coworkers to establish the critical features of sol–gel bioencapsulation: - the sol–gel technique permits the encapsulation of biomolecules; - the biocomposites display the characteristic activities of the entrapped species; - the polymer matrices are permeable enough to enable the diffusion of low molecular weight species, but not enough to permit the captured biologicals to leak [56, 59]. The studies of Glad group were recognized in 1990, when the technique was applied independently to the doping of transparent silica glasses with alkaline phosphatase, chitinase, aspartase and β-glucosidase [60].

In recent years, several reviews on sol–gel technology have appeared with specific applied areas [50, 51, 53, 56, 61 - 66]. This section of the chapter presents the main advantages, developments, applications and perspectives of sol–gel immobilized biomolecules, which includes enzymes, antibodies and protein. Because of already published comprehensive studies in the field, we didn't propose to discuss here bioencapsulation of microorganisms, plant and animal cells.

Sol-Gel Process and Bioencapsulation

The preparation of inorganic silica gels has been extended to novel hybrid nanomaterials in which organic and inorganic species are mixed at the molecular level [66].

The two classes of materials (unmodified and organically-modified silica glasses) are good hosts for the incorporation (using the sol-gel process) of different additive such as pigments, organic dyes, metal particles and a variety of chemical and biological compounds. They can all be combined with the silica sol in solution before gelling and incorporated in silica-based organic-inorganic hybrid matrices for different applications (e.g., photonics, optics, etc.) depending notably on the properties and the functionality induced by the surface modification, for instance [66 - 72].

As with traditional sol–gel fabrication, the starting point for bioencapsulation is the precursor (Figure 10). This is typically an alkyl silicate, an alkoxymetallate or an alkoxysilane, or a mixture of them [56, 71]. The precursor is hydrolysed by water, either spontaneously or under acid or base catalysis, to form hydroxy derivatives (silicic acids, hydroxometallates, hydroxysilanes, etc.). A cascade of condensation reactions gives rise to soluble, colloidal and ultimately phase-separated polymers (polysilicates, hydrous metal oxides, polysiloxanes etc.), which produce the final matrices (silica, metallosilicate, metal oxide and siloxane) [56, 71].

Using this basic technique and specific fabrication processes (e.g. block casting, reverse emulsion polymerization, screen or contact printing, fluid-bed coating and dip- or spin-coating), one can obtain bio-doped hydrogels or xerogels in various configurations (e.g. monoliths, sheets, granulates, microparticles and thick and thin films) [56, 71].

The resulted sol–gel is an interconnected rigid network with pores of sub-micrometer dimensions and polymeric chains whose average length is greater than a micrometer. When the liquid in the pore is removed at or near ambient pressure by thermal evaporation, drying and shrinkage occurs, the resulted monolith is termed as xerogel (Figure 10). If the liquid is primarily alcohol, the monolith is termed as an "alcogel." Usually, xerogels are superior in mechanical properties and chemical resistance to hydrogels in view of cross-linking and densification [50]. But the drying of hydrogels inevitably reduces porosity, increases steric compression and diffusional limitations, and results in a reduced bioactivity, especially for inorganic sol-gels.

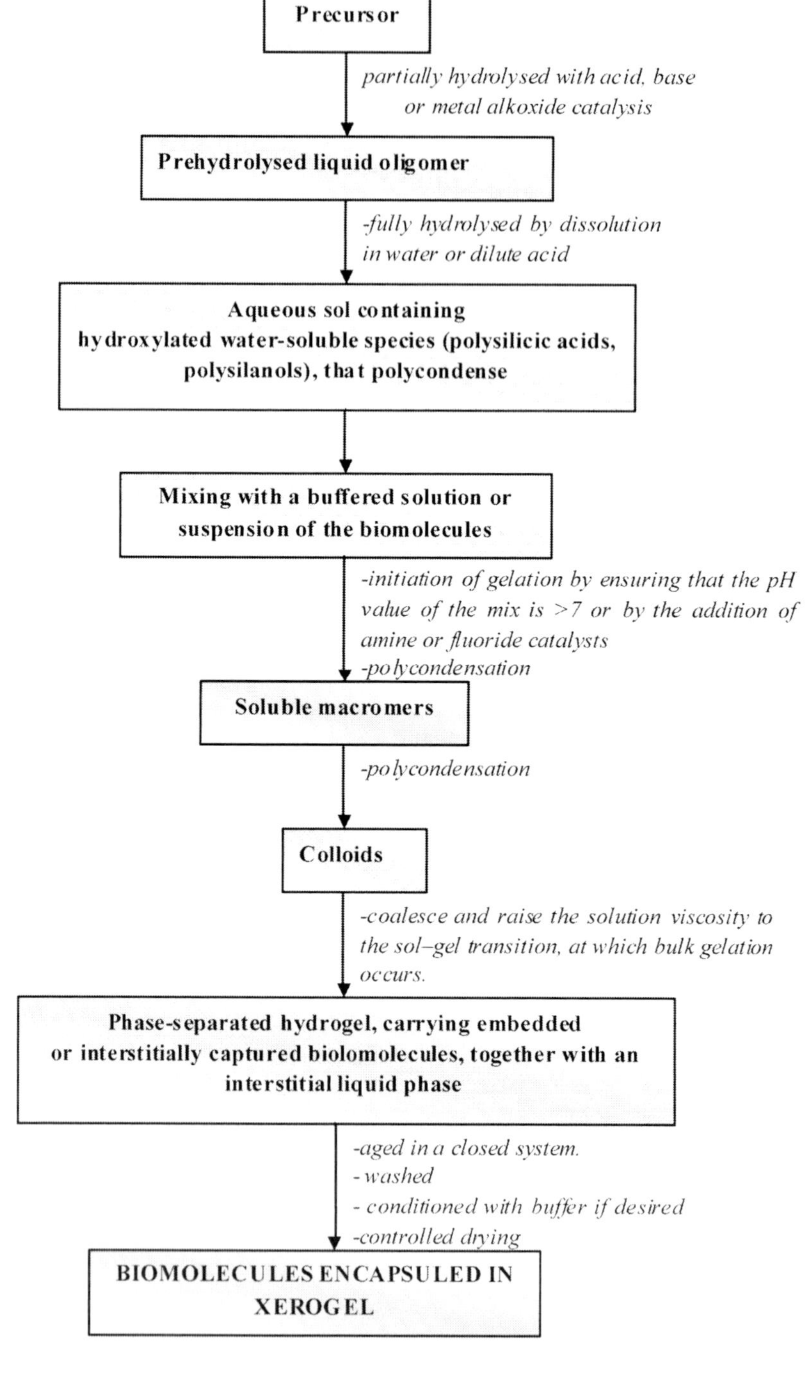

Figure 10. The sol-gel bioencapsulation technique (adapted by Gill) [56].

The studies point out that, depending on the precursors and the protocol used, proteins with molecular weights of 8000–15000 (~1.3–1.7 nm ellipsoids) can be irreversibly encapsulated in sol–gels [50] The trapping of such small molecules results in the close interpenetration of the biological and polymer surfaces and the establishment of specific binding interactions [56]. Investigations suggest that there is enough conformity between the matrix and the surface of the biological, to capture biomolecules and to hinder their diffusion and global motion but sufficient latitude for local conformational transitions [56]. Gill et al. [56] suggest in the review from 2000, the following model for the encapsulation of biomolecules:

- The biological entity resides within a polymer cavity whose interior is templated to conform physically and chemically to the surface features of the protein.
- There is substantial interpenetration of exposed biomolecule segments with the sol–gel polymer framework, resulting in varying degrees of embedding.
- A restricted solvent shell is trapped between the protein and polymer surfaces.
- Protein rotation and conformational transitions are restrained according to the conformity of the polymer surface, the composition of the sol–gel framework, the amount and mobility of the trapped solvent, and biomolecule – polymer interactions (ionic, hydrogen bonding, hydrophobic).
- There is sufficient accessibility between the binding or catalytic site of the protein and the pore structure, and enough freedom for local conformational transitions, to enable the entry, recognition and processing of substrates in a similar way to the free, soluble protein.

Types of Sol–Gel Precursor and Matrix

Much effort on bioencapsulation has focused on silica, metallosilicates and titanium, zirconium and aluminum oxides, which form hard transparent glasses that are micro- to mesoporous [72 - 75]. Since the first example of sol-gel bioencapsulation introduced by Avnir and co-workers [61], several types of sol-gel matrices have been developed to be used as substrates (Figure 11):

Inorganic sol-gels: The pure inorganic xerogels, such as aluminum, titanium, zirconium and tin oxides as well as their mixed oxides with silica, are always hard, transparent glasses with microporous structure. They are chemically robust, but limited by their brittleness and too small pore size, which prevents small molecule diffusion through the matrix [76].

Organically modified silica sol-gels *(Ormosils)*: In this category, organic groups, from simple alkyl, alkenyl, and aryl to those additionally bearing amino, amido, carboxy, hydroxy, thiol, and mixed functionalities as well as nicotinamides, flavins and quiniones, can be grafted on precursor silanes. Thus after sol-gel reactions, those organic functional groups are attached on the silica matrix by stable Si-C bonds. Because of those groups, tailorable properties, such as hydrophilic, hydrophobic, ionic as well as H-bonding capacities can be achieved in the silica matrix. However, the optical transparency and stability are lower than inorganic sol-gels [76].

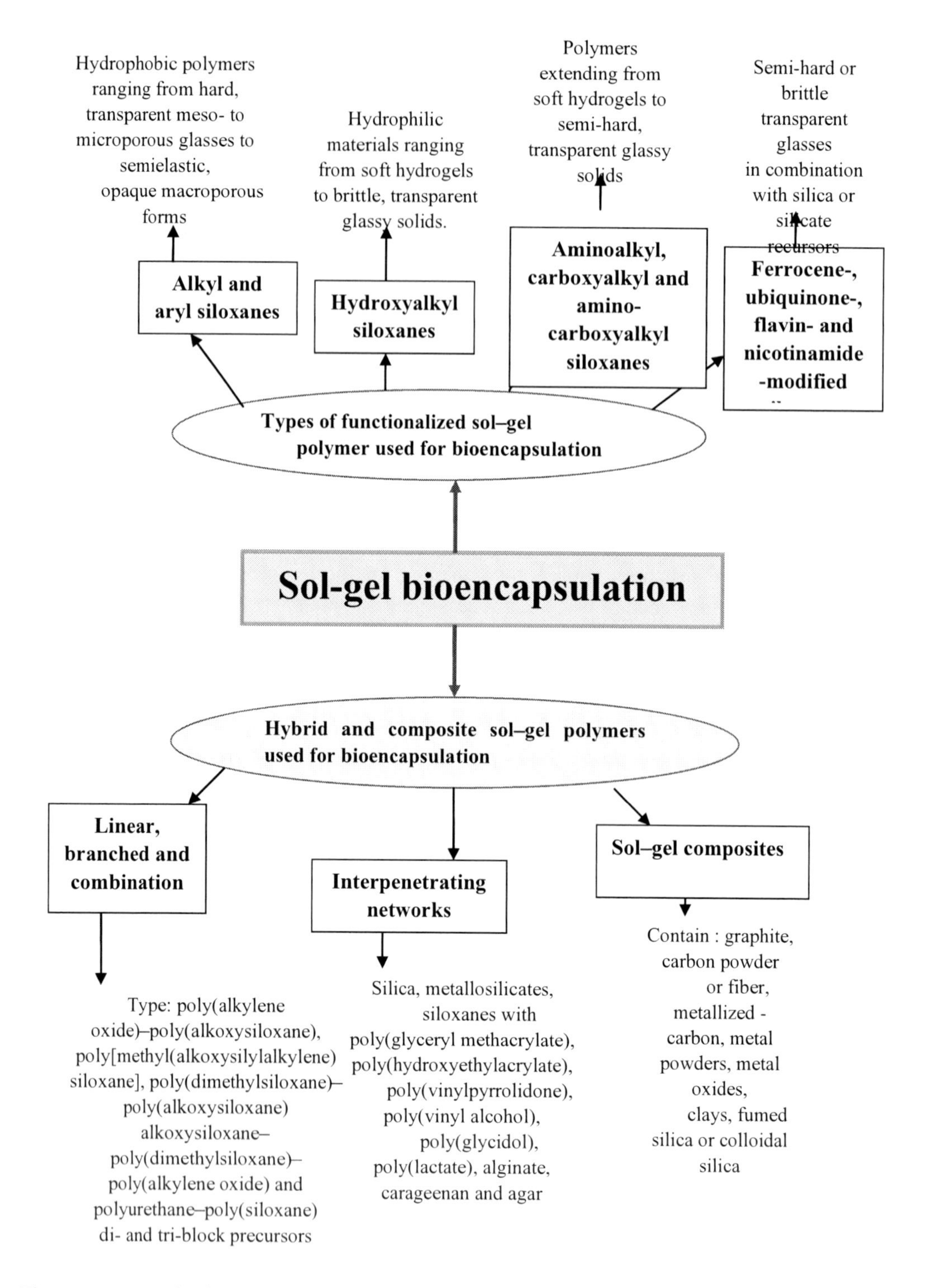

Figure 11. Type of sol-gel precursors and matrix used in sol gel bioencapsulation (adapted by Gill) [56].

Hybrid sol-gels: Amino- or hydroxyl- functional polymers, such as polymethy silane, polyurethane, polyacrylate, and polyphosphazene are mixed with alkoxysilane during the sol-gel reaction. After polymerization, hybrid organic-inorganic structure can be formed in the silica matrix to provide good mechanical properties and variable hydrophilic-hydrophobic balances. But they are often not optically transparent and in some case only are available as hydrogels [77, 78].

Reinforced/filled composite sol-gels: To improve the mechanical properties and processing behavior of the sol-gel materials, some nano-or micro-particles, such as graphite powder, fume silica, clays, cellulose and so on, can be incorporated inside the sol-gel silica. In addition, some active metal filler like gold, palladium, platinum can be used when conducting and redox-active materials are desired [79, 80].

Enzyme (Biocatalysts) Immobilization

The excellent specificity and selectivity (including enantioselectivity) properties of enzymes have been employed to carry out processes of high complexity in less harmful experimental and environmental conditions. There are several tools, as protein engineering, available to improve the enzyme characteristics, but some drawbacks associated with use of soluble enzymes like high price, product contamination, difficulty of separation from the reaction mixture, and insufficient operational stability may reduce the industrial application possibilities. Among the immobilization methods described in the literature, bioencapsulation looks particularly valuable while direct linkage of the enzyme and support, frequently associated with activity loss, is avoided. A large diversity of natural or synthetic organic polymers, possessing a broad variety of available functional features, has been already studied for entrapment of biomolecules. They can fulfill several requirements needed for a highly efficient biocatalyst, but mechanical strength and chemical stability are often not appropriate [81]. The approach to sol-gel silica loaded with biological systems, particularly enzymes, has produced important and valuable implementations in biotechnology and open tremendous possibilities to understand the protein folding process [56, 61, 82, 83]. Immobilization has long been used to improve enzyme activity and stability in aqueous media, and, more recently, also in non-aqueous media [84], once it became clear that in those media enzymes can catalyze reactions that are difficult to perform in water, become more stable and exhibit altered selectivity [85, 86].

Silica has been widely used as an inert and stable matrix for enzyme immobilization owing to its high specific surface areas and controllable pore diameters, which can be tailored to the dimension of a specific enzyme: that is, microporous (<2 nm pore size), mesoporous (2–50 nm pore size) or macroporous (>50 nm pore size) silica. Because most enzymes are of the order of 3 to 6 nm in diameter, mesoporous materials are most commonly used [87]. One of the primary limitations of the sol-gel technique, however, is poor loading efficiency and enzyme leakage. The problem has in some instances been addressed by designing protocols for the preparation of matrices with a pore size that is adequate to allow the flow of substrates and products but small enough to prevent the elution of the entrapped biomolecules [88]. Recent interest in nanotechnology has provided a wealth of diverse nanoscaffolds that could potentially support enzyme immobilization and enzymes immobilized to nanosized scaffolds, such as spheres, fibers and tubes, have recently been reported [89 - 91]. The premise of using nanoscale structures for immobilization is to reduce diffusion limitations and maximize the functional surface area to increase enzyme loading. In addition, the physical characteristics of nanoparticles, such as enhanced diffusion characteristics and particle mobility, can impact the inherent catalytic activity of attached enzymes. Betancor et al [88] present in a recent review the advantages of biomimetic silica for enzyme immobilization: (i) *Inexpensive* (no need for chemical reagents for laborious synthesis); (ii) *Rapid* (immobilization occurs within seconds); (iii) *Mild conditions* (formation of the particles occurs at room temperature and neutral pH); (iv) *Nanosized* (Lower diffusion limitations and higher volumetric activities); (v) *Smart* (Matrix can be dissolved to release the entrapped enzyme); (vi) *Robust* (Physical properties suitable for flow-through applications); (vii) *Stabilizing* (Numerous enzymes have been stabilized by entrapment in this support); (viii) *Polymorphous* (Shapes can be tailored by varying the conditions of silica deposition).

A very large number of enzymes have been trapped within sol–gel glasses showing that they usually retain their catalytic activity and can even be protected against degradation. Trapped enzymes usually exhibit better activity and longer lifetimes than free enzymes. During encapsulation, they remain trapped within a silica cage tailored to their size. Mobility within this confined space is restricted. Unfolding is prevented avoiding denaturation even when exposed to harsh conditions (temperature, pH, solvents) [92].

In 1955 Dickey [57] demonstrated for the first time bioencapsulation with partial activity of urease and catalase, but muscle adenylic acid deaminase completely lost its activity. After 3 decades, considerable attention has been paid again towards the bioencapsulation using sol–gel glasses. Avnir's group successfully encapsulated alkanie phosphatase in silica gel, which retained its activity up to 2 months (30% of initial) with improved thermal stability [93].

Braun et al. reported encapsulation of a purified enzyme alkaline phosphatase in a TMOS-derived sol–gel material, which exhibited only 30% activity inside silica glasses [60].

Bovine liver catalase (BLC) was immobilized in TEOS based sol–gels. TEOS sol–gels were found to be suitable for entrapping BLC with a high degree of immobilization. Kinetic constants for the catalatic mode of action of the immobilized enzyme were determined. Immobilized BLC retained less than 1% catalatic but 50% peroxidatic activity compared to its free counterpart. Addition of various additives did not improve the catalatic performance. Sol–gel immobilized BLC was found to be stable for 17 consecutive reactor batch cycles with no apparent loss in activity [94].

Ormosils and polymers are shown to be good to modulate enzyme activity. The ormosil methyltrimethoxysilane (MTMS) doped with chitosan shows good biocompatibility for immobilisation of glucose oxidase (GOD) [95]. Horseradish peroxidase (HRP) and acetylcholinesterase can also be encapsulated in Ormosils [66, 96].

Improvement of enzyme characteristics as activity and enantioselectivity is the most important aim for every biocatalytic process. Selecting the appropriate precursors, additives and template compounds is fundamental to obtain solid-phase biocatalysts with higher efficiency and operational stability. Ionic liquids could have an important function in this issue, as they contribute to the formation of the sol-gel framework and can also facilitate the enzymatic reaction by their hydrophilic or hydrophobic nature. The results provide that fine tuning of ionic liquid and silane precursor structure can result in entrapped lipase composites with higher activity and enantioselectivity than the native enzyme. However, several other aspects are yet to be improved, as avoiding sol-gel matrix shrinkage and enzyme aggregation during the immobilization process [81].

A number of reports indicate that sol-gel glass encapsulation retains the activity of a wide variety of enzymes. Most of the studies to date concerned with bio-encapsulation in unmodified tetramethoxysilane-based glass or in the organically-modified TMOS glass system with R-Si(OCH3)$_3$ (R = functional organic group). Many reported the catalytic activity of the proteins encapsulated that associate with lipid interfaces or proteins that bind nonpolar ligands, such as lipases [65, 97, 98].

Enzyme interactions with the sol-gel matrices lead to conformational changes of the lid that result in an open, substrate-accessible form of the enzyme, and to the formation of the oxyanion hole. This phenomenon is found with most lipases and also with cutinase. Cutinase is a versatile enzyme that catalyzes synthetic and hydrolytic reactions on a wide range of substrates in aqueous and non-aqueous media [99]. Cutinase from *Fusarium solani pisi* was encapsulated in sol–gel matrices prepared with a combination of alkyl-alkoxysilane precursors of different chain-lengths. The specific activity of cutinase in a model transesterification reaction at fixed water activity in *n*-hexane was highest for the precursor combination tetramethoxysilane/*n*-butyltrimetoxysilane (TMOS/BTMS) in a 1:5 ratio, lower and higher chain lengths of the mono-alkylated precursor or decreasing proportions of the latter relative to TMOS leading to lower enzyme activity. Results obtained using combinations of three precursors confirmed the beneficial effect of the presence of BTMS in the preparations. The impact that fictionalization of some of the additives had on cutinase activity indicates that the enzyme/matrix interactions must play an important role. Some of the best additives from the standpoint of enzyme activity were also the best from the standpoint of its operational stability (ca. 80% retention of enzyme activity at the tenth reutilization cycle). None of the additives that proved effective for cutinase could improve the catalytic activity of sol–gel encapsulated *Pseudomonas cepacia* lipase [100].

Another example refers to sol–gel encapsulated creatine kinase that is protected from thermal denaturation. It retains 50% of its activity ten times longer at 47° C (125 hours) than the free enzyme (13 hours) at the same temperature. Surprisingly, a 4-fold increase in activity is even observed upon heating for a short time. This may be explained by structural changes in both enzyme and the sol–gel matrix. Actually the structure of the creatine kinase may be slightly modified upon encapsulation, but reverts to the right conformation upon heating. Moreover, short heating could also increase the pore size of the silica matrix allowing easier diffusion and decreasing the stresses arising from encapsulation [92].

In another example tripsin inhibitor was chosen as the model protein, because its size (21 kD) is similar to that of bone growth factors. The viability of using sol-gels intended to serve as both substrates for bone growth as well as to allow incorporated proteins such as growth factors to diffuse out and stimulate cell function and tissue healing was under investigation. The data documented that the in vitro release of tripsin inhibitor was dose and time dependent during immersion up to 9 weeks [101].

Sugar (sorbitol) and amino acid N-methylglycine additives can be used to stabilize enzymes within sol–gel matrices. Chymotrypsin and ribonuclease T1 have been trapped in the presence of sorbitol and N-methylglycine. Both osmolytes significantly increase the thermal stability and biological activity of the proteins by altering their hydration and increasing the pore size of the silica matrix [92, 102].

D-glucolactone and D-maltolactone have also been covalently bound via a coupling reagent (APTS) to the silica network giving non hydrolyzable sugar moieties. Firefly Luciferase, trapped in such matrices, has been used for the ultra sensitive detection of ATP via bioluminescent reactions [92]. The silica matrix forms around the trapped biomolecules, but some shrinkage always occurs during the condensation process and the drying of the gel. Stresses can then lead to some partial denaturation of the enzymes. Polymers have been used as additives to form hybrid organic–inorganic gels in order to reduce shrinkage via a 'pore filling' effect. Nice results have been obtained with hydroxyethyl carboxymethyl cellulose (HECMC). Shrinkage is significantly reduced, decreasing stresses and preserving the porous structure of the xerogel. Glucose oxidase trapped within such hybrid HECMC–silica xerogels retains its high biocatalytic activity and remains accessible whilst being stored up to 3 years under ambient conditions [92].

Bioconversion Using Sol–Gel Entrapped Enzymes

In the chemical industry a variety of catalysts for well-defined chemical transformations are required. These reactions do not or only do unspecifically occur under normal conditions, so that the desired product is available only in a small concentration or as a mixture with by-products. This often results in the need of costly purification processes, which makes the chemical synthesis economically non-efficient. An alternative possibility is the use of enzymes. These biocatalysts enable highly specific transformations under moderate reaction conditions. Because of the high stereo- or regio-selectivity, the application of enzymes may be superior to chemical synthesis. Sometimes enzymatic processes can be used for the production of compounds, which are difficult to synthesize chemically. Especially enzymatic methods are suitable for the synthesis of optically active compounds. Often only one enantiomer of a compound has a pharmacological effect, whereas the other enantiomere has no, or an unwelcome, effect. Therefore, biotechnical syntheses gain importance in the production of compounds requiring high enantiomeric purity for pharmaceuticals or commercially important products [50].

Lipases provide a good example showing how a chemical control of the sol–gel process can be used to improve enzymatic activity. These enzymes are involved in hydrolysis and esterification reactions. In aqueous media they hydrolyze fats and oils into fatty acids and glycerol whereas esterification reactions occur in organic media. Actually most lipases are interfacial activated enzymes. In an aqueous solution, an amphiphilic peptidic loop over the active site just likes a lid. At a lipid/water interface, this lid undergoes a conformational rearrangement which renders the active site accessible to the substrate [92, 103].

Reetz *et al.* established the practical utility of sol–gel immobilized biocatalysts with the encapsulation of lipases in Ormosils [104]. They reported a technique of sol gel bioencapsulation which induce lid-displacement activation of lipases during sol–gel formation and leads to the surface encapsulation of the activated lipases in micro-phase separated poly(alkylsiloxanes). Their method appears to be generic for lipases and produces highly active particulate and thick-film immobilization, which mediate regio-, chemo- and enantioselective transformations in aqueous–organic media and organic solvents [103, 104]. The promise of the technique is indicated by the high immobilization efficiencies, increased resistance to denaturation and enhanced storage and operating stabilities of hydrolases, oxidoreductases, lyases and transferases entrapped in silica, metallosilicate, Ormosil and composite sol–gels [56, 103, 104].

One of the most powerful applications of sol–gel biocatalysts is the co-entrapment of multi-enzyme systems, in which the nanoconfinement of the catalysts and reactions increases the effective concentrations of reaction intermediates, thereby enhancing overall efficiencies. Thus, sialic-acid aldolase, myokinase, pyruvate kinase, pyrophosphatase, CMP–sialate synthase and (2,6)-sialyl transferase can be trapped in poly(3-aminopropylsiloxane)-zirconosilicate to enable the continuous synthesis of a(2,6)-sialyl-*N*-acetyllactosamine [56]. Also, formate dehydrogenase, formaldehyde dehydrogenase and alcohol dehydrogenase have been co-encapsulated in silica to allow the high-yield conversion of carbon dioxide to methanol [105].

Co-immobilization of enzymes is a convenient approach for the multi-enzyme processes. By co-immobilizing formate dehydrogenase, formaldehyde dehydrogenase and alcohol dehydrogenase in silica gels, together with the electron donor NADH (Nicotinamide adenine dinucleotide), synthesis of methanol can be performed at low temperatures and pressures [106, 107].

Kuncova et al. [108] studied the effect of different organic solvents on steric acid esterification using immobilized lipase. Acetone and toluene impaired the activity of free lipase, whereas sol–gel immobilized enzyme partially lost the activity. The half-life of immobilized lipase was 4 months at 40°C in hexane with efficient esterification capacity. This stability was not only due to the physical entrapment but also to additional multiple interactions with SiO_2 through hydrogen, ionic, or hydrophobic interaction, and through multi-point attachment to the support [108].

Pierre [65] in his review of the sol–gel encapsulated enzymes and their applications concluded that even though the sol–gel entrapped enzymes show improved stabilities against physicochemical changes around their vicinity and activity; still, they are facing diffusion limitations, which lead to the high *K*m value and have great impact on conversion at large scale use of immobilized enzymes in industrial applications. Hence more efforts are needed for the improved diffusion properties of sol–gel matrices for efficient biotransformation/ synthetic reactions.

Antibodies Immobilization

The period of 2000–2005 has seen a very intensive activity in the study of sol–gel entrapped antibodies and their applications: more than 100 papers have been published. Because these topic it only a part of the review we followed to show only a briefly description of this activity, although the very volume of it attests to the attractiveness of this specific application, which was introduced only about a decade ago.

Antibodies (Ab) are immune system-related proteins called immunoglobulins. Each antibody consists of four polypeptides with two heavy chains and two light chains joined to form a "Y" shaped molecule. The amino acid sequence in the tips of the "Y" varies greatly among different antibodies. This variable region, composed of 110–130 amino acids, gives the antibody its specificity for binding antigen. Because of their high specificity and sensitivity they have been widely employed in number of modern diagnostic and therapeutic technologies such as in vitro diagnostics, like radioimmunoassay (RIA), immunoradiometric assay (IRMA), enzyme-linked immunosorbent assay (ELISA), or blotting techniques [50].

Similarly to the other bioactive molecules, antibodies are also not much stable in aqueous solutions even though affinity is good towards the antigen and functionality. By attachment to an inert support material, bioactive molecules may be rendered insoluble, retaining catalytic activity, thereby extending their useful life [50].

Immobilization of Abs on to a solid support was first reported in 1967 and the technology has widespread application in affinity chromatography (AC) and other areas [109]. However the major problem associated with covalent immobilization of antibody on solid surface is partial loss of biological activity due to the random orientation of the asymmetric macromolecules, sometimes steric hindrance caused by the neighboring antibody molecules.

Three decades ago, taking into account the inertness and biocompatibility of the sol–gel entrapment technique antibodies have been successfully encapsulated in the sol–gel materials for different applications. The Abs encapsulated in sol–gel-derived glasses can interact with target molecules with a high degree of specificity as in solution, and the signal can be detect using an appropriate sensing scheme. Generally antibodies are high molecular weight proteins, if antigen is also high molecular weight compound the interactions between antibody and antigen is difficult through the small pores of the matrix [110]. In such a case, the antigen can be tagged with small signaling compounds such as ferrocene or their derivatives, depending upon the detection mode [111]. Wang et al. [112] encapsulated firstly antifluorescein antibodies in TMOS sol–gel. Yang et al. [113] reported that the addition of poly(ethylene glycol)-PEG did not affect the encapsulation efficiencies of the antigentamicin antibody but determined a strong effect on the binding activity of the encapsulated antibody . When the antibody was encapsulated in sol–gel along with PEG, 95% of the gentamicin was bound to the column comparative with the case of sol gel without PEG when only 42% was bounded to the column. Other studies also reported that the addition of PEG stabilized the urease and prevented the fouling and adhesion of unwanted protein to the surface [114].

Shabat et al. [115] successfully encapsulated catalytic antibodies through sol–gel method and used them in the transformation reactions. Antibody 14D9 is an effective catalyst for various hydrolytic reactions including the hydrolysis of a cyclic acetal, ketals, epoxides and enolethors. The catalyst has been homogeneously doped inside the gel matrix and shows a catalysis followed Michaelis–Menten kinetics. The entrapped antibody is more stable than that immobilized through surface attachment [50]. The antigentamicin Mab entrapped in mesoporous TMOS so-gel monolith was employed for the development of flow injection fluorescence immunoassay for the quantitative analysis of the gentamicin [112]. Gentamicin is a broad spectrum antibiotic and at higher dosages it causes impaired renal function, hence accurate monitoring of the drug in serum of patients is mandatory. Anticortisol antibodies were encapsulated in optically transparent sol–gel silica matrices and competitive immunoassays for cortisol were conducted using the antibody doped silica material as sensing elements. Between the monolith and thin-film, thin films showed good accessibility of antigen to the encapsulated antibody and significant reduction in assay time. The encapsulated antibodies were able to detect cortisol in the range of 1–100 μg dl−1 with controlled non-specific binding to the sol–gel silica matrix. More importantly the encapsulated antibody fluorescence signal was 10 times higher than the surface-adsorbed antibody. It was attributed to the possibility of more antibody encasulation per unit volume of sol–gel [50, 110].

Antigentamicin antibody was also immobilized in a mesoporous sol gel material using tetramethoxysilane as a precursor and poly(ethylene glycol) as a template. The sol-gel glass was used to develop an immunoaffinity column for the flow injection immunoassay of gentamicin. The immunoassay was based on the competition between gentamicin and fluorescein isothiocyanate-labeled gentamicin for a limited number of encapsulated antibody binding sites. NaOH solution was used for regeneration of encapsulated antibody binding sites after each measurement, which allowed the immunoreactor to be used for up to 20 times without any loss of reactivity. It can be concluded that the described method of biochemical analysis seems to be simple, rapid, stable, sensitive, and renewable [101]. The development of the novel hepatitis B surface antigen (HBsAg) immumosensor was achieved by self-assembling gold nanoparticles to a thiolcontaining sol–gel network. Thus, a gold electrode was first derivatized with mercaptopropyl-trimethoxysilane sol–gel solution, forming a mercapto-silica gel layer, then gold nanoparticles were chemisorbed onto the thiol groups. Finally, hepatitis B surface antibodies (HBsAb) were adsorbed onto the surface of the gold nanoparticles. The electrochemical ferricyanide redox process was used as a probe to determinate HBsAg. It was found that this approach is superior to the glutaraldehyde binding approach, in terms of the larger amount of adsorbed antibodies immobilized by this method and in terms of higher immunoactivity [116].

Liu et al. [117] prepared luminescent sol–gel based silica nano-particles (20 nm), doped with dibromofluorescein (D–SiO2), placed on a polyamide membrane. The particles phosphoresce intensively and the phosphorescence can be quenched with Pb(Ac)2. This system was used to detect human IgG, through the interaction with a goat-anti-human IgG antibody labeled with D–SiO2, by following quantitatively the regained phosphorescence intensity of the particles due to the immunoreaction. The limit of detection is quite low (0.018 pg per spot), and the reproducibility is very good, with about 4% deviation over 11 measurements.

Brennan et al [118] developed a sol–gel-based method for the preparation of protein microarrays that has the potential to allow pin-spotting of active proteins for high-throughput multianalyte biosensing and screening of protein–small molecule interactions, using various antibodies. Microarrays were printed and a wide variety of factors that control the production and the optimization of the microarrays were studied. Such factors included the ability to pin-print without clogging of the pins, the adhesion of the sol–gel spot to the substrate, the dimensions of the microspot, and the stability of both the microspot and the entrapped protein. The arrays were shown to have higher signal-to-background levels than conventional arrays formed by covalent immobilization of antibodies on chemically derived surfaces [92]. In the work of Pierre et al [119], TMOS-based silica gels, as well as organically-modified matrices were investigated as hosts for low molecular weight salmon sperm DNA. Only a limited amount (up to 45%) of the encapsulated material could be extracted in a buffer solution. In order to explain the low rate of DNA recovery, 31P NMR spectroscopy was used to investigate silica–DNA interactions, revealing the presence of a P–O–Si link. More recently, hybrid matrices obtained from TEOS and aminopropyl-triethoxysilane (APTES) precursors were used to entrap DNA. The increase in APTES content of the gel correlates with a decrease in DNA leaching upon rinsing. Ethidium bromide and dibenzofuran could be successfully intercalated in encapsulated DNA, although to a limited extent when compared to surface immobilized DNA[92].

Protein Immobilization

A modified sol–gel procedure for entrapment of proteins was reported by Ellerby et al. [120] without alcohol and at higher pH between 5.0 and 8.0 of the precursor solution. Use of buffer (containing the dopants and biomolecule) raises the pH to physiological range but accelerate the gelation process.

The major advantages of sol–gel derived silicate materials for immobilisation of proteins are [52]: (i) they can be made to be optically transparent, making them ideal for the development of chemical and biochemical sensors that rely on changes in an absorbance or fluorescence signal, (ii) they are open to a wide variety of chemical modifications based on the inclusion of various polymer additives, redox modifiers and organically modified silanes (Ormosils) (iii) they have a tunable pore size and pore distribution, which allows small molecules and ions to diffuse into the matrix while large biomolecules remain trapped in the pores, allowing size-dependent bioanalysis.

Entrapment of proteins via the sol–gel process

The preparation of biologically-doped materials via the sol–gel process requires that several parameters be accommodated simultaneously [52]: (a) the method must be amenable to aqueous solutions, since these are generally required to maintain the biological function of biomolecules; (b) the polymerization reaction must be compatible with the pH and ionic strength ranges that are required for protein function (pH 4–10, ionic strength ranging from 0.01 to 1.00 M); (c) the process must occur at or near room temperature to maintain proteins in their native conformation; (d) the material must have a pore size that is sufficiently small to prevent leaching of the protein, but large enough to allow smaller analytes to enter the matrix with ease; (e) the material properties should be tuneable to allow for the modification of the internal environment so as to maximize the activity of the entrapped protein; (f) it should be possible to form the final material such that it is optically transparent or electrically conductive to allow spectroscopic or electrochemical measurements; (g) the fabrication of the material should be straight forward and reproducible, and should be amenable to a variety of formats including bulk glasses, thin films, columns, fibres, powders, and arrays.

Braun and co-workers [60] described the entrapment of proteins into alkoxysilane-derived silicate materials using the sol-gel method. Two years after, Dunn's and Zink's groups [120], demonstrated that other proteins, such as cytochrome *c* and myoglobin, could be entrapped into TMOS derived silicates with retention of O_2 binding ability [66, 119].

Protein conformation and dynamics in sol–gel matrices

The study on protein conformation in biocompatible materials is very important in materials sciences for the development of new and efficient silica-based biomaterials (e.g., sensors, drug delivery systems or implanted devices) but also in medicine as it is well-known that misfolded/unfolded proteins and disturbed protein-protein interactions are the cause of devastating diseases (e.g., Alzheimer, Huntington, diabetes) [66, 121 - 125].

The protein encapsulation in porous sol–gel glasses [53, 56, 92, 126] has been widely applied to heterogeneous biocatalysis applications and to the development of solid state optical, electrochemical biosensors [66].

Apomyoglobin (apoMb) is also a model protein that has been used for decades for the study of protein folding-unfolding process in solution. The holoprotein is very stable, but once its heme has been removed to form apomyoglobin, it is easily unfolded and several partially folded states can be populated depending on its surrounding environment. The protein is then ideal to probe the change of the protein conformation as function of the different properties of the host matrix. In the recent studies, the protein was encapsulated in different glass systems [(100-x) TMOS: (x) RnSi(OCH3)4-n, n = 1, 2, 3, R = alkyl, vinyl, fluorine] obtained for different molar composition and including also the unmodified TMOS glass ($x = 0$) as control [66].

Entrapped proteins typically reside in pores that are of a similar size to the protein, thus, it is important to ascertain whether entrapped proteins maintain their native conformation during and after entrapment, and whether they are able to undergo changes in conformation once entrapped. The latter point is particularly important since in many cases the binding of an analyte to a protein requires that the protein be able to undergo conformational changes, which may in turn be used to derive an analyte-dependent fluorescent signal, as is the case for a number of fluorescent allosteric signal transduction proteins [52, 127].

The conformational motions of entrapped proteins have been examined using absorbance, fluorescence, resonance Raman and dipolar relaxation measurements [52, 128, 129].

Conformational studies of a variety of entrapped proteins (monellin, parvalbumin, oncomodulin, human serum albumin (HSA) and bovine serum albumin (BSA) have indicated that such proteins tend to retain a native conformation immediately upon entrapment, although some proteins, such as myoglobin, may undergo substantial conformational changes during entrapment [130].

Another aspect of entrapped proteins that has been widely studied is the conformational stability of the protein in presence of denaturing stresses. In these studies, large scale conformational motions of proteins are induced by introducing a denaturing compound such as guanidine hydrochloride or by increasing the temperature of the entrapped protein. These studies have indicated that the conformational motions of large proteins, such as BSA, can be substantially restricted in sol–gel media and that in some cases inactive conformations may end up “trapped” in the sol–gel matrix [131, 132]. On the other hand, small proteins, such as cytochrome c and parvalbumin, appear to be able to retain full conformational flexibility upon entrapment and are only moderately affected by aging of the matrix [52]. The enhanced stability of entrapped proteins has been related to a “molecular confinement” process, wherein the protein is restricted in its ability to undergo conformational changes, and thus, requires a higher energy input before it can unfold.

Overall, studies of conformational motions of entrapped proteins suggest that the conformational stability of the native form of a protein determines to a large degree the conformational changes that are induced by sol–gel entrapment and the degree of stabilization that is imparted upon entrapment. That is “hard” proteins (such as cytochrome c and antibodies, which are rigid and do not easily undergo conformational changes more easily) tend to maintain their native conformation in the glass, while on the other hand, “soft” proteins (such as HSA or BSA, which are more flexible and undergo conformational change more easily) tend to denature. Furthermore, such studies indicate that large scale motions (e.g. folding and unfolding) in the sol–gel is greatly restricted, especially in dry-aged gels, but segmental motions (including those required for substrate binding) are largely unaffected.

Protein function is intimately linked to the ability of the protein to perform structural fluctuations among many different conformational substrates. Various electrochemical and spectroscopic studies on a number of proteins in silica hydrogels and xerogels showed the existence of native conformation of protein along with restricted rotations and global conformational changes inside tight silica cages and still the possibility of local motions required for binding and catalysis [56] Thus, these studies showed unambiguously that large-scale dynamics of biomolecules is strongly hindered in the glassy cage. Various physical forces, e.g., specific electrostatic interactions between silicate sites and protein surface residues and mechanical forces have been implicated to reduce flexibility of the entrapped protein. Few studies have focused on aging effect on dynamics of proteins in sol–gel. Flora and Brennan [133] studied conformation and dynamics of protein HSA, entrapped in TEOS-derived monoliths using time-resolved anisotropy decay measurements. HSA in solution showed unhindered rotation ($r_\infty = 0$) along with two rotational reorientation times, viz. global motion (20 ns) and local motion (0.44 ns). Global reorientation time of HSA decreased rapidly as compared to local motion when kept in denaturing agent guanidium hydrochloride (GdHCl). The most significant change was the high value of the residual anisotropy ($r\infty >$ 0.11 in all cases) for entrapped HSA. This was due to adsorption of the probe on the surface of the glass, causing restriction in the global rotational motion of the protein [72].

Boltan and Scherer have shown, still 1980, that the degree of protein hydration and/or local solvent composition can affect a protein's structure and dynamics [134].

They have shown that the structure of BSA, cast as a thin film was affected by the relative humidity. Thus, any attempt towards exploiting biomolecule, as chemical recognition element should be carried out with attention on the hydration of the biomolecule-reporter group. [72].

Perhaps the most important issue regarding entrapped proteins is whether they remain functional, and to what degree. The function of an entrapped protein depends on a number of factors, including the protein location (related to accessibility of the protein, as described above), protein structure (native versus unfolded), and the charge and polarity of the local environment, which can affect the binding properties of the protein [52].

The development of biotechnology is based on the immobilization of biomolecules or micro-organisms onto solid substrates. For obvious reasons, natural and synthetic polymers have been currently used for bio-immobilisation via grafting or entrapment. Using inorganic materials such as glasses or ceramics would offer many advantages, but the harsh conditions associated with their processing are not compatible with fragile biomolecules. Sol–gel science is well developed and applications of sol–gel glass as a porous matrix for chemical and biological molecules are growing. Various industrial applications of sol–gel technology are very much established. Sol–gel glasses for the entrapment of sensing agents have potential advantages over other methods however, the diffusional limitations inside the porous network (in case of monoliths), reproducibility of results and sensitivity remains to be achieved for entrapped biomolecules. In recent years, a number of new sol–gel-derived materials have been designed with the purpose of making the matrix more compatible with entrapped biological molecules. New biocompatible silane precursors and processing methods based on glycerated silanes, sodium silicate, or aqueous processing conditions were primarily directed towards removal of alcohol byproducts by evaporation before the addition of proteins. Other approaches includes the use of protein stabilizing additives such organosilanes, polymers, sugars and amino acids (osmolytes) to silica to improve the protein stability. This review emphasises the importance of functionalized organosilane precursors for the preparation of silica-based sol-gel glasses employed for bioencapsulation. The nature and choice of organosilanes can modulate the properties and the functionality of the biomaterials (hydrophobicity, crowding effects, and so on). The biocompatibility of the silica-based materials makes them suitable for the study of the protein folding process or for the developpment of new bionanodevices. The materials are highly porous (mesopores) materials and had the ability to absorb easily solutes such as phosphate ions that can also play a role to enhance the folding of the proteins. The materials are obtained at room temperature that is convenient to host the proteins; they are synthesised using the sol-gel route *via* hydrolysis and polycondensation from alkoxysilane precursors with addition of the aqueous sol composed of the acid catalyst HCl and distilled water. The addition of alcohols as organic solvents is not necessary for the hydrolysis and polycondensation of the silica network because of the sensitivity and possible denaturation of the protein.

THE APPLICATION OF SOL-GEL TECHNOLOGY IN DRUG DELIVERY

The production of very pure materials with high biocompatibility and/or bioerodible, obtained at low temperatures, by including in the synthesis inorganic compounds is an interest in medical field. This versatile technique, called sol-gel technique, has recently been investigated in diagnostic field, as well as in pharmaceutical applications.

The research on sol-gel technique has been made to have application in different fields and recently have strong attention for controlled drug-delivery. Porous materials, xerogels, organic-inorganic hybrids and nanocomposites can be part for pharmaceutics studies.

The synthesis of inorganic material by a sol-gel technique through the formation of colloidal suspension and gelation of the sol into the gel is enables to incorporate heat-sensitive active substances into the material during processing. Recent studies aimed at obtained hybrid materials with advanced properties. Incorporation of organic part into the inorganic network combines advantages of organic and inorganic materials, which results hybrids materials with good thermal stability, high resistance, satisfying hardness as well as very good flexibility and workability of the materials.

The organic part are represented by the biodegradable polymers, which are used with inorganic part due to they can be degraded to non-toxic monomers inside the body and have minim inflammatory side effects. Polymers have a particular advantage, they represent important drug reservoir and protect the integrity of the drug against physiological environment. The polymeric material chosen in dual system should be pharmaceutically acceptable, soluble in a variety of suitable solvents, biodegradale and available in different grades to enable the release rate of the active agent to be controlled. Same suitable synthetic polymers for drug encapsulation include homopolymers or copolymers such as aliphatic or aromatic polyolefins, polyvinyls, polyamide, polyester, polyurethane, polystyrene, etc. Preferable biopolymers are represented by polyamino acids, hyaluronic acid, starch, celluloses, etc. The active agents may be encapsulated in polymeric sheet before or during the start of the polymerization reaction. The initiators, starters or catalysts must be removed from polymeric systems if they are included in the reaction[135].

Drug carrier systems are defined as reservoir systems in which a reservoir of therapeutic agent is surrounded in an organic/inorganic network. The biologically active agent can be any organic or inorganic agent that is biologically active in crystalline, polymorphous or amorphous form. This type of drug carriers includes the drug dose needed for a period of time, which is administered at one time and released in a controlled manner. This leads to increase both comfort and confidence in the treatment of the patient.

The therapeutically agent can be added as a function of the application, respectively the condition to be treated, the age, sex, and condition of the patient, the nature of the therapeutic agent, the nature of the sol-gel system, and/or the nature of the medical device, among other factors.

The synthesis of controlled release sol-gels system involves several steps: an acid-catalyzed hydrolysis to form a sol with the bioactive molecules included, followed by casting, aging and drying. To produced porous sol-gel particles with desirable size are requested additional steps, like grinding or sieving. Other important parameters of synthesis are pH and time of gelatinizing, drug concentration in the sol, the rotational speed and time of drug release system.[136] The drugs are chemically (by covalent bonding) or physically (by adsorption) incorporated into the sol-gel system.

Table 2. The non-genetic therapeutic agents.

Class: non-genetic therapeutic agents	
1. Anti-thrombotic agents	heparin, synthetic heparin analogs, urokinase, dextrophenylalanine, proline, arginine;
2. Antibiotics	vancomycin, doxorubicin, cefoxitin, tetracyclines, chloram-phenicol, neomycin, gramicidin, kanamycin, amikacin, sismicin
3. Anti-inflammatory steroids agents (SAIDs)	cortisone, hydrocortisone, estrogen, dexamethasone, fluocortolone, prednisone, triamcinolone, budesonide, sulfasalazine, mesalamine, fluocinolone
4. Non-steroidal anti-inflammatory drugs (NSAIDs)	flurbiprofen, ibuprofen, indomethacin, piroxicam, naproxen, antipyrine, phenylbutazone, aspirin, diclofenac, fenoprofen, ketoprofen, mefenamic acid
5. Antineoplastic /antiproliferative / antimiotic agents	paclitaxel, 5-fluorouracil, cisplatin, vinblastine, vincristine, epothilones, endostatin, angiostatin, angiopeptin
6. Anesthetic agents	lidocaine, bupivacaine, ropivacaine, procaine, benzocaine, xylocaine
7. Anti-coagulants	heparin, hirudin, antithrombin compounds, platelet receptor antagonists, anti-thrombin antibodies, anti-platelet receptor antibodies, aspirin, prostaglandin inhibitors, platelet inhibitors
7. Vascular cell growth promoters	growth factors, transcriptional activators, translational promotors
8. Vascular cell growth inhibitors	growth factor inhibitors or receptor antagonists, transcriptional or translational repressors, replication inhibitors, inhibitory antibodies, antibodies directed against growth factors

Table 2. (Continued)

Class: non-genetic therapeutic agents	
9. Protein kinase and tyrosine kinase inhibitors	tyrphostins, genistein, quinoxalines
10. Antimicrobial agents	triclosan, thimerosal, chloramine, boric acid, phenol, cephalosporins, aminoglycosides, nitrofurantoin;
11. cytotoxic agents, cytostatic agents, cell proliferation affectors	

The therapeutically agents are selected depending on the application. The more required drugs to be included in a sol-gel system embrace: anti-thrombotic agents, anti-proliferative agents, anti-inflammatory agents, anti-migratory agents, anesthetic agents, anti-coagulants, cell growth promoters, cell growth inhibitors and combinations thereof. Depending on the genetic or non-genetic action, the therapeutically agents are classified into two classes.

In the Tables 2 and 3 are summarized the types of non-genetic and genetic therapeutic agents.

Table 3. The genetic therapeutic agents.

Class: genetic therapeutic agents	
1. Anti-sense RNA	
2. tRNA or rRNA to replace defective or deficient endogenous molecules	
3. Angiogenic and other factors	acidic and basic fibroblast growth factors, epidermal growth factor, transforming growth factor α and β, platelet-derived endothelial growth factor, platelet-derived growth factor, tumor necrosis factor, hepatocyte growth factor, insulin-like growth factor
4. Cell cycle inhibitors	

Genetic therapeutic agents are including anti-sense DNA and RNA as well as DNA coding for the various proteins[137].

Drug delivery systems should be checked for local and systemic toxicity. By using targeted drug-delivery systems based on polymers, liposome or microsphere, the toxic effects of certain drugs types can be mastered without decreasing the drug potency. In this regard have formed an encapsulation of doxorubicin in a polysiloxane by sol-gel method and the good results are correlated with the short time of gelation and high efficiency of drug encapsulation, which was realized by pre-doping method, but with low drug release. The pre-doping method refers to the fact that the drug loading was done with the gel formation[138].

The drugs encapsulations in hybrid organic-inorganic matrices are influenced by few processing parameters, such as porosity degree and specific surface of the pores. These parameters affect also the rate of the drug release. But even dense gels may be employed as drug delivery systems, because can form thin film protective with anti-inflammatory agents.

Sustained release of low molecular compounds entrapped into sol-gel derived xerogels were intended to have oral or localized parenteral administration, specially for antibiotics and anticancer drugs. The quantity and duration of release can vary due to the processing parameters. High concentration of drugs can be incorporated in the gel, which after the drying becomes a porous matrix.

In the next studies will present some sol-gel systems and their use as drug carriers system and controlled delivery, where both the load and drug release are influenced by the geometric shape of the final product or the principal components of hybrids systems. The systems have different area of applications.

Drug Delivery from Nanoparticles Synthesized by Sol-Gel Method

Polymeric nanoparticles, named nanocapsules, have also been widely studied as drug carriers due to their manly advantages: drug sustained release, increase of drug selectivity and effectiveness, higher bioavailability and decrease of drug toxicity.

The most candidates for controlled drug-delivery particles synthesized via sol gel route are derived silica particles, which have the capacity to encapsulate biologically and therapeutically active molecules and release to the application. The sol-gel derived silica particles show a very slow degree both for drug loading and release of adsorbed bioactive agents then the soft silica gels [139].

To avoid the fast release rate of drugs, in an extended period, from the gel is based on the use of polymeric nanocapsules as coating material. A major advancement is that the coatings can influence the release proprieties and the coating and degradation rate determines the large time until the onset of release.

Two methods for drug incorporation in the nanoparticles are applied, the drug incubation and the drug *in situ* incorporation [140]. Additional benefits of using nanoparticles drug carriers are reduced drug toxicity and have more efficient drug distribution.

A series of nanoparticles based on organic-anorganic materials synthesized by co-hydrolysis and co-polycondensation reactions, led to the formation of a core–shell structure with a hybrid core of silica-polymer containing hydroxyl groups and a silica-polymer containing poly(ethylene-glycol) groups monolayer shell. The hydroxyl methyl triethoxy silane and ω-methoxy (polyethyleneoxy) propyl trimethoxy silane can form by sol-gel method, water-soluble organo-silica particles with controllable size and molecular weight. These organico-silica nanoparticles are expected to be useful in drug delivery application [141]. The efficiency of drug delivery is directly affected by particle size, the geometric shape of the particles, the main compounds, the natural/synthetic type of the polymer, etc. [142].

It was reported in the literature that porous silica particles have the capability to incorporate drug and release his through a degradation process upon silica. The biodegradability of silica particles investigated, in vitro, in simulated physiological buffer, show a linear behavior for the dissolution rate; this aspect is important when the drug-silica system is used for a implantable therapeutically system. When added the serum proteins in the test solution, the degradation rate was slower by 20-30% [143, 144].

Drug Delivery from Porous systems Synthesized by Sol-Gel Method

The interest for nanostructured materials, which are synthesised from biopolymers-inorganic hybrids, has been growing in the last decades. The interest has been stimulated by the large variety of applications and due their capacity to insert different bioactive agents.

An advantage in drug release phenomena is to have a porous morphological aspect, obtained by controlled drying or be modified catalyst molar ratio during the synthesis. [93] The pore surface is significant because they serve as the absorption for different type of molecules, such as bioactive agents, enzymes, proteins and growth factors. In addition such a drug carrier improves the therapeutic efficiency, change pharmacokinetics and provide protection from degradation, which is necessary in certain circumstances. The influence of synthesis parameters were studied upon different sol-gel / drugs matrix.

For example, a crosslinked network can be formed through catalyzed hydrolysis and condensation of alkoxysilanes. Few general methods relying for drug incorporation in the matrix can been considered: drug incubation, drug *in situ* incorporation [140], by combining sol-gel polymerization with spray-drying or emulsion chemistry [145], spray-drying. The drug incorporation after the sol-gel system was synthesized has the advantage to avoid the permanent bonding of drug into organic-inorganic system [146].

Nanostructured amorphous microporous silica, synthetized under acid-catalyzed sol-gel conditions, represents a support for drug incorporation with oral controlled release. To confirm these assumptions, ibuprofen it was dispersed in the silica matrix and the release profile was viewed in a medium simulating of gastrointestinal tract buffer. It is concluded that the drug release rate can be tailored by adjusting the pore diameter [147].

Using a porous matrix as a host for magnetic nanoparticles has attractive application in drug delivery. The magnetic particles have unique magnetic properties and can bind and transport drugs to their structure. Advantage of including nanoparticles in silica matrices is that it can protect the metallic particles against air-oxidation. Usually the particles are embedded inside the pores by in situ sol-gel synthesis or by impregnation of mesoporous silica gels [148].

Drug Delivery from Gels by Sol-Gel Method for Orthopedic Application

Ambrosio and others were synthesized hybrid organic-inorganic amouphous materials to control release mechanism for a broad-spectrum antibiotic in simulate body fluid. The hybrid materials based on silica/polycaprolactone form a wet gel with sodium ampicillin by sol-gel method. This system was designed to treat bone infection in orthopaedic surgery and have a good rate of bone integration and regeneration [149].

Pro-drug Delivery by Sol-Gel Method

Special attention has been shown for the conversion of a pro-drug into a biologically active agent by sol-gel method. The enzyme L-amino acid decarboxylase is incorporated into a biocompatible silica-organic carrier sol-gel matrix. This complex is transferred in the brain, where converts L-dopa to dopamine and help in treatment of Parkinson's disease. The pro-drug may be formulated for administration by, for example, injection, inhalation or insufflations (either through the mouth or the nose) or oral, buccal, parenteral, rectal administration or implanted subcutaneous form [150].

Molecular Imprinting of Drugs as Drug Delivery Systems

Even molecular imprinting the drugs can be attached to the hybrid system. The molecular imprinting technique refers to the fact that the association between the imprint molecule and hybrid material is non-covalent or covalent type. The covalent imprinting over the non-covalent bond leads to strong interaction and the drug is released after these bonds are cleaved to liberate by degradation. Combining sol-gel process with molecular imprinting enables to produce ideal drug carriers with different medical application fields [151]. Use indomethacin, naproxen and ketoprofen as model drugs, which were incorporated into silica derivate system by molecular imprinting technique, the results shows selectivity for type of drug blind in the system to the functional silane and the specific geometric cavity. [152].

Drug Delivery from Thin Films Obtained from Sol-Gel Method

A thin film prepared under acid-catalysis, a sol-gel method, for a derived from tetraethyl orthosilicate precursor synthesized for coating Ti-6Al-4V wires show remarkable properties to prevent and treat the bone infection. The thin film acts as antibacterial film when vancomycin-antibiotic is incorporated [153].

To improved sol-gel barrier films, a tetrafunctional alkoxide silicate and Lewis acid or metal chelat like catalyst was investigated to form a crosslinked sol-gel polymer composition, where the polymer composition is uniformly coating a substrate and dried. The application of this kind of research is for oxygen barrier films [154].

The thickness of these sol-gel films due to the synthesis process is capable of regulating the drug release rate by diffusion mechanism [155]. Also organic-inorganic hybrids can be used to cover the compressed tablets which include therapeutic agents, for example a polydimethylsiloxane-sol-gel silica hybrid was tested for coating of tablets with hydrochlorothiazide [93].

Drug delivery by Sol-Gel Method for Tissue Engineering Applications

Great attention should be given to biomaterials used like wound dressings, because they must include both growth factors and anti-inflammatory drug. Once applied, the wound dressing remains in the body where it gradually degrades the absence of residue.

The scaffolds should provide space and substrate for barrier against bacteria, cell differentiation and proliferation. The dextran and Vancomycin in a sol-gel/copolymer composite was synthetized such as a film for wound dressing. Tyrosine-polyethylene glycol (Mw = 1000 Da)-derived polycarbonates composite with silica xerogel shows good mechanical properties and have the ability to bind and release gradually bioactive molecules, controlled synthesis parameters [156].

The Drug Release from Sol-Gel Systems

The controlled release of therapeutic agents from a hybrid matrix has become increasingly important for oral, transdermal or implantable therapeutic systems, due to they advantages of safety, efficacy and patient convenience.

As perhaps expected, the experimental results suggest that the release rate of drug from the hybrid sol-gel materials can be controlled by manufacturing parameters (increase the organic part in the synthesis, aging/drying conditions) [157]. Was been observed after the tests that adding organic precursors fine down the release of the drug.

The drug-release behaviors can be explained by considering the effect of textural properties of the xerogels (matrix swelling, matrix dissolution) and the bioavailability and/or solubility of the drug in aqueous media, the delivery pH, temperature, type or concentration of the drug, the addition of acids, and aging and drying times [158].

However, although the release of bioactive agents may be delayed and the release phenomena from sol-gel systems may exhibit large fluctuations, which can lead to unfortunate side effects. This shortcoming can be adjusted using water–soluble substance to accelerate the release rate of drug or modifying agents that increase the permeability of the matrix by including in synthesis polyanionic compounds.

The drug release is generally known to be diffusion controlled with relatively fast rate. Inclusion of a dopant, in synthesis, improves the drug-release behaviour, because the matrix/dopant interactions which change the solubility and diffusivity of the drug. There can be three types of interaction between drug and sol-gel system that can affect the release rate: electrostatic interaction, hydrogen bonding and hydrophobic interaction. The strength of these interactions is given by the functional groups of the main component of the system and the nature of dissolution medium [158].

The influence of component proportion on drug release properties of poly methylmethacrylate/3-(trimethoxysilyl) propyl methacrylate/silica composite has been evaluated using aspirin as a model drug. Mei et al. [159], conclude that the drug release fitted with the Fickian diffusion model and is influenced by the coupling agent content and the interface between the polymer matrix and the silica. Thereby with increase of coupling agent content, 3-(trimethoxysilyl) propyl methacrylate can obtain a low degree of drug released in vivo and can reduce the toxic levels, into the body, due to a sudden drug release.

For a matrix system obtained by sol-gel technique, in almost all cases the release kinetics obeys to Higuchi model [160]. In 1961 Higuchi propose the first mathematical model to understand the drug release for a planar matrix-drug system, which can happen through a pure diffusion of drug out an encapsulating matrix, without erosion. The Higuchi model is used both for water-soluble drug and low soluble drugs incorporated in solid/semi-solid matrix:

$$Q = \mathrm{A}\sqrt{t},$$

where, ***Q*** represent the amount of the drug release in time(***t***), per unit aria (***A***). When the drug solubility is consider, the formula becomes:

$$Q=\sqrt{(D\ (2C - C_S)\ C_S t}$$

where ***C*** is the initial drug concentration, ***C_S*** is the drug solubility in the matrix media and ***D*** represent the drug load in matrix. In this case, the drug release follows square root of time kinetics until the concentration in the matrix falls below the saturation value (C>> Cs).

The release profile is obtained by plotting the graph between amounts of drug release into a buffer solution versus time, which match to first–order kinetic, with a steady increase of the drug release into the external medium.

The therapeutic agent release from a sol-gel controlled system can be achieved by several mechanisms, such as diffusion, by osmosis, ion exchange or by degradation of the system components. Also the therapeutic agent in the porous hybrid material is released as the material dissolves and/or slowly diffuses from the pores.

If in synthesis were used biodegradable polymers in biological medium, governed by the local enzymes, the drug release occur. The therapeutic agent is covalently linked to the polymer backbone or physical inserted. The drug is released as the bond between drug and polymer chain or breaking of polymer chain.

When the drug is released due to osmotic swelling, the kinetics release fit to Fick´s law. In this case the water penetrates into drug carrier system, inducing relaxation of polymer chain and the drug is released outside.

The liberation rate for nanoparticles is influenced by the grain size of the composites, by modification of the silica matrix with methyl-triethoxysilane and by low or high molecular weight sol-soluble additives (sorbitol, polyethyleneglycol) [161].

The sol-gel products have a promising future due to the successful results in the medical field. The sol-gel tehnologies was accepted and considered safe for human use, thus the U.S. Food and Drug Administration has acknowledged receipt for a product based on therapeutic agent encapsulating in silica microcapsule to treat rosacea [162].

Other commercial application are for dressing burn and diabetic wounds with silica gel fibers designed to heal wounds caused by burns and diabetes. The fibers are produced by means of wet-chemical material synthesis and support matrix for newly growing skin cells. The dressings are absorbed to the burn area contact during the healing process. The fiber provides structural support for the healthy cells around the wound that is needed for a proper supply of growth-supporting nutrients.

The Application of Sol-Gel Technique to Realize Medical Devices

Other application of sol-gel technique is to realize medical diagnostic tool, which requiring biocompatibility, non-toxic, bioactivity and/or drug delivery properties. The medical devices must be without inflammatory tissue responses or immune reactions, if the medical devices it's used like implant materials such as surgical and/or orthopedic devices, artificial heart-valves, stents covered with polymeric thin film. It has also been found that the silica or titania gels can be used for implantable medical devices, but one challenge in the field of implantable medical devices has been release, to used polymeric compound in synthesis.

This type of medical apparatus is accompanied by selective sensing elements in the form of thin film, porous structures, which are sensible system in non-invasive monitoring and different diagnosis of diseases [163].

Polymeric part can hold the drug onto the surface of implantable medical devices, and controlled drug release via degradation of the polymer or diffusion into liquid or living tissue [164]. The problems is that the polymeric coatings can be used to adhere bioactive materials to implanted medical devices, but the adherence of a polymeric coating to a substantially different substrate, such as a stent's metallic substrate, is difficult due to differing characteristics of the materials. Further, most inorganic solids are covered with a hydrophilic native surface oxide that is characterized by the presence of surface hydroxyl groups.

Same implantable or insertable medical device comprising a sol-gel derived ceramic region with a porous structure and different geometries were realized. Ceramic region was moulded to comprise metal oxides, semi-metal oxide and polymers. Results showed that use of polymers as substrates to improve properties of devices. Preferred polymers were selected, that can be dissolved preferential relative to sol-gel regions, such as polymethylmethacrylate, polytetrafluoroethylene, various polyvinyl polymers and polyurethanes. A polymer can be introduced to the sol gel derived ceramic by infiltrating the polymer into the pores of the template or the monomer may be polymerized in the presence of the template. For polymers in fluid form may be introduced to the template by spin coating, spray coating, dip coating, ink jet printing, coating with an applicator such as a roller brush or blade and solidified. This ceramic region corresponds to the medical device in its entirety or to a discrete component of the medical device and can incorporate a therapeutic agent. These implantable or insertable medical devices are susceptible for non-planar, tubular medical device and stent [137].

Prior to several studies and Patent Application Number was used sol-gel compositions as drug reservoirs on implantable medical devices and also was used such sol-gel compositions to improve adhesion between organic and inorganic surfaces.

Examples for suitable implantes are tooth-implants, hip-implants, knee-implants, mini plates, external fixation pins, stents (e.g. for use in repair of blood vessels) or any other metallic, polymeric, ceramic or organic implants can be coated with a layer of sols and gels, and/or sol-gel derived materials and a therapeutic agent may be incorporated into this coating. The coating dissolves in the tissue and releases the active substance locally [165]. Titanium, alloys of nickel and titanium (NiTi-alloys), other memory-shape metals, Al_2O_3 or other ceramic materials can be coved with organic thin films to increase the properties such as biocompatibility.

Promising for future medical devices are those covered with an organosilane thin film obtained by sol-gel technique. Such equipments are resistant to oxidative or corrosive medium, especially resistant to iodine [166]. The coating process can be obtained by partly hydrolytically condensing an organosilane compound, thereby forming a sol, which sol is subsequently provided on a substrate. After the hydrolytic condensation is finished by curing at an elevated temperature, there form a network with drug loading possibility.

Materials produced by sol-gel technique shows real advantages and application in various areas of medical studies.

SOL-GEL TECHNIQUE FOR BIOSENSOR PREPARATION

The organic–inorganic sol–gels have received significant interest because the incorporation of organic polymers in the inorganic sol–gel can lead to newcomposite materials possessing the properties of each component that would be useful in particular applications. Incorporation of organic polymers, especially those with amino or amide groups, allows the formation of molecular hybrids often stabilized by strong hydrogen bonding. Thus, the sol–gel technique provides a unique method to prepare the three-dimensional network suited for the encapsulation of a variety of biomolecules – enzymes or proteins – as well as sensors [167-169]. Also, the sol–gel-derived inorganic materials are particularly attractive because they can be prepared under ambient conditions, can possess and exhibit tunable porosity, high thermal stability, and chemical inertness, and can experience negligible swelling in aqueous solution.

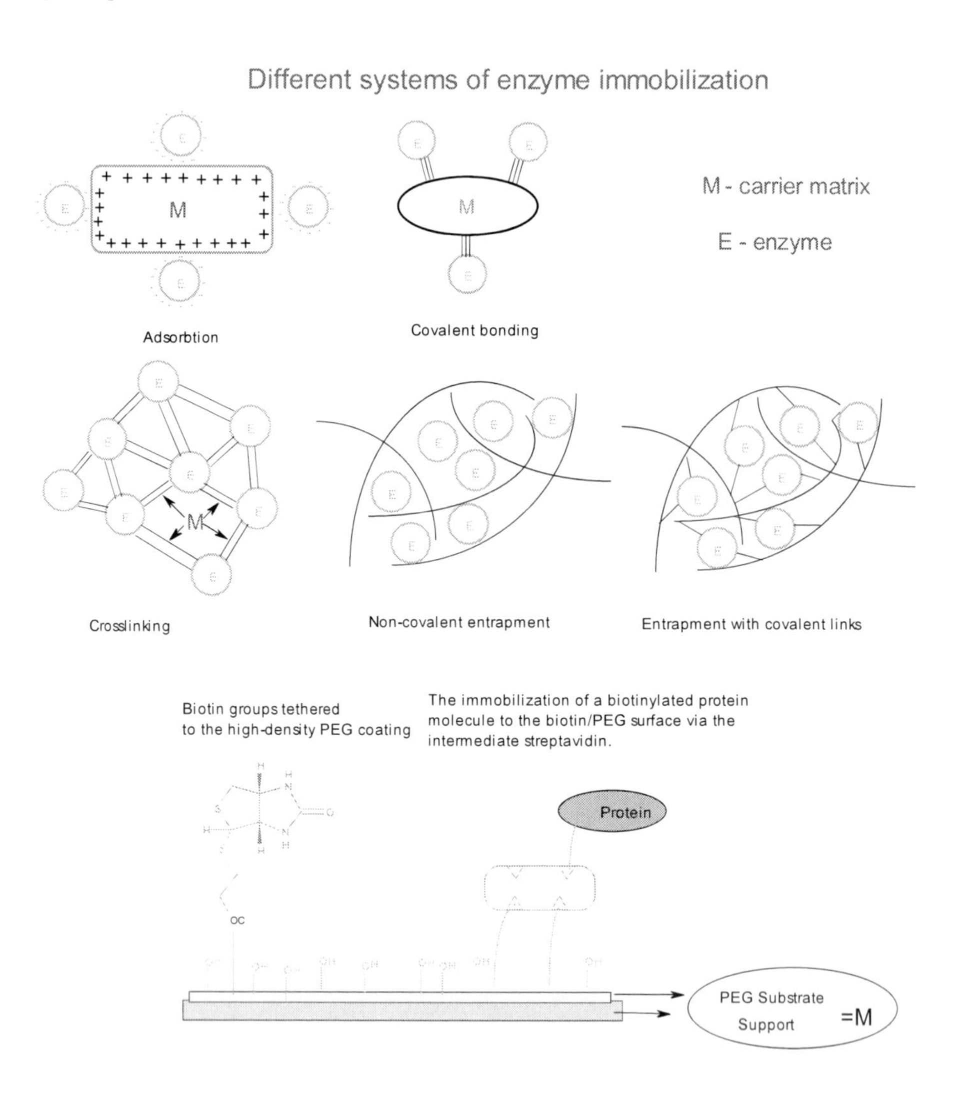

Figure 12. Immobilization techniques: adsorption to solid supports, covalent attachment, tethering via an intermediate linker molecule and entrapment in polymer networks.

The incorporation of sensing molecules in suitable matrix and monitoring/quantitating the interactions between the analytes and these molecules represent the principal tasks for performing suitable biosensors. Many biological macromolecules are highly efficient at recognizing specific analytes or catalyzing reactions in aqueous biological media. These characteristics make biomolecules desirable reagents [170]. Immobilization of biomolecules has become the most important area of research in development of biosensor. Various immobilization techniques have been applied, including adsorption to solid supports, covalent attachment, tethering via an intermediate linker molecule and entrapment in polymers (Figure 12) [171].

For the electrochemical biosensor, the effective immobilization of an enzyme on an electrode surface with a high retention of its biological activity is also a crucial point for the development of biosensors [172-174]. Among various approaches, the sol–gel method of processing is particularly advantageous to the immobilization of biomolecules [175].

Immobilization of enzymes with low temperature sol–gel process became a topic of ongoing research owing to its advantages as well as to the porous and open frame works in the sol–gel silica materials which provide effective access of the analyte into the active functionalities and could result in high sensitivity for the detection of analyte. The various sol–gel systems such as silica, alumina, and titania have been reported to be suitable materials for immobilizing glucose oxidase (GOD), horseradish peroxidase (HRP), tyrosinase and other biomolecules [175 -182]. Braun et al. firstly reported the attempt to encapsulate proteins inside SiO_2 glasses in 1990 [178]. Since then, these new kinds of inorganic materials are particularly attractive to the development of electrochemical biosensor [183].

However, besides complicated procedures and fragility, the shrinkage and cracks of the sol–gel silicate modified membranes greatly decreased the stability of the biosensors and limited their applications. Other non-silica sol–gel matrices have been developed to immobilize enzymes for fabrication of the biosensors. Deng and co-workers immobilized tyrosinase in a positively charged Al_2O_3 sol–gel matrix instead of negatively charged silicate matrix, which was more suitable for negatively charged tyrosinase in neutral solution [176].

Another alternative substrate for biomolecule immobilization is nanocrystalline TiO_2. Durrant and co-workers investigated the use of nanoporous TiO_2 film as substrates for protein adsorption. They proved that the nanoporous structure of the titania film greatly enhanced the active surface area available for protein bonding. In another paper, they compared the use of nano-TiO_2 and ZnO as substrates for protein immobilization and as electrode probes for electrochemical reduction of the absorbed substrates [184, 185]. Wang et al. illustrated that titania sol–gel carbon composite electrodes could offer a substantial decrease in the over voltage required for nicotinamide adenine dinucleotide (NADH) oxidation, as well as minimization of surface-fouling effects [186].

As was mentioned before the enzymes and proteins immobilized within sol–gel matrices maintain their native properties and reactivates, which assure the development potential for biosensors. Also, the sufficient amount of trapped interstitial water in gels allows the retention of the tertiary structure and active reactivity of encapsulated biomolecules. At the same time, the pore size in sol–gel can be controlled into an appropriate size for the diffusion of the analyte to the redox active sites, as well as preventing enzyme leakage. Even so, some papers mention that some gels lose their activities after the first few cycles, primarily ascribed to leaching of enzymes from the sol–gel matrix [187].

Sol-Gel Organic Hybrid Composite Biosensor

As a consequence a suitable method for immobilization/entrapment is still a challenge in the development of the biosensor field. Sol–gel glass offers a better way to immobilize biomolecules within its porous optically transparent matrix and demonstrated functional activity of encapsulated biomolecules. This is due to simple sol–gel processing conditions and possibility of tailoring for specific requirements. Due to this inherent versatility, sol–gel-derived glasses are excellent host matrix for chemical sensing and biosensing. This approach is unique compared to the conventional methods involving adsorption on glass surfaces, entrapment in polymer matrices or impregnation in porous glass powders because entrapment is based on the growing of siloxane polymer chains around the biomolecule within an inorganic oxide network. Also, because of the porous nature of the sol–gel network, entrapped species remains accessible and can interact with other external chemical species or analytes. At the same time, sol–gel-based sensors suffer from some disadvantages, e.g., entrapment in sol–gel glass may change chemical and biological properties of the entrapped species, due to reduced degrees of freedom and interactions with the inner surface of the pores. The entrapping environment of the biomolecules must be careful chosen since this will affect both stability and functionality of the sensor. The analytes must also have the capacity to enter the glass by diffusion to the site of entrapped biomolecules so that measurable signals to be generated from the interaction process for sensing. Research studies present and underline the effect of various physical-chemical properties, e.g., pH, polarity and microviscosity of the environment within sol–gel-derived matrices as well as the structure, dynamics, activity and function of the biomolecules [188 – 192]. Some reports also, focused on the effect of long-term sol–gel matrix aging on dynamics of protein and the detailed kinetics of the interaction between analytes and entrapped proteins. Occurrence of appropriate environment around entrapped biomolecules and its long-term stability is one of the important factors determining the functionality of entrapped proteins and enzymes [193 – 195]. The characterization of the internal environment of the sol–gel matrix is one of the focuses of research activities to design appropriate matrix for sensing applications. Thus, it have been made attempts to characterize the internal environment of the bulk gel and the thin gel films prepared from different sol compositions using various fluorescent molecules as a function of long-term storage. Fluorescence spectroscopy was used in this order.

A fundamental component of amperometric biosensors and biofuel cells are enzyme-modified electrodes. The selection of appropriate combinations of materials, such as: enzyme, electron transport mediator, binding and encapsulation materials, conductive support matrix and solid support, for construction of enzyme-modified electrodes also governs the efficiency of the electrodes in terms of electron transfer kinetics, mass transport, stability, and reproducibility [196].

The inorganic M–O–M (M-metal, O-organic) or M–OH–M bridges into the sol–gel materials form a continuous network able for bioimmobilization as well as includes a liquid phase which can then be dried out to form a solid, porous polymeric matrix. In biosensors, this technique allowed the development of novel strategies for the immobilization of biological receptors within silica, metal oxide, organosiloxane, and hybrid sol–gel polymers, which finally induces advanced materials [197].

Silica sol–gel material prepared under ambient conditions is well biocompatible and can retain the catalytic activities of enzymes to a large extent. Thus, it has emerged as one matrix well suited for the immobilization of enzyme, the construction of hydrogen peroxide biosensor.

Lim et al. fabricated nanostructured electrodes for enzymatic glucose–oxygen biofuel cells [198]. To provide enhanced electronic conduction the electrodes were based on enzyme encapsulation in sol–gel SiO_2 matrixes and incorporated carbon nanotubes (CNTs) in the matrix. The SiO_2 matrix was designed to be sufficiently porous that both glucose and O_2 have access to the enzymes and yet provides a protective cage for immobilizing the bio-molecules without affecting biological function. Voltammetry indicated that the effect of the SiO_2 matrix on mediator diffusion was minimal, although for one mediator, 2, 2'- azino-bis (3-ethylbenzothiazoline-6-sulfonic acid) diammonium salt, chemical modification of the solvent phase with polyethylene glycol was necessary. The polyethylene glycol addition resulted in a more uniform dispersion of the CNTs. The enzymes maintained their biocatalytic activity in the sol–gel matrix. A glucose–O_2 biofuel cell based on nanostructured SiO_2 sol–gel/CNT composite electrodes generated approximately $120\mu Wcm^{-2}$ at 0.24V at room temperature.

Another type of sol–gel/organic hybrid composite material based on the crosslinking of the natural polymer – chitosan – with (3-acryloxypropyl) dimethoxymethylsilane was developed for the fabrication of an amperometric H_2O_2 biosensor [199]. The composite film was used to immobilize horse radish peroxidase (HRP) on a gold disk electrode. With the aid of a catechol mediator, the biosensor had a fast response of less than 2 s with linear range of 5.0×10^{-9} to 1.0×10^{-7} mol L^{-1} and a detection limit of 2×10^{-9} mol L^{-1}. The biosensor retained approximately 75% of its original activity after about 60 days of storage in a phosphate buffer at 4°C.

The preliminary results of immobilization of the glucose oxidase (GOx) in silica gel on an oxygen electrode were also reported. The influence of the components of the casting solution (gel precursor, pH of the enzyme solution, sol–water ratio) on the electrode response was investigated [200].

The interest for organic–inorganic sol–gels develops into the incorporation of organic polymers in the inorganic sol–gel which can lead to new composite materials possessing the properties of each component that would be useful in particular applications. Incorporation of organic polymers, especially those with amino or amide groups, allows the formation of molecular hybrids often stabilized by strong hydrogen bonding.

The preparation of a platinum/multi-walled carbon nanotube (Pt/MWNTs) nanocomposite as glucose biosensor it was in this order mentioned [201]. Platinum nanoparticles were grown by electrodeposition onto MWNTs directly with the average diameter of the nanoparticles about 30–40 nm. The resulting Pt/MWNTs material presents new capabilities for electrochemical devices by using the synergistic action of Pt nanoparticles and CNTs. The immobilization of glucose oxidase onto electrode surfaces was carried out by chitosan-SiO_2 gel. Thus, the synergistic action of Pt and MWNTs and the biocompatibility of chitosan-SiO_2 sol–gel induce excellent electrocatalytic activity and high stability. The resulting biosensor exhibits good response performance to glucose with a wide linear range from 1 μM to 23mM and a low detection limit 1 μM. The biosensor also shows a short response time (within 5 s), and a high sensitivity (58.9 μAmM^{-1} cm^{-2}). For this hybrid nanocomposite it was provided a new electrochemical platform for designing a variety of bioelectrochemical devices with high sensitivity and good stability.

Biosensors for Total Cholesterol Estimation

As it is well known total cholesterol estimation is very important for patients suffering from heart diseases, hypertension, arteriosclerosis, cerebral thrombosis and other disorders that require continuous monitoring of total cholesterol. In this context, biosensors have recently gained much attention as they provide easy operation, accuracy, sensitivity and selectivity for cholesterol estimation. Several matrices such as nanomaterials, conducting polymers, sol–gel films and self-assembled monolayers have been used for fabricating a cholesterol biosensor via layer-by-layer technique, physical adsorption, hydrogel or sol–gel entrapment, electro-polymerization entrapment, cross-linking and covalent techniques, etc.[202 – 212].

The sol–gel-derived silica prepared using tetra-ethyl-orthosilicate (TEOS) precursor has been found to be more reactive towards the condensation reaction and has high affinity towards enzymes as compared to other precursors such as tetra-methyl-orthosilicate (TMOS), etc. [213]

Cholesterol esterase (ChEt) and cholesterol oxidase (ChOx) have been immobilized via glutaraldehyde as a cross-linker onto sol–gel-derived silica (SiO_2)/chitosan (CHIT)/multi-walled carbon nanotubes (MWCNT) - based nanobiocomposite film deposited onto indium-tin-oxide (ITO) glass for estimation of esterified cholesterol [214]. The realized ChEt–ChOx/MWCNT/SiO_2–CHIT/ITO bioelectrode, characterized using Fourier transform infrared (FTIR), scanning electron microscopy (SEM) and electrochemical techniques, shows response time of 10 s, linearity as 10–500 mg/dL for esterified cholesterol (cholesterol oleate), sensitivity as 3.8μA/mM and shelf-life of about 10 weeks under refrigerated conditions. The value of Michaelis–Menten constant (*K*m) estimated as 0.052mM using Lineweaver–Burke plot indicates high affinity of ChEt and ChOx to cholesterol oleate.

Sol–gel-derived silica-CHIT (CHIT–SiO_2) hybrid biocomposite exhibits interesting biosensor characteristics but very small response current were observed [215, 216]. The incorporation of multiwalled carbon nanotubes (MWCNT) into CHIT–SiO_2 hybrid film is likely to result in improved sensing characteristics. MWCNT enhance electro-catalytic activity due to presence of edge-plane-like sites located at both ends and in the defect region. Also, chitosan (CHIT) as a natural cationic biopolymer attracted much interest owing to its interesting properties such as biocompatibility, non-toxicity, low-cost, good film forming ability, high mechanical strength and high hydrophilicity. Moreover, presence of amino and hydroxyl groups in CHIT facilitates immobilization of enzymes via covalent binding for biosensor application.

There are reports on integration of MWCNT with a biopolymer such as CHIT to fabricate biosensors to gain synergistic action using organic–inorganic hybrid nanobiocomposites. Tan et al. have fabricated a free cholesterol biosensor based on TEOS sol–gel CHIT–silica and MWCNT [217].

Other results report on studies relating to the immobilization of ChEt and ChOx immobilized onto SiO_2–CHIT/MWCNT bionanocomposite film deposited onto ITO coated glass for detection of total cholesterol using differential pulse voltammetry (DPV). DPV technique is more advantageous because it yields both peak shaped response and considerable reduction of undesirable currents contribution [218].

The use of such hybrid matrices provides a good microenvironment for the enzyme immobilization and help to avoid cracking of the gel macrostructure because the stabilizing organic groups in the developing matrix enable structural relaxation as well as a large extension of the stable pH region [219]. The study presents the formation of micropatterned sol-gel structures containing active proteins by patterning with polydimethylsiloxane (PDMS) microchannels. To transport sol solution efficiently into the hydrophobic PDMS microchannels, a hydrophilic-hydrophobic block copolymer was used to impart hydrophilicity to the PDMS microchannels. Poor adhesion of the micropatterned gel structure onto glass slides was improved by treating the glass surface with a polymeric substrate. To minimize cracks in the gel microstructure, hybrid matrices of interpenetrating organic and inorganic networks were prepared containing the reactive organic moieties polyvinylalcohol or polyvinylpyrrolidone. Retention of biochemical activity within the micropatterned gel was demonstrated by performing immunobinding assays with immobilized immunoglobulin G (IgG) antibody. The potential application of microfluidics technology to immobilized-enzyme biocatalysis was demonstrated using PDMS-patterned microchannels filled with trypsin-containing sol-gels. The work also provides a foundation for the microfabrication of functional protein chips using sol-gel processes.

Redox Enzymes Based Biosensors

Direct electrochemistry of redox enzymes has attracted increasing interest both for the mechanism study of electron exchange of among enzymes in biological system as well as for practical application involving bioelectrochemistry. As it is well known, when enzymes are immobilized on the electrode surface denaturation and loss of electrochemical activities and bioactivities generally occur. Also, it is usual that the active sites of enzymes are deeply embedded in the insulated protein shells, resulting in a too long distance between the active sites and the underlying electrode, which is not suitable for electron exchange. As a result, direct electron transfer between redox enzymes and electrodes remains a great challenge and extensive efforts have been made to achieve the direct electrochemistry of redox enzymes. Most developed methods focus on utilization of functional organic or inorganic materials which are expected to meet the requirements. Firstly, the enzymes must be immobilized firmly on the electrode surfaces without altering their original conformations and activities. Secondly, they must be capable of facilitating the electron transfer between the active sites of enzymes and underlying electrodes. These materials mainly include polymer, nano-scale materials (carbon nanotubes, Au nanoparticles, metal oxide nanoparticles) [220 – 226].

The enzyme-containing hybrid film is formed on an electrode by casting method, layer-by-layer assembly and Langmuir–Blodgett film technique [227, 228]. Uniformity and stability of the formed hybrid film are necessary to guarantee good performances to the modified electrode. Thus, it's a challenge to obtain uniform hybrid film when various materials are put together, especially when using solid particles which tend to aggregate in suspension system and fall off the electrode surface. A one-step method to enzyme immobilization process and to achieve the direct electron transfer of glucose oxidase (GOx) on screen-printed carbon electrodes by immobilizing the enzyme in silica sol–gel and polyvinyl alcohol hybrid film, it was reported [229]. The favorable results were attributed to both the porous structure of the electrode surface and silica sol–gel/PVA hybrid material. The screen-printed electrodes were considered inexpensive and easy to prepare, they having porous surfaces, and providing more conductive sites to interact with enzymes. At the same time, the screen-printing technology it is considered the most economical mean for mass production of disposable electrodes [230].

Glucose oxidase is considered a model enzyme due to its wide range of applications in many areas such as glucose detection or biofuel cells [231, 232].

The hybrid material performed to encapsulate GOx consists into silica sol–gel and polyvinyl alcohol (PVA). The GOx-containing hybrid film was formed on the surface of screen printed electrodes (SPE) through one-step casting procedure. Thus, the direct electrochemistry of GOx was achieved just using cheap and easy-prepared materials and simple process and disposable screen-printed electrodes and the proposed modified SPE was developed as a novel mediator - free biosensor for glucose detection. The immobilized glucose oxidase displays a couple of stable and well-defined redox peaks with an electron transfer rate constant of 8.38 s^{-1} and a formal potential of −460mV in phosphate buffer (0.05 M, pH 7.0) at a scan rate of 300mVs^{-1}. The presented results suggested that conformation and bioactivity of glucose oxidase could be retained efficiently using the proposed immobilization method and the porous structure of screen-printed electrode surface was helpful for electron communication of glucose oxidase and the electrode. The modified electrode was used as a glucose biosensor, exhibiting a linear response to glucose concentration ranging from 0 to 4.13mM and a sensitivity of 3.47μAmM^{-1} cm^{-2} at an applied potential of −0.5 V. The detection limit of the biosensor was 9.8μM, based on a signal-to-noise ratio of 3.

Sol-Gel Method in Manufacture Surface Plasmon Resonance Biosensor

Various surface plasmon resonance (SPR) biosensors have been developed and applied for the detection and identification of specific analyte in the areas of biotechnology, environmental protection, medical diagnostic, drug screening, food safety and security [233].

A SPR biosensor system with magnetic beads was developed to detect the combination of antigen with antibody [234]. Owing to the difficulties to achieve complete dissociation of magnetic beads from the Au film due to the direct contact between magnetic beads and the Au film the titania sol–gel matrix was constructed on Au thin films to prevent the direct contact of the magnetic beads and the Au film. This determines also the easy regeneration and the sensitivity enhancement of the SPR biosensor.

The silica sol–gel process is usually carried out in acidic condition, which is hostile to the activities of biomolecules. At the same time, the silica sol–gel-derived matrix is fragile and easy to shrink, crack and desquamate from the electrode surface. As a result new sol–gel materials were thought for biosensor construction.

Thin films of crystalline TiO_2 nanoparticles are potential candidates for sensing applications having extremely large surface-to-volume ratio and peculiar surface properties of nanocrystals, which are expected to be beneficial for gas sensing purposes for example. TiO_2 thin films have also demonstrated good sensing properties towards humidity, oxygen and organic vapor, as well as for interfering gases such as CO, NO_2 and benzene.

The titania sol–gel matrix has been reported to immobilize enzymes in conductive polymers for electrochemical biosensor constructions. Titania is a kind of nonsilica material easily obtained from the sol–gel process being an alternate support material to silica for column packing in high performance liquid chromatography (HPLC) because of its high chemical stability, enough rigidity, and amphoteric ion-exchange properties [235, 236].

Usually, TiO_2 sol–gel, which was hydrolyzed in acidic conditions, was also not suitable for immobilizing enzymes because of the denaturation of enzymes under such low pH surroundings. Dry fracture of the gel membranes was also an obvious drawback for retaining the biosensor stability. Additionally, some membranes calcined at high temperature in the process of sol–gel fabrication before biomolecules immobilization prevented the biomolecules to be entrapped in the gel directly.

The achievement of an immunoassay of heat shock protein 70 (Hsp 70) by surface plasmon resonance (SPR) biosensor with magnetic beads immobilized on the Au film, it was reported [237]. The sol–gel-derived titania-modified Au film retained the bioactivity and provided for long-term stability of the biomaterials in storage. The presented vapor deposition method simplified the traditional sol–gel process and prevented the cracking of conventional sol–gel-derived glasses. In the presence of titania sol–gel matrix as the safeguard, the SPR biosensor with magnetic beads exhibited a satisfactory response concentration range of Hsp 70 from 0.20 to 30.00μgml^{-1}. Compared to the measurement without modification of titania sol–gel matrix, larger resonant wavelength shift and sensitivity enhancement of the biosensor were obtained simultaneously.

Another titania sol–gel inorganic–organic hybrid nanocomposite film was prepared to fabricate a sensitive tyrosinase biosensor for the amperometric detection of trace phenolic compounds without additional electron mediators [238]. The used acetylacetone worked as a complexing ligand to chelate with Ti atom and to prevent sol from agglomerating during the hydrolysis procedure, then the $Ti(OBu)_3$ (acac) was used as precursor to synthesize TiO_2 sol–gel. Hence, the pH of the titania solution could be adjusted over a wide range by varying the buffer solution in the hydrolysis process to the value which was optimum for retaining tyrosinase activity and such a membrane was stably attached onto the surface of a glassy carbon electrode (GCE). Amperometry coupling tyrosinase with mediators is one of the most sensitive measurements with rapid response. The results showed that TiO_2 sol–gel matrix could also act as an effective promoter for the electron transfer between quinones and the electrode as well as co-immobilized mediator. This performed titania matrix supplies a good environment for enzyme loading which resulted in a high sensitivity of 15.78$\mu A \mu M^{-1} cm^{-2}$ for monitoring phenols with a detection limit of 1×10^{-8}M at a signal-to-noise ratio of 3. The TiO_2 sol–gel derived biosensor exhibited a fast response less than 10 s and a good stability for more than 2 months.

Sol-Gel Technique Performing Fluorescence-based Glucose Sensors

Problems with existing devices based on electrochemistry have encouraged alternative approaches to glucose sensing in recent years, and those based on fluorescence intensity and lifetime have special advantages, including sensitivity and the potential for non-invasive measurement when near infrared light is used. Several receptors have been employed to detect glucose in fluorescence sensors, and these include the lectin concanavalin A (Con A), enzymes such as glucose oxidase, glucose dehydrogenase and hexokinase/glucokinase, bacterial glucose-binding protein, and boronic acid derivatives (which bind the diols of sugars). Techniques include measuring changes in fluorescence resonance energy transfer (FRET) between a fluorescent donor and an acceptor either within a protein which undergoes glucose-induced changes in conformation or because of competitive displacement; measurement of glucose-induced changes in intrinsic fluorescence of enzymes (e.g. due to tryptophan residues in hexokinase) or extrinsic fluorophores (e.g. using environmentally sensitive fluorophores to signal protein conformation). Noninvasive glucose monitoring can be accomplished by measurement of cell autofluorescence due to *nicotinamide adenine dinucleotide* (the NAD(P)H coenzyme with two nucleotides joined through their phosphate groups, with one nucleotide containing an adenine base and the other containing nicotinamide – used as substrate of enzymes that add or remove chemical groups from proteins)(Figure 13), and fluorescent markers of mitochondrial metabolism can signal changes in extracellular glucose concentration.

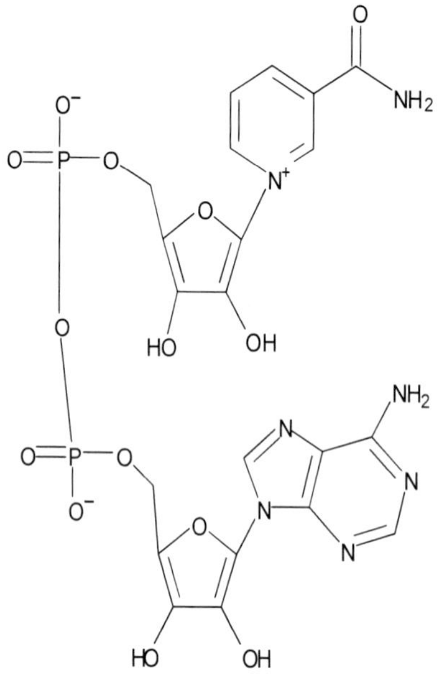

Diphosphopyridine nucleotide (DPN$^+$), Coenzyme I

Figure 13. Nicotinamide adenine dinucleotide Coenzime.

The problem of configuration of hexokinase-based glucose sensors for in vivo use by studying its encapsulation in porous sol–gels based on tetramethyl orthosilicate (TMOS), with or without covering membranes, it was studied [239].

Sol–gels have several potential advantages for glucose sensors intended for implantation in the body, including the provision of a matrix for immobilizing and containment of the enzyme, protection and excluding interfering substances such as fluorescence quenchers from the biological medium, favourably altering the enzyme kinetics (e.g. extending the linear range) and providing a biocompatible medium. Sol–gel immobilized yeast hexokinase retains its activity, showing an approximately 20% decrease in intrinsic fluorescence in response to saturating glucose concentrations. The maximal response of hexokinase in a monolithic sol–gel matrix was increased to about 40mM glucose (cf. 0.8mM for enzyme in solution) and the *K*m to about 12.5mM (cf. 0.3mM for free hexokinase). Application of an outer membrane consisting of 5% poly(2-hydroxyethyl methacrylate) extends the linear range to up to 100mM and the *K*m to about 57mM glucose. Unlike hexokinase in solution, sol–gel encapsulated hexokinase responds to glucose in serum with a decrease in intrinsic fluorescence comparable to that observed in serum free buffer, whether or not coated with an outer membrane. Techniques for encapsulation of in vivo fluorescence sensors are also in their infancy, at least as far as testing in humans or animals are concerned. Also, sol–gel immobilization and implanted fibre-optic probes are amongst the approaches that seem to have promise. More improved fluorophores for use in biological systems are in study, and in this respect quantum dots are already making an impact. Several fluorescent nanocrystal variants of quantum dots will emerge in the next few years and are likely to be incorporated in sensors. There is no doubt that fluorescence technologies have considerable promise for glucose sensing.

Sol-Gel Technique and Carbon Nanotube in Performing Biosensor

Extensive research studies on carbon nanotubes (CNTs) revealed their unique properties [240]. Their electrical and other properties have been exploited for a variety of applications that include electrochemical sensors [241,242]. CNTs have been employed as an electrode modifying material on the surface of glassy carbon, graphite, carbon fiber, gold, and platinum electrode towards fabrication of electrochemical sensors. Modified electrodes (MEs) with CNTs and supporting molecules as a binder or ion-exchanger have been used for the detection of many biomolecules, such as glucose, dihydronicotinamide adenine dinucleotide, hydrogen peroxide and amino acids [243 – 246]. However, the surface hydrophobicity and poor solubility of CNTs in common organic solvents limit their utility for applications. Functionalization of CNTs on their side walls as well in open ends makes them useful for the fabrication of MEs. Various MEs have been fabricated with functionalized CNTs and used for the electrochemical determination of bioanalytes. CNTs were functionalized with a conducting polymer and used as electrochemical sensors [247]. Polyaniline (PANI) was grafted onto multiwalled carbon nanotubes (MWNT) and used for the adsorptive reduction and determination of celecoxib [248]. MWNTs were non-covalently wrapped by a cationic polymer and loaded into electrospun polymer nanofibers and a glucose sensor was fabricated [249]. Functionalized CNTs were also incorporated with metal nanoparticles (MNPs) towards fabrication of sensors. MNPs have attracted significant attention because of their unusual size-dependent optical and electronic properties. Among MNPs, gold nanoparticles (AuNPs) possess good catalytic activity for a number of reactions. The hydrophobic surface of CNTs can be modified or functionalized with adequate groups and subsequently MNPs could be loaded onto the surface of CNTs. AuNPs have been widely used for the construction of biosensors due to their excellent ability to immobilize biomolecules and at the same time retain the biocatalytic activities of the tested biomolecules. AuNPs could be anchored onto the surface of functionalized CNTs. AuNPs were self-assembled on a thiol-terminated sol–gel derived silicate network and a sensor was developed for the selective and sensitive detection of NADH etc. [250]. As was mentioned before natural polymers such as chitosan can be mixed with sol–gel material and such composites are less brittle and more biocompatible [251, 252]. These composites facilitate the electron transfer process between the biomolecules and the electrodes.

In general, sol–gel derived materials are electrical insulators with low charge transfer efficiency, poor mechanical stability and low resistance that limit their electrochemical applications [253, 254].

While silica imparts useful properties for the sensors, an additional material is required to have adequate electron transduction to the electrode. In this order MWNTs have been incorporated into sol–gel matrix to improve the electrical conductivity [255].

Considering the advantages of sol–gel based organic–inorganic hybrids, electrical properties of CNTs and catalytic properties of AuNPs, composites comprising of CNTs, silica and AuNPs were considered to have improved electrochemical sensor characteristics.

The preparation of CNT–silica–AuNP composite and its utility for electrochemical sensing of ascorbic acid (AA), it was reported [256-258]. AA is one of the most important water soluble components present in fruits and vegetables. AA is used in large scales as an antioxidant in food, animal feed, beverages, pharmaceutical formulations and cosmetic applications. Because of the biological and technological importance of AA, studies have been pursued to establish rapid and sensitive methods for the reliable determination of AA. The determination of the AA content was performed with several methods, such as titrimetry, fluorimetry, flow injection analysis, spectrophotometry and liquid chromatography.

Electrocatalytic oxidation at modified electrodes has been proved to be an important approach for the determination of AA. Electrochemical methods are simple and inexpensive. In the electrochemical process (Figure 14), AA can be easily oxidized to dehydroascorbic acid through a two-electron and one-proton process.

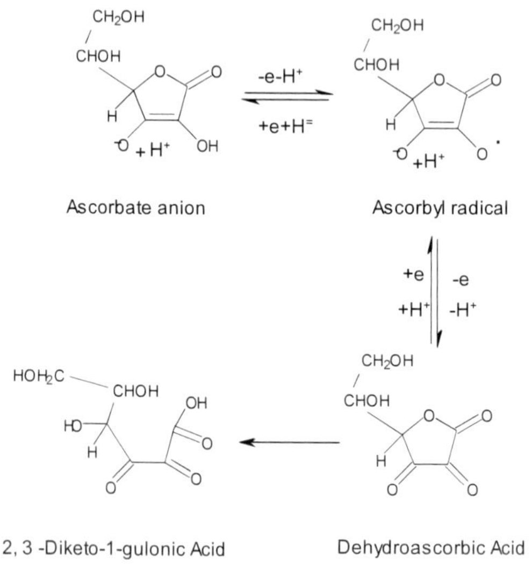

Figure 14. Oxidative metabolism of ascorbic acid.

Subsequently, the irreversible hydrolysis could result in an electro inactive product, 2, 3-diketogluconic acid. The usefulness of various modified electrodes has been demonstrated for the electrochemical determination of AA.

The literature mentions a screen-printed electrode modification with o-amino phenol film for the fabrication of an amperometric sensor for AA [259]. It was expected that the combined presence of CNTs, sol–gel silica and AuNPs to provide synergistic influences on the sensor performances. The preparation of a nanocomposite, comprising of MWNTs, AuNPs and silica and its use as an electrode modifier for the fabrication of AA sensor have been reported [260]. The modified electrode, GC/MWNT–silica–AuNPs, was fabricated by modifying the surface of multiwalled carbon nanotubes with silica networks and subsequent electrodeposition of gold nanoparticles (AuNPs). The morphology of the GC/MWNT–silica-NW–AuNPs modified electrode (ME) was analyzed by field emission scanning electron microscopy (FE-SEM). Cyclic voltammetry (CV) and electrochemical impedance spectroscopy (EIS) were used to evaluate the electrochemical performances of GC/MWNT–silica-NW–AuNPs-ME. Ascorbic acid (AA) was found to be electrocatalytically oxidized at GC/MWNT–silica-NW–AuNPs-ME due to the synergistic influence of MWNTs, silica-NW and AuNPs. The MWNT–silica-NW–AuNPs-ME sensor electrode presents a high sensitivity (8.59μA/mM) and selectivity for AA in the presence of dopamine (DA). Differential pulse voltammetry revealed that the sensor electrode exhibited excellent selectivity for AA in the presence of large excess of DA with a large potential separation of ~0.26 V. The effective electrocatalytic activity, excellent peak resolution and ability for independent determination of DA in the presence of AA reveal that MWNT–silica-NW–AuNPs-ME is suitable for simultaneous and selective determination of DA and AA.

It is obvious that the enzyme-modified electrode is the fundamental component of amperometric biosensors and biofuel cells [261]. The selection of appropriate combinations of materials, such as: enzyme, electron transport mediator, binding and encapsulation materials, conductive support matrix and solid support, for construction of enzyme-modified electrodes governs the efficiency of the electrodes in terms of electron transfer kinetics, mass transport, stability, and reproducibility. The electrodes based on enzyme encapsulation in sol–gel SiO_2 matrixes and incorporated CNTs in the matrix provide enhanced electronic conduction. The SiO_2 matrix was designed to be sufficiently porous that both glucose and O_2 have access to the enzymes and yet provides a protective cage for immobilizing the bio-molecules without affecting biological function.

Voltammetry indicated that the effect of the SiO_2 matrix on mediator diffusion was minimal, although for one mediator, 2, 2 γ-azino-bis (3-ethylbenzothiazoline-6-sulfonic acid) diammonium salt, chemical modification of the solvent phase with polyethylene glycol was necessary. The polyethylene glycol addition resulted in a more uniform dispersion of the CNTs. The enzymes maintained their biocatalytic activity in the sol–gel matrix. A glucose–O_2 biofuel cell based on nanostructured SiO_2 sol–gel/CNT composite electrodes generated approximately $120\mu Wcm^{-2}$ at 0.24V at room temperature. Another new type of sol–gel/organic hybrid composite material based on the crosslinking of the natural polymer chitosan with (3-acryloxypropyl) dimethoxymethylsilane was developed for the fabrication of an amperometric H_2O_2 biosensor [262].

The composite film was used to immobilize horseradish polymerase on a gold disk electrode. With the aid of a catechol mediator, the biosensor presented a fast response of less than 2 s with linear range of 5.0×10^{-9} to 1.0×10^{-7} mol L^{-1} and a detection limit of 2×10^{-9} mol L^{-1}. The biosensor retained approximately 75% of its original activity after about 60 days of storage in a phosphate buffer at 4°C.

REFERENCES

[1] Brinker, C. J. & Scherer, G. W. (1990). *Sol-Gel Science: The Physics and Chemistry of SoI-Gel Processing.* New York: Academic Press, Inc.

[2] Brinker, C. J. (1994). *The Colloid Chemistry of Silica.* Columbus, OH: ACS Publications, ch. 18.

[3] Brinker, C. J. (1988). Hydrolysis and Condensation of Silicates: Effects on Structure. *J Non-Cryst Solids*, *100*, 31-50.

[4] Gupta R. & Kumar, A. (2008) Bioactive materials for biomedical applications using sol–gel technology. *Biomed Mater*, *3*, 034005, 1-15.

[5] Gupta R. & Chaudhury N. K. (2007). Entrapment of biomolecules in sol–gel matrix for applications in biosensors: Problems and future prospects. *Biosens Bioelectron*, *22*, 2387–2399.

[6] Sakka, S. (2003). Sol-Gel Technology as Reflected in Journal of Sol-Gel. Science and technology. *J Sol-Gel Sci Technol*, *26,* 29-33.

[7] Uhlmann, D. R. & Teowee, G. (1998). Sol-Gel Science and Technology: Current State and Future Prospects. *J Sol-Gel Sci Technol*, *13*, 153–162.

[8] Wei, Y., Jin, D., Yang, C. & Wei, G. (1996). A fast convenient method to prepare hybrid sol-gel materials with low v&olume-shrinkages. *J Sol-Gel Sci Technol*, *7*, 191-201.

[9] Mackenzie, J. D. (1982). Glasses From Melts and Glasses From Gels: A Comparison. *J Non-Cryst Solids*, *48,* 1-10.

[10] Mascia, L., Zhang, Z. & Shaw. S. J. (1996). Carbon Fibre Composites Based on Polyimide/Silica Ceramers: Aspects of Structure-Property Relationship. *Composites Part* A, 27A (12), 12ll-1221.

[11] Sakka, S., Tanaka, Y. & Kokubo, T. (1986). Hydrolysis and Polycondensation of Dimethyldiethoxysilane and Methyltriethoxysilane as Materials for the *Sol-Gel Process. J Non-Cryst Solids*, *82,* 24-30.

[12] Brinker, C. J., Shegal, R., Hietala, S. L., Deshpande, R., Smith, D. M., Loy, D. & Ashley, C. S. (1994). Sol-Gel Strategies for Controlled Porosity Inorganic Materials. *J Membrane Sci*, *94,* 85-102. 54 Aurica P. Chiriac, Iordana Neamtu, Loredana E. Nita et al.

[13] Schmidt, H. (1989). Organic modification of glass structure new glasses or new polymers. *J Non-Cryst Solids*, *112,* 419-423.

[14] Nogami, M. & Moriya, Y. (1980). Glass formation through hydrolysis of $Si(OC_2H_5)_4$ with NH_4OH and HCl solution. *J Non-Cryst Solids*, *37,* 191-201.

[15] Zusman, R. D., Beckman, A., Zusman, I. & Brent, R. L. (1992). Purification of sheep immunoglobin G using protein A trapped in sol-gel glass. *Anal Biochem*, *201,* 103-106.

[16] Yakubovich, T. N., Zub, Y. L. & Leboda, R. (1994). Metal Coordination Compounds Built Into Polyorganosiloxane Matrices. Russ. *J Coord Chem*, *20 (10)*, 761-765.

[17] Hobson, S. T. & Shea, K. J. (1997). Bridged Bisimide Polysilsesquioxane Xerogels: New Hybrid Organic-Inorganic Networks. *Chem Mater, 9(2)*, 616-623.

[18] Mulhaupt, R., Buchholz, U., Rosch, J. & Steinhauser, N. (1994). Progress in Synthesis and Application of Segmented Polymers. *Angew Makromol Chem, 223,* 47-60.

[19] Hashim, A., Kawabata, N. & Kohjiya, S. (1995). Silica reinforcement of epoxidized natural rubber by the sol-gel method. *J. Sol-Gel Sci. Technol.*, 5:211{*218,* 1995.*J Non-Cryst Solids., Vol. 5*, 211-218.

[20] Mark, J. E. (1992). Novel Reinforcement Techniques for Elastomers. *J Appl Pol Sci: Appl Polym Symp, 50,* 273-282.

[21] Baumann, F., Deubzer, B., Geck, M., Dauth, J. & Schmidt, M. (1997). Elastomeric Organosilicon Micronetworks. *Macromolecules, 30(24)*, 7568-7573.

[22] Wen, J. & Wilkes, G. L. (1996). Organic-Inorganic Hybrid Network Materials by the Sol-Gel Approach. *Chem Mater, 8*, 1667-1681.

[23] Amberg-Schwab, S., Weber, U., Burger, A.,Nique, S. & Xalter, R. (2006). Development of Passive and Active Barrier Coatings on the *Basis of Inorganic–Organic Polymers. Monatsch Chemie, 137,* 657–666.

[24] Reisfeld, R. (2001). Prospects of sol-gel technology towards luminescent materials. *Opt Mater, 16,* 1–7.

[25] Chaudhury, N. K., Gupta, R. & Gulia, S. (2007). Sol-gel technology for sensor applications. *Def Sci J, 57,* 241-253.

[26] Livage, J., Henry, M. & Sanchez, C. (1988). Sol-gel chemistry of transition metal oxides. *Prog Solid State Chem, 18 (4)*, 259-341.

[27] Livage J. (1997), Sol-gel processes, *Curr. Opin. Solid State Mater. Sci.*, 2, 132-1 36.

[28] Hench, L. L. & West, J. K. (1990). The Sol-Gel Process. *Chem Rev, 90,* 33-72.

[29] Dimitriev, Y., Ivanova, Y. & Iordanova R. (2008), History of Sol-Gel Science and technology (Review). *Journal of the University of Chemical Technology and Metallurgy, 43(2)*, 181-192.

[30] Kozhukharov, S. (2009), Relationship between the Conditions of Preparation by the Sol-Gel Route and the Properties of the Obtained Products, *Journal of the University of Chemical Technology and Metallurgy , 44(2)*, 143-150.

[31] Sakka, S. (2003), Sol-Gel Technology as Reflected in Journal of Sol-Gel Science and Technology. *J Sol-Gel SciTechnol, 26,* 29–33.

[32] Coradin, T., Boissière, M. & Livage, J. (2006). Sol-gel Chemistry in Medicinal Science. *Curr Med Chem, 13,* 99-108.

[33] Zarzycki, J. (1997). Past and Present of Sol-Gel Science and Technology. *J Sol-Gel Sci Technol, 8*, 17–22.

[34] Gvishi, R. (2009). Fast sol–gel technology: from fabrication to applications, *J Sol-Gel Sci Technol, 50,* 241–253.

[35] Mackenzie J. D. (2003). Sol-Gel Research—Achievements Since 1981 and Prospects for the Future. *J Sol-Gel Sci Technol, 26,* 23–27.

[36] Schmidt, H. (1985) New type of non-crystalline solids between inorganic and organic materials *J Non-Cryst Solids, 73,* 681-691.

[37] Schmidt, H., Jonschker, G., Goedicke, S. & Mennig, M. (2000). The Sol-Gel Process as a Basic Technology for Nanoparticle-Dispersed Inorganic-Organic Composites. *J Sol-Gel Sci Technol, 19,* 39-51.

[38] Philipp, G. & Schmidt, H. (1984). New materials for contact lenses prepared from Siand Ti-alkoxides by the sol-gel process. *J Non-Cryst Solids*, 63(1-2), 283-292.

[39] Avnir, D., Levy, D. & Reisfeld, R. (1984). The nature of the silica cage as reflected by spectral changes and enhanced photostability of trapped Rhodamine 6GJ. *Phys Chem, 88 (24)*, 5956-5959.

[40] Reisfeld, R., Saraidarov, T. S. & Levchenko, V. (2009). Strong emitting sol–gel material based on interaction of luminescence dyes and lanthanide complexes with Silver Nanoparticles, *J Sol-Gel Sci Technol, 50,* 194 – 200.

[41] Young, S. K. *Overview of Sol-Gel Science and Technology*. January 2002. Available from: http://handle.dtic.mil/100.2/ADA398036.

[42] Yoldas, B. E. (1984). Modification of Polymer-Gel Structures. *J Non-Cryst Solids, 63,* 145-154.

[43] Ng, L. V., Thompson, P., Sanchez, J., Macosko, C. W. & McCormick, A. V. (1995). Formation of Cagelike Intermediates From Nonrandom Cyclization During Acid-Catalyzed *Sol-Gel Polymerization of Tetraethyl Orthosilicate. Macromolecules, 28 (19)*, 6471- 6476.

[44] Julbe, A., Balzer, C., Barthez, J. M., Guizard, C., Larbot, A. & Cot, L. (1995). Effect of Non-Ionic Surface Active Agents on TEOS-Derived Sols, Gels and Materials. *J Sol-Gel Sci Technol , 4 (2)*, 89-97.

[45] Schmidt, H., Scholze, H. & Kaiser, A.(1984). Principles of Hydrolysis and Condensation Reaction of Alkoxysilanes. *J Non-Cryst Solids*, 63(1-2),1-11.

[46] Schmidt, H., Kaiser, A., Patzelt, H. & Scholze, H. (1982). Mechanical and Physical Properties of Amorphous Solids Based on $(CH_3)_2SiO$-SiO_2 *Gels. J Phys, 12(C9), 43,* 275-278.

[47] Schmidt, H., Seiferling, B. (1986). Chemistry and Applications of Inorganic-Organic Polymers. *Mat Res Soc Symp Proc, 73,* 739-750.

[48] Schmidt, H. (1984). Organically modified silicates by the sol-gel process. *Mat Res Soc Symp Proc, 32,* 327-335.

[49] Sanchez, C., Julian, B., Belleville, P. & Popall, M.(2005). Applications of hybrid organic–inorganic nanocomposites. *J Mater Chem, 15,* 35-36, 3559-3592.

[50] Kandimalla, V. B., Tripathi, V. S. & Ju, H. (2006). Immobilization of Biomolecules in Sol–Gels: Biological and Analytical *Applications. Crit Rev Anal Chem, 36,* 73–106.

[51] Gill, I., (2001). Biodoped nanocomposite polymers: Sol-gel bioencapsulates. *Chem Mater, 13,* 3404-3421.

[52] Gill, I. (2001). Biodoped nanocomposite polymers: Sol-gel bioencapsulates. *Chem Mater, 13,* 3404–3421.

[53] Gill, I., (2001). Biodoped nanocomposite polymers: Sol-gel bioencapsulates. *Chem Mater, 13,* 3404–3421. 56 Aurica P. Chiriac, Iordana Neamtu, Loredana E. Nita et al.

[54] Carturan, G., Toso, R. D., Boninsegna, S. & Monte, R. D. (2004). Encapsulation of functional cells by sol–gel silica: actual progress and perspectives for cell therapy. *J Mater Chem, 14,* 2087–2098.

[55] Smith, K., Silvernail, N. J., Rodgers, K. R., Elgren, T. E., Castro, M. & Parker, R. M. (2002). Sol-gel encapsulated horseradish peroxidase: A catalytic material for peroxidation. *J Sol-Gel SciTechnol,, 124,* 4247–4252.

[56] Gill, I & Ballesteros, A. (2000). Bioencapsulation within synthetic polymers (Part 1): sol–gel encapsulated biologicals. *TIBTECH, 18,* 282-296.

[57] Dickey, F. H. (1955). The exchange properties given are mean values for standard deviations for all resins. *J Phys Chem*, *58,* 695-700.

[58] Venton, D. L., Cheeseman, K. L., Chatterton, R. T. & Anderson, T. L. (1984). Entrapment of a highly specific antiprogesterone antiserum using polysiloxane copolymers. *Biochim Biophys Acta, 797,* 343–347.

[59] Glad, M., Norrlow, O., Sellergren, B., Siegbahn, N. & Mosbach, K. (1985). Use of silane monomers for molecular imprinting and enzyme entrapment in polysiloxanecoated porous silica. *J Chromatogr, 347,* 11–23.

[60] Braun, S., Rappoport, *S.*, Zusman, R., Avnir, D. & Ottolenghi, M. (1990) Biochemically active sol–gel glasses: the trapping of enzymes. *Mater Lett*, *10,* 1–8.

[61] Avnir, D., Braun, S., Ovadia, L. & Ottolengthi, M. (1994). Enzymes and other proteins entrapped in sol–gel materials. *Chem Mater*, *6*, 1605–1614.

[62] Dave, B. C., Dunn, B., Valentine, J. S. & Zink, J. I. (1994), Sol–gel encapsulation method for biosensors. *Anal Chem, 66,* 1120A–1127A.

[63] Armon, R., Dosoretz, C., Starosvetsky, J., Orshansky, F. & Saadi, I. (1996). Sol–gel applications in environmental biotechnology. *J Sol-Gel SciTechnol,, 51,* 279–285.

[64] Coradin, T., Nassif, N. & Livage, J. (2003). Silica–alginate composites for microencapsulation. *Appl Microbiol Biotechnol*, *61,* 429–434.

[65] Pierre, A. C. (2004). The sol–gel encapsulation of enzymes. *Biocatal Biotransform*, *22,* 145–170.

[66] Menaa, B., Menaa, F. (2010). Silica-based nanoporous sol-gel glasses: from bioencapsulation to protein folding studies. *Int J Nanotechnol*, *7(1)*, 1-45.

[67] Menaa, B., Takahashi, M., Tokuda, Y. & Yoko, T. (2008). Dispersion and photoluminescence of free-metal phtalocyanine doped in sol-gel polyphenylsiloxane glass films. *J Photochem Photobiol A*, 194 (2–3), 362–366.

[68] Menaa, B., Takahashi, M., Tokuda, Y., Yoko, T. (2006). High dispersion and fluorescence of anthracene doped in polyphenylsiloxane films. *J Sol-Gel Sci Technol*, *39 (2)*, 85–194.

[69] Dunn, B., Zink, J.I. (1991). Optical properties of sol-gel glasses doped with organic molecules. *J Mater Chem*, *1(6)*, 903–913.

[70] Avnir, D. (1995). Organic chemistry within ceramic matrixes: doped sol-gel materials, *Acc Chem Res*, *28(8)*, 328–334.

[71] Avnir, D., Klein, L.C., Levy, D., Schubert, U., Wojcik, A.B. (1998). Organo-silica solgel materials, in Rappoport, Z. and Apeloig, Y.J. (Eds.): In the *Chemistry of Organic Silicon Compounds*, Wiley and Sons, New York, *Vol. 2*, 2317–2362. [72] Gupta, R. & Chaudhury N. K. (2007). Entrapment of biomolecules in sol–gel matrix for applications in biosensors: Problems and future prospects. *Biosens Bioelectron*, *22,* 2387–2399.

[73] Rao, M. S. & Dave, B. C. (1998). Selective intake and release of proteins by organically modified silica sol–gels. *J Am. Chem Soc, 120,* 13270–13271.

[74] Brennan, J. D. (1999). Using intrinsic fluorescence to investigate proteins entrapped in sol–gel derived materials. *Appl Spectrosc, 53,* 106A–121A.

[75] Miller, J. M., Dunn, B. & Valentine, J. S. (1996). Synthesis conditions for encapsulating cytochrome c and catalase in SiO2 sol–gel materials. *J Non-Cryst Solids 202,* 279–289.

[76] Kuenzelmann, U. & Boettcher, H. (1997). Biosensor properties of glucose oxidase immobilized within silica gels. *Sens Actuat* B, *39,* 222–228.

[77] Huang, W., Kim, J. B., Bruening, M. L. & Baker, G. L. (2002). Functionalization of surfaces by water-accelerated atom-transfer radical polymerization of hydroxyethyl methacrylate and subsequent derivatization. *Macromolecules*, *35,* 1175-1179.

[78] Bontempo, D., Tirelli, N., Masci, G., Crescenzi, V. & Hubbell, J. A. (2002). Thick oating and functionalization of organic surfaces via ATRP in water. *Macromol Rapid Commun*, *23,* 417-422.

[79] Flickinger, M. C., Mullick A. & Ollis D. F. (1998). Method for construction of a simple laboratory-scale nonwoven filament biocatalytic filter. *Biotechnol Progr*, *14,* 664-666.

[80] Havens, P. L. & Rase, H. F. (1993). Reusable immobilized enzyme/polyurethane sponge for removal and detoxification of localized organophosphate pesticide spills. Ind *Eng Chem Res*, *32,* 2254-2258.

[81] LeJeun, K. E., Wild, J. R. & Russell, A. J. (1998). Nerve agent degraded by enzymatic foams. *Nature*, *395,* 27-28.

[82] Zarcula, C., Croitoru, R., Corîci, L., Csunderlik, C. & Peter, F. (2009). Improvement of Lipase Catalytic Properties by Immobilization in Hybrid Matrices. *World Academy of Science. Eng Technol*, *52,* 179-184.

[83] Livage, J. (1996). Bioactivity in sol-gel glasses. CR *Acad Sci Paris*, *322 (5)*, 417–427.

[84] Rao, M. S. & Dave, B. C. (2003). Thermally-regulated molecular selectivity of organosilica sol-gels. *J Sol-Gel SciTechnol,*, *125(39)*, 11826–11827.

[85] Bornscheuer, U. T. (2003). Immobilizing enzymes: how to create more suitable biocatalysts. *Angew Chem Int Ed.*, *42,* 3336–3337.

[86] Klibanov, A. M. (2001). Improving enzymes by using them in organic solvents. *Nature 409,* 241–246.

[87] Krishna, S. H. (2002). Developments and trends in enzyme catalysis in nonconventional media. *Biotechnol Adv, 20,* 239–266.

[88] Betancor, L. & Luckarift, H. R. (2008). Bioinspired enzyme encapsulation for biocatalysis *Trends Biotechnol*, *26,*10, 566-572.

[89] Berne, C., Betancor, L., Luckarift, H. R. & Spain, J. C. (2006). Application of a microfluidic reactor for screening cancer prodrug activation using silica-immobilized nitrobenzene nitroreductase. *Biomacromolecules*, *7*, 2631–2636.

[90] Kim, J., Grate, J. W. & Wang, P. (2006). Nanostructures for enzyme stabilization. *Chem Eng Sci*, *61,* 1017–1026.

[91] Yim, T. J., Kim, D. Y., Karajanagi, S. S., Lu, T. M., Kane, R. & Dordick, J. S. (2003). Si nanocolumns as novel nanostructured supports for enzyme immobilization. *J Nanosci Nanotechnol*, *3*, 479-482.

[92] Martin, C. R. & Kohli, P. (2003). The emerging field of nanotube biotechnology. Nat *Rev Drug Discov*, *2*, 29–37. 58 Aurica P. Chiriac, Iordana Neamtu, Loredana E. Nita et al.

[93] Avnir, D., Coradin, T., Levc, O. & Livage J. (2006).Recent bio-applications of sol–gel materials. *Mater Chem*, *16,* 1013–1030.

[94] Shtelzer, S., Rappoport, S., Avnir, D., Ottolenghi, M. & Braun, S. (1992). Properties of trypsin and of acid phosphatase immobilized in sol–gel glass matrices. *BiotechnolAppl Biochem*, *15,* 227–235.

[95] Jurgen-Lohmann, D. L. & Legge, R. L. (2006). Immobilization of bovine catalase in sol–gels. *Enzyme Microb Technol*, *39,* 626–633.

[96] Chen, X., Jia, J. & Dong, S. (2003). Organically modified sol–gel chitosan composite based glucose biosensor. *Electroanalysis, 15 (7)*, 608–612.

[97] Pandey, P. C., Upadhyay, S., Tiwari, I. & Tripathi, V. S. (2001). An ormosil based peroxide biosensor – a comparative study on direct electron transport from horseradish peroxidase, *Sens Actuators B*, *72,* 3, 224–232.

[98] Reetz, M. T., Tielmann, P., Wiesenhofer, W., Konen, W. & Zonta, A. (2003). Second generation sol–gel encapsulated lipases: robust heterogeneous biocatalysts, *Adv Synth Catal*, 345 (6–7), 717–728.

[99] Rassy, E. l, Maury, H., Buisson, S. & Pierre, A. C. (2004). Hydrophobic silica aerogel–lipase biocatalysts. Possible interactions between the enzyme and the gel. *J Non-Cryst Solids*, *350,* 23–30.

[100] Carvalho, C. M. L., Aires-Barros, M. R. & Cabral, J. M. S. (1999). Cutinase: from molecular level to bioprocess development. *Biotechnol Bioeng*, *66,* 17–34.

[101] Vidinha, P., Augusto, V., Almeida, M., Fonseca, I., Fidalgo, A., Ilharco, L., Cabral, J. M. S. & Barreiros, S. (2006). Sol–gel encapsulation: An efficient and versatile immobilization technique for cutinase in non-aqueous media, *J Sol-Gel Sci Technol,*, *121,* 23–33.

[102] Podbielska, H. & Ulatowska-Jarza A. (2005). Sol-gel technology for biomedical engineering, *Bull Polish Academy Sci*, *53(3)*, 261-271.

[103] Brennan, J. D., Benjamin, D., DiBattista E. & Gulcev, M. D. (2003). Using Sugar and Amino Acid Additives to Stabilize Enzymes within Sol- Gel Derived Silica, *Chem Mater*, *15,* 737-743.

[104] Reetz, M. T., Zonta, A. & Simpelkamp, J. (1996). Efficient immobilization of lipases by entrapment in hydrophobic sol–gel materials. *Biotechnol Bioeng*, *49,* 527–534.

[105] Reetz, M. T. (1997) Entrapment of biocatalysts in hydrophobic sol–gel materials for use in organic chemistry. *Adv Mater*, 9, 943–954.

[106] Obert, R. & Dave, B. C. (1999) Enzymatic conversion of carbon dioxide to methanol: enhanced methanol production in silica sol–gel matrices. *J Am Chem Soc*, *121,* 12192–12193.

[107] Wu, H. Z., Jiang, Y., Xu, S. W. & Huang S. F. (2003). A new biochemical way for conversion of CO2 to methanol via dehydrogenases encapsulated in SiO2 matrix. *Chin Chem Lett*, *14,* 423–425.

[108] Kuncova, G., Szilva, J., Hetflejs, J. & Sabata, S. (2003) Catalysis in organic solvents with lipase immobilized by sol–gel technique. *J Sol–Gel Sci Technol*, *26,* 1183–1187.

[109] Chen, J. P. & Lin, W. S. (2003). Sol–gel powders and supported sol–gel polymers for immobilization of lipase in ester synthesis. *Enzyme Microb Technol*, *32,* 801–811.

[110] Catt, K. & Niall, H. D. (1967). Solid phase radioimmunoassay. *Nature*, *213,* 825–827.

[111] Zhou, J. C., Chuang, M. H., Lan, E. H., Dunn, B., Gillman, P. L. & Smith, S. M. (2004). Immunoassays for cortisol using antibody-doped sol–gel silica. *J Mater Chem*, *14,* 2311–2316.

[112] Wang, R., Narang, U., Prasad, P. N. & Bright, F. V. (1993). Affinity of antifluorescein antibodies encapsulated within a transparent sol–gel glass. *Anal Chem*, *65,* 2671–2675.

[113] Yang, H. H., Zhu, Q. Z., Qu, H. Y., Chen, X. L., Ding, M. T. & Xua, J. G. (2002). Flow injection fluorescence immunoassay for gentamicin using sol–gel derived mesoporous biomaterial. *Anal Biochem*, *308,* 71–76.

[114] Yadavalli, V. K., Koh, W. G., Lauzer, G. J. & Pishko, M. V. (2004). Micro fabricated protein containing poly (ethylene glycol) hydrogel arrays for biosensing. *Sens Actuators B*, *97,* 290–297.

[115] Shabat, D., Grynszpan, F., Saphier, S., Turniansky, A., Avnir, D. & Keinan, E. (1997). An efficient sol–gel reactor for antibody-catalyzed transformations. *Chem Mater*, *9*, 2258–2260.

[116] Liang, R. P., Qiu, H. D. & Cai, P. X. (2005). A novel amperometric immunosensor based on three-dimensional sol-gel network and nanoparticle self-assemble technique. *Anal Chim Acta*, *534,* 223-230.

[117] Liu, J. M., Zhu, G. H., Rao, Z. M., Wei, C. J., Li, L. D., Chen, C. L. & Li, Z. M. (2005). Determination of human IgG by solid substrate. *Anal Chim Acta*, *528,* 29-36.

[118] Rupcich, N., Goldstein A. & Brennan, J. D. (2003). Optimization of Sol – Gel technique for Protein. Microarrays *Chem Mater*, *15,* 1803 – 1810.

[119] Pierre, A., Bonnet, J., Vekris A. & Portier, J. (2001). Encapsulation of deoxyribonucleic acid molecules in silica. *J Mater Sci:Mater Med*, *12,* 51-60.

[120] Ellerby, L., Nishida, C. R., Nishida, F., Yamanaka, S. A., Dunn, B., Valentile, J. S. & Zink, J. I. (1992). Encapsulation of proteins in transparent porous silicate glasses prepared by the sol-gel method. *Science*, *255(5048)*, 1113–1115.

[121] Keeling-Tucker, T. & Brennan, J. D. (2001). Fluorescent probes as reporters. *Chem Mater*, *13 (10)*, 3331–3350.

[122] Menaa, B., Herrero, M., Rives, V., Lavrenko, M. & Eggers, D. K. (2008). Favourable influence of hydrophobic surfaces on protein structure in porous organically-modified silica glasses. *Biomaterials*, *29(18)*, 2710–2718.

[123] Menaa, B., Torres, C., Herrero, M., Rives, V., Gilbert, A. R. W. & Eggers, D. K. (2008). Protein adsorption onto organically modified silica glass leads to a different structure than sol-gel encapsulation. *Biophys J*, *95(8)*, 51– 53.

[124] Hungerford, G., Rei, A., Ferreira, M. I. C., Suhling, K. & Tregidgo, C. (2007). Diffusion in a sol–gel derived medium with a view toward biosensor applications. *J Phys Chem B*, *111 (13)*, 3558–3562.

[125] Tsai, H. & Doong, R. (2007). Preparation and characterization of urease-encapsulated, biosensors in poly(vinyl alcohol)-modified silica sol–gel materials. *Biosens Bioelectron*, *23 (1)*, 66–73.

[126] Murphy, R. M., Kendrick, B. S., Chiti, F. & Dobson, C. M. (2006). Protein misfolding, functional amyloid, and human disease. *Annu Rev Biochem*, *75,* 333–366.

[127] Bettati, S., Pioselli, B., Campanini, B., Viappiani, C. & Mozzarelli, A. (2004). Proteindoped nanoporous silica gel. Encyclopedia of Nanoscience and Nanotechnology, American Scientific, *Stevenson Ranch*, *9*, 81–103. 60 Aurica P. Chiriac, Iordana Neamtu, Loredana E. Nita et al.

[128] Brennan, D. (1999). Using intrinsic fluorescence to investigate proteins entrapped in sol-gel derived materials, *Appl Spectrosc*, *53,* 106A-121A.

[129] Dave, B. C., Soyez, H., Miller, J. M., Dunn, B., Valentine, J. S. & Zink, J. I. (1995). Sol-Gel encapsulated heme proteins with chemical sensing properties. *Chem Mater*, *7*, 1431-1438.

[130] Das, T. K., Khan, I., Rousseau, D. L. & Friedman, J. M. (1998). Preservation of the native structure in myoglobin at low pH by sol-gel encapsulation. *J Sol-Gel SciTechnol,*, *120,* 10268-10269.

[131] Edmiston, P. L., Wambolt, C. L., Smith, M. K. & Saavedra, S. S. (1994). Spectroscopic characterization of albumin. *J Colloid Interf Sci*, *163,* 395 -400.

[132] Wamboldt, C. L. & Saavedra, S. S. (1996). Iodide fluorescence quenching of sol-gel immobilized BSA. *J Sol-Gel Sci Technol*, *7*, 53-57.

[133] Flora, K. K. & Brennan, J. D. (2001). Characterization of the Microenvironments of PRODAN Entrapped in Tetraethyl Orthosilicate Derived Glasses. *J Phys Chem B*, *105 (48)*, 12003–12010.

[134] Boltan, B. A. & Scherer, J. R. (1989). Raman spectra and water absorption of bovine serum albumin. *J Phys Chem*, *93 (22)*, 7635–7640.

[135] Asgari, S. (2006). Drug delivery materials made by sol/gel technology. *US Patent Application* , 20060171990.

[136] Radin, S., Chen, T. L. & Ducheyne, P. (2007). Emulsified sol-gel microspheres for controlled drug delivery, *Key Eng Mater*, 330-*332,* 1025-1028.

[137] Weber, J., Atanasoska, L., Zoromski, M. & Robert, W. (2009). Medical devices having sol-gel derived ceramic regions with molded submicron surface features. US Patent Application 20090048659.

[138] Prokopowicz, M., Lukasiak, J. & Przyjazn, A. (2004). Utilization of a sol–gel method for encapsulation of doxorubicin. *J Biomater Sci Polym Ed*, *15,* 343–356.

[139] Gulaim, S., Kessler, V. G., Unell, M. & Håkansson, S. (2009). *Self-Assembled Titania Micelles for Drug Delivery and Bio-Control Applications.* Proceeding of 1st International Conference on Nanostructured Materials and Nanocomposites, Kottayam, India. ICNM – *2009,* 135 – 139.

[140] Fonseca, L. S., Silveira, R. P., Debonia, A. M., Benvenuttia, E. V., Costa, T. M. H., Guterres, S. S. & Pohlmanna, A. R. (2008), Nanocapsule@xerogel microparticles containing sodium diclofenac: a new strategy to control the release of drugs. *Int J Pharm*, *358,* 292–295.

[141] Du, H., Hamilton, P. D. , Reilly M. A., Avignon, A., Biswas, P. & Ravi, N. (2009), A facile synthesis of highly water-soluble, core–shell organo-silica nanoparticles with controllable size via sol–gel process. *J Colloid Interface Sci*, *340,* 202–208. [142] Saeed Arayne, M., Sultana, N. & Noor-us-Sabah, (2007). Fabrication of solid nanoparticles for drug delivery. *Pak J Pharm Sci*, *20,* 251-259.

[143] Finnie, K. S., Waller, J. D., Perret, F. L. , Krause-Heuer, A. M., Lin, H. Q., Hanna, J. V. & Barbe, C. V. (2009). Biodegradability of sol–gel silica microparticles for drug delivery. *J Sol-Gel Sci Technol*, *49,* 12–18.

[144] Bottcher, H., Slowik, P. & Sub, W. (1998). Sol-gel carrier for controlled drug delivery. *J Sol-Gel Sci Technol*, *13,* 277-281.

[145] Barbé, C., Bartlett, J., Kong, L., Finnie, K., Lin, H. Q., Larkin, M., Calleja, S., Bush, A. & Calleja, G. (2004). Silica particles: a novel drug-delivery system. *Adv Mater*, *16,* 1959-1966.

[146] Jin, W. & Brennan, J. D. (2002). Properties and applications of proteins encapsulated within sol–gel derived materials. *Anal Chim Acta*, *461,* 1–36.

[147] Aerts, A. C., Verraedt, E., Mellaerts, R., Depla, A., Augustijns, P., Humbeeck, J., Van den Mooter, G. & Martens, J. A. (2007). Tunability of pore diameter and particle size of amorphous microporous silica for diffusive controlled release of drug compounds. *J Phys Chem, 111 (36)*, 13404–13409.

[148] Popovici, M., Gich, M. & Savu, C. (2006). Ultra-light sol-gel derived magnetic nanostructured materials. *Rom Rep Phys*, *58*, 369–378.

[149] Gaetano, F. De, Ambrosio, L., Raucci, M. G., Marotta, A. & Catauro M. (2005). Sol-gel processing of drug delivery materials and release kinetics. *J Mater Sci: Mater Med*, *16*, 261– 265.

[150] Babich, J. W., Zubieta, J. & Bonavia, G. (2002). Matrices for drug delivery and methods for making and using the same. *US Patent*, 6395299.

[151] Gupta, R. & Kumar, A. (2008). Molecular imprinting in sol–gel matrix. *Biotechnol Adv*, *26*, 533–547.

[152] Farrington, K. & Regan, F. (2009). Molecularly imprinted sol gel for ibuprofen: An analytical study of the factors influencing selectivity. *Talanta*, *78(3)*, 653–659

[153] Radin, S., Antoci, V., Hickok, N., Adams, C. S., Parvizi, J., Shapiro, I. & Ducheyne, P. (2007). In Vitro and In Vivo Bactericidal Effect of Sol-Gel/Antibiotic Thin Films on Fixation Devices. *Key Eng Mater*, 330-332, 1323-1326.

[154] Volpe, R. A. (1997). *Sol gel barrier films*. US Patent 5618628.

[155] Katstra, W. E., Palazzolo, R. D., Rowe, C. W., Giritlioglu, B., Teung, P. & Cima, M. J. (2000). Oral dosage forms fabricated by Three Dimensional Printing™. *J Controlled Release, 66,* 1-9.

[156] Kim, J., Radin, S., Qu, H., Ducheyne, P., Knabe , C., Costache, M. & Devore D. (2008). Early stage treatment of compartment syndrome using polymer sol-gel composite growth factor delivery wound dressings. Dec. Available from: http://handle.dtic.mil/100.2/ADA505821.

[157] Jin, W. & Brennan, J. D. (2002). Properties and applications of proteins encapsulated within sol–gel derived materials. *Anal Chim Acta*, *461,* 1–36.

[158] Wu, Z., Joo, H., Lee, T. G. & Lee, K. (2005). Controlled release of lidocaine hydrochloride from the surfactant-doped hybrid xerogels. *J Controlled Release*, *104,* 497–505.

[159] Mei, L., Wang, H., Meng, S., Zhong, W., LI, Z., Cai, R., Chen, Z., Zhou, X. & Du, Q. (2007). Structure and release behavior of PMMA/Silica composite drug delivery system. *J Pharm Sci*, *96,* 1518-1526.

[160] Grassi, M. & Grassi, G. (2005). Mathematical Modelling and Controlled Drug Delivery: Matrix Systems. *Curr Drug Delivery*, 2, 97-116.

[161] Böttcher, H., Slowik, P. & Süß, W. (1998). Sol-Gel carrier systems for controlled drug delivery. *J Sol-Gel Sci Technol*, *13,* 277–281.

[162] FDA accepts Sol-Gel's Investigational New Drug Application for DER45-EV Gel. (2009). Available from: http://www.news-medical.net/news/20091111/FDA-accepts-Sol-Gele28099s-Investigational-New-Drug-Application-for-DER45-EV-Gel.aspx 62 Aurica P. Chiriac, Iordana Neamtu, Loredana E. Nita et al.

[163] Iyer, K. K., Prasad, A. K. & Gouma, P. I. (2006). A smart medical diagnostic tool using resistive sensor technology. *Mater. Res Soc Symp Proc*, *888,* 211-218.

[164] Pantelidis, D., Bravman, J. C., Rothbard, J. & Klein, R. L. (2007) *Bioactive material delivery systems comprising sol-gel compositions*. US Patent Application 20070071789.

[165] Moritz, N., Kangasniemi, I., Yli-urpo, A., Peltola, T. & Jokinen, M. (2004), Treatment of sol, gels and mixtures thereof. *US Patent Application*, 20040121451.

[166] Kole, H., Bouwkampwijnoltz, A. L., Legierse, P. E. J., Visser, C. G. & Boehmer, M. R. (2008). Medical equipment provided with a coating, *US Patent 7416787*.

[167] Jia, J., Wang, B., Wu, A., Cheng, G., Li, Z. & Dong, S. (2002). A Method to Construct a Third-Generation Horseradish Peroxidase Biosensor: Self-Assembling Gold Nanoparticles to Three-Dimensional Sol-Gel Network. *Anal Chem*, *74,* 2217-2223.

[168] Jia, N., Zhou, Q., Liu, L., Yan, M. & Jiang, Z. (2005). Direct electrochemistry and electrocatalysis of horseradish peroxidase immobilized in sol–gel-derived tin oxide/gelatin composite films. *J Electroanal Chem*, *580,* 213–221.

[169] Wang, Q., Lu, G. & Yang , B. (2004). Hydrogen peroxide biosensor based on direct electrochemistry of hemoglobin immobilized on carbon paste electrode by a silica sol–gel film. *Sens Actuators B*, *99,* 50–57.

[170] Dave, B. C., Dunn, B., Valentine, J. S. & Zink, J. I. (1994). Sol-Gel Encapsulation Methods for Biosensors. *Anal Chem*, *66 (22)*, 1120A-1127A.

[171] Gupta, R. & Chaudhury, N. K. (2007).Entrapment of biomolecules in sol–gel matrix for applications in biosensors: Problems and future prospects. *Biosens Bioelectron*, *22,* 2387–2399.

[172] Yang, M., Yang, Y., Liu, Y., Shen, G. & Yu, R. (2006). Platinum nanoparticles-doped sol–gel/carbon nanotubes composite electrochemical sensors and biosensors. *Biosens Bioelectron*, *21,* 1125–1131.

[173] Tang, H., Chen, J., Yao, S., Nie, L., Deng, G. & Kuang, Y. (2004), Amperometric glucose biosensor based on adsorption of glucose oxidase at platinum nanoparticlemodified carbon nanotube electrode. *Anal Biochem*, *331,* 89–97.

[174] Ricci, F., Amine, A., Palleschi, G. & Moscone, D. (2003). Biosens Bioelectron, *18,* 165–Prussian Blue based screen printed biosensors with improved characteristics of long-term lifetime and pH stability. *Biosens Bioelectron*, *8*, 165- 174.

[175] Wang, B., Zhang, J. & Dong, S. (2000). Silica sol–gel composite film as an encapsulation matrix for the construction of an amperometric tyrosinase-based biosensor. *Biosens Bioelectron*, *15,* 397–402.

[176] Liu, Z. J., Liu, B. H., Kong, J. L. & Deng, J. Q. (2000). Probing Trace Phenols Based on Mediator-Free Alumina Sol- Gel-Derived Tyrosinase Biosensor. *Anal Chem*, *72,* 4707-4712.

[177] Wang, J. (1999). Sol–gel materials for electrochemical biosensors. *Anal Chim Acta*, *399(1)*, 21-27.

[178] Braun, S., Rappoport, S., Zusman, R., Avnir, D. & Ottolenghi, M. (1990). Biochemically active sol-gel glasses: the trapping of enzymes. *Mater. Lett.* 10, 1-5.

[179] Avnir, D. (1995). Organic Chemistry within Ceramic Matrixes: Doped Sol-Gel Materials. *Acc Chem Res*, *28,* 328 - 334.

[180] Petit-Dominguez, M. D., Shen, H., Heineman, W. R. & Seliskar, C. J. (1997). Electrochemical Behavior of Graphite Electrodes Modified by Spin-Coating with Sol-Gel-Entrapped Ionomers. *Anal Chem*, *69,* 703-710.

[181] Liu, Z. J., Liu, B. H., Zhang, M., Kong, J. L. & Deng, J.Q. (1999). Al_2O_3 sol–gel derived amperometric biosensor for glucose. *Anal Chim Acta*, *392,* 135 - 141.

[182] Wang, B., Zhang, J. & Dong, S. (2000). Silica sol–gel composite film as an encapsulation matrix for the construction of an amperometric tyrosinase-based biosensor. *Biosens Bioelectron*, *15,* 397 - 402.

[183] Yang, Y. M., Wang, J. W. & Tan, R. X. (2004). Immobilization of glucose oxidase on chitosan–SiO_2 gel. *Enzyme Microb Technol*, *34,* 126–131.

[184] Topoglidis, E., Cass, A. E. G., Gilardi, G., Sadeghi, S., Beaumont, N. & Durrant, J. R. (1998). Protein Adsorbtion on Nanocrystalline TiO_2 Films: An Immobilization Strategy for Bioanalytical *Devices. Anal Chem*, *70,* 5111 - 5113.

[185] Topoglidis, E., Cass, A. E. G., O'Regan, B. & Durrant, J. R. (2001). Immobilisation and bioelectrochemistry of proteins on nanoporous TiO_2 and ZnO films. *J Electroanal Chem*, *517,* 20 - 27.

[186] Wang, J., Pamidi, P. V. & Jiang, M. (1998). Low-potential stable detection of ß-NADH at sol–gel derived carbon composite electrodes. *Anal Chim Acta*, *360,* 171 - 178.

[187] Shi, G., Sun, Z., Liu, M., Zhang, L., Liu, Y., Qu, Y. & Jin, L. (2007). Electrochemistry and Electrocatalytic Properties of Hemoglobin in Layer-by-Layer Films of SiO_2 with Vapor- Surface Sol- Gel Deposition. *Anal Chem*, *79,* 3581–3588.

[188] Narang, U., Prasad, P. N., Ramanathan, K., Kumar, N. D., Malhotra, B. D., Kamalasanan, M. N., Chandra, S. & Bright, F.V. (1994). Glucose Biosensor Based on a Sol-Gel-Derived Platform. *Anal Chem*, *66 (19)*, 3139–3144.

[189] Narang, U., Rahman, M. H., Wang, J. H., Prasad, P. N. & Bright, F. V. (1995). Removal of ribonucleases from solution using an inhibitor-based sol-gel-derived Biogel. *Anal Chem*, *67(13)*, 1935–1939.

[190] Lev, O., Tsionsky, L., Rabinovich, L., Glezer, V., Sampath, S., Pankratov, I. & Gun, J. (1995). Organically modified sol-gel sensors. *Anal Chem*, *67 (1)*, 22A–30A [191] Jordan, J. D., Dunbar, R. A. & Bright, F. V. (1995). Dynamics of Acrylodan-Labeled Bovine and Human Serum Albumin Entrapped in a Sol-Gel-Derived Biogel. *Anal Chem*, *67 (14)*, 2436–2443.

[192] Besanger, T. R., Chen, Y., Deisingh, A. K., Hodgson, R., Jin, W., Mayer, S., Brook, M. A. & Brennan, J. D. (2003). Ion Sensing and Inhibition Studies Using the Transmembrane Ion Channel Peptide Gramicidin A Entrapped in Sol- Gel-Derived Silica. *Anal Chem*, *75(10)*, 2382–2391.

[193] Gupta, R., Mozumdar, S. & Chaudhury, N. K. (2005). Effect of ethanol variation on the internal environment of sol–gel bulk and thin films with aging. *Biosens Bioelectron*, *21 (4)*, 549–556.

[194] Gupta, R., Mozumdar, S. & Chaudhury, N. K. (2005). Fluorescence spectroscopic studies to characterize the internal environment of tetraethyl-orthosilicate derived sol–gel bulk and thin films with aging. *Biosens Bioelectron*, *20 (7)*, 1358–1365.

[195] Bhaskar M. M., Anand, S., Nivedita, K. G. & Chaudhury, N. K. (2007). Fluorescence spectroscopic study of dip coated sol-gel thin film internal environment using fluorescent probes Hoechst33258 and Pyranine. *J Sol-Gel Sci Technol*, *41,* 147–155.

[196] Sarma, A. K., Vatsyayan, P., Goswami, P. & Minteer S. D. (2009).Recent advances in material science for developing enzyme electrodes. *Biosens Bioelectron*, *24,* 2313–2322. 64 Aurica P. Chiriac, Iordana Neamtu, Loredana E. Nita et al.

[197] Gorton, L. (2005). Comprehensive Analytical Chemistry Volume XLIV: Biosensors and Modern Biospecific *Analytical Techniques. Elsevier Science*, 285–327.

[198] Lim, J., Malati, P., Bonet, F. & Dunn, B. (2007). Nanostructured Sol–Gel Electrodes for Biofuel Cells. *J Electrochem Soc*, *154,* A140–A145.

[199] Wang, G., Xu, J. J., Chen, H. Y. & Lu, Z. H. (2003). Amperometric hydrogen peroxide biosensor with sol–gel/chitosan network-like film as immobilization matrix. *Biosens Bioelectron*, *18,* 335–343.

[200] Przybyt, M. & Bialkowska, B. (2002). Template-directed lattices of nanostructures; preparation and physical properties. In : Frontiers of Multifunctional Nanosystems. II. Mathematics, *Physics and Chemistry*, *57,* 91–108.

[201] Zou, Y., Xiang, C. & Sun, Li-Xian, Xu, F. (2008). Glucose biosensor based on electrodeposition of platinum nanoparticles onto carbon nanotubes and immobilizing enzyme with chitosan-SiO2 sol–gel. *Biosens Bioelectron*, *23,* 1010–1016

[202] Wang, S. G., Zhang, Q., Wang, R. & Yoon, S. F. (2003). A novel multi-walled carbon nanotubebased biosensor for glucose detection. *Biochem Biophys Res Commun*, *311,* 572–576. [203] Khan, R., Kaushik, A., Solanki, P. R., Ansari, A. A., Pandey, M. K. & Malhotra, B. D. (2008). Zinc oxide nanoparticles-chitosan composite film for cholesterol biosensor. *Anal Chim Acta*, *616,* 207 – 213.

[204] Vidal, J. C., Espuelas, J., Ruiz, E. G. & Castillo, J. R. (2004). Amperometric cholesterol biosensors based on the electropolymerization of pyrrole and the electrocatalytic effect of Prussian-Blue layers helped with self-assembled monolayers. *Talanta*, *64,* 655–664.

[205] Singh, S., Solanki, P. R., Pandey, M. K. & Malhotra, B. D. (2006). Cholesterol biosensor based on cholesterol esterase, cholesterol oxidase and peroxidase immobilized onto conducting polyaniline films. *Sens Actuators B*, *115,* 534–541.

[206] Li, J., Peng, T. & Peng, Y. (2003). A cholesterol biosensor based on entrapment of cholesterol oxidase in a silicic sol–gel matrix at a Prussian blue modified electrode. *Electroanalysis*, *15,* 1031–1037.

[207] Kumar, A., Pandey, R. R. & Brantley, B. (2006). Tetraethylorthosilicate film modified with protein to fabricate cholesterol biosensor. *Talanta*, *69,* 700–705.

[208] Solanki, P. R., Arya, S. K., Nishimura, Y., Iwamoto, M. & Malhotra, B. D. (2007). Cholesterol biosensor based on amino-undecanethiol self-assembled monolayer using surface plasmon resonance technique *Langmuir*, *23*, 7398–7403.

[209] Ram, M. K., Bertoncello, P., Ding, H., Paddeu, S. & Nicolini, C. (2001). Cholesterol biosensors prepared by layer-by-layer technique. *Biosens Bioelectron*, *16,* 849–856.

[210] Malhotra, B. D., Chaubey, A. & Singh, S. P. (2006). Prospects of conducting polymers in biosensors. *Anal Chim Acta*, *578,* 59–74.

[211] Singh, S., Singhal, R. & Malhotra, B. D. (2007). Immobilization of cholesterol esterase and cholesterol oxidase onto sol–gel films for application to cholesterol biosensor. *Anal Chim Acta*, *582,* 335–343.

[212] Arya, S. K., Solanki, P. R., Singh, R. P., Pandey, M. K., Datta, M. & Malhotra, B. D. (2006). Application of octadecanethiol self-assembled monolayer to cholesterol biosensor based on surface plasmon resonance technique. *Talanta*, *69,* 918–926.

[213] Narang, U., Prasad, P. N., Bright, F. V., Ramanathan, K., Kumar, N. D., Malhotra, B. D., Kamalasanan, M. N. & Chandra, S. (1994). Glucose biosensor based on a sol–gelderived platform. *Anal Chem*, *66,* 3139–3144.

[214] Solanki, P. R., Kaushik, A., Ansari, A. A., Tiwari, A. & Malhotra, B. D. (2009). Multiwalled carbon nanotubes/sol–gel-derived silica/chitosan nanobiocomposite for total cholesterol sensor. *Sens Actuators B*, *137,* 727–735.

[215] Miao, Y. & Tan, S. N. (2000). Amperometric hydrogen peroxide biosensor based on immobilization of peroxidase in chitosan matrix crosslinked with glutaraldehyde. *Analyst*, *125,* 1591–1594.

[216] Xu, C., Cai, H., He, P. & Fang, Y. (2001). Electrochemical detection of sequencespecific DNA using a DNA probe labeled with aminoferrocene and chitosan modified electrode immobilized with ss DNA, *Analyst*, *126,* 62–65.

[217] Tan, X., Li, M., Cai, Luo, P. L. & Zou, X. (2005). An amperometric cholesterol biosensor based on multiwalled carbon nanotubes and organically modified sol–gel/chitosan hybrid composite film. *Anal Biochem*, *337,* 111–120.

[218] Ortuno, J. A., Serna, C., Molina, A. & Gil, A. (2006). Differential pulse voltammetry and additive differential pulse voltammetry with solvent polymeric membrane ion sensors. *Anal Chem*, *78,* 8129–8813

[219] Kim, Y., Park, C. & Clark, D. (2001). Stable sol-gel microstructured and microfluidic networks for protein patterning. *Biotechnol Bioeng*, *73,* 331-337.

[220] Hong, J., Moosavi-Movahedi, A. A., Ghourchian, H., Rad, A. M. & Rezaei-Zarchi, S. (2007). Direct electron transfer of horseradish peroxidase on Nafion-cysteine modified gold electrode. *Electrochim Acta*, *52,* 6261–6267.

[221] Kafi, A. K. M., Lee, D. Y., Park, S. H. & Kwon, Y. S. (2007). Amperometric biosensor based on direct electrochemistry of hemoglobin in poly-allylamine (PAA) film. *Thin Solid Films*, *515,* 5179–5183.

[222] Yin, Y., Lü, Y., Wu, P. & Cai, C. (2005). Direct electrochemistry of redox proteins and enzymes promoted by carbon nanotubes. *Sensors*, *5*, 220–234.

[223] Yin, Y., Wu, P., Lü, Y., Du, P., Shi, Y. & Cai, C. (2007). Immobilization and direct electrochemistry of cytochrome *c* at a single-walled carbon nanotube-modified electrode. *J Solid State Electrochem*, *11,* 390–397.

[224] Xu, S., Peng, B. & Han, X. (2007). A third-generation H_2O_2 biosensor based on horseradish peroxidase-labeled Au nanoparticles self-assembled to hollow porous polymeric nanopheres. *Biosens Bioelectron*, *22,* 1807–1810.

[225] Zong, S., Cao, Y., Zhou, Y. & Ju, H. (2006). Zirconia nanoparticles enhanced grafted collagen tri-helix scaffold for unmediated biosensing of hydrogen peroxide. *Langmuir*, *22,* 8915–8919.

[226] Salimi, A., Sharifi, E., Noorbakhsh, A. & Soltanian, S. (2006). Direct voltammetry and electrocatalytic properties of hemoglobin immobilized on a glassy carbon electrode modified with nickel oxide nanoparticles. *Electrochem Commun*, *8*, 1499–1508.

[227] Wu, B. Y., Hou, S. H., Yin, F., Li, J., Zhao, Z. X., Huang, J. D. & Chen, Q. (2007). Amperometric glucose biosensor based on layer-by-layer assembly ofmultilayer films composed of chitosan, gold nanoparticles and glucose oxidase modified Pt electrode. *Biosens Bioelectron*, *22,* 838–844.

[228] Yin, F., Shin, H. K. & Kwon, Y. S. (2005). A hydrogen peroxide biosensor based on Langmuir–Blodgett technique: direct electron transfer of hemoglobin in octadecylamine *layer. Talanta*, *67,* 221–226. 66 Aurica P. Chiriac, Iordana Neamtu, Loredana E. Nita et al.

[229] Zuo, S., Teng, Y., Yuan, H. & Lan, M. (2008). Direct electrochemistry of glucose oxidase on screen-printed electrodes through one-step enzyme immobilization process with silica sol–gel/polyvinyl alcohol hybrid film. *Sens Actuators B*, *133,* 555–560.

[230] Tudorache, M. & Bala, C. (2007). Biosensors based on screen-printing technology, and their applications in environmental and food analysis. *Anal Bioanal Chem*, *388,* 565–578.

[231] Wilson, R. & Turner, A. P. F. (1992). Glucose oxidase: an ideal enzyme. *Biosens Bioelectron*, *7*, 165–185.

[232] Chen, T., Barton, S. C., Binyamin, G., Gao, Z., Zhang, Y., Kim, H. H. & Heller, A. (2001). A miniature biofuel cell. *J Sol-Gel SciTechnol,*, *123,* 8630–8631.

[233] Karlsson, R. (2004). SPR formolecular interaction analysis: a review of emerging application areas. *J Mol Recogn*, *17,* 151–161.

[234] Sun, Y., Bai, Y., Song, D. Q., Li, X., Wang, L. & Zhang, H. Q. (2007). Design and performances of immunoassay based on SPR biosensor with magnetic microbeads. *Biosens Bioelectron*, *23,* 473–478.

[235] Choi, H. N., Lyu, Y. K., Han, J. H. & Lee, W. Y. (2007). Amperometric ethanol biosensor based on carbon nanotubes dispersed in sol–gel-derived titania–nafion composite film. *Electroanalysis*, *19,* 1524–1530.

[236] Zheng, M. P., Jin, M. Y., Wang, H. H., Zu, P. F., Tao, P. & He, J. B. (2001). Effects of PVP on structure of TiO2 prepared by the sol–gel process. *Mater Sci Eng B*, *87,* 197–201.

[237] Sun, Y., Bi, N., Song, D., Bai, Y., Wang, L. & Zhang, H. (2008). Preparation of titania sol–gel matrix for the immunoassay by SPR biosensor with magnetic beads. *Sens Actuators B*, *134,* 566–572.

[238] Zhang, T., Tian, B., Kong, J., Yang, P. & Liu, B. (2003). A sensitive mediator-free tyrosinase biosensor based on an inorganic–organic hybrid titania sol–gel matrix. *Anal Chim Acta*, *489,* 199–206.

[239] Pickup, J. C., Hussain, F., Evans, N. D., Rolinski, O. J. & Birch, D. J. S. (2005). Fluorescence-based glucose sensors. Biosens *Bioelectron*, *20,* 2555–2565.

[240] Iijima, S. (1991). Helical microtubules of graphitic carbon. *Nature*, *354,* 56–58.

[241] Gooding, J. J., Wibowo, R., Liu, J., Yang, W., Losic, D., Orbons, S., Mearns, F. J., Shapter, J. G. & Hibbert, D. B. (2003). Protein electrochemistry using aligned carbon nanotube arrays. *J Sol-Gel SciTechnol,*, *125,* 9006–9007.

[242] Wang, J., Hocevar, S. B. & Ogorevc, B. (2004). Carbon nanotube-modified glassy carbon electrode for adsorptive stripping voltammetric detection of ultratrace levels of 2,4,6-trinitrotoluene. *Electrochem Commun*, *6*, 176–179.

[243] Gopalan, A. I., Lee, K. P., Ragupathy, D., Lee, S. H. & Lee, J. W. (2009). An electrochemical glucose biosensor exploiting a polyaniline grafted multiwalled carbon nanotube/perfluorosulfonate ionomer–silica nanocomposite. *Biomaterials*, *30,* 5999–6005.

[244] Zeng, J., Gao, X., Wei, W., Zhai, X., Yin, J., Wu, L., Liu, X., Liu, K. & Gong, S. (2007). Fabrication of carbon nanotubes/poly(1,2-diaminobenzene) nanoporous composite via multipulse chronoamperometric electropolymerization process and its electrocatalytic property toward oxidation of NADH. Sens *Actuators B*, *120,* 595–602.

[245] Zou, Y., Sun, L. & Xu, F. (2007). Prussian blue electrodeposited on MWNTs–PANI hybrid composites for H_2O_2 detection. *Talanta*, *72,* 437–442.

[246] Luque, G. L., Ferreyra, N. F. & Rivas, G. A. (2007). Electrochemical sensor for amino acids and albumin based on composites containing carbon nanotubes and copper microparticles. *Talanta*, *71,* 1282–1287.

[247] Balasubramanian, K. & Burghard, M. (2008). Electrochemically functionalized carbon nanotubes for device applications. *J Mater Chem*, *18,* 3071–3083.

[248] Manesh, K. M, Santhosh, P., Komathi, S., Kim, N. H., Park, J. W., Gopalan, A. I. & Lee, K. P. (2008). Electrochemical detection of celecoxib at a polyaniline grafted multiwall carbon nanotubes modified electrode. *Anal Chim Acta*, *626,* 1–9.

[249] Manesh, K. M., Kim, H. T., Santhosh, P., Gopalan, A. I. & Lee, K. P. (2008). A novel glucose biosensor based on immobilization of glucose oxidase into multiwall carbon nanotubes–polyelectrolyte-loaded electrospun nanofibrous membrane. *Biosens Bioelectron*, *23,* 771–779.

[250] Jena, B. K. & Raj, C. R. (2006). Electrochemical biosensor based on integrated assembly of dehydrogenase enzymes and gold nanoparticles. *Anal Chem*, *78,* 6332–6339.

[251] Chen, H. & Dong, S. (2007). Direct electrochemistry and electrocatalysis of horseradish peroxidase immobilized in sol–gel-derived ceramic–carbon nanotube nanocomposite film. *Biosens Bioelectron*, *22,* 1811–1815.

[252] Zhao, G. C., Yin, Z. Z., Zhang, L. & Wei, X. W. (2005). Direct electrochemistry of cytochrome c on a multi-walled carbon nanotubes modified electrode and its electrocatalytic activity for the reduction of H_2O_2. *Electrochem Commun*, 7, 256–260.

[253] Walcarius, A., Mandler, D., Cox, J. A., Collinson, M. M. & Lev, O. (2005). Exciting new directions in the intersection of functionalized sol–gel materials with electrochemistry. *J Mater Chem*, *15,* 3663–3689.

[254] Lev, O., Wu, Z., Bharathi, S., Glezer, V., Modestov, A., Gun, J., Rabinovich, L. & Sampath, S. (1997). Sol–gel materials in electrochemistry. *Chem Mater*, 9, 2354– 2375.

[255] Gong, K., Zhang, M., Yan, Y., Su, L., Mao, L., Xiong, S. & Chen, Y. (2004). Sol–gelderived ceramic–carbon nanotube nanocomposite electrodes: tunable electrode dimension and potential electrochemical applications. *Anal Chem*, *76,* 6500–6505.

[256] Ragupathy, D., Gopalan, A.I., Lee, K. P. & Manesh, K. M. (2008). Electro-assisted fabrication of layer-by-layer assembled poly(2,5-dimethoxyaniline)/phosphotungstic acid modified electrode and electrocatalytic oxidation of ascorbic acid. *Electrochem Commun*, *10,* 527–530.

[257] Qian, L., Gao, Q., Song, Y., Li, Z. & Yang, X. (2005). Layer-by-layer assembled multilayer films of redox polymers for electrocatalytic oxidation of ascorbic acid. *Sens Actuators B*, *107,* 303–310.

[258] Qu, L. Y., Lu, R. Q., Peng, J., Chen, Y. G. & Dai, Z. M. (1997). H_3P $W11MoO_{40}x \cdot 2H_2O$ protonated polyaniline-synthesis, characterization and catalytic conversion of isopropanol. *Synth Met*, *84,* 135–136.

[259] Nassef, H. M., Civit, L., Fragoso, A. & O'Sullivan, C.K. (2008). Amperometric sensing of ascorbic acid using a disposable screen-printed electrode modified with electrografted o-aminophenol film. *Analyst*, *133,* 1736–1741.

[260] Ragupathy, D., Gopalan, A. I. & Lee, K. P. (2010) Electrocatalytic oxidation and determination of ascorbic acid in the presence of dopamine at multiwalled carbon 68 Aurica P. Chiriac, Iordana Neamtu, Loredana E. Nita et al. nanotube–silica network–gold nanoparticles based nanohybrid modified electrode. *Sens Actuators B*, *143*, 696–703.

[261] Sarma, A., Vatsyayan, K. P., Goswami, P. & Minteer, S. D. (2009). Recent advances in material science for developing enzyme electrodes. *Biosens Bioelectron*, *24,* 2313–2322.

[262] Wang, G., Xu, J. J., Chen, H. Y. & Lu, Z. H. (2003). Amperometric hydrogen peroxide biosensor with sol–gel/chitosan network-like film as immobilization matrix. *Biosens Bioelectron*, *18,* 335–343.

In: The Sol-Gel Process
Editor: Rachel E. Morris
ISBN 978-1-61761-321-0

Chapter 3

DYE-REMOVAL CHARACTERISTICS OF PURE AND MODIFIED NANOCRYSTALLINE SOL-GEL TITANIA

Satyajit Shukla*
Ceramic Technology Department, Materials and Minerals Division (MMD), National Institute for Interdisciplinary Science and Technology (NIIST), Council of Scientific and Industrial Research (CSIR), Thiruvananthapuram, Kerala, India

ABSTRACT

In this chapter, significant results obtained (past 5 years) in our laboratory related to the development of sol-gel processed nanocrystalline anatase-titania and its use as a photocatalyst for the industrial dye-removal application has been reviewed in detail. The pure nanocrystalline sol-gel anatase-titania has been processed using an alkoxide-precursor and utilized for the removal of methylene blue, a model catalytic dye-agent, from an aqueous solution via photocatalytic degradation mechanism under the ultraviolet- and solar-radiation exposure. The effect of various processing (conventional and modified sol-gel) and material parameters, such as '*R*' (the ratio of molar concentration of water to that of the alkoxide precursor), calcination temperature, polymer concentration, specific surface-area, average nanocrystallite size, surface-purity, and volume fraction of the rutile phase on the photocatalytic activity of the nanocrystalline sol-gel anatase-titania has been taken into account. The role of surface-functionalization of the nanocrystalline sol-gel anatase-titania in enhancing its photocatalytic activity has been discussed by considering the deposition of silver via ultraviolet- and chemical-reduction techniques. The use of sol-gel titania photocatalyst as a surface-sensitizer for depositing the conducting-metals, such as copper and silver, on the non-conducting flyash particle surface has been demonstrated. In order to ease the separation of nanocrystalline sol-gel anatase-titania, from the treated effluent by applying an external magnetic field, the processing and the use of a conventional magnetic photocatalyst have

* Corresponding author: Phone: +91-471-2515282; Fax: +91-471-2491712; E-Mail: satyajit_shukla@niist.res.in, Phone: (630) 252-4658; Fax: (630) 252-3604; E-Mail: sshukla@anl.gov,
Present Address: Indo-US Science and Technology Forum (IUSSTF) Research Fellow, Ceramics Section, Energy Systems Division, Argonne National Laboratory, Argonne, Illinois 60439-4838

been systematically described, in which a ceramic magnet has been introduced in between the nanocrystalline particles of the sol-gel anatase-titania. It has been shown that, the major limitations associated with the conventional magnetic photocatalyst can be overcome using a newly developed magnetic dye-adsorbent catalyst. It has been revealed that, the new magnetic dye-adsorbent catalyst removes an organic dye from an aqueous solution much quickly via surface-adsorption mechanism under the dark-condition, which is in contrast to the photocatalytic degradation mechanism, under the ultraviolet-radiation exposure, associated with the conventional magnetic photocatalyst. It appears that, the morphology of the nanocrystalline sol-gel titania plays a significant role in governing the predominant dye-removal mechanism, and hence, the overall efficiency and the cost of the dye-removal process.

1. Introduction

Organic synthetic dyes are extensively used in various industrial sectors such as textile, leather tanning, paper production, food technology, agricultural research, light-harvesting arrays, photo-electrochemical cells, and hair-coloring. Due to the large-scale production, extensive use, and subsequent discharge of colored waste-waters containing these toxic and non-biodegradable pollutants, the organic synthetic dyes cause considerable environmental pollution and health-risk factors. Moreover, this also affects the sunlight penetration and the oxygen solubility in the water-bodies, which in turn affect the under-water photosynthetic activity and life-sustainability. In addition to this, due to their strong color even at lower concentrations, the organic synthetic dyes generate serious aesthetic issues in the waste-water disposal [1-5].

For this reason, powerful oxidation/reduction methods are needed to be applied to ensure the complete decolorization and degradation of the organic synthetic dyes and their metabolites present in the waste-water effluents. The technologies such as adsorption on inorganic or organic matrices and microbiological or enzymatic decomposition have been developed for the removal of organic synthetic dyes from the waste-water to decrease their impact on the environment [5-7]. However, the treatment of waste-water containing organic synthetic dyes and their decolorization using these techniques is very costly and has lower efficiency in the color removal and mineralization. Therefore, alternative effective methods of waste-water treatment capable of removing the color and degrading toxic organic compounds from the industrial effluents are needed to be developed.

Photocatalysis has been an area of rapidly growing interest over the last two decades for the removal organic synthetic dyes from the industrial effluents. The use of semiconductor particles as photocatalyst for the initiation of the redox chemical reactions continues to be an active area of research [8-11]. When the semiconductor oxide particle is illuminated with the radiation having the energy comparable to its band-gap energy, it generates highly active oxidizing/reducing sites, which can potentially oxidize/reduce large number of organic-wastes. Photocatalytic decolorization of dyes has been proposed as an efficient method for the removal of color from the industrial effluents. Metal-oxide and metal-sulfide semiconductors, such as titania (TiO_2) [8-11], zinc oxide (ZnO) [12], tin oxide (SnO_2) [13], zinc sulfide (ZnS) [14], cadmium sulfide (CdS) [15] have been successfully applied as photocatalyst for the removal of highly toxic and non-biodegradable pollutants commonly present in air and wastewater. Among them, TiO_2 is believed to be the most promising one since it is cheaper, environmentally friendly, non-toxic, highly photocatalytically active, and stable to chemical/photo-corrosion. However, its effective application as a photocatalyst is hindered due to some of its serious limitations. First, TiO_2 nanocrystallites tend to aggregate (or agglomerate) into large-sized nanoparticles, thus affecting its performance as a photocatalyst due to a decreased specific surface-area. Secondly, it has lower absorption in the visible-region, which makes its less effective in using the readily available solar-energy. Third, the separation and recovery of the photocatalyst is difficult and time consuming. Nevertheless, extensive research has been conducted to process the doped and surface-modified nanocrystalline TiO_2 in different forms such as powders, thin/thick films, and one-dimensional nanostructures using variety of techniques including hydrothermal [16], precipitation [17], sol-gel [18], microemulsion [19], chemical vapor deposition [20], and plasma spraying [21] to overcome its major limitations.

In the present chapter, we review our own significant efforts directed towards the dye-removal using the nanocrystalline pure and modified TiO_2 processed via sol-gel [22-30]. For this purpose, the methylene blue (MB) dye has been considered as a model catalytic dye-agent, which is a brightly colored blue cationic thiazine dye. It has been used as an antidote for cyanide poisoning in humans, antiseptic in veterinary medicine, and most commonly in vitro diagnostic in biology, cytology, hematology, and histology. It has various harmful effects on human being; for example, on inhalation, it may cause breathing problem; while, ingestion produces burning sensation, nausea, vomiting, diarrhea and gasestrics, painful micturation, and methemoglabinemia-like syndromes [31].

Typically, in the section-2, 3 and 4 of this chapter, various experimental details and the characterization results obtained using the different microscopic and spectroscopic techniques are summarized. In the section-5, the key results of the photocatalytic activity measurements are presented, which are correlated with the characterization results. Lastly, in the section-6, some of the new emerging trends in the application of nanocrystalline sol-gel TiO_2 photocatalyst are described.

2. Nanocrystalline Titania via Sol-Gel

2.1. Conventional Method

The nanocrystalline TiO_2 was synthesized via conventional sol-gel using the hydrolysis and condensation of titanium(IV) iso-propoxide ($Ti(OC_3H_7)_4$) in an anhydrous alcohol medium [30]. A measured quantity of water was first dissolved in an anhydrous 2-propanol. A second solution was prepared in which $Ti(OC_3H_7)_4$ was dissolved completely in an anhydrous 2-propanol. Both the solutions were sealed immediately and stirred rapidly using the magnetic stirrer to obtain the homogeneous solutions. The water part of the solution was then added drop-wise to the alkoxide-part under the continuous stirring. As a result of hydrolysis and condensation reactions due to the reaction of $Ti(OC_3H_7)_4$ with water, change in the color of the solution from colorless to white was visible. The time required for the observable color change was, however, different depending on the processing conditions. After the complete addition of water part of the solution to that of the alkoxide part, the resulting solution was stirred overnight before drying in the furnace at 80 °C for the complete removal of solvent and residual water. The dried powders were then calcined at higher temperature within the range of 400-800 °C for 2 h for the crystallization of amorphous TiO_2 powders. Different powders were prepared by varying the '*R*' (the ratio of molar concentration of water to that of the alkoxide precursor) values within the range of 5-150 [29,30].

The hydrolysis and condensation reactions, which are responsible for the formation of TiO_2 particles, can be summarized as [30],

$$\textit{Hydrolysis: } Ti(OC_3H_7)_4 + 4H_2O \rightarrow Ti(OH)_4 + 4C_3H_7OH \tag{1}$$

$$\textit{Condensation: } Ti(OH)_4 \rightarrow TiO_2 + 2H_2O \tag{2}$$

$$\textit{Net Reaction: } Ti(OC_3H_7)_4 + 2H_2O \rightarrow TiO_2 + 4C_3H_7OH \tag{3}$$

The variation in the morphology of the nanocrystalline sol-gel TiO_2, processed with different '*R*' values within the range of 5-90, is shown in Figure 1. At the lowest '*R*' value, Figure 1(a), the TiO_2 powder is in the form of large aggregates with an average size of ~5 μm having near-spherical shape. With increasing '*R*', Figure 1(b)-(e), progressive decrease in the average aggregate size is noted. The smallest aggregate size (or the nanoparticle size) as small as ~100-200 nm is noted for the '*R*' value of 90. Hence, the aggregation tendency is noted to decrease with increasing '*R*'. The variation in the average nanocrystallite size (determined using the X-ray diffraction (XRD) analysis) and the average nanoparticle size (determined using the Brunauer-Emmett-Teller (BET) method) as a function of '*R*' is presented in Figure 2 for two different calcination temperatures. After the calcination at lower temperature (400 °C), although marginally, the average nanocrystallite size and the average nanoparticle size increase with '*R*'. However, this trend is observed to reverse after the calcination at higher temperature (600 °C), where the average sizes decrease with increasing '*R*'. This has been primarily attributed to the reaction mechanism, Eqs. 1-3, which form the TiO_2 particles with different degree of aggregation. With increasing '*R*', although the hydrolysis reaction, Eq. 1, is driven in the forward direction, the condensation reaction, Eq. 2, is driven in the reverse direction, which suggests more dissolution of TiO_2 particles, which effectively reduces the average size of the nucleated TiO_2 particles within the sol. Hence, from the net reaction presented in Eq. 3, it appears that, the nucleation rate of TiO_2 particles increases and the average nucleus size decreases with increasing '*R*'. However, due to the formation of large number of small-size nuclei within the sol, the growth rate would tend to increase for larger '*R*' values. Hence, after calcination at lower temperature (400 °C), the average nanocrystallite size and average nanoparticle size are observed to increase with '*R*', which possibly reflects the condition which is likely to be present just after the drying step. Interestingly, the nature of the variation in the average nanocrystallite size and the average nanoparticle size as a function '*R*' is reversed after the calcination at 600 °C. It is noted that, at lower '*R*' values, the average aggregate size is much larger due to lower reaction kinetics, which effectively increases the co-ordination number (defined here as the number of the nearest neighbors surrounding a nanoparticle within an aggregate) of the nanoparticles within the aggregates.

This is highly conducive in enhancing the diffusion kinetics in the large-sized aggregates. As a result, the growth rate is possibly much higher during the calcination in these aggregates, which leads to larger average nanocrystallite size and average nanoparticle size for lower '*R*' values. Thus, the nucleation/growth rates and the degree of aggregation strongly govern the nature of the variation in the average nanocrystallite size and average nanoparticle size as a function of '*R*' at lower and higher calcination temperatures. These variations have a drastic effect when the photocatalytic activity of nanocrystalline sol-gel TiO_2 is investigated as a function of average nanocrystallite size, which is discussed in detail in the section-5.1 and 5.2.2.

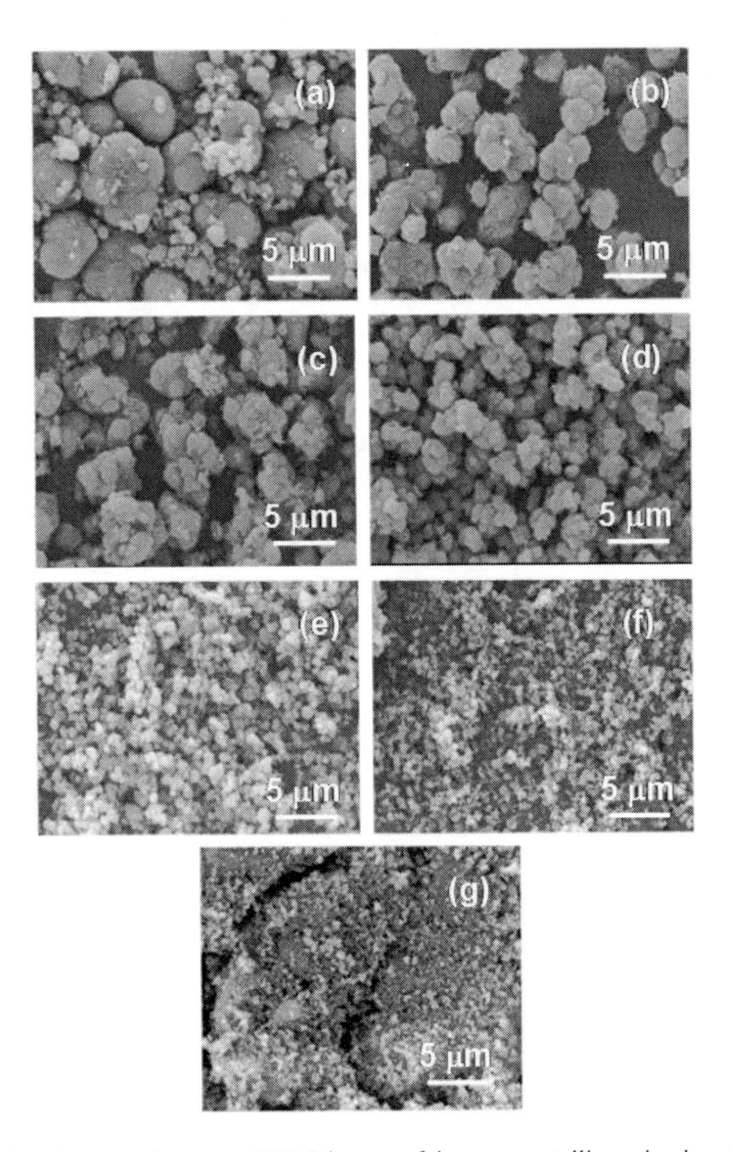

Figure 1. Scanning electron microscope (SEM) images of the nanocrystalline sol-gel anatase-TiO_2 processed with different '*R*' values. (a) 5, (b) 15, (c) 30, (d) 60, (e)-(g) 90. Powders (a)-(e) and (f)-(g) are processed via conventional and modified sol-gel methods. The HPC concentration for the powders presented in (f) and (g) are 1.0 and 2.0 $g{\bullet}L^{-1}$. All powders are calcined at 400 °C for 2 h [29,30]. Copyright 2007 American Chemical Society ((a)-(e)); Copyright 2008 Springer ((f) and (g)).

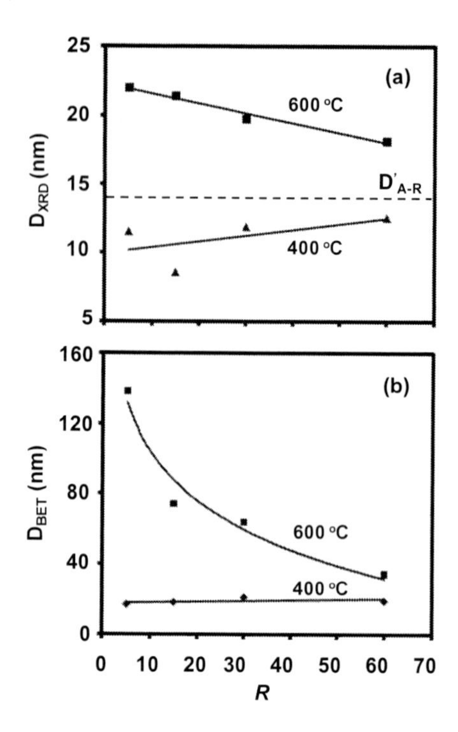

Figure 2. Variation in the average nanocrystallite size (D_{XRD}) (a) and average nanoparticles size (D_{BET}) (b) as a function of '*R*' obtained for the nanocrystalline sol-gel anatase-TiO_2 calcined at two different temperatures [30]. D'_{A-R} represents the theoretical critical size for the anatase-to-rutile phase transformation, which is comparable with the critical size for observing the maximum photocatalytic activity (D^*), as discussed later in the section-5.2.2. Copyright 2007 American Chemical Society.

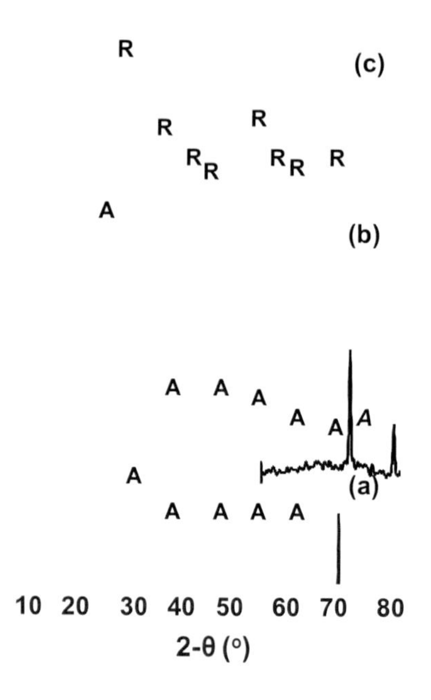

Figure 3. Broad-scan XRD patterns obtained using the nanocrystalline sol-gel TiO_2, processed with the '*R*' value of 5. The powders are calcined at different temperatures: (a) 400 °C, (b) 600 °C, and (c) 800 °C. A-anatase and R-rutile [30]. Copyright 2007 American Chemical Society.

Typical XRD patterns obtained using the nanocrystalline sol-gel TiO_2 have been presented in Figure 3 at different calcination temperatures. The sol-gel TiO_2 in the as-synthesized condition is amorphous; however, it gets crystallized into the anatase-TiO_2 after the calcination at 400 °C, Figure 3(a), as identified by the comparison of the XRD pattern with the JCPDS card # 21-1272. With increasing calcination temperature, the peak intensity increases with the reduction in the full-width at the half-maximum (FWHM) intensity of the main peak, Figure 3(b), which suggests an improved crystallinity of the powder and increase in an average nanocrystallite size. At higher calcination temperature of 800 °C, anatase-to-rutile phase transformation is observed, which results in the formation of rutile-TiO_2 (JCPDS card # 21-1276). Thus, during the calcination treatment, the metastable anatase-TiO_2 is stabilized first at room temperature which gets transformed to more stable rutile-TiO_2 after the calcination at higher temperature. It is known that, for the bulk-TiO_2 at room temperature, the rutile-phase is more stable than the anatase-phase. The stable rutile-TiO_2, however, gets transformed to the metastable anatase-TiO_2 when the nanocrystallite is reduced below a critical size at room temperature. Based on the thermodynamic considerations, the critical size ($D'_{A\rightarrow R}$) for the stabilization of the metastable anatase-phase in a single isolated TiO_2 nanocrystallite is given by the relationship of the form [32],

$$D'_{A\rightarrow R} = \frac{(2t+3)M}{\left(\Delta G_f^o\right)_{A\rightarrow R}} \times \left(\frac{\gamma_R}{\rho_R} - \frac{\gamma_A}{\rho_A}\right) \quad (4)$$

where, 'M' is the molecular weight of TiO_2 (80 g/mol), '$(\Delta G^0_f)_{A\rightarrow R}$' the change in the volume free-energy associated with the anatase-to-rutile phase transformation for the bulk-TiO_2 at room temperature (~6 kJ/mol), 't' the proportionality constant between the surface stress and the surface free-energy for the bulk-TiO_2 (~3.5), 'γ_R' and 'γ_A' the surface free-energies of the rutile-TiO_2 (1.91 J/m^2) and anatase-TiO_2 (1.32 J/m^2) respectively, and 'ρ_R' and 'ρ_A' the densities of the rutile-TiO_2 (4.26 g/cm^3) and anatase-TiO_2 (3.84 g/cm^3) respectively. Substituting these values in Eq. 4, the critical size for the anatase-to-rutile phase transformation, at room temperature, is calculated to be ~14 nm, which is marked as a dotted-line in Figure 2(a). It appears that, the metastable anatase-TiO_2 is stabilized at room temperature after calcining the amorphous-TiO_2 in between the temperature range of 400-600 °C. The transmission electron microscope (TEM) and high-resolution TEM (HRETM) images of anatase-TiO_2 and rutile-TiO_2 are shown in Figure 4(a)-(c) and Figure 4(d) respectively. The corresponding selected-area electron diffraction (SAED) patterns are shown as insets in Figure 4(c) and (d). After the calcination at 400 °C, Figure 4(a), the nanocrystallite size is observed to be within the range of ~8-12 nm, which is below the critical size of ~14 nm. Hence, in this case, the stabilization of the metastable anatase-phase in the nanocrystalline sol-gel TiO_2 is in accordance with the thermodynamic considerations. However, after the calcination at 600 °C, the nanocrystallite size increases to ~18-22 nm, Figure 4(c), which is above the critical size of ~14 nm. The stabilization of metastable anatase-TiO_2, at room temperature, within the TiO_2 nanocrystallites of size greater than 14 nm has also been reported earlier by others [33,34]. It is to be noted that, the critical size of ~14 nm is calculated only for a single isolated TiO_2 nanocrystallite. In practice, the nanocrystallites form aggregates, which modify their interfacial energies than those considered in Eq. 4. As reported for other system such as zirconia (ZrO_2), the change in the interfacial energies due to aggregation may increase the critical size for the metastable phase stabilization at room temperature [35]. Hence, the room temperature stabilization of the metastable anatase-phase within the TiO_2 nanocrystallites of size greater than 14 nm is attributed to the possible modification in the interfacial energies due to the aggregation of TiO_2 nanocrystallites. In Figure 4(d), the nanocrystallite size associated with rutile-TiO_2, as obtained after the calcination at 800° C, is observed to be as large as ~100-200 nm. The anatase-to-rutile phase transformation, observed here at higher calcination temperature, is mainly attributed to a significant amount of growth in the nanocrystallite size above the critical size. The stabilization of metastable anatase-phase in the nanocrystalline sol-gel TiO_2 and its transformation to more stable rutile-TiO_2 at higher calcination temperature have significant influence on the dependence of the photocatalytic activity as a function of calcination temperature, which is discussed in detail in the section-5.2.3.

The nitrogen (N_2) adsorption/desorption isotherms and the differential pore volume curve (Barret-Joyner-Halenda (BJH) plot) as obtained for the nanocrystalline sol-gel anatase-TiO_2 are presented in Figure 5(a) and 5(b). The isotherms are of type IV and exhibit typical hysteresis behavior of type H3, which suggest that the nanocrystalline sol-gel anatase-TiO_2 is mesoporous having the pore size within the range of ~2-20 nm. The presence of mesoporosity is of a major significance in enhancing the photocatalytic activity of nanocrystalline sol-gel anatase-TiO_2. Moreover, an excess oxygen-ion vacancy concentration is known to be created at room temperature within the nanocrystallites of ceramic oxides below a critical size [35,36]. Higher concentration of oxygen-ion vacancies (V_o), produced as a result of the size-effect, is in turn responsible for the formation of excess surface-concentration of the super-oxide ions (O^{2-}, O^-) via oxygen-spillover reactions, in which the conduction band electrons are picked-up by oxygen from the surrounding atmosphere [37].

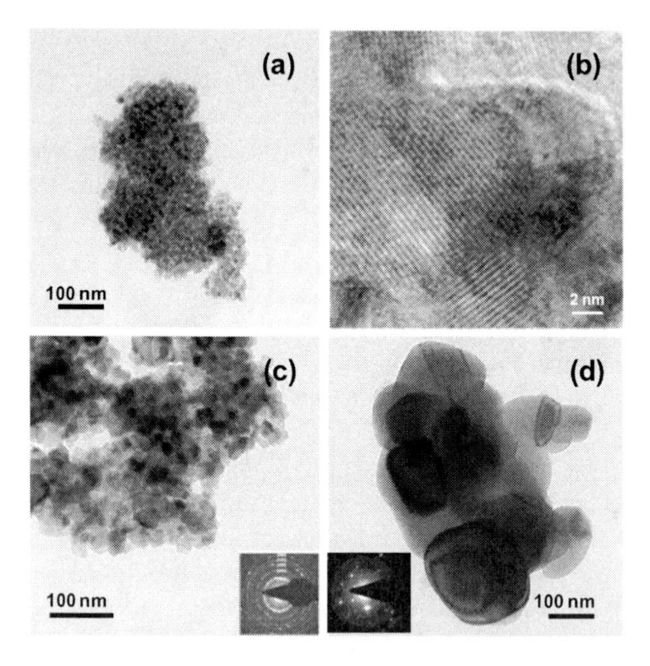

Figure 4. TEM/HRTEM images of the nanocrystalline sol-gel anatase-TiO_2 (a)-(c) and rutile-TiO_2 (d). The powders are processed with the '*R*' value of 90 and calcined at 400 °C (a)-(b), 600 °C (c), and 800 °C (d) for 2 h [24,28]. Copyright 2008 (a,b) and 2009 (c) American Chemical Society.

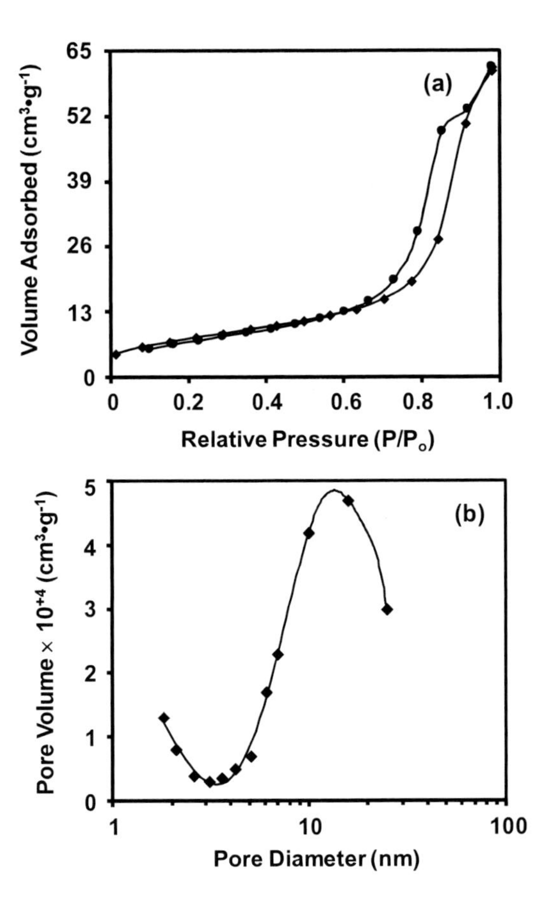

Figure 5. N_2 adsorption/desorption isotherms (a) and BJH pore-size distribution curve (b) as obtained for the nanocrystalline sol-gel anatase-TiO_2 processed with the '*R*' value of 90 and calcined at 600 °C for 2 h [24]. Copyright 2009 American Chemical Society.

$$O_{2(gas)} + V_o + e^- \xrightarrow{Ag,TiO_2} O^-_{2(ads)} \tag{5}$$

$$\frac{1}{2}O_{2(gas)} + V_o + e^- \xrightarrow{Ag,TiO_2} O^-_{(ads)} \tag{6}$$

The double-layer consisting the negatively charged superoxide-ions on the particle-surface, with the positive metal-ions just below it, is often termed as the "space-charge-layer". The variation in the space-charge-layer thickness significantly affects the amount of the dye-adsorption under the dark-condition, which in turn affects the photocatalytic activity of the nanocrystalline sol-gel anatase-TiO_2 as discussed later in detail in the section-5.2.6.

2.2. Modified Method

Nanocrystalline TiO_2 powders were also synthesized via modified sol-gel in which the conventional process was modified using the hydroxypropyl cellulose (HPC) polymer

[38,39]. In the modified sol-gel, measured quantities of water and HPC were first dissolved in anhydrous 2-propanol. A second solution was prepared in which $Ti(OC_3H_7)_4$ was dissolved completely in anhydrous 2-propanol. Both the solutions were sealed immediately and stirred rapidly using the magnetic stirrer to obtain the homogeneous solutions. The water part of solution was then added drop-wise to the alkoxide-part under the continuous stirring. All other steps were similar to those described earlier for the conventional method. Different powders were prepared by varying the HPC concentration within the range of 0-2 $g \cdot L^{-1}$.

The chemical structure of HPC is shown in Figure 6. It consists of -OH and ether groups in its structure, and hence, gets readily adsorbed on the surface of oxide particles via hydrogen bonding. The HPC adsorption has been confirmed earlier for the nanocrystalline sol-gel ZrO_2 via X-ray photoelectron spectroscope (XPS) analysis [39]. As indicated by arrows in Figure 7(a), the adsorption of HPC on the surface of TiO_2 nanoparticles has been confirmed via fourier transform infrared (FTIR) analysis, where typical absorption peaks corresponding to the different functional groups preset within the HPC structure have been detected on the surface of the nanocrystalline anatase-TiO_2 synthesized via modified sol-gel. In addition to this, the thermal gravimetric analysis (TGA) has shown that [29], the total weight-loss of dried TiO_2 powder increases linearly with increasing HPC concentration, which further supports the surface-adsorption of HPC. This was in good agreement with the report [39], where [C]:[Zr] atomic ratio, as determined via XPS analysis, was found to increase linearly with the HPC concentration. This strongly suggests that, modified sol-gel produces TiO_2 nanoparticles with the surface-adsorbed HPC.

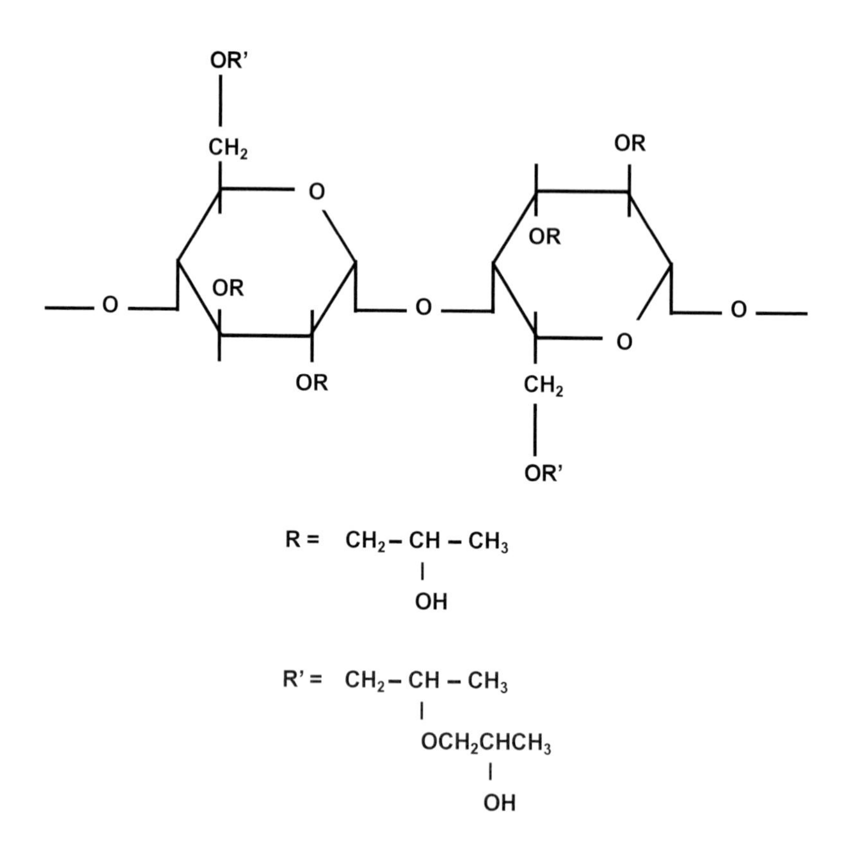

Figure 6. The chemical structure of HPC, an organic polymer [29]. Copyright 2008 Springer.

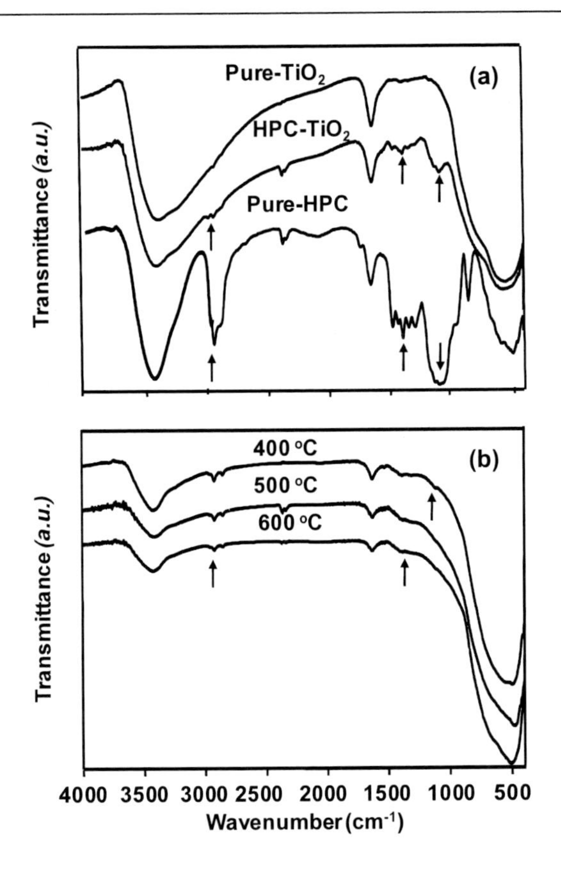

Figure 7. (a) FTIR spectra obtained for dried TiO_2 powder synthesized via conventional sol-gel with the '*R*' value of 90, pure-HPC, and dried TiO_2 powder synthesized via modified sol-gel with the '*R*' value and HPC concentration of 90 and 2.0 $g{\bullet}L^{-1}$. (b) Typical FTIR spectra obtained for the nanocrystalline anatase-TiO_2 synthesized via modified sol-gel with the '*R*' value and HPC concentration of 90 and 2.0 $g{\bullet}L^{-1}$. The powders are calcined at different temperatures [29]. Copyright 2008 Springer.

The adsorption of HPC on the surface of TiO_2 nanoparticles creates a steric hindrance to the particle-particle aggregation, which limits the particle growth within the sol and during the drying. Comparison of Figure 1(f) and 1(g) with Figure 1(e) clearly shows that the average nanoparticle size decreases with the addition of HPC during the sol-gel process. The average nanocrystallite size, as determined using XRD, also decreases with increasing HPC concentration, Table 1. HPC, being an organic polymer, gets decomposed when it is heated at higher temperature leaving behind the pores within the nanocrystalline ceramic powder. The BJH pore size distribution data suggest that, the removal of HPC from the surface of nanocrystalline sol-gel TiO_2 results in the formation of intra-aggregate pores of size ~10-30 nm, which are larger than the pore size of ~1-10 nm as obtained for the nanocrystalline TiO_2 processed via conventional sol-gel. As a result, the pore size distribution is unimodial and bimodal for the nanocrystalline TiO_2 processed without and with the addition of HPC [29]. Hence, the specific surface-area of the nanocrystalline sol-gel TiO_2 increases with the HPC concentration, Table 1.

Table 1. Average nanocrystallite size and specific surface-area as calculated for the nanocrystalline sol-gel anatase-TiO_2 processed under different conditions [29]. Copyright 2008 Springer.

R	Calcination Temperature (°C)	D_{XRD} (nm)			Specific Surface-Area ($m^2 \cdot g^{-1}$)		
		HPC Concentration ($g \cdot L^{-1}$)			HPC Concentration ($g \cdot L^{-1}$)		
		0.0	1.0	2.0	0.0	1.0	2.0
90	400	9.9	8.8	8.5	84	90	112
	600	21.2	21.5	18.9	30	31	37

Typical FTIR spectra, obtained for the nanocrystalline sol-gel TiO_2 processed using the HPC and calcined at different temperatures, are presented in Figure 7(b). Comparison of Figure 7(a) and 7(b) shows that, the calcination of dried TiO_2 powders at 400 °C results in a drastic dehydroxilation and the removal of chemisorbed/physisorbed H_2O as indicated by the reduced absorbance by the –OH (3400 cm^{-1}) and H_2O (1620 cm^{-1}) groups. Comparison further shows that, the absorbance by the C–O–C group (1076 and 1122 cm^{-1}) is also reduced after the calcination treatment, which suggests the decomposition of HPC at higher temperatures. The presence of C–O–C group along with the other functional groups such as –CH_2 (1420 and 2919 cm^{-1}) and –CH_3 (1382 cm^{-1}) after the calcination at 400 °C indicates that, at least a monolayer of HPC exists on the surface of nanocrystalline sol-gel TiO_2 after the calcination at relatively lower temperature. The absorbance peak due to the C–O–C group, however, disappears completely after the calcination at relatively higher temperatures (500 and 600 °C), which suggests the complete decomposition of HPC under these processing conditions. In Figure 7(b), the absorbance by the functional groups –CH_2 and –CH_3 is seen to decrease gradually with increasing calcination temperature. However, the presence of these peaks suggests that, even after the calcination at relatively higher temperatures (500 and 600 °C), small amount of carbon still remains on the powder surface in the form of –CH_2 and/or –CH_3 groups as surface-contaminants. Hence, it appears that, the modified sol-gel method produces the nanocrystalline anatase-TiO_2 with reduced average nanocrystallite size, higher specific surface-area, and increased surface-contaminants. The combined effect of these parameters on the photocatalytic activity of the nanocrystalline sol-gel anatase-TiO_2 is discussed in detail in the section-5.2.4.

3. Mixed-Phase Nanocrystalline Sol-Gel Titania

The nanocrystalline sol-gel anatase-TiO_2 (Figure 4(a)) and the rutile-TiO_2 (Figure 4(d)) were utilized to prepare the mixed phase nanocrystalline TiO_2 via solvent mixing and calcination (SMC) treatment [28]. In SMC, the rutile-TiO_2 was first mechanically mixed in different weight-ratios with the anatase-TiO_2. The mechanically mixed precursors were then dispersed in an anhydrous 2-propanol under the continuous magnetic stirring to achieve the homogenous mixing. The suspension was then dried in an oven till the solvent was evaporated completely. The nanocrystalline TiO_2, with the homogeneously mixed anatase-TiO_2 and rutile-TiO_2, was then calcined at higher temperature for establishing an electronic coupling between the two mixed-phases. Typical electrical contact between the anatase-TiO_2

and rutile-TiO_2 nanocrystallites, as obtained for the sample with 40 wt.% rutile, is shown in Figure 8.

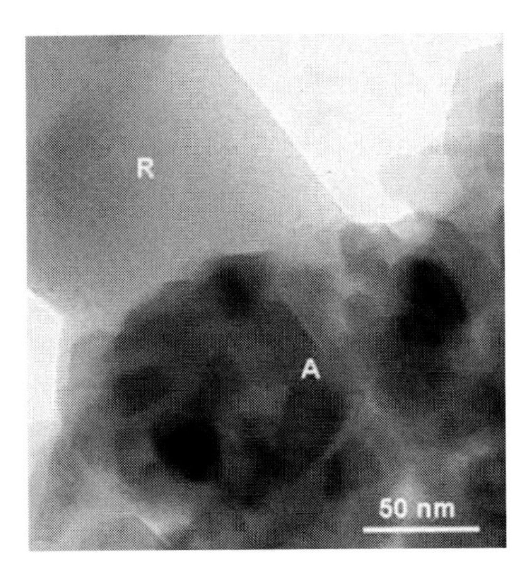

Figure 8. TEM image of the mixed-phase nanocrystalline TiO_2, with an optimum rutile-content of 40 wt.%, processed via sol-gel SMC. 'A' and 'R' represent the anatase-TiO_2 and the rutile-TiO_2 [28]. Copyright 2008 American Chemical Society.

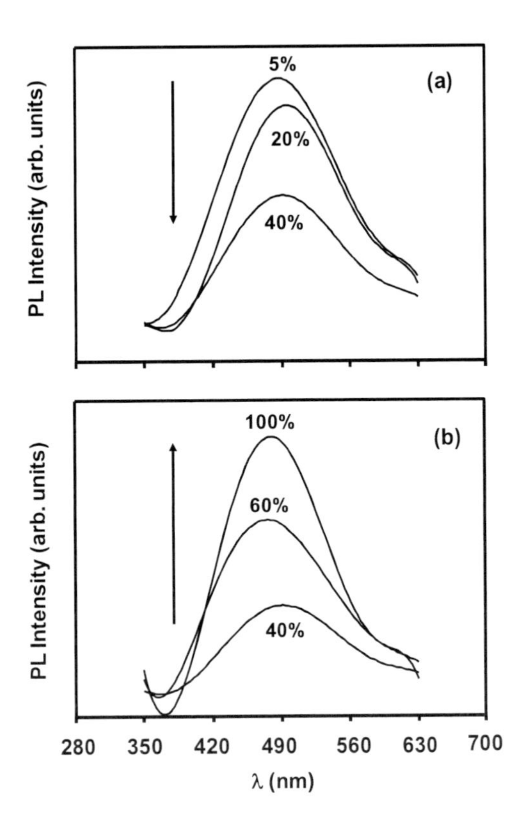

Figure 9. PL spectra obtained for the mixed-phase nanocrystalline sol-gel TiO_2 processed with different rutile-contents. Excitation wavelength used is 327 nm [23]. Copyright 2009 Springer.

The photoluminescence (PL) spectra of the mixed-phase nanocrystalline TiO_2 with varying rutile-content, obtained at an excitation wavelength of 327 nm, are presented in Figure 9. It is noted that, all the powders show a broad PL peaking at 486 nm in the visible-region, which is an excitonic PL attributed to the defects present in the samples, namely the oxygen-ion vacancies, which provide acceptor levels near the conduction-band edge, and hence, mediate the transfer of photo-induced electrons from the conduction-band to the valence-band [40]. As indicated by an arrow in Figure 9(a), the PL intensity is seen to decrease progressively as the amount of rutile increases within the range of 5-40 wt.%. However, as indicated by an arrow in Figure 9(b), the PL intensity increases with the amount of rutile within the range of 40-100 wt.%. Thus, the trend in the variation of the PL intensity of mixed-phase nanocrystalline sol-gel TiO_2 as a function of amount of rutile is different below and above 40 wt.% rutile. Since all the powders have been processed in pure form under the similar processing conditions, variation in the PL intensity cannot be attributed to the changes in the oxygen-ion vacancy concentration in different powders. However, such variation can be very well explained using the "charge-separation mechanism" [41] and a newly proposed model based on variation in the band-gap energy of mixed-phase nanocrystalline TiO_2 as a function of size-distribution and phases involved [28]. According to these mechanisms, when the anatase-TiO_2 and rutile-TiO_2 are in contact with each other, the photo-induced holes in anatase-TiO_2 accumulate in rutile-TiO_2; while, the photo-induced electrons remain in anatase-TiO_2 due to the band-bending at the interface. This results in an effective charge-separation and increased photo-induced e^-/h^+ life-time. The PL intensity is, hence, expected to reduce for the mixed-phase nanocrystalline TiO_2 having the anatase-TiO_2 and rutile-TiO_2 in an electronic contact with each other. As a consequence, the decreasing PL intensity as observed in Figure 9(a) is attributed to an increased number of anatase-rutile contacts with increasing rutile-content within the range of 5-40 wt.%. However, for the amount of rutile above 40 wt.%, more rutile-rutile contacts are established, which dominate the anatase-rutile contacts. Since the former are not as effective as the latter for an effective charge-separation, the photo-induced e^-/h^+ life-time decreases with increasing rutile-content within the range of 40-100 wt.%; which in turn increases the PL intensity. Thus, the PL analysis suggests the maximum photo-induced e^-/h^+ life-time for the mixed-phase nanocrystalline TiO_2 with 40 wt.% rutile. This also suggests that, for the mixed phase nanocrystalline TiO_2 processed via sol-gel SMC, the maximum photocatalytic activity would be for the sample with 40 wt.% rutile content, which is discussed in further details in the section-5.2.5.

4. Surface-Functionalized Nanocrystalline Sol-Gel Titania

The sol-gel derived nanocrystalline anatase-TiO_2 processed via conventional method was utilized to surface-deposit nano-sized clusters of silver oxide-silver (Ag_2O/Ag^0) using two different techniques, namely the UV-reduction and chemical-reduction (using Sn^{2+}-ions) [24,27].

4.1. UV-Reduction Method

In this technique, the nanocrystalline sol-gel anatase-TiO_2 was dispersed in an aqueous silver nitrate ($AgNO_3$) solution under continuous magnetic stirring. A proper concentration range of $AgNO_3$ was chosen so as to obtain Ag/Ti ratio within the range of 10^{-4}-10^{-1}. The solution-pH was adjusted to ~10-12 by slowly adding an aqueous ammonium hydroxide (NH_4OH) solution to an aqueous $AgNO_3$ solution containing the nanocrystalline sol-gel anatase-TiO_2. The resulting suspension was then exposed to the UV-radiation in a photoreactor containing tubes which emitted the UV-radiation having the wavelength within the range of 200–400 nm peaking at 360 nm. Ag_2O/Ag^0-deposited nanocrystalline sol-gel anatase-TiO_2 was then separated using a centrifuge and dried in an oven overnight. Change in the color of nanocrystalline sol-gel anatase-TiO_2/Ag composite powder was noted from initial white to dark-grey with increasing Ag-concentration within the investigated range.

The addition of NH_4OH to an aqueous $AgNO_3$ solution results in the formation of silver-ammonia complex-ions via following reactions [24,27],

$$NH_4OH_{(aq)} \rightarrow NH_{3(aq)} + H_2O \tag{7}$$

$$Ag^+(aq) + 2NH_{3(aq)} \xrightarrow{TiO_2} [Ag(NH_3)_2]^+_{(ads)} \tag{8}$$

The net reaction can be written as,

$$Ag^+(aq) + 2NH_4OH_{(aq)} \xrightarrow{TiO_2} [Ag(NH_3)_2]^+_{(ads)} + 2H_2O \tag{9}$$

In the presence of nanocrystalline sol-gel anatase-TiO_2, silver-ammonia complex-ions are surface-adsorbed due to the presence of the space-charge-layer on the surface of TiO_2 nanoparticles. When the suspension is illuminated with the UV-radiation, e^-/h^+ pairs are created within the nanocrystalline sol-gel anatase-TiO_2. The photo-induced electrons then reduce the surface-adsorbed silver-ammonia complex-ions to metallic Ag^0.

$$[Ag(NH_3)_2]^+_{(ads)} + e^- \xrightarrow{TiO_2,UV} Ag^0 + 2NH_{3(aq)} \tag{10}$$

The overall reaction for the Ag^0-deposition can be written as,

$$Ag^+ + 2NH_4OH + e^- \xrightarrow{TiO_2,UV} Ag^0 + 2NH_3 + 2H_2O \tag{11}$$

In the presence of the space-charge-layer on the surface of nanocrystalline sol-gel anatase-TiO_2, the deposited Ag^0 gets oxidized immediately to Ag_2O via "reverse-spillover-effect" [42].

$$4Ag^0 + O^-_{2(ads)} \xrightarrow{TiO_2} 2Ag_2O + e^- \tag{12}$$

$$2Ag^0 + O^-_{(ads)} \xrightarrow{TiO_2} Ag_2O + e^- \tag{13}$$

The deposition of Ag_2O/Ag^0 on the surface of nanocrystalline sol-gel anatase-TiO_2 via UV-reduction method (and also via chemical-reduction method as discussed below) is confirmed using the narrow-scan XPS analysis, Figure 10, where the amount of Ag_2O (~367.4-367.7 eV) is found to be relatively larger than that of Ag^0 (~368.5-368.7 eV). Hence, the reverse-spillover effect appears to be a dominant oxidation mechanism of deposited-Ag^0. The narrow-scan XPS analysis further reveals a shift in Ag $3d_{5/2}$ binding energy (BE) level by +0.3-0.5 eV relative to that (~368.2 eV) of the bulk-Ag^0, which suggests the presence of Ag^0 nanoparticles with average size less than 10 nm. Such small Ag^0 nanocrystallites may exhibit a positive core-level BE shift as a result of the "cluster-size-effect" involving the initial and final-state effects [43].

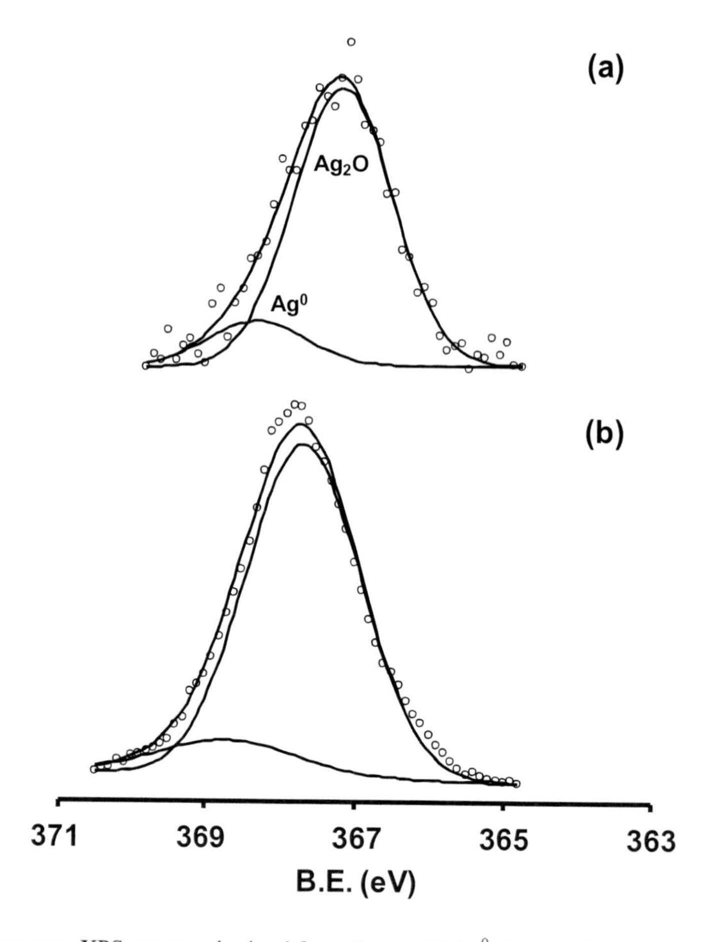

Figure 10. Narrow-scan XPS spectra obtained from the Ag_2O/Ag^0-deposited nanocrystalline sol-gel anatase-TiO_2 obtained via chemical-reduction (1.0% Ag/Ti) (a) and UV-reduction (10% Ag/Ti) (b) methods [24]. Copyright 2009 American Chemical Society.

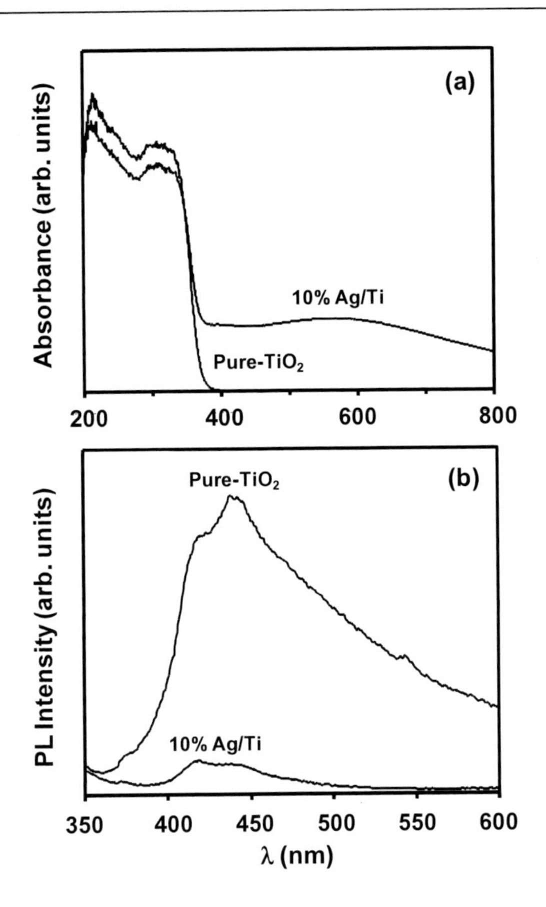

Figure 11. UV-visible absorption (a) and PL spectra (b) obtained for the nanocrystalline sol-gel anatase-TiO_2 processed with the '*R*' value of 90, calcined at 600 °C-2 h, and subsequently deposited with Ag_2O/Ag^0 using the UV-reduction method. The excitation wavelength is 327 nm [27]. Copyright 2009 Springer

The UV-visible diffuse reflectance (DR) absorption spectra obtained for the pure and surface-modified nanocrystalline sol-gel anatase-TiO_2, Figure 11(a), shows that the pure nanocrystalline sol-gel anatase-TiO_2 absorbs the radiation in the UV-region below 400 nm. However, the absorption spectrum of the Ag_2O/Ag^0-deposited nanocrystalline sol-gel anatase-TiO_2 is different than that of the former and shows a broad absorption in the visible-region as well, which peaks at ~575 nm. In the literature, the absorption in the visible-region for the Ag-deposited nanocrystalline TiO_2 has been attributed to the surface plasmon absorption due to the spatially confined electrons in the Ag^0 nanoparticles [44]. Since, the band-gap energy of Ag_2O is ~1.3-1.5 eV, the absorption in the visible-region (λ=400-800 nm) is also likely to be contributed by Ag_2O. Moreover, more intense and broader absorbance peak in the visible-region has been ascribed to relatively larger average-size and broader size-distribution of Ag^0 nanoparticles, high refractive index of TiO_2, and the interaction between Ag^0 and TiO_2 [44]. Hence, from the absorption spectrum obtained for the Ag_2O/Ag^0-deposited nanocrystalline sol-gel anatase-TiO_2 processed via UV-reduction method, it is inferred that the average nanoparticle size, its distribution, and surface-coverage by Ag^0 and Ag_2O are relatively larger.

The PL spectra obtained from the pure and Ag_2O/Ag^0-deposited nanocrystalline sol-gel anatase-TiO_2 are presented in Figure 11(b). A broad PL spectrum in the visible-region peaking at ~433 nm is observed for the former, which is an excitonic PL as discussed earlier. Comparison shows that, the PL intensity of the Ag_2O/Ag^0-deposited nanocrystalline sol-gel anatase-TiO_2 is considerably quenched compared with that of the pure sample. Such PL quenching has been attributed to an enhanced photo-induced e^-/h^+ life-time caused by an effective trapping of the photo-induced electrons by the deposited Ag_2O and Ag^0 [40,45]. As a result, the photo-induced e^-/h^+ life-time appears to be larger for the surface-modified nanocrystalline sol-gel anatase-TiO_2 relative to that for the pure one. Thus, the DR and PL analyses suggest that, the surface-modified nanocrystalline anatase-TiO_2 not only absorbs the visible-radiation but also exhibits higher photo-induced e^-/h^+ life-time compared with those of pure nanocrystalline sol-gel anatase-TiO_2. Both of these factors have a strong effect on the photocatalytic activity under the solar-radiation, which would be discussed in the section-5.2.6.

4.2. Chemical-Reduction Method

In this technique, the nanocrystalline sol-gel anatase-TiO_2 was dispersed in an acidic aqueous solution containing tin(II) chloride ($SnCl_2$) under continuous magnetic stirring. The solution-pH was adjusted by adding an aqueous hydrochloric acid (HCl) solution. The nanocrystalline sol-gel anatase-TiO_2, surface-sensitized with Sn^{2+}-ions, was then separated using a centrifuge after stirring the suspension for sufficient amount of time. The surface-sensitized anatase-TiO_2 was dispersed in an aqueous $AgNO_3$ solution under continuous magnetic stirring. Similar to the previous case, a proper concentration of $AgNO_3$ was chosen to obtain Ag/Ti ratio in a specific range. The solution-pH was then adjusted to ~10-12 by slowly adding an aqueous NH_4OH solution to the above suspension under continuous magnetic stirring. The Ag_2O/Ag^0-deposited nanocrystalline sol-gel anatase-TiO_2 was then separated using a centrifuge and dried in an oven overnight. Change in the color of nanocrystalline sol-gel anatase-TiO_2/Ag was noted from initial white to light-grey with increasing Ag-concentration [24].

When the pure nanocrystalline sol-gel anatase-TiO_2 is added to an aqueous solution of $SnCl_2$, it adsorbs Sn^{2+}-ions on the surface (surface-sensitization) due to the presence of the space-charge-layer. When the surface-sensitized nanocrystalline sol-gel anatase-TiO_2 is stirred in an aqueous solution containing the silver-ammonia complex-ions, it results in the deposition of metallic Ag^0 on the surface via following chemical reaction,

$$2[Ag(NH_3)_2]^+_{(ads)} + Sn^{2+}_{(ads)} \xrightarrow{TiO_2} 2Ag^0 + 4NH_{3(aq)} + Sn^{4+}_{(ads)} \tag{14}$$

As mentioned before, large amount of deposited-Ag^0 gets oxidized due to the reaction with the super-oxide ions via reverse-spillover-effect, Eqs. 12-13. The deposition of Ag_2O/Ag^0 on the surface of nanocrystalline sol-gel anatase-TiO_2 via chemical-reduction method is confirmed using the narrow-scan XPS analysis, Figure 10. For larger Ag-concentrations (1 and 10 mol%), the formation of SnO_2 was also detected via XPS analysis [24], possibly via reactions presented in Eq. 15 and 16.

$$Sn^{4+} + 4O^{-}_{2(ads)} \xrightarrow{TiO_2} SnO_2 + 3O_{2(gas)} \quad (15)$$

$$Sn^{4+} + 4O^{-}_{(ads)} \xrightarrow{TiO_2} SnO_2 + O_{2(gas)} \quad (16)$$

The formation of Ag_2O and SnO_2, specifically at higher Ag-concentrations, drastically reduce the surface-concentration of the super-oxide ions, which may in turn reduce the amount of surface-adsorbed dye under the dark-condition, thus affecting the photocatalytic activity as discussed in detail in the section-5.2.6.

5. Industrial Dye-Removal Using Nanocrystalline Sol-Gel Titania

5.1. Photocatalysis Mechanism

In this section, the industrial dye-removal characteristics of the nanocrystalline sol-gel TiO_2, via photocatalysis mechanism, is reviewed for the methylene blue (MB) dye as a model catalytic dye-agent. The chemical structure of the MB dye is schematically shown in Figure 12(a). The MB dye has a cationic configuration in an aqueous solution [30,46], which results in its adsorption through the coulombic interaction with the OH^- or the superoxide-ions present on the surface of the nanocrystallites sol-gel TiO_2.

As described schematically in Figure 12(b), when an aqueous solution containing the nanocrystalline sol-gel TiO_2 is irradiated with the UV-radiation having the energy greater than its band-gap energy, e^-/h^+ pair is created within the nanocrystallites due to the ejection of an electron from the valence-band into the conduction-band leaving behind a hole in the former (*Charge Carrier Generation*). The generated holes may react with the surface-adsorbed OH^- ions forming the -$OH^•$ radicals, which may also be formed by the reaction of dissolved oxygen (O_2) with the generated electrons forming the hydrogen peroxide (H_2O_2). This intermediate product subsequently gets decomposed to -$OH^•$ radical. The overall reaction may be summarized as (*Interfacial Charge Transfer*) [30,47]

$$TiO_2 + h\nu \rightarrow TiO_2 + e^- + h^+ \quad (17)$$

$$OH^- + h^+ \rightarrow OH^• \quad (18)$$

$$O_2^- + H^+ \rightarrow HO_2 \quad (19)$$

$$2HO_2 \rightarrow H_2O_2 + O_2 \quad (20)$$

$$H_2O_2 + e^- \rightarrow OH^• + OH^- \quad (21)$$

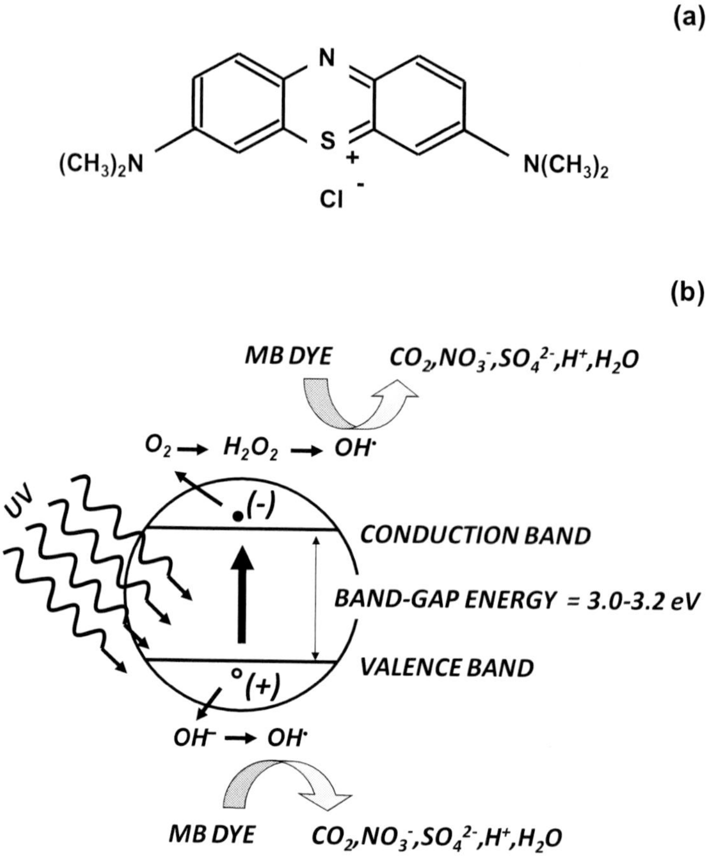

Figure 12. (a) Schematic representation of the molecular structure of the MB dye. (b) Photocatalysis mechanism using the nanocrystalline sol-gel TiO_2. In (b), the MB dye is considered as a model catalytic dye-agent [30]. Copyright 2007 American Chemical Society.

The -$OH^{\bullet}$ radicals, thus, formed are mainly responsible for the degradation of the MB dye through its successive attacks via formation of several other intermediate products. The degradation of the MB dye mainly begins with the cleavage of the $C\text{-}S^{+}\text{=}C$ functional group since this group is responsible for the adsorption of the MB dye on the surface of nanocrystalline sol-gel TiO_2. The overall reaction, which results in the decomposition of the MB dye into carbon dioxide (CO_2), nitrate (NO_3^-) ions, sulfate (SO_4^-) ions, protons, and water may be summarized as [30,46],

$$C_{16}H_{18}N_3S^+ + 102OH^{\bullet} \xrightarrow{UV, TiO_2} 16CO_2 + 3NO_3^- + SO_4^{2-} + 6H^+ + 57H_2O \quad (22)$$

The efficacy of the above mechanism in decomposing the MB dye depends on the effectiveness of the photocatalytic process in transferring the photo-induced e^-/h^+ pair from the particle volume to the particle surface and subsequently to the surface-adsorbed species. The generated e^-/h^+ pair, hence, must migrate to the particle surface as a separate entity; however, if the TiO_2 nanocrystallite size is relatively larger, which increases the travel distance for the e^-/h^+ pair, then they may get recombined within the particle volume before reaching the particle surface (*Volume Charge Carrier Recombination*) [30,47].

$$h^+ + e^- \rightarrow Heat \tag{23}$$

On the other hand, if the nanocrystallite size is relatively smaller, then the generated e^-/h^+ pair may escape to the particle surface and get trapped at the active surface sites, before undergoing the volume charge carrier recombination process (*Surface Charge Carrier Trapping*).

$$e^- + > Ti^{IV}OH \rightarrow \left(> Ti^{III}OH\right) \tag{24}$$

$$h^+ + > Ti^{IV}OH \rightarrow \left(> Ti^{IV}OH^\bullet\right)^+ \tag{25}$$

where, >TiOH is the hydrated surface functional group, ($>Ti^{+3}OH$) the surface-trapped conduction-band electron, and $(>Ti^{+4}OH^\bullet)^+$ the surface-trapped valence-band hole. The surface-trapped charge carriers may get transferred to the surface-adsorbed species via following reactions (*Interfacial Charge Transfer*),

$$\left(> Ti^{III}OH\right) + oxd \rightarrow \left(> Ti^{IV}OH\right) + oxd^{\bullet -} \tag{26}$$

$$\left(> Ti^{IV}OH^\bullet\right)^+ + red \rightarrow \left(> Ti^{IV}OH\right) + red^{\bullet +} \tag{27}$$

where, '*oxd*' and '*red*' are the surface-adsorbed oxidant and reductant species respectively. However, if the nanocrystallite size is too small, the surface-trapped charge carriers may get annihilated by the subsequent photo-induced e^-/h^+ pair reaching the surface, before the interfacial charge transfer process takes place (*Surface Charge Carrier Recombination*).

$$e^- + \left(> Ti^{IV}OH^\bullet\right)^+ \rightarrow \left(> Ti^{IV}OH\right) \tag{28}$$

$$h^+ + > Ti^{III}OH \rightarrow \left(> Ti^{IV}OH\right) \tag{29}$$

It appears that, for an optimum photocatalytic activity, the rate of volume and surface charge carrier recombination processes should be minimum; while, that of the interfacial charge transfer process should be maximum. For enhancing the photocatalytic dye-degradation rate, the adsorption of dye molecules on the surface of photocatalyst particle should also be higher along with the high rate of interfacial charge transfer process since these processes occur in series.

5.2. Effect of Processing and Materials Parameters

5.2.1. 'R'

Typical kinetics of the MB dye degradation using the nanocrystalline sol-gel anatase-TiO_2, under the UV-radiation exposure, is presented in Figure 13(a) for two different '*R*' values. The corresponding plots for determining the apparent first-order reaction rate-constant (k_{app}) are presented in Figure 13(b). Higher MB dye degradation kinetics is observed for larger '*R*' value. The obtained variation in k_{app} as a function of '*R*' within the investigated range of 5-150 is shown in Figure 14(a). The maximum k_{app} value is, thus, observed at the '*R*' value of 90, which is comparable with that of the commercial Degussa P25. Comparison of Figure 14(a) and (b) shows that, the variation in k_{app} as a function '*R*' follows almost similar trend as that exhibited by the variation in the specific surface-area as a function of '*R*'. Higher the specific surface-area, larger would be the potential sites for the surface-adsorption of MB dye molecules, which results in higher photocatalytic activity. However, the maximum MB adsorption may block the radiation reaching the particle volume; hence, the maximum photocatalytic activity may not be observed under this typical condition [29,30].

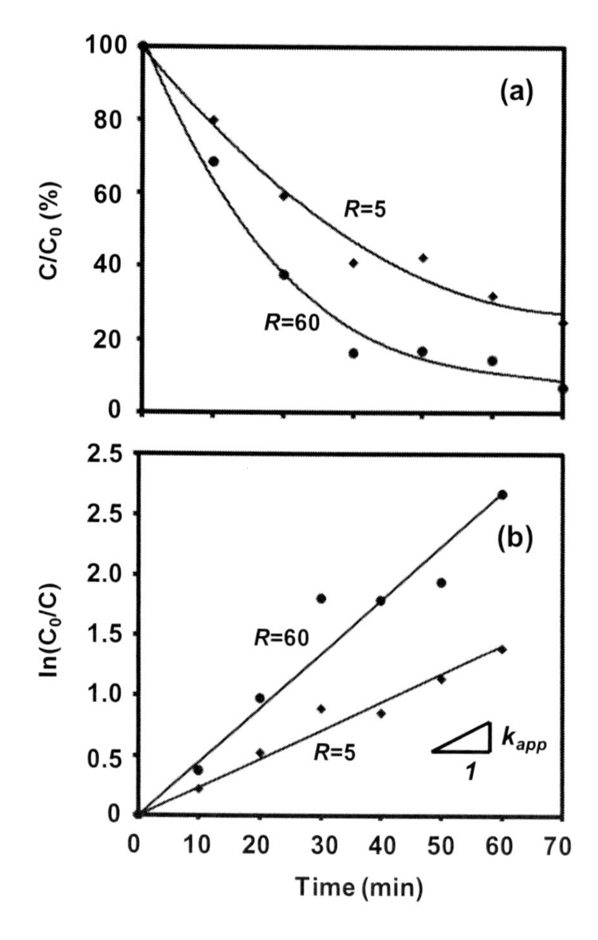

Figure 13. (a) Variation in the normalized residual MB dye concentration as a function of UV-radiation exposure time for two different '*R*' values after calcination at 600 °C-2h. (b) Corresponding plots for determining the k_{app} [30]. Copyright 2007 American Chemical Society.

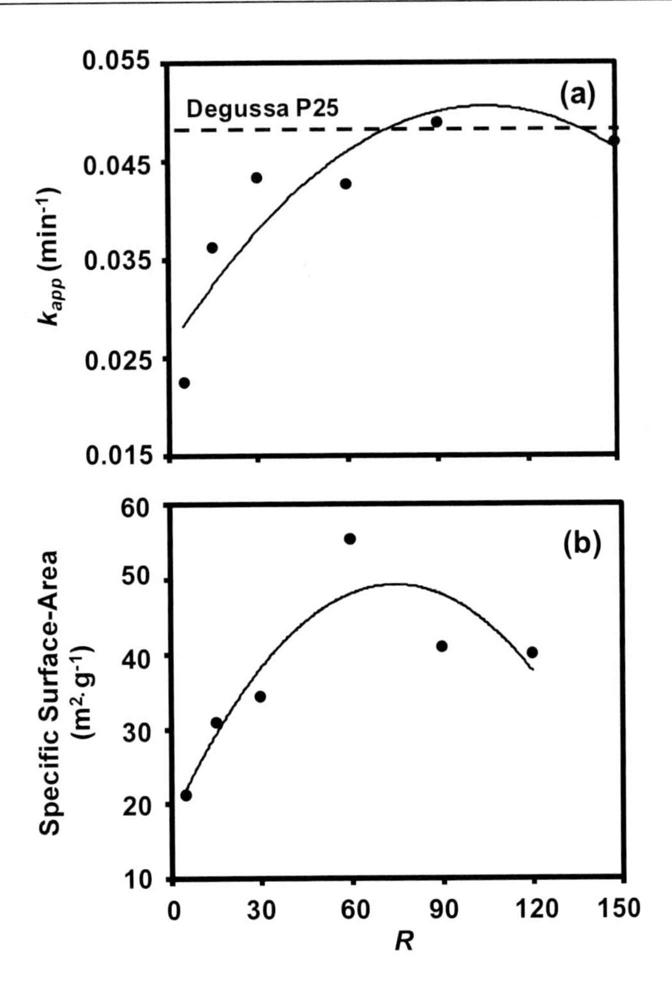

Figure 14. Variation in k_{app} (a) and specific surface-area (b) as a function of 'R' obtained for the nanocrystalline sol-gel anatase-TiO_2 calcined at 600 °C-2 h. In (a), the dotted horizontal line represents the k_{app} value corresponding to the commercial Degussa-P25 [29]. Copyright 2008 Springer.

5.2.2. Nanocrystallite size

The variation in k_{app} as a function of average nanocrystallite is presented in Figure 15. The average nanocrystallite size determined using the XRD, Figure 2(a), is utilized for this purpose. It appears that, k_{app} increases first with decreasing average nanocrystallite size of the nanocrystalline sol-gel anatase-TiO_2. It reaches the maximum value for an optimum average nanocrystallite size (D^*) of ~12 nm (experimental value) and then decreases with average nanocrystallite size below 12 nm. This optimum average nanocrystallite size is almost comparable with the theoretical critical size for the anatase-to-rutile phase transformation. Thus, an inversion phenomenon is clearly observed for the variation in the k_{app} as a function of average nanocrystallite size. This can be very well explained based on the mechanism of photocatalysis discussed in the section-5.1. An increase in the photocatalytic activity is contributed by an increase in the specific surface area, which increases the number of active surface sites for the dye-adsorption. Moreover, this is further aided by an increase in an interfacial charge transfer process due to the decrease in the rate of volume charge carrier recombination with the decreasing average nanocrystallite size. Nevertheless, the photocatalytic activity does not increase continuously with decreasing average nanocrystallite size. A critical size is reached below which the photocatalytic activity begins to decrease [30,47]. It appears that, below the critical size, although the rate of volume charge carrier recombination is less effective in annihilating the photo-induced e^-/h^+ pair, the rate of surface charge carrier recombination becomes a dominant process. This reduces the rate of interfacial charge transfer, and hence, the photocatalytic activity with decreasing average nanocrystallite size below the critical size. The existence of a critical size of ~10-12 nm has also been experimentally demonstrated in the literature [46,48].

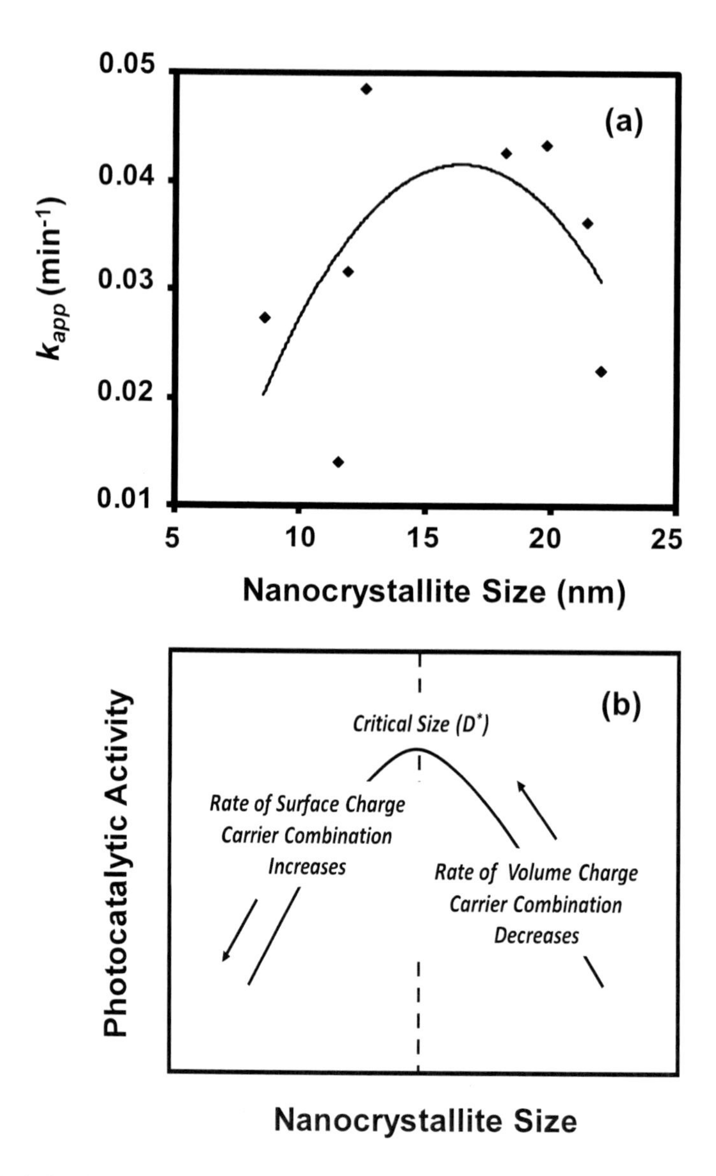

Figure 15. (a) Variation in k_{app} as a function of average nanocrystallite size as obtained for the nanocrystalline sol-gel anatase-TiO_2. All powders are processed with the 'R' value of 90. (b) Schematic representation of the variation in the photocatalytic activity as a function of average nanocrystallite size [30]. Copyright 2007 American Chemical Society.

5.2.3. Calcination temperature

The variation in k_{app} as a function of '*R*', for three different calcination temperatures, is presented in Figure 16. As mentioned previously, the as-synthesized sol-gel TiO_2 is amorphous; however, it gets transformed into nanocrystalline anatase-TiO_2 after calcination at 400 °C ((Figure 3(a)). The crystallinity of anatase-TiO_2 increases with increasing calcination temperature above 400 °C. As a result, the photocatalytic activity is observed to be higher at 600 °C relative to that observed at 400 °C within the investigated range of '*R*' values. Interestingly, as demonstrated earlier in Figure 2(a), at 400 °C the average nanocrystallite size increases with increasing '*R*'; while, within this size range which is below the critical size (D^*), Figure 15(a), the photocatalytic activity increases with average nanocrystallite size. Hence, at 400 °C, the photocatalytic activity is observed to increase with '*R*'. On the other hand, at 600 °C, the average nanocrystallite size decreases with increasing '*R*', and within this size range which is above the critical value (D^*), Figure 15(a), the photocatalytic activity increases with decreasing average nanocrystallite size. Hence, even at 600 °C, the photocatalytic activity increases with increasing '*R*'. The specific surface-area, however, decreases with increasing calcination temperature due to an increase in an average nanocrystallite size and the reduced pore volume [29], which in turn reduces both the amount of surface-adsorbed MB under the dark-condition and the photocatalytic activity of nanocrystalline sol-gel anatase-TiO_2. The increase in an average nanocrystallite size with increasing calcination temperature also leads to the phase transformation from the anatase-to-rutile, which is invariably followed by a rapid growth in the average nanocrystallite size [30]. As a consequence, the band-gap energy variation in the connected nanocrystallites of different sizes is not observed for the rutile-TiO_2 since the nanocrystallite size range is often larger than the critical size required for the band-gap energy enhancement [28]. The electron mobility in rutile-TiO_2 is, hence, lower than that in anatase-TiO_2, which results in higher rate of e^-/h^+ recombination in the former. This reduces the photo-induced e^-/h^+ life-time, and hence, the photocatalytic activity of the rutile-TiO_2. As a result, the photocatalytic activity of nanocrystalline sol-gel TiO_2 calcined at 800 °C is noted to be considerably lower than that calcined at 600 °C, Figure 16.

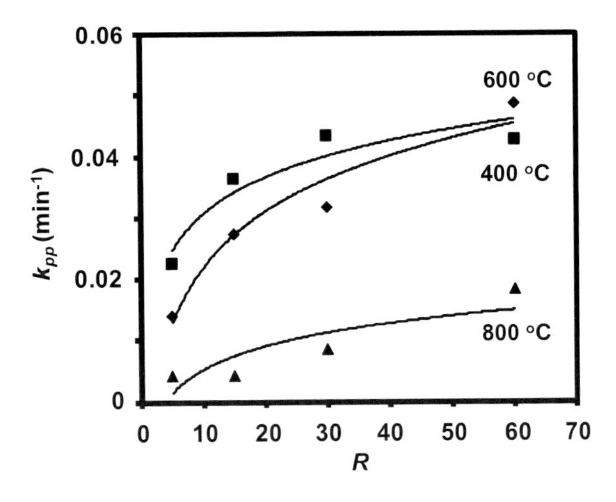

Figure 16. Variation in k_{app} as a function of '*R*' as obtained for the nanocrystalline sol-gel TiO_2 calcined at different temperatures [30]. Copyright 2007 American Chemical Society.

Table 2. k_{app} and amount of surface-adsorbed MB after stirring under the dark-condition as obtained for the nanocrystalline sol-gel anatase-TiO_2 processed under different conditions [29]. Copyright 2008 Springer.

R	Calcination Temperature (°C)	k_{app} (min^{-1})			MB Adsorbed (%)		
		HPC Concentration (g•L^{-1})			HPC Concentration (g•L^{-1})		
		0.0	1.0	2.0	0.0	1.0	2.0
90	400	0.024	0.015	0.019	16.9	19.4	15.6
	500	0.022	0.023	0.029	21.4	20.4	22.5
	600	0.049	0.035	0.034	27.2	20.6	17.7

5.2.4. Surface-purity

As discussed in the section-2.2, the nanocrystalline sol-gel anatase-TiO_2, synthesized using the HPC, exhibits higher specific surface-area and smaller average nanocrystallite size relative to those processed without using the HPC (Table 1). This strongly suggests that, the nanocrystalline sol-gel anatase-TiO_2 processed via modified approach should have higher photocatalytic activity than that processed via conventional approach. Nevertheless, the data tabulated in Table 2 suggests that, the nanocrystalline sol-gel anatase-TiO_2 is deactivated due to the addition of HPC to the conventional process. With the addition of HPC within the concentration range of 0-2 g•L^{-1}, the range of k_{app} is observed to be reduced from 0.022-0.049 min^{-1} to 0.015-0.035 min^{-1}. This is in contrary to the recent report [49], where an enhanced photocatalytic activity of the sol-gel derived nanocrystalline TiO_2 thin films has been reported due to an enhanced specific surface-area caused by the addition of HPC. It is also noted that, with the addition of HPC, the range in the amount of surface-adsorbed MB is also reduced from 16.9-27.2 % to 15.6-22.5 %, Table 2. Hence, it appears that, the photocatalytic process has been interfered by the HPC addition by affecting the amount of MB adsorbed on the powder-surface under the dark-condition. As mentioned previously, the surface-adsorption of MB is one of the key steps in enhancing the kinetics of the photocatalytic degradation of the dye. The reduced photocatalytic activity has been reported earlier due to the presence of a polymer coating on the surface of TiO_2 particles [50]. Hence, it appears that, the presence of a small amount of carbon remaining on the powder-surface after the removal of HPC via calcination, is possibly responsible for reducing the MB dye-adsorption under the dark-condition. It is to be noted that, the form of carbon (amorphous or crystalline) plays an important role in the surface-adsorption of MB dye. The crystalline form of the carbon (that is, the activated carbon) has been known to enhance the photocatalytic activity of nanocrystalline TiO_2 via enhancing the surface-adsorption of the dye molecules [51-54]. Since the photocatalytic activity has been reduced here due to the addition of HPC, it is deduced that, the residual carbon deposited on the surface of nanocrystalline sol-gel anatase-TiO_2, as a result of the HPC decomposition, is possibly amorphous in nature. The crystallinity of nanocrystalline TiO_2 powders has also been reported to reduce when they are synthesized using the HPC possibly due to the effect of residual surface-carbon, which retards the amorphous-to-anatase phase transformation [29]. This reduced crystallinity is an additional contributing factor for the decreased photocatalytic activity of the nanocrystalline anatase-TiO_2 processed via modified sol-gel. Hence, although higher specific surface-area and smaller average nanocrystallite size have been obtained through the HPC addition, the reduced photocatalytic activity is mainly attributed to both the reduced crystallinity and the decreased surface-adsorption of the MB dye, under the dark-condition, caused by the presence of the residual amorphous carbon which possibly covers the active surface-sites responsible for the MB dye adsorption.

5.2.5. Synergy effect

The experimental variation in the k_{app} as a function of amount of rutile is presented in Figure 17(a). The k_{app} is noted to increase first with the amount of rutile within the range of 0-40 wt. %. It reaches the maximum value for 40 wt.% rutile, beyond which it is seen to decrease with further increase in the rutile-content. This is in consonance with the maximum photo-induced e^-/h^+ life-time as observed using the variation in the PL intensity as a function of amount of rutile-content, Figure 8. The dotted-line joining the k_{app} values of the pure nanocrystalline sol-gel anatase-TiO_2 and pure rutile-TiO_2 represents the hypothetical path which would be followed in the absence of any strong electronic interaction between the two phases involved. The various mechanisms, operating in different regions of the experimentally determined variation in the k_{app} as a function of amount of rutile, are shown in Figure 17(b) and are discussed in detail below.

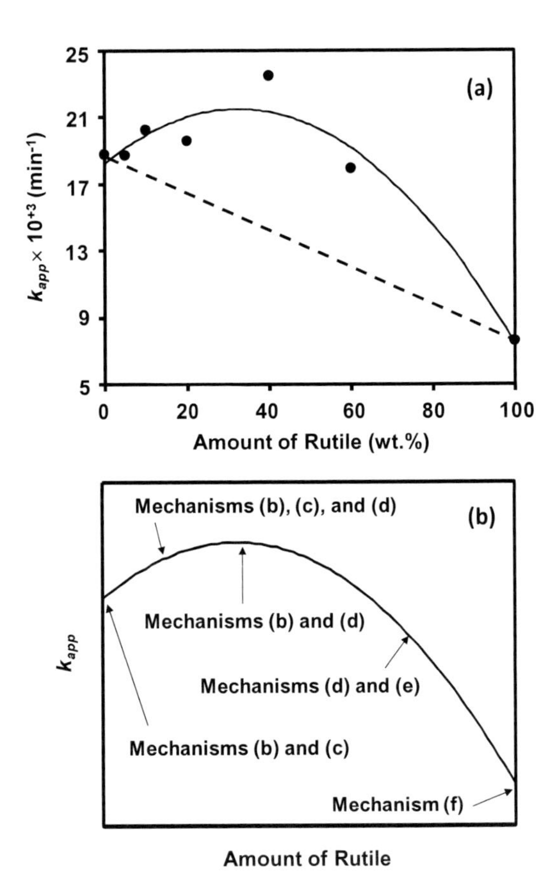

Figure 17. (a) Variation in k_{app} as a function of amount of rutile as obtained for the mixed-phase nanocrystalline TiO_2 processed via sol-gel SMC. The dotted-line represents a hypothetical variation in the absence of the synergy effect between the anatase-TiO_2 and rutile-TiO_2. (b) Schematic diagram showing the different mechanisms, as proposed in the new model (Figure 19), operating in the different regions of the experimentally determined graph presented in (a) [28]. Copyright 2008 Springer.

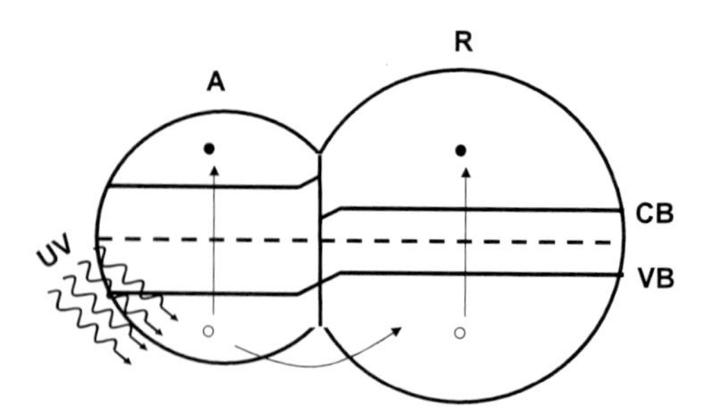

Figure 18. Schematic representation of the "charge-separation mechanism" as proposed in the literature for an enhanced photo-induced e^-/h^+ life-time in the mixed-phase nanocrystalline TiO_2. 'A', 'R', 'CB', and 'VB' represent the anatase-TiO_2, rutile-TiO_2, conduction-band, and valence-band respectively [28,41]. Copyright 2003 American Chemical Society.

The maximum photocatalytic activity of the mixed-phase nanocrystalline sol-gel TiO_2, observed at an intermediate rutile-content, can be partially explained using different models proposed earlier in the literature [41,55]. One of the proposed mechanisms, known as the "charge-separation mechanism" [41] which was considered earlier in the section-3, is described schematically in Figure 18. According to this mechanism, when the anatase-TiO_2 and rutile-TiO_2 nanocrystallites are in contact with each other, the resulting band bending at the interface does not allow the electrons generated in the anatase-TiO_2 to escape into the rutile-TiO_2. On the other hand, the same band bending allows the holes generated in the anatase-TiO_2 to escape into the rutile-TiO_2. This effectively separates the charges with the electrons accumulating in the anatase-TiO_2 and the holes in the rutile-TiO_2, which in turn enhance the photo-induced e^-/h^+ life-time and the photocatalytic activity, as demonstrated experimentally by others [56]. The second mechanism, known as the "antenna mechanism" [28,55], considers the connected anatase-TiO_2 nanocrystallites having the same lattice-orientation, which allows the photo-induced e^-/h^+ pair to move via multi-crystallite transfer process, without the recombination, from the site of their origin to the nanocrystallite where the dye molecule is adsorbed. As a result, according to this mechanism, the nanocrystallite in which the e^-/h^+ pair is created and the nanocrystallite over which the dye molecule is adsorbed may be separated by a large distance. Thus, an entire chain of aligned nanocrystallites acts as an antenna to transfer the incoming photon-energy from the light-absorption site to the reaction-site. The charge-separation ultimately takes place when the photo-induced e^-/h^+ pair reaches the trap-sites; thus, leading to an increased separation life-time.

Both of the proposed mechanisms are, however, associated with the major limitations [28]. For example, it is well-known that, the nanocrystalline sol-gel TiO_2 invariably contains a size distribution with a large half-width; secondly, the band-gap energy of the anatase-TiO_2 is dependent on the nanocrystallite size; third, the anatase-to-rutile phase transformation is also dependent on the nanocrystallite size. Both of the mechanisms discussed above do not take these factors into consideration; and hence, these mechanisms do not provide a complete description of the synergy effect associated with the connected nanocrystallites of anatase-TiO_2 and ruitle-TiO_2 or with those of anatase-TiO_2. As a result, the existence of an optimum rutile-content for observing the maximum photocatalytic activity in the mixed-phase nanocrystalline sol-gel TiO_2 cannot be satisfactorily predicted using these mechanisms. To overcome these limitations associated with the existing mechanisms, a new model has been recently proposed, Figure 19, based on the band-gap energy variation in the connected nanocrystallites as a function of the size distribution and the phases involved [28]. The "charge-separation mechanism" and the "antenna mechanism" are appropriately combined in a single model.

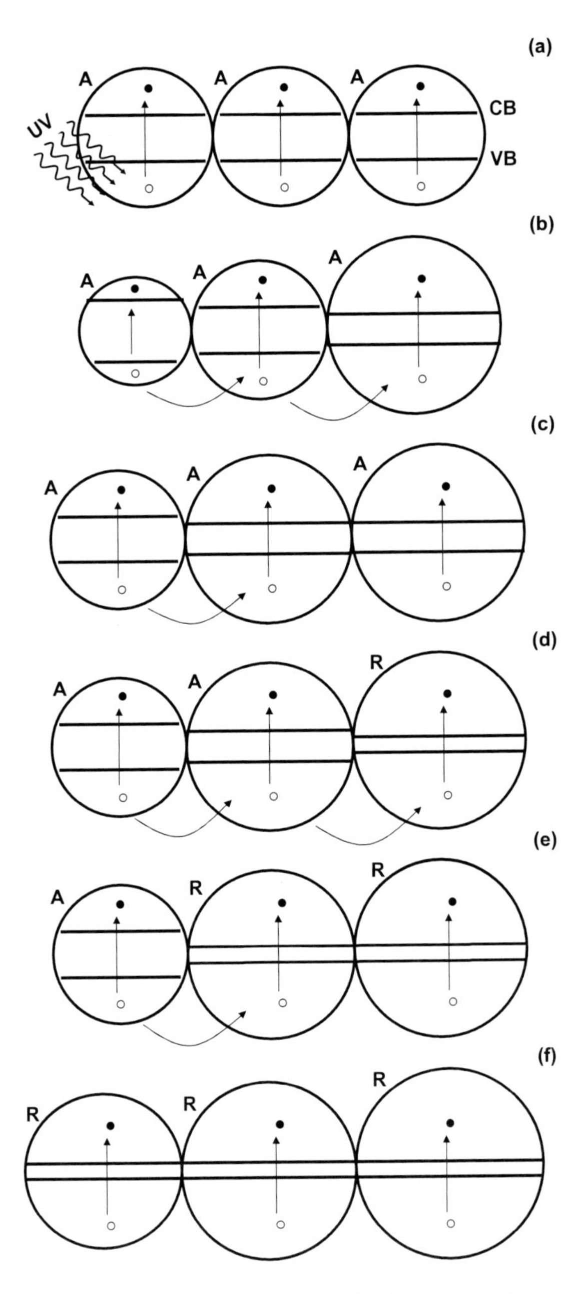

Figure 19. Model based on the band-gap energy variation in the connected nanocrystallites as a function of the size distribution and phases involved, which satisfactorily explains the existence of an optimum rutile-content in the mixed-phase nanocrystalline TiO_2 corresponding to the maximum photocatalytic activity. 'A', 'R', 'CB', and 'VB' represent the anatase-TiO_2, rutile-TiO_2, conduction-band, and valence-band respectively [28]. Copyright 2008 American Chemical Society.

In Figure 19(a), the connected nanocrystallites of pure anatase-TiO_2 (as proposed in the "antenna mechanism" but not necessarily having the same lattice orientation) with a monodispersed nanocrystallite size distribution, is schematically shown. Since the band-gap energies of these nanocrystallites would be the same due to their identical size and phase, the driving force for the transfer of photo-induced hole from one nanocrystallite to the neighboring nanocrystallite would be minimum. As a result, the photo-induced e^-/h^+ life-time in pure anatase-TiO_2 with a monodispersed nanocrystallite size distribution would be very low leading to reduced photocatalytic activity. The photocatalytic activity of the pure anatase-TiO_2 with a monodispersed nanocrystallite size distribution has not been reported yet, which may be due to the issues associated with processing such powders. In practice, the anatase-TiO_2 synthesized via wet-chemical routes (including sol-gel) contains a broad nanocrystallite size distribution with larger half-width and the band-gap energy of pure anatase-TiO_2 increases below a critical size (~17 nm) [57,58]. Hence, if the nanocrystallite size distribution is below this critical size, then the band-gap energies of the connected nanocrystallites would be different depending on their size, Figure 19(b). As a result, the photo-induced hole produced in one nanocrystallite can easily escape into the other leading to an effective charge-separation. Due to the variation in the band-gap energies, the nanocrystallite located in the center acts as a sink as well as a source of holes; thus, serving a dual function in this band-gap energy configuration. Hence, the anatase-TiO_2 having such band-gap energy configuration of the connected nanocrystallites would have relatively longer photo-induced e^-/h^+ life-time leading to its higher photocatalytic activity. This also suggests the significance of the nanocrystallite size distribution with larger half-width to achieve higher photocatalytic activity using the nanocrystalline sol-gel anatase-TiO_2.

If a part of the nanocrystallite size distribution lies above the critical size for the band-gap energy enhancement, then that part of the size distribution would not exhibit the variation in the band-gap energy as a function of size, Figure 19(c). This is possible since the critical size for the anatase-to-rutile phase transformation may be higher than ~14 nm due to an aggregation of the nanocrystallites, as discussed in the section-2.1. Since the critical size for an enhancement in the band-gap energy is ~17 nm, under these circumstances, the two connected nanocrystallites of anatase-TiO_2 of size greater than 17 nm would have the same band-gap energy and the transfer of holes from one nanocrystallite to its neighbor would be restricted. The present model, hence, assumes that the nanocrystalline sol-gel anatase-TiO_2 primarily contains a mixture of band-gap energy configurations of the connected nanocrystallites as shown in Figure 19(b) and (c). The photocatalytic activity of pure anantase-TiO_2 would be then governed by the relative fractions of these band-gap energy configurations. The band-gap energy configuration in the connected nanocrystallites, as shown in Figure 19(c), may further change to that shown in Figure 19(d) provided a part of the nanocrystallite size distribution crosses the critical size for the phase transformation. The mixed-phase nanocrystalline sol-gel TiO_2 thus formed would now contain a mixture of band-gap energy configurations shown in Figure 19(b) and (d) instead of those shown Figure 19(b) and (c). Although the phases involved are different, the band-gap energy configuration shown in Figure 19(d) appears to be similar to that shown in Figure 19(b). In both the band-gap energy configurations, the nanocrystallite located in the center serves a dual function. Hence, due to the change in the relative fractions of the band-gap energy configurations in the connected nanocrystallites, the formation of rutile-TiO_2 would enhance the photo-induced e^-/h^+ life-time in the mixed-phase nanocrystalline sol-gel TiO_2 relative to that in the pure anatase-TiO_2. Hence, the present model suggests that, at the initial stage, the photocatalytic activity would tend to increase with increasing rutile-content, which justifies the experimental observation made in Figure 17(a).

However, with further increase in the amount of rutile-TiO_2, more number of contacts are established in between the rutile-rutile nanocrystallites having an identical band-gap energy, Figure 19(e). As mentioned earlier, the rutile phase is normally formed as a result of the anatase-to-rutile phase transformation, which occurs at relatively higher temperatures (>600 ^oC) and is invariably accompanied by an excessive growth in the nanocrystallite size of the rutile-TiO_2 [28]. Hence, in Figure 19(e) and (f), the nanocrystallites of rutile-TiO_2 of different sizes having identical band-gap energies are considered. As the fraction of such band-gap energy configuration increases, the photo-induced e^-/h^+ life-time would be reduced, which in turn would tend to decrease the photocatalytic activity of the mixed-phase nanocrystalline TiO_2. Hence, it appears that, the maximum photocatalytic activity would be observed for an optimum rutile-content in the mixed-phase nanocrystalline TiO_2. Above this optimum value, the effect of the band-gap energy configuration shown in Figure 19(e) would be more pronounced than that of the band-gap energy configuration shown in Figure 19(d) due to the variation in their relative fractions. Hence, increasing the rutile-content above the optimum value would decrease both the photo-induced e^-/h^+ life-time and the photocatalytic activity of the mixed-phase nanocrystalline sol-gel TiO_2. According to the present model, the mixed-phase nanocrystalline sol-gel TiO_2 with an optimum rutile-content would ideally contain a mixture of band-gap energy configurations shown in Figure 19(b) and (d). For the pure rutile-TiO_2, Figure 19(f), the band-gap energy would not change from one nanocrystallite to the other as their size, although different, would be well above the critical size required for the band-gap energy variation. This may drastically reduce the photo-induced e^-/h^+ life-time for the pure rutile-TiO_2, as experimentally demonstrated by others [56], which in turn would significantly reduce its photocatalytic activity. Smaller specific surface area of the pure rutile-TiO_2 would be another contributing factor for its lower photocatalytic activity.

Thus, the present model suggests that, the maximum photocatalytic activity would be obtained for the mixed-phase nanocrystalline TiO_2 having an optimum rutile-content. The various mechanisms discussed above, based on the band-gap energy variation in the connected nanocrystallites as a function of the size distribution and the phases involved, operate for different rutile-content and have been summarized in Figure 17(b), which is an experimentally observed variation in the k_{app} as a function of amount of rutile. Thus, the new model proposed here satisfactorily explains the existence of an optimum rutile-content in the mixed-phase nanocrystalline TiO_2, for observing the maximum photocatalytic activity, by overcoming the limitations of the models proposed earlier in the literature. It is believed that, the present model is valid irrespective of the processing method used for obtaining the mixed-phase nanocrystalline TiO_2 with varying rutile-content.

5.2.6. Surface-functionalization

As discussed earlier in the section-2.1, the formation of negatively charged super-oxide ions on the surface of pure nanocrystalline sol-gel anatase-TiO_2 results in the creation of a space-charge-layer. This space-charge-layer is responsible for the surface-adsorption of MB dye molecules, which are cationic in an aqueous solution [46]. Hence, it is obvious that, the variation in the surface-concentration of super-oxide ions would affect the surface-adsorption of MB, and hence, the photocatalytic activity of nanocrystalline sol-gel anatase-TiO_2.

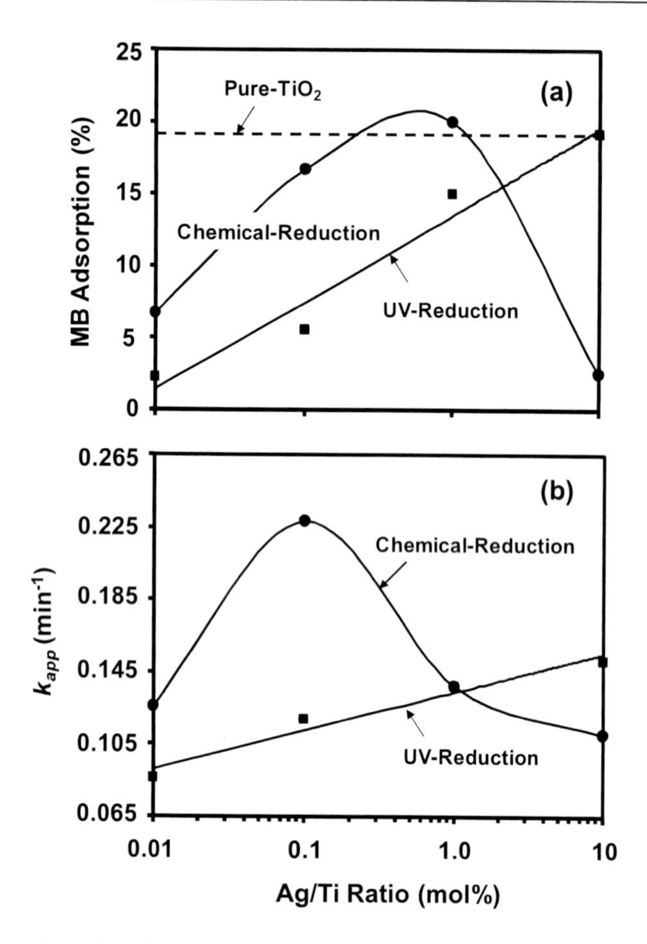

Figure 20. Variation in k_{app} (a) and amount of surface-adsorbed MB (b) as a function of Ag/Ti ratio as obtained for Ag_2O/Ag^0-deposited nanocrystalline anatase-TiO_2 processed via two different methods. In (a), the k_{app} value for the pure nanocrystalline sol-gel anatase-TiO_2, processed with the 'R' value of 90 and calcined at 600 °C-2 h, is 0.065 min^{-1} [24]. Copyright 2009 American Chemical Society.

As shown by the dotted line in Figure 20(a), the nanocrystalline sol-gel anatase-TiO_2 adsorbs 19.2% MB on the surface under the dark-condition before an exposure to the UV-radiation. When Ag_2O/Ag^0 are surface-deposited via UV-reduction method, with the lowest Ag-concentration of 0.01%, a drastic reduction in the amount of surface-adsorbed MB is noted as the concentration of potential surface-adsorption sites for MB, namely the super-oxide ions, is possibly reduced due to the formation of significant amount of Ag_2O via reverse-spillover effect, Eqs. 12-13. The MB adsorption is, however, noted to increase almost linearly with increasing Ag/Ti ratio within the range of 0.01-10 mol%. It is well known that, Ag^0 acts as a catalyst for the spillover of oxygen on the surface of nanocrystalline sol-gel anatase-TiO_2, Eqs. 5 and 6 [59]. Larger the surface-coverage of Ag^0-catalyst, stronger would be the oxygen-spillover effect. It appears that, with increasing Ag-concentration, the loss in the concentration of super-oxide ions due to the formation of significant amount of Ag_2O is gradually compensated by an enhanced oxygen-spillover effect of Ag^0 and the super-oxide formation on the surface of Ag_2O [24]. As a result, the amount of MB adsorbed on the powder-surface, increases with Ag/Ti ratio within the investigated range of 0.01-10 mol%, which is highly favorable in enhancing the photocatalytic activity. In the case of chemical-reduction method as well, for the lowest Ag/Ti ratio of 0.01 mol%, decrease in the amount of surface-adsorbed MB is noted due to the Ag_2O/Ag^0-deposition, which consumes the potential sites for the MB adsorption. However, the amount of MB adsorbed is relatively larger than that observed in the previous case. According to Eqs. 5 and 6, an excess surface-concentration of oxygen-ion vacancies, produced as a result of the charge-balance due to the presence of Sn^{2+}-ions, possibly may lead to higher surface-concentration of super-oxide ions, which in turn may increase the amount of surface-adsorbed MB. Within the Ag/Ti range of 0.01-1.0 mol%, higher amount of surface-adsorbed MB relative to that observed for the UV-reduction method is, hence, attributed to higher surface-concentration of oxygen-ion vacancies caused by the presence of Sn^{2+}-ions. This effect of Sn^{2+}-ions is absent for the UV-reduction method as Sn^{2+}-ions are not involved in this process.

For the chemical-reduction method, at the highest Ag/Ti ratio of 10 mol%, a sudden decrease in the amount of surface-adsorbed MB is noted, Figure 20(a), which is not observed for the UV-reduction method. There are three possible major contributing factors to this effect [24]. First, at the highest Ag/Ti ratio of 10 mol%, most of the Sn^{2+}-ions present on the surface may have been utilized for reducing the silver-ammonium complex-ions to Ag_2O/Ag^0. As a result, an excess surface-concentration of oxygen-ion vacancies, which was generated due the presence of Sn^{2+}-ions is possibly eliminated, which in turn may reduce the surface-concentration of super-oxide ions. Second, the formation of SnO_2 was detected for the highest Ag-concentration, which may also reduce the surface-concentration of super-oxide ions via reverse-spillover effect, Eq. 15 and 16. Third, at the highest Ag-concentration, the XPS analysis indicated the absence of Ag^0 on the powder-surface for the present method. This may relatively reduce the oxygen-spillover effect, thus reducing the surface-concentration of super-oxide ions. The net effect of these three possible contributing factors is reflected in a drastic reduction in the amount of surface-adsorbed MB as observed for the highest Ag/Ti ratio of 10 mol%.

The experimental variation in k_{app} as a function of Ag/Ti ratio as obtained for the Ag_2O/Ag^0 deposited nanocrystalline sol-gel anatase-TiO_2 processed via two different techniques is presented in Figure 20(b). For both the reduction techniques, an enhanced photocatalytic activity has been observed for the Ag_2O/Ag^0-deposited nanocrystalline sol-gel anatase-TiO_2, relative to that (0.065 min^{-1}) of pure nanocrystalline sol-gel anatase-TiO_2. This has been attributed to the result of an increased photo-induced e^-/h^+ life-time caused by an effective trapping of the photo-induced electrons by the deposited-Ag_2O/Ag^0 [45]. Moreover, the nature of the variation in the photocatalytic activity is primarily governed by the corresponding variation in the amount of surface-adsorbed MB as a function of Ag/Ti ratio.

The variation in the normalized residual MB concentration as a function of solar-radiation exposure time, as obtained for the pure and surface-modified nanocrystalline sol-gel anatase-TiO_2, is presented in Figure 21(a); while, the corresponding plots for obtaining an apparent first-order reaction rate-constant (k_{app}) is presented in Figure 21(b). The k_{app} values for the pure and surface-modified nanocrystalline sol-gel anatase-TiO_2 have been estimated to be 0.147 min^{-1} and 0.310 min^{-1}. Thus, the surface-deposition of Ag_2O and Ag^0 (predominantly Ag_2O) is observed to enhance the photocatalytic activity of the nanocrystalline sol-gel anatase-TiO_2 under the solar-radiation by the factor of 2. Since the solar-radiation contains both the UV- and visible-radiations, an enhanced solar-radiation induced photocatalytic activity has been attributed to the absorption of the visible-radiation and higher life-time of the photo-induced e^-/h^+ pairs (Figure 11), caused by the presence of

Ag_2O and Ag^0 (predominantly Ag_2O) on the surface-modified nanocrystalline sol-gel anatase-TiO_2. Thus, it is shown that, the surface-modified nanocrystalline sol-gel anatase-TiO_2 is suitable for the dye-removal application under the solar-radiation.

6. Recent Trends in Application of Nanocrystalline Sol-Gel Titania as a Photocatalyst

In the previous section, the nanocrystalline sol-gel anatase-TiO_2 is shown to be photocatalytically active for the degradation of MB dye in an aqueous solution under the UV- or solar-radiation exposure. In this section, some of the emerging trends in the novel applications of the nanocrystalline sol-gel TiO_2 as a photocatalyst are described.

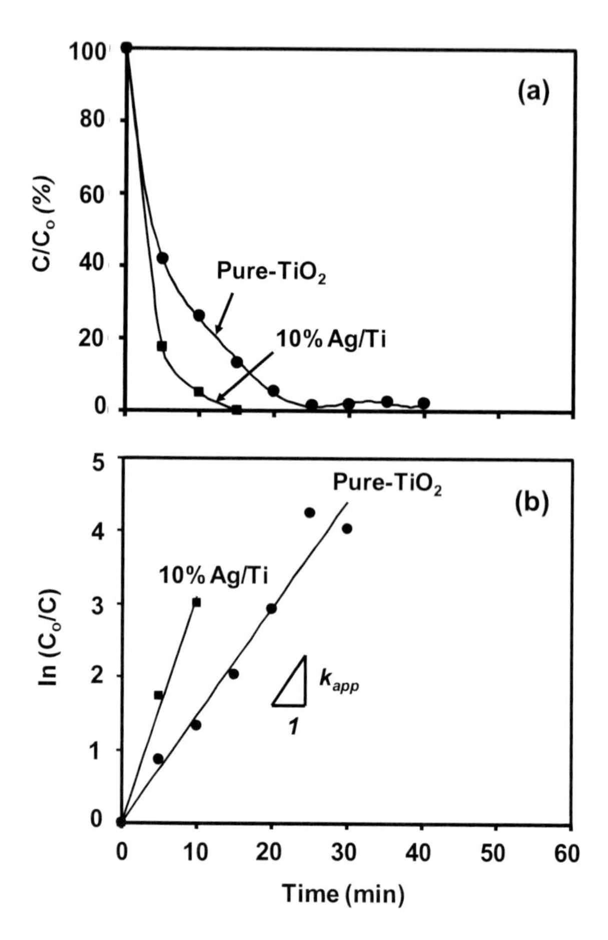

Figure 21. (a) Variation in the normalized residual MB dye concentration as a function of solar-radiation exposure time as obtained for the nanocrystalline sol-gel anatase-TiO_2, processed with the '*R*' value of 90 and calcined at 600 °C-2 h. (b) Corresponding plots for determining the k_{app} values [27]. Copyright 2009 Springer.

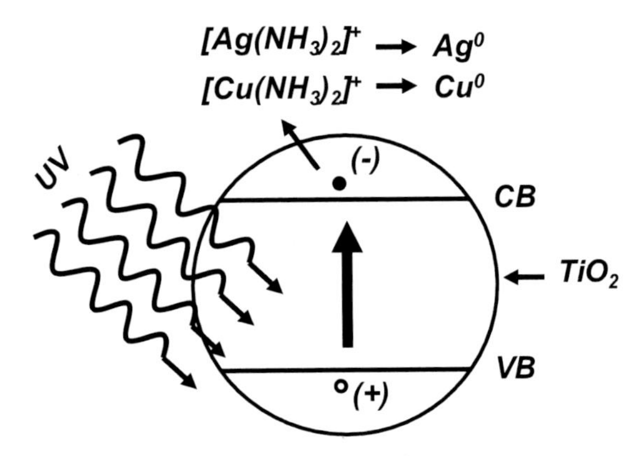

Figure 22. Proposed mechanism of using the sol-gel TiO_2 photocatalyst as a surface-sensitizer for the subsequent surface-activation and Cu or Ag coating of flyash via novel electroless process [25]. Copyright 2009 Springer.

6.1. Surface-Sensitizer for Metal Deposition

As discussed earlier in the section-5.1, photocatalysis using the nanocrystalline sol-gel TiO_2 involves the generation of e^-/h^+ pairs via exposure to the UV- or solar-radiation, which subsequently take part in the $–OH^•$ radical formation on the particle surface, which attack and degrade the MB dye molecules. As described schematically in Figure 22, the photo-induced electrons can also be utilized to reduce the metal-cations such as copper- or silver-ions (Cu^+ or Ag^+) from an aqueous solution to deposit the conducting metal such as Cu and Ag on the non-conducting ceramic substrates [25].

Flyash, a ceramic byproduct of thermal power plants, is a waste and environmentally hazardous material, which pose major disposal problems. Efforts are ongoing to recycle this byproduct via surface-modifications to reduce the associated environmental issues. Typically, due to their lower density, the Cu or Ag-coated flyash find application in manufacturing the conducting polymers for the electromagnetic interference (EMI) shielding application. Flyash is primarily composed of a mixture of silica (SiO_2) and alumina (Al_2O_3) with the trace amount of calcium oxide (CaO) and TiO_2. Due to the presence of small amount of TiO_2 on the surface, the flyash particles do exhibit some amount of photocatalytic activity, which can be further enhanced by depositing TiO_2 on its surface via sol-gel.

In order to deposit Cu or Ag on the flyash particle surface, the conventional electroless process has been recently modified [25]. In Figure 23, the mechanisms of Cu-coating of flyash particles via conventional electroless process, which utilizes the Sn-Pd catalyst-system [60] is compared with that of the modified process, which utilizes the TiO_2/UV-radiation/metal catalyst-system, where the metal is Cu or Ag. Both the processes begin with the as-received flyash and end with the Cu-coated flyash. However, these processes significantly differ in the intermediate techniques utilized to surface-sensitize and surface-activate the flyash particles. In the conventional electroless process, the flyash particles are surface-sensitized by adsorbing Sn^{2+}-ions by stirring the particles in an acidic aqueous bath of tin chloride ($SnCl_2$), Figure 23(a). On the other hand, in the modified electroless process, Figure 23(b), the flyash particles are surface-sensitized by coating the particles with TiO_2 via sol-gel by using an alkoxide-precursor followed by calcination at higher temperature for converting amorphous-TiO_2 into crystalline-TiO_2. Secondly, in the conventional electroless process, the flyash particles are surface-activated by stirring the surface-sensitized particles in an acidic aqueous palladium chloride ($PdCl_2$) bath to surface-deposit Pd-clusters as surface-catalyst, Figure 23(a). On the other hand, in the modified electroless process, the flyash particles are surface-activated by stirring the surface-sensitized particles in a basic (pH is adjusted using an aqueous ammonium hydroxide (NH_4OH) solution) aqueous bath, under the UV-radiation exposure, containing Cu- or Ag-ions so as to deposit Cu- or Ag-clusters as surface-catalyst, Figure 22 and 23(b). In both the processes, the surface-activated flyash particles are then stirred in a conventional electroless bath to deposit thicker Cu- or Ag-coating.

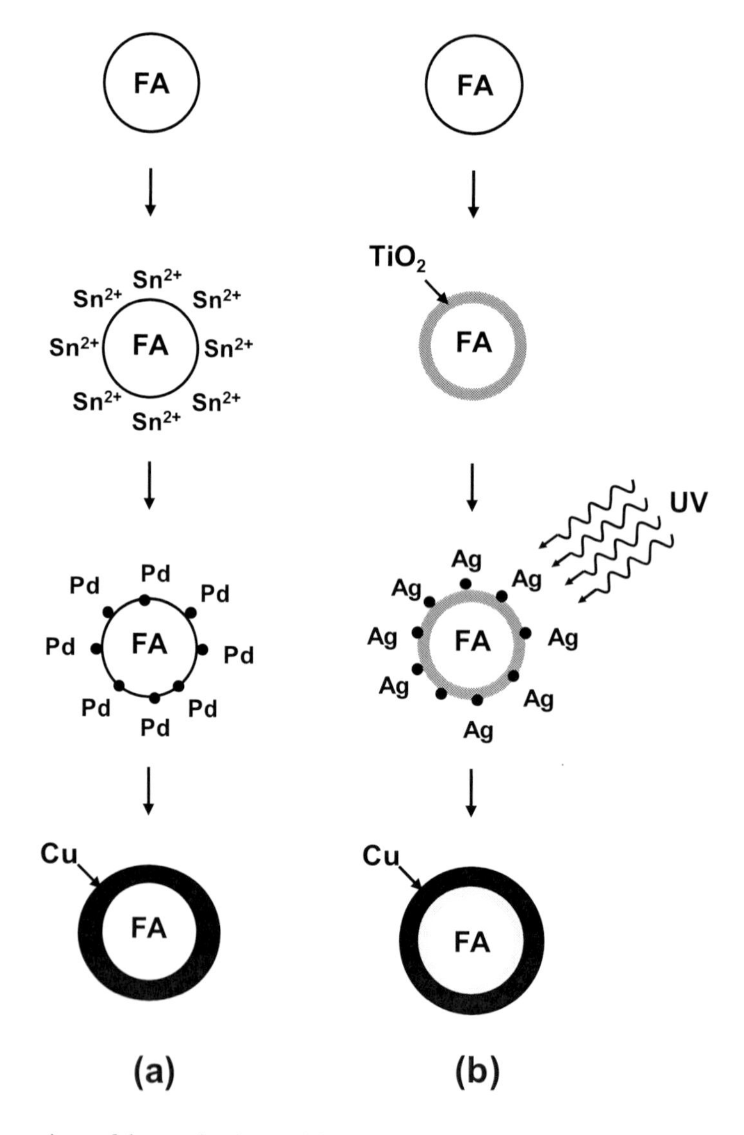

Figure 23. Comparison of the mechanisms of Cu-coating of flyash via conventional (a) and novel (b) electroless processes. FA represents the flyash particle [25]. Copyright 2009 Springer.

The SEM images of the as-received and Cu-coated flyash particles, as obtained via modified approach, are presented in Figure 24(a) and (c). The corresponding energy dispersive X-ray (EDX) analysis spectra are shown in Figure 24(b) and (d). It is noted that, the as-received flyash possesses a featureless surface and primarily shows the presence of Si, Al, Ca, and Ti near the surface. On the other hand, a dense coating of small particles is clearly visible on the surface of Cu-coated flyash particle. The EDX spectrum clearly identifies an additional presence of Cu on the surface after the modified electroless process. Comparison of the XRD spectra of the as-received and Cu-coated flyash particles, Figure 25, further confirms the crystalline nature of as-deposited Cu^0. Thus, for the modified electroless process, the quality of the Cu-coating obtained is comparable with that reported earlier for the conventional electroless process [60]. Thus, the nanocrystalline sol-gel TiO_2 is shown to be useful as a surface-sensitizer to deposit the conducting metals on the surface of non-conducting ceramic particles.

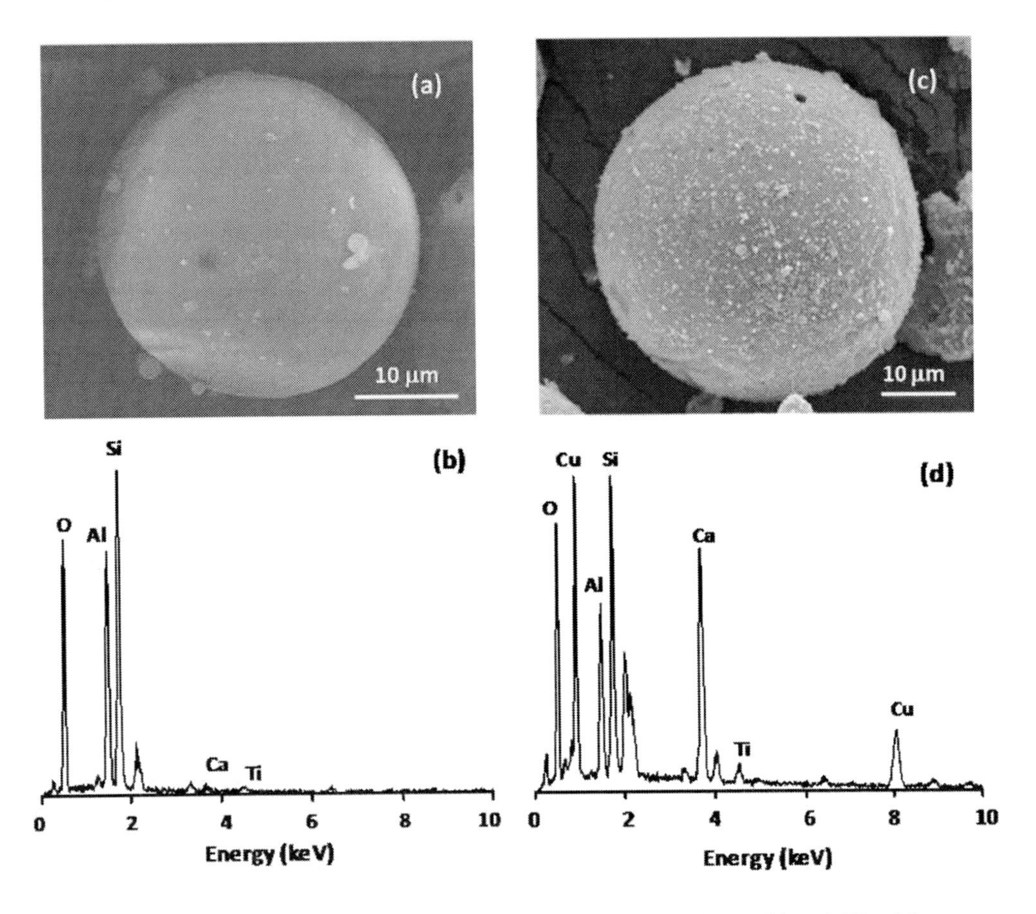

Figure 24. SEM micrographs, (a) and (c), and corresponding EDX analyses, (b) and (d), of the spherical-shaped as-received (a) and Cu-coated (b) flyash particles obtained using the '*R*' value of 2 for the surface-sensitization and Ag as a surface-activator and [25]. Copyright 2009 Springer.

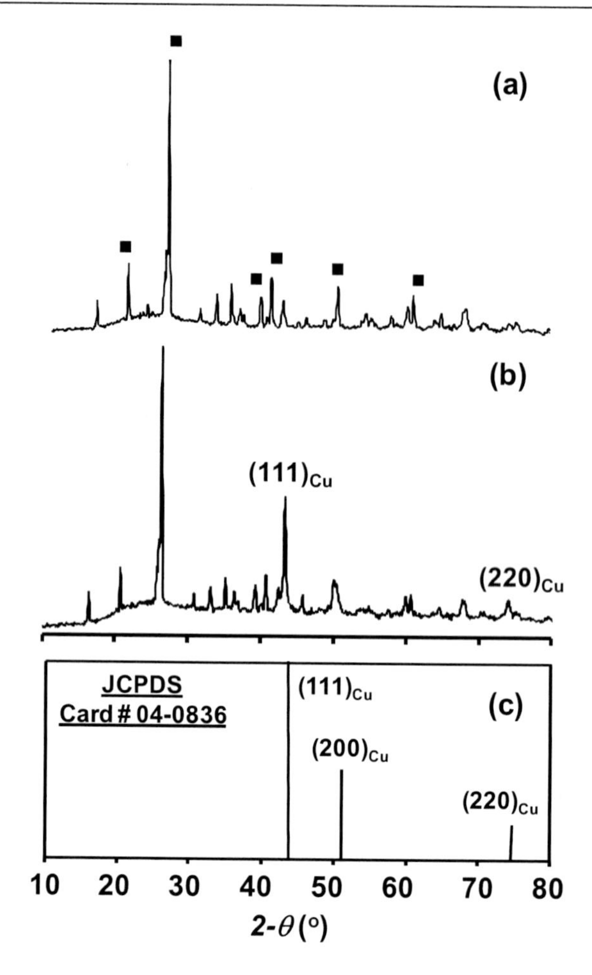

Figure 25. XRD patterns obtained for the as-received flyash calcined at 600 °C-2 h (a) and Cu-coated flyash (b) obtained using the '*R*' value of 2 for the surface-sensitization and Ag as a surface-activator. The diffraction peaks in (a), which are labeled using '■', correspond to those of silica (quartz) as identified using the JCPDS card # 83-0539. In (c), the standard diffraction pattern of pure-Cu^0, JCPDS card # 04-0836, is shown as a reference [25]. Copyright 2009 Springer.

6.2. Magnetic Photocatalyst

Although the semiconductor TiO_2 has been the most commonly applied photocatalyst, several issues have been associated in removing the photocatalyst from the treated effluents. The traditional methods for the separation such as coagulation, flocculation, and sedimentation are tedious. Moreover, the requirements of additional chemicals and purification stage to wash the coagulant from the photocatalyst make these processes expensive. The approach to overcome these problems has been to facilitate the photocatalyst removal using an external magnetic field. However, the TiO_2-based photocatalysts are inherently non-magnetic and cannot be separated using this technique. Hence, the TiO_2-based photocatalyst have been conventionally modified into a "core-shell" composite particles, comprising the core of a magnetic ceramic and the shell of nanocrystalline TiO_2 particles [61-

68]. Such core-shell composite structure, which is termed here as a "conventional magnetic photocatalyst", Figure 26(a), possesses both the magnetic and photocatalytic properties.

In a conventional magnetic photocatalyst, various magnetic materials including magnetite [61,62] hematite (Fe_2O_3) [63], manganese ferrite [64], nickel ferrite [65], barium ferrite [66], cobalt ferrite ($CoFe_2O_4$) [67], and nickel [68], have been used as a core; while, the coating of nanocrystalline TiO_2 has been popular as a shell. The latter has been developed using different techniques including sol-gel [61-64,66,67], and hydrolysis/precipitation [65]. In order to avoid an electrical contact between the TiO_2 shell and the magnetic core, an insulating layer of silica (SiO_2) or a polymer is deposited in between the two [67,69]. This intermediate layer acts as a barrier for the diffusion of core magnetic material into the photocatalyst layer during the calcination treatment and also for the photo-dissolution of the core magnetic material during the photocatalysis experiment [69]. The Stober and the microwave techniques have been commonly employed for obtaining an intermediate SiO_2 layer [66,70]. The deposition of metal-catalyst over the TiO_2-shell to increase the photocatalytic activity of the conventional magnetic photocatalyst has also been reported [71]. A typical "core-shell" morphology of the conventional magnetic photocatalyst with a mixture of $CoFe_2O_4$ and Fe_2O_3 as a core magnetic particle and sol-gel deposited nanocrystalline anatase-TiO_2 as a shell is shown in Figure 26(b).

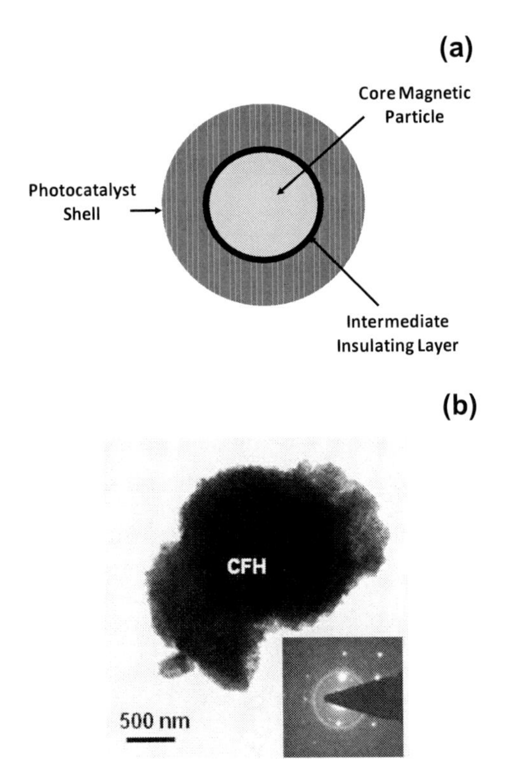

Figure 26. Schematic diagram (a) (Copyright 2010 John Wiley and Sons) and TEM image (b) of the conventional magnetic photocatalyst particle having a "core-shell" composite structure. The SAED pattern of the core magnetic ceramic particle is shown as an inset in (b). CFH represents a mixed cobalt ferrite and hematite magnetic particle.

The variation in the induced magnetization (M) as a function of applied magnetic field strength (H), as obtained for the conventional magnetic photocatalyst (calcined-sample), is presented in Figure 27(a). The presence of a hysteresis loop suggests the ferromagnetic nature of the conventional magnetic photocatalyst. The measured values of the saturation magnetization, remenance magnetization, and coercivity are 59 emu•g^{-1}, 24 emu•g^{-1}, 1410 Oe respectively. The variation in the residual MB dye concentration as a function of stirring time, under the UV-radiation exposure, is presented in Figure 27(b) for the conventional magnetic photocatalyst without and with the surface-metal (Pd) catalyst, which is obtained via UV-reduction method as described previously in the section-4.1. It is noted that, only ~45-60 % of MB dye is degraded after 70 min of UV-radiation exposure. Thus, the conventional magnetic photocatalyst, in the form of a composite particle, exhibits both the photocatalytic as well as the magnetic characteristics. The photocatalytic property is used for the dye-removal and the magnetic property is used for the separation of the photocatalyst from an aqueous solution after the dye-removal process.

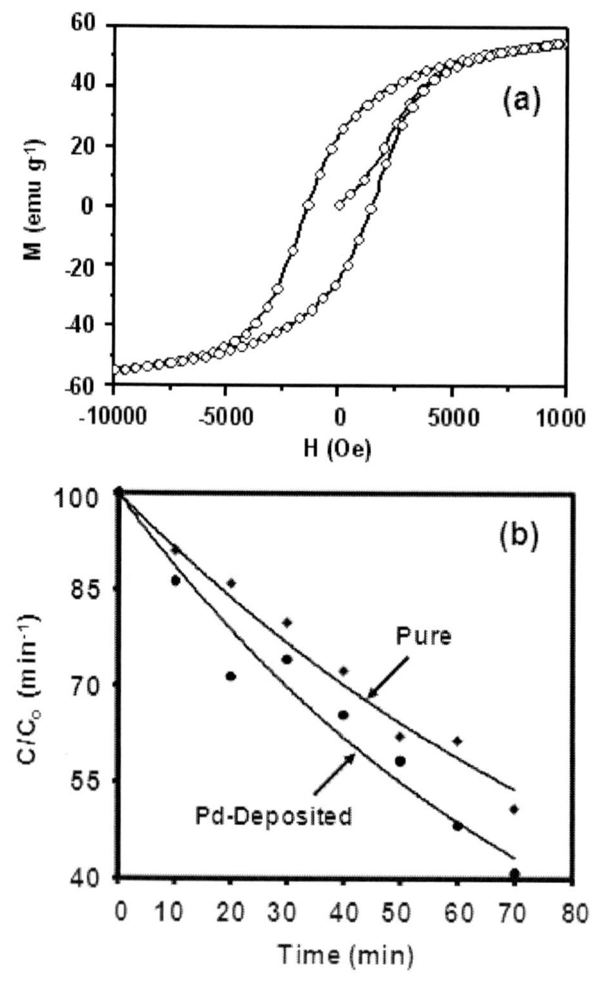

Figure 27. (a) Variation in the induced magnetization (M) as a function of applied magnetic field strength (H) as obtained for the conventional magnetic photocatalyst. Copyright 2010 John Wiley and Sons. (b) Variation in the normalized residual MB dye concentration as a function of UV-radiation exposure time as obtained under the different processing conditions.

However, the conventional magnetic photocatalyst has been associated with several drawbacks. First, as noted above, they show limited photocatalytic activity due to the presence of a core magnetic particle, which reduces the volume fraction of the photocatalyst available for the dye-removal. Second, the total time for a complete dye-removal using the conventional magnetic photocatalyst is substantially higher (few hours). Third, the dye-removal using the conventional magnetic photocatalyst is predominantly via photocatalytic degradation mechanism. Forth, being an energy-dependent mechanism (that is, requiring an exposure to the UV, visible-, or solar-radiation), the photocatalytic degradation is a relatively expensive process. Fifth, the dye-removal via other mechanism(s) such as the surface-adsorption, which is not an energy-dependent process (that is, it can be carried out in the dark) has never been utilized for the conventional magnetic photocatalyst. This has been mainly due to the non-suitability of the conventional magnetic photocatalyst for the surface-adsorption mechanism as a result of its lower specific surface-area. Sixth, the techniques to enhance the specific surface-area of the conventional magnetic photocatalyst have not been reported yet.

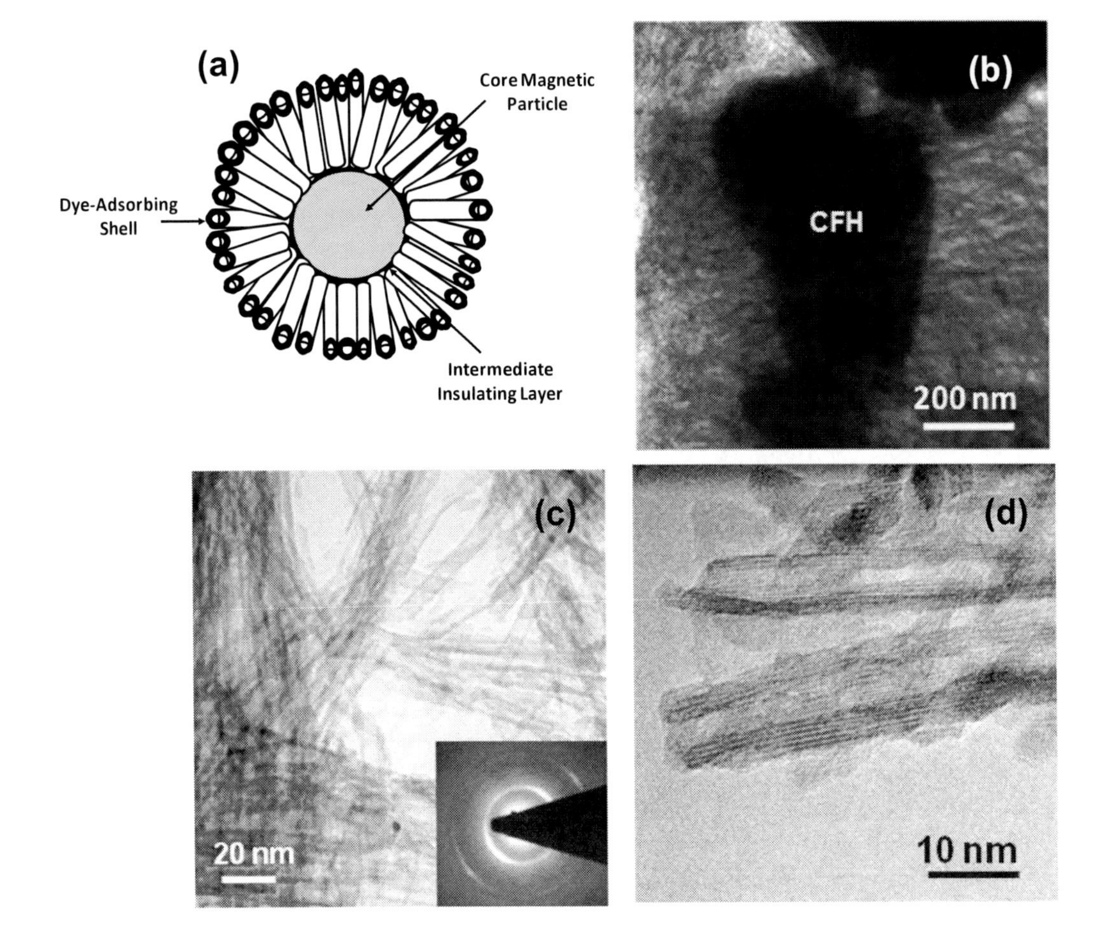

Figure 28. Schematic diagram (a) (Copyright 2010 John Wiley and Sons) and TEM/HRTEM images (b-d) of the magnetic dye-adsorbent catalyst having a "core-shell" composite structure. The SAED pattern of the nanotubes is shown as an inset in (c). CFH represents mixed cobalt ferrite and hematite magnetic particle.

6.3. Magnetic Dye-Adsorbent Catalyst

In order to overcome the major limitations of the conventional magnetic photocatalyst, having lower specific surface-area, a "magnetic dye-adsorbent catalyst", having higher specific surface-area, has been developed [72,73]. It consists of a composite structure with the core of a magnetic ceramic particle and the shell of nanotubes of a dye-adsorbent catalyst, as schematically described in Figure 28(a). For this, the conventional magnetic photocatalyst has been subjected to a hydrothermal treatment (high temperature and pressure conditions in an autoclave) under highly alkaline condition (10 M sodium hydroxide (NaOH) solution). Such treatment is then followed by a typical washing cycle using 1 M HCl acid solution and distilled-water, which results in the formation of nanotubes of hydrogen titanate ($H_2Ti_3O_7$) on the surface of core ceramic particles. Such conversion is accompanied by a concurrent change in the mechanism of organic dye-removal from an aqueous solution from the photocatalytic degradation to the surface-adsorption due to an enhanced specific surface-area of the product. Typical TEM/HRTEM images of the "magnetic dye-adsorbent catalyst" are presented in Figure 28(b)-(d). In Figure 28(b), the dark area corresponds to the core magnetic ceramic particle; while, the surrounding area with lower contrast corresponds to the $H_2Ti_3O_7$ nanotubes, which are clearly revealed at higher magnifications, Figure 28(c) and (d).

The variation in the induced magnetization (M) as a function of applied magnetic field strength (H), as obtained for the magnetic dye-adsorbent catalyst (calcined-sample), is presented in Figure 29(a). Similar to the previous case, the presence of a hysteresis loop suggests the ferromagnetic nature of the magnetic dye-adsorbent catalyst. The measured values of the saturation magnetization, remenance magnetization, and coercivity are 45 emu•g^{-1}, 15 emu•g^{-1}, 578 Oe respectively. Comparison shows that, the obtained values are lower than those obtained for the conventional magnetic photocatalyst. This has been attributed to the formation of nanotubes, which reduce the volume fraction of the core magnetic ceramic particle, and an increase in the average particle size of the latter following the hydrothermal and calcination treatments. Surprisingly, the magnetic dye-adsorbent catalyst possesses very high MB dye adsorption capacity under the dark-condition. Typical variation in the normalized concentration of the surface-adsorbed MB as a function of stirring time in the dark, as obtained for the conventional magnetic photocatalyst (calcined-sample) and the magnetic dye-adsorbent catalyst (dried-sample and calcined-sample), is presented and compared in Figure 29(b). It is noted that, in general, the surface-adsorption increases rapidly within the first 30 min of stirring time. A near-saturation level is reached with further increase in stirring time within the time interval of 30-180 min. The conventional magnetic photocatalyst shows lower surface-adsorption (~50-55 %) of MB dye under the dark-condition. On the other hand, the magnetic dye-adsorbent catalyst shows very high surface-adsorption capacity (~90-100 %) in just 30 min depending on the processing conditions. Slight decrease in the maximum surface-adsorption capacity of the calcined-sample relative to that of the dried-sample is noted possibly due to the loss in the specific surface-area due to the thermal treatment. Comparison of Figure 27(b) and 29(b), thus, shows that the conventional magnetic photocatalyst removes the MB dye predominantly via photocatalytic degradation mechanism under the UV-radiation exposure; while, the magnetic dye-adsorbent catalyst removes the dye predominantly via surface-adsorption mechanism under the dark-condition. Hence, the morphology of the coating (that is, nanocrystalline particles or nanotubes) on the surface of a core magnetic ceramic particles plays a significant role in governing the dye-removal mechanism. The qualitative variation in the initial blue color of an aqueous MB dye solution, for the above three different catalysts, is shown in Figure 30. It is seen that, only small change in the initial blue color is noted for the conventional magnetic photocatalyst for the total stirring time of 180 min under the dark-condition; whereas, almost colorless solution is obtained in just 30 min when the magnetic dye-adsorbent catalysts are used, Figure 30(a). As demonstrated in Figure 30(b), the initial white-colored nanotubes-based photocatalyst powder (without the core magnetic ceramic particle) becomes blue after the surface-adsorption of MB, which can be removed from the powder-surface, under the dark-condition, for the reuse of catalyst powder for the next cycle of dye-adsorption [73]. Hence, due to its greater efficiency and cost-effectiveness, the magnetic dye-adsorbent catalyst provides a new and promising approach for treating the textile effluents than that provided by the conventional magnetic photocatalyst.

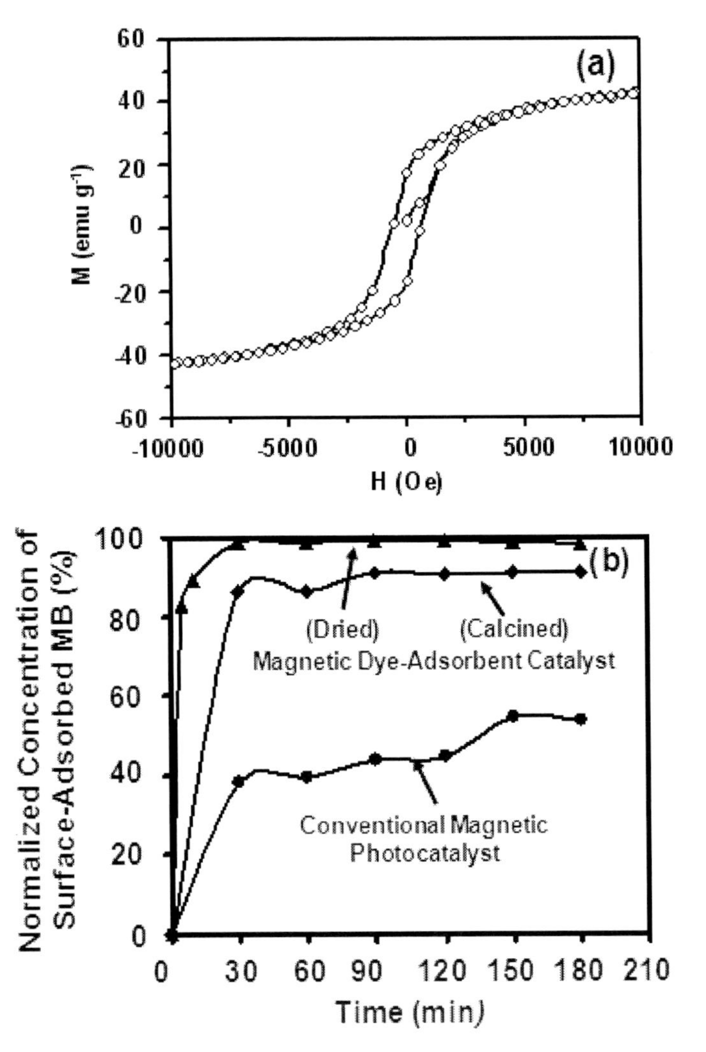

Figure 29. (a) Variation in the induced magnetization (M) as a function of applied magnetic field strength (H), as obtained for the magnetic dye-adsorbent catalyst (calcined-sample). (b) Variation in the normalized concentration of surface-adsorbed MB as a function of stirring time, under the dark-condition, as obtained for different samples. Copyright 2010 John Wiley and Sons.

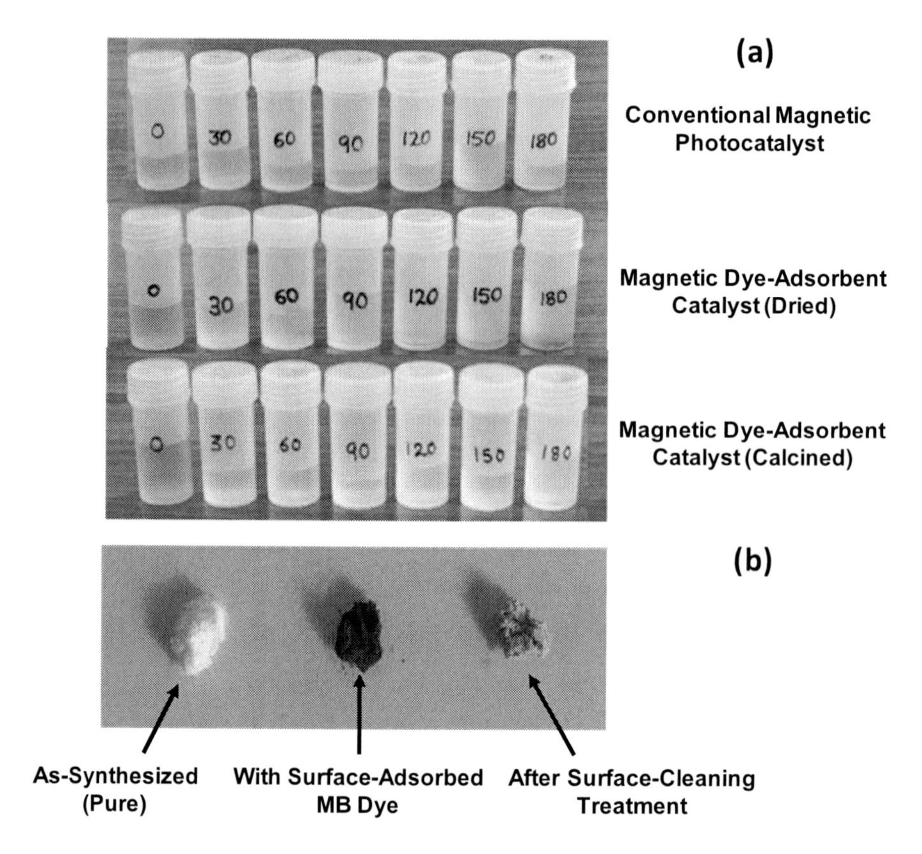

Figure 30. (a) Digital photographs showing the variation in the initial blue color of the MB dye solution as a function of stirring time (in min), under the dark-condition, as obtained for different samples. (b) Digital photographs showing a typical variation in the color of hydrogen titanate nanotubes (without the core magnetic particle) under different conditions.

7. Summary

Sol-gel is a versatile wet-chemical approach to process the nanocrystalline TiO_2 for a typical industrial organic dye-removal application via photocatalysis mechanism. The kinetics of an organic dye-removal using the nanocrystalline sol-gel TiO_2 can be controlled by tuning the major processing parameters such as '*R*' and calcination temperature, which in turn control the various material parameters such as an average nanocrystallite size, specific surface-area, porosity, crystallinity, amount and nature of the evolving phases. The processing of nanocrystalline sol-gel TiO_2 in the presence of an organic polymer can effectively control the specific surface-area and average nanocrystallite size of the photocatalyst; however, it also affects the surface-purity of the powders during its removal via high temperature calcination, which interferes with the dye-removal process by reducing the amount surface-adsorbed catalyst. The metal-deposition on the surface of nanocrystalline sol-gel TiO_2 not only improves its photocatalytic activity under the UV-radiation exposure, but also leads to higher rates of dye-removal under an exposure to the solar-radiation. Nanocrystalline sol-gel TiO_2 is demonstrated to be useful as a surface-sensitizer in the modified electroless process for depositing the conducting metals such as Cu or Ag on the surface of non-conducting ceramic-substrates. The conventional magnetic photocatalyst provides a convenient way for the separation of nanocrystalline sol-gel TiO_2 photocatalyst, from an aqueous solution, using an external magnetic field after the dye-degradation process. However, it seriously suffers from several demerits, which are recently overcome via the development of magnetic dye-adsorbent catalyst, which primarily relies on the surface-adsorption mechanism for an organic dye-removal, under the dark condition. This mechanism is in contrast to the photocatalytic degradation mechanism, under the UV-radiation exposure, associated with the conventional magnetic photocatalyst, and hence, it provides a cost-effective and efficient approach for treating the industrial effluents than that offered by the existing ones.

ACKNOWLEDGMENTS

Author thanks CSIR, India for funding the photocatalysis, nanotechnology, and ceramic research at the NIIST-CSIR (Project # NWP0010) and the Indo-US Science and Technology Forum (IUSSTF) Research Fellowship (Grant # P81113) at the Argonne National Laboratory (2009-2010).

REFERENCES

[1] Forgacsa, E; Cserhati, T; Oros, G. *Environ. International,* 2004, 30, 953-971.
[2] Gupssssta, GS; Shukla, SP; Prasad, G; Singh, VN. *Environ. Technol.,* 1992, 13, 925-936.
[3] Shukla, SP; Gupta, GS. Ecotoxicol. *Environ. Saf.,* 1992, 24,155-163.
[4] Sokolowska-Gajda, J; Freeman, HS; Reife, A. *Dyes Pigments,* 1996, 30, 1-20.
[5] Robinson, T; McMullan, G; Marchant, R; Nigam, P. *Bioresource Technol.,* 2001, 77, 247-255.
[6] Shaul, GM; Holdsworth, TJ; Dempsey, CR. Dostal KA. *Chemosphere,* 1991, 22, 107-119.
[7] Gupta, VK; Suhas *J. Environ. Manage.,* 2009, 90, 2313–2342.
[8] Carp, O; Huisman, CL; Reller, A. *Prog. Solid State Ch.,* 2004, 32, 33-177.
[9] Tachikawa, T; Fujitsuka, M; Majima, T. *J. Phys. Chem., C* 2007, 111, 5259-5275.
[10] Chen, X; Mao, SS. *Chem. Rev.,* 2007, 107, 2891-2959.
[11] Fujishima, A; Zhang, X; Tryk, DA. *Surf. Sci. Rep.,* 2008, 63, 515-582.
[12] Martoa, J; Marcosa, PS; Trindadeb, T; Labrinchaa, JA. *J. Haz. Mat.,* 2009, 163, 36-42.
[13] Pan, SS; Shen, YD; Teng, XM; Zhang, YX; Li, L; Chu, ZQ; Zhang, JP; Li, GH; Hub, X. *Mater. Res. Bull.,* 2009, 44, 2092-2098.
[14] Feng, S; Zhaoa, J; Zhua, Z. *Mater. Sci. Eng., B* 2008, 150, 116-120.
[15] Datta, A; Priyama, A; Bhattacharyya, SN; Mukherjea, KK; Saha, A. *J. Colloid Interf Sci.,* 2008, 322, 128-135.
[16] Mills, A; Hill, G; Bhopal, S; Parkin, IP; O'Neil, SA. *J. Photoch. Photobio., A* 2003, 160, 185-194.

[17] Ambrus, Z; Bala'zs, N; Alapi, T; Wittmann, G; Sipos, P; Dombi, A; Mogyorosi, K. *Appl. Catal., B* 2008, 81, 27-37.

[18] Su, C; Hong, BY; Tseng, CM. *Catal. Today,* 2004, 96, 119-126.

[19] Mohapatra, P; Mishra, T; Parida, KM. *Appl. Catal., A* 2006, 310, 183-189.

[20] Mills, A; Elliott, N; Parkin, IP; O'Neill, SA; Clark, RJ. *J. Photoch. Photobio.,* 2002, 151, 171-179.

[21] Burlacov, I; Jirkovský, J; Müller, M; Heimann, RB. *Surf. Coat. Tech.,* 2006, 201, 255-264.

[22] Baiju, KV; Shukla, S; Biju, S; Reddy, MLP; Warrier, KGK. *Catal. Lett.,* 2009, 131, 663-671.

[23] Baiju, KV; Zachariah, A; Shukla, S; Biju, S; Reddy, MLP; Warrier, KGK. *Catal. Lett.,* 2009, 130, 130-136.

[24] Priya, R; Baiju, KV; Shukla, S; Biju, S; Reddy, MLP; Patil, KR; Warrier, KGK. *J. Phys. Chem., C* 2009, 113, 6243-6255.

[25] Shijitha, T; Baiju, KV; Shukla, S; Patil, K; Warrier, KGK. *Appl. Surf. Sci.,* 2009, 255, 6696-6704.

[26] Baiju, KV; Shukla, S; Biju, S; Reddy, MLP; Warrier, KGK. *Mater. Lett.,* 2009, 63, 923-926.

[27] Priya, R; Baiju, KV; Shukla, S; Biju, S; Reddy, MLP; Patil, KR; Warrier, KGK. *Catal. Lett.,* 2009, 128, 137-143.

[28] Zachariah, A; Baiju, KV; Shukla, S; Deepa, KS; James, J; Warrier, KGK. *J. Phys. Chem. C,* 2008, 112, 11345-11356.

[29] Baiju, KV; Shukla, S; Sandhya, KS; James, J; Warrier, KGK. *J. Sol-Gel Sci. Technol.,* 2008, 45, 165-178.

[30] Baiju, KV; Shukla, S; Sandhya, KS; James, J; Warrier, KGK. *J. Phys. Chem., C* 2007, 111, 7612-7622.

[31] Mohapatra, P; Parida, KM. *J. Mol. Catal.,* 2006, 258, 118-123.

[32] Zhang, H; Banfiled, JF. *J. Mater. Chem.,* 1998, 8, 2073-2076.

[33] Reidy, DJ; Holmes, JD; Morris, MA. *J. Europ. Ceram. Soc.,* 2006, 26, 1527–534.

[34] Qi, JQ; Wanga, Y; Chen, WP; Tian, Li, LT; Chan, HL. W. *J. Alloy Compd,* 2006, 413, 307-311.

[35] Shukla, S; Seal, S. *Inter. Mater. Rev.,* 2005, 50, 45-64.

[36] Zhou, XD; Huebner, W. *Appl. Phys. Lett.,* 2001, 79, 3512-3514.

[37] Okumura, M; Coronado, JM; Soria, J; Haruta, M; Conesay, JC. *J. Catal.,* 2001, 203, 168-174.

[38] Seal, S; Shukla, S. U.S. Patent Numbers 7, 288,324; 7,572,431; 7,595,036; 7,758,977.

[39] Shukla, S; Seal, S; Vanfleet, R. *J. Sol-Gel Sci. Technol.,* 2003, 27, 119-136.

[40] Jing, L; Qu, Y; Wang, B; Li, S; Jiang, B; Yang, L; Fu, W; Fu, H; Sun, J. *Sol. Energ. Mat. Sol. C.,* 2006, 90, 1773–1787.

[41] Sun, B; Vorontsov, AV; Smirniotis, PG. *Langmuir,* 2003, 19, 3151-3156.

[42] Nicole, J; Tsiplakides, D; Pliangos, C; Verikios, XE; Comninellis, Ch; Vayenas, CG. *J. Catal.,* 2001, 204, 23-34.

[43] Shukla, S; Seal, S. *Nanostruct. Mater.,* 1999, 11, 1181-1193.

[44] Yu, J; Xiong, J; Cheng, B; Liu, S. *Appl. Catal., B* 2005, 60, 211-221.

[45] Kuo, YL; Chen, HW; Ku, Y. *Thin Solid Films,* 2007, 515, 3461-3468.

[46] Lachheb, H; Puzenat, E; Houas, A; Ksibi, M; Elaloui, E; Guillard, C; Herrmann, JM. *Appl. Catal. B,* 2002, 39, 75-90.
[47] Zhang, A; Wang, CC; Zakaria, R; Ying, JY. *J. Phys. Chem. B,* 1998, 102, 10871-10878.
[48] Inagaki, M; Imai, T; Yoshikawa, T; Tryba , B. *Appl. Catal., B* 2004, 51, 247-254.
[49] Zhao, G; Tian, Q; Liu, Q; Han, G. *Surf. Coat. Tech.,* 2005, 98, 55-58.
[50] Manorama, SV; Reddy, KM; Reddy, CVG; Narayanan, S; Raja, PR; Chatterji, PR. *J. Phys. Chem. Solids,* 2002, 63, 135-143.
[51] Wang, W; Silva, CG; Faria, JL. *Appl. Catal., B* 2007, 70, 470-478.
[52] Gao, Y; Liu, H. *Mater. Chem. Phys.,* 2005, 92, 604-608.
[53] Qourzal, S; Assabbane, A; Ichou, YA. *J. Photchem. Photobiol., A* 2004, 163, 317-321.
[54] Lin, L; Lin, W; Zhu, YX; Zhao, BY; Xie, YC; He, Y; Zhu, YF. *J. Mol. Catal., A* 2005, 236, 46-53.
[55] Wang, CY; Pagel, R; Dohrmann, JK; Bahnemann, DW*C. R. Chimie,* 2006, 9, 761-773.
[56] Kolen'ko YV; Churagulov, BR; Kunst, M; Mazerolles, L; Justin, CC. *Appl. Catal., B* 2004, 54, 51-58.
[57] Venktatachalam, N; Palanichamy, M; Murugesan, V. *Mater. Chem. Phys.,* 2007, 104, 454-459.
[58] Lin, H; Huang, CP; Li, W; Ni, C; Shah, SI; Tseng, YH. *Appl. Catal., B* 2006, 68, 1-11.
[59] Yang, X; Xu, L; Yu, X; Guo, Y. *Catal. Commun.,* 2008, 9, 1224-1229.
[60] Shijitha, T; Baiju, KV., Shukla, S; Patil, K; Warrier, KGK. *Appl. Surf. Sci.,* 2009, 255, 6696–6704.
[61] Beydoun, D; Amal, R; Scott, J; Low, G; Evoy, SM. *Chem. Eng. Technol.,* 2001, 24, 745-748.
[62] Song, X; Gao, L. *J. Am. Ceram. Soc.,* 2007, 90, 4015-4019.
[63] Gao, Y; Chen, B; Li, H; Ma, Y. *Mater. Chem. Phys.,* 2003, 80, 348-355.
[64] Xiao, HM; Liu, XM; Fu, SY. *Compos. Sci. Technol.,* 2006, 66, 2003-2008.
[65] Rana, S; Rawat, J; Sorensson, MM; Misra, RDK. *Acta Biomater.,* 2006, 2, 421-432.
[66] Lee, SW; Drwiega, J; Mazyckb, D; Wu, CY; Sigmunda, WM. Mater. *Chem. Phys.,* 2006, 96, 483-488.
[67] Fu, W; Yang, H; Li, M; Li, M; Yang, N; Zou, G. *Mater. Lett.,* 2005, 59, 3530-3534.
[68] Jiang, J; Gao, Q; Chen, Z; Hu, J; Wu, C. *Mater. Lett.,* 2006, 60, 3803-3808.
[69] Beydoun, D; Amal, R; Low, G; McEvoy, S. *J. Mol. Catal., A* 2002, 180, 193-200.
[70] Siddiquey, A; Furusawa, T; Sato, M; Suzuki, N. *Mater. Res. Bull.,* 2008, 43, 3416-3424.
[71] Xu, MW; Bao, SJ; Zhang, XG. *Mater. Lett.,* 2005, 59, 2194-2198.
[72] Thazhe, L; Shereef, A; Shukla, S; Reshmi, CP; Varma, MR; Suresh, KG; Patil, K; Warrier, KGK. *J. Am. Ceram. Soc.,* No. DOI: 10.1111/j.1551-2916.2010.03949.x.
[73] Shukla, S; Warrier, KGK; Varma, MR; Lajina, MT; Harsha, N; Reshmi, CP. PCT Application No. PCT/IN2010/000198.

In: The Sol-Gel Process
Editor: Rachel E. Morris

ISBN 978-1-61761-321-0

Chapter 4

SOL-GEL SYNTHESIS OF MESOPOROUS ORDERED NANOMATERIALS ON THE BASIS OF ALUMINA AND TITANIA

A. V. Agafonov[a*]***, V. V. Vinogradov***[a] ***and A. V. Vinogradov***[b]

[a]Institute of Solution Chemistry, Russian Academy of Sciences,
ul. Akademischeskaya 1, Ivanovo, 153045 Russia
[b]Ivanovo State University of Chemistry and Technology, pr.
Engel'sa 7, Ivanovo, 153000 Russia

ABSTRACT

Ordered nanoporous alumina and titania represent very interesting molecular sieves exhibiting a narrow pore size distribution with higher surface areas compared to conventional alumina and titania used as a support for catalytically active species and photocatalyst in numerous large-scale industrial processes. This contribution encompasses various synthesis approaches to ordered nanoporous alumina and titania, description of their structures and properties, and characterization by various experimental techniques. Self-assembly mechanism, micelle formation in solution is described as well. Unlike siliceous materials with ordered nanoporosity that are commonly prepared through the sol-gel process with surfactants as structure-directing agents or by utilizing the nanocasting method with silica or carbon materials as hard templates, synthesis of ordered and thermal stable alumina and titania represents a much more complex problem due to its susceptibility for hydrolysis as well as to the phase transitions accompanying the thermal breakdown of the ordered structure. Effects of synthesis conditions, calcination temperature, removal of surfactants and drying is discussed. The potential of the ordered alumina and titania with respect to use in catalysis and photocatalysis is considered.

[*] Corresponding author: e-mail: ava@isc-ras.ru

INTRODUCTION

The first successful synthesis of mesoporous molecular sieves opened a new era in studies of this kind of materials all over the world [1, 2]. This discovery offered new possibilities for synthesis of new types of molecular sieves with larger pore size compared to zeolites and narrow pore size distribution that are applicable not only in catalysis but also in other fields of chemistry. The mesoporous molecular sieves were first obtained by Mobil Scientific Research Corporation on the basis of silicas and aluminosilicates with fixed pore sizes ranging from 1.5 to 10 nm and had surface area of more than 1000 m^2/g. For this purpose a new approach was used making it possible to form material structure through self-assembly on the molecular template surfaces. Since then, a huge number of published works has been devoted to the synthesis aspects, internal structure characteristics and practical application of mesoporous molecular sieves [3-7].

The first successful synthesis of ordered mesoporous alumina and titania was produced slightly later. In the most cases the molecular sieve synthesis studies were focused on obtaining the mesoporous silicas or aluminosilicates that were stable in various chemical processes [8] and seemed to be the most perspective for modification of its surfaces by catalytically active phases. This is probably due to the fact that the synthesis of ordered mesoporous alumina and titania is more complex compared to the synthesis of mesoporous silica. The first synthesis of mesoporous alumina is described by Landau et al. [9]. The synthesis process was carried out on the surface of MCM-41 mesoporous silica that was subsequently covered with alumina forming an extra layer of final material.

Alumina is very perspective material with a wide spectrum of application as a medium modified by different catalytically active phases that is used in various technological industrial processes [10, 11]. Traditional alumina obtained by precipitation possesses specific surface area of 50–300 m^2/g. A special attention in literature is paid to the description of properties of alumina-based materials [12]. Different phase transitions of alumina (α, γ, δ, η and θ) during the precursor (boehmite, pseudoboehmite, bayerite et al.) thermal treatment process as well as the internal structure changes are well described in Ref. [13]. Traditional alumina possesses surface area as small as 350–400 m^2/g, and its main disadvantage is wide pore size distribution.

On the other hand, ordered mesoporous titania with high crystallinity was obtained for the improvement of its adsorption characteristics and semiconductor properties [14]. Such an approach has led to substantial development of photoactivity of similar materials, they have gained wide application as photocells, membranes, biomedical materials, however, the perspective of application of these materials as photocatalysts is still equivocal because of low crystallinity of formed samples. The successful results on structurization of ordered titania-based materials include 2D hexagonal with mesochannels running parallel to substrates [15], 3D cubic [16], and 3D hexagonal structures [17]. Evidently, the synthesis of such titania-based materials requires the use of stabilizing agent that, in its turn, would facilitate the increase in catalytic activity as well. That is why the newest approaches to obtaining the highly photoactive ordered TiO_2-based materials involve the polyvalent and rare-earth metal (Ce, Fe, In et al.) ions.

For this cause, synthesis of ordered mesoporous alumina and titania with surface area of more than 500 m^2/g and narrow pore size distribution is a highly perspective scientific direction.

For the production of ordered mesoporous alumina and titania synthesis, it was necessary to change and optimize processes that are well described and studied for the mesoporous silica synthesis. The hydrolysis behavior of alumina and titania is very complicated and strongly affected by acid, water, temperature, relative humidity, and other factors, giving rise to rather strict synthetic conditions for ordered mesoporous aluminas and titanias.

The main purpose of this chapter is to describe the regularities of synthesis and the prospects of ordered mesoporous alumina and titania application. Various variants of synthesis of ordered and organized mesoporous alumina and titania that are currently known, characteristics of molecular sieves, their main properties as well as the prospects of application in catalysis and photocatalysis are reflected in the present chapter.

1. Characterization of the Structure of Ordered Mesoporous Materials

In a similar way to mesoporous silicas and aluminosilicates, small angle X-ray scattering, sorption isotherms of nitrogen and transmission electron microscopy were mainly employed to characterize organized mesoporous aluminas.

1.1. Small Angle X-Ray Scattering (SAXS)

Small-angle X-ray scattering (SAXS) is a technique where the elastic scattering of X-rays by a sample which has inhomogeneities in the nanometer range, is recorded at very low angles (typically 0.1–10°). SAXS measures spatial correlations in the scattering density, averaged over the time scale of the measurement. Structures with electron density fluctuations on a length scale from several ten nanometers up to the micrometer regime give rise to this scattering. In a SAXS instrument a monochromatic beam of X-rays is brought to a sample from which some of the X-rays scatter, while most simply go through the sample without interacting with it. The scattered X-rays form a scattering pattern which is then detected at a detector which is typically a 2-dimensional flat X-ray detector situated behind the sample perpendicular to the direction of the primary beam that initially hit the sample. The scattering pattern contains the information on the structure of the sample. Reliable experimental methods are essential for characterizing tailored pore sizes in a reproducible manner. SAXS measurements typically are concerned even with scattering angles < 1°. As dictated by Bragg's Law, the diffraction information about structures with large *d*-spacings lies in the region. Therefore the SAXS technique is commonly used for probing large length scale structures such as high molecular weight polymers, biological macromolecules (proteins, nucleic acids, etc.), and self-assembled superstructures (e.g. surfactant templated mesoporous materials). In specific, SAXS is used for the determination of the microscale or nanoscale structure of particle systems in terms of such parameters as averaged particle sizes, shapes, distribution, and surface-to-volume ratio.

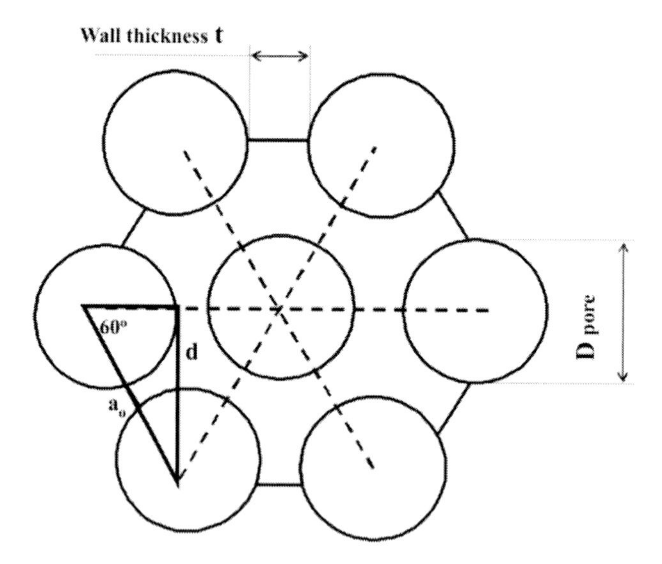

Figure 1. Evaluation of lattice parameter and wall thickness from the hexagonal pore arrangement.

Mesoporous materials usually have long-range order; however, the pore walls are amorphous. In the absence of such short-range order, they can simply be interpreted as semicrystalline solids. SAXS pattern provide valuable information on a scale of sizes larger than those covered by the more widely applied X-ray diffraction (XRD) at normal angles, and is ideally suited for studying porous structures in the nanometer range [18].

Micelle templated materials exhibit regular hexagonal pattern of cylindrical pores in the case of ordered SiO_2, Al_2O_3, and TiO_2. The unit-cell parameter, a_0, is calculated from the *d*-spacing as expressed by Bragg's equation:

$$n = 2 \cdot \lambda \cdot d \cdot \sin\theta,$$

$$a_0 = d \cdot 2/\sqrt{3}$$

where n is the integer determined by the order given; λ is the wavelength of X-rays; d is the spacing between the planes in the atomic lattice; and θ is the angle between the incident ray and the scattering planes.

The wall thickness of the mesoporous materials, t, can be calculated using the unit cell parameter a_0 and the pore diameter value obtained from the N_2 sorption measurements and is given in equation: $t = a_0 - D_{pore}$

1.2. Nitrogen Adsorption/Desorption Isotherms

High surface area and narrow pore size distribution are important attributes of mesoporous materials. Hence the ability to quantify these values is essential when characterizing any mesoporous material. The most common method of evaluating the surface area and pore size distribution of mesoporous materials involves studying the adsorption and desorption of an inert gas such as nitrogen onto the surface of the material at liquid nitrogen temperature and relative pressures (P/P_0, where P_0 = 1 atm) ranging from 0.05 to 1. Nitrogen adsorption studies are widely reported in the literature for a range of nanoporous materials [19,20].

Nitrogen adsorption isotherms are extremely useful to determine structural details of nanoporous materials. Initially at very low relative pressures, a monolayer of nitrogen molecules forms on the surface of the solid. As the limit of monolayer coverage is reached, nitrogen begins to condense and fill in the very smallest pores of the sample. Further increases in the relative pressure result in the filling of steadily larger pores. IUPAC distinguished six types of adsorption isotherms accounting for the majority of isotherms and these are shown below in Figure 2.

The reversible Type I isotherms are given by microporous solids having relatively small external surface areas, the initial rise in adsorption corresponds to nitrogen filling the micropores after which multilayer growth gives a steady increase in adsorption. The reversible Type II isotherm is characteristic of a non-porous or macroporous adsorbent. It represents unrestricted monolayer-multilayer adsorption. The reversible Type III isotherms are not uncommon, but there are a number of systems (e.g. nitrogen on polyethylene) which give isotherms with gradual curvature. Type IV isotherms are characteristic of mesoporous materials, the sharp jump in adsorption corresponding to the pressure at which nitrogen condenses into the mesopores. Another characteristic feature of the Type IV isotherm is the hysteresis loop, which is associated with capillary condensation taking place in the mesopores. The Type V isotherm is related to the Type III isotherm in which the adsorbent-adsorbent interaction is weak, but is obtained with certain porous adsorbents. Finally the Type VI isotherm, in which the sharpness of the steps depends on the system and the temperature, represents stepwise multiplayer adsorption on a uniform non-porous surface.

Isotherms of type IV and V above both have what is known as a hysteresis loop. This occurs when the isotherm is irreversible (i.e. The desorption isotherm does not retrace the adsorption isotherm but rather lies above it over a range of relative pressure before eventually rejoining the adsorption isotherms). Hysteresis loops can take different shapes according to the properties of the adsorbate and IUPAC has distinguished four main categories (1-4) as outlined below in figure.

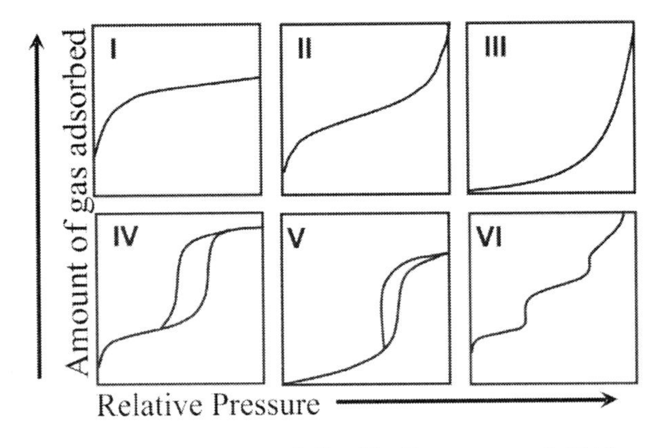

Figure 2. The six types of sorption isotherm as defined by Brunauer et al. [21].

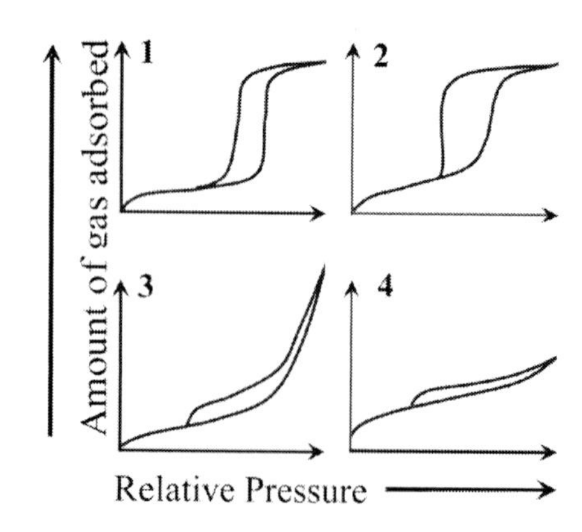

Figure 3. The different types of hysteresis loop.

Type 1 is often associated with porous materials with narrow pore size distributions; Type 2 is attributed to a difference between condensation and evaporation processes occurring in pores with narrow necks and wide bodies; 3 is observed with aggregates of plate-like particles giving rise to slit-shaped pores; type 4 loop is often associated with narrow slit-like pores[22].

Hysteresis appearing in the multilayer range of physisorption isotherms and is usually associated with capillary condensation in mesoporous structures. The observed hysteresis may be the result of two basic mechanisms, namely, "the single pore mechanism" and "the network mechanism". In "the single pore mechanism", a metastable phase may persist beyond the vapor liquid coexistence pressure during the adsorption and desorption processes where a vapor phase may be present at pressure above the condensation pressure, and a liquid phase below the condensation pressure. The second mechanism is related to the topology of the pore network. During the adsorption process, the vapors needed to fill the pore can be transported either through the liquid or through the vapor phases. However, during the desorption process, the desorbed vapors must be transported to outside only through the vapor phase. Vaporization therefore occurs only in pores connected to the bulk vapor phase, not in pores surrounded by other liquid filled pores. Once vaporization has occurred in some of the pores near the external surface, the adjacent pores now have contact with the vapor phases, and will vaporize when it is thermodynamically favorable. As a result, clusters of vapor-filled pores grow from the surface until enough pores are opened.

1.3. Transmission Electron Microscopy

Transmission electron microscopy is widely used to examine mesoporous materials and provides another important technique to characterize the pore size diameter and long-range channel ordering. The results should agree with the results of X-ray powder diffraction and nitrogen adsorption. The technique involves passing a parallel beam of electrons through a thin slice of specimen to form an image from the transmitted electrons. The technique is analogous to a standard upright or inverted light microscope. Dense areas of the sample absorb or scatter the beam, producing dark spots; whereas less dense regions such as pores do not adsorb or scatter the beam as strongly and as such appear lighter.

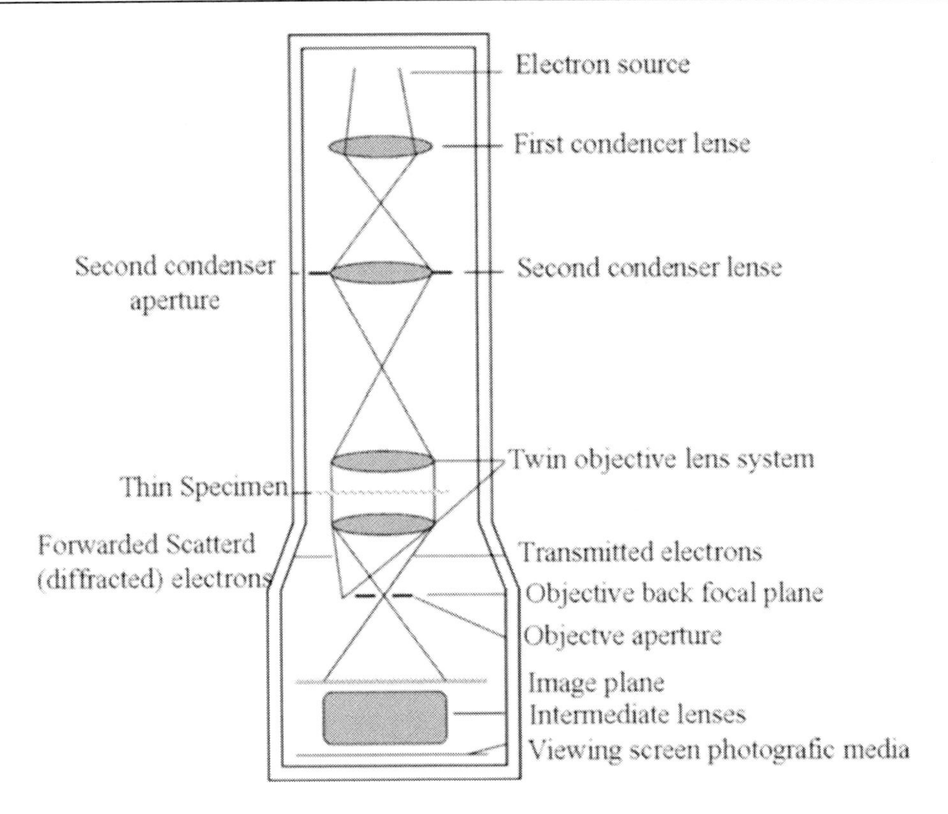

Figure 4. Schematic representation of a transmission electron microscope.

The electrons are emitted by an electron gun which accelerates the electrons produced from a filament. The accelerating voltage is typically between 100 and 300 kV allowing the beam to penetrate the specimen. Below the electron gun are two or more condenser lenses, which both demagnify the beam emitted by the gun and control the beam diameter as it hits the specimen. Next the objective lens serves to focus the electron beam onto the surface of the specimen. Beneath the specimen chamber a series of lenses serve to magnify and focus the transmitted image before the image is projected onto a fluorescent screen [23].

2. Alumina and Titania Polymorphs

2.1. Alumina Polymorphs

Alumina has many appealing properties which makes the material interesting for applications in many different areas. For example, it is hard, stable, insulating, transparent, beautiful, etc. This section describes briefly the complexity of alumina crystalline phases and reviews the properties of the three phases.

Alumina exists in a number of crystalline phases (polymorphs), three of the most important being γ, θ, and α. The α structure is thermodynamically stable at all temperatures up to its melting point at 2051°C, but the metastable phases (e.g., γ and θ) still appear frequently in alumina growth studies. The common alumina polymorphs can all be formed within typical synthesis temperatures, i.e., from room temperature up to about 1000°C. This complicates the study and growth of alumina, since it becomes difficult to control the process so that the desired phase is achieved. But the polymorphism also opens many opportunities for applications in various areas of technical science, since the properties of one alumina phase in some respects differ from the properties of another[24]. All alumina phases are involved in transformation sequences, which all have in common that they end in the α phase at high temperature. Below phase transition relations for the alumina phases is shown. The transformations to the α phase are irreversible and typically take place at above 1000°C[25].

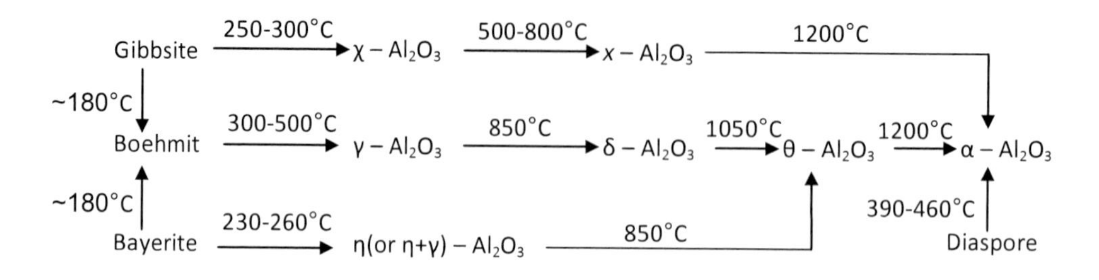

Alumina is formed through the thermal dehydration of aluminum hydroxides and oxyhydroxides. Among the alumina precursors, boehmite (γ-AlOOH) is of special interest because its transformation into $\gamma - Al_2O_3$ is pseudomorphic which preserves the boehmite texture. It was reported that $\gamma - Al_2O_3$ was formed through the dehydration of the boehmite at temperatures ranging from 400 to 700°C. During heating, boehmite nanostructures undergo an isomorphism transformation to nanocrystalline alumina.

It is well known that the properties of nanostructured materials depend on their size and morphology. Therefore the investigation of morphology and size control of boehmite is of fundamental importance in designing the properties of alumina nanomaterials.

Properties of α-alumina

The α form of alumina is also known as *corundum* (the name comes from the naturally occurring mineral corundum, which consists of pure α-Al_2O_3). It is transparent and uncolored and is known in its single crystal form as *sapphire*. It is used not only in materials science, but occurs also as gemstones. The gem known as ruby is α-alumina doped with small amounts of chromium, while the gemstone sapphire is actually α-alumina doped with iron and titanium. Like all alumina phases the α phase is highly ionic with calculated valences of $+2.63e$ and $-1.75e$ for aluminum and oxygen, respectively [26]. Thus the chemical bonds between ions are almost purely ionic (or electrostatic). The corundum structure is also formed by a number of other metal sesquioxides, such as Cr_2O_3, Ti_2O_3, and Fe_2O_3. The structure is rhombohedral with two formula units (10 atoms) in the primitive unit cell. However, a more often used unit cell is the hexagonal representation containing six formula units. The corundum structure can be described as a hexagonal close-packed (hcp) oxygen sublattice, in which the aluminum atoms, or ions, occupy two thirds of the octahedral interstices, i.e., they have six oxygen nearest neighbors. There is thus only one coordination (octahedral) for aluminum and one for oxygen (with four surrounding aluminum ions).

The thermodynamic stability of α-alumina makes it the most suited phase for use in many high-temperature applications, although also the κ phase is used due to its high transformation temperature. Other important characteristics of α-alumina are chemical inertness and high hardness. The elastic modulus and hardness are measured to be ~440 and ~28 GPa,[27] respectively. Combined, these properties have made α-alumina thin films important as, e.g., wear-resistant[28] and high-temperature diffusion barrier [29] coatings. Other uses of α-alumina is in electronics, where it is used, e.g., as an insulator due to the wide band gap of 8.8 eV,[30] and in optics, since it is completely transparent and stable at high temperature.

Properties of θ-alumina

The θ-phase of alumina is metastable and transforms into the α-phase at about 1050°C.[24] It is less dense than the α-phase with a density of about 3600 kg/m^3 compared to 4000 kg/m^3 for α-alumina [24]. The structures of all alumina phases are built up around (slightly distorted) close-packed oxygen lattices and while the α-phase has an hcp framework, the θ-structure is based on an oxygen lattice. Within this oxygen framework, half the aluminum ions occupy octahedral interstitial sites and half occupy tetrahedral (with four oxygen neighbors) sites. This is also in contrast to the α-phase. The oxygen ions have three different possible surroundings, each of which is occupied by one third of the oxygen ions. Two of these oxygen sites there have three aluminum nearest neighbors and the third has four. The structure is monoclinic and the unit cell contains four formula units (20 atoms).

There are not as many investigations made on the θ-phase as on α-alumina. It is clear, though, that it is highly ionic[31] and insulating with a band gap of 7.4 eV θ-phase is often chosen as a representative of the metastable alumina phases. The reason is the well defined crystal structure, in contrast to, e.g., γ, and the structural similarities between the metastable phases.

Properties of γ-alumina

γ – Al_2O_3 is widely used in catalysis as an active phase and is characterized by having acidic sites which determines the activity and selectivity of the catalyst for specific catalytic reactions . Due to its excellent thermal stability and chemical properties, γ-alumina has also been extensively used as carrier and support for a variety of industrial catalysts in many chemical processes including cracking, hydrocracking, and hydrodesulphurization of petroleum feedstock. Since nanomaterials possess large surface areas, research into the nanochemistry of alumina-based nanomaterials will lead to new catalysts. The low surface energy also means that the γ-phase is surface energy stabilized when the surface area is high relative to the bulk volume, e.g., for small grain sizes. In high-temperature applications a problem with the use of the γ-phase is that it transforms into θ at 700–800°C [24]. This has led to the experimental research on doping of alumina to increase its thermal stability.

The γ-alumina structure has two main similarities with the θ-phase, the oxygen lattice and the mixture of octa- and tetrahedrally coordinated aluminum ions. However, the exact structure is not well defined. It is commonly believed that the structure can be described as a defect cubic spinel with the aluminum ions more or less randomly distributed between octa- and tetrahedral sites [32].

2.2. Titania Polymorphs

Freshly precipitated hydrated titania represents amorphous phase with variable composition. However, ordered material morphology forming in the drying stage undergoes substantial changes during the calcination processes due to phase transitions. Thus titanium dioxide mainly exists in three crystalline polymorphs namely, anatase, rutile and brookite. These three polymorphs have different crystalline structure. Anatase and rutile have tetragonal structure, whereas brookite is orthorhombic [33-35]. Anatase and brookite are metastable phases, whereas rutile is the most stable phase. Brookite and anatase convert to rutile when they are calcined at higher temperatures (≥300°C). The phase transition temperatures varying with the method of preparation of the powders are presented below.

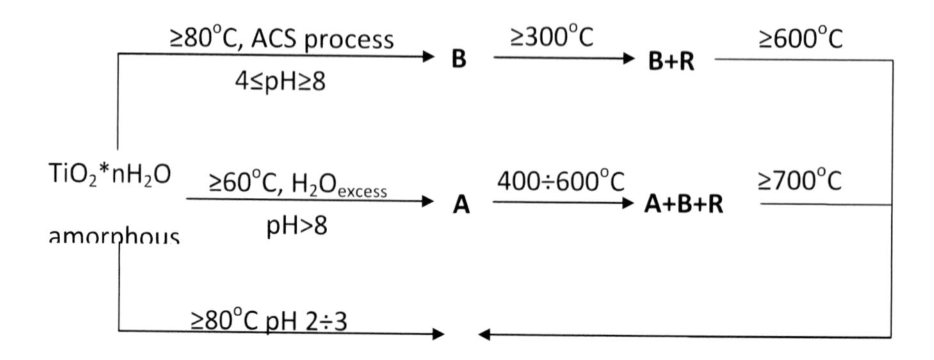

All of these phases consist of TiO_6^{2-} octahedra. Figure 5 above shows the structure of TiO_6^{2-} octahedra. The octahedron has center atom of titanium surrounded by six oxygen atoms.

For the formation of titanium dioxide crystal, at first two octahedra condense together to form a bond as shown in Figure 6. Then position of the third octahedron determines the phase that will be formed.

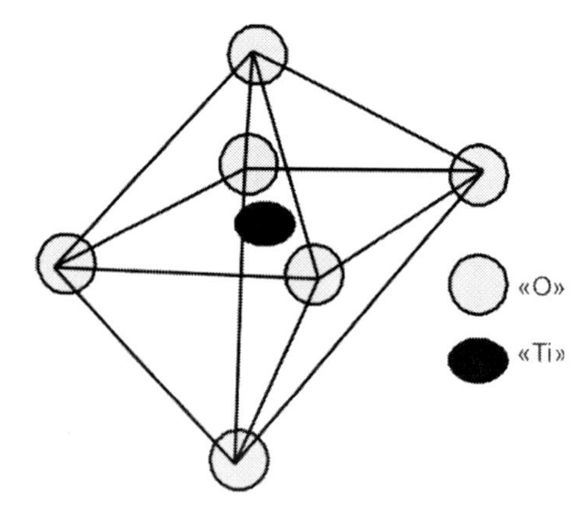

Figure 5. Structure of TiO_6^{2-} octahedron.

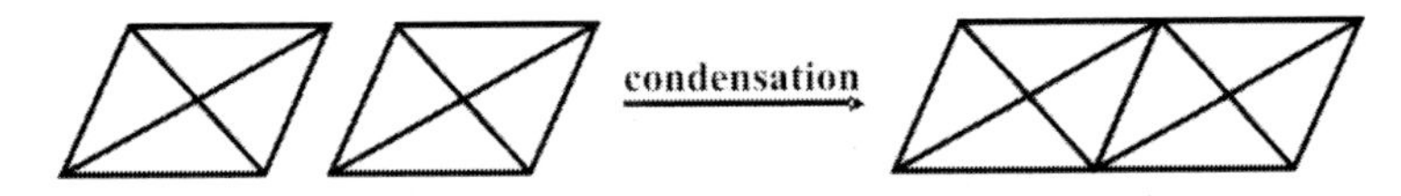

Figure 6. Formation of bond by two octahedral.

Rutile has a tetragonal structure, the octahedra join in such a way that they form a linear chain and so just two of the twelve edges of the octahedra are connected. The linear chain is joined by sharing of corner oxygen atoms. Figure 7 shows the formation of rutile by joining of two edges of the octahedra. Figure 8 shows the crystal structure of rutile. Anatase also has a tetragonal structure. In this case, there is no corner oxygen sharing. Four edges are shared per octahedra. Figure 9 shows the formation of anatase. Figure 10 shows the crystal structure of anatase. Brookite has orthorhombic crystal structure. In the brookite formation there is sharing of three edges of the octahedra. Figure 11 shows the formation of brookite structure. Figure 12 shows the crystal structure of brookite.

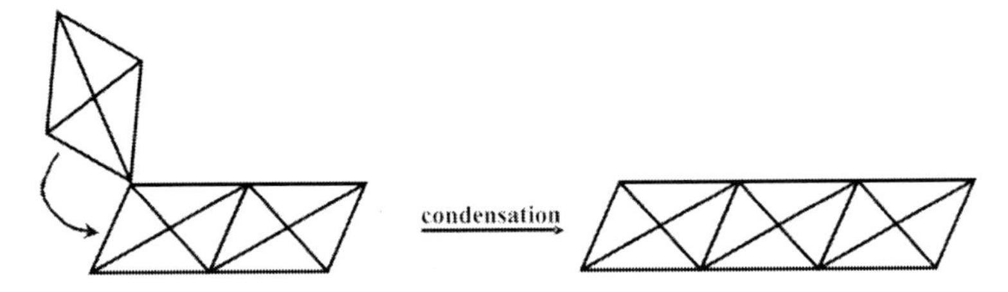

Figure 7. Rutile formation.

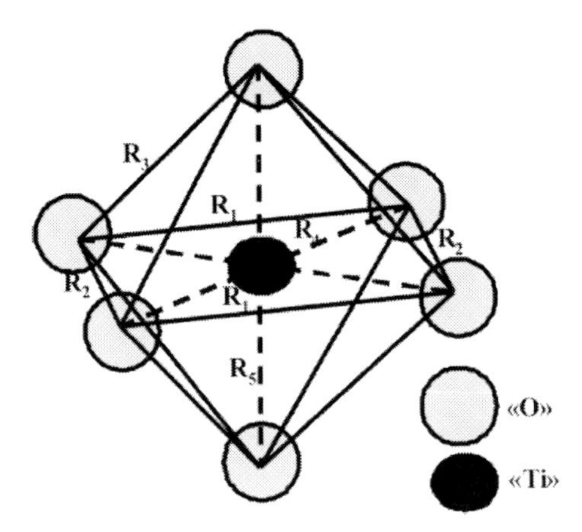

Figure 8. Crystal structure of rutile.

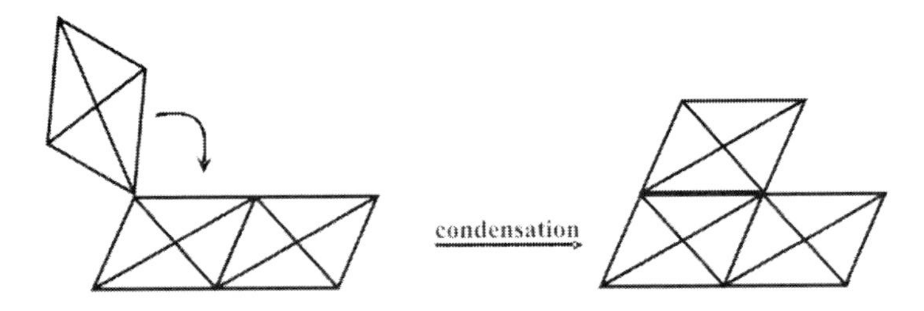

Figure 9. Anatase formation.

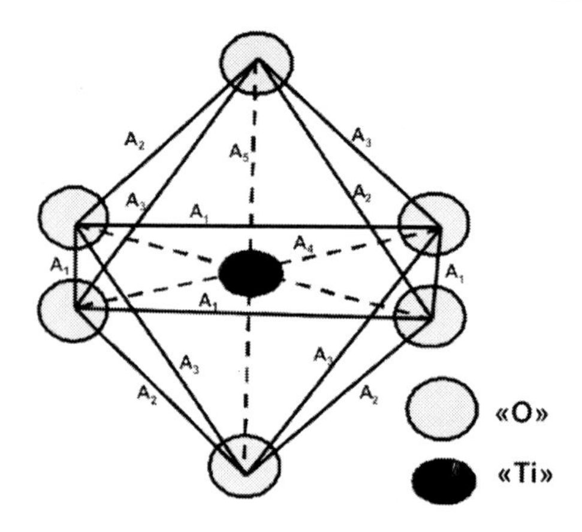

Figure 10. Anatase crystal structure.

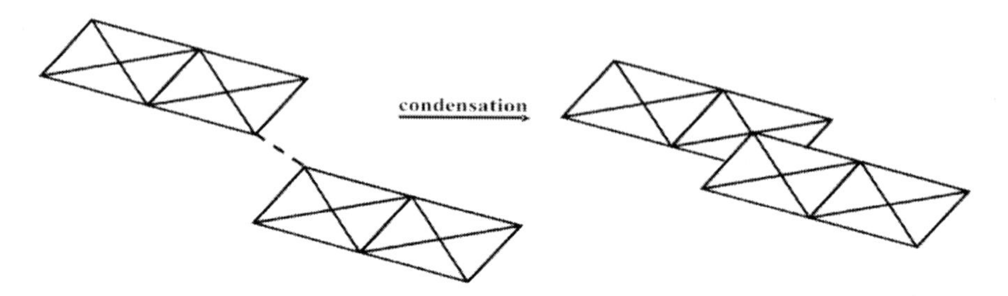

Figure 11. Brookite formation.

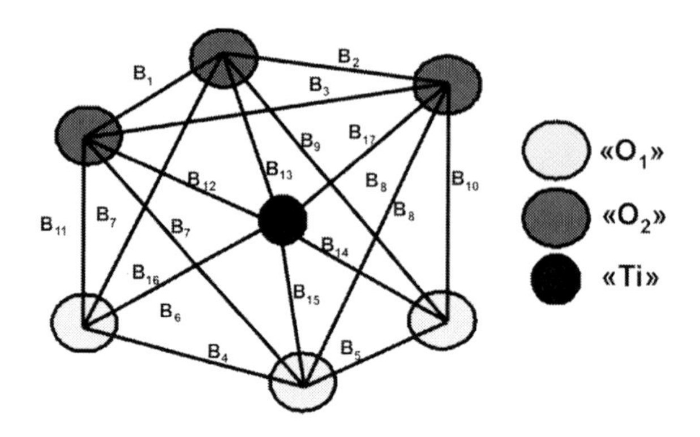

Figure 12. Crystal structure of brookite.

Table 1. Phase composition* of TiO_2 calcined samples versus pH value of titanium isopropoxide acetic acid solution during hydrolysis process.

Calcination temperature, °C	pH of hydrolysis			
	3	4	5	6
1000	A+R	A+R	R	R
800	A+R(trace)	A+R(trace)	A+R	R
600	A	A	A	A
400	A	A	A	A

*: A — anatase, R — rutile.

A substantial effect on phase composition of final product is exercised by pH value. For sol-gel synthesis of titania-based materials, the hydrolysis in the presence of glacial acetic acid, hydrochloric or nitric acids is the most popular. Properties of salts with variable structure formed at the course of hydrolysis and condensation reactions have a considerable effect on phase transformations during calcination process.

According to literature data, from a practical point of view the TiO_2 anatase and rutile forms are more preferable compared to brookite. They are commonly used in photocatalysis, with anatase showing a higher photocatalytic activity. The differences in lattice structures cause different mass densities and electronic band structure between the two forms of TiO_2 (anatase and rutile). The band gap energies are approximately 3.2 eV for anatase and 3.0 eV for rutile. Anatase is thermodynamically less stable than rutile, but its formation is kinetically favored at low temperatures (< 600°C), which could explain its higher surface area, and its higher surface density of active sites for adsorption and for catalysis.

3. Hydrolysis of Aluminum and Titanium Alkoxides

The processes of self-organization for inorganic precursor and organic template are determined in many respects by the rates of hydrolysis and condensation for titanium and aluminum alkoxides. The precisely controlled conditions allow slowing down the hydrolysis/ condensation reactions and preventing the uncontrolled phase separation between the inorganic and organic components so that a higher degree of cross linking between the precursor molecules occurs during the formation of the mesoscopic assemblies. For this reason, the regularities of course and controlling the processes of hydrolysis and condensation for the main representatives of titanium and aluminum alkoxides will be considered in this section.

The hydrolysis for aluminum and titanium alkoxides proceeds fairly quickly. High coordination number of metals in its alkoxy compounds leads to increased oligomerization and polymerization, which prevents the substitution of OH^- groups for alkoxy ones during the process of interaction with water molecules. This leads to the formation of structurally complex partially substituted oligomeric or polymeric aluminum and titanium alkoxides.

The process of formation of alumina and titania from their alkoxides can be schematically represented as follows[36,37]:

$$M(OR)_n \xrightarrow{+H_2O} \underset{\text{sols/gels(-ROH)}}{M_2O_n{\cdot}xH_2O + \text{solvent}} \xrightarrow{\text{drying}} M_2O_n{\cdot}xH_2O \xrightarrow{t} \underset{\text{xerogels}}{M_2O_n}$$

As a rule, during the first stage of alkoxide hydrolysis there are formed amorphous hydroxides (xerogels) transforming into crystalline oxides upon drying and calcination. It is necessary to note that the properties of the final products, oxides, are mainly conditioned by the first stages of hydrolysis. Molecular structure of metal alkoxides in solution and their ability to form oxoalkocomplexes play an important role in the course of hydrolysis. Oxobridges in molecules of oxoalkoxide complexes are the centers of formation of the future crystal structures. Stability of colloids obtained by hydrolysis of aluminum alkoxide is usually much higher compared to colloids obtained from inorganic salts (in the absence of destabilizing anions). Formation of M-O-M' bonds in the course of hydrolysis can occur via two directions [38]. The first one takes place as a result of hydroxide ageing; the central atoms of metals are coordinated through hydroxo - oxo - and H_2O groups. Dehydration of these products leads to the formation of amorphous oxide structures, such as:

Ì —OH + H O—Ì ⟶ Ì —O—Ì +Í 2Î

The second direction involves the interaction only between the molecules of metal alkoxides and includes the production of the corresponding ether:

Ì —OR + R O—Ì ⟶ Ì —O—Ì Ι +R2Î

Water presence catalyzes this reaction, especially upon stirring the mixture[38]. As a result, crystal phases are formed at very low temperatures (compared to dehydration and crystallization of ordinary gels). Upon addition of water or an aqueous-alcoholic mixture to an alkoxide solution in organic solvent there takes place a nucleophilic attack of molecules with positively charged aluminum or titanium atom (a) that immediately leads to formation (b):

H, H O: + Ì —OR ⟶ H, H O:→Ì —OR ⟶ OH—Ì ← :O(R, H) →Ì —OH + ROH

a á â ã

Proton shift from water molecule to negatively charged oxygen atom of neighboring OR-group leads to the state (c) and further production of the corresponding alcohol. That is why the whole process occurs by mechanism of nucleophilic substitution. The thermodynamics of this reaction is determined first of all by nucleophility of attacking group and electrophility of metal atom [$\delta(O)<<0$ и $\delta(Al)>>0$] [38]. The rate of substitution process depends on the coordination saturation of metal in its alkoxide N-z (N is the maximum coordination number of metal in its oxide, z is the oxidation state of metal). The higher the N-z value, the less the activation energy for nucleophilic attack is.

Right after hydroxide ions enter the coordination sphere of aluminum or titanium atom, there takes place the process of condensation proceeding according to one of three possible ways:

(a) Alkoxylation with production of an alcohol:

$$H_2O: + M{-}OR \rightarrow (H)(M)O{:}\rightarrow M{-}OR \rightarrow M{-}O{-}M \leftarrow :O(R)(H) \rightarrow M{-}O{-}M + ROH$$

(b) Oxylation with production of water:

$$(M)(H)O: + Al{-}OH \rightarrow (M)(H)O{:}\rightarrow M{-}OH \rightarrow M{-}O{-}M \leftarrow :O(H)(H) \rightarrow M{-}O{-}M + H_2O$$

(c) Olation with production of solvent (water or alcohol) molecules:

$$M{-}OH + M \leftarrow :O(H)(R) \rightarrow M{-}O{-}M + ROH$$

$$M{-}OH + M \leftarrow :O(H)(H) \rightarrow M{-}O{-}M + H_2O$$

Condensation process proceeds much more slowly compared to the reaction of nucleophilic substitution of OH^- group for alkoxide one. Kinetic study of aluminum and titanium alkoxide hydrolysis has shown that condensation starts to occur after substitution of hydroxide groups for 25 to 50% alkoxide ones in the course of hydrolysis (approximately after 80 milliseconds after mixing the reagents). The rate of alkoxide hydrolysis also depends on the alkoxide group structure. For example, for aluminum and titanium butoxides the hydrolysis rate decreases in the following row: tertiary > secondary > primary[39]. Nature of OR-groups has an effect on condensation as well. Specifically, hydrolysis of $Ti(OR)_4$ proceeds very quickly in the case of R=Et, Pr^i and is slowed down in the case of R=Bu^n, Am^n.

The rate of hydrolysis and condensation processes depends also on pH of a solution (on addition of acids or bases). The role of acid catalysis is in fast protonation of negatively charged OR groups by H_3O^+ groups and theoretically allows substituting all OR-groups. However, OH-groups are mainly formed in a chain's end that leads to the formation of linear polymers. In strongly acidic environment ($[H^+] \sim [Al]$) the condensation processes are substantially slowed down, the protonation of hydroxide groups becomes possible and leads to the formation of hydrated hydroxocomplexes. The aluminum alkoxide hydrolysis in acidic environment at 20°C leads to the formation of stable sols with mixed octahedral and tetrahedral coordination of aluminum atoms. Acidity reduction promotes the formation of a precipitate of amorphous aluminum hydroxide containing a considerable quantity of alkoxide that has not entered the reaction. Hydrolysis at 80°C leads to the formation of a sol with octahedral coordination of all aluminum atoms. Boehmite is formed only in the neutral environment in the form of a precipitate that can be easily peptized upon addition of a small amount of acid.

In the case of base catalysis, condensation process activation occurs through formation of highly nucleophilic fragments, such as M-O⁻ [38]:

$$\rangle M - OH + :D \rightarrow \rangle M - O^{-} + DH^{+} (D = OH^{-}, NH_3)$$

This very reactive precursor attacks positively charged aluminum or titanium atom. As a result of this process, a solid polymer framework is formed.

The aluminum and titanium alkoxide condensation occurring as a result of uncontrollable hydrolysis affects substantially the further hydrolysis of reagents as well. Therefore properties of hydrolysis products strongly depend on the water content in dehydrated solvent used for preparation of an alkoxide solution, as well as on time and conditions of solvent storage before use.

At the same time, the key parameter affecting the hydrolysis rate is the molar ratio of reagents (the water – alkoxide $[H_2O]:[M(OR)_n]$ ratio). This ratio defines the structure and the properties of hydrolysis products, which allows obtaining powders, films, glasses, and fibers of the corresponding oxide and hydroxide [38].

4. Sol-Gel Synthesis of Ordered Mesoporous Al_2O_3 and TiO_2 and Their Potential for Catalytic and Photocatalytic Applications

Ordered mesoporous materials are defined by pore sizes within the range of 2 – 50 nm and are a relatively new field of materials. Research on mesoporous materials was originally motivated by the need for ordered alumina/silica materials with pores larger than those found in zeolitic materials for use in petrochemical catalysis. The first successful synthesis of ordered mesoporous materials was in 1992 when Beck and coworkers reported a range of mesoporous amorphous silicate materials known as M41S [1,2]. The M41S family [40] of materials consists of three structure types, each of which has a highly ordered channel systems with narrow pore size distributions (pore diameters > 2 nm). Figure 13a-c shows TEM images highlighting the pore morphologies observed for the M41S family. The first material known as MCM-41 (Mobil Corporation Material 41) contains a hexagonal arrangement of channels. Figure 13a shows a cross section through this channel system, clearly highlighting the highly ordered hexagonal arrangement of the channels. The second material, MCM-48, has a complex three dimensional channel system (FIgure 13b) whilst the third material, MCM-50, consists of stacked silica sheets (FIgure 13c).

The preparation of ordered mesoporous materials focuses on using self-assembled surfactant aggregates as structure directing agents, and these materials are labeled according to the liquid crystal phase from which they were formed (hexagonal (HI), cubic (V1) and lamellar (La)). The high surface area and tightly controlled pore sizes were among the many desirable properties that made these materials the focus of a great deal of research over the following years. The mobile workers also reported that partial exchange of silica for alumina, in M41S materials, resulted in mesoporous aluminosilicate materials. This initial effort was soon followed by the synthesis of other mixed oxides, such as titanosilicates and borosilicates [41-43], which in turn led to the development of a variety of other non-silica inorganic oxides [44,45,4]. Recently, new mesoporous materials based on TiO_2 and Al_2O_3 with ordered structure were also obtained, for the synthesis of ordered mesoporous alumina and titania, it was necessary to modify and optimize procedures that have been well described and understood for the synthesis of mesoporous silicas. Ordered mesoporous alumina and titania with continuously adjustable pore size can be synthesized by carefully changing the synthesis parameters.

In this section the principles of self-assembly and liquid crystal phase formation will be also discussed.

4.1. Surfactants and Liquid Crystal Phases

As already mentioned, the ordered mesoporous materials described above were synthesized using surfactant aggregates as structure directing agents in a sol-gel process. This is known as a liquid crystal templating approach and is the predominant approach for the production of mesoporous materials. In order to better understand how these ordered structures are formed it is necessary to have an understanding of how surfactants behave in aqueous solutions. Figure 14 shows a schematic representation of a surfactant molecule. The molecule consists of a water soluble hydrophilic headgroup and a water insoluble hydrophobic tailgroup.

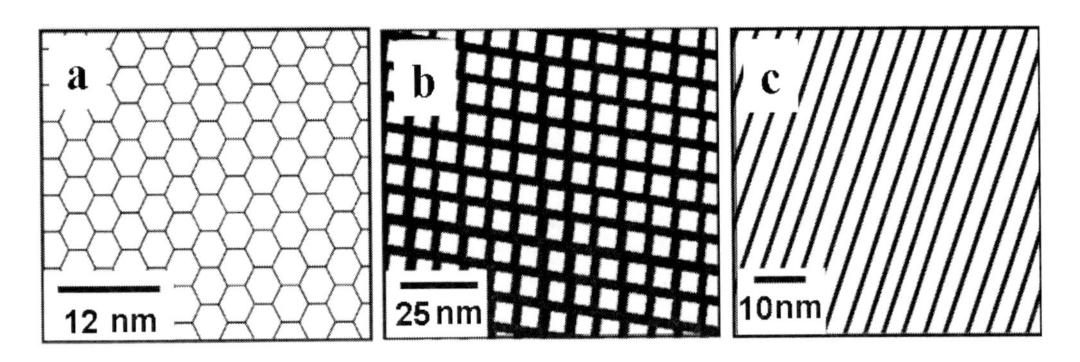

Figure 13. Images of mesoporous silicas prepared by the Mobil method. a)Hexagonal pore morphology (hexagonal array of cylindrical holes); b) Cubic pore morphology; c) Stacked silica sheets.

Due to the amphilic nature of surfactant molecules when they are added to water, they aggregate together to form micelles (Figure 15) with the polar head groups on the outside and the hydrocarbon tails towards the centre. This aggregation only occurs after a certain concentration is reached (called the critical micelle concentration or cmc) which is typically about 1 wt. % of surfactant. It is easy to see why micelles are stable structures because the hydrophobic components of the molecules are shielded from the water by the hydrophilic headgroups.

Further increases in the surfactant concentration result in the formation of more micelles, until eventually a point is reached when the micelles reorganize to form larger structures (Figure 16). One such structure is the hexagonal phase which consists of a hexagonal arrangement of micellar rods. Once again this structure forms so that the hydrophilic headgroups are in contact with the water whilst the hydrophobic tailgroups are tucked away inside the columns. Another common structure is known as the lamellar structure. This forms at even higher concentrations than the hexagonal structure and consists of bilayers separated from each other by water. At the extremes where the surfactant concentration is very high, inverse phases are formed such as the inverse hexagonal and inverse micellar phase. As with the micellar phase, the inverse micellar also consists of spherical micelles, however, in this case we have the hydrophobic tailgroups projected outwards whilst the hydrophilic headgroups surround small pockets of water.

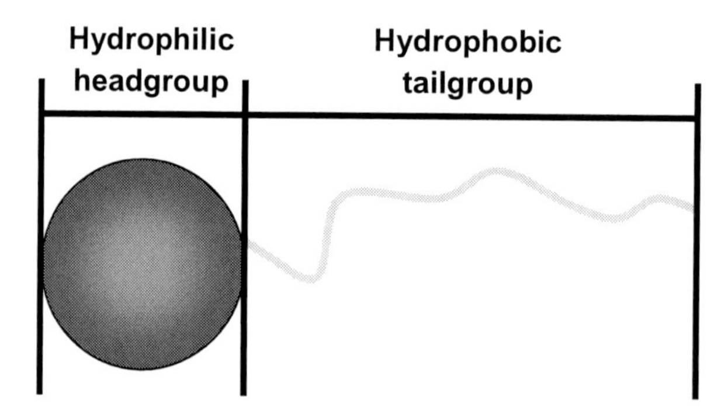

Figure 14. Schematic representation of a typical surfactant molecule.

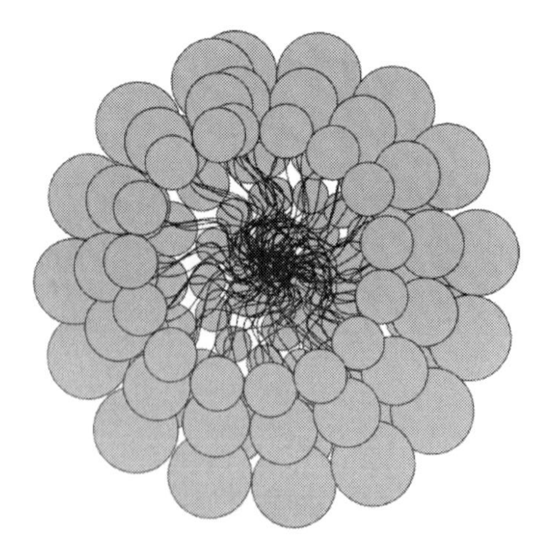

Figure 15. Schematic representation of a spherical micelle.

Amphilic molecules can form a diverse range of mesostructures and those shown in figure 16 represent only a small number of the possibilities.

The larger structures, formed beyond the concentrations of surfactant required for micelles, are said to be liquid crystalline. This is because the molecules are oriented on average along some preferred direction, producing fluid but structurally anisotropic phases. The molecules in a liquid crystal phase do not have the positional and orientational order of a solid nor do they have the total disorder associated with a liquid but tend to lie somewhere between the two extremes. The liquid crystal phases formed by mixtures of surfactant and water are termed lyotropic phases. As with thermotropic liquid crystals, altering the temperature can alter the phase. However, in lyotropic systems, altering the water content can also change the structure of the phase.

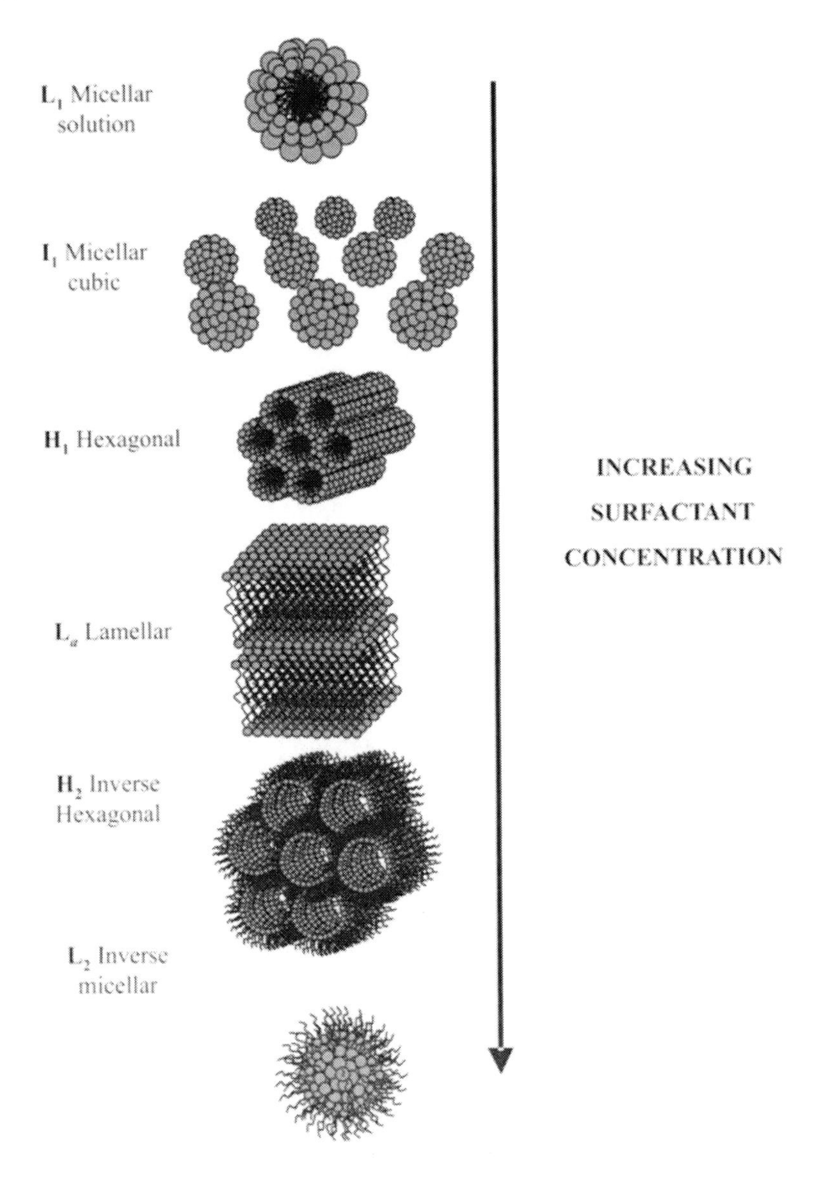

Figure 16. Examples of surfactant aggregate structures formed in lyotropic systems.

General rule: surfactants with one hydrophobic chain form micelles and other straight structures while surfactants with two hydrophobic chains mainly form lamellar phases and inverse structures. It is easy to understand if we try to construct spatial models of surfactant aggregates with various structures. We will thus find out that because of great volume of hydrophobic "tails" molecules with two chains cannot be packed into spherical micelles. Upon introducing the second alkyl chain and leaving other parameters constant, the other geometrical forms become preferable. The critical packing parameter (CPP) can be considered in another way, namely as the ratio of the cross-section areas of hydrocarbon chain and polar group.

As the interrelation of chemical structure and structure of aggregates is much more complex than this simple geometrical analysis, it is given only for illustration and serves as a starting point, in particular, for the analysis of trends in phase behavior. The formed structure is a result of balance between polar and non-polar parts of a surfactant molecule.

There are two important factors affecting greatly the structure of aggregates that are not considered by simple geometrical model. First of all, this is an interaction of polar groups in the aggregate. Clearly, the strong repulsion between polar heads of a surfactant will lead to a shift of aggregates to the left whereas shift in the opposite direction requires the action of attractive forces. The problem can be solved by estimating the "effective" area of polar group. So, for ionic surfactant, the interaction of polar groups is strongly affected by electrolyte concentration. The interaction of polar groups of non-ionic surfactants is affected more strongly by temperature, instead of electrostatic interactions. Temperature change appears to be the principal factor defining structure of an aggregate.

Answering a question, which surfactants belong to various categories, spherical micelles are characteristic of surfactants with one non-polar chain and highly polar "head", for example, ionogenic group in the electrolyte absence. Non-ionogenic surfactants with the large polar groups fall under this category as well. Rod-like aggregates are corresponded by ionogenic single-chained surfactants in the electrolyte presence or with strongly bonded counterions as well as non-ionogenic surfactants with medium-sized polar groups. Higher CPP values are characteristic of surfactants with two non-polar chains or of non-ionogenic surfactants with small polar groups. In the case of ionogenic surfactants the addition of an electrolyte can cause transition from lamellar structures to inverse structures.

4.2. Mechanisms of Self-Assembly

The similarities between liquid crystal mesophases and the M41S family of materials led researchers at Mobil to suggest a liquid crystal templating (LCT) mechanism of formation. This suggests that the structure of the channel system in these mesoporous materials is determined by the surfactant aggregation behavior. The Mobil researchers suggested two alternative mechanisms by which the surfactants might self-assemble into the appropriate mesophases. These mechanisms are discussed below along with the evaporation induced self-assembly mechanism which has been extensively used to fabricate mesoporous films.

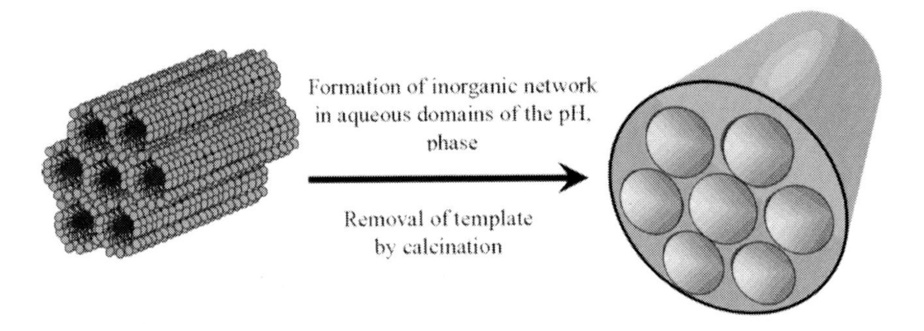

Figure 17. Schematic representation of the direct liquid crystal templating approach.

4.2.1. Direct liquid crystal templating

The first mechanism postulated that there was a hexagonally ordered surfactant phase already formed prior to the addition of silicate source. Upon addition of this silica source the anions diffuse into the water regions of the mesophase, subsequent condensation in the interstitial voids results in the mesostructured silica.

This mechanistic pathway received little support in the literature because it was noted that the Mobil method utilized surfactant concentrations of between 1–10 wt. % which is far lower than required to form surfactant mesophases [46] (typically greater than 20 wt. %). Further research suggested that, provided the concentration of surfactant was greater than the CMC, MCM-41 could be successfully synthesized, which further suggests problems with the direct liquid crystal model.

Direct liquid crystal templating of mesoporous materials was demonstrated in 1995 by the use of high surfactant concentrations [40]. The characteristic phase of the surfactant water system can be used as a direct mould for the forming oxide, ceramic or metal.

The preparation of mesoporous silica directly from a liquid crystal phase involves addition of a silicon alkoxide (typically TEOS or TMOS) to a surfactant (> 50 wt. %), water and acid mixture. Initially, a large amount of methanol is produced which destroys the liquid crystal phase, but removing the methanol under a gentle vacuum can regenerate the phase. After leaving the mixture to gel, the surfactant is removed by calcination. The aqueous domains of the surfactant phases determine the regions in which condensation of the silica network occurs. Consequently, the structure of the calcined silica is a cast of the supra-molecular architecture of the liquid crystal phase.

4.2.2. Co-operative self-assembly

The second mechanism proposed by the Mobil researchers suggested that the addition of silica resulted in the ordering of silicate encased surfactant micelles. In this case prior to the addition of silica micelles are present in solution but no hexagonal phase is observed. Upon addition of the silica source to the system silicate anions coat the surface of the micellar rods. These silicas encased micelles then self-assemble themselves before condensation of the silicate occurs resulting in the observed mesostructured silica (Figure 18).

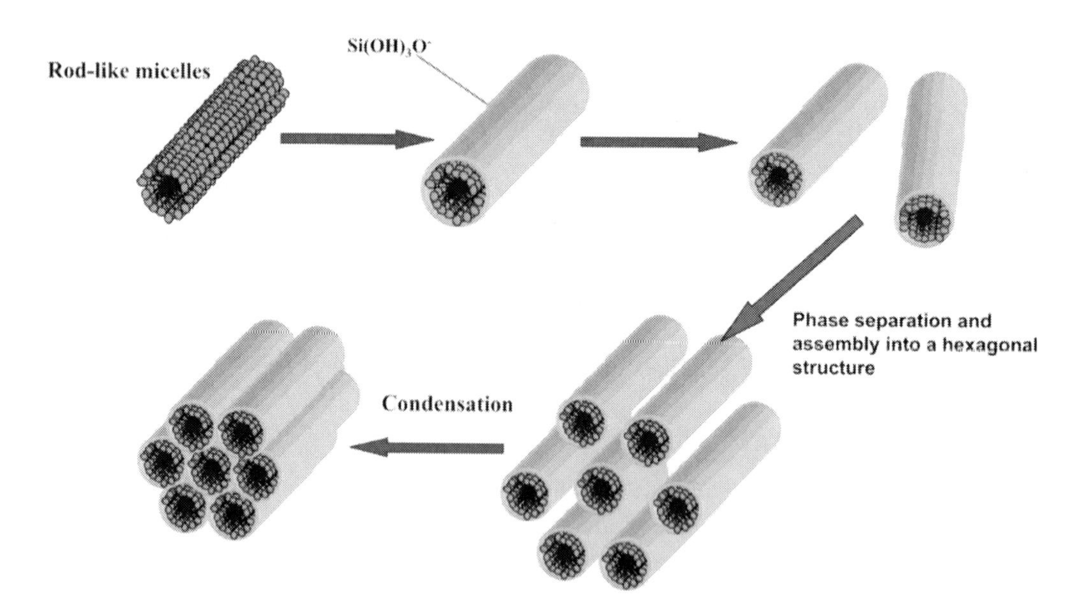

Figure 18. Schematic representation of the cooperative self-assembly mechanism.

It has been well established that no preformed liquid crystal phase is required for the formation of MCM-41, however, there is still some debate over the precise details with various models being suggested [7,47,48]. All these models share one fundamental principle however, which is that silicate species promotes the formation of a liquid crystal phase at low surfactant concentrations. This observation is consistent with the effect of polyvalent ionic species on the phase behavior of ionic surfactant solutions. This consists of macroscopic aggregates of micelles dispersed in water. Structural transitions in the micelle-rich regions, driven by the incipient polyvalent inorganic network lead to micron-sized domains of liquid crystalline phases.

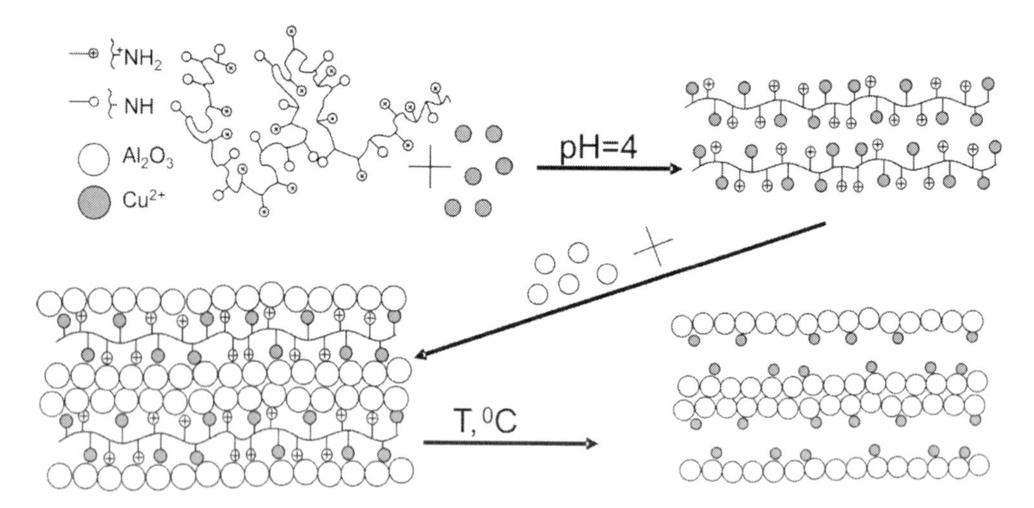

Figure 19. Scheme of $CuO–Al_2O_3$ formation.

4.2.3. Self-assembly with polyelectrolytes

Another mechanism proposed by Agafonov et al. [49] for the synthesis of ordered mesoporous materials is based on application of polyethylenimine permolecular formations as a structure directing agents (Figure 19).

Addition of acid results, on the one hand, in transition of the aluminum hydroxide precipitate to a sol and thus formation of positively charged micelles, and on the other hand, in uncoiling of polyethylenimine macromolecules. Donor–acceptor bonds are formed between imine groups and copper ions, and, as a result, a polymer-colloid complex is formed. Subsequently, alumina precursor particles are adsorbed on its surface. This approach should produce an ordered mesoporous material under conditions of layer-by-layer self-assembly, where first copper oxide precursor layers and then alumina precursor layers are formed in succession on the surface of supramolecular polyethylenimine.

4.2.4. Self-assembling at temperature treatment

The authors of [50] describe peculiarities of formation of ordered titanium dioxide film structures during the process of sol-gel synthesis using PEG as a template and analyze the regularities of phase change of synthesized material when exposed to temperature, Figure 20. It is reported that the structural conversion of titanium dioxide – monooleate PEG hybrid organic-inorganic material thin film applied from an alcohol sol to the glass substrate surface into ordered periodic lattice framework constructed by the titanium dioxide nanosized channels is induced by the temperature effect.

The revealed structural change in the films may be induced by various reasons. For example, the initial film formed by sol-gel method may be originally structured due to the appearing of convection currents while drying and removing the solvent. Thus obtained ordered channels formed by titanium dioxide hydroxoforms are filled with the surfactant and remaining solvent molecules and are not identified by scanning probe microscopy. Upon the film calcination the organic phase is removed and the channel structure develops. On the other hand, the structural change observed may occur in originally non-structured film and be due to the fact that heating upon the calcination of glass substrate with the film takes place under non-isothermal conditions favoring the convection transfer of the film substance.

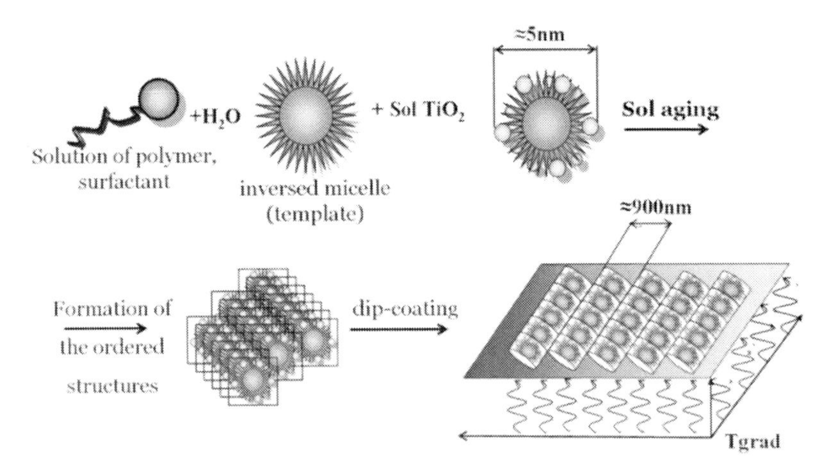

Figure 20. Mechanism of formation of the mesoporous macroordered film structures.

All of these mechanisms were used for the synthesis of mesoporous TiO_2 and Al_2O_3. Due to some problems related with different behavior of materials during calcination firstly only organized mesoporous TiO_2 and Al_2O_3 were obtained.

4.3. From Organized to Ordered Mesoporous Structure of Al_2O_3

For the synthesis of ordered mesoporous alumina and titania, it was necessary to modify and optimize the procedures that were well-described and understood for the synthesis of mesoporous silicas. However, the synthesis of ordered and thermal stable mesoporous alumina and titania represents a much more complex problem due to its susceptibility for hydrolysis as well as to the phase transitions accompanying the thermal breakdown of the ordered structure. Although particular attentions have been devoted to the synthesis of mesoporous alumina and titania, unfortunately disordered structures with amorphous walls were fabricated in most cases.

4.3.1. Sol – gel synthesis of organized mesoporous alumina

"Neutral" way of synthesis based on interaction of electrostatically neutral polyethylene oxide and aluminum alkoxide as inorganic precursor was used by group of Pinnavaia and coworkers [3,51]. The mesoporous alumina obtained possessed wormhole-like pore structure with surface area of about 500 m^2/g. In the process of neutral synthesis various templates, namely, Tergitols, Tritons and Pluronics, were used. In particular, dependence between the size of the Tergitols supramolecular formations of polyethylene oxide (PEO) and the formed pore size [51] was found. Introduction of a small amount of the Ce^{3+} or La^{3+} ions into system allowed to considerably increase thermostability of mesoporous alumina [52]. The given fact confirms the stabilization of metastable phases of alumina preventing coagulation and further structural transformations upon introduction of rare-earth metals. Formation of mesoporous structure of alumina by the neutral way of synthesis using triblock copolymers also was confirmed by Luo et al. and Deng et al. [53]. Gonzalez-Pena et al. have shown that 1,4-dioxane can be effectively used as a medium for synthesis of mesoporous alumina by the neutral way. Diameter of final material pores is in certain dependence on the size of micellar structures of a surfactant.

"Anionic" method for synthesis of mesoporous alumina was described by Vaudry et al. [54]. Capric, lauric and stearic acids were used as structure forming substances, and ethyl alcohol, formamide, chloroform and diethyl ether were chosen as media for synthesis. Aluminum alkoxide was used as alumina precursor. The calcinated alumina surface area values were within the range of 500 and 700 m^2/g with the pore size of about 2.0 nm. At the same time, Cejka et al. [55] have shown that diameter of a pore of mesoporous alumina obtained with stearic acid is more than one obtained with lauric. The presence of decane in the solution leads to the increase in pore size distribution spectrum.

Controllable synthesis of mesoporous alumina by means of aluminum hydroxide precipitation in the presence of urea and sodium dodecyl sulphate is described by Yada et al. [56]. The authors of the cited works have shown that micellar structures of sodium dodecyl sulphate originally form layered mesophases with the adjustable size of interlayer space. Surfactants form a space that bonds the layers of alumina precursor. As a result of the further hydrolysis of urea, layered mesophases are transformed into hexagonal packing [57], the process of dodecyl sulphate removal being the most important in the synthesis of mesoporous alumina. The surfactant full decomposition occurs at temperatures less than 200°C while sulphate groupings are removed in the temperature range of 400 to 550°C. The data obtained give evidence to the strong interaction between SO_4^{2-} groups and alumina surface. The mechanism of formation of alumina hexagonal packing was studied by a fluorescence method in Ref. [58]. It was shown that the process of polymerization occurring on the surface of dodecyl sulphate micelle is carried out at the expense of formation of hydrogen bonds between hydroxyl ions formed upon decomposition of urea. Thus, the analysis of data has shown that complex formation between alumina and dodecyl sulphate micelles occurs during the thermal treatment of a system. Yada et al. [56,57]did not study the calcination of the mesostructured materials. At the same time, Sicard et al. [59] has shown that mesoporous alumina is transformed upon thermal treatment into microporous with pore diameter of 1.5 to 2.0 nm. Therefore, the structure of non-calcinated mesoporous materials possessing hexagonal or lamellar packing stabilized by strong interaction between nanotubes or layers of alumina and sodium dodecyl sulphate sulphatic groups is destroyed during thermal treatment that leads to the collapse of alumina mesopores. At the same time, the presence of small quantity of yttrium in alumina structure stabilizes structural changes during thermal treatment process, which prevents pore collapse [60,61]. The process of sodium dodecylsulphate removal from non-calcinated alumina is described by Valange et al. [62]. Upon calcination of a material in a nitrogen current at 450°C, the specific surface area of mesoporous alumina amounted to 450 m^2/g, the average pore size being 3,4 nm. Thus, for the first time, it has been shown that mesoporous alumina can be obtained in the water environment. Change in system porosity was studied, depending on the nature of alumina precursor and surfactant micellar structure (hexadecyl trimethyl ammonium bromide, sodium dodecyl sulphate, long-chained carboxylic acids, Triton X or its combinations) [62].

"Cationic" method for synthesis of mesoporous alumina has been described by Cabrera et al. [63], using hexadecyl trimethyl ammonium bromide in the water environment with triethanolamine in water as a surfactant. Authors have shown that ratio change between quantity of surfactant and triethanolamine allows changing the pore size from 3.3 to 6.0 nm. The cited approach has been applied for the description of the other oxide systems [64]. The cationic method seemed to be the most convenient for regulation of the pore size, however, reproducibility of experiments was bad. Lemon acid and aluminum isobutoxide have been used by Liu et al. [65]. Depending on surfactant concentration, the specific surface area increased from 380 to 430 m^2/g, and the pore size changed from 3.8 to 5.0 nm.

4.3.2. Synthesis of ordered mesoporous alumina

According to previous reports, the hydrolysis behavior of alumina is very complicated and strongly affected by acid, water, temperature, relative humidity, and other factors, giving rise to rather strict synthetic conditions for ordered mesoporous aluminas. Pinnavaia et al. [66] obtained pseudolamellar mesostructured γ-alumina with crystalline framework walls. Employing aluminum tri-*tert*-butoxide as the main inorganic precursor and anhydrous aluminum chloride as the pH adjustor and hydrolysis-condensation controller, Zhao et al. [67] fabricated partly ordered mesoporous alumina. Somorjai et al. [68] first reported the synthesis of ordered mesoporous alumina with amorphous walls through a sol-gel route under strict control of the hydrolysis procedure as well as the condensation of reagents. By utilizing the dip-coating method, Sanchez et al. synthesized ordered nanocrystalline γ-Al_2O_3 films with contracted *fcc* mesoporosity [69]. Following the above work, Grosso et al. improved the synthesis procedure and fabricated ordered nanocrystalline mesoporous γ-alumina powders by aerosol generation of the initial solution using an ITS atomizer [70]. After treatment at 700°C γ-alumina was obtained and it could be stable up to 900°C. Recently, Zhang and co-workers [71] developed an ordered crystalline mesoporous alumina molecular sieve with CMK-3 as hard template, which presented a new route to obtain ordered mesoporous alumina. However, this synthesis procedure requires multiple steps and is time-consuming.

From the viewpoint of synthesis, it is still a significant challenge to obtain γ-alumina with highly ordered mesostructures via a one-step, convenient, and economic approach. Moreover, the thermal stability and catalysis properties of ordered mesoporous alumina have not been studied in detail yet. Recently Quan Yuan et al. [72] presented an easily accessible, reproducible, and high-throughput method to synthesize highly ordered mesoporous aluminas with amorphous and/or crystalline γ-phase framework walls through a simple sol-gel route with block copolymers as the soft templates (Figure 21). With his strategy, a series of ordered mesoporous aluminas with 2D hexagonal structure were readily obtained, and partly ordered mesoporous alumina was also fabricated with hydrous aluminum nitrate as the precursor. More important, these mesoporous aluminas exhibit a high thermal stability up to 1000°C, possess high surface areas, tunable pore sizes, and large amount of surface Lewis acid sites.

The ordered mesoporous aluminas synthesized with different aluminum precursors, surfactants, and acids. Ordered mesoporous aluminas are achieved with whichever aluminum isopropoxide, aluminum sec-butoxide, or hydrous aluminum nitrate used as the precursor. Different acids are added as the pH adjustors for the hydrolysis of aluminum precursors, and the results demonstrate that it is flexible to obtain the ordered mesostructure in a wide range of acidity and water content.

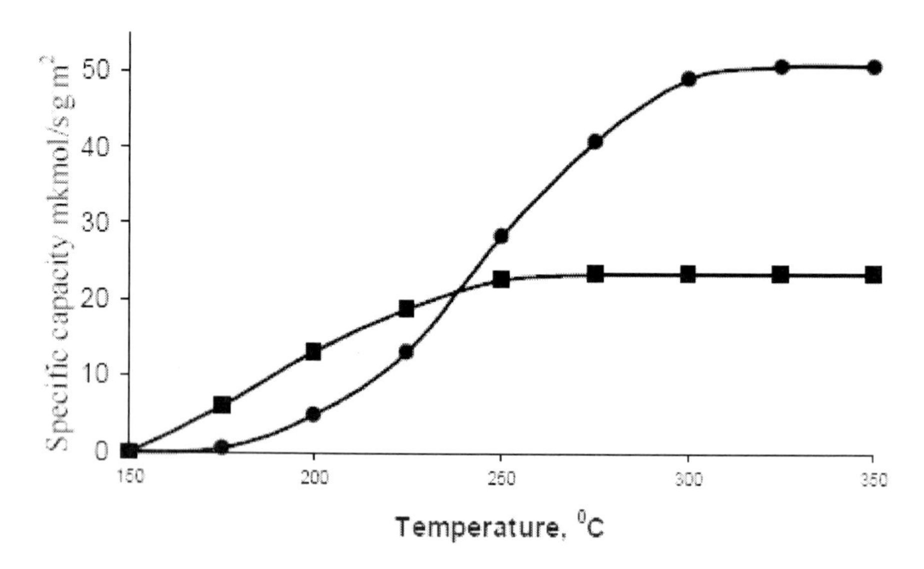

Figure 21. Specific capacity of obtained samples for methanol conversion:
•- meso – Al_2O_3, ■-commercial Al_2O_3.[77]

According to the "liquid crystal templating" mechanism for the formation of mesoporous structure proposed by Beck et al.,1 the ultimate liquid crystalline phases may be related to the ionic strength, counterion polarizability, surfactant concentration, counterion charge, and other factors.

Authors [72] assume that volatilization behavior of the acids is an important factor that should be taken into account. The reactivity of the alumina precursors can be efficiently regulated through various methods such as careful adjustment of the pH and dilution of solutions or the use of condensation inhibitors. A subsequent slowing step is necessary to irreversibly freeze a half-condensed structure and hence avoids the fast formation of an inorganic network. Inorganic hydrolysis and condensation have to be controlled to fabricate a robust mesostructure instead of a hard solid. The starting solutions are relatively dilute, and the inorganic polymerization can be readily dominated by an acid, which is subsequently removed by evaporation. The volatility of HCl is higher than that of HNO_3, the acidity of the whole system added with HCl in evaporation process reduces more quickly and thus consequently affects the hydrolysis behavior of aluminum species and the self-assembly process. When non-volatile hydrocarboxylic acids are introduced, they act as sustained-release agents to maintain an acidic equilibrium environment for precursors. In the beginning, their effect is not obvious, whereas at the end of the evaporation process this effect becomes more important as hydrocarboxylic acid concentrations increase in the whole system.

Besides, the synthesis conditions, particularly the amount of water, pH, temperature, and relative humidity, have a great impact on the final mesophase. The water quantity has crucial effects on the formation kinetics of the condensed phase and on the chemical compatibility at the hybrid interface. The fine-tuning of the water/precursor ratios should place the systems in the right position to obtain the desired mesostructure. A low water quantity is necessary. The species and dose of acid do directly determine not only the sort of cations that affect the self-assembly process through coordination interaction with the aluminum precursor but also the pH of the medium. A small amount of H^+ slows down the hydrolysis rate of the aluminum alkoxide, making the dissolution of the precursor molecules difficult in ethanol, whereas higher H^+ concentration results in uncontrollable fast hydrolysis of aluminum alkoxide. Therefore, the amount of H^+ is restricted to a relatively narrow range. At the same time, it is also necessary to control proper ambient factors, like temperature and relative humidity. The formation of a well-defined mesostructure depends critically on these factors, which controls two competitive processes: solvent evaporation and inorganic polymerization. High temperature and low relative humidity ($< 20\%$) are adopted by placing the reaction vessels in a drying oven with the temperature set at 60°C. After evaporating for 2 days, the product has been totally condensed to a foam-like product without fluidity. Varying the evaporation time from 2 to 8 days, the same 2D hexagonal mesostructures are obtained, implying that it is the most stable mesostructure under this synthesis condition. Two days is a proper choice for the sake of synthesis efficiency.

The mesostructures are stable at a temperature as high as 1000°C, which is an excellent advantage in high-temperature catalytic reactions. Moreover, tunable pore sizes from 3 to 7 nm make the shape-selective catalysis become possible. Strong Lewis acidity, provided by aluminum atoms in tetrahedral and octahedral coordination sites, proves these ordered mesoporous aluminas to be ideal supports for heterogeneous catalysis.

4.3.3. Catalytic properties of mesoporous alumina

Wieland et al. [73] tested organized mesoporous alumina, synthesized according to Vaudry et al. [54], using carboxylic acids as structure-directing agents, and modified by cesium and boron, in toluene alkylation with methanol. It was found that the catalyst was inactive in toluene methylation but methanol was decomposed to carbon monoxide. The authors proposed that the physical constraints imposed by mesopores together with the less than optimum proximity of acid–base sites within molecular sieve environment did not facilitate the side-chain toluene alkylation reaction [73]. Mesoporous alumina modified with copper was tested as a catalyst for selective hydrogenation of cinnamaldehyde. The catalyst was prepared by direct synthesis using aluminum Keggin polycations, copper nitrate, palmitic acid and hexadecyl trimethyl ammonium bromide. The remarkable selectivity of this catalyst towards formation of unsaturated alcohol was observed compared to conventional alumina. This finding was attributed to the particularly strong interaction of the nanometer-sized Cu^0 particles with the mesoporous alumina walls, while the conjugated C=C bond was more readily hydrogenated on larger Cu^0 clusters exhibiting weaker interaction with the support [74]. Hydrodesulphurization reactions represent one of the most important applications of conventional alumina as a support for various Co–Mo or Ni–Mo phases in sulfide forms. Cejka et al. [75] tested organized mesoporous alumina for hydrodesulphurization of thiophene, and compared it to the commercial catalyst. Organized mesoporous alumina prepared by the "anionic" route with long-chain carboxylic acids was modified by conventional impregnation with a solution of ammonium heptamolybdate and by thermal spreading of molybdenum oxide. Due to the significantly larger surface area of different mesoporous aluminas compared to commercial Mo catalyst (BASF M8-30), it was possible to spread about 30 wt.% of MoO_3 on mesoporous alumina. This resulted in a significantly higher conversion compared to the conventional catalyst possessing only 15 wt.% of MoO_3.

Royer et al. [76] studied a set of mesoporous aluminas with different morphologies and physical properties for thiophene hydrodesulphurization. The most active solid exhibited a fine fibrillar morphology and the largest water pore volume (3.1 cm^3/g).

We have shown that application of mesoporous alumina catalysts with pore size at 6.3nm for obtaining dimethyl ether allows increasing transformation degree of methanol from 80 % to 88 %. Mesoporous Cu-based alumina catalysts allow increasing transformation degree of methanol up to 100 % [77]

The authors [72] examined well-ordered mesoporous aluminas as catalyst supports in the hydrogenation of acetone, glucose and cellobiose with different molecular sizes as test reactions. The conversion rates (normalized by the total ruthenium content) depend largely on the size of the support pores and reactants. The two catalysts are active and give identical rates (4207–4208 mol/mol·ruthenium·h,) for the hydrogenation of acetone, the smallest molecule chosen, demonstrating their same intrinsic hydrogenation activities. For glucose and cellobiose, their hydrogenation rates on ruthenium/meso-Al_2O_3 are 173 and 54 mol/ mol·ruthenium·h, respectively. In contrast, glucose and cellobiose convert at similar rates (510 and 492 mol/mol·ruthenium·h) on ruthenium/commercial-Al_2O_3, which are approxi-mately three and nine times greater than those on ruthenium/meso-Al_2O_3, respectively. Such differences in the conversion rates are apparently related to the effects of steric hindrance imposed by the molecular sizes of the reactants relative to the pore sizes of the Al_2O_3 supports on the diffusion and accessibility of the reactants to the ruthenium domains confined inside the pores [72].

4.4. From Organized to Ordered Mesoporous Structure of Tio$_2$

4.4.1. Sol-gel synthesis of organized mesoporous titania

As shown, the problem of synthesis of nanosized titania stable forms is successfully solved, however, one can obtain materials with predefined structural and morphological parameters only by thorough selection of synthesis conditions or by applying new approaches[78-80]. One such method is successfully used in practice and defined by use of organic templates as coordinators of pores and particles formed in final product. For titania-based materials, such approach promotes emergence of unique characteristics due to high photocatalytic properties that are developed upon structurization of nanolevel (20–100 Å) materials.

The first such materials (Ti–TMS1) were obtained in 1995 using alkylphosphate surfactant and titanium isopropoxide when it was shown that the key feature in obtaining the mesophase is the precise control of hydrolysis and condensation processes for titanium isopropylate as its alkoxides are instantly hydrolyzed in the presence of water. Because of this, non-stabilized process leads to the formation of insoluble titanium hydroxide with dense packing.

Authors of Ref. [81] obtained hexagonal packing of titania while performing hydrolysis in the presence of phosphatic surfactant, however, because of high thermal stability of such templates, they could not be completely removed during the calcination process and remained in the form of an admixture in a mesoporous material, according to IR spectroscopy data. Such samples were characterized by the surface area of about 200m^2/g and average mesopore distribution of 32 Å. The functional groups of surfactant were supposed to have reacted with alkoxides, leading to the formation of mesophases, however, because of the large residual quantities in the calcinated samples it was accepted to call them titanium oxophosphates, instead of pure titania, and after two years it has been shown that it is not hexagonal packing but a powder with scaly structure [81]. Later, mesoporous titania (712 m^2/g) was obtained at lower calcination temperature (200°C), with further extraction of templates by an acid method. [82]. However, it was not possible to completely remove phosphatic groups of samples what resulted in sharp decrease in photocatalytic activity of materials, as amorphous pore walls of titania hardly displayed photoactivity [82] for obtaining the highly active materials, considerable thermal treatment is required for complete transformation from an amorphous phase into crystal one [83] However, there is a number of the complexities caused by infringement of generated mesostructures in the course of crystallite growth. In Ref. [84], the generalized synthesis of mesoporous metal-oxide materials with polycrystalline structure including obtaining the mesoporous titania with large pore diameter (≈15 nm) is given. In this approach, block copolymers were used as predecessors of mesostructures, and $TiCl_4$ was used as the inorganic predecessor instead of alkoxides. The samples obtained possessed high crystallinity after calcination at 400°C for 5 hours and also hexagonal or cubic packing of mesophases.

The similar method was also used to obtain thin transparent films of anatase modification [84], with average pore diameter of 5 Å and the surface area of about 200m^2/g, the thickness of pore wall amounting to 51 Å. According to authors, the mechanism of formation of such materials represents self-assembly of a block copolymer with simultaneous course of complexation reactions. One of advantages of such materials is the thick crystal wall formed by anatase modification due to which pores do not collapse during calcination process. For the mesostructured materials with thinner walls, such as, for example, M41S group, crystallization of an inorganic part, after course of organo-inorganic self-assembly reactions, leads to a macroscopical phase separation.

Another example of mesoporous titania formation was the use of non-ionogenic triblock copolymer with the subsequent hydrothermal treatment [85], thus crystallization of an amorphous phase proceeded at low temperature, leaving the generated structure unbroken. Such materials possessed the high surface area of 250 m^2/g, with average pore diameter of about 80 Å, what promoted high thermostability up to 500°C, the X-ray electronic microscopy data indicating the formation of chaotic wormhole-like channels. In Ref. [86], the synthesis of mesoporous titania using non-ionogenic surfactants with amino groups as templates was performed. According to this method, the first stage was supposed to be the complexation reaction of metal-ligand and surfactant, leading to obtaining organo-inorganic hybrid materials, and then the product obtained was subjected to controllable hydrolysis. It was established that the given technique allows performing precise control of mesostructure parameters at the expense of metal/surfactant stoichiometric variations. According to the given method, the mesostructured materials with the surface areas under 700 m^2/g have been obtained using acid method for removal of template. Then the series of works on obtaining the mesoporous TiO_2 followed, first of which consisted in the use of cationic surfactant slowing down the hydrolysis process, then mesoporous titania was obtained using tetrabutylammonium hydroxide, and finally synthesis of a group of Ti-TMS mesoporous materials using the surfactant containing amino groups etc. was performed. However, because surfactant concentrations were low, they affected too weakly the formation of inorganic materials, so it was suggested using molecules of different nature organic templates as predecessors of the mesostructures, leading to the formation organo-inorganic hybrid materials. Along with this, the use of some organic acids, for example, stearic acid, tartaric acid, for obtaining mesoporous alumina and silica [87], as well as the use of some enzymes (for example, alkaline phosphatase 45) in mesostructured silica, was reported. However, initial attempts to use typical non-surfactants (for example, dibenzoyl tartaric acid (DBTA), D-glucose etc.) with titanium isopropylate have not led to obtaining homogeneous sol-gel system. (Note: such templates as glycerol, pentaerythritol etc. were successfully used for obtaining mesoporous TiO_2, whereas titanium tetrabutylate (TBT) was used as inorganic precursor. Then studies on synthesis of the mesostructured titania matrix using non-ionogenic PEG with possibility of pore structure management by selection of glycols with various molecular weights have been performed. All subsequent studies on obtaining mesoporous titania have been focused on the mechanism of formation of the ordered structures, in various packings, and self-assembly processes as well as on attempts to establish an interrelation between defined textural properties of highly structured materials and photoactivity.

4.4.2. Synthesis for ordered mesoporous titania

The key problem in preparation of the ordered titania is that titania has a rather low nucleation-to-growth rate, making bulk nanoparticles grow to a diameter of 10–20 nm. If such a process starts in much thinner titania walls of a mesoporous system, it is evident that the primary structure is lost throughout the crystallization process, partly due to the mass transport in the walls and partly due to the missing connectivity between the nanoparticles. There are two complementing solutions for this problem. Either the conditions of sol chemistry are changed in a way that the preformed "amorphous" titania structures are better interlinked and show a molecular pre-organization which supports faster nucleation. This can be done by reducing the water content in the recipes, for instance, by addition of tert-butyl alcohol or in ionic liquids. The other possibility is the development of new templates promoting larger pore sizes which self-organize sufficiently early in the process to support a robust titania gel structure. Note that also the first solution would strongly benefit from more robust templates, as the addition of alcohols and organic solvents usually interferes with the template organization process. As from 1998 several groups have successfully synthesized ordered mesoporous titania powders and thin films with well-ordered 2D hexagonal with mesochannels running parallel to substrates [15] and 3D cubic structures.[16,88]. Tian et al. [89] developed the "acid-base pair" strategy to fast prepare highly ordered mesoporous titania employing titanium alkoxide as the main inorganic precursor and $TiCl_4$ as the pH "adjustor" and hydrolysis-condensation "controller" [67], and the synthetic period was only 1–2 days. Unfortunately, mesoporous titania particles obtained through this method was only thermally stable to 623 K, resulting in the incomplete removal of block copolymer templates and the semi-crystalline anatase walls that could limit their wider application in nanotechnology and catalysis. Calcinations of the synthesized mesoporous TiO_2 at the higher temperature would induce a collapse of the ordered mesostructure.

Authors [90] presented a first step into this direction and employ a new block copolymer template (PHB-PEO block copolymers (based on the commercial Kraton liquid block)) for the preparation of mesoporous titania with a well-defined contracted Im3m mesostructure and a crystalline anatase framework between the mesopores.

Although these mesostructures exhibit ordered pore size and high surface area, the lack of perpendicular porosity greatly reduces the merit of the films. Therefore, the crystalline titania films with perpendicular porosity have been highly demanded because this pore configuration should increase the accessibility and enhance the efficiencies of adsorption and catalysis.

Chia-Wen Wu and et. al. [91] studied structural transformation of mesoporous titania films at the calcination to produce new structures because contraction of the mesostructure along the direction perpendicular to substrates preferentially occurs during the thermal processes. For example, rectangular and grid-like structures can be derived from 2D hexagonal and 3D cubic structures, respectively. The mesostructure of the as-synthesized titania film (heat-treatment at 100–400°C) was determined as a 3D hexagonal structure with an ABAB stacking on the basis of the HR-SEM and TEM results, [91]. After calcination at 400°C, the 3D hexagonal structure was transformed to the arrays of titania pillars [91].

T.Sakai et al. [17] presented a formation mechanism for hexagonal-structured self-assemblies of nanocrystalline titania. It was formed after mixing of aqueous solutions, containing CTAB spherical micelles and titanium oxysulfate acid hydrate ($TiOSO_4 \cdot H_2SO_4 \cdot xH_2O$) as a titania precursor. They founded that crystal growth of crystalline (anatase) titania (polymorphic crystallization) was promoted with higher temperature and lower titania precursor content in aqueous solutions, and hexagonally structure was formed in only cationic surfactant micellar solutions, but not in polycation solutions and formamide, because small molecules and polymers containing nitrogen can promote the hydrolysis reaction initiation and polymorphic crystallization, while the hexagonally structure was not formed in the absence of micelles.

4.4.3. Catalytic properties of ordered mesoporous titania

Titanium dioxide (TiO_2) is a promising photocatalyst owing to its peculiarity of chemical inertness, no photo-corrosion, low cost and non-toxicity. Titanium dioxide with highly ordered mesoporous structure was reported to improve the photocatalytic activity efficiently since the ordered mesoporous channels facilitate reactant molecular movement, and the large surface area of the mesoporous materials can improve the adsorption of the reactant and enhance the light harvesting. Anatase is excellent catalyst for the photocatalytic oxidation of a-methylstyrene under UV light, and acetophenone as a major product often was obtained [92]. Table 2 shows the yield change of acetophenone with UV illumination time using commercial anatase, P25 titania, and mesoporous titania calcined at 733 K as catalysts, according to [93].

The partial oxidation of a-methylstyrene is negligible without the light irradiation. The yields of acetophenone are 81 mol% for mesoporous titania and 63 mol% for commercial anatase under the 6 h irradiation time, which indicate that ordered mesoporous titania shows the higher photocatalytic activity than commercial anatase probably because of more active centers of high surface mesoporous titania or the higher quantum yield of the photocatalytic process [94]. Besides, P25 titania shows a lower photocatalytic activity in this photo oxidation reaction, suggesting that the anatase may be the active component in this photocatalytic reaction rather than rutile [93].

Table 2. Comparison the photocatalytic activities of commercial anatase, P25 titania and mesoporous titania calcined at 733 K for 2 h.

time, min	yield of acetophenone, %		
	p-25	commercial anatase	mesoporous titania
50	4	4	5
100	10	11	13
150	19	22	25
200	26	34	38
250	34	42	51
300	40	51	70
350	50	60	80

Table 3. Photocatalytic degradation of Rhodamine B in the presence of mesoporous ordered TiO_2 calcined at 450, 600, and 700°C.

time, min	C/C_0 of Rhodamine B		
	TiO_2 calcined at 450°C	TiO_2 calcined at 600°C	TiO_2 calcined at 700°C
25	0,84	0,84	0,74
50	0,74	0,71	0,55
75	0,65	0,58	0,4
100	0,57	0,47	0,32
125	0,5	0,39	0,24
150	0,42	0,28	0,19
175	0,36	0,2	0,15

On the other hand, in [95] the photocatalytic activity of the carbon-titania nanocomposites obtained at different temperatures was examined by measuring the photodegradation of Rhodamine B in an aqueous suspension following the reported procedure. They showed that the nanocomposite calcined at 450°C exhibits the lowest photocatalytic activity, whereas the nanocomposite TiO_2 calcined at 700°C exhibits the highest photocatalytic activity, table 3.

It was interpreted in terms of a compromise from the crystallinity and the surface area. Although the mesostructural regularity of the titania decreases with the increase in temperature, large surface area, and highly crystalline anatase obtained at high temperature contributes high adsorptive capacity, resulting in high photocatalytic activity. For the comparison, the photodegradation of Rhodamine B was performed in the presence of the commercial Degussa P25 TiO_2 as a reference sample under the same condition. The rate constant for the sample P25 is measured to be 4.7 × 10-2 min-1, which is a little higher than that of the obtained ordered nanocomposites calcined at different temperatures. This phenomenon has been observed in the previous report [96], which may be due to the smaller domain size of the crystalline TiO_2 in the nanocomposites compared with that of the powdered P25 sample. However, mesoporous ordered titania can be formed as membranes, which may be easy to handle and considered as promising candidate for useful photocatalysts.

CONCLUSION

The field of ordered mesoporous titania and alumina represents a great opportunity for materials to have an impact on modem technology. Substantial progress has been made in the inorganic-organic and organic-organic interactions that have been reflected in new types of molecular sieves with larger pore size compared to zeolites and narrow pore size distribution that are applicable not only in catalysis but also in other fields of chemistry. Using organic templates (surfactants, polymers, microspheres) in combination with hydrolysis by sol-gel technology, with the precise control of intermolecular interactions, has led to obtaining ordered materials that, depending on nature of modificating additives, produce mesostructures of predefined forms.

This chapter describes the precise control of the interactions inside the inorganic or organic species and organic template/precursor species, effects of condensation, pH and etc. for understanding the surfactant self-assembly approach in the fabrication of ordered mesoporous titania and alumina materials. Valuable applications emerge in the fields of biology, photoelectronic materials, sensors, electrodes and membranes. Although the number of studies on non-siliceous mesoporous materials has increased because of the high potential impact on practical applications, there is still a significant need to design, synthesize, and evaluate new materials, since the general methods for controlling the interactions between template and precursor leading to the formation of ordered materials have not yet been defined. Bulk production of non-siliceous mesoporous materials is also imperative under the situation. The present syntheses are limited by used templates. Once this is amplified, it will be possible to fabricate ordered mesoporous metal oxides and sulfides with crystallite walls, high melting point semiconductors, as well as mesoporous SiO_2, with open frameworks, together with achieving direct organic functional walls or channels of the mesoporous materials. The challenges are inevitably great in exploiting their functions in catalysis, sensors, microelectrodes and dielectric materials.

REFERENCES

[1] Kresge, CT; Leonowicz, ME; Roth, WJ; Vartuli, JC; Beck, JS. Ordered Mesoporous Molecular Sieves Synthesized by a Liquid-crystal Template Mechanism. *Nature,* 1992, 359, 710-712.

[2] Beck, JS; Vartuli, JC; Roth, WJ; Leonowicz, ME; Kresge, CT; Schmidt, KD; Chu, CT-W; Olson, DH; Sheppard, EW; McCullen, SB; Higgins, JB; Schlenker, JL. A New Family of Mesoporous Molecular Sieves Prepared with Liquid Crystal Templates. *J. Am. Chem. Soc.*, 1992, 114, 10834-10843.

[3] Bagshaw, SA; Prouzet, E; Pinnavaia, TJ. Templating of Mesoporous Molecular Sieves by Nonionic Polyethylene Oxide Surfactants. *Science*, 1995, 269, 1242-1244.

[4] Huo, Q; Margolese, DI; Ciesla, U; Feng, P; Gier, TE; Sieger, P; Leon, R; Petroff, PM; Schuth, F; Stucky, GD. Generalized Synthesis of Periodic Surfactant/Inorganic Composite Materials. *Nature*, 1994, 368, 317-321.

[5] Yang, H; Kuperman, A; Coombs, N; Mamiche-Afara, S; Ozin, GA. Synthesis of Oriented Films of Mesoporous Silica on Mica. *Nature,* 1996, 379, 703-705.

[6] Firouzi, AD; Kumar, D; Bull, LM; Besier, T; Sieger, P; Huo, Q; Walker, SA; Zasadzinski, JA; Glinka, C; Nicol, J; Margolese, D; Stucky, GD; Chmelka, BF. Cooperative Organization of Inorganic-Surfactant and Biomimetic Assemblies. *Science*, 1995, 267, 1138-1143.

[7] Monnier, A; Schuth, F; Huo, Q; Kumar, D; Margolese, D; Maxwell, RS; Stucky, GD; Kirshnamurty, M; Petroff, PM; Firouzi, A; Janicke, M; Chmelka, BF. Cooperative Formation of Inorganic-Organic Interfaces in the Synthesis of Silicate Mesostructures. *Science*, 1993, 261, 1299-1303.

[8] Huo, Q; Leon, R; Petroff, PM; Stucky, GD. Mesostructure Design with Gemini Surfacants: Supercage Formation in a Three-Dimensional Hexagonal Array. *Science*, 1995, 268, 1324-1327.

[9] Landau, MV; Dafa, E; Kaliya, ML; Sen, T; Herskowitz, M. Mesoporous alumina catalytic material prepared by grafting wide-pore MCM-41 with an alumina multilayer. *Microporous Mesoporous Mater,* 2001, 49, 65-81.

[10] Misra, C. *Industrial Alumina Chemicals*, Washington: ACS Monograph, 1986.

[11] Topsoe, H; Clausen, BS; Massoth, FE. *Hydrotreating Catalysis*, Berlin: Springer, 1996.

[12] Rouquerol, F; Rouquerol, J; Sing, K. *Adsorption by Powders & Porous Solids*, Academic Press: San Diego, 1999.

[13] Zhou, RS; Snyder, RL. Structures and Transformations mechanisms of the η, γ and θ transition Aluminas. *Acta Crystallogr*, 1991, 47, 617.

[14] Hagfeldt, A.; Gratzel, M. Light induced redox reactions in nanocrystalline systems. *Chem. ReV.*, 1995, 95, 49-68.

[15] Yun, H. S.; Miyazawa, K.; Zhou, H. S.; Honma, I.; Kuwabara, M. Synthesis of mesoporous thin TiO_2 films with hexagonal pore structures using triblock copolymer templates. *AdV. Mater.* 2001, 13, 1377-1380.

[16] Alberius, P. C. A.; Frindell, K. F.; Hayward, R. C.; Kramer, E. J.; Stucky, G. D.; Chmelka, B.F.General predictive syntheses of cubic, hexagonal, and lamellar silica and titania mesostructured thin films. *Chem. Mater.* 2002, 14, 3284-3294.

[17] Toshio, S; Hanae, Y; Mitsuri, O; Hirobumi, S; Kanjiro, T; Shigenori, U; Kazutami, S; Naokiyo, K; Satoshi, A; Hideki, S; Masahiko,A. Formation mechanism for hexagonal-Strucrured Self-Assemblies of Nonocrystallite titania Templated by cetyltrimethylammonium bromide. *Journal of Oleo Science.* 2008 57, (11), 629-637.

[18] Selvam, P; Bhatia, SK; Sonwane, CG. Recent Advances in Processing and Characterization of Periodic Mesoporous MCM-41 Silicate Molecular Sieves, *Ind. Eng. Chem. Res.*, 2001, 40, 3237-3261.

[19] Sing, KSW. The use of nitrogen adsorption for the characterization of porous materials. *Colloids and Surfaces, A: Physicochemical and Engineering Aspects*, 2001, 187-188, 3-9.

[20] Sing, KSW. The use of gas adsorption for the characterization of porous solids. *Int. Conf. Fundam. Adsorpt., 3rd*, 1991, 69-83.

[21] Brunauer, S; Deming, LS; Deming, WE; Teller, E. A theory of the van der Waals adsorption of gases. *Journal of the American Chemical Society* 1940, 62, 1723-1732.

[22] Brunauer, S; Emmett, PH; Teller, E. Adsorption of gases in multimolecular layers. *Journal of the American Chemical Society*, 1938, 60, 309-319.

[23] Williams, DB; Carter, B. *Transmission Electron Microscopy – A Textbook for Materials Science*, Published by Springer, 1996.

[24] Levin, I; Brandon, D. Metastable Alumina Polymorphs: Crystal Structures and Transition Sequences. *J. Am. Ceram. Soc.*, 1998, 81, 1995–2012.

[25] Gitzen, WH. *Alumina as a ceramic material.* The American Ceramic Society: Westerville, 1970.

[26] Xu, YN; Ching, WY. Selfconsistent Band Structures, Charge Distributions and Optical Properties of MgO, α-Al_2O_3 and $MgAl_2O_4$, *Phys. Rev. B,* 1991 ,43, 4461-4472.

[27] Oliver, WC; Pharr, GM. An improved technique for determining hardness and elastic modulus using load and displacement sensing indentation experiments. *J. Mater. Res.* 1992, 7, 1564-1583.

[28] Schneider, JM; Sproul, WD; Voevodin, AA; Matthews, A. Crystalline Alumina Deposited at Low Temperatures by Ionised Magnetron Sputtering. *J. Vac. Sci. Technol.*, 1997, 15, 1084-1088.

[29] Müller, J; Schierling, M; Zimmermann, E; Neuschütz, D. Chemical vapor deposition of smooth a-Al_2O_3 films on nickel base superalloys as diffusion barriers. *Surf. Coat. Technol.* 1999, 120-121, 16-21.

[30] French, RH. Electronic Structure of Al_2O_3, with comparison to AlON and AlN. *J. Am. Ceram. Soc.*, 1990, 73, 477-489.

[31] Borosy, AP; Silvi, B; Allavena, M; Nortier, P. Structure and Bonding of Bulk and Surface θ-Alumina from Periodic Hartree-Fock Calculations. *J. Phys. Chem.*, 1994, 98, 13189-13194.

[32] Mo, S-D; Xu, Y-N; Ching, W-Y. Electronic and Structural Properties of Bulk γ-Al_2O_3. *J. Am. Ceram. Soc.* 1997, 80, 1193-1197.

[33] Kim, SJ; Park, SD; Jeong, YH. Synthesis of brookite TiO_2 nanoparticles by ambient condition sol process. *J. Am. Ceram. Soc.,* 1999, 82, 927–932.

[34] Wang, CC; Ying, JY. The effects of different acids on the preparation of TiO2 nanostructure in liquid media at low temperature.*Chem. Mater.*, 1999, 11, 3113–31203.

[35] Ozawa, T; Iwasaki, M; Tada, H; Akita, T; Tanaka, K. Thermally induced phase and photocatalytic activity evolution of polymorphous titania. *J. of Colloid and Interface Science,* 2005, 281, 510-513.

[36] Brinker, CJ; Scherer, GW. Sol–Gel–Glass: I. Gelation and Gel Structure. *J. Non-Cryst. Solids*, 1985, 70, 301-322.

[37] Brinker, CJ; Scherer, GW. *Sol-Gel Science: The Physics and Chemistry of Sol-Gel Processing*, Academic Press: San Diego, 1990.

[38] Turova, NY; Kessler, VG; Turevskaya, EP; Yanovskaya MI. *The chemistry of metal alkoxide*. Springer, 2002.

[39] Winter G. Esters of Titanium and their use in Paint. Part I: Preparation of Polymeric Butyl Titanates. *J. Oil and Colour Chem.*, 1953, 36, 689-695.

[40] Attard, GS; Glyde, JC; Goltner, CG. Liquid-crystalline phases as templates for the synthesis of mesoporous silica. *Nature,* 1995, 378, 366-368.

[41] Das, TK; Chaudhari, K; Chandwadkar, A J; Sivasanker, S. Synthesis and catalytic properties of mesoporous tin silicate molecular sieves. *Chemical Communications*, 1995, 24, 2495-2496.

[42] Gontier, S; Tuel, A.. Synthesis and characterization of Ti-containing mesoporous silicas. *Zeolites*, 1995, 15, 601-610.

[43] Reddy, KM; Moudrakovski, I; Sayari, A. Synthesis of mesoporous vanadium silicate molecular sieves. *Chemical Communications* 1994, 9, 1059-1560.

[44] Ciesla, U; Demuth, D; Leon, R; Petroff, P; Stucky, G; Unger, K; Schueth, F. Surfactant controlled preparation of mesostructured transition-metal oxide compounds. *Chemical Communications,* 1994, 11, 1387-1388.

[45] Huo, Q; Margolese, DI; Ciesla, U; Demuth, DG; Feng, P; Gier, TE; Sieger, P; Firouzi, A; Chmelka, BF. Organization of Organic Molecules with Inorganic Molecular Species into Nanocomposite Biphase Arrays. *Chemistry of Materials,* 1994, 6, 1176-1191.

[46] Vartuli, JC; Kresge, CT; Roth, WJ; McCullen, SB; Beck, JS; Schmitt, KD; Leonowicz, ME; Lutner, JD; Sheppard, EW. Designed synthesis of mesoporous molecular sieve systems using surfactant-directing agents. *Advanced Catalysts and Nanostructured Materials,* 1996, 1-19.

[47] Chen, CY; Burkett, SL; Li, HX; Davis, ME. Studies on mesoporous materials. II. Synthesis mechanism of MCM-41. *Microporous Materials* 1993, 2, 27-34.

[48] Stucky, GD; Monnier, A; Schueth, F; Huo, Q; Margolese, D; Kumar, D; Krishnamurty, M; Petroff, P; Firouzi, A. Molecular and atomic arrays in nano and mesoporous materials synthesis. *Molecular Crystals and Liquid Crystals Science and Technology, Section A: Molecular Crystals and Liquid Crystals,* 1994, 240, 187- 200.

[49] Vinogradov, VV; Agafonov, AV; Vinogradov, AV; Application of polyethyleneimine to obtain a mesoporous CuO–Al_2O_3 composite. *Mendeleev Communications*, 2009, 19, 222-223.

[50] Vinogradov, A.V; Agafonov, A.V; Vinogradov, V.V. Sol-gel synthesis of titanium dioxide based films possessing highly ordered channel structure. *J. Mendeleev Comm.* 2009, 19, 340-341.

[51] Bagshaw, SA; Pinnavaia, TJ. Mesoporous alumina molecular sieves. *Angew Chem. Int. Ed. Engl.*, 1996, 35, 1102-1105.

[52] Zhang, W; Pinnavaia, TJ. Rare Earth Stabilization of Mesoporous Alumina Molecular Sieves Assembled through an NoIo Pathway. *Chem. Commun.,* 1998, 1185-1186.

[53] Deng, W; Bodart, P; Pruski, M; Shanks, BH. Characterization of Mesoporous Alumina Molecular Sieves Synthesized by Nonionic Templating. *Microporous Mesoporous Mater.,*2002, 52, 169-177.

[54] Vaudry, F; Khodabandeh, S; Davis, ME. Synthesis of Pure Alumina Mesoporous Materials. *Chem. Mater.*, 1996, 8, 1451-1464.

[55] Cejka, J; Žilkova, N; Rathousky, J; Zukal, A. Nitrogen adsorption study of organised mesoporous alumina. *Phys. Chem. Chem. Phys.*, 2001, 3, 5076-5081.

[56] Yada, M; Kitamura, H; Machida, M; Kijima, T. Biomimetic Surface Patterns of Layered Aluminium Oxide Mesophases Templated by Mixed Surfactant Assemblies. *Langmuir,* 1997, 13, 5252-5257.

[57] Yada, M; Hiyoshi, H; Ohe, K; Machida, M; Kijima, T. Synthesis of Aluminium-based Surfactant Mesophases Morphologically Controlled through a Layer to Hexagonal Transition. *Inorg. Chem.*, 1997, 36, 5565-5569.

[58] Sicard, L; Lebeau, B; Patarin, J; Zana, R. Study of the Mechanism of Formation of a Mesostructured Hexagonal Alumina Means of Fluorescence Probing Techniques. *Langmuir,*2002, 18, 74.

[59] Sicard, L; Llewellyn, PL; Patarin, J; Kolenda, F. Investigation of the mechanism of the surfactant removal from a mesoporous alumina prepared in the presence of sodium dodecylsulfate. *Microporous Mesoporous Mater.*, 2001, 44–45, 195-201.

[60] Yada, M; Okya, M; Machida, M; Kijima, T. Synthesis of porous yttrium aluminium oxide templated by dodecyl sulfate assemblies. *Chem. Commun.,*1998, 1941-1942

[61] Yada, M; Okya, M; Ohe, K; Machida, M; Kijima, T. Porous Yttrium Aluminum Oxide Templated by Alkyl Sulfate Assemblies. *Langmuir*, 2000, 16, 1535-1541.

[62] Valange, S; Guth, J-L; Kolenda, F; Lacombe, S; Gabelica, Z. Synthesis strategies leading to surfactant-assisted aluminas with controlled mesoporosity in aqueous media. *Microporous Mesoporous Mater.*, 2000, 35–36, 597-607.

[63] Cabrera, S; Haskouri, JEl; Alamo, J; Beltran, A; Beltran, D; Mendioroz, S; Dolores Marcos, M; Amoros, P. Surfactant Assisted Synthesis of Mesoporous Alumina with Continuously Adjustable Pore Sizes. *Adv. Mater.*, 1999, 11, 379-382.

[64] Cabrera, S; Haskouri, JEl; Guillem, C; Latorre, J; Beltran-Porter, A; Beltran-Porter, D; Dolores Marcos, M; Amoros, P. Generalized Synthesis of Ordered Mesoporous Oxides:The Atrane Route. *Solid State Sci.*, 2000, 2, 405-420.

[65] Liu, X; Wei, Y; Jin, D; Shih, W-H. Synthesis of mesoporous aluminum oxide with aluminum alkoxide and tartaric acid. *Mater. Lett.,* 2000, 42, 143-149.

[66] Zhang, ZR; Hicks, RW; Pauly, TR; Pinnavaia, TJ. Mesostructured forms of gamma-Al_2O_3. *J. Am. Chem. Soc.,* 2002, 124, 1592-1593.

[67] Tian, BZ; Yang, HF; Liu, XY; Xie, SH; Yu, CZ; Fan, J; Tu, B; Zhao, DY. Fast preparation of highly ordered nonsiliceous mesoporous materials via mixed inorganic precursors, *Chem. Commun.,* 2002, 1824-

[68] Niesz, K; Yang, P; Somorjai, GA. Sol-gel synthesis of ordered mesoporous alumina. *Chem. Commun.* 2005, 1986-1987.

[69] Kuemmel, M; Grosso, D; Boissie`re, C; Smarsly, B; Brezesinski, T; Albouy, PA; Amenitsch, H; Sanchez, C. Thermally stable nanocrystalline gamma-alumina layers with highly ordered 3D mesoporosity. *Angew. Chem.*, *Int. Ed.* 2005, 44, 4589-4592.

[70] Boissie`re, C; Nicole, L; Gervais, C; Babonneau, F; Antonietti, M; Amenitsch, H; Sanchez, C; Grosso, D. Nanocrystalline mesoporous gamma-alumina powders "UPMC1 Material" gathers thermal and chemical stabiity with high surface area. *Chem. Mater.* 2006, 18, 5238-5243.
[71] Liu, Q.; Wang, A. Q.; Wang, X. D.; Zhang, T. Ordered Crystalline Alumina Molecular Sieves Synthesized via a Nanocasting Route *Chem. Mater.* 2006, 18, 5153-5155.
[72] Yuan, Q; Yin, A-X; Luo, C; Sun, L-D; Zhang, Y-W; Duan, W-T; Liu, H-C; Yan C-H. Facile Synthesis for Ordered Mesoporous γ-Aluminas with High Thermal Stability. *J. Am. Chem. Soc.*, 2008, 130, 3465-3472.
[73] Wieland, WS; Davis, RJ; Garces, JM. Side-Chain Alkylation of Toluene with Methanol over Alkali-Exchanged Zeolites X, Y, L, and β. *J. Catal.*, 1998, 173, 490-500.
[74] Valange, S; Barrault, J; Derouault, A; Gabelica, Z. Binary Cu-Al mesophases precursors to uniformly sized copper particles highly dispersed on mesoporous alumina. *Microporous Mesoporous Mater.*, 2001, 44–45, 211-220.
[75] Kaluža, L; Zdražil, M; Žilkova, N; Cejka, J. High activity of highly loaded MoS_2 hydrodesulfurization catalysts supported on organised mesoporous alumina. *Catal. Commun.*, 2002, 3, 151-157.
[76] Bejenaru, N; Lancelot, C; Blanchard, P; Lamonier, C; Rouleau, L; Payen, E; Dumeignil, F; Royer, S. Synthesis, Characterization, and Catalytic Performances of Novel CoMo Hydrodesulphurization Catalysts Supported on Mesoporous Aluminas. *Chem. Mater.* 2009, 21, 522–533.
[77] Vinogradov, VV; Agafonov, AV; *Kat. Ind.*, 2008, **5**, 17-21 (in Russian).
[78] Agafonov, AV; Vinogradov, AV. Sol–gel synthesis, preparation and characterization of photoactive TIO2 with ultrasound treatment. *J Sol-Gel Sci Technol.*, 2009, 49, 180–185.
[79] Vinogradov, VV; Agafonov, AV; Vinogradov, AV. Superhydrofobic effect of hybrid organo-inorganic materials. *J Sol-Gel Sci Technol.,* 2010, 53, 312–315.
[80] Agafonov, AV; Vinogradov, AV. Catalytically Active Materials Based on Titanium Dioxide: Ways of Enhancement of Photocatalytic Activity. *High Energy Chemistry.* 2008, 42, 70–72.
[81] Putnam, RL; Nakagawa, N; McGrath, KM; Yao, N; Aksay, IA; Grunner, SM; Navrotsky, A.. Titanium Dioxide−Surfactant Mesophases and Ti-TMS1. *Chem. Mater.* 1997, 9, 2690-2697.
[82] Stone, VF; Davis, RJ. Synthesis, Characterization and Photocatalytic Activity of Titania and Niobia Mesoporous Molecular Sieves. *Chem. Mater.* 1998, 10, 1468-1474.
[83] Yang, P; Zhao, D; Margolese, DI; Chmelka, BF; Stucky, GD. Block Copolymer Templating Syntheses of Mesoporous Metal Oxides with Large Ordering Lengths and Semicrystalline Framework. *Chem. Mater.* 1999, 11, 2813-2826.
[84] Yang, P; Zhao, D; Margolese, DI; Chmelka, BF; Stucky, GD. Generalized syntheses of large-pore mesoporous metal oxides with semicrystalline frameworks. *Nature.* 1998, 396, 152-155.
[85] Yue, Y; Gao, Z. Synthesis of mesoporous TiO2 with a crystal- line framework. *Chem. Commun.* 2000, 1755-1756.
[86] Antonelli, DM. Synthesis of Phosphorous-Free Mesoporous Titania via Templating with Amine Surfactants. *Microporous Mesoporous Mater.* 1999, 30, 315-319.

[87] Izutsu, H; Mizukami, F; Kiyozumi, Y; Maeda, K. Structure and Properties of Silica Derived from Silicon Alkoxide Reacted with Tartaric Acid. *J. Am. Ceram. Soc.* 1997, 80, 2581-2589.

[88] Crepaldi, EL; Soler-Illia, GJ. de AA.; Grosso, D.; Cagnol, F; Ribot, F; Sanchez, C. Controlled formation of highly organized mesoporous titania thin films: From mesostructured hybrids to mesoporous nanoanatase TiO2. *J. Am. Chem. Soc.* 2003, 125, 9770-9786.

[89] Tian, B; Liu, X; Tu, B; Yu, C; Fan. J; Wang, L; Xie, S; Stucky, G; Zhao. D. Self-adjusted synthesis of ordered stable mesoporous minerals by acid-base pairs. *Nature Mater.* 2003, 2, 159-163.

[90] Smarsly, B; Grosso, D; Brezesinski, T; Pinna, N; Boissière, C; Antonietti, M; Sanchez, C. Highly Crystalline Cubic Mesoporous TiO2 with 10-nm Pore Diameter Made with a New Block Copolymer Template. *Chem. Mater.* 2004, 16, 2948–2952.

[91] Wu, C; Ohsuna, T; Kuwabara, M; Kuroda, K. Formation of Highly Ordered Mesoporous Titania Films Consisting of Crystalline Nanopillars with Inverse Mesospace by Structural Transformation. *J. Am. Chem. Soc.* 2006, 128, 4544-4545.

[92] Yahiro, H; Miyamoto, T; Watanabe, N; Yamaura, H. Photocatalytic partial oxidation of α-methylstyrene over TiO2 supported on zeolites. *Catal. Today.* 2007, 120, 158-162.

[93] Aiguo, K.; Jiang, L; Xin, Y; Hanming, D; Yongkui, S. Fast preparation of ordered crystalline mesoporous titania. *J Porous Mater.* 2009, 16, 9–12.

[94] Yuan, S; Sheng, Q; Zhang, J; Chen, F; Anpo, M; Zhang, Q. Synthesis of La3+ doped mesoporous titania with highly crystallized walls. *J. Microporous Mesoporous Mater.* 79, 93-99 (2005)

[95] Liu, R; Ren, Y; Shi, Y; Zhang, F; Zhang, L; Tu, B; Zhao, B. Controlled Synthesis of Ordered Mesoporous C-TiO2 Nanocomposites with Crystalline Titania Frameworks from Organic-Inorganic-Amphiphilic Coassembly. *Chem. Mater.* 2008, 20, 1140–1146.

[96] Zhang, DY; Yang, D; Zhang HJ; Lu, CH; Qi, L. Synthesis and Photocatalytic Properties of Hollow Microparticles of Titania and Titania/Carbon Composites Templated by Sephadex G-100. M. Chem. *Mater.* 2006, 18, 3477–3485.

In: The Sol-Gel Process
Editor: Rachel E. Morris

ISBN 978-1-61761-321-0

Chapter 5

NANOCOMPOSITES WITH CONTROLLED PROPERTIES OBTAINED BY THE THERMAL TREATMENT OF SOME TETRAETHYL ORTHOSILICATE-DIOLS- METAL NITRATES GELS

Mircea Stefanescu, Marcela Stoia and Oana Stefanescu
Industrial Chemistry and Environmental Engineering,
University 'Politehnica' of Timisoara, Timisoara, Romania

ABSTRACT

The preparation of nanoparticles and nanocrystallites by conventional synthesis methods is difficult to achieve due to their strong tendency to agglomerate into bigger aggregates. This fact leads to the loss of unique properties generated by the large number of atoms located at the surface of the nanomaterial compared to the bulk material. The embedding of the nanocrystallites in amorphous matrices represents one of the possibilities to limitate the nanocrystallite's growth and agglomeration and to achieve a narrow size distribution. The sol-gel method is one of the few synthesis methods which lead to nanocomposites at low temperatures, by embedding or formation of oxidic particles within amorphous matrices. Using inorganic matrices (silica dioxide, titanium dioxide) as host materials, represents an efficient preparation way of oxidic nanoparticles with controlled dimensions and shapes, i.e. of nanocomposites with pre-defined properties. Since the properties of the nanocomposites significantly depend on the interface interaction between the oxidic particle and matrix, an important role is played by the nature and morphology of the matrix. This chapter will focus on a study regarding the obtaining of nanocomposites metal oxides/silica with controlled magnetic properties, using an original modified sol-gel method. The synthesis method used consists in the obtaining of some gels in the system tetraethyl orthosilicate-diol-metal nitrates, using a controlled thermal treatment. The diol present in the system interacts with both tetraethyl orthosilicate (TEOS) and metal nitrates. The interaction of the diol with TEOS during the gelation process influences the formation of the gels and finally the morphology of the resulted silica matrix. During the thermal treatment of the gels, the diol interacts with the metal nitrates by a redox reaction, leading to the formation of some carboxylate type compounds embedded in the silica matrices pores (precursors for nanocomposites). The controlled thermal treatment applied to these precursors leads to the formation of nanocrystalline oxidic phases embedded within the silica matrix. The first part of this chapter will present a systematic study on the interaction of the diols with the hydrolysis products of TEOS and the effect of this interaction on the silica matrices morphology. The second part will focus on the preparation of different nanocomposites with magnetic properties. The synthesis method involves particular conditions (the diols nature, molar ratio of reactants, concentration of metal oxide within the matrix, thermal treatment) for every studied oxidic system, especially in the cases of those that contain ions with variable valences.

INTRODUCTION

Nanocomposites are materials consisting of at least two phases with one dispersed in the other, the so-called matrix which forms a three-dimensional network, in which the dispersed nanograined materials are single phased and polycrystalline [1].

The composite systems containing metal oxide nanoparticles embedded in a mesoporous polymeric matrix represent an important issue in the field of nanoscience and nanotechnology, as the nanoparticle incorporation allows imparting of unique properties to polymeric materials: catalytic, optical, magnetic, sensing. If the polymeric matrix is nanostructured, with pores of sizes and shapes controlled before the formation of the nanoparticles, then the matrix contains domains of different chemical natures, divided by interfaces, which reflects a high degree of nanostructural organization. The presence of these kinds of interfaces in polymer systems allows a subtle control of the nanoparticle growth of their shapes and size distribution [2].

The nanoparticles can be embedded by using some organic or inorganic matrices. The organic matrix is usually an easy to shape polymer. The inorganic matrices, of which the most used is silica, are generally made of oxides and prepared by a hydrolytic polycondensation of sol-gel type. The choice between organic and inorganic polymers for the attainment of the matrix is essentially determined by the intended application. The role of the matrix is very important because it is the matrix that transforms the nanoparticle into nanomaterial, thus permitting it to display its special properties [3]. For example, the structure of the matrices pores and the interactions between the magnetic particles and the matrix can be used to control their magnetic properties [4]. The use of inorganic matrices as host materials for the nanoparticles represents an efficient way to prepare particles with uniform sizes and controlled morphology. A porous matrix provides enough nucleation sites for nanoparticle formation and a way to avoid their aggregation. One has to keep in mind that the properties of the SiO_2 matrix can be modified by molecular additives.

One of the actual tendencies for the obtaining of nanocomposites with controlled properties is the encasement of oxidic nanoparticles or of nanoparticles precursors, in hybrid organic-inorganic matrices. These matrices synergistically combine the advantages of organic matrices with those of inorganic matrices. By embedding oxidic precurors in hybrid matrices, the thermal treatment necessary for the decomposition of these precursors simultaneously induces the conversion of the hybrid organic-inorganic matrix in an inorganic matrix with modified morphology dependent on the nature of the organic component.

Processes and Factors in the Synthesis of Silica Gels

The sol-gel process is one of the most used ways for the preparation of nanocomposites of oxide nanoparticles/ inert matrix (organic, inorganic or hybrid) type. The silica gels are ideal for the obtaining of this type of nanocomposites due to their unreactiveness, their high specific surface and their high porosity and transparency. The obtaining of the silica gels consists in the formation of an amorphous inorganic network starting from a molecular precursor (≡Si-OR, R – alkyl group), by hydrolysis – polycondensation reactions, in solution, at low temperatures ($t < 100^{o}C$), when the transition from the liquid (sol) to the gel phase (solid capilar network which contains within the pores the liquid from the system) takes place. The alkoxides $Si(OR)_4$ are sensitive to the action of water [5] giving facile hydrolysis with the formation of hydroxoderivatives $Si(OH)_x(OR)_{4-x}$.These compounds can be obtained as real or colloidal (sol) solutions which under certain conditions can be transformed to half-solid or solid gels. Complete dehydrating and thermal treatment of the gels can lead to oxidic materials at much lower temperatures compared to those used in traditional technology. The gelation is based on two types of chemical reactions.

*1. **Hydrolysis***– the alkoxidic precursor reacts with water, giving ≡Si-OH groups. Depending on the nature of the alkoxide, the hydrolysis may take place with different rates according to the reaction:

$$Si(OR)_4 + x\ H_2O \Rightarrow Si(OH)_x(OR)_{4-x} + x\ R\text{-}OH$$

When the hydrolysis reaction takes place with a lower water amount than necessary for the stoechiometry of the reaction, the condensation takes place before the completion of the hydrolysis and the partial hydrolized species begin to condense [6].

*2. **Condensation***– Si-O-Si units are formed and the reaction products are water or alcohol. The condensation reaction can be regressed by the solubilization of the alkoxide in alcohol, because the alcohol is a reactant which can participate to esterification and alcoholysis.

The formation of the silica gels, as a result of hydrolysis and condensation, is influenced by multiple parameters. The most important are:

Organic radical of the alkoxide precursor

The nature of the alkyl radical influences the rate of the hydrolysis reaction. The main characteristics of the silica matrix (specific surface, porosity) obtained by hydrolysis and condensation of tetraalkoxysilicates depend on the nature of the alkyl group [7, 8]. The majority of the researchers use as raw materials for the obtaining of silica gels the tetraethyl orthosilicate (TEOS) [9-19] or the tetramethyl orthosilicate (TMOS) [20-24].

The ratio water : TEOS

The ratio water: TEOS significantly influences the hydrolysis and implicitly, the condensation processes. Several studies have reported the use of molar ratios $r = H_2O/TEOS$ in the range 1-50. For $r < 1$ the sol-gel synthesis leads to the attainment of linear siloxane polymers, for $1 < r < 2$ one can obtain a viscoelastic gel. An increase of the molar ratio $H_2O/TEOS$ leads to a more advanced hydrolysis although the reaction rate decreases. Because water is a condensation product, the water excess favorizes the inverse reaction, the hydrolysis of siloxane bonds. Independent on the used catalyst, even for $r > 4$, the hydrolysis and condensation reactions are not complete (quantitative), residual carbon remains in the gels matrix [25]. According to the literature, for a complete hydrolysis of the initial alkoxysilanes it is necessary to use a molar ratio H_2O: TEOS = 25 : 1.

There are different possibilities for a sol-gel synthesis without water in the initial system, which opens new opportunities in the sol-gel chemistry. Thus, by using a mixture of alcohol and carboxyllic acid in the initial system, the water is generated in situ by the acids esterification (e. g. citric acid) [26] or by means of other condensation reactions (heterofunctional condensation of silicon alkoxides) [27].

Effect of the solvent

The solvent is necessary in order to homogenize the two components, water and alkoxide, which are non-miscible. The alkoxide is solubilized in alcohol in order to lower the rate of the condensation reaction as compared to that of the hydrolysis. The alcohol is not only a solvent, but it appears in the ecuations of hydrolysis-polycondensation as a reaction product. The concentration of the alcohol is involved in the hydrolysis equilibrum [5].

The solvents can be classified in polar and non-polar, protic and aprotic. Protic solvents can form H bonds increasing the electrophylicity of the H^+ ions, in case of the sol-gel processes in acid catalysis. Aprotic solvents can not form hydrogen bonds with HO^- groups. Thus, the nucleophile character of the HO^- ions, in aprotic environment is substantially increased. Aprotic solvents hinder the hydrolysis and condensation processes in acid catalysis, while protic solvents have an opposite effect [6].

The catalyst

Generally, the used catalysts are acids (mineral acids: HCl, HNO_3, HF) or bases (ammonia and hydroxides of alkaline metals) [28, 29]. The catalysts initiate and accelerate the hydrolysis. The low reactivity of $Si(OR)_4$ at hydrolysis and condensation can be due to the low electrophile character of silicon in tetrahedral coordination. The hydrolysis rate is minimal at pH = 7 and increases with the increase of the $[H_3O^+]$ and $[HO^-]$ ions concentration.

The condensation degree partly depends on the catalyst. Thus, the use of acid catalysts leads to a lower condensation degree (~70%) compared to the use of base catalysts (~90%) [30].

Temperature

The temperature significantly influences the gelation process. The increase of the temperature leads to a shorter gelation time, but at the same time, the temperature can not exceed a maximum value, when the thermal agitation of the particles blocks the formation of the network. Many of the sol-gel syntheses are achieved in the temperature range 40-80 ^{o}C.

The gelation time is a specific parameter of the sol-gel method and it depends on many parameters: the nature of the alkoxide, pH value, catalyst, solvent, molar ratio water-alkoxide, temperature. Because the gelation takes place through several dependent processes, it is difficult to determine the exact gelation time [6].

After gelation, the network becomes flexible and there is the possibility for subsequent condensation reactions to develop. Moreover, the sol still exists in the interior of the pores and the corresponding oligomers continue to get attached to the network, causing the consolidation and strengthening of the system. This process is called gel aging. The gel suffers syneresis (contraction) when the liquid dispersion medium is spontaneously eliminated from the matrix gel (disperse phase). During the maturation, the system composition remains practically the same [25].

Drying is one of the most important factors in the sol-gel process. It starts with the evaporation of the dispersion medium, which generally consists of water and solvents. The actual drying starts when the interface approaches the monoliths surface. The drying process takes place under the influence of capillary forces which assure the transfer of the liquid phase to the surface of the monolith, for evaporation. The movement of the interface liquid-gas, in the interior of the monolith, is attended by the formation of the pores, so that the number of the latter and the specific surface increases considerably. The monolith tends to lower the surface energy which increases and that causes cracks. In case of drying under normal conditions (room temperature, atmospheric pressure), when the wet gels are kept, after gelation, in covered recipients, they start to contract and crack, and they get a glass- like aspect.

The gels structure is correlated with hydrolysis and condensation parameters. Acid catalysis leads to the formation of linear polymers transversally bonded with fine, condensed structures. These polymers overlap and form additional chains which lead to gelation. In basic catalysis, clusters more branched are formed due to the fast condensation. The binding of the clusters lead to the formation of gels with granular texture, which retain less organic rest. Generally, SiO_2 materials prepared in acid conditions are microporous with narrow size distribution of the pores and high superficial surface while those prepared in basic conditions present a large mesopore distribution and relatively low superficial area [31].

Further thermal treatment of the gels leads to sintering. Sintering is a process of densification determined by the high interfacial surface of the system. The less the particles dimension, the higher the specific surface and superficial energy. Polymeric gels with a specific surface of 600 m^2/g are sintered at temperatures lower than 1100 ^{o}C and silicate gels with a specific surface up to 100 m^2/g are completely sintered at temperatures higher than 1400 ^{o}C. The reason for this big difference between the sintering temperatures is the presence of a high number of hydroxyl groups at the gels surface. The higher the surface, the higher the hydroxyl groups number and their condensation by heating favorizes the sintering [25].

By determination of the exact drying and sintering conditions, one can develop different obtaining methods for materials with directed properties using the sol-gel process [32-34].

The influence of molecular organic compound on the formation of silica gels

The possibility of using organic compounds in order to modify the inorganic framework has long been recognized as an interesting instrument for the development of new composite materials. One of the possibilities to synthesize hybrid materials was to modify the inorganic network by introducing selected organic groups leading to organic modified silicates (Ormosils) [35, 36]. The use of polysiloxanes modified with SiO_2 or TiO_2 leads to hybrid materials called by Wilkes: Ceramers (CERAmic PolyMERS). These materials contain Si-C bonds and very small SiO_2 particles.

Organic compounds with small molecules such as formamide, dimethylformamide, ethylene glycol, glycerol, oxalic acid, etc. are used in the sol-gel synthesis as drying control chemical additives (DCCAs) in order to achieve a faster drying without cracks. The use of DCCAs leads to the attainment of uniform pores and particle sizes, which reduce the differential drying stress by minimizing the variation of the eveporation rate. The presence of organic compounds with small molecules, especially diols (ethylene glycol, glycerol, oligomers of ethylene glycol), significantly influences the structure and morphology of the silica matrix, by leading to organic-inorganic hybrid materials [37].

From the analysis of the alumino-silicate gels synthesized in the presence of diols as co-solvents, a mass loss at temperatures lower than 471 K (boiling point of ethylene glycol, but also at higher temperatures) has been recorded. These mass losses at high temperaturas were attributed to the elimination of organic residues derived from the diols. It was thus concluded that the diols greatly affect the inorganic network of alumino-silicates without knowing precisely how these additives influence the formation of the inorganic matrix [38].

It was established that polyethylene glycol can be used to influence the dimensions of fractals and the particle size distribution during and after gelation [39]. The studies of the textural properties of the gels synthesized from tetrametyl orthosilicate and ethylene glycol as organic additive have evidenced the fact that these species interact with tetrametyl orthosilicate leading to pore shrinkage, although the mechanism is not known. Higgibotham et al. [40] have observed that small molecule additives, such as ethylene glycol, do not significantly influence the pore size during gelation. If the molecular mass of the modifier increases, the pores are formed in the presence of bigger molecules and are larger. The pore size increases with the concentration of the organic additive.

Ethylene glycol, as well as water and alcohol, is amphoteric and it is capable to form hydrogen bonds with Si-OH which are stronger and produce a sterical shield around the silicon atom. The strong hydrogen bond network shields the reactive centers ($\equiv$Si-O^-) of the incomplete condensed chains. This prevents the efficient condensation which leads to the formation of a branched network and produces larger and uniformly distributed micropores within the polymeric network.

Glycerol is considered a special DCCA because of the three hydroxyl groups from its molecule. Thus, glycerol is strongly adsorbed on the surface of the gel, reducing the capillar tension in two different ways: it reduces the contact angle by forming a film on the surface of the gel and does not evaporate, due to the low vapor pression. The main disadvantage of glycerol is that it remains retained in the pores and it is difficult to remove. Even at high temperatures, it tends to decompose to carbonate than to evaporate [6, 41]. It has been recorded that gelation time increases due to the adsorption of the glycerol molecules to the alkoxide molecules and to the surface of colloidal particles, slowing down the condensation between siloxane chains. The gelation rate for the gels with glycerol as additive is higher than for the gels with polyethylene glycol. This can be explained by the high number of – OH groups. Thus, the interaction with the hydrolysis products of tetramethyl orthosilicate and the formation of a three-dimensional network, takes place with higher rates.

In literature, it is usually considered that the presence of primary – OH groups is responsible for the interaction of organic compounds with the hydrolysis products of tetraethyl orthosilicate during the sol-gel synthesis [42]. This interaction between organic compounds and tetraethyl orthosilicate and its hydrolysis products influences the morphology of the silica matrix and thus, the dispersion and size of the embedded oxidic nanoparticles.

In this chapter, we present a study regarding the preparation of Fe_2O_3/SiO_2 and $Ni_{0.65}Zn_{0.35}Fe_2O_4/SiO_2$ nanocomposites with magnetic properties, obtained from some carboxylate coordination compounds embedded in silica gels.

EXPERIMENTAL

Our research has focused on the obtaining of oxide nanoparticles embedded in SiO_2 matrix using an original modified sol-gel method. This method consists of the obtaining of hybrid gels TEOS – diol – metal nitrates, where the diol and the metal nitrates are reactants for the formation of oxide precursors within the gels pores. The precursors are coordinative compounds of the corresponding metal cations having as ligands the products resulted at oxidation of diols by the NO_3^- anion. In order to obtain nanocomposites of metal oxides/SiO_2 type we have used for synthesis the following diols: ethane-1,2-diol (ethylene glycol, EG), 1, 2 propane diol (1,2 PG), 1, 3 propane diol (1,3 PG) and 1, 4 buthane diol (1,4 BG) mixed with metal nitrates and tetraethyl orthosilicate as a silica precursor. The presence of the diol in the system TEOS – H_2O, influences the formation and the final properties of the silica matrix. The participation of diols in the formation process of the matrix can influence the synthesis of oxide precursors as well as the dispersion and dimensions of the oxide particles embedded in the matrix. These experimental observations require a detailed study of the diols effect on the formation and evolution of silica gels.

1. Study on the Interaction of Diols with $Si(OC_2H_5)_4$ and Its Hydrolysis Products

The particularity of the study presented in this chapter is the high value of the ratio diol:TEOS compared to those reported in the literature (diols are usually used as additives) [40, 42, 43] due to the fact that the diol is used, as a reactant together with metal nitrates, in order to obtain nanocomposites. To elucidate the interaction between diol (D), tetraethyl orthosilicate (TEOS) and the hydrolysis products of TEOS, during the gelation process, were synthesized gels of different compositions: D:TEOS and H_2O:TEOS.

The synthesis method consists of slow adding, at room temperature, under magnetic stirring, of an ethanolic TEOS solution to the hydroalcoholic diol solution acidified with HNO_3 (c_{ac}=0.001 mol/L). The clear solution was stirred for 30 min and left for gelation at room temperature. After gelation, the obtained gels were grinded and dried at 60^oC for 3 hours. Table 1 presents the quantities of reactants used for the synthesis of the gels and the gelation time.

In order to evidence the influence of the diols on the system TEOS-H_2O at formation of the silica matrix, we have studied the silica gels synthesized with and without diol, in identical conditions (Table 1).

1.1. The Formation of Silica Gels

The processes which take place at formation of the silica matrix (hydrolysis, condensation, evaporation of the solvent and gel contraction) are temperature dependent processes with mass variation. Therefore, thermal analysis represents an excellent technique for the study of the silica gels evolution with temperature. Thus, the gels with and without water have been first characterized by this technique.

Figure 1 presents the TG and DTA curves of the gels G4 (with water) and G0 (without water).

Table 1. Characteristics of the synthesized gels.

Sample	Diol	Quantity (mols)				Molar ratio TEOS:Diol:H_2O	t_g (hours)
		TEOS	Diol	H_2O	EtOH		
G0	-	0.05	-	-	0.12	1:0:0	240
G4	-	0.05	-	0.2	0.12	1:0:4	120
G8	-	0.05	-	0.4	0.12	1:0:8	84
$G4_{EG}^{0.25}$	EG	0.05	0.0125	0.2	0.12	1:0.25:4	144
$G4_{EG}^{0.5}$	EG	0.05	0.0250	0.2	0.12	1:0.50:4	132
$G0_{EG}^{1}$	EG	0.05	0.0500	-	0.12	1:1:0	114
$G4_{EG}^{1}$	EG	0.05	0.0500	0.2	0.12	1:1:4	168
$G8_{EG}^{1}$	EG	0.05	0.0500	0.4	0.12	1:1:8	200
$G4_{EG}^{1.5}$	EG	0.05	0.0750	0.2	0.12	1:1.5:4	180
$G4_{12PG}^{0.2}$	1,2 PG	0.05	0.0100	0.2	0.12	1:0.2:4	140
$G4_{12PG}^{0.5}$	1,2 PG	0.05	0.0250	0.2	0.12	1:0.5:4	144
$G4_{12PG}^{1}$	1,2 PG	0.05	0.0500	0.2	0.12	1:1:4	154
$G4_{12PG}^{1.5}$	1,2 PG	0.05	0.0750	0.2	0.12	1:1.5:4	168
$G4_{13PG}^{0.25}$	1,3 PG	0.05	0.0125	0.2	0.12	1:0.25:4	132
$G4_{13PG}^{0.5}$	1,3 PG	0.05	0.0250	0.2	0.12	1:0.5:4	228
$G4_{13PG}^{1}$	1,3 PG	0.05	0.0500	0.2	0.12	1:1:4	216
$G4_{13PG}^{1.5}$	1,3 PG	0.05	0.0750	0.2	0.12	1:1.5:4	312
$G4_{14BG}^{0.2}$	1,4 BG	0.05	0.0100	0.2	0.12	1:0.2:4	148
$G4_{14BG}^{0.5}$	1,4 BG	0.05	0.0250	0.2	0.12	1:0.5:4	130
$G0_{14BG}^{1}$	1,4 BG	0.05	0.0500	-	0.12	1:1:0	312
$G4_{14BG}^{1}$	1,4 BG	0.05	0.0500	0.2	0.12	1:1:4	264
$G8_{14BG}^{1}$	1,4 BG	0.05	0.0500	0.4	0.12	1:1:8	216
$G4_{14BG}^{1.5}$	1,4 BG	0.05	0.0750	0.2	0.12	1:1.5:4	288

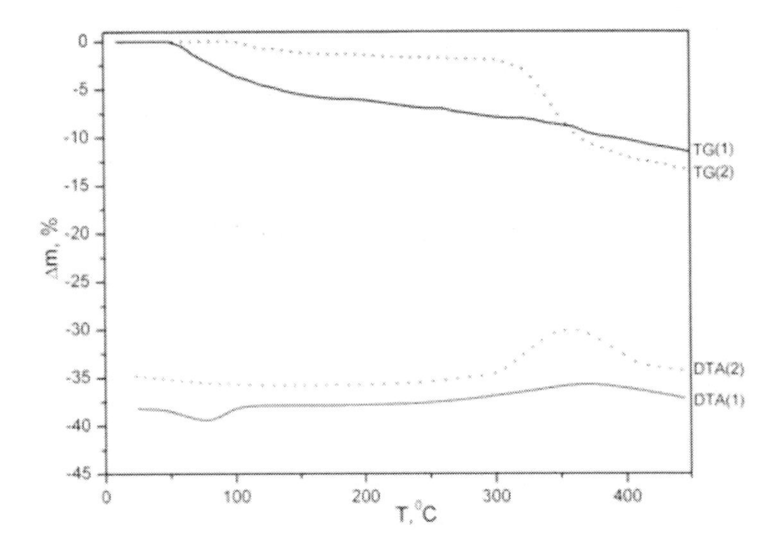

Figure 1. TG and DTA thermal curves for gels G4 (1) and G0 (2) dried at 200°C.

In the case of the gel with water (G4) (1) up to ~150°C appears a mass loss of ~ 7%, due to the elimination of adsorbed water and volatile products resulted at condensation (C_2H_5OH, H_2O). This process is evidenced on the DTA curve by an endothermic effect. Up to 500°C is registered a slow mass loss attributed to the progress of the polycondensation reaction. In this temperature range, on DTA there are no thermal effects. Similar results were obtained in the case of gel G8 thermally analyzed under the same conditions.

In the case of the gel without water (G0) (2) on TG appears a mass loss of ~9%, in the temperature range 300-400 °C. The large exotherm at 350°C, on DTA, is due to the combustion of residual organic groups ($-OCH_2CH_3$) from the silica matrix [44]. We can conclude that in this case, an ethoxy type silica matrix is formed (with a high number of $-OC_2H_5$ groups on the surface).

From thermal analysis, results are that G0 (without water) is an ethoxy type gel (with $\equiv Si-OC_2H_5$ surface groups) while G4 and G8 (with water) are predominant silanolic gels (with $\equiv Si-OH$ surface groups).

Figure 2 presents the FT-IR spectra in the range 400-4000 cm^{-1} for gels G4 (spectrum 1) and G0 (spectrum 2). Spectrum (1) corresponding to gel G4 presents bands characteristic for the silica matrix: 480 cm^{-1} (attributed to Si-O vibrations), the shoulder from 580 cm^{-1} attributed by some authors [45] to cyclic structures, 798 cm^{-1} (attributed to tetrahedron SiO_4), 945 cm^{-1} (attributed to Si-OH groups), 1080 cm^{-1} with shoulder at 1200 cm^{-1} (attributed to stretching vibrations of Si-O-Si bonds). The bands from 1650 cm^{-1} (attributed to deformation vibrations of H-O-H bond) and in the range 3400-3500 cm^{-1} (attributed to –OH groups from water and matrix) [46] confirms the silanolic character of the gels synthesized with a ratio of H_2O: TEOS $\geq$ 4, corresponding to the results of thermal analysis.

Spectrum (2) of gel G0, without water, presents, along with the bands characteristic for the silica matrix, a series of bands characteristic for the ethoxy groups: at 2850 cm^{-1} and 2993 cm^{-1} (attributed to the stretching vibrations of the C-H bond) between 1400-1500 cm^{-1} (attributed to the deformation vibrations of the C-H bond) [47].

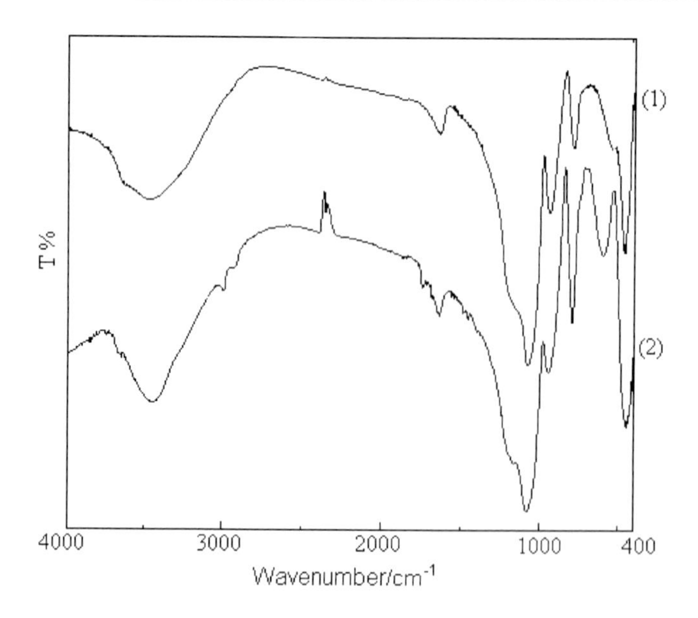

Figure 2. FT-IR spectra of the gels without ethylene glycol: G4(1) and G0(2).

The intense band from 580 cm^{-1} is due to the cyclic structures Si-O-Si; the gelation of the gels with low water content is achieved through linear and cyclic intermediates [48]. The low intensity band from 1165 cm^{-1} can be attributed to the vibrations of the Si-O-C bonds [49, 50] being a proof for the presence of ethoxy groups in the gels structure. The band from 1730 cm^{-1} was attributed by some authors [51] to C=O bonds formed during thermal treatment.

1.2. The formation of hybrid gels silica-diol

Thermal and FT-IR analysis

In order to study the interaction effect between the four diols (EG, 1,2 PG, 1,3 PG si 1,4BG), tetraethyl orthosilicate (TEOS) and the hydrolysis products of TEOS on the obtaining of the silica matrix and its morphology, we have studied the formation and thermal evolution of the gels (TEOS: D:H_2O = 1: 1: 4): $G4^1_{EG}$, $G4^1_{12PG}$, $G4^1_{13PG}$ and $G4^1_{14BG}$, respectively. During the gelation process (hydrolysis and polycondensation), the diols present in the system TEOS-H_2O interact with the hydrolysis products of tetraethyl orthosilicate depending on their structure, influencing both hydrolysis and condensation. The thermal evolution of hybrid gels obtained in this system allows us to establish how diols interact with the hydrolysis products of TEOS.

Figure 3 presents the TG curves of gel $G4^1_{EG}$ thermally treated at different temperatures: 60°C, 130°C, 200°C corresponding to the thermal processes.

From the evolution of the TG curves, one can distinguish the following stages:

(i) up to ~100°C takes place the elimination of volatile products formed at condensation and the elimination of solvent (EtOH, H_2O);
The loss in this range is diminished by treating the gels at 130°C and 200°C due to the advanced evaporation of volatiles.

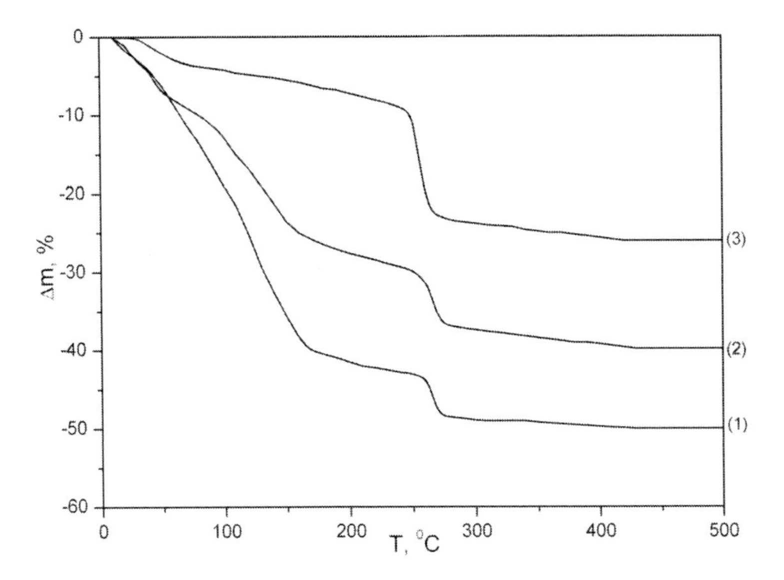

Figure 3. TG curves of gel $G4^1_{EG}$ at 60 °C (1), 130 °C (2) and 200 °C (3).

(ii) in the range 100°C – 200°C, the mass loss can be attributed to the evaporation of the free EG from the pores, to a part of the EG bonded by H bonds with Si-OH groups, as well as to the chemical interaction (condensation) of ethylene glycol bonded by hydrogen bonds with the silanol groups of the matrix;
(iii) between 200°C and ~ 250°C takes place a slow mass loss due to the evolution of the poly-condensation reaction;
(iv) the mass loss, with a high rate, in the range 250-280°C, is attributed to the oxidative decomposition of the organic chains bonded in the siloxane network as a result of the interaction between ethylene glycol and Si-OH groups;
(v) up to 500°C takes place a slow mass loss corresponding to the evolution of the polycondensation reaction of the silanol groups.

On the TG curve (3) of gel $G4^1_{EG}$ thermally treated at 200°C (the temperature is higher than the boling point of EG) there is a significant mass loss registered only in the temperature range 250-280°C, corresponding to the burning of the (-O-CH_2-CH_2-O-) groups interposed in the matrices network.

Figure 4 presents the FT-IR spectra of gel $G4_{EG}{}^1$ thermally treated at different temperatures. The spectra of gel $G4^1_{EG}$,thermally treated at 60°C (1) and 200°C (2), present absorption bands in the ranges 2800-3000 cm^{-1}, 1300-1400 cm^{-1} and 880 cm^{-1}, characteristic for ethylene glycol present in the gel and for the organic chains (-O-CH_2-CH_2-O-) bonded in the matrices network. Spectrum (3) obtained for $G4^1_{EG}$ thermally treated at 300°C (after elimination of the organic groups bonded within the matrix), differs from spectra (1) and (2) by the absence of the bands from 880 cm^{-1}, 1300-1400 cm^{-1} and 2800-3000 cm^{-1}. Thus, the results of FT-IR analysis sustain the proposed stages for the evolution of the gels with diols.

In order to elucidate the interaction mechanism between diols and TEOS or the hydrolysis products of TEOS, the gels $G4^1_{EG}$, $G4^1_{12PG}$, $G4^1_{13PG}$ and $G4^1_{14BG}$ were thermally treated at 200°C (when the majority of the unbonded diol is eliminated) and thermally analyzed in air and nitrogen.

Figures 5 and 6 present the derivatograms of gels $G4^{1}_{EG}$ and $G4^{1}_{13PG}$, in air, up to 800°C. It was observed that all four gels have similar thermal behavior.

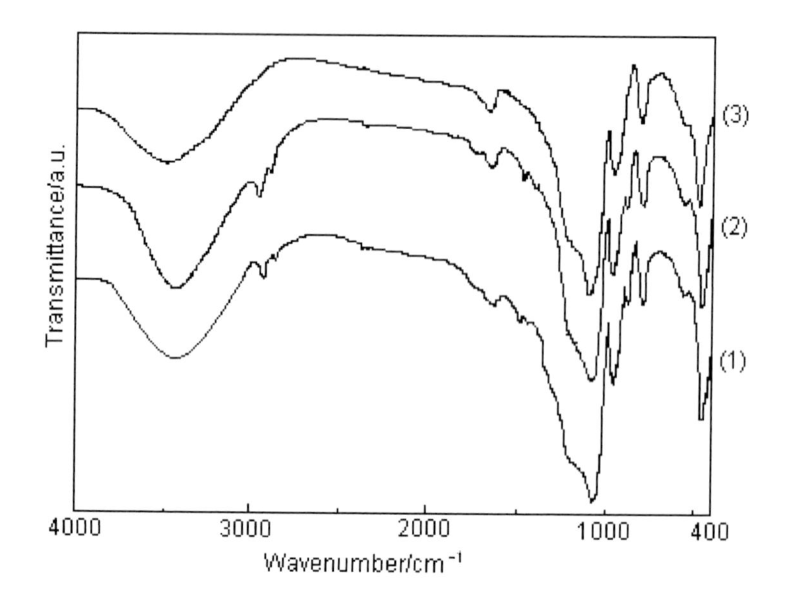

Figure 4. FT-IR spectra of gel $G4^{1}_{EG}$ thermally treated at different temperatures: (1) 60°C, (2) 200 °C, (3) 300 °C.

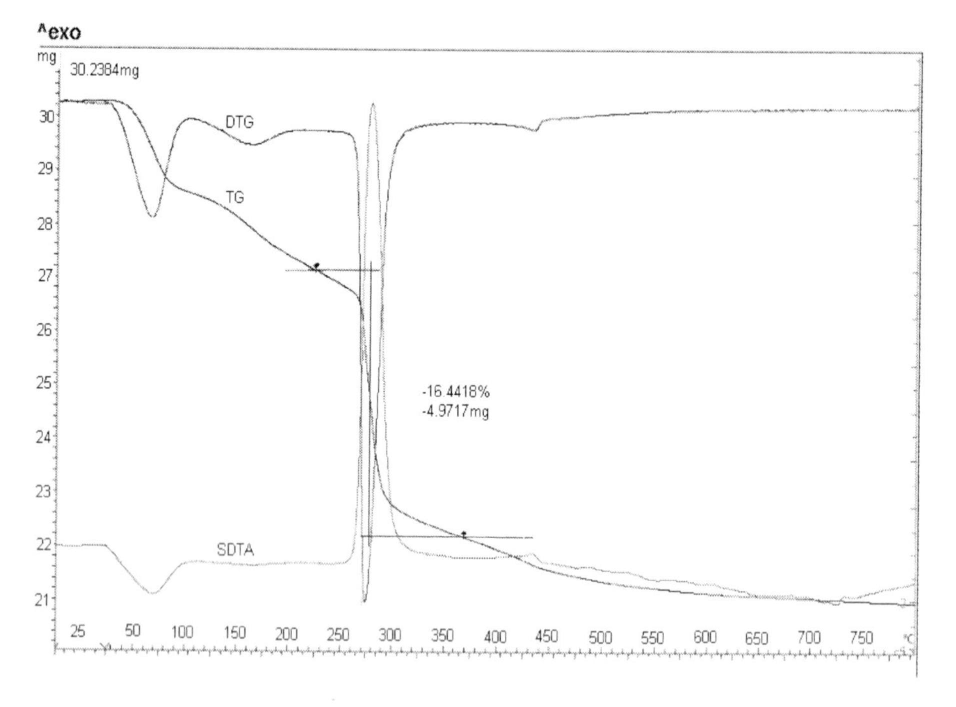

Figure 5. Derivatogram obtained at air heating of the gel $G4^{1}_{EG}$.

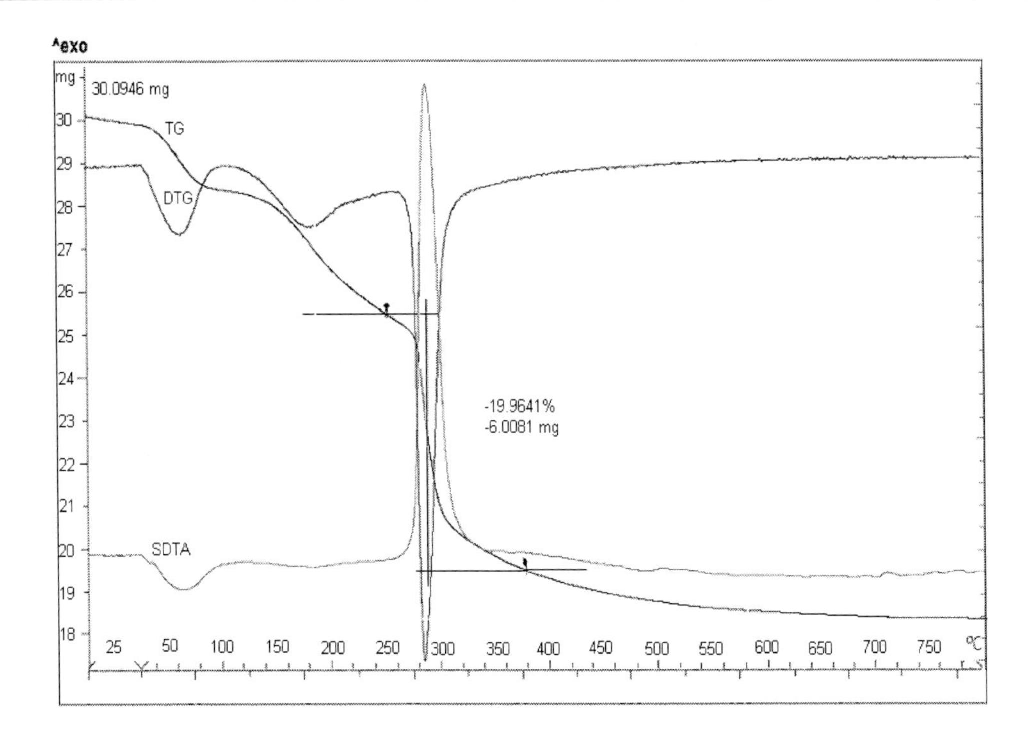

Figure 6. Derivatogram obtained at air heating of gel $G4^{1}_{13PG}$

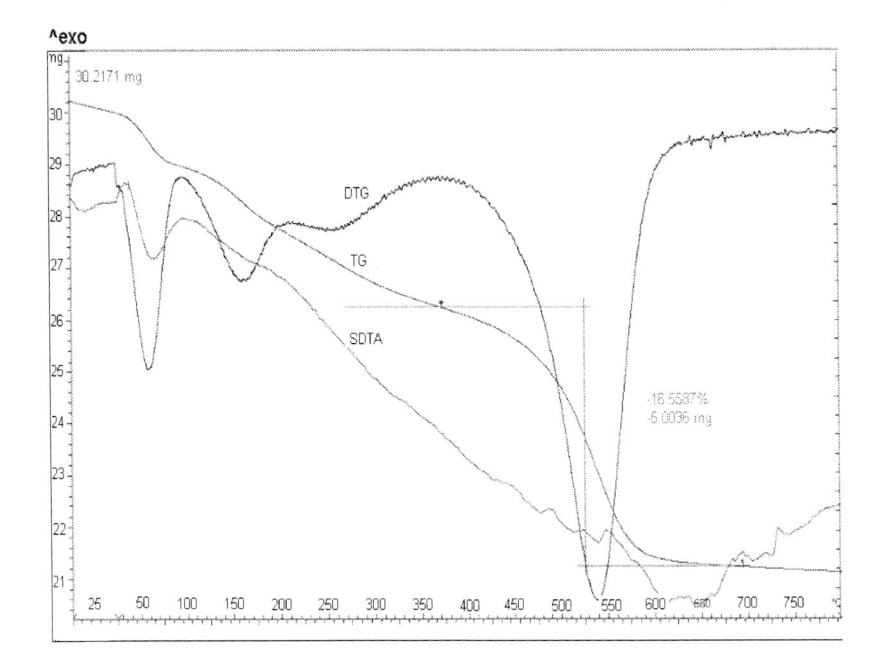

Figure 7. Derivatogram obtained at nitrogen heating of gel $G4^{1}_{EG}$

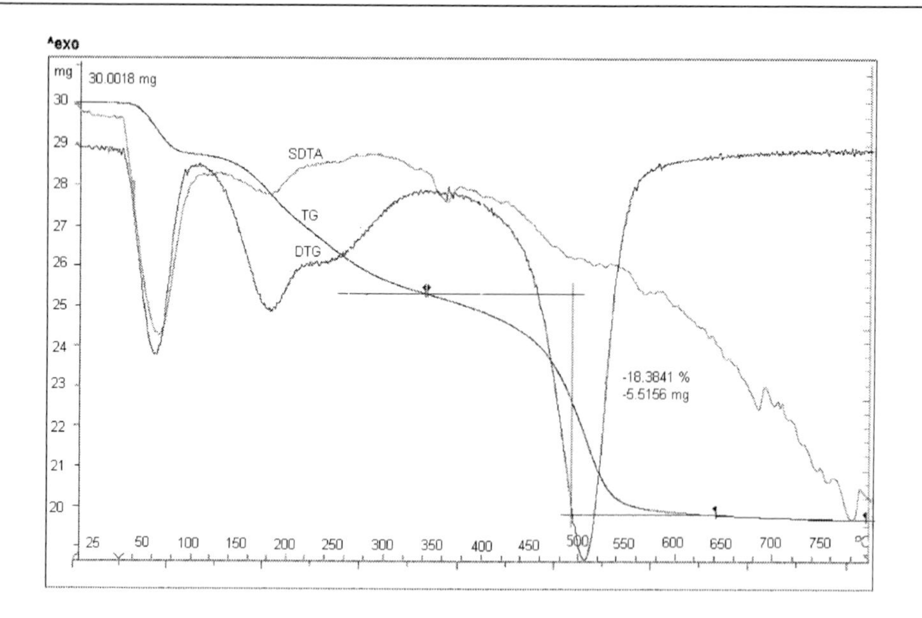

Figure 8. Derivatogram obtained at nitrogen heating of gel $G4^{1}_{13PG}$

The mass loss up to 225°C is attributed to the elimination of adsorbed water and volatile products from the polycondensation (EtOH, H_2O). The process with mass loss and strong exothermic effect (DTA) in the range 225-350°C is attributed to the oxidative decomposition of the organic chains corresponding to the diols bonded within the silica matrix. The slow mass loss up to 800°C corresponds to dehydroxylation of the silica matrices surface.

The presence of the organic chains chemically bonded within the silica matrix is confirmed by thermal analysis in dynamic nitrogen atmosphere, up to 800°C. The derivatograms obtained under these conditions for the gels $G4^{1}_{EG}$ and $G4^{1}_{13PG}$ thermally treated at 200°C are presented in Figures 7 and 8.

The mass loss up to 360°C is due to the elimination of adsorbed water, to the evolution of the matrices polycondensation and to the evaporation of the residual diols from the matrices pores. The mass loss with high rate in the temperature range 350-600°C corresponds to the decomposition of the diols organic chains chemically bonded within the network. There are no thermal effects on DTA in this temperature range. The shift of the temperature range, for the decomposition of the organic chains from the matrix, to higher temperatures, at thermal analysis in nitrogen atmosphere, confirms the fact that the chains are chemically bonded.

The close values of the mass losses registered for the four gels, in air, in the temperature range 250-400°C and in nitrogen in the range 350-600°C, presented in Table 2, confirm the fact that it is the same decomposition stage.

Based on the thermal analysis studies, we can conclude that the studied diols chemically interact with the silanol groups of the hydrolysis products of TEOS leading to the formation of a hybrid matrix stable up to ~250°C.

The gels $G4^{1}_{EG}$, $G4^{1}_{12PG}$, $G4^{1}_{13PG}$ and $G4^{1}_{14BG}$, thermally treated at 200°C, were analyzed by FT-IR (Figure 9). At this temperature, the gels contain the organic chains corresponding to the diols chemically bonded in the matrix. The bands that confirm their presence are in the ranges 3000-2800 cm^{-1} and 1400-1300 cm^{-1} and correspond to the vibrations characteristic for

C-H bonds (from $-CH_2-$ and $-CH_3$). The other bands overlap with the bands of the silica matrix.

Table 2. Mass losses registered at termal analysis in air and nitrogen of the gels with diols, thermally treated at 200°C.

Sample (200°C)	Mass losses, wt%			
	Air		Nitrogen	
	250-400 °C	20-800 °C	350-600 °C	20-800 °C
$G4^1_{EG}$	16.4	29.7	16.6	28.9
$G4^1_{12PG}$	20.8	32.4	19.1	31.8
$G4^1_{13PG}$	19.9	39.2	18.4	35.6
$G4^1_{14BG}$	21.9	52.7	19.6	51.3

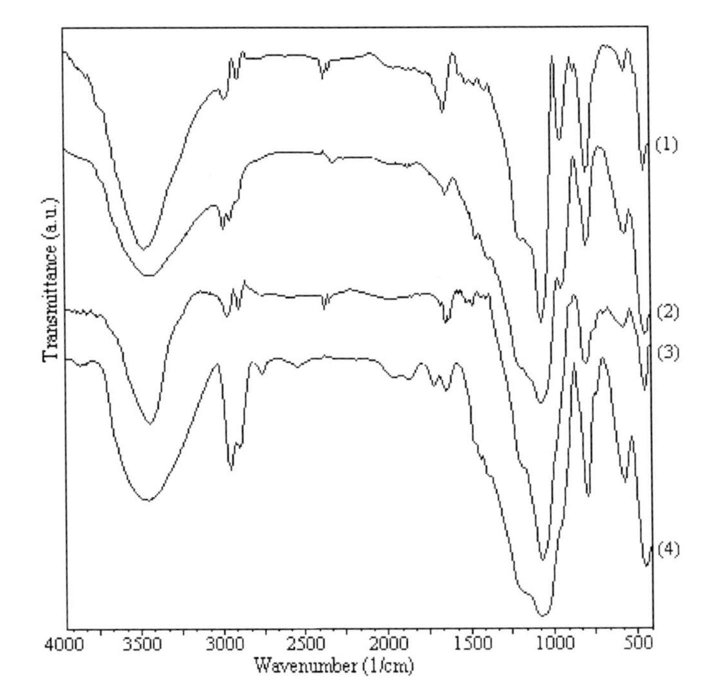

Figure 9. FT-IR spectra of the gels thermally treated at 200°C: (1) $G4^1_{EG}$; (2) $G4^1_{12PG}$; (3) $G4^1_{13PG}$; (4) $G4^1_{14BG}$.

The FT-IR spectra were processed by deconvolution and the obtained bands are presented in Table 3.

In conclusion, the FT-IR study of the gels synthesized with diols, thermally treated at 200°C, confirms the presence in the matrix of the organic chains $(-CH_2-CH_2-)$, $(-CH_2-CH(CH_3)-)$, $(-CH_2-CH_2-CH_2-CH_2-)$ interposed in the matrices network, by chemical interaction of the diols with silanol groups.

Solid state ^{29}Si-NMR

The gels $G4^1_{EG}$, $G4^1_{12PG}$, $G4^1_{13PG}$, $G4^1_{14BG}$,thermally treated at 200°C, were characterized by ^{29}Si-NMR on solids, in order to describe the local environment (vicinities) of the Si atom within the network. Figure 10 presents the ^{29}Si-NMR spectra for gels $G4^1_{EG}$ and $G4^1_{13PG}$ thermally treated at 200°C.

Table 3. FT-IR band assignment for the gels thermally treated at 200°C.

Wave number [cm^{-1}]				Assignment
$G4^1_{EG}$	$G4^1_{12PG}$	$G4^1_{13PG}$	$G4^1_{14BG}$	
3442	3446	3440	3443	H_2O assoc.
2960,2893	2980,2942,2889	2968,2897	2950,2890	CH [52]
			1727	C=O [51]
1640	1643	1644	1639	H_2O [45, 51]
1460	1455,1418, 1390	1452,1420,1390		C-H (CH_2, CH_3) [45, 51]
1202	1205	1207	1207	Si-O-Si assym. [45]
		1166		Si-O-C [49]
1081	1087	1086	1078	Si-O-Si symm[45, 51]
960	946	950	960	Si-O(H) [51]
880	840	889	860	C-C [53]
790	794	796	796	Si-O-Si assym[45, 51]
565	570	570	568	Si-O-Si cyclic [45]
460	460	450	450	Si-O-Si [45, 51]

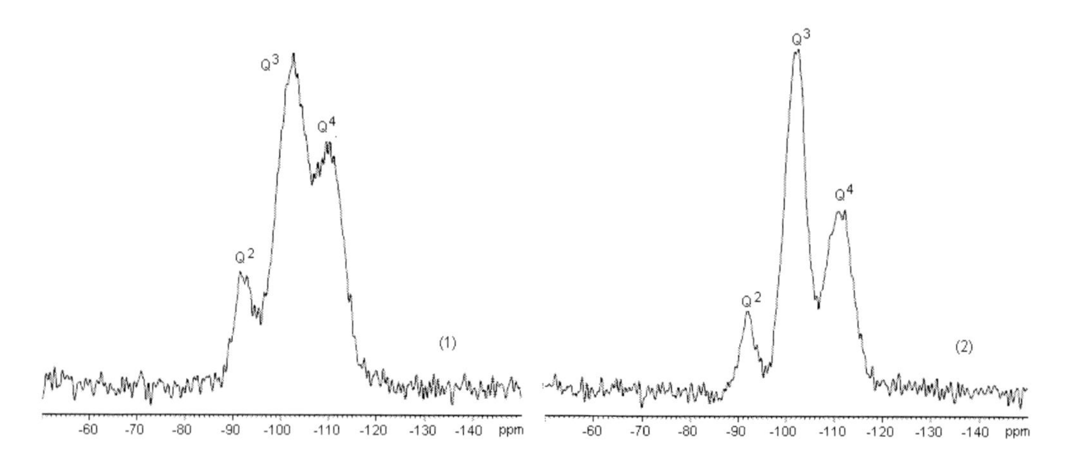

Figure 10. ^{29}Si-NMR spectra of the gels $G4^1_{EG}$ (1) and $G4^1_{13PG}$ (2) thermally treated at 200°C.

From the presented spectra one can observe that in both cases, the species Q^4 [≡ Si-(OSi)$_4$] (-110 ppm), Q^3 [XO-Si-(OSi)$_3$] (-100 ppm) and Q^2 [(XO)2-Si-(OSi)$_2$] (-92 ppm) are present, which indicates a low condensation degree as a result of the chemical bonds formed between diols and silanol groups.

By thermal treatment of the gels at 400°C, the silica matrices present a higher condensation degree, as a result of the elimination of the organic chain from the matrices network. Figure 11 presents the NMR spectra for $G4^{1}_{EG}$ (1) and $G4^{1}_{13PG}$ (2), which evidence only the presence of Q^4 species.

The results of Si-NMR analysis sustain the conclusions obtained by thermal and FT-IR analyses about the interactions of the studied diols with tetraethyl ortosilicate and the hydrolysis products of TEOS. The schemes from Figure 12 present the possible interactions of the studied diols with the silanol groups, based on the results of thermal and FT-IR analyses.

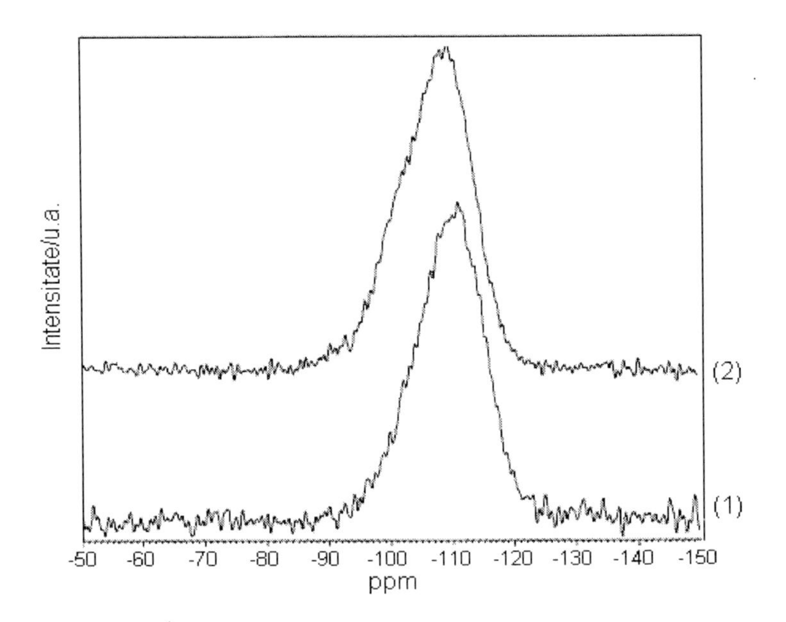

Figure 11. Si-NMR spectra for gels: (1) $G4^{1}_{EG}$ and (2) $G4^{1}_{13PG}$ annealed at 400°C.

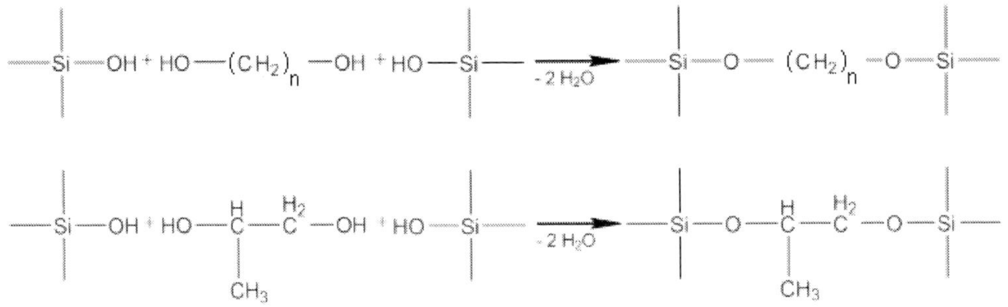

$$\equiv Si-OH + HO-(CH_2)_n-OH + HO-Si\equiv \xrightarrow[-2H_2O]{} \equiv Si-O-(CH_2)_n-O-Si\equiv$$

$$\equiv Si-OH + HO-CH(CH_3)-CH_2-OH + HO-Si\equiv \xrightarrow[-2H_2O]{} \equiv Si-O-CH(CH_3)-CH_2-O-Si\equiv$$

Figure 12. Interaction of the diols with silanol groups.

These chemical interactions of the diols with the silanol groups of the matrix may lead, by appropriate thermal treatments, to the obtaining of silica matrices with modified morphology.

1.2.1. Influence of the ratio H_2O – TEOS on the interaction diol-tetraethyl orthosilicate

In order to study the influence of the ratio H_2O-TEOS on the interaction of the diol with tetraethyl orthosilicate or its hydrolysis products, we have synthesized gels (TEOS: EG: H_2O) with different water content (TEOS: H_2O = 0; 4; 8). Figures 13 and 14 present the TG curves of the gels $G0^1_{EG}$, $G4^1_{EG}$, $G8^1_{EG}$ synthesized with ethylene glycol and the gels $G0^1_{14BG}$, $G4^1_{14BG}$, $G8^1_{14BG}$ synthesized with 1, 4 buthane diol thermally treated at 200°C, for one hour. According to earlier statements, the gels contain the diol chemically bonded within the network.

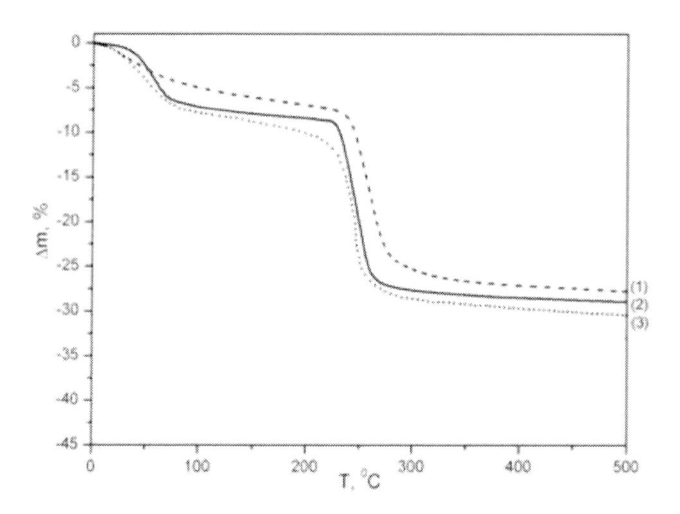

Figure 13. TG curves of gels $G0^1_{EG}$, $G4^1_{EG}$, $G8^1_{EG}$ thermally treated at 200°C.

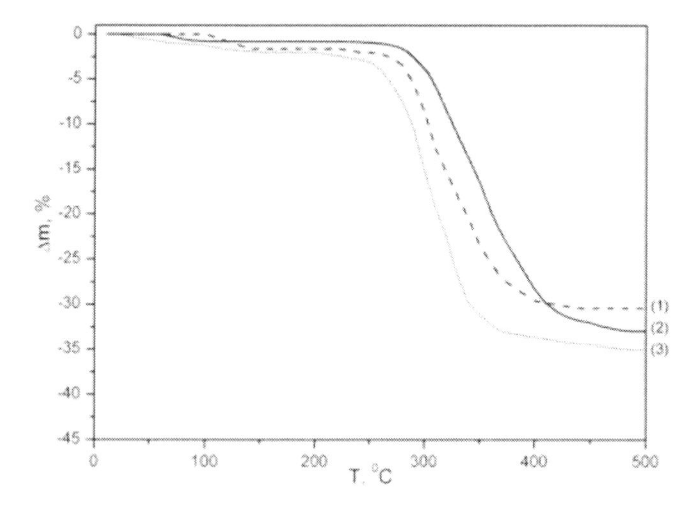

Figure 14. TG curves of gels $G0^1_{14BG}$, $G4^1_{14BG}$, $G8^1_{14BG}$ thermally treated at 200°C.

One can notice that all TG curves (Figs. 13 and 14) present mass losses in the temperature range 250-380°C. The mass losses are attributed to the combustion of organic groups (-O-$(CH_2)_n$-O-) interposed in the matrices network. From the presented ones, results show that in the systems (TEOS – D – H_2O) and (TEOS-D), the diols chemically interact with the hydrolysis products of TEOS and with TEOS, interposing within the matrices network.

In the case of the gels synthesized with water, the diol interacts with tetraethyl orthosilicate according to mechanism from Figure 12.In the case of the gels synthesized without water ($G0^1_{EG}$ and $G0^1_{14BG}$), the diol interacts with $\equiv$Si-O-C_2H_5 groups. According to the literature, the –OC_2H_5 groups of tetraethyl orthosilicate can be substituted by the diols leading to species with different substitution degree [37, 42]. Thus, the interaction of the diol with tetraethyl orthosilicate is given by the equations:

$$\equiv Si-O-C_2H_5 + HO\text{-}(CH_2)_n\text{-}OH <=> \equiv Si-O-(CH_2)_n-OH + C_2H_5OH$$

$$\equiv Si-O-(CH_2)_n-OH + \equiv Si-O-C_2H_5 <=> \equiv Si-O(CH_2)_nO-Si\equiv + C_2H_5OH$$

From the presented ones, it results that the condensation degree of the diol within the hybrid gels does not depend, significantly, on the ratio H_2O: TEOS introduced in synthesis.

1.2.2. Influence of the ratio Diol: TEOS on the formation of the hybrid gels

In order to study the influence of the inital ratio D: TEOS on the polycondensation reaction, the gels with the four diols: EG, 12 PG, 13PG and 14BG for different D: TEOS molar ratios, thermally treated at 200 ^{0}C, have been characterized by thermal analysis. All gels have presented similar thermal behavior, with a significant mass loss registered in the range 250-350°C corresponding to the burning of the organic chaines bonded within the silica network. These mass losses depend on the molar ratio D: TEOS introduced in synthesis and on the diols nature. Figures 15 and 16 present the thermal behavior of the gels synthesized at 200°C with 1,2 PG and 1,4 BG.

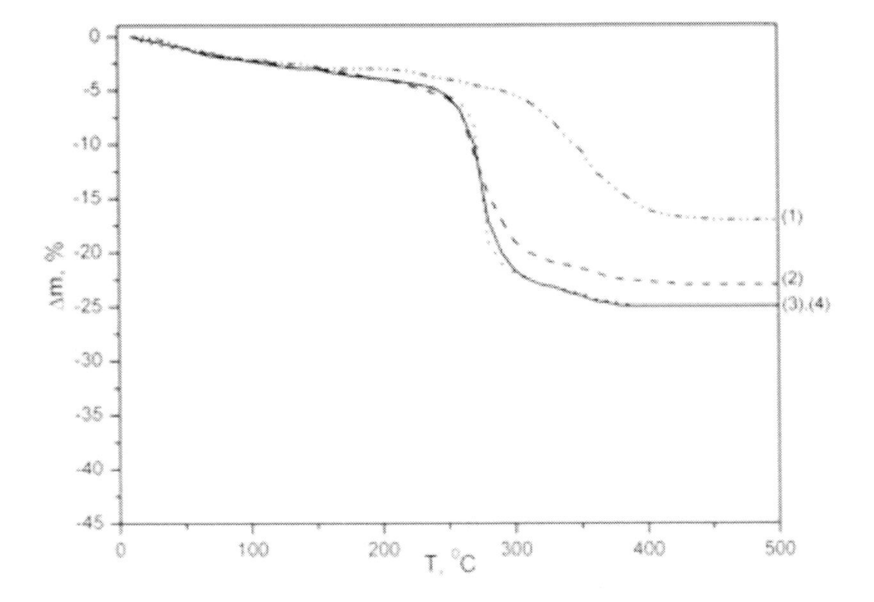

Figure 15. TG curves of the gels $G4_{12PG}$ with molar ratio D:TEOS: (1) 0.2; (2) 0.5; (3) 1; (4) 1.5 .

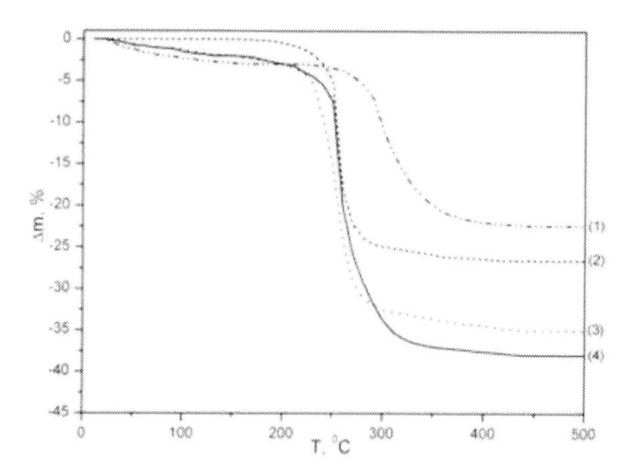

Figure 16. TG curves of the gels $G4_{14BG}$ with molar ratio D:TEOS: (1) 0.2; (2) 0.5; (3) 1; (4) 1.5.

Table 4. Interpretation of the results of thermal analysis in air of the synthesized gels.

Sample	Initial molar ratio Diol:TEOS	Δm_t% 20-500°C	Δm % 250-400°C	$m_{residue}$ $_{SiO2}$ (%) 500°C	mols diol chemically bounded	mols TEOS	Experimental molar ratio Diol:TEOS
$G4^{0.25}_{EG}$	0.25	23	10	77	0.23	1.28	0.18
$G4^{0.5}_{EG}$	0.50	27	11	73	0.25	1.22	0.20
$G4^{1}_{EG}$	1.0	31	12	69	0.27	1.15	0.24
$G4^{1.5}_{EG}$	1.5	31	13	69	0.29	1.15	0.25
$G4^{0.2}_{12PG}$	0.20	16	14	84	0.24	1.40	0.17
$G4^{0.5}_{12PG}$	0.50	23	18	77	0.31	1.28	0.24
$G4^{1}_{12PG}$	1.0	25	20	75	0.34	1.25	0.27
$G4^{1.5}_{12PG}$	1.5	25	20	75	0.34	1.25	0.27
$G4^{0.25}_{13PG}$	0.25	27	14	73	0.24	1.22	0.20
$G4^{0.5}_{13PG}$	0.50	32	16	68	0.28	1.13	0.25
$G4^{1}_{13PG}$	1.0	33	19	67	0.33	1.12	0.29
$G4^{1.5}_{13PG}$	1.5	35	20	65	0.34	1.10	0.31
$G4^{0.2}_{14BG}$	0.20	22	17	78	0.23	1.3	0.18
$G4^{0.5}_{14BG}$	0.50	27	25	73	0.34	1.21	0.28
$G4^{1}_{14BG}$	1.0	35	31	65	0.43	1.08	0.40
$G4^{1.5}_{14BG}$	1.5	38	34	62	0.47	1.03	0.46

Based on the mass losses from the range 300-400°C, corresponding to the organic chains bonded within the matrix and the residue from 600°C (SiO_2), we have estimated the molar ratio diol (bounded): TEOS, for each gel. Table 4 presents the results obtained for the gels of different compositions synthesized with the four diols. From the tabled data, results show that the quantity of the chemically bounded diol within the matrix increases with the increase of the initial molar ratio diol:TEOS up to 1:1, for every diol. For higher initial molar ratio, the quantity of the bounded diol remains constant. An exception represents the gels with 1,4BG, where the bounded quantity significantly increases with the initial molar ratio 1,4BG:TEOS. An explanation of this feature might by the increasing of the gelation time, which provides a longer interaction leading to an advanced degree of condensation.

1.3. Study of the Silica Matrix Obtained by Annealing of Hybrid Gels

In order to evidence the effect of the diols condensation within the silica matrix on the final properties of the silica matrix, we have studied the textural properties of the silica matrices obtained by annealing of gel $G4_{12PG}$ at 600°C, for different initial 1,2PG: TEOS molar ratios. The textural properties of the silica-matrix obtained by thermal treatment of the hybrid gels have been studied by N_2 adsorption isotherm (BET) measurements. The results showed the same type of isotherms for all compositions (Figure 17) but differences in the values of the textural parameters (table 5).

The shapes of the isotherms are close to type 1, characteristic for microporous materials. The pore size distributions for the gels synthesized with 1,2PG ($G4^{0.5}_{12PG}$, $G4^{1}_{12PG}$, $G4^{1.5}_{12PG}$) are presented in Figure 18. The corresponding textural properties of these gels are listed in Table 5.

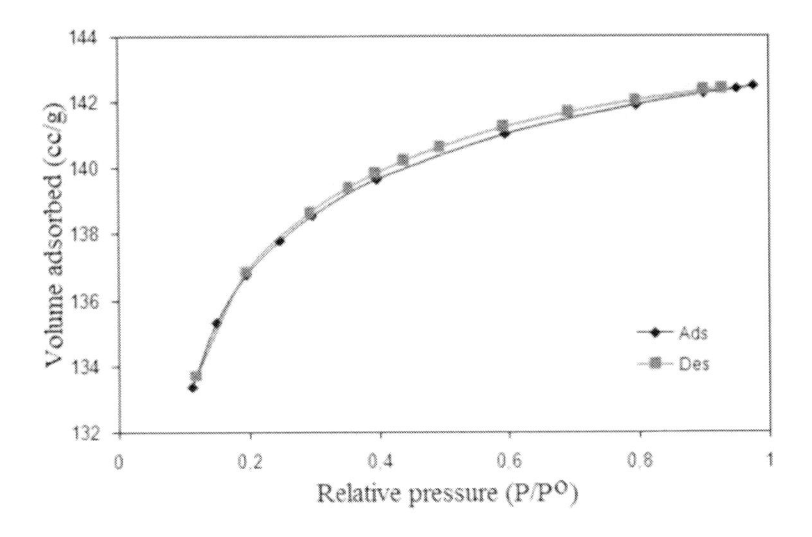

Figure 17. Nitrogen adsorption – desorption isotherms for gel $G4^{1}_{12PG}$.

Table 5. Textural characteristics of the SiO_2 matrices from the BET adsorption isotherm.

Sample	Surface area S_{BET} (m^2/g)	Pore volume BJH_{des} (cm^3/g)	Average pore diameter BJH_{des} D_p (nm)
PG0.5	284	0.161	1.97
PG1.0	396	0.220	3.05
PG1.5	420	0.241	3.40

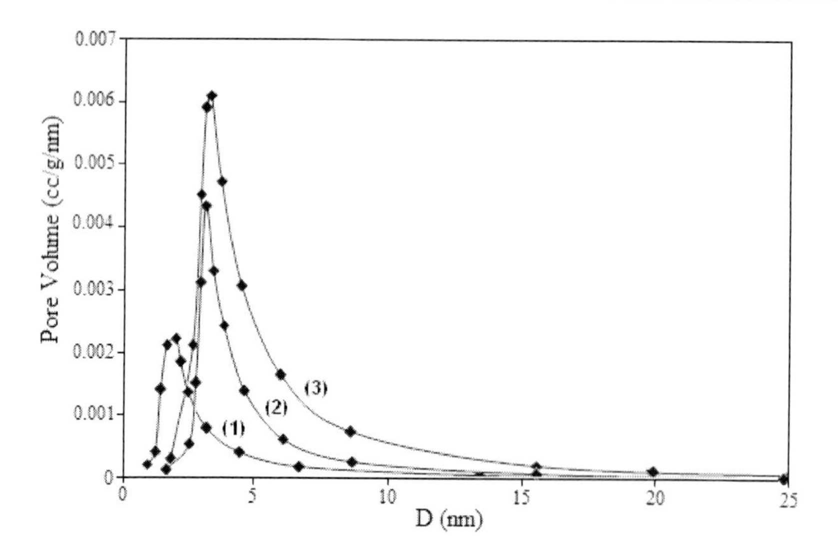

Figure 18. Pore size distribution for the gels: (1) $G4^{0.5}_{12PG}$, (2) $G4^{1}_{12PG}$, (3) $G4^{1.5}_{12PG}$.

The results of N_2 adsorption-desorption analysis show that the surface area, the total pore volume and the pores diameter are influenced by the molar ratio diol: TEOS, by the different amount of bounded diol.

From this study we have established that for the synthesis of nanocomposites we have to take into account the interaction diol-TEOS, which consumes a part of the diol introduced as a reactant for the synthesis of the oxides precursors. Also, this interaction influences the morphology and structure of the silica matrix and the morphology of the embedded oxide nanoparticles.

2. Obtaining of Fe_2O_3/SiO_2 and $Ni_{0.65}Zn_{0.35}Fe_2O_4/SiO_2$ Nanocomposites with Magnetic Properties

The formation of simple or mixed oxide nanoparticles is based on the synthesis of some metal carboxylate complex combinations, which by decomposition and appropriate thermal treatments lead to nanomaterials with directed properties [54]. The complex combinations are formed in the redox reaction between metal nitrates and diols [17]. The redox reaction takes place between the NO_3^- ion and the –OH group bonded to the primary C atom from the structure of the diol, according to the general equations:

$$3\,HO-\overset{H_2}{C}-(CH_2)_n-\overset{H_2}{C}-OH + 8\,NO_3^- + 2\,H^+ \rightleftharpoons 3\,^-OOC-(CH_2)_n-COO^- + 8\,NO + 10H_2O$$

$$3\,H_3C-\underset{OH}{\overset{H}{C}}-\overset{H_2}{C}-OH + 4\,NO_3^- + H^+ \rightleftharpoons 3\,H_3C-\underset{OH}{\overset{H}{C}}-COO^- + 4\,NO + 5\,H_2O$$

The method introduced by us for the synthesis of the nanocomposites metal oxides/SiO_2 is a modified sol-gel method which presents the particularity of the formation of carboxylate type complex combinations within the pores of the silica matrix.

The modified sol-gel method combines the advantages of the carboxylate type complex combinations decomposition with those of the sol-gel method. The formation and decomposition of the carboxylate type compounds within the pores of the silica matrix may lead to the obtaining of oxide nanoparticles uniformly dispersed in silica matrix at low temperatures. Under the conditions of the obtaining of the carboxylate type precursors starting from the mixture (TEOS - metal nitrates - diol), the diol interacts with tetraethyl orthosilicate with the formation of silica matrices with modified morphology. At the same time, the diol reacts with the metal nitrates leading to complex combinations formed within the pores of the hybrid gel.

At synthesis of the carboxylate type compounds, one has to take into account the mixing molar ratio TEOS: diol: nitrate, corresponding to the stoechiometry of the redox reaction diol-NO_3^-, as well as to the polycondensation reaction diol:TEOS.

2.1. Obtaining of γ-Fe_2O_3/SiO_2 nanocomposites by the modified sol-gel method

From all iron oxides the most difficult to obtain are γ-Fe_2O_3 (maghemite) and ε-Fe_2O_3 in comparison with α-Fe_2O_3 (hematite) or Fe_3O_4 (magnetite). It is important to obtain γ-Fe_2O_3 as well crystallized sole phase, , as powder or dispersed in silica matrix. Therefore, one has to use the corresponding precursor and synthesis method in order to obtain the desired final material γ-Fe_2O_3 or γ-Fe_2O_3/SiO_2, as nanoparticles with well controlled morphology and directed properties. A characteristic feature of iron oxides is the variety of possible transitions between the different crystalline phases. The obtaining of undispersed (bulk) maghemite as sole phase is difficult due to the transition of γ- to α-Fe_2O_3 at low temperatures (400 °C), which is hard to control. Within the silica matrix, γ-Fe_2O_3 stabilizes up to higher temperatures, after that it turns to α-Fe_2O_3 via ε-Fe_2O_3 as intermediary phase [55, 56]. The silica matrix has a stabilizing effect on the maghemite nanoparticles with dimensions lower than 10 nm having a superparamagnetic behavior [57, 58]. γ-Fe_2O_3 nanoparticles are not only an important industrial material but also a preparation source of the rare polymorph ε-Fe_2O_3 [55]. Recently, the ε-Fe_2O_3 phase attracted much attention due to its enormous room-temperature coercivity of about 2T and coupling of magnetic and dielectric properties (so-called magneto-electric response) [59]. Despite the increasing number of scientific papers focused on this direction [60], the mechanism of the transition γ-Fe_2O_3 → ε-Fe_2O_3 → α-Fe_2O_3 is not completely elucidated.

There are many methods which can be used for the obtaining of the oxidic system γ-Fe_2O_3 as nanoparticles but their complexity, the high cost and purity of the obtained material limits their use at large scale. The researches from the last years on the preparation of γ-Fe_2O_3 were focused on the obtaining of some iron precursors by unconventional methods, especially chemical ones [62-65].

The aim of our research is to obtain γ-Fe_2O_3 as sole stable phase as powder or dispersed in silica matrix, by the thermal decomposition of some particular precursors: carboxylate type coordination compounds of Fe (III), obtained by the redox reaction between iron (III) nitrate and some diols.

2.1.1. Obtaining of γ-hFe_2O_3 nanoparticles by thermal decomposition of the carboxylate type complex combinations

The obtaining of iron oxide nanoparticles is based on the synthesis of iron carboxylate type complex combinations in the redox reaction between $Fe(NO_3)_3 \cdot 9\ H_2O$ and the diols: EG, 1,2 PG, 1,3 PG and 1,4 BG.

The synthesis method consists of dissolving $Fe(NO_3)_3{\cdot}9H_2O$ in the corresponding diol amount, followed by controlled heating, until the redox reaction starts, visible by evolving of nitrogen oxides. The redox reaction takes place between NO_3^- ion and the –OH groups from the primary C atom [66, 67] simultaneous with the isolation of the Fe (III) carboxylate type complex combination. The high acidity of the reaction medium due to the aquacation $[Fe(H_2O)_9]^{3+}$ (pK_a = 2.22), leads to the starting of the redox reaction at low temperatures (~ 70°C).

Table 6. Characteristics of the synthesized samples.

Sample	Diol	Quantity (mole)				Molar ratio
		$Fe(NO_3)_3{\cdot}H_2O$	NO_3^-	Diol		NO_3^- : diol
				Stoechiometric	50 % excess	
B1	EG	0.0375	0.1125	0.0563	0.0844	1 : 0.750
B2	1,2 PG	0.0375	0.1125	0.0844	0.1266	1 : 1.125
B3	1,3 PG	0.0375	0.1125	0.0422	0.0633	1 : 0.563
B4	1,4 BG	0.0375	0.1125	0.0422	0.0633	1 : 0.563

There were synthesized samples $Fe(NO_3)_3$- diol, with 50% diol excess reported to NO_3^- according to the stoechiometry of the redox reaction (Table 6).

The evolution of the redox reaction and the formation of the complex combinations, in the solution mixture $Fe(NO_3)_3$-diol was established by thermal analysis and FT-IR spectrometry. The reaction products were heated at 130°C, 3h, until the finalization of the redox reaction.

Figure 19 presents the TG and DTA curves obtained in air, up to 500 °C, for the solution B4 ($Fe(NO_3)_3$-1,4 BG), deposed in thin film on Pt plates. The DTA curve consists of two exothermic effects. The first exotherm at ~60 °C corresponds to the redox reaction between the anion NO_3^- and diol, simultaneous with the isolation of the Fe(III) complex combination. The mass loss in the temperature range 40-100°C is due to the water evaporation and the elimination of the nitrogen oxides resulted in the redox reaction. The second exotherm, in the temperature range 220-300 °C, accompanied by a mass loss on the TG curve, corresponds to the oxidative thermal decomposition of the complex combination. In the case of the other three solutions (B1, B2, B3), there were similar results obtained on the evolution of the thermal curves during the redox reaction. Thus, 130 °C was established as optimal synthesis temperature for the obtaining of the organic precursors.

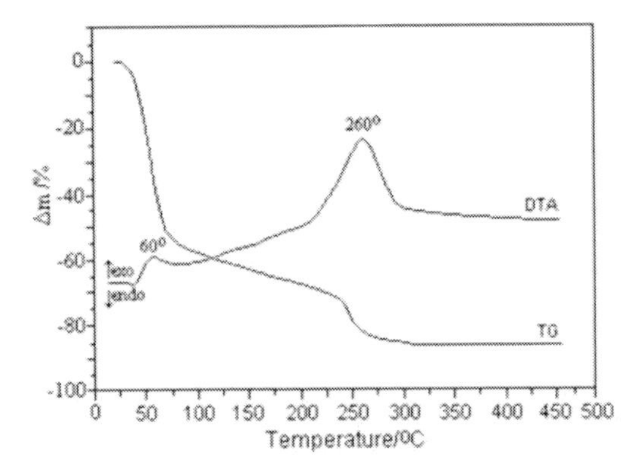

Figure 19. Thermal curves of the solution B4 ($Fe(NO_3)_3$-1,4 BG).

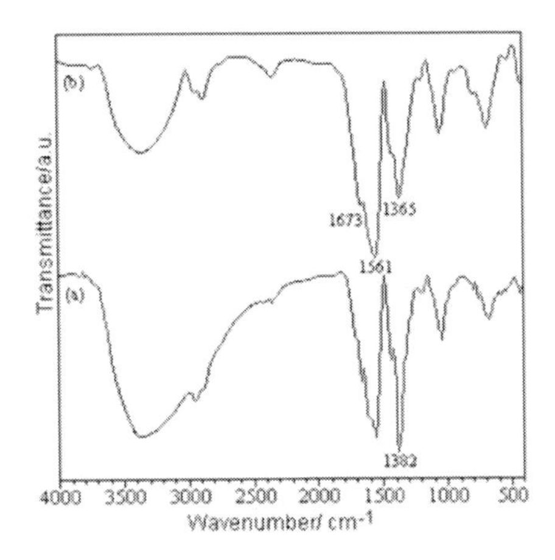

Figure 20. FT-IR spectra of the sample B4 heated at 60 °C (a) and 130 °C (b).

FT-IR spectrometry (Figure 20) has evidenced the evolution of the redox reaction (spectrum a) and the formation of the complex combination (spectrum b). Spectrum (a) of sample B4 heated at 60 °C, when the redox reaction took place, partially presents the characteristic band of the ion NO_3^- at 1382 cm^{-1} along with bands of the carboxylate ion, partially formed at this temperature. Spectrum (b) of sample B4 heated at 130 °C, 2h (when the redox reaction is finished) does not present the band of NO_3^- (1382 cm^{-1}) anymore. The spectrum presents the characteristic bands of the carboxylate type compunds at: 1673-1561 cm^{-1} - $\upsilon_{as}(COO^-)$ [68], 1365 cm^{-1} - $\upsilon_s(COO^-)$ [69] and 1320 cm^{-1} - $\upsilon_s(CO)+\delta(OCO)$. There are also bands at 3368 cm^{-1} attributed to υ (H_2O) and υ ($OH_{assoc.}$), at 2948 and 2880 cm^{-1} assigned to υ (CH). The bands from 1048 cm^{-1} are attributed to υ (C-OH) + δ(OH) and bridged υ (OH), while the band from 681 cm^{-1} is assigned to $\rho(H_2O)$ and υ (M-O). Similar results were also obtained for the other three complex combinations.

Figures 21 and 22 present the thermal behavior, in air, of the compounds synthesized at 130 °C with EG (B1) and 1,4 BG (B4).

The mass losses in the range 200-270 °C, with strong exothermic effects on DTA, correspond to the oxidative decomposition of the Fe(III) carboxylate compunds (glyoxylate and succinate [66]). At thermal conversion, a reducing environment (CO, C) is generated which influences the formation of the final decomposition product (Fe_2O_3). The exothermic effect from ~350 °C (Figure 22) can be attributed to the burning of the residual carbon of the organic ligand with a higher carbon content.

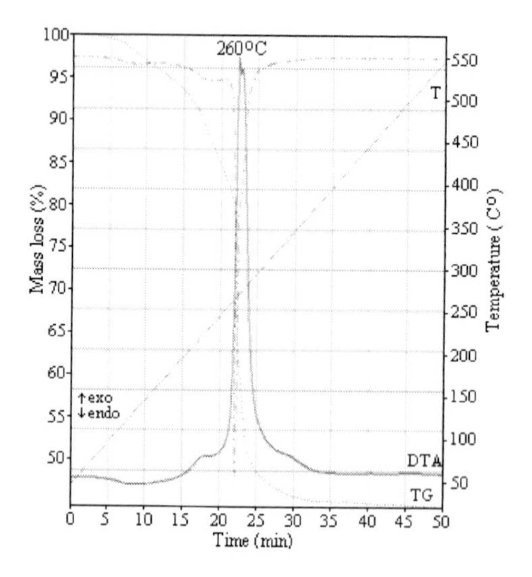

Figure 21. Thermal curves of gel B1 (with EG) thermally treated at 130 °C.

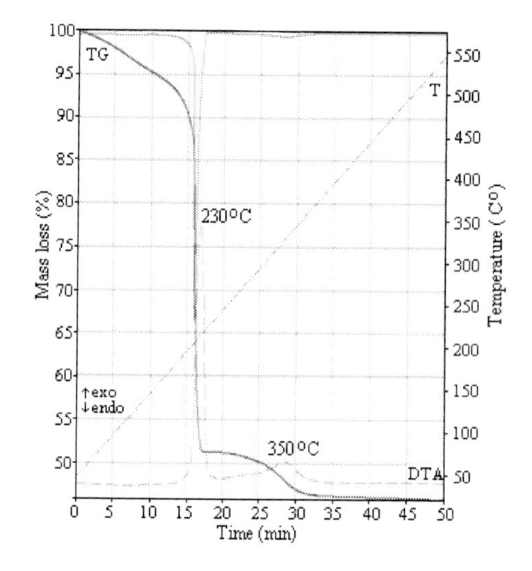

Figure 22. Thermal curves of gel B4 (with 1,4 BG) thermally treated at 130 °C.

The decomposition product from 300 °C presents magnetic properties and corresponds to the phase γ-Fe_2O_3. Figure 23 presents the XRD patterns of sample (B1) synthesized with EG, annealed at 300 °C, 400 °C and 500 °C, 3 hours. From the XRD spectra, γ-Fe_2O_3 was identified at 300 °C as sole crystalline phase, with traces of α-Fe_2O_3 at 400 °C, while at 500 °C α-Fe_2O_3 appears well crystallized.. The Figure 24 presents the magnetization curve of the sample with γ-Fe_2O_3 nanoparticles (ferrimagnetic) obtained by annealing, in air at 300 °C, of the precursor (B1).

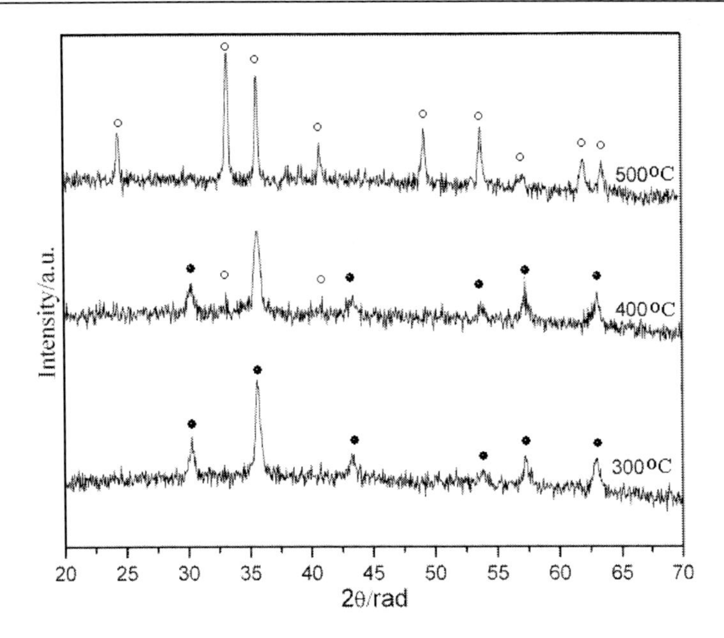

Figure 23. XRD patterns of sample B1 annealed at different temperatures.

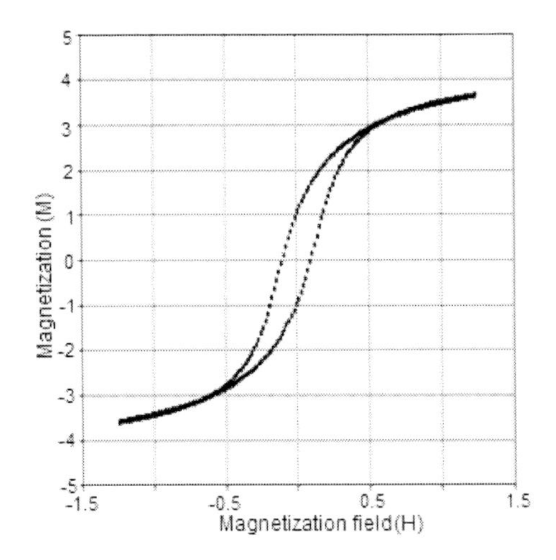

Figure 24. Magnetization curve of sample B1 annealed at 300 °C.

The XRD study is sustained by Mössbauer spectroscopy. Mössbauer spectra were carried out at room temperature for sample B1 annealed at 300 °C (Figure 25) and 500 °C (Figures 26). The isomer shifts are consistent with Fe^{3+} , which means that no Fe^{2+} ions are present in the samples. The Mössbauer spectrum of sample B1 annealed at 300 °C (Figure 25) can be fitted with three sextets and one doublet. Two sextets are ascribed to γ-Fe_2O_3 and one to α-Fe_2O_3; the doublet could be attributed to paramagnetic (or magnetically ordered phase in superparamagnetic state) in the base of large quadrupole splitting values which are known to increase with decreasing particle size and the absence of hyperfine magnetic field. The spectral parameters are presented in Table 7.

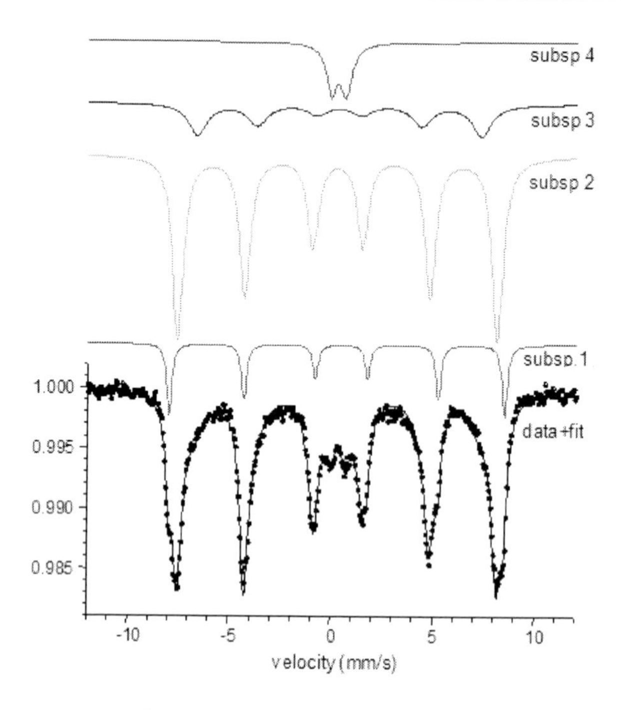

Figure 25. Mössbauer spectrum of sample B1 annealed at 300 °C.

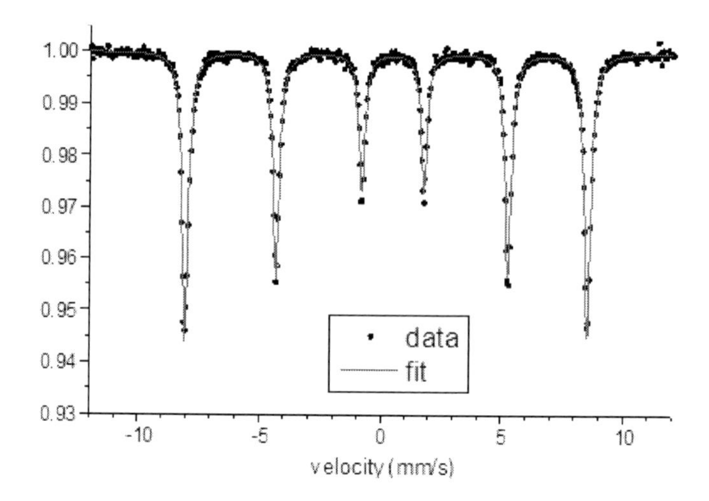

Figure 26. Mössbauer spectrum of sample B1 annealed at 500 °C.

The Mössbauer spectrum of sample B1 annealed at 500 °C (Figure 26) is composed from one sextet with a quadrupole splitting of – 0.21 mm/s and a hyperfine field of 51.5 T, both spectral parameters being characteristic to pure hematite (100 % Fe_2O_3).

It is notable that by thermal decomposition of these precursors (Fe(III) carboxylates) at 300 °C in air, γ-Fe_2O_3 is obtained as the major stable phase.

Table 7. Spectral parameters of sample B1 annealed at 300 °C.

	Isomer shift (δ)	Quadrupole splitting (ΔEQ)	Hyperfine field (BHf)	Relative area (%)	Interpretation
Subsp. 1	0,38 mm/s	-0,21 mm/s	51,2	12	α-Fe2O3 - hematite
Subsp. 2	0,32 mm/s	0,01 mm/s	48,7	63	γ-Fe2O3 - maghemite
Subsp. 3	0,38 mm/s	0,01 mm/s	43,3	17	γ-Fe2O3 - maghemite
Subsp. 4	0,35 mm/s	0,75 mm/s	-	8	(super)paramagnetic Fe3+

2.1.2. Obtaining of Fe_2O_3/SiO_2 nanocomposites by the modified sol-gel method

The nanocomposites of Fe_2O_3/SiO_2 type were obtained from Fe(III) carboxylate type complex combinations embedded within the pores of the hybrid gel TEOS-diol. There were synthesized gels starting from tetraethyl orthosilicate (TEOS), $Fe(NO_3)_3 \cdot 9\ H_2O$, ethylene glycol (EG), 1,2 propane diol (1,2 PG), 1,3 propane diol (1,3 PG) and 1,4 buthane diol (1,4 BG). At preparation of the gels, the necessary amount of the diol was used for the interaction with TEOS and its hydrolysis products and for the redox reaction, with an excess of 50% diol compared to the stoechiometry (according to the general equations).

The synthesis method consists of mixing, under magnetic stirring of the solutions $Fe(NO_3)_3$ – diol with an ethanolic TEOS solution, in acid catalysis ($[Fe(H_2O)_9]^{3+}$; pK_a = 2.22). During the gelation time, at room temperature, the interaction between the diol and TEOS or its hydrolysis products takes place, leading to a hybrid gel. The solutions of Fe(III) nitrate and diol are present in the pores of the hybrid gel. The synthesized gels were dried at 40 °C and grinded, then thermally treated at temperatures between 70-100 °C when the redox reaction takes place between the NO_3^- ion and the diol with formation of the Fe(III) carboxylate type complex combinations within the pores of the hybrid gels. After grinding, the synthesized gels are very fine brown powders.

The gels can be thermally treated, at 300 °C, when the oxidative decomposition of the complex combinations takes place with the formation of γ-Fe_2O_3. Studies on the formation of the nanocomposites were achieved for the gels synthesized with all four diols for different compositions 20, 30, 50 and 70 wt% Fe_2O_3/SiO_2, noted with $S_{diol}^{wt\%}$ (diol = EG, 1,2PG, 1,3 PG, 1,4 BG; wt% = 30, 50, 70).

The formation of the Fe(III) carboxylate type complex combinations within the gels pores was studied by thermal analysis and FT-IR spectrometry. Figure 27 presents the TG and DTA curves of gel $S_{1,4BG}^{30}$ heated at 40 °C. The DTA curve presents an exothermic effect at 70 °C attributed to the redox reaction between $Fe(NO_3)_3$ and 1,4 BG with the formation of the Fe(III) complex combination within the gels pores. The second exothermic effect at 250 °C corresponds to the oxidative decomposition of the formed complex combination. The mass losses on TG, in the first stage, correspond to the elimination of volatile products: H_2O, NO_x, and in the second stage, to the elimination of the oxidative decomposition products CO, CO_2 and condensation products of the matrix.

According to the thermal analysis data, we have established 130 °C as an optimal temperature for the formation of the complex combinations within the gels pores.

Figure 28 presents the FT-IR spectra of gel $S_{1,4BG}^{30}$ at 40 °C (spectrum a) which evidences a clear band at 1381 cm^{-1} attributed to NO_3^-, free in the gels pores. Spectrum (b) corresponding to the gel thermally treated at 130 °C shows the disappearance of the band from 1381 cm^{-1} and the appearance of bands characteristic for the formed complex combination from the gels pores: $\nu_{as}(COO^-)$ at 1677 cm^{-1} and $\nu_s(COO^-)$ at1364 cm^{-1}. In the range 2800-3000 cm^{-1} (spectra a and b) the bands characteristic for the groups $-CH_2-$ are registered corresponding to diols chemically bonded within the matrices network. The intense band from 1062 cm^{-1} is attributed to the asymmetric stretching vibration ν_{as}(Si-O-Si) [51].

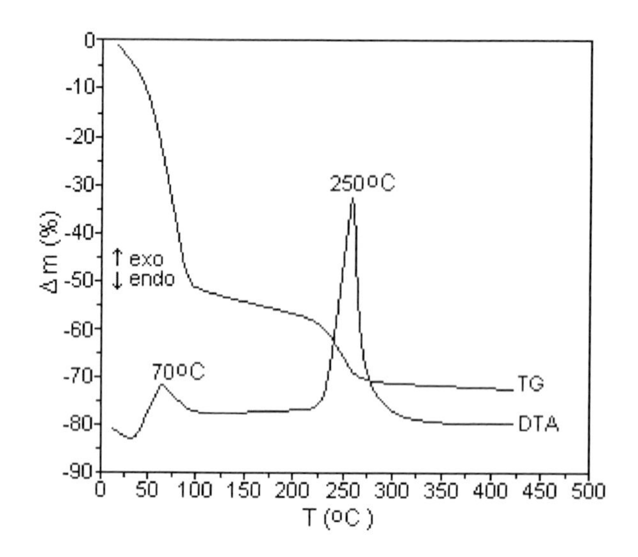

Figure 27. TG and DTA curves of gel $S_{1,4BG}^{30}$ heated at 40 °C.

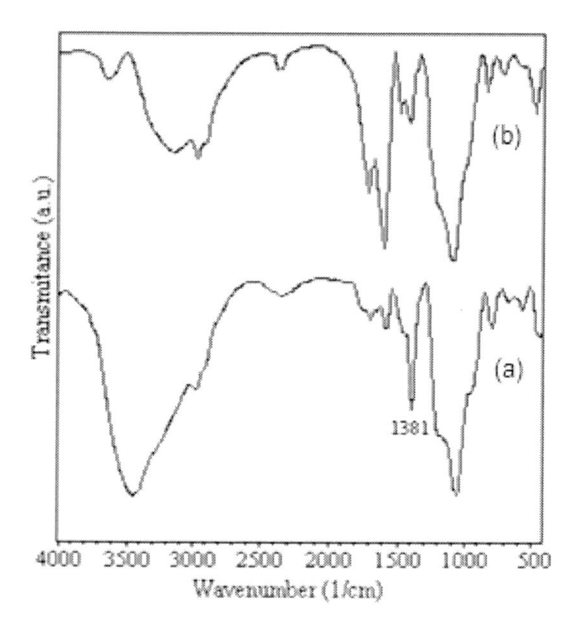

Figure 28. FT-IR spectra of gel $S_{1,4BG}^{30}$ thermally treated at 40 °C (spectrum a) and 130 °C (spectrum b).

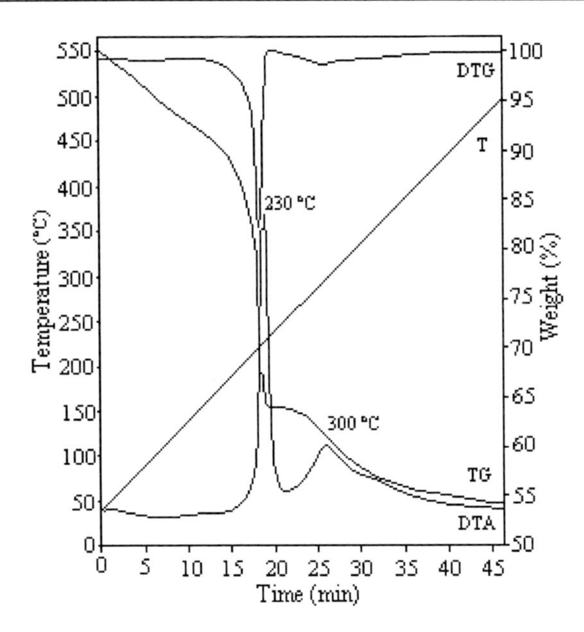

Figure 29. Thermal curves of gel $S_{1,4BG}^{50}$ thermally treated at 130 °C.

The thermal behavior of the gels thermally treated at 130°C confirms the presence of the Fe(III) carboxylate compounds within the xerogels pores. Figure 29 exhibits the thermal curves obtained for gel $S_{1,4BG}^{50}$ up to 500°C in air.The mass loss registered up to 180 °C corresponds to the elimination of adsorbed water and to the on-going poly-condensation process. The mass loss in a single step from the range 180-270 °C, associated with a strong exothermic effect on DTA. corresponds to the oxidative decomposition of the Fe(III) carboxylate precursor within the pores of silica matrix. The decomposition of Fe(III) carboxylate generates a reducing environment(CO, C) in the pores of the matrix. Thus, Fe(III) is reduced to Fe(II) which re-oxidizes to Fe_2O_3 [18]. The mass loss with exothermic effect at 300°C is attributed to the burning of residual C resulted at the oxidative decomposition of the organic precursor.

The gels synthesized at 130°C, S_{EG}^{50} and $S_{1,4BG}^{50}$, were annealed at 700 °C, 3 hours (Figure 30). From the XRD patterns presented in Figure 30 , one can notice that in the case of gel S_{EG}^{50} (spectrum 1), the majoritary crystalline phase is fayalite (Fe_2SiO_4) with traces of α-Fe_2O_3, and in the case of sample $S_{1,4BG}^{50}$ (spectrum 2), γ-Fe_2O_3 is obtained as unique crystalline phase in the amorphous silica matrix. In the case of gels with 1,2 and 1,3 PG, thermally treated in the same conditions, the evolution is similar to the gel with EG. The formation of fayalite in these cases, is due to the lower porosity of the gels which hinders the air diffusion in the pores of the matrix. This fact makes the re-oxidation of Fe(II) to Fe(III) harder, favorizing the oxidation state Fe(II), which interacts with SiO_2 from the matrix with formation of fayalite (Fe_2SiO_4). In the case of the gel with 1,4BG, due to its high porosity, the air diffusion within the gels pores is facilitated leading to the re-oxidation of Fe(II) to Fe(III) with formation of γ-Fe_2O_3, which is stable within the pores of the silica matrix.

In the case of a pre-treatment of the samples at 300 °C, γ-Fe_2O_3 is formed as sole phase and it is stable in the pores of the silica matrix up to 700°C for all gels. Figure 31 presents the XRD patterns of gels S_{EG}^{50} and $S_{1,4BG}^{50}$ pre-treated at 300°C and annealed for 3 hours at 700°C. The crystallization degree of the phase γ-Fe_2O_3 depends on the distribution of the particles within the silica matrix, which are determined by the modified morphology of the matrix, depending on the diol used for synthesis.

Figure 32 presents the evolution of the phase γ-Fe_2O_3 obtained at 300 °C, within the silica matrix, by annealing the gel $S_{1.4BG}^{50}$ up to 1200 °C. The XRD patterns present γ-Fe_2O_3 as sole phase up to 800 °C. At 900 °C, γ-Fe_2O_3 turns to the sole phase ε-Fe_2O_3 [70]. According to Tronc et al. [71], the structural transformation of γ-Fe_2O_3 nanoparticles to ε-Fe_2O_3 takes place by direct thermal transformation in the case of ultrafine particles (<10 nm) of γ-Fe_2O_3, well dispersed in a silica matrix. By heating at 1200 °C, ε-Fe_2O_3 turns to the stable phase α-Fe_2O_3.

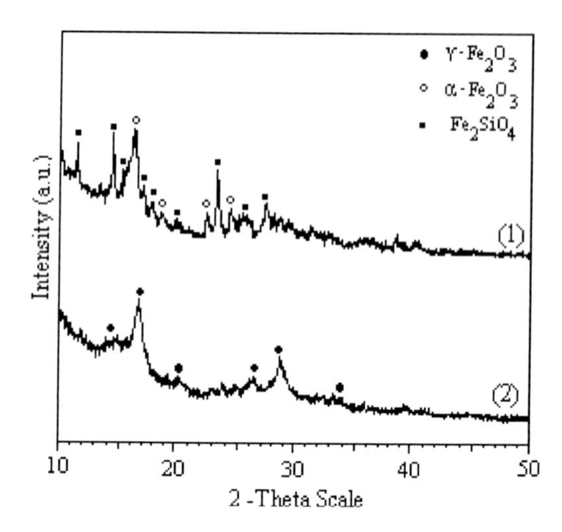

Figure 30. XRD patterns of sample S_{EG}^{50}, $S_{1.4BG}^{50}$ thermally treated at 130 →700 °C.

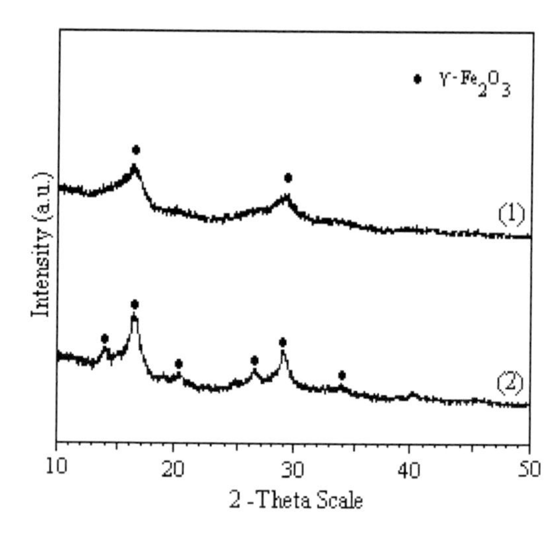

Figure 31. XRD patterns of sample S_{EG}^{50}, $S_{1.4BG}^{50}$ thermally treated at 300 →700 °C.

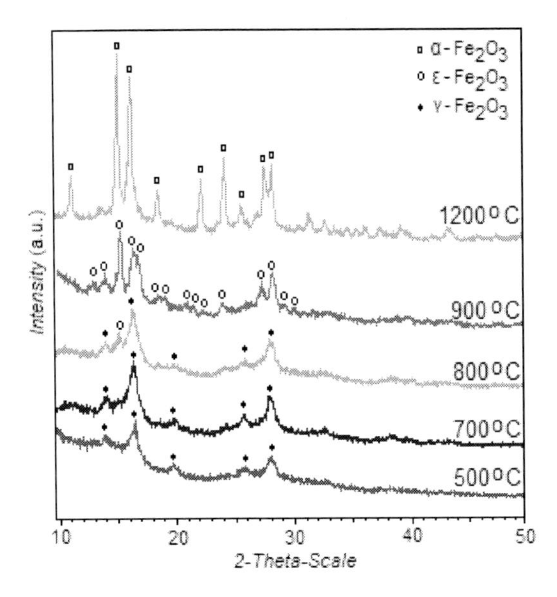

Figure 32. XRD patterns of sample $S_{1,4BG}^{50}$ annealed at different temperatures.

In order to study the influence of the iron oxide content inside the silica matrix on the evolution of crystalline phases, the samples $S_{1,4BG}^{20}$, $S_{1,4BG}^{30}$, $S_{1,4BG}^{50}$, $S_{1,4BG}^{70}$ were pre-treated at 300°C and annealed at 800°C. From the corresponding XRD patterns presented in Figure 33, one can observe that for compositions lower than 50 wt% Fe_2O_3/SiO_2, the unique phase is γ-Fe_2O_3, while for a 70 wt% composition the sole phase is α-Fe_2O_3, well crystallized. The difference in the phase evolution in the case of the sample with 70% Fe_2O_3 it can be explained, according to the previous studies reported in the literature [55], to the extended agglomeration of γ-Fe_2O_3 particles (present in silica matrix up to 500°C) that leads to the straight transition of γ-Fe_2O_3 to α-Fe_2O_3.

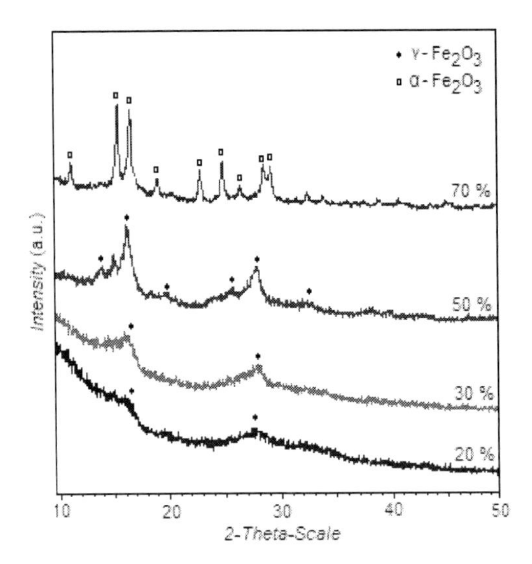

Figure 33. XRD patterns of sample $S_{1,4BG}^{wt\%}$ annealed at 800 °C for different compositions of Fe_2O_3/SiO_2.

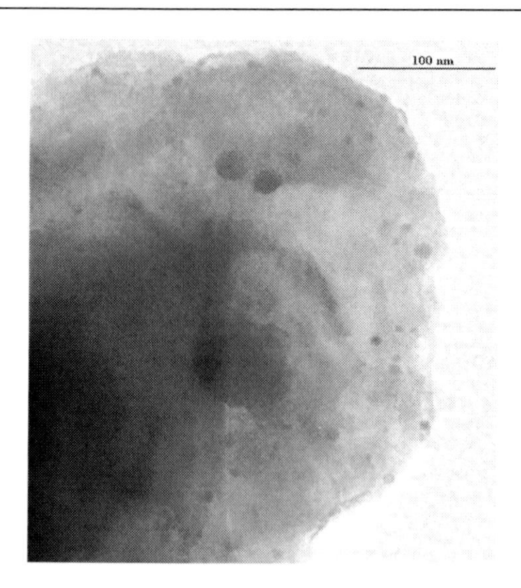

Figure 34. TEM image of sample $S_{1,4BG}^{50}$ annealed at 700 °C.

The dimensions of the nanocrystallites estimated from the XRD patterns using the Scherrer formula [72] are between 5 and 7 nm, independent on their concentration in the matrix or on the annealing temperature. The TEM image for gel $S_{1,4BG}^{50}$ annealed at 800 °C (Figure 34) shows that the γ-Fe_2O_3 nanoparticles are spherical and homogenous dispersed within the silica matrix with diameters of < 10 nm, confirming the dimensions obtained from XRD data.

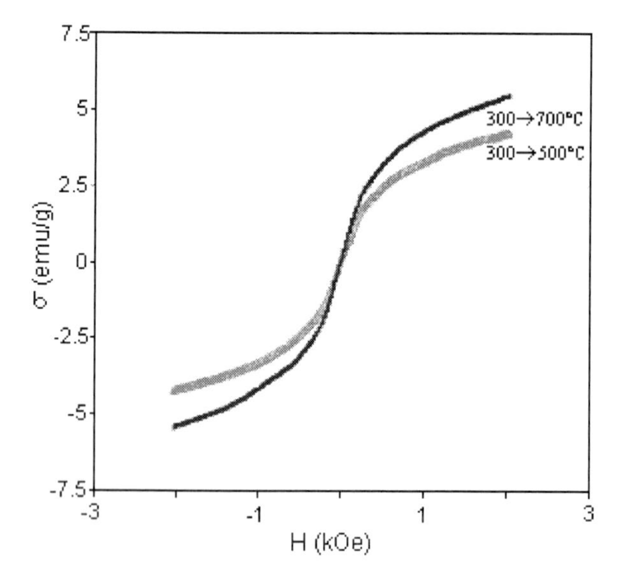

Figure 35. Magnetization curves of sample $S_{1,4BG}^{50}$ annealed at 500 and 700 °C.

The γ-Fe_2O_3/SiO_2 nanocomposites, which present magnetic properties, were also characterized by magnetic measurements in an external magnetic field. Figure 35 presents the magnetization curves of sample $S_{1,4BG}^{50}$ annealed at 500 and 700 °C. We can notice that the coercive field is $H_C = 0$, showing a superparamagnetic behavior of these nanocomposites due to the reduced size of the magnetic nanoparticles dispersed in silica matrix. The value of the saturation magnetization slightly increases with temperature due to the increased crystallinity of γ-Fe_2O_3. The magnetic behavior remains superparamagnetic because of the fine nature of the magnetic nanoparticles.

2.2. The obtaining of $Ni_{0.65}Zn_{0.35}Fe_2O_4/SiO_2$ nanocomposites by the modified sol-gel method

The Ni-Zn ferrite represents a material of high technological importance and it has been the subject of many scientific papers. The Ni-Zn ferrite belongs to the class of soft ferrimagnetic materials, with low magnetic coercivity and resistivity, properties that make it an excellent material with different applications in telecommunications, electronics, microwaves devices, magnetic heads for reading digital equipments of high speed [73-75]. The nanocomposites obtained by embedding within inorganic matrices of Ni, Zn ferrite nanoparticles, present interest due to their performant chemical, electric and magnetic properties, as a result of their particulate structure compared to the bulk material.

In order to obtain Ni,Zn ferrite embedded in silica matrix, by thermal decomposition of the carboxylate type precursors formed within the matrixes pores, the same experimental procedure was used as in chapter 2.1.2.. In the first step, gels from tetraethyl orthosilicate - metal nitrates – diol were prepared for two different compositions corresponding to the nanocomposites x%($Ni_{0.65}Zn_{0.35}Fe_2O_4$)/(100-x)%$SiO_2$, where x = 35% ($F^{35}$) and x = 65% ($F^{65}$). There were three different diols used: ethylene glycol (EG), 1,2 propane diol (12PG) and 1,3 propane diol (1,3PG). The compositions of the prepared gels are presented in Table 8.

Table 8. Compositions of the synthesized gels.

Sample	Diol	Quantity (mole)							Molar ratio	
		TEOS	Diol	$Fe(NO_3)_3 \cdot 9H_2O$	$Ni(NO_3)_2 \cdot 6H_2O$	$Zn(NO_3)_2 \cdot 6H_2O$	EtOH	H_2O	TEOS:Diol:NO_3^-:H_2O	x% ferrite
F^{65}_{EG}	EG	0.030	0.120	0.030	0.009	0.005	0.120	0.12	1 : 4 : 4 : 12	65
F^{65}_{12PG}	1,2PG	0.030	0.120	0.030	0.009	0.005	0.120	0.12	1 : 4 : 4 : 12	65
F^{65}_{13PG}	1,3PG	0.030	0.120	0.030	0.009	0.005	0.120	0.12	1 : 4 : 4 : 12	65
F^{35}_{EG}	EG	0.110	0.120	0.030	0.009	0.005	0.100	0.44	1 : 1 : 1 : 3	35
F^{35}_{12PG}	1,2PG	0.110	0.120	0.030	0.009	0.005	0.100	0.44	1 : 1 : 1 : 3	35
F^{35}_{13PG}	1,3PG	0.110	0.120	0.030	0.009	0.005	0.100	0.44	1 : 1 : 1 : 3	35

The obtained gels, dried at 40 ^{0}C, were thermally analyzed in order to study the evolution of the redox reaction between the nitrate ion (from the mixture of Fe(III), Ni(II) and Zn(II) nitrates) and diols. The diols are oxidized to the corresponding carboxylate anions (gyoxylate, lactate, malonate) which form with the metal ions existent within the matrixes pores, carboxylate type coordination compounds [67].

Figure 36 presents the DTA curves for gels F^{35} synthesized with EG (curve1), 1,2 PG (curve 2), 1,3PG (curve 3) dried at 40°C. In all cases, in the temperature range 100-120°C, the exothermic effect on DTA corresponds to the redox reaction between diol and Fe(III), Ni(II) and Zn(II) nitrates with formation within the gels pores of the carboxylate type compounds. These coordinative compounds thermally decompose in the range 220-350°C, generating a pronounced exothermic effect on DTA. In the case of gels, F_{EG}^{65}, F_{12PG}^{65} and F_{13PG}^{65} with a higher metal nitrates content, dried at 40 °C, similar thermal effects were obtained on DTA curves, corresponding to the two processes: the redox reaction nitrate-diol with formation of carboxylate type complex combinations and their thermal decomposition. In the case of these gels, the redox reaction was initiated during gelation with evolving of nitrogen oxides. Based on thermal analysis data for the gels dried at 40°C, the optimal temperature forobtaining the carboxylate type ferrite precursors was established at 130°C.

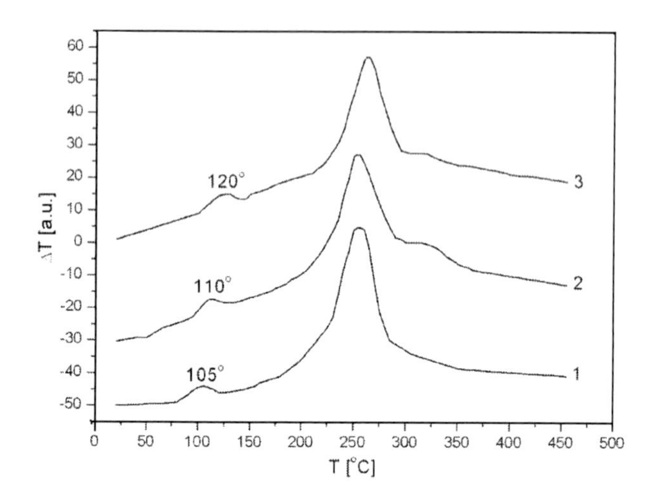

Figure 36. DTA curves corresponding to gels F^{35} with different diols, dried at 40°C: 1) EG; 2) 12PG; 3) 13PG.

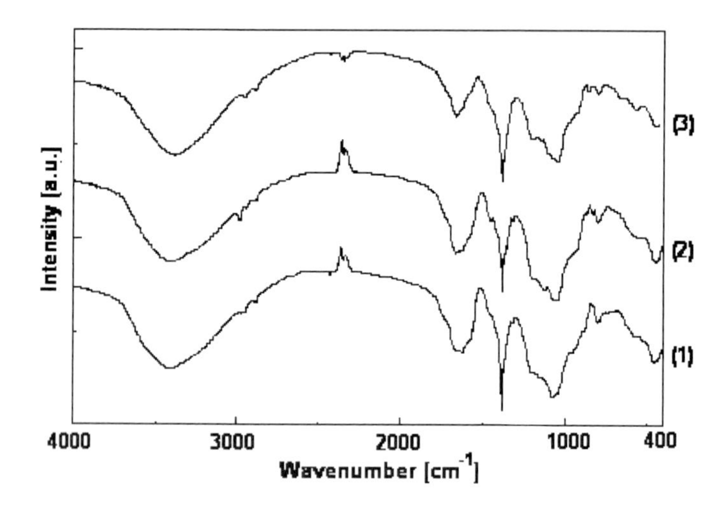

Figure 37. FT-IR spectra of gels, F_{EG}^{65}, F_{12PG}^{65} and F_{13PG}^{65} dried at 40°C.

The formation within the matrix of the coordination compounds of Fe(III), Ni(II) and Zn(II) cations with the carboxylate anions was confirmed by FT-IR analysis of the synthesized gels, dried at 40°C and thermally treated at 130°C. Figure 37 presents the FT-IR spectra of gels, F_{EG}^{65}, F_{12PG}^{65} and F_{13PG}^{65} dried at 40°C. In all spectra, the intense band from 1380 cm^{-1}, characteristic for the nitrate ions vibration is evidenced, coming from the metal nitrates existent in the pores, along with the bands characteristic for the silica matrix (at 580 cm^{-1}, 800 cm^{-1}, 960 cm^{-1}, 1000-1200 cm^{-1} and 3420 cm^{-1} [51]).

Figure 38 presents the FT-IR spectra corresponding to gels, F_{EG}^{65}, F_{12PG}^{65} si F_{13PG}^{65} thermally treated at 130°C. In all spectra, the disappearence of the band from 1380 cm^{-1} due to the integral consumption of the nitrate ions in the reaction with the diol is observed. Due to the formation within the matrixes pores of the carboxylate type complex combinations, one can notice the appearance of the bands characteristic for the coordinated carboxylate group: $\nu_s(CO)+\delta(OCO)$ at 1311 cm^{-1}, $\nu_s(COO^-)$ at 1360 cm^{-1}[69]. The band characteristic for the asymmetric vibration $\nu_{as}(COO^-)$ from 1620 cm^{-1} overlaps with the one characteristic for the vibrations of H-O-H bonds, and the band characteristic for the vibration ν(C-OH) from the carboxylic group, which appears at 1060 cm^{-1}, overlaps with the characteristic band ν(Si-O-Si), leading to an increase in intensity of these bands.

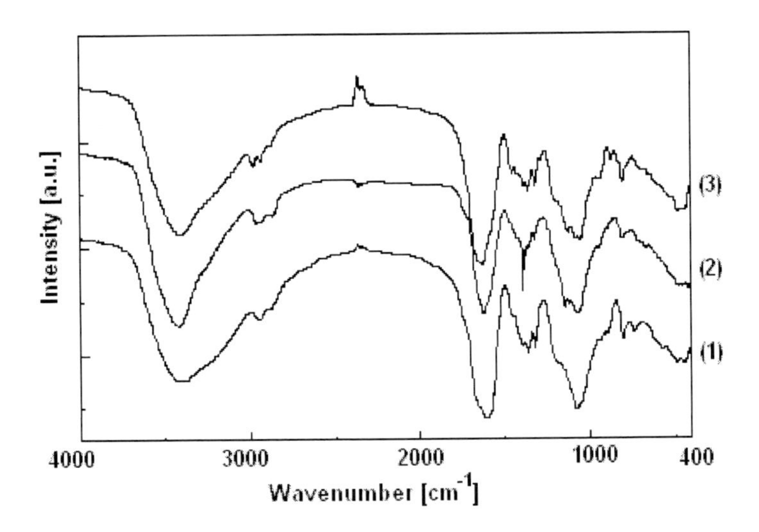

Figure 38. FT-IR spectra of gels F_{EG}^{65}, F_{12PG}^{65} and F_{13PG}^{65} thermally treated at 130°C.

The thermal analysis and FT-IR spectrometry data confirm the evolution within the pores of the silica matrix of the redox reaction between the studied diols (EG, 12PG, 13PG) and Fe(III), Ni(II), Zn (II) metal nitrates with the formation of Fe(III), Ni(II), Zn (II) carboxylate type compounds, embedded in the matrix, which are precursors for Ni, Zn ferrite.

The thermal behavior of gels F^{35} and F^{65}, thermally treated at 130°C, which contain the Fe(III), Ni(II), Zn (II) carboxylate type complex combinations, was studied by thermal analysis in order to establish the optimal obtaining conditions of the Ni, Zn ferrite embedded in silica matrix. Figures 39-41 present the thermal curves TG and DTA of gels F_{EG}^{65}, F_{12PG}^{65} and F_{13PG}^{65}.

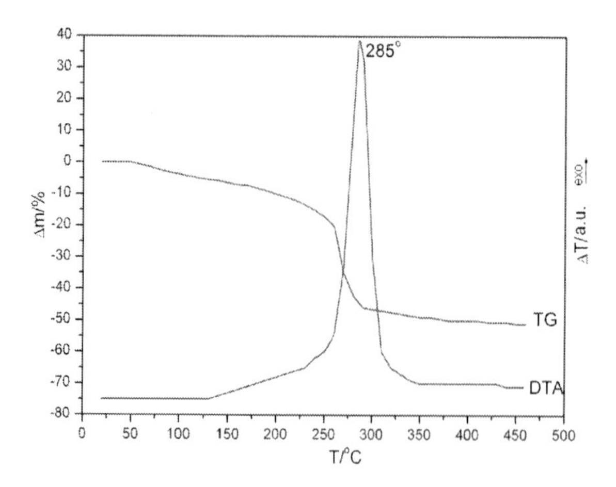

Figure 39. TG and DTA curves of gel F_{EG}^{65} thermally treated at 130°C.

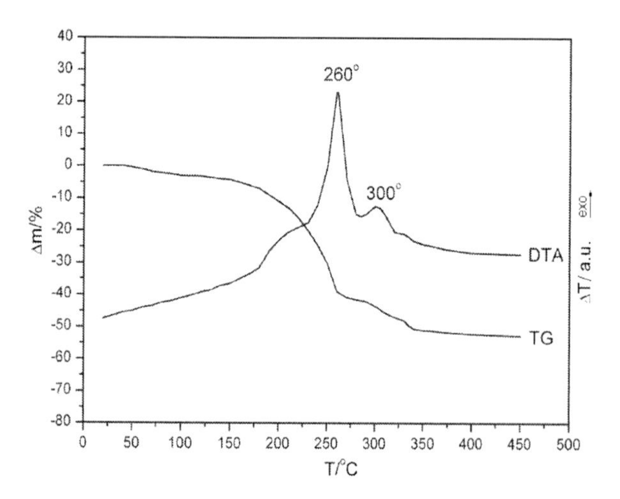

Figure 40. TG and DTA curves of gel F_{12PG}^{65} thermally treated at 130°C.

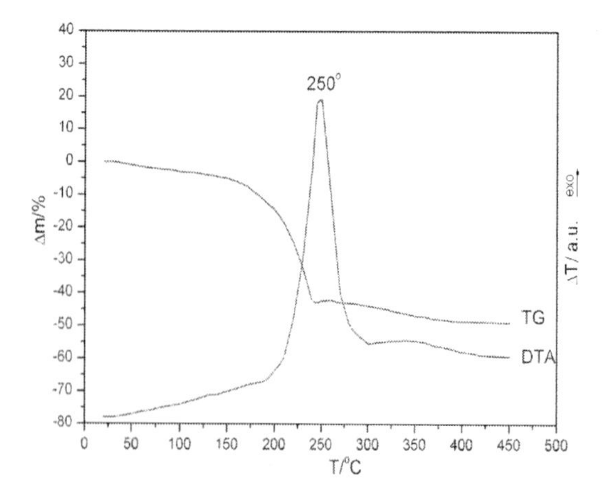

Figure 41. TG and DTA curves of gel F^{65}_{13PG} thermally treated at 130°C.

The mass loss registered in each case up to 200°C is due to the elimination of water adsorbed at the surface and eventually to the elimination of the coordination water from the complex. The mass loss in the 200-350°C can be attributed to the oxidative thermal decomposition of the carboxylate type compounds and to the decomposition of the organic chains from the matrix. In the case of gel with 1,2 PG, on DTA (Figure 40), in this temperature range appear many individualized exothermal effects. This behavior and the mass loss in steps, from the range 175-350°C, can be due to the formation of a homogenous mixture of Fe(III), Ni(II) and Zn(II) homopolynuclear complex combinations. In the case of gels F^{65}_{EG} (Figure 39) and F^{65}_{13PG} (Figure 41), the thermal decomposition in the range 200-350°C develops unitary, with generation on DTA of a single pronounced exothermic effect. In this case, there is the probability of the formation of some heteropolynuclear complex combinations by functionalizing of the ligands (glyoxylate and malonate) as bridged ligands [67].

One can observe from the presented thermal curves that, independent on the nature of the precursors, their oxidative decomposition within the matrix is finished at ~350°C, with formation within the matrixes pores of the corresponding oxide mixture. Up to 500°C another slow mass loss due to the elimination of the mai are loc –OH groups of the silica matrix takes place.

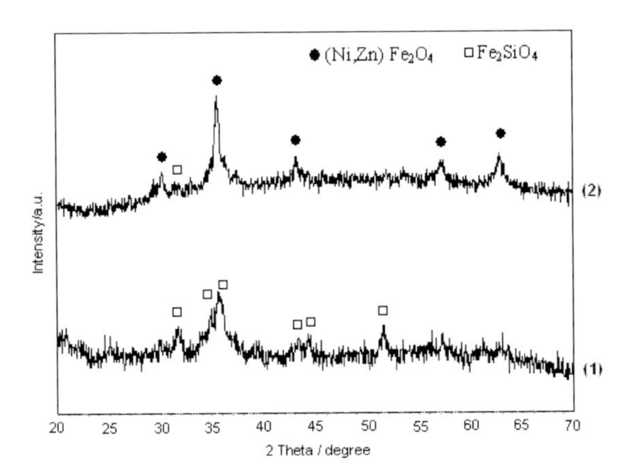

Figure 42. XRD patterns of samples F^{35}_{EG}(1) and F^{65}_{EG}(2) annealed at 1000°C.

In order to obtain Ni,Zn ferrite embedded in silica matrix, the xerogels thermally treated at 130°C, were annealed at 600°C, 800°C and 1000°C in order to study the formation of the Ni, Zn ferrite. The XRD patterns of gels, F^{35}_{EG}, F^{35}_{12PG} and F^{35}_{13PG} annealed at 600°C had an amorphous aspect. In the case of annealing at 1000°C, the XRD patterns have evidenced besides Ni, Zn ferrite of Fe_2SiO_4 (fayalite) as secondary crystalline phases. This is due to the fact that, in case of the gels synthesized with a lower diol content, the pore volume is smaller and the air diffusion within the pores is slower, which makes the reducing atmosphere (CO) created at decomposition of metal carboxylates to favorize the oxidation state (II) for Fe, with the formation of fayalite. In the case of gels, F^{65}_{EG}, F^{65}_{12PG} and F^{65}_{13PG} annealed at 1000°C, the

XRD patterns have evidenced only the diffraction lines characteristic for Ni,Zn ferrite. This different behavior may be due to the higher porosity of the gels with a higher metal nitrates and diol content which favorize the air diffusion within the matrixes pores. Figure 42 presents the XRD patterns of gels, F_{EG}^{35} and F_{EG}^{65} annealed at 1000°C.

In order to obtain Ni,Zn ferrite as sole phase within the silica matrix, a controlled thermal treatment was applied to the gels similarly with the case of Fe_2O_3/SiO_2 nanocomposites (chapter 2.1.2.). Thus, the gels were thermally treated for 3 hours at 300°C, when the decomposition of metal carboxylates takes place with the reduction and re-oxidation of iron with the formation of Ni, Zn ferrite nuclei. The ulterior annealing at higher temperatures (600°C ,800°C, 1000°C) of these pre-treated samples leads to crystallization of Ni,Zn ferrite as sole phase within the silica matrix.

Figures 43 and 44 present the XRD patterns of samples with 35% ferrite in silica matrix (Figure 43) and with 65% ferrite in silica matrix (Figure 44) thermally pre-treated at 300°C for 3 hours, and then annealed at 1000°C, for 3 hours. According to the XRD patterns, all samples present Ni, Zn ferrite as sole crystalline phase inside the amorphous silica matrix.

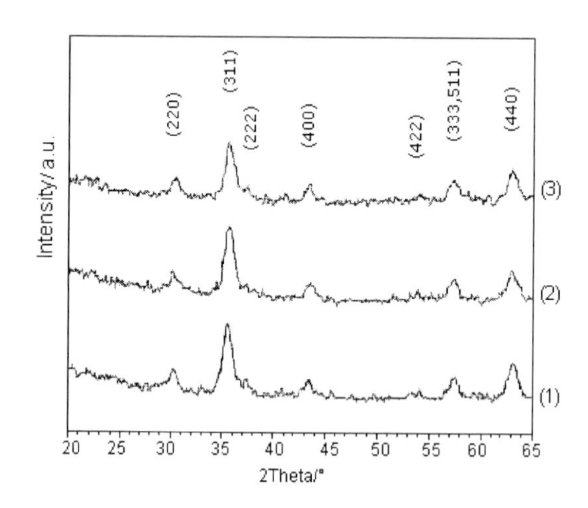

Figure 43. XRD patterns of samples F_{EG}^{35} (1), F_{12PG}^{35} (2) and F_{13PG}^{35} (3) pre-treated at 300°C and annealed at 1000°C.

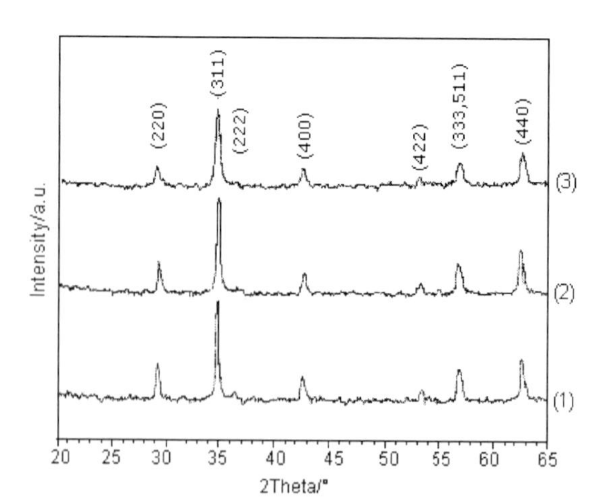

Figure 44. XRD patterns of samples F_{EG}^{65}, F_{12PG}^{65} and F_{13PG}^{65} pre-treated at 300°C and annealed at 1000°C.

Tabel 9. Mean diameter of the particles estimated from XRD data at 1000°C.

Sample	Mean diameter		
	EG	1,2 PG	1,3 PG
F35	15	15	15
F65	39	40	41

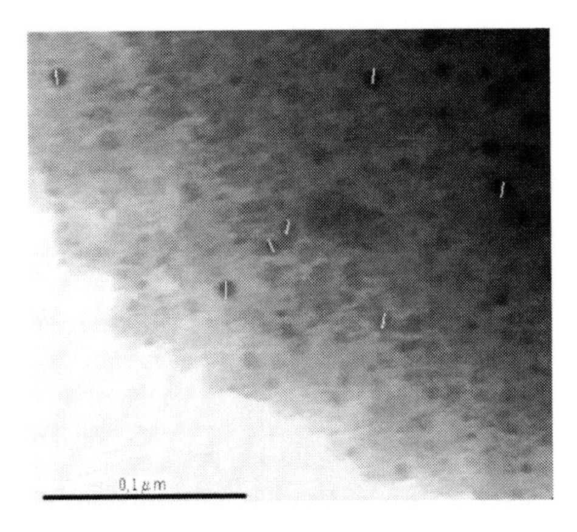

Figure 45. TEM image of sample F_{12PG}^{35} (800°C).

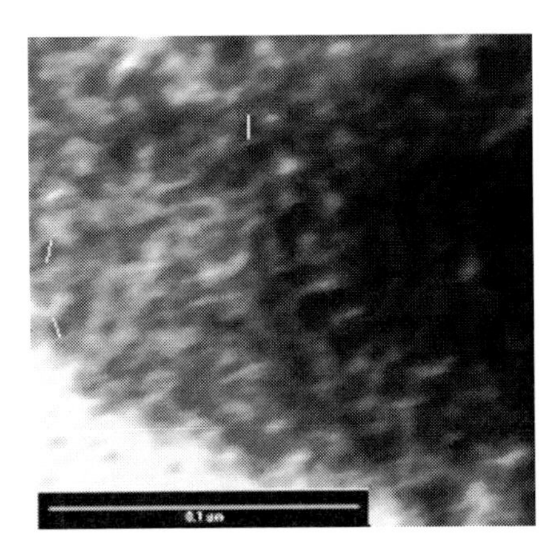

Figure 46. TEM image of sample F_{12PG}^{65} (800°C).

The influence of the used diols is reflected on the mean diameter of the Ni,Zn ferrite nanocrystallites from the silica matrix and on the magnetic properties of these nanocomposites. Thus, at 800°C the smallest mean diameter (7,2 nm) is obtained in the case of using 1,2 PG. At higher annealing temperatures (1000°C), the diameter of ferrite nanocrystallites increases in all four cases and the differences are insignificant (Table 9).

In figures 45 and 46, the TEM images of the nanocomposites synthesized with 1,2-propandiol are presented, with a content of 35% ferrite in SiO_2, respectively 65% ferrite, thermally treated at 800°C for 3 hours. One can observe that, in the case of sample F_{12PG}^{35} (35%

ferrite în SiO_2), the ferrite nanoparticles are spherical, uniformly dispersed within the silica matrix, with diameters in the range 4 - 8 nm. In the case of the sample with 65% ferrite in SiO_2, the nanoparticles are more agglomerated, of approximately spherical shape, with diameters in the range 6-10 nm. It is notable that the diameters of the nanoparticles established from the TEM images are in accord with the values calculated from XRD data.

The histogram achieved from the TEM image for sample F_{EG}^{35} (Figure 47) shows a narrow size distribution of the ferrite nanoparticles. The TEM image confirms the uniform distribution of the Ni, Zn ferrite nanoparticles embedded in silica matrix in this case.

Magnetic properties of the nanocomposites $Ni_{0.65}Zn_{0.35}Fe_2O_4/SiO_2$

The interest for granular magnetic solids formed out of ultrafine nanometric particles with magnetic properties, inserted in an unmiscible isolating or metal matrix has considerably increased in the last years due to the magnetic properties of these particular structures [75]. The magnetic properties of the Ni,Zn ferrite /SiO_2 nanocomposites are different from those of bulk Ni-Zn ferrite [76, 77]. The magnetic properties of Ni,Zn ferrite nanoparticles embedded in SiO_2 matrix are dictated by the superexchange between magnetic ions and oxygen atoms. Moreover, the properties of the composites depend on microstructure and measurable parameters of the particles as: dimension, porosity and magnetization frequency.

In composite materials with high ferrite content, the homogenous distribution of magnetic nanoparticles within the non-magnetic matrix insures the continuity of the magnetic flux. The ferrite particles are in contact with each other and the activation energy of the mass transfer is low. The particles increase takes place easily, resulting in higher grains and higher μ_i. In contrast, in composites with low ferrite content, there exists much non-magnetic SiO_2 and many pores which create holes around the magnetic particle. The holes determine not only magnetic dilution but also the interruption in the magnetic flux within the composite material. Ferrite particles are isolated in the non-magnetic matrix and the contact between each other is difficult, resulting in low grains dimensions and a low μ_i value [58].

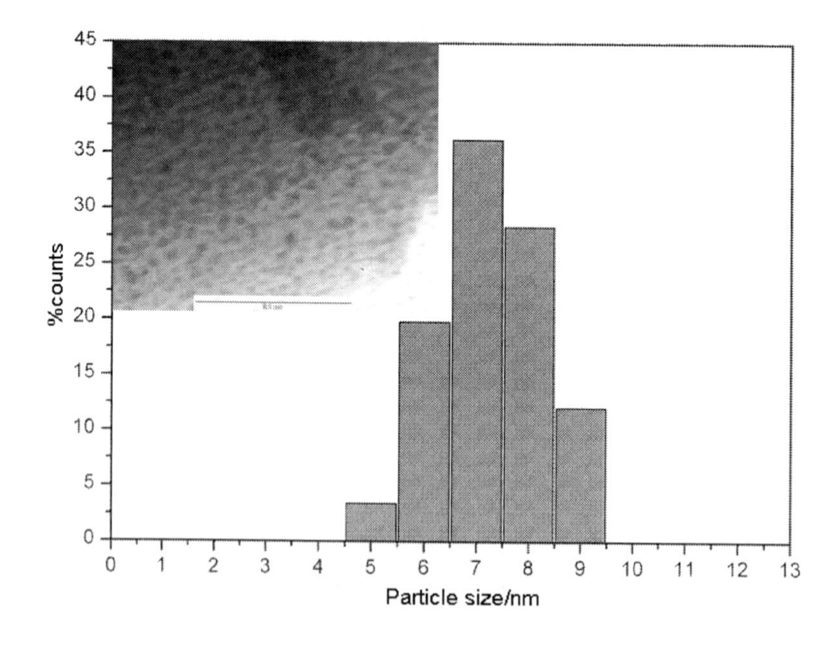

Figure 47. Histogram of the ferrite nanoparticles diameter for sample F_{EG}^{35} (400°C – 1000°C).

Table 10. Values of the magnetic parameters of the samples with 65% ferrite in SiO_2,at 800°C.

Sample	F_{EG}^{65}	F_{12PG}^{65}	F_{13PG}^{65}
σ (emu/g)	23.40	19.76	24.34
Hc (Oe)	32	20	0

In the case of the synthesized nanocomposites, the magnetic behavior modifies from superparamagnetic (SPM) to ferrimagnetic (FM), depending on the ferrite concentration and the nature of the diol used in synthesis, which determines the nanoparticles dimensions. The magnetic parameters established based on the magnetization curves registered for the synthesized nanocomposites are presented in Table 10.

The specific saturation magnetization of 50-65 emu/g in a field of 1,7 kOe corresponds to ferromagnetic nanoparticles domain. With a temperature increase of the thermal treatment at 1000°C, the hysteresis cycle becomes larger in all cases and the saturation magnetization values increase significantly, due to the increase of the Ni,Zn ferrite nanocrystallites dimensions. At higer temperatures (1000°C), all samples are ferromagnetic and at lower (600°C) temperatures, the behavior is (F_{13PG}^{65}) or close to superparamagnetic ($F_{EG}^{65}, F_{12PG}^{65}$). Figure 48 presents the magnetization curves registered for the samples F_{13PG}^{65} thermally pretreated at 400°C and then annealed at 600°C, 800°C and 1000°C.

From the registered curves, one can notice that the temperature influences significantly the magnetic behavior of the nanocomposites obtained from the synthesized precursors. The increase of the specific saturation magnetization with temperature is due, in the first place, to the variation of diameters of the Ni,Zn ferrite nanoparticles embedded in the matrix, which at 1000°C are higher (agglomerated). In the temperature range 800-1000°C, the magnetic parameters modify significantly, as a result of the transition from a monodomenial magnetic structure (d<20nm) to a multidomenial structure (d > 30 nm) of the nanoparticles, which explains the significant increase of the saturation magnetization values.

In the case of the samples with 35% ferrite in SiO_2, the magnetic behavior is different. Thus, at the magnetic field used there is no hysteresis cycle, the magnetizations are reduced and the shapes of the curves turns from superparamagnetic (high temperatures) to approximately linear (low temperatures). This unusual behavior of the obtained nanocomposites can be attributed to the very small sizes (2 ÷ 9 nm) of Ni,Zn ferrite formed within the silica matrix. At these diameters, the ferrite nanoparticles are totally monodomenial (from magnetic structure point of view), being situated in the superparamagnetic domain, which leads to the absence of hysteresis and coercive field [17]. Also, in the temperature range 800°C – 1000°C, the restructuration of the silica matrix takes place, which influences the magnetic behavior of the ferrite nanoparticles, contributing to the significant modification of the nanocomposites magnetic properties. In conclusion, the magnetic behavior of the obtained nanocomposites is significantly influenced by the procentual content of Ni,Zn ferrite within the matrix and by the thermal treatment temperature.

From our studies, it results that it is possible to control the size of the Ni,Zn ferrite nanoparticles embedded in the silica matrix by the nature of the diol, the temperature of the thermal treatment and the ferrite concentration within the matrix.

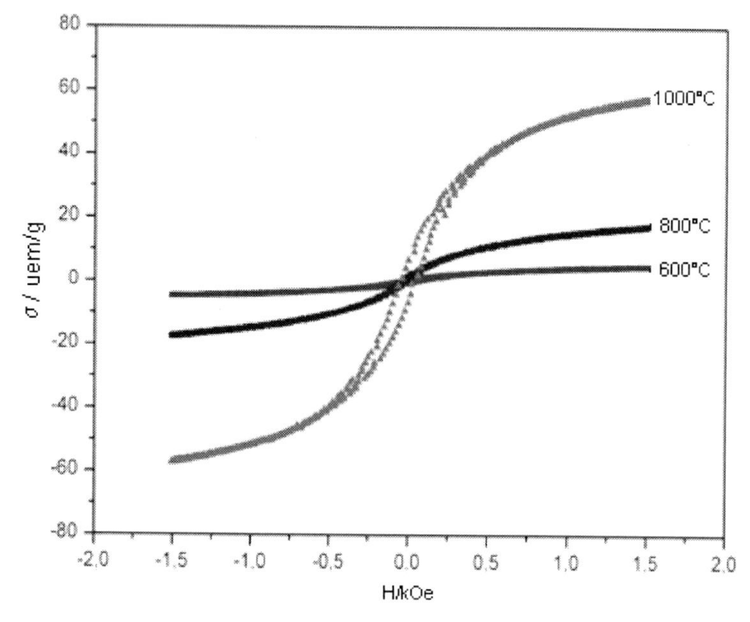

Figure 48. Magnetization curves of simple F^{65}_{13PG} , thermally pre-treated and annealed at different temperaturas.

This aspect is very important from theoretical and practical point of view, because it allows the obtaining of nanoparticles with controlled diamaters in a large value range and with good energetic efficiency (lower temperatures).

CONCLUSION

In this chapter, we have reported some of our results regarding the preparation of oxidic nanoparticles/SiO_2 nanocomposites with magnetic properties, using an original version of sol gel synthesis method. The particularity of this modified sol-gel method is the in-situ formation in the pores of silica matrix of some precursors – coordination compounds of the involved metal ions with carboxylate anions – resulted in the redox reaction between metal nitrates and diols. This redox reaction takes place by the controlled heating of tetraethyl ortosilicate – diol – metal nitrates gels. By thermal decomposition of the precursors in the pores of the gels, followed by appropriate thermal treatments, we have obtained fine oxides nanoparticles homogenously dispersed in silica matrix. The excellent results of the synthesis method are also due to the formation of a organic/inorganic hybrid matrix by the chemical interaction of the diol with the functional groups of the silica network. During the thermal treatment of the gels, the hybrid matrix turns to a silica matrix with modified morphology.

Using the modified sol-gel method, we have successfully synthesized Fe_2O_3/SiO_2 and $Ni_{0.65}Zn_{0.35}Fe_2O_4/SiO_2$ nanocomposites with controlled structure and properties. An important feature of this method is the reduction of iron (III) ions as a result of the reducing atmosphere generated in the thermal decomposition of the precursors. In the case of the bulk materials, the reducing process is followed by the re-oxidation of Fe(II) to Fe(III) with formation of γ-Fe_2O_3 (300°C). At thermal decomposition of the precursors in the pores of silica matrix, the reoxidation process is slower due to the difficult diffusion of the air (O_2) in the pores. A controlled thermal treatment insures the re-oxidation of Fe(II) to γ-Fe_2O_3 in the pores of the silica matrix, as sole crystalline phase inside the amorphous silica matrix, stabilized up to 800°C. For the nanocomposites with Fe_2O_3 content in silica matrix lower than 50% above 800°C, γ-Fe_2O_3 turns to ε -Fe_2O_3 as sole crystalline phase, which is difficult to obtain by other synthesis methods.

The same synthesis method was used also for the preparation of $Ni_{0.65}Zn_{0.35}Fe_2O_4/SiO_2$ nanocomposites with controlled magnetic properties. The mixture of the metal nitrates interacts with the diol in the pores of silica matrix, forming heteropolynuclear/homonuclear metal carboxylates. By controlled thermal decomposition and annealing, pure Ni,Zn ferrite as fine spherical nanoparticles is obtained homogenously dispersed in silica matrix.

All synthesized γ-Fe_2O_3/SiO_2 and $Ni_{0.65}Zn_{0.35}Fe_2O_4/SiO_2$ nanocomposites presented superparamagnetic behavior, with variable saturation magnetization, depending on the magnetic phase content inside the silica matrix and on the annealing temperatures.

REFERENCES

[1] Cao, G. (2004). Nanostructures & Nanomaterials: Synthesis, *Properties & Applications*. London, Imperial College Press.

[2] Predoi, D., Crisan, O., Jitianu, A., Valsangiacom, M. C., Raileanu, M., Crisan, M. & Zaharescu, M. (2007). Iron oxide in a silica matrix prepared by the sol-gel method. *Thin Solid Films*, *515*, 6319-6323.

[3] Corriu, R. & Anh, N. T. (2009). *Molecular Chemistry of Sol-Gel Derived Nanomaterials*. West Sussex, United Kingdom, John Wiley & Sons Ltd.

[4] Zhang, L., Papaefthymiou, G. C. & Ziolo, R. F. (1997). Novel γ-Fe_2O_3/SiO_2 magnetic nanocomposites via sol-gel matrix-mediated synthesis. *Nanostruct Mater, 9*, 185-188.

[5] Bradley, D. C., Mehrotra, R. C. & Gaur, D. P. (1978). *Metal Alkoxides*, London, Academic Press.

[6] Brinker, C. J. & Scherer, G. W. (1990). Sol-Gel Science. *The Physics and Chemistry of Sol-Gel Processing*. Boston–San Diego–New York–London–Sydney–Tokyo–Toronto, Academic Press.

[7] Springer, L. & Yan, M. F. (1984). *Ultrastructure Processing of Ceramics, Glasses and Composites*. Ed. In: L. L., Hench, D. R., Ulrich, New York, Wiley.

[8] Vallet-Regi, M., Veiga-Blanco, M. L. & Mata-Arjona, A. (1980). Textura de geles de TiO_2: I. Caracterizacion de materiales obtenidos en diferentes condiciones de preparacion. *Ann Quim*, *76B*, 172-176.

[9] Venkateswara Rao, A. & Kalesh, R. R. (2004). Organic Surface Modification of TEOS Based Silica Aerogels Synthesized by Co-Precursor and Derivatization Methods. *J Sol-Gel Sci Technol, 30(3)*, 141-147.

[10] Rubio, F., Rubio, J. & Oteo, J. L. (1997). Further Insights into the Porous Structure of TEOS Derived Silica Gels. *J Sol-Gel Sci Technol*, *8(1-3)*, 159-163.

[11] Muşat, V., Budrugeac, P. & Gheorghieş, C. (2008). Effect of reagents mixing on the Thermal behavior of sol-gel precursors for silica-based nanocomposite thin films, *J Therm Anal Cal, 94(2)*, 373-377.

[12] Zaharescu, M., Vasilescu, A., Badescu, V. & Radu, M. (1997). Hydrolysis-polycondensation in binary phosphorus alkoxides - TEOS system studied by GC- MS. *J Sol-Gel Sci Technol, 8(1-3)*, 59-63.

[13] Sakka, S. & and Kozuka, H. (1998). Sol-Gel Preparation of Coating Films Containing Noble Metal Colloids. *J Sol-Gel Sci Technol, 13(1-3)*, 701-705.

[14] Poddenezhnyi, E. N., Kravchenko, I. P., Mel'nichenko, I. M., Kapshai, M. N. & Boiko, A. A. (2000). Drying of bulk silica gels molded from tetraethyl orthosilicate and aerosil. *Glass and Ceram, 57(5-6)*, 158-161.

[15] Silva, R. F. & Vasconcelos, W. L. (2000). Influence of Processing Variables on the Pore Structure of Silica Gels Obtained with Tetraethyl orthosilicate. *Mater Res, 2(3)*, 197-200.

[16] Ferreira, R. V., Pereira, I. L. S., Cavalcante, L. C. D., Gamarra, L. F., Carneiro, S. M., Amaro Jr., E., Fabris, J. D., Domingues, R. Z. & Andrade, A. L. (2010). Synthesis and characterization of silica-coated nanoparticles of magnetite. *Hyperfine Interact, 195(1-3)*, 265-274.

[17] Stefanescu, M., Stoia, M., Caizer, C. & Stefanescu, O. (2009). Preparation of $x(Ni_{0.65}Zn_{0.35}Fe_2O_4)/$ (100-x) SiO_2 nanocomposite powders by a modified sol-gel method, *Mater Chem Phys, 113*, 342-348.

[18] Stefanescu, O., Stoia, M., Stefanescu, M. & Vlase, T. (2009). Study on the influence of TEOS-diol molar ratio on their chemical interaction during the gelation process. *J Therm Anal Cal, 97*, 251-256.

[19] Stoia, M., Stefanescu, M., Dippong, T., Stefanescu, O., Barvinschi, P. (2010). Low temperature synthesis of Co_2SiO_4/SiO_2 nanocomposite using a modified sol-gel method. *J Sol-Gel Sci Technol, 54(1)*, 49-56.

[20] Mazúr, M., Mlynárik, V., Valko, M. & Pelikán, P. (1999). The time evolution of the sol-gel process: ^{29}Si NMR study of hydrolysis and condensation reactions of tetramethoxysilane. *Appl Magn Reson, 16 (4)*, 547-557.

[21] Dahmouche, K., Bovier, C., Dumas, J., Serughetti, J. & Mai, C. (1995). Densification of acid- catalysed silica xerogels. *J Mat Sci, 30(16)*, 4149-4154.

[22] Li, L. P. & Li, G. S. (2000). Hyperfine characterization of the crystallization of nanocrystalline $NiFe_2O_4$ from the amorphous silica gel. *Hyperfine Interact, 128(4)*, 437-442.

[23] Bouajaj, A., Ferrari, M. & Montagna, M. (1997). Crystallization of Silica Xerogels: A Study by Raman and Fluorescence Spectroscopy. *J Sol-Gel Sci Technol, 8(1-3)*, 391-395.

[24] Khimich, N. N. (2003). Dependence of the Pore Size of Silica Gels on the Acidity of the Medium. *Glass Phys Chem, 29(6)*, 596-598.

[25] Khimich, N. N. (2004). Synthesis of silica gels and organic – inorganic hybrids on their base, *Glass Phys Chem, 30(5)*, 430-442.

[26] Campero, A., Cardoso, J. & Pacheco, S. (1997). Ethylene glycol-citric acid-silica Hybrid organic-inorganic materials obtained by the sol-gel method. *J Sol-Gel Sci Technol*, 8, 535-539.

[27] Sharp, K. G. (1994). A two-component, non-aqueous route to silica gel. *J Sol-Gel Sci Technol*, *2*, 35-41.

[28] Schmidt, H., Scholze, H. & Kaiser, A. (1984). Principles of hydrolysis and condensation reaction of alkoxysilanes. J Non-Cryst Solids, *63(1-2)*, 1-11.

[29] Brinker, C. J., Keeffer, K. D., Schaeffer, D. W., Assiuk, R. A., Kay, B. D. & Ashley, C. S. (1984). Sol-gel transition in simple silicates II. *J Non-Cryst Solids*, *63*, 45-59.

[30] Nassar, E. J., Neri, C. R., Calefi, P. S. & Serra, O. A. (1999). Functionalized silica synthesized by sol–gel process. *J Non-Cryst Solids*, *247*, 124-128.

[31] D'Adamo, A. F. & Kienle, R. H. (1955). The Mode of Hydrolysis of Tetraalkyl Titanates. *Amer Chem Soc*, *77(16)*, 4408.

[32] Lu, Y., Ganguli, R., Drewien, C. A., Anderson, M. T., Brinker, C. J., Gong, W., Guo, Y., Soyez, H., Dunn, B., Huang, M. H. & Zink, J. I. (1997) – Continous formation of supported cubic and hexagonal mesoporous films by sol-gel dip coating, *Nature, 389 (6649)*, 364-368.

[33] Brinker, C. J. & Dunphy, D. R. (2006). Morphological control of surfactant – templated metal oxide films, *Curr Opin Colloid Interface Sci, 11(2-3)*, 126-132.

[34] Brinker, C. J., Hurd, A. J., Schunk, P. R., Frye, G. C. & Ashley, C. S. (1992). Review of sol-gel thin film formation. *J Non-Cryst Solids*, *147(48)*, 424-436.

[35] Pena-Alonso, R., Rubio, F. & Rubio, J. (2005). The Role of γ – Aminopropyltriethoxysilane (γ-APS) on Thermal Stability of TEOS-PDMS Ormosils. *J Sol-Gel Sci Technol*, *36*, 77-85.

[36] Ravaine, D., Seminel, A., Charbouillot, V. & Vincens, M. (1986). A new family of organically-modified silicates prepared from gels. *J Non-Cryst Solids*, *82*, 210- 219.

[37] Touati, F., Gharbi, N. & Zarrouk, H. (1997). Synthesis of New Hybrid Organic-Inorganic Alumina Gels by the Sol-Gel Method. *J Sol-Gel Sci Technol*, *8*, 595-598.

[38] Mizukami, F., Kiyozumi, Y., Sano, T., Niwa, S. I., Toba, M. & Shin, S. (1998). Effect of Solvent Diols and Ligands on the Properties of Sol-Gel Alumina-Silicas. *J Sol- Gel Sci Technol, 13*, 1027-1031.

[39] Agren, P., Counter, J. & Laggner, P. (2000). A light and X-ray scattering study of the acid catalyzed silica synthesis in the presence of polyethylene glycol. *J Non-Cryst Solids*, *261*, 195-203.

[40] Higginbotham, C. P., Browner, R. F., Jenkins, J. D. & Rice, J. K. (2003). Dependence of drying technique on surface area and pore size for polyethylene glycol/tetramethoxysilicate hybrid gel. *Mat Lett, 57*, 3970-3975.

[41] Rao, V. A. & Kulkarni, M. M. (2002). Effect of glycerol additive on physical properties of hydrophobic silica aerogels. *Mat Chem Phys, 77*, 819-825.

[42] Shilova, O. A., Terasyuk, E.V., Shevchenko,V. V., Klimenko, N. S., Movchan, T. G., Hashkovsy, S. V. & Shilov, V. V. (2003). The Influence of Low- and High-Molecular Hydroxyl-Containing Additives on the Stability of Sol–Gel Tetraethoxysilane-Based Systems and on the Structure of Hybrid Organic–Inorganic *Coatings. Glass Phys Chem*, *29*, 378-389.

[43] Hinic, I., Stanisic, G. & Popovic, Z. (2003). Influence of the synthesis conditions on the photoluminescence of silica gels. *J Serb Chem Soc*, *68*, 953-959.

[44] Parashar,V. K., Raman,V. & Bahl, O. P.(1996). The role of N, N, Dimethylformamide and glycol in the preparation and properties of sol-gel derived silica. *J Mat Sci Lett*, *16*, 1403-1407.

[45] Lenza, R. F. S. & Vasconcelos, W. L. (2003). Study of the influence of some DCCAs on the structure of sol–gel silica membranes. *J Non-Cryst Solids*, *330*, 216-225.

[46] Schlottig, F., Textor, M., Georgi, U. & Roewer, G. (1999). Template Synthesis of SiO2 Nanostructures. *J Mater Sci Lett*, *18*, 599-601.

[47] Mondragon, M. A., Castano, V. M., Garcia, M. J. & Tellez, S. C. A. (1995). Vibrational analysis of $Si(OC_2H_5)_4$ and spectroscopic studies on the formation of glasses via silica gels. *Vibr. Spectrosc.*, *9*, 293-304.

[48] Wagh, P. B., Venkateswara Rao, A. & Haranath, D. (1998). Influence of molar ratios of precursor, solvent and water on physical properties of citric acid catalyzed TEOS silica aerogels. *Mat Chem Phys*, *53*, 41-47.

[49] Voulgaris, C., Amanatides, E., Mataras, D. & Rapakoulias, D. E. (2005). On the effect of the substrate pretreatment parameters on the composition and structure of plasma deposited SiO_2 thin films. *J Phys: Conference Series*, *10*, 206-209.

[50] Vasconcelos, D. C. L., Campos, W. R., Vasconcelos, V. & Vasconcelos, W. L. (2002). Influence of process parameters on the morphological evolution and fractal dimension of sol–gel colloidal silica particles. *Mat Sci Eng A*, *334*, 53-58.

[51] Lenza, R .F. S. & Vasconcelos, W. L. (2001). Preparation of silica by sol-gel method using formamide. *Mater Res*, *4*, 189-194.

[52] Jitchum, V., Chivin, S., Wongkasemjit, S. & Ishida, H. (2001). Synthesis of spirosilicates directly from silica and ethylene glycol/ethylene glycol derivatives. *Tetrahedr*, *57*, 3997- 4003.

[53] Maroni, V. A. & Epperson, S. J. (2001). An in situ infrared spectroscopic investigation of the pyrolysis of ethylene glycol encapsulated in silica sodalite. *Vibr Spectrosc*, *27*, 43-51.

[54] Stefanescu, M., Dippong, T., Stoia, M. & Stefanescu, O. (2008). Study on the obtaining of cobalt oxides by thermal decomposition of some complex combinations, undispersed and dispersed in SiO_2 matrix. *J Therm Anal Cal*, *94*, 389-393.

[55] Zboril, R., Mashlan, M., Barcova, K. & Vujtek, M. (2002). Thermally induced solid-state syntheses of γ- Fe_2O_3 nanoparticles and their transformation to α-Fe_2O_3 via ε-Fe_2O_3. *Hyperf Interact*, 139/140, 597-606.

[56] Nakamura, T., Yamada, Y. & Yano, K. (2006). Novel synthesis of highly monodispersed γ- Fe_2O_3/SiO_2 and ε-Fe_2O_3/SiO_2 nanocomposite spheres. *J Mat Chem*, *16*, 2417-2419.

[57] Cannas, C., Concas, G., Gatteschi, D., Falqui, A., Musinu, A., Piccaluga, G. & Sangregorio, C. (2001). Superparamagnetic behavior of gamma-Fe_2O_3 nanoparticles dispersed in a silica matrix. *Phys Chem Chem Phys*, *3*, 832-838.

[58] Caizer, C., Savii, C. & Popovici, M. (2003). Magnetic behavior of iron oxide nanoparticles dispersed in a silica matrix. *Mater Sci Eng B*, *97*, 129-134.

[59] Brazda, P., Niznansky, D., Rehspringer, J. L. & Vejpravova, J. P. (2009). Novel sol- gel method for preparation of high concentration ε-Fe_2O_3/SiO_2 nanocomposite. *J Sol-Gel Sci Technol*, *51*, 78-83.

[60] Chaneac, C., Tronc, E. & Jolivet, J. P. (1995). Thermal behavior of spinel iron oxide-silica composites. *Nanostruct Mater*, *6*, 715-718.

[61] Popovici, M., Gich, M., NIznansky, D., Roig, A., Savii, C., Cassas, L., Molins, E., Zaveta, K., Enache, C., Sort, J., de Brion, S., Chouteau, G. & Nogues, J. (2004).

Optimized Synthesis of the Elusive ε-Fe_2O_3 Phase via Sol–Gel *Chemistry. Chem Mater, 16*, 5542-5548.

[62] Kang, Y. S., Risbud, S., Rabolt, J. F. & Stroeve, P. (1996). Synthesis and characterization of nanometer – size Fe_3O_4 and γ- Fe_2O_3 particles. *Chem Mater, 8*, 2209-2211.

[63] Hyeon, T., Lee, S. S., Park, J., Chung, Y. & Na, H. B. (2001). Synthesis of Highly Crystalline and Monodisperse Maghemite Nanocrystallites without a Size-Selection Process. *J Am Chem Soc, 123*, 12798-12801.

[64] Ida, T., Tsuiki, H., Ueno, A., Tohji, K., Udagawa, Y., Iwai, K. & Sano, H. (1987). (1987). Characterization of iron. *J Catal, 106*, 428-439.

[65] Bentivegna, F., Nyvlt, M., Ferré, J., Jamet, J., Brun, P., Visnovsky, S. & Urban, R. (1999). Magnetically textured γ-Fe_2O_3 nanoparticles. *J Appl Phys, 85*, 2270-2279.

[66] Stefanescu, M., Stefanescu, O. & Stoia, M., Lazau, C. (2007). Thermal decomposition of some metal-organic precursors. Fe_2O_3 nanoparticles. J *Therm Anal Cal, 88*, 27-32.

[67] Niculescu, M., Vaszilcsin, N., Birzescu, M., Budrugeac, P. & Segal, E. (2001). Thermal and Structural Investigation of the Reaction between 1,2-propanediol and $Co(NO_3)_2 \cdot 6H_2O$. *J Therm Anal Cal, 65*, 881-889.

[68] Nakamoto, K. (1970). *Infrared spectra of inorganic and coordination compounds*. New York, John Wiley and Sons.

[69] Prasad, R. & Sulaxna, Kumar, A. (2005). Kinetics of thermal decomposition of iron(III) dicarboxylate complexes. *J Therm Anal Cal, 81*, 441-450.

[70] Powder diffraction file No.16-653, JCPDS-International Center for Diffraction Data *Swarthmore*, PA, (1993).

[71] Tronc, E., Chaneac, C. & Jolivet, J. P. (1998). Structural and Magnetic Characterization of ε-Fe_2O_3. *J Solid State Chem, 139*, 93-104.

[72] Jenkins, R. & Snyder, R. L. (1996). *Introduction to X-ray Powder Diffractometry*. New York, John Wiley & Sons Inc.

[73] Deka, S. & Joy, P. A. (2006). Characterization of nano-sized Ni,Zn ferrite powders synthesized by an autocombustion method. *Mater Chem Phys, 100*, 98-100.

[74] Anil Kumar, P. S., Shrotri, J. J. & Kulkarni, S. D. (1996). Low temperature synthesis of $Ni_{0.8}Zn_{0.2}Fe_2O_4$ powder and its characterization. *Mater Lett, 27*, 293-296.

[75] Wu, K. H., Chang, Y. C. & Chang, T. C. (2004). Effects of SiO_2 content and solution pH in raw materials on Ni-Zn ferrite magnetic properties. *J Magn Magn Mater, 283*, 380-384.

[76] Sedlář, M., Matějec, V., Grygar, T. & Kadlecová, J. (2000). Sol-gel processing and magnetic properties of nickel zinc ferrite thick films. *Ceramics International, 26*, 507-512.

[77] Arshed, M., Siddique, M., Anwar-ul-Islam, M., Butt, N.M., Abbas, T. & Ahmed, M. (1995). Site occupancy dependence of magnetism in Ni-Zn ferrites by the Mössabuer effect. Solid State *Commun, 93*, 599-602.

In: The Sol-Gel Process
Editor: Rachel E. Morris

ISBN 978-1-61761-321-0

Chapter 6

METHODS FOR PREVENTION, DIAGNOSIS AND TREATMENT ACHIEVED WITH THE AID OF SOL-GEL CHEMISTRY

Martin F. Desimone*[*], *Gisela S. Alvarez, Guillermo J. Copello, Maria L. Foglia, Maria V. Tuttolomondo and Luis E. Diaz
IQUIMEFA-CONICET, Facultad de Farmacia y Bioquímica,
Universidad de Buenos Aires, Buenos Aires, Argentina

ABSTRACT

The sol-gel process has several well-known advantages such as high purity precursors, homogeneity of the obtained material and especially the possibility of making hybrids and composite materials with new chemical and mechanical properties. This emerging field of material science has generated considerable and increasing interest from researchers of various disciplines. Recent advances due to the generation of multidisciplinary investigations have allowed real interesting advances in biology and medicine applications. The relation with the biomedical applications involves principally the development of methods for treatment, diagnosis, monitoring and control of biological systems. The first part of this chapter is dedicated to illustrate latest development in the immobilization of bactericide ions such as silver, copper or zinc and organic compounds with proven antimicrobial properties to obtain antimicrobial coatings, since microbial surface contamination control is an important tool for the prevention of infectious diseases. Secondly, due to the fact that early detection of infectious diseases is of great importance for the implementation of efficient treatments the entrapment of proteins, bacteria and parasites for the detection of infectious diseases is presented. The final section is an overview of the state of the art in materials for the treatment of human diseases including drug delivery systems and materials for tissue engineering achieved with the aid of sol-gel chemistry.

[*] Corresponding author: E-mail: desimone@ffyb.uba.ar

INTRODUCTION

The sol-gel process has several well-known advantages such as choice of high purity precursors (monomers or condensed species), homogeneity of the obtained material with different shapes (i.e., gels, films, particles) and especially the possibility of making hybrids and composite materials with new chemical and mechanical properties, conductivity and permeability (Iler, 1979; Brinker *et al.*, 1990). The general chemical pathways to prepare hybrid organic-inorganic nanocomposites and their applications have been summarized in an interesting review article by Sanchez *et al.*, (Sanchez *et al.*, 2005). This emerging field of material science has generated considerable and increasing interest from researchers of various disciplines (Sanchez, 2010). Recent advances due to the generation of multidisciplinary investigations have allowed real interesting advances in biology and medicine applications with improvements in the quantity and quality of commercialized products (Avnir *et al.*, 2006; Coradin *et al.*, 2006; Desimone *et al.*, 2009). Moreover, sol-gel chemistry is an interesting domain because it was early identified as an eco-friendly process compared to traditional synthesis routes to ceramics and glasses, thus improving sustainability in product development (Baccile *et al.*, 2009). Additionally, these materials can be employed in the development of new eco-friendly processes compared to conventional chemical ones (Meunier *et al.*, 2010). It is noteworthy to mention that there is also an increased interest in the development of biohybrid and biomimetic materials that exhibit improved structural and functional properties that offer valuable opportunities for applications that involve disciplines dealing with engineering, biotechnology, medicine, pharmacy, agriculture and nanotechnology among others (Ruiz-Hitzky *et al.*, 2010). The relation of sol-gel chemistry with biomedical applications involves principally the development of methods for prevention, treatment, diagnosis, monitoring and control of biological systems.

Over the last years, new strategies have been added to the challenge of control and prevention of microbial contamination. One was the development of materials with antimicrobial action that could be applied to critical surfaces to limit microbial colonization. The sol-gel chemistry has proved to be an efficient alternative to make antimicrobials coatings on various types of materials with proven activity against a wide spectrum of microorganisms (Gupta *et al.*, 2008). The first part of this chapter is dedicated to illustrate latest developments in the immobilization of bactericide ions such as silver, copper or zinc and organic compounds with proven antimicrobial properties to obtain antimicrobial coatings, since microbial surface contamination control aims to remove or reduce disease risk factors that is an important tool for the prevention of infectious diseases.

Secondly in this chapter, due to the fact that early detection of infectious diseases is of great importance for the implementation of efficient treatments, the entrapment of proteins, bacteria and parasites for the diagnosis of infectious diseases is presented. Considerable research work has been focused on the sol-gel encapsulation of labile biological materials with catalytic and recognition functions within robust polymer matrices (Gill *et al.*, 1998). The development of highly specific detection and quantification methods is a challenging task for the analytical and bioanalytical scientist (Kandimalla *et al.*, 2006). For this purpose, the molecular recognition with high selectivity and affinity is one of the most interesting properties of biological systems. In this sense, the detection of antibodies elicited in response to an infectious agent is usually employed for the diagnosis of a broad range of diseases, thanks to the high affinity and selectivity of the antigen-antibody recognition system. In order to detect serum antibodies against pathogens, various techniques employ immobilization of cells (i.e. bacteria or parasites) or antigens to different supports. Then the patient's serum is incubated in contact with fixed antigens and specific antibodies are recognized with species specific conjugated antibodies. The syntheses conditions that can be associated with sol-gel chemistry allow the obtaining of mineral phases under biocompatible conditions necessaries to maintain the antigen-antibody recognition sites. Soft chemistry based synthesis conditions involve low temperature, coexistence of inorganic, organic and biological systems, adequate pH and ionic strength. These synthesis strategies allowed the immobilization of biological systems without denaturation or deleterious conditions (Nassif *et al.*, 2002; Desimone *et al.*, 2006; Livage *et al.*, 2006; Amoura *et al.*, 2007; Coradin *et al.*, 2007; Perullini *et al.*, 2007). Moreover, the chemical strategies to design this materials must be carefully selected, since slight changes of experimental parameters (i.e., pH, temperature, concentrations) can lead to substantial modifications in the morphology and structure of the materials (Sanchez *et al.*, 2001; Soler-Illia *et al.*, 2002) and thus in the bioactivity (Alvarez *et al.*, 2007; Perullini *et al.*, 2008; Alvarez *et al.*, 2009).

The final section of the chapter is an overview of the state of the art in materials for the treatment of human diseases including drug delivery systems and materials for tissue engineering achieved with the aid of sol-gel chemistry. Several exciting biotechnological approaches have been developed to overcome the present difficulties related to organ failure or wound healing and drug delivery. The *in vivo* biological response of these sol-gel materials has been analyzed, showing benign local biocompatibility (Hudson *et al.*, 2008).

Cell encapsulation for tissue and organ replacement, as well as for continuous and controlled release of therapeutic agents to the host has a potentially significant future in medicine (Orive *et al.*, 2003a; Orive *et al.*, 2004a). It was demonstrated that specific properties of the materials may be beneficial to promote cell adhesion, proliferation and healing. The fact that the response of cells to biomaterials is critical in medical devices has attracted considerable attention and opened a land of opportunities for the designed construction of functional materials (Sanchez *et al.*, 2010).

The process by which a drug is delivered can significantly affect the pharmacological activity of the drug. Therefore, it is essential to develop suitable drug delivery systems (Orive *et al.*, 2003b; Orive *et al.*, 2004b). Moreover, silica particles have been studied as potential drug delivery systems because of their ordered pore structure, large surface areas and possibility to be easily modified (Gadre *et al.*, 2006). Although, significant challenges must be addressed, including the optimization of pore size, morphology, surface topography, elemental ratio, hydrophobic/hydrophilic balance, water activity, erosion rate and diffusion kinetics, among others, to understand the nature of the inherently complicated interactions between the inorganic, organic and biological components, which will determine the success of resulting material (Desimone *et al.*, 2009).

This chapter highlights, as mentioned above, recent successfully applications of sol-gel materials in the biomedical and pharmaceutical field. The huge diversity of materials shape that can be achieve depending on the desired function is presented with emphasis in the broad variety of applications of sol-gel chemistry.

Prevention

The purpose of every antimicrobial system is the prevention of cross-contamination. Massive contamination is expected to be controlled by means of disinfection procedures which utilizes the traditional chemical compound solutions, e.g. sodium hypochlorite, ethanol, iodine (Gardner *et al.*, 1986). Therefore the objective of the development of an antimicrobial system is not to prevent high number of microorganisms to interact with a patient but to avoid the interaction of few microbial cells or specific pathogens.

With this aim in mind in the last decade the benefits of sol-gel chemistry were used in the development of antimicrobial systems. In this field the main advantage of using sol-gel chemistry is the possibility to obtain materials with inorganic ions or organic molecules trapped within a metal or silicon polymeric network. By controlling monomers polymerization conditions materials of various shapes, sizes, tunable pore diameters or functionalities can be obtained (Brinker *et al.*, 1990). For the generation of antimicrobial systems the immobilization of several compounds with proven antimicrobial activity was described.

Reviewing literature and patents of sol-gel antimicrobial systems it is easy to notice that there are assorted ways to achieve this target. Thus, the strategies employed by authors can be classified in two general groups: the *ex-vivo* systems and the *in-vivo* systems.

The *ex-vivo* systems could be defined as the ones that accomplish their objective outside the patient. This group includes every antimicrobial system which activity is effective mainly in devices that could function as pathogen vehicles or reservoirs. Examples of these systems are SiO_2 micro and nanoparticles that have been endowed with antimicrobial activity by adding an active agent (Kawashita *et al.*, 2003; Li *et al.*, 2006). Literature also reported the obtaining of biocidal coatings in the form of thin films on different surfaces, such as glass, stainless steel, etc (Jeon *et al.*, 2003; Copello *et al.*, 2006; Evans *et al.*, 2007). There has been a great interest in the development of antimicrobial fabrics. The strategies of textile functionalization involve sol coatings, nanoparticles and doping among others (Haufe *et al.*, 2005; Haufe *et al.*, 2008; Jasiorski *et al.*, 2009; Wang *et al.*, 2009).

On the other hand, many developments have been proposed to achieve their action *in-vivo*. In this area there have been also studies of materials in the form of particles and surfaces (Gupta *et al.*, 2008). Silica xerogel powders containing gentamicin were obtained and proved its activity against intracellular pathogens. The silica xerogel particles were inoculated intraperitoneally to mice and no significant inflammatory reaction was observed (Munusamy *et al.*, 2009). Other studies focus on the prevention or treatment of periprosthetic infections. With this aim vancomycin was immobilized in thin sol–gel films for coating titanium alloy fracture plate materials (Radin *et al.*, 2007). Similarly, there has been a great interest in the obtaining of nitric oxide releasing coatings for medical implants (Frost *et al.*, 2005; Nablo *et al.*, 2005a).

From another point of view, immobilization can be carried out by many methods, e.g. entrapment, electrostatic interaction, adsorption, covalent binding, vapor deposition (Iler, 1979). IN respect of the antimicrobial chosen accounts, the chemical properties of the compound not only will set the antimicrobial spectrum of the final material but it will also limit the strategies for its entrapment and applications. By revising developments reported in the literature it could be analyzed the strategies of immobilization of the different biocidal agents as shown in the following section.

Immobilized metal ions as antimicrobial agents

This kind of antimicrobial was the first studied for their immobilization in a sol-gel network. This was probably due to the simplicity of the immobilization methodology. Most of the systems reported employed the technique of immobilization by entrapment, which usually is achieved by adding the agent during the polymerization process. This is the case for cations such as silver, copper and zinc (Gupta *et al.*, 2008). The most frequently immobilized cation is silver by far, a wide spectrum antimicrobial, which has been incorporated in films, micro and nanoparticles, fabrics, powders, etc (Bellantone *et al.*, 2000; Kawashita *et al.*, 2000; Jeon *et al.*, 2003; Kawashita *et al.*, 2003; Mahltig *et al.*, 2004; Mahltig *et al.*, 2007; Page *et al.*, 2007; Xing *et al.*, 2007; Zhang *et al.*, 2009). In addition this cation has been incorporated to sol-gel material in the form of nanoparticles, profiting from its property of silver slow release (Li *et al.*, 2006; Tomšič *et al.*, 2008; Jasiorski *et al.*, 2009). For silver, zinc and copper antibacterial activity have been known for many years but have been less studied for its incorporation in sol-gel materials. It can be mentioned the immobilization of zinc ions onto hydroxyapapatite coatings and copper in silicon oxide mesoporous materials among other studies (Trapalis *et al.*, 2003; Ren-Jei *et al.*, 2006; Wu *et al.*, 2009).

Spetial attention should be given to titanium derived materials. Depending on experimental conditions Ti itself could show antimicrobial activity or be biologically inactive. In recent studies on TiO_2 materials photoactive biocidal properties have been demonstrated (Evans *et al.*, 2007; Mahltig *et al.*, 2007; Ditta *et al.*, 2008). Still other reports on TiO_2 materials proved that Ti bioactivity was absent or insignificant compared to Ag activity (Page *et al.*, 2007; Zhang *et al.*, 2009). Other inorganic compounds with biocidal activity have been not so thoroughly studied as the ones mentioned before. For example literature reports the development of hydroxyapatite nanoparticles containing immobilized cerium ions, or the immobilization of boric acid in SiO_2 matrices (Bottcher *et al.*, 1999; Haufe *et al.*, 2005; Lin *et al.*, 2007).

Immobilized organic compounds as antimicrobial agents

Organic antimicrobials are probably the most promising biocidals since they are available in a wide range of synthetic and natural compounds which have specific antibacterial or antifungal spectrum. Moreover there are molecules against target pathogens that can be used with biomedical purposes. In addition organic molecules have the possibility of being immobilized by non-covalent or covalent methodologies.

The use of sol-gel chemistry avoids molecular stability drawbacks. Here is when the soft conditions of sol-gel chemistry allow the immobilization of unstable molecules, such as peroxides (Zeglinski *et al.*, 2007). Sol-gel precursors could be polymerized at low temperatures, desired pH and without caustic reagents.

Non-covalent immobilized compounds are usually encapsulated by entrapment. Among them it can be mentioned benzoic and sorbic acids which have proved to have both antibacterial and antifungal effect combined with boric acid (Bottcher *et al.*, 1999). Other kind of wide spectrum organic biocidals are the surface active agents such as quaternary ammonium salts or amphoteric compounds (Block, 1983; D'Aquino *et al.*, 1995). By embedding coatings with the quaternary ammonium salts cetyltrimethylammoniumbromide (CTAB) and octenidine, Mahtlig *et al.*, achieved the inhibition of growth of both fungi and bacteria (Mahltig *et al.*, 2004). CTAB has been also used by Wang *et al.*, together with octadecyl dimethyl benzyl ammonium chloride and ethylene-bis (octadecyl trimethyl ammonium chloride) in order to obtain an antibacterial cotton finish (Wang *et al.*, 2009). In other studies bactericidal surfaces with bioactivity against hospital and food-borne pathogens were developed by immobilizing the amphoteric surfactant dodecyl-di-(aminoethyl)-glycine within a silicon oxide network (Copello *et al.*, 2006; Copello *et al.*, 2008b).

The immobilization of several antimicrobials has lead to the development of systems with narrower spectra. As mentioned above, by being added within the gellification mixture gentamicin was immobilized in order to obtain a silicon oxide powder with the capability of targeting intracellular pathogens (Munusamy *et al.*, 2009). On the other hand, vancomycin has been immobilized within tetraethylorthosilicate (TEOS) multilayers in an attempt to control Gram-positive pathogens (Radin *et al.*, 2007). On a different approach various bioactive liquids such as primrose and perilla oil have been incorporated within modified silica coatings for antimicrobial and antiallergic effects (Haufe *et al.*, 2008).

When covalent immobilization is chosen, the mechanism of action of the biocidal had to be considered. Not all antimicrobials would conserver their activity if they are bounded to the sol-gel matrix. Therefore the compounds typically covalently immobilized are the ones that have surface activity or the ones in which the bioactive compound is generated *in situ* by matrix degradation. These systems usually involve one or more synthesis steps in order to incorporate the antimicrobial agent or proagent to the polymeric network.

A surface can be endowed with contact bacteria-killing capacity through chemical modification with tethered bactericidal functionalities. Quaternary ammonium compounds, known as surface active agents, have been covalently immobilized or even synthesized in sol-gel matrices in order to achieve non-leaching systems (Ferreira *et al.*, 2009). Saif *et al.*, and Marini *et al.*, synthesized bioactive materials containing quaternary ammonium compounds holding in their structures an alkoxysilane together with N-alkylated aniline (Marini *et al.*, 2007; Saif *et al.*, 2008). Li *et al.*, also remarked the advantages of modifying a surface with a permanent agent to retain the antibacterial activity once a releasable biocide, such as silver, is eluted from the matrix (Li *et al.*, 2006). This dual action system contains octadecyldimethyl(3-trimethoxysilylpropyl)-ammonium chloride (ODDMAC), which bears a quaternary ammonium salt structural unit coupled with a long hydrophobic alkyl chain (C18). ODDMAC is commercially available as a monomer and has also been used in the obtaining of permanent antimicrobial films based on surface-modified microfibrillated cellulose (Andresen *et al.*, 2007).

A different perspective of non-leaching antibacterial systems was taken in the finishing of cotton fabrics with fluoroalkoxysilane (FAS) among other compounds (Vilcnik *et al.*, 2009). In this study authors observed that the presence of FAS prevents or at least hinders the adhesion of bacteria and their consequent growth and the formation of a biofilm on the finished fabrics. The latter effect was called "passive antimicrobial activity" because no component of the fabric had intrinsic biocidal activity.

Alternatively, several studies reported developments for bioactive agent releasing materials. In those systems the initial material has the antimicrobial proagent anchored to a mother structure which degradation will free the active compound. Polymeric nitric oxide (NO) release represents one of the most studied alternatives for actively preventing bacterial adhesion. The cellular synthesis of NO mediates much of the antimicrobial activity of macrophages against some fungal, helminthic, protozoal and bacterial pathogens (Nathan *et al.*, 1991). Additionally, there have been reports suggesting that surface-localized NO release may be antibacterial (Nablo *et al.*, 2001). Thus, properties of NO are though to be beneficial for designing improved biomedical coatings. Nablo *et al.,* have extensively studied the synthesis of surfaces holding diazeniumdiolates as NO precursors (Nablo *et al.*, 2001; Nablo *et al.*, 2003; Nablo *et al.*, 2005a; Nablo *et al.*, 2005b). S-nitrosothiols (RSNOs) are compounds that are formed endogenously via the nitrosation of free thiols by reactive nitrogen species and are also promising NO donors that have been studied in the development of antimicrobial systems (Frost *et al.*, 2005; Riccio *et al.*, 2009).

Diagnosis

The immobilization of biological molecules including enzymes, antibodies, non-catalytic proteins, oligonucleotides as wll as animal, vegetal and bacterial cells in matrices prepared by the sol–gel process has made of this technique a potential tool for the development of biosensors and bioreactors (Livage, 2001; Jin *et al.*, 2002; Du *et al.*, 2003; Zhong *et al.*, 2004; Lee *et al.*, 2005); . Sol–gel chemistry allows the coating of different materials because of the mild conditions of the polymerization reaction, which include room temperature operation and polymerization at physiological compatible pH (Iler, 1979). When functional groups are required, an organically modified silicate (ORMOSIL) precursor containing the desired group can be used. One of the most used ORMOSILS is APTES (3- amino-propyltriethoxysilane), which provides the silica network with an amine terminal group that can be easily derivatized with, for example, glutaraldehyde. Magnetic particles can be included within silica micro and nanoparticles in order to further facilitate laboratory diagnostics (Yang *et al.*, 2004; Gupta *et al.*, 2005; Neuberger *et al.*, 2005)

Cell Immobilization

Immobilizing whole cells may be the method of choice when the enzymes, proteins or target molecules required for a specific determination are difficult or expensive to extract, very unstable, or if a series of enzymes are required for a certain reaction to occur. Livage *et al.,* have encapsulated parasites within sol-gel silica matrices. It was demonstrated that pore size was dependant on experimental conditions, and that they were large enough to allow large biomolecules diffusion. They could also see that cellular organization and integrity of the plasma membrane of entrapped parasites were preserved and they retained their antigenic activity. Two examples were described in this work, one with parasitic protozoa, *Leishmania*, and the other one with the cystic hydatid stage of tapeworm parasites, *Echinococcus granulosus* (Livage *et al.*, 1996). Based in this premises, G.J. Copello *et al.*, describe a suitable alternative method for the covalent attachment of parasites with a homogeneous distribution for the detection of specific antibodies against *T. cruzi* or *L. guyanenesis*. The TEOS–APTES network in the coating solution allows the homogeneous covalent attachment of parasites for antibody detection by immunofluorescence (IFA) and immunoperoxidase (IPA) assays, without matrix background. The sensitivity and specificity of the system is comparable to heat fixation systems for IFA and IPA detection. In addition, parasites remain attached even after several washing steps. The generation of these slides is simple and they do not require extreme storage conditions (Copello *et al.*, 2008a).

Whole cell immobilization can also be used in biosensors. A biosensor is a device for the detection of an analyte that combines a biological component with a physicochemical detector component. Gavlasova *et al.*, describe a whole cell biosensor for the analysis of polychlorinated biphenyl. They encapsulated cells of *Pseudomonas* sp. P2 into pre-polymerized tetramethoxysilane (TMOS) in the presence of dissolved biphenyl. The advantages of silica entrapped *Pseudomonas* sp. P2 cells in comparison with physically adsorbed cells on porous glasses are a shorter time for the immobilization process, accurate biphenyl dosage, and higher and controlled cell concentration. Bioassays based on *Pseudomonas* sp. P2 cells grown on glass spheres take 6 days, while assays based on silica entrapment are completed within a few hours after 1-day cultivation of free cells. (Gavlasova *et al.*, 2008).

Other bacterial strains such as genetically modified *E. coli* have been immobilized by Premkumar *et al.*, in oder to make sol gel luminescent biosensors in thick silicate films. They used TEOS in order to immobilize the bacteria onto glass slides. An important advantage of the solid silicate–*E. coli* sensing elements compared to suspended bacterial cultures is their ability to operate under continuous flow conditions (Premkumar *et al.*, 2002).

Not only bacteria and parasites can be immobilized in sol-gel matrices. Zolkov *et al.*, described a way of detecting polio virus using immobilized mammalian cells with methyltrimethoxysilane (MTMOS) and tetraethoxysilane (TEOS) as silica matrix components. The immobilized Buffalo Green Monkey kidney cell tissues were grown in a monolayer. Then the virus is inoculated and "plaques" (clear areas when cells are stained) can be seen where infective virus were present. Cell dye was entrapped in the silica matrix as well. With this new immobilization technique viral detection analysis time was reduced from 48 to 24 hours. (Zolkov *et al.*, 2009).

Biomolecules Immobilization

Antibodies

Immunoassays combining specific antigen–antibody (Ag–Ab) recognition for analytical purposes have been successfully applied to many fields including food industry, environmental protection and clinical control. Analyses of food-borne pathogens are of great importance in order to minimize the health risk for customers. Thus, very sensitive and rapid detection methods are required. Current conventional culture techniques are time consuming. Modern immunoassays and biochemical analysis also require pre-enrichment steps resulting in a turnaround time of at least 24 h. Biomagnetic separation (BMS) is a promising more rapid pre enrichment and detection method, without the need of using complex and expensive reagent and devices such as the ones used by molecular biology (Steingroewer *et al.*, 2005). Tuttolomondo *et al.*, describe a way of using immobilized antibodies to separate *Salmonella spp* from a mixed culture using magnetic silica beads obtained by inverse microemulsion by the aqueous sol-gel process and using APTES and glutaraldehyde as the antibody ligand. The beads were incubated with the desired sample and then separated by a permanent magnet (Tuttolomondo *et al.*, 2009).

Grant and Glass describe the development of a biosensor based on fluorescein-labeled D dimmer antibodies immobilized on the tip of an optical fiber by dip coating from a silica sol-gel solution. It is used for the detection of fibrinolytic products produced during lysis of "soft" blood clots. The presence of D dimer antigens above a threshold level is a clinical diagnose used to determine the presence of such occlusions following a stroke. When D dimer antigens combine with the antibodies, fluorescence intensity decreases. The D dimer antibodies remain viable for at least 4 weeks while encapsulated in the sol-gel network and it is being developed for use with other catheter-based microtools to treat stroke resulting from occlusion in the vascular system (Glass, 1999). Electrochemical immunoassays and immunosensors are drawing more attention in a wide range of uses, especially for determination of clinically important substances, due to their advantages such as simple pretreatment procedure, fast analytical time, precise and sensitive current measurements, and inexpensive and miniaturizable instrumentation. In the development of electrochemical immunosensing strategies, stability of the immobilized biocomponents and signal amplification of the immunoconjugates are two key factors. The latter has been achieved by using enzymes to label immunocomponents on transducer surface. Colloidal gold nanoparticles have been extensively used as an immobilizing matrix for retaining the bioactivity of macromolecules such as proteins, enzymes and antibodies and promoting the direct electron transfer of the immobilized proteins; combined with sol–gel technique to form a composite structure, Chen *et al.*, proposed a novel strategy to construct the reagentless immunosensor by encapsulating immunocomponent-adsorbed gold nanoparticles in titania sol-gel matrices. This composite architecture formed via a vapor deposition method provided a very hydrophilic interface for retaining the bioactivity and improved the stability of the immobilized enzyme labels and immunocomponents. It also promoted the electrical communication between redox sites of enzyme labels and sensing surface for direct electrochemical immunoassay of protein or antigen analytes. A novel immunosensor for human chorionic gonadotrophin (hCG) was thus been constructed based on this promising approach. hCG, a 37 kDa glycoprotein hormone, is an important diagnostic marker of pregnancy and one of the most important carbohydrate tumor markers. By utilization of sol-gel chemistry and nanotechnology, the immunosensor proposed in this work showed good sensitivity and stability, and could be prepared in mass-production (Chen *et al.*, 2006).

Liu *et al.*, using the same vapor deposition method, developed a highly hydrophilic and nontoxic colloidal silica nanoparticle/titania sol-gel composite membrane. With carcinoembryonic antigen (CEA) as a model antigen and encapsulation of carcinoembryonic antibody (anti-CEA) in the composite architecture, this membrane could be used for reagentless electrochemical immunoassay. The formation of an immunoconjugate by a simple one-step immunoreaction between CEA in sample solution and the immobilized anti-CEA introduced the change in the potential. Compared to the commonly applied immobilization methods, this strategy allowed higher loading of immobilized antibodies with a better retained immunoactivity. Analytical results of clinical samples show that the developed immunoassay is comparable with the enzyme-linked immunosorbent assays (ELISAs) method, implying a promising alternative approach for detecting CEA in the clinical diagnosis (Liu *et al.*, 2006).

Doping a sol-gel matrix with an antibody that retains its ability to bind free antigen from an aqueous solution was described by Turniansky et al., when evaluating a monoclonal anti-atrazine mouse antibody. The silica matrix was prepared from tetramethoxysilane by several methods. Atrazine was selected as a model compound for this study, within the framework of the development of immunochemical-based methods for monitoring pesticide residues and other organo-synthetic environmental contaminants. The combination of the properties of the sol-gel matrix (e.g., stability, inertness, high porosity, high surface area and optical clarity), together with the selectivity and sensitivity of the antibodies lead to the development of a novel group of immunosensors which could be used for purification, concentration and monitoring of a variety of residues from different sources *(Turniansky* et al., *1996).*

Lan *et al.*, immobilized anti-TNT antibodies by entrapment in a silica network. Porous sol-gel glasses with immobilized antibodies, therefore, represent solid-state probes for analytes of interest. The need for reliable, specific, and sensitive detectors for trinitrotoluene (TNT), a commonly used explosive, is escalating due to the need to detect landmines, soil, and groundwater contamination. Using antibody-antigen reactions to detect TNT has been demonstrated. The sol-gel approach is robust and rugged since the antibodies are immobilized in the interior of a porous silica material. The use of the sol-gel probes in both displacement and competitive assays, showed that the fluorescence signal as a function of TNT level exhibits the expected behaviour in standard immunoassays. The portability of such method is a considerable advantage over instrumental detection methods such as HPLC, GC, and MS whose field usage is limited (Lan *et al.*, 2000).

Enzymes

Due to the inherent low temperature process, sol–gel technology provides an attractive way for the immobilization of heat-sensitive biological entities (enzymes, proteins, and antibodies). This class of sol–gel silicate matrix possesses chemical inertness, physical rigidity, negligible swelling in aqueous solution, tuneable porosity, high photochemical and thermal stability, and optical transparency. These attractive features have led to intense research in this area mainly based on optical and electrochemical detection methods. In particular, most of the activities on sol–gel-derived electrochemical biosensors have focused on the design of amperometric enzyme electrodes. The analogous immobilization of enzymes for the design of conductometric biosensors is limited (Lee *et al.*, 2000).

Another way of immobilizing enzymes is through entrapment in the silica matrix. Voss *et al.*, have developed a way of entrapping horseradish peroxidase in silica particles in a one step process. The enzyme remained active for at least three months, demonstrating the biocompatibility of the entrapment process (Rebecca Voss *et al.*, 2007). This method could lead to the entrapment of other enzymes of clinical or environmental significance. The determination of urea nitrogen in human blood (BUN) is one of the most frequent analyses in a routine clinical laboratory. An abnormally high level of urea in blood is a strong indication of a kidney function impairment or failure. Some causes of decreased values for urea nitrogen include pregnancy, severe liver insufficiency, overhydration and malnutrition. The interest in the conductometric biosensor stems, to a large extent, from the relative simplicity (no reference electrode needed) and ease of fabrication of the biosensor. A very important factor in the development of enzyme-based biosensors is the immobilization of enzymes on some type of transducers. Lee *et al.*, reported a new conductometric biosensor by immobilizing urease in a silica sol–gel matrix derived from tetramethyl orthosilicate (TMOS) on a screen-printed platinum IDA electrode. The conductometric sensor was applied to the determination of urea levels in control human serum samples. (Lee *et al.*, 2000).

Nucleic Acids

Recent advances have accelerated the development of biosensors for the analysis of specific gene sequences. In this kind of biosensor, a DNA probe is immobilized on a transducer and the hybridization with the target DNA is monitored. Riccardi *et al.*, have developed an amperometric biosensor for the detection and genotyping of hepatits C virus, based on the immobilization of biotinilated DNA onto encapsulated streptavidin. The streptavidin was encapsulated in thin films siloxane–poly (propylene oxide) hybrids prepared by sol–gel method and deposited on the graphite electrode surface by dip-coating process. Biotinylated DNA probes were immobilized through streptavidin. (Riccardi *et al.*, 2006)

Other Molecules

Nivens *et al.*, have developed an organo-silica sol-gel membrane in a single layer format for pH measurement and multiple-layer format for both carbon dioxide and ammonia. These sensors employ hydroxypyrenetrisulfonic acid (HPTS) as the fluorescent pH indicator electrostatically attached to an APTES and TEOS containing sol-gel matrix. Gel preparation is simple. CO_2 sensors are also very fast and are highly stable, with shelf lives of at least 12 months storage in the laboratory. The key component of the gas sensor is the highly porous hydrophobic ORMOSIL matrix that allows the reversible diffusion of CO_2 gas. The porous hydrophobic sol-gel membrane allows dry sensor storage and reduces cross reactivity to pH when either CO_2 or NH_3 are examined. (Nivens *et al.*, 2002). Copper is an essential trace element for the catalytic activity of many enzymes involved in biological processes. Under normal conditions the urinary excretion of copper is very low. However, its value is increased in several pathologies related to abnormalities in copper metabolism. The most important is Wilson's disease, or hepatolenticular degeneration, that results in excessive accumulation of copper in the liver, brain, cornea and kidneys. Hence, the determination of copper in urine is of particular interest in clinical chemistry for purposes of diagnosis, for monitoring Wilson's patients under chelation therapy, for detection of environmental or occupational exposure, and for nutritional studies. Determination of copper in urine is usually performed by relatively expensive techniques, such as graphite furnace atomic absorption spectrometry and inductively coupled plasma mass spectrometry. There are some colorimetric methods for measurement of urine copper levels but they are time-consuming and usually require previous ashing and/or chelation followed by extraction. PAR (4-(2-pyridylazo) resorcinol) is a non-selective azo dye widely used as a colorimetric reagent for metal ions because it forms very stable water-soluble chelates with most transition metals. At pH values above 5.0, PAR forms a 2:1 complex with Cu (II). Sol-gel is a suitable host matrix for development of viable chemical sensors, since sensing agents can be readily entrapped in the porous glass matrix during the steps of the process, preserving their selectivity and activity. Jerónimo *et al.*, immobilised PAR in sol–gel thin films to obtain a Cu(II) optical sensor. The proposed sensor was then coupled to a continuous flow system to accomplish the direct determination of urinary copper, in order to achieve a simpler, cleaner, and biohazard protected handling of samples. (Jerónimo *et al.*, 2004)

Drug Delivery Systems

Drug delivery systems consist of a vehicle, usually a natural or synthetic polymer, where drugs or therapeutic components such as proteins are included to achieve their release from the material in a controlled manner. These systems can be designed either to release drugs at a sustained rate for a long period or to release them only in certain environments or conditions. The advantages of using delievery systems for medical therapy include greater efficacy, lower toxic effects, higher effectiveness in the treatment of chronic conditions and the need for fewer administrations which leads to improved patient compliance.

Nowadays there are a great variety of drug delivery systems which can be classified into three different groups:

Enteral drug delivery systems: they are introduced into the digestive tract through the mouth or rectum.

Topical drug delivery systems: the introduction of the drug to the body is obtained by absorption through the skin or mucus membranes. Skin patches are an example of topical drug delivery systems as well as sprays, inhalation aerosols, eye drops or creams.

Parenteral systems: they involve the injection or infusion of the formulation into the body and are specially used for chemotherapy.

According to Brannon-Peppas the ideal drug delivery system should be inert, biocompatible, mechanically strong, comfortable for the patient, capable of achieving high drug loading, safe from accidental release, simple to administer and remove, and easy to fabricate and sterilize (Branno-Peppas, 2002). A wide range of polymeric materials have been used for drug delivery systems like polyurethanes, polysiloxanes, polymethyl methacrylate, polyvinyl alcohol, polyethylene or polyvinyl pyrrolidone. In recent years, new materials have been tested for drug controlled release, which have the ability to degrade within the body. These new biodegradable materials include polymers such as polylactides, polyglycolides, poly (lactide-co-glycolides), polyanhydrides and polyorthoesters.

In addition, other experimental drug delivery systems have been evaluated including dendrimers, fullerenes, liposomes, liquid crystals, hydrogels or solid lipid nanoparticles.

The mechanisms by which the therapeutic agent is released from the vehicle depend on the material used for the construction of the delivery system and include:

Dissolution. The drug is released over time as the polymer dissolves in the gastro-intestinal tract

Diffusion. The release of the drug is controlled by its rate of diffusion out of the polymer.

Osmotic drug deliverys systems. The simplest design consists of an osmotically active core surrounded by a semipermeable membrane, with one or more orifices through which the therapeutic agent is delivered when water gets into the core.

Silica for Drug Delivery

Silica gels are promising materials for controlled drug delivery of therapeutic agents. Sol-gel chemistry is used to engineer and optimize silica vehicles for a wide variety of drug delivery formulations. The physical characteristics of the matrices such as density, pore size or area depend on the sol-gel reaction conditions. By controlling the rates of hydrolysis and condensation of the silica precursors, the physical characteristics of the material can be standardized as well as the drug release kinetics.

Silica vehicles offer a number of advantages over organic polymers as silica is highly inert and stable, hydrophilic and therefore easy to disperse in aqueous medium and suitable for many water insoluble therapeutic agents. Unlike organic polymers, silica is not subject to microbial attack and there is no porosity change occurring in these particles with media variations.

Hollow Silica Nanoparticles

Among the different silica formulations investigated at the moment, hollow silica nanoparticles (HSNP), are one of the most popular. They are hollow spherical particles with sizes below 200 nm and a core shell of silica. These hollow spheres provide some advantages over solid ones because of their low densities and large surface area. However, their main disadvantage is the low stability in aqueous suspensions as HSNPs tend to agglomerate. A common strategy used to stabilize nanoparticles is the use of surface active agents that are adsorbed to the particle's surface to form a physical barrier. Wen *et al.*, found that the adsorption of polyethylenimine on the surface of HSNP increased the repulsive energy among particles reducing their agglomeration (Wen *et al.*, 2007). while Bagwe *et al.*, indicated that the addition of appropriate ratios of inert functional groups (e.g. methyl phosphonate) to active functional groups (e.g. amino groups) to the surface of silica nanoparticles results in a highly negative zeta potential, which is necessary to keep the particles well dispersed (Bagwe *et al.*, 2006).

There are different methods to prepare nanoparticles with a hollow interior. One of them consists in coating the surfaces of colloidal particles (templates) with layers of silica followed by the removal of the colloidal templates. Porous hollow silica nanoparticles are usually fabricated in a suspension containing calcium carbonate as the sacrificial nanoscale template. Silica precursors, such as sodium silicate, are added into the suspension, which is then dried and calcinated creating a core of the template material coated with a porous silica shell. The template material is then removed (Hughes, 2005) (Chen *et al.*, 2004). Organic compounds, such as polystyrene (PS) (Caruso *et al.*, 1998) and polymethylmethacrylate (PMMA) (Stein, 2001) have also been employed as templates. The incorporation of drugs to hollow nanoparticles is commonly achieved by mixing the drugs with the carriers and then drying the mixture to coalesce the drug to the surface of the nanoparticles.

Micro and Nano-Sized Solid Silica Particles

In contrast to nonporous particles, these mesoporous silica particles offer the possibility of drug incorporation inside the pores or in the exterior surface for molecules with different characteristics. The mechanism of release in this case is a combination of diffusion and dissolution processes. Unger *et al.*, (Unger *et al.*, 2008) have shown that basic drugs were released in a sustained manner from silica particles whereas neutral drugs were released very quickly from silica xerogels. Yilmaz and Bengisu compared the release kinetics of pure anti-inflammatory drugs (ibuprofen and naproxen) with their entrapped counterparts and found that pure drugs exhibited linear release with time, while sol-gel silica entrapped drugs were released exhibiting a logarithmic time dependence starting with an initial burst effect followed by a gradual decrease (Elvan *et al.*, 2006)

Electrospray is a common technique used to prepare polymeric micro or nanoparticles and consists in the production of fine droplets from a solution of the pre-hydrolyzed sol-gel precursor by means of an electric field. Solid particles are then formed by solvent evaporation from the droplets travelling through the electrical field. Kortesuo *et al.*, obtained sol–gel-derived spray dried silica gel microspheres with a low specific surface, a smooth surface and without pores on the external surface and studied them as a carrier material for dexmedetomidine HCl and toremifene citrate (Kortesuo *et al.*, 2000a). According to Barbé *et al.*, (Barbé *et al.*, 2004), the main problems associated to this method are:

- The size of the particles is usually limited to the micrometer range
- Low yields because of particle adhesion to the reactor walls
- High temperatures applied during the drying step can damage temperature-sensitive molecules.
- Spray drying of less water-soluble drugs leads to low incorporation into de particles and to drug aggregation on their surface.

Another method used to obtain solid silica particles involves the preparation of water-in-oil microemulsions. Microemulsions are macroscopically homogeneous, and thermo-dynamically stable solutions containing at least three components: a polar phase like water, a nonpolar phase, usually oil, and a surfactant. The surfactant molecules form an interfacial film separating the polar water droplets from non-polar domains. The droplets of polar water are dispersed in a continuous oil phase which can be used as nanoreactors for the synthesis of silica nanoparticles. Alchoxides like tetramethylorthosilicate (TMOS) (Finnie *et al.*, 2007) or tetraethyl orthosilicate (TEOS) (Bagwe *et al.*, 2004) as well as aqueous precursors (sodium orthosilicate) (Gan *et al.*, 1996) have been used for the synthesis of silica particles through inverse microemulsion technique.

The Stöber synthesis of silica nanoparticles (Stöber *et al.*, 1968), on the other hand, utilizes the ammonia-catalyzed reaction of tetraethylorthosilicate (TEOS) with water in low-molecular-weight alcohols to obtain monodisperse, spherical silica nanoparticles that range in size from 5 to 2000 nm (Bogush *et al.*, 1988; Nozawa *et al.*, 2004).

Silica Xerogels for Implants

The gel drying process consists in the removal of water from the gel system, with simultaneous collapse of the gel structure, under conditions of constant temperature, pressure and humidity. The result is a dense solid with a great surface area and a very small pore size called xerogel. In contrast, when solvent removal occurs under hypercritical conditions, the network does not shrink and a highly porous, low-density material called aerogel is produced.

Xerogels have been used for drug delivery in animal implants. Kortesuo *et al.*, examined the controlled delivery of the drug toremifene from silica xerogel implants which had an approximated diameter of 4.7 mm and weighted about 15 mg (Kortesuo *et al.*, 2000b). Furthermore, Czarnobaj utilized the sol-gel method to synthesize different forms of xerogel matrices in order to obtain drug delivery systems and studied the influence of synthesis conditions and drug solubility on the drug release profile. Two model compounds, diclofenac diethylamine (DD), a water-soluble drug and ibuprofen (IB), a water insoluble drug, were used for the kinetics studies and it was found that, in all the cases studied, the released amount of DD was higher and the released time was shorter compared with IB for the same type of matrices (Czarnobaj, 2008). Another work that can be mentioned on the same thematic is the one by Roveri *et al.*, where they investigated a sol–gel derived silica matrix as a delivery system for the prolonged release of different molecular weight heparins and found that the surface area of the matrix was a determinant parameter affecting drug release kinetic as xerogels were obtained by embedding heparins into matrices (Roveri *et al.*, 2005).

Bioactive agents can be incorporated into silica xerogels either by adsorbing the drug on the surface of the silica xerogel, as in the previously mentioned work, or by adding the drug during the sol-gel manufacturing process (Ahola *et al.*, 2001). The release of a drug from these matrices occurs according to a combined process of diffusion through solvent filled capillarity channels and dissolution of the matrices. Therefore chemical and structural characteristics of the silica xerogel strongly affect drug release kinetics. In this sense, the chemical nature of the silica pore walls can be modified to modulate the confinement and delivery kinetics of the drug (Balas *et al.*, 2006).

Silica Aerogels

Supercritical drying is the key step in making aerogels. In this final step, the liquid within the gel is removed, leaving only the linked silica network. The process can be performed in two different ways: venting the ethanol above its critical point or by prior solvent exchange with CO_2 followed by supercritical venting.

While standard aerogels are prepared from precursors such as tetramethoxysilane (TMOS) or tetraethyl orthosilicate (TEOS), hydrophobic aerogels are prepared using the previously mentioned agents and one of the following organosilane co-precursors: methytrimethoxysilane (MTMS), ethyltrimethoxysilane (ETMS), or propyltrimeth-oxysilane

(PTMS) (Anderson *et al.*). Another way to obtain hydrophobic matrices is to modified the surfaces of monolithic silica aerogels using surface modification agents like hexamethyldisilazane (HMDS) (Kartal *et al.*). CO_2 is then used as an effective solvent to deliver the therapheutic compounds to the inner surfaces of the aerogels.

There are two types of silica aerogels and therefore, two mechanisms of drug release from them (Smirnova *et al.*, 2004):

- Hydrophilic aerogels which collapse in the aqueous media promoting the dissolution on the compound. In this case, therapeutic agents are adsorbed on the aerogels surface and they show a fast release kinetic.
- Hydrophobic aerogels which are more stable and have a slower drug release controlled by the diffusion of the drug from the pores.

Some works have been published concerning silica aerogels as oral drug delivery systems. It has been demonstrated that the dissolution rate of poorly water soluble drugs can be significantly enhanced by adsorption on highly porous hydrophilic silica aerogels (Smirnova *et al.*, 2004).

On the other hand, Guenther *et al.*, studied the implementation of drug loaded silica aerogels as dermal delivery systems. The authors studied several formulations of dithranol and showed that this drug adsorbed on hydrophilic silica aerogels exhibited superior penetration behaviour into human stratum corneum compared to that of the standard ointment which is the drug in white soft paraffin (Guenther *et al.*, 2008).

Silica Coated Liposomes

Silica nanoparticle-liposomes hybrid materials are promising as delivery vehicles for proteins and peptides. A hybrid silica-liposome system containing insulin has been developed. The formulation strategy is based on using insulin-loaded liposomes as a template for the deposition of inert silica nanoparticles which protect liposomes against degradation by digestive enzymes (Mohanraj *et al.*). Liposomes loaded with insulin having an outer layer of silica have also been prepared to enhance the stability and improve the encapsulation efficiency within the liposome by inhibiting leakage of insulin (Dwivedi *et al.*, 2010). Begú *et al.*, described the synthesis and characterization of hybrid silica nanospheres in which the trapped unilamellar liposomes maintain their fundamental properties and called these systems liposils. These silica-based particles were obtained via liposome templating where the non-porous amorphous silica cladding protected the liposomes used for drug storage and delivery (Bégu *et al.*, 2007).

Targeted Drug Delivery

The concept of targeted drug delivery implies selective accumulation of a drug within the affected tissue after administration, with the minimal possible side effects on other organs and tissues.

Magnetic silica drug delivery systems

The basic premise is that therapeutic agents are attached to, or encapsulated within, a magnetic micro- or nanoparticle. These particles may have magnetic cores with a polymer or metal coating or may consist of porous polymers that contain magnetic nanoparticles precipitated within the pores. Magnetic fields are focused over the target site and the forces on the particles as they enter the field allow them to be captured and extravasated at the target. However, while this may be effective for targets close to the body's surface, as the magnetic field strength falls off rapidly with distance, deeper sites within the body are still difficult to access (McBain *et al.*, 2008). Some groups have recently proposed a way around this problem by implanting magnets near the target site, within the body (Yellen *et al.*, 2005).

Magnetic hollow silica nanocomposites have been successfully synthesized using a coating of Fe_3O_4 magnetic nanoparticles and silica on nanosized spherical and nanoneedle-like forms (Zhou *et al.*, 2005). Iron-loaded hollow silica microcapsules have also been obtained by several cycles of metal impregnation under vacuum over hollow silica microcapsules (Arruebo *et al.*, 2006). Furthermore, Kim *et al.*, synthesized monodisperse magnetite (Fe_3O_4) nanocrystals embedded in mesoporous silica spheres and used them to study the uptake and controlled release of drugs (Kim *et al.*, 2005) as well as Zhao and co-workers who studied the release of ibuprofen from particles consisting of an inner magnetic core and an outer mesoporous silica shell (Zhao *et al.*, 2005). Despite the many works that exist on magnetic silica nanoparticle synthesis and their potential as target drug delivery systems, there is still little information on their in vivo behaviour.

Surface-functionalized silica particles

Active targeting of chemotherapy drugs is of great interest in the medical field, especially when treating different types of cancer. Using the expression of specific recognition markers by the tumour, bioconjugation of the nanoparticles with antibodies directed against such tumour markers improves localisation of the particles specifically at the cancer cells. Two tumour markers most commonly used as targets for directed therapy are the folic acid receptor and the EGFR-2 (erbB2/HER2), as their implication in tumorigenesis results in their overexpression on the cancer cell surface. (Van Vlerken *et al.*, 2006). Different groups have been working with folic silica functionalized- particles. Folic acid, in principle able to be recognized from specific tissues, has been covalently coupled on the external function of mesoporous silica particles thanks to the presence of aminopropyl groups as a potential tool for drug targeting (Pasqua *et al.*, 2007) Rosenholm *et al.*, showed that the total number of folic silica functionalized particles internalized by cancer cells was about an order of magnitude higher than the total number of particles internalized by normal cells, a difference high enough to be of significant biological importance. (Rosenholm *et al.*, 2008).

Scientists have also made use of Her-2 overexpression in tumors for the synthesis of functionalized-silica particles. Tsai *et al.*, conjugated anti-HER2/neu mAb (monoclonal antibody) to green fluorescent dye loaded mesoporous silica nanoparticles through a polyethylene glycol spacer and examined their targeting properties toward HER2/neu over-expressing breast cancer cells and their internalization into the cells (Tsai *et al.*, 2009).

Silica biodegradability and toxicity

In vitro and *in vivo* studies have been done to know the biodegradability and toxicity of silica matrices. Wilson *et al.*, performed a series of both *in vitro* and *in vivo* tests to evaluate the toxicity of matrices containing silica and showed silica to be non-toxic and biocompatible (Wilson *et al.*, 1981). Palumbo *et al.*, showed that implanted sol-gel glasses did not elicited a marked inflammatory response by polymorphonuclear leukocytes.(Palumbo *et al.*, 1997) while Gerritsen *et al.*, inserted sol–gel coated discs subcutaneously in the back of rabbits and, after 4 and 12 weeks, processed the implants histologically finding only a thin fibrous capsule containing a few inflammatory cells at the tissue implant interface (Gerritsen *et al.*, 2000).

According to Finnie *et al.*, it is of a great importance the study of the fate of silica particles in the body after the delivery of a drug in order to achieve the acceptance of regulatory authorities. Nanoparticles taken up into cells must degrade effectively to avoid the toxicity risk associated with accumulation of non-degradable fine particles. They have conducted in vitro tests for silica degradability and found initial linear dissolution rates which increased with sample surface area and have also showed that the addition of serum proteins to the dissolution media acted to slow dissolution rates by 20–30%, suggesting a slower degradation *in vivo* (Finnie *et al.*, 2009). Furthermore, Lai *et al.*, implanted bioactive glass granules in rabbits in order to determine the pathway of the silicon released from bioactive glass *in vivo*. No significant increase in silicon was found in any of the organs including the kidney, and they concluded that the silica gel was harmlessly excreted in a soluble form through the urine. The metabolic pathway for silica biodegradation includes erosion of the nanoparticle surface with formation of plasma-soluble silicic acid and its subsequent excretion by the kidney (Lai *et al.*, 2002).

On the other hand, Nishimori *et al.*, investigated the relationship between particle size and toxicity using silica particles with diameters of 70, 300 and 1000 nm and found that small silica particles of 70 nm induced liver injury at 30 mg/kg body weight, while particles of 300 and 1000 nm had no effect even at 100 mg/kg when administered intravenously (Nishimori *et al.*, 2009).

APPLICATIONS OF THE SOL-GEL TECHNOLOGY IN THE TISSUE ENGINEERING FIELD

Sol-gel technology has proved to be suitable for the elaboration of acellularized bioceramics such as apatites or bioglasses (Vallet-Regi *et al.*, 2008), mainly used in the bone engineering field. However, immobilization in inorganic matrices is possible for bacteria and yeasts but for mammalian cells hybrid sol-gel materials are preferred.

Bone Tissue Engineering

Bone regeneration has attracted special attention in the recent years due to the requirements of new technologies to treat loss of tissue caused by trauma, disease or tumor extraction. In the field of bone engineering the material chosen should provide sufficient mechanical strength while enabling osteogenic cells to survive and differentiate into osteoblasts and osteocites. Ideally the scaffold should consist of a 3D interconnected pore network with a pore size of at least 100 μm, in order to allow cell penetration and proper vascularization of the implanted tissue. An efficient and hierarchically 3D mesoporous-giantporous composite obtained by a combination of the sol-gel, polymer templating and rapid prototyping techniques with a gantry robotic deposition apparatus was reported (Yun *et al.*, 2007). The obtained scaffold proved to be efficient in terms of biocompatibility and mecanichal properties.

The chosen material for osteogenic cells immobilization should be biocompatible, have a controlled biodegradation, allow nutrient diffusion and not less important, promote cell adhesion. Osteoblasts are an anchorage-dependent cell population that relies on support for survival. Proteins present in the extracellular matrix (ECM) such as vitronectin, fibronectin and type I collagen interact with integrins present in the cell membrane and allow cell adhesion which is of vital importance for future events such as osteoblast survival, proliferation, differentiation, and matrix mineralization, as well as bone formation (LeBaron *et al.*, 2000).

Silicon is an essential element in connective tissue, especially in bone and cartilage, being its primary effect the formation of an ECM composed of collagen and glycosaminoglycans. Silica matrices obtained by the sol-gel process are chemically inert, optically transparent, resistant to microbial attack and thermally stable(Meunier *et al.*, 2010). Moreover, silica hydrogels could be obtained under very mild conditions, maintaining the viability of the immobilized cells during the encapsulation process (Desimone *et al.*, 2010).

Gil *et al.*, evaluated the in vivo behaviour of a sol–gel glass and a glass-ceramic during critical diaphyseal bone defects healing (Gil-Albarova *et al.*, 2005). The morphometric study showed minimum degradation or resorption of glass-ceramic cylinders in comparison to sol-gel cylinders, which showed abundant fragmentation and surface resorption. Due to the better mechanical properties of glass-ceramics in comparison to sol-gel cylinders it was suggested their application in critical bone defects locations of transmission forces or load bearing, whereas sol–gel glass cylinders could be useful in locations where a quick resorption should be preferable, considering the possibility of serving as drug or cells vehicle for both of them.

Type I collagen, which is the most abundant protein in vertebrates, is present in many organs and is a major constituent of bone, tendons, ligaments, and skin. The study of naturally occurring silica – collagen based biocomposites, showed that high collagen contents give origin to the higher flexibility observed in Hexactinellida skeletons, which represent an example of a biological material in which a collagenous organic matrix serves as a scaffold for the deposition of a reinforcing silica mineral phase (Ehrlich *et al.*, 2008).

Bioactive silica-collagen composite xerogels modified by calcium phosphate phases have been presented for bone replacement (Heinemann *et al.*, 2009). Silicon-collagen matrices have proved to be efficient in stimulating bone remodeling by demonstrating differentiation of human monocytes into osteoclasts-like cells. Furthermore, the incorporation of calcium phosphate cement allowed a better control of the material's mechanical properties.

Dermal substitutes

As life expectancy increases, the need of treatments for chronic wounds has increased in the past years. Dermal substitutes should be able to provide protection to the wound from infection and fluid loss, a stable and biodegradable extracellular matrix able to stimulate neodermal tissue synthesis without arousing an immune response, a biocompatible matrix to enable dermal cells proliferation and ease of handling and resistance to shear forces (van der Veen *et al.*, 2009)

The use of fibroblast encapsulation in hybrid silica-collagen hydrogels for chronic ulcer treatment has been reported (Desimone *et al.*, 2010). Hybrid silica-collagen scaffolds were obtained by the simultaneous polymerization of aqueous silicates and self-assembly of collagen triple helix in the presence of living human dermal fibroblasts.

Recently it was reported the immobilization 3T3 mouse fibroblasts and CRL-epithelial cells in a silica matrix. Cells survived encapsulation conditions inside gels in a wet state, supporting the non toxicity of the material. Furthermore, the cell clumping performed proved to be an efficient method to control cell attachment and protect the encapsulated cells, while preserving its viability (Nieto *et al.*, 2009).

Pancreatic tissue substitutes

Insulin-dependent diabetes is a condition which requires daily blood glucose monitoring and insulin injections. A pancreatic tissue substitute should provide the patient with a more physiological and less invasive treatment of the disease. The encapsulation of murine islet cells as the first mammalian material in silica ceramics was also reported. Results showed that capsules allowed diffusion of insulin and cytokines, while preventing the passage of antibodies. Furthermore, after one month of implantation no fibrosis was observed and insulin secretion was maintained (Peterson *et al.*, 1998).

Pancreatic rat islets were encapsulated by a siliceous layer deposited on its surface upon reaction with gaseous siliceous precursors(Boninsegna *et al.*, 2003). Viability as well as islets dimensions were conserved. Dynamic perfusion experiments with glucose stimulation were carried out in both encapsulated and non-encapsulated islets. Results showed that siliceous deposits are biocompatible and do not prevent islet feeding or maintenance, while providing immunological protection. It was also reported the encapsulation of rat islets in an alginate/aminopropyl-silicate/alginate (Alg/AS/Alg) membrane prepared on Ca-alginate gel beads by a sol–gel process. A molecular permeability study for glucose, insulin, BSA and g-globulin showed that the membrane had an optimal balance between permeability for low-molecular-weight substances and molecular weight cut-off (Sakai *et al.*, 2002).

Transplantation of the encapsulated rat islets into mice with streptozotocin-induced diabetes showed maintenance periods of normoglycemia from 2 to 15 weeks showing the feasibility of the Alg/AS/Alg membrane for a microcapsule-shaped bioartificial pancreas.

One major drawback in tissue substitutes engineering is the elicited immune response when the host´s immune system detects the implanted tissue, leading to rejection. Autologous cells are immunologically accepted, but as an insulin dependent patient has no β cells, then non β cells should be genetically engineered to be capable of insulin secretion. An hybrid pancreatic tissue substitute consisting of recombinant insulin-secreting cells and glucose-responsive material was reported (Cheng *et al.*, 2004). The glucose-responsive material provided better control of the insulin secretion profile.

Liver tissue substitutes

Hepatocyte-based bioartificial livers should be able to provide support for liver patients in the clinic or be useful as a bioreactor for drug metabolites synthesis. The resulting bioartificial liver will need to contain a high cell density culture of hepatocytes, maintaining their viability and preventing loss of liver-specific functions in cultured cells for a large period of time. Metabolic requirements of hepatocytes are particularly high, which implies the necessity to develop matrices with optimum nutrient and oxygen diffusion rates (Jasmund *et al.*, 2002).

Muraca *et al.*,investigated liver-specific metabolic activities in silica-overlaid hepatocytes obtained by the Biosil method. Bilirubin conjugation, ammonia removal, urea synthesis, and diazepam metabolism were evaluated. Results showed maintained cell viability and increased ammonia removal rate and diazepam metabolism after 48 hr, whereas urea synthesis was unaffected. The enhancement of metabolic activities 48 hr after the entrapment suggests that favorable changes in the hepatocyte microenvironment may have occurred (Muraca *et al.*, 2002). Tissue-derived buffalo green monkey kidney cells, were successfully grown on hydrophobically modified silica sol–gel thin films and on such films in which poly-L-lysine (PLL) was entrapped in order to affect surface positive charge density (Zolkov *et al.*, 2004).

ACKNOWLEDGMENTS

G.S.A. and M.L.F. are grateful for their doctoral fellowship granted by the National Research Council (CONICET) and the University of Buenos Aires, respectively. The authors would like to acknowledge the support of grants from the University of Buenos Aires UBACYT B049 (L.E.D.) and B407 (M.F.D.) and Agencia Nacional de Investigaciones Científicas y Técnicas BID 1728/OC-AR PICT 32310 (M.F.D.).

REFERENCES

[1] Ahola, MS; Säilynoja, ES; Raitavuo, MH; Vaahtio, MM; Salonen, JI and Yli-Urpo, AUO. In vitro release of heparin from silica xerogels. *Biomaterials*, 2001, 22, 2163-2170.

[2] Alvarez, GS; Desimone, MF and Diaz, LE. Immobilization of bacteria in silica matrices using citric acid in the sol-gel process. *Appl Microbiol Biotechnol*, 2007, 73, 1059-1064.

[3] Alvarez, GS; Foglia, ML; Copello, GJ; Desimone, MF and Diaz, LE. Effect of various parameters on viability and growth of bacteria immobilized in sol-gel-derived silica matrices. *Appl Microbiol Biotechnol*, 2009, 82, 639-646.

[4] Amoura, M; Nassif, N; Roux, C; Livage, J and Coradin, T. Sol-gel encapsulation of cells is not limited to silica: long-term viability of bacteria in alumina matrices. *Chem Commun* (Camb), 2007,, 4015-4017.

[5] Anderson, A; Carroll, M; Green, E; Melville, J; Bono, M. Hydrophobic silica aerogels prepared via rapid supercritical extraction. *Journal of Sol-Gel Science and Technology* 53, 199-207.

[6] Andresen, M; Stenstad, P; Moretro, T; Langsrud, S; Syverud, K; Johansson, LS; Stenius, P. Nonleaching antimicrobial films prepared from surface-modified microfibrillated cellulose. *Biomacromolecules*, 2007, 8, 2149-2155.

[7] Arruebo, M; GalÃ¡n, M; NavascuÃ©s, N; TÃ©llez, C; Marquina, C; Ibarra, MR;

SantamarÃ-a, Js. Development of Magnetic Nanostructured Silica-Based Materials as Potential Vectors for Drug-Delivery Applications. *Chemistry of Materials*, 2006, 18, 1911-1919.

[8] Avnir, D; Coradin, T; Lev, O; Livage, J. Recent bio-applications of sol-gel materials. *Journal of Materials Chemistry*, 2006, 16, 1013-1030.

[9] Baccile, N; Babonneau, F; Thomas, B; Coradin, T. Introducing ecodesign in silica sol-gel materials. *Journal of Materials Chemistry*, 2009, 19, 8537-8559.

[10] Bagwe, RP; Yang, C; Hilliard, LR; Tan, W. Optimization of Dye-Doped Silica Nanoparticles Prepared Using a Reverse Microemulsion Method. *Langmuir,* 2004, 20, 8336-8342.

[11] Bagwe, RP; Hilliard, LR; Tan, W. Surface Modification of Silica Nanoparticles to Reduce Aggregation and Nonspecific Binding. *Langmuir,* 2006, 22, 4357-4362.

[12] Balas, F; Manzano, M; Horcajada, P; Vallet-RegÃ-, Ma. Confinement and Controlled Release of Bisphosphonates on Ordered Mesoporous Silica-Based Materials. *Journal of the American Chemical Society*, 2006, 128, 8116-8117.

[13] Barbé, C; Bartlett, J; Kong, L; Finnie, K; Lin, HQ; Larkin, M; Calleja, S; Bush, A; Calleja, G. (2004) *Silica Particles: A Novel Drug-Delivery System*. In, Vol. 16, p. 1959-1966.

[14] Bégu, S; Pouëssel, AA; Lerner, DA; Tourné-Péteilh, C; Devoisselle, JM. Liposil, a promising composite material for drug storage and release. *Journal of Controlled Release*, 2007, 118, 1-6.

[15] Bellantone, M; Coleman, NJ; Hench, LL. Bacteriostatic action of a novel four-component bioactive glass. *J Biomed Mat Res A*, 2000, 51, 484-490.

[16] Block, S. Quaternary ammonium compounds in disinfectants and antiseptics. In: *Surface active agents*, 1983, . Lea & Febiger, Philadelphia, USA, p. 263-273.

[17] Bogush, GH; Tracy, MA; Zukoski Iv, CF. Preparation of monodisperse silica particles: Control of size and mass fraction. *Journal of Non-Crystalline Solids*, 1988, 104, 95-106.

[18] Boninsegna, S; Bosetti, P; Carturan, G; Dellagiacoma, G; Dal Monte, R; Rossi, M. Encapsulation of individual pancreatic islets by sol-gel SiO2:: A novel procedure for perspective cellular grafts. *Journal of Biotechnology*, 2003, 100, 277-286.

[19] Bottcher, H; Jagota, C; Trepte, J; Kallies, KH; Haufe, H. Sol-gel composite films with controlled release of biocides. *J Control Release*, 1999, 60, 57-65.

[20] Branno-Peppas, L. *Polymers in Controlled Release*. John Wiley & Sons, 2002, .

[21] Brinker, C; Scherer, G. *Sol-Gel science*, 1990, . Academic Press, San Diego, USA.

[22] Caruso, F; Caruso, RA; ouml; hwald, H. (1998) *Nanoengineering of Inorganic and Hybrid Hollow Spheres by Colloidal Templating*. In, Vol. 282, p. 1111-1114.

[23] Chen, J-F; Ding, H-M; Wang, J-X; Shao, L. Preparation and characterization of porous hollow silica nanoparticles for drug delivery application. *Biomaterials*, 2004, 25, 723-727.

[24] Chen, J; Tang, J; Yan, F; Ju, H. A gold nanoparticles/sol-gel composite architecture for encapsulation of immunoconjugate for reagentless electrochemical immunoassay. *Biomaterials*, 2006, 27, 2313-2321.

[25] Cheng, SY; Gross, J; Sambanis, A. Hybrid pancreatic tissue substitute consisting of recombinant insulin-secreting cells and glucose-responsive material. *Biotechnology and Bioengineering*, 2004, 87, 863-873.

[26] Copello, GJ; Teves, S; Degrossi, J; D'Aquino, M; Desimone, MF; Diaz, LE. Antimicrobial activity on glass materials subject to disinfectant xerogel coating. *J Ind Microbiol Biotechnol*, 2006, 33, 343-348.

[27] Copello, GJ; De Marzi, MC; Desimone, MF; Malchiodi, EL; Díaz, LE. Antibody detection employing sol-gel immobilized parasites. *Journal of Immunological Methods*, 2008a, 335, 65-70.

[28] Copello, GJ; Teves, S; Degrossi, J; D'Aquino, M; Desimone, MF; Díaz, LE. Proving the antimicrobial spectrum of an amphoteric surfactant-sol-gel coating: a food-borne pathogen study. *J Ind Microbiol Biotechnol*, 2008b, 35, 1041-1046.

[29] Coradin, T; Boissiere, M; Livage, J. Sol-gel chemistry in medicinal science. *Curr Med Chem*, 2006, 13, 99-108.

[30] Coradin, T; Livage, J. Aqueous silicates in biological sol-gel applications: new perspectives for old precursors. *Acc Chem Res*, 2007, 40, 819-826.

[31] Czarnobaj, K. (2008) Preparation and Characterization of Silica Xerogels as Carriers for Drugs. In, Vol. 15. *Informa Healthcare*, 485-492.

[32] D'Aquino, M; Rezk, R. Desinfección: Desinfectantes, Desinfestantes, Limpieza. In: Características de los agentes químicos desinfectantes, 1995, . E.U.DE.B.A, Buenos Aires, *ARG,* 67-76.

[33] Desimone, MF; De Marzi, MC; Copello, GJ; Fernández, MM; Pieckenstain, FL; Malchiodi, EL; Diaz, LE. Production of recombinant proteins by sol-gel immobilized Escherichia coli. *Enzyme and Microbial Technology*, 2006, 40, 168-171.

[34] Desimone, MF; Alvarez, GS; Foglia, ML; Diaz, LE. Development of sol-gel hybrid materials for whole cell immobilization. *Recent Pat Biotechnol*, 2009, 3, 55-60.

[35] Desimone, MF; Hélary, C; Mosser, G; Giraud-Guille, M-M; Livage, J; Coradin, T. Fibroblast encapsulation in hybrid silica–collagen hydrogels. *Journal of Materials Chemistry*, 2010, 20, 666-668.

[36] Ditta, IB; Steele, A; Liptrot, C; Tobin, J; Tyler, H; Yates, HM; Sheel, DW; Foster, HA. Photocatalytic antimicrobial activity of thin surface films of TiO_2, CuO and TiO_2/CuO dual layers on Escherichia coli and bacteriophage T4. *Applied Microbiology and Biotechnology*, 2008, 79, 127-133.

[37] Du, D; Yan, F; Liu, S; Ju, H. Immunological assay for carbohydrate antigen 19-9 using an electrochemical immunosensor and antigen immobilization in titania sol-gel matrix. *Journal of Immunological Methods*, 2003, 283, 67-75.

[38] Dwivedi, N; Arunagirinathan, MA; Sharma, S; Bellare, JR. Silica Coated Liposomes for Insulin Delivery. *Journal of Nanomaterials*, 2010.

[39] Ehrlich, H; Heinemann, S; Heinemann, C; Simon, P; Bazhenov, VV; Shapkin, NP; Born, R; R, K; Tabachnick; Hanke, T; Worch, H. Nanostructural Organization of Naturally Occurring Composites—Part I: Silica-Collagen-Based Biocomposites. *Journal of Nanomaterials*, 2008, 8

[40] Elvan, Y; Murat, B. (2006) *Drug entrapment in silica microspheres through a single step sol-gel process and <I>in vitro</I> release behavior.* In, Vol. 77B, 149-155.

[41] Evans, P; Sheel, DW. Photoactive and antibacterial TiO_2 thin films on stainless steel. *Surface and Coatings Technology*, 2007, 201, 9319-9324.

[42] Ferreira, L; Zumbuehl, A. Non-leaching surfaces capable of killing microorganisms on contact. *J. Mater. Chem.*, 2009, 19, 7796-7806.

[43] Finnie, K; Waller, D; Perret, F; Krause-Heuer, A; Lin, H; Hanna, J; Barbé, C. Biodegradability of sol–gel silica microparticles for drug delivery. *Journal of Sol-Gel Science and Technology*, 2009, 49, 12-18.

[44] Finnie, KS; Bartlett, JR; BarbÃ©, CJA; Kong, L. Formation of Silica Nanoparticles in Microemulsions. *Langmuir*, 2007, 23, 3017-3024.

[45] Frost, MC; Meyerhoff, ME. Synthesis, characterization, and controlled nitric oxide release from S-nitrosothiol-derivatized fumed silica polymer filler particles. *J Biomed Mater Res*, 2005, 72A, 409-419.

[46] Gadre, SY; Gouma, PI. Biodoped ceramics: Synthesis, properties, and applications. *Journal of the American Ceramic Society*, 2006, 89, 2987-3002.

[47] Gan, LM; Zhang, K; Chew, CH. Preparation of silica nanoparticles from sodium orthosilicate in inverse microemulsions. Colloids and Surfaces A: *Physicochemical and Engineering Aspects*, 1996, 110, 199-206.

[48] Gardner, J; Peel, M. Introduction to sterilization and disinfection, 1986, *Churchill Livington,* Melbourne, AUS.

[49] Gavlasova, P; Kuncova, G; Kochankova, L; Mackova, M. Whole cell biosensor for polychlorinated biphenyl analysis based on optical detection. *International Biodeterioration & Biodegradation*, 2008, 62, 304-312.

[50] Gerritsen, M; Kros, A; Sprakel, V; Lutterman, JA; Nolte, RJM; Jansen, JA. Biocompatibility evaluation of sol-gel coatings for subcutaneously implantable glucose sensors. *Biomaterials*, 2000, 21, 71-78.

[51] Gil-Albarova, J; Salinas, AJ; Bueno-Lozano, AL; Román, J; Aldini-Nicolo, N; García-Barea, A; Giavaresi, G; Fini, M; Giardino, R; Vallet-Regí, M. The in vivo behaviour of a sol-gel glass and a glass-ceramic during critical diaphyseal bone defects healing. *Biomaterials*, 2005, 26, 4374-4382.

[52] Gill, I; Ballesteros, A. Encapsulation of biologicals within silicate, siloxane, and hybrid sol- gel polymers: An efficient and generic approach. *Journal of the American Chemical Society*, 1998, 120, 8587-8598.

[53] Glass, SAGaRS. Sol-Gel-Based Biosensor for Use in Stroke Treatment. *IEEE Transactions on Biomedical Engineering*, 1999, 46.

[54] Guenther, U; Smirnova, I; Neubert, RHH. Hydrophilic silica aerogels as dermal drug delivery systems-Dithranol as a model drug. *European Journal of Pharmaceutics and Biopharmaceutics,* 2008, 69, 935-942.

[55] Gupta, AK; Gupta, M. Synthesis and surface engineering of iron oxide nanoparticles for biomedical applications. *Biomaterials*, 2005, 26, 3995-4021.

[56] Gupta, R; Kumar, A. Bioactive materials for biomedical applications using sol-gel technology *Biomedical Materials*, 2008, 3, doi: 10.1088/1748-6041/1083/1083/034005.

[57] Haufe, H; Thron, A; Fiedler, D; Mahltig, B; Böttcher, H. Biocidal nanosol coatings. Surface Coatings International Part B: *Coatings Transactions*, 2005, 88, 55-60.

[58] Haufe, H; Muschter, K; Siegert, J; Böttcher, H. Bioactive textiles by sol–gel immobilised natural active agents. *Journal of Sol-Gel Science and Technology*, 2008, 45, 97-101.

[59] Heinemann, S; Christiane, H; Ricardo, B; Antje, R; Berthold, N; Michael, M; Hartmut, W; Thomas, H. Bioactive silica collagen composite xerogels modified by calcium phosphate phases with adjustable mechanical properties for bone replacement. *Acta biomaterialia*, 2009, 5, 1979-1990.

[60] Hudson, SP; Padera, RF; Langer, R; Kohane, DS. The biocompatibility of mesoporous silicates. *Biomaterials*, 2008, 29, 4045-4055.

[61] Hughes, GA. Nanostructure-mediated drug delivery. Nanomedicine: Nanotechnology, Biology; *Medicine*, 2005, 1, 22-30.

[62] Iler, R. *The Chemistry of silica,* 1979, J Wiley, New York, USA.

[63] Jasiorski, M; Leszkiewicz, A; Brzeziński, S; Bugla-Płoskońska, G; Malinowska, G; Borak, B; Karbownik, I; Baszczuk, A; Stręk, W; Doroszkiewicz, W. Textile with silver silica spheres: its antimicrobial activity against *Escherichia coli* and *Staphylococcus aureus*. *Journal of Sol-Gel Science and Technology*, 2009, 51, 330-334.

[64] Jasmund, I; Bader, A. Bioreactor Developments for Tissue Engineering Applications by the Example of the Bioartificial Liver. *Advances in Biochemical Engineering*, 2002, 74, 100-108.

[65] Jeon, HJ; Yi, SC; Oh, SG. Preparation and antibacterial effects of Ag-SiO_2 thin films by sol-gel method. *Biomaterials*, 2003, 24, 4921-4928.

[66] Jerónimo, PCA; Araújo, AN; Montenegro, MCBSM; Pasquini, C; Raimundo, IM. Direct determination of copper in urine using a sol–gel optical sensor coupled to a multicommutated flow system. *Analytical and Bioanalytical Chemistry*, 2004, 380, 108-114.

[67] Jin, W; Brennan, JD. Properties and applications of proteins encapsulated within sol-gel derived materials. *Analytica Chimica Acta*, 2002, 461, 1-36.

[68] Kandimalla, V; Tripathi, VS; Ju, H. Immobilization of biomolecules in sol-gels: Biological and analytical applications. *Critical Reviews in Analytical Chemistry*, 2006, 36, 73-106.

[69] Kartal, AM; Erkey, C. Surface modification of silica aerogels by hexamethyldisilazane-carbon dioxide mixtures and their phase behavior. *The Journal of Supercritical Fluids* In Press, Corrected Proof.

[70] Kawashita, M; Tsuneyama, S; Miyaji, F; Kokubo, T; Kozuka, H; Yamamoto, K. Antibacterial silver-containing silica glass prepared by sol-gel method. *Biomaterials*, 2000, 21, 393-398.

[71] Kawashita, M; Toda, S; Kim, HM; Kokubo, T; Masuda, N. Preparation of antibacterial silver-doped silica glass microspheres. *J Biomed Mat Res A*, 2003, 66, 266-274.

[72] Kim, J; Lee, JE; Lee, J; Yu, JH; Kim, BC; An, K; Hwang, Y; Shin, C-H; Park, J-G; Kim, J; Hyeon, T. Magnetic Fluorescent Delivery Vehicle Using Uniform Mesoporous Silica Spheres Embedded with Monodisperse Magnetic and Semiconductor Nanocrystals. *Journal of the American Chemical Society*, 2005, 128, 688-689.

[73] Kortesuo, P; Ahola, M; Kangas, M; Kangasniemi, I; Yli-Urpo, A; Kiesvaara, J. In vitro evaluation of sol-gel processed spray dried silica gel microspheres as carrier in controlled drug delivery. *International Journal of Pharmaceutics*, 2000a, 200, 223-229.

[74] Kortesuo, P; Ahola, M; Karlsson, S; Kangasniemi, I; Yli-Urpo, A; Kiesvaara, J. Silica xerogel as an implantable carrier for controlled drug delivery--evaluation of drug distribution and tissue effects after implantation. *Biomaterials*, 2000b, 21, 193-198.

[75] Lai, W; Garino, J; Ducheyne, P. Silicon excretion from bioactive glass implanted in rabbit bone. *Biomaterials*, 2002, 23, 213-217.

[76] Lan, EH; Dunn, B; Zink, JI. Sol–Gel Encapsulated Anti-Trinitrotoluene Antibodies in Immunoassays for TNT. *Chemistry of Materials,* 2000, 12, 1874-1878.

[77] LeBaron, RG; Athanasiou, KA. *Extracellular matrix cell adhesion peptides: functional applications in orthopedic materials*, Vol. 6, 2000, .

[78] Lee, W-Y; Kim, S-R; Kim, T-H; Lee, KS; Shin, M-C; Park, J-K. Sol-gel-derived thick-film conductometric biosensor for urea determination in serum. *Analytica Chimica Acta*, 2000, 404, 195-203.

[79] Lee, W; Park, K-S; Kim, Y-W; Lee, WH; Choi, J-W. Protein array consisting of sol-gel bioactive platform for detection of E. coli O157:H7. *Biosensors and Bioelectronics*, 2005, 20, 2292-2299.

[80] Li, Z; Lee, D; Sheng, X; Cohen, RE; Rubner, MF. Two-Level Antibacterial Coating with Both Release-Killing and Contact-Killing Capabilities. *Langmuir,* 2006, 22, 9820-9823.

[81] Lin, Y; Yang, Z; Cheng, J. Preparation, Characterization and Antibacterial Property of Cerium Substituted Hydroxyapatite Nanoparticles. *Journal of Rare Earths*, 2007, 25, 452-456.

[82] Liu, Y; Jiang, H. Electroanalytical Determination of Carcinoembryonic Antigen at a Silica Nanoparticles/Titania Sol-Gel Composite Membrane-Modified Gold Electrode. *Electroanalysis,* 2006, 18, 1007-1013.

[83] Livage. Sol–gel encapsulation of bacteria: a comparison between alkoxide and aqueous routes. *J. Mater. Chem*, 2001, 11.

[84] Livage, J; Roux, C; Costa, JM; Quinson, JF; Desportes, I. Immunoassays in sol-gel matrices. *Journal of Sol-Gel Science and Technology*, 1996, 7, 45-51.

[85] Livage, J; Coradin, T. (2006) Living cells in oxide glasses. In: *Reviews in Mineralogy and Geochemistry*, Vol. 64, p. 315-332.

[86] Mahltig, B; Fiedler, D; Böttcher, H. Antimicrobial Sol–Gel Coatings. *Journal of Sol-Gel Science and Technology*, 2004, 32, 219-222.

[87] Mahltig, B; Gutmann, E; Meyer, DC; Reibold, M; Dresler, B; Günther, K; Faßler, D and Böttcher, H. Solvothermal preparation of metallized titania sols for photocatalytic and antimicrobial coatings. *J. Mater. Chem.*, 2007, 17, 2367 - 2374.

[88] Marini, M; Bondi, M; Iseppi, R; Toselli, M; Pilati, F. Preparation and antibacterial activity of hybrid materials containing quaternary ammonium salts via sol-gel process. *European Polymer Journal*, 2007, 43, 3621-3628.

[89] McBain, SC; HP Yiu, H; Dobson, J. Magnetic nanoparticles for gene and drug delivery. *Int J Nanomedicine*, 2008, 3, 169-180.

[90] Meunier, CF; Dandoy, P; Su, B-L. Encapsulation of cells within silica matrixes: Towards a new advance in the conception of living hybrid materials. *Journal of Colloid and Interface Science*, 2010, 342, 211-224.

[91] Mohanraj, VJ; Barnes, TJ; Prestidge, CA. Silica nanoparticle coated liposomes: A new type of hybrid nanocapsule for proteins. *International Journal of Pharmaceutics* In Press, Uncorrected Proof.

[92] Munusamy, P; Seleem, MN; Alqublan, H; Tyler Jr, R; Sriranganathan, N; Pickrell, G. Targeted drug delivery using silica xerogel systems to treat diseases due to intracellular pathogens. *Materials Science and Engineering*: C, 2009, 29, 2313-2318.

[93] Muraca, M; Vilei, MT; Zanusso, GE; Ferraresso, C; Boninsegna, S; Monte, RD; Carraro, P; Carturan, G. SiO2 Entrapment of Animal Cells: Liver-Specific Metabolic Activities in Silica-Overlaid Hepatocytes. *Artificial Organs*, 2002, 26, 664 -669.

[94] Nablo, BJ; Chen, T-Y; Schoenfisch, MH. Sol-Gel Derived Nitric-Oxide Releasing Materials that Reduce Bacterial Adhesion. *Journal of the American Chemical Society*, 2001, 123, 9712-9713.
[95] Nablo, BJ; Schoenfisch, MH. Antibacterial properties of nitric oxide-releasing sol-gels. *J Biomed Mat Res A*, 2003, 67A, 1276-1283.
[96] Nablo, BJ; Prichard, HL; Butler, RD; Klitzman, B; Schoenfisch, MH. Inhibition of implant-associated infections via nitric oxide release. *Biomaterials*, 2005a, 26, 6984-6990.
[97] Nablo, BJ; Rothrock, AR; Schoenfisch, MH. Nitric oxide-releasing sol-gels as antibacterial coatings for orthopedic implants. *Biomaterials*, 2005b, 26, 917-924.
[98] Nassif, N; Bouvet, O; Noelle Rager, M; Roux, C; Coradin, T; Livage, J. Living bacteria in silica gels. *Nat Mater,* 2002, 1, 42-44.
[99] Nathan, CF; Hibbs, JB. Role of nitric oxide synthesis in macrophage antimicrobial activity. *Current Opinion in Immunology*, 1991, 3, 65-70.
[100] Neuberger, T; Schöpf, B; Hofmann, H; Hofmann, M; von Rechenberg, B. Superparamagnetic nanoparticles for biomedical applications: Possibilities and limitations of a new drug delivery system. *Journal of Magnetism and Magnetic Materials*, 2005, 293, 483-496.
[101] Nieto, A; Areva, S; Wilson, T; Viitala, R; Vallet-Regi, M. Cell viability in a wet silica gel. *Acta biomaterialia*, 2009, 5, 3478-3487.
[102] Nishimori, H; Kondoh, M; Isoda, K; Tsunoda, S-i; Tsutsumi, Y; Yagi, K. Silica nanoparticles as hepatotoxicants. European *Journal of Pharmaceutics and Biopharmaceutics*, 2009, 72, 496-501.
[103] Nivens, DA; Schiza, MV; Angel, SM. Multilayer sol-gel membranes for optical sensing applications: single layer pH and dual layer CO2 and NH3 sensors. *Talanta*, 2002, 58, 543-550.
[104] Nozawa, K; Gailhanou, H; Raison, L; Panizza, P; Ushiki, H; Sellier, E; Delville, JP; Delville, MH. Smart Control of Monodisperse Stober Silica Particles: Effect of Reactant Addition Rate on Growth Process. *Langmuir,* 2004, 21, 1516-1523.
[105] Orive, G; Hernandez, RM; Gascon, AR; Calafiore, R; Chang, TM; De Vos, P; Hortelano, G; Hunkeler, D; Lacik, I; Shapiro, AM; Pedraz, JL. Cell encapsulation: promise and progress. *Nat Med,* 2003a, 9, 104-107.
[106] Orive, G; Hernandez, RM; Rodriguez Gascon, A; Dominguez-Gil, A; Pedraz, JL. Drug delivery in biotechnology: present and future. *Curr Opin Biotechnol*, 2003b, 14, 659-664.
[107] Orive, G; Hernandez, RM; Rodriguez Gascon, A; Calafiore, R; Chang, TM; de Vos, P; Hortelano, G; Hunkeler, D; Lacik, I; Pedraz, JL. History, challenges and perspectives of cell microencapsulation. *Trends Biotechnol*, 2004a, 22, 87-92.
[108] Orive, G; Gascon, AR; Hernandez, RM; Dominguez-Gil, A; Pedraz, JL. Techniques: new approaches to the delivery of biopharmaceuticals. *Trends Pharmacol Sci*, 2004b, 25, 382-387.
[109] Page, K; Palgrave, RG; Parkin, IP; Wilson, M; Savin, SLP; Chadwick, AV. Titania and silver-titania composite films on glass - Potent antimicrobial coatings. *Journal of Materials Chemistry*, 2007, 17, 95-104.
[110] Palumbo, G; Avigliano, L; Strukul, G; Pinna, F; Principe, DD; D'Angelo, I; Annicchiarico-Petruzzelli, M; Locardi, B; Rosato, N. Fibroblast growth and

polymorphonuclear granulocyte activation in the presence of a new biologically active sol–gel glass. *Journal of Materials Science: Materials in Medicine*, 1997, 8, 417-421.
[111] Pasqua, L; Testa, F; Aiello, R; Cundari, S; Nagy, JB. Preparation of bifunctional hybrid mesoporous silica potentially useful for drug targeting. *Microporous and Mesoporous Materials*, 2007, 103, 166-173.
[112] Perullini, M; Rivero, MM; Jobbagy, M; Mentaberry, A; Bilmes, SA. Plant cell proliferation inside an inorganic host. *J Biotechnol,* 2007, 127, 542-548.
[113] Perullini, M; Jobbagy, M; Moretti, MB; García, SC; Bilmes, SA. Optimizing silica encapsulation of living cells: In situ evaluation of cellular stress. *Chemistry of Materials*, 2008, 20, 3015-3021.
[114] Peterson, KP; Peterson, CM; Pope, EJA. Silica sol-gel encapsulation of pancreatic islets. *Proc Soc Exp Biol Med.*, 1998, 218, 365-369.
[115] Premkumar, R; Rachel, R; Shimshon, B; Ovadia, L. Sol-gel luminescence biosensors: Encapsulation of recombinant E. coli reporters in thick silicate films. *Analytica Chimica Acta*, 2002, 462, 11-23.
[116] Radin, S; Ducheyne, P. Controlled release of vancomycin from thin sol-gel films on titanium alloy fracture plate material. *Biomaterials*, 2007, 28, 1721-1729.
[117] Rebecca Voss; Michael A. Brook; Jordan Thompson; Yang Chen; Pelton, RH; Brennan, JD. Non-destructive horseradish peroxidase immobilization in porous silica nanoparticles. *J. Mater. Chem*, 2007, 17, 4854 - 4863.
[118] Ren-Jei, C; Ming-Fa, H; Chine-Wen, H; Li-Hsiang, P; Hsiao-Wei, W; Tsung-Shune, C. Antimicrobial effects and human gingival biocompatibility of hydroxyapatite sol-gel coatings. *Journal of Biomedical Materials Research Part* B: *Applie*d *Biomaterials*, 2006, 76B, 169-178.
[119] Riccardi, CdS; Dahmouche, K; Santilli, CV; da Costa, PI; Yamanaka, H. Immobilization of streptavidin in sol-gel films: Application on the diagnosis of hepatitis C virus. *Talanta*, 2006, 70, 637-643.
[120] Riccio, DA; Dobmeier, KP; Hetrick, EM; Privett, BJ; Paul, HS; Schoenfisch, MH. Nitric oxide-releasing S-nitrosothiol-modified xerogels. *Biomaterials*, 2009, 30, 4494-4502.
[121] Rosenholm, JM; Meinander, A; Peuhu, E; Niemi, R; Eriksson, JE; Sahlgren, C; Lindeǐ• n, M. Targeting of Porous Hybrid Silica Nanoparticles to Cancer Cells. *ACS Nano,* 2008, 3, 197-206.
[122] Roveri, N; Morpurgo, M; Palazzo, B; Parma, B; Vivi, L. Silica xerogels as a delivery system for the controlled release of different molecular weight heparins. *Analytical and Bioanalytical Chemistry*, 2005, 381, 601-606.
[123] Ruiz-Hitzky, E; Darder, M; Aranda, P; Ariga, K. Advances in biomimetic and nanostructured biohybrid materials. *Advanced Materials*, 2010, 22, 323-336.
[124] Saif, MJ; Anwar, J; Munawar, MA. A Novel Application of Quaternary Ammonium Compounds as Antibacterial Hybrid Coating on Glass Surfaces. *Langmuir*, 2008, 25, 377-379.
[125] Sakai, S; Ono, T; Ijima, H; Kawakami, K. In vitro and in vivo evaluation of alginate/sol-gel synthesized aminopropyl-silicate/alginate membrane for bioartificial pancreas. *Biomaterials*, 2002, 23, 4177-4183.
[126] Sanchez, C; Soler-Illia, GJDAA; Ribot, F; Lalot, T; Mayer, CR; Cabuil, V. Designed hybrid organic-inorganic nanocomposites from functional nanobuilding blocks.

Chemistry of Materials, 2001, 13, 3061-3083.

[127] Sanchez, C; Julián, B; Belleville, P; Popall, M. Applications of hybrid organic-inorganic nanocomposites. *Journal of Materials Chemistry*, 2005, 15, 3559-3592.

[128] Sanchez, C. Advanced nanomaterials: A domain where chemistry, physics and biology meet. *Comptes Rendus Chimie*, 2010, 13, 1-2.

[129] Sanchez, C; Rozes, L; Ribot, F; Laberty-Robert, C; Grosso, D; Sassoye, C; Boissiere, C; Nicole, L. "Chimie douce": A land of opportunities for the designed construction of functional inorganic and hybrid organic-inorganic nanomaterials. *Comptes Rendus Chimie*, 2010, 13, 3-39.

[130] Smirnova, I; Arlt, W. Supercritical fluids as solvents and reaction media. Edited by G. Brunner. *Elsiever*, 2004, Chapter 3.1.

[131] Smirnova, I; Suttiruengwong, S; Seiler, M; Arlt, W. Dissolution rate enhancement by adsorption of poorly soluble drugs on hydrophilic silica aerogels. *Pharm Dev Technol.*, 2004, 9, 443-452.

[132] Soler-Illia, GJ; Sanchez, C; Lebeau, B; Patarin, J. Chemical strategies to design textured materials: from microporous and mesoporous oxides to nanonetworks and hierarchical structures. *Chem Rev*, 2002, 102, 4093-4138.

[133] Stein, A. Sphere templating methods for periodic porous solids. Microporous and *Mesoporous Materials*, 2001, 44-45, 227-239.

[134] Steingroewer, J; Knaus, H; Bley, T; Boschke, E. A Rapid Method for the Pre-Enrichment and Detection of <I>Salmonella</I> Typhimurium by Immunomagnetic Separation and Subsequent Fluorescence Microscopical Techniques. *Engineering in Life Sciences*, 2005, 5, 267-272.

[135] Stöber, W; Fink, A; Bohn, E. Controlled growth of monodisperse silica spheres in the micron size range. *Journal of Colloid and Interface Science*, 1968, 26, 62-69.

[136] Tomšič, B; Simončič, B; Orel, B; Černe, L; Tavčer, P; Zorko, M; Jerman, I; Vilčnik, A; Kovač, J. Sol–gel coating of cellulose fibres with antimicrobial and repellent properties. *Journal of Sol-Gel Science and Technology*, 2008, 47, 44-57.

[137] Trapalis, CC; Kokkoris, M; Perdikakis, G; Kordas, G. Study of Antibacterial Composite Cu/SiO_2 Thin Coatings. *Journal of Sol-Gel Science and Technology*, 2003, 26, 1213-1218.

[138] Tsai, C; Chen, C; Hung, Y; Chang, F; Mou, C. Monoclonal antibody-functionalized mesoporous silica nanoparticles (MSN) for selective targeting breast cancer cells. *J. Mater. Chem.*, 2009, 19, 5737-5743.

[139] Turniansky, A; Avnir, D; Bronshtein, A; Aharonson, N; Altstein, M. Sol-gel entrapment of monoclonal anti-atrazine antibodies. *Journal of Sol-Gel Science and Technology*, 1996, 7, 135-143.

[140] Tuttolomondo, V; Teves, S; Copello, G; Villanueva, M; Gerber, FP; Díaz, L. Isolation of pathogen microorganism using immobilized antibodies onto magnetic silicate beads. In: 45th Annual Meeting Argentine Society for *Biochemistry and Molecular Biology*, 2009, Vol. 33, 60.

[141] Unger, K; Rupprecht, H; Valentin, B; Kircher, W. The use of porous and surface modified silicas as drug delivery and stabilizing agents. *Drug Development and Industrial Pharmacy*, 2008, 9, 69-91.

[142] Vallet-Regi, M; Arcos, D. *Biomimetic nanoceramics in Clinical Use*. In: Cambridge., 2008, RSCPublishing.

[143] van der Veen, VC; van der Wal, MBA; van Leeuwen, MCE; Ulrich, MMW; Middelkoop, E. *Biological background of dermal substitutes. Burns*, 2009, 36, 305-321.
[144] Van Vlerken, LE; Amiji, MM. Multi-functional polymeric nanoparticles for tumour-targeted drug delivery. *Expert Opinion on Drug Delivery*, 2006, 3, 205-216.
[145] Vilcnik, A; Jerman, I; Surca Vuk, A; Kozelj, M; Orel, B; Tomsic, B; Simoncic, B; Kovac, J. Structural Properties and Antibacterial Effects of Hydrophobic and Oleophobic Sol-Gel Coatings for *Cotton Fabrics. Langmuir*, 2009, 25, 5869-5880.
[146] Wang, X; Wang, C. The antibacterial finish of cotton via sols containing quaternary ammonium salts. *Journal of Sol-Gel Science and Technology*, 2009, 50, 15-21.
[147] Wen, L; Wang, Q; Zheng, T; Chen, J. Effects of polyethylenimine on the dispersibility of hollow silica nanoparticles. *Frontiers of Chemical Engineering in China*, 2007, 1, 277-282.
[148] Wilson, J; Pigott, GH; Schoen, FJ; Hench, LL. Toxicology and biocompatibility of bioglasses. *J. Biomed. Mater. Res*, 1981, 15, 805-817.
[149] Wu, X; Ye, L; Liu, K; Wang, W; Wei, J; Chen, F; Liu, C. Antibacterial properties of mesoporous copper-doped silica xerogels *Biomedical Materials*, 2009, 4, doi: 10.1088/1748-6041/1084/1084/045008.
[150] Xing, Y; Yang, X; Dai, J. Antimicrobial finishing of cotton textile based on water glass by sol–gel method. *Journal of Sol-Gel Science and Technology*, 2007, 43, 187-192.
[151] Yang, HH; Zhang, SQ; Chen, XL; Zhuang, ZX; Xu, JG; Wang, XR. Magnetite-Containing Spherical Silica Nanoparticles for Biocatalysis and Bioseparations. *Analytical Chemistry*, 2004, 76, 1316-1321.
[152] Yellen, BB; Forbes, ZG; Halverson, DS; Fridman, G; Barbee, KA; Chorny, M; Levy, R; Friedman, G. Targeted drug delivery to magnetic implants for therapeutic applications. *Journal of Magnetism and Magnetic Materials*, 2005, 293, 647-654.
[153] Yun, HS; Kim, SE; Hyun, YT; Heo, SJ; Shin, JW. Three-Dimensional Mesoporous-Giantporous Inorganic/Organic *Composite Scaffolds for Tissue Engineering. Chemistry of Materials*, 2007, 19, 6363-6366.
[154] Zeglinski, J; Cabaj, A; Strankowski, M; Czerniak, J; Haponiuk, JT. Silica xerogel-hydrogen peroxide composites: Their morphology, stability, and antimicrobial activity. Colloids and Surfaces B: *Biointerfaces*, 2007, 54, 165-172.
[155] Zhang, H; Chen, G. Potent Antibacterial Activities of Ag/TiO_2 Nanocomposite Powders Synthesized by a One-Pot Sol-Gel Method. *Environmental Science & Technology*, 2009, 43, 2905-2910.
[156] Zhao, W; Gu, J; Zhang, L; Chen, H; Shi, J. Fabrication of Uniform Magnetic Nanocomposite Spheres with a Magnetic Core/Mesoporous Silica Shell Structure. *Journal of the American Chemical Society*, 2005, 127, 8916-8917.
[157] Zhong, T-S; Liu, G. Silica Sol-gel Amperometric Immunosensor for Schistosoma Japonicum Antibody Assay. *Analytical Sciences*, 2004, 20, 537.
[158] Zhou, W; Gao, P; Shao, L; Caruntu, D; Yu, M; Chen, J; O'Connor, CJ. Drug-loaded, magnetic, hollow silica nanocomposites for nanomedicine. Nanomedicine: Nanotechnology, *Biology and Medicine*, 2005, 1, 233-237.
[159] Zolkov, C; Avnir, D; Armon, R. Tissue-derived cell growth on hybrid sol–gel films. *Journal of Materials Chemistry*, 2004, 14, 2200-2205.
[160] Zolkov, C; Avnir, D; Armon, R. Sol-gel-based poliovirus-1 detector. *Journal of Virological Methods*, 2009, 155, 132-135.

In: The Sol-Gel Process
Editor: Rachel E. Morris

ISBN 978-1-61761-321-0

Chapter 7

DESCRIPTION OF SOLVENT-DEPENDENT AND NANOSCALE DIFFUSION PROCESS IN SILICA BASED SOL–GEL MATRIX

Kazushige Yokoyama
Department of Chemistry, The State University of New York,
Geneseo College, Geneseo, New York, USA

ABSTRACT

The solvent diffusion process in silica based sol-gel nano-thin film was examined as the film was immersed in water, methanol, ethanol, 2-propanol, and glycerol by observing the luminescence decay time constant of the photo-excited ^{3}MLCT (metal-ligand charge-transfer) state of the $[Ru(bpy)_3]^{2+}$ ion dopant in the film. The luminescence decays of the films in the methanol, 2-propanol, and glycerol were better explained by the KWW model, while the luminescence decay of film immersed in water and ethanol were both well explained by a single, exponential decay. The water and ethanol were the only two solvents that exhibited solvent dependence in the aging effect of the film. The resulting photodynamics were influenced by the interaction between dopant and residual water/ethanol solvents co-encapsulated inside the cavity. We further investigated the effect of the diffusion speed of aqueous acid solution as a function of sizes of encapsulated gold nanocolloidal particles ranging from 5 nm to 100 nm by observing the color change of gold colloids as a function of time. While the maximum rate was found for silica sol-gel matrix encapsulating 15 nm gold colloid, 60 nm gold colloid doped silica sol-gel showed the smallest acid interaction. It was speculated that the surface of these gold colloids was homogeneously covered by the silica gel layer, thus avoiding direct contact of the acid with the surface of the gold colloid. The close match of the average size of the cavity with the size of the gold colloids was considered to prevent acid penetration. This study revealed that the nanoscale dopant size affects the rate of solvent penetration into a sol-gel cavity.

1. INTRODUCTION

Modern drug delivery technology uses sophisticated systems that allow controlled release of drug substances at a targeted object.[1, 2] It is desired to have biomaterial polymers which control the rate of drug release as the drug diffuses out from the polymer matrix. Sol-gel derived silica gel plays a very important role in biomedical applications for drug delivery, since sol-gel based materials are bioactive and bond to bone.[3-7] One of the challenges in drug delivery technology is to administer through the gastrointestinal tract. For example, general drugs show poor physicochemical properties under a high first-pass metabolism in the liver and drugs are easily degraded under the acidic environment of the stomach. As a promising solution to this problem, the injection of drugs encapsulated in sol-gel matrix is proposed. A transparent sol-gel host matrix is highly desirable compared to the many existent carrier materials due to its high thermo-resistance, transparency, and relatively robust photostability,[8-14] especially suitable for preparing inorganic oxide thin films doped with functional organic molecules and organometalic complexes.[8, 12, 13, 15] The advantage of encapsulating materials in sol–gels is their capacity to modify entrapment environments through external liquids, which may penetrate through the three dimensional silica gel network into finite amorphous-like cavities. A general requirement of sol–gel material for a practical use is the stability of the dopants under given external conditions. With this respect, a host material mainly determines the physical properties of an entire material and modifies the physical properties of a dopant. The encapsulation environment for the dopants can be controlled by thermal energy, dielectric constant, pH, and even bonding states. As a practical form for a device aimed at a biological applications, the versatile usage of such devices under exposure to a liquid media has important applications in the diffusion across cellular or membrane-like environments[16-18] and kidney dialysis.[19-28] Diffusion across nanoporous structured membranes involves numerous physical and chemical systems.[19, 22, 29] As seen in the phenomena of crossing cell membranes, the diffusion of solvent into a sol-gel matrix involves complicated changes in the physical environment involving the transport of different molecules, the entry of metabolite, and the exit of substances.[17, 18] By nature, porous materials exposed to fluid materials (gas or liquid) have less control in their diffusion rate, while the rates of their transportation are considered to be dominated by their own diffusion property. So the drugs encapsulated inside the porous materials (dopants) interact with an external solvent which reaches inside the cavities, then is delivered to the outside of the pores by the diffusion of the solvent. Thus, it has been a great challenge to systematically and externally control delivery time of the dopants.

Over the last few decades, the elucidation of deviations from ordinary diffusion under the influence of confinement have been extensively discussed in theoretical physics and material sciences.[30] Primarily, a vast variety of microdynamic and microstructural models with respect to the resulting patterns of particle migration were established.[31-33] Experimental techniques[34] have contributed to the development of a model with the availability of advanced nanoporous materials. However, in order to achieve a rate controlled diffusion process *in vivo*, further experiments *in vitro* are still needed. Metal compound encapsulated in a nanoscale microcavity in a host material provides useful information for rationally designing biochemical -sensing devices.[35-40]

While most host materials immobilize the guest species, the immobilization of the guest in a sol–gel matrix is considered to provide a relatively higher degree of intramolecular freedom for a dopant. Sometimes dopants even behave as an individual mobile complex within a silica gel matrix, thus they are considered to retain their properties as if in the solution states.[15, 41-44] However, our study of organic dye molecules entrapped in a solvent-exposed silica sol–gel matrix revealed a rather different view of dopant mobility, where encapsulation is acting more rigidly and does not allow a free solution-like environment. Under the exposure of water solvent, the R6G dye dopant became less mobile and more rigidly encapsulated by the sol–gel matrix.[45, 46] In order to understand the ability of the sol–gel cavity to withstand the intrusion of external media (especially fluid), it is important to understand the role of externally introduced solvents to a host-guest interaction and to the stability of a dopant encapsulated in the sol–gel matrix. Up to now, few studies have described the change in the encapsulation environment for a dopant under the influence of liquid media. The permeability of solvents through a sol–gel network is an important index for material binding to an interfacial environment. Currently, there are no additional details available about permeability at this interface region or boundary environment. However, the degree of diffusional motion undertaken at the border region between liquid media, and the permeability of a porous surface requires detailed characterization of the inter/intra-molecular interactions occurring. By utilizing a dopant whose physical properties are sensitive to the encapsulation environment, we were able to probe matrix-solvent interactions at this key interfacial region of interest.

Here we report two major studies on solvent diffusion property in silica gel based sol-gel matrix.

(a) **Solvent dependent diffusion rate of silica sol-gel matrix.** The selectivity of solvent intrusion over the silica sol–gel surface can be monitored by a dopant's luminescence lifetime. The silica- based sol–gel matrix is expected to exhibit a selective permeability to aqueous solutions whereas hydrophobic solvents may have low permeability. Since the interfacial region of the silica gel cavity possesses a highly hydrophilic group (possibly $-SiO^{-}$), it may result in a high interaction with only aqueous media and thereby allowing penetration of the water solvent.[47]

(b) **Controlling the solvent diffusion rate by nanoscale dopants.** Variation of the dopant's size must naively affect the interaction between dopants and the silica gel. A physical matching of dopant sizes with cavity sizes of silica gel creates a strong direct contact between the silica gel surface and the dopant. This condition can be used to measure the degree of stabilization of the dopants against external solvents. In order to systematically investigate interactions between the dopant, solvent, and silica gel cavity surface, the probe dopants were prepared in various sizes around the sizes of silica sol–gel cavities. For this purpose, different diameters of gold colloidal nanoparticles were utilized, and corresponding solvent diffusion rate was observed.[48]

2. Solvent- Dependent Diffusion Rate of Silica Sol–Gel Matrix

The optical properties of thin films provide a unique host system to study the interaction and dynamics of guest molecules. Surface adsorption can alter the physical and chemical properties of an adsorbate, and this has been advantageously exploited in silica sol-gel chemistry to synthesize optical and biocomposite materials with a higher photostability.[49-51] The photodynamics of the Ru(II)-complex ion dopant demonstrates a strong dependence on the encapsulation environment[12, 52, 53]; thus change of the dopant's dynamics will reveal the entrapment environment of the sol-gel matrix under the influence of the solvent. With this respect, our study took an approach of using a luminescence decay time constant of the $[Ru(bpy)_3]^{2+}$ ion dopant entrapped in a silica-based sol–gel pore to indicate the sol–gel encapsulation environment as the degree of surrounding solvent's penetration progress. The degree of solvent penetration is dependent on the selective permeability of the sol-gel matrix for an external media. With photo-physical properties of the $[Ru(bpy)_3]^{2+}$ ion investigated in solution and sol–gel bulk,[54-57] the photodynamics of the $[Ru(bpy)_3]^{2+}$ ion dopant upon solvent intrusion in a sol-gel matrix clarifies the encapsulation environment and revealed significant solvent dependent permeability.

2.a. Sample Handling and Method

A gel precursor was prepared by a sol-gel method which was utilized in our previous studies of single molecule optical microscopy.[45, 46] The 200 proof ethanol, deionized H_2O, tetraethyl orthosilicate (TEOS) from Aldrich Co., and 0.1 mM HCl were mixed in a volume ratio of 85:40:12:1, respectively. The solution was then vortexed for 20 s and sonicated in an ice bath for two hours at 5 °C. The aqueous solution of 4 mM $[Ru(bpy)_3]^{2+}$ $Cl_2 \bullet 6H_2O$ (tris-2;2′-bypyridal-dichloro-ruthenium (II) hexahydrate) from Aldrich Co. was then added with the same volume amount of deionized water. The final concentration of the $[Ru(bpy)_3]^{2+}$ ion was 0.4 mM and the OD of the thin film sample at 452 nm was estimated about 0.01 based on a known extinction coefficient at 452 nm ($\varepsilon_{452\ nm} = 1.46 \times 10^4\ M^{-1}\ cm^{-1}$) [58] and film thickness of 600 nm. The solution was then vortexed again for 20 seconds and stored in the dark at room temperature. A lime-glassed microscope cover glass was sequentially rinsed with 1 M NaOH, acetone, and deionized water for one hour each and kept in a deionized water bath. These slides were dried with nitrogen gas before their usage. A sol–gel thin film was prepared by casting 20 μL of sol–gel solution on the cover glass plate using a home-made spin coater with a speed of 550 revolutions per minute. The thickness of the thin film was found to be 601 ± 32 nm by a Spectroscopic Ellipsometer M-2000. The average pore size inside a sol–gel film can be controlled by varying the composition of the precursor solution with TEOS/H_2O,[59-61] and 30 Å is estimated to be an average pore size of our sol–gel matrix. A prepared thin film was placed in a home-made flow cell made of Pyrex glass which was temperature controlled by an external water bath, ranging from 10 to 65 °C with an accuracy of ±0.05 °C (See Figure 1). Each solvent tested in this study was purged with nitrogen gas for one hour before the measurements were taken. In order to conduct comparative studies, the aqueous solution and sol-gel bulk solution were both prepared with a concentration of 20 μM of $[Ru(bpy)_3]^{2+}$ ion. The solution sample was placed in a quartz flow cell and the temperature was controlled by the external thermostat bath. Bulk sol-gel samples (mixture of TEOS sol-gel precursor and$[Ru(bpy)_3]^{2+}$ ion dopants), were mixed in a plastic cuvette, and placed in a temperature controlled flow cell bath.

The excitation light source was the output of a computer controlled Nd:YAG pumped OPO laser system (Polytech–Spectron Laser SGL450-GW OPO system), wavelength 450 nm with 0.5 mJ/pulse, 20 $\pm$ 1 ns pulse duration, and 10 Hz repetition rate. The thin film was placed at around a Brewster angle to the plane of laser light injection in order to minimize the reflection of the beam from the glass surface. The signal was collimated and sent to a 1/8 m monochromator (ORIEL UV-VIS Cornerstone 130), and was detected by a photomultiplier tube (Hamamatsu HC120-08) and stored by a 500 MHz digital oscilloscope (LeCroy LT322) with 4000 sample averaging. (See Figure 1) The luminescence decay data and the temperature dependence of the luminescence decay signals were analyzed using ORIGIN program (version 7.0). For some cases, the luminescence decay constants were found to deviate by ±10% depending upon a sampling spot on the film surface; indicating that the film conditions prepared in this study possessed an inhomogeneous environment within the gel network. Therefore, the final time constant for a film sample was determined by averaging over 10 different sampling spots.

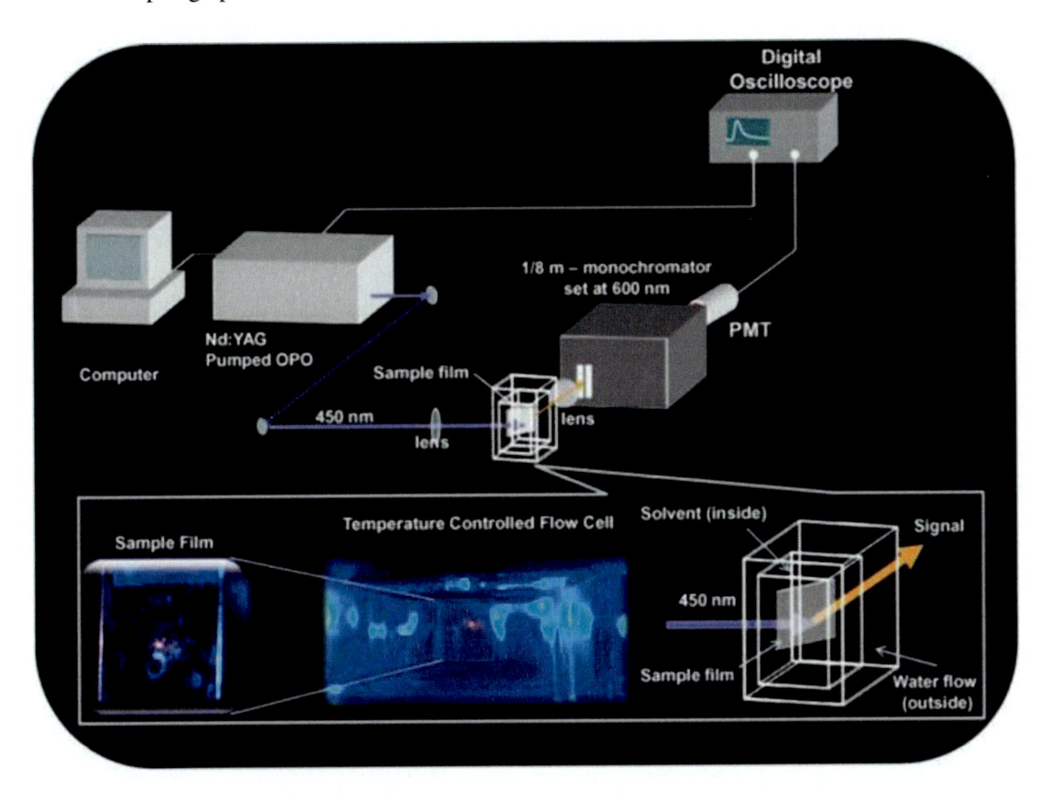

Figure 1. Schematic diagram of an arrangement of luminescence measurement from the $[Ru(bpy)_3]^{2+}$ ion encapsulated sol–gel thin film situated in a home-made temperature controlled flow cell. The set up is also given in the picture. The orange color spot is the luminescence from the sample film.

2.b. Effect of an External Solvent to a Dopant's Photodynamics

Every luminescence decay curve observed in this experiment was first analyzed with a single exponential component: i.e.,

$$I(t) = Ae^{-t/\tau} \tag{1}$$

where the luminescence intensity at time t is given by I(t), A is an exponential pre-factor which was normalized to be one for all data, and the lifetime was given by τ (See Figure 2). The data was also analyzed by the KWW model shown in Eq. (2).

$$I(t) = \alpha e^{-(t/\tau)^{\beta}} \tag{2}$$

Here, α is a pre-exponential factor and β is related to the width of an underlying Levy distribution of relaxation rates and τ is the lifetime at the maximum amplitude of the distribution. The KWW model can be considered an approximation of multiple components of exponential functions, i.e., $I(t) = \sum_{i=1}^{n} e^{-(t/\tau_i)}$, where β reflects *n*.[62] For a single exponential component, where *n* = 1 the β value is 1 as seen for water and ethanol solvents at 25 °C (See Table 1). As *n* increases, the relaxation process become more diverse and β decreases from 1 as seen for methanol, propanol, glycerol as well as sol–gel bulk. The photoluminescence decay times, τ, measured at 25 °C, are tabulated in Table 1 along with previously reported values.[63, 64] In single exponential analysis, the values of τ of $[Ru(bpy)_3]^{2+}$ ion in solution were significantly increased when it was contained in sol–gel films except for glycerol. This was true for the KWW model except for the film in methanol where the decay constant remained the same between solution and the film in the solvent. This increase in τ indicates that the sol-gel matrix provides a different relaxation path from excited MLCT (Metal Ligand Charge Transfer) states of a dopant than that of solvent by preventing or decreasing direct contact. From the fact that the values of the luminescence decay constants in films showed similar values within their experimental errors, the entrapment environment of a dopant was speculated to be dominated by silica gel-$[Ru(bpy)_3]^{2+}$ electrostatic interaction despite the presence of polar solvents.

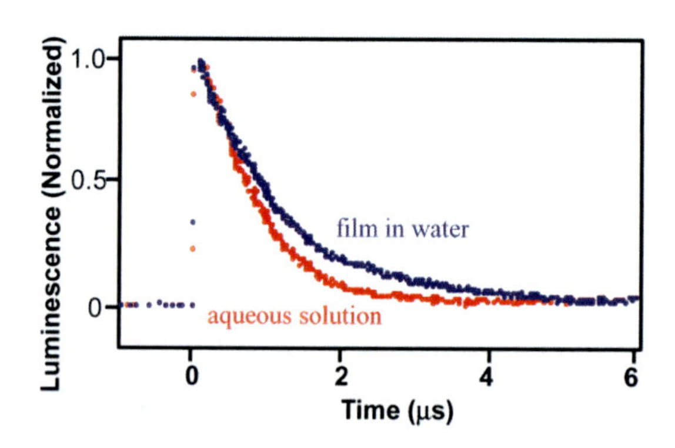

Figure 2. Photoluminescence decay profiles of $[Ru(bpy)_3]^{2+}$ ion encapsulated in a sol–gel film in water and aqueous solution of $[Ru(bpy)_3]^{2+}$ ion.

Table 1. Summary of lifetime, τ, of dopant in a film immersed in various solvents at 25 °C. The available literature values are shown in parentheses. ([a] Juris et al. [64], [b] Maruszewski et al. [63]).

Solution	Single exponential model /μs	KWW Model (β)
Water	0.740 ± 0.010 (0.576 ~ 0.685 [a])	
Methanol	0.543 ± 0.004	
Ethanol	0.703 ± 0.003 (0.680 ~ 0.900 [a])	
2-propanol	0.874 ± 0.004	
Glycerol	1.064 ± 0.002	
Sol-gel bulk	1.21 ± 0.02 (2.145[b])	1.12 ± 0.04 (0.89 ± 0.01)
Film in		
Air	1.58 ± 0.12 (1.776[b])	
Water	1.33 ± 0.03	1.43 ± 0.03(1.05 ± 0.01)
Methanol	1.10 ± 0.09	0.87 ± 0.12(0.78 ± 0.04)
Ethanol	1.58 ± 0.01	1.58 ± 0.01(0.92 ± 0.01)
2-propanol	1.10 ± 0.08	0.84 ± 0.09(0.77 ± 0.02)
Glycerol	1.05 ± 0.10	0.81 ± 0.17(0.72 ± 0.08)

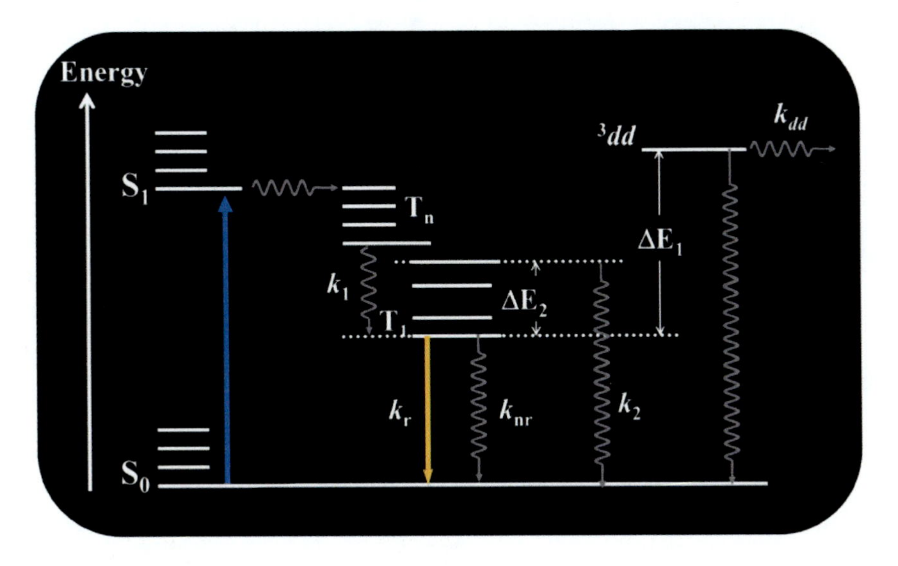

Figure 3. Schematic Jablonski energy level diagram and photophysical relaxation processes of $[Ru(bpy)_3]^{2+}$.

In our KWW analysis, the β value was almost one for the film samples in water and ethanol confirming that the dynamics under those solvents are best described by a single exponential curve at 25 °C. Therefore, the encapsulation environment created in our film in water and ethanol did not suggest as strong an inhomogeneity at least at 25 °C as was observed in other studies. On the other hand, slight non-exponential behaviors were observed for the dynamics of the dopants in the films in methanol, propanol and glycerol at 25 °C. The relaxation process after a photoexcitation is described in Figure 3. The lowest triplet state (T_1) can emit a photon with the rate k_r, or return to the electronic ground state via the non-radiative process(es) with the rate k_{nr}. On the other hand, there is the possibility of thermal population of the metal-localized (^3dd) states as well as the fourth ligand-localized ^{3}MLCT state. Thermal population of the ^3dd states (approximately 3,000 cm^{-1} above the lowest ^{3}MLCT state, i.e., approximately 18,000 cm^{-1} above the ground state [65]) leads to the fast non-radiative decay (k_{dd}) and might result in the possible photochemical processes (k_{rx}) occurring from the metal-localized ^3dd states (possessing largely σ* antibonding character). The fourth ^{3}MLCT state decay process, k_2, has non-radiative character [65]. The measured excited state lifetime is equal to the reciprocal of the total deactivation rate constant, k_{tot}, shown in Eq. (3).

$$1/\tau = k_{tot} = k_r + k_{nr} + k_{dd}e^{-\frac{\Delta E_1}{k_B T}} + k_2 e^{-\frac{\Delta E_2}{k_B T}} \tag{3}$$

where k_B is the Boltzmann constant, T is temperature, ΔE_1 is the energy gap between the ^3dd states and the cluster of the ^{3}MLCT emitting states, and ΔE_2 is associated with the additional ^{3}MLCT state separated from the cluster of the three emitting states by several hundredths of a wavenumber [65-68]. Our data suggests that the temperature dependence is explained by a single exponential component, thus, we approximated ΔE_2 to be negligible,

$$k_{tot} = k_o + k_{dd}e^{-\frac{\Delta E}{k_B T}} \tag{3'}$$

where the non-radiative constant k_o is given by $k_r + k_{nr} + k_2$. Our experimental results were best explained by the Eq. (3′), and ΔE in Eq. (3′) corresponds to ΔE_1 in Eq. (3). Therefore, the present study probed the relaxation process from the metal centered ^3dd atomic state to the ground state after the photoexcitation and intersystem crossing from the singlet excited state.

The extracted energy gap and the rate constants are tabulated in Table 2 for solution, films in the solvents, and the sol–gel bulk. The variation observed in the energy gap for various solvents are not prominent for the energy gap observed in films exposed to different solvents. The energy gaps obtained for the films in methanol, propanol, and glycerol gave relatively large deviations. We observed a strong resemblance among values of ΔE between sol–gel bulk and films in water or ethanol with relatively small deviations. Significant deviations seen in the values of ΔE of the films inserted in methanol, propanol and glycerol are considered to originate directly from the deviations found in the decay lifetimes, which were strongly dependent on the location of sample spot. On the other hand, the location dependence was small for the sol-gel bulk and the dopants in water or ethanol. For the films inserted in water, ethanol, and sol–gel bulk, the temperature dependence was found in the power of the exponent term β given in the KWW model in Eq. (2). (See Table 3 and Figure 4) The value increased as the temperature decreased and approached the regime where the dynamics is well explained by a single exponential.

The dynamics of the dopant in the film in methanol, propanol, and glycerol exhibited signs of non-exponential behavior. From the fact that β values in glycerol, propanol, and methanol were significantly deviating from unity and these are not the solvents pre-trapped in the sol-gel film, the sol-gel cavity created a solvent-controlled inhomogeneous environment for a dopant where the ethanol/water (presolvent) condition and non-presolvent-like condition were locally coexisting. If the ethanol or water co-exist in a prepared sol-gel matrix with a dopant, the intruding solvent can create a drastically different solvent environment than those pre-trapped. Thus, the water or ethanol may not create a significantly different entrapment environment for the dopant, whereas the other solvents may easily create localized environment depending upon the amount of that solvent interacting with a dopant. The hydrophobicity of the solvents can be interacting somewhat repulsively especially for glycerol and propanol.

A temperature dependence observed in β parameter in the KWW model used for interpreting the dopants dynamics in the film immersed in water and ethanol provided a similar dependence observed in the sol–gel bulk. This observation implies that the environment of the dopant in sol-gel bulk can be somewhat similar to the mixture of water and ethanol. Since a deviation of β value from 1 corresponds to an inhomogeneous dopants' environment, a comparable magnitude of temperature dependence, the value of A given in Table 3, between sol-gel bulk and ethanol indicates that the ethanol plays a large role to determine the inhomogeneity of the cavity environment. The negative slope indicates the deviation of β from 1 was enhanced as the temperature increased. This temperature dependence is thought to be due to thermal energy enabling more diverse population across energy states and resulting in increased inhomogeneity in the dopants' environments.

Table 2. The relaxation rates and energy gaps, ΔE, obtained in solutions and films from Eq. (3′). The values given in the brackets are based on the rate constants extracted from the KWW model.

	Solutions			Films		
Solvent	k_o/s^{-1}	k_{dd}/s^{-1}	ΔE/kJ mol^{-1}	k_o/s^{-1}	k_{dd}/s^{-1}	ΔE/kJ mol^{-1}
Water	9.4(9) ×10^{5}	9(15) ×10^{10}	31 ± 5	1(2) ×10^{5} [2(2) ×10^{5}]	4(5) ×10^{7} [8(9) ×10^{7}]	11 ± 4 [12 ± 4]
Methanol	7(5) ×10^{5}	4(17) ×10^{12}	37 ± 12	7.0(3) ×10^{5} [7(2) ×10^{5}]	4.5(5) ×10^{12} [2(20) ×10^{13}]	41 ± 20 [44 ± 25]
Ethanol	1(2) ×10^{5}	3.9(4) ×10^{13}	42 ± 5	1.4(7) ×10^{5} [2.3(5) ×10^{5}]	1.2(8) ×10^{8} [2(2) ×10^{9}]	14 ± 2 [21 ± 3]
2-propanol	3(1) ×10^{5}	4(2) ×10^{13}	44 ± 12	7(1) ×10^{5} [9(1) ×10^{5}]	6(50) ×10^{12} [2(25) ×10^{15}]	43 ± 23 [58 ± 37]
Glycerol	5.0(6) ×10^{5}	1(2) ×10^{11}	31 ± 4	7(2) ×10^{5} [8(2) ×10^{5}]	6 (50) ×10^{10} [2(10) ×10^{11}]	30 ± 23 [33 ± 15]
Sol-gel Bulk	3(1) ×10^{5} [4.5(5) ×10^{5}]	2(2) ×10^{8} [2(2) ×10^{10}]	15 ± 3 [26 ± 3]			

Table 3. Temperature dependence of parameter, β, in the KWW fit for all films immersed in each solvent and sol-gel bulk. The value change was fit with a linear form: β = A•T + B.

Solvent	A/K^{-1} ×10^{3}	B	(r^2)
Water	–3(2)	2.13(8)	0.92
Ethanol	–6.0 (3)	2.7(1)	0.95
Sol-gel bulk	–5.7(4)	2.6(1)	0.93

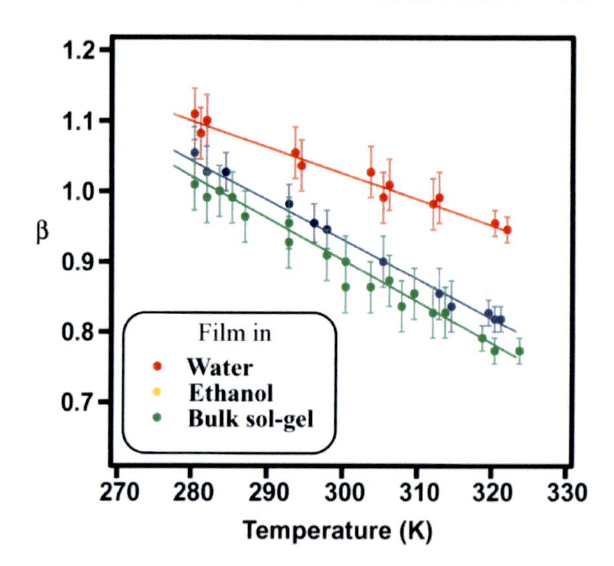

Figure 4. The beta value, β, obtained in the KWW model plotted as a function of temperature.

The sol–gel matrix environment cannot be described as the same as that of any of the solutions examined in this study. Therefore, the sol–gel matrix is considered to play a role in modifying a direct solvent contact with the dopant. Only the ΔE values for water and ethanol in this study gave reliable data, while the rest of the solvents possessed a relatively large deviation. This study showed that the encapsulation by a sol-gel film slightly lowered the energy gap when the film was inserted into solvents (Table 2). While the energy gap for the solvents ranged from 31 to 44 kJ/mol, energy gap values observed in films under all solvents examined in this experiment were similar to that of sol-gel bulk (i.e., ΔE = 15 ± 3 kJ/mol or 26 ± 3 kJ/mol in the KWW model). Since the photodynamics of a dopant in a film observed at 25°C exhibited negligible dependence on immersed solvent, the entrapment environment must be dominated by the interaction between dopant and a component of the sol-gel network. The sol-gel network however, may have significant residual solvents but minimal amounts of external solvent, and this may play a key role in determining the trapping.

Entrapments in zeolites or cellulose acetates were reported to increase the energy level of the low-lying ^{3}MLCT states.[65] This means that the energy gap ΔE_1 decreased, if the ^3dd state was assumed to be insensitive to an encapsulation. Thus, it is consistent with the observed energy gap which showed a decrease of ΔE_1 by encapsulation. In a rigid matrix, because the solvent cannot re-orient, the excited state is not completely relaxed within its lifetime, and hence the emission occurs from a higher energy level than that in a fluid solution. This restriction in the freedom of movement of the solvent molecules is explained by the rigid-chromism of the $[Ru(bpy)_3]^{2+}$ complexes, resulting in a blue shift of the emission peak on sol-gel film.[43] For example, Maruszewski et al., reported the 21 nm blue shift (588 $cm^{-1} \approx$ 7 kJ/mol) [63] as the dopant was placed from solution to sol–gel bulk matrix, while our study shows about a 15 kJ/mol change in ΔE from aqueous solution to sol–gel bulk. If the lowest triplet state was raised about 7 kJ/mol, the lowering of the ^3dd state is considered to be 8 kJ/mol in sol–gel bulk. We speculate that the metal localized state can be significantly lowered through silanol-Ru^{II} center interaction due to stabilization through bonding, analogous to R6G dye described in the work by Gilliland et al.[45, 46]

Table 4. Aging day dependence of photoluminescence rate, k, for a film immersed in water and ethanol measured at 20 °C. The rate change as a function of date was fitted with a linear form: k = A•(days) + B.

Solvent	$A/s^{-1} \times 10^4$	$B/s^{-1}\,day^{-1}$	(r^2)
Water	5.1 (1)	$7.3(2)\times10^5$	0.99
Ethanol	1.0 (2)	$6.2(2)\times10^5$	0.92

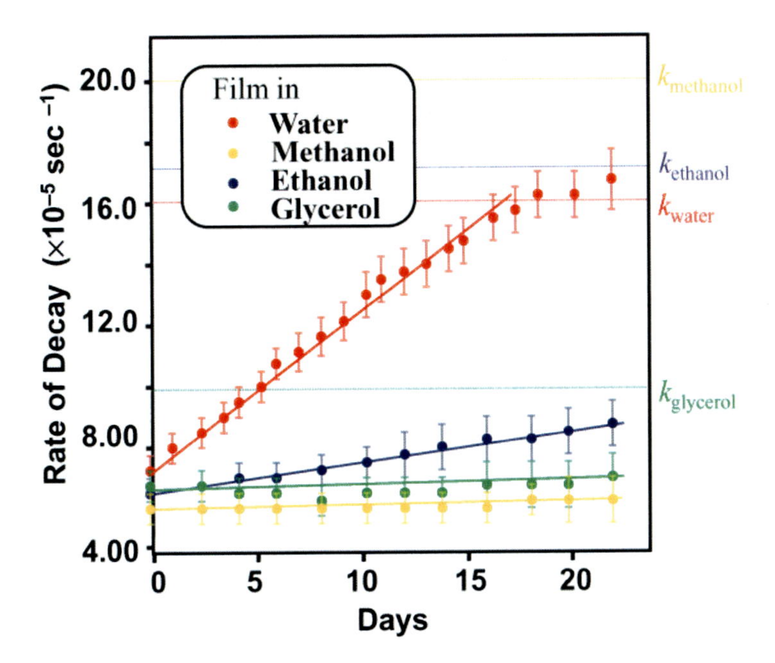

Figure 5. The change of luminescence decay rates as a function of immersed days in various solvents. Each horizontal dotted line indicates the rate of a $[Ru(bpy)_3]^{2+}$ ion in solution, k_i (from Table 1), and encapsulated in sol-gel immersed in solvent i, i = water, ethanol, methanol, and glycerol.

2.c. Solvent Dependent Intrusion Rate into the Aged Sol–Gel Film

The luminescence decay time constants observed in the thin films at constant temperature (25.0 ± 0.2 °C) were monitored for over three weeks in various solvents. Figure 5 shows the inverse of the luminescence decay time, $1/\tau = k$, versus immersed days of the films in each solvent. A monotonic increase of decay rate as a function of immersed days was observed for water and ethanol. Those responses were found to be linearly dependent for a certain duration. A clear example was seen in water, where a monotonic increase of the rate was observed over 21 days, then the rate approached a certain asymptote and stayed approximately constant. The asymptotic value is close to the value of $[Ru(bpy)_3]^{2+}$ ion in water. The rate reached close to the value corresponding to aqueous solution. The rate change as a function of the immersed days was fitted with a linear formula for the duration of the range exhibiting a monotonic change (See Table 4). The change of the rate observed in ethanol was smaller than that of the water by a factor of five. The aging day dependence of the rate seen in water and ethanol indicates that the sol–gel matrix has selective permeability to certain solvents. The sol–gel environment prepared in our study showed that they have positive attraction to those solvents to intrude to inner sites where dopants are located. The specific interaction can be either that dipolar character or hydrogen bonding is significantly interacting with the silanol groups of the sol-gel surface. The domination of the hydrophilic character of the sol-gel matrix then fully explains the case of glycerol and propanol which showed no difference between the solution and the film condition. This fits the fact that the glycerol and propanol are inert to the entrapment of a dopant and that water or ethanol can be preferential in intrusion. It is speculated either that the hydrophilic part of the sol-gel pore is avoiding the intrusion of the hydrophobic portion of the solvent or that the diffusional speed of the glycerol is slow enough not to intrude into the dopant's site when the photodynamics were measured. Different intrusion speeds among solvents can be due to viscosity, which dominates the diffusional movement of the solvent. The viscosity at 20 °C of the solvents tested in this study are in increasing order of methanol (0.597 cP), water (1.002 cP), ethanol (1.2 cP), propanol (2.256 cP), and glycerol (1490 cP). While those with less aging dependence can be verified by these values, the observed order seen in water and methanol are inconsistent. This implies that the observed aging effect cannot be explained solely by the physical property of an interface but it requires intermolecular interactions. The silica sol-gel has an isoelectronic point of pH = 2, and the $SiOH_2^+$ is a predominant form below pH = 2 and SiO^- is the predominant form above pH = 2.[15] The internal pH value of the prepared sol–gel matrix was speculated to be 4.9 from our previous study,[46] thus, the SiO^- is the predominant form prepared for the sample surface. The presence of SiO^- group on the silica gel surface implies a consideration of electrostatic interaction not only between a dopant and matrix but also between matrix and external solvent. Therefore, a polar solvent can be more diffusive through the matrix compared to non-polar solvents. The single molecule optical microscopy study of the positively charged R6G dye molecule indicated that the majority of molecules were fixed (firmly bound to the gel matrix) [45]. This bonding character may explain the unexpected minimal solvent dependence in observed photodynamics and energy gaps.

It is speculated that the amount of the quencher in the air dissolved in each solvent plays a key role in dominating photodynamics, mainly oxygen gas in the air. The solvent dependence may reflect the solubility of oxygen in a solvent. For example, the Ostwald coefficient of liquid water for O_2 is almost three orders of magnitude larger than those of the rest of solvents for O_2.[69] In order to distinguish the effect by the oxygen dissolved in water, the fresh water was replaced and purged with N_2 gas after 10 days, and the values did not deviate from the trend of the previous measurements. This indicates that the time constant was not determined by the dissolved quencher (in this case, oxygen gas molecule) but by the amount of water; or it could be that quenchers already dissolved in water intruded in to the gel pores up to a given time. However, the most puzzling result was that there was no sign of solvent intrusion for the film immersed in methanol. Since the permeability of methanol and ethanol are expected to be similar, rejection of methanol penetration through the silica gel matrix is unexpected. For some reasons, the sol-gel cavity prepared in our method was selectively permeable only to the solvents used in the preparation of a gel precursor. It can be interpreted that the observed aging dependence reflects an on-going hydrolysis of TEOS, where only water and ethanol were involved in our sol-gel method. Our result implies that an encapsulated sensing dopant can be used in most solvents when direct contact with water and ethanol is avoided. However, at this point, we are not able to provide a satisfactory explanation for the cause of this selectivity. This work clearly demonstrates the use of photosensitive heavy metal complexes to probe an optically inactive carrier material. Based on our study, we propose that the silica sol-gel matrix relationship with the surrounding solvent environment is being dominated not only by electrostatic interactions with the negatively charged silanol groups, SiO^-, of the sol-gel but also by hydrophilic interactions dominated, at least early on, by solvents trapped during the fabrication process.

This study shows that the solvents used in fabrication remain within the sol-gel network in fairly significant quantities for some initial period, which has serious implications for applications such as tissue engineering and sensor development if fabrication solvents are toxic or damaging to the dopant or embedded ''device''. Intriguingly, the solvent selectivity demonstrated by the silica sol-gel material might be tunable for use as a molecular sieve. Or, if solvents trapped during fabrication are extruded into the surrounding environment, this material might be applicable as a carrier and a delayed delivery mechanism for low molecular weight fluids.

3. The Description of Acid Penetration to the Gold Nano Colloids Encapsulated Silica Sol-Gel Matrix

Metal compounds encapsulated in nanoscale pores of silica sol-gel have been suggested for application to optical or sensing devices.[70] Since gold's surface plasmon resonance (SPR) is influenced by the refractive index of the surrounding environment, it possesses a significant potential for biosensors as demonstrated by the application of nanoparticles to labeling DNA molecules.[36, 71-73] This medium is cost effective owing to its stable physical and chemical properties, useful catalytic activities and small dimensional size.[74-76]

By embedding various sizes of gold colloids in the silica gel, the interaction rate of the guest dopant with the external acid solvent media was investigated. Differing from the solution condition, the acid needs to interact with the silica sol-gel layer as it intrudes and meets with the gold colloid dopants in silica sol-gel matrix cavities. While the size of the dopants embedded in silica gel cavities can be systematically prepared with only a few nm deviations, the size of the sol-gel cavities can deviate by a few 100 nm. Thus, a specific size dependence for an interaction between dopants and external solvent acid may be overshadowed by the large inhomogeneous distributions of the cavity sizes.

This study intends to capture the major interaction that takes place between solvent intrusion and nanosize dopants in nanoscale cavities. It also aims to investigate if dopant size and solvent interaction can play a major role in the solvent intrusion rate into sol-gel cavities. In most studies, the dopant size is relatively small compared to the size of the cavity created in the sol-gel matrix, and a guest molecule entrapped inside a silica sol-gel matrix exhibits a wide range of molecular mobility originating from different extents of surface interaction.[77-81] However, in the case of encapsulation of gold colloid nanoparticles, the size of the gold colloid dopants can be comparable in size to the cavity resulting in a huge restriction in mobility and a more severe form of interaction from the cavity wall of the sol-gel matrix. This strong host-guest interaction is expected to affect the diffusion rate of an external solvent.

3.a. Sample Handling and Method

A gel precursor was prepared from a solution with 18 mL of TEOS (tetraethyl orthosilicate) from Sigma-Aldrich Co. (St. Louise, Missouri, USA) 0.4 mL of 0.1 M HCl solution, and 5.6 mL distilled and deionized H_2O.[48]This mixture was sonicated for 1 h at room temperature, and stored at 4 °C. Gold colloidal nanoparticles were all purchased from Ted Pella Inc. (Redding, California, USA) with the following diameters (estimated number of particles under experimental condition of OD observed at the peak of the SPR band of each solution for path length 0.5 cm) 5 ± 0.4 nm (2.5×10^{13} particles/mL), 9.9 ± 0.5 nm (2.9×10^{12} particles/mL), 15.2 ± 0.8 nm (7.0×10^{11} particles/mL), 19.8 ± 0.8 nm (2.8×10^{11} particles/mL), 30.7 ± 1.2 nm (8.0×10^{10} particles/mL), 40.6 ± 1.6 nm (3.6×10^{10} particles/mL), 51.5 ± 2.1 nm (1.6×10^{10} particles/mL), 61.3 ± 2.5 nm (8.7×10^{9} particles/mL), 78.9 ± 3.2 nm (4.4×10^{9} particles/mL), and 99.5 ± 4.0 nm (3.2×10^{9} particles/mL). All solutions were reported to contain citric acid (0.1% in volume).

The gold colloid was mixed with sodium tetra borate buffer (pH 9.18) from VWR International Co. (West Chester, Pennsylvania, USA) to catalyze both condensation and polymerization reactions of the TEOS. The volume ratio between the gold colloid solution, sodium tetra borate buffer, and TEOS precursor was 1:2:2. The gel formation was confirmed within 24 h. Whereas we have tested various buffers ranging from pH 3 to pH 10 to optimize storage of gold colloid in silica sol-gel, only sodium tetraborate buffer (pH 9.18) preserved the feature of the SPR band. Other buffers decomposed the colloidal particles or aggregates of the particles, causing precipitation or destabilization of gold colloid and resulted in bluish color.(See Figure 6) This observation clearly indicated that a chemical interaction between gold colloidal surfaces and the contents of pH 9.18, i.e., sodium tetra borate decahydrate ($Na_2B_4O_7 \cdot 10H_2O$), buffer was responsible for stable gel formation. Sodium tetra borate is commonly used to reduce gold colloid particles from $Au^{III}Cl_4^-$ to $Au^{I}Cl_2^-$.

In order to monitor the acid interaction of gold colloid as a function of time, the sol-gel sample was exposed to external solvent and the corresponding absorption spectrum was monitored as the solvent progressively intruded into the sol-gel cavity. The gel precursor and gold colloid were mixed with buffer solution obtaining 2.35 mL volume and placed in a 4.7 mL fluorescence plastic cuvette as shown in step 1 of Figure 7. The cuvette was capped and placed on the flat surface of an optical bench with one of the cuvette faces laid horizontally down for 1 day (step 2 in Figure 7) after which the sol-gel containing the gold colloid was formed. The sample cuvette had a half space vertically filled with silica gel containing gold colloid as shown in step 3 of Figure 7. The empty space was used to contain the solvent (in this experiment it was 0.1 M HCl solution). The volume transported into the cuvette determines the vertical gel thickness (ca. 0.5 cm) and corresponds to the sampling path length of the gel. The thickness of the gel layer was found to vary within ±0.02 mm around the optical path and was consistent within different samples (±0.05 mm). Since the sample path length was 0.5 cm, the concentration of the gold colloid was increased by a factor of two to gain a signal level comparable with the solution condition.

The 0.1 M HCl solution was introduced into the empty side of the cuvette at room temperature (18 °C), and the absorption spectrum was monitored (Hewlett Packard Diode Array Spectrophotometer Model 8452A) as the acid intruded into the gel side of the sample cell. A corresponding color change from red to blue was observed for up to 2 h. The collected spectra were analyzed with the ORIGIN program (version 7.0) to extract components of the spectra, peak averages, and the rate of time dependent peak shifts. Nitrogen porosimetry was conducted by utilizing Micromeritics ASAP2020 gas sorption analyzer, which measures porosity with diameters of ~1.5 nm and above by the N_2 absorption method. The aero gels were prepared from sol-gel samples containing no gold colloids, 20, 40, 60, 80, and 100 nm gold colloids by dehydrating at 120 °C for 3 h. (See Figure 8)

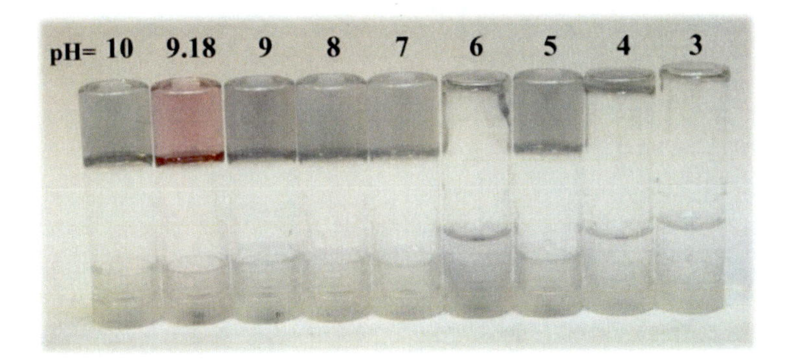

Figure 6. The silica-based sol-gel prepared under various buffer of pH ranging from pH 3 to pH 10. The gel container was placed up-side down in order to demonstrate a formation of sol-gel under each pH condition.

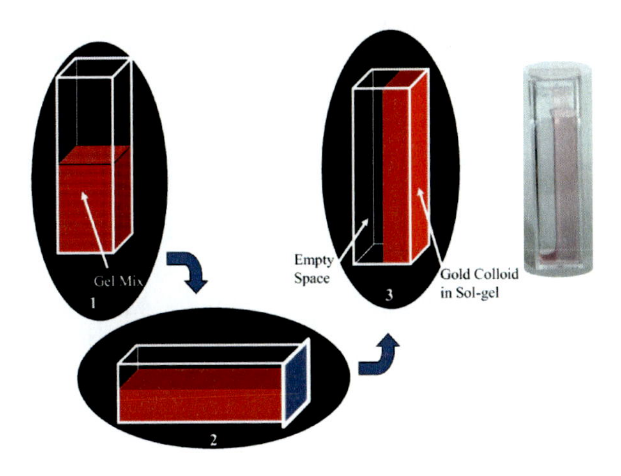

Figure 7. Sketches showing the sample preparation process. 1) The silica gel precursor and a gold colloid mixture are placed in a plastic cuvette with sodium tetra-borate buffer. 2) The cuvette was laid with one of the faces down and the gel formed overnight. 3) The half volume of the cuvette was filled by the sol-gel sample and the other half was an open space. The picture shows an example of a prepared sample cuvette containing gold colloid encapsulated sol-gel.

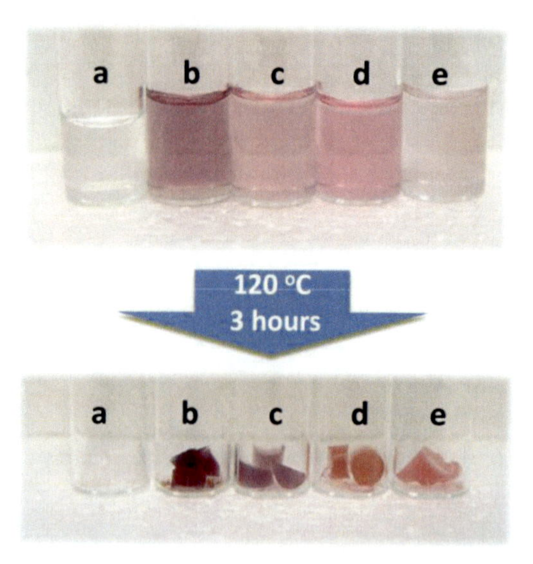

Figure 8. (Top) The sol-gel solutions containing a) no, b) 5 nm, c) 15nm , d) 60 nm and e) 100 nm gold colloids. (Bottom) The samples heated under 120 °C for 3 hours.

3.b. Characterization of Acid Intrusion Rate

As a control, the interaction of the gold with acid solution was investigated in solution for gold colloid of sizes 5 to 100 nm. The strong acid (pH<2) condition was found to provide a color change from red to blue for all tested sizes of gold colloid in the solution.[82-84] Therefore, for all sizes of gold colloid tested in this study (5–100 nm), it was confirmed that the acid of pH 1 caused a color change in the colloid. Based on these results, the interaction of gold colloid with acid of pH 1 was tested on the gold colloid encapsulated in the silica sol-gel environment. Most of the gold colloid, originally red, clearly turned bluish in silica sol-gel after the acid (0.1 M HCl) intrusion was completed. The colorimetric change of the gold colloid was monitored by absorption spectroscopy of the SPR band, which featured a prominent absorption peak around 534 ± 6 nm. The absorption bands were fit with a Gaussian profile for the range between 400 and 800 nm.

When the band component consisted of multi parts the peak position, λ_{peak}, was given by

$$\lambda_{peak} = \sum_{i=1}^{n} a_i \lambda_i \tag{4}$$

where λ_i and a_i represent the peak position and fraction of the ith component band. Most of the bands observed in our study were fully analyzed with two components or one component and a large background band with maximum of 350 ± 50 nm. The component a_i was determined by the fraction of the band's area to the total sum of the entire bands' areas, e.g., $a_1 = A_1/\sum_i A_i$. The background band was excluded in this analysis.

In Figure 9, the feature of the SPR band of gold colloid in silica sol-gel is displayed for (a) before adding the acid and (b) 1 h after adding the acid to the empty side of the cuvette. Since the gel sample contained a broad background band originating from light scattering in the silica sol-gel, the spectra are shown after the background band was subtracted. As for the spectrum observed before adding acid (spectrum shown in Figure 9a with black line), the analysis indicated that it contained two components of λ_1 = 530 nm (a_1 = 0.38) and λ_2 = 592 nm (a_2 = 0.62) with the average peak position λ_{peak} = 568 nm, where the number in the parenthesis indicates the fraction of the area of each component. While the detail features did not exactly match each other, this spectrum closely corresponds to the aqueous solution under pH 3.7 (red line) but not under pH 7 (blue dotted line) as indicated in Figure 9a. The spectrum of 20 nm gold aqueous solution at pH 3.7 showed an average peak position of λ_{peak} = 576 nm with two components of λ_1 = 526 nm (a_1 = 0.39) and λ_2 = 607 nm (a_2 = 0.61). Therefore, the pH condition of the silica gel environment (even before adding acid) was acidic, if we assume that the spectrum feature could only be attributed from pH condition. The spectrum of 20 nm gold colloid in silica gel after an addition of acid (pH 1) showed a clear sign of color change in its spectrum as shown in Figure 9b. The spectrum width was broadened and the average peak position was calculated to be 618 nm with two components λ_1 = 544 nm (a_1 = 0.22) and λ_2 = 640 nm (a_2 = 0.78). While no exact match was found, the closest was the spectrum of aqueous solution under pH 2.75 (red line), which contains λ_1 = 532 nm (a_1 = 0.34) and λ_2 = 660 nm (a_2 = 0.66) band components resulting average peak of λ_{peak} = 618 nm.

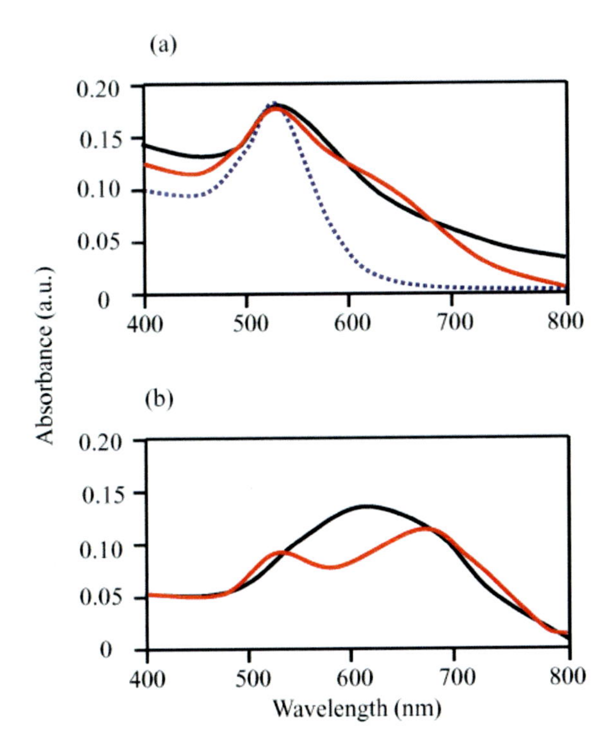

Figure 9. The spectra of gold colloidal nanoparticles of 20 nm size in aqueous solution and encapsulated in a silica sol-gel for (a) before addition of the acid and (b) 1 h after addition of the acid. (a) After the base line was subtracted, the gold colloid in silica gel (black line) is plotted together with aqueous solution of 20 nm gold colloid under pH 7 (blue dotted line) and pH 3.7 (red line). (b) After the base line was subtracted, the gold colloid in silica gel (black line) is plotted together with aqueous solution of 20 nm gold colloid under pH 2.75 (red line).

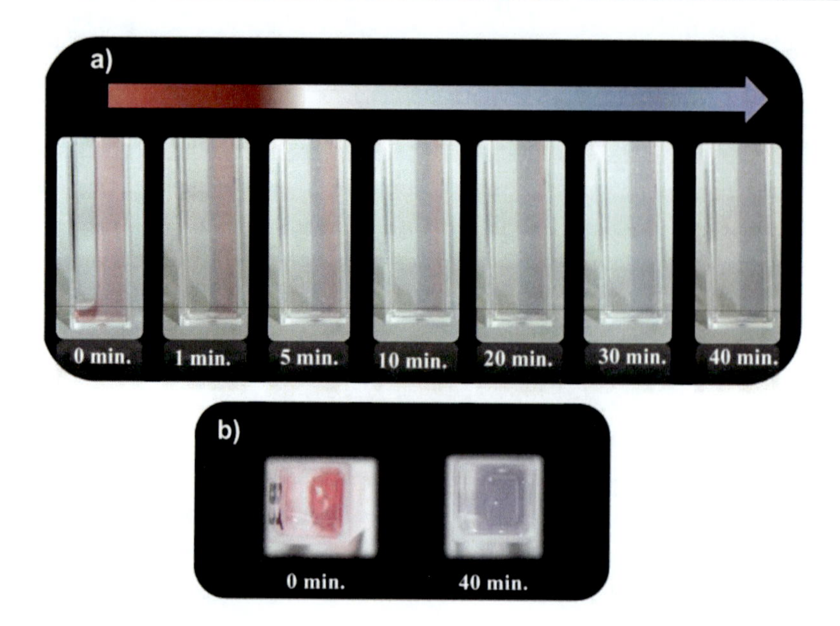

Figure 10. The color change observed for 20 nm gold colloids encapsulated silica sol-gel. a) Side view of sample cell observed at 0, 1, 5, 10, 20, 30, and 40 min. after acid was inserted. b) Top view of sample cell before (0 min.) and 40 min. after acid was inserted.

For most of the sizes tested in our study, the color change of the gold colloid did not happen instantaneously. For example, the process of color change in 20 nm gold colloid was completed within 40 min after the acid was introduced. (See Figure 10) We observed a clear boundary line between different colors. A shift from the solvent side to the deeper silica gel side indicated the direction of solvent intrusion. Since a vertical boundary was almost parallel to the surface plane of the cuvette, the speed of acid intrusion was homogenous within the sample. The thickness of the sol-gel sample and the condition of interaction with an external solvent differ at the top edge area of the cuvette. Since the spectroscopic data was obtained from the middle area of the cuvette, the edge effect was ignored in the analysis. For each time interval, the corresponding absorption spectrum, indicating color change as a function of time, is shown in Figure 11 for the case of the gold colloid size of 20 nm encapsulated in a silica sol-gel. Each spectrum exhibited two components where λ_1= 531 ± 13 nm and λ_2 = 585 ± 27 nm or a longer wavelength. It is interpreted that the spectra has an isosbestic point around 590 nm. The component given as λ_1 was identified as the original SPR band of gold colloid before acid insertion. The spectral feature of this component remained constant until all of the gold colloid in the silica gel interacted with the acid solvent. This then caused a gradual shift toward a longer wavelength and a decreased the spectral area. The peak position of λ_2 also shifted toward a longer wavelength and its spectral area grew as time progressed. The component represented by λ_2 must correspond to the SPR band of gold colloid in the region affected by the acid.

The shift of average peak position as a function of time, $\lambda_{peak}(t)$, implies the diffusional motion of the acid in silica gel and the resulting interaction. Since the $\lambda_{peak}(t)$ is the average peak position weighted by the area of the band, it represents the concentration of gold colloid that interacted with acid and turned bluish in color. While the diffusion model of the acid solution in a silica gel was not established at this point, among all tested analytical formulas, we found that the observed time dependent peak shift, $\lambda_{peak}(t)$, could be best explained by a single exponential function:

$$\lambda_{peak}(t) = \lambda_A - \lambda_B e^{-kt} \tag{5}$$

where λ_A indicates the final peak position, λ_{final}, and λ_B is the difference between final and initial peak positions, $\lambda_{final} - \lambda_{initial}$, i.e., $\lambda_{peak}(0) = \lambda_{initial}$ and $\lambda_{peak}(\infty) = \lambda_{final}$. The time constant, k, indicates the rate of acid intrusion into the gel cavity. The average peak shift was plotted as shown in Figure 11 and fitted with the model described in Eq. 5.

The time constants were obtained for all sizes of the gold colloid tested in this study and plotted as a function of size in Figure 12, with values recorded in Table 5. As the size of gold colloid increased, the rate reached a maximum at 15 nm and drastically slowed down as the size approached 100 nm. We found that the gold colloid sizes of 60 nm almost remained the original color, as confirmed by the smallest values of λ_B in Table 5.

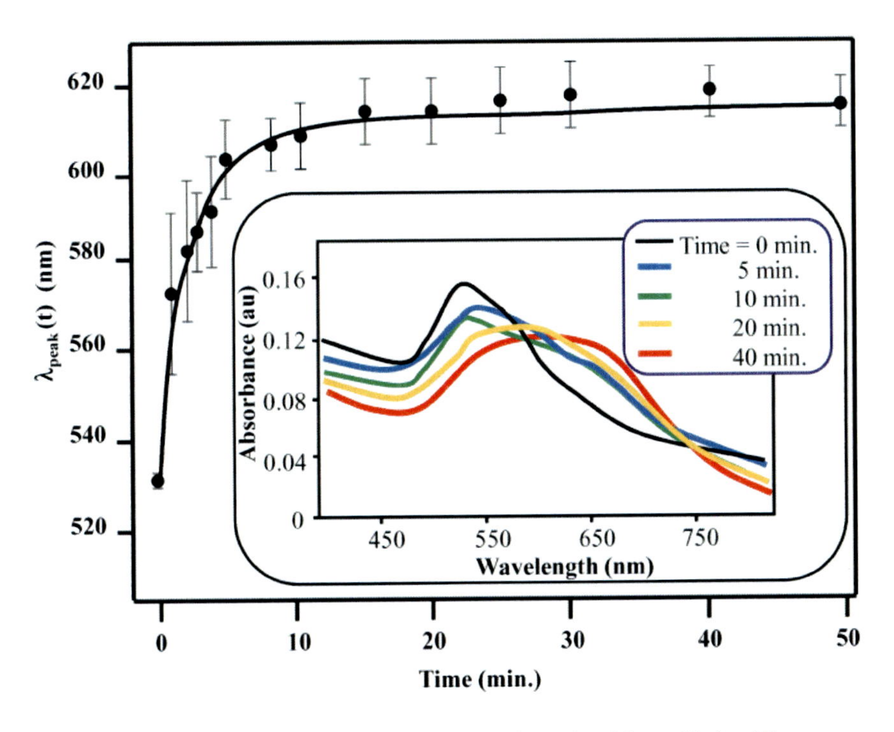

Figure 11. Time dependence of the average spectral peak, $\lambda_{peak}(t)$, of the gold size 20 nm encapsulated in a silica sol-gel after an exposition to the acid solution (pH 1). The average peak position (black circle with an error bar) of each time was calculated from Eq. 4. The data was fit by the model shown in Eq. 5, and the best fit is given by the solid line in this figure. (Inset) The absorption spectra of the gold colloid size of 20 nm encapsulated in a silica sol–gel is shown. The range from 400 to 800 nm is displayed as a function of time after it was exposed to the acid (pH 1) solution.

Table 5. The summary of parameters extracted for $\lambda_{peak}(t)$ given in Eq. 5 of various sizes of gold colloid.

Size/nm	λ_A/nm	λ_B/nm	k/min^{-1}
5	556 ±1	30 ± 1	0.15 ± 0.01
10	585 ±1	51 ± 2	0.18 ± 0.01
15	588 ±4	69 ± 10	0.7 ± 0.2
20	611 ±4	70 ± 8	0.37 ± 0.09
30	614 ±3	92 ± 5	0.19 ± 0.02
40	624 ± 2	115 ± 5	0.22 ± 0.01
50	557 ±2	24 ± 2	0.11 ± 0.02
60	550 ±1	15 ± 1	0.08 ± 0.01
80	555 ±1	21 ± 1	0.14 ± 0.01
100	607 ±4	46 ± 3	0.07 ± 0.02

Generally, the silica gels may not possess the same microstructure (pore size, pore size distribution, pore volume, and pore connectivity) from one sample to another. If the microstructures are considerably different from one gel to another, the observed phenomenon may be due to the different diffusion rates of the gels themselves, not the sizes of nanoparticles. Therefore, it is critical to confirm deviation of the cavity conditions via the inclusion of nitrogen porosimetry data of the resultant aerogels. Since aerogels have very low shrinkage, measurements on aero gels provide similar pore size, pore size distribution, and pore volume as their corresponding wet gels. According to the nitrogen porosimetry measurements, the adsorption average pore width and desorption average pore width for all five samples shown in Figure 8 were obtained as 46 Å ± 2 and 40 Å ± 3, respectively. We conclude that microstructures are considerably equivalent from one gel to another, and that the different diffusion rates shown in Figure 12 were due to the sizes of gold nanoparticles.

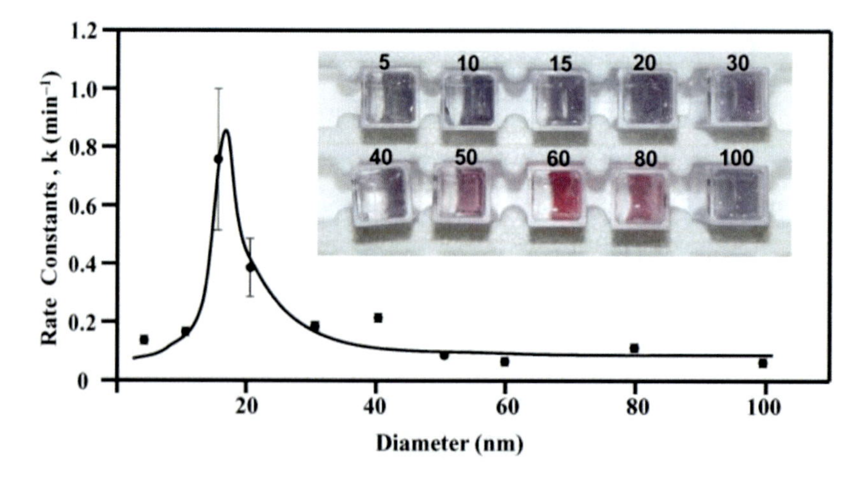

Figure 12. The reaction rate constant, k, extracted with the use of Eq. 5 for each gold colloidal size. The solid curve indicates the predicted value by an analytical formula (Eq. 6) adopted from the Beutler-Fano profile. (Inset) The top views of the sample cells encapsulating 5, 10, 15, 20 30, 40, 50, 60, 80, and 100 nm gold colloids are shown.

3.c. Nanosize Dependence in Acid Intrusion Rate

The degree of solvent intrusion was quantitatively extracted from the spectral feature indicating the intensity of color change in the gold colloid. Since the sampling path of the absorption spectrum goes perpendicular across the gel layer, the observed spectrum was an integration of absorbance originating from the regions both affected and unaffected by the acid before the completion of the reaction. The average spectral shift for a given time, $\lambda_{peak}(t)$, was weighted by the area of both intruded and non-intruded regions. Blue regions indicated the completion of the acid reaction with the gold; and thus, the degree of spectral shift is proportional to the amount of the gold reacted with the acid. The rate of color change in a given location exhibits the diffusion rate of the solvent and the rate of interaction between acid and the gold encapsulated in the cavity. The rate roughly decreased by 10 when the pH of the solvent was increased by one (i.e., from pH 1 to 2), which is a signature of first order reactions. The rate did not show a significant change as the number of gold colloids increased by a factor of two. Thus, the rate of the reaction must be first order in acid for a gold colloid.

Remarkably, the reaction rate constant, k, was found to have a narrow size range dependency on the dopant size as shown in Figure 12. The selective maximization of the reaction rate with 15 nm gold colloid may be explained by the relative nanoscale size relationship between the cavity and the gold colloid. We think this can be subdivided into five categories: a) 5-10 nm gold colloids, b) 15-30 nm gold colloid, c) 40-50 nm gold colloids, d) 60-80 nm gold colloids, and e) 100 nm gold colloid. The corresponding conditions are all given as sketch in Figure 13, and they show the rough relationship between the size of the dopant and the size of the cavity. In the figure, the channels are drawn as paths that connect each cavity. Thus, the solvent can be transported from cavity to cavity. In order to simplify the argument, each cavity and gold colloid is represented by a sphere. Based on the nitrogen porosimetry, the average cavity size was estimated to be around 5 nm and was consistent from one sample to another. This implies that all tested gold colloid sizes are larger than the average cavity size. Therefore, the silica sol-gel matrix needs to be physically arranged in order to create larger cavity sizes to encapsulate the gold colloids.

(a) If the dopant (5-10 nm) is relatively small, the gold colloid does not require a new cavity of silica sol-gel matrix to be formed and maintains the position and path of the original channels. Since most of the surface area of the gold colloid could be covered by the silica sol-gel, the diffusion rate of the acid into the colloid would be relatively slow yet the channels would allow the acid to penetrate. (See Figure 13a)

(b) As the dopant size gets bigger, the original cavity needs to be stretched further while widening the diameter of the channels to contain the gold colloid. This constraint allows the channels to gather around the gold colloids and creates easier access for the external acid solvent. According to our observation, we think that the optimum condition for creating channels around the colloidal surfaces was achieved at 15 nm of gold colloid as shown in Figure 13b.

(c) As the dopant size increases (40-50 nm), the sol-gel matrix requires further rearrangement and requires the space of other cavities, i.e., combinations of multiple cavity spaces. Therefore, the condition of the channels around the gold colloidal surface is the same as case of a), except for the size of the gold colloids. This condition, shown in Figure 13c, reduces the surface exposed to the external solvent compared to the case shown in Figure 13b. Thus, the rate of interaction is less than that at 15 nm gold colloid.

(d) At an optimum condition, the size of the dopant fortuitously matches and fits the separating multiple silica gel cavities without changing the configuration of the channels as shown in Figure 13d. This particular situation may correspond to the dopant size of 60 nm, which clearly exhibited red color even after it was fully immersed in acid. It also implies that this constraint may cause the channels to be closed due to the shifting of the silica sol–gel matrix. This would prevent the acid penetration.

(e) Since the dopant size of 100 nm is far larger than the cavity size, multiple cavities cannot be arranged because the size of the dopant may exceed the flexibility of the silica sol–gel matrix. In this situation, the cavities surrounding the gold dopant act like a channel, thereby making the external solvent more accessible to the dopant surface as sketched in Figure 13e. However, at this dopant size, a significant percentage of the gold colloids were still covered by the silica sol–gel matrix resulting in a relatively slower reaction rate.

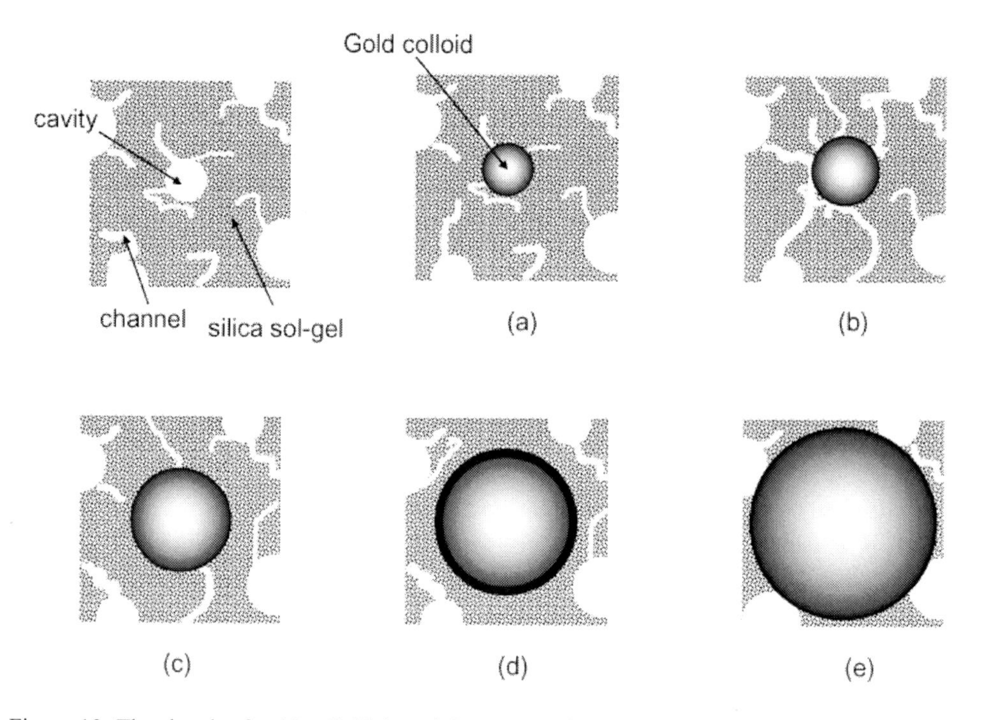

Figure 13. The sketch of gold colloidal particles encapsulated in a cavity of silica sol gel. a) The size of the gold colloidal particle is relatively small. b) The dopant's size causes channels to be created around the dopant's surface. c) The size of the dopant is large enough to contact multiple cavities and keeps the channels disconnected from one another. d) The dopant size reaches an optimum condition fortuitously matching the exact volume size of the existing multiple cavities. The dopant's surface is tightly covered by the silica-sol matrix and the channels are disconnected from each other. e) The dopant size is well over the combinations of multiple cavity sizes. This causes the original cavity to be damaged and to be turned into multiple channels for an external solvent.

The observed trend in diffusion rate shown in Figure 12 had an asymmetric profile centering on the peak at 15 nm, none of the popularly used peak functions reproduced the asymmetric profile. To the best of our knowledge, Beutler-Fano formula [85] is considered to be the most appropriate analytical formula with which was able to reproduce the observed trend. The Beutler-Fano formula provides an absorption cross section, $\sigma(\varepsilon)$, at an auto-ionization process by

$$\sigma(\varepsilon) = \sigma_o \frac{(q+\varepsilon)^2}{1+\varepsilon^2} + \sigma_c \tag{6}$$

where σ_o or σ_c is the dissociation or continuum dissociation cross section and q is the line shape factor. The reduced energy, ε, at an energy, E, is given by $(E - E_r)/(\Gamma/2)$ with the width of resonance, Γ, and the resonance energy, E_r. While the notations and the meanings of the parameters used in Eq. 6 do not physically match the parameters characterized in Figure 12, we converted $\sigma(\varepsilon)$ and E to the rate constant, $k(\varepsilon)$, and the size of gold nanocolloid (diameter of the gold colloid in the unit of nm), d, respectively. The value given by E_r was interpreted to be noted as d_r, which is the size at which the rate of interaction is maximized. We have arranged an equation given by Eq. 6 where ε is given by $(d - d_r)/(\Gamma/2)$,

$$k(\varepsilon) = k_o \frac{(q+\varepsilon)^2}{1+\varepsilon^2} + k_c \tag{6$'$}$$

and k_o implies a normalization factor for the reaction rate constant at the peak. In this case, the k_c corresponds to the rate of diffusion of acid when no dopant is present. The best fit curve is shown in Figure 12 as a solid line, and the best fit parameters are shown in Table 6. The predicted values from Eq. 6 best explain the overall asymmetric trend observed in our study. Some data points around the tail regions do not match these predicted values, especially the local maximum seen around 40 nm.

The Beutler-Fano formula describes the absorption cross section (or a spectral line shape) when a discrete ionization energy (resonance energy) level exists over the continuous ionization levels. In the present system, the acid interaction of a given dopant takes place with various cavity sizes, and the inhomogeneous distribution of the cavity sizes resembles the situation of continuous ionization energy levels. The resonance energy (or a discrete energy level) corresponds to a specific dopant size, which creates an optimum interaction (maximum exposure to the acid) and a maximum rate. The value Γ in $\varepsilon = (d-d_r)/(\Gamma/2)$ implies the inhomogeneity of diameter d_r, and parameter q is the line shape factor which particularly controls an asymmetric portion of the profile. The maximum rate for acid interaction in our sol-gel system will be achieved at the dopant size of $d_r = 16$ nm with an accuracy of ±2 nm. The width of this size condition, however, was predicted to be $\Gamma = 5$ nm which is relatively large compared to the accuracy of the d_r. This value must be attributed to a relatively broad asymmetric tail observed at sizes larger than d_r, and it indicates that the average cavity size is larger than d_r with broad deviation, whereas the optimum condition of the rate was given at d_r with relatively narrow area. The largest deviation seen at 40 nm may indicate that there is a secondary optimizing size existing around the 40 nm region. However, peak height is relatively small compared to the primary feature shown around 15 nm, and we did not attempt to reproduce this feature by analytical formula.

Table 6. The parameters used to best explain the dopants' size dependence in the acid reaction rate constant by the Beutler- Fano formula given in Eq. 6′.

Parameters	
k_o/min^{-1}	0.01 ± 0.02
k_c/min^{-1}	0.10 ± 0.04
q	7± 8
d_r/nm	16 ± 2
Γ/nm	5 ± 4

CONCLUSION

The luminescence decay time constants of the photo-excited ^{3}MLCT state of a $[Ru(bpy)_3]^{2+}$ ion dopant encapsulated in a silica based sol-gel thin film was examined while immersed in water, methanol, ethanol, 2-propanol, and glycerol. As temperature increased, the dynamics of the dopants immersed in water, ethanol, as well as in sol-gel bulk were better explained by the KWW model. Generally, the ΔE(sol- gel film) values were reduced from ΔE(solution). The films immersed in water for three weeks presented the most remarkable increase in relaxation rates finally approaching the asymptotic value observed in the water solution. This phenomenon must be heavily due to a hydrophilic interaction through the sequential intrusion of water or ethanol solvent into sol-gel pores. Water and ethanol were the only two solvents that exhibited solvent dependence in the aging effect of the film. From these facts, the encapsulated environment of the dopants entrapped in sol-gel film prepared in our method was dominated by silica gel network as well as the water and ethanol solvent co-encapsulated inside the cavity, and the resulting photodynamics must be somewhat influenced by the interaction between dopant and residual water/ethanol solvents.

The color change of the gold colloid as a sign of interaction with acid was monitored by an absorption spectrum. As the size of encapsulated gold was varied between 5 nm and 100 nm, we were able to investigate the rate of acid intrusion into the dopant's site and measure the interaction with the gold colloid nanoparticles in real time. The interaction rate of the gold colloid with acid was controlled by the intrusion rate of the solvent and nanoscale dopants' size. The degree of the interaction between the gold colloid and the acid was considered to be dependent on the relative size between the gold colloids and the silica gel cavity. This study confirmed that the nanoscale dopant size affects the rate of solvent penetration into a sol-gel cavity, and the maximum rate was found for 15 nm gold colloid. On the other hand, the smallest acid interaction and color change was observed for 60 nm gold colloid. The surface of these gold colloids was speculated to be homogeneously covered by the silica gel layer avoiding direct contact of the acid with the surface of the gold colloid. The closest match of the average size of the cavity was observed with the size of the gold colloids of 60 nm diameter.

ACKNOWLEDGMENTS

This work is supported by the National Science Foundation under grant number NSF-NER 0508240. A generous contribution from The SUNY-Geneseo Foundation at an initial stage of this project is greatly acknowledged. I thank The Analytical Technology Division, Kodak Co. (Rochester, NY) for providing the analysis of our film with M-2000 V. Nitrogen porosimetry measurements were kindly supported by Professor Hong Yang of the University of Rochester. The following individuals also should be recognized for their contribution to the project described here: Bradley E. Johnson, Jonathan W. Bourne, Jeffrey R. Swana, Tonya M. Gilbert, Duo D. Chen, Liwen Chen, and Paul Kogan. With kind permission from Springer Science & Business Media, the contents of the following articles are displayed in this document:

"The Photodynamics of $[Ru(bpy)_3]^{2+}$ Ion Dopant in a Solvent Exposed Silica Sol-gel Thin Film, Journal of Sol-Gel Science and Technology", 43, 2007, 259-268, by K. Yokoyama, B. E. Johnson and J. W. Bourne.

"The Nanoscale Description of Acid Penetration to the Gold Colloids Encapsulated in Silica Sol-Gel Matrix", Journal of Sol-Gel Science and Technology, 50, 2009, 48-57 by K. Yokoyama, J. R. Swana, T. M. Gilbert, D. D. Chen, L. Chen and P. Kogan.

REFERENCES

[1] Langer, R. *Chem. Eng. Commun.*, 1980, 6.

[2] Santini, JT; Cima, MJ; Langer, R. *Nature*, 1999, 397, 335-338.

[3] Unger, K; Rupprecht, H; Valentin, B; Kircher, W. *Drug Dev. Ind. Pharm.*, 1983, 9, 69-91.

[4] Klein, CPAT; Li, P; Blieck-Hogervorst, JMA; de Groot, K. *Biomaterials*, 1995, 16, 715-719.

[5] Wilson, J; Douek, E; Rust, KD. In *Bioceramics,* J. Wilson; L. Hench; D. Greenspan, eds; Alden Press: Oxford, 1995, Vol. 8, 239-245.

[6] Suominen, E; Kinnunen, J. *J. Plast. Reconstr. Surg. Hand Surg.*, 1996, 30, 281-289.

[7] Stoor, P; Söderling, E; Grenman, R. *J. Biomed. Mater. Res.*, 1999, 48, 869-874.

[8] Dvorak, O; De Armond, MK. *J. Phys. Chem.*, 1993, 97, 2646 - 2648.

[9] Novak, BM. *Adv Mater.*, 1993, 5, 422.

[10] Lee, GR; Crayston, JA. *Advanced Materials*, 1993, 5, 434-442.

[11] MacCraith, BD; McDonagh, C; O'Keefe, G; Keyes, ET; Vos, JG; O'Kelly, B; McGlip, JF. *Analyst*, 1993, 118, 385.

[12] Castellano, FN; Heimer, TA; Tandhasetti, MT; Meyer, GJ. *Chem. Mater.*, 1994, 6, 1041-1048.

[13] Kiernan, P; McDonagh, C; MacCraith, BD; Mongey, K. *Sol-Gel Sci. Technol.*, 1994, 2, 513-517.

[14] MacCraith, BD; McDonagh, C; O'Keefe, G; McEvoy, AK; Butler, T; Sheridan, FR. *Proc. SPIE-Int. Soc. Opt. Eng.*, 1994, 118, 2288 (Sol-Gel Optics III), 518.

[15] Matsui, K; Sasaki, K; Takahashi, N. *Langmuir*, 1991, 7, 2866 - 2868.

[16] Valiullin, R; Kortunov, P; Karger, J; Timoshenko, V. *J Chem Phys*, 2004, 120, 11804.
[17] Alberts, B; Alberts, B; Bray, D; Lewis, J; Raff, M; Roberts, K; Watson, JD. *Molecular Biology of the Cell*, Garland: New York, 1994.
[18] Crick, F. *Nature (London)*, 1970, 225, 420.
[19] Mitra, PP; Sen, PN; Schwartz, LM; Le Doussal, P. *Phys Rev Lett*, 1992, 68, 3555.
[20] Mitra, PP; Sen, PN; Schwartz, LM. *Phys Rev B*, 1993, 47, 8565.
[21] Callaghan, P. *Principles of Nuclear Magnetic Microscopy*, Clarendon Press: Oxford 1993.
[22] Latour, LL; Svoboda, K; Mitra, PP; Sotak, CH. *Proc Natl Acad Sci,* USA, 1994, 91, 1229.
[23] Eigen, M; Rigler, R. *Proc. Natl. Acad .Sci.,* U S A, 1994, 91, 5740-5747.
[24] Berne, BJ; Pecora, R. *Dynamic Light Scattering: with Applications to Chemistry, Biology, and Physics*, Dover: New York 2000.
[25] Potma, EO; de Boeji, WP; van Haastert, PJM; Wiersma, DA. *Proc Natl Acad Sci,* USA, 2001, 98, 1577.
[26] Stallmach, F; Vogt, C; Karger, J; Helbig, K; Jacobs, F. *Phys Rev Lett*, 2002, 88, 105505.
[27] Darqui, A; Poline, JB; Poupon, C; Saint-Jalmes, H; Le Bihan, D. *Proc Natl Acad Sci,* USA, 2001, 98, 9391.
[28] Pfeuffer, J; Flogel, U; Dreher, W; Leibfritz, D. *NMR Biomed*, 1998, 11, 19.
[29] Zientara, GP; Freed, JH. *J. Chem. Phys.*, 1980, 72, 1285.
[30] Kutner, R; Pekalski, A; Sznajd-Weron, K. *Anomalous Diffusion: From Basics to Applications*, Springer: Berlin 1999.
[31] Weiss, GH; Rubin, RJ. *Adv. Chem. Phys.*, 1983, 52, 363.
[32] Binder, K; Heerman, DW. *Monte Carlo Simulations in Statistical Physics*, Springer: Berlin 1992.
[33] Ben-Avraham, D; Havlin, S. *Diffusion and Reaction in Fractals and Disordered Systems*, Cambridge University Press: Cambridge 2000.
[34] Kärger, J; Heitjans, P; Haberlandt, R. *Diffusion in Condensed Matter*, Vieweg: Braunschweig/Wiesbaden 1998.
[35] Haes, AJ; Van Duyne, RP. *J, Ame. Chem. Soc,*, 2002, 124, 10596 -10604.
[36] Mirkin, CA; Letsinger, RL; Mucic, RC; Storhoff, JJ. *Nature*, 1996, 382, 607-609.
[37] Xu, XH. N; Patel, R. In *Encyclopedia of Nanoscience and Nanotechnology*, HS. Nalwa, ed; American Scientific Publisher: CA, 2004, Vol. 1, 181-192.
[38] Kyriacou, SV; Brownlow, WJ; Xu, XHN. *Biochemistry*, 2004, 43, 140-147.
[39] Xu, XHN; Brownlow, WJ; Kyriacou, SV; Wan, Q; Viola, JJ. *Biochemistry*, 2004, 43, 10400-10413.
[40] Lyon, LA; Pena, DJ; Natan, MJ. *J. Phys. Chem. B*, 1999, 103, 5826-5831.
[41] Matsui, K; Momose, F. *Chem. Mater.*, 1997, 9, 2588 - 2591.
[42] Castellano, FN; Meyer, GJ. *J. Phys. Chem.*, 1995, 99, 14742 - 14748.
[43] Innocenzi, P; Kozuka, H; Yoko, T. *J. Phys. Chem. B*, 1997, 101, 2285-2291.
[44] Momose, F; Maeda, K; Matsui, K. *J. Non-Cryst. Solids*, 1999, 244, 74-80.
[45] Gilliland, JW; Yokoyama, K; Yip, WT. *Chem Mater*, 2004, 16, 3949.
[46] Gilliland, JW; Yokoyama, K; Yip, WT. *J Phys. Chem B*, 2005, 109, 4816.
[47] Yokoyama, K; Johnson, BE; Bourne, JW. *J Sol-Gel Sci Technol*, 2007, 43, 259.

[48] Yokoyama, K; Swana, JR; Gilbert, TM; Chen, DD; Chen, L; Kogan, P. *J. Sol-Gel Sci. Technol.*, 2009, 50, 48-57.
[49] Avnir, D; Levy, D; Reisfeld, R. *J. Phys. Chem.*, 1984, 88, 5956-5959.
[50] Dubois, A; Canva, M; Brun, A; Chaput, F; Boilot, JP. *Syn. Metals*, 1996, 81, 305-308.
[51] Suratwala, T; Gardlund, Z; Davidson, K; Uhlmann, DR; Watson, J; Bonilla, S; Peyghambarian, N. *Chem. Mater.*, 1998, 10, 199-209.
[52] Murtagh, MT; Shahriari, MR; Krihak, M. *Chem. Mater.*, 1998, 10, 3862-3869.
[53] Murtagh, MT; Kwon, HC; Shahriari, MR; Krihak, M; Ackley, DE. *J. Mater. Res.*, 1998, 13, 3326-3331.
[54] Handy, ES; Pal, AJ; Rubner, MF. *J. Am. Chem. Soc.*, 1999, 121, 3525-3528.
[55] Rudmann, H; Shimada, S; Rubner, MF. *J. Am. Chem. Soc.*, 2001, 124, 4918.
[56] Bernhard, S; Barron, JA; Houston, PL; Abruna, HD; Ruglovsky, JL; Gao, X; Malliara, GG. *J. Am. Chem. Soc.*, 2002, 124, 13624-13628.
[57] Bernhard, S; Gao, X; Malliaras, GG; Abruna, HD. *Advanced Materials*, 2002, 14, 433-436.
[58] Lin, CT; Bottcher, W; Chou, M; Creutz, C; Sutin, N. *Journal of the American Chemical Society*, 1976, 98, 6536-6544.
[59] Hench, LL. *Sol-Gel Silica*, Westwood, Noyes Publications 1998.
[60] Martínez, JR; Ruiz, F; Vorobiev, YV; Pérez-Robles, F; Gonález-Hernández, J. *J. Chem. Phys.*, 1998, 109, 7511-7514.
[61] Dunn, B; Zink, JI. *Chem. Mater.*, 1997, 9, 2280-2291.
[62] Wong JAC. *Glass Structure by Spectroscopy*, Dekker: New York 1976.
[63] Maruszewski, K; Jasiorski, M; Salamon, M; Strek, W. *Chemical Physics Letters*, 1999, 314, 83-90.
[64] Juris, A; Balzani, V; Barigelletti, F; Campagna, S; Belser, P; Von Zelewsky, A. *Coord. Chem. Rev.*, 1988, 84, 85-277.
[65] Maruszewski, K; Strommen, DP; Kincaid, JR. *J. Am. Chem. Soc.*, 1993, 115, 8345 - 8350.
[66] Kober, EM; Meyer, TJ. *Inorg. Chem.*, 1984, 23, 3877 - 3886.
[67] Maruszewski, K; Bajdor, K; Strommen, DP; Kincaid, JR. *J. Phys. Chem.*, 1995, 99, 6286 - 6293.
[68] Lumpkin, RS; Kober, EM; Worl, LA; Murtaza, Z; Meyer, TJ. *J. Phys. Chem.*, 1990, 94, 239 - 243.
[69] Battino, R. *Oxygen and Ozone*, Pergamon Press: Oxford 1981.
[70] Liz-Marzan, LM; Kamat, PV. eds; Springer: US, 2003, 227.
[71] Alivisatos, AP; Johnson, KP; Peng, X; Wilson, TE; Loweth, CJ; Bruchez, MP; Jr; Schultz, PG. *Nature*, 1996, 382, 609-611.
[72] Taton, TA; Mirkin, CA; Letsinger, RL. *Science*, 2000, 289, 1757-1760.
[73] Reichert , J; Csaki, A; Kohler, JM; Fritzsche, W. *Anal. Chem.*, 2000, 72, 6025-6029.
[74] Katz, E; Shipway, AN; Willner, I. In *Nanoparticles from Theory to Application*; Wiley-VCH: Germany, 2004, 368-421.
[75] Daniel, MC; Astruc, D. *Chem. Rev.*, 2004, 104, 293-346.
[76] Verma, A; Simard, JM; Rotello, VM. *Langmuir*, 2004, 20, 4178-4181.
[77] Dave, BC; Soyez, H; Miller, JM; Dunn, B; Valentine, JS; Zink, JI. *Chem. Mater.*, 1995, 7, 1431 - 1434.
[78] Zhang, Z; Suo, J; Zhang, X; Li, S. *Chem. Commun.*, 1998, 241 - 242.

[79] Jordan, JD; Dunbar, RA; Bright, FV. *Anal. Chem.*, 1995, 67, 2436-2443.
[80] Doody, MA; Baker, GA; Pandey, S; Bright, FV. *Chem. Mater.*, 2000, 12, 1142-1147.
[81] Flora, KK; Brennan, JD. *Chem. Mater.*, 2001, 13, 4170-4179.
[82] Yokoyama, K; Welchons, DR. *Nanotechnology*, 2007, 18, 105101-105107.
[83] Yokoyama, K; Briglio, NM; Sri Hartati, D; Tsang, SMW; MacCormac, JE; Welchons, DR. *Nanotechnology*, 2008, 19, 375101-375108.
[84] Yokoyama, K. In *Advances in Nanotechnology*, EJ. Chen; N. Peng, eds; Nova Science Publishers; NY, 2010, Vol. 1, 65-104.
[85] Fano, U. *Phys Rev B*, 1961, 124, 1866.

In: The Sol-Gel Process
Editor: Rachel E. Morris

ISBN 978-1-61761-321-0

Chapter 8

SOL-GEL PROCESSES ON CHEMICALLY INERT SUBSTRATES TO FABRICATE FUNCTIONAL HYBRID ARCHITECTURES AND THEIR APPLICATIONS

Haiqing Li and Il Kim*
The WCU Center for Synthetic Polymer Bioconjugate Hybrid Materials,
Department of Polymer Science and Engineering,
Pusan National University, Pusan, Korea

ABSTRACT

The precise organization of functional matters on the chemically inert surfaces of varied substrates for the fabrication of uniform hybrid architectures in the meso- and nanoscale is one of the new challenging frontiers in materials chemistry. Sol-gel process provides a promising synthesis route towards such fascinating materials because it not only endows the resulting hybrids with cooperatively distinct and superior properties, but also offers the refreshed surfaces to facilitate linking diverse functionalities targeted to specific applications. In this chapter, we mainly focused on the sol-gel processes on the chemically inert surfaces of carbon nanotubes and polystyrene colloids. The newly developed preparation protocols and tentative application explorations of such novel hybrid materials were reviewed. Three sections are involved in the context of this chapter: 1) an overview of the relative research background; 2) recent advances in the typical fabrication protocols, the correspondingly fascinating properties and representative applications for these novel hybrid materials (carbon nanotube- and PS-inorganic oxides) with well-defined architectures; 3) an account on the correspondingly recent researches, the perspectives and outlook for the future research directions.

* Corresponding author: Email: ilkim@pusan.ac.kr, Tel.: +82-51-510-2466; Fax: +82-51-513-7720.

1. INTRODUCTION

A variety of conventional materials such as metals, ceramics and plastics cannot fulfill all requirements for new technologies seeking to solve the world's most immediate problems related to energy and environment. In many cases, however, the hybrid materials consisting of multiple components can merge the properties of the constituencies in a way that creates new properties distinct from those of either building block. Nowadays, one of the key issues in the field of hybrid materials is to synthesize micro- and nano-scaled hybrid materials with monodisperse sizes, uniform morphologies and functionalized surfaces owing to the following reasons: 1) The utilization of such micro- or nano-scaled hybrids as building blocks for devices not only helps downscale conventional technologies by at least an order of magnitude but also offers a cheaper and more environmentally friendly production route, since a drastic reduction in the necessary amount of raw materials, thus mimicking nature's efficient ways of managing with less when it comes to chemical and physical processing; 2) Scaling down the hybrid substance size to micro and nanometer dimensions can increase the specific surface area of the materials, which benefits the applications with reactions at the gas-solid or liquid-solid interface. This makes these hybrids well-suited to be utilized as catalysis, energy conversion, electrochemistry and environmental chemistry, where the use of such hybrids can increases response time, efficiency and sensitivity; 3) When the hybrid materials exist in a micro- or nano-scaled size, the chemical and physical properties of the substances can be considerably altered and finely tuned. For instance, in metal nanoparticles, there is a gap between the valence band and conduction band, unlike in the bulk metals. Also the size-induced metal-insulator transition and size-dependent quantization effects generally occurs when the metal particles are small enough.[1] 4) The uniform and fine anisotropic morphologies of particles are of significant for some specific applications of these materials. Especially, for the potential electronic applications such as nanocapacitors or nanotransitors, addressing and assembling arrays of particles are quite necessary. In those alignment and functionalization procedures, particles with well-defined morphologies are very advantageous owing to their ability to shear align, form porous structures and percolate at low concentrations. Additionally, a great number of facts, such as the morphology-dependant photonic properties of metal nanoparticles,[2] the shape-induced electron distribution of oxides nanocrystallines [3] and morphology-determined catalytic capacity of alloy nanoparticles [4] *etc.*, set representative examples for the significance of morphology-tunability of materials.

In the last decades, enormous efforts have been focused on the exploration of various synthetic strategies for micro- and nano-scaled hybrid materials with uniform size and morphologies. The typical protocols include the hydrothermal synthetic strategy, self-assembling procedures, sol-gel processes and their combinations, which have been well documented.[5,6] Among all those pioneer developed methodologies, the sol-gel process is a versatile, solution-based process for producing various ceramic and glass materials in the form of nanoparticles, thin film coatings, fibers, aerogels and involves the transition of a liquid, colloidal "sol" into a solid "gel" phase. Furthermore, it is a cheap and low-temperature technique that allows the fine control of chemical composition and the introduction of lowest concentration of finely dispersed dopants, such as organic dyes and rare earth metals. Although one of the major drawbacks of sol-gel process is that the product typically consists of an amorphous phase rather than defined crystals and thus requires crystallization and post-annealing steps for some certain applications, sol-gel process provides a facile, low-cost, environmentally friendly and energy efficient protocol for the fabrication of a wide range of hybrids. Indeed the mild synthetic conditions offered by the sol-gel processes meet the requirements for the future development of manmade materials which are produced by using environmentally benign, reliability, and less energy consumption protocols.

To date, numerous hybrids with well-defined architectures have been developed on the basis of sol-gel processes. These hybrid materials can be classified into two categories based on the strength of interaction between the two components. Category I hybrids are typically constructed by weak interactions between two phases, such as van der Waals, hydrogen bonding or weak electrostatic interactions. The representative examples of these hybrids include the organic compounds that are trapped in sol-gel matrix or inorganic oxide particles that are embedded into polymer networks. Category II hybrids are formed by strong chemical bonding of two components. These types of hybrid materials are generally fabricated by covalently bonding the discrete inorganic oxide building blocks to organic articles, modifying sol-gel networks with organic functional materials, or covalently linking the inorganic oxides and organic polymers by capping or linking agents.

Both types of hybrids described above are fabricated by a template-assisted process, where the templates can be either organic or inorganic materials in the form of one-, two or three-dimensional structures such as carbon nanotubes (CNTs), polymer fibers, inorganic wires, grapheme sheets, polymer networks and assembled architectures of organic molecules, *etc.* Especially, if the employed templates possess well-defined architectures, the resulting inorganic oxide-based hybrids with uniform hierarchical structures could be expected *via* an appropriate template-assisted process. To design and fabricate such fascinating hybrid materials, the surface properties of templates are the most important factor to be considered because that they provide interface for attaching the other functional component. If the template surfaces intrinsically contain functional moieties which also can be further modified to link diverse functionalities, versatile second components are allowed to be attached on these functional surfaces, yielding a wide range of hybrids. For example, the surface of monodisperse silica spheres obtained by Stöber sol-gel method bearing numerous silanol groups, which offers perfect interface for depositing various inorganic oxides layer *via* traditional sol-gel processes or directly grafting polymer networks.[7,8] In addition, by suitable modifying these silanol groups, more functional silica-based surface can be obtained aiming at the deposition of some certain components. For instance, our group has developed a facile protocol to grow crosslinked polystyrene/divinylbenzene layer on the C=C bond-contained-coupling-agent modified silica surface by disperse polymerization, yielding monodisperse silica/polymer hybrid spheres.[9] Following by the removal of silica core with HF acid, the rigid hollow polymer spheres were obtained. Further utilizing these polymer capsules as micro reactors, varied metal nanoparticles were in-situ generated, yielding the novel polymer capsules encapsulating single metal nanoparticle hybrids.

However, many fascinating template materials such as graphene materials tend to lack functional groups on their surfaces, which make them with bad solubility or dispersity in solvents, difficult surface modification and poor incompatibility with other components. These shortcomings greatly hinder the extensive utilization of these materials as template to create diverse hybrid materials. Therefore, to explore the facile and effective protocols to obtain versatile hybrids with well-defined topological structures based on chemically inert templates is still a challenging task in the material fields. Recently, numerous efforts have been paid targeted to those fascinating materials. Herein, in this chapter, we mainly focus on the recently developed promising protocols for deposition of inorganic oxides onto the chemically inert surfaces of CNTs and monodisperse polystyrene (PS) spheres in sol-gel processes to fabricate versatile inorganic oxide-based hybrid materials with well-defined architectures. The representative applications of such hybrid materials are also briefly described in this chapter.

2. CNT/Inorganic Oxide Nanohybrids

2.1. Synthesis of CNT/Inorganic Oxide Nanohybrids

CNTs are allotropes of carbon with a cylindrical nanostructure. These cylindrical carbon molecules have many novel properties including high aspect ratio and tubular geometry, which provides ready gas access to a large specific surface area and percolation at very low volume fractions. They also exhibit extraordinary mechanical, thermal, electrical and optical properties, which support CNTs as ideal building blocks in hybrid materials with potentially useful in many applications in nanotechnology, electronics and optics. By templating against CNTs, a variety of compounds including polymers, oxides, ceramics and nanoparticles have been successfully used to decorate CNTs sidewalls or fill CNTs matrix.[10] Among them, the inorganic oxides derived from sol-gel processes are the most commonly explored species.

However, CNTs generally exist in the form of solid bundles, which are entangled together giving rise to a highly complex network. Together with the chemically inert surfaces, pristine CNTs tend to lack of solubility and be difficult manipulated in any solvents, which have imposed great limitations to the use of CNTs as templates to assemble varied of functional components. To efficiently synthesize CNT-based nanohybrids, it is necessary to activate the graphitic surface of CNT. In this direction, two main approaches including covalent and non-covalent methodologies have been extensively explored in recent years. In this section, the CNT/inorganic oxides nanohybrids fabricated by templating against surface modified CNTs in sol-gel processes are mainly considered.

2.1.1. Covalent processes

The end caps of CNTs (when not closed by the catalyst particles) tend to be composed of highly curved fullerene-like hemispheres, which are therefore highly reactive, as compared with the sidewalls.[11] The sidewalls themselves contain defective sites such as pentagon-heptagon pairs called Stone-Wales defects, sp^3-hybridized defects and vacancies in the nanotube lattice (Figure 1).[12] Frequently, these intrinsic defects are supplemented by oxidative damage to the nanotube framework by strong acids which leave holes function-nalized with oxygenate functional groups such as carboxylic acid, ketone, alcohol, and ester groups.[13] In particular, the treatment of CNTs with strong acids such as nitric acid or with other strong oxidizing agents including $KMnO_4/H_2SO_4$, oxygen gas, $K_2Cr_2O_7$/ H_2SO_4 and OsO_4,[14-16] tends to open these tubes and to subsequently generate oxygenated functional moieties that serve to tether many different types of chemical functionalities onto the ends and defect sites of CNTs. By means of appropriate functionalities on these surface modified CNTs, the CNT/inorganic oxides nanohybrids can be generated *via* sol-gel processes.

Bottini *et al.* [17] grafted tetraethyl or tetramethyl- orthosilicate (TEOS or TMOS) onto carboxylic acid groups containing CNTs obtained under concentrated HNO_3 oxidizing conditions, forming coupling aninopropyltriethyoxysiane functionalized CNTs through a carboxamide bond. On the basis of these surface-modified CNTs, silica beads were generated and decorated along the CNTs by a sol-gel process in the presence of ammonia water. Based on the strong acid treated CNTs, the silica nanobeads also can be directly fabricated onto the sidewalls of CNTs in the presence of small molecular surfactants. For example, mediated by anionic sodium dodecyl sulfate, Fan *et al.* [18] grew the silica nanobeads onto the concentrated acid treated CNTs by means of vapor-phase method. Another typical example was reported by Liu et al, [19] who used cationic surfactant cetyltrimethyl ammonium bromide to suspend oxidized CNTs in aqueous solution. Followed by a sol-gel process catalyzed by ammonia water, silica decorated CNTs nanohybrids were successfully synthesized.

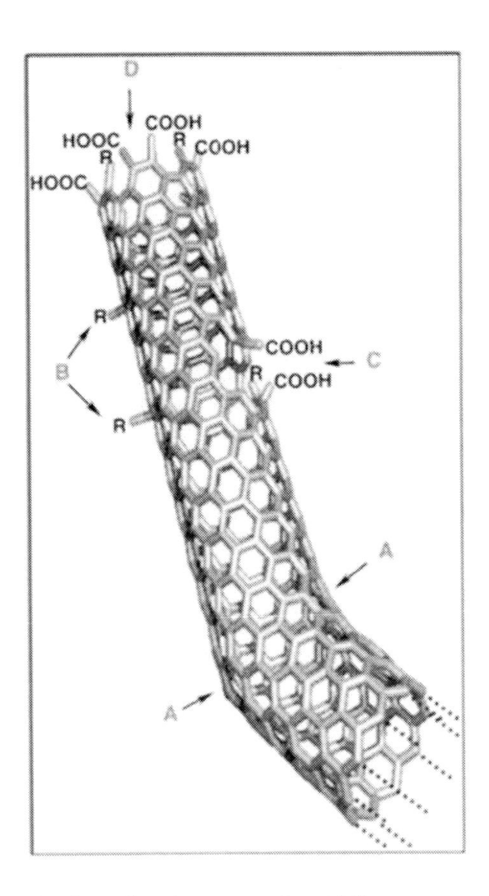

Figure 1. Typical defects in a single wall carbon nanotube: A) five- or seven-membered rings in the C framework, instead of the normal six-membered ring, leads to a bend in the tube, B) sp3-hybridized defects (R=H and OH). C) Carbon framework damaged by oxidative conditions, which leaves a hole lined with –COOH groups, and D) open end of the carbon nanotubes, terminated with -COOH groups. Besides carboxyl termini, the existence of which has been unambiguously demonstrated, other terminal groups such as $-NO_2$, -OH, -H, and =O are possible. Reprinted with permission from Ref 12. Copyright 2002 Wiley-VCH.

The above described pioneering works are very interesting but unfortunately, during all those processes for modifying CNT surfaces, more chemicals such as modifier agents, surfactants, organic solvents, amphiphilic polymers or other additives are indispensable. These would inevitably increase the hazard to environment; enhance the preparation cost and complex the functionalization processes. Therefore, it is still a challenging work to develop a facile, low-cost and green CNT surface-modification method for fabricating CNT-based nanocomposites. Recently, we have demonstrated an effective route to introduce hydroxyl groups onto the side walls of pristine SWCNTs by means of plasma treatment technique (Figure 2).[20] Followed by a co-condensation process between these hydroxyl groups bearing on the SWCNTs and TEOS (or together with MPTO), a uniform SiO_2 and thiol groups-functionalized SiO_2 coating on the CNTs can be fabricated effectively. Utilizing SWCNT@SiO_2-SH, a stable SWCNT@SiO_2/Ag heterogeneous hybrid has been generated *via in-situ* growth process in the absence of any additional reducing agents. Particularly, in this synthetic strategy, no more solvent and chemicals are involved, which simplifies the modification process, reduces the preparation cost and decreases the hazard to environment. Moreover, this facile procedure could offer a promising alternative to create varied SWCNT/inorganic oxide (TiO_2, GeO_2 *et al.*) composites and the corresponding SWCNT/i norganic oxide/metal nanoparticles hybrids.

2.1.2. Non-covalent processes

Although the strong oxidizing acids treatments can introduce a variety of organic groups on the CNTs surfaces, these functional groups tend to be with limited control over their number, type and location. Moreover, such treatment processes generally cause the surface etching and shortening of CNTs, resulting in the compromise of the electronic and mechanical properties thus suppress their extensive applications. In addition, for the sol-gel processes on these surface modified CNTs, the incomplete functionalization of CNT surfaces also often leads to the non-uniform oxide coatings on the acid-treated CNTs. Therefore, the acid-treated CNTs are less efficient templates for deposition of inorganic oxide coatings in a sol-gel process. To achieve uniform inorganic oxide coatings on CNTs sidewalls, recently developed non-covalent (non-destructive) protocols provide facile and efficient ways.

By means of non-covalent attractions, π-π stacking and/or wrapping interactions in the presence of surfactants and/or polymers, aqueous-based CNT sols can be achieved. These surface modified CNTs can be further assembled with a variety of nanoparticles or ceramic materials by means of *in-situ* synthesis techniques. The resulting CNT-based hybrids exhibit tailored properties while still reserving nearly all the intrinsic properties of CNTs. For example, Bourlinos *et al.* wetted pristine CNTs with vinyl silane molecules *via* non-covalent interactions between the vinyl groups and CNT surface.[21] After condensation to an oligomeric siloxane network and subsequent calcinations, the authors obtained silica nanoparticles (5-12 nm), which were well-dispersed on the CNT surface. Another approach was explored by Cao *et al.*, who modified CNTs with surfactants such as sodium dodecylsulfate.[22] The hydrophobic aliphatic chain interacted with the sidewalls of CNTs, while the hydrophilic end attracted the metal ions of the $RuCl_3$ precursor, which then reacted to form RuO_2 in a sol-gel process.

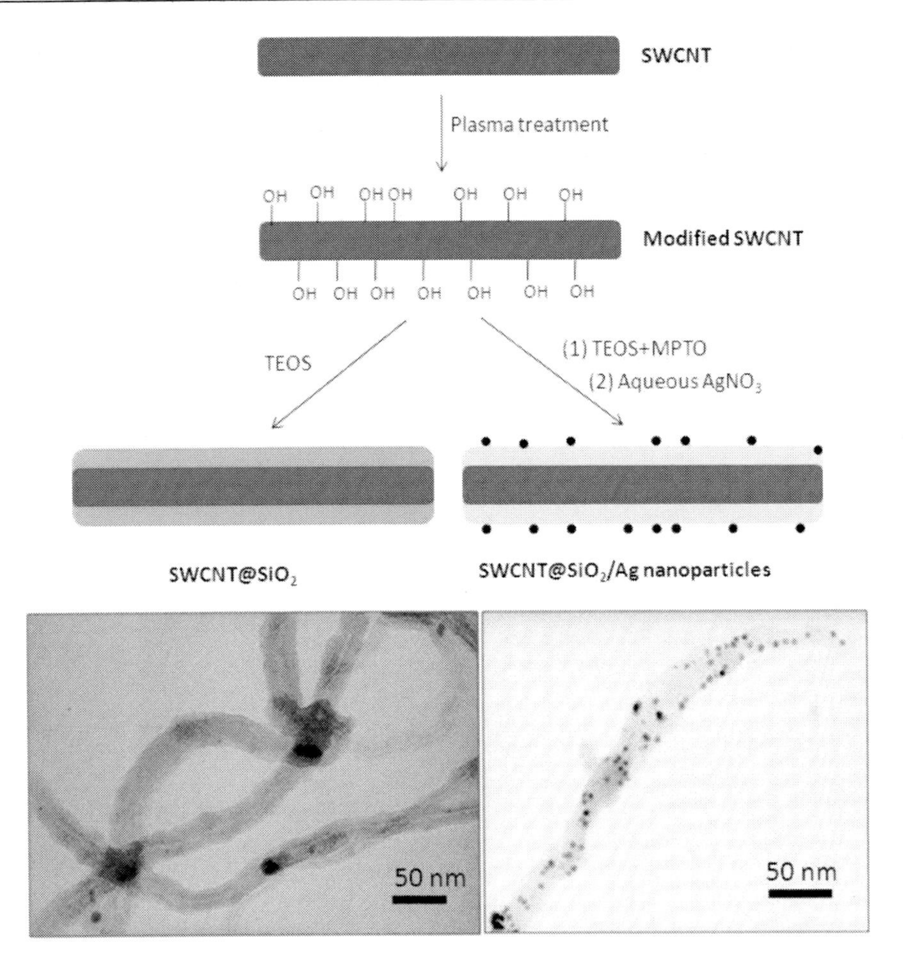

Figure 2. Schematic illustration of fabrication of SiO_2 coated single wall carbon nanotubes (SWCNT@SiO_2) and Ag nanoparticles immobilized SWCNT@SiO_2 (SWCNT@SiO_2/Ag nanoparticles) based on the plasma treated SWCNTs. TEOS and MPTO in scheme are tetraethoxysilane and 3-mercaptopropyl-triethoxysilane, respectively. Reprinted with permission from Ref 20.

More recently, Li *et al.* have developed a simple and efficient non-destructive [23] approach to decorate pristine CNTs with silica nanoparticles mediated by 1-aminopyrene surfactant. The inherently hydrophobic surfaces of pristine CNTs are first irreversibly modified with bifunctional 1-aminopyrene molecules using the non-covalent interactions between the pyrenyl groups of 1-aminopyrene and the six-membered carbon rings of the nanotubes. The resulting amino-modified CNTs specially adsorb *in-situ* formed silica nanoparticles *via* preferential affinity in a sol-gel process in the presence of amino water. Another promising non-covalent protocol has been developed to deposit titania coatings on pristine CNTs by using benzyl alcohol as a surfactant (Figure 3).[24] The benzyl alcohol was adsorbed onto the CNTs' sidewalls *via* π-π interactions, while simultaneously providing hydrophilic hydroxyl groups for the hydrolysis of the titanium precursor. In comparison with the sample in the absence of surfactant, the addition of appropriate amount of titania precursor resulted in a very uniform titania coating on the whole CNT surface. This work also showed that benzyl alcohol strongly affected the phase transition from anatase to rutile, providing very high specific surface areas.

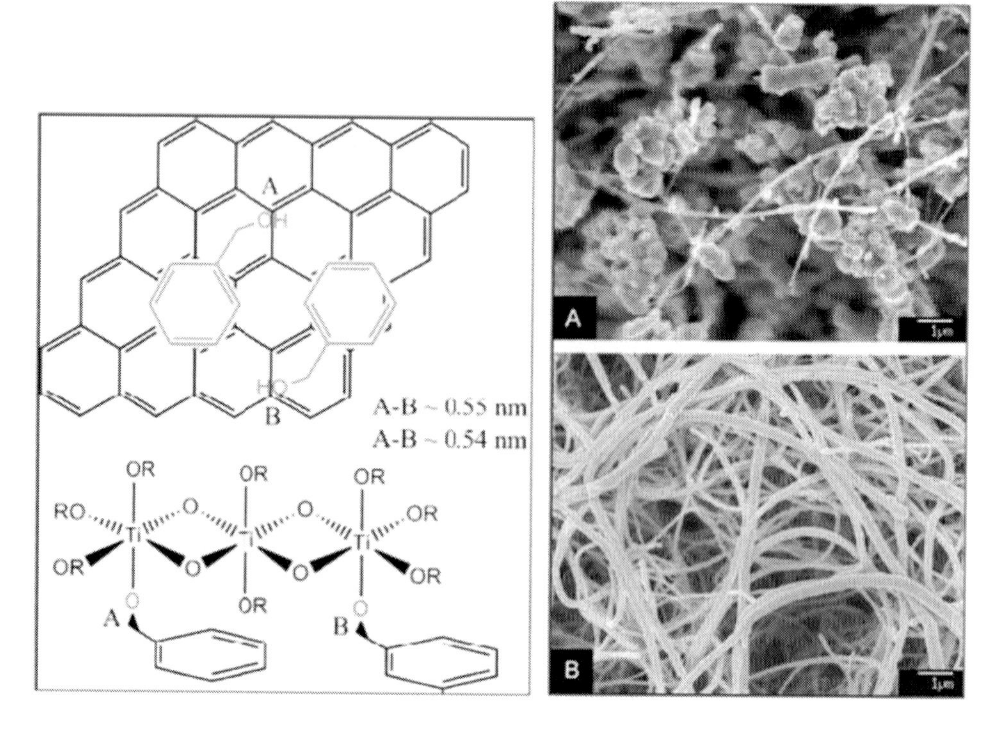

Figure 3. (Left) Scheme of the beneficial role of benzyl alcohol in the *in situ* coating of pristine CNTs with TiO_2. The benzene rings of the alcohol adsorb onto the CNT surface *via* π-π interactions and at the same time provide a high density of hydroxyl groups for the hydrolysis of the titanium precursor directly on the CNT surface. (Right) SEM images of partial and irregular coating of TiO_2 on acid-treated CNTs without benzyl alcohol (A) and uniform TiO_2 coating of pristine CNTs with benzyl alcohol as surfactants. Reprinted with permission from Ref 24. Copyright 2008 Wiley-VCH.

Besides the traditional non-destructive process mentioned above, the use of electrostatic interactions for the *in-situ* sol-gel processes has also been demonstrated for the facile fabrication of CNT/inorganic oxide nanohybrids. Seeger *et al.* have functionalized the CNTs with the polyelectrolyte PEI which provided positive charges on the CNT surface.[25] Using TEOS as precursor, the negatively charged SiO_x colloids can be easily attached onto CNT surface to form an amorphous silica coating *via* electrostatic interactions. However, it took rather long time (exceeded 100 h) to complete reaction since the reaction was carried out at room temperature. Similarly, by using different metal halides as precursors, alumina, silica and titania coatings on the CNT surface have also been successfully addressed in the presence of SDS surfactant.[26] On the other hand, the uses of metal organic precursors such as aluminum isopropoxide and zinc acetate in the presence of surfactants, the nanoparticles were generally formed in the solution rather than form uniform coatings on the CNT surface. Therefore, the applied surfactant had a repressive role in the adhesion of the inorganic precursor due to the different polarities between surfactant-treated CNTs (ionic) and the alkyl groups of the precursors.

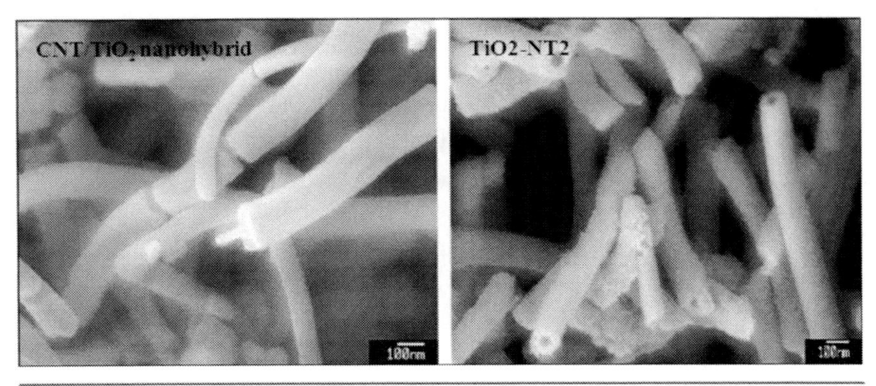

Samples	Conversion at 7.5 ml/min	Conversion at 30 ml/min	Conversion at 60 ml/min
P25	100	78	50
CNT–TiO_2	71	41	38
TiO_2-NT1	65	21	14
TiO_2-NT2	69	25	16
TiO_2-NT3	45	13	7

Figure 4. SEM images (Top) of CNT/TiO_2 nanohybridand TiO_2-NT2; The table (Bottom) shows the propene conversion by P25 (commercially obtained, mixture of anatase and rutile), CNT/TiO_2 nanohybrid and TiO_2 nanotubes with different crystal structures (TiO_2-NT1: anatase; TiO_2-NT2: mixture of anatase and rutile; TiO_2-NT3: rutile) photocatalysts using different 100 ppmv propene stream flow rates. Reprinted with permission from Ref 31. Copyright 2009 Elsevier Publishing.

2.2. Applications of CNT/Inorganic Oxide Nanohybrids

The CNT/inorganic oxides hybrids with well defined architectures exhibit the combined properties of both oxide and CNTs components, which paves the promising avenue for the extensive applications of these fascinating hybrids. In this section, several exciting examples of the improved performance of CNT/inorganic oxides in the applications such as catalysts in the photoeletronic, environmental field as well as sensors were presented.

2.2.1. Catalytic applications

As is well known, titania is the most suitable material for industrial use in photo-electrochemical and photocatalytic applications due to its efficient photoactivity, chemical and biological inertness, nontoxicity, higher photostability, cost effectiveness and easy production.[27] Therefore, quite a few studies have observed that CNT/titania hybrids exhibited superior photocatalytic performance towards the oxidation degradation of organic compounds. For examples, Wang *et al.* coated acid-treated MWCNTs with titania *via* sol-gel process and investigated the degradation of phenol under UV light and visible light.[28,29] In both cases, they observed a considerable acceleration of the oxidation reaction, with most of the phenol being transformed after 4 h (6 h in visible), while pure titania needed 6 h (9 h) in visible and pristine MWCNTs only converted up to 5%. The similar catalytic behavior has been shown by Yu *et al.* for the oxidation of acetone.[30] The authors believed that the higher photocatalytic activity was attributed to the enhanced interface of CNTs and titania. Another typical example for the photocatalytic application has been demonstrated by Bouazza and coworkers (Figure 4),[31] who synthesized the CNT/titania hybrids and applied them to the photocatalytic oxidation of propene at low concentration (1000 ppmv) in gaseous phase. They have found that the reaction activity heavily depended on the crystalline composition of titania components in the order of rutile < anatase < rutile/anatase mixture. The key result of this work is the exceptional performance of the CNT/titania hybrid, which yielded the highest observed photocatalytic activity. These improved performances were attributed to the synergistic effects derived from the hybrid nature of the materials. Moreover, the formed anatase crystalline with small sizes (CNTs act as heat sinks) and a reduced electron-hole pair recombination rate (CNTs act as electron traps) also play key roles for the significantly improved phtocatalytic performance.

The well dispersed oxide components on the CNTs surface is beneficial to the higher specific surface area of the corresponding CNT/inorganic oxide hybrids, which make them well suited to be applied as heterogeneous catalysts. There have been many reports on CNT hybrids with transition metal oxides for heterogeneous catalysts, such as RuO_2,[32] V_xO_5,[33] WO_x,[34] and ZrO_2,[35] as well as electrocatalysis such as SnO_2.[36] The typical examples include that the CNT/SnO_2 nanohybrids demonstrated improved activities towards the aerobic oxidation of various aromatic, saturated, and cyclic alcohols and quite selective to the corresponding aldehydes or ketones in liquid phase under mild conditions.[36] The CNT/inorganic oxide hybrids also have been utilized as solid acid catalysts in the petrochemical industry. For example, CNT/WO_3 hybrids exhibited very high skeletal isomerization activity and selectivity exclusively toward olefin reactions.[34] In this case, the improved performance of CNT/WO_3 catalysts was not only attributed to the uniform distribution of oxides on CNT surfaces, CNTs also prevented the complete reduction of oxides to metal materials, a process that typically cause the complete deactivation of the catalysts. In addition, CNT/inorganic oxide hybrids also provide substrates for depositing various metal nanoparticles. These complex hybrids are especially used in the oxidation of CO to CO_2. For instance, the poisoning of the surface of Pt electrode with CO always result in the slow kinetics of electro-oxidation of alcohols in the direct alcohol fuel cells systems. One of the promising solutions is the involvement of transition metal oxide. CNT/oxide hybrids can assist in the oxidation of CO to CO_2 by the dissociation of water, which was favorable for the significant improvement of eletrocatalytic activities and CO tolerances of the catalysts.[37]

2.2.2. Photovoltaic applications

For the dye-sensitive solar cell, the recent studies have demonstrated that photo-conversion efficiency was mainly limited by the transportation of photogenerated electrons across grain boundaries in the semiconductor particle network which often results in a random transit path for the electrons and thus increasing the probability of their recombination with oxidized sensitizers.[38] The use of CNT/titania in such photovoltaic devices have provided a promising alternative to improve the performance of dye-sensitive solar cell owing to the following reasons: 1) the titania networks especially with nanotubes and nanowires morphology can effectively direct the flow of photogenerated charge and facilitate the electron transport by rapid regeneration of oxidized sensitizers;[39] 2) CNTs can serve as media to accept photogenerated electrons and transport them to the collecting electrode and thus increased the photoconversion efficiency of the solar cell; 3) the use of

CNTs were also favorable for the strong increase in charge lifetimes.[40] On the basis of these great findings, the use of SWCNT/titania in the photovoltaic devices has been investigated by Kamat *et al.*, who found that the photoconversion efficiency for the hybrid materials was almost doubled compared with that of pure titania.[41] They also proved the electron transfer between the titania and CNTs. Note that the future research direction for the improvement of performance of solar cells will be focused on the shape-control of titania nanoparticles and the improvement of interface of titania and CNTs.

2.2.3. Sensors

Some metal oxides such as WO_3 and SnO_2 are promising sensor materials, since their electrical properties are highly sensitive to the surrounding environments. In addition, CNTs generally show excellent absorption capacity owing to their high specific area, which offers a large number of active surface sites. Therefore, the combination of such oxides materials and CNTs are promising sensor candidate materials. For example, SnO_2 is a conventional gas sensor for NO_x, CO and C_2H_4, which typically operated at high temperatures (200~500 ^oC). However, by depositing SnO_2 onto the CNT surface *via* a sol-gel process, the resulting CNT/SnO_2 hybrids showed the excellent gas-sensing capacity for NO_2 gas even at room temperature.[42] Such considerably enhanced sensitivities were attributed to the common interface with CNTs. In the CNT/SnO_2 hybrid sensor, the electric properties of the oxide are strongly enhanced by the highly conducting CNTs and thus the sensor resistance is dominated by the Schottky barrier at the interface of n-type oxide grains and p-type CNTs, resulting in the formation of additional depletion layers. This fact amplified the increase in resistance upon NO_2 absorption and enabled the operation of the gas sensor at room temperature. The CNT/SnO_2 hybrids also exhibited excellent sensitivities to the ethanol,[43] NH_3,[44] and acetylene gases.[45] Moreover, the sensors showed shorter response times, increased recovery properties and better stability as well as reproductivity. More importantly, they can be operated at room temperature. Therefore, the CNT/metal oxides nanohybrids with combined advantages of oxides and CNTs provide a type of very useful sensor materials, which could be expected to extend to the other promising applications. For instance, if depositing WO_3 onto the CNT surface, the resulting hybrids could be expected to exhibit excellent sensitivity toward the pollutants such as SO_2, H_2S and NO.

In addition, Gavalas *et al.* have tested the feasibility of CNT/silica nanohybrids for the development of biosensors.[46] Aimed to that, they used aqueous sol-gel process to fabricate enzyme-friendly composite materials by integration of enzymes and CNTs within a sol-gel matrix. The authors have found that the involvement of CNTs in the composite facilitated the fast electron transfer rates. By using L-amino acid oxidase as model enzyme, the feasibility of this type of composite for the development of biosensors was demonstrated. The corresponding results showed that the composites exhibited enhanced stability to the enzyme without adverse effects in their activity compared with the bare oxides, which make them suitable for the development of stable biosensors. More surprisingly, the sensor retained more than 50% of its response after 1 month of testing.

2.2.4. Synthesis of inorganic nanotubes

The inorganic oxides components in CNT/inorganic oxide nanohybrids obtained by sol-gel processes are generally amorphous, which is not useful in some specific applications where the crystalline oxides are necessary. Upon sintering treatment of the CNT/inorganic oxide nanohybrids, CNTs can act as support during crystallization and phase transformation of oxide, which lead to the inorganic oxides in crystalline phase. Simultaneously, the inorganic oxide nanotubes are generated. The advantage of such oxides in nanotubular morphology is their three-dimensionally mechanical coherent architecture, which allows ready gas access to a very high specific surface area. Therefore, most of these oxide nanotubes showed excellent performance in gas sensing or photocatalysis.[47,48] The typically reported examples include the nanotubular SiO_2,[49] TiO_2,[50] RuO_2,[51] Co_3O_4,[52] *etc.* This template-directed synthetic strategy provides a simple, versatile and cost-effective approach for pure and modified inorganic nanotubes. Moreover, since the CNTs utilized in those processes are with lengths of millimeters, this protocol also is adapted for fabrication of very long nanotubes from the rich pool of materials. However, the involvement of very high temperature treatments in those synthetic procedures tend to induce oxides grain growth and cause collapse of the tubes' structure, thus the specific surface area are considerably reduced. Consequently, the phase transformation of oxides need to carry out in an inert atmosphere, the preserved CNTs could act as a support so that the stresses associated with the reconstruction of the phase did not destroy the nanotubular structures. The correspondingly representative example has been demonstrated by Eder and coworkers (Figure 5).[10,53] The authors have found that the quality of the TiO_2 coating was strongly affected by various reaction conditions, including the order of mixing, the choice of drying method, and the water concentration. Also the calcination treatment had great effect on the crystallization of amorphous TiO_2 to anatase and the phase transformation from anatase to rutile. The key achievement of that work was the control of morphology and structure of the TiO_2 coating and the rutile nanotubes, which also allowed the production of the ideal CNT/TiO_2 hybrid material for the desired photochemical, catalytic and sensor applications.

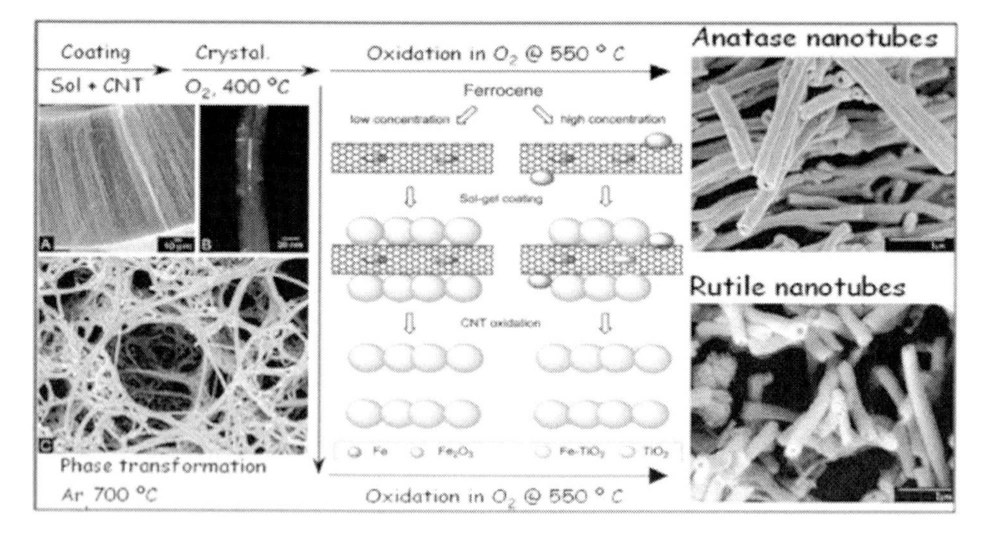

Figure 5. Scheme for the synthesis of iron-doped anatase and rutile nanotubes, using the catalyst residues within CNTs as iron source (A, B) involving coating of pristine CNTs with TiO2 *via* sol-gel process (C), heat treatment in argon to either anatase or rutile, and oxidation of the CNT template in air at 550 °C, during which the iron oxidizes and enters the TiO2 lattice as substituents (Fe^{3+}) for Ti^{4+} ions. Reprinted with permission from Ref 10. Copyright 2010 American Chemical Society.

2.2.5. Other applications

Besides the promising applications of CNT/inorganic oxide nanohybrids mentioned above, other fascinating applications have been also extensively explored. For example, on the basis of the exceptional electronic properties of CNTs, the combination of electroactive oxides and CNTs can form electrochemical capacitors, which generally exhibit considerably enhanced capacitance values compared with pure CNTs.[54] Also, the CNT/inorganic oxide nanohybrids with good cyclability and capacity can be applied as anode material for lithium battery.[55] By means of the special photonic properties of CNTs, the CNT/inorganic oxide nanohybrids have also been successfully utilized as photonic devices such as transistors and field emission devices.[56,57]

3. PS/Inorganic Oxide Hybrids

3.1. Synthesis of PS/Inorganic Oxide Hybrids

In the last two decades, highly monodisperse Stöber silica spheres and PS colloidal have received the most intensive research in the fabrication and application of uniform hybrids and the corresponding functional architectures, where they generally were utilized as powerful templates. Stöber silica spheres possess numerous silanol groups onto their surface, which provides functional platforms for the integration of the second component to yield hybrids with well-defined morphologies. After removal of silica core from these hybrids, the hollow particles can be generated. As competitively powerful template materials, PS spheres with micrometer or nanometer scale also have attracted extensive research interests in recent years. Monodisperse PS spheres are typically synthesized by surfactant-free emulsion polymerization.[58] These PS colloids can be well dispersed in water and alcohol solutions. The highly uniform size, regularly spherical morphology and good disperse in solvent make PS spheres well suited to be used as templates to generate a variety of organic/inorganic hybrids and the corresponding inorganic functional materials. However, in comparison with silica spheres, PS colloids possess chemically inert surfaces bearing a large number of benzyl groups, which greatly limit the facile deposition of functional components onto their surfaces. Especially, for our current topic, the sol-gel process cannot directly occur on such chemically inert surface. Consequently, in order to deposit inorganic oxides onto PS surface to form uniform PS/inorganic oxide hybrids, the suitable surface modification of PS spheres are quite necessary.

So far, many successful protocols have been developed to use PS colloids as powerful templates for fabrication of various PS/oxide spherical hybrids and hollow oxide spheres. In a typical procedure, surface-modified PS spheres are coated in solution either by controlled surface precipitation of inorganic molecule precursors or by direct surface reaction utilizing specific functional groups on the cores to create core/shell composites. The template spheres are subsequently removed by selective dissolution in an appropriate solvent or calcinations at elevated temperature to generate hollow structures. In this process, one of the most crucial steps is to modify the hydrophobic and chemically inert surface of PS sphere with tunable chemical and physical environment, which is favorable for the growth of oxide through the specific interaction. According to different protocols for surface modification of PS spheres, the synthetic approaches of PS/inorganic oxide can be classified into two types, chemical and non-chemical strategy.

3.1.1. Chemical strategy

There have been reported two methods to chemically modify the surface of PS sphere for fabrication of hollow oxide spheres. One is termed as prior-modification method, in which the surface-modified PS spheres can be obtained by one-pot reaction. For example, Tissot and coworkers synthesized PS beads bearing silanol groups on the surface *via* emulsion polymerization using 3-(trimethoxysily)propyl methacrylate as a comonomer.[59] Using this functionalized PS beads as template, the hollow silica spheres were synthesized by co-condensation between the silanol groups bearing on the surface of PS latex and TEOS, followed by thermal degradation of PS core. Another successful example has been demonstrated by Agrawal and coworkers, who synthesized β-diketone-functionalized PS beads by emulsion copolymerization of styrene and acetoacetoxyethyl methacrylate.[60] They used these surface modified PS colloids as sacrificial templates to fabricate sub-micrometer-size PS/tantalum oxide composites and hollow tantalum oxide spheres with tailored shell thickness by controllable depositing tantalum oxide nanoparticles on the functionalized template surface.

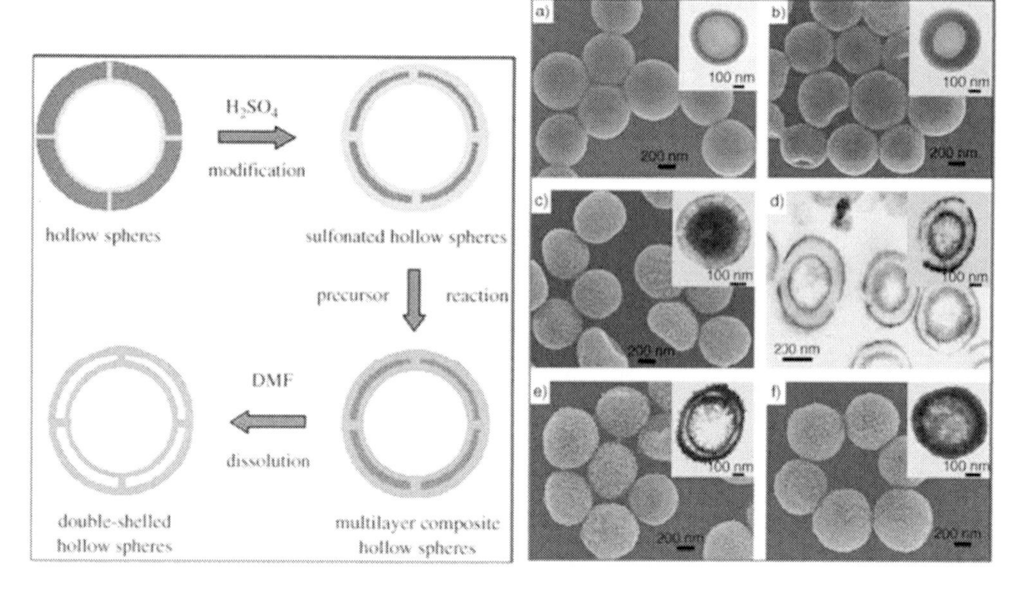

Figure 6. Schematic illustration of the formation of double-shelled hollow spheres; Morphologies of representative templates and titania hollow spheres. a) SEM and TEM (inset) images of unsulfonated polymer hollow-sphere templates. b) SEM and TEM (inset) images of the sulfonated S1 templates. c) SEM image of titania composite hollow spheres templated by S1; the inset shows a TEM image of the corresponding double-shelled titania hollow spheres after treatment with DMF. d) Cross-sectional TEM images of ultramicrotomed titania hollow spheres before and after (inset) treatment with DMF. e) SEM image of titania composite hollow spheres templated by S2; the inset shows a cross-sectional TEM image of the composite hollow spheres after treatment with DMF. f) SEM image of titania composite hollow spheres templated by S3; the inset shows a TEM image of the corresponding titania hollow spheres after treatment with DMF. Reprinted with permission from Ref 62. Copyright 2005 Wiley-VCH.

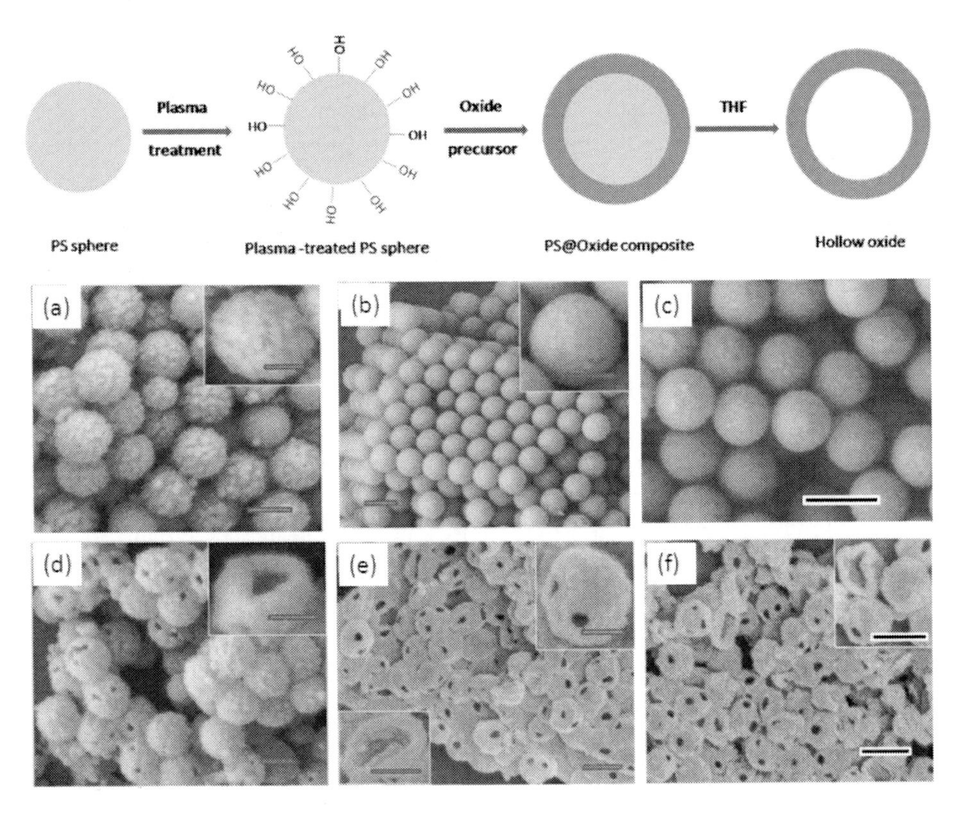

Figure 7. Schematic illustration fir fabrication of hollow oxides using plasma-treated PS spheres as sacrificial templates; SEM images of PS/SiO_2 (a), PS/TiO_2 (b) and PS/GeO_2 (c) hybrids obtained by templating against plasma-treated PS spheres *via* sol-gel processes; SEM images of the hollow SiO_2 (d), TiO_2 (e) and GeO_2 (f) obtained by removal of PS core from the corresponding PS/oxide hybrids using THF. Reprinted with permission from Ref 63.

The other method can be defined as post-modification, which allows to covalently grafting the functional groups onto the surface of preformed PS spheres by means of chemical approaches. For instance, Yang *et al.* used the sulfonating reaction technique to introduce the sulfonate groups onto the surface of bare PS colloids.[61] By templating against these sulfonated PS spheres followed by a sol-gel process, the titania/PS capsules and hollow titania spheres were created. And the thickness of the titania shell and the cavity size were controllable in the entire particle radius range. The authors also reported a one-step approach to the synthesis of hollow spheres with a double-shelled complex structure by using commercial polymer capsules as templates (Figure 6).[62] In this process, a PS hollow sphere containing a thin hydrophilic inner layer and transverse channels of poly(methyl methacrylate)-poly(methacrylic acid) is treated with sulfuric acid, and the sulfonation takes place in three locations: the exterior shell surface, the inner shell surface and the transverse channels. This process creates a hollow sphere that contains a hydrophilic inner layer, outer layer and channels of sulfonated PS gel. The newly formed surface bears sulfonic acid groups that are capable of adsorption or of forming complexes with metal oxide precursors. For example, using titania alkoxides as precursor, preferential growth of titania layer within the hydrophilic sulfnated regions followed by removal of the middle PS layers creates double-shelled titania hollow spheres.

The above described pioneering works are very interesting but unfortunately, during all those processes for modifying PS surface, more chemicals such as the co-monomers, organic solvents, modifier agents or other additives are indispensable. These would inevitably increase the environmental imperilment, augment the preparation cost and complex the preparation processes. Therefore, it is still a challenging work to develop a facile, low-cost and environment-benign PS surface-modification method for fabricating hollow inorganic materials. Recently, we have firstly demonstrated a novel post modification approach, plasma-treatment technique, to modify the PS surface (Figure 7).[63,64] The plasma treatment process can easily introduce hydroxyl groups onto the surface of PS spheres. Moreover, in comparison to the previous modification approaches mentioned above, no more solvent and chemicals are involved in this process, which can simplify the modification process, reduce the preparation cost and decrease the hazard to environment. By using these plasma-treated PS spheres as sacrificial template, silica- titania- and germania-coated PS composites have been successfully fabricated by the co-condensation between hydroxyl groups with TEOS, titanium (IV) isopropoxide and tetra-n-butoxygermane in a sol-gel process, respectively. Followed by the removal of PS cores with tetrahydrofuran (THF), the hollow silica, titania and germania spheres were generated. Moreover, the shell thickness of the resulting hollow oxide spheres is controllable by simply changing the employed concentration of precursors.

3.1.2. Non-chemical strategy

For the non-covalent strategy for the fabrication of PS-based hybrids, one of the most typical examples is layer-by-layer (LBL) self-assembly technique developed by Caruso and coworkers (Figure 8).[65] In this process, the polyelectrolytes were absorbed on the surface of PS spheres through electrostatic interaction to reverse the surface charge. A novel and intriguing result arose from the subsequent addition of a second solution of oppositely charged polyelectrolyte to the polymer-coated PS spheres; adsorption of a second layer on the particle surface occurred through electrostatic self-assembly in the same way that multilayered polymer films have been assembled on planer substrates. Again, a reversal in surface charge was observed. Repetition of this process resulted in the formation of multiple bilayers on the particle surface in a controlled fashion. The main advantages of LBL technique including: 1) The thickness of polymer coatings can be fine tuned by altering the number of layers deposited and the solution conditions from which the polymer are absorbed; 2) The multicomposite polymer films can be assembled through choice of a large variety of polymers; 3) Colloids of different size, shapes and composition can be employed as templates since polyelectrolyte self-assemble onto numerous surface. On the other hand, the biggest limitation of the LBL method is the time-consuming sequential polyelectrolyte deposition cycles and purification steps.

Using these polyelectrolyte coated PS spheres as template, a varied of small oxide nanoparticles and inorganic molecule precursors can be allowed to consecutively assemble onto PS colloids by a sol-gel process, yielding various PS/inorganic oxide hybrids. By removing the PS template, the resulting composite can be converted into hollow oxide spheres. The relative promising works have been well summarized in the review report.[66] For example, the authors reported on the construction of composite multilayers of silica nanoparticles and poly(diallyldimethylammonium chloride) (PDADMAC) on submicrometer-sized PS latex particles *via* the sequential electrostatic adsorption of silica and PDADMAC from dilute solution.[67] Alternating silica-PDADMAC multilayers with thickness ranging from tens to hundreds of nanometers have been fabricated. Following the similar procedures, ZrO_2 and Y_2O_3 nanoparticles obtained by sol-gel processes also have been assembled onto the PS surface *via* LBL methods.[68]

Mediated by a suitable polymer layer deposited onto the PS surfaces through weak non-covalent interactions, the inorganic oxides also can be in-situ produced and deposited on the modified PS surface *via* sol-gel processes, yielding PS/inorganic oxide hybrids. This protocol presents a typically protocol to non-covalently tailor PS surface environments which facilitate the occurrence of sol-gel reaction. Yang *et al.* have explored to in-situ grow a controllable poly-L-lysine coating onto the amine functionalized PS beads.[69] Mediated by these poly-L-lysine layers, the silica and titania networks were deposited onto the modified PS surfaces *via* a sol-gel process. After the removal of PS core by calcinations or dissolution, hollow silica and titania spheres were generated. In addition, another more facile method to fabricate hollow oxide spheres have been reported by Deng and coworkers.[70] They physically absorbed an amphiphilic poly(vinyl pyrilidone) (PVP) layer onto the PS latex surface. Subsequently, ammonia water catalyzed sol-gel processes took place and in-situ formed dense silica and titania shells. More interestingly, the use of higher concentration of ammonia water during the sol-gel process, the hollow oxides spheres can be directly created without any core-removal processes.

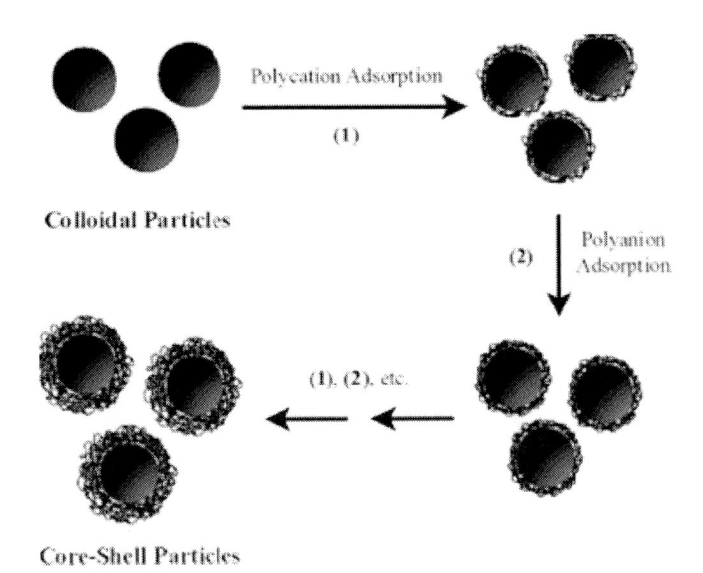

Figure 8. Schematic illustration of the LbL process for forming polyelectrolyte multilayers on particles. The scheme is shown for negatively charged particles. The process entails the sequential deposition of oppositely charged polyelectrolytes onto colloidal particles, exploiting primarily electrostatic interactions for polymer multilayer build-up. Following deposition of each polymer layer, excess polyelectrolyte is removed by centrifugation or filtration, with intermediate water washings. The key to the formation of the polyelectrolyte multilayers is that not all of the cationic (or anionic) groups of the deposited polyelectrolyte interact with the particle surface (or the underlying polymer layer beneath). Hence, non-utilized charged groups, which cause charge overcompensation, facilitate the electrostatic binding of the subsequently adsorbed layer. Finally, a polyelectrolyte multilayer film of tailored hickness is obtained on the colloidal template *via* this strategy. Reprinted with permission from Ref 65. Copyright 2001 Wiley-VCH.

3.2. Applications of PS/Inorganic Oxide Hybrids

As described above, PS spheres have provided powerful templates for the fabrication of a variety of PS/inorganic oxide hybrids and the corresponding hollow oxide architectures. These resultant hollow spheres generally possess uniform size, porous structures and high specific surface area. Together with the intrinsic properties of oxides, the hollow oxide structures are very promising for the applications in the field of drug delivery, catalysts, energy storage, low dielectric constant materials and piezoelectric materials. Aimed to various applications, although the hollow oxides have not been sufficiently fabricated so far by PS-template-assisted sol-gel processes, it is believed that most of hollow oxides spheres can be created by templating against PS spheres. Even though for specific applications where the crystalline oxides are necessary, the amorphous oxide layer on the PS/oxide composites can be easily tranformed into crystalline phase by calcinations treatments, while the core/shell structure of hybrids were converted into hollow architectures. In this section, we mainly focused on the recently developed application cases of these fascinating materials but not limited the hollow oxides obtained by using PS spheres as templates.

3.2.1. Application as Sensors

Oxide semiconductor gas sensors such as SnO_2, ZnO, In_2O_3 and WO_3 show a significant resistance change upon exposure to a trace concentration of reducing or oxidizing gases. However, these bulk oxidizes tend to exhibit relatively lower sensitivity owing to their lower specific surface area and lack of porous structure. In contrast, the hollow structures of these oxides generated by templating against PS spheres generally show thin shell thickness, high specific surface area, porous structures and good gas permeability, which offer attractive platform for gas sensing reactions.[71] For instance, Martinez *et al.* prepared Sb-doped SnO_2 hollow spheres by LBL coating on PS templates and utilized them as gas sensors for CH_3OH.[72] The R_a/R_g ratio of Sb:SnO_2 hollow spheres to 0.4-1 ppm gas at 400 ^{o}C were around 3- and 5- fold higher than those of SnO_2 polycrystalline chemical vapor deposition film and Sb:SnO_2 microporous nanoparticle films. Li *et al.* synthesized hollow WO_3 spheres with around 400 nm of average diameter and about 30 nm of thin shells composed of numerous small nanocrystals.[73] The as-synthesized spheres exhibited good sensitivity to alcohol, acetone, CS_2 and other organic gases. The In_2O_3 hollow spheres have also been fabricated by Li and coworkers,[74] who demonstrated these hollow spheres performed well as a gas-sensing material in response to both ethanol and formaldehyde gases. In addition, Zhang *et al.* fabricated hollow zinc oxide spheres and applied them as gas sensors for efficiently detecting NO_2 gas.[75]

The applications of PS/oxides hybrid materials as chemical sensors have also been explored by Wang *et al.*[76] They developed a simple method to prepare colloidal crystal arrays based on PS/ZrO_2 core shell spheres sensing microgel that detects the organ-ophosphorus compound paraoxon at ppm concentrations in aqueous solution. The molecular recognition agent for this sensor is ZrO_2, which has a chemical affinity for paraoxon. Colloidal crystal arrays based on these core-shell particles were prepared by a simple gasket method at room temperature. Reflectance measurements were carried out for varying concentrations of paraoxon from 0.024 to 0.096 ppm, showing a characteristic shift in the reflectance peak from 468 to 488 nm. The shift in the peak wavelength from right to left suggested that the distance between the crystals or lattice spacing increased due to addition of paraoxon. Thus, it acts as a chemical sensor for paraoxon since zirconia has an affinity for the phosphate group of paraoxon. It was demonstrated that by increasing paraoxon concentration, the reflectance peak increased linearly, making colloidal crystal array a potential candidate for future application as chemical sensors.

3.2.2. Catalytic applications

The hollow oxide spheres also have attracted tremendous interest as catalytic materials owing to their high specific surface area, low density and better permeation. A typical example has been demonstrated by Huang *et al.* (Figure 9).[77] The authors have synthesized hollow Ta_2O_5 spheres *via* LBL techniques by templating against PS spheres. These hollow oxides showed remarkably effective photocatalytic properties for hydrogen evolution from a methanol solution upon exposure to UV light irradiation. The rate of hydrogen generation over crystalline hollow Ta_2O_5 spheres exceeded 4 times and 20 times that of commercial amorphous Ta_2O_5 hollow spheres and Ta_2O_5 powder, respectively. The authors addressed that the significant enhanced photocatalytic performance of hollow Ta_2O_5 spheres were attributed to their crystalline structures and high specific surface area. Although an amorphous hollow sphere had a slightly higher surface area, it may have more surface defects because of its amorphous nature, which will result in a higher recombination rate of electrons and holes. The number of surface defects is expected to decrease upon crystallization. High crystallinity ensures the lattice to the surface. Thus the crystallized hollow spheres show considerable enhancement of photocatalytic activity. Another excellent examples have been demonstrated by Li *et al.*[78] who using the as-prepared amorphous Fe_2O_3 hollow spheres with mesoporous structures as photocatalytic materials. More interestingly, they performed better than the nanocrystal samples. In addition, Yu *et al.* have demonstrated that polycrystalline $WO_3 \cdot 1/3$ H_2O hollow microspheres with hierarchically porous wall structures showed a continuous absorption band below 470 nm, consistent with a semiconductor material with a 2.86 eV bandgap.[79] Moreover, these hollow $WO_3 \cdot 1/3$ H_2O exhibited highly efficient photocatalytic activity with regard to the degradation of methyl orange. These results also suggested that hollow $WO_3 \cdot 1/3$ H_2O spheres could have uses in environmental applications concerning air purification, water disinfection and purification, and hazardous waste remediation. More recently, the CeO_2 hollow spheres with shell thickness in around 20 nm have been synthesized and utilized as catalysts towards the CO oxidation reaction.[80] These hollow spheres exhibited a higher catalytic properties compared to the commercial CeO_2 powders.

3.2.3. Photoelectronic applications

The photoluminescence properties of hollow hybrid microspheres have been also successfully demonstrated by Gao and coworkers.[81] These well-defined nitrogen-doped, hollow SiO_2/TiO_2 hybrid spheres were successfully prepared through a PS template assisted two-step sol-gel synthesis combined calcination process. A stronger visible absorption of N-doped hollow SiO_2/TiO_2 hybrid spheres was observed with increasing amounts of nitrogen doping. The PL bands showed spectral lines at about 421, 472, and 529 nm, which were attributed to the self-trapped excitons (i.e., F and F^+ centers). In particular, the doping of nitrogen into hollow SiO_2/TiO_2 hybrid spheres led to the drastic quenching of photoluminescence, indicating that nitrogen doping increased the separation efficiency of the photoinduced electron and hole pairs. This work further demonstrated that photoluminescence is a powerful technique for investigating the mechanism of photocatalytic reaction and provided a practical guide.

Wang *et al.* have addressed that the hollow SnO_2 microspheres with 200-300 nm in average diameter can be utilized as reversible Li-ion storage.[82] They have found that such hollow materials exhibited noticeable improvement in a number of performance areas such as specific capacity, rate capability, and cyclability. These improvements could be attributed to a high degree of crystallinity, which increases the electronic conductivity, and the facile transport of Li ions in a hollow shell with nanoscale thickness, as well as significantly shortens the solid-state diffusion length. Han *et al.* has also demonstrated the hollow SnO_2 microspheres can be used as lithium battery anode.[83] They exhibited extraordinarily high discharge capacities and columbic efficiency.

4. Challenges and Outlook

Aiming at the specific applications of CNT-based nanohybrids, the recently developed synthetic strategies for a variety of CNT/inorganic oxides nanohybrids by means of sol-gel processes on the surface-modified CNTs have been reviewed in this chapter. It is quite obvious that the suitable modification of CNT surface not only endows the CNTs with good dispersity in solvents, but also creates functional surfaces for deposition varied functionally organic or inorganic components. Although the chemical modification of CNTs have provided powerful alternatives for building CNT-based nanohybrids, the intrinsic properties of CNTs are greatly destroyed owing to the involvement of structural defects caused by the chemical treatment processes. Moreover, the chemically introduced moieties are generally anchored on the defective sites and un-uniformly distributed onto CNT sidewalls, directly resulting in the un-uniform coating of second functional components. Therefore, the developments of novel approaches for effectively non-destructive (non-covalent) modification CNTs are still desperate. Even though varied of non-covalent protocols have been explored, the interaction between the CNT and modifiers are rather weak, which generally results in the less improvement of modified CNTs solubility and unstable hierarchically structures of the corresponding CNT-based hybrids. Especially for single wall carbon nanotubes, the effective non-covalent approach for well-dispersing the single nanotubes in solvents has not been achieved so far. Consequently, how to get well-dispersed and stable CNTs in solvents and CNT-based nanohybrids *via* non-destructive protocols is still a challenging task in the future years. Particularly, the corresponding research for modifying single wall carbon nanotubes will become a hot issue and receive intense attraction. Therefore, the synthesis of novel non-covalent modifier and the exploration of newly non- or less-destructive protocol of CNTs offer the promising alternatives to improve the interaction of CNTs and the other functional components, leading to the formation of the stable and versatile CNT-based nanohybrids.

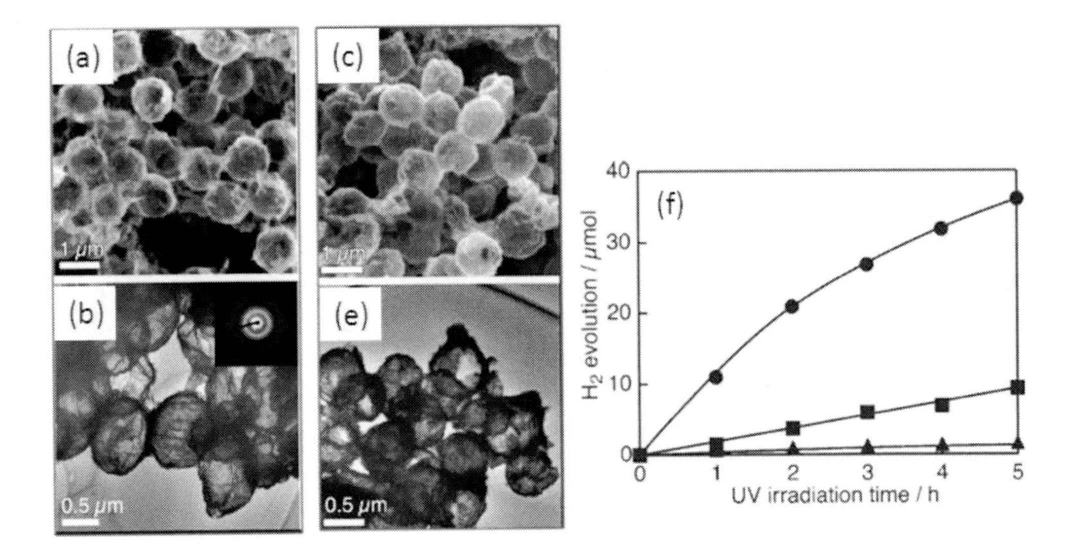

Figure 9. SEM (a) and TEM (b) images of amorphous Ta_2O_5 hollow spheres obtained by calcination at 600 °C. Inset of (b) shows the corresponding SAED pattern. SEM (d) and TEM (e) of crystalline Ta_2O_5 hollow spheres obtained after calcination at 650 °C. (f) Time course of hydrogen evolution over Ta_2O_5 hollow spheres (●); amorphous hollow spheres (■); commercial Ta_2O_5 powder (▲). Reaction conditions: 1 mg of catalyst, 2 cm^3 of aqueous 10 vol % methanol solution. Reprinted with permission from Ref 77. Copyright 2010 American Chemical Society.

The surface-modified PS colloids serve as powerful templates for generating various functional inorganic materials. However, from the recent developments described in our current chapter, the effective modification protocols are less versatile compared with the other common template materials such as silica and CNTs. Indeed, some promising surface modification protocols such as for CNTs also could be adaptable for PS spheres. Unfortunately, few reports have been involved. From this point of view, more versatile functionalization protocols for PS colloids might be expected targeted to the generation of a wide range of novel materials with well-defined architectures.

Besides the CNTs and PS spheres we discussed here, other materials with chemically inert surfaces, such as grapheme sheets, nanodiamonds, fullerene, *etc.* possess fascinating mechanical, photonic, electronic, thermal and biofunctional properties. By suitable surface modification of these materials, a wide range of functional hybrids are highly expected by incorporation of various multiple functional components. Although the corresponding researches have not been mentioned in this chapter, the cases of CNTs and PS latex have provided valuable references for the surface modification and integration of functional components for such fascinating materials. In more broad eyesight, such novel materials will pave a broad avenue for the versatile applications in catalyses, bioengineering, nanoscience and nonotechnology, as well as photoelectronic devices.

ACKNOWLEDGMENTS

This work was supported by grants-in-aid for the *World Class University Program* (No. R32-2008-000-10174-0).

REFERENCES

[1] Daniel, MC; Astruc, D. *Chem. Rev.,* 2004, 104, 293.
[2] Sun, Y; Xia, Y. *Science,* 2002, 298, 2176.
[3] Cozzoli, PD; Kornowski, A; Weller, H. *J. Am. Chem. Soc.,* 2003, 125, 14539.
[4] Park, JY; Zhang, Y; Grass, M; Zhang, T; Somorjai, GA. *Nano Lett.,* 2008, 8, 673.
[5] Sanchez, C; Julian, B; Belleville, P; Popall, M. *J. Mater. Chem.,* 2005, 15, 3559.
[6] Schubert, U; Huesing, N; Lorenz, A. *Chem. Mater.,* 1995, 7, 2010.
[7] Yu, M; Lin, J; Fang, J. *Chem. Mater.,* 2005, 17, 1783.
[8] Achilleos, DS; Vamvakaki, M. *Materials,* 2010, 3, 1981.
[9] Li, H; Ha, CS; Kim, I. *Macromol. Rapid Commun.,* 2009, 30, 188.
[10] Eder, D. *Chem. Rev.,* 2010, 110, 1348.
[11] Niyogi, S; Hamon, MA; Hu, H; Zhao, B; Bhowmik, P; Sen, R; Itkis, ME; Haddon, RC. *Acc. Chem. Res.,* 2002, 35, 1105.
[12] Hirsch, A. *Angew. Chem. Int. Ed.,* 2002, 41, 1853.
[13] Chen, J; Hamon, MA; Hu, H; Chen, Y; Rao, AM; Eklund, PC; Huddon, RC. *Sience,* 1998, 282, 95.
[14] Liu, J; Rinzler, AG; Dai, H; Hafner, JH; Bradley, RK; Boul, PJ; Lu, A; Iverson, T; Shelimov, K; Huffman, CB; Rodriguez-Macias, F; Shon, YS; Lee, TR; Colbert, DT; Smalley, RE. *Science,* 1998, 280, 1253.
[15] Hiura, H; Ebbesen, TW; Tanigaki, K. *Adv. Mater.,* 1995, 7, 275.
[16] Ajayan, PM; Ebbesen, TW; Ichihashi, T; Iijima, S; Tangigaki, K; Hiura, H. *Nature,* 1993, 361, 333.
[17] Bottini, M; Tautz, L; Huynh, H; Monosov, E; Bottini, N; Dawson, MI; Bellucci, S; Mustelin, T. *Chem. Commun.,* 2005, 785.
[18] Fan, W; Gao, L. *Chem. Lett.,* 2005, 34, 954.
[19] Liu, Y; Tang, J; Chen, X; Wang, R; Pang, GKH; Zhangm, Y; Xin, JH. *Carbon,* 2006, 44, 158.
[20] Li, H; Ha, CS; Kim, I. *Nanoscale Res. Lett.,* 2009, 4, 1384.
[21] Bourlinos, AB; Georgakilas, V; Zboril, R; Dallas, P. *Carbon,* 2007, 45, 2136.
[22] Cao, L; Scheiba, F; Roth, C; Schweiger, F; Cremers, C; Stimming, U; Fuess, H; Chen, L; Zhu, W; Qiu, X. *Angew. Chem. Int. Ed.,* 2006, 45, 5315.
[23] Li, X; Liu, Y; Fu, L; Cao, L; Wei, D; Wang, Y. *Adv. Funct. Mater.,* 2006, 16, 2431.
[24] Eder, D; Windle, AH. *Adv. Mater.,* 2008, 20, 1787.
[25] Seeger, T; Kohler, T; Frauenheim, T; Grobert, N; Ruhle, M; Terrones, M; Seifert, G. *Chem. Commun.,* 2002, 34.
[26] Hernadi, K; Ljubovic, E; Seo, JW; Forro, L. *Acta Mater.,* 2003, 51, 1447.
[27] Carp, O; Huisman, CL; Reller, A. *Prog. Solid State Chem.,* 2004, 32, 33.
[28] Wang, WD; Serp, P; Kalck, P; Faria, JL. *J. Mol. Catal. A: Chem.,* 2005, 235, 194.

[29] Wang, WD; Serp, P; Kalck, P; Faria, JL. *Appl. Catal., B* 2005, 56, 305.
[30] Yu, Y; Yu, JC; Yu, JG; Kwok, YC; Che, YK; Zhao, JC; Ding, L; Ge, WK; Wong, PK. *Appl. Catal., A* 2005, 289, 186.
[31] Bouazza, N; Ouzzine, M; Lillo-Ródenas, MA; Eder, D; Linares-Solano, A. *Appl. Catal. B: Environ.,* 2009, 92, 377.
[32] Fu, X; Yu, H; Peng, F; Wang, H; Qian, Y. *Appl. Catal., A* 2007, 321, 190.
[33] Chen, XW; Zhu, Z; Haevecker, M; Su, DS; Schloegl, R. *Mater. Res. Bull.,* 2007, 42, 354.
[34] Pietruszka, B; DiGregorio, F; Keller, N; Keller, V. *Catal. Today,* 2005, 102-103, 94.
[35] Juan, JC; Jiang, Y; Meng, X; Cao, W; Yarmo, MA; Zhang, J. *Mater. Res. Bull.,* 2007, 42, 1278.
[36] Zhang, D; Pan, C; Shi, L; Mai, H. X. Gao, *Appl. Surf. Sci.,* 2009, 255, 4907.
[37] Jusys, Z; Behm, RJ. *J. Phys. Chem. B,* 2001, 105, 10874.
[38] Baxter, JB; Walker, AM; Ommering, KV; Aydil, ES. *Nanotechnology,* 2006, 17, S304.
[39] Kamat, PV. *J. Phys. Chem., C* 2007, 111, 2834.
[40] Guldi, DM; Rahman, GMA; Sgobba, V; Kotov, NA; Bonifazi, D; Prato, M. *J. Am. Chem. Soc.,* 2006, 128, 2315.
[41] Kongkanand, A; Dominguez, RM; Kamat, PV. *Nano Lett.,* 2007, 7, 676.
[42] Wei, BY; Hsu, MC; Su, PG; Lin, HM; Wu, RJ; Lai, HJ. *Sens. Actuators, B,* 2004, 101, 81.
[43] Chen, Y; Zhu, C; Wang, T. *Nanotechnology,* 2006, 17, 3012.
[44] Van Hieu, N; Thuy, LTB; Chien, ND. *Sens. Actuators, B* 2008, 129, 888.
[45] Liang, YX; Chen, YJ; Wanga, TH. *Appl. Phys. Lett.,* 2004, 85, 666.
[46] Gavalas, VG; Law, SA; Ball, JC; Andrews, R; Bachas, LG. *Anal. Biochem.,* 2004, 329, 247.
[47] Adachi, M; Murata, Y; Okada, I; Yoshikawa, S. *J. Electrochem. Soc.,* 2003, 150, G488.
[48] Lin, CH; Lee, CH; Chao, JH; Kuo, CY; Cheng, YC; Huang, WN; Chang, HW; Huang, YM; Skih, MK. *Catal. Lett.,* 2004, 98, 61.
[49] Satishkumar, BC; Govindaraj, A; Vogl, EM; Baumallick, L; Rao, CNR. *J. Mater. Res.,* 1997, 12, 604.
[50] Eder, D; Kinloch, IA; Windle, AH. *Chem. Commun.,* 2006, 13, 1448.
[51] Satishkumar, BC; Govindaraj, A; Nath, M; Rao, CNR. *J. Mater. Chem.,* 2000, 10, 2115.
[52] Du, N; Zhang, H; Chen, B; Wu, JB; Ma, XY; Liu, ZH; Zhang, YQ; Yang, D; Huang, X. HJ; Tu, P. *AdV. Mater.,* 2007, 19, 4505.
[53] Eder, D; Windle, AH. *J. Mater. Chem.,* 2008, 18, 2036.
[54] Sivakkumar, SR; Ko, JM; Kim, DY; Kim, BC; Wallace, GG. *Electrochim. Acta,* 2007, 52, 7377.
[55] Chen, MH; Huang, ZC; Wu, GT; Zhu, GM; You, JK; Lin, ZG. *Mater. Res. Bull.,* 2003, 38, 831.
[56] Lin, YM; Tsang, JC; Freitag, M; Avouris, P. *Nanotechnology,* 2007, 18, 295202.
[57] Yi, WK; Jeong, T; Yu, SG; Heo, J; Lee, C; Lee, J; Kim, W; Yoo, JB; Kim, J. *Adv. Mater.,* 2002, 14, 1464.
[58] Gu, Z; Chen, H; Zhang, S; Sun, L; Xie, Z; Ge, Y. *Colloids Surf. A: Physicochem. Eng. Aspects,* 2007, 302, 312.
[59] Tissot, I; Reymond, JP; Lefebvre, F; Bourgeat-Lami, E. *Chem. Mater.,* 2002, 14, 1325.

[60] Agrawal, M; Pich, A; Gupta, S; Zafeiropoulos, NE; Simon, P; Stamm, M. *Langmuir,* 2008, 24, 1013.
[61] Yang, Z; Niu, Z; Lu, Y; Hu, Z; Han, CC. *Angew. Chem. Int. Ed.,* 2003, 42, 1943.
[62] Yang, M; Ma, J; Zhang, C; Yang, Z; Lu, Y. *Angew. Chem. Int. Ed.,* 2005, 44, 6727.
[63] Li, H; Ha, CS; Kim, I. *Langmuir,* 2008, 24, 10552.
[64] Li, H; Ha, CS; Kim, I. *J. Sol-gel Sci. Technol.,* 2010, 53, 232.
[65] Caruso, F. *Adv. Mater.,* 2001, 13, 11.
[66] Caruso, F. *Chem. Eur. J.,* 2000, 6, 413.
[67] Caruso, F; Lichtenfeld, H; Giersig, M; Möhwald, H. *J. Am. Chem. Soc.,* 1998, 120, 8523.
[68] Hotta, Y; Jia, Y; Kawamura, M; Omura, N; Tsunekawa, K; Sato, K; Watari, K. *J. Mater. Sci.,* 2006, 41, 2779.
[69] Yang, J; Lind, JU; Trogler, WC. *Chem. Mater.,* 2008, 20, 2875.
[70] Deng, Z; Chen, M; Zhou, S; You, B; Wu, L. *Langmuir*, 2006, 22, 6403.
[71] Lee, JH. *Sens. Actuators B,* 2009, 140, 319.
[72] Martinez, CJ; Hockey, B; Montgomery, CB; Semancik, S. *Langmuir,* 2005, 21, 7937.
[73] Li, XL; Lou, TJ; Sun, XM; Li, YD. *Inorg. Chem.,* 2004, 43, 5442.
[74] Li, B; Xie, Y; Jing, M; Rong, G; Tang, Y; Zhang, G. *Langmuir,* 2006, 22, 9380.
[75] Zhang, J; Wang, S; Wang, Y; Xu, M; Xia, H; Zhang, S; Huang, W; Guo, X; Wu, S. *Sens. Actuator B,* 2009, 139, 411.
[76] Wang, D; Song, C; Hu, Z; Fu, X. *J. Phys. Chem., B* 2005, 109, 1125.
[77] Huang, J; Ma, R; Ebina, Y; Fukuda, K; Takada, K; Sasaki, T. *Chem. Mater.,* 2010, DOI: 10.1021/cm903733s.
[78] Li, L; Chu, Y; Liu, Y; Dong, L. *J. Phys. Chem., C* 2007, 111, 2123.
[79] Yu, J; Yu, H; Guo, H; Li, M; Mann, S. *Small,* 2008, 4, 87.
[80] Yang, Z; Han, D; Ma, D; Liang, H; Liu, L; Yang, Y. *Cryst. Growth Des.*, 2010, 10, 291.
[81] Song, L; Gao, X. *Langmuir,* 2007, 23, 11850.
[82] Wang, Y; Su, F; Lee, JY; Zhao, XS. *Chem. Mater.,* 2006, 18, 1347.
[83] Han, S; Jang, B; Kim, T; Oh, SM; Hyeon, T. *Adv. Funct. Mater.,* 2005, 15, 1845.

In: The Sol-Gel Process
Editor: Rachel E. Morris
ISBN 978-1-61761-321-0

Chapter 9

PROTEIN SOL-GEL ENCAPSULATION WITH POLYMER ADDITIVES

Monika Sommerhalter*
Department of Chemistry and Biochemistry,
CSU East Bay, Hayward, California, USA

ABSTRACT

This review article describes how the sol-gel process can be modified to encapsulate proteins in a silica sol-gel matrix. The entrapment of intact and functional proteins leads to intriguing applications in biocatalysis and biosensing. The main challenges are protein deactivation and limited material transport. Proteins are prone to deactivation by the release of alcohol during the sol-gel process or by steric or chemical interactions with the sol-gel cage. The fine pores of the sol-gel matrix result in a very efficient encapsulation of the protein molecules but also restrict the transport of substrate or analyte molecules to the proteins' active sites. A major goal in the further development of sol-gel based biocatalysts and biosensors is to achieve better pore size control in more uniform sol-gel matrices. This review article focuses on recent efforts to improve the sol-gel process with the help of polymer additives for applications in biocatalysis and biosensing. The effect of the polymer additive on the encapsulated protein and the sol-gel host will be discussed. As more sophisticated spectroscopic, physisorption, and imaging tools are employed to characterize these novel hybrid materials, our understanding about the effect of polymer additives increases. Together with the systematic exploration of polymers with different chemical and physical properties, these technological advancements will result in the emergence of improved sol-gel based biocatalysts and biosensors.

* Corresponding author: monika.sommerhalter@csueastbay.edu

INTRODUCTION

The sol-gel process can be used to prepare glass-like materials or ceramics for numerous applications, including optical coatings [1], thermal insulators [2], or chromatography materials [3]. Silica based sol-gels also have medical applications as bioceramic implants [4] or drug release systems [5, 6]. Most sol-gel procedures yield optically transparent materials that can be doped to generate lasers [7]. Doping with conductive molecules facilitates electrochemical applications [8]. Sol-gel technology can also be employed to create diverse nanomaterials [9, 10].

The formation of sol-gels is a room-temperature process and is therefore compatible with the encapsulation of biomolecules. The incorporation of biomolecules, in particular proteins, broadens the range of possible applications even further [11-14]. Proteins have highly selective binding properties and catalytic functions. Sol-gel encapsulated proteins can thus be used as biosensors or biocatalysts. Their immobilization enables continuous reuse which results in more economical and also more practical process development. Silica based sol-gel glasses are particularly well suited as immobilization matrices. They are chemically inert, mechanically robust, and stable in a broad range of solvents and a broad pH range (pH 2-9). Further, sol-gel materials are inexpensive and easy to prepare. They can be cast into different shapes, such as fibers, thin film coatings, beads, disks, or other bulk shapes (monoliths). Alternatively, sol-gels can be ground into fine powders. The sol-gel process can be used to create organic-inorganic hybrid materials, also called ormosils [15, 16]. These ormosils can house diverse chemical groups, such as alkyl, alkenyl, or aryl chains with or without additional functional groups (thiol, amino, or hydroxyl groups). Additives, including sugar molecules, ionic liquids, surfactants, or polymers can be employed to influence the pore structure and other properties of the sol-gel material. Mesoporous materials with pore sizes ranging from 2-50 nm and microporous materials with pore sizes smaller than 2 nm can be created [17]. In sum, sol-gel materials are impressively versatile.

The optical transparency of the silica sol-gel glass and the possibility to dope sol-gels with conducting molecules are major advantages for using silica sol-gels in biosensor applications. The sol-gel entrapment of glucose-oxidase, for example, led to the development of photometric and electrochemical biosensors to detect glucose [18]. A prime example for the development of a reusable biocatalyst is the sol-gel entrapment of lipase in an ormosil [19]. It is also possible to co-immobilize several different enzymes in one sol-gel matrix. Such a multienzyme scheme with cholesterol oxidase and cholesterol esterase led to the development of a cholesterol biosensor [20]. The co-encapsulation of formate dehydrogenase and alcohol dehydrogenase in the presence of NADH resulted in a biocatalyst that is able to produce the fuel methanol directly from carbon dioxide at ambient temperature and pressure [21].

Despite many successful developments in biocatalysis and biosensing, challenges still remain. The catalytic performance of sol-gel encapsulated enzymes is typically diminished in comparison to the catalytic performance of the free enzyme in solution. Relative activity values for sol-gel encapsulated enzymes can range between 2% and 100% [22]. This loss of activity has two main causes. First, the encapsulated protein might be damaged or functionally compromised. Second, the material transport properties of the sol-gel host are limited. Protein damage can be caused by the presence of residual alcohol, steric or chemical interactions with the sol-gel cage, or slow dehydration of the protein as the sol-gel material ages. Substrate or analyte diffusion is limited by the pore size distribution and the pore connectivity of the silica network. Furthermore, the silica surface carries negative charges, which results in a repulsion of negatively charged substrate or analyte molecules. The addition of polymers, ionic liquids, surfactants, or sugar molecules into the sol-gel procedure can help to improve the performance of sol-gel encapsulated proteins [22-27]. This review article will focus on the effect that polymer additives exhibit on the sol-gel host and the encapsulated protein. Selected examples for the successful creation of biocatalysts and biosensors will be highlighted.

The Sol-Gel Process

Sol-gels are derived from main group and transition metal-oxide or metal-alkoxide precursor molecules. Silica-based sol-gels are most often used for bioencapsulations. The most common precursors for silica sol-gels are tetraethoxysilane (TEOS) and tetramethoxysilane (TMOS). TEOS is less expensive and less hazardous than TMOS [28]. The traditional sol-gel precursors ($Si(OR)_4$) can be modified by substituting a hydrolysable Si-OR bond with a non-hydrolysable Si-C bond. Various organic functional groups can thereby be incorporated in the sol-gel material resulting in the production of organically modified silica glasses, also called ormosils [16]. The sol-gel precursor molecules are first hydrolyzed in the presence of a catalyst, such as hydrochloric acid, sodium hydroxide, or sodium fluoride. Depending on the catalyst used during hydrolysis, colloidal particles (base catalysis) or polymeric networks (acid catalysis) are obtained [29, 30]. A polycondensation reaction follows after hydrolysis. A silica network evolves and the interpenetrating solid and liquid phases form a gel. The gel strengthens over time and "ages" as the condensation reactions progress and the network matures. Sol-gels can be prepared as wet hydrogels, dried xerogels, or supercritically dried aerogels [31]. The available surface area and the morphology of the sol-gel (i.e. the pore geometry and pore size distribution of the sol-gel network) are very dependent on the drying process, the composition of the sol-gel mixture, and other details of the sample preparation. Furthermore, the average pore size, pore geometry, and pore size distribution of sol-gels can be influenced profoundly with the use of templating agents, notably surfactants. If the silica sol is prepared in the presence of a water soluble surfactant, and if this templating agent is removed via calcination, a mesoporous silicate is obtained [32, 33].

Protein Properties

Proteins are composed of amino acids that are linked to each other in long polypeptide chains typically comprised of 100 to 300 amino acids. The 20 naturally occurring amino acids can be grouped according to their side-chains into uncharged polar, charged polar, or non-polar amino acids. Several amino acids have acidic or basic side chains. This chemical versatility results in an impressive array of different proteins with unique properties, structures, and functions. These functions include catalysis (enzymes), immune defense (antibodies), structural support (keratin), muscle contraction (myosin), transport and storage of small molecules (hemoglobin, myoglobin), or signal transduction (hormone receptors). The polypeptide chains of proteins are organized in well defined structures that can be characterized based on their secondary, tertiary, or quaternary structure. The secondary structure describes the presence of alpha-helices, beta-strands, and loops. The spatial arrangement of these secondary structure elements defines the tertiary structure. Some proteins are assembled from several polypeptide chains. The quaternary structure describes the orientation and position of these subunits. The overall shape of a protein and the position of individual amino acids are critical for the function of a protein. It is therefore crucial that proteins retain their structure after sol-gel encapsulation. It is also important that the catalytically active site of enzymes or the epitope binding patch of antibodies remain accessible. In addition, proteins require a certain degree of conformational freedom to remain functional.

Proteins are unique. They differ in size, shape amino acid composition, surface charges, polarity, and cofactor requirements. Many enzymes depend on metal centers, metalloporphyrins (heme, for example), NADH, or other redox active cofactors. Proteins are found in different cellular locations. Integral membrane proteins have large hydrophobic patches so that they can be embedded in the lipid bilayer of biological membranes. Cytosolic or secreted proteins have mostly hydrophilic surfaces. The sol-gel encapsulation procedure is based on a very generic protocol. However, the unique characteristics of proteins often require very case dependent modifications and a good understanding of the properties that are characteristic for the target protein.

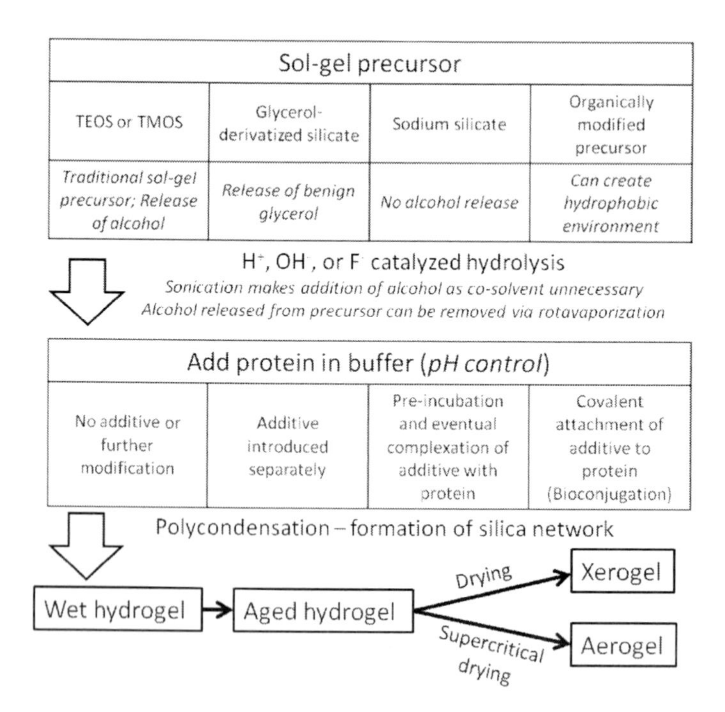

Figure 1. Different strategies for the sol-gel encapsulation of proteins.

Modifications to the Sol-Gel Procedure for Protein Encapsulation

To make it possible to introduce proteins into sol-gels, several precautions are necessary (see Figure 1). Proteins are sensitive to changes in pH, ionic strength, and the presence of organic solvents (for example, alcohol) or chaotropic agents. In the traditional sol-gel process alcohol is used as a co-solvent. Ellerby et al. [34] modified the sol-gel process by introducing a sonication step. Sonication ensures that the sol-gel precursor is well distributed in the hydrolysis mixture and renders the addition of alcohol as a co-solvent unnecessary. Nevertheless, some alcohol is still released from the precursor molecule. The hydrolysis of TEOS generates ethanol and the hydrolysis of TMOS generates methanol. Ellerby et al. [34] added a buffer after the acid-catalyzed hydrolysis of TMOS to raise the pH above 5.0 so that proteins are protected from very harsh pH conditions. The proteins are added during the polycondensation step. As the silica network develops the proteins are encapsulated into the nascent mesh. The procedure developed by Ellerby et al. [34] was further modified by removing the residual alcohol by rotavaporization before a buffered enzyme solution was added [35]. Alternatively, other sol-gel precursors can be used, such as glycerol-derivatized silicates [36] or sodium silicate [37, 38]. The release of glycerol instead of alcohol can have a beneficial effect on the protein. However, the glycerol-derivatized silicates are not commercially available and need to be synthesized. The aqueous route with sodium silicate as a precursor is the best choice according to the principles of Green Chemistry [28, 38]. Less environmentally friendly compounds, such as organosilane precursors, however, enable the design of more versatile organic-inorganic hybrid materials (ormosils). Another technique for protein sol-gel encapsulation involves liposomes that carry protein molecules as cargo [39, 40]. The liposome is incorporated into the sol-gel during the polycondensation phase and later destroyed with a strong electric pulse. This procedure results in very large cages for the encapsulated proteins. Liposomes also have other intriguing applications in sol-gel technology. Liposomes can be used to generate cerasomes. Cerasomes are organic-inorganic hybrid nanomaterials composed of a spherical lipid-bilayer membrane with an internal aqueous compartment and covered with a silicate framework [41, 42]. A reverse construct, with a lipid bilayer on the outside and a porous silica nanoparticle in the inside can also be made [43, 44].

How Does the Protein Influence the Sol-Gel Host?

The size of protein molecules is large enough to influence the pore size distribution of sol-gel materials. The molecular dimensions of ovalbumin, for example, are 7 nm × 4.5 nm × 5 nm [45]. The addition of ovalbumin to a TEOS-based sol-gel created mesopores and also some larger macropores [46]. The additive free control had only micropores. The templating effect of proteins was mostly studied with nitrogen sorption techniques [46, 47]. This technique requires that the sample be dried and evacuated. Already dried xerogels are therefore more suitable than wet hydrogels. To gain insight into the pore size distribution of hydrogels a supercritical drying protocol can be employed to convert the hydrogel into an aerogel after the stepwise replacement of the wet buffer inside the pores with liquid CO_2, followed by a temperature raise and slow gas venting [37, 48]. A recent alternative technique is small angle neutron scattering (SANS). In contrast to nitrogen sorption experiments, SANS also assesses closed pore structures [49]. The two techniques therefore yield different results.

In addition, the presence of proteins also influences the maturation process of the silica host. Proteins carry acidic and basic groups that can catalyze hydrolysis and polycondensation reactions and thus influence aging of the silica matrix. Gao et al. [49] encapsulated bovine serum albumin (BSA) in sol-gels derived from TMOS and MTMS (methyltrimethoxysilane) mixtures. They monitored the maturation of the silica matrix with ATR-FTIR (Attenuated Total Reflection – Fourier Transform InfraRed Spectroscopy) and solid-state ^{29}Si NMR spectroscopy. ATR-FTIR is useful to characterize the surface of sol-gel materials. Spectroscopic fingerprints indicative for a condensed silica network with Si-O-Si bonds can be distinguished from free Si-OH groups. As the BSA content increased, more Si-O-Si species were observed which is typical for a more mature silica network. The NMR spectroscopic analysis confirmed that the protein influenced the evolution of the matrix network.

The process of biosilicification strikingly illustrates that proteins can have a profound structure-directing role in the formation of silica materials. Diatoms, a major class of phytoplankton, are encased in a wall of silica, called frustule. These frustules show beautiful silica-based nanostructures. The formation of the silica cage is guided by proteins called silaffins [50]. Polyamines and short peptide sequences can also be used to trigger the precipitation of silica from dilute sodium silicate solutions [51, 52]. Diverse polypeptide-based biopolymers are now employed to create silica nanoparticles from aqueous silica [52]. Biosilicification and its application in nanotechnology is a rapidly growing research field [38].

Assessing the Structural Integrity of Sol-Gel Encapsulated Proteins

How do we know whether an encapsulated protein has retained its structural integrity? Sol-gel glasses are optically transparent. Therefore many spectroscopic tools, such as circular dichroism (CD), absorption spectroscopy, and fluorescence spectroscopy, can be used to assess potential protein damage [53]. In addition, NMR spectroscopy can also be employed [54]. CD can be used to assess the presence and integrity of secondary structure elements. Studies on apomyoglobin in different types of sol-gel glasses with covalent organic groups revealed an increase in helical content of apomyoglobin with increasing hydrophobicity of the sol-gel glass [55, 56]. The emission wavelength of the amino acid tryptophan is indicative for the polarity of the local tryptophan environment. This intrinsic spectroscopic fingerprint can be used to study the effect of the sol-gel encapsulation procedures on protein folding [25, 57]. The attachment of fluorophores to proteins makes it possible to assess the conformational flexibility and reorientation dynamics of the protein inside the sol-gel glass [58]. UV/VIS spectroscopy, Resonance Raman spectroscopy and EPR spectroscopy can be used to study the coordination of heme cofactors and thereby reveal details on the active site of hemeproteins that are encapsulated in sol-gel glasses [59]. In most cases, it was observed that encapsulated proteins retain their structural integrity very well upon sol-gel encapsulation. Furthermore, in many cases, an increase in stability with respect to heat or organic solvents was observed [60]. It was rationalized that the confinement of proteins into the sol-gel cages hinders large conformational movements that result in unfolding [61, 62]. How stable a protein remains after sol-gel encapsulation, nevertheless, is still dependent on the individual protein and details of the sol-gel procedure.

Limited Material Transport in Sol-Gel Materials Restricts the Performance of Encapsulated Proteins

Overall, the major common obstacle for optimum catalytic performance is associated with the limited material transport properties of the silica sol-gel materials. The smaller the pores, the more hindered the material transport. However, if the pore size is too large, the protein is poorly retained. The connectivity of the individual sol-gel pores is also important. Enzyme molecules entrapped in closed pores or enzymes with concealed active sites cannot participate in biocatalysis or biosensing. In the quest for optimum poor size distribution and optimum pore connectivity the consequences of aging and drying sol-gel samples also need to be carefully considered. As the sol-gel network matures, sol-gels have a tendency to shrink. The problem of small pore sizes is compounded further by pore collapse upon sample drying. As a sol-gel sample is air dried, the removal of the wet phase should be gentle and slow, so that the pore collapse is minimized. Gentle drying procedures are controlled desiccation or pinhole drying [14]. The co-immobilization of polymer additives or the use of ormosil precursors can protect the silica network from pore collapse.

Material transport can also be hindered by electrostatic interactions. The silica host carries negative surface charges. The size, polarity, and charge of potential substrate molecules impact their diffusion in the sol-gel host. Kadnikova et al. [63] encapsulated horse radish peroxidase (HRP) in silica prepared from TMOS and methylated silica prepared from methyltrimethoxysilane. In both matrices HRP catalyzed the oxidation of ABTS with hydrogen peroxide. However, the catalytic performance of HRP encapsulated in the methylated silica was superior, because of diminished electrostatic repulsion between the negatively charged ABTS substrate and the methylated silica surface. Smith et al. [59] encapsulated HRP in TMOS and TEOS derived sol-gels and tested the peroxidation activity for different substrates, including I^- and guaiacol. The smaller substrate (I^-) was oxidized at much higher rates. Limited mass transport was further demonstrated by casting the sol-gel materials into different shapes (pellets versus monoliths). As surface:volume ratios decreased from pellets to monoliths, the catalytic performance of encapsulated HRP also decreased.

Incorporation of Additives into the Sol-Gel Procedure

A large variety of additives, including sugar molecules [25, 40, 64-66], ionic liquids [67-71], surfactants [26, 72], and diverse polymers [22, 23, 27, 73-78] have been incorporated into the sol-gel procedure. These additives can influence the pore size distribution of the sol-gel host and/or stabilize the encapsulated protein. Additives can improve pore size distributions by acting as templating agents that increase the average pore size during the formation of the silica network and by helping to prevent pore collapse by pore filling during the drying stage. Additives can also help to improve the catalytic performance of sol-gel encapsulated enzymes by protecting the enzyme from damage caused by steric or chemical interactions with the sol-gel cage. It was also proposed that ionic liquid additives protect the enzyme from alcohol that is released during the hydrolysis step in sol-gel processes that employ TMOS or TEOS [70, 71]. Other additives, such as polyethylene glycol (PEG), can protect proteins from dehydration [22]. In some cases, the additive can alter the properties of the encapsulated enzyme. The incorporation of surfactant molecules in TMOS-based sol-gels drastically changed the working pH range for alkaline and acid phosphatase [72].

Molecular additives can be introduced in different ways into the sol-gel procedure. The use of organically modified silica precursor molecules results in the covalent attachment of organic groups into the silica sol-gel cage. The creation of such ormosils requires the synthesis of a specific silica precursor molecule. However, the introduced organic group is firmly embedded in the sol-gel material and cannot be lost by leaching. If molecular additives are simply added to the sol-gel mixture, no extra steps are required. Thus this type of modification is not labor intensive. However, the additive is either washed out on purpose or is eventually leaching out over time. Wei et al. [66], for example, modified the pore size distribution of a TMOS-based sol-gel by adding glucose. The sol-gel also contained the enzyme acid phosphatase. The addition of high glucose concentrations (42-60% by weight) during the sol-gel preparation step followed by subsequent glucose removal with washing resulted in a three-fold increase of enzymatic activity relative to an additive-free control. The average pore size of the samples prepared in the presence of high glucose concentrations increased to 3.4 nm. Low concentrations of glucose only yielded microporous materials similar to the additive-free control. In other studies, the additive is not washed out on purpose. The additive is supposed to stabilize the encapsulated protein inside the sol-gel cage. Keeling-Tucker et al. [22], for example, incorporated low-molecular weight PEG into ormosils that housed the enzyme lipase. The addition of PEG increased the catalytic performance of the encapsulated lipase significantly. However, as the researchers continued to use these samples in several catalytic cycles, they noticed that the PEG additive was leaching out. The decrease in catalytic performance of the encapsulated lipase was more pronounced with extensive washing between the catalytic cycles. A similar observation was made by Lee et al. who used ionic liquid additives to increase the performance of sol-gel encapsulated lipase [70, 71]. An alternative approach involves the covalent attachment of additives to the protein. Dattelbaum et al. [23] modified a periplasmic maltose-binding protein by covalent conjugation with PEG 5000. The maltose binding protein also carried an NBDamide group, to create a fluorescence based biosensor for maltose. The silica sol-gel host was created from diglycerylsilane, so that benign glycerol molecules, instead of alcohol, were released during the encapsulation step. The bioconjugation of the protein with PEG increased the signaling response of the maltose biosensor.

Mesoporous Silica

Another variation to use additives in the sol-gel process is the creation of mesoporous silica. In this process the silica sol is mixed with a surfactant. A surfactant is a wetting agent that lowers the surface tension of liquids. The surfactant and the silicate form supermolecular structures that result in large, uniform pores with characteristic geometry (for example, hexagonal or cubic). The templating agent is removed by calcination. In the calcination step the dried silica sol-gel is heated up to 500^0C or 800^0C. Large open structures result and thus these materials can be used as molecular sieves. Chiral surfactants can be used to imprint the sol-gel host to confer enantioselectivity [79, 80]. The most studied ordered mesoporous silica systems are MCM-41, MCM-48, and SBA-15. MCM-41 and MCM-48 have pore sizes in the order of 2-10 nm. SBA-15 has larger pore sizes in the order of 5-30 nm [81]. The pore size is tunable and depends on the alkyl chain length of the surfactant, the silica to water ratio, and the temperature setting of the calcination step. Mesoporous silica can also be employed for protein immobilization. However, the protein is not entrapped, but adsorbed to the silica surface. To increase storage and operational stability of these biocatalysts or biosensors, covalent cross-linkers can be introduced [82, 83]. The enzyme chloroperoxidase has been immobilized in different mesoporous silica hosts [83-87]. Chloroperoxidase is one of the most versatile heme proteins and catalyzes peroxidation, oxygenation, epoxidation, halogenations, and sulfoxidation reactions with high substrate promiscuity but high stereoselectivity. Thus, CPO is an attractive biocatalyst for oxidative transformations that are often employed to synthesize drug molecules [88, 89].

Polymer Additives

In the following, the effect of polymer additives on aged hydrogels and xerogels will be discussed. Due to their large size, polymers are less likely to be washed out by accident. Continuous reusability and long term stability of the sol-gel material are important parameters for applications in biocatalysis and biosensing. A gradual change in the composition of a biosensor would be detrimental. However, as mentioned earlier, low molecular weight polymers can still leak from sol-gels [22]. In some cases, covalent attachment of the additive to the sol-gel host or the protein might be warranted. Due to the growing importance of bioconjugation for medical applications, in particular cell targeted protein delivery, many chemical tools have been developed for the attachment of polymers to proteins [90]. Notably, the conjugation of enzymes with polymers often leads to an enhancement of the enzyme's catalytic performance in organic solvents. A synergistic effect of polymer conjugation and sol-gel encapsulation is therefore possible.

There are numerous examples for the successful co-immobilization of polymers and enzymes in silica-based sol-gels. Shtelzer et al. [75], for example, observed that the addition of PEG 6000 to TMOS based sol-gels increased the proteolytic activity of the encapsulated trypsin from 21% to 47% relative to an additive-free control. The entrapment of two membrane proteins, the nicotinic acetylcholine receptor and the dopamine D2 receptor, required the addition of PEG (PEG 600 or PEG 10000) to maintain an active state for both ligand-binding receptors [91]. However, there are also examples for which the incorporation of a polymer additive decreased the catalytic performance of the encapsulated enzyme. The activity of sol-gel encapsulated chlorophyllase, for example, decreased upon co-immobilization with PEG 1500, polyethyleneimine (PEI), or polyvinyl alcohol (PVA) [27].

Some very common polymer additives are shown in Figure 2. The properties of a polymer are mostly determined by the properties of the monomeric unit that builds up the polymer. Most polymers contain only a single type of unit (for example PEG), but other polymers are composed of different units that have similar functional groups in common. Proteins are excellent examples for such complex polymers because they are built from a diverse pool of amino acids. Polymers are distinguished by their chain length or molecular weight. PEG 400, for example, has an average molecular mass of 400 Da, which means that an individual polymer molecule contains approximately 8 to 9 monomeric units. Very common polymers can be purchased in different molecular weight ranges. The typical molecular weight range for PEG 400 is 380-420 Da. The molecular weight range typically increases with the chain length. PEG 10000, for example, is offered in a molecular weight range of 8,500 to 11,500 by Hampton Research or other vendors. Some polymers are linear, but others can have branches. PEI can be obtained in a linear or a branched version. Unfortunately, complete specifications (average molecular weight, molecular weight range, and the presence of branches) are not always provided in the studies that present an improved biocatalyst or biosensor.

Effect of Polymer Additives

The combined incorporation of polymers and proteins in sol-gels will have multiple effects on the properties of the composite material. In the following, individual aspects of these effects will be discussed. However, it is important to keep in mind that some effects might counterbalance each other and that all three components (protein, polymer, and sol-gel host) will influence each other.

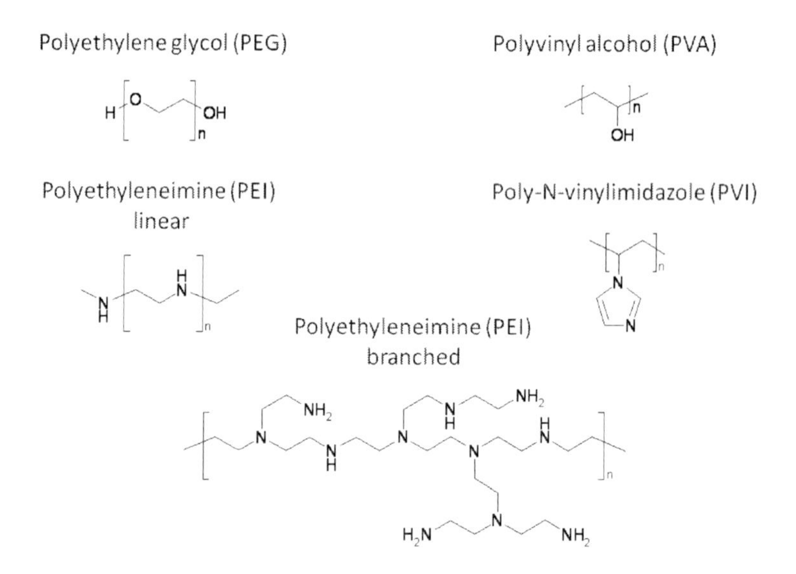

Figure 2. Some polymers that are often used as additives in protein sol-gel encapsulation.

Pore size distribution in silica sol-gels

The addition of water soluble polymers into the sol-gel procedure can have a profound influence on the pore size distribution of silica sol-gels. Sato et al. [46] prepared TEOS-based sol-gels in the presence of uncharged polymers (polyvinylalcohol and polyethylenglycol) and charged polymers (anionic polyacrylic acid and cationic polyethyleneimine). The sol-gels were calcined after drying to remove the polymer additive. The pore size distribution was analyzed with nitrogen sorption isotherms and SEM imaging. Without any polymer additive the calcined TEOS sol-gels were mostly microporous with a few mesopores with diameters up to 10 nm. With uncharged polymer additives the calcined TEOS sol-gels are more mesoporous, but the mesopores were still smaller than 20 nm in diameter. The addition of charged polymers increased the number of large mesopores significantly. Some of these pores were larger than 20 nm in diameter. The properties of the added polymer (charged versus uncharged) are therefore crucial for the pore size distribution. Charged polymers can have electrostatic interactions with the silica sol-gel particles as the sol-gel particles grow, thus tailoring their size. The size of the sol-gel particles determines the pore size dimensions as the silica network forms during the gelation phase of the sol-gel process. As mentioned previously, the presence of the protein itself also influences the pore size distribution of the sol-gel host. Upon addition of lipase and ovalbumin, larger mesopores and some macropores were formed in a TMOS-based sol-gel [46]. Proteins have large dimensions of several nm across and carry charged amino acids on their surface. Glutamate and aspartate have negatively charged carboxylate groups. Histidine, lysine, and arginine carry positive charges. The sheer bulk of the protein and the electrostatic interactions with the charged silica particles influence the final pore dimensions of the silica network.

If the polymer-doped sol-gels are not calcined and the polymer remains inside the cages of a sol-gel, polymer doping typically results in a decrease of the surface area and overall pore volume [27, 92]. The polymers fill the pores. In a wet hydrogel this is typically a disadvantage, since the polymer might hinder substrate or analyte diffusion. Pore filling, however, can protect the silica sol-gel cage from pore collapse upon drying of the sol-gel. This in turn, has a positive effect on the performance of enzymes encapsulated in xerogels.

Physical appearance and water retention

TMOS or TEOS based sol-gels are optically transparent as wet hydrogels and also as xerogels, provided that the drying process proceeds in a slow and careful manner. Xerogels that develop cracks and are brittle tend to become opaque. The incorporation of polymers can help to prevent cracking and can improve the dehydration and rehydration stability of the sol-gels [22]. PEG, notably, has a strong tendency to be hydrated by water molecules [93]. PEG's ability to retain water is also very beneficial for the encapsulated protein.

Sato et al. [46] state that the incorporation of uncharged polymers results in transparent sol solutions and transparent gels (whether they were wet, dried, or calcined). In their study, PVA and PEG with molecular weights of 1000, 3000, 8500, and 50000 were added in amounts up to 37.5% w/w to TEOS based sol-gels. Brennan et al., however, note that the addition of PEG with molecular weights of 1000 or greater resulted in opaque and very brittle materials [22]. The addition of PVA also compromised the optical transparency. Brennan et al. used TEOS as well as organosilane sol-gel precursors.

Polarity of the internal sol-gel environment

The entrapment of polarity sensitive probes, such as pyrene, 7-azaindol, or prodan, makes it possible to study the local chemical environment inside the sol-gel glasses. Baker et al. [92] used the fluorophore pyrene to assess the local polarity of PEG doped xerogels prepared from

TMOS. The addition of PEG 8000 (with final concentrations of 3, 5, and 10% by weight) resulted in a significant enhancement in the polarity inside the sol-gel host. In another study, Baker et al. [94] used PEG 200, PEG 300, and PEG 400 (with final concentrations of 1, 3, and 5% by weight) to make TEOS/PEG xerogel composites. In contrast to their previous study, the local polarity of the pyrene probe decreased with increasing PEG content. This decrease was most pronounced for the PEG with the lowest molecular weight. Apparently, the molecular weight of the PEG additive has a significant influence on the local polarity inside the sol-gel host. Alternatively, the location of the probe might depend on the molecular weight of the PEG additive. The probe might be located in different microenvironments, either closer to the surface of the sol-gel cage or surrounded by polymer molecules. The assessment of the rotational mobility of the fluorophore rhodamine 6 G (R6G) also showed a strong dependence on the molecular weight of the PEG additive. Inclusion of PEG 8000 increased R6G mobility [92]. In contrast, inclusion of low molecular weight PEG additives (PEG 200-PEG 400) decreased R6G mobility [94].

Brennan et al. [22] used the fluorophores 7-azaindol and prodan to assess the polarity of PEG and PVA doped TEOS-derived sol-gels. PEG 400 and PEG 600 were included in total amounts of 5% and 10% weight per volume. PVA was added in an amount of 5% w/v. PVA enhanced the local polarity within the sol-gels, whereas PEG decreased the local polarity. This trend is in agreement with the different amounts of polar groups present in PEG versus PVA. PEG houses only one terminal alcohol group, whereas PVA has one alcohol group per monomeric unit. The decrease in polarity with the addition of low-molecular weight PEG is in agreement with the second study by Baker et al. [94].

Electrostatic interactions

Proteins with positive surface charges will have electrostatic interactions with negatively charged silanol groups on the surface of the sol-gel pores. These interactions can result in protein inactivation. If proteins are co-immobilized with a charged polymer, these electrostatic interactions can be altered. Polycations, for example, should be able to shield positively charged proteins from the negatively charged silanol groups in the sol-gel matrix. Heller et al. [76, 77] tested this hypothesis with three oxidase enzymes that have different isoelectric points (IP) and therefore different surface charges. The enzymes glucose oxidase (IP=3.8), lactate oxidase (IP=4.6), and glycolate oxidase (IP=9.6) were encapsulated in TMOS-based gels after vacuum evaporation of methanol. The gels were air dried. The resulting xerogels were ground into a fine powder and later dispersed in aqueous solutions for measuring the enzymatic activity. Without the incorporation of polymer additives, only glucose oxidase retained its enzymatic activity in the silica xerogels. Notably, the heat stability of sol-gel encapsulated glucose oxidase increased considerably. In contrast, glycolate oxidase and lactate oxidase were inactivated upon sol-gel encapsulation. Glucose oxidase has the most acidic IP and carries mostly negative surface charges at neutral pH. Glycolate oxidase with the most basic IP carries positive charges at neutral pH, but lactate oxidase with an acidic IP should carry mostly negative charges at neutral pH and therefore behave more like glucose oxidase. However, both glycolate and lactate oxidase have positively charged arginine residues at the opening of a channel that guides substrate molecules into the enzyme's active site. Glucose oxidase, in contrast, has positively and negatively charged residues at the channel's opening. The inactivation of glycolate and lactate oxidase can therefore be explained with unfavorable electrostatic interactions between negatively charged silanol groups at the sol-gel surface and positive surface charges located at the opening of a channel that is essential for catalysis. A polycation additive should be able to provide a protective shield. As expected, the electrostatic complexation of lactate oxidase and glycolate oxidase with the strong base polyethylenimine (PEI) stabilized both proteins and they retained their activity in the silica host. The weak base poly-N-vinylimidazole (PVI) was only able to stabilize sol-gel encapsulated lactate oxidase (with acidic IP), but not glycolate oxidase (with basic IP).

Electrostatic interactions between polymer and protein can also be detrimental, however. The polycation PEI and the uncharged polymeric additives PEG and PVA were employed for the encapsulation of chlorophyllase in TMOS-based, dried sol-gels [27]. Even though PEI-doped sol-gels had larger pore volumes than PEG and PVA doped sol-gels, the lowest enzymatic activity was determined for the PEI-doped sol-gel. The relative activity for PEI-containing samples was 30% with respect to an additive-free control. PEG and PVA containing samples showed relative activities of 40%. Chlorophyllase has an acidic IP and therefore carries negative surface charges. The authors suggested that electrostatic interactions between the polycation PEI and chlorophyllase might partially denature the enzyme [27]. Notably, chlorophyllase is a membrane protein. Yi and coworkers therefore also introduced lipid additives [27]. With increasing amounts of the lipid monogalactosyldiglyceride the activity of the sol-gel entrapped chlorophyllase increased up to a value of 140% in comparison to a TMOS-based gel prepared without any additive.

O'Neill et al. [95] reported that negatively charged polymers (blue dextran and alginic acid) enhanced the activity of sol-gel encapsulated invertase more than positively charged polymers (PEI). The invertase from *Candida utilis* was encapsulated in TMOS-based hydrogel beads. The IP of invertase from *Candida utilis* is located at pH 3.35 [96]. Thus invertase carries negative surface charges at neutral pH. In addition, this large protein (300kDa in mass) is glycosylated. It is conceivable that the sugar molecules on the enzyme surface provide additional shielding. O'Neill et al. proposed that the repulsion between negatively charged polymers and the silica surface increases the pore size [95]. This results in a more efficient substrate diffusion to the encapsulated enzyme molecules. A pore size determination was not performed, however.

The above examples illustrate the uniqueness of every case of protein sol-gel encapsulation. There are multiple interactions between protein, polymer, and sol-gel host with beneficial or adverse effects. Some of these effects might counterbalance each other. For example, a polymer can hinder substrate diffusion by filling out the channels of a sol-gel network and thus diminish the catalytic efficiency of the composite material. Yet at the same time, the polymer can improve the stability of the encapsulated enzyme in the sol-gel cage and thus increase the catalytic efficiency of the composite material.

Biocatalysis

Using protein sol-gel encapsulation for biocatalytical applications has several advantages. The chemical inertness of the silica host ensures that the matrix used for the immobilization does not interfere with the catalyzed reaction. The sol-gel material can be manufactured in different shapes. For example, small sol-gel beads that fill a bioreactor can be produced [95].

The sol-gel material can also be cast into a monolithic column, and solvents and reactants can be applied with the help of a pump. Practically zero leaching of the protein from the silica matrix ensures that the biocatalyst can be re-used extensively. Often, an enhanced stability of the entrapped protein is observed, so that the reaction can be performed at elevated temperature or in organic solvents. Many chemicals that are of interest for industrial transformations (such as intermediates for the synthesis of pharmaceuticals) are better soluble in organic solvents. Also, the encapsulation process is very simple and cost efficient, thus rendering the biocatalyst economical. Notably, sol-gel encapsulation can be used to entrap several different enzymes in one sol-gel host [30]. Thereby, several sequential catalytic steps can be performed with one multi-enzyme biocatalyst.

Most studies on biocatalysis with sol-gel encapsulated enzymes use the enzyme lipase. Lipases can be used for the synthesis of biopolymers, biodiesel, and fine chemicals (therapeutics, agrochemicals, cosmetics, and flavor compounds) [97]. Lipases catalyze the esterification of alcohols with carboxylic acids as well as the reverse hydrolysis reaction. The active site of lipase is covered with a loop, also called the inactivation lid. When lipase approaches a non-polar interface, the inactivation lid moves out of the way and the substrate can approach the active site. This mechanism is known as interfacial activation [98]. The immobilization of lipase in sol-gel matrices is of great industrial interest, but the first attempts to encapsulate lipase in TMOS or TEOS based sol-gels resulted in a very poor catalytic performance [99, 100]. For example, relative activity values of less than 5% in comparison to non-immobilized lipase were obtained for the esterification of lauric acid with n-octanol in iso-octane [99]. As Reetz et al. started to use hydrophobic ormosils, the catalytic performance of the encapsulated lipase increased dramatically [99]. The encapsulation of lipase in an ormosil derived from methyltrimethoxysilane (MTMS) resulted in relative catalytic activity values of 1300% in comparison to non-immobilized lipase. In concord with the previous example, the esterification reaction of lauric acid with n-octanol in iso-octane was employed to monitor lipase activity [99]. It is conceivable that the hydrophobic environment of the ormosil triggers a similar conformational change (i.e., the opening of the inactivation lid) as expected for the interfacial activation of lipase. Due to the commercial interest in immobilized lipase, further optimization of the sol-gel encapsulation procedure is a very active research area. These optimizations include the incorporation of various additives, such as PEG [22], ionic liquids [70, 71], crown ether [101], and calixarenes [102]. In addition, the usage of different ormosil precursors is becoming more elaborate [103]. For example, mixtures of alkylalkoxysilanes with short chain and long chain alkyl substituents are used [102, 104, 105]. The deposition of the sol-gel onto another immobilization support, for example celite, proved to be a valid strategy to increase the thermostability of the encapsulated lipase [106-110]. Notably, lipases are very abundant enzymes and can be obtained from various organisms. Sol-gel encapsulated lipases from *Candida antarctica, Candida rugosa,* and *Pseudomonas cepacia* are already commercially available (Fluka). Notably, sol-gel encapsulated lipase could also serve as a biosensor for triacylglycerides [22]. A high level of triacylglycerides is a risk factor for heart disease. The detection of triacylglycerides in blood samples is therefore of interest.

Biosensing

Protein encapsulation in silica sol-gels provides many opportunities for the development of biosensors. Many sol-gel glasses are transparent. Transparency is a prerequisite for optical detection methods. In addition, sol-gel materials can be doped with conductive molecules for amperometric (electrochemical) detection. A prime advantage of the sol-gel process is the ability to cast sol-gels into various different shapes, notably, thin film coatings on electrodes. The sol-gel with the entrapped biosensor can also be integrated into an optical fiber. Practically zero leaching of the protein from the silica matrix ensures that the biosensor can be stored. Long-term storage and extended usage without losing sensitivity are very important characteristics for successful biosensor development.

In the past, an impressive range of different biosensors has been developed [11, 12, 20, 23, 30, 53, 111-119]. In this section, three selected examples that all make use of the enzyme cholinesterase, but differ in the format of the sensor, are highlighted. Cholinesterase enzymes catalyze the hydrolysis of the neurotransmitter acetylcholine. This deactivation of the neurotransmitter is important for the return of cholinergic neurons to the resting state. Cholinesterases are inhibited by organophosphorous compounds and carbamates [120]. These inhibitors are used as pesticides, therapeutic compounds in Alzheimer's treatment, or chemical warfare agents. A sol-gel material with encapsulated cholinesterase could be used to detect pesticides, drugs, or chemical warfare agents in a sample by comparing the cholinesterase activity of this sample to a non-inhibited cholinesterase standard.

Altstein et al. [74] encapsulated different cholinesterases, acetylcholinesterase from electric eel, acetylcholinesterase from bovine erythrocytes, and butyrylcholinesterase from horse serum in TMOS-based gels that were cast in microtiterplates. The composition of the sol-gel was optimized by varying the ratio of TMOS/hydrochloric acid, including the additive PEG 400, and by altering the amount of encapsulated cholinesterase. A mutual dependence of all optimization parameters was detected. Best results were achieved for a combination of low enzyme concentration, low TMOS/hydrochloric acid ratios, and PEG 400 incorporation. The microtiterplate format enables the simultaneous analysis of different samples and is convenient for re-use after washing the whole plate.

The next biosensor is an example for an immobilized enzyme reactor (IMER) [121]. This example is particularly enticing, because it employs a biomimetic silification process. IMER systems contain enzymes that are immobilized to packed matrices within flow-through devices (i.e., columns that can be attached to a liquid chromatography system). A commercially available column, pre-packed with a metal chelating resin, was charged with cobalt ions. A silica-condensing peptide (R5) was synthesized with and without a histidine tag. The R5-peptide with the histidine tag was loaded onto the column. Addition of the R5-peptide, TMOS, and the cholinesterase enzyme (equine butyrylcholinesterase) triggered a silification process that resulted in the successful immobilization of butyrylcholinesterase in a column. This column was integrated into a liquid chromatography system to screen for cholinesterase inhibitors.

Brennan et al. [122] developed a paper-based solid-phase biosensor that uses inkjet printing of enzyme-doped, sol-gel based inks to create colorimetric sensor strips. To create a biosensor in this novel format of a bioactive paper strip, Brennan et al. developed a multilayer deposition method [122]. First, polyvinylamine was printed on paper, followed by a sol-gel mixture containing diglyceryl silane and sodium silicate. The enzyme acetylcholinesterase from electric eel was printed next and overprinted by another layer of the biocompatible sol-gel mixture. Ellman's colorimetric assay was used to measure cholinesterase activity. This assay results in the generation of the yellow colored 5-thio-2-nitrobenzoate anion. This anion was captured on the strip on the polyvinylamine layer. The inkjet printing technique resulted in a very cost effective, simple, and scalable deposition of the biosensor material on paper. The development of a simple and portable bioassay is of importance to enable rapid screening for toxic compounds in emergency situations, remote settings, or developing countries [122].

Conclusion

The sol-gel process is a very versatile technique. The properties of the sol-gel can be tuned by using different precursors or by including additives. The sol-gel material can vary profoundly in pore structure, available surface area, and chemical character. Ormosils can create a hydrophobic encapsulation environment, whereas the traditional sol-gel is hydrophilic and polar in character. Dopants can render a sol-gel conductive, colorful or introduce other new functionalities. Of particular interest are biomolecule dopants due to their impressive range of highly specialized functionalities. Sol-gels can be cast into various forms and prepared as wet hydrogels, dried xerogels, or very light and porous aerogels. This versatility of the sol-gel process has inspired many creative scientists. A flurry of diverse applications evolved in recent years. Sol-gel science is thus evolving into a highly multidisciplinary field. Analytical chemists who aim to develop new separation media in chromatography or new sensors to detect and quantify small molecules are working together with inorganic or organic chemists, who specialize in the fabrication of new sol-gel precursors or new additives. Physical chemists and material scientists bring in their expertise to characterize sol-gel materials with new techniques. Biochemists will characterize and identify more proteins that are good candidates for sol-gel based applications. As more applications enter the field of medical science, medical chemists, toxicologists, and pharmacologists are also entering the sol-gel arena. The combined efforts of experts with different backgrounds will propel this field forward.

References

[1] Penard, AL; Gacoin, T; Boilot, JP. *Acc. Chem. Res.*, 2007, 40, 895-902.

[2] Takahashi, R; Sato, S; Sodesawa, T; Tomita, Y. *J. Ceram. Soc. Jpn.*, 2005, 113, 92-96.

[3] Ishizuka, N; Kobayashi, H; Minakuchi, H; Nakanishi, K; Hirao, K; Hosoya, K; Ikegami, T; Tanaka, N. *J. Chromatogr.*, A 2002, 960, 85-96.

[4] Vallet-Regi, M; Balas, F; Colilla, M; Manzano, M. *Prog. Solid State Chem.*, 2008, 36, 163-191.

[5] Coradin, T; Boissiere, M; Livage, J. *Curr. Med. Chem.*, 2006, 13, 99-108.

[6] Quintanar-Guerrero, D; Ganem-Quintanar, A; Nava-Arzaluz, MG; Pinon-Segundo, E. *Expert Opin. Drug Del.*, 2009, 6, 485-498.

[7] Rahn, MD; King, TA. *Appl. Optics,* 1995, 34, 8260-8271.

[8] Collinson, MM. *Crit. Rev. Anal. Chem.*, 1999, 29, 289-311.

[9] Knopp, D; Tang, DP; Niessner, R. *Anal. Chim. Acta,* 2009, 647, 14-30.
[10] Ruiz-Hitzky, E; Darder, M; Aranda, P; Ariga, K. *Adv. Mater.*, 2010, 22, 323-336.
[11] Avnir, D; Coradin, T; Lev, O; Livage, J. *J. Mater. Chem.*, 2006, 16, 1013-1030.
[12] Gadre, SY; Gouma, PI. *J. Am. Ceram. Soc.*, 2006, 89, 2987-3002.
[13] Gill, I; Ballesteros, A. *Trends Biotechnol.*, 2000, 18, 469-479.
[14] Gill, I. *Chem. Mater.*, 2001, 13, 3404-3421.
[15] Collinson, MM. *Mikrochim. Acta,* 1998, 129, 149-165.
[16] Mackenzie, JD; Bescher, EP. *J. Sol-Gel Sci. Techn.*, 1998, 13, 371-377.
[17] Walcarius, A; Collinson, MM. *Annu. Rev. Anal. Chem.*, 2009, 2, 121-143.
[18] Avnir, D; Braun, S; Lev, O; Ottolenghi, M. *Chem. Mater.*, 1994, 6, 1605-1614.
[19] Reetz, MT; Wenkel, R; Avnir, D. *Synthesis-Stuttgart,* 2000, 781-783.
[20] Singh, S; Singhal, R; Malhotra, BD. *Anal. Chim. Acta,* 2007, 582, 335-343.
[21] Jiang, ZY; Wu, H; Xu, SW; Huang, SF. *ACS Sym. Ser.*, 2003, 852, 212-218.
[22] Keeling-Tucker, T; Rakic, M; Spong, C; Brennan, JD. *Chem. Mater.*, 2000, 12, 3695-3704.
[23] Dattelbaum, AM; Baker, GA; Fox, JM; Iyer, S; Dattelbaum, JD. *Bioconjugate Chem.*, 2009, 20, 2381-2384.
[24] Gulcev, MD; Goring, GLG; Rakic, M; Brennan, JD. *Anal. Chim. Acta,* 2002, 457, 47-59.
[25] Brennan, JD; Benjamin, D; DiBattista, E; Gulcev, MD. *Chem. Mater.*, 2003, 15, 737-745.
[26] Nadzhafova, OY; Zaitsev, VN; Drozdova, MV; Vaze, A; Rusling, JF. *Electrochem. Commun.*, 2004, 6, 205-209.
[27] Yi, YY; Neufeld, R; Kermasha, S. *J. Sol-Gel Sci. Techn.*, 2007, 43, 161-170.
[28] Baccile, N; Babonneau, F; Thomas, B; Coradin, T. *J. Mater. Chem.*, 2009, 19, 8537-8559.
[29] Buckley, AM; Greenblatt, M. *J. Chem. Edu.*, 1994, 71, 599-602.
[30] Pierre, AC. *Biocatal. Biotransfor.*, 2004, 22, 145-170.
[31] Pierre, M; Buisson, P; Fache, F; Pierre, A. *Biocatal. Biotransfor.*, 2000, 18, 237-251.
[32] Kresge, CT; Vartuli, JC; Roth, WJ; Leonowicz, ME. *Stud. Surf. Sci. Catal.*, 2004, 148, 53-72.
[33] Vartuli, JC; Schmitt, KD; Kresge, CT; Roth, WJ; Leonowicz, ME; Mccullen, SB; Hellring, SD; Beck, JS; Schlenker, JL; Olson, DH; Sheppard, EW. *Chem. Mater.*, 1994, 6, 2317-2326.
[34] Ellerby, LM; Nishida, CR; Nishida, F; Yamanaka, SA; Dunn, B; Valentine, JS; Zink, J. *I. Science,* 1992, 255, 1113-1115.
[35] Ferrer, ML; Del Monte, F; Levy, D. *Chem. Mater.*, 2002, 14, 3619-3621.
[36] Gill, I; Ballesteros, A. *J. Am. Chem. Soc.*, 1998, 120, 8587-8598.
[37] Bhatia, RB; Brinker, CJ; Gupta, AK; Singh, AK. *Chem. Mater.*, 2000, 12, 2434-2441.
[38] Coradin, T; Livage, J. *Acc. Chem. Res.*, 2007, 40, 819-826.
[39] Li, Y; Yip, WT. *J. Am. Chem. Soc.*, 2005, 127, 12756-12757.
[40] Brennan, JD. *Acc. Chem. Res.*, 2007, 40, 827-835.
[41] Katagiri, K. *J. Sol-Gel Sci. Techn.*, 2008, 46, 251-257.
[42] Katagiri, K; Hashizume, M; Ariga, K; Terashima, T; Kikuchi, J. *Chem. Eur. J.*, 2007, 13, 5272-5281.

[43] Carnes, EC; Harper, JC; Ashley, CE; Lopez, DM; Brinker, LM; Liu, JW; Singh, S; Brozik, SM; Brinker, CJ. *J. Am. Chem. Soc.*, 2009, 131, 14255-14257.
[44] Liu, JW; Stace-Naughton, A; Jiang, XM; Brinker, CJ. *J. Am. Chem. Soc.*, 2009, 131, 1354-1355.
[45] Stein, PE; Leslie, AG; Finch, JT; Carrell, RW. *J. Mol. Biol.*, 1991, 221, 941-959.
[46] Sato, S; Murakata, T; Suzuki, T; Ohgawara, T. *J. Mater. Sci.*, 1990, 25, 4880-4885.
[47] Maury, S; Buisson, P; Pierre, AC. *Langmuir,* 2001, 17, 6443-6446.
[48] Nguyen, DT; Smit, M; Dunn, B; Zink, J. *I. Chem. Mater.*, 2002, 14, 4300-4306.
[49] Gao, Y; Heinemann, A; Knott, R; Bartlett, J. *Langmuir*, 2010, 26, 1239-1246.
[50] Kroger, N; Lorenz, S; Brunner, E; Sumper, M. *Science*, 2002, 298, 584-586.
[51] Luckarift, HRJ. Liq. *Chromatogr. R T*, 2008, 31, 1568-1592.
[52] Li, L; Jiang, ZY; Wu, H; Feng, YN; Li, J. *Mat. Sci. Eng. C-Mater.*, 2009, 29, 2029-2035.
[53] Gupta, R; Chaudhury, NK. *Biosens. Bioelectron.*, 2007, 22, 2387-2399.
[54] Brennan, JD; Hartman, JS; Ilnicki, EI; Rakic, M. *Chem. Mater.*, 1999, 11, 1853-1864.
[55] Menaa, B; Miyagawa, Y; Takahashi, M; Herrero, M; Rives, V; Menaa, F; Eggers, DK. *Biopolymers*, 2009, 91, 895-906.
[56] Rocha, VA; Eggers, DK. *Chem. Commun.*, 2007, 1266-1268.
[57] Zheng, LL; Reid, WR; Brennan, JD. *Anal. Chem.*, 1997, 69, 3940-3949.
[58] Doody, MA; Baker, GA; Pandey, S; Bright, FV. *Chem. Mater.*, 2000, 12, 1142-1147.
[59] Smith, K; Silvernail, NJ; Rodgers, KR; Elgren, TE; Castro, M; Parker, RM. *J. Am. Chem. Soc.*, 2002, 124, 4247-4252.
[60] Menaa, B; Menaa, F; Aiolfi-Guimaraes, C; Sharts, O. *Int. J. Nanotechnol.*, 2010, 7, 1-45.
[61] Eggers, DK; Valentine, JS. *Protein Sci.*, 2001, 10, 250-261.
[62] Ping, G; Yuan, JM; Sun, ZF; Wei, Y. *J. Mol. Recognit.*, 2004, 17, 433-440.
[63] Kadnikova, EN; Kostic, NM. *J. Mol. Catal. B-Enzym.*, 2002, 18, 39-48.
[64] Brook, MA; Chen, Y; Guo, K; Zhang, Z; Brennan, JD. *J Mater. Chem.*, 2004, 14, 1469-1479.
[65] Cruz-Aguado, JA; Chen, Y; Zhang, Z; Brook, MA; Brennan, JD. *Anal. Chem.*, 2004, 76, 4182-4188.
[66] Wei, Y; Xu, JG; Feng, QW; Lin, MD; Dong, H; Zhang, WJ; Wang, C. *J. Nanosci. Nanotechno.*, 2001, 1, 83-93.
[67] Zhu, WL; Zhou, Y; Zhang, JR. *Talanta*, 2009, 80, 224-230.
[68] Liu, Y; Shi, LH; Wang, MJ; Li, ZY; Liu, HT; Li, JH. *Green. Chem.*, 2005, 7, 655-658.
[69] Liu, Y; Wang, MJ; Li, J; Li, ZY; He, P; Liu, HT; Li, JH. *Chem. Commun.*, 2005, 1778-1780.
[70] Lee, SH; Doan, TTN; Ha, SH; Chang, WJ; Koo, YM. *J. Mol. Catal. B-Enzym.*, 2007, 47, 129-134.
[71] Lee, SH; Doan, TT. N; Ha, SH; Koo, YM. *J. Mol. Catal. B-Enzym.*, 2007, 45, 57-61.
[72] Frenkel-Mullerad, H; Avnir, D. *J. Am. Chem. Soc.*, 2005, 127, 8077-8081.
[73] Shchipunov, YA; Karpenko, TY; Bakunina, LY; Burtseva, YV; Zvyagintseva, TN. *J. Biochem. Bioph. Meth.*, 2004, 58, 25-38.
[74] Altstein, M; Segev, G; Aharonson, N; Ben-Aziz, O; Turniansky, A; Avnir, D. *J. Agric. Food Chem.*, 1998, 46, 3318-3324.

[75] Shtelzer, S; Rappoport, S; Avnir, D; Ottolenghi, M; Braun, S. *Biotechnol. Appl. Biochem.*, 1992, 15, 227-235.
[76] Chen, Q; Kenausis, GL; Heller, A. *J. Am. Chem. Soc.*, 1998, 120, 4582-4585.
[77] Heller, J; Heller, A. *J. Am. Chem. Soc.*, 1998, 120, 4586-4590.
[78] Eggers, DK; Valentine, JS. *J. Mol. Biol.*, 2001, 314, 911-922.
[79] Fireman-Shoresh, S; Marx, S; Avnir, D. *Adv. Mater.*, 2007, 19, 2145-+.
[80] Fireman-Shoresh, S; Marx, S; Avnir, D. *J. Mater. Chem.*, 2007, 17, 536-544.
[81] Ispas, C; Sokolov, I; Andreescu, S. *Anal Bioanal. Chem.*, 2009, 393, 543-554.
[82] Aburto, J; Ayala, M; Bustos-Jaimes, I; Montiel, C; Terres, E; Dominguez, JM; Torres, E. *Micropor. Mesopor. Mat.*, 2005, 83, 193-200.
[83] Jung, D; Streb, C; Hartmann, M. *Int. J. Mol. Sci.*, 2010, 11, 762-778.
[84] Hartmann, M; Streb, C. *J. Porous Mat.*, 2006, 13, 347-352.
[85] Hudson, S; Cooney, J; Hodnett, BK; Magner, E. *Chem. Mater.*, 2007, 19, 2049-2055.
[86] Borole, A; Dai, S; Cheng, CL; Rodriguez, M; Davison, BH. *Appl. Biochem. Biotech.* 2004, 113-16, 273-285.
[87] Jung, D; Streb, C; Hartmann, M. *Micropor. Mesopor. Mat.*, 2008, 113, 523-529.
[88] Kim, JB; Grate, JW; Wang, P. *Trends Biotechnol.*, 2008, 26, 639-646.
[89] Conesa, A; Punt, PJ; van den Hondel, CAMJJ. *J Biotechnol.*, 2002, 93, 143-158.
[90] Veronese, FM; Morpurgo, M. *Farmaco*, 1999, 54, 497-516.
[91] Besanger, TR; Easwaramoorthy, B; Brennan, JD. *Anal. Chem.*, 2004, 76, 6470-6475.
[92] Baker, GA; Jordan, JD; Bright, FV. *J. Sol-Gel Sci. Techn.*, 1998, 11, 43-54.
[93] Oesterhelt, F; Rief, M; Gaub, HE. New *Journal of Physics,* 1999, 1, 1-11.
[94] Baker, GA; Pandey, S; Maziarz, EP; Bright, FV. *J. Sol-Gel Sci. Techn.*, 1999, 15, 37-48.
[95] O'Neill, H; Angley, CV; Hemery, I; Evans, BR; Dai, S; Woodward, J. *Biotechnol. Lett.* 2002, 24, 783-790.
[96] Belcarz, A; Ginalska, G; Lobarzewski, J; Penel, C. *Bba-Protein Struct.,* M 2002, 1594, 40-53.
[97] Jaeger, KE; Eggert, T. *Curr. Opin. Biotech.*, 2002, 13, 390-397.
[98] Kim, KK; Song, HK; Shin, DH; Hwang, KY; Suh, SW. *Structure*, 1997, 5, 173-185.
[99] Reetz, MT. Adv. *Mater.*, 1997, 9, 943-&.
[100] Kawakami, K; Yoshida, S. *Biotechnol. Tech.*, 1995, 9, 701-704.
[101] Reetz, MT; Tielmann, P; Wiesenhofer, W; Konen, W; Zonta, A. *Adv. Synth. Catal.* 2003, 345, 717-728.
[102] Sahin, O; Erdemir, S; Uyanik, A; Yilmaz, M. *Appl. Catal. A-Gen.*, 2009, 369, 36-41.
[103] Yang, G; Wu, JP; Xu, G; Yang, LR. *Bioresource Technol.*, 2009, 100, 4311-4316.
[104] Yang, JK; Liu, LY; Cao, XW. *Enzyme Microb. Tech.*, 2010, 46, 257-261.
[105] Menaa, B; Herrero, M; Rives, V; Lavrenko, M; Eggers, DK. *Biomaterials*, 2008, 29, 2710-2718.
[106] Koszelewski, D; Muller, N; Schrittwieser, JH; Faber, K; Kroutil, WJ. *Mol. Catal. B-Enzym*. 2010, 63, 39-44.
[107] Kawakami, K; Yoshida, S. *J Ferment Bioeng.*, 1996, 82, 239-245.
[108] Zarcula, C; Kiss, C; Corici, L; Croitoru, R; Csunderlik, C; Peter, F. *Rev. Chim.-Bucharest,* 2009, 60, 922-927.
[109] Meunier, SM; Legge, RL. *J. Mol. Catal. B-Enzym.*, 2010, 62, 54-58.
[110] Kawakami, K. Biotechnol. *Tech.*, 1996, 10, 491-494.

[111] Ivnitski, D; Artyushkova, K; Rincon, RA; Atanassov, P; Luckarift, HR; Johnson, GR. *Small*, 2008, 4, 357-364.

[112] Jia, WZ; Wang, K; Zhu, ZJ; Song, HT; Xia, XH. *Langmuir*, 2007, 23, 11896-11900.

[113] Nadzhafova, O; Etienne, M; Walcarius, A. *Electrochem. Commun.*, 2007, 9, 1189-1195.

[114] Hungerford, G; Rei, A; Ferreira, MIC; Suhling, K; Tregidgo, C. *J. Phys. Chem. B,* 2007, 111, 3558-3562.

[115] Altstein, M; Aharonson, N; Segev, G; Ben-Aziz, O; Avnir, D; Turniansky, A; Bronshtein, A. Ital. *J. Food. Sci.*, 2000, 12, 191-206.

[116] Hreniak, A; Rybka, J; Gamian, A; Hermanowicz, K; Hanuza, J; Maruszewski, K. *J. Lumin.*, 2007, 122, 987-989.

[117] Kim, J; Grate, JW; Wang, P. *Chem. Eng. Sci.*, 2006, 61, 1017-1026.

[118] Lee, CW; Yi, SS; Kim, J; Lee, YS; Kim, BG. *Biotechnol. Bioproc. E.*, 2006, 11, 277-281.

[119] Rupcich, N; Nutiu, R; Li, YF; Brennan, JD. *Anal. Chem.*, 2005, 77, 4300-4307.

[120] Andreescu, S; Marty, JL. *Biomol. Eng.*, 2006, 23, 1-15.

[121] Luckarift, HR; Johnson, GR; Spain, JC. *J Chromatogr.*, B 2006, 843, 310-316.

[122] Hossain, SMZ; Luckham, RE; Smith, AM; Lebert, JM; Davies, LM; Pelton, RH; Filipe, CDM; Brennan, JD. *Anal. Chem.*, 2009, 81, 5474-5483.

In: The Sol-Gel Process
Editor: Rachel E. Morris
ISBN 978-1-61761-321-0

Chapter 10

FABRICATION AND PROPERTY OPTIMIZATION OF MULTIFERROIC COMPOSITE THICK FILMS

W. Chen[1], *W. Zhu*[1]* *and Z. H. Wang*[2]
[1]Microelectronics Center, School of Electrical and Electronic Engineering, Nanyang Technological University, Singapore 639798
[2]Advanced Nanofabrication Core Laboratory, King Abdullah University of Science and Technology, Thuwal 23955-6900, Saudi Arabia

ABSTRACT

Combining ferromagnetism and ferroelectricity in one material and coupling between ferromagnetism and ferroelectricity are two big challenges in Multiferroics. Hybrid sol-gel processing inherently has advantages in making biphasic nanocomposites which exhibit coexisting ferroelectric and ferromagnetic properties. In this chapter, multiferroic $CoFe_2O_4$-$Pb(Zr_{0.53}Ti_{0.47})O_3$ (CFO-PZT) composite thick films have been fabricated onto $Pt/Ti/SiO_2/Si$ by the hybrid sol-gel processing for their multifunctional applications. This chapter includes the development of deposition technique of thick films on silicon substrate as well as the continuous optimization for multiferroic properties. In the first part, a suitable route for sol-gel deposition of CFO-PZT composite thick films on silicon substrate has been developed, and then deposition of dense, crack-free CFO-PZT thick films with desired thickness range (3~5 μm) along with a coexistence of ferromagnetic and ferroelectric properties have been achieved. In the second part, we study the effect of polyvinylpyrrolidone (PVP) on microstructures and ferromagnetic, ferroelectric and dielectric properties of CFO-PZT composite thick films. It is observed that PVP is beneficial for enhancing the film thickness remarkably and avoids the cracks probably induced by different thermal expansion coefficients or lattice mismatch between different materials. However, the disadvantage of PVP exists in diluting the ferroelectric and dielectric properties of composite thick films. To deal with this issue, the third part of this chapter introduces the sol infiltration to enhance the dielectric and ferroelectric properties of the composite thick films. It is found that enough sol infiltration between slurry layers, the maximum layers investigated here are four

* Corresponding author: Email: ewgzhu@gmail.com.

layers, dramatically improve the multiferroic properties as well as the density of the microstructure. Based on the obtained microstructures and multiferroic properties, the last part optimizes the deposition sequence and ferromagnetic phase content. Experimental results indicate that 8 wt% of CFO content along with an optimized deposition sequence gives satisfied coexisting ferromagnetic and ferroelectric properties, which is a precondition for exploring multiferroic devices.

1. Introduction

The past few years have seen a tremendous flurry of research interest in multiferroic materials, which are proving to be a rich source for exploring the fundamental science of phase control and magnetoelectric interactions. Huge magnetoelectric effects have been observed in the form of ferroelectric phase transitions induced by magnetic fields in perovskite manganites [1] and ferromagnetism induced by electric fields in hexagonal manganites [2]. Magnetoelectric memory effects and magnetic switching of ferroelectric domains (and the converse process) have been demonstrated. An optical technique has been developed that allows separate access to both coexisting domain structures of a multiferroic in the same experimental setup; with this method, coupling between magnetic and electric domains has been observed [3]. However, these single-phase multiferroics are not very attractive for applications in the short term, because none of the existing materials combines large and robust electric and magnetic polarizations at room temperature. The difficulties associated with uniting electrical and magnetic ordering in a single phase have been circumvented by forming two-phase composite multiferroics that consist of a ferroelectric constituent [such as $PbZr_{1-x}Ti_xO_3$ (PZT)] and a ferromagnetic constituent [such as $Tb_{1-x}Dy_xFe_2$ (Terfenol-D)] [4]. In such composites, the magnetoelectric effect arises from the interaction of the elastic components of the ferromagnetic and ferroelectric constituents. An electric field induces strain in the ferroelectric; this strain is passed on to the ferromagnet, where it causes magnetization (vice versa). The magnetoelectric effect will be large if the coupling at the interface is large; therefore, composites with large surface area and strong ferroelastic constituents are particularly effective.

Recently, $CoFe_2O_4$-$Pb(Zr_{0.53}Ti_{0.47})O_3$ (abbreviated as CFO-PZT) multiferroic bulk ceramics and thin film have attracted much attention as a typical composite multiferroic material due to the large magnetostrictive coefficient of CFO and large piezoelectric coefficient of PZT [5-10], which are able to achieve extrinsic magnetoelectricity depending on the composite microstructure and coupling interaction across magnetic-piezoelectric interfaces [11]. Furthermore, the achievements of multiferroic materials suggest potential applications in novel devices based on the mutual controls of magnetic and electric fields [12-13]. However, the current researches are mainly focusing on the multiferroic thin film and bulk ceramics with little attention on the multiferroic thick films. It is known that the piezoelectric thick films, in particular in the range of 3~10 μm, are still the important candidates in piezoelectric-MEMS devices [14-15]. The success in fabricating multiferroic thick films would stimulate the potential applications on multifunctional devices similarly. However, due to the trap of spurious ferroelectric behavior in multiferroic materials [16], most of them were demonstrated as typical lossy dielectrics mainly caused by space charges, which should be eliminated further. Fortunately, it is noticed that the typical ferroelectric behavior have been achieved in CFO-PZT multiferroic thin films continually with a good dielectric property [17-18], which would be of particular interest to fabricate the multiferroic CFO-PZT films with thickness over micrometer for device applications.

In the present chapter, multiferroic thick films process is initially developed. Then polyvinylpyrrolidone (PVP) and sol infiltration are employed to optimize the microstructures and multiferroic properties of the thick films. The former is beneficial for improving the thick films quality, while the latter remarkably enhances the ferroelectric and dielectric properties. By properly combining both of them, high quality thick films with promising properties are achieved. Based on this, the influence of deposit sequence and magnetic phase content on multiferroic properties is further studied. Detailed mechanisms are also discussed.

2. Fabrication of Multiferroic Thick Films

2.1. Preparation of PZT and CFO Nanoparticles

PZT commercial powder (APC-850) was firstly dry milled for 20 hours using high energy ball milling (HEBM) and then milled for another 4 h with addition of 5 wt% organic vehicle (ESL400) as dispersion agent. CFO powder was prepared from solid state reaction using Fe_2O_3 and Co_3O_4 as raw materials. Both powders were first mixed together and milled for 1 h by HEBM, then annealed at 1100 °C for 2 h after being pressed as a bulk disk. The calcined body ceramics were crashed into powder using roll-milling and subsequently were high-energy ball milled for 20 h to obtain nanoparticle. Five weight percent organic vehicle (ESL400) was added into the resultant nanoparticle and milled for another 4 h to adjust powder dispersion.

X-ray diffraction (XRD) results of these nanoparticles are shown in Figure 1. It is clearly seen from the figure a broader peak and change of their respective preferential orientations by the high-energy ball milling. The powder size was calculated from half-peak-width according to the equation below:

$$d = \frac{K * \lambda}{B * \cos\theta_B} \tag{1}$$

where d is the size of crystallite, K is constant dependent on crystal shape (0.89), λ is the x-ray wavelength, B is the full width at half maximum (FWHM) and θ_B is the Bragg angle. The average crystal size is calculated as about 203 nm for CFO nanoparticle and 460 nm for PZT nano-particle using the equation (1).

To decrease the aggregation in milled green powder, ESL400 was used to make the particle dispersion. Firstly, it was added into PZT powder and CFO powder respectively with the ratio of 5 wt%. The dispersed particles were then dissolved into ethanol with a ratio of 5 mg: 30 ml and ultrasonically agitated for 30 minutes for good dispersion. After kept statically for 4~6 h, the dispersed particles were measured by Zeta Sizer (Nano-ZS), and the particle size distributions of both PZT and CFO nanoparticles are given in Figure 2. It is seen that PZT particles are distributed in a wide range from 70 to 600 nm with its peak location at around 200 nm. On the other hand, the CFO nanoparticles are distributed from 90 to 400 nm, with its narrower range than that of PZT particle and the distribution peak at less than 200 nm, which are consistent with the results calculated from FWHM.

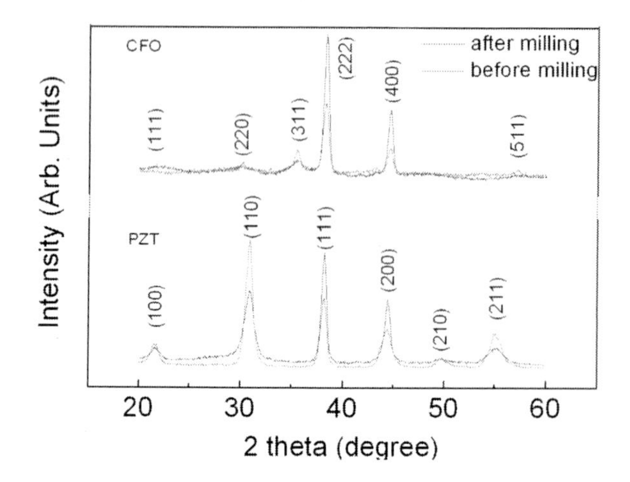

Figure 1. Effect of high energy ball milling on the x-ray diffraction patterns of PZT and CFO powder.

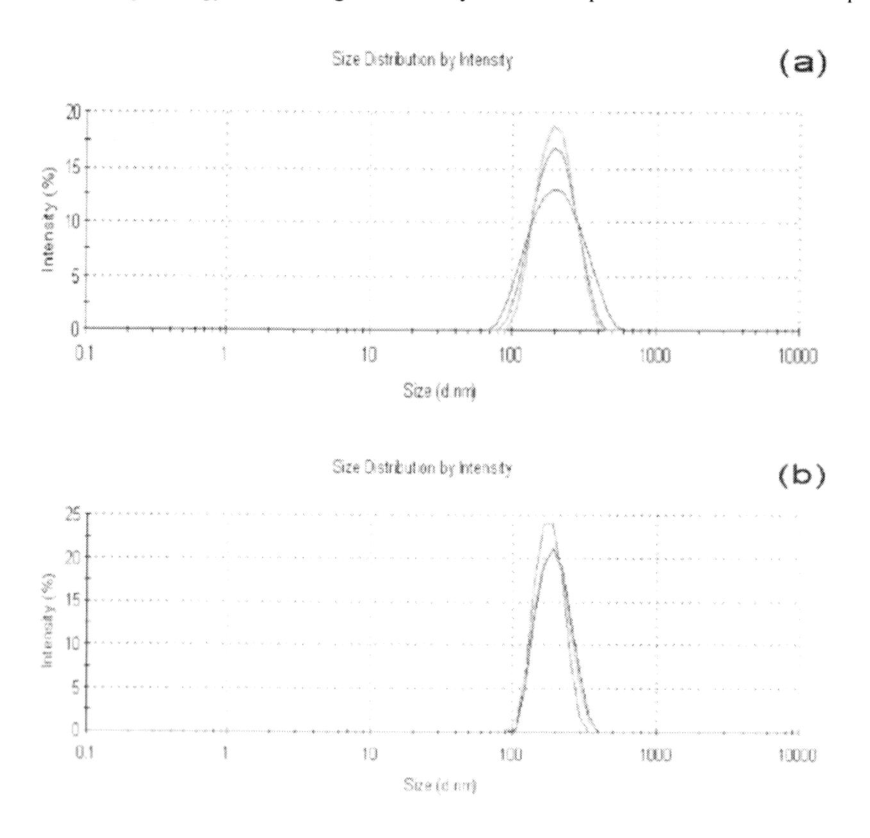

Figure 2. Size distribution of PZT and CFO particles with ethanol as the medium.

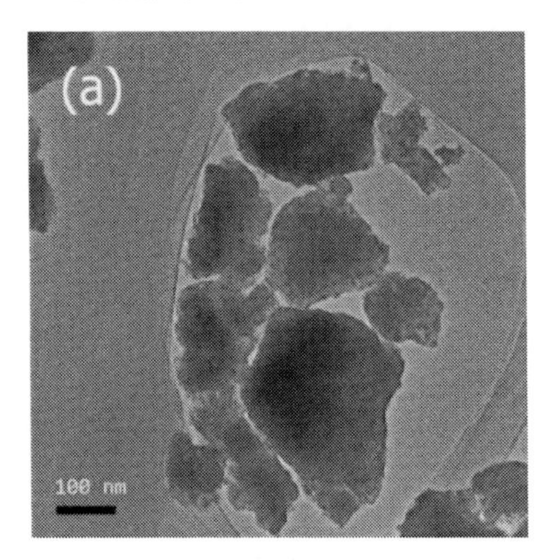

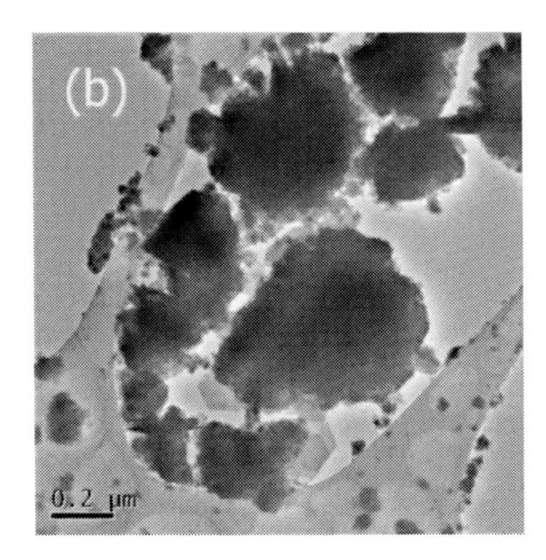

Figure 3. Transmission electron microscope pictures of PZT and CFO particles.

In addition, Transmission Electron Microscope pictures of both particles are shown in Figure 3 after they are dispersed in the ethanol, which are conducted using Transmission Electron Microscope (TEM-JEOL2010). It can be seen that both of them belongs to sub-micro-meter level, which is corresponding to the calculated size from XRD and measured ones from size distributions.

2.2. Preparation of PZT Sol-Gel Solution

Tetraisopropyl titanate ($[(CH_3)_2CHO]_4Ti$) and acetylaceton ($CH_3COCH_2COCH_3$) were firstly mixed and stirred at 80 ^oC for 10 min. Lead acetate ($(CH_3COO)_2Pb{\cdot}3H_2O$) with additional 10 mol% was added into the solution to compensate the lead loss during annealing. After stirring for another 10 min at 120 ^oC, zirconium acetylacetone ($[CH_3COCH{=}C(O-)CH_3]_4Zr$) was added into the above solution and stirred at 220 ^oC for 10 min. The resultant solution was yellowish and clear. The solution was transferred into the vacuum oven to keep for 4 h at 100 ^oC to obtain the xerogel. Subsequently, xerogel was dissolved into proper 2-methoxyethanol ($CH_3OC_2H_4OH$) to get 0.3 M of PZT sol-gel solution.

2.3. Preparation of PZT and CFO Slurry

To prepare PZT slurry, PZT nanoparticles were mixed with PZT sol-gel solution with the mass ratio of 2: 3 and milled for 15 h in an agate bowl, and the yellowish sticky PZT slurry was then collected. By using the PZT slurry, it can improve the thickness of films to several dozens of micrometers [18-19]. The PZT slurry was further utilized to make CFO-PZT multiferroic thick films on account of the lower resistivity of CFO nanoparticles, which is responsible for the conducting behavior in films, to avoid negative effect on ferroelectric and dielectric properties.

CFO nanoparticles were dispersed into PZT sol-gel solution with the mass ratio of 1: 15 and milled for 15 h in the agate bowl. The ratio of CFO was kept low to avoid possible negative effect to dielectric and ferroelectricity. On the other hand, the selected CFO percentage should be enough to produce detectable magnetic signal. After milling, black CFO slurry was obtained. In this process, PZT sol-gel acted as the matrix to disperse and carry CFO nanoparticles.

2.4. Deposition of Layered Thick Films

The flow-chat of CFO-PZT multiferroic thick films is given in Figure 4. Three kinds of solutions and slurries were spin-coated on the Pt/Ti/SiO_2/Si substrate according to the sequence in the schematic diagram depicted in Figure 5. Two CFO layers and three PZT slurry layers were alternatively deposited as well as two PZT thin films layers as the initial buffer layer and the top capping layer. The spin speed and time were 3000 rpm and 30 s followed by baking at 250 °C and pyrolyszing at 500 °C for 2 minutes, respectively.

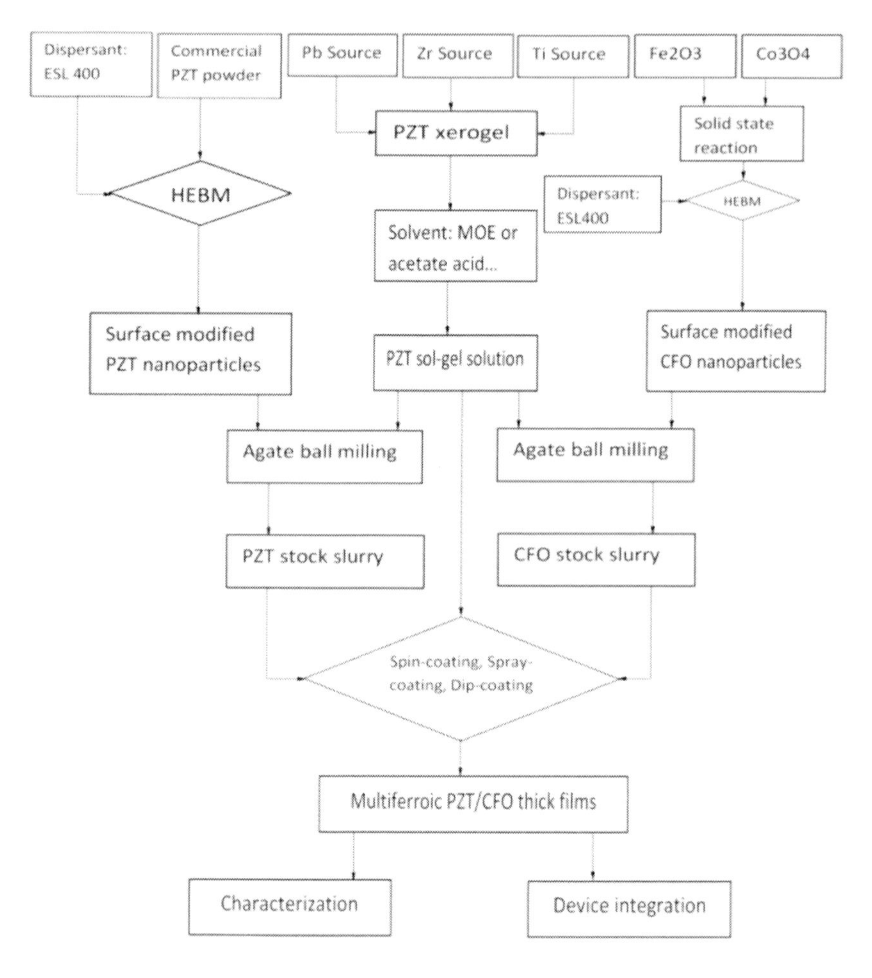

Figure 4. Flow-chart of the hybrid process of fabricating multiferroic CFO-PZT thick films.

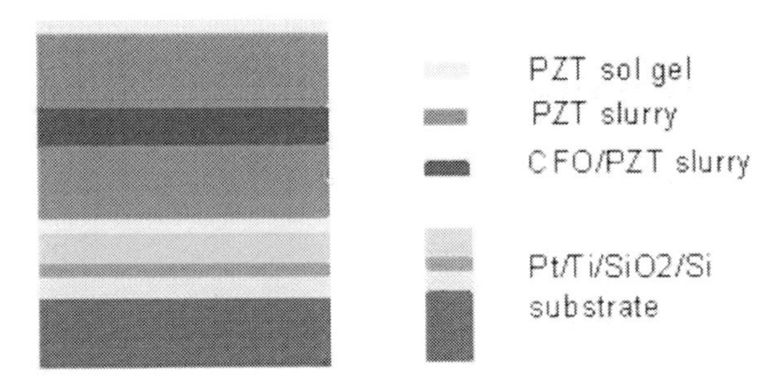

Figure 5. Schematic diagram of CFO-PZT layered structures.

2.5. Annealing and Characterizations

The thick films were annealed from 600 to 750 °C for 1 hour in air, and it was found that the thick film was peeled off from the substrate when the temperature reached 700 °C or higher. Hence, strictly controlling of annealing temperature below 700 °C is conducted in the following part. Gold was then sputtered onto the thick films as top electrodes with each size of 0.5 mm × 0.5 mm. It is noticed that the surface of thick films is not as smooth as that of thin films. As such, the gold electrode was deposited to 200 nm for proper measurements of electric properties. In addition, Ti layer, as the adhesive one between our CFO-PZT thick films and Au electrode, was also deposited to 20 nm for effective adhesion.

TGA and DTA were performed using a Thermal Analyzer (TA-60WS) with a heating rate of 2 °C/min. X-ray diffraction was carried out using a Rigaku X-RAY Diffractometer with a scanning speed of 2 °/min with current of 40 mA and voltage of 40 kV. The cross-sections of thick films were observed using a field emission scanning electronic microscope (Jeol-6340F). P-E electric hysteresis loop was measured using Precision Pro ferroelectric analyzer (Radiant Technology Inc.) with high voltage interfaces. The duration was set as 10 ms for P-E hysteresis measurement. Ferromagnetic behavior was characterized by a Vibration Sample Magnetometer (Lakeshore VSM 7404). Dielectric measurement was made by an Agilent 4294A precision impedance analyzer and the ac oscillation level was set as 100 mV.

The thickness of one layer sol-gel PZT film is of about 60 nm measured by surface profiler, much less than the size of CFO nanopowder of ~200 nm. As the direct consequence, the spin-coated single CFO film would not be smooth due to the larger CFO particle size than thickness of sol-gel formed thin film, resulting in large film surface roughness. For this reason, spin-coat PZT slurry on top of it was adopted to reduce the interfacial roughness. Since the thickness of PZT slurry formed one-layer film was about 600 nm, which was about 3 times of that of CFO slurry formed film, the CFO layer lies between two PZT slurry layers is thus largely submerged. In addition, PZT slurry was also used for adjusting thickness of resultant films. In the previous work on hybrid techniques [20-21], adjusting the ratio of sol-gel solution and powder could achieve proper thick films to meet different requirements. Finally, PZT sol-gel solution was coated as a top capping layer to reduce the pores in CFO-PZT multiferroic thick films. Lee reported [22] the effect of PZT sol-gel solution on microstructure and ferroelectric properties of screen-printed PZT thick films, the sol infiltration to fill the pores in the thick films and thus improving the ferroelectrics. It is therefore expected that sol-gel solution is penetrated into the multiferroic thick films with similar effects, and also reducing the film surface roughness.

DTA and TGA curves of PZT xerogel are shown in Figure 6. TGA yields a weight loss of 37% before 500 °C, and then not much weight loss up to 900 °C. In the DTA curve, the peak before 100 °C was attributed to the organic solvent evaporation. The decomposition and combustion of the bound organic species in the xerogel contributed to the exothermic peaks before 450 °C. The other peak around 670 °C without significant weight loss indicated the formation of the perovskite phase, whereas the formation of pyrochlore phase occurred in the range from 427 to 670 °C. Based on the DTA and TGA results, it was noted that the main weight loss occurred at 227 °C and 427 °C, and thus drying steps at 250 °C and pyrolyzing step at 500 °C were chosen.

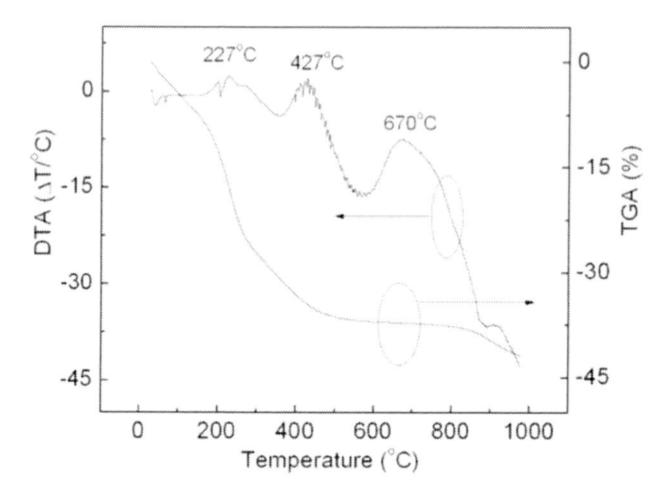

Figure 6. TGA and DTA results of PZT xerogel from room temperature to 1000 °C with a heating rate of 2 °C/min.

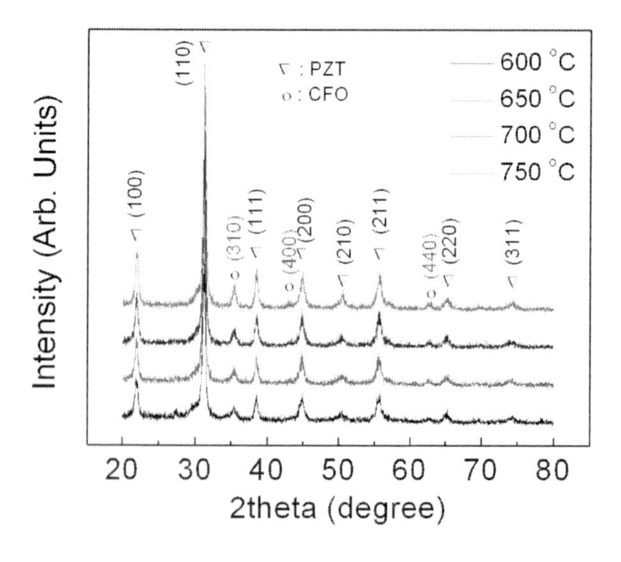

Figure 7. Typical XRD result of a CFO-PZT multiferroic thick film.

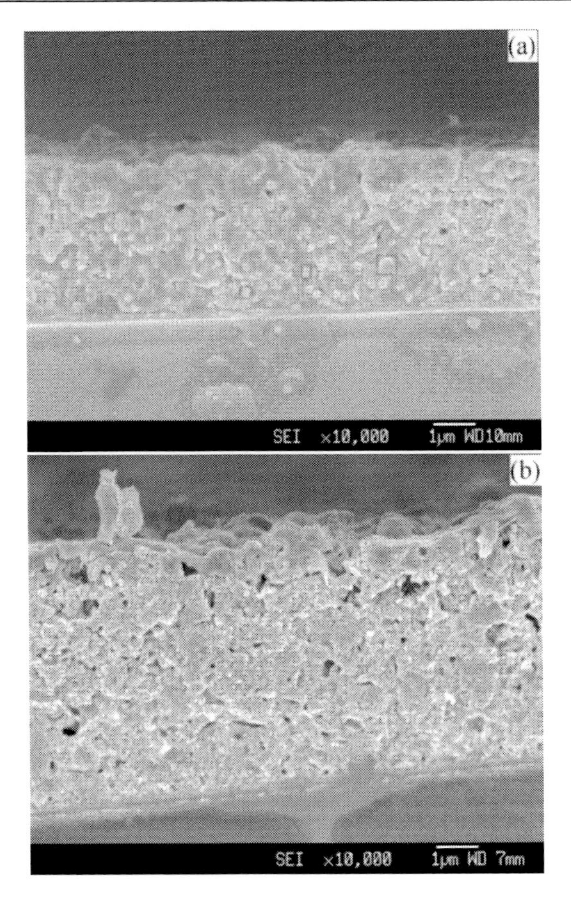

Figure 8. SEM pictures of cross-sections of CFO-PZT multiferroic thick film in (a) and PZT thick film in (b).

Figure 7 shows the XRD patterns of the CFO-PZT multiferroic thick film annealed at different temperatures. Both of the PZT and CFO phase are observed from Figure 7. SEM cross section micrographs of multiferroic composite thick film and PZT thick film prepared from the same processing with an annealing temperature of 650 ^{o}C were compared in Figure 8. Compared to PZT thick films in Figure 8 (b), some big grains with the size of 150 to 400 nm which are not found in (b) are obviously seen in Fig 8(a) [marked ones], CFO particles may play a role in this phenomenon. In addition, it is worth to note that CFO slurry in Figure 5 is formed by dispersing CFO particles in PZT sol solution, thus the CFO particles are scattered in the PZT matrix. Therefore, there is no obvious boundary like what depicted in Figure 5.

Magnetic hysteresis loops are shown in Figure 9. The CFO-PZT thick film reaches a magnetized saturation in both in-plane and out-of-plane directions. The *Hc* of 831.4 Oe along the in-plane direction is smaller than that of out-of-plane direction (941.5 Oe). The possible reason is that it is easier to magnetize the film along the in-plane direction than out-of-plane direction due to the demagnetization effect. The remnant magnetizations (*Mr*) and *Ms* are 4.3 emu/cm^3 and 13.4 emu/cm^3 along the in-plane direction; while in the out-of-plane direction, they are 3.7 emu/cm^3 and 12.7 emu/cm^3, respectively. The similar magnetization values with little different coercive fields in two directions indicate that CFO nanoparticles in PZT matrix have isotropic magnetic property. Normally, strain or stress induced by lattice mismatch between substrate and film is the main contribution to magnetic anisotropy in thin film [17,23]. However, in the present CFO-PZT composite thick film, the CFO magnetic nanoparticles are distributed far away from the substrate, so the aforementioned effect could be negligible. The isolated CFO nanograin islands with a 0-3 connection in the PZT matrix, such as the situation described in [24], could be the distributed form in this study. In order to obtain the mass fraction of CFO in the PZT matrix for a comparison, the effective *Ms* (defined as *Mse=measured Ms/mass fraction of CFO in composite thick films*) was described below. Pure CFO thin film from sol-gel processing yields a *Ms* of 320 emu/cm^3 [25], because the *Ms* is directly related to the ratio of magnetic content in samples, the value of 320 dividing 13.4 (the *Ms* along in-plane in the present work) gives the ratio of 23.9. It means that the volume fraction of CFO in this film is about 0.04 (1/23.9). According to Wan [18], the *Ms* and the volume fraction of CFO in their CFO-PZT film is 28.4 emu/cm^3 and 0.21, respectively. It is hence obviously that *Mse* of 13.4/0.04 in this paper is far larger than that of 28.4/0.21 reported by Wan. According to [18], the low *Hc* of 330 Oe should be attributed to the large residual stress in CFO due to the lattice mismatch and thermal expansion coefficient difference between PZT and CFO phase. Because of the enhanced film thickness and the reduced CFO ratio, the above mentioned reasons could not be fitted for our mediate *Hc*. Generally speaking, high coercivity can be obtained in the system with nano-structure or preferential orientation, such as thin films with preferred crystal textured, or nano-particles with single domain diameter [26]. Similar shapes of the hysteresis loops with different directions of the applied magnetic field confirm that there is no clear in-plane or out-of-plane magnetic anisotropy in the thick film. This implies the random orientation of the CFO crystalline, which agrees well with the XRD result. Furthermore, the single domain diameter of CFO is about 40 nm [26], which is much smaller than the diameters of the CFO particles in our composites (~300 nm). Therefore, the observed *Hc* in our composite thick film is mainly attributed to the magnetic multi-domain configuration of the CFO particles in the composites [27], which is also evidenced by TEM results of CFO particles. As can be seen in Figure 10, many fingerprints-like nano-domains are observed. The domains sizes are below 10 nm and their directions are totally random.

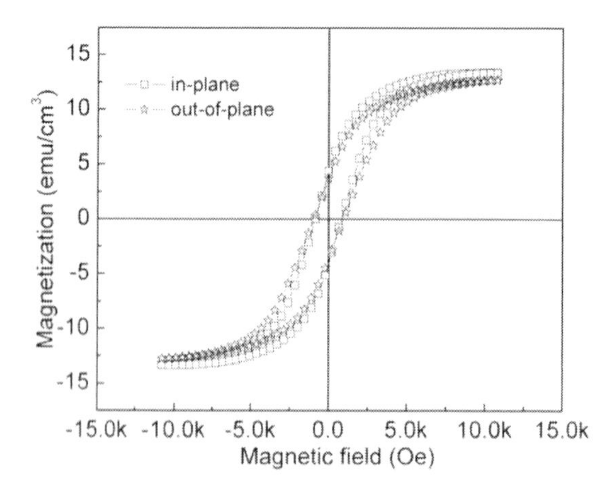

Figure 9. Magnetic hysteresis loops of CFO-PZT multiferroic thick films along in-plane and out-of-plane directions.

Figure 10. Multi-domains state of CFO particles characterized by TEM.

Ferroelectric behavior of the CFO-PZT thick films is shown in Figure 11 (a), which is measured at room temperature. Remnant polarization of 12.5 μC/cm^2 and coercive electric field of 252.8 kV/cm were measured but it did not reach the polarization saturation. Figure 11 (b) shows the ferroelectric hysteresis loops of CFO-PZT thick films under different frequencies. With increasing frequency, the loop becomes slim due to the effect of capacitance at high frequency, whereas the leakage current which yields a fat Lissajous-like plot does not increase. Therefore, the hysteresis loops are not affected by loss or leakage current, but they are reliable ferroelectric loops. Compared with the CFO-PZT film assisted by PVP [18], which showed a P_r of 14.6 μC/cm^2 and E_c of 43 kV/cm, the present CFO-PZT multiferroic thick film has smaller remnant polarization and larger coercive electric field. It is known that the addition of PVP is beneficial for avoiding film cracks, promoting film thickness, decreasing residual stress in films, and then results in a good ferroelectric property [28-30]. It is the lack of PVP that causes the possible cracks in the composite film, and the cracks could be one of the reasons leading to the reduced *Pr* and enhanced *Ec*. The further optimization on composition and processing procedure will be introduced later in this chapter. On the other hand, the nature of non-ferroelectric phase CFO is to dilute the ferroelectric remnant polarization, and meanwhile, the hindered and pinned domain wall motion of ferroelectric regions due to the existence of CFO leads to the increase of coercive electric field. This is why there are different ferroelectric behaviors between CFO-PZT and PZT thick films shown in Figure 11 (a). Furthermore, compared with bi-layer CFO-PZT thin films [8], not only the remnant polarization of our CFO-PZT thick film is smaller, but also the coercive field, since that there exists the clamping effect in the thin films from the substrate and the top CFO layer during the switching of ferroelectric domains. It also could be seen that the switching of ferroelectric domains and domain walls motion are more difficult in thin film than that in composite thick films.

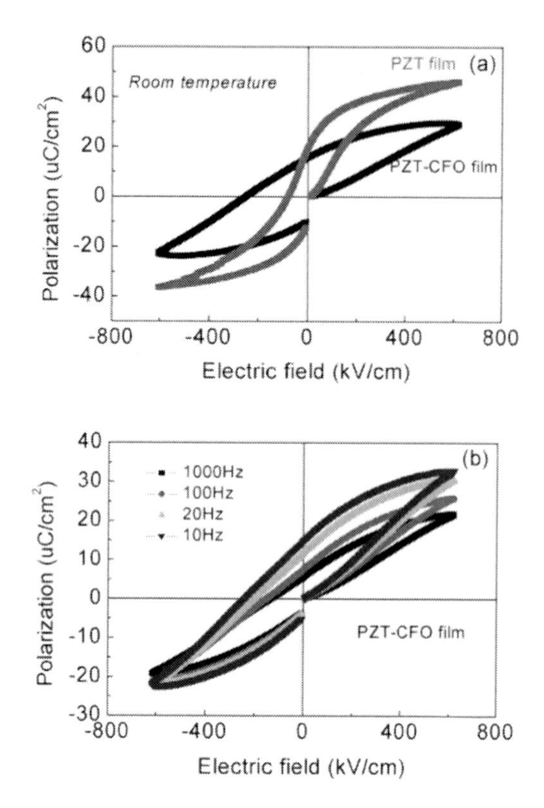

Figure 11. Ferroelectric hysteresis loop of CFO-PZT thick film and pure PZT thick film measured at room temperature under 20 Hz (a); ferroelectric hysteresis loops of CFO-PZT film under different frequencies (b).

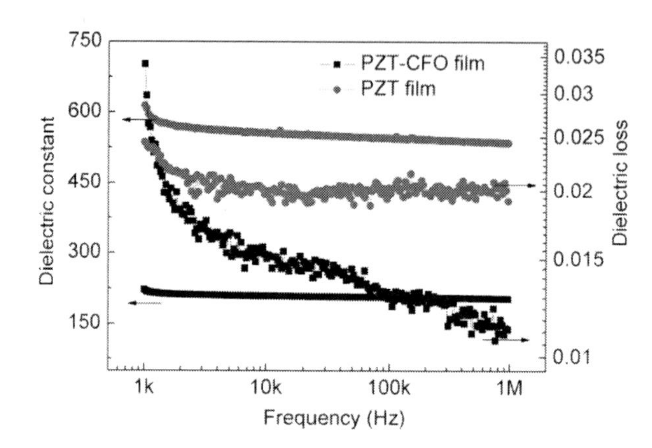

Figure 12. Dielectric permittivity and dielectric loss of CFO-PZT thick film and pure PZT thick film vs. frequency.

In addition, the dielectric permittivity and loss were characterized from 100 Hz to 1 MHz at room temperature, and they are shown in Figure 12. Compared with pure PZT thick films, the dielectric constants of our CFO-PZT thick film is low, which should be attributed to the introduction of CFO phase with a low dielectric constant [31-32]. According to the modified equation [33],

$$\log\varepsilon_m = log\varepsilon_1 + v_2(1-k)\log(\varepsilon_2/\varepsilon_1) \quad (2)$$

where ε_m is the dielectric permittivity of the mixture, ε_1 and ε_2 are the dielectric permittivity of matrix and embedded powder, respectively, v_2 is the volume fraction of embedded powder in the matrix, and k is a fitting constant. Considering that ε_2 of CFO is much smaller than ε_1 of PZT, the second part in the equation is negative, and thus the overall dielectric permittivity is decreased from ~550 to ~220. Because the density of the composite film is also a key factor to affect dielectric constant, further optimization methods on it, such as introducing PVP and enhancing sol infiltration [22], have also been carried out. Moreover, the dielectric loss of CFO-PZT decreases monotonously with increasing frequency and the lowest value can reach 0.01 at 1 MHz, which is lower than that of pure PZT thick films with a stable value of ~0.02, which indicates that our CFO-PZT multiferroic thick film is a possible candidate in high frequency electronic device applications.

In summary, CFO-PZT thick films were obtained by a hybrid sol-gel processing. XRD results indicated the coexistence of PZT and CFO phases, and the SEM cross-sectional microphotos of thick films exhibited the dense nano-compositional structure with less pores. Both ferromagnetic and ferroelectric behaviors were measured. The lower dielectric loss at 1 MHz offers possible applications of our CFO-PZT multiferroic composite thick films for high frequency electronic devices. Ferroelectric and ferromagnetic properties can be further improved to achieve a large magnetoelectric effect. The following sections are attempts of optimizing process through different routes.

3. Influence of PVP on CFO-PZT Composite Thick Films

Although alkoxide-based sol-gel method is a popular and widespread technique for preparing ceramic coating films in laboratories, people in industry are not encouraged enough to employ this technique for mass production. One of the factors discouraging is the low critical thickness, i.e. the maximum thickness achievable without crack formation via non-repetitive and single-step deposition, especially in non-silicate crystalline films. Films over submicrometer in thickness are favored in general for achieving better crystallinity, and especially in piezoelectric devices, of which the displacement induced by electric field is proportional to the film thickness. In laboratories, cycles of deposition and heat-treatment of gel films are usually employed, which is, however, time-consuming and impracticable in industrial production.

It has been reported that incorporating PVP (Figure 13) in precursor solutions could increase the critical thickness of gel-derived ceramic coating films. The introduced PVP would be hybridized with metalloxane polymers through hydrogen bonding as demonstrated by Saegusa and Chujo, retarding condensation and promoting structural relaxation in films subjected to heat-treatment.

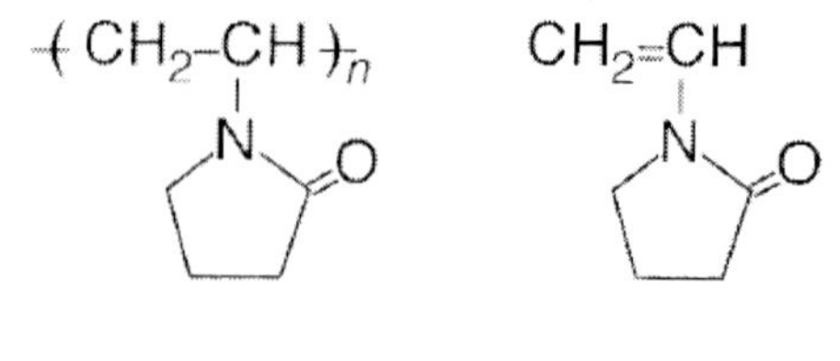

$-(CH_2-CH)_n-$ $CH_2=CH$

PVP VP

Figure 13. PVP and VP.

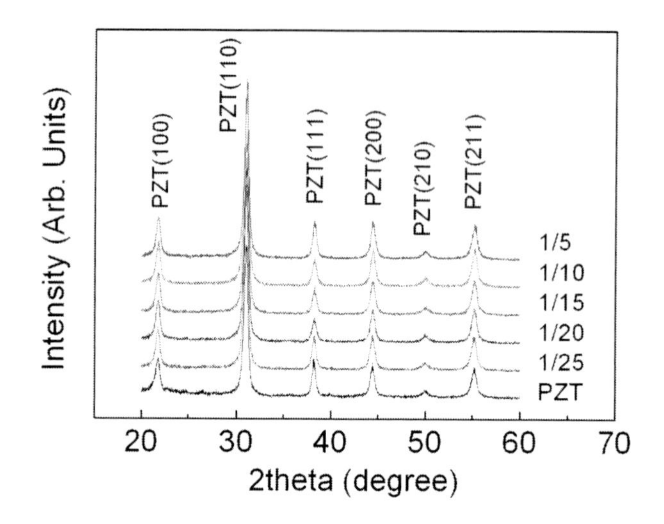

Figure 14. XRD patterns of CFO/PZT multiferroic composite thick films.

In this Section, PVP is introduced to adjust the viscosity and deposit process of PZT sol and deposit process of thick films is the same with Figure 5. The mass ratio of PVP and PZT xerogel 2:5 is chosen. CFO-PZT slurry layer with mass ratio of 1/5, 1/10, 1/15, 1/20 and 1/25 and plus pure PZT as well were prepared, to study the effect of magnetic content on the properties of the films. All of the films were annealed at 650 °C for 1 h in air.

As shown in Figure 14, PZT pervoskite phase is detected in all films without any spinel phase and other phase observed. Lack of CFO spinel phase in XRD measurement is due to its very low ratio and concealed positions where are distributed in the center of the composite thick films (the measured magnetic behavior in the thick films demonstrates the existence of CFO phase, which will be discussed later). Furthermore, the introducing of PVP promotes the thickness of films, which would cause the detection of CFO phase more difficult. The previous study on thick films without PVP has demonstrated this phenomenon. The fact that the peak positions of all film are the same as that of pure PZT thick film, further demonstrates that PZT and CFO phases maintain their own lattice structures respectively.

Cross-sectional SEM micrograph of a typical film, for example 1/15, is shown in Figure 15 (a), to compare with previous film without PVP added. It can be seen from the figure that the composite film has a dense microstructure with less pores scattered. Moreover, compared with the results without PVP, the film thickness increased to 7 μm, promoted by 100%. In addition, the films also showed reasonable smooth surface without any cracks, seen in Figure 15 (b). The enhancement of film thickness and the suppression of micro-crack in the films were strongly related to the introducing of PVP. Moreover, the color of film is also changed from dark-grey to light-grey, which is more suitable for piezoelectric displacement characterization using a scanning homodyne interferometer. The reason is that the smooth surface would improve the reflection with enhanced signal. Furthermore, with PVP added the sol solution or slurry would be more stable and could be stored for a long period of time without any precipitates observed, offering great convenience for industry or laboratory processing. However, the pores appeared in the film surface indicates that a further optimization on its densification is still needed.

Figure 16 shows the magnetic hysteresis loops under the same condition as described in section 2. It is seen that saturated magnetization (*Ms*) along in-plane is enhanced with increasing CFO content as expected. The maximum *Ms* in sample 1/5 is 14.7 emu/cm^3 whereas the minimum value in sample 1/25 is just 1.15 emu/cm^3. Compared with the previous one without PVP introduced, the film 1/15 shows a *Ms* of 5.76 emu/cm^3 as well as a *Hc* of 854 Oe. The *Ms* is reduced dramatically by more than 100% from 13.5 emu/cm^3 but with *Hc* unchanged, indicating that the increase of film volume thus the film thickness is the root cause of the Ms reduction.

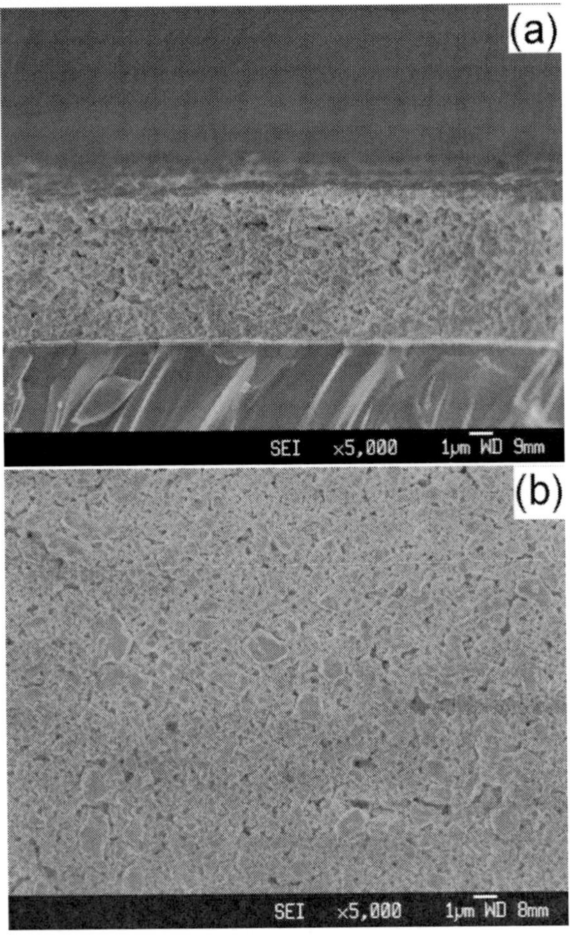

Figure 15. Typical SEM cross-sectional pictures of CFO/PZT (1/15) composite thick films (a) and its surface morphology (b).

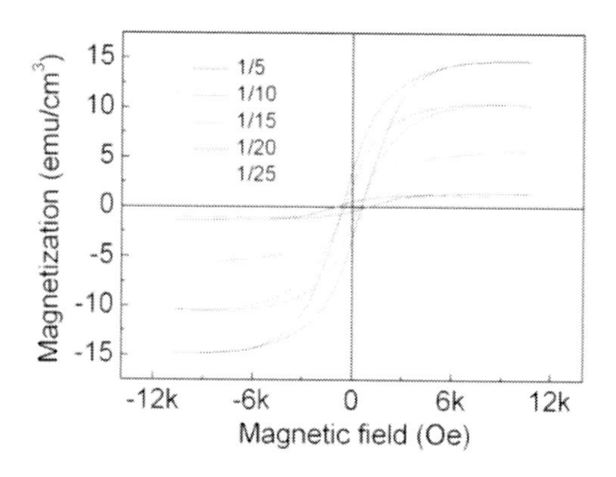

Figure 16. Magnetic hysteresis loops of all PZT/CFO composite thick films along in-plane direction.

Ferroelectric hysteresis loops of the typical films are shown in Figure 17. Compared to pure PZT thick film, the ferroelectric behavior of 1/15 film is degenerated, which is due to the dilute effect of non-ferroelectric phase CFO particles. Moreover, the other CFO-PZT thick films have the similar situation. In addition, with PVP introduced, remnant polarization is degraded but with little change on coercive electric field while compared with the previous results detailed in section 2.

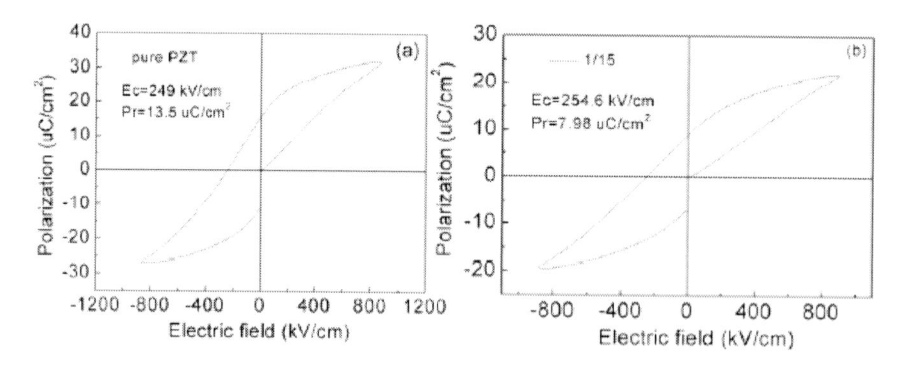

Figure 17. Typical ferroelectric hysteresis loops of pure PZT thick film and 1/15 film assisted by PVP.

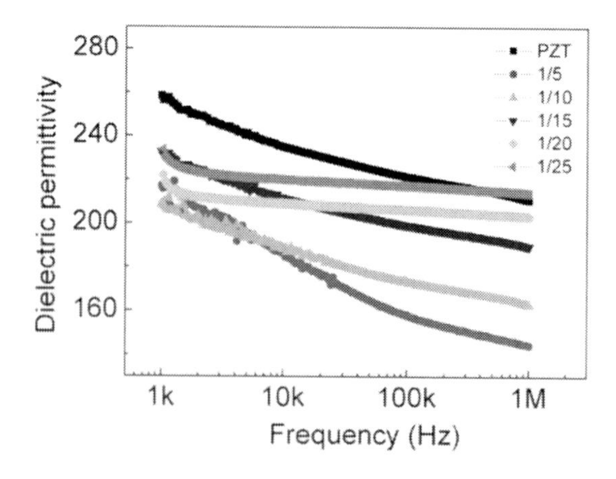

Figure 18. Dielectric permittivity of CFO/PZT composite thick films assisted by PVP.

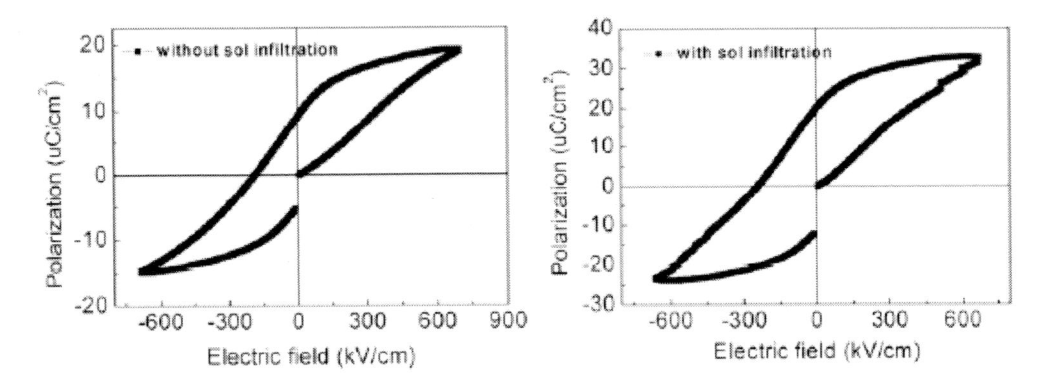

Figure 19. Comparison between PZT thick films without and with sol infiltration. Microstructures indicated an improved densification and a smoother surface; ferroelectric properties were also enhanced due to sol infiltration introducing.

The effect of PVP addition on dielectric permittivity of CFO-PZT composite thick films is shown in Figure 18. PZT has the maximum dielectric permittivity but the 1/5 film has the minimum value due to its largest CFO content. With CFO content decreasing, dielectric permittivity of composite thick films is increasing. However, the overall dielectric permittivity is still below the value expected [34]. In further process optimization, sol infiltration would be considered to improve the densification of thick films [22].

In summary, the influence of PVP on phase structure, microstructure, ferromagnetic, ferroelectric, dielectric properties has been studied. The results indicate that PVP is beneficial for increasing the film thickness and avoiding cracks, finally providing a smooth and uniform film for piezoelectric characterization. However, ferroelectric domain wall motion and domain rotation becomes difficult due to much pore introduced, resulting in a smaller polarization and a larger coercive electric field. In comparison, PZT thick films with or without sol infiltration were prepared as shown in Figure 19. It is obvious that densification of PZT composite thick films could be improved dramatically and its ferroelectric properties are also promoted remarkably with sol infiltration introduced. In the next Section, the effect of sol infiltration on multiferroic CFO-PZT composite thick films will be discussed in detail.

4. Influence of Sol Infiltration on CFO-PZT Composite Thick Films

Sol infiltration is adopted to improve the densification of composite thick films, based on the same sample as section 2. After each coating layer, PZT sol infiltration was deposited to infiltrate into the pores. Between every two layer of slurries, there are 1, 2, 3, 4 layers of sol applied for infiltration, respectively (denote PZT1, PZT2, PZT3, and PZT4). The final thick films are annealed at 600 °C for 1 hour in air.

Figure 20 (a) shows the ferroelectric hysteresis loops of multiferroic thick films measured at room temperature and under 100 Hz. All the loops show the slim shape and reach the saturation at a low electric field, indicating a typical soft ferroelectric behavior. With increasing sol infiltration layer from PZT1 to PZT4, the saturated polarization (*Ps*) and remnant polarization (*Pr*) of composite thick films are enhanced from 25.99 μC/cm^2 and 9.03 μC/cm^2 to 42.6 μC/cm^2 and 17.22 μC/cm^2, respectively, as shown in Figure 20 (b). Similar improvement through sol infiltration was also reported on PZT thick film by Perez [35]. In addition, the coercive electric field (*Ec*) was reduced from 91 kV/cm to 73 kV/cm with increasing sol layers due to the depinning effect of defect dipoles in the composite thick films [36], and hence the ferroelectric domain wall motion and domain rotation would be easier.

Moreover, duration dependence of *P-E* curves of typical PZT1 and PZT4 under 100 V of external voltage was investigated to further evaluate the influence of sol infiltration layers on the ferroelectric property of the films, the results are depicted in Figure 20 (c) and (d). It is seen that the polarization and coercive electric field were increased with prolonging duration for both of them. It is usually attributed to the space charges in the thick films, which would be pronounced at low frequency. In this work, due to different polarization mechanisms of CFO and PZT, a potential barrier might be formed at their interface with accumulated interfacial charges. It is known that interfacial charge could enhance or degrade polarization depending on the polarization difference between neighboring layers [37]. When these charges were accumulated and active at a longer duration, the dipolar effect will give way to them and give rise to a dilated hysteresis loop. The increased coercive electric field is also related to these charges due to their causing large leakage current in composite thick films. Quantitatively analyzing, when 50 ms of duration was applied on the PZT1, the hysteresis loop became distorted and as the duration reached 100 ms, the dielectric behavior changed into leaky behavior, as can be seen in inset of Figure 20 (c), such rounded hysteresis loops are always indicative of leakage currents. However, as can be seen in the Figure 20 (d), the same duration could not affect the loops shape of PZT4. Even at duration of 1000 ms, the slim loop is still able to be taken on, indicative of a dipolar polarize predominated state. Their comparison hence reveals that more sol infiltration could reduce the interfacial charges and improving ferroelectric behavior.

Figure 21 (a) shows the leakage current density (*J*) of PZT1, PZT2, PZT3, and PZT4 measured at room temperature. With increasing sol infiltration layer from 1 to 4, the average *J* value is decreased from 5×10^{-6}A/cm^2 to 6×10^{-7}A/cm^2. Besides the contribution of exchange between Fe^{2+} and Fe^{3+} to leakage, another main contribution to a large leakage is caused by porous structure in the composite thick films. By introducing enough PZT sol layers in the thick films, the porous structure could be densified along with a reduced leakage current for MEMS device applications [15]. On the other hand, the reduced leakage current could also suppress the contribution of conductive loss to the ferroelectric hysteresis loops that may result in pseudo-ferroelectricity in multiferroic materials [16], and achieve the true multiferroic property coupling with its ferromagnetic property.

In addition, the improvement of sol infiltration on dielectric properties of composite thick films is also obvious. Figure 21(b) and (c) show the frequency dependence of their dielectric constant (ε) and loss tangent (*tan* δ) at room temperature. It is seen clearly that the dielectric constants nearly keep stable in the whole frequency range except for some small dielectric dispersion at low frequency similar to that of dielectric loss, in which large dc conductivity might be induced by mobile charge carriers existed [38]. The movement of these mobile charges is suppressed at high frequency, and hence the decreased dielectric loss is observed. For quantitatively clarifying the influence of sol infiltration on dielectric properties of the films, the dielectric constant at 100 kHz is observed to be enhanced from 517 to 896 as the sol infiltration layer is increased from 1 to 4, along with a reduced loss value from 0.025 to

0.017. These results are also in agreement with the enhanced ferroelectric property since the dielectric constant is mainly contributed by ferroelectric polarization.

Besides the improvement of ferroelectric and dielectric properties, sol infiltrations show a slight effect on the magnetic properties of multiferroic thick films, as can be seen in Figure 22. The thick films exhibit an obvious magnetic hysteresis loop and their saturated magnetizations are decreased slightly with increasing sol infiltration from 1 to 4 layers due to the reduced mass fraction of CFO ferrite particles in the PZT matrix. However, the coercive magnetic field remains unchanged since it is mainly dependent on the embedded CFO and not related to its mass fraction in the films.

From the results above, we can see that through introducing sufficient PZT sol infiltration in CFO-PZT multiferroic composite thick films, their ferroelectric and dielectric properties were enhanced as well as a longer endurance to the external voltage. *Ps* and *Pr* could reach 42.6 $\mu C/cm^2$ and 17.22 $\mu C/cm^2$ with 4 layers of PZT sol layers. Dielectric constant was also enhanced to 869 at 100 kHz with a reduced loss of 0.017. Magnetic property kept a controlled decrease with increasing sol infiltration. All of them were attributed to the optimized film densifications.

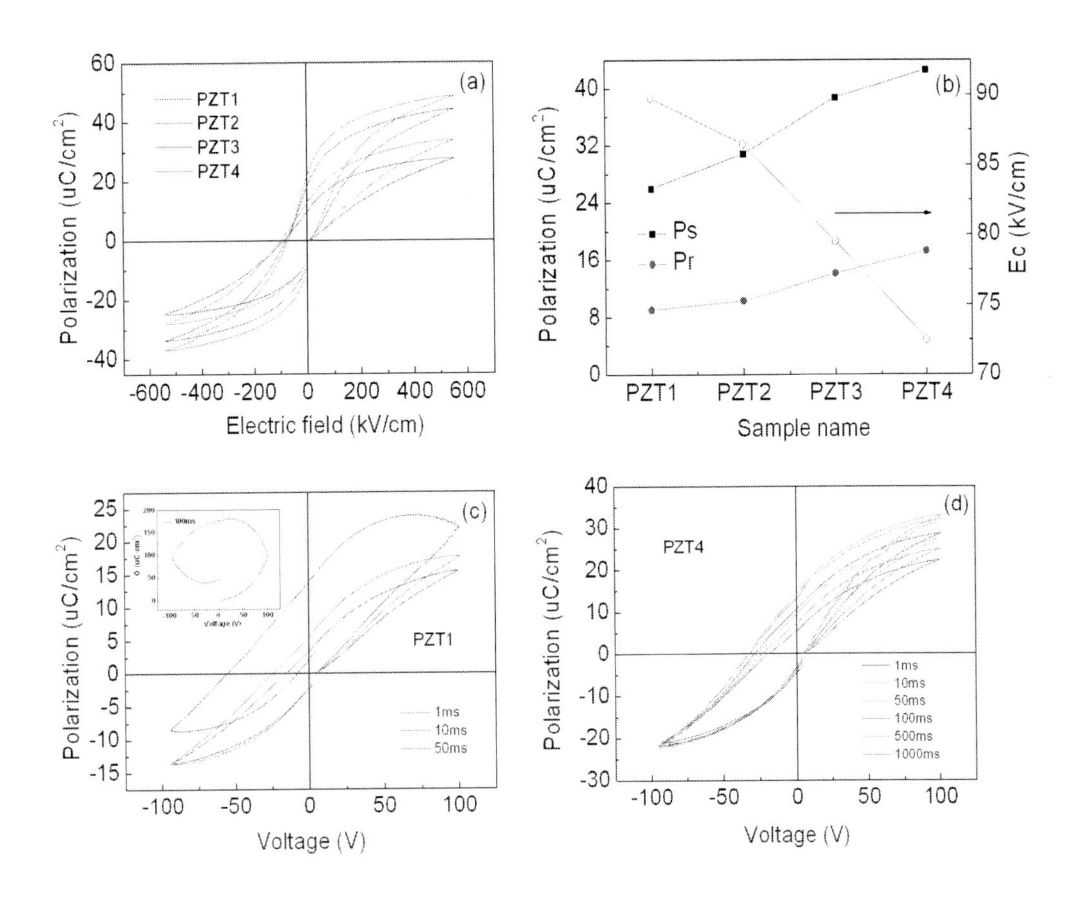

Figure 20. Ferroelectric hysteresis loops of PZT1, PZT2, PZT3 and PZT4 at room temperature under 100 Hz (a), and their ferroelectric parameters (b). Duration dependence of P-E curves of typical PZT1 (c) and PZT4 (d) composite thick films under 100 V of external voltage.

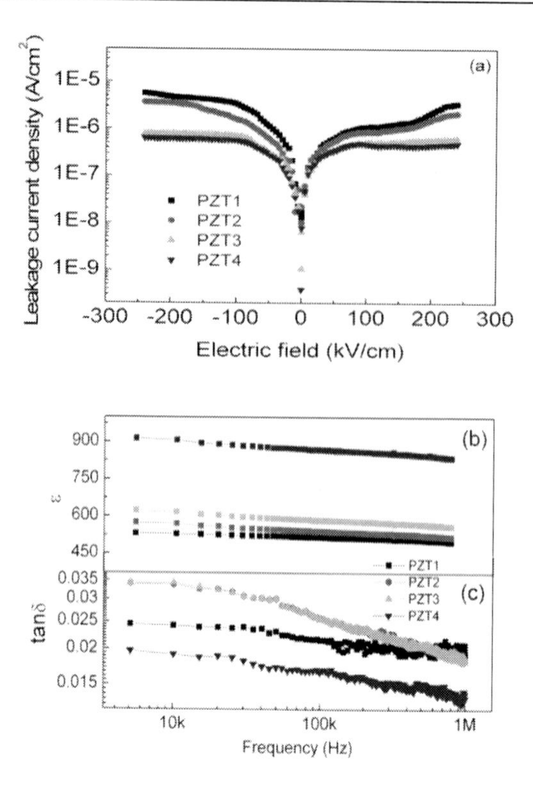

Figure 21. Leakage current densities of PZT1, PZT2, PZT3 and PZT4 at room temperature (a) as well as their dielectric constants (b) and losses (c) under different frequencies.

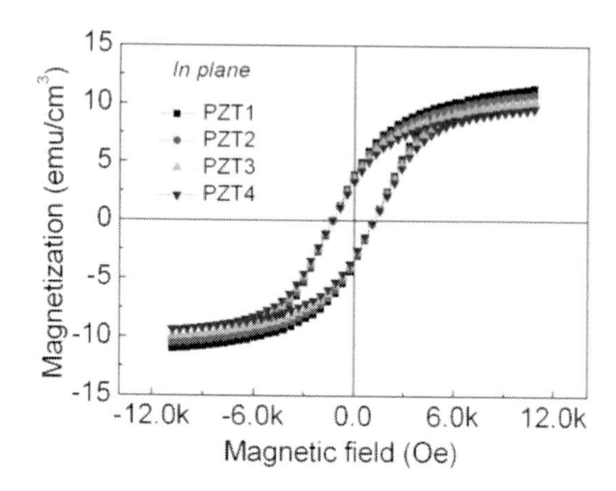

Figure 22. Magnetic hysteresis loops of PZT1 through PZT4 measured at room temperature.

5. Co-Effect of PVP and Sol Infiltration on Composite Thick Films

Since PVP improves the uniformity of the microstructure, while sol infiltration promotes the densification of the thick films, the present section combines both of them together to get the further optimization on multiferroic thick films. Based on this optimization, magnetic content percentage is hence increased to balance the ferroelectric and ferromagnetic properties.

PZT sol-gel solution is modified by PVP with a proper weight fraction and PZT slurry (Marked as P) keeps the same ratio of PZT sol-gel solution and modified PZT particles as before. Modified CFO and PZT nanoparticles mixed as a mass ratio of 1: 1 and then dissolved into PZT sol-gel solution with a mass ratio of 2: 3 to form heterogeneous slurry (marked as C). Both slurries (P and C) were deposited onto $Pt/Ti/SiO_2/Si$ substrate as PPPP, PPCC, and CCCC, see Figure 23. Additionally PCPC, CPCP, and CCPP were also prepared using both slurries. Between every two slurry layers several layers of PZT sol were coated to compensate the porosities on thick films and improve their densities. The resultant composite thick films are annealed at 600 °C for 1 hour in air, and their thicknesses measured by surface profiler are above 5 μm.

Figure 24 (a) shows the x-ray diffraction (XRD) of the three thick films. The well-defined peaks of the pervoskite PZT and spinel CFO phase without any other impurity phase are identified in sample CCCC and PPCC. For the samples PPPP, only the peaks of typical pervoskite phase are observed. The difference is mainly determined by the CFO phase in PZT matrix. However, as can be seen in Figure 24 (b), even if CFO phase existed, no spinel peak is observed in CCPP. More specifically, the closer of CFO location to the surface, the stronger its spinel main peak, revealing an important effect of deposit sequence on magnetic phase detection.

Figure 25 exhibits a typical cross-sectional image and surface morphologies of the sample PPCC. It can be seen from the cross-sectional picture that the deposited films are uniform and no apparent projectors appeared on its surface, indicating a well composite structure for PZT and CFO. This is reasonable because "C" layer is composed of dispersed CFO and PZT particles into PZT slurry to avoid peeling off between pure CFO and PZT layer due to their lattice mismatch and different thermal expansion coefficients. The film thicknesses are estimated as 5 μm, in agreement with the measurements from surface profiler. In addition, the surface morphologies of thick film suggest a smooth and dense microstructure without any cracks, which is beneficial for the top electrode deposition and efficient electric measurements.

Figure 26 shows the magnetic hysteresis loops of the three composite thick films at room temperature. The field-dependent magnetization was measured by applying magnetic fields parallel to the plane of the films. The magnetic hysteresis loops of the thick films demonstrated the effect of the different CFO contents on the coercivity as well as saturation magnetization. As can be seen in Figure 26, no magnetic response was observed in sample PPPP. The coercive fields of sample CCCC and CCPP were the same as 1288 Oe, no dependence on magnetic phase content. This value is closing to that of pure CFO thin film [39], also indicative of a less effect from film thickness. However, the saturation magnetizations (*Ms*) of in-plane loops yielded 64.9 emu/cm^3 of CCCC and 34.5 emu/cm^3 of PPCC. Because *Ms* value is mainly determined by magnetic content in composites, and pure CFO thin films reported by sol gel processing yielded a *Ms* value of 320 emu/cm^3 [25], therefore, in the above mentioned two films, the mass fractions of CFO in PZT matrix are approximately estimated as 20 wt% and 8 wt% [40]. Additionally, the *Hc* and *Ms* values of CCPP, CPCP, PCPC, and PPCC are listed in Table 1. Based on the same composition, no change is observed on *Hc* value with different deposit sequences. However, when the "C" layer is more closing to the film surface, the *Ms* value of film becomes large. This is in agreement with the case of multilayered CZFO-PZT thin films [41].

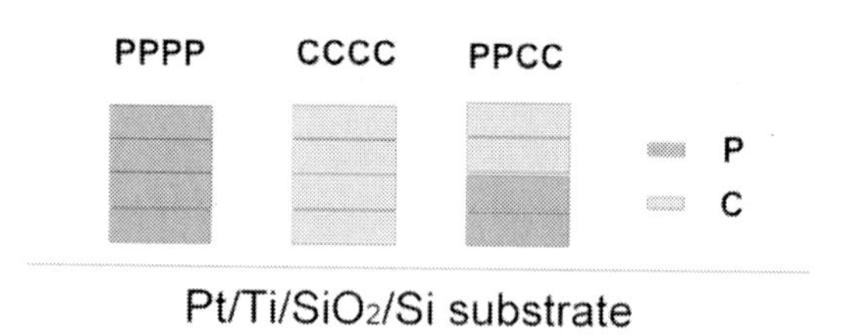

Figure 23. Deposited schematic illustration of multiferroic CFO-PZT composite thick films: PPPP, CCCC, and PPCC.

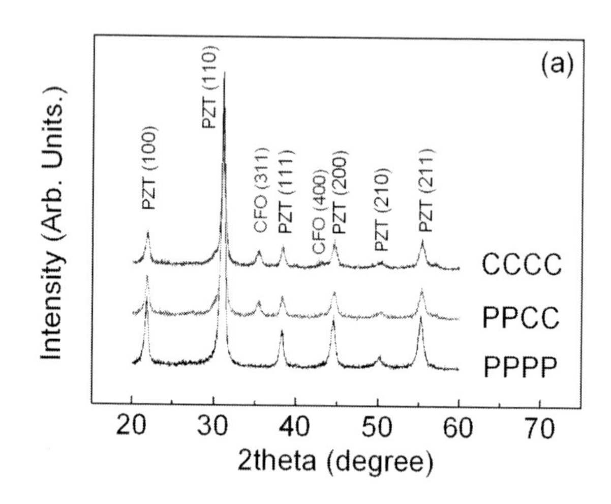

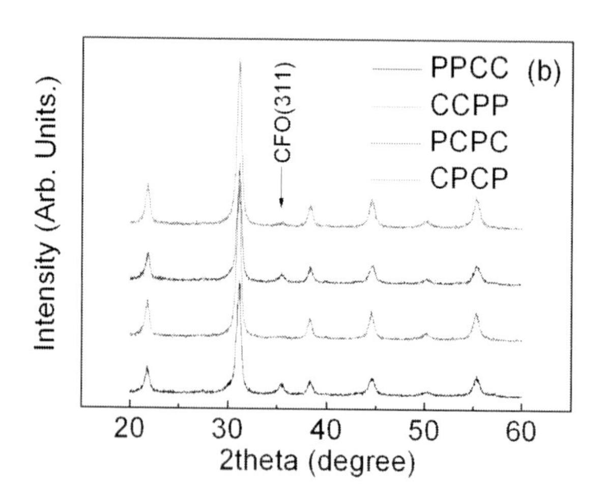

Figure 24. Influence of magnetic phase content (a) and deposit sequence (b) reflected from XRD patterns.

Figure 25. A typical cross-sectional picture and morphology of sample PPCC.

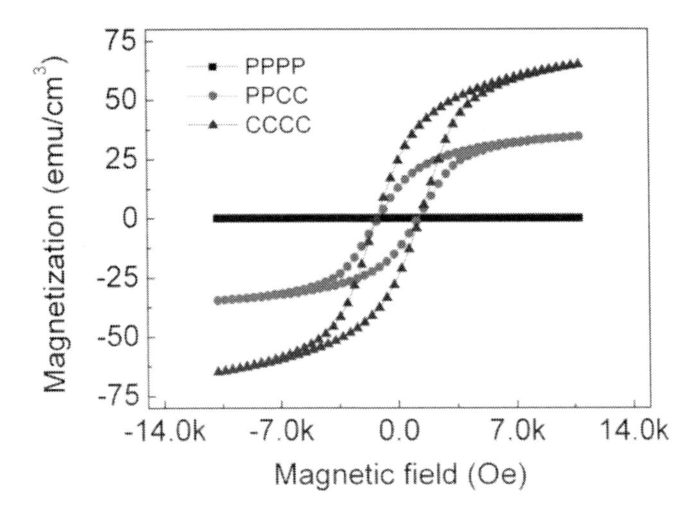

Figure 26. In-plane magnetic hysteresis loops of typical multiferroic CFO-PZT composite thick films measured at room temperature.

Table 1. Saturated magnetization, saturated polarization and remanent polarization of PPCC, PCPC, CPCP and CCPP.

	Ms (emu/cm^3)	*Hc* (Oe)	*Ps* (μC/cm^2)	*Pr* (μC/cm^2)
PPCC	34.5	1288	33.3	17.5
PCPC	29.7	1288	32.3	17.9
CPCP	27.5	1288	32.4	18.9
CCPP	26.1	1288	31.8	19.1

Figure 27 (a) presents ferroelectric hysteresis loops of the sample PPPP, PPCC and CCCC. A typical soft ferroelectric hysteresis loop is observed in PPPP. As for the PPCC, it is interesting that its saturated polarization and remnant polarization are larger than that of PPPP as well as a larger coercive electric field under similar external electric field, which is different from the diluted polarization caused by magnetic phase [7,42]. The main reason is that low resistivity of CFO particles increases the accumulation of space charges in the PPCC, this part of charges contribute partially to the polarization of thick film but sacrifice the number of discharging charges, evidenced by the red shadow area in this figure compared with black shadow area of PPPP. While lossy ferroelectric loop is observed in CCCC due to nearly 20 wt% of CFO particles, abundant space charges are hence formed and distributed along with the vertical direction between top and bottom electrodes. A low barrier is then induced in the whole thick film and leakage current is naturally increased, causing the spurious ferroelectric hystereis loop. In addition, due to different CFO content in three samples, their maximum endured voltages are hence different. That is also why the electric fields applied on them are different even though they have the same film thickness. The influence of deposit sequence on ferroelectric properties of CFO-PZT thick films can be seen in Figure 27 (b), where complete hysteresis loops of PPCC, PCPC, CPCP, and CCCC with an approximate CFO content of 8 wt% are presented for detailed comparison. Compared with PPPP, the loops of these four samples become fattened, and their polarization values read from the loops are very closing, which is shown in Table 1. However, for cases of CCPP and CPCP, their polarization values shown in the loops begin to decline before reaching the maximum applied electric field, indicative of a typically leaky mechanism. While for the case of PPCC and PCPC, this kind of phenomenon becomes weak, especially when the first layer to the substrate is of more ferroelectric phase, as can be seen in the loop of PPCC, where the polarization keeps rising trend at maximum electric field. This is corresponding to the effect of deposit sequence on ferroelectric properties of multilayered CZFO-PZT thin films [41]. According to literatures [16], the fattened loop is generally attributed to the effect of space charges and the typical loop of this case is usually like the one of CCPP and CPCP. As we known, introducing low resistivity materials into ferroelectric materials will bring with space charge carriers inevitably, which is undesirable to ferroelectric or dielectric properties. However, through adjusting the deposit sequence, like the situation reported in CZFO-PZT thin film and our CFO-PZT thick film, the effect of space charges can be suppressed, and meanwhile, enhanced polarization values can be achieved, just like the situation of PPCC and PCPC.

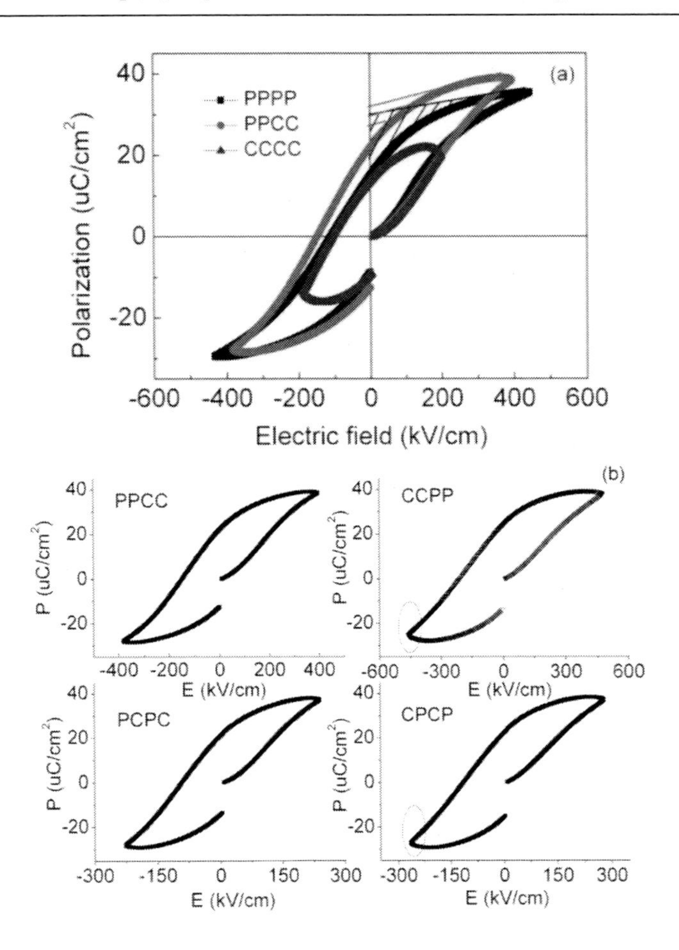

Figure 27. Influence of magnetic phase content (a) and deposit sequence (b) reflected from room temperature ferroelectric hysteresis loops.

Figure 28 (a) firstly exhibits the leakage current of PPPP, PPCC and CCCC. The leakage behavior of PPPP is stable with increasing external electric field, indicating low energy loss during ferroelectric domain motion and rotation. The situation of PPCC is similar to that of PPPP except enhanced leakage value in the whole measured range. But the leakage behavior of CCCC is totally different, where leakage current is nearly increasing linearly with the increase of electric field. This is attributed to the abundant charges induced by low resistivity of CFO particles. On the other hand, the leakage comparison of CCPP, CPCP, PCPC, and PPCC can be seen in Figure 28 (b). CCPP and CPCP have similar leakage curves to that of CCCC. Whereas for the case of PCPC and PPCC, their leakage behaviors are much similar to that of PPPP, stable leakage is obtained as increasing electric field. Apparent contrast between CCPP, PPCC and CPCP, PCPC reveals that a proper deposit sequence enables us to suppress the leakage of composite thick films. Additionally, suppressed leakage behavior in reverse demonstrates the true ferroelectric behavior of PPCC and PCPC.

From the results above, we can see that introducing magnetic CFO particles into composite thick films brings with the free electrons due to unfilled d-orbits of Fe^{3+} and Co^{2+} ions. These free electrons form spin moments in the thick films and on one hand, contributed to the magnetization with their ordered sequence under magnetic field, producing the magnetic storage ability of the film. On the other hand, it is these free electrons that lower the barrier in the thick film and an internal electric field which has the same direction with external electric field is hence formed, indicating that a larger electric field have to be endured by the original ferroelectric materials. That is also why the ferroelectric PZT is easier to be breakdown after introducing more CFO particles.

However, with the approximate CFO content of 8 wt% in the PZT matrix, the ferromagnetic and ferroelectric properties are still full of varieties due to different deposit sequence. For the ferromagnetic part, this influence has been found in multilayered CZFO-PZT thin films [41], in which the closer the thinner CZFO layer is to substrate, the lower the saturation magnetization. This is the same with the present case in composite thick films. However, the higher coercivity is also observed in the thin films [41], which is not found here. Both decreased *Ms* and increased *Hc* are attributed to the strong constraint effect in the multilayered CZFO-PZT thin films. In our composite thick films, the constraint effect mainly induced by lattice mismatch between CFO and PZT has been reduced smartly. That is also why the observed *Hc* values among four of same composites don't show apparently difference. While as for the different *Ms* values among them, we are more inclining to believe that the accumulated space positive charges nearby the substrate neutralize the free electrons around Fe^{3+} and Co^{2+} ions, then reducing the number of magnetic spin moments, finally decreasing the magnetic storage ability of CCPP and CPCP.

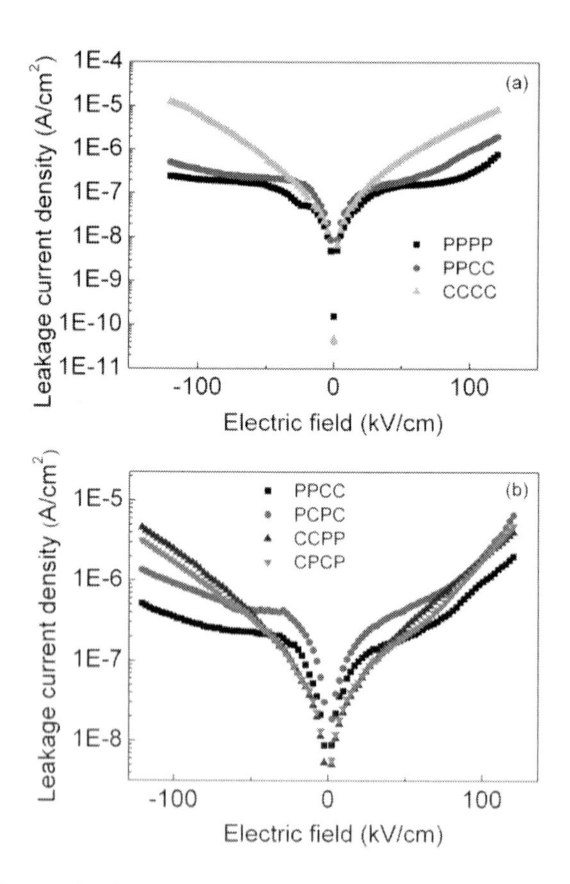

Figure 28. Influence of magnetic phase content (a) and deposit sequence (b) observed from leakage current densities.

For the ferroelectric part, ferroelectric hysteresis loops of most CFO-PZT multilayered thin film exhibited a 'melon' shape [43-44] and nonlinear behavior between polarization and electric field is degraded severely, like the case of present CCPP and CPCP. Space charge effects induced by various low resistivity mediums, such as grain boundaries, air porosities and even some ferromagnetic particles, are believed widely to be attributed to them [43-44]. In order to reduce this effect, organic PVP and ferroelectric sol infiltration were introduced to improve ferroelectric behavior of CFO-PZT thick films, as clarified above. The present thick films are also optimized by PZT sol infiltration and their leakage behavior could be partially controlled. However, the main reason causing different ferroelectric behavior is still attributed to its sensitive to the first layer to the substrate. Overall, through well controlling deposit sequence and CFO content, below 10 wt%, promising ferromagnetic and ferroelectric properties are able to be achieved and their magnetoelectric coupling effect could hence be predicted.

CONCLUSION

A study on hybrid sol-gel routes was carried out to specify the preparation of modified CFO and PZT nanoparticles, PZT sol solution and its uniform slurry. Thermal analyses (DTA/TGA) were carried out for the PZT xerogel and the baking and pyrolizing temperatures of coated films were hence confirmed. CFO-PZT multiferroic composite thick films were successfully prepared on $Pt/Ti/SiO_2/Si$ substrate by spin-coating method. After annealed at 600 °C or 650 °C for 1 hour, their ferromagnetic and ferroelectric properties were observed simultaneously at room temperature.

Through introducing PVP into the sol-gel solution, the viscosity was adjusted and uniform and stable slurry was hence made through suspending PZT and CFO particles in it. These slurries were used to coat PZT composite thick films and crack-free and uniform CFO-PZT multiferroic composite thick films were achieved along with a smooth and reflected surface. Moreover, the thickness of the resultant CFO-PZT composite film was promoted by 100%. However, PVP also brought with much pores, which degenerated the multiferroic properties. HIP (Hot Isostatic Press) process could be able to improve the quality of films by reducing its porosity.

Sol infiltration was further performed for process optimization of thick films. It is observed that enough sol infiltration can promote the densification of CFO-PZT composite thick films remarkably and hence enhance their dielectric and ferroelectric properties along with the slight effect on magnetic properties.

Based on the optimization of CFO-PZT composite thick films by introducing sol infiltration and proper PVP, CFO magnetic content was enhanced in the PZT matrix, leakage current density indicates that 20 wt% CFO can still keep the CFO-PZT composite thick film in an acceptable dielectric level. However, its ferroelectric property is severely degraded. Good ferroelectric and ferromagnetic properties were achieved simultaneously with a promising dielectric permittivity at 8 wt% of CFO content. It is hence predicted that 8 wt% of CFO in PZT matrix may be a critical point to have a maximum magnetoelectric effect in CFO-PZT composite thick film due to its promising multiferroic behaviors. The resultant thick films could also be a good candidate for future multifunctional application.

References

[1] Kimura, T; Goto, T; Shintani, H; Ishizaka, K; Arima, T; Tokura, Y. *Nature*, 2003, 426, 55-58.

[2] Lottermoser, T; Lonkai, T; Amann, U; Hohlwein, D; Ihringer, J; Fiebig, M. *Nature*, 2004, 430, 541-544.

[3] Fiebig, M. *J. Phys. D,* 2005, 38, R123-152.

[4] Nan, CW; Bichurin, MI; Dong, SX; Viehland, D; Srinivasan, G. *J. Appl. Phys*., 2008, 103, 031101-35.

[5] Zhou, JP; Qiu, ZC; Liu, P. *Mater. Res. Bul*., 2008, 43, 3514-20.

[6] Lin, R; Wu, TB; Chu, YH. *Scr. Mater*., 2008, 59, 897-900.

[7] Sim, CH; Pan, AZZ; Wang, J. *J. Appl. Phys*., 2008, 103, 124109-7.

[8] He, HC; Ma, J; Wang, J; Nan, CW. *J. Appl. Phys*., 2008, 103, 034103-5.

[9] Ortega, N; Bhattacharya, P; Katiyar, RS; Dutta, P; Manivannan, A; Seehra, S; Takeuchi, I; Majumder, SB. *J. Appl. Phys*., 2006, 100, 126105-3.

[10] Zhou, JP; He, HC; Shi, Z; Gang, L; Nan, CW. *J. Appl. Phys*., 2006, 100, 094106-6.

[11] Eerenstein, W; Mathur, ND; Scott, JF. *Nature*, 2006, 442, 759-765.

[12] Hur, N; Park, S; Sharma, PA; Ahn, JS; Guha, S; Cheong, SW. *Nature*, 2004, 429, 392-395.

[13] Kimura, T; Kwawmoto, S; Yamada, I; Azuma, M; Takano, M; Tokura, Y. *Phys. Rev*. B, 2003, 67, 180401-4.

[14] Wang, ZH; Miao, JM; Zhu, WG. *Appl. Phys*., A 2008, 91, 107-117.

[15] Wang, ZH; Miao, JM; Zhu, WG. *J. Eur. Ceram. Soc*., 2007, 27, 3759-64.

[16] Scott JF. *J. Phys.: Condens. Matter,* 2008, 20, 021001-2.

[17] Zhang, JX; Dai, JY; Chow, CK; Sun, CL; Lo, VC; Chan, HLW. *Appl. Phys. Lett*., 2008, 92, 022901-3.

[18] Wan, JX; Zhang, H; Wang, XW; Pan, DY; Liu, JM; Wang, GH. *Appl. Phys. Lett*., 2006, 89, 122914-3.

[19] Ahn, BL; Lee, J; Park, SM; Lee, SG. *J. Mater. Sci*., 2008, 43, 3408-11.

[20] Zhu, W; Wang, ZH; Zhao, CL; Tan, OK; Hng, HH. *Jpn. J. Appl. Phys*., 2002, 41, 6969-75.

[21] Haertling, GH. *J. Am. Ceram. Soc*., 1999, 82, 797-818.

[22] Lee, SG. *Mater. Lett*., 2007, 61, 1982-85.

[23] Huang, W; Zhu, J; Zeng, HZ; Wei, XH; Zhang, Y; Li, YR. *Appl. Phys. Lett*., 2006, 89, 262506-3.

[24] Liu, M; Li, X; Lou, J; Zheng, SJ; Du, K; Sun, NX. *J. Appl. Phys*., 2007, 102, 083911-3.

[25] Sathaye, SD; Patil, KR; Kulkarni, SD; Bakre, PP; Pradhan, SD; Sarwade, BD; Shintre, SN. *J. Mater. Sci*., 2003, 38, 29-33.

[26] Yin, JH; Ding, J; Liu, BH; Miao, XS; Chen, JS. *Appl. Phys. Lett*., 2006, 88, 162502-3.

[27] Lee, JG; Lee, HM; Kim, CS; Jei, OY. *J. Magn. Magn. Mater*., 1998, 177, 900-902.

[28] Kozuka, H; Kajimura, M; Hirano, T; Katayama, K. *J. Sol-Gel. Sci. Tech*., 2000, 19, 205-209.

[29] Park, GT; Choi, JJ; Park, CS; Lee, JW; Kim, HE. *Appl. Phys. Lett*., 2004, 85, 2322-24.

[30] Yamano, A; Kozuka, H. *J. Am. Ceram. Soc*., 2007, 90, 3882-89.

[31] Selim, MS; Turky, G; Shouman, MA; El-Shobaky, GA. *Solid State Ionics,* 1999, 120, 173-181.

[32] Garg, A; Agrawal, DC. *J. Mater. Sci.: Mater. Electron,* 1999, 10, 649-652.

[33] Abraham, R; Guo, RY; Bhalla, AS. *Ferroelectrics*, 2005, 315, 1-15.

[34] Wang, ZH; Zhu, W; Zhu, H; Miao, JM; Chao, C; Zhao, CL; Tan, OK. *IEEE Trans.* Ultrason. Ferroelectr. Freq. Control 2005, 52, 2289-97.

[35] Perez, J; Vyshatko, P; Vilarinho, PM; Kholkin, AL. *Mater. Chem. Phys.*, 2007, 101, 280-284.

[36] Li, BS; Li, GR; Yin, QR; Zhu, ZG; Ding, AL; Cao, WW. *J. Phys. D,* 2005, 38, 1107-11.

[37] Misirlioglu, IB; Alexe, M; Pintilie, L; Hesse, D. *Appl. Phys. Lett.*, 2007, 91, 022911-3.

[38] Khodoroy, A; Rodrigues, SAS; Pereira, M; Gomes, MJM. *J. Appl. Phys.*, 2007, 102, 114109-7.

[39] Lee, JG; Park, JY; Oh, YJ; Kim, CS. *J. Appl. Phys.*, 1998, 84, 2801-04.

[40] Chen, W; Wang, ZH; Zhu, W; Tan, OK. *J. Phys. D,* 2009, 42, 075421-5.

[41] He, HC; Wang, J; Zhou, BP; Nan, CW. *Adv. Func. Mater.*, 2007, 17, 1333-38.

[42] Tan, SY; Shannigrahi, SR; Tan, SH; Tay, FEH. *J. Appl. Phys.*, 2008, 103, 094105-4.

[43] Ortega, N; Kumar A; Katiyar, RS; Scott, JF. *Appl. Phys. Lett.*, 2007, 91, 102902-3.

[44] Deng, CY; Zhang, Y; Ma, J; Lin, YH; Nan, CW; *J. Appl. Phys.*, 2007, 102, 074114-5.

In: The Sol-Gel Process
Editor: Rachel E. Morris

ISBN 978-1-61761-321-0

Chapter 11

APPLICATION OF A SOL-GEL BASED NANOSTRUCTURED CERAMIC MEMBRANE FOR HYDROGEN SEPARATION IN CO_2 CAPTURE PURPOSES

***T. H. Y. Tran*[1], *W. G. Haije*[1,2] *and J. Schoonman*[1]**
[1]Delft University of Technology, Faculty of Applied Sciences, Department ChemE, Delft, The Netherlands
[2]Energy research Centre of the Netherlands, Hydrogen and Clean Fossil Fuels, Petten, The Netherlands

ABSTRACT

The two most significant industrial processes for hydrogen production from fossil fuels are coal gasification and steam reforming of natural gas. A membrane process for CO_2-H_2 separation is an attractive alternative to conventional technologies such as solvent scrubbing, pressure swing adsorption, and cryogenic distillation for capturing the CO_2. The advantages of using membrane reactors, *i.e.*, the combination of a chemical reactor and membrane separation as one unit operation, are highly evident, *e.g.*, the possibility to shift the chemical equilibrium towards the product side, the improvement of hydrogen production, along with high fossil fuel conversion at milder operating conditions and thus a lower efficiency penalty and lower cost. In principle, two different types of membranes are under development, dense Pd-alloy based and porous inorganic ceramic ones. The non-metallic inorganic membranes are in principle more robust, thermally and mechanically, but suffer from the drawback that the pore size has to be tuned to the kinetic diameter of hydrogen in order to obtain sufficient selectivity, for instance by Atomic Layer Deposition (ALD). The present contribution provides a literature overview of ceramic membranes and their performance which are currently under investigation for hydrogen separation at high temperatures, as well as recent research results by the authors. The scope of materials presented here is limited to those that are primarily based on sol-gel synthesis.

INTRODUCTION

The present energy system which utilizes fossil fuels like coal, oil, and natural gas has several impacts on modern society such as the degradation of the natural environment, the current consequences of global warming, the exhaustion of fossil fuels reserves, economic dependence, and political tensions in the oil-rich region - Middle East [1-3]. Therefore, it is urgent to switch to alternative sustainable energy sources. The potential of hydrogen (H_2) as an environmentally clean energy carrier in the near future, *i.e.*, the only combustion product being water, has been attracting a lot of attention of a vast number of researchers. Meanwhile, until the transition to a sustainable energy system fossil fuels will be used but in a clean carbon free way. Therefore, H_2 needs to be separated from its carbon carrier, the fuel. The carbon containing fuels are transformed into reaction mixtures via different processes such as coal (gasification, carbonization), natural gas (steam reforming, partial oxidation, autothermal reforming, plasma reforming), biomass (gasification, steam reforming, biological conversion) from which H_2 is extracted. The most common energy feedstocks and existing commercial processes are H_2 production from coal gasification and from steam reforming of natural gas (methane (CH_4) is its principal constituent) [3, 4]. Generally, the production involves two main steps. The first step is the steam reforming of the feedstocks to a syngas, mainly consisting of carbon monoxide (CO) and H_2 as shown in the following chemical reactions:

(a) Coal gasification (CG)

$$C + H_2O \underset{40\text{ bar}}{\overset{900°C}{\rightleftarrows}} CO + H_2 \qquad \Delta H^0_{298} = 119\text{ kJ/mol} \tag{1}$$

(b) Steam-methane reforming (SMR)

$$CH_4 + H_2O \underset{30\text{ bar}}{\overset{850°C}{\rightleftarrows}} CO + 3H_2 \qquad \Delta H^0_{298} = 206\text{ kJ/mol} \tag{2}$$

In the second step, CO in the syngas is subsequently converted to carbon dioxide (CO_2) and H_2 via the water-gas-shift (WGS) reaction:

(c) Water-gas-shift (WGS)

$$CO + H_2O \underset{20\text{ bar}}{\overset{350°C}{\rightleftarrows}} CO_2 + H_2 \qquad \Delta H^0_{298} = -41\text{ kJ/mol} \tag{3}$$

With associated CO_2 capture and storage (CCS) technology [5-7] at minimum energy penalty, H_2 produced via coal gasification and steam methane reforming can be a key factor to secure a carbon free H_2 supply and to commercial H_2-based technologies, such as fuel cells [8]. In order to be both economically competitive and environmentally sustainable, various alternative methods such as pressure-swing adsorption (PSA) [9], cryogenic distillation [10], and membrane separation [11, 12] can be used to separate H_2. Membrane separation is considered to be an energy-efficient continuous process alternative to the commercial PSA and cryogenic distillation methods, which are quite energy demanding for the separation and purification of H_2. Under normal conditions, the steam-methane reforming (SMR) is an equilibrium endothermic reaction (reaction (2)). If H_2 is removed from the reaction system by using H_2-selective membranes, the driving force of the chemical equilibrium favours the product side, which yields higher conversion of CH_4 to CO or CO_2 and H_2, at even lower reaction temperature. Membrane reactors, in which the reaction and separation happen simultaneously, can provide such opportunity for minimising the energy penalty for carbon capture. There are various types of membranes which hold the promise for separating H_2, including metal- [13, 14], polymer- [15], silica- [16-18], and zeolite-based [18-20] ones. The most investigated metallic membranes are dense palladium (Pd)-based alloys, intrinsically having 100% selectivity. However, a pure Pd membrane suffers from H_2 embrittlement due to the phase transition between the α- and β-hydride form in a H_2-containing environment and from poisoning effects of sulphur (H_2S) and CO, which lead to membrane degradation or reduced membrane performance. The chemical stability of the Pd membranes can be improved by alloying the Pd with other metals such as silver (Ag) [14]. Currently, Pd and Pd-alloy membranes are used only when ultra-pure H_2 is needed and only as a separation step. Polymeric membranes are commercially used for H_2 separation only at low temperatures (<250°C) with H_2/CO selectivity in the range 21-30 [15], and have therefore no interest to high temperature processes as SMR and WGS. In comparison to their Pd and polymeric counterparts, ceramic membranes are not adversely affected by elevated operating temperatures and have higher chemical resistance, which make them good candidates for H_2 separation in reforming processes, provided they exhibit high (hydro)thermal stability.

Many reviews have been published [11, 12, 21] recently, which have focused on particular aspects of the above mentioned membrane classes. In the present contribution, membrane materials solely related to sol-gel based production will be considered in the present discussion. Sol-gel technology has gained great interest as a suitable method for making thin porous membranes with desired properties [22], mainly due to the relative simplicity, it being a low-temperature process, and the possibility of good control over the process parameters. The following presentation focuses on the recent results of the sol-gel preparation of membranes with an emphasis on three key parameters, i.e., the choice of materials, the tailoring/design of the pore structure for improving the gas separation factor, and the characterization of membranes. A full understanding of these parameters will put us in a better position to develop sol-gel derived materials as a potential application in porous membrane based-H_2 separation.

Choice of Materials

The state-of-the-art ceramic membranes have been a subject of research for several decades and have been made from a variety of inorganic materials. Usually, a sol-gel preparation of a membrane involves first the formation of a sol, which is a suspension of solid particles in a liquid, then the deposition of the sol on the surface of a porous support where a gel layer is formed. The solvent can be removed from the gel layer by either conventional drying in order to obtain a product known as a xerogel, or drying with supercritical extraction in order to have an aerogel. Calcination of the gel at a temperature higher than the drying temperature causes further removal of the solvent and results in a porous ceramic layer. It can be a microporous (pore size $d_p < 2$ nm) or mesoporous ($2 < d_p < 50$ nm) membrane. Figure 1 shows all sequences of a conventional membrane preparation as described.

Although each step can be critical for the success of the membrane preparation, which is discussed in the excellent review of Brinker et al. [22], it is considered that the resulting membrane characteristics are mostly determined by the structure and chemistry of the sol. The formation of a sol is usually involved in the hydrolysis-condensation of metal alkoxide precursor $M\text{-}(OR)_x$ (M=Si, Ti, Zr, Al; $OR=OC_nH_{2n+1}$; x is the valence state of the metal). The overall hydrolysis-condensation reactions of the sol-gel routes can be presented by the following simplified ones:

Hydrolysis:

$$M\text{-}(OR)_x + H_2O \longrightarrow M\text{-}(OH)(OR)_{x-1} + R\text{-}OH \tag{4}$$

Condensation:

$$M\text{-}(OH)(OR)_{x-1} + M\text{-}(OR)_x \xrightarrow[\text{Alcoxolation}]{} (OR)_{x-1}\text{-}M\text{-}O\text{-}M\text{-}(OR)_{x-1} + ROH \tag{5}$$

$$M\text{-}(OH)(OR)_{x-1} + M\text{-}(OH)(OR)_{x-1} \xrightarrow[\text{Oxolation}]{} (OR)_{x-1}\text{-}M\text{-}O\text{-}M\text{-}(OR)_{x-1} + H_2O \tag{6}$$

It is well known that silicon alkoxides $Si\text{-}(OR)_4$, such as tetraethylorthosilicate (TEOS/ $Si(OCH_2CH_3)_4$), are excellent precursors in the sol-gel process. These precursors can be utilized in synthesizing microporous silica via the polymeric sol-gel route under acid-catalysed conditions (HCl or HNO_3) [23-25]. During acid-catalysed hydrolysis the alkoxy OR is substituted by water according to a nucleophilic substitution reaction, S_N2, accompanied by inversion of the silicon tetrahedron (reaction (7a)). Condensation of the silanols is directed preferentially towards the ends of siliceous oligomers which result in the formation of an oxo-bridge (Si-O-Si) as shown in the reaction (7b) [26]:

$$H_3O^+ + (RO)_3Si\text{—}OR \xrightarrow[\text{Hydrolysis}]{\text{fast}} H\text{—}\overset{+}{\underset{H}{O}}\text{—}Si(OR)_3 + ROH \tag{7a}$$

$$(RO)_3Si\text{—}OH + H\text{—}\overset{+}{\underset{H}{O}}\text{—}Si(OR)_3 \xrightarrow[\text{Condensation}]{\text{slow}} (RO)_3Si\text{—}O\text{—}Si(OR)_3 + H_3O^+ \tag{7b}$$

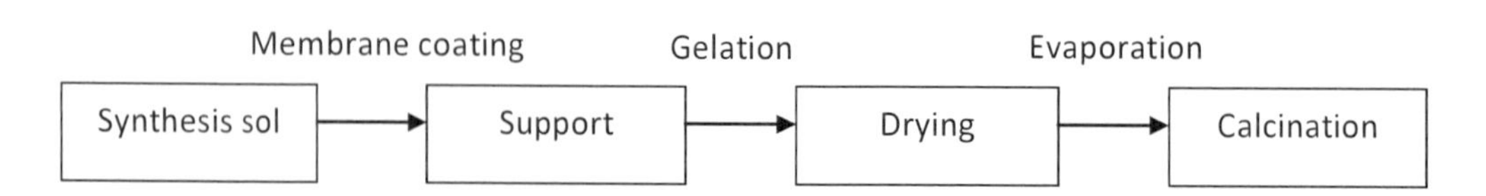

Figure 1. Sol-gel ceramic membrane preparation.

Table 1. Summary of reported gas separation results of H_2 membranes.

T (°C)	Support	ΔP [a)] (Pa)	H_2 permeance (mol. $m^{-2} Pa^{-1} s^{-1}$)	Gas A	Selectivity H_2/gas A	Ref.
Summary of reported gas separation results of sol-gel-derived silica membranes						
300	γ-alumina	10^5	$2*10^{-7}$ - $7*10^{-7}$	CO_2	11-36	[36]
100	γ-alumina	200	$2.1*10^{-7}$	CO_2	9	[37]
200	α-alumina	$3*10^5$	$4.8 * 10^{-8}$	CO_2	16	[17]
Summary of reported gas separation results of palladium-based membranes						
200	porous glass	10^4	$2.2* 10^{-5}$ [b)]	N_2	7	[40]
373	α-alumina	$3*10^4$	$6.7* 10^{-6}$ [b)]	N_2	5000	[41]
445	α-alumina	10^5	$5* 10^{-7}$	N_2	1000	[42]

[a)] pressure drop across the membrane.
[b)] calculated from reference data.

Contrary to silicon alkoxides, the sol-gel chemistry of transition metal alkoxides is more complex, because transition metal precursors are highly reactive due to their low electronegativity and existence of various coordination numbers [27-31]. This in turn makes the metal ion very susceptible to nucleophilic attack, which makes it difficult to obtain a straightforward and controlled mechanism for hydrolysis-condensation reactions of these alkoxides. Dedicated chelating schemes have to be developed to prevent direct precipitation of the metal hydroxides.

In any case, both for silicon and transition metal alkoxides, it is necessary to control hydrolysis and condensation reaction rates. The fast hydrolysis and condensation rates can result in a precipitated product instead of a stable sol. By controlling a number of factors such as the nature of the alkoxy groups, solvent, molar ratio of water/alkoxide (h), pH, aging temperature and time, and drying, it is possible to change the rate of hydrolysis and condensation reactions. The hydrolysis rate can be influenced by the nature of the alkyl group of metal alkoxides. It is decreased in the order tertiary > secondary > primary group. This is attributed to steric hindrance effects and the decreased positive partial charge of the metal ion with increased alkyl chain length. Substitution of the OR ligands by a chelating bidentate ligand such as acetylacetone (AcAcH) [32, 33] leads to the formation of a new molecular precursor which exhibits a different molecular structure. Here AcAc acts as a bidentate, ligand on titanium isopropoxide ($Ti(OPr^i)_4$ with Pr^i=$OCH(CH_3)_2$), i.e.,

$$Ti(OPr^i)_4 + AcAcH \longrightarrow Ti(OPr^i)_3AcAc + Pr^iOH \quad (8)$$

These complexing ligands are much more difficult to hydrolyse than alkoxy groups. They restrain the fast hydrolysis of the precursor and consequently slow down the condensation process. Selecting an appropriate solvent such as an aprotic solvent (tetrahydrofurane, dioxane) instead of a protic sovent (H_2O, alcohols) [22] can also slow down the hydrolysis rate. In general, the solvents most often used are parent alcohols, which have the same number of carbon atoms in the alkyl group as in the alkoxy group of the precursor. The precursor can be hydrolyzed with H_2O released in-situ by the esterification reaction [34] or the oxolation reaction (Eq. 6). The molar ratio of water/alkoxide (h) can control nucleation and growth of clusters in the sol. Wang et al. [35] prepared ultrafine titania particles (6nm) by carrying out hydrolysis of titanium isopropoxide with a high water/alcohol ratio (h>50). The pH is also an important parameter to prepare a sol. The peptization phenomenon with a mineral acid such as HNO_3, used to prepare a colloidal sol [27], is based on repulsive forces which prevent particle aggregation in the sol. When the pH value is decreased or increased away from the isoelectric point (IEP) of metal oxides, high repulsive forces exist in the sol stage. Near the IEP a flocculation phenomenon occurs due to absence of repulsive forces between particles, which results in precipitate formation.

Given its microporous structure and good permeability, sol-gel derived silica is one of the most studied candidates for H_2 separation [17, 36, 37]. The silica synthesis from TEOS under acid catalyzed conditions results in the formation of an amorphous structure after sintering at T>400°C, with pore sizes which can be smaller than 1nm. State-of-the-art microporous silica membranes often consist of a silica layer supported on a mesoporous γ-Al_2O_3 or α- Al_2O_3, or both, asymmetric membrane. The support provides mechanical strength to the selective silica layer, which is typically in the order of 100 nm thick. The H_2 permeance of these microporous silica membranes was reported to be in the range 10^{-8} to 10^{-7} $mol \cdot m^{-2} \cdot Pa^{-1} \cdot s^{-1}$ as shown in Table 1, with limited information on their H_2 selectivity over CO_2, though. However, the use of steam in the reforming processes has a serious impact on the resulting performance and longevity of the microporous silica membranes. The hydrophilic nature of the silica surface leads to the degradation of the silica microstructure in the presence of water vapour which involves the rupture and hydrolysis of siloxane bridges (Si-O-Si) at high temperature [38]. De Vos et al. [23] incorporated hydrophobic methyl groups which covalently bind to silica in the sol–gel process in order to obtain hydrophobic selective silica membranes, but only studied the sorption of water vapour of the membrane. Upon exposure to water in pervaporation experiments, this type of membrane did not perform stable at temperatures higher than 95°C [31]. More recently, the research group of Vente at the Energy Research Centre of The Netherlands (ECN) [39] successfully developed a microporous hybrid membrane based on silica HybSi®. They introduced organic bridges to the inorganic network structure of the membrane by using 1,2-bis(triethoxysilyl)ethane ($(EtO)_3Si–CH_2CH_2–Si(OEt)_3$). The role of ethyl links was suggested to provide additional shielding of the Si–O–Si bonds and reduce the number of hydrolysable groups per Si. After two years of testing at 150°C in pervaporation, the performance of the membrane remained unaltered [39].

The hydrothermal stability of amorphous silica membranes is said to be improved by doping the starting silica sol with a second metal oxide such as Al_2O_3, TiO_2, ZrO_2 , or NiO [25, 38]. For example, the alumina-doped silica membrane exhibited separation properties similar to the pure silica membranes and the hydrothermal stability was improved after doping with 3% alumina [38]. Upon exposure to 50 mol% steam/air at 600°C for 30h, the 3% alumina-doped silica had 64% reduction in the surface area and a loss of 86% micropore volume, compared to 85% and 94%, respectively, for the pure silica. It remains to be seen however if these are really doped materials or physical mixtures of the two compounds. Pure Al_2O_3, TiO_2, and ZrO_2 membranes [27-31] are considered more stable alternatives to replace silica membranes under hydrothermal conditions. However, the sol-gel chemistry of transition metal oxides is more complex than that of silica, because transition metals are highly reactive to form hydroxides and exhibit various coordination numbers. These characteristics lead to the difficult control of hydrolysis and condensation rates in the sol-gel process. Most of the reported porous membranes have pore sizes in the mesopore range [43-45]. The mesoporous membranes have lower selectivity for H_2 than the silica membranes because Knudsen diffusion is dominant. The ideal selectivity, $S_{i/j}$, of Knudsen diffusion is determined by the molecular mass M of the components i and j, which is given by:

$$S_{i/j}=\sqrt{\frac{M_j}{M_i}} \quad (9)$$

The Knudsen selectivity of H_2 over CO_2 is about 4.69 and, therefore, insufficient for high purity H_2 separation. A high H_2 selectivity is one of the key factors for membrane performance. The presence of pinholes in the sol-gel-derived membranes may also be responsible for the relatively lower selectivity. In general, mesoporous membranes are used to provide a homogenous layer with narrower pore size distribution and sufficient smoothness suitable for the deposition of a thin microporous top layer such as silica. In order to produce a structure of defect-free layers with continuously decreasing pore sizes, it is necessary to repeat the coating and sintering steps (Figure 1) several times. This large number of processing steps is a disadvantage of such asymmetric ceramic membranes, which result in a time consuming, and expensive process, and an increase in membrane thickness, leading to a decrease in the permeability of the membrane [44]. The permeability F ($mol\cdot m^{-1}s^{-1}Pa^{-1}$) of a gas through a membrane is usually defined as:

$$F=\frac{Jl\mu}{\Delta PA} \quad (10)$$

where J is the gas flow through the membrane ($mol\cdot s^{-1}$), l is the membrane thickness (m), μ is the pore tortuosity (dimensionless, number in between 2 and 13), ΔP is the pressure drop across the membrane (Pa), and A is the exposed area of the membrane (m^2). Due to the fact that membrane thickness and pore tortuosity can not be obtained unambiguously, the permeance P_m ($mol\cdot m^{-2}s^{-1}Pa^{-1}$) can often be used to evaluate the membrane, and is given by Eq. (11):

$$P_m=\frac{J}{\Delta PA} \quad (11)$$

Typically, the colloidal sol-gel route is used in the fabrication of mesoporous membranes which results in a wide pore size distribution, whereas uniform pore sizes with small width of the distribution are required for high performance (=selectivity and flux) application of the membrane reactor. Both α- and γ-Al_2O_3 are known for exhibiting a wide pore distribution. Even the γ-Al_2O_3 has a pore size distribution centered around 6 nm and tailing up to 20 nm (see section CHARACTERISATION OF SOL-GEL DERIVED MEMBRANES), clearly not an ideal starting point for pore size tailoring. The mesoporous silicate molecular sieves, known as MCMs, were synthesized using the sol-gel route of silica precursors in the presence of a surfactant assembly [46]. In the evaporation-induced self-assembly (EISA) process, the arrays of self-assembled surfactant molecules are used as structural directing templates for the pore structure, around which siliceous species are polymerized. The organic surfactant is removed at the end of the synthesis by calcination in order to create pores which have the shape and size of the surfactant core filling the silica framework. These MCMs have very narrow pore size distributions that can be tuned ranging from 2 to 100 nm and low tortuosity [47]. This provides a bottom-up method for creating new porous materials and controlling their pore size through varying surfactants with different molecular sizes. Several groups have reported alternative sol-gel synthesis methods based on the surfactant assembly mechanism for the preparation of ordered mesoporous materials based on different metal oxides with similar MCM-like symmetries. For example, SBA-15 silica [47] exhibits a two-dimensional hexagonal structure with a well-ordered hexagonal array and one-dimensional channel structure. The characteristics such as uniform pore size, narrow pore size distribution make MCM-like material a suitable one for other approaches such as pore size reduction via Atomic Layer Deposition (ALD). Figure 2 gives an illustration of two types of pore size distribution which can be obtained by sol-gel process with and without EISA.

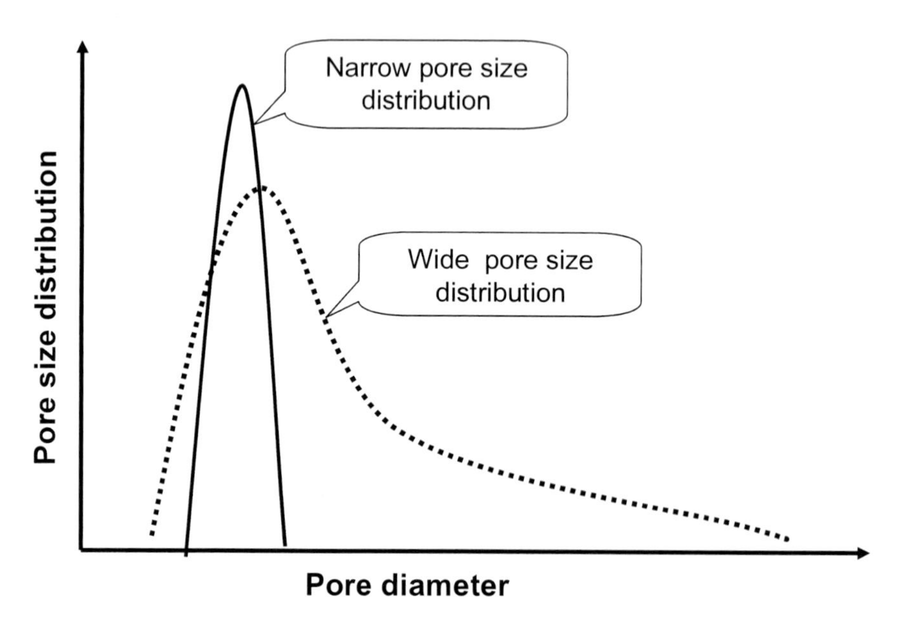

Figure 2. Schematic of pore size distribution as a function of pore size of two sol-gel derived materials, prepared with (solid line) and without the evaporation-induced self-assembly (EISA) process (dotted line).

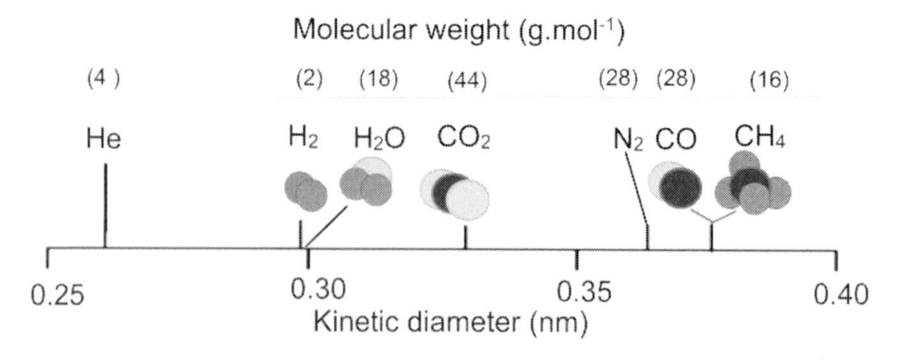

Figure 3. Selected properties of some small gases.

Modification of Sol-Gel Derived Membranes

It must be noted that in order to enhance the selectivity of H_2 membranes, the mentioned mesoporous membranes must be modified to reduce the mesopore size to close to the size of the kinetic diameter of H_2. As shown in Figure 3 if the pore size of membrane is between the molecular sizes of H_2 and CO_2, it can result in extremely high H_2 selectivity based on the molecular sieving mechanism. However, it is a challenge to develop such a membrane because the kinetic diameter of H_2 is very close to the other gases involved in the steam-methane reforming and water gas-shift processes.

According to literature data [48], the separation of H_2 from other small molecular size gases (CO_2, CO, H_2O) relies on the competition between molecular diffusivity and/or the molecular sieving effect when the pore size is in the range of 0.5 nm. A variety of surface modification techniques has been explored to reduce the pore size in order to improve the performance of membranes, including Chemical Vapour Deposition (CVD), and Atomic Layer Deposition (ALD). CVD can graft silica on mesoporous membranes such as γ-alumina [48]. The resulting silica-alumina composite membrane exhibits a high H_2 permeance in the order of 10^{-7} mol. m^{-2}·Pa^{-1}·s^{-1} with expected higher hydrothermal stability compared with the sol-gel derived silica membranes as mentioned earlier [37]. Similar to CVD, ALD is a promising strategy to the modification of membranes. In fact, ALD has been investigated to deposit thin metal-oxide films onto complex porous substrates with a highly conformal film coverage [49-52]. The use of ALD is based on the fact that the deposition of metal-oxide films consists of two reactions in which a metal precursor first attaches to available surface sites such as surface hydroxyls, followed by a separate reaction of the metal precursor with an oxygen source such as H_2O to form a monolayer film and regenerate the surface sites. These sequential saturating reactions are referred to an ALD cycle. The review paper by Puurunen [53] described most aspects about this process with ALD of trimethylaluminum ($Al(CH_3)_3$) and H_2O as a case study. The reaction conditions lead to a self-limiting growth mechanism which is ALD's most typical characteristic. In addition, since this process uniformly deposits the film onto any exposed surface that contains active sites, including internal walls of pores as well as steps and corners, it is highly conformal with atomic precision. Conformal coatings with monolayer precision of ALD have been used to deposit metal oxide layers of Al_2O_3, TiO_2, and ZrO_2 either on the internal pore surface or on the entrance of the pores in order to reduce the initial pore sizes of the membranes. Figure 4 depicts the overall scheme of this work in our laboratory. The mesoporous membrane is prepared by a sol-gel process combined by the evaporation-induced self-assembly (EISA) process. The EISA through cooperative self-assembly of inorganic species and organic surfactant is a synthesis approach to fabricate mesostructured materials with well-defined pore structures and uniform pore sizes. This multi-layer system with the tailored top layer serves as a starting point for pore size tuning. It can be further exposed to a metal halide precursor and an oxygen source in a typical type of ALD apparatus as developed in our laboratory [54] and leads to the deposition of metal oxide layers. By adjusting the metal oxide thickness, the initial mesopore size can significantly be reduced to the desired dimension (micropores) and, therefore, improve H_2 selectivity of the membranes.

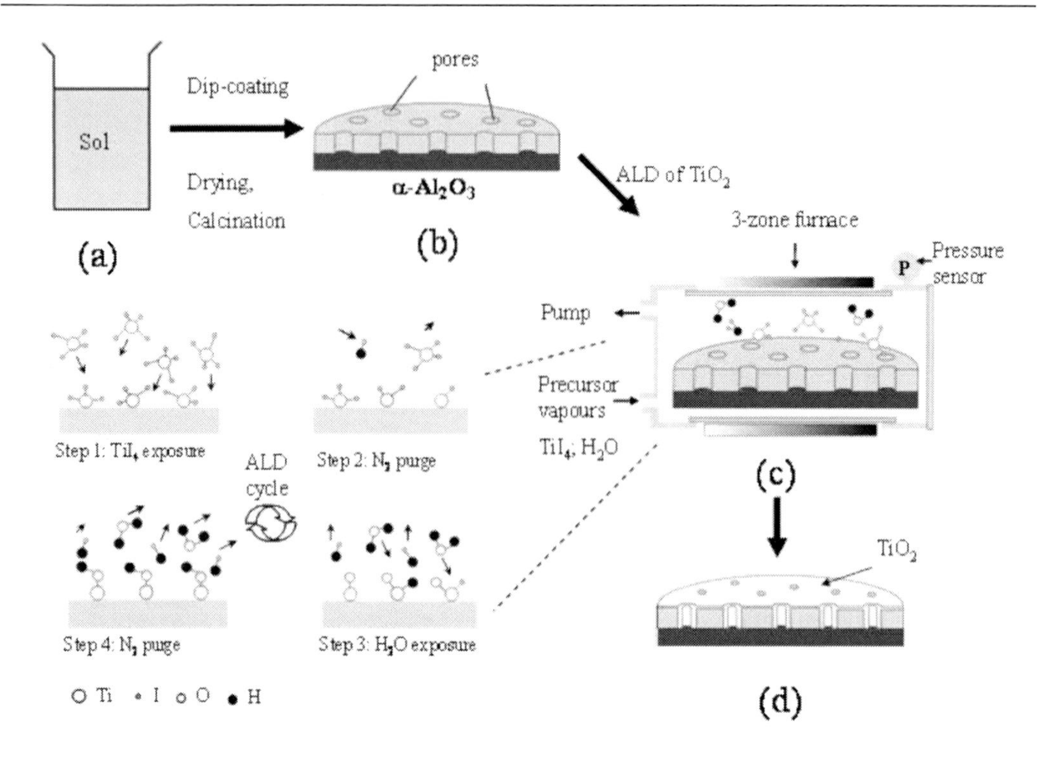

Figure 4. Schematic of the structural and chemical modification of ideal sol-gel derived membranes which includes (a) sol prepared with the evaporation-induced self-assembly (EISA) process, (b) sol-gel derived membrane, (c) atomic layer deposition (ALD) of TiO_2, (d) TiO_2 modified sol-gel membrane.

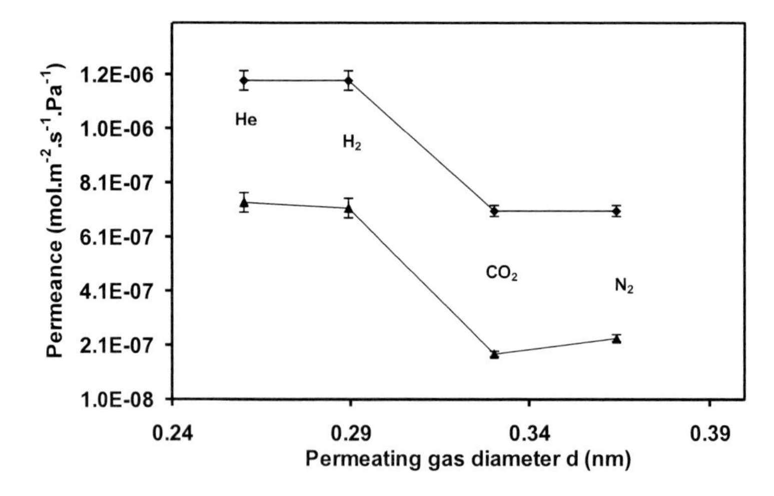

Figure 5. Relationship between permeance and the gas molecule's kinetic diameter d from single gas permeation experiments at 25 °C and a pressure drop of 10^5 Pa of bare alumina (diamond symbols) and TiO_2-deposited alumina (triangle symbols) membranes.

Figure 5 shows the permeance of single component gases (He, H_2, CO_2, and N_2) through the mesoporous alumina and modified alumina membranes with ALD of TiO_2 by using TiI_4 and H_2O as precursors. The alumina membrane used for this experiment is 10 nm in pore diameter, prepared by the EISA process with tri-block copolymer Pluronic© P123 ($EO_{20}PO_{70}EO_{20}$, EO=ethylene oxide, PO=propylene oxide) as template agent and aluminium tri-secbutoxide as aluminium source, which is further coated with 15 ALD cycles of TiO_2 deposition (~ 3 nm)[55]. The gas permeances decrease with increasing the molecular size of the permeating gas, which are quite different as well in both cases. The modified alumina membrane exhibits a permeance one order of magnitude smaller than that of the unmodified membrane. This result is a strong indication that the pore structure of the membrane is modified by the ALD process. The permeance of H_2 is approximately two times higher than the value of CO_2 for the mesoporous alumina. In the case of TiO_2-deposited alumina membrane, the permeance of H_2 is approximately four times higher than the value of CO_2. These results clearly show that with ALD it is possible to reduce large pore diameters of alumina and proves that the ALD technique can improve the H_2 selectivity. Knudsen diffusion takes place for micropore transport of H_2 and CO_2 in this membrane.

CHARACTERISATION OF SOL-GEL DERIVED MEMBRANES

The information of the pore size distribution and of the pore volume of a sol-gel derived membrane is particularly interesting for the fundamental membrane properties, which allow for predicting membrane performance such as the selectivity, the flux, or the permeability. A variety of techniques is available with both advantages and disadvantages, and certain techniques depend strongly on the membrane material, morphology, and other specific properties. Characterization techniques can be different for "unsupported" or "supported" thin membrane films, depending on the state in which the membrane samples are prepared. This distinction is meaningful, because the membrane preparation can either involve the sol-gel layer without or with the meso- or macroporous support. A schematic overview of characterization techniques for sol-gel derived membranes is given in Table 2.

The conventional physisorption technique [56], employing gases like argon (Ar), nitrogen (N_2), or krypton (Kr) is predominantly used to analyze surface areas and porosities of unsupported membrane films. The characteristics of physisorption isotherms can be grouped into six types, among which type I (for micropore) and type IV (for mesopore) isotherms according to the IUPAC classification are of interest for this review. The specific surface area can be deduced from the isotherm data of the relative pressure between 0.05 and 0.30 using the Brunauer, Emmett, and Teller (BET) equation [57]. Figure 6 shows the N_2 physisorption isotherm of an unsupported silica membrane prepared in our laboratory with TEOS ($Si(OCH_2CH_3)_4$)with a typical sol molar composition of 1(TEOS)-6.4(H_2O) –0.085(HNO_3) – 3.8 (C_2H_5OH) (a). The isotherm of pure SiO_2 is type I without hysteresis, due to the absence of liquid phase formation as a result of capillary condensation but by micropore filling at low relative pressure. Figure 6 shows the N_2 physisorption isotherm of an unsupported γ-Al_2O_3 membrane prepared with a typical sol molar composition of 1 (Aluminium tri-secbutoxide)-55.6 (H_2O) –0.07 (HNO_3) (c). Most of the γ-Al_2O_3 in literature is synthesized using Yoldas's recipe [27]. The N_2 isotherm is type IV with the typical hysteresis between the adsorption and desorption branches due to the capillary condensation of N_2 inside the mesopores. For the alumina/silica hybrid membrane with 36 mole % of SiO_2 (Figure 6b), the isotherm is similar to that of γ-Al_2O_3, but the volume of adsorbed N_2 is almost doubled at low relative pressures compared to that of pure γ-Al_2O_3. This result shows fingerprints of both the meso- and microporous nature of its constituents [58].

Table 2. Overview of some characterisations for sol-gel derived membranes.

Characterization method	Determined membrane property	Membrane state during characterization
N_2 physisorption	Surface area, pore size distribution	Unsupported
Permporometry	Gas flux, pore size , crack, defects	Supported
Gas permeation	Selectivity, separation	Supported
Low- and wide-angle X-ray diffractions	Composition, structure, particle size, pore size	Unsupported
SEM, TEM, AFM	Morphology	Unsupported Supported
Spectroscopic Ellipsometry (SE)	pore size distribution, thickness	Supported

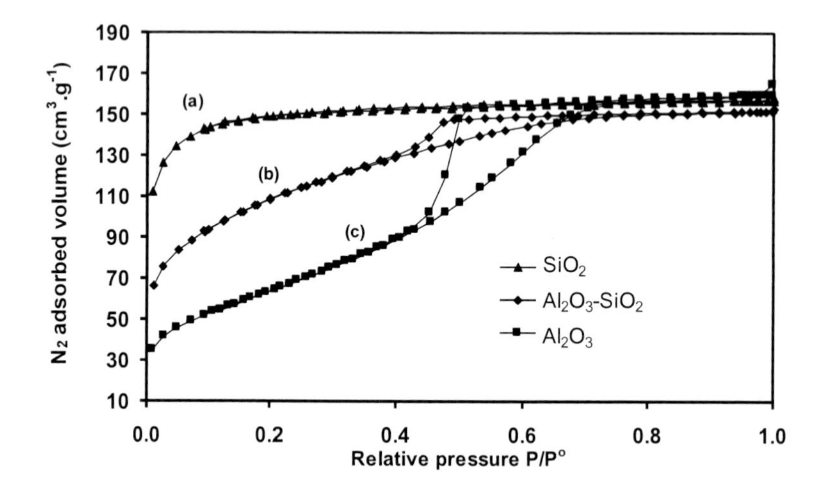

Figure 6. N_2 physisorption isotherms at 77K for SiO_2 (triangle symbols), alumina/silica hybrid (diamond symbols), and Al_2O_3 (rectangle symbols) calcined at 600 °C.

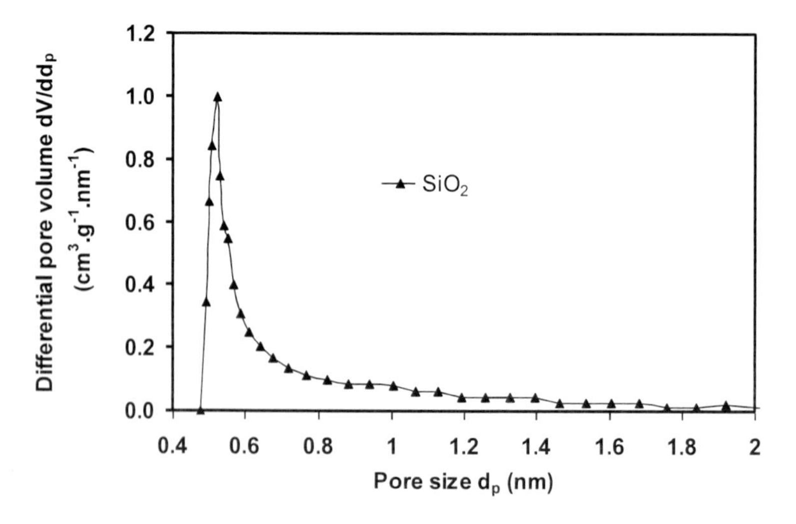

Figure 7. Pore size distribution at 77K of unsupported SiO_2 calcined at 600 °C calculated using the Horvath-Kawazoe method.

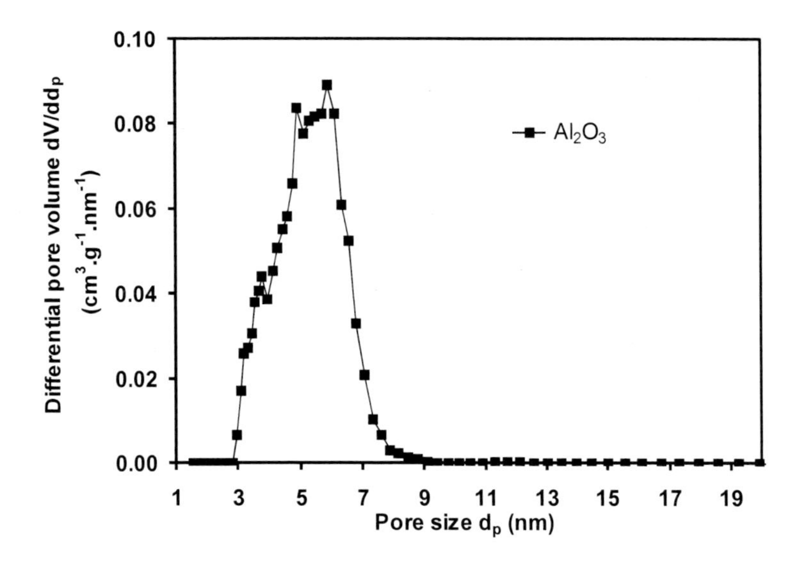

Figure 8. Pore size distribution at 77K of unsupported Al_2O_3 calcined at 600 °C calculated using the non-local density functional theory method.

The corresponding pore size distributions of SiO_2 and Al_2O_3 are shown in Figures 7 and 8, respectively. The average pore size of the SiO_2 is in the microporous regime while the pore size of the γ-Al_2O_3 centers around 6 nm in the mesoporous regime.

In order to determine the pore size distribution of a mesoporous layer on a macroporous support, which really contributes to the permeability of a membrane, permporometry is an effective technique [59]. It is supposed that unsupported sol-gel materials have a similar structure as a processed sol-gel layer on the support, even while in reality the forces developing during the drying process of both materials are different, *i.e.*, not only evaporation and setting but evaporation, infiltration, and setting. The N_2 physisorption technique is commonly used for the measurement of the pore size of unsupported membranes but cannot distinguish dead-end pores from active pores. In addition, it cannot differentiate between the thin separative (sol-gel) layer and the macroporous support. Hence, permporometry is a more adequate technique for studying the whole assembly, because active pore sizes in the separating layer on a supported membrane can be detected in-situ. In short, a condensable vapour (H_2O) and a non-condensable gas (He) are passed through the membrane. During the measurement the pores are stepwise blocked by the condensation of H_2O vapour at increasing partial water vapour pressure. The pore size in which capillary condensation of H_2O can still occur is expressed as a function of the relative H_2O pressure by the Kelvin equation [59]:

$$\ln\left(\frac{P}{P_0}\right)=\frac{-2\gamma V_M}{rRT}\cos(\theta) \tag{12}$$

where P/P_o the relative vapour pressure of the condensable gas, γ the surface tension ($N{\cdot}m^{-1}$), V_M the molar volume ($m^3{\cdot}mol^{-1}$), R the gas constant ($J{\cdot}K^{-1}{\cdot}mol^{-1}$), T the absolute temperature (K), r the Kelvin radius of the pore (m), and θ is the contact angle of the liquid on the solid (rad).

For each relative pressure all pores with radius smaller than the threshold value r are blocked by the condensed H_2O vapour, the He permeation passes through the larger than r pores. At a relative pressure of 1 all pores of the membrane are filled and He transport through the membrane is not possible, provided no large defects and cracks are present. When the partial water vapour pressure is reduced, pores with larger size than r are emptied and become available for He diffusion. The Kelvin radius r of the pore does not correspond exactly to the actual pore radius. In order to obtain the real pore radius, it must be corrected by adding the thickness of one or more H_2O monolayers (t-layer) [59], which forms on the surface of the pores before capillary condensation occurs. As an example, the resulting He permeance – Kelvin radius r plot for a γ-Al_2O_3 membrane with increasing and decreasing partial pressure of H_2O vapour is shown in Figure 9. Considering the existence of a t-layer, the pore size might be larger than 2r (~ 6 nm), which is in good agreement with that obtained by the N_2 physisorption technique.

Gas permeation measurements can be used to determine the permeability of membranes (see Eq. (10)). This method is the main tool to evaluate the membrane performance. Comparison of the permeations of different gases (He, H_2, CO_2, N_2) can provide an estimate of the membrane pore size and the selectivity. For a binary gas mixture, i.e., H_2/CO_2 the separation factor α_{H_2/CO_2} is expressed as a function of their molar fractions as is given in the following equation:

$$\alpha_{H_2/CO_2} = \frac{y_{p,H_2}}{x_{r,H_2}} \frac{x_{r,CO_2}}{y_{p,CO_2}} \tag{13}$$

where y_{p,H_2}, y_{p,CO_2} are the molar fractions of H_2 ,CO_2 at the permeate side and x_{r,H_2}, x_{r,CO_2} are the molar fractions of H_2, CO_2 at the retentate side.

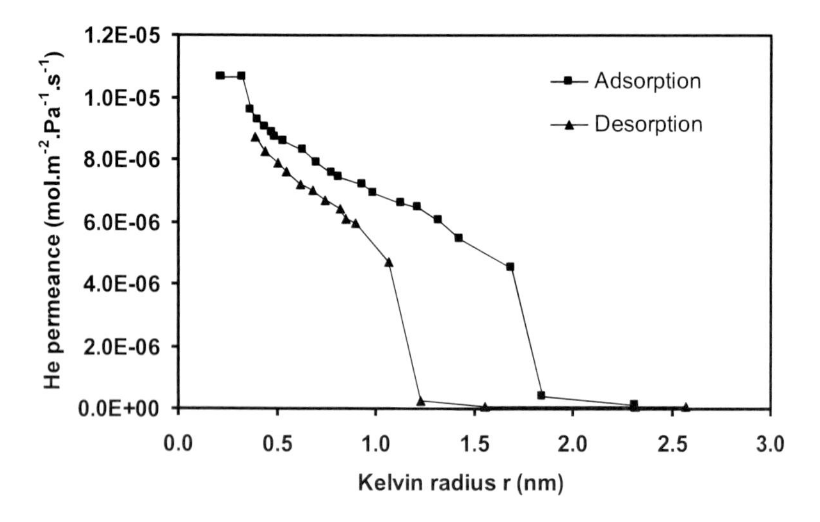

Figure 9. He permeance as a function of the Kelvin radius measured with H_2O used as a condensable gas in permporometry for a γ-Al_2O_3 membrane.

Low- and wide-angle X-ray diffraction (XRD) studies are performed to characterize the structural properties of the porous systems. The spatial resolution of low-angle XRD fits well with the dimensions of a porous system such as MCM. If there is an ordered mesoporous structure, the peaks in the low-angle range of 1-10° can be observed with the d spacing superior to 1.8 nm. Figure 10 represents the low-angle XRD diffraction pattern of mesoporous (templated) silica after calcination at 500°C and displays one peak at $2\theta=2.85°$, the (1 0 0) reflection or more precisely the (1 0) reflection since a 2D array of cylinders is concerned. This result indicates good long-range hexagonal ordering with a d-spacing of 3.15 nm.

Wide-angle XRD is routinely used to identify crystalline phases, quantitative phase analysis, and estimate of crystallite size of sol-gel derived materials. Ceramic membranes prepared by sol–gel processing can be categorized into amorphous materials such as silica, and crystalline materials such as alumina and titania. Figure 11 shows the XRD pattern of γ-Al_2O_3, which is calcined at 600°C in air. Depending on the calcination temperature, structural evolution in sol-gel derived materials gives other, metastable, phases of metal oxides, for example, in the case of alumina it can be summarized in the following reaction scheme:

$$\text{Boehmite}\ (\gamma\text{-AlOOH}) \xrightarrow{500°C} \gamma\text{-}Al_2O_3 \xrightarrow{800°C} \delta\text{-}Al_2O_3 \xrightarrow{1000°C} \theta\text{-}Al_2O_3 \xrightarrow{1200°C} \alpha\text{-}Al_2O_3 \quad (14)$$

Microscopy methods are widely used to investigate the morphology of membranes. Atomic Force Microscopy (AFM) [60] can give information about the membrane roughness, morphology, and even pore size, provided the probing tip is carefully chosen. Scanning Electron Microscopy (SEM) can only give qualitative information about the membrane morphology and an estimation of the "separative layer" thickness, because the resolution is not high enough to visualize the pores in the "separative layer" of a membrane. In addition, any asymmetry of the structure and defects (cracks, pinholes) can be observed. A ZrO_2 membrane is synthesized with an asymmetric structure as shown in Figure 12. The top ZrO_2 layer has a controlled pore size in the region of 10 nm which is responsible for the separation properties of the membrane, while the α-Al_2O_3 porous support (pore size > 100 nm) provides the mechanical strength required for the membrane to operate at high pressures.

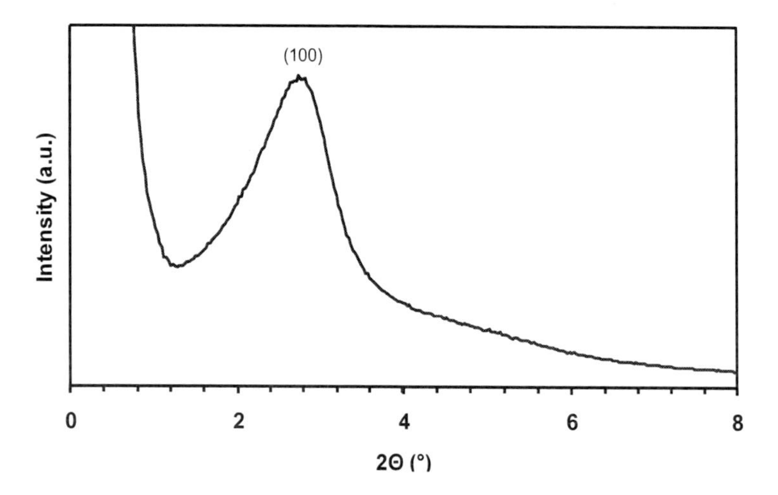

Figure 10. Low-angle XRD diffraction pattern for a templated silica sample calcined at 500°C in air.

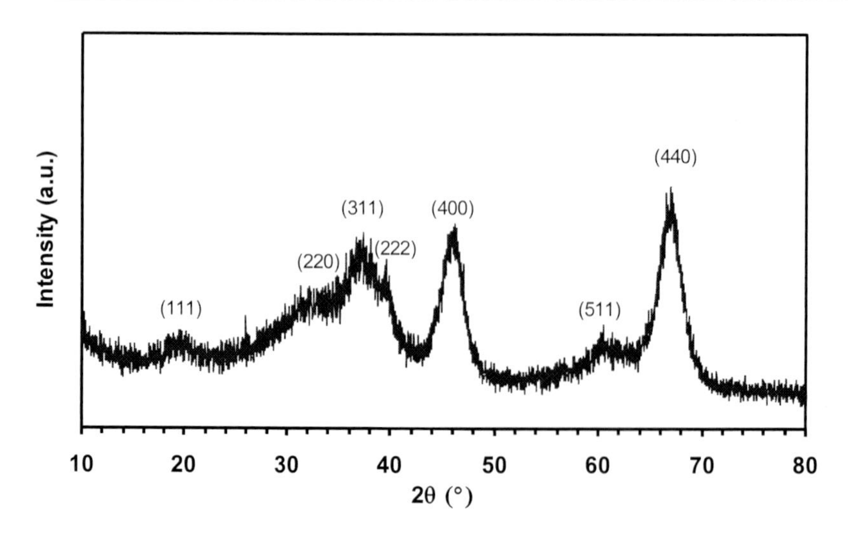

Figure 11. Wide-angle XRD pattern of γ-Al_2O_3 calcined at 600°C in air.

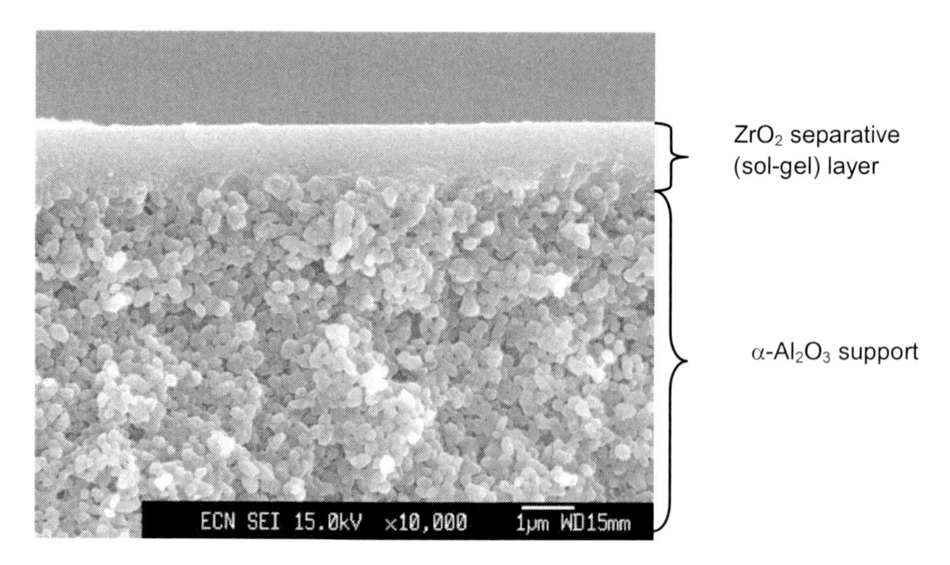

Figure 12. Scanning electronic micrograph of a cross section of a typical zirconia membrane prepared by a sol-gel synthesis method. The membrane consists of a top ZrO_2 layer (thickness of ~ 1 μm), followed by a support with larger, coarser particles of α-Al_2O_3.

For Transmission Electron Microscopy (TEM) a thin slice has to be cut from the membrane which is embedded in a non-conducting resin. TEM has a very high resolution (< 1 nm), but often small cracks can be seen in the samples, caused by the cutting procedure. Also the influence of the embedding resin on the membrane structure is an issue. Figure 13 shows a TEM micrograph of a cross section of a typical γ-Al_2O_3 membrane prepared as previously described. The membrane consists of a top γ-Al_2O_3 layer, followed by a support with larger, coarser particles of α-Al_2O_3.

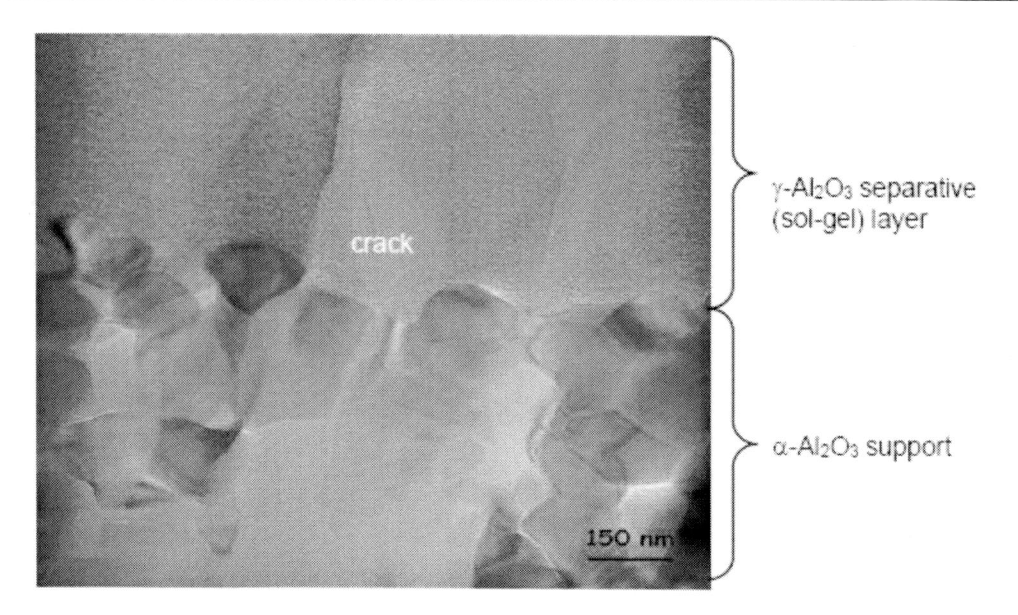

Figure 13. Transmission electronic micrograph of a cross section of a typical γ-Al_2O_3 membrane prepared by a sol-gel synthesis method. The membrane consists of a top γ-Al_2O_3 layer, on top of a support with larger, coarser particles of α-Al_2O_3.

An optical technique such as Spectroscopic Ellipsometry (SE) has also proven to be useful for membrane characterization. Ellipsometry measurements are routinely used to determine thickness and optical constants of dielectric and semiconductor films [61]. SE measures the change of polarization in light intensity reflected from or transmitted through a sample. The polarization change is described by an amplitude ratio, tan (Ψ), and phase difference, Δ, between light oriented in the p- and s- directions relative to the sample surface:

$$\rho=\tan(\psi)e^{i\Delta}=\frac{R_p}{R_s} \tag{15}$$

where R_p and R_s represent the complex Fresnel coefficients for p- and s- polarizations, respectively. From the obtained ellipsometric data Ψ and Δ, the film thickness and the optical constants, e.g., the refractive index (n) and extinction coefficient (k) can be deduced from the Fresnel equations. The application of SE for studying thin film synthesis reveals a fast and reliable technique, especially in-situ during ALD where the film thickness and the growth rate per ALD cycle can be monitored [62]. Combining ES with gas adsorption (ellipsometric porosimetry) can be used for measuring porosity and pore size distribution of thin porous membranes. The method is based on the ability of the physisorption of a solvent vapor into the porous network, in the similar way of N_2 physisorption. It consists in measuring the change of the optical characteristics of the porous film during solvent vapour adsorption and desorption in order to calculate porosity and pore size distribution of the film [63].

MEMBRANE APPLICATION IN THE WGS REACTOR

Since the relatively low H_2 selectivity in other sol-gel derived membranes than those based on silica, there have predominantly been studies on WGS and SMR with H_2-selective microporous silica [64-66]. Giessler et al. [64] studied hydrophobic and hydrophilic silica membranes for WGS. The hydrophobic membranes presented very high hydrogen permeation fluxes at 250°C, which reinforces the idea that higher flow rates and high conversion can be obtained with such membrane reactors as compared to a conventional reactor. At higher temperature (280°C) another silica membrane [65] showed 95% CO conversion which was achieved at 4 bar. Silica membranes show potential benefits in WGS, though steam presence in the process is a serious issue for longevity of a membrane as previously described.

OUTLOOK

This chapter focused on the recent progress in the research and development of sol-gel ceramic membranes for potential application in high-temperature H_2 separation. The advantages of the sol-gel coating are that the sol-coating methods are simple and do not require high investment costs. The sol-gel technique can be combined with other techniques such as CVD and ALD in order to develop hybrid membranes. Amorphous microporous silica membranes prepared by the sol-gel technique or modified with CVD are the most notable of inorganic membranes with the capability of separating H_2. They exhibit very high H_2 fluxes due to their thin thickness. However, the major problem is the hydrothermal instability of these membranes. They only show excellent separation performance for H_2 in dry applications. While in steam-reforming processes hydrothermal conditions are apparent, it is necessary to choose alternative materials with higher stability. From the point of view of hydrothermal stability, non-silica materials such as Al_2O_3, TiO_2, and ZrO_2 promise a high potential in H_2 separation for membrane application in steam-reforming processes. However, most of these are mesoporous membranes, but also here the selectivity of the membrane can be increased by tuning the pore size as in the silica case. Current membranes do not have high H_2 selectivity. Strategy towards the synthesis of membranes exhibiting high transport and selectivity for H_2 can be possible by combining sol-gel with templating molecules and ALD in the membrane fabrication which has proven functional in tailoring the pore size. Membrane materials with high H_2 selectivities have been studied on the laboratory scale, which can be scaled-up for the industrial applications. Moreover, more attention to defects (pinholes), high fluxes, and reasonable H_2 selectivity will be beneficial for the successful application of ceramic membrane reactors.

ACKNOWLEDGMENTS

The authors whish to thank Stanford University for supporting the research through the Global Climate and Energy Project (GCEP) and sponsors of GCEP. We gratefully acknowledge Prof. F. Kapteijn (TU Delft University) and his co-workers for providing access to equipments for membrane characterizations.

REFERENCES

[1] Salameh, MG. Quest for Middle East oil: the US versus the Asia-Pacific region. *Energy Policy* 2003, 31 (11), 1085-1091.

[2] Farrell, AE; Zerriffi, H; Dowlatabadi, H. Energy infrastructure and security. *Annual Review of Environment and Resources,* 2004, 29, 421-469.

[3] BP *Statistical Review of World Energy,* 2009.

[4] World Coal Institute *The coal resource: a comprehensive overview of coal* London, UK 2005.

[5] Yang, HQ; Xu, ZH; Fan, MH; Gupta, R; Slimane, RB; Bland, AE; Wright, I. Progress in carbon dioxide separation and capture: A review. *Journal of Environmental Sciences-China,* 2008, 20(1), 14-27.

[6] Ebner, AD; Ritter, JA. State-of-the-art Adsorption and Membrane Separation Processes for Carbon Dioxide Production from Carbon Dioxide Emitting Industries. *Separation Science and Technology,* 2009, 44 (6), 1273-1421.

[7] Davison, J; Thambimuthu, K. An overview of technologies and costs of carbon dioxide capture in power generation. *Proceedings of the Institution of Mechanical Engineers Part A-Journal of Power and Energy,* 2009, 223 (A3), 201-212.

[8] Goto, S; Tagawa, T; Assabumrungrat, S; Praserthdam, P. Simulation of membrane microreactor for fuel cell with methane feed. *Catalysis Today,* 2003, 82(1-4), 223-232.

[9] Sircar, S; Golden, TC. Purification of hydrogen by pressure swing adsorption. *Separation Science and Technology,* 2000, 35(5), 667-687.

[10] Huang, CP; Raissi, A. Analyses of one-step liquid hydrogen production from methane and landfill gas. *Journal of Power Sources,* 2007, 173 (2), 950-958.

[11] Ritter, JA; Ebner, AD. State-of-the-art adsorption and membrane separation processes for hydrogen production in the chemical and petrochemical industries. *Separation Science and Technology,* 2007, 42 (6), 1123-1193.

[12] Ockwig, NW; Nenoff, TM. Membranes for hydrogen separation. *Chemical Reviews,* 2007, 107 (10), 4078-4110.

[13] Paglieri, SN; Way, JD. Innovations in palladium membrane research. *Separation and Purification Methods,* 2002, 31 (1), 1-169.

[14] Ma, YH; Mardilovich, IP; Engwall, EE. Thin composite palladium and palladium/alloy membranes for hydrogen separation. *Advanced Membrane Technology,* 2003, 984, 346-360.

[15] Perry, JD; Nagai, K; Koros, WJ. Polymer membranes for hydrogen separations. *MRS Bulletin,* 2006, 31 (10), 745-749.

[16] Prabhu, AK; Oyama, ST. Highly hydrogen selective ceramic membranes: application to the transformation of greenhouse gases. *Journal of Membrane Science,* 2000, 176 (2), 233-248.

[17] Duke, MC; da Costa, JCD; Lu, GQ; Petch, M; Gray, P. Carbonised template molecular sieve silica membranes in fuel processing systems: permeation, hydrostability and regeneration. *Journal of Membrane Science,* 2004, 241(2), 325-333.

[18] Dong, J; Lin, YS; Kanezashi, M; Tang, Z. Microporous inorganic membranes for high temperature hydrogen purification. *Journal of Applied Physics,* 2008, 104 (12).

[19] McLeary, EE; Jansen, JC; Kapteijn, F. Zeolite based films, membranes and membrane reactors: Progress and prospects. *Microporous and Mesoporous Materials,* 2006, 90 (1-3), 198-220.

[20] Caro, J; Noack, M. Zeolite membranes-Recent developments and progress. *Microporous and Mesoporous Materials,* 2008, 115(3), 215-233.

[21] Barelli, L; Bidini, G; Gallorini, F; Servili, S. Hydrogen production through sorption-enhanced steam methane reforming and membrane technology: A review. *Energy,* 2008, 33 (4), 554-570.

[22] Brinker, CJ; Hurd, AJ; Schunk, PR; Frye, GC; Ashley, CS. Review of Sol-Gel Thin-Film Formation. *Journal of Non-Crystalline Solids,* 1992, 147, 424-436.

[23] de Vos, RM; Verweij, H. Improved performance of silica membranes for gas separation. *Journal of Membrane Science,* 1998, 143 (1-2), 37-51.

[24] Duke, MC; da Costa, JCD; Do, DD; Gray, PG; Lu, GQ. Hydrothermally robust molecular sieve silica for wet gas separation. *Advanced Functional Materials,* 2006, 16 (9), 1215-1220.

[25] Kanezashi, M; Asaeda, M. Hydrogen permeation characteristics and stability of Ni-doped silica membranes in steam at high temperature. *Journal of Membrane Science,* 2006, 271 (1-2), 86-93.

[26] Brinker, CJ; Scherer, GW. *Sol-Gel Science: The Physics and Chemistry of Sol-Gel Processing;* Academic Press ed; Boston, 1990.

[27] Yoldas, BE. Alumina Gels That Form Porous Transparent Al2O3. *J. Mater. Sci.,* 1975, 10 (11), 1856-1860.

[28] Leenaars, AFM; Keizer, K; Burggraaf, AJ. The Preparation and Characterization of Alumina Membranes with Ultra-Fine Pores. *J. Mater. Sci.,* 1984, 19 (4), 1077-1088.

[29] Guizard, CG; Julbe, AC; Ayral, A. Design of nanosized structures in sol-gel derived porous solids. Applications in catalyst and inorganic membrane preparation. *J. Mater. Chem.,* 1999, 9 (1), 55-65.

[30] Kikkinides, ES; Stoitsas, KA; Zaspalis, V. Correlation of structural and permeation properties in sol-gel-made nanoporous membranes. *J. Colloid Interface Sci.,* 2003, 259 (2), 322-330.

[31] Campaniello, J; Engelen, CWR; Haije, WG; Pex, PPAC; Vente, JF. Long-term pervaporation performance of microporous methylated silica membranes. *Chemical Communications,* 2004, (7), 834-835.

[32] Putnam, RL; Nakagawa, N; McGrath, KM; Yao, N; Aksay, IA; Gruner, SM; Navrotsky, A. Titanium Dioxide-Surfactant Mesophases and Ti-TMS1. *Chemistry of Materials,* 1997, 9 (12), 2690-2693.

[33] Kessler, VG. Geometrical molecular structure design concept in approach to homo- and heterometallic precursors of advanced materials in sol-gel technology. *Journal of Sol-Gel Science and Technology,* 2004, 32 (1-3), 11-17.

[34] Wu, JCS; Cheng, LC. An improved synthesis of ultrafiltration zirconia membranes via the sol-gel route using alkoxide precursor. *J. Membr. Sci.,* 2000, 167 (2), 253-261.

[35] Wang, CC; Ying, JY. Sol-gel synthesis and hydrothermal processing of anatase and rutile titania nanocrystals. *Chemistry of Materials,* 1999, 11 (11), 3113-3120.

[36] Kim, YS; Kusakabe, K; Morooka, S; Yang, SM. Preparation of microporous silica membranes for gas separation. *Korean Journal of Chemical Engineering,* 2001, 18 (1), 106-112.

[37] Kusakabe, K; Shibao, F; Zhao, GB; Sotowa, KI; Watanabe, K; Saito, T. Surface modification of silica membranes in a tubular-type module. *Journal of Membrane Science,* 2003, 215 (1-2), 321-326.

[38] Fotou, GP; Lin, YS; Pratsinis, SE. Hydrothermal Stability of Pure and Modified Microporous Silica Membranes. *Journal of Materials Science,* 1995, 30 (11), 2803-2808.

[39] Castricum, HL; Sah, A; Kreiter, R; Blank, DHA; Vente, JF; ten Elshof, JE. Hydrothermally stable molecular separation membranes from organically linked silica. *Journal of Materials Chemistry,* 2008, 18 (18), 2150-2158.

[40] Altinisik, O; Dogan, M; Dogu, G. Preparation and characterization of palladium-plated, porous glass for hydrogen enrichment. *Catalysis Today,* 2005, 105 (3-4), 641-646.

[41] Itoh, N; Akiha, T; Sato, T. Preparation of thin palladium composite membrane tube by a CVD technique and its hydrogen permselectivity. *Catalysis Today,* 2005, 104 (2-4), 231-237.

[42] Yan, SC; Maeda, H; Kusakabe, K; Morooka, S. Thin Palladium Membrane Formed in Support Pores by Metal-Organic Chemical-Vapor-Deposition Method and Application to Hydrogen Separation. *Industrial & Engineering Chemistry Research,* 1994, 33(3), 616-622.

[43] Kessler, VG; Seisenbaeva, GA; Werndrup, P; Parola, S; Spijksma, GI. Design of molecular structure and synthetic approaches to single-source precursors in the sol-gel technology. *Materials Science-Poland,* 2005, 23 (1), 69-78.

[44] Kreiter, R; Rietkerk, MDA; Bonekamp, BC; van Veen, HM; Kessler, VG; Vente, JF. Sol-gel routes for microporous zirconia and titania membranes. *Journal of Sol-Gel Science and Technology,* 2008, 48 (1-2), 203-211.

[45] Spijksma, GI. Modification of zirconium and hafnium alkoxides : the effect of molecular structure on derived materials. University of Twente, *The Netherlands*, 2006.

[46] Kresge, CT; Leonowicz, ME; Roth, WJ; Vartuli, JC; Beck, JS. Ordered Mesoporous Molecular-Sieves Synthesized by A Liquid-Crystal Template Mechanism. *Nature,* 1992, 359 (6397), 710-712.

[47] Zhao, DY; Feng, JL; Huo, QS; Melosh, N; Fredrickson, GH; Chmelka, BF; Stucky, GD. Triblock copolymer syntheses of mesoporous silica with periodic 50 to 300 angstrom pores. *Science,* 1998, 279 (5350), 548-552.

[48] Gu, YF; Hacarlioglu, P; Oyama, ST. Hydrothermally stable silica-alumina composite membranes for hydrogen separation. *Journal of Membrane Science,* 2008, 310(1-2), 28-37.

[49] Elam, JW; Routkevitch, D; Mardilovich, PP; George, SM. Conformal coating on ultrahigh-aspect-ratio nanopores of anodic alumina by atomic layer deposition. *Chemistry of Materials,* 2003, 15 (18), 3507-3517.

[50] Xiong, G; Elam, JW; Feng, H; Han, CY; Wang, HH; Iton, LE; Curtiss, LA; Pellin, MJ; Kung, M; Kung, H; Stair, PC. Effect of atomic layer deposition coatings on the surface structure of anodic aluminum oxide membranes. *Journal of Physical Chemistry B,* 2005, 109 (29), 14059-14063.

[51] Mccool, BA; DeSisto, WJ. Self-limited pore size reduction of mesoporous silica membranes via pyridine-catalyzed silicon dioxide ALD. *Chemical Vapor Deposition,* 2004, 10 (4), 190-194.

[52] Mahurin, S; Bao, LL; Yan, WF; Liang, CD; Dai, S. Atomic layer deposition of TiO2 on mesoporous silica. *Journal of Non-Crystalline Solids,* 2006, 352(30-31), 3280-3284.

[53] Puurunen, RL. Surface chemistry of atomic layer deposition: A case study for the trimethylaluminum/water process. *Journal of Applied Physics,* 2005, 97(12).

[54] Reijnen, L; Meester, B; Goossens, A; Schoonman, J. Atomic layer deposition of CuxS for solar energy conversion. *Chemical Vapor Deposition,* 2003, 9 (1), 15-20.

[55] Tran, THY; Haije, WG; Schoonman, J. Pore size control of ceramic membranes for hydrogen separation by atomic layer deposition *International Conference on Inorganic Membranes (ICIMs)*, Tokyo 2008.

[56] Sing, KSW; Everett, DH; Haul, RAW; Moscou, L; Pierotti, RA; Rouquerol, J; Siemieniewska, T. Reporting Physisorption Data for Gas Solid Systems with Special Reference to the Determination of Surface-Area and Porosity (Recommendations 1984). *Pure and Applied Chemistry,* 1985, 57 (4), 603-619.

[57] Brunauer, S; Emmett, PH; Teller, E. Adsorption of gases in multimolecular layers. *Journal of the American Chemical Society,* 1938, 60, 309-319.

[58] Tran, THY; Stoitsas, K; Haije, WG; Schoonman, J. *Advanced membrane reactors in energy systems: a carbon-free conversion of fossil fuels* Global Climate & Energy Project (GCEP) technical report 2008.

[59] Cuperus, FP; Bargeman, D; Smolders, CA. Permporometry-the Determination of the Size Distribution of Active Pores in Uf Membranes. *Journal of Membrane Science,* 1992, 71 (1-2), 57-67.

[60] Boussu, K; Van der Bruggen, B; Volodin, A; Snauwaert, J; Van Haesendonck, C; Vandecasteele, C. Roughness and hydrophobicity studies of nanofiltration membranes using different modes of AFM. *Journal of Colloid and Interface Science,* 2005, 286 (2), 632-638.

[61] Tompkins, HG; Irene, EA. *Handbook of Ellipsometry;* William Andrew Publishing: Norwich NY, 2005.

[62] Langereis, E; Heil, SBS; Knoops, HCM; Keuning, W; van de Sanden, MCM; Kessels, WMM. In situ spectroscopic ellipsometry as a versatile tool for studying atomic layer deposition. *Journal of Physics D-Applied Physics,* 2009, 42(7).

[63] Baklanov, MR; Mogilnikov, KP; Polovinkin, VG; Dultsev, FN. Determination of pore size distribution in thin films by ellipsometric porosimetry. *Journal of Vacuum Science & Technology B,* 2000, 18 (3), 1385-1391.

[64] Giessler, S; Jordan, L; da Costa, JCD; Lu, GQ. Performance of hydrophobic and hydrophilic silica membrane reactors for the water gas shift reaction. *Separation and Purification Technology,* 2003, 32 (1-3), 255-264.

[65] Brunetti, A; Barbieri, G; Drioli, E; Lee, KH; Sea, B; Lee, DW. WGS reaction in a membrane reactor using a porous stainless steel supported silica membrane. *Chemical Engineering and Processing,* 2007, 46(2), 119-126.

[66] Moon, JH; Bae, JH; Bae, YS; Chung, JT; Lee, CH. Hydrogen separation from reforming gas using organic templating silica/alumina composite membrane. *Journal of Membrane Science,* 2008, 318(1-2), 45-55.

In: The Sol-Gel Process
Editor: Rachel E. Morris
ISBN 978-1-61761-321-0

Chapter 12

ISOCYANATE MODIFIED SILANES AS A NEW GENERATION PRECURSORS IN THE SOL-GEL TECHNOLOGY: FROM MATERIALS DESIGN TO APPLICATION

Ts. Gerganova*[1]*, Y. Ivanova*[1]*, T. Gerganov*[1]*,
***I. M. Miranda Salvado*[2] *and M. H. V. Fernandes*[2]**
[1]University of Chemical Technology and Metallurgy, Department of Silicates, 8 Kliment Ohridski blvd., 1756 Sofia, Bulgaria
[2]Ceramic and Glass Engineering Department CICECO, Aveiro University, 3810-193 Aveiro, Portugal

ABSTRACT

This work is devoted to the development of a new generation of sol-gel hybrid materials, their driving forces on the material science and technology and especially on their impact on possible advanced applications. The motivation of the carried work was to point out that the choice of the appropriate precursors can facilitate the connection between application and science, the future of the sol-gel technology. In this context, our group decided to combine the properties of new generation polymeric products commercialized in the last few years: the isocyanate modified silanes with traditional sol-gel products such as tetramethoxysilane in order to synthesize several attractive inorganic/organic networks. Among the group of isocyanate modified silanes, a trimethylsilyl isocyanate (TMSI) has not been used before in the sol-gel technology. The obtained nanostructured gel siliceous backbone with urea (-NHC(-O)NH-) or urethane (-NH(C-O)O-) linkages could be classified as ureasiles and urethanesiles, respectively. This kind of hybrid products has application in the domains of optics, magnetism and electrochemistry. Here we show that they possess unique behavior when modified by Ti, Zr, B and Al. The coordination ability of the designed hybrid Si-O-C-N host matrix extends its well-known applications to the optic field and makes urea and urethanesile sol-gel materials applicable in the area of thin films, adsorbents and hybrid membranes.

When the obtained nanostructured gel products were subjected to heat treatment from 60 up to 1100°C in an inert atmosphere, the structural changes such as glass formation and crystal nucleation proved that they can be suitable for high temperature use. These features were established with the help of different traditional and advanced methods such as XRD, FTIR, ^{29}Si, ^{27}Al and ^{11}B MAS NMR, SEM, AFM, TG and BET. In general, the obtained results provide conclusive evidence that the nature of the organic group incorporated in the siloxane structure plays a major role in the understanding of the relationships between design of new sol-gel materials and their application. This study is very interesting since it refers to the application of sol-gel process as synthesis route to advanced materials development and engineering.

Introduction

The advance in the field of high performance materials with structural and functional applications depend on the materials processing, which allows the control of structures at atomic, molecular and nanoscale level. Sol–gel processing for such materials provides tremendous opportunities for the development of novel high technology products [1]. This process that is mainly based on inorganic polymerization reactions is a chemical synthesis method initially used for the preparation of inorganic materials such as glasses and ceramics. Its unique low temperature processing characteristic also provides unique opportunities to make pure and well-controlled organic/inorganic compositions named hybrid materials through the incorporation of low molecular weight and oligomeric/polymeric organic molecules with appropriate inorganic moieties [2]. The organic/inorganic hybrid materials made in this way, have been termed normal nanocomposites (if the organic and inorganic components are connected to each other by weak chemical bonds) or nanostructures (if the organic and inorganic components are connected to each other by covalent chemical bond) [2-4]. The final hybrid materials have the potential for providing interesting combinations of properties which cannot be achieved by other materials. The preparation, characterization, and applications of organic/inorganic hybrid materials have become a fast expanding area of research in materials science. The major driving forces behind the intense activities in this area are the new and different organometallic compounds (based on alkoxides of silicon, titanium, aluminum, zirconium etc.) that appeared in the past few decades. This class of potential precursors is very suitable for the sol-gel technology and is characterized by the presence of –M-O-R (metal-organic bonds) in a metal oxo-network. The formation of Si-O-Si bonds by hydrolysis/polycondensation of a Si-OR group is the key chemical step [3]. Starting from a homogeneous solution of the precursor in a solvent, the formation of the Si-O-Si network leads to the formation of growing species - oligomers, polymers, crosslinked chains, and colloids, until a gel finally forms. Moreover, the mixture of various precursors makes the number of different organic-inorganic materials that can be designed and prepared by sol-gel technology practically unlimited [4]. The selection of molecular or polymeric species of quite different natures and properties that could also be incorporated into these organic-inorganic systems offers also a great number of alternatives.

Finally, the use of precursors containing reactive groups such as alcohol, mercapto, amino, carboxyl, carbonyl phenyl or unsaturated functions, presents the possibility of developing further reactions of the resulting polysiloxane networks with appropriate molecular species [5-10]. This opportunity additionally extends the application area of the sol-gel products especially in the catalysis and membrane technologies [11-18].

Due to the tremendous number of sol-gel methods and to the more precise knowledge of the physical properties of molecules, the chemistry is now an unavoidable partner in the development of innovative materials with advanced applications.

One interesting feature of the sol-gel processes carried out is the preparation of metastable phases such as silicon oxycarbide or oxynitride amorphous, nanostructured and nanocomposite materials, glasses and ceramics [19]. The replacement of a part of the divalent oxygen atoms of silica by tetravalent carbon atoms and trivalent nitrogen atoms leads to an improvement of the thermo mechanical properties of the technologically important ceramics [20-22]. Despite the long history of the sol–gel process, as a clean and environmentally friendly process, the transformation of the carbon and nitrogen containing metalloxanes to oxide ceramics is increasingly attractive for investigation because it features a new route to control the ceramic microstructure [23]. The first practical transformation of organo-modified silica compounds to ceramic material was developed by Verbeek in the early 1970's [24]. Until then, significant improvements were made in the development of novel pre-ceramic precursors in order to control the microstructure and processing behavior [25]. In recent years, many examples of polysilanes [26], polycarbosilanes [27], polyorganosilazanes [28], and polysiloxanes [29] as preceramic polymer precursors have been reported. Besides the great variety of organically modified siloxanes, there are additional metallic Ti, Al, Zr etc. alkoxides that can give quaternary Si-(M)-C-N or five-component Si-(M)-O-C-N systems [30]. The additional incorporation of nitrogen atoms in the above mentioned systems can usually be made by ammonolysis, amminolysis, hydrazinolysis and silazanolysis [31]. The main disadvantages of these precursor routes are the difficult separation of the main reaction products from the solid byproducts such as NH_4Cl, H_3NRCl, which additionally leads to nitrogen losses. Moreover, the nitridation reactions very often lead to formation of volatile compounds and additionally decrease the ceramic yields. In order to avoid this problem the nitridation process usually could occur at low temperatures [31].

As an alternative approach, olygomeric silazanes can also be cross-linked into non-volatile polysilazanes when urea or urethane units are present in the starting precursors [32]. However, the hydrolysis and condensation reactions of urea or urethane precursors during the sol-gel process are possible only under heating up to 90-100°C [33].

Nowadays, the new products that were commercialized in the last few years as isocyanate-modified silanes and siloxanes appear as promising precursors that can be used in the sol-gel synthesis at room temperature. In combination with the traditional cross-linking agents such as tetraethyl- or tetramethylortosilicate, the urea or urethane groups can be made during hydrolysis and condensation reactions [34-36].

Among the group of isocyanate-modified silanes and siloxanes, a trimethylsilyl isocyanate (TMSI) has not been used in the sol-gel technology before. In this context, we wish to show that TMSI is much more attractive than the conventionally used polysilazanes for sol-gel synthesis of nitrogen-containing organosilicas owing to its low cost, acceptable synthetic temperature and chemical properties close to the most useful cross-linking agent (TMOS and TEOS) in the sol-gel route. During the hydrolysis and condensation reactions, when trimethylsilyl isocyanate takes a part in the sol-gel synthesis, two possible networks can be prepared – urethanesil or ureasil, depending on the used synthesis conditions. The obtained hybrid products found application in the domains of optics, magnetism and electrochemistry.

Here we report, that the synthesized di-urethanesiles possess unique behavior and can be much more atractive in the area of thin film coatings, adsorbents and hybrid membranes when they have been additionaly modified by Ti, Zr, B and Al.

Experimental Part

Five series of urethane-functionalized xerogels were prepared using two organosilane precursors [tetramethoxysilane (TMOS) (Fluka), and trimethylsilyl isocyanate (TMSI) (Fluka)] and four kinds of metal-organic alkoxydes [titanium tetraisopropoxyde (TTP); zirconium tetraisopropoxide (ZTP); aluminum sec-butoxide (ASB); and trimethylborate (TMB)] (Aldrich)] that have been used as received.

Sol-Gel Synthesis of Urethane-Functionalized Hybrid Materials in the TMOS-TMSI System

In order to investigate the optimal TMOS/TMSI ratio that will provide an homogeneous hybrid material, three gels (Table 1) were prepared in the proportions 50/50; 70/30 and 80/20 as a first step of the experiment. They have been prepared as follows: 50, 70 or 80 wt. % tetramethoxysilane (TMOS) were dissolved in tetrahidrofurane (THF in ratio 1:3) and hydrolyzed with acidified water (H_2O: silane 2:1 and pH=1.5) for 10 min. Then the proper amount (50, 30 or 20 wt. %) of trimethylsilyl isocyanate (TMSI) firstly dissolved in THF (1:1) was added. After 30 min stirring, the obtained sols were left for gelation at room temperature for 12h.

The obtained hybrid gels from the non-modified TMOS/TMSI system with different contents of TMSI were investigated by means of XRD. The results are presented on Figure 1 and clearly indicate that the most homogeneous and amorphous gel was produced when 20 wt. % of TMSI was used during the sol-gel synthesis. Therefore, the TMOS/TMSI ratio that will be used in further studies is the 80/20 ratio.

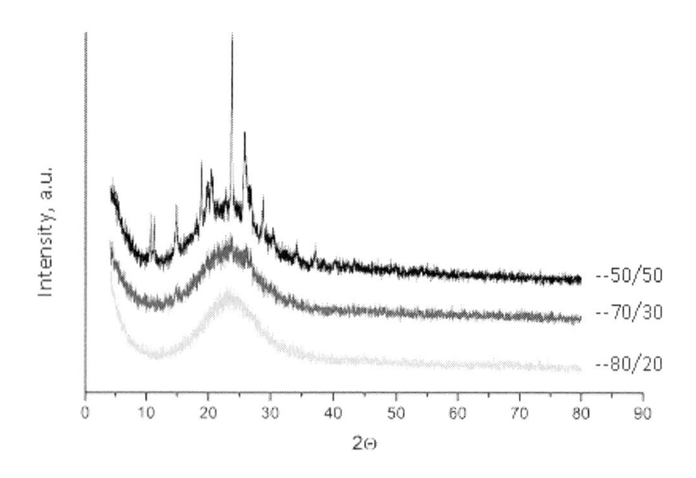

Figure 1. XRD results of gels prepared with different TMSI content.

Table 1. Composition of the prepared gels from different TMOS/TMSI ratios.

label	TMOS	TMSI	TMOS/THF	TMSI/THF	pH	aspect
50/50	50	50	1:3	1:1	1, 5	white powder
70/30	70	30	1:3	1:1	1, 5	white flake
80/20	80	20	1:3	1:1	1, 5	transparent gel

Sol-Gel Synthesis of Urethane-Functionalized Hybrid Materials in the TMOS-TMSI-Mt(Or)$_X$ (Mt=Ti, Zr, B, Al) System

Ti, Zr, B and Al-modified urethanesil hybrid xerogels from the TMOS-TMSI-M(OR)x system with 70/20/10; 60/20/20 and 50/20/30 molar ratio (M10, M20 and M30, M=Ti, Zr, B, Al), have been successfully prepared as follows: 80 wt. % TMOS was partially hydrolyzed using a H_2O: silane ratio 2:1 at room temperature for 10 minutes under acidic conditions (pH = 2) with tetrahydrofurane (THF) as solvent. The 20 wt. % TMSI which has been first dissolved in THF was added for further co-hydrolysis. After stirring for more 30 minutes at room temperature, the prepared sols were modified by a mixture of the required amount (10, 20 and 30 wt. % at TMOS instead) of titanium tetraisopropoxyde (TTP); zirconium tetraisopropoxide (ZTP); aluminum sec-butoxide (ASB); and trimethylborate (TMB) for further hydrolysis and co-condensation. In this way, the number of Si–C and Si–N bonds per Si is kept constant, while the number of Si–O–M (M=Ti, Zr, B, Al) bonds varies. The used Ti, Zr, Al and B alkoxydes were first dissolved in a specific complexing agent (ethylacetoacetat - EtAcAc). The influence of this complex-forming agent from a structural point of view is become apparent on the control of the strict effect of the used alkoxides during the condensation process and contributes to their homogeneous distribution in the final organic–inorganic network.

After one hour of stirring, the sols with composition TMOS-TMSI-M(OR)$_x$ were cast and gelled for12h at room temperature under air atmosphere.

Heat Treatment of the Ti, Zr, Al and B Modified Urethanesil Hybrid Materials

All synthesized xerogels were placed in an electric furnace under nitrogen atmosphere and pyrolyzed up to 1100 ^oC. The temperature was raised at a heating rate of 10 ^oC/min and then held for 2 hours followed by cooling to room temperature.

Tools for Materials Characterization

All materials were studied by XRD (Rigaku /New X- Ray Diffractometer System "Geigetflex" D/Max- C Series with CuKα radiation in the range from 10 to 80 2Θ at scan rate of 3-20V/min), FTIR (Mattson 7000 by the KBr disk methods over the wave number range 4000–400 cm^{-1}), ^{29}Si CP MAS NMR (^{29}Si solid state NMR spectra were recorded at 79.79

MHz respectively on a (9.4 T) Bruker Avance 400 spectrometer, measured with 40° pulses); ^{27}Al NMR (Bruker Avance 500 spectrometer); ^{11}B NMR (Bruker Avance 500 spectrometer, measured at 128.4 MHz respectively, on a (9.4 T)) SEM (Hitachi S-4100), AFM (Atomic Force Microscope Multimode, Nanoscope IIIA, Digital Instrument) and BET (*BET-Analysis-Geminy 2370 V5.00*)

Results and Discussion

1. X-Ray Diffraction

XRD patterns of the xerogels with different modifying ions are presented on Figure 2. The obtained hybrid products are amorphous independently of the kind and concentration of the modifying ions (Figure 2a). Some differences have been observed in the XRD pattern of the sample prepared with 30 wt. % TMB. A reflection was detected at $2\Theta = 27^{o}$ and it was attributed to the crystallization of $B(OH)_3$ (Figure 2a). The behavior of boron-containing gels completely changes with the heat treatment. $B(OH)_3$ crystallites disappear and at 200°C the gel is amorphous.

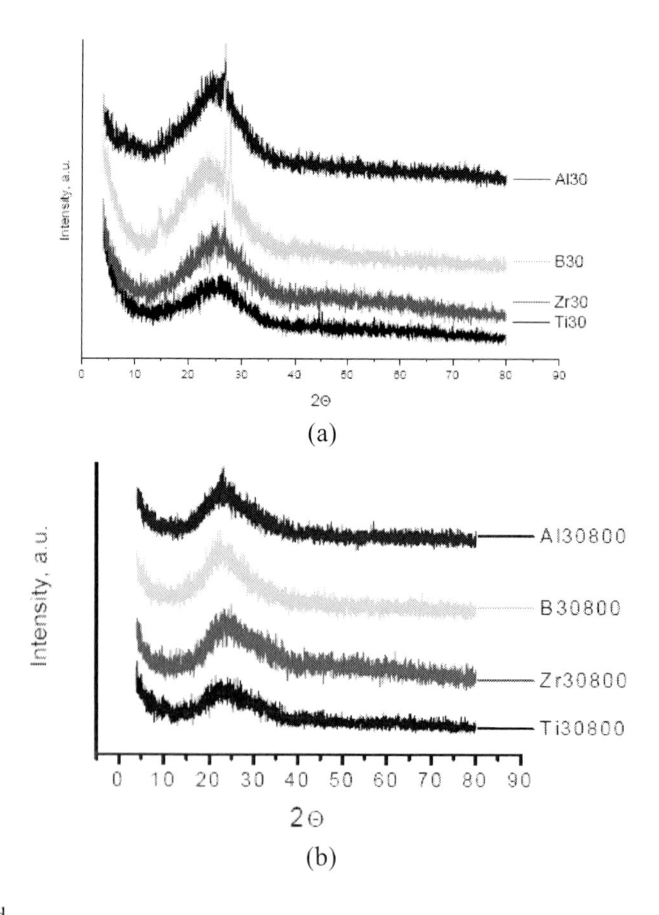

Figure 2. Continued.

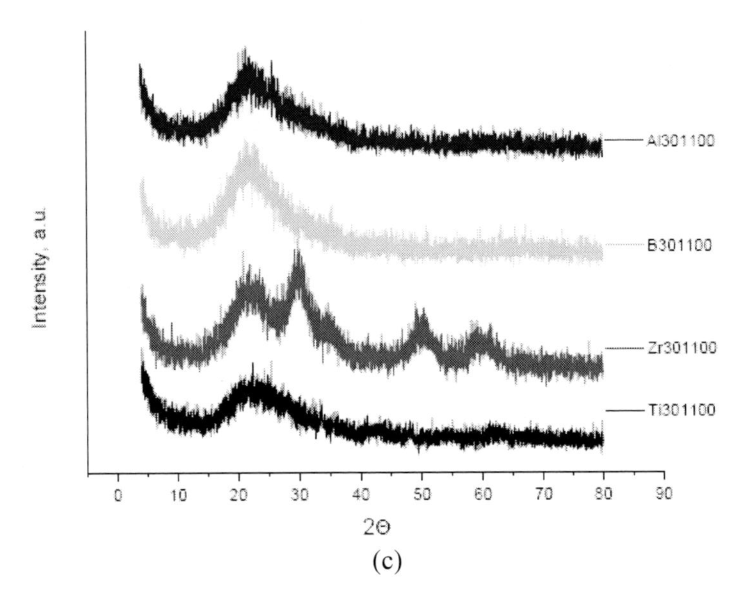

(c)

Figure 2. XRD of the samples modified by 30wt. % Ti, Zr, B and Al alkoxyde at a) xerogel state; b) after gel pyrolysis at 800°C and c) after gel pyrolysis at 1100°C.

From the obtained XRD results of the materials after pyrolysis up to 1100 °C (Figure 2b and c), it can be seen that the compositions modified by Al, Ti and B retain their amorphous structure but in the xerogels modified by more than 10 wt.% Zr the process of crystallization of ZrO_2 (tetragonal) is favored. This statement finds confirmation in the appearance of peaks at $2\Theta=30^{\circ}$; $2\Theta=50^{\circ}$ and $2\Theta=60^{\circ}$ in the XRD diffractograms.

The observed broad peaks indicate the presence of nanosized crystallites (<20nm) The average crystallite size, calculated from the width of the diffraction peaks using the Scherer equation [37, 38] was about 10 to 18 nm (Figure 2c). By means of proper software (Origin 5 PeackFit™), the degree of cristallinity was determined from the obtained XRD patterns as a ratio between the sum of integrated intensity of all crystal peaks and total integrated intensity in spectra as in Ref.[39].The calculations for the pyrolyzed zirconosiloxanes showed that the cristallinity degree is around 18%.

2. Scanning Electron and Atomic Force Microscopy

The obtained XRD results are in accordance to the SEM observations (Figure 3) and these two analyses confirm that the prepared hybrid urethanesiles are amorphous materials (Figure 3 a and b) with exception of boron-modified gels (Figure 3c). On Figure 3a and Figure 3b, it can be observed that the surface of the titanium and zirconium- modified samples show no crystals, cracks and pores. In the SEM pictures obtained for the xerogels modified with Al (Figure 3d), the low content of the modifying ion leads to the highest degree of open porosity with a characteristic pore size of 240 nm.

The cross-section surface of this sample was investigated by AFM in order to assess the interconnectivity of pores (Figure 4).

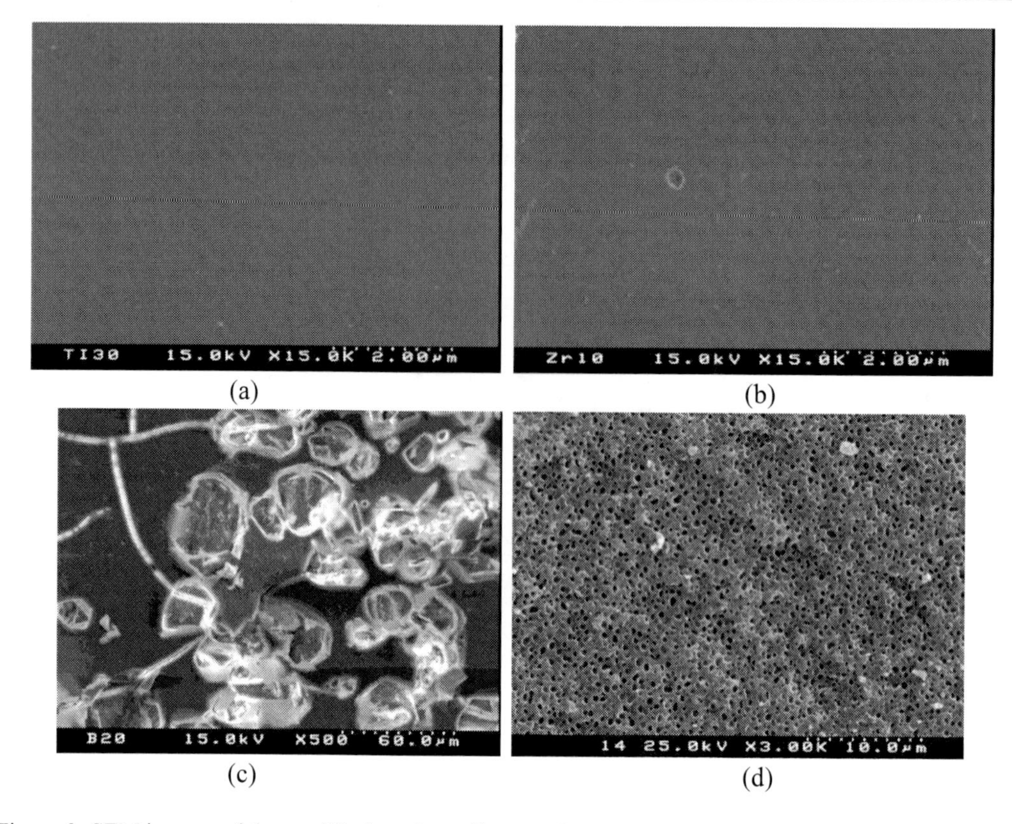

Figure 3. SEM images of the modified urethanesil xerogels: a) 30 wt. % TTP; b) 30 wt. % ZTP; c) 30 wt. % TMB and d) 30 wt. % ASB.

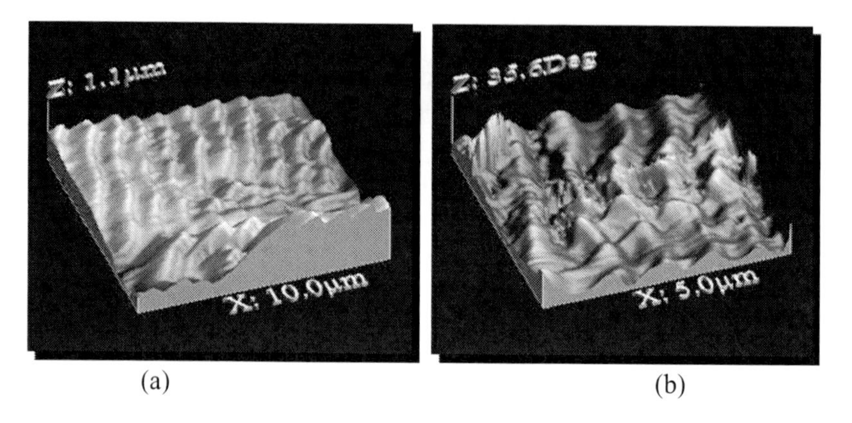

Figure 4. AFM images of samples with 30 wt.% ASB before pyrolysis (a) and after pyrolysis at 1100°C.

From the obtained results, it was observed that the Al modified urethanesiles are characterized by a canal structure as a result of pores connected to each other and uniformly distributed within the body of the sample (Figure 4a). The RMS (Root Mean Square) surface roughness was calculated, and it is of the order of 7 nm .

Figure 5 and Figure 6 show for comparison SEM images of xerogels modified by maximum concentration of Ti, Zr, B and Al after pyrolysis at 1100°C, where the microstructural changes of the sample surface with temperature can be observed. According to SEM observation, Ti-modified urethanesil is a non-porous solid with a typical flat, vitreous and defect-free surface (Figure 3(a)). These features are maintained up to 800°C. Above this temperature (Figure 5 a) open porosity is observed and it is uniformly distributed representative for an inorganic materials. The minimal pore size is about 30 up to 60 nm after heating at 1100°C (Figure 5(a)).

SEM images of the Al and B modified samples after pyrolysis at 1100°C are presented in Figure 6a) and Figure 6b) . The surface of the boron modified materials remain vitreous and without any visible defects as cracks or pores (Figure 6b). The hybrid materials modified by Al show a porous structure at gel state that is retained after pyrolysis up to 1100°C (Figure 6a). At this temperature, the pore size increases up to 600 nm, which could be a result of the combustion and evaporation of organic components from the structure of the materials.

Figure 5. SEM images of modified urethanesil xerogels after pyrolysis at 1100°C: a) 30 wt. % TTP; b) 30 wt. % ZTP.

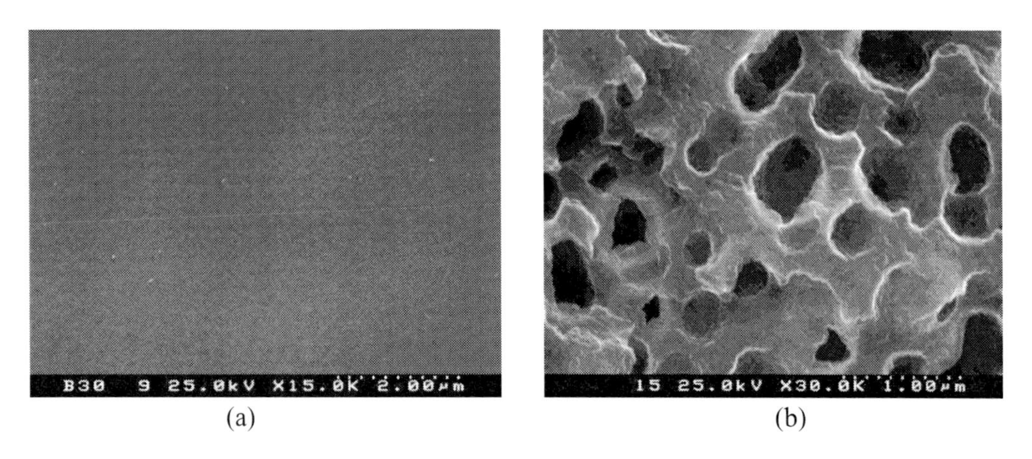

Figure 6. SEM images of modified urethanesil xerogels after pyrolysis at 1100°C: a) 30 wt. % ASB and b) 30 wt. % TMB.

3. FTIR Investigations

The structural changes of the urethanesil hybrid material after modification with Ti, Zr, B and Al have been studied by FTIR and the obtained results are presented on Figure 7.

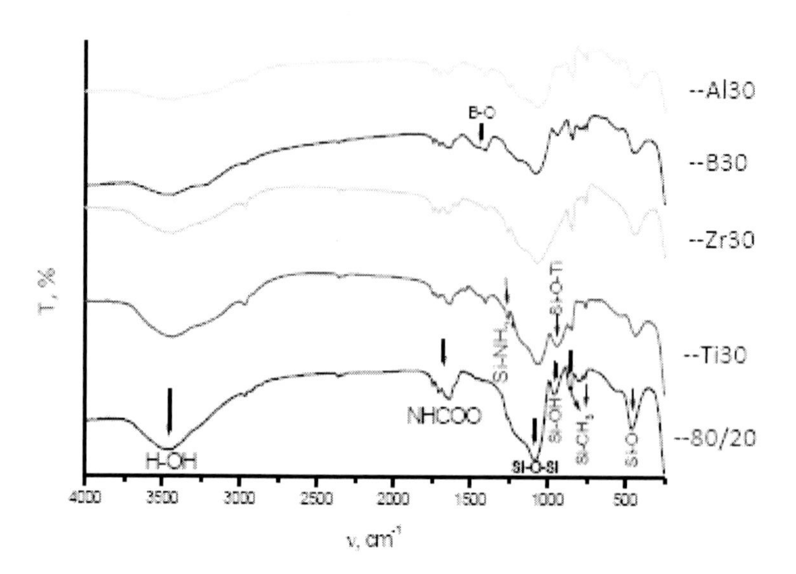

Figure 7. FTIR spectra of non modified urethanesil xerogel (80/20) and xerogels with different modifying ions (Ti, Zr, B and Al).

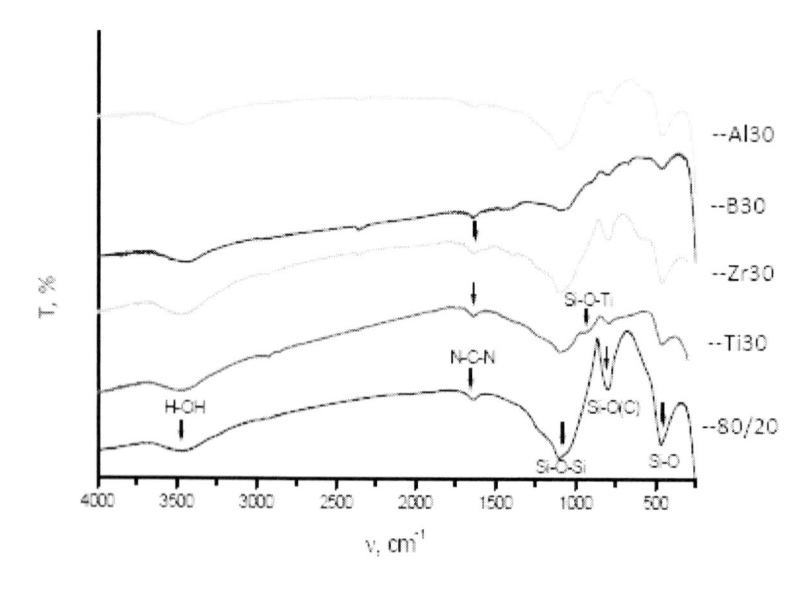

Figure 8. FTIR spectra of non modified urethanesil xerogel (80/20) and xerogels with different modifying ions (Ti, Zr, B and Al) pyrolyzed at 1100°C.

The FTIR spectra obtained for all synthesized hybrid materials show that there is no band corresponding to the free -NCO- groups at 2300 cm^{-1} which provides an indirect evidence for the lack of unreacted trimethylsilyl isocyanate during the sol-gel synthesis. The combination of bands between 1600 and 1750 cm^{-1} corresponds to the vibrations of the urethane linkages formed during the polycondensation processes between prehydrolized TMOS and added TMSI [40, 41]. The bands detected at 1746cm^{-1} are due to the C=O bond and at 1707 cm^{-1} are due to the NH-CO-O groups. The contribution of NH-CO bonds results in bands found at 1660cm^{-1} and 1632cm^{-1}. The peaks that appear at 1080 cm^{-1} and 480 $cm^{-1}$are due to stretching vibration and deformation vibration of Si-O-Si bonds, while the deformation vibration corresponding to Si-$(CH)_3$ groups for all materials appeared at 845 cm^{-1} and 765 cm^{-1}. The vibrations of the Si-NH_3 groups results in a band at 1210-1230 cm^{-1} [42]. The fact that the peak at 1080 cm^{-1} for composition 80/20 is shifted to 1050 cm^{-1} when the samples were modified by Zr, Ti, B and Al, prove the formation of mixed Si–O–M bonds (M=Ti, Zr, B and Al) and their influence on the Si–O–Si vibration. The main signals due to Si–O–M bonds are visible in the FTIR spectra only for the Ti-modified xerogels at 940–920 cm^{-1}. In the spectra obtained for the hybrids modified by Zr, B and Al, the Si-O-M bonds are not observed due to the broad peak for Si-O-Si linkage centered at 1050 cm^{-1}.

Moreover, from the obtained FTIR patterns, it was established that in the synthesized hybrid gels the organic and inorganic building blocks are linked to each other by covalent bonds (Si–NH_2 and Si–$(CH_3)_3$) forming a single homogeneous nanostructured material [43]. During the pyrolysis process, the final structure of the materials is formed at 1100 °C (Figure 8). According to the obtained results from Figure 8 the nanostructured gel materials undergo several structural changes during pyrolysis process. The intensity of the signal due to adsorbed water (3300 cm^{-1}) decreases with the increasing temperature, suggesting the reduction of the surface area and the decrease of Si–OH terminal groups, respectively. A wide and broad Si–O–Si band centered at 1080 cm^{-1} in the spectrum of initial gels (80/20) is shifted to higher wave lengths to 1092 cm^{-1} at 1100 °C. This result is consistent with length and energy modification of Si–O–Si bonds [44]. The transformation of the Si–O–Si peaks and the consumption of Si–CH_3 groups are related to the loss of the initial arrangement between siloxane units in the gel network as a consequence of Si–O–Si bond redistribution reactions.

According to the presence of the mixed Si–O–Ti bonds at 920 cm^{-1} , they are still present in the FTIR spectra at the maximum pyrolysis temperature. Moreover, a new band around 790 cm^{-1} that corresponds to SiX_4 (X = O; C) is observed related to the typical oxycarbide network [45]. In addition, the formed urethane groups connected to the siloxane network are transformed in crebodiimid groups visible at 1645 cm^{-1} and stable up to 1100°C.

4. Solid State NMR Investigations

^{29}Si CP MAS NMR

Silicon-29 MAS NMR can often resolve resonances from five different types of silicate tetrahedra having varying numbers (n) of bridging oxygen, commonly described as Q(n) species [46]. Systematic variations of ^{29}Si chemical shifts with structure occur, with the ^{29}Si chemical shift becoming increasingly negative as the number of bridging oxygen linkages increases. This occurs due to a greater degree of electronic shielding of the central Si-atom.

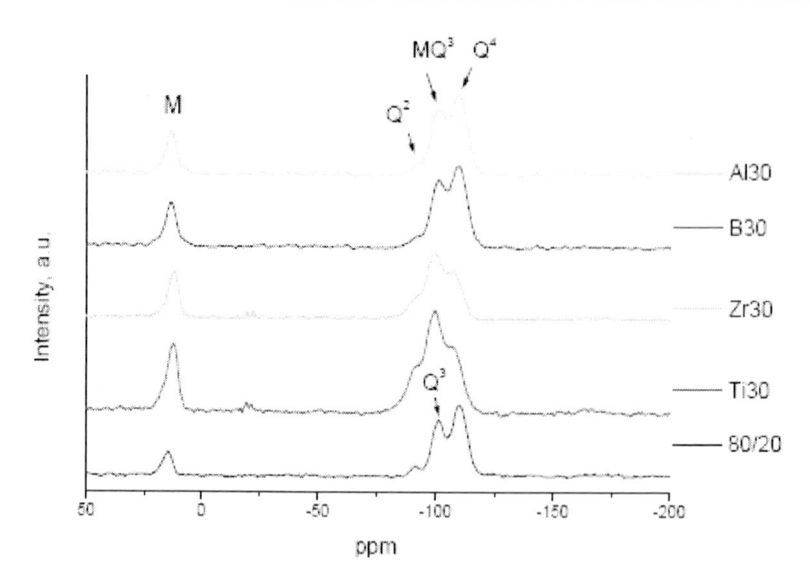

Figure 9. [29] Si CP MAS NMR spectra of non modified urethanesil xerogel (80/20) and xerogels with different modifying ions (Ti, Zr, B and Al).

The variation of ^{29}Si CP MAS NMR spectra of the hybrid materials with different modifying ions and pyrolysis temperature has been studied. The obtained results are presented on Figure 9. The obtained NMR spectra independent of the concentration and the type of modifying ions (Ti, Zr, B and Al) are characterized by signals at −91, −101 and −111 ppm attributed to different Si units namely Q^2 [$Si(OSi)_2(OCH_3)_2$], Q^3 [$Si(OSi)_3OCH_3$] and Q^4 [$Si(OSi)_4$], respectively [46].

Moreover, a signal at +14 ppm is observed in all NMR spectra on Figure 9. From the bibliographic data, this signal is due to M [$(CH)_3SiN$] structural units [47], which allow to conclude that during the hydrolysis –condensation reactions, the Si–N and Si–C bonds, typical of TMSI as a precursor, remain stable.

In addition to the above discussed chemical shifts in all NMR spectra, it can be pointed out that no signals are detected as a contribution of T- structural units at -40.56 ppm (3OH); -40.90 ppm (2OH, 1OCH_3) and -41.40 ppm (1OH $_2OCH_3$). This is evidence of the lack of unreacted methoxy groups of TMOS during the hydrolysis and the presence of unreacted OH groups during the process of polycondensation. The absence of such groups means that the reactions of hydrolysis and condensation have been completely accomplished.

The presence of trivalent modifying ions (Al and B) in the urethanesil network does not influence the NMR spectra in Q^n region in contrast to the presence of tetravalent modificators. When Ti and Zr are used for the synthesis of the hybrid urethanesiles, a decreasing of the intensity of Q^4 (all four oxygen atoms of the SiO_4-tetrahedra linked to other polyhedra) structural units is observed at the expense of Q^3 units (three oxygen linkages to silicon, one OH group). For more precise understanding of the modified urethansiles, the Q^n region from the NMR spectra of Ti and Zr modified samples have been deconvoluted by using PeakFit™ software (Figyre 10). It was proved that the lower intensity of the signal due to Q^4 units (−111 ppm) is due to the formation of Ti–O–Si and Zr-O-Si bonds. Evidence for this statement is the appearance of an unassigned peak along with Q^2, Q^3 and Q^4 signals. For Ti-modified urethanesiles, this peak appeared at −107 ppm (Figure 10 a) and for Zr-modified, it was detected at -105 ppm (Figure 10b).These unassigned peaks could be ascribed to a titanosiloxane $(Si{\equiv}O)_3Si(OTi)$ and zirconosiloxane $(Si{\equiv}O)_3Si(OZr)$ units, denoted as TiQ^3 and ZrQ^3, respectively [48-53]. All of these results again confirm the very high homogeneity of the siloxane network at the gel state.

When the samples were pyrolyzed at 1100 °C, the NMR spectra were strongly changed (Figure 11). The NMR spectra of all samples are built mainly from Q^4, C_2SiON, and C_3SiO structural units. C_4Si and N_2SiO_2 spaces are visible only for the hybrid urethanesiles modified by boron alkoxyde. After deconvolution of the Qn region (Figure 12) of the NMR spectra of the Ti and Zr modified urethanesiles after their pyrolysis, it was proved that the TiQ^3 structural units are still present in the Ti-modified sample (Figure 12(a)), but with lower intensity than in the spectrum of gel state. In comparison, on Figure 12 (b) the deconvoluted ^{29}Si NMR spectrum of Zr-modified urethanesiles is presented. The obtained result showed that the material has Q^4 and Q^3 silica units and no Si-O-Zr bonds are retained in the structure after the pyrolysis process.

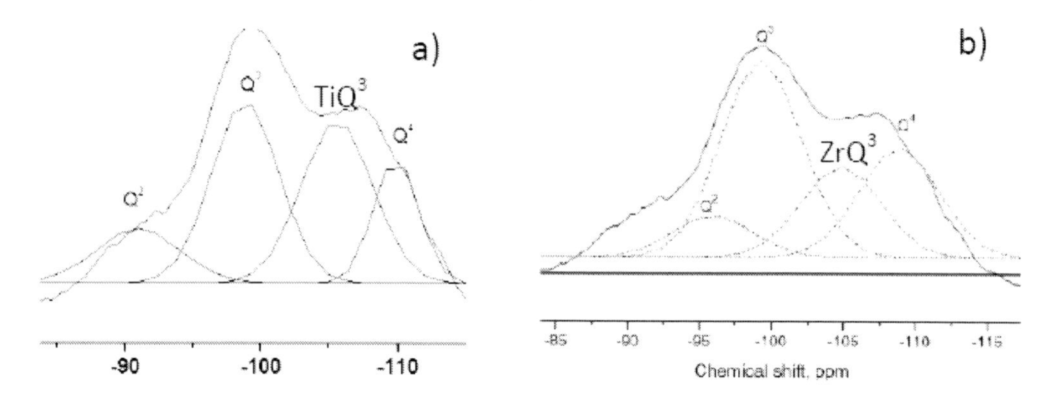

Figure 10. ^{29}Si CP MAS NMR spectra of the xerogels modified by Ti (a) and by Zr (b) deconvoluted by using PeakFittingi software.

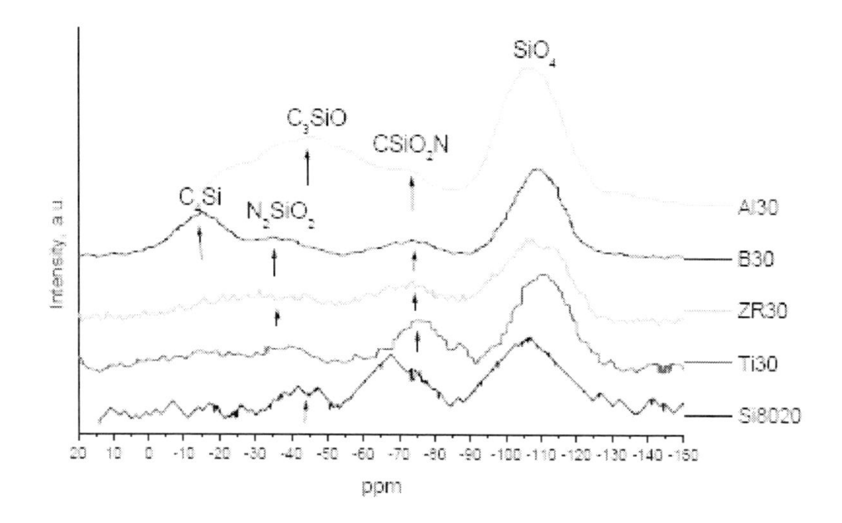

Figure 11. ^{29}Si CP MAS NMR spectra of non modified urethanesil xerogel (80/20) and xerogels with different modifying ions (Ti, Zr, B and Al) after pyrolysis at 1100°C.

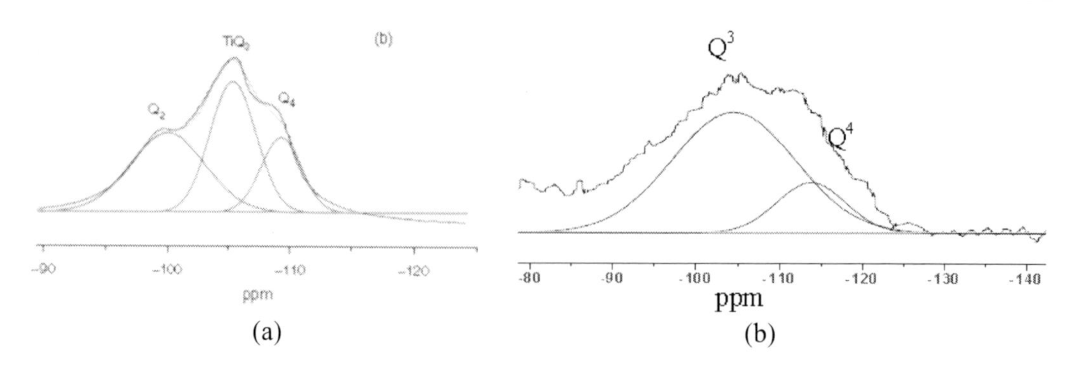

Figure 12. ^{29}Si CP MAS NMR spectra of the pyrolyzed xerogels modified by Ti (a) and Zr (b) deconvoluted by using PeakFittingi software.

^{27}*Al NMR spectra*

In order to establish the presence of Al and B ions in the hybrid urethanesiles, ^{27}Al and ^{11}B NMR were used as additional characterization tools. The obtained results from ^{27}Al NMR investigation for the sample modified by 30 wt. % ASB is presented on Figure 13.

From the obtained results in Figure 13(a), it was established that the ^{27}Al NMR spectrum is consistent with the presence of only octahedral Al species (resonance centered at 0 ppm) tentatively assigned to higher-molecular-weight species containing asymmetrical Al sites. No peaks near 60 ppm due to tetrahedrally coordinated Al species resolved in the sample at the gel state. The resonance centered at 0 ppm is remarkably stable with respect to the temperature (implying a fairly rigid structure) and pyrolysis time. A heat treatment at 1100°C led to a decrease in intensity of the 0-ppm resonance with a corresponding increase in the intensity of a new peak visible at 62 ppm (Figure 13b). This new resonance is attributed to the presence of tetrahedral aluminum atoms, an evidence for rapid conversion of the gel state to the high-temperature form [54].

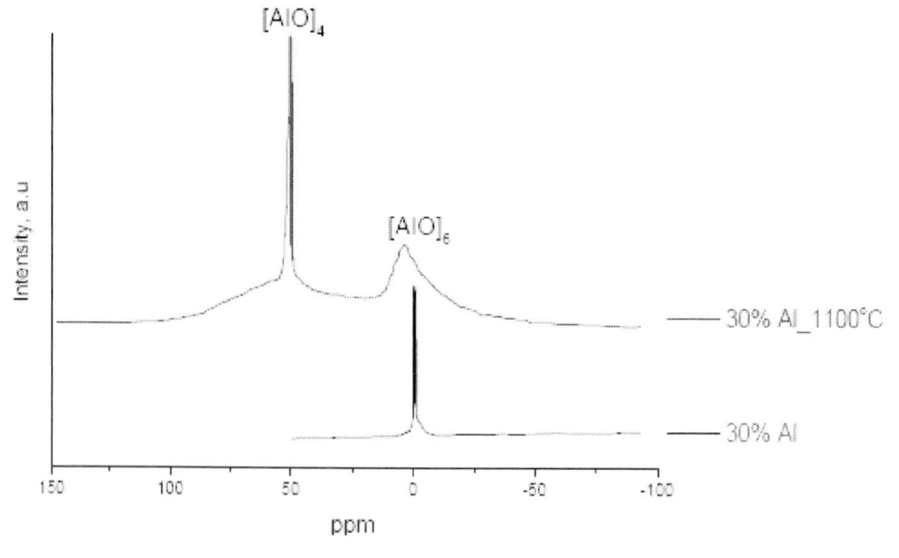

Figure 13. ^{27}Al NMR spectra of gel modified by 30 wt.% ASB before and after pyrolysis at 1100°C.

^{11}B NMR spectra

There have been several recent reports of ^{11}B MAS NMR spectroscopic studies on borates and borosilicates [57-64]. These studies have shown that the BO_4 sites have isotropic chemical shifts in the range of 1.5 to -2.5 ppm. These sites are bonded either to 2, 3, or 4 other boron atoms consisting of B–O–B or B–O–Si linkages. In contrast, the isotropic shifts of the BO_3 sites depending on the number of non-bridging oxygen their isotropic shifts fall in the range of 12–24 ppm.

Figure 14 shows ^{11}B NMR for boron-modified urethanesiles before and after pyrolysis. In ^{11}B NMR spectrum for as synthesized material (Figure 14a), the symmetric sharp peak at around – 10 ppm is attributed to B3 in BO_3 units, and the asymmetric broad peak at - 20 ppm is assigned to B4 in BO_4 units [55, 56]. This result reveals that the trimethylborate is well incorporated in the silica network of the xerogel dried at room temperature.

After deconvolution of the mentioned ^{11}B MAS NMR spectrum (Figure 14 a), it can be seen that the peak characterized for BO_4 units is much more intense than the peak assigned to BO_3 units. This result indicates that most of the TMB quantity used for boron modification of the derived urethanesiles, is co-polymerized with the silica network.

Figure 14 (b) shows the ^{11}B spectrum of boron-modified urethanesil pyrolized at 1100°C. Deconvoluted Boron-^{11} NMR measurement resolve four distinct boron environments at this pyrolysis temperature. As we have discussed, the resonance with chemical shift values of -1.0 ppm is assigned to tetrahedral boron atoms [59]. The next two resonances, with a chemical shift of -18 ppm and -22 ppm, arises from a Q^4B (3 Si, 1 B) site and a Q^4B (4 Si) site, respectively [63]. In the ^{11}B NMR spectrum of the pyrolized sample after deconvolution a small peak was detected at 17 ppm. According to the bibliographic data, it corresponds to the symmetric BO_3 (BO3S) sites having zero or 3 non-bridging oxygen [64]. Usually, the BO_3S sites can be further classified into $BO_{3(ring)}$ and $BO_{3(non\text{-}ring)}$ sites with isotropic shifts around 18 ppm and 13 ppm, respectively. Thus, the BO3S sites in the synthesized urethanesiles after pyrolysis process are present predominantly in ring structures due to the isotropic chemical shift centered at 17 ppm.

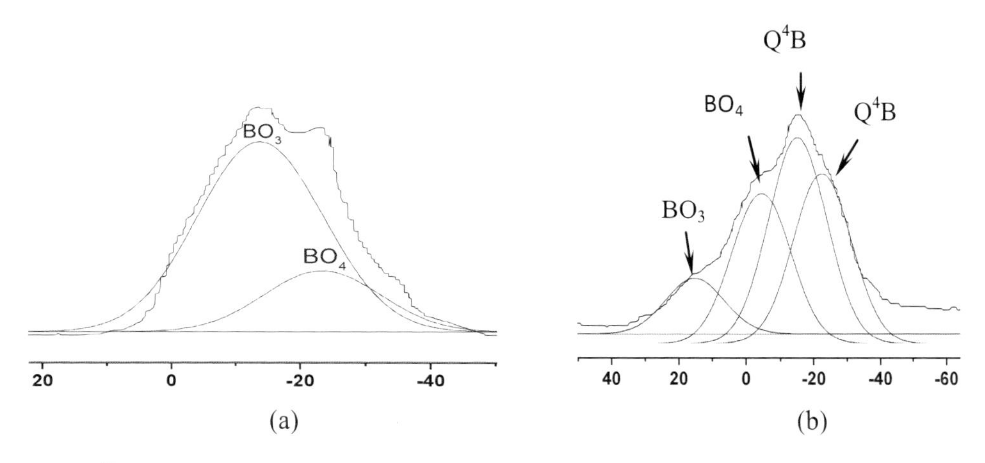

Figure 14. ^{11}B NMR spectra of gel modified by 30 wt.% ASB before (a) and after pyrolysis at 1100°C (b).

Briefly, from the obtained FTIR, XRD and ^{29}Si, ^{27}Al and ^{11}B MAS NMR patterns it was established that the synthesized hybrid gels are amorphous materials and the organic and inorganic building blocks are linked to each other by a covalent bond forming a single homogeneous phase in the Si–O–C–N–M (M=Ti, Zr. B, Al) system. According to the hybrid materials classification, this type of hybrid products belongs to the group of the nanostructured materials.

Design of the Synthesized New Hybrid Materials

Based on the obtained spectral results, a suggestion has been made for the most probable cross-linking mechanism that occurs during the gel formation in the synthesized hybrid materials. An example will be given for the gels modified by titanium ions:

- Hydrolysis and partially polycondensation of the TMOS are processes that can be considered as a first stage of a mechanism discussed. The reactions describing this mechanism first stage, which determine the sol formation are presented in the following reactions (1) and (2):

$$(H_3C\text{-}O)_2\text{-}Si\text{-}(O\text{-}CH_3)_2 + HOH \rightarrow (HO)_2\text{-}Si\text{-}(OH)_2 + CH_3OH \quad (1)$$

silanol

$$\sim Si\text{-}OH + HO\text{-}Si \sim \rightarrow HO\text{-}Si(OH)_2\text{-}O\text{-}Si(OH)_2\text{-}OH + H_2O \quad (2)$$

di-silanol

- The interaction between the isocyanate groups from the TMSI with silanol groups from reaction (2) is the second stage of the proposed cross-linking mechanism that leads to the formation of urethane bridges or so-called urethanesil (see reaction (3)):

$$H_3C\text{-}Si(CH_3)_2\text{-}N{=}C{=}O + HO\text{-}Si(OH)_2\text{-}OH \rightarrow CH_3\text{-}Si(CH_3)_2\text{-}NHCOO\text{-}Si(OH)_2\text{-}OH \quad (3)$$

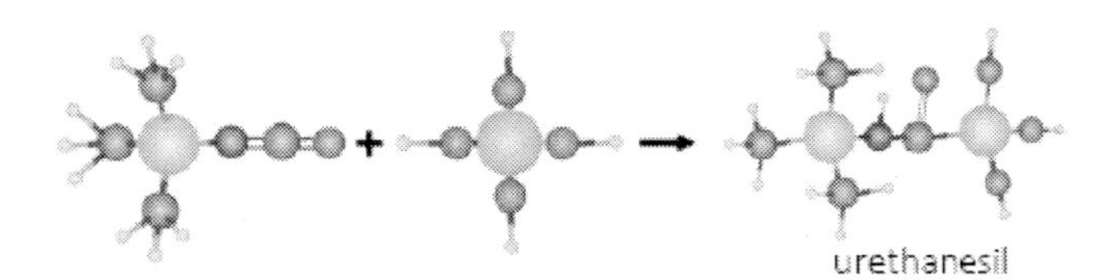

- The next step is polycondensation process, which includes interaction between the urethanesil monomers with each other, by means of their functional OH groups (reaction (4)).

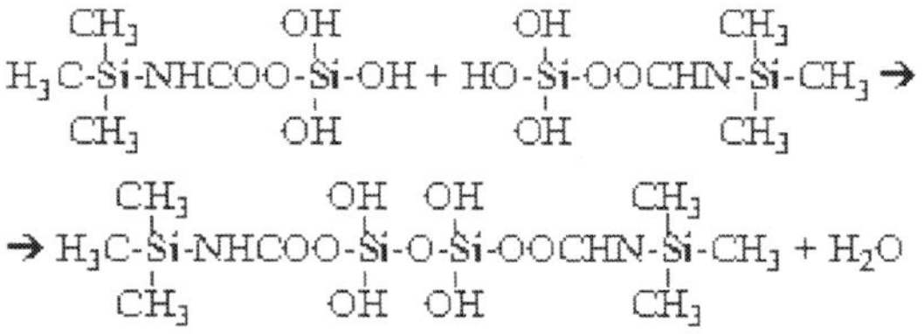

$$H_3C\text{-}Si(CH_3)_2\text{-}NHCOO\text{-}Si(OH)_2\text{-}OH + HO\text{-}Si(OH)_2\text{-}OOCHN\text{-}Si(CH_3)_2\text{-}CH_3 \rightarrow$$
$$\rightarrow H_3C\text{-}Si(CH_3)_2\text{-}NHCOO\text{-}Si(OH)_2\text{-}O\text{-}Si(OH)_2\text{-}OOCHN\text{-}Si(CH_3)_2\text{-}CH_3 + H_2O \quad (4)$$

di-urethanesil

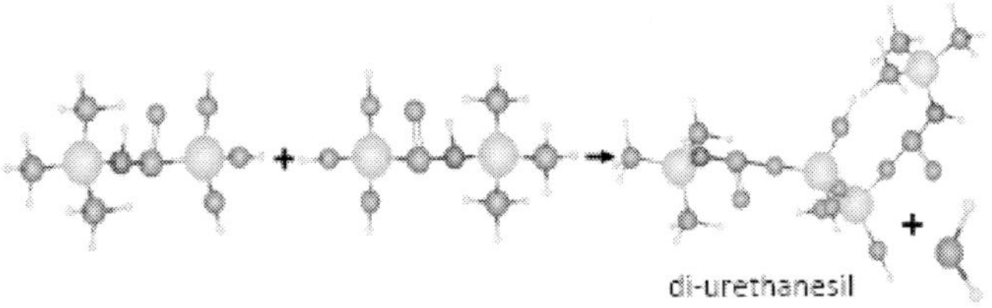

- The reaction (4) was followed by a co-condensation process between the formed di-urethanesiles (reaction 5):

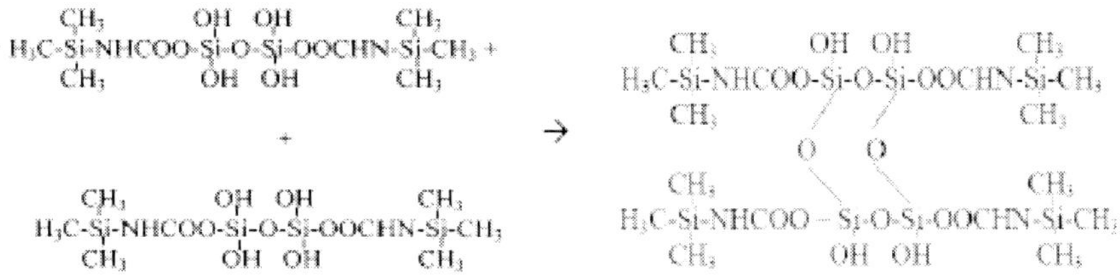

$$H_3C\text{-}Si(CH_3)_2\text{-}NHCOO\text{-}Si(OH)_2\text{-}O\text{-}Si(OH)_2\text{-}OOCHN\text{-}Si(CH_3)_2\text{-}CH_3 + H_3C\text{-}Si(CH_3)_2\text{-}NHCOO\text{-}Si(OH)_2\text{-}O\text{-}Si(OH)_2\text{-}OOCHN\text{-}Si(CH_3)_2\text{-}CH_3 \rightarrow$$
$$\begin{array}{l} H_3C\text{-}Si(CH_3)_2\text{-}NHCOO\text{-}Si(OH)\text{-}O\text{-}Si(OH)\text{-}OOCHN\text{-}Si(CH_3)_2\text{-}CH_3 \\ \qquad\qquad\qquad\qquad\quad O \qquad\;\; O \\ H_3C\text{-}Si(CH_3)_2\text{-}NHCOO\text{-}Si(OH)\text{-}O\text{-}Si(OH)\text{-}OOCHN\text{-}Si(CH_3)_2\text{-}CH_3 \end{array} \quad (5)$$

cross-linked di-urethanesil

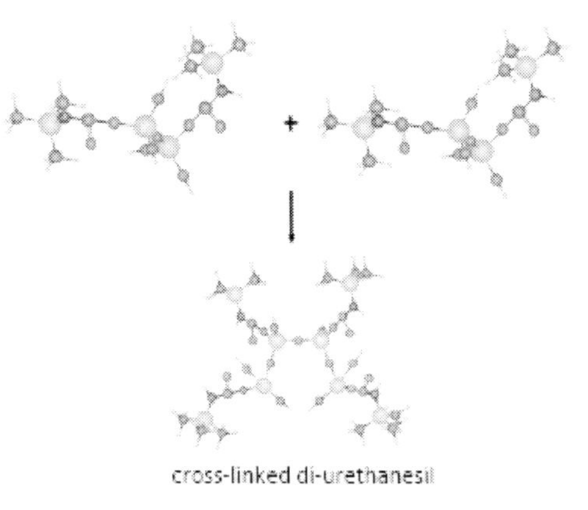

- According to the synthesis scheme described in Experimental Part, the required amount of titanium alkoxide is added during the polycondensaton. Its presence in the sol leads to exchange reactions between the propyl groups of the titanium isopropoxide and OH groups of the silanol. Due to these exchange reactions, the heterometallic Si–O–Ti bonds (reaction (6)) are formed simultaneously with methyl alcohol as by-product:

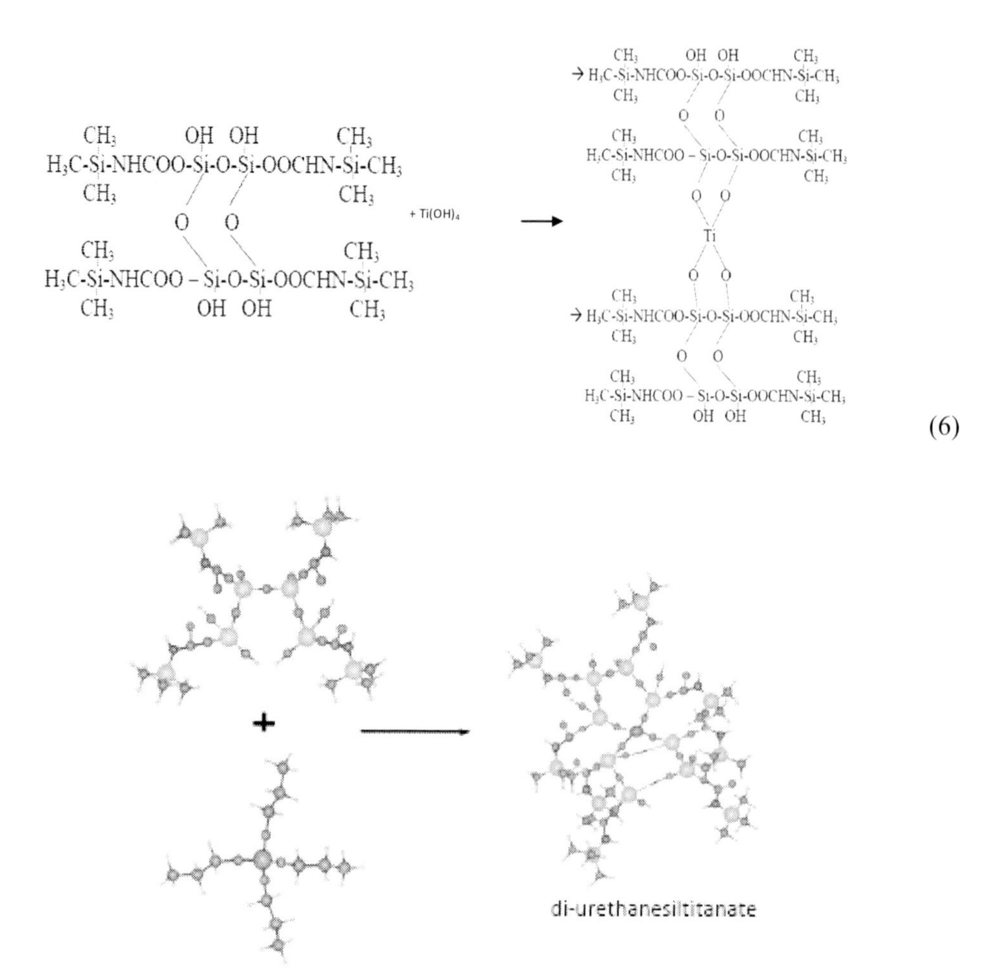

(6)

Therefore, from the discussed cross-linking mechanism it could be concluded that the synthesized new nanostructured hybrid materials are built from Si–CH_3 and Si–O–M (M=Ti, Zr, B, Al) structural units covalently bonded to the siloxane network by urethane (NH-COO) bridges. Moreover, during the sol–gel process discussed above, it was established (from the spectral analysis) that the Si–N and Si–C bonds are still present in the xerogel network. In this way, the trimethysilyl isocyanate (TMSI) can be used as a precursor for sol–gel synthesis of hybrid materials in Si–O–C–N system.

POSSIBLE TRANSFORMATION OF THE MODIFIED NANOSTRUCTURED DI-URETHANESILES TO NANOCOMPOSITES

From the obtained XRD, SEM, AFM, FTIR and NMR results, it was established that the sol-gel synthesized hybrid di-urethanesiles are nanostructured materials built from Si-CH_3 and Si-O-M structural units bonded to the silica network by urethane bridges forming an homogeneous Si-O-C-N-M phase. This means that the organic and inorganic building blocks in the gel network are linked to each other by strong covalent bonds. During pyrolysis process up to 1100°C in an inert atmosphere, it was established that the Si-CH_3 and NHCOO bonds are transformed into SiO(C) and N-C-N linkages. Moreover, according to the ^{29}Si NMR spectra it was additionally proved that the incorporated carbon and nitrogen atoms in the gel network by using of trimethylsilyl isocyantate as a new silica source are still present in the siloxane network. Therefore, this transformation process is related to pyrolyzed materials that possess Si oxycarbonitride network modified by Ti, Zr, B or Al ions. Regarding the obtained XRD results at 1100°C, it was proved that the pyrolyzed samples modified by Ti, B and Al, as well as non-modified material, are still amorphous. The gel modified by zirconium alkoxyde up to 30 wt. % and pyrolyzed at 1100°C results in the presence of Zr–O–Zr linkages by means of tetragonal ZrO_2 nano-crystals and an evidence for this statement was given by the obtained XRD results and their analysis. Consequently, two phases are odentified—amorphous (built of Si–O–C–N network) and crystalline (built of ZrO_2 tetragonal nano-crystals). Because of them, the materials after pyrolysis can be considered as nanocomposites built of Si–O–C–N vitreous matrix and ZrO_2 nano-crystals. Consequently, the sol–gel synthesized zirconium modified nanostructurred hybrid xerogel materials are transformed into nanocomposite after pyrolysis process at 1100 °C.

POTENTIAL APPLICATIONS OF THE SOL-GEL DERIVED TI, ZR, B AND AL- MODIFIED DI-URETHANESILES

Zirconium-Modified Di-Urethanesiles

Based on the results obtained on this study, practical application can be suggested. The results from the preliminary experiments carried out motivate our research team to continue the study in the direction of the obtained materials deposition on metal substrates in order to improve their resistance toward oxidation and corrosion or to modify their surface properties.

The synthetic sol–gel method, based on preparation of silica–zirconium–urethane gel matrix, gives a possibility to obtain ZrO_2 nano-crystals with an average size in the range of 10–20 nm dispersed in the glass matrix. The size of the particles and crystal growth rate could be controlled by heat treatment, as well as the concentration of $Zr(OPr)_4$ in the precursor solution. Due to the strong chemical bonding between the organic and inorganic parts, the hybrid materials offer superior mechanical properties (elasticity, flexibility) and higher chemical stability [65]. The presence of nitrogen and carbon atoms in the siloxane network makes the obtained hybrid materials applicable for the preparation of oxidation resistant coatings or films in electronic and optoelectronic devices [66]. The participation of ZrO_2 in silicon oxicarbonitride materials makes them applicable as barrier against oxidation since ZrO_2 has the advantage to have thermal expansion coefficient close to that of stainless steel. On the other hand, amorphous silica is a much better barrier to oxygen diffusion [67]. Further investigation of the synthesized zirconium modified hybrid materials and nanocomposites should be performed as part of our future work.

Titanium-Modified Di-Urethanesiles

The sol-gel synthesized titanium modified hybrid organic-inorganic materials could be applicable for advanced functional materials such as VOC (Volatile Organic Compound) or gas-separation membranes. Based on the performed FTIR investigation, it was proved that the structure of the hybrid materials obtained by using TMSI are built from OH, C=O, COO, NH_2, NH-CO-groups linked to the siloxane network. These results show that TMSI as a precursor for preparation of hybrid membranes results in urethane copolymers. Because of this, the synthesized materials could exhibit excellent gas-separation, high selectivity and heat stability. These properties make them applicable especially for the separation of gas mixture containing carbon dioxide and methane, carbon dioxide and hydrogen, carbon dioxide and nitrogen [68, 69].

The participation of di- or trifunctional methylsilanes may additionaly lead to different spatial location of Si-CH_3 groups and thus can result in different levels of porosity and distribution of pore size during the pyrolysis process [70]. Gas–transport properties of these kinds of materials could be dependent of the concentration of siloxane that is incorporated in the matrix [69]. The permeability coefficients of CO_2, O_2 and N_2 may be increased proportionally with increasing of SiO_2 content as it has been reported in [5, 6]. According to Seshadri [73-76], the extent of modifying ions distribution, as well as the conditions of pyrolysis affect the permeability and selectivity of the hybrid materials applicable as membranes.

SEM observation of the sol-gel synthesized titanium modified di-urethanesiles showed that the samples are amorphous at room temperature independently of titanium concentration. They possess a vitreous surface and a nonporous and defect-free structure. This is an indirect evidence of a high homogeneity of the obtained gel materials and suggests valuable properties of the hybrid materials as membranes [69]. According to the IUPAC nomenclature, the synthesized materials applicable as membranes could be classified as dense class of membranes [75].

A nonporous structure is retained during pyrolysis up to 800°C. Above this temperature, open porosity uniformly distributed is observed on the obtained SEM images . The porosity is a result of the combustion and evaporation of organic groups within the bulk of the material during the pyrolysis stage. The minimal pore size is about 3 up to 6 nm and it remains in this range up to 1100°C. In this case, the pyrolyzed samples belong to the group of the meso-porous materials, which may act as intermediate supporting layers for synthesis of micro-porous membranes [76] . These titanium modified di-urethanesile seem to be very attractive as gas-separative membrane, when their selectivity is based on the presence of space charges in the pore, usually induced by the adsorbtion of ions at the membrane surface [76].

Aluminum-Modified Di-Urethanesiles

A modified silica containing Al can be used for technological applications such as extracting metallic cations from aqueous and non-aqueous solutions [77–12], for high performance liquid chromatography and gas chromatography in many different investigations [13–15], in catalysis [16, 17], for treatment of waste and toxic effluents produced by a variety of chemical processes [18], ion exchange [19, 20], and so on. Moreover, the hybrid materials based on silicas are mesoporous structures, and more importantly, they posses abundant number of surface hydroxyl groups. These properties make them ideally suited for chemical grafting. The grafted sorbents are widely used for a number of applications. The most important application is sorbent for selective adsorption. The trimethylsilyl isocyanate that was used for the first time in this study has been established as a very good grafting compound. From the obtained spectral results it was proved that the products from the hydrolysis and condensation reactions using TMSI are urethane and amine grafted silicas. This kind of hybrid materials could be promising sorbents for selective adsorption, especially for applications in environmental control and catalysis due to their high surface areas, high silanol numbers, as well as thermal stability. In order to confirm this statement a, nitrogen adsorption isotherm of aluminum modified sample was done (Figure 15). According to the IUPAC classification, the isotherm is of type IV (Figure 15a). A characteristic feature of this type is the hysteresis that is normally attributed to the existence of pore cavities larger in diameter than the opening leading into them (so-called inkbottle pores) [89]. The obtained isotherm shows low adsorption at low pressure and a sharp increase in adsorption with increasing relative pressure at P/P_0=0.25-0.45 due to capillary condensation. This indicates that the material has low or insignificant microporosity and a broader distribution of larger pores (200-400 nm). Usually for these kind of materials, the microporosity results from the compaction of the individual globular structure. The surface area, pore volume, and average pore size determined from the adsorption branches are 764 m^2/g; 60 m^2/g and 350 nm, respectively. The related BJH pore size distribution curves for aluminum modified urethanesil is shown on Figure 15 (b).

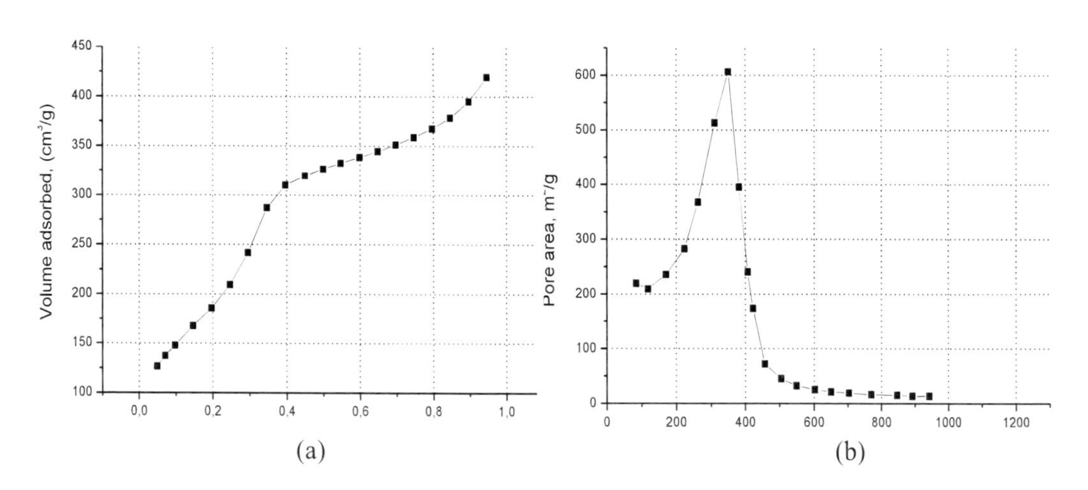

Figure 15. Nitrogen adsorption isotherm (a) and BJD pore-size distribution curves (b).

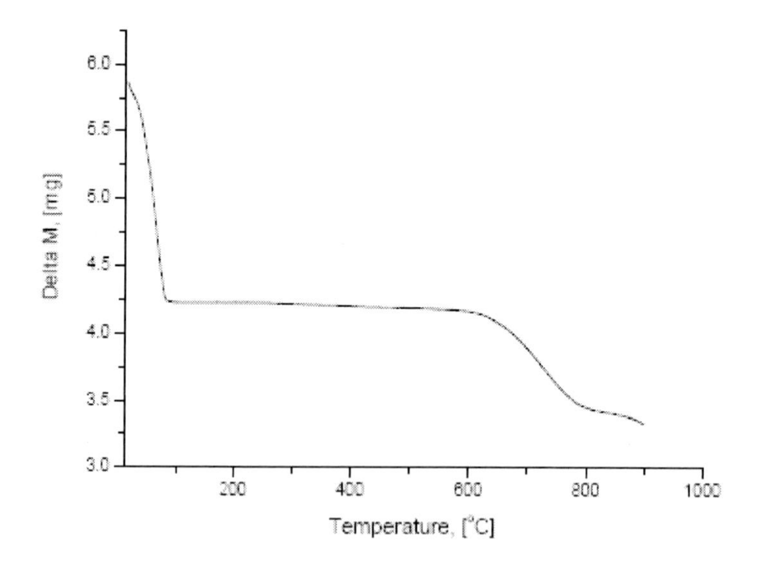

Figure 16. TGA curve of aluminum modified di-urethanesil.

In order to investigate the thermal stability of the synthesized new aluminum modified urethanesiles, a thermogravimetric analysis was performed and the obtained curve is shown in Figure 16. The obtained TG result shows no significant weight loss in the lower temperature region (100-500 °C) clearly indicating that the synthesized material has strong thermal stability. The weight loss below 100 °C is due to the residual water adsorbed in the sample and it is typical for the silica hybrid materials.

From the obtained preliminary results for urethane-functionalized alumosilica xerogel, it was established that textural properties (BET surface area, pore diameter, pore volume), and thermal stability are suitable for adsorbents for removing heavy metal ions such as Pb^{2+}, Hg^{2+}, Cd^{2+}, Zn^{2+}, Cu^{2+}, and Cr^{3+} from wastewater. With urethane moieties grafted to the alumino-silica porous framework, the stability of the mesostructure is improved, which makes the material structurally stable and a promising adsorbent for removing heavy metal ions from polluted water.

Conclusion

Within the context of sol-gel science and technology, well-controlled design and preparation of new organic-inorganic hybrid materials provided with suitable functions is now realistic. Using this technology, Ti, Zr, B and Al-modified di-urethanesiles hybrid materials have been fabricated through the hydrolysis and polycondensation of tetramethylortosilicate (TMOS) and trimethylsilyl isocyanate (TMSI). A considerable set of experimental data has established that the structure of the derived xerogels can be described as nanostructured di-uretanesiles built from Si–O–Si, Si–O–M (M=Ti, Zr, B, Al), Si–CH_3 and Si–NHCOO structural units. During the pyrolysis process, the materials can be fabricated to a variety of forms that range from nonporous nanostructures and nanocomposites to high surface area porous monolithic structures. More importantly, this process does not affect the covalent bonds between the organic and the inorganic parts of the obtained new materials. These features, together with the ability to prepare materials with organic functionality as a repeated unit in the gel network, without disturbing the gel formation or with adverse effects to the porosity, make these materials attractive not only in optics, magnetism and electrochemistry but also in the area of thin films, adsorbents and hybrid membranes.

REFERENCES

[1] Ivanova, Y; Gerganova, Ts. I; Dimitriev, Y; Salvado, IM; Miranda; Fernandes, MHV. *J. Thin Solid Films,* 2006, 515, 271.

[2] Wilkes, GL; Orler, B; Huang, H. *J. Polym. Prepr.*, 1985, 26, 300.

[3] Soraru, GD; Colombo, SM; Egan, EPJ; Pontano, C. *J.Am. Ceram. Soc.*, 2002, 85, 1530.

[4] Schmidt, H. *Mater. Res. Soc. Symp. Proc.*, 1990, 171, 3.

[5] Babonneau, F; Maquet, JL. *J. Chem. Mater.*, 1995, 7, 1050.

[6] Komarneni, S. *J. Sol-Gel Sci. Techno,.* 1996, 6, 127

[7] Mackenzie, JD. *Ceramic Trans,* 1995, 31, 5525

[8] Klein L; Sanchez , CJ. *J. Sol-Gel Sci Technol,* 1995, 5, 75

[9] Babonneau, F; Maquet, JL. *J Chem Mater,* 1995, 7, 1050

[10] Hasegawa, I. *J. Sol-Gel Sci. Techno,* 1996, 5, 93

[11] Zhang , GC; Chen, YF; Xie, YS. *J. Eng. Chem. Metall.* 2000, 21, 225

[12] Smaihi, M; Jermoumi, T; Mariganan, J; Noble, RD. *J. Membrane Sci.*, 1996, 116, 211

[13] Hirata, K; Matsuda, A; Hirata, T; Tatsumisago, M; Minami, T. *J. Sol–Gel Sci. Technol.*, 2000, 17, 61

[14] Matsuda, A; Kanzaki, T; Tatsuminago, M; Minami, T. *Solid State Ionics,* 2001, 145, 161.

[15] Hoebbel, D; Nacken, M; Schmidt, H. *J. Sol–Gel Sci. Technol.*, 1998, 12, 169.

[16] Park, Y; Nagai, M. *Solid State Ionics,* 2001, 145, 149.

[17] Kreuer, KD. *J. Membr. Sci.*, 2001, 185, 29.

[18] Li, S; Liu, M. *Electrochimica Acta,* 2003, 48, 4271.

[19] Chen, Y; Xie, LJY. *J. Sol– Gel Sci. Technol.*, 1998, 13, 735.

[20] Mutin, PH. *J. Sol– Gel Sci. Technol.*, 1999, 14 , 27.

[21] Iwamoto, Y; Kroke, WVE; Reidel, R. *J. Am. Ceram. Soc.*, 2000, 184, 2170.

[22] Roux, S; Audebert, P; Pegetti, J; Roche, M. *Sol– gel Sci. Technol.*, 2003, 26, 435.

[23] Corriu, R. *Eur. J. Inorg. Chem.*, 2001, 11, 09.

[24] Verbeek, W; Ger. Pat. No. 2218960 (Bayer AG) 1973, Nov. 8, (U.S. Pat. No. 3853567)

[25] Wynne, KJ; Rice, RW. *Annu. Rev. Mater. Sci.*, 1984, 14, 297.

[26] Yajima, S; Hayashi, J; Omori, M. *Chem. Lett.*, 1975, 931.

[27] Shilling, CL; Jr., Wesson, JP; Williams, TC. *Am. Ceram. Soc. Bull.*, 1983, 62, 912.

[28] Seyferth, D; Tasi, M; Woo, HG. *Chem.Mater.*, 1995, 7, 236.

[29] Bahloul, D; Pereira, M; Goursat, P; Choong Kwet Yive, NS; Corriu, RJP. *J. Am. Ceram. Soc.*, 1993, 76, 1156.

[30] Kroll, P. *J. Eur.Ceram. Soc.*, 2005, 25, 163.

[31] Kroke, Ed. ;Ya-Li Li, Konetschny, Ch; Lecomte, E; Fasel, C; Riedel, R. *Materials Science and Engineering*, 2000, 26, 97.

[32] Reucket, M; Vaahs, T; BruÈck, M. *Adv. Mater.*, 1990, 2, 398

[33] Galusek, D; Reschke, S; Riedel, R; Dresser, W; Sajgalik, P; Lences, Z; Majling, J. *J. Eur. Ceram. Soc.*, 1999, 19, 1911.

[34] Goncalves, MC; Bermudez, V; de Z., Ostrovskii, D; Carlos, LD. *Electrochimica Acta,* 2003, 48, 1977.

[35] Silva, MM; Nunes, SC; Barbosa, PC; Evansa, A; Bermudez, V.de Zea Smith, MJ; Ostrovskii, D. *Electrochimica Acta*, 2006, 52, 1542.

[36] Widmaier, JM; Bonilla, G. *Polym. Adv. Technol.*, 2006, 17, 634.

[37] Xue, BPC; Hong, Qi; Lin, J; Tan, KL. *J. Mater. Chem.*, 2001, 11, 2378.

[38] Alexander, LE. X-ray Diffraction Methods in Polymer Science, *Wiley- Interscience*, 1969.

[39] Bryaskova, R; Mateva, R; Djourelov, N; Krasteva, M. *Cent. Eur. J. Chem.*, 2008, 6, 575

[40] Colombo, P; Griffoni, M; Modesti, M. *J. Sol-Gel Science and Technology*, 1998, 13, 195

[41] Nyquist, RA. Interpreting Infrared, Raman, and Nuclear Magnetic Resonance Spectra, 0-12-523355-8, *San Diego, Academic Press*, 2001, 46-47.

[42] Dire, S; Ceccato, R; Babonneau, F. *J. Sol–Gel Sci. Technol.*, 2005, 34, 53.

[43] Ivanova, Y; Gerganova, Ts.I; Vueva, Y; Salvado, Is. MM; Fernandes, MH. *J. Nanosci. Nanotechnol.*, 2010, 10, 2444.

[44] Hatfield, G; Carduner, K. *J. Mater. Sci.*, 1989, 24, 4209.

[45] Mutin, P. *J. Sol–Gel Sci. Technol.*, 1999, 14, 27.

[46] Mackenzie, KJD; Smith, ME. Multinuclear Solid-state NMR of Inorganic Materials, ISBN 0-08-043787-7, Elsevier Science Ltd, New York, 2002, 205-206.

[47] Corriu, R. *Eur. J. Inorg. Chem.*, 2001, 1109.

[48] Gunji, T; Okogawa MY; Hongo, M; Abe, Y. *J. Plym. Sci.*, 2005, 43, 763.

[49] Kasgoz, A; Oshimura, KY; Misono, T; Abe, Y. *J. Sol–Gel Sci. Technol.*, 1994, 1, 185.

[50] Akamura, NT; Aguchi, KT; Gunji, T; Abe, Y. *J. Sol–Gel Sci. Technol.*, 1999, 16, 227.

[51] Riedel, RAG; Miehe, G; Dressler, W; Fuess, H; Bill, J; Aldinger, F. *Angew. Chem.,* 1997, 36, 603

[52] Chen, RXS. *J. Chin. Chem. Soc.*, 2004, 51, 945

[53] Mocaer, DRP; Naslain, R; Richard, C; Pillot, JP. *J. Mater. Sci.*, 1993, 28, 2615.

[54] Krawietz, TR; Lin, P; Lotterhos, KE; Torres, PD; Barich, D; Clearfield, A; Haw, JF. *J. Am. Chem. Soc.*, 1998, 120, 8502.

[55] Martens, R; Müller-Warmuth, W. J. *Non-Cryst. Solids,* 2000, 265, 167.

[56] Zhao, P; Kroeker, S; Stebbins, JF. *J. Non-Cryst. Solids,* 2000, 276, 122.

[57] Du, LS; Stebbins, JF. *J. Phys. Chem.*, B 2003, 107 (37), 10063.

[58] Martens, R; Muller-Warmuth, W. *J. Non-Cryst. Solids,* 2000, 265(1–2), 167.

[59] Youngman, RE; Zwanziger, JW. *J. Am. Chem. Soc.*, 1995, 117(4), 1397.

[60] Youngman, RE; Zwanziger, JW. *J. Phys. Chem.*, 1996, 100(41), 16720.

[61] S. Sen, Z. Xu, JF. *Stebbins, J. Non-Cryst. Solids,* 226 (1–2) (1998) 29.

[62] Yazawa, T; Kuraoka, K; Akai, T; Umesaki, N; Du, WF. *J. Phys.Chem.*, B 2000, 104(9), 2109

[63] Stebbins, JF; Zhao, PD; Kroeker, S. *Solid State Nucl. Magn. Reson.*, 2000, 16(1–2), 9.

[64] Kroeker, S; Stebbins, JF. *Inorg. Chem.*, 2001, 40(24), 6239.

[65] Saraidarov, TRR; Saschiuk, A; Lifshitz, E. *J. Sol– Gel Sci. Technol.*, 2003, 26, 533.

[66] Colombo, P. *J. Sol– Gel Sci. Technol.*, 1994, 2, 601.
[67] Guglielmi, M. *J. Sol– Gel Sci. Technol.*, 1997, 8, 443.
[68] Siwen Li, ML. *Electrochimica Acta,* 2003, 48, 4271.
[69] Smaihi, MJCS; Lesimple, C. *Prevost, Iswabelle Christian Guizard*, 1999, 161, 157.
[70] Dire, SEP; Babonneau, F; Ceccato R; Garturan, G. *Journal of Material Chemistry.*, 1997, 7 (1), 67.
[71] Kickelbick, G. *Progress in Polymer Science,* 2003, 28, 83.
[72] Razdan, U. *Current science.*, 2003, 85(6), 761.
[73] Kagramanov, GG. *Glass and Ceramics.*, 2001, 58 (5-6), 166.
[74] Seshadri, KS. *Bull. Mater. Sci.*, 2002, 26 (2), 221.
[75] Byron Gates, YY. *Younan Xia Chemistry of Materials,* 1999, 11, 2827.
[76] Verweij, H. *Journal Of Materials Science,* 2003, 38, 4677.
[77] Arakaki, LNH; Airoldi, C. *Polyhedron,* 2000, 19, 367.
[78] Arakaki, LNH; Sousa, AN; Esp´ınola, JGP; Oliveira, SF; Airoldi, C. *J. Colloid Surf. Sci.*, 2002, 249, 290.
[79] Fonseca, MG; Esp´ınola, JGP; Oliveira, SF; Ramos, LC; Souza, AG; Airoldi, C. *Colloids Surf.*, 1998, A 133, 205.
[80] Pavan, FA; Lucho, AMS; Gonçalves, RS; Costa, TMH; Benvenutti, EV. *J. Colloid Interface Sci.*, 2003, 263, 688.
[81] Silva, CR; Jardim, ICSF; Airoldi, C. *J. High Res. Chromatogr.*, 1999, 22, 103.
[82] Silva, CR; Jardim, ICSF; Airoldi, C. *J. Chromatogr.*, 2003, A 987, 127.
[83] Buszewski, B; Jezierska, M; Welniak, M; Berek, D. *J. High Res. Chromatogr.*, 1998, 21, 267.
[84] Lygin, VL. *Kinet. Katal.*, 1994, 35, 480.
[85] Oliveira SF; Esp´ınola, JGP; Lemus, WES; Souza, AG; Airoldi, C. *Colloids Surf.*, 1998, A 136, 151.
[86] Prado, AGS; Arakaki, LNH; Airoldi, C. *J. Chem. Soc., Dalton Trans,* 2001, 2206.
[87] Collman, JP; Belmont, JA; Brauman, JI. *J. Am. Chem. Soc.*, 1983, 105 7288.
[88] Vieira, EFS; Simoni, JA; Airoldi C. *J. Mater. Chem.*, 1997, 7 , 2249.

In: The Sol-Gel Process
Editor: Rachel E. Morris
ISBN 978-1-61761-321-0

Chapter 13

EX-SITU XPS ANALYSIS OF SURFACES APPLIED IN ADVANCED OXIDATION PROCESSES (AOPS)

Jens Muff, Morten E. Simonsen and Erik G. Søgaard
Department of Chemistry, Biotechnology, and Environmental Engineering,
Aalborg University, Esbjerg, Denmark

ABSTRACT

Advanced oxidation processes are used for treatment of recalcitrant and non-biocompatible pollutants in soil, air, and water. Two of the key AOPs are heterogeneous photocatalysis and electrochemical oxidation where interactions between the polluted media and a surface play the main role. In this chapter, research are presented where XPS has been applied for characterization of differently prepared photocatalytic TiO_2 films before and after UV irradiation for specific determination of the amount of OH groups on the surface. When investigated under different humid conditions, the spectral analyses have been applied for studies concerning the mechanisms for photocatalysis and photoinduced super hydrophilicity. Platinum based metal alloys are a widely used electrode material also considered for water treatment by electrochemical oxidation due to its inert and catalytic properties. XPS has in this context been applied for studying the extent of oxidation of a Ti/Pt_{90}-Ir_{10} anode surface before and after electrolysis under oxygen and chlorine evolution conditions. The results provide insight in both the risk of corrosion of the surface in these oxidative environments and support a suggested mechanism for the oxidation of organics in electrolytic water treatment.

INTRODUCTION

Advances in chemical water and wastewater treatment have led to the development of methods termed advanced oxidation processes (AOPs). AOPs can be broadly defined as aqueous phase oxidation methods based on the intermediacy of highly reactive species such as (primarily but not exclusively) hydroxyl radicals in the mechanisms leading to the destruction of the target pollutant (including organics, inorganics, and pathogens). The list of AOPs that have been considered for treating wastewater is long and increasing as new processes are developed.[1,2] In this chapter two of the key AOPs are considered; heterogeneous photocatalysis and electrochemical oxidation, both advanced oxidation processes which require the interaction of the aqueous pollutant with a surface for the electron transfer process to occur. X-ray photoelectron spectroscopy (XPS) has in this area of environmental engineering been shown to be a valuable tool for the development and evaluation of the surface materials and the study of the mechanisms of the oxidation processes.

Heterogeneous Photocatalysis

Heterogeneous photocatalysis is one of the AOPs that couples light activation with semiconductors acting as photocatalysts. In this process, completely mineralization of organic pollutants to carbon dioxide and inorganic acids can be achieved. In the field of photocatalysis, most applications are concerned with water and air purification or self-cleaning properties of different materials. Titanium dioxide (TiO_2) is the most often used photocatalyst due to its high photocatalytic activity, non-toxic properties, chemical stability over a wide pH range, and that it is not subject to photocorrosion.[3]

The initial step of photocatalysis is the generation of electron-hole pairs within the TiO_2 particle.

$$TiO_2 \xrightarrow{hv < 388nm} TiO_2\,(e^- + h^+)$$

The photogenerated holes have been suggested directly to oxidize adsorbed molecules or to react with surface hydroxyl groups or water molecules to produce hydroxyl radicals. The produced hydroxyl radicals then react with the organic substances/pollutants and thereby enhance the rate of photocatalytic degradation.

$$h^+ + OH^-_{ad} \rightarrow \cdot OH_{ad}$$

$$h^+ + H_2O \rightarrow \cdot OH_{ad} + H^+$$

The photogenerated electrons (e^-) have been suggested to be trapped by oxygen or surface defect sites, which in turn diminish the recombination of the photogenerated electron-hole pairs and hence enhance the photocatalytic activity. The reduction potential for the conduction band electrons is -0.52 V, however the electrons may become trapped and lose some of their reducing power. A significant number of the photogenerated electrons will even after the trapping process keep their reducing power and be able to produce superoxide from dioxygen.[4] The species generated from the electron-hole pair may include hydroxyl radicals ($HO\cdot$), superoxide (O_2^-), and hydrogen peroxide (H_2O_2). Depending on the exact conditions the photogenerated reactive species can play an important role in the photocatalytic reaction mechanisms.

Liu et al. (2003)[5] has by mean of XPS analysis shown the formation of Ti(III) defect sites on the TiO_2 surface through a surface reduction process of Ti(IV) to Ti(III) during UV irradiation. It was assumed that the surface Ti(III) defects were created by the removal of surface oxygen atoms, or most likely bridging oxygen atoms and thus generate oxygen vacancies on the surface. Yu et al. (2002)[6] observed an increase in the amount of hydroxyl groups on TiO_2 films after UV irradiation using high resolution XPS analysis. The observed increase in hydroxyl groups after UV irradiation supports the idea that UV irradiation promotes the generation of reactive species that will mediate the photocatalytical destruction of pollutants.

In addition to the ability of titanium dioxide to photodegrade organic contaminants, it was found that UV irradiation of the titanium dioxide surface will induce superhydrophilicity, which changes the nature of a surface from hydrophobic to hydrophilic. The mechanism for the photocatalytic mineralization feature of TiO_2 films is well recognized, however different models have been proposed for the mechanism leading to superhydrophilicity. One model for the photoinduced superhydrophilicity is based on surface structural changes caused by the photogenerated electron-hole pair. The group of Fujishima[7, 8] has suggested the following mechanism for the photoinduced superhydrophilicity. The photogenerated holes are assumed to be trapped by lattice oxygen on the surface of the TiO_2 film. The trapped holes are thought to weaken the bond between the titanium atoms and the lattice oxygen. As a result, such oxygen is liberated to create oxygen vacancies. At these defect sites water molecules are dissociatively absorbed creating a more hydroxylated surface. The photogenerated electrons are assumed to be trapped by Ti(IV) sites, as Ti(III), which are subsequently oxidized by oxygen. After the superhydrophilic state has been obtained through UV irradiation prolonged irradiation will not lead to any changes in this property. However if the TiO_2 film is stored in the dark, the surface of the TiO_2 film will convert back to its initial more hydrophobic state. The recovery of the film is thought to be due to relaxation of the surface where the weakly bonded hydroxyl groups desorb and the initial surface is recreated.[9] A second model suggesting the superhydrophilicity being obtained by photocatalytically removal of organic contaminates from the surface has also been proposed.[10, 11] This model suggests a direct relation between photocatalysis and the photoinduced hydrophilicity, where the regeneration of the hydrophobic surface in the dark is thought to be caused by re-contamination of air-borne hydrophobic organics.

Electrochemical Oxidation

Electrochemical advanced oxidation (EAOP) is one of the new emerging processes within the AOP family. In the electrochemical cell pollutants are oxidized directly at the surface of the anode or indirectly by powerful oxidants generated in the process from dissolved salts or other natural occurring water constituents.[12] Several produced oxidant have been identified in electrochemical oxidation of polluted water dependent on the operating parameters and materials used.[12-14]

While oxidation and regeneration of selected inorganic ionic pollutants by electrochemical means is a well known technique already established on the market[15], electrochemical oxidation of organic pollutants is a relatively new method in water treatment.

Since the early 1990'ties EAOP has in laboratory and pilot scale studies showed to be a versatile and non-selective oxidation technique and has been applied for the treatment of a wide range of polluted water bodies as landfill leachate, municipal sewage, pesticide polluted groundwater, different kinds of industrial waste waters etc.[1, 12, 16]

Traditionally, expensive platinum based electrodes have been widely used in the electrochemical and analytical industry due to the inert, non-reactive and high electrocatalytic properties of the metals.[16] Through the 1970'ties, the dimensionally stable mixed oxide anodes (DSA) were developed for use especially in the chlor-alkali industry.[17, 18] Since the late 1990'ties boron-doped diamond (BDD) film electrodes have been developed as an electrode material specifically for use in the water treatment industry with highly non-specific oxidation properties.[1,19] The BDD electrodes are characterized by a very efficient production of highly reactive hydroxyl radicals originating from a high over-potential of 1.3 V for the molecular oxygen evolution reaction.[1] In order to prevent inhibition of the anode surface by organic polymerization, which can occur at low anodic potentials, electrochemical oxidation for water treatment purposes are usually carried out in the water oxidation region.[1, 20] According to the generally accepted models, the first step in the electrochemical oxidation of organics is the oxidation of water generating reactive oxygen species on or very close to the anode surface.[20, 21] The specific strong oxidation power of the BDD electrodes compared to the platinum and DSA type electrodes are due to a very weak interaction of the oxidized water molecule and the electrode surface, which generates physisorbed hydroxyl radicals immediately ready for further organic oxidation.[20] At the Pt and DSA electrodes, the interaction is stronger and results in the formation of higher chemisorbed oxides performing a more selective organic oxidation.[20, 22] The model is among other studies based on in-situ DEMS (Differential Electrochemical Mass Spectroscopy) studies of IrO_2 anodes, where the formation of IrO_3 has been showed as an intermediate product in the oxygen evolution reaction by atomic oxygen labeling.[23, 24]

Due to its inert properties metallic platinum is generally considered as resistant to corrosion and chemical attacks which provides platinum based electrodes with high durability in corrosive environments. However, the existence of PtO_x has been demonstrated in numerous studies[25-29] and the oxidative environments present in electrolytic water treatment processes are very likely to promote oxidation of the platinum surface as well as proposed in the models for electrochemical oxidation of organics. In this context, XPS has been applied as a powerful tool in a study of the extent of oxidation of a Ti/Pt_{90}-Ir_{10} anode surface during oxygen and chlorine evolution, respectively, in search of higher oxidative states of especially Pt.

Characterization of TiO_2 by XPS

This section presents a comparison study between the surface properties and the photocatalytic activity of four differently produced TiO_2 powder; two commercially available powders (Degussa P25 and Hombikat UV100) and in-house produced powders by a microwave assisted sol-gel method and a supercritical carbon dioxide procedure (SC 134). Some particle properties of the differently prepared TiO_2 catalyst particles are shown in Table 1.

Table 1. Properties of the comparable commercial and in-house prepared TiO_2 catalytic materials.

	Degussa P25	Hombikat UV100	SC 134	Sol-gel film
Crystallinity				
Anatase (%)	75.6	86.2	65.6	69.1
Rutile (%)	21.6	-	-	-
Amorphous (%)	2.8	13.8	34.4	30.9
Crystallite size determined by XRD				
Crystallite size powder (nm)	18	12	7	10
Crystallite size film (nm)	23	10	7	11
AFM				
Surface roughness (nm)	59.9	131.7	198.7	24.4
BET				
Surface area powder (m2/g)	50	360	221	117
DLS				
Particle size of powder in suspension (nm)	600 ± 15	818 ± 24	1274 ± 36	270 ± 2

Source: Simonsen et al. (2009) *J. Photochem. Photobiology A: Chemistry*, 200, 192-200, Reprinted with permission from Elsevier.

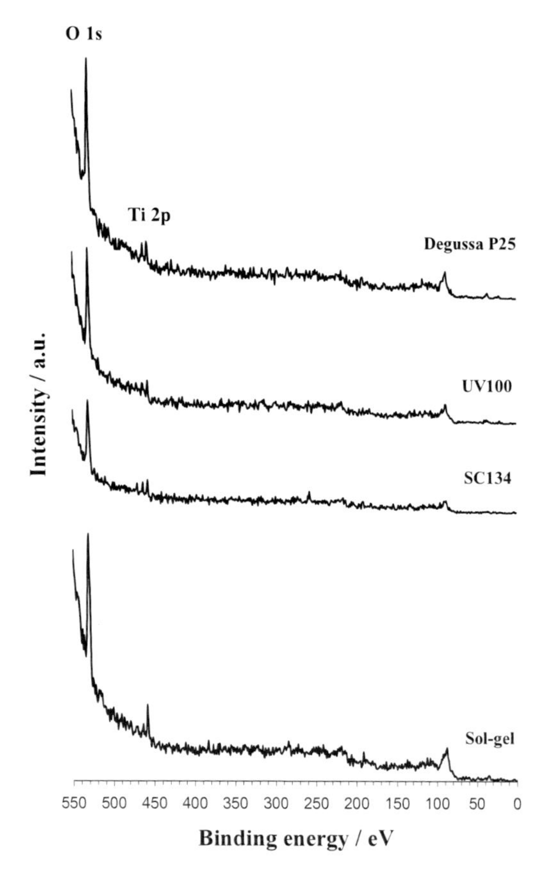

Figure 1. Survey spectra of the commercial (Degussa P25, UV100) and in-house prepared (SC134, Sol-gel) TiO_2 films obtained at 600 eV.

Prior to the spectral analysis, the differently prepared TiO_2 powder samples were immobilized as films on microscope slides as described in Simonsen et al. (2009)[30], and the dried films were placed in a furnace at 450°C for one hour to secure adhesion between the TiO_2 film and the glass.

Synchrotron XPS was used to investigate the surface changes occurring during UV irradiation of the different TiO_2 films. Survey spectra of the different TiO_2 films were obtained at a photon energy of 600 eV. The spectra are shown in Figure 1, where the Ti 2p and O 1s peaks are identified together with auger peaks. A small carbon signal may be observed for the SC134 film.

Analysis of the Titanium 2p Peak

Representative Ti 2p spectra of the TiO_2 films before and after UV irradiation are shown in Figure 2. The sharp and strong peak located at 458.7 eV in the Ti 2p spectrum indicated that the elemental Ti mainly existed as Ti(IV).[5, 31] In photocatalysis the photogenerated electrons are assumed to be trapped by Ti(IV) sites, as Ti(III), which are subsequently oxidized by oxygen. In this case it should be possible to observe Ti(III) as a small shoulder at 1.6 eV lower binding energies. However, the Ti 2p XPS spectra obtained in this work did not show any significant difference indicating that UV irradiation under these conditions induce a significant change in the Ti chemical oxidation states resulting in the formation of Ti(III) and Ti(II), which have been suggested by other groups.[32, 33]

Analysis of the Oxygen 1s Peak

The spectra of the O 1s peak of a typical TiO_2 film before and after UV irradiation are shown in Figure 3. From the spectra it is seen that the oxygen peak was asymmetric indicating that at least two different chemical states of oxygen were present. In literature, the O 1s peak has been proposed to consist of 4-5 contributors such as Ti-O in TiO_2 and Ti_2O_3, hydroxyl groups, C-O bonds, and adsorbed H_2O.[31] Although some H_2O is easily adsorbed on the surface of TiO_2 films during the deposition process, the physically adsorbed H_2O on TiO_2 is easily desorbed under the ultrahigh vacuum condition of the XPS system.[34] From the survey spectra of the TiO_2 films, Figure 1, it is seen that carbon is not present to a greater extent due to the calcination at 450 °C before analysis. The spectra were deconvoluted by addition of two Gaussian and/or Lorentzian shaped peaks centret at 529.9 and 531.9 eV corresponding to Ti–O in TiO_2 and hydroxyl groups (–OH), respectively.[6]

From the O 1s spectrum of the TiO_2 film before UV irradiation it is seen that most of the oxygen was present as lattice oxygen. In comparison, the XPS spectrum after one hour of UV irradiation showed a significant change in the amount of surface oxygen. After UV illumination, the area of the peak due to surface oxygen (hydroxyl groups) increased, indicating that the chemisorption of water molecules on the surface of the TiO_2 films was enhanced by UV irradiation. The modeled peak data are shown in Table 2 for all four TiO_2 films before and after UV illumination.

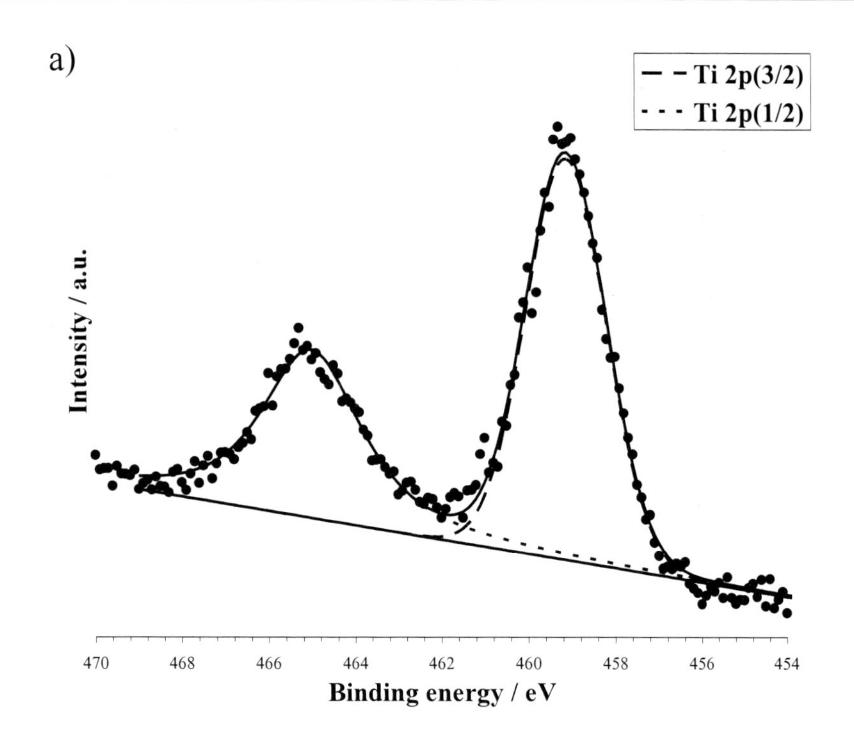

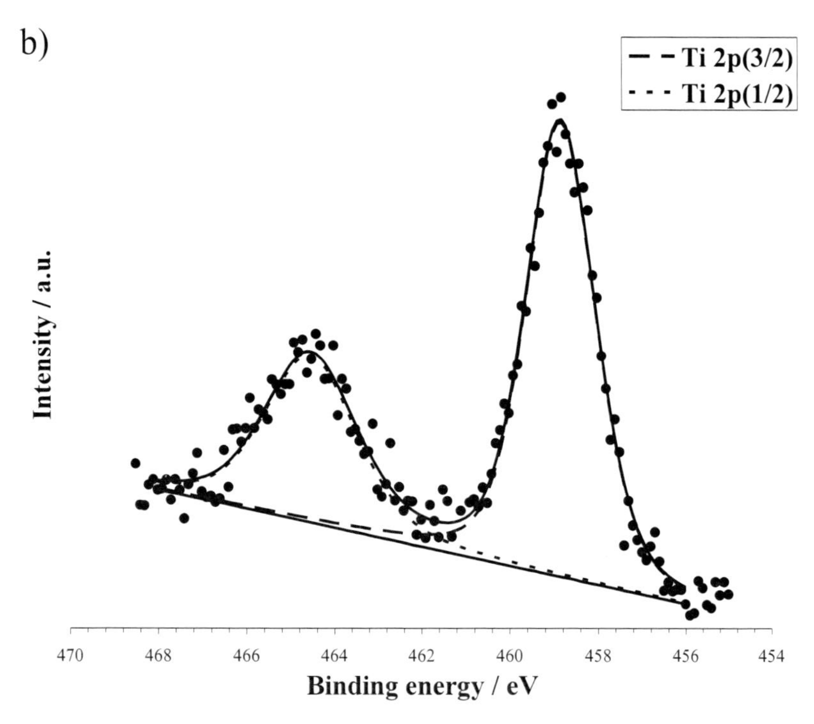

Figure 2. Representative Ti spectra for the TiO_2 film before (a) and after (b) UV irradiation in a climate chamber at 50% relative humidity (RH). The TiO_2 films were UV irradiated using UVC light with an intensity of 10 mW/cm^2 measured at the surface of the TiO_2 films. The spectra were obtained at 600 eV. Reprinted with permission from Elsevier.[30]

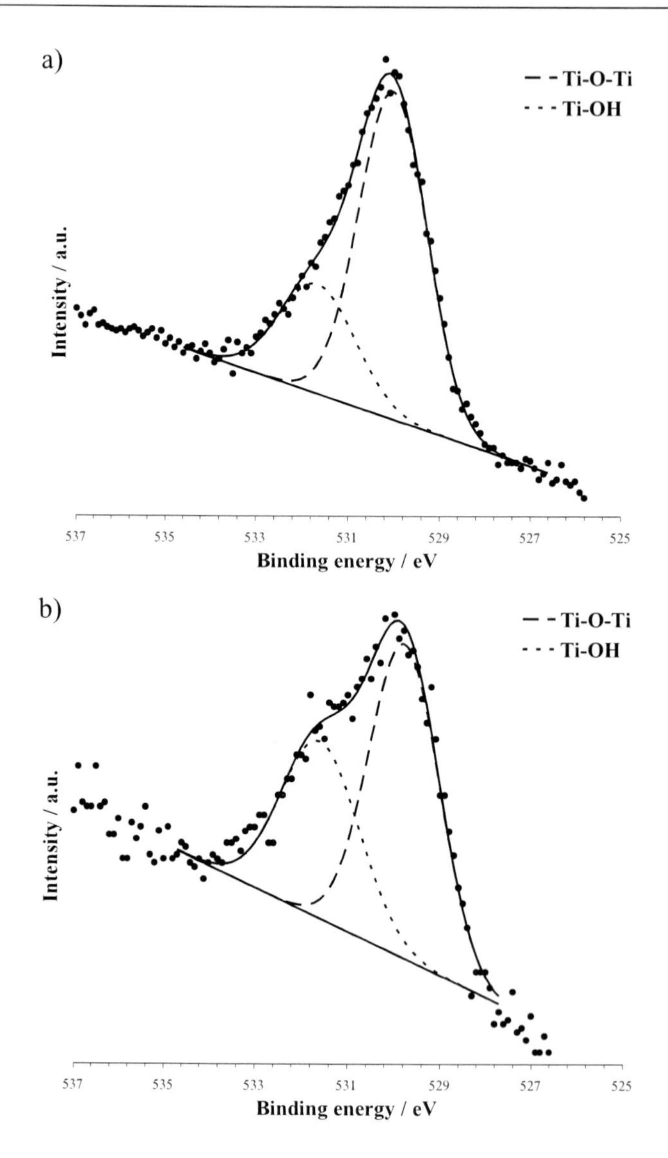

Figure 3. O 1s spectra obtained from XPS analysis of sol-gel films before (a) and after (b) UV irradiation at 50 % RH (25 °C). The TiO_2 films were UV irradiated using UVC light with an intensity of 10 mW/cm^2 measured at the surface of the TiO_2 films. Reprinted with permission from Elsevier.[30].

The peak data in Table 2 show that 24 – 28.8 % of the oxygen was present in form of OH groups. The difference in the amount of OH groups between the films can be ascribed to the difference in surface roughness and surface areas. The films prepared from the smallest particles or aggregates (Table 1), in this case the MW220 film, can be assumed to have the largest surface area. After one hour of UV irradiation the percentage of OH groups was increased by up to 32 % of the amount prior to UV irradiation, see Table 2. The change in the amount of OH groups observed before and after UV irradiation suggested that UV irradiation of the TiO_2 films enhanced the chemisorption of water molecules on the surface of the TiO_2 films.

Table 2. Fitted parameters for the O 1s peak of the TiO_2 films before and after UV irradiation under ambient conditions (254 nm, 6 mW cm^{-2}).

	P25	UV100	SC 134	MW220
O 1s peak data				
$Area_{surface}$ (%)	28.0	25.7	24.0	28.8
$Area_{lattice}$ (%)	72.0	74.3	76.0	71.2
E_b (eV) Ti – O	529.9	529.9	529.9	529.9
E_b (eV) OH	531.8	531.9	531.8	531.8
O 1s peak data after UV				
$Area_{surface}$ (%)	34.4	34.0	30.1	35.8
$Area_{lattice}$ (%)	65.6	66.0	68.9	64.2
E_b (eV) Ti – O	529.9	529.9	529.9	529.9
E_b (eV) OH	531.9	531.7	531.9	531.9
Relative increase in surface oxygen (%)	23.9	32.1	25.4	24.5

Source: Simonsen et al. (2009) *J. Photochem. Photobiology A: Chemistry*, 200, 192-200, Reprinted with permission from Elsevier.

Table 3. Fitted parameters for the O 1s peak of the TiO_2 films before and after UV irradiation under ambient conditions (254 nm, 6 mW cm^{-2}).

	Atomic ratio of $O_{lattice}$-Ti	Atomic ratio of O_{total}-Ti
TiO_2 films		
Degussa P25 film	2.00	2.64
Hombikat UV100 film	1.82	2.45
SC 134 film	1.82	2.52
Sol-gel film	1.87	2.62
TiO_2 films after UV irradiation		
Degussa P25 film	1.88	2.87
Hombikat UV100 film	1.82	2.76
SC 134 film	1.96	2.85
Sol-gel film	1.91	2.97

Source: Simonsen et al. (2009) *J. Photochem. Photobiology A: Chemistry*, 200, 192-200. Reprinted with permission from Elsevier.

In order to compare the amount of OH groups on the TiO_2 films under the different experimental conditions, the atomic ratio between the crystal lattice oxygen and Ti was calculated (Table 3). The atomic concentration was determined by dividing the integrated intensities by the relative sensitivity factors, which for titanium $2p_{3/2}$ is 1.1 and 0.63 for oxygen 1s.[35] However, without a standard TiO_2 reference surface, it was not possible to use these ratios quantitatively, and the ratios were therefore only used semi-quantitatively to obtain knowledge about changes in surface (Ti-OH) and lattice (Ti-O-Ti) oxygen.

From Table 3 it is seen that the atomic ratios between crystal lattice oxygen and Ti were found to be around 1.9 both before and after UV irradiation. In comparison, the ratios between total oxygen and Ti were found to vary significantly before and after UV irradiation. After UV illumination the total oxygen/Ti ratio has increased indicating an increase in the oxygen present at the surface. These results were similar to the results reported by Yu et al. who also observed an increase in the hydroxyl group content after UV irradiation using high resolution XPS analysis.[6] Usually, an increase in the hydroxyl content on the surface of TiO_2 films enhances the photo-induced super-hydrophilicity and photocatalytic activity.[6]

Amount of OH Groups versus Photocatalytic Activity

The photocatalytic activity of the different TiO_2 films was investigated from the degradation profiles of stearic acid (SA), see Figure 4b, where the measured activity of the TiO_2 films were in the order of P25 > MW220 > UV100 > SC 134.

Based on the previous discussion concerning the amount of OH groups on the surface of TiO_2 films before and after UV irradiation, it was suggested that the photocatalytic destruction of organic matter on TiO_2 films proceeded partly through formation of hydroxyl radicals which were formed from adsorbed hydroxyl groups. In Figure 4a the photocatalytic activity of the TiO_2 is plotted against the amount of surface OH groups of the different films after UV irradiation determined from Table 3. The resulting plot showed that there was a good correlation between the amount of surface OH groups and the photocatalytic activity.

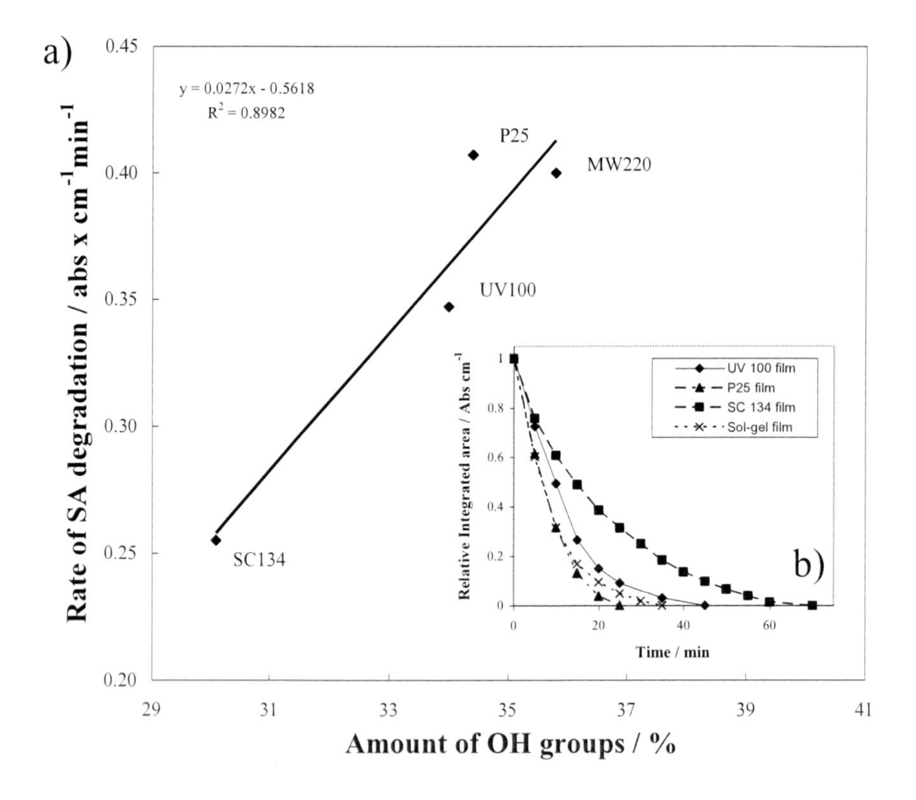

Figure 4. (a) Correlation between the initial rates of degradation of stearic acid (SA) obtained from (b) and the amount of OH groups on the surface of the different TiO_2 films. Subplot (b) Degradation of SA as a function of time. The experiments were conducted under ambient conditions using a UVC light source (254 nm) with an intensity of 6 mW cm^{-2} measured at the surface of the film. Reprinted after permission from Elsevier.[36]

Other factors affecting the photocatalytic activity are the crystallinity and the surface area of the different TiO_2 film (Table 1). The P25 film was the only film found to consist of both anatase and rutile. The anatase/rutile structure of titania has been suggested by different research groups to be beneficial for suppressing the recombination of photogenerated electrons and holes and thus enhance the photocatalytic activity.[34, 37] The UV100 film was found to consist of 86.2 % anatase with the rest being amorphous. In comparison, the in-house prepared MW220 and SC134 films were both found to consist of about 70 % anatase. Based on AFM imaging, the sol-gel and P25 films have the largest surface area available for reaction as they are assembled from the smallest particles. In addition, XPS analysis of the films showed that the MW220 film and the P25 film have the highest amount of OH groups consisting with these films having the largest surface area.

Fundamental Study of the Surface Structural Changes during UV Irradiation – Mechanisms for Photocatalysis and Photoinduced Super Hydrophilicity

In the previous section of this chapter it was shown how XPS is a valuable tool for quantification of OH groups on the surface of TiO_2 films. In this section, a study is presented of the changes in surface structure of the in-house prepared sol-gel TiO_2 films during UV irradiation under different controlled humid conditions. Furthermore, the influence of the surface OH groups on the photocatalytic activity and the photoinduced hydrophilicity was investigated.

From Table 4 it is seen that the atomic ratios between crystal lattice oxygen and Ti were found to be around 2 both before and after UV irradiation. In comparison, the ratios between the surface oxygen (Ti-OH) and Ti were found to vary significantly between TiO_2 films UV irradiated under different humidities. Comparison of the atomic ratios before UV irradiation in Table 4 shows that an increase in the relative humidity (RH) leads to an increase in the amount of adsorbed OH groups on the TiO_2 surface (the atomic ratio increases from 0.7 to 1.5). Although H_2O is easily adsorbed on the surface of TiO_2 films the physically adsorbed H_2O on TiO_2 desorb under the ultrahigh vacuum condition of the XPS system.[34] Thus, this change should be assigned to chemically bonded OH groups. In comparison an even higher increase in the amount of OH groups was observed when the TiO_2 films were UV irradiated under the same conditions (the atomic ratio increased from 1.1 to 2.1). This difference between the atomic ratios suggested that chemisorption of H_2O occurred during UV irradiation leading to formation of OH groups. The high increase in the amount of OH groups at high RH may be due to adsorption of H_2O in the surface of the TiO_2 film.

The increase in the amount of OH groups was suggested to be due to dissociative adsorption of water at oxygen vacancies. The photogenerated holes were assumed to be trapped at lattice oxygen at the surface of the TiO_2 film. The trapped holes were thought to weaken the bond between the titanium atoms and the lattice oxygen. As a result such oxygen was liberated to create oxygen vacancies. At these defect sites water molecules were dissociatively adsorbed creating a more hydroxylated surface. The formation of oxygen vacancies during UV irradiation of TiO_2 films under ultra high vacuum (UHV) conditions has been reported by Shultz et al. (1995).[38] Subsequent water exposia was found to regenerate the TiO_2 surface by production of surface hydroxyl groups.[39] Water has been reported to be easily adsorbed on the TiO_2 films and interact with the vacancies even if it is present in almost negligible amounts.[40] The interaction between the vacancy and the water molecule was assumed to result in diffusion of a proton to a neighboring oxygen atom thereby creating two hydroxyl groups. According to density functional theory (DFT) calculation conducted by Wendt et al. (2005)[40] the system was stabilized by the diffusion of the proton to a more distant oxygen atom. Moreover, DFT calculations have shown that the adsorption of molecular oxygen can stabilize the system even further.[40] However, this process was not believed to occur to a greater extent if the principle part of the vacancies was already occupied by hydroxyl groups. That molecular oxygen did not occupy these vacancies immediately after creation was suggested to be a result of the difference in kinetics between the two competing molecules (molecular oxygen and water) during interaction with the oxygen vacancies.

Effect of Humidity on the Photocatalytic Activity and the Photoinduced Hydrophilicity

Increasing humidity was found to have a positive effect on the photocatalytic activity determined by stearic acid (SA) degradation, and the increase in activity was suggested to be caused by the increased amount of OH groups on the surface of the TiO_2 films observed by XPS analysis. As Figure 5a shows there was a very good correlation between the photocatalytic activity and the amount of OH groups on the TiO_2 film. The photocatalytic acitivity was given by the initial SA degradation rates obtained from the degradation profiles obtained under different humidities in Figure 5b.

Table 4. Fitted XPS peak data for the sol-gel TiO_2 film at different relative humidities (RH).

	Atomic ratio $O_{lattice}$-Ti	Atomic ratio O_{OH}-Ti	Atomic ratio O_{total}-Ti
TiO_2 films before UV irradiation			
35 % RH	1.9	0.7	2.6
50 % RH	2.1	1.0	3.1
75 % RH	1.8	1.4	3.2
95 % RH	1.9	1.5	3.4
TiO_2 films after UV irradiation			
35 % RH	1.9	1.1	3.0
50 % RH	1.9	1.6	3.5
75 % RH	2.0	2.1	4.1
95 % RH	2.2	2.0	4.2

Source: Simonsen et al. (2009) *Applied Surface Science*, 255, 8054-8062. Reprinted with permission from Elsevier.

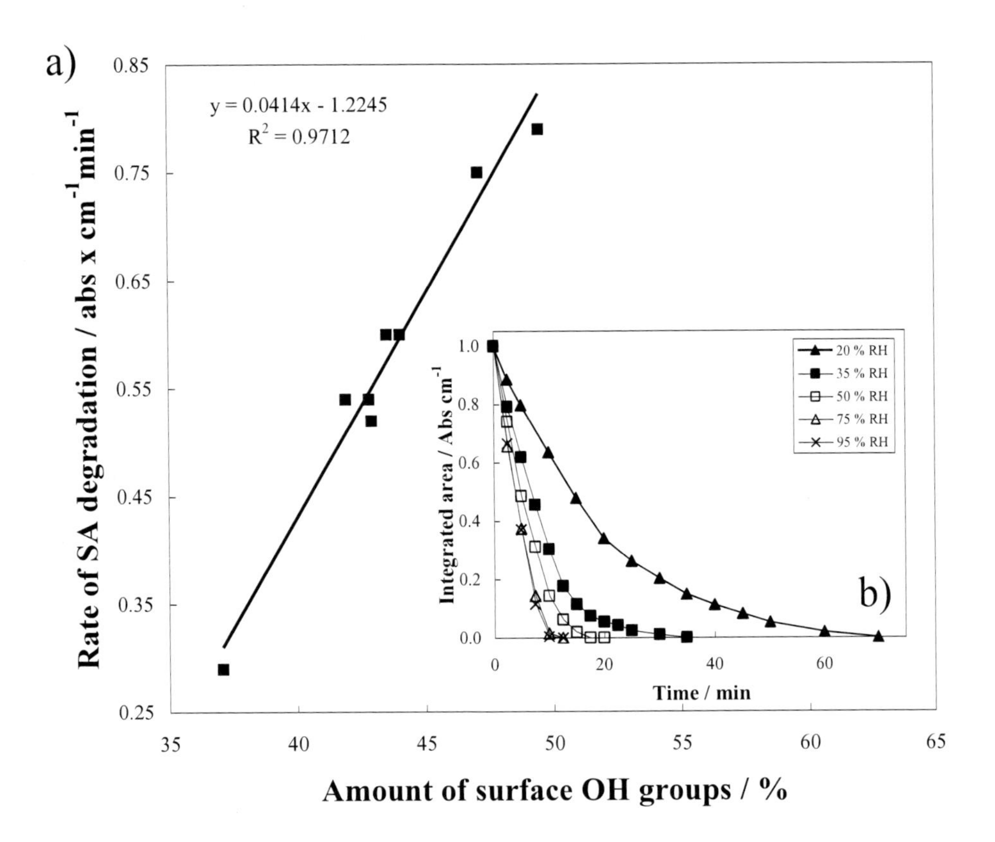

Figure 5. (a) Correlation between the photocatalytic activity (initial rate of SA degradation) and the percentage of OH groups on the surface of the TiO_2 films. (b) Degradation profiles of SA under varying humidity (20 – 95 % RH). In the experiments the initial area of SA varied between 6.2 – 7 Abs cm^{-1}. The films were UV irradiated using UVC light with an intensity of 10 mW cm^{-2}. Reprinted with permission from Elsevier.[30]

In addition, the effect of the humidity on the photoinduced superhydrophilicity has been investigated through contact angle (CA) measurements. The initial CA of a water droplet on pristine TiO_2 films under the different humid conditions was found to be in an interval between 28° and 40°. The difference in the initial values of CA was considered to be due to the difference in the amount of OH groups on the surface of the TiO_2 films at different humidity (Table 4). After UV irradiation was initiated, the CA dropped to a value close to zero within 30 min. In comparison, the dark recovery of the TiO_2 films was found to proceed over a period of up to 60 days (Figure 6b). The rate of conversion is given by the empirical conversion rate obtained from the plot of CA^{-1} versus UV irradiation time as reported by Sakai et al. (2003).[9] For the superhydrophilic conversion under different humid conditions (20 – 95 % RH), the empirical conversion rate is given by the plot in Figure 6a. The results of this study showed that there was a linearly relationship between the rate of hydrophilic conversion given by the reciprocal CA and RH (Figure 7a). Additionally, it was found that there was a linear relation between the rate of conversion and the amount of OH groups on the TiO_2 film surface (Figure 7b).

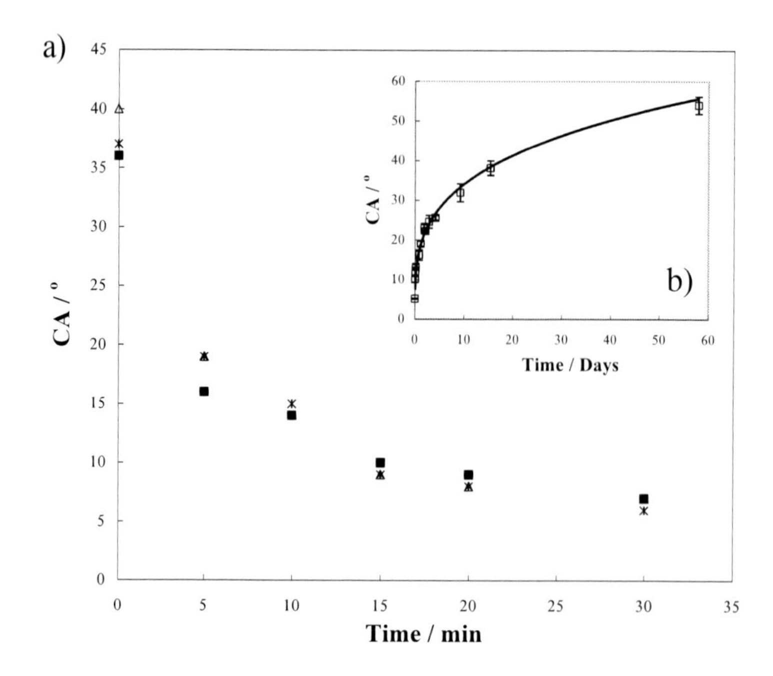

Figure 6. (a) Contact angle (CA) of water as a function of UV irradiation time. The films were UV irradiated using UVC light with an intensity of 10 mW cm^{-2}. (b) The TiO_2 films measured by the water CA as a function of time during dark storage. Reprinted with permission from Elsevier.[30]

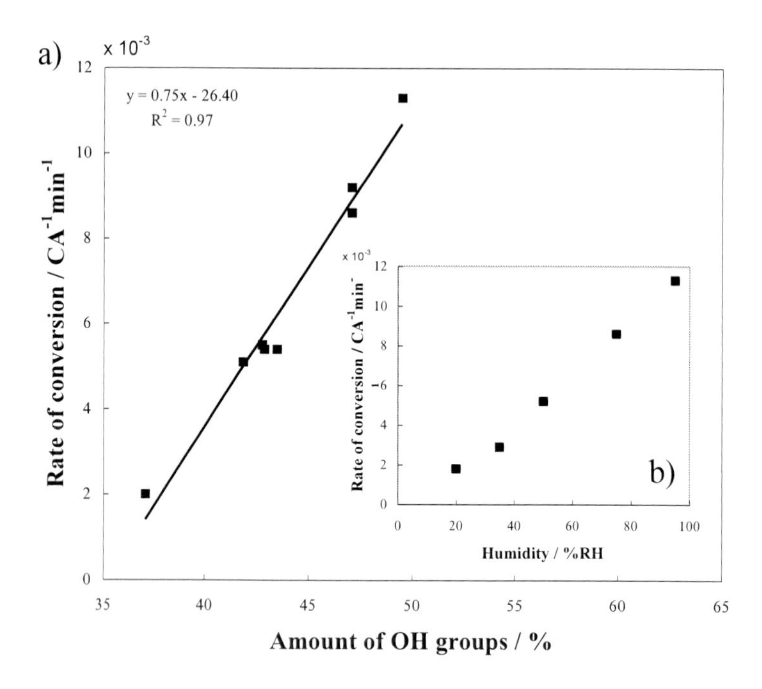

Figure 7. (a) Hydrophilic conversion rate ($CA^{-1} \times min^{-1}$) as a function of relative humidity. (b) Correlation between the amount of surface OH groups of the TiO_2 films and the hydrophilic conversion rate ($CA^{-1} \times min^{-1}$). Reprinted with permission from Elsevier.[30]

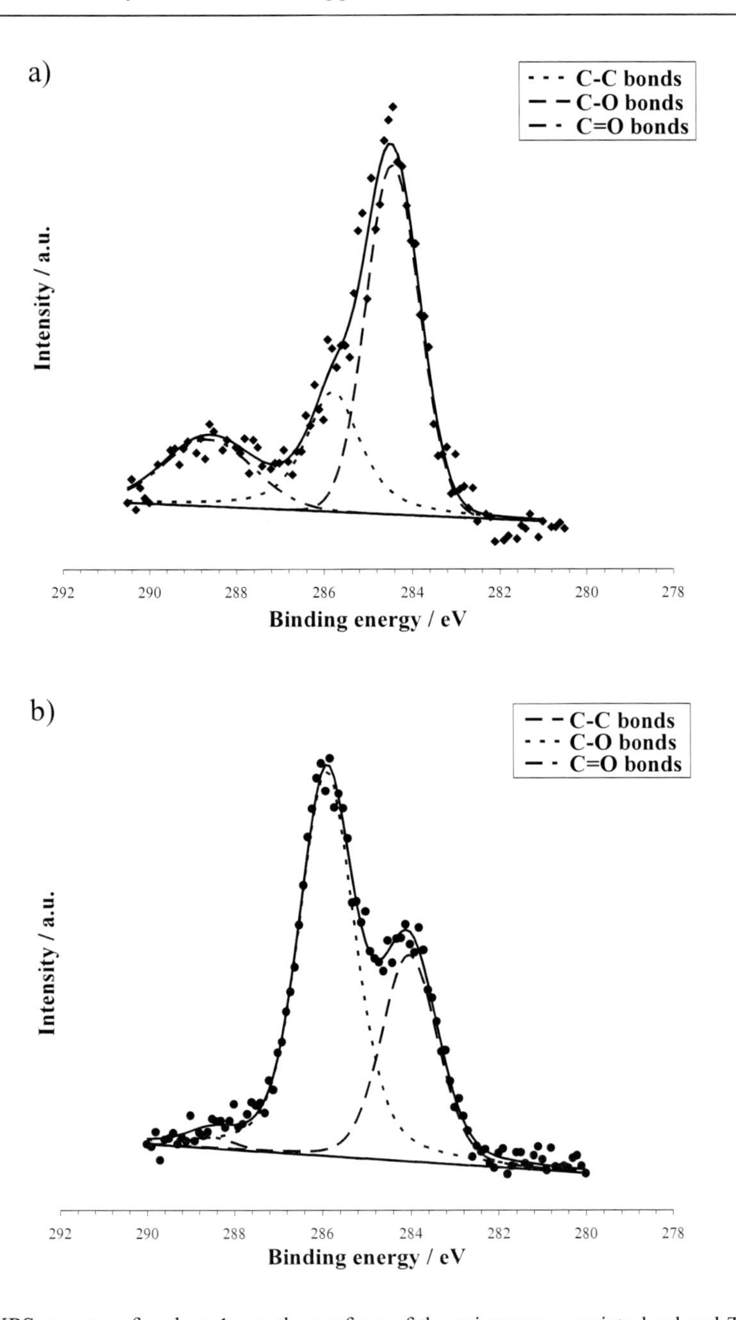

Figure 8. The XPS spectra of carbon 1s on the surface of the microwave assisted sol-gel TiO_2 film (MW220) before (a) and after (b) UV irradiation. The C 1s signal was deconvoluted into three peaks. The first peak at 288.4 eV is due to C=O bonds, the second peak at 285.5 eV is due to C-O bonds and the third peak at 284.3 eV is due to C-C bonds. The film was UV irradiated under ambient conditions using UVC light (8 mW cm^{-2}).

In regard to the two models suggested for the photoinduced superhydrophilicity the results of the presented study supported the surface structural change model in the sense that UV irradiation was found to increase the amount of hydroxyl groups on the surface of the investigated TiO_2 films creating a more hydrophilic surface. However, it was found that despite the samples showing superhydrophilic properties (small CA), the amount of OH groups on the surface varied significantly. This result may suggest that superhydrophilicity was obtained when a certain amount of OH groups are present on the TiO_2 surface and that a further increase will not lead to a change in the hydrophilic property. The difference in time required to obtain superhydrophilicity may however also be described by the difference in the photomineralization rate of organic contaminants on the TiO_2 surface, since it was found that increasing amounts of OH groups on the TiO_2 surface enhanced the photocatalytic activity.

XPS analysis of the surface of an initial hydrophobic TiO_2 films before and after UV irradiation confirmed this explanation. A significant change in the appearance of the C 1s peak was observed after one hour of UV irradiation (Figure 8). From the detailed scan of the C 1s region obtained at 380 eV it is seen that the ratio between the area of peak 2, assigned to C-O, and the area of peak 3, assigned to C-C bonds, is reversed. This provided a clear indication that the chained carbon molecules present on the surface was oxidized during the UV irradiation.

Based on the conducted experiments it was found that the amount of OH groups on the TiO_2 surface highly influenced the photocatalytic activity and the photoinduced superhydrophilicity, which may suggest that the two mechanisms are closely related. It is suggested that UV irradiation of the TiO_2 films may lead to the formation of vacancies, which in turn will react with H_2O and lead to the formation of OH groups. These OH groups are then thought to react with photogenerated holes leading to the formation of hydroxyl radicals which will mediate the destruction of residual hydrophobic organics on the TiO_2. This mechanism suggests that the superhydrophilicity was obtained through a combination of the photomineralization and surface structural change model. Lee et al. (2007)[41] has in their investigation of the role of water adsorption on the photoinduced superhydrophilicity on TiO_2 films arrived to similar conclusions, in which the formation of hydroxyl groups and the removal of pollutants by photocatalysis jointly form the necessary conditions for superhydrophilicity.

The recovery of the TiO_2 film is ascribed to the fact that the surface is contaminated by adsorbing gaseous pollutants from the air and that the vacancies are healed or replaced gradually by oxygen atoms creating a more sTable surface (DFT calculations conducted by Wendt et al., 2005)[40], which changes the surface wettability from hydrophilic to hydrophobic.

XPS Analysis of Ti/Pt_{90}-Ir_{10} Anode before and after Electrolysis under Oxygen and Chlorine Evolution Conditions

Regarding the other advanced oxidation process considered in this chapter; electrochemical oxidation of organics, the minor study presented in this section focused on the extent of oxidation of the platinum anode after electrolysis in two different chemical domains; the oxygen evolution reaction studied in a 0.10 M inert sodium sulphate electrolyte and the chlorine evolution reaction studied in 0.10 M sodium chloride.

The spectral analysis was initiated by survey scans in order to get insight into the diversity of contaminating elements present on the surface. The survey spectra were obtained at a photon beam energy of 600 eV. In addition, a cleaning procedure including sonolytic treatment in acetone/methanol mixture followed by sonolytic rinsing in demineralised water was tested in order to test how strongly adsorbed the contaminants were attached to the surface. Three examples of spectra of the raw electrode before and after sputtering and the cleaning procedure are shown in Figure 9.

The dominant high intensity peaks were the O 1s and the Pt 4f doublets located at binding energies around 530 and 71 eV respectively. The signal for the Ir 4f doublets around 61 eV were in the spectrum for the raw and cleaned surface to weak, but the peaks were more pronounced when the surface was sputtered and the bulk composition of the material showed up. Like most other surfaces, the survey spectra of the raw electrode revealed the presence of several surface contaminants with carbon, chlorine and sodium as the most distinguished. The cleaning procedure was especially efficient in removing adsorbed sodium chlorine and some unidentified peaks in the 80-90 eV region of the spectra. The adsorbed layer of carbon containing molecules was jointly removed with the main part of the oxygen by the sputtering procedure, indicating that oxygen almost exclusively was bounded in surface chemical species.

Detailed scans of the Pt 4f region were obtained using a photon energy of 160 eV and the Pt 4f 7/2 and 5/2 doublets were seen very clearly in the sputtered sample (Figure 10).

The spectrum was deconvoluted with mixed Gaussian-Lorentzian functions into four peaks labeled 1, 2 (7/2), 1', and 2' (5/2). It is clearly demonstrated throughout the literature that the dominant peaks 1 and 1' at 71.1 and 74.4 eV respectively (Pt1), can be ascribed to surface metallic Pt(0)[42-47], whereas discrepancies were found regarding the origin of peaks 2, and 2' (Pt2). Most papers refer to the peaks as being assigned to Pt(II) chemical states such as PtO and $Pt(OH)_2$[26, 43, 45, 48, 49], whereas contradictory suggestions assign the Pt2 doublet to bulk metallic Pt(0).[42] The latter proposal was supported by theoretical calculations on platinum nanoparticles where Kua and Goddard[50, 51] used nonlocal density functional methods, based on the interstitial electron model, and found that bulk Pt has a $5d^8 6s^2$ electronic configuration, while the surface atoms at the Pt(111) face have a $5d^9 6s^1$ configuration. In addition, the deconvoluted spectrum in Figure 10 was obtained after sputtering and hence presented an elemental analysis of the bulk anode material. In this case no oxidized states of Pt could be expected and the present results in this way supported the ascription of the Pt2 doublet as representing bulk Pt(0).

The structural change of the anode surface before and after the oxygen and chlorine evolution reaction was studied by galvanostatic electrolysis at 200 mA cm^{-2} current density in 0.10 M sodium sulphate and 0.10 M sodium chloride electrolyte, respectively. At platinum based electrodes sodium sulphate can be regarded as an inert supporting electrolyte with no significant anodic formation of peroxodisulphate under the applied conditions. In both electrolytes, hydrogen evolution was the main cathodic reaction. The samples were studied after 30 and 60 min of electrolysis using two separate (but otherwise identical) sample objects. The deconvoluted spectra for the detailed scans in the Pt 4f region are seen in Figure 11. The reference samples were submerged in the respective electrolytes for 60 min with no applied voltage. The scans of the samples electrolyzed in 0.10 M sodium sulphate are presented in a, b, and c and the samples electrolyzed in 0.10 M sodium chloride are presented to the right in d, e, and f.

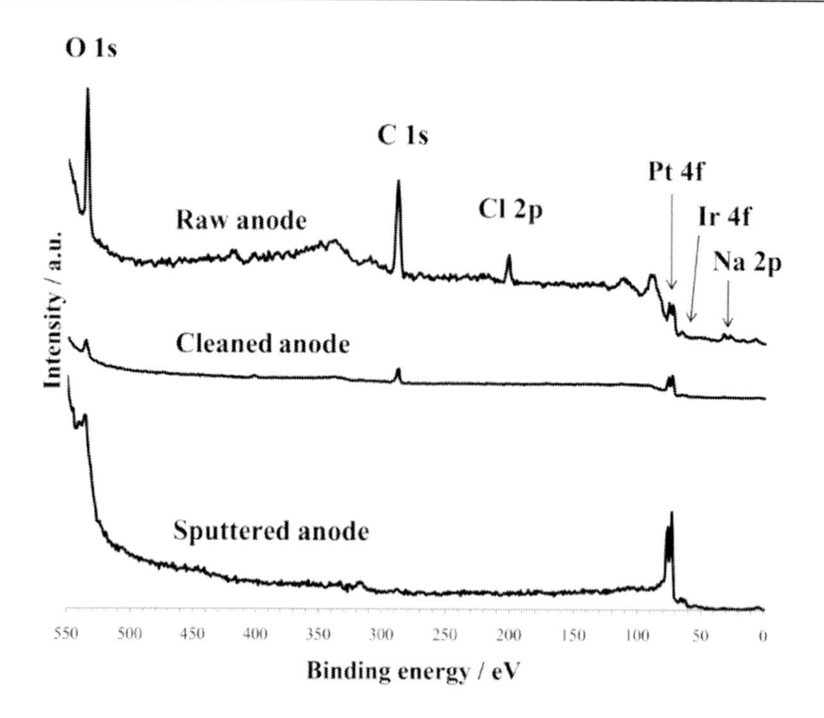

Figure 9. Examples of survey spectra of the Ti/Pt_{90}-Ir_{10} anode. The spectra were obtained at a photon energy of 600 Ev.

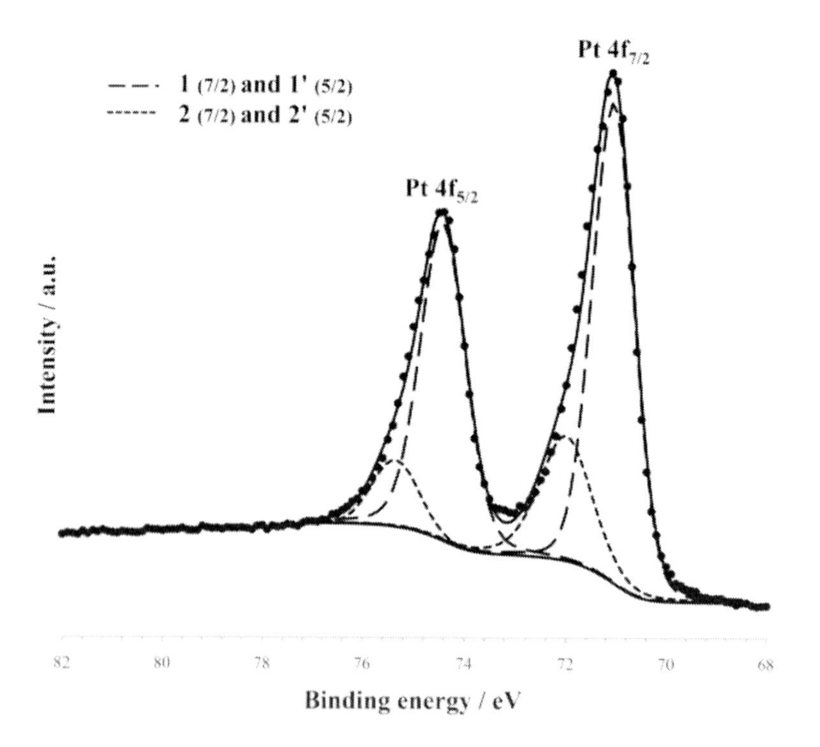

Figure 10. The XPS spectra of platinum 4f clearly showing the 7/2 and 5/2 doublet. The spectrum was deconvoluted into two doublets with the peaks 1 and 1' (Pt1) being ascribed to surface Pt(0) and 2 and 2' being ascribed to bulk Pt(0) (Pt2). The scan was obtained at a photon energy of 160 eV.

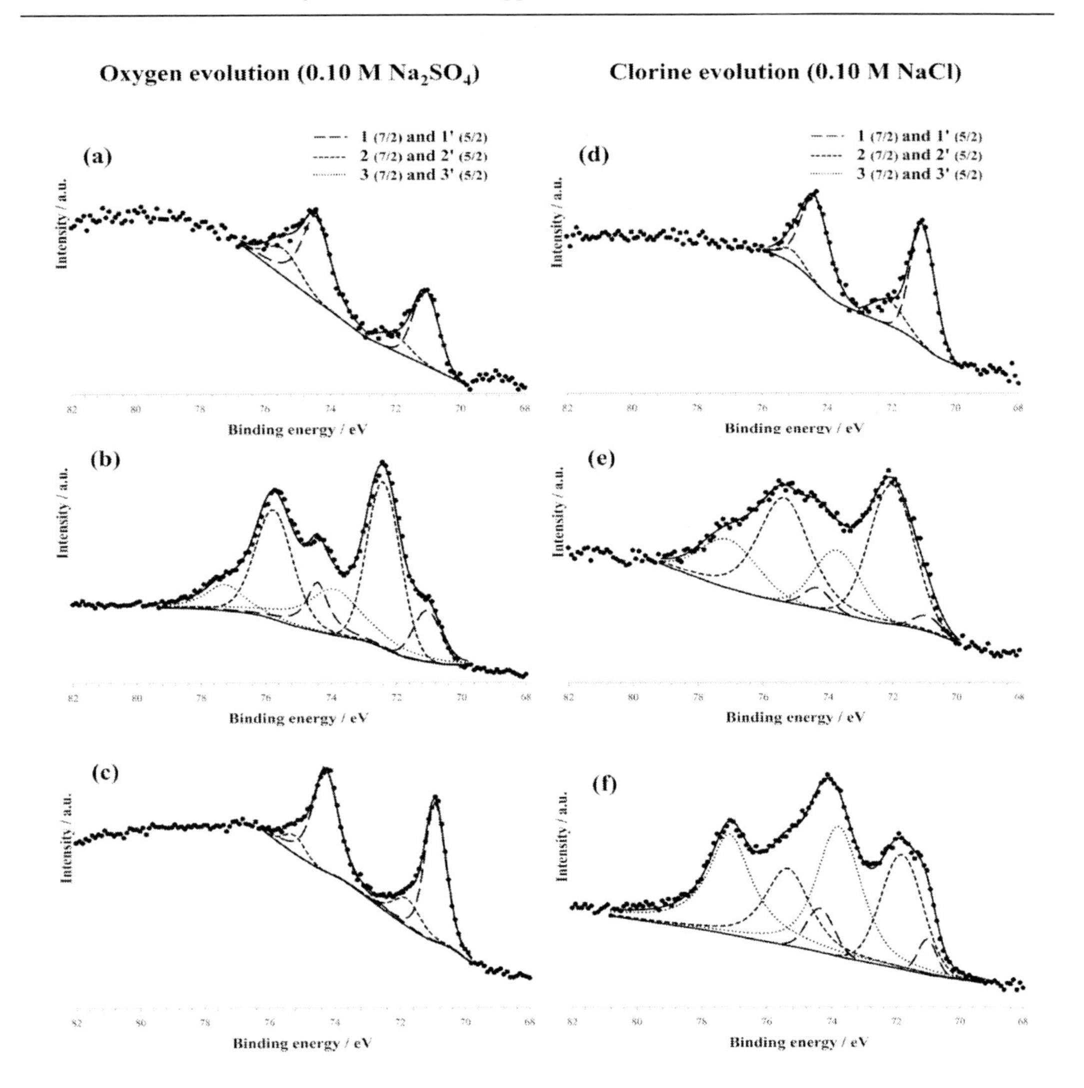

Figure 11. The spectra from the detailed scans of the Pt 4f region of the samples after electrolysis. The samples in (a), (b), and (c) have been electrolyzed under oxygen evolution conditions in 0.10 M sodium sulphate and (d), (e), and (f) have been electrolyzed under chlorine evolution conditions in 0.10 M sodium chloride. (a) and (d) are reference samples submerged in the elecotrolytes with no applied voltage for 60 min, (b) and (e) have been electrolyzed for 30 min, and (c) and (f) have been electrolyzed for 60 min at a constant current density of 200 mA cm^{-2}.

The spectra of the reference samples could in both electrolytes be deconvoluted in the same manner as Figure 10 with the Pt1 and Pt2 doublets assigned to surface and bulk Pt(0). Hence, the anode surface was not significantly oxidized in any of the two investigated electrolytes when no voltage was applied, showing the noble nature of the Pt surface not amenable to oxidation at ambient conditions. For both the oxygen and the chlorine evolution study, the spectra broadened significantly during the electrolysis process and introduction of a Pt3 doublet 3 (7/2) and 3' (5/2) peak was necessary in order to obtain a proper fit to the experimental values. The Pt3 doublet can be assigned to the formation of oxidized chemical states of platinum such as Pt(II) in PtO and $Pt(OH)_2$[43] or Pt(IV) in PtO_2[42] considering the discrepancies found in literature. Since Pt(IV) has a greater binding energy than Pt(II), the chemical shift can be used to specify valence. The chemical shift between Pt(0) and Pt(II) (in PtO) is ≈3 eV and ≈1.6 eV (in $Pt(OH)_2$) and between Pt(0) and Pt(IV) (in PtO_2) ≈ 4 eV.[44] The data presented in Figure 11 indicate chemical shifts between Pt1 and Pt3 of 2.6-3 eV suggesting that platinum in the formed oxide layer was in the +2 chemical state, however strongly implying that the oxide layer was composed of mixed platinum oxide and hydroxide.[42]

The spectral analysis of the samples electrolyzed in sodium sulphate under oxygen evolution conditions showed the development of an oxidized platinum layer seen by the presence of the Pt3 doublet after 30 min. However, the Pt3 peaks were not present in analysis of the 60 min sample. This indicates that the oxidized layer formed under these conditions is unstable in nature. It is suggested that with increased time of electrolysis the system reached steady-state conditions, where the oxidized layer itself participated in the oxygen evolution reaction during water electrolysis. This lead to a return into the initial composition of the surface chemical states as seen in Figure 11c. Under the chlorine evolution conditions, the development of the oxide layer was even more predominant with the Pt3 doublet increasing its area from the 30 min to the 60 min sample. This was in fine agreement with the decreasing area of the Pt1 doublet (surface metallic Pt(0)) as the electrolysis time was increased and platinum was oxidized to higher chemical oxidation states (Pt3). The difference observed in the samples electrolyzed in the two different electrolyte compositions demonstrated the harsh oxidative environment under especially the chlorine evolution process. When chlorine is produced from the combination of two oxidized chloride ions, it rapidly hydrolyses to the conjugate acid-base pair hypochlorous acid/hypochlorite, both strong oxidants forming a more stable oxide layer on the platinum surface compared to the oxide layer formed in the water electrolysis process. The oxide layer will over time change the electrocatalytic properties of the metallic platinum, however creating a higher surface area due a higher roughness. The demonstrated formation of an oxidized layer of platinum during electrolysis supported the model proposed for electrochemical oxidation presented in the introductory section, where platinum based electrodes were characterized as active anodes utilizing chemisorbed oxygen for the oxidation of organic pollutants. The next successive XPS study will focus on the change of the surface with the presence of an organic pollutant in the solution, which might promote a regeneration of the initial dominating Pt(0) surface.

CONCLUSION

Research presented in this chapter demonstrate the strength of the X-ray photoelectron spectroscopy technique, when it comes to increasing the understanding of surface chemical reactions and properties. In the area of environmental engineering significant effort in further research within advanced oxidation processes are needed in order to increase efficiencies of the processes and durability of the materials in order to decrease especially operating costs. An still increasing shortage of water in both developed and undeveloped countries force us to think of waste water as a valuable resource, and AOPs are key solutions in the possibilities to reuse water. XPS has with great success been applied in elucidating the mechanisms in photocatalytic degradation of organics and the concept of super hydrophilicity through detailed investigations of the amount of OH groups at the surfaces of TiO_2 films under different conditions. In addition to process investigations, XPS is a powerful tool for product analysis through fundamental studies of the TiO_2 films and comparisons of production methods, coating techniques etc. In electrochemical oxidation both processes and properties under different environmental conditions has been investigated as well. For both heterogeneous oxidation processes XPS will remain an important analytical tool for maturing the techniques for larger scale use within treatment of polluted air and water.

REFERENCES

[1] Comninellis, C; Kapalka, A; Malato, S; Parsons, S. A; Poulios, L; Mantzavinos, D. *Journal of Chemical Technology and Biotechnology*, 2008, 83, 769-776.
[2] Mantzavinos, D; Kassinos, D; Parsons, S. A. *Water Res.*, 2009, 43, 3901-3901.
[3] Mills, A; LeHunte, S. *Journal of Photochemistry and Photobiology A-Chemistry,* 1997, 108, 1-35.
[4] Fujishima, A; Rao, T; Tryk, D. *Journal of Photochemistry and Photobiology C: Photochemistry Reviews,* 2000, 1, 1-21.
[5] Liu, HM; Yang, WS; Ma, Y; Cao, Y; Yao, JN; Zhang, J; Hu, TD. *Langmuir,* 2003, 19, 3001-3005.
[6] Yu, J; Yu, JC; Ho, W; Jiang, Z. *New Journal of Chemistry,* 2002, 26, 607-613.
[7] Fujishima, A; Rao, TN. *Pure and Applied Chemistry,* 1998, 70, 2177-2187.
[8] Sakai, N; Fujishima, A; Watanabe, T; Hashimoto, K. *J Phys Chem B,* 2001, 105, 3023-3026.
[9] Sakai, N; Fujishima, A; Watanabe, T; Hashimoto, K. *J Phys Chem B,* 2003, 107, 1028-1035.
[10] Zubkov, T; Stahl, D; Thompson, TL; Panayotov, D; Diwald, O; Yates, JT. *J Phys Chem B,* 2005, 109, 15454-15462.
[11] Mills, A; Crow, M. *Journal of Physical Chemistry C,* 2007, 111, 6009-6016.
[12] Panizza, M; Cerisola, G. *Chem. Rev.*, 2009, 109, 6541-6569.
[13] Kraft, A. *Platinum Metals Review,* 2008, 52, 177-185.
[14] Serrano, K; Michaud, PA; Comninellis, C; Savall, A. *Electrochim. Acta,* 2002, 48, 431-436.
[15] Chen, GH. *Separation and Purification Technology,* 2004, 38, 11-41.
[16] Martinez-Huitle, CA; Ferro, S. *Chem. Soc. Rev.,* 2006, 35, 1324-1340.
[17] Comninellis, C; Vercesi, GP. *J. Appl. Electrochem.,* 1991, 21, 335-345.
[18] Vercesi, GP; Rolewicz, J; Comninellis, C; Hinden, J. *Thermochimica Acta,* 1991, 176, 31-47.
[19] Kraft, A. *International Journal of Electrochemical Science,* 2007, 2, 355-385.
[20] Kapalka, A; Foti, G; Comninellis, C. *J. Appl. Electrochem.,* 2008, 38, 7-16.
[21] Comninellis, C. *Electrochim. Acta,* 1994, 39, 1857-1862.
[22] Kapalka, A; Foti, G; Comninellis, C. *Electrochemistry Communications,* 2008, 10, 607-610.
[23] Fierro, S; Nagel, T; Baltruschat, H; Comninellis, C. *Electrochemical and Solid State Letters,* 2008, 11, E20-E23.

[24] Fierro, S; Nagel, T; Baltruschat, H; Comninellis, C. *Electrochemistry Communications,* 2007, 9, 1969-1974.
[25] Derry, GN; Ross, PN. *Surf. Sci.,* 1984, 140, 165-180.
[26] Puglia, C; Nilsson, A; Hernnas, B; Karis, O; Bennich, P; Martensson, N. *Surf. Sci.,* 1995, 342, 119-133.
[27] Saliba, N; Tsai, YL; Panja, C; Koel, BE. *Surf. Sci.,* 1999, 419, 79-88.
[28] Gambardella, P; Sljivancanin, Z; Hammer, B; Blanc, M; Kuhnke, K; Kern, K. *Phys. Rev. Lett.,* 2001, 87.
[29] Kodera, F; Kuwahara, Y; Nakazawa, A; Umeda, M. *J. Power Sources,* 2007, 172, 698-703.
[30] Simonsen, ME; Li, ZS; Sogaard, EG. *Appl. Surf. Sci.,* 2009, 255, 8054-8062.
[31] Que, W; Zhou, Y; Lam, YL; Chan, YC; Kam, CH. *Applied Physics A-Materials Science & Processing,* 2001, 73, 171-176.
[32] Yu, JG; Zhao, XJ; Zhao, QN. *Thin Solid Films,* 2000, 379, 7-14.
[33] Liqiang, J; Xiaojuna, S; Weimina, C; Zilic, X; Yaoguoc, D; Honggang, F. *Journal of Physics and Chemistry of Solids,* 2003, 64, 615-623.
[34] Yu, J; Yu, H; Cheng, B; Zhao, X; Yu, JC; Ho, W. *The Journal of Physical Chemistry B,* 2003, 107, 13871-13879.
[35] Wagner, CD; Riggs, WM; Davis, LE; Moulder, JF; Muilenberg, GE. Eds; In *Handbook of X-Ray Photoelectron Spectroscopy;* Perkin-Elmer Corporation: Eden Prairie, 1979.
[36] Simonsen, ME; Jensen, H; Li, ZS; Sogaard, EG. *Journal of Photochemistry and Photobiology A-Chemistry,* 2008, 200, 192-200.
[37] Bickley, RI; Gonzalezcarreno, T; Lees, JS; Palmisano, L; Tilley, RJD. *Journal of Solid State Chemistry,* 1991, 92, 178-190.
[38] Shultz, AN; Jang, W; Hetherington, WM; Baer, DR; Wang, LQ; Engelhard, MH. *Surf. Sci.,* 1995, 339, 114-124.
[39] Wang, LQ; Baer, DR; Engelhard, MH. *Structure and Properties of Interfaces in Ceramics,* 1995, 357, 97-102.
[40] Wendt, S; Schaub, R; Matthiesen, J; Vestergaard, EK; Wahlstrom, E; Rasmussen, MD; Thostrup, P; Molina, LM; Laegsgaard, E; Stensgaard, I; Hammer, B; Besenbacher, F. *Surf. Sci.,* 2005, 598, 226-245.
[41] Lee, FK; Andreatta, G; Benattar, J. *J. Appl. Phys. Lett.* 2007, 90.
[42] Zhang, GX; Yang, DQ; Sacher, E. *Journal of Physical Chemistry C,* 2007, 111, 565-570.
[43] Acharya, CK; Li, W; Liu, ZF; Kwon, G; Turner, CH; Lane, AM; Nikles, D; Klein, T; Weaver, M. *J. Power Sources,* 2009, 192, 324-329.
[44] Wagner, CD; Naumkin, AV; Kraut-Vass, A; Allison, JW; Powell, CJ; Rumble, JR. NIST X-ray Photoelectron Spectroscopy (XPS) Database, Version 3.5. http://srdata.nist.gov/xps/ (accessed 12/16/2009, 2009).
[45] Wang, GQ; Lin, Y; Xiao, XR; Li, XP; Wang, WB. *Surf. Interface Anal.,* 2004, 36, 1437-1440.
[46] da Silva, L; Alvesb, VA; de Castroc, SC; Boodts, JFC. *Colloids Surf. Physicochem. Eng. Aspects,* 2000, 170, 119-126.
[47] Delgass, WN; Hughes, TR; Fadley, CS. *Catalysis Reviews,* 1970, 4, 179-&.
[48] Liu, Z; Gan, L; Hong, L; Chen, W; Lee, J. *J. Power Sources,* 2005, 139, 73-78.

[49] Shukla, AK; Ravikumar, MK; Roy, A; Barman, SR; Sarma, DD; Arico, AS; Antonucci, V; Pino, L; Giordano, N. *J. Electrochem. Soc.,* 1994, 141, 1517-1522.
[50] Kua, J; Goddard, WA. *J Phys Chem B,* 1998, 102, 9481-9491.
[51] Kua, J; Goddard, WA. *J Phys Chem B,* 1998, 102, 9492-9500.

In: The Sol-Gel Process
Editor: Rachel E. Morris

ISBN 978-1-61761-321-0

Chapter 14

SOL-GEL CHEMISTRY OF TRANSITIONAL METALS IN AQUEOUS MEDIUM

Luminita Predoana and Maria Zaharescu
Ilie Murgulescu Institute of Physical Chemistry, Romanian Academy,
Splaiul Independentei, Bucharest, Romania

ABSTRACT

The chemistry of the sol-gel METHOD in aqueous medium in systems containing transition metals can be very complicated as several molecular species depending on the oxidation state of the metal, the pH or the concentration could be formed. In the aqueous medium during the reaction of hydrolysis of inorganic salts, new ionic species or precipitates can occur. To prevent this precipitation along the metal salts dissolved in water, organic ligands are used. In this synthesis pathway, the metal ions are coordinated by organic ligands with a strong electron donor character which prevents their precipitation in the solution. By water evaporation, an amorphous inorganic-organic network is formed. In this network, the metal ions are homogeneously dispersed and linked by coordinative bonds. The sol-gel method in aqueous medium is a low cost procedure and is an environmental friendly alternative compared to the classical sol-gel technique that uses organic solvents and metal alkoxides.

In the present work, the sol-gel chemistry in aqueous medium of cobalt-based system (Li-Co and La-Co) with citric acid as chelating agent is presented. Information concerning the influence of the precursors on the sol-gel method and the properties of the resulted gels and powders are discussed.

1. INTRODUCTION

To obtain oxide materials, conventional ceramic methods based on solid phase reactions at high temperatures are commonly used. However, in recent years, the interest in the synthesis of oxide materials by unconventional methods significantly increased. Some of those methods such as precipitation/coprecipitation methods [1, 2], hydrothermal method [3], the method of decomposition of complex combinations [2], sol-gel method [4, 5] play an important role.

Among these un-conventional methods in solution, the sol-gel process based on alkoxide or the sol-gel process in aqueous medium is most intensively studied and used [6].

Sol-gel alkoxide route presents a major interest in the field of ceramics. The processing of oxide compounds by alkoxide route is based on the hydrolysis and polycondensation of metallic alkoxides $M(OR)_n$. The starting materials, the metallic alkoxide (with general formula $M(OR)_n$), are commercially available for a large variety of metals. As shown by Brinker and Scherer [7] by changing the reaction conditions of the synthesis, particles, films, fibers, bulk materials can be obtained. By reactions in the presence of organic compounds, hybrid materials are synthesized [4].

Sol-gel process in aqueous medium uses inorganic salts and chelating agents of carboxylic acids or polyol type as precursors. The citrate route using citric acid as a chelating agent compared to other methods, has the advantage of not only a good mixing of the components at the molecular scale, but also offers the possibility of obtaining films and fibers which have technological importance [5]. Carboxylic precursors are used in the sol-gel route for obtaining oxide materials especially in systems containing transition metals with the role of chelating agents in order to avoid their precipitation. The chemistry of the sol-gel process in aqueous medium, in systems containing transition metals can be very complicated as several molecular species depending on the oxidation state of the metal, the pH or the concentration could be formed. Moreover, in the case of non-tetravalent cations oxides, hydroxides or oxo-hydroxides are obtained. Hydrolysis of salts may involve formation of hydrated cations, anions or both types of ions. The hydrolysis of metal cations was first studied by Bjerrum in the early twentieth century [8]. Previous to the studies conducted by Sillen [9], the formation of the polynuclear hydrated compounds was almost ignored. Sillen proposes a mechanism of hydrolysis in which the hydroxyl groups are linked to cations and lead to the formation of condensed species. Now, the iso and heteropoly oxometalate are well known and in the specialized literature, there is detailed data about cation hydrolysis [10].

The main reactions that take place during the sol-gel process in aqueous medium were discussed in detail by Livage [10].

When metal salts MX_Z are dissolved in water, they dissociate in M^{Z+} and X^- ions and are solvated by dipolar water molecules leading to the formation of $[M(OH_2)_N]^{Z+}$ species

$$MX_Z + H_2O \rightleftarrows [M(OH_2)_N]^{Z+} + X^-$$

Aquo-cations behave as Brønsted acids and a spontaneous deprotonation may take place as follows:

$$[M(OH_2)_N]^{Z+} + hH_2O \rightleftarrows [M(OH)_h(OH_2)_{N-h}]^{(Z-h)+} + hH_3O^+,$$

h = hydrolysis ratio as it corresponds to the number of protons that have been removed from the solvatation sphere of aquo-cation:

$$M^{Z+}\text{-}OH_2 \rightleftarrows M^{Z+}\text{-}OH^- + H^+ \rightleftarrows M^{Z+}\text{-}O^{2-} + 2H^+.$$

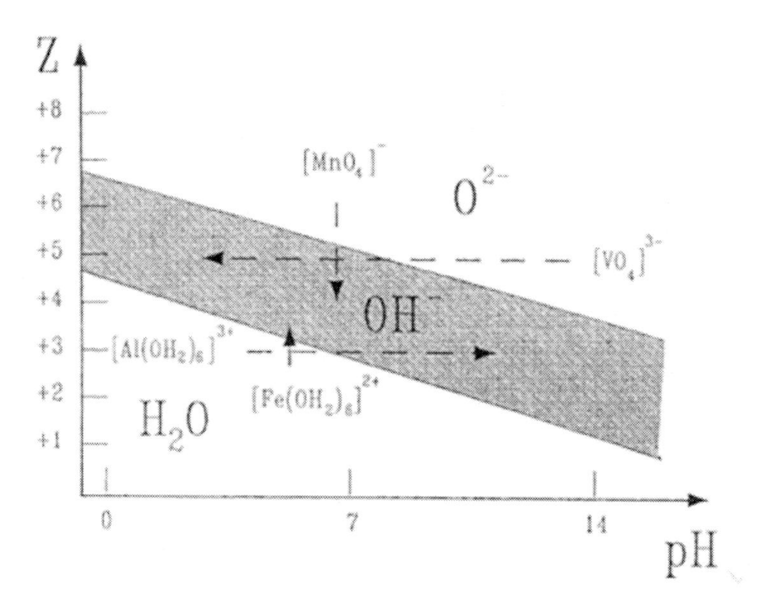

Figure 1. Charge-pH diagram showing the domains in which H_2O, OH^- or O^{2-} ligands are formed.

In Figure 1, the charge-pH diagram showing the domains in which H_2O, OH^- or O^{2-} ligands are formed is presented [11].

Condensation reactions of the hydorxylated precursors proceed via nucleophilic attack of $HO^{\delta-}$ groups on metal cations $M^{\delta+}$ by olation or oxolation

Olation represents the nucleophilic addition of a negatively charged OH^- group to a positively charged hydrated metal cation by the following mechanism:

$$\text{-M-OH} + \text{H}_2\text{O-M-} \Rightarrow \text{M-OH-M-} + \text{H}_2\text{O}$$

Oxolation involves the condensation of two OH groups to form one water molecule, which is then removed giving rise to an "oxo" bridge:

$$\text{-M-O-H} + \text{H-O-M-} \rightarrow \text{-M-O-M-} + \text{HOH}$$

$$\text{-M-O(H)} + \text{(H)O-M-} \rightarrow \text{-M-O}\begin{matrix}\text{H}\\ \\ \text{H}\end{matrix}\text{O-M-} \rightarrow \text{-M-O-M-} + \text{HO}$$

The morphology, structure and even chemical composition of the resulting solid particles can be modified via complexation [11].

Replacement of coordinated water molecules in aqua species by organic complexants, A^{m-} leads to new precursors whose chemical reactivity can be noticeably modified:

$$[M(OH_2)_n]^{z}+ + a\, A^{m-} \Leftrightarrow [M(H_2O)_w(A)_a]^{(z-am)} + (n-w)H_2O,$$

The intention of the complexation of metal cations is to prevent the rapid hydrolysis of the coordinated water molecules in the following deprotonated reaction:

$$[M(H_2O)_w(A)_a]^{(z-am)} + (n-w)H_2O \Leftrightarrow [M(OH)(H_2O)_{w-1}(A)_a]^{(z-am-1)} + H_3O^+$$

Based on the considerations presented above, in the present work the sol-gel chemistry in aqueous medium of cobalt-based system (Li-Co and La-Co) with citric acid as chelating agent is presented. Information concerning the influence of the precursors on the sol-gel method and the properties of the resulted gels and crystallized powders are discussed.

2. Synthesis of $LiCoO_2$ by Sol-Gel Method in Aqueous Medium

Layered oxide compounds containing lithium and transition metals such as $LiMO_2$ (where M is a 3d transition metals like Ni, Co, Mn, V,) were studied as active compounds for high capacity cathode materials or for lithium rechargeable batteries [12].

Among other compounds mentioned above, $LiCoO_2$ is the most used material for the commercial lithium batteries due to its excellent electrochemical stability after a long period of cycling. The excellent stability comes from the material structural stability, where the layered cation ordering is extremely well preserved even after a repeated process of insertion and removal of Li^+ ions.

In the literature, two phases for $LiCoO_2$ depending of the temperature of formation, namely: LT-$LiCoO_2$ low temperature phase (400^{0}C) and HT-$LiCoO_2$ high temperature phase (800^{0}C) have been mentioned. High temperature phase presents an excellent electrochemical stability at cyclization and it crystallizes in an isotype layered network of α-$NaFeO_2$ type structure [13]. Low temperature phase has partially disordered plans of lithium and cobalt, the structure is different from the high temperature phase. The two phases have different electrochemical property.

The literature data indicates two types of preparation methods for lithium cobaltite.

The first type of synthesis is represented by solid phase reactions, which consist of mixing by grinding of hydroxides, oxides and carbonates, followed by heat treatment at high temperatures. In the literature, using solid phase reactions [14-24], it can be seen that the molar ratio between precursors is 1:1 or a small excess of Li is used knowing its high volatility at high temperatures. Heat treatment is done for long periods, for 12-24 h at 400°C, to obtain low-temperature form (LT) and at temperatures > 650 ° C to form high temperature form (HT). .

The second type of synthesis is performed in solution, where several methods of preparation are known. Amatucci et al. [25] used the hydrothermal method using $LiOH.H_2O$ and CoOOH, with an excess of Li, as precursors,

The sol–gel method is a widely spread technique for both powders and films preparation due to some well known advantages: an easy way to obtain homogenous distribution of precursors, the possibility to introduce controlled amounts of dopants, chemical methods of reaction control, viscosity advanced control as well as low processing temperatures [7]. Several sol–gel routes were approached in order to prepare $LiCoO_2$. The sol–gel method starting with alkoxides was used by Fey et al. [26], Quinlan et al. [27], Szatvanyi et al.[28]. The carboxylic route was experienced by Yazami et al. [14], Yoon and Kim [29], Shlyakhtin et al. [30] and Cho et al.[31] Some combined inorganic-carboxylic routes were also taken into consideration by Zhecheva et al.[32], Kang et al. [33], Kushida and Kuriyama [34] and Uchihida et al. [35].

Zhecheva et al. [32] have used Li_2CO_3, $Co(NO_3)_2.6H_2O$ as metal precursors and citric acid as an chelating agent.

The chemical methods in solutions where acetates are used as precursors [14, 29, 30] or nitrates with a carboxylic acid as chelating agent [28, 32-40] are also mentioned. It is noted that heat treatment temperatures are similar to the case summaries by solid phase reaction but the time taken by thermal forming of compounds is shorter. Kang et al. [33] studied $LiCoO_2$ obtained using the "citrate" method. Precursors, Li_2CO_3, $Co(NO_3)_2.6H_2O$ and citric acid were dissolved in distilled water in a 1:1:2 molar ratio. Depending on the temperature of heat treatment, LT-$LiCoO_2$ or HT-$LiCoO_2$ was obtained. Kim et al. [38] have achieved $LiCoO_2$ thin films. The most difficult problems encountered in the manufacturing of thin film was the adherence of the prepared solution to the substrate. Lithium acetate, lithium acetyl acetonate and lithium nitrate were tested as sources of lithium and cobalt acetate, cobalt acetyl acetonate and cobalt nitrate as sources of cobalt. Ethanol, 2-propanol, 1-butanol, 2-methoxyethanol were used as solvents. Many combinations of these materials had low solubility or poor wetting property. Films treated at high temperatures showed a small discharge capacity and good recharging capacity. In terms of results, $LiCoO_2$ films prepared by sol-gel method can be used for making all microbattery semiconductor cathodes.

Park et al. [36] have prepared $LiCoO_2$ by sol-gel process from lithium acetate and cobalt acetate in cationic ratio Li: Co = 1:1. Precursors were dissolved in distilled water and homogenized at 50-60°C. An aqueous solution of glycolic acid, introduced as chelating agent was added to the mixture of solutions leading to gel. The ratio of glycolic acid to total metal ions was fixed at 1.5. Tao et al. [41] have prepared $LiMeO_2$ (Me = Co, Ni) using lithium nitrate and cobalt acetate as precursors in the molar ratio of 1:1. Precursors were dissolved in glycol and distilled water to form sol. For preparation of layered $LiCoO_2$ powders with small particle size at low temperatures (400-600 °C), Yazami and Schleich [14, 42] have used solutions of acetate and glycolate as precursors.

Yamaki et al. [16] have developed a new method for synthesis of $LiCoO_2$ powders with high specific surfaces area, characterized by a layered crystalline structure. This method is the thermal decomposition of an intimate mixture of lithium acetate and cobalt acetate at 550°C in air for at least 2 hours. By the "citrate" method, several types of synthesis were developed, emphasizing that it has great potential in the preparation of multicomponent oxides (perovskite, spinel, ferrites, polititanate) with a high chemical homogeneity and small particle [13]. Another method is based on the average solubility of citrate salts in ethanol compared with water. In the sol-gel citrate route, amorphous citrate precursors are prepared by adding citric acid to the solution of metal salts (usually nitrates) followed by evaporation in vacuum [43-45]. By Fu et al., lithium cobaltites were prepared from aqueous solutions of lithium and cobalt salts (nitrates or acetates) mixed with an aqueous solution of PAA (polyacrylic acid) [46]. Son et al. obtained a transparent sol from a mixture of $Co(CH_3COO)_2.4H_2O$ and $CH_3COOLi.2H_2O$ dissolved in distilled water and homogenized continuously at 80^0C for 4h

[47]. Zhu et al. [48] and Yang et al. [49] obtained ultrafine $LiCoO_2$ powders using a novel sol-gel method with citric acid as a chelating agent, hydroxypropyl cellulose as a dispersant agent and $LiNO_3$, $Co(NO_3)_2.6H_2O$ as starting materials. Zeng et al. [50] prepared $LiCoO_2$ powder using $LiOH.H_2O$, $Co(OAC)_2.4H_2O$ and L-apple acid as starting materials.

Using a carboxylic route, the bonding of the metallic ions in a chelating complex is assumed to lead to a highly homogeneous precursor gel and should diminish the particle size of the resulting oxides due to their formation in an organic matrix. In the same time during the thermal treatment, the organic species evolved by matrix decomposition could be retained on the particle surface, hindering again the particle growth [28].

2.1. Sol-Gel Chemistry of the $LiCoO_2$ Formation

The sol-gel chemistry of the $LiCoO_2$ formation in aqueos medium was approached using inorganic salts and carboxylic acids (citric acid) as chelating agents. Two different reagents, namely Li and Co nitrates and acetates, respectively and citric acid in 1.1:1:1 or 1.1:1:2, molar ratio in aqueous solutions were used. The solutions, the resulting gels and the powders, starting with nitrate were labeled (N), while the similar solutions and gels obtained starting with acetates were labeled (A). In the conditions presented above, when nitrates were used as precursors, the pH of the starting solutions was 5, while in the case of the acetates precursor, the pH of the starting solutions was 6.

The gelling process in the studied solutions was approached by spectroscopic methods. The positions of the absorbance bands of the UV-VIS spectra for all studied samples are summarized in Table 2, while in Figure 2, the evolution of the UV-Vis spectra during the sol-gel process is shown for solution in which 1 mol citric acid was used as chelating agent.

Table 1. The absorption bands of the UV-VIS spectra of the sol-gel solutions starting with nitrates Li-Co-CA (N1and N2) and acetates (A1 and A2).

Sample	Absorption bands	Assignement
Li-Co-CA (N1) nitrate	301 nm 461 - 476 nm 508 nm ~620 nm	CT $^4T_{1g} \rightarrow {}^4T_{1g}(P)$ splitting $^4T_{1g} \rightarrow {}^2T_{1g}$
Li-Co-CA (N2) nitrate	301 nm 461 - 476 nm 508 nm	CT $^4T_{1g} \rightarrow {}^4T_{1g}(P)$ splitting
Li-Co-CA (A1) acetate	466 - 480 nm 509 - 511 nm ~625 nm	$^4T_{1g} \rightarrow {}^4T_{1g}(P)$ splitting $^4T_{1g} \rightarrow {}^2T_{1g}$
Li-Co-CA (A2) acetate	466 - 480 nm 509 - 511 nm ~625nm	$^4T_{1g} \rightarrow {}^4T_{1g}(P)$ splitting $^4T_{1g} \rightarrow {}^2T_{1g}$

Legend : (N1) and (A1) corresponds to synthesis with 1 mol citric acid and (N2) and (A2) corresponds to synthesis with 2 mols citric acid

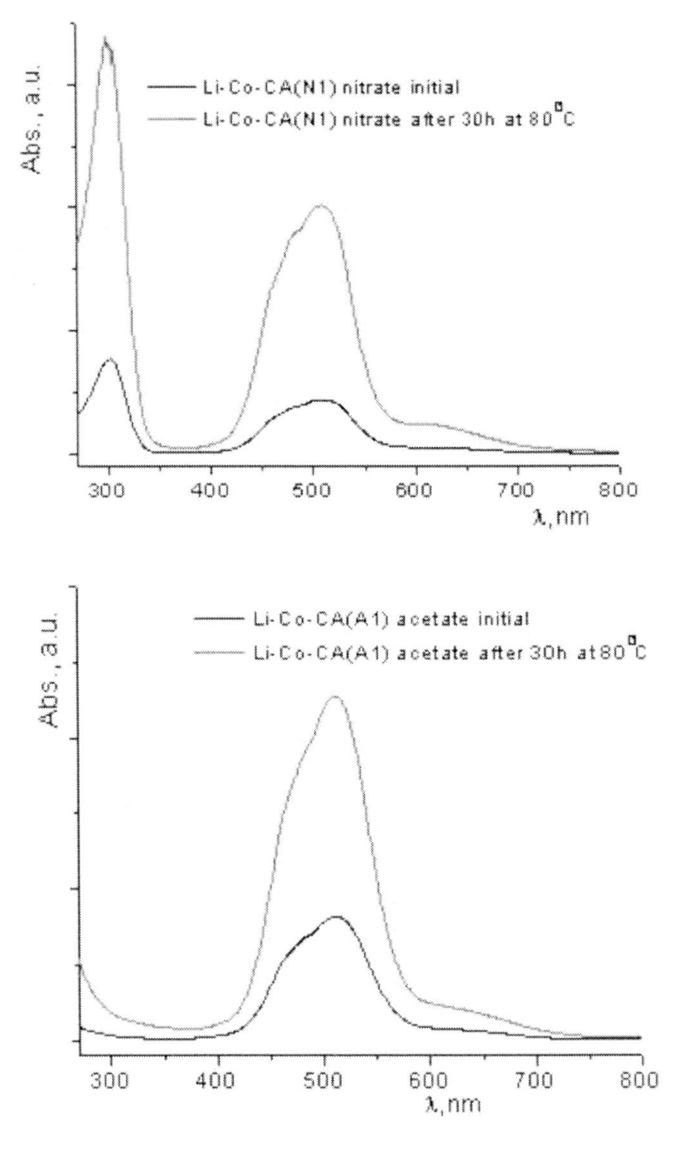

Figure 2. Evolution of the UV-VIS spectra of Li-Co-Citric acid solutions based on nitrate and acetate during the sol-gel process.

In both cases, an asymmetric band with a maximum at 510 nm was identified in the solutions and it was assigned to $^4T_{1g} \rightarrow {}^4T_{1g}$ (P) transition. The deconvolution of the spectrum points out that the band with a principal maximum at 510 nm covers two split transitions with maxima at 510 nm and 468 nm. For the split structure of this band, several explanations could be done. The splitting could occur due to the admixture of a spin forbidden transition to doublet states, or due to spin-orbit coupling, or due to the presence of the components with a low symmetry [51]. The band with the maximum at 468 nm has a lower intensity compared to the band placed at 510 nm and this is consistent with spin forbidden transitions. The $^4T_{1g} \rightarrow {}^4T_{1g}$ (P) transition identified for this system is characteristic for a Co(II) ion in an octahedral coordination.

Table 2. The absorption bands of the UV-VIS spectra of the gels obtained starting with nitrates Li-Co-CA (N1) and acetates (A1).

Gels	Absorption bands	Assigmenent
Li-Co-CA (N1)	346 nm 510 nm 588 nm	CT $^4T_{1g} \rightarrow {}^4T_{1g}(P)$ $^4T_{1g} \rightarrow {}^2T_{1g}$
Li-Co-AC (A1)	536 nm 596 nm	$^4T_{1g} \rightarrow {}^4T_{1g}(P)$ $^4T_{1g} \rightarrow {}^2T_{1g}$

Table 3. The assignment of the vibration bands in IR spectra of the solution and gels.

Gel		Gel		Vibration mode	Assignments
N1	N2	A1	A2		
3194	3350	3276	3326 3200	ν OH	structural OH group
1717	1727	1736	1717	ν_{asym} (COO)	carbonyl asymmetric stretching
1614	1623	1607	1626	δ_{H2O}	adsorption of water
1575	1574	-	1571	ν_{asym} (COO)	carbonyl asymmetric stretching
1429	1412	1425	1435	ν_{sym} (COO)	carbonyl symmetric stretching
1382	1386	-	-	$\nu_{NO_3^-}$	introduced by the precursors
1291		1283		ν_{C-O}	
1265	1260		1247		
1203	1198	1188	1205	ν -COOH	introduced by the citric acid
1138		1156			
1076	1078	1075	1078	π_{CH}	
1033	1033	1036	1036	ν_{C-OH}	

In the case of the solutions starting with nitrates, a supplementary band at 300 nm was identified that could be assigned to the charge transfer from 2p orbitals of nitrogen or oxygen from NO_3^- to the 3d orbitals of Co^{2+}. (Figure 2).

During the sol-gel process, the position of the UV-VIS bands do not change, only their intensity increase due to an ordering of the ligands around to Co^{2+} ions.

As a result of the sol-gel process, red gels were obtained. The results of the UV-Vis spectroscopic investigation of the gels presented in Table 2 have shown the presence of the same band as those identified in the solutions.

The assignment of the vibration bands in IR spectra of the resulted gels are presented in Table 3.

In the case of the gels obtained with nitrates, the vibration of the carbonyl groups are situated between 1727-1412 cm^{-1}, showing the ionisation of all carboxil groups that leads to the complex formation. According to the literature data, the separation frequences between the vibration of the asymmetric ν_{asym}(COO) la 1717 cm^{-1} and ν_{sym}(COO) stretching of carbonyl ion at 1429 cm^{-1} for N1 gel of $\Delta\nu$ = 288 cm^{-1} and ν_{asym}(COO) la 1727 cm^{-1} and ν_{sym}(COO) stretching of carbonyl ion at a 1412 cm^{-1} for N2 gel of $\Delta\nu$ = 315 cm^{-1}, indicate the presence of a unidentate ligands.

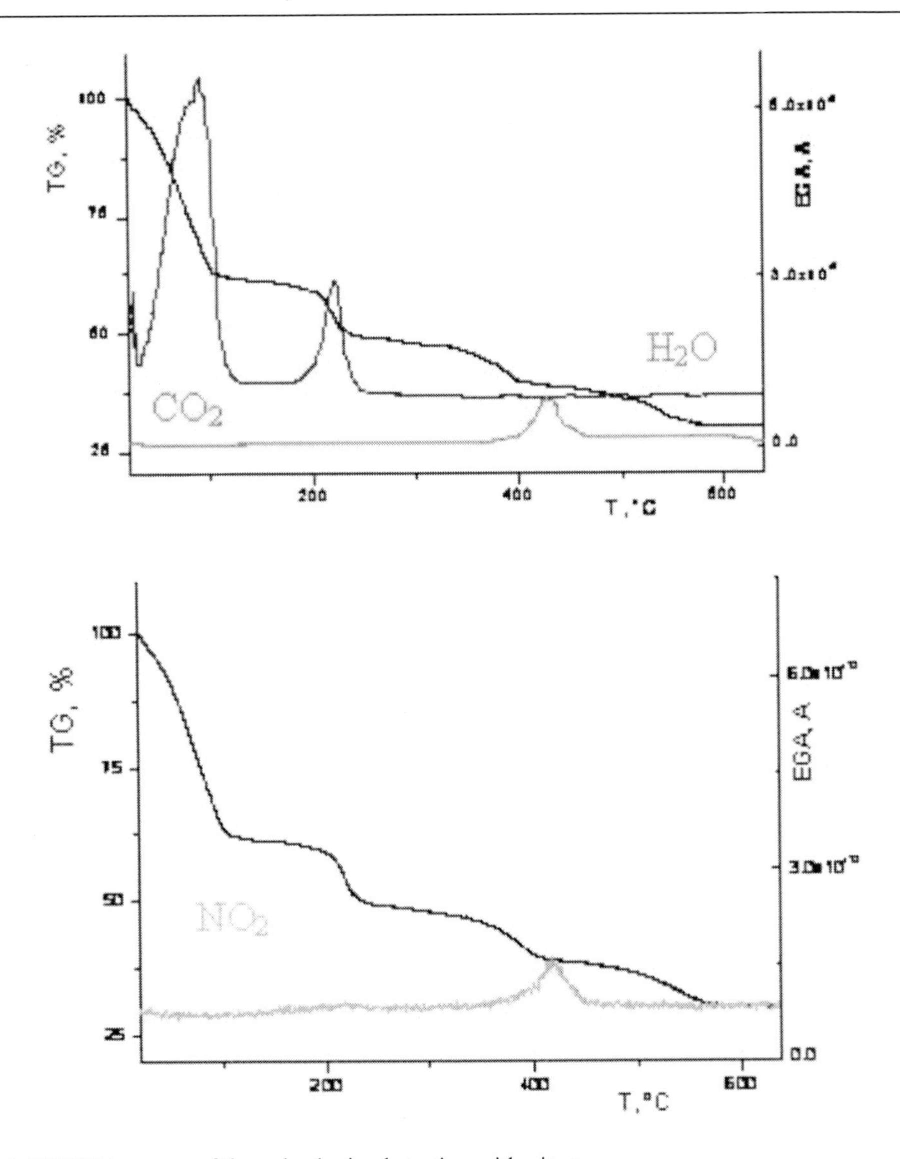

Figure 3. TG/EGA curves of the gels obtained starting with nitrates.

The band at ~1380 cm^{-1}corresponds to the vibration of $-NO_3$ introduced by the metal nitrates used as precursors, while the bands located at about ~1265 cm^{-1} corresponds to the vibrations of ν -COOH from the citric acid.

The band present at about 1620 cm^{-1} is assigned to the stretching vibration of the molecular water and its presence in the IR spectrum is correlated to the chemical method in aqueous medium used for the gels preparations

In the case of the gel samples starting with acetates, the vibration of the asymmetric and symmetric bands occur in the 1736-1425 cm^{-1}. According to the literature data, the frequency separation of $\Delta\nu = 311$ cm^{-1} for A1 and 282 cm^{-1} for A2, between the two type of vibrations suggests the presence of a unidentate ligands.

The band situated at about 1607 could be assigned to the stretching vibration of the molecular water, while the bands form the 1100-1075 cm^{-1} correspond to the $-M-O-COO^-$ bonds.

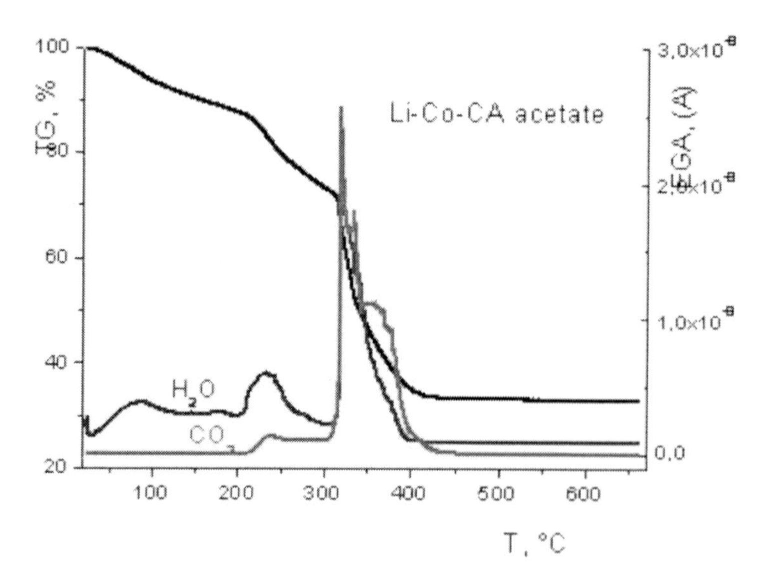

Figure 4. TG/EGA of the gels obtained starting with acetates.

The TG/AEG curves of the gels obtained starting with nitrate are presented in Figure 3 and for the gels obtained with acetates in Figure 6.

In both cases, a stepwise themal decomposition was noticed. In the case of gels obtained starting with nitrates, in the first step, under 100^0C, the adsorbed molecular water is evolved. Around 200^0C, the coordinated water is eliminated, while the decomposition of the chelating agents thakes place at the same temperature around 400^0C.

In the case of the gels obtained starting with acetates, a very low amount of adsorbed water is noticed while the coordinated water is eliminated at two temperatures, around 200 and about 350^0C.

Based on spectroscopic results obtained, on the mechanism proposed by Livage [11] and on the charge-pH diagram (see Figure 1), in the studied cases the following reactions could be proposed

1. Dissolution of metal salts in water

   In the nitrate-salt solution with pH = 5, aquo-cations are formed:

$$LiNO_3 \xrightarrow{dissociation} Li^+ + NO_3^- \xrightarrow{solvation} [Li(H_2O)_4]^+ + (NO_3)^-$$

$$Co(NO_3)_2 \xrightarrow{dissociation} Co^{2+} + 2(NO_3)^- \xrightarrow{solvation} [Co(H_2O)_6]^{2+} + [(NO_3)^-]_2$$

In the acetate-salt solution with pH = 6, aquo-cations are formed:

$$CH_3COOLi \xrightarrow{dissociation} Li^+ + CH_3COO^- \xrightarrow{solvation} [Li(H_2O)_4]^+ + (CH_3COO)^-$$

$$Co(CH_3COO)_2 \xrightarrow{dissociation} Co^{2+} + 2(CH_3COO)^- \xrightarrow{solvation}$$
$$[Co(H_2O)_6]^{2+} + [(CH_3COO)^{1-}]_2^{2-}$$

Negatively charged anions X^{x-} can react with cationic precursors to give metal complexes. For monovalent anions, the reaction can be written as follows:

$$[M(H_2O)_X]^{z+} + x\,X^- \Leftrightarrow [M(X)_x(H_2O)_{N-\alpha X}]^{(z-x)+} + \alpha x\,H_2O$$

where α corresponds to the number of water molecules which are replaced by X^-.

2. Complexation and chemically controlled condensation in the presence of the citric acid in the studied system could be the following;
   For the nitrate solution:

$$[Li(H_2O)_4]^+ + (NO_3)^- \rightarrow [Li(NO_3)_x(H_2O)_{3-x}]^{(1-x)} + xH_2O$$
$$[Li(NO_3)(H_2O)_3] + x(C_6H_5O_7)^{3-} \rightarrow [Li(NO_3)_x(C_6H_5O_7)_y(H_2O)_{3-x-y}]^{(1-x-y)+} + (x+y)H_2O$$

$$[Co(H_2O)_6]^{2+} + 2(NO_3)^- \rightarrow [Co(NO_3)_m(H_2O)_{6-m}]^{(2-m)+} + mH_2O$$
$$[Co(NO_3)_m(H_2O)_{4-m}] + n(C_6H_5O_7)^{3-} \rightarrow [Co(NO_3)_m(C_6H_5O_7)_n(H_2O)_{4-m-n}]^{(2-m-n)+} + (m+n)H_2O$$

For the acetate solution:

$$[Li(OH_2)_4]^+ + x(CH_3COO)^- \rightarrow [Li(CH_3COO)_x(OH_2)_{4-x}]^{(1-x)+} + xH_2O$$

$$[Li(CH_3COO)_x(OH_2)_{3-x}] + y(C_6H_5O_7)^{3-} \rightarrow [Li(CH_3COO)_x(C_6H_5O_7)_y(OH_2)_{3-x-y}]^{(1-x-y)+} + (x+y)H_2O$$

$$[Co(OH_2)_6]^{2+} + m(CH_3COO)^- \rightarrow [Co(CH_3COO)_m(OH_2)_{4-m}]^{(2-m)+} + mH_2O$$

$$[Co(CH_3COO)_m(OH_2)_{4-m}] + n(C_6H_5O_7)^{3-} \rightarrow [Co(CH_3COO)_m(C_6H_5O_7)_n(OH_2)_{4-m-n}]^{(2-m-n)+} + (m+n)H_2O$$

3. Formation of metals-citric acid chelate:
   For the nitrate solution:

$$[Li(NO_3)_x(C_6H_5O_7)_y(H_2O)_{3-x-y}]^{(1-x-y)+} + (x+y)H_2O + [Co(CH_3COO)_m(C_6H_5O_7)_n(OH_2)_{4-m-n}]^{(2-m-n)+} + (m+n)H_2O \rightarrow [LiCo(NO_3)_{x+m}(C_6H_5O_7)_{y+n}(H_2O)_{7-x-y-m-n}]^z.(x+y+m+n)H_2O$$

For the acetate solution:

$$[Li(CH_3COO)_x(C_6H_5O_7)_y(OH_2)_{3-x-y}]^{(1-x-y)+} + (x+y)H_2O + [Co(CH_3COO)_m(C_6H_5O_7)_n(OH_2)_{4-m-n}]^{(2-m-n)+} + (m+n)H_2O \rightarrow [LiCo(CH_3COO)_{x+m}(C_6H_5O_7)_{y+n}(H_2O)_{7x-y-m-n}]^z.(x+y+m+n)(H_2O)$$

(Note: $z=7-x-y-m-n$)

In the experimental conditions presented above, amorphous polinuclear complex gels have been obtained.

The SEM images of the samples obtained with different starting precursors (nitrate N and acetate A), show pieces of gel. The sizes of the gels' fragments are different and depend on the starting precursors.

In Figure 5, SEM images of $LiCoO_2$ powders obtained by heat treatment at different temperatures are shown. It can observe that heat treatment affects the structure and morphology of powders, causing particle growth due to their crystallization and decreasing their level of aggregation.

The SEM images indicate that by heat treatment, crystallization occurs with increasing temperature the crystal size increases.

Based on thermal behavior of the complex gel, the thermal treatment to obtaine single pure perovskite rhombohedral phase of $LiCoO_2$ (JCPDS 77-1370) (Figure 6) was the following: heating rate of 1^0C/min up to 400^0C; 1 hour plateau and heating rate of 5^0C/min up to 600^0C or 800^0C; 6 hours plateau.

The results presented in Figure 6 have shown the same rhombohedral perovskite structure for both samples.

X-ray diffraction patterns show that the synthesis at 600°C the process the $LiCoO_2$ phases is formed and it is well crystallized. However, un-reacted of Co_2O_3 are noticed in the samples obtained starting with nitrates. By heat treatment at 800^0C, in both cases single phase rhombohedral perovskite $LiCoO_2$ structure is put in evidence.

Figure 5. SEM images of the $LiCoO_2$ sample thermally treated at different temperatures, starting with nitrates (a. 600^0C-6h, b. 800^0C-6h) and starting with acetates (c.600^0C-6h, d. 800^0C-6h).

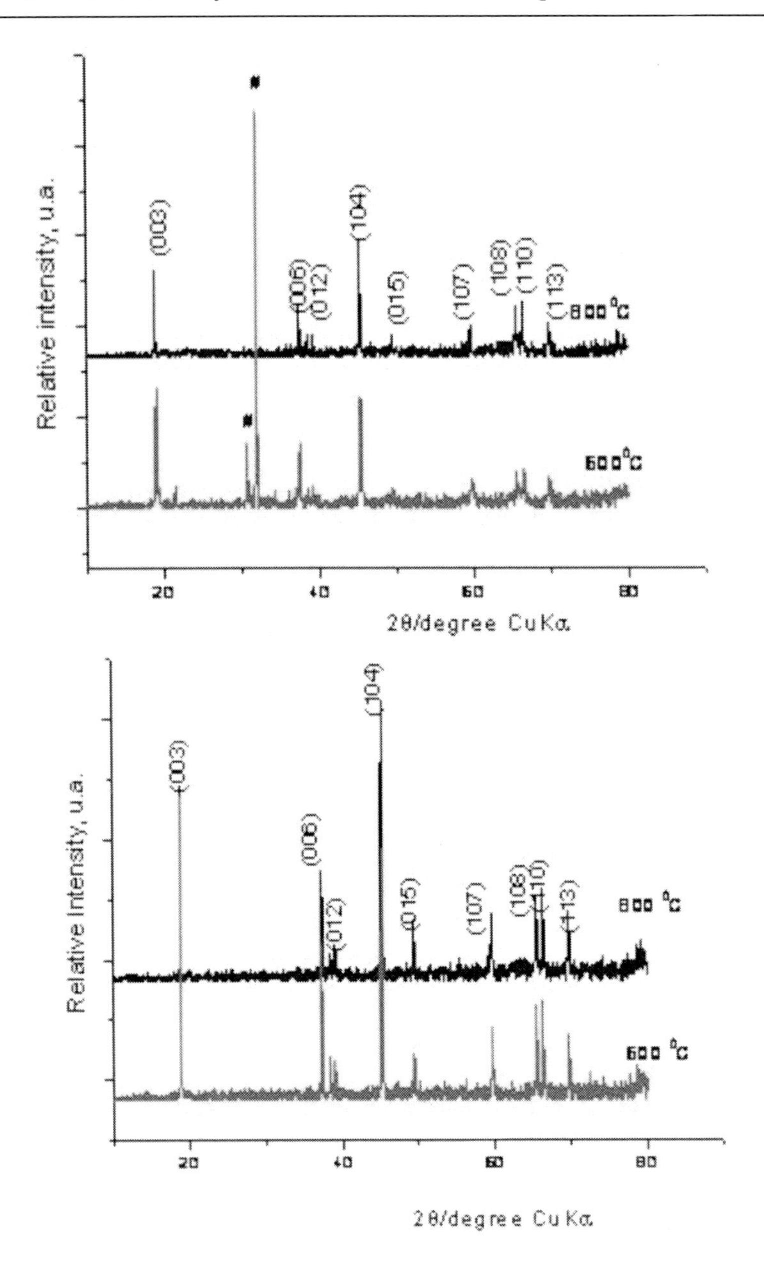

Figure 6. XRD patterns of $LiCoO_2$ powders thermally treated at different temperatures (a) powders obtained starting with nitrates, (b) powders obtained starting with acetates (# Co_3O_4).

2.2. Electrochemical Characteristics of $LiCoO_3$ Powders

The electrochemical characteristics of the $LiCoO_2$ investigated samples were performed and a typical charge–discharge curves are obtained.[52] The capacity at the first charge is 155 mAh g^{-1}, a value very close to the capacity recorded on the slow voltammetric test (SVT) and the discharge capacity is relatively low only 109 mAh g^{-1}, but again, very close to the data pointed out from the SVT. From the second charge- discharge curves, the displayed capacities are more reasonable 112 mAh g^{-1} for charge and 103 mAh g^{-1} for discharge. The results are satisfactory and show the potential for improvement.

3. Synthesis of $LaCoO_3$ by Sol-Gel Method in Aqueous Medium

The mixed oxides type perovskite with general formula ABO_3 can be considered strategic materials due to the fact that it presents magnetic, electrical and catalytic properties [53]. Special attention was given to $LnMeO_3$ type oxides with perovskite structure, where Ln belongs to the lanthanides group and Me is an element of transition metal groups, due to their applications as materials for electrodes in solid electrolyte cells (SOFC) or materials for chemical sensors. $LaCoO_3$ compound belongs to this class of materials showing interesting electrical and catalytic properties, due to a high ionic and electronic conductivity.

It is also known that $LaCoO_3$ electrode offers much better performance than the cathode materials based on $LaMnO_{3+\delta}$. [54, 55]. To get good performance and functional properties, dense materials with a well defined microstructure are desired. For this purpose, it is important to prepare high quality powders with controlled microstructure and stoechiometry. Single phase materials are preferred because in most cases, the presence of secondary phases decreases the functional properties. Many authors were interested to synthesize powders or films in this system [2, 54-65].

$LaCoO_3$ compound can be obtained through several methods of preparation. As conventional preparation techniques, solid state reactions based on metal oxides, carbonates or oxalate followed by a thermal treatment around 1000°C, with intermediate grinding were used [55]. This method leads to materials with low specific surface areas, with low activity, requiring high temperatures and long thermal treatment periods.

To avoid the disadvantages described above, several new methods, especially techniques in solution, leading to improved conditions for obtaining pure phases were developed. Such syntheses require low reaction temperatures and leads to homogeneous and fine powders [2].

These methods include the sol-gel process (alkoxide route) and (aqueous route) in the presence of citrate or tartrate [7], hydrothermal and coprecipitation [66-68].

Kleveland et al. [60] used a chemical route starting with nitrates and ethylendiamine tetracetic acid, while Faaland at al. [69] applied a spray pyrolysis method starting with the same precursors. Berger at al. [70] proposed a combustion method, while Li et al. [71] presented a combined method of milling nitrates with nitric acid followed by co-precipitation with potassium hydroxide.

The sol-gel method, using alkoxides as precursors $La(O_iPr)_3$ and $Co(NO_3)_2$, was used by Hwang [59]. The polymerizable complex method was used by Popa et al. [72] and Guo et al. [73]. The carboxylic route was applied by Taguchi and Ajami using lanthanum and cobalt nitrate and citric acid as chelating agent [74, 75].

Sol-gel method is often used as a technique for preparation of powders, because of well known advantages such as an easy way to obtain homogenous distribution of precursors with the possibility of controlled introduction of a given quantity of dopant, control by chemical reaction, control viscosity and low temperature of processing. Citrate sol-gel process route presents advantages over other methods not only of good mixing of the components at the atomic scale, but also offers the possibility of obtaining films and fibers which have technological importance [9].

3.1. Sol-Gel Chemistry of the $LaCoO_3$ Formation

The studies of the $LaCoO_3$ powder formation started with the investigation of the sol-gel process in aqueous medium using inorganic salts and carboxylic acids (citric acid) as chelating agents [76]. Two different reagents, namely La- and Co nitrates and acetates, respectively and citric acid in 1:1:1 molar ratio in aqueous solutions were used [77-79]. The solutions, the resulting gels and the powders, starting with nitrate, were labeled (N), while the similar solutions and gels obtained starting with acetates, were labeled (A).

The gelling process in the studied solutions was approached by spectroscopic methods.

In the UV-VIS spectra of the La-Co solutions based on nitrates and acetates immediately after their preparation and after 30 hours of storage at $80^{0}C$ the Co^{2+} ions in octahedral coordination were observed [80]. The positions of the absorbance bands of the UV-VIS spectra are presented in Table 4.

Table 4. The maxima of absorbance identified in the UV-VIS spectra recorded on the studied solutions with their assignments.

Sample	Maxima of absorbance nm (cm^{-1})	Assignments
La-Co-CA (N) solution	300 nm (33231 cm^{-1}) 468 nm (21368 cm^{-1}) 510 nm (19531 cm^{-1})	CT Splitting $^4T_{1g} \rightarrow {}^4T_{1g}(P)$
La-Co-CA (A) solution	467 nm (20217 cm^{-1}) 512 nm (18587 cm^{-1})	Splitting $^4T_{1g} \rightarrow {}^4T_{1g}(P)$

Table 5. The assignment of the vibration bands in IR spectra of the solution and gels The table should not be broken. Please move the entire table on the same page.

N		A		Vibration mode	Assignments
solution	gel	solution	gel		
3300	3350	3276	3326 3200	ν OH	structural OH group
1718	1720	-	-	ν_{asym} (COO)	carbonyl asymmetric stretching
1635	1616	1636	1616	δ_{H2O}	adsorption of water
-	-	1552	1568	ν_{asym} (COO)	carbonyl asymmetric stretching
1404	1432	1410	1415	ν_{sym} (COO)	carbonyl symmetric stretching
1333	1383	-	-	$\nu_{NO_3^-}$	introduced by the precursors
1230		1229	1263	ν_{C-O}	introduced by the citric acid
-	1078	-	1078	π_{CH}	
-	1033	-	1036	ν_{C-OH}	
-	813	-	912	ν_{CH2}	

Table 6. The maxima of absorbance identified in the UV-VIS spectra recorded on the obtained gels with their assignments.

Sample	Maxima of absorbance nm (cm^{-1})	Assignments
La-Co-AC (N) gel	356 nm (28089 cm^{-1}) 406 nm (24639 cm^{-1}) 532 nm (18797 cm^{-1}) 671 nm (14903 cm^{-1})	CT $^4T_{1g} \rightarrow {}^4A_{2g}$ $^4T_{1g} \rightarrow {}^4T_{1g}(P)$ $^4T_{1g} \rightarrow {}^4B_{1g}$
La-Co-AC (A) gel	372 nm (26881 cm^{-1}) 540 nm (18518 cm^{-1}) 685 nm (14598 cm^{-1})	CT $^4T_{1g} \rightarrow {}^4T_{1g}(P)$ $^4A_{2g} \rightarrow {}^4T_{1g}(P)$

In Table 5, the assignment of the vibration mode in the FT-IR spectra of the same solutions is presented. In the case of the La-Co solutions based on nitrates, the frequency separation between the antisymmetric stretching and symmetric stretching of the carbonyl ion ($\Delta\nu = 314\ cm^{-1}$) showing that carboxylic groups of the citrate act as unidentate ligands [81]. On the other hand, in the case of La-Co solutions based on acetates, the same separation is only, $\Delta\nu = 142\ cm^{-1}$, leading to the conclusion that carboxylic groups of citrate ligands act as bridging ligands.

Under the above presented experimental conditions, amorphous red gels were obtained.

The positions of the absorbance bands of the UV-VIS spectra, of gels obtained starting with nitrates and acetates are presented in Table 6. In the spectrum of the nitrate-derived gel, the d-d transitions correspond to the Co(II) ion in the tetragonal distorted geometry. In the acetate-derived gel, the d-d transitions correspond to the formation of complex gels in which Co(II) ion has an octahedral symmetry.

The thermal decomposition of the gels obtained starting with nitrates La-Co-CA(N1) and with acetates La-Co-CA(A1), as compared with that of the gels obtained in monocomponent La-CA and Co-CA systems has shown that the main decomposition effects of the binary La-Co complex (La-Co-CA) gels lies between the decomposition effects of the monocomponent complexes (La-CA and Co-CA). This observation is leading to the conclusion that the binary systems are not mixtures of phases but polynuclear complex gels are formed [80]. Moreover, the TG/EGA results of the gel obtained with nitrates show the decomposition at two different temperatures (200°C and 350°C), with the evolution of the same gases (NO_2, CO_2 and H_2O) that means that the water, citric acid and nitric reagents are bound in two different ways in the complex.

Based on experimental results obtained and on the mechanism proposed Livage [11], the following mechanism of formation was proposed for the polynuclear gelic complexes [80]:

1. Dissolution of metal salts in water

   In the nitrate-salt solution with pH = 3, aquo-cations are formed:

$$La(NO_3)_3 \xrightarrow{dissociation} La^{3+} + 3(NO_3)^- \xrightarrow{solvation} [La(H_2O)_6]^{3+} + 3(NO_3)^-$$

$$Co(NO_3)_2 \xrightarrow{dissociation} Co^{2+} + 2(NO_3)^- \xrightarrow{solvation} [Co(H_2O)_6]^{2+} + 2(NO_3)^-$$

In the acetate-salt solution with pH = 8, Co aquo-cations and La neutral hydroxyl species are formed:

$$La(CH_3COO)_3 \xrightarrow{dissociation} La^{3+} + 3(CH_3COO)^- \xrightarrow{solvation} [La(OH)^-_3]^0 + 3(CH_3COO)^-$$

$$Co(CH_3COO)_2 \xrightarrow{dissociation} Co^{2+} + 2(CH_3COO)^- \xrightarrow{solvation} [Co(H_2O)_6]^{2+} + 2(CH_3COO)^-$$

2. Complexation and chemically controlled condensation in the presence of the citric acid
   For the nitrate solution:

$$[La(H_2O)_6]^{3+} + 3(NO_3)^- + x(C_6H_5O_7)^{3-} \rightarrow [La(C_6H_5O_7)_x(H_2O)_{6-x}]^{3-x}\ [(NO_3)^-_3\ (H_2O)_x]$$
$$[Co(H_2O)_6]^{2+} + 2(NO_3)^- + y(C_6H_5O_7)^{3-} \rightarrow [Co(C_6H_5O_7)_y(H_2O)_{6-y}]^{2-y}\ [(NO_3)^-_2\ (H_2O)_y]$$

   For the acetate solution:

$$[La(OH)^-_3]^0 + 3(CH_3COO)^- + x(C_6H_5O_7)^{3-} \rightarrow [La(C_6H_5O_7)_x(OH)^-_{3-x}]^{3-x}\ [(CH_3COO)^-_3\ (H_2O)_x]$$

$$[Co(OH_2)_6]^{2+} + 2(CH_3COO)^- + y(C_6H_5O_7)^{3-} \rightarrow [Co(C_6H_5O_7)_y(OH_2)_{6-y}]^{2-y}\ [(CH_3COO)^-_2\ (H_2O)_y]$$

3. Formation of metals-citric acid chelate:
   For the nitrate solution:

$$[La(C_6H_5O_7)_x(H_2O)_{6-x}]^{3-x}[(NO_3)^-_3\ (H_2O)_x] + [Co(C_6H_5O_7)_y(OH_2)_{6-y}]^{2-y}\ [(NO_3)^-_2(H_2O)_y] \rightarrow [LaCo(C_6H_5O_7)_{y+x-z}(OH_2)_{6-y-x}\ (NO_3)^-_{3z}]\ [(C_6H_5O_7)_z(NO_3)^-_2] \cdot (H_2O)_{x+y}$$

   For the acetate solution:

$$[La(C_6H_5O_7)_x(OH)^-_{3-x}]^{3-x}[(CH_3COO)^-_3(H_2O)_x] + [Co(C_6H_5O_7)_y(OH_2)_{6-y}]^{2-y}\ [(CH_3COO)^-_2\ (H_2O)_y] \rightarrow [LaCo(C_6H_5O_7)_{y+x}(OH_2)_{6-y}(OH)^-_{3-x}\ (CH_3COO)^-_5] \cdot (H_2O)_{x+y}$$

Based on the mechanism of formation proposed above, in the case of the gels obtained starting with nitrates the following formula was proposed.

$$[LaCo(C_6H_5O_7)_{y+x}(H_2O)_{6-y-x}]\ [(NO_3)^{1-}]_5 \cdot x+yH_2O$$

In this case, the citric acid is bound in the position of a coordinated ligand and nitric anion as weakly bonded ligand. However, as the TG/EGA results show the decomposition of the polynuclear complex gel in two steps at two different temperatures (200°C and 350°C), with the evolution of the same gases (NO_2, CO_2 and H_2O), one may conclude that the same ligands are bound in two different ways in the complex and the real formula could be the following:

$$[LaCo(C_6H_5O_7)_{y+x-z}(H_2O)_{6-y-x}\,(NO_3)_{3Z}{}^{1-}]\,[(C_6H_5O_7)_Z(NO_3)_2{}^{1-}]\cdot x+yH_2O$$

In this formula, citric acid and the nitrate are bound in two positions as weakly bound ligands and as coordinated ligands.

The SEM images of the samples obtained with different starting precursors (nitrate N and acetate A), are presented in Figure 7 and show pieces of gel. The sizes of the gels' fragments are different and depend on the starting precursors.

Based on the results obtained by DTA/TG analysis, the gels were thermally treated at 600 ^0C, for 6 hours leading to the formation of the single pure perovskite rhombohedral phase of $LaCoO_3$ (JCPDS 84-0848) (Figure 8).

In Table 7, the XRD evaluation of the crystallographic parameters, particle size and internal stress of the thermally treated powders is presented.

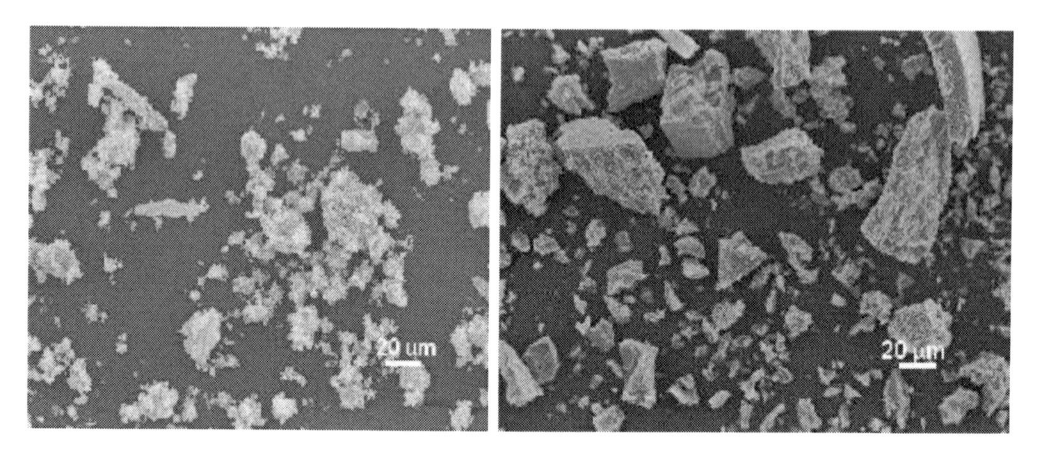

Figure 7. SEM images of the $LaCoO_3$ gel sample.

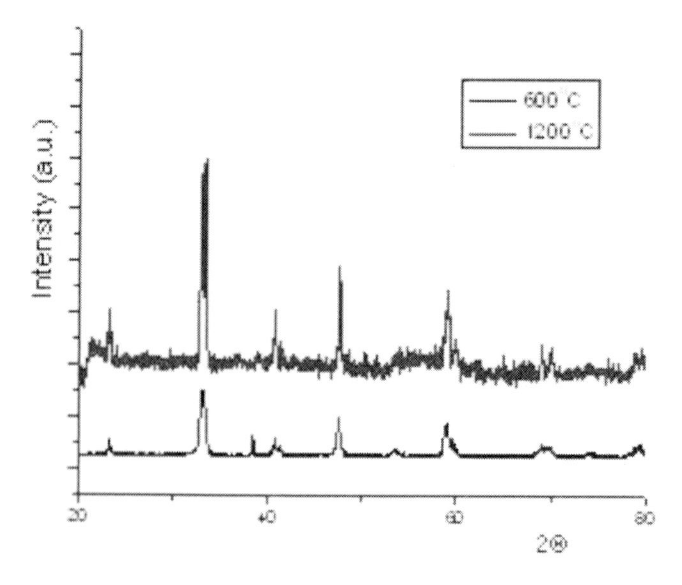

Figure 8. XRD patterns of $LaCoO_3$ powders thermally treated.

Table 7. Crystallographic parameters, particle size and internal stress of the thermally treated powders at 600^{0}C [82].

Sample	a(Å)	c(Å)	c/a	V (Å^3)	S (E^{-3})	D (Å)	BET S.area (m^2/g)
$LaCoO_3$ (N)	5.4390	13.1616	2.4199	337,19	0.87	537	8.82
$LaCoO_3$ (A)	5.4455	13.2098	2.4258	339.24	1.04	777	13.30

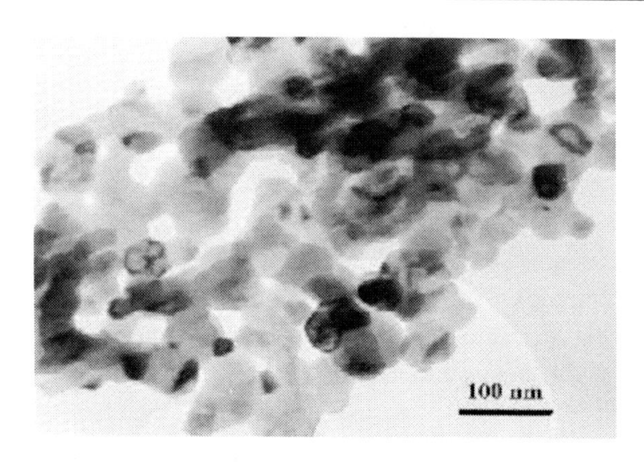

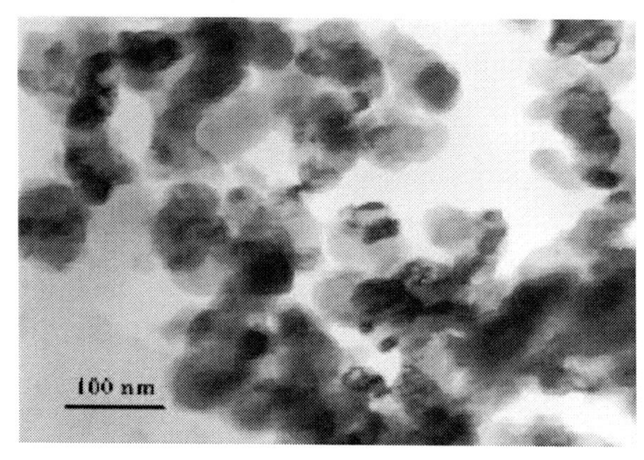

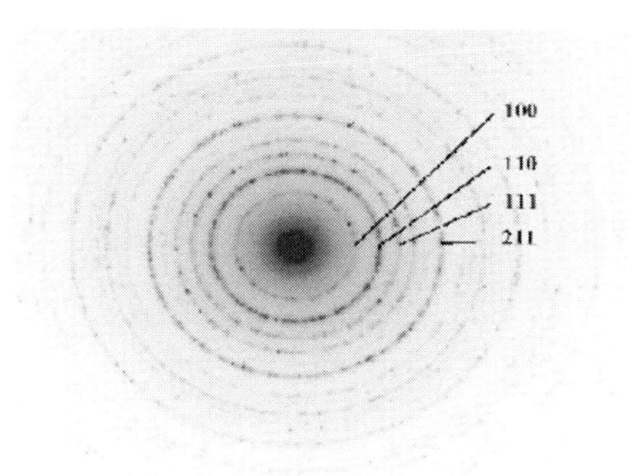

Figure 9. TEM and SAED images of the $LaCoO_3$ sample.

The presented results have shown the same rhombohedral perovskite structure for both samples, but with higher values of the crystallographic parameters, higher internal stress of the crystals, higher particles size and higher surface area when acetates were used as precursors. Due to the fact that in the case of the powders obtained from acetates both particles size and specific surface area are higher than in the case of the powders obtained from nitrates, one may assume that in this later case, an intragranular porosity is present.

The results obtained by XRD are confirmed by TEM and SAED measurements that confirm the formation of the rhomboedral perovskite structure of the $LaCoO_3$ (Figure 9).

3.2. $LaCoO_3$ Ceramic Samples

According to the literature data, no matter of precursor powders used the sintering temperature and time required for obtaining dens ceramic samples were high. Schmidt at al. [82] obtained sintered samples by thermal treatment at 1200^0C for one day, starting with powders obtained by classical method, while Kleveland at al. [60] used the same procedure using powders obtained by chemical methods.

Predoana et al. [83] studied the sintering properties of the pressed powders characterized above in the temperature range between 800-1200^0C. The ceramic properties of the pressed and sintered samples are presented in Table 8. One may notice that the variation with the thermal treatment of the open porosity of the samples obtained from both powders presents close values. However, the ceramics obtained from the powders prepared from acetates have shown much lower densities as compared with the samples obtained form powders prepared from nitrates. This behavior could be correlated to the presence of an internal porosity of the grains obtained starting with acetates.

The structure of the sintered powders, as determined by XRD, was the same as in the initial powders, but with a higher intensity. In the case of the powders obtained starting with acetates, Co_2O_3 occurred during the sintering process at highest temperature (1200^0C), as a secondary phase in low amount.

Table 8. Ceramic properties of the sintered samples at different temperatures.

Sintering temperature (^{0}C)	Scrinkage Δl/l, %	Open porosity %	Experim. density, g/cm^3	Relative density, %
Sintered samples from powders obtained from nitrates				
800	1.65	7.12	4.92	73
900	6.00	5.29	6.14	91
1000	10.60	4.77	6.34	94
1100	15.50	1.29	6.50	96
1200	18.85	0.12	6.71	99
Sintered samples from powders obtained form acetates				
800	6.9	6.09	3.88	58
900	10.45	9.36	5.11	76
1000	13.00	5.68	5.66	84
1100	19.25	2.58	5.81	86
1200	24.30	1.31	6.21	92

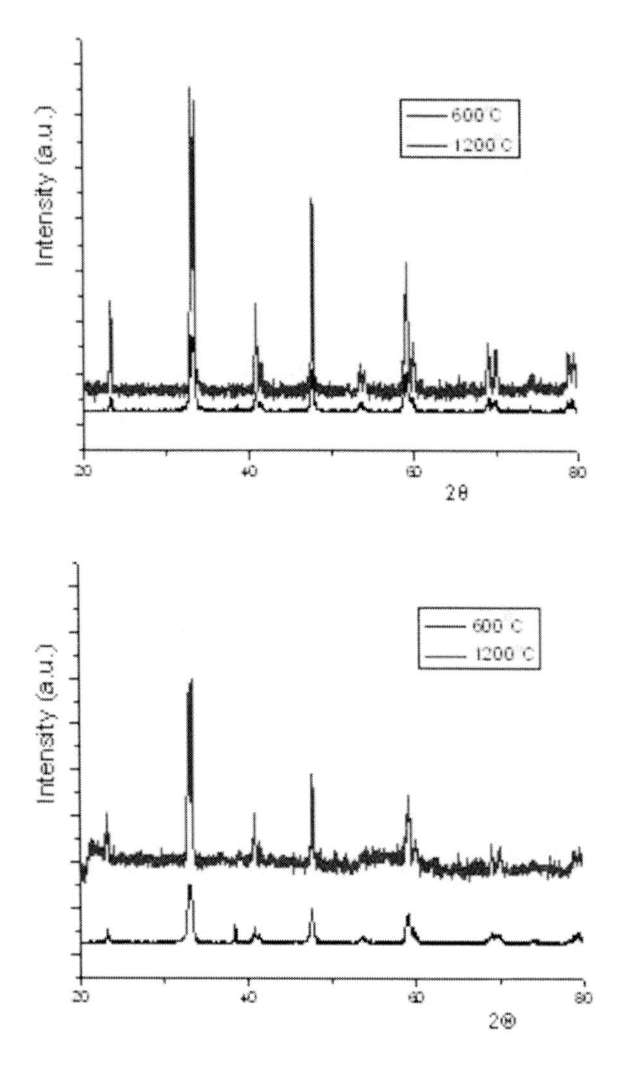

Figure 10. XRD patterns of $LaCoO_3$.

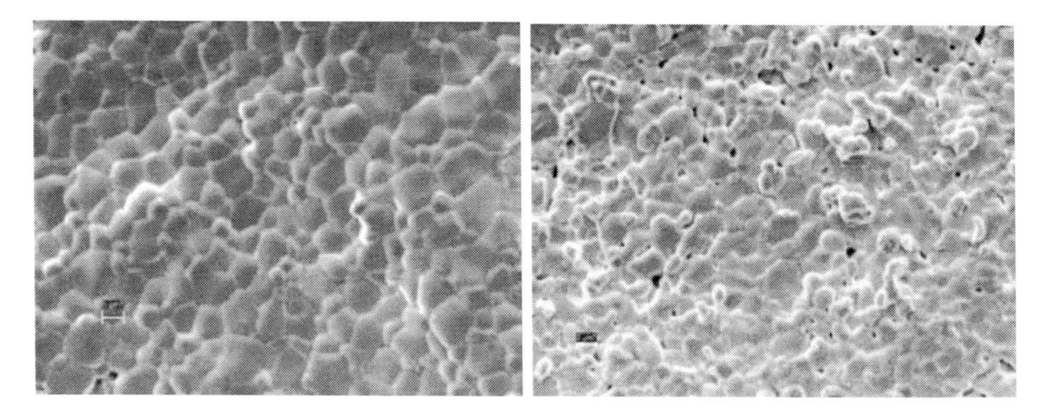

Figure 11. SEM images of the $LaCoO_3$ sintered sample.

The SEM images (Figure 11) have shown a lower tendency for sintering and a less homogeneous morphology for the ceramics prepared from powders obtained from acetates.

The ceramics sintered at 1100^{0}C from powders obtained from nitrates present uniform grains size distribution and practically no open porosity.

Comparing our data with results published before, one may underline that the formation of $LaCoO_3$ powder, as pure phase, takes place at lower temperature as the previously published [60, 83]

The powders have increased sintering ability and led to ceramic bodies with good ceramic properties and very interesting electrical characteristics at lower temperature than that was reported before (1100, as compared to 1250 ^{0}C).

Conclusions

General considerations on the sol-gel chemistry in aqueous system are presented.

The aqueous sol-gel method in the case of transition metals in the presence of the chelating agent is described for the case of Li-Co-citric acid (CA) and La-Co-citric acid (CA) systems.

The chemistry of the sol-gel process was followed by spectroscopic methods (UV-Vis and FT-IR) and a corresponding gelling mechanism was proposed. In both cases, polynuclear gelic complexes are formed

In both cases, Co^{2+} ion presents an octahedral symmetry during the gelling process and the presence of the secondary ions do not influence the UV-Vis spectra.

The FT-IR spectra depend on the secondary ions present in the system, showing that the same ligands are bound in different ways in the polynuclear gelic complexes formed when different type of secondary ion are present.

The resulted $LiCoO_2$ powders with nanosized dimensions present very promising electrochemical properties, while the pure $LaCoO_3$ perovskite powder showed improved sintering ability .

References

[1] Kakihana, M. *J.Sol-Gel Sci.Tehnol.*, 1996, 6, 7-55.

[2] Lazău, I; Ecsedi, Z; Păcurariu, C; Ianoş, R. "Metode conventionale utilizate în sinteza compuşilor oxidici"; *Editura Politehnica,Timişoara*, RO, 2006.

[3] Suchanek, WL; Riman, RE. *Advances Sci. Technol.*, 2006, 45, 184-193.

[4] Schmidt, H; Seferling, A. *Mat. Res. Soc., Pittsburgh*, 1986, 739.

[5] Chen, CC; Shen, CH; Liu, RS; Lin, JG., Huang, CY. *Mater. Res. Bull.*, 2002, 37, 235-246.

[6] Campagnoli, E; Tavares, A; Fabbrinu, L; Rossetti, I; Dubitsky, YA; Zaopo, A; Forni, L. *Appl. Catal. B: Environ.*, 2005, 55,133 -139.

[7] Brinker, CJ; Scherer, GW. *Sol–Gel Science: The Physics and Chemistry of Sol–Gel Processing.* Academic Press, San Diego, 1990.

[8] Bjerrum, NZ. *Phys. Chem.*, 1907, 59, 336.

[9] Sillen, LG. *Quart. Rev.*, 1959, 13, 146.
[10] Livage, J; Henry, M; Sanchez, C. *Progr. Solid State Chem.*, 1988,18, 259-341.
[11] Livage, J. *Catalysis Today,* 1998, 41, 3-19.
[12] Landschoot, NV; Ionica, CM; Kelder, EM; Schoonman, J. Proceedings on CD ROM of the Second Conference on *Advanced Materials and Technologies Euro-TECHMAT,* Bucharest, 9-13 September 2001
[13] Cho, PJ; Jeong, ED; Shim, JB. *Bull. Korean. Chem. Soc.*, 1998, 19, 39.
[14] Yazami, R; Lebrun, N; Bonneau, M; Molteni, M. *J. Power Sources,* 1995, 54, 389-392.
[15] Cho, J; Kim, G. *Electrochem. Solid-State Lett.*, 1999, 2 (6), 253.
[16] Yamaki, JI; Baba, Y; Katayama, N; Takatsuji, H; Egashira, M; Okada, S. *J. Power Sources,* 2003, 119-121, 789-793.
[17] Rossen, E; Reimers, JN; Dahn, JR. *Solid State Ionics*, 1993, 62, 53-60.
[18] Wang, B; Bates, JB; Harts, FX; Sales, BC; Zuhr, RA; Robertson, JD. *J.Electrochem. Soc.*, 1996, 143, 3203.
[19] Poulsen, JM; Mueller-Neuhaus, JR; Dahn, JR. *J.Electrochem. Soc.*, 2000, 147(2), 508.
[20] Cho, J.*Chem. Mater.*, 2000, 12, 3788.
[21] Endo, E; Yasuda, T; Kita, A; Yamaura, K; Sekai, K. *J.Electrochem. Soc.*, 2000, 147(4), 1291.
[22] Cho, J; Kim, YJ; Park, B. *J.Electrochem. Soc.*, 2001, 148(10), A1110.
[23] Jang, SW; Lee, HY; Lee, SJ; Baik, HK; Lee, SM. *Mater. Res. Bull.*, 2003, 38,
[24] Gopukumar, S; Jeong, Y; Kim, KB. *Solid State Ionics,* 2003, 159, 223-232.
[25] Amatucci, GG; Tarascon, JM; Larcher, D; Klein, LC. *Solid State Ionics,* 1996, 84, 169-180.
[26] Fey, GT; Huang, DL. *Electrochim. Acta*, 1999, 45, 295–314.
[27] Quinlan, FT; Vidu, R; Predoana, L; Zaharescu, M; Gartner, M; Groza, J; Stroeve, P. *Ind. Eng. Chem. Res.*, 2004, 43, 2468–2477.
[28] Szatvanyi, A; Crişan, M; Crişan, D; Jitianu, A; Stanciu, L; Zaharescu, M. Revue *Roumaine de Chimie,* 2002, 47(12), 1255-1259.
[29] Yoon, WS; Kim, KB. *J. Power Sources,* 1999, 81-82, 517-523.
[30] Shlyakhtin, OA; Yoon, YS; Oh, YJ. *J. European Ceram.Society*, 2003, 23, 1893.
[31] Cho, J; Kim, GB; Lim, HS; Kim, CS; Yoo, SI. *Electrochem. Solid State Lett.*, 1999, **2**(12), 607-609.
[32] Zhecheva, ER; Stoyanova, Gorova, M; Alcantara, R; Morales, J; Tirado, JL. *Chem. Mater.*, 1996, 8, 1429-1440.
[33] Kang, SG; Kang, SY; Ryu, KS; Chang, SH. *Solid State Ionics,* 1999, 120, 155-161.
[34] Kushida, K; Kuriyama, K. *J. Crystal Growth,* 2002, 237-239, 612-615.
[35] Uchida, I; Fujiyoshi, H; Waki, S. *J.Power Sources,* 1997, 68, 139-144.
[36] Park, SC; Kim, YM; Kang, YM; Kim, KT; Lee, PS; Lee, JY. *J.Power Sources,* 2001, 103, 86-92.
[37] Park, SC; Han, YS; Y.S.Kang, Lee, P; Ahn, S; Lee, HM; Lee, JY. *J.Electrochem. Soc.*, 2001, 148 (7), A680.
[38] Kim, MK; Chung, HT; Park, YJ; Kim, JG; Son, JT; Park, KS; Kim, HG. *J.Power Sources,* 2001, 99, 34-40.
[39] Cho, SI; Yoon, SG. *J.Electrochem. Soc.,* 2002, 149(12), A1584.
[40] Liu, J; Wen, Z; Gu, Z; Wu, M; Lin, Z. *J.Electrochem. Soc.*, 2002, 149(11), A1405.
[41] Tao, S; Wu, Q; Zhan, Z; Meng, G. *Solid State Ionics,* 1999, 124, 53-59.

[42] Schleich, DM. *Solid State Ionics*, 1994, 70, 407-411.
[43] Chu, CT; Dunn, B. *J. Am. Ceram. Soc.*, 1987, 70, C375.
[44] Baythoun, MSG; Sale, FR. *J. Mater.Sci.*, 1982, 17, 2757.
[45] Gummow, RJ; Thackeray, MM; David, WIF; Hull, S. *Mat. Res. Bull.*, 1992, 27, 327.
[46] Fu, LJ; Liu, H; Li, C; Wu, YP; Rahm, E; Holze, R; Wu, HQ. *Progress in Materials Science,* 2005, 50, 881.
[47] Son, JT; Cairns, EJ. *Journal of Power Sources,* 2007, 166, 343-347.
[48] Zhu, C; Yang, C; Yang, WD; Hsieh, CY; Ysai, HM; Chen, YS. *J. Alloys Compounds* , 2010, 496, 703-709.
[49] Yang, WD; Hsieh, CY; Chuang, HJ; Chen, YS. *Ceram. International*, 2010, 36, 135-140.
[50] Zeng, XL; Huang, YY; Luo, FL; He, YB; Tong, DG. *J Sol-Gel Sci Technol,* 2010, 54, 139-146.
[51] Lever ABP; Inorganic Electronic Spectroscopy" Elsevir Publ. Comp., Amsterdam, London, New York, 1984.
[52] Predoanã, L; Barau, A; Zaharescu, M; Vassilchina, H; Velinova, N; Banov, B; Momchilov, A. *J.Eur.Ceram.Soc.*, 2007, 27, 1137-1142.
[53] Fernandes, JDG; Melo, DMA; Ziner, LB; Salustiano, CM; Silva, ZR; Martinelli, AE; Cerqueira, M; Alves Junior, C; Longo, E; Bernardi, MIB. *Mater. Lett.*, 2002, 53, 122-125.
[54] Ohno, Y; Nagata, S; Sato, H. *Solid State Ionics*, 1981, 3-4, 439-442.
[55] Figueiredo, FM; Frade, JR; Marques, FMB. *Solid State Ionics,* 1999, 118, 81-87.
[56] Traversa, E; Nunziante, P; Sakamoto, M; Sadaoka, Y; Montanari, R. *Mater. Res. Bull.* 1998, 33 (5), 673-681.
[57] Teraoka, Y; Zhang, HM; Okamoto, K; Yamazoe, N. *Mater. Res. Bull.*, 1988, 23, 51-58.
[58] Figueiredo, FM; Marques, FMB; Frade, JR. *Solid State Ionics,* 1998, 111, 273-281.
[59] Hwang, HJ; Moon, J; Awano, M; Maeda, K. *J. Am. Ceram. Soc.*, 2000, 83, 2852-2854.
[60] Kleveland, K; Orlovskaya, N; Grande, T; Mardal Moe, AM; Einarsrud, MA; Breder, K; Gogotsi, G. *J. Am. Ceram. Soc.*, 2001, 84 [9], 2029–2033.
[61] Orlovskaya, N; Kleveland, K; Grande, T; Einarsrud, MA. *J.Eur. Ceram. Soc.*, 2000, 20, 51-56.
[62] De Souza, RA; Kilner, JA. *Solid State Ionics,* 1998, 106(3-4), 175-187.
[63] Aruna, ST; Muthuraman, M; Patil, KC. *Mater. Res. Bull.*, 2000, 35, 289-296.
[64] Popa , M; Calderon-Moreno, JM. *Thin Solid Films,* 2009, 517, 1530-1533.
[65] Popa , M; Calderon-Moreno, JM. *J.Eur. Ceram. Soc.*, 2009, 29, 2281-2287.
[66] Berger, D; Matei, C; Papa, F; Voicu, G; Fruth, V. "Pure and doped lanthanum cobaltites obtained by combustion method", *Progress in Solid State Chemistry*, 35 (2007) 183 – 191.
[67] Campagnoli, E; Tavares, A; Fabbrinu, L; Rossetti, I; Dubitsky, YA; Zaopo, A; Forni, L. Appl. Catal. B: Environ., 2005, 55, 133 -139.
[68] Yao, T; Uchimoto, Y; Sugiyama, T; Nagai, Y. *Solid State Ionics* , 2000, 135 [1-4], 359–364.
[69] Faaland, S; Grande, T; Einarsrudw, MA; Vullum, PE; Holmestad, R. *J. Am. Ceram. Soc.*, 2005, 88, 726–730.
[70] Berger, D; van Landschoot, N; Ionica, C; Papa, F; Fruth, V. *J. Optoelectron. Adv. Mater.*, 2003, 5, 719-724.

[71] Li, F; Yu, X; Chen, L; Pan, H; Xin, X. *J.Am.Ceram.Soc.,* 2002, 85, 2177–2180.
[72] Popa, M; Frantti, J; Kakihana, M. *Solid State Ionics,* 2002, 154-155, 135-141.
[73] Guo, J; Lou, H; Zhu, Y; Zheng, X. *Mater Letters,* 2003, 57, 4450-4455.
[74] Taguchi, H; Yamada, S; Nagao, M; Ichikawa, Y; Tabata, K. *Materials Research Bulletin,* 2002, 37, 69-76.
[75] Ajami, S; Mortazavi, Y; Khodadadi, A; Pourfayaz, F; Mohajerzadeh, S. *Sensors and Actuators,* 2006, B 117, 420-425.
[76] Chen, CC; Shen, CH; Liu, RS; Lin, JG; Huang, CY. *Mater. Res. Bull.*, 2002, 37, 235-246.
[77] Predoana, L; Jitianu, A; Malic, B; Zaharescu, M. *J. Sol-Gel Sci.Technol.*, to be published
[78] Predoana, L; Malic, B; Kosec, M; Carata, M; Caldararu, M; Zaharescu, M. *J. Eur. Ceram. Soc.*, 2007, 27, 4407–4411.
[79] Predoana, L; Malic, B; Kosec, M; Scurtu, M; Caldararu, M; Zaharescu, M. *Processing and Application of Ceramics,* 2009, 3 [1-2], 39–42.
[80] Predoana, L; Malic, B; Zaharescu, M. *J. Thermal Analysis Calorimetry,* 2009, 98(2), 361-366.
[81] Nakamoto, K. Infrared and Raman Spectra of Inorganic and Coordination Compounds'' Part B., *Applications in Coordination, Organometallic and Bioinorganic Chemistry,* Sixth Edition, John Wiley & Sons Inc., New-York, 2009.
[82] Predoana, L; Malic, B; Crisan, D; Dragan, N; Gartner, M; Anastasescu, M; Calderon-Moreno, J; Zaharescu, M. Proceedings of the 11th *ECERS Conference, Krakow*, 2009, 22-24.
[83] Schmidt, R; Wu, J; Leighton, C; Terry, I. *Phys.Rev.B.*, 2009, 79, 125105 -1-8.

In: The Sol-Gel Process
Editor: Rachel E. Morris
ISBN 978-1-61761-321-0

Chapter 15

APPLICATION OF SOL-GEL PROCESS IN THE POLYMER MATERIALS MODIFICATION

Nika Veronovski[*]
University of Maribor, Characterization and Processing of Polymers Laboratory, Faculty of Mechanical Engineering, University of Maribor, Smetanova, Maribor, Slovenia

ABSTRACT

Nanotechnology, which employs physical or chemical techniques to construct materials, devices or systems on a nanometre scale, is recently being introduced in preparation of functional textiles. It is a technology that can significantly improve properties of materials compared with those of conventional ones. Various functionalities can be enhanced or imparted to a textile material by applying different coatings of nano particles; e.g. self-cleaning properties and antimicrobial activity. Numerous techniques for the production of nanocoatings exist. Sol-gel process is one of the most applicable techniques to manufacture thin films. One of the most advantageous and attractive characteristics of the sol-gel process is the ability to control and manipulate composition and microstructure of the material at the molecular level and at room temperature.

Two basic coating processes will be explained; (i) process of direct (in situ) formation of nanocoatings on fibres' surfaces and (ii) process of nanocoatings' preparation starting from TiO_2 P25 powder. Conditions, under which the processes for nanocoating can give rise to self-cleaning cellulose materials, were studied. In the first phase, the preparation technique and properties of nano titanium dioxide (TiO_2) coatings generated directly on the surface of materials via sol-gel process have been studied extensively. Process was optimized in order to obtain coatings with desired structure and properties. The influence of sol-gel process conditions on the particle size and effectiveness was examined. When applying the procedure to the fibres some problems could occur, i.e. high temperature treatment conditions, problems connected with the formation of required polymorph TiO_2 form, fibre damage risks, problems connected with durability of the modification, etc. For that reason, in the second part of research,

[*] Corresponding author: nikaveronovski@gmail.com

already formed TiO_2 P25 nanoparticles were used for obtaining self-cleaning modified surfaces. In addition to that, the process for attaching composite TiO_2-SiO_2 nanoparticles, where SiO_2 acts as a binding agent was applied; furthermore SiO_2 protects the fibres against possible photocatalytic influence of photocatalyst, since TiO_2 P25 is known as a photocatalyst with high photocatalitic activity. In spite of the fact, that investigations indicated self-cleaning effectiveness of all TiO_2 coated samples, the self-cleaning effectiveness was the highest in the case of samples, treated using stable dispersions of TiO_2 P25 nanoparticles. According to functional effectiveness and yielded technologically-applicable properties, the process of preparing composite TiO_2-SiO_2 nanocoatings was the most appropriate.

1. Introduction

Nanotechnology and Its Role in the Field of Technical Textiles

Nanotechnology, which employs physical or chemical techniques to construct materials, devices or systems on a nanometre scale, is recently being introduced in preparation of functional textiles. It is a technology that can significantly improve properties of materials compared with those of conventional ones. This remarkable properties improvement is attributed to the significant increase in the fibre's boundary area or its functional surface. In the textile sector, nanotechnology provides opportunities by creating new materials such as nanofibres, nanocomposites and nanocoated textiles [1]. Various functionalities can be enhanced or imparted to a textile material by applying different coatings of nano particles [2, 3]; e.g. self-cleaning properties [4-6] and antimicrobial activity. [7-9]. Such treatments of textiles is especially interesting for the technical textile sector [10]. In the practice, cleaning surfaces of different materials is energy and water consuming process, where huge amounts of chemical agents are used, which results in high costs and environmental pollution. Environmental friendly process results in highly efficient TiO_2 nanocoatings, which will contribute to cleaner environment, since photocatalysis is promising method for elimination of most problems related to increased pollution.

Textile materials for technical application must fulfil special requirements due to their exposure to different weather and climatic conditions (rain, sun, snow, hail, etc.). In addition, environmental contents such as dust, pollen and other impurities and particles can bind themselves on textile material therefore special care is needed to enhance the appropriate appearance. Consequently, having a self-cleaning surface ensures a special advantage for these materials. This chapter presents a nano-modification of textile materials for technical application in order to obtain special self-cleaning properties. The properties of the nanostructured material show remarkable improvement when compared to those of conventional items. These unique properties are attributed to the significant increase in the fibre's boundary area or its functional surface.

Cellulose fibres have a large active surface area and are therefore suitable starting material for various modification procedures that lead to improved consumer qualities of the material. [11]. The solvent spun cellulose fibre type, produced by the amine oxide process (NMMO fibre or lyocell fibre-CLY) were used for modification, with characteristic round cross-section and smooth surface. The special properties of Lyocell fibres, high strength and comparatively low elongation are arising from their high degree of crystallinity and molecular orientation.

TiO_2, Photocatalysis and Self-Cleaning

Among semiconductors, TiO_2 has been chosen, since it is biologically and chemically inert and relatively low-cost product [12]. TiO2 usually exists in three common crystal structures: rutile, brookite and anatase. Most studies in TiO_2 photocatalysis have been carried out using either pure anatase form or pure rutile form (natural) or a mixture of the two as photocatalysts [13]. Only rutile and anatase are commercially important, often the anatase form of TiO_2 is believed to be more photocatalytically active than rutile form [14-18]. Interestingly, P-25, a commercial TiO2 catalyst containing a mixture of rutile and anatase crystalline forms, is the most widely used photocatalyst and has proven to be the best photocatalyst towards a broad range of organic pollutants [19, 20].

Recently, titanium dioxide (TiO_2) has attracted much interest for the good photocatalytic and hydrophilic properties of its surface [21-24].

TiO_2 is a nontoxic inorganic semiconductor with a band gap of ca. 3.2 eV (anatase), corresponding to a photonic energy of 388 nm [25, 26]. Upon absorption of the photons in the UV range, free electrons (e^-) and holes (h^+) are generated in the TiO_2 crystal (electron-hole pair). The charge carriers are trapped at the crystal's surface and act as spatially separated redox centers. The electrons then lead to the formation of superoxide anions (from atmospheric oxygen) and finally hydrogen peroxide and hydroxyl radicals, which are highly oxidising species. On the other hand, the holes participate in oxidative processes by direct oxidation of pollutants or by the generation of hydroxyl radicals from the present water.

Photocatalysis of TiO_2 in the presence of light proceed as indicated in the following Eqs. (1), (2) and (3) [27, 28]:

$$TiO_2 + h\nu \rightarrow e_{cb}^- + h_{vb}^+ \quad (1)$$

$$O_2 + e_{cb}^- \rightarrow O_2^{-}\bullet \quad (2)$$

$$H_2O + h_{vb}^+ \rightarrow OH\bullet + H^+ \quad (3)$$

The electronic structure of TiO_2 and the basic processes are shown (in simplified form) in Figure 1.

The oxidative processes are the basis of the photoinduced mineralization of organic pollutants [30, 31], eventually leading to the formation of CO_2 and H_2O. Photocatalysis has been shown to act efficiently against water pollutants [32] and air contaminants [33] as well as against bacteria [34]. By use of photocatalytic coatings, soil and microorganisms present on the surfaces are photooxidized (i.e. mineralized) and self-cleaning or hygienic surfaces become effective. In addition, TiO_2 shows superhydrophilicity, leading to enhanced wettability under (and also after) UV irradiation.

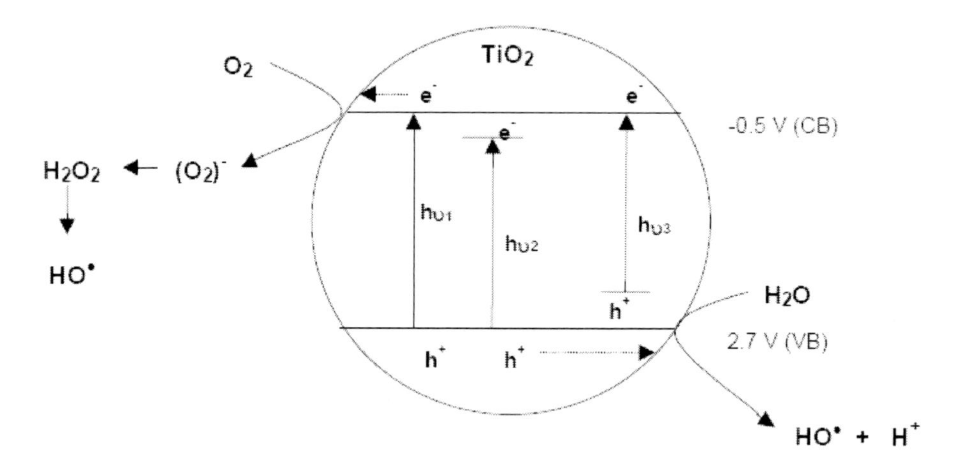

Figure 1. Mechanism of photoxidative degradation of organic pollutants at a TiO_2 surface [29].

According to the literature, different factors seem to affect the photocatalytic properties of TiO_2, i.e.: particle size, phase composition, level of crystallinity, crystallite size, morphology, specific surface area, surface state, etc. [35]. From a catalysis point of view, a high surface area is an advantage in terms of a greater concentration of active sites per square metre, and hence generally leads to higher reactivity [36, 37]. These properties are affected by the production techniques and by the preparative conditions. For that reason this research was focused on achieving the desired properties of TiO_2 and the processes that allow getting these properties.

2. Techniques for Nanocoatings Formation

Numerous techniques for the production of nanocoatings exist. Different procedures for the preparation of TiO_2 nanocoatings on fibre's surfaces were used in the research and will be presented hereafter, i.e.:

1. Process of direct (in situ) formation of nanocoatings on fibres surfaces starting from precursor Titanium isopropoxide (TIP) under various sol-gel process conditions;
2. Process of nanocoatigs preparation starting from TiO_2 P25 powder using colloid suspensions of different stability;
3. The process for attaching composite TiO_2-SiO_2 nanoparticles.

2.1. The Process of Direct (in Situ) Formation of Nanocoatings on Fibres Surfaces Starting from Precursor (TIP) under Various Sol-Gel Process Conditions

Sol-gel process is one of the most applicable techniques to manufacture thin films [38, 39]. One of the most advantageous and attractive characteristics of the sol-gel process is the ability to control and manipulate composition and microstructure of the material at the molecular level and at room temperature. Since a *sol* is defined as a colloidal dispersion of particles in a liquid and *gel* as a substance that contains a continuous solid skeleton, which serves as an entrapment for continuous liquid phase, than the definition for sol-gel processing is the growth of colloidal particles and their linkage together to form a gel. Formation of films, fibres and particles are all considered sol-gel processes, even though gelation may not occur, as in the case of particles formation [40, 41].

The sol-gel process involves the hydrolysis (1) and condensation (2) and (3) of metal alkoxides [42-45]:

$$\text{M-OR} + \text{HOH} \rightarrow \text{M-OH} + \text{ROH} \tag{1}$$

$$\text{M-OH} + \text{RO-M} \rightarrow \text{M-O-M} + \text{ROH} \tag{2}$$

$$\text{M-OH} + \text{HO-M} \rightarrow \text{M-O-M} + \text{HOH} \tag{3}$$

Sol-gel process is presented at Figure 2.

According to Iler [47], sol-gel polymerization occurs in three stages:

1. Monomers polymerization and particles formation
2. Particles growth
3. Particles linkage into chains and thickening of networks, which extend throughout the liquid medium, into a gel.

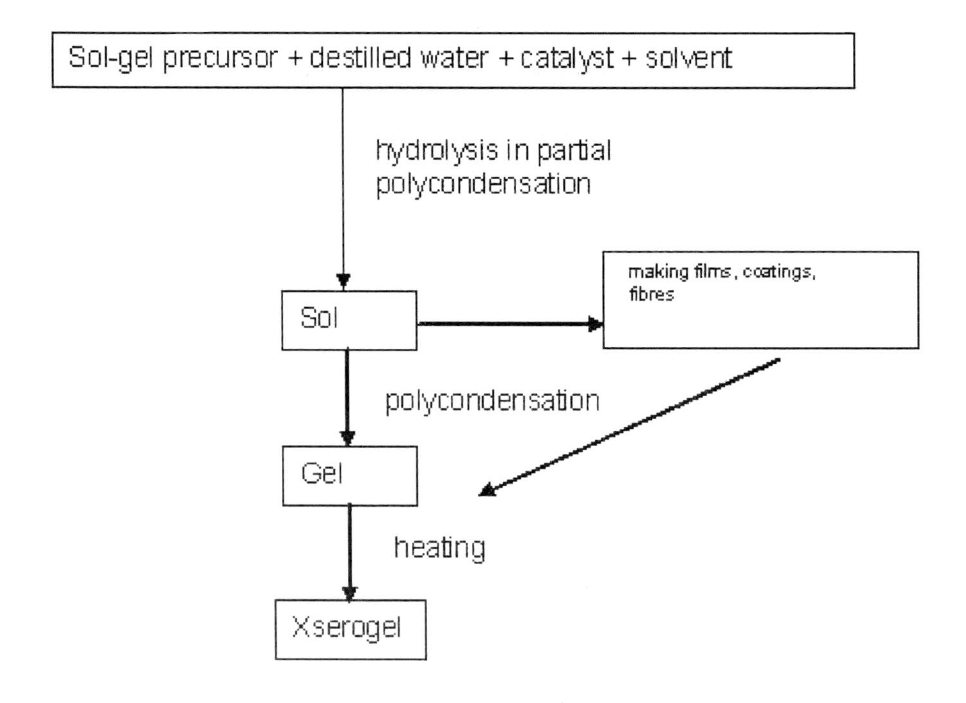

Figure 2. Sol-gel process [46].

Within the context of these stages, many factors affect the resulting Titania network, such as: pH, temperature and time of reaction, reagent concentrations, catalyst nature and concentration, H_2O/Ti molar ratio (R), aging temperature and time. Sol particle size and cross-linking inside the particles (density) is pH and R-value dependant [41]. By controlling the synthesis conditions carefully, these reactions may lead to a variety of structures, and to different final states of the material [43, 44].

Self-cleaning, anti-microbial materials were developed by coating fibres with the semiconducting photocatalyst titanium dioxide (TiO_2) via sol-gel process. In the process, following chemicals were used as received: Ethanoic acid (CH_3COOH), 100 %, or Ammonia solution (NH_3), 25 %, as a catalyst; Isopropanol (C_3H_7OH), 99%, as a solvent; and precursor Titanium isopropoxide ($Ti[OCH(CH_3)_2]_4$), 97 %. Superior coatings were generated when TiO_2 nanoparticles were synthesized directly onto the material under contolled synthesis conditions. The cellulose fabrics were soaked in an ethanoic acid/ammonia solution (10 mL 100 % CH_3COOH or 10 mL 25 % NH_3 in 40 mL 100 % isopropanol) for approximately 1 h before placing them in between two filter papers soaked in Titanium isopropoxide (TIP) (10 vol % in isopropanol). To allow the reagents to be homogeneously absorbed, the fabrics were carefully blotted and left for 4 hrs in a covered Petri dish. After that, the fabrics were immersed in pure isopropanol for one day in order to remove any free TiO_2 nanoparticles. The fabrics were dried at room temperature and thermally treated for 2 hrs at 90 °C.

2.2. The Process of Nanocoatigs Preparation Starting from TiO_2 P25 Powder Using Colloid Suspensions of Different Stability

Another technique for the production of nanocoatings is the process using pre-formed TiO_2 nanoparticles. TiO_2 P25 with average diameter of 21 nm and specific surface area close to 55±15 m^2/g and refractive index above 2.5 [48, 49] is a commercial photocatalyst produced by Degussa containing a mixture of rutile and anatase crystalline forms which has proven to be the best photocatalyst towards a broad range of organic pollutants owing to special crystalline structure [19, 20, 50-52]. Titanium dioxid TiO_2 P25 consists of 80 percent anatase and only 20 percent rutile [53]. Rutile and anatase have different band gaps. The band gap is the energy difference between the valence band and the conduction band in a semiconductor and is the major determinant of the absorption behavior. For rutile this band-gap is 3.05 eV, corresponding to absorption at 420 nm. Anatase on the other hand has a band gap of 3.20 eV and absorbs at 385 nm. The rutile absorption starts at 420 nm and the quantum yield, with the resulting reactivity of the electrons at the surface of TiO_2 P25, is particularly high because the rutile contained in the mixed crystal initially functions as an antenna. This conducts absorbed quanta to the anatase system, thus increasing the probability of generating reactive electrons at the surface. Coupling of the two absorption processes results in improved utilization of the incident light [53].

TiO_2 P25 pigments are made of extremely small TiO_2 particles, which are characterized by many important properties. TiO_2 P25 nanoparticles if separated into smaller particle size populations possess large surface area; unfortunately its dispersions are intrinsically unstable. Such particles tend to agglomerate and form clusters (Figure 3) due to attractive interactions between nanoparticles [54-57]. Such drawbacks may drastically reduce their performances.

Surface of TiO_2 nanoparticles is covered with hydroxyl groups of amphoteric character, formed upon adsorption of water [58]. By controlling nanoparticles aggregation the applicability of TiO_2 nanoparticles suspensions would be considerably more effective. That is why the conditions for an efficient dispersion are the critical step in preparing such materials. The influence of pH varying and/or surfactants addition on stability of TiO_2 P25 aqueous suspensions was determined. For the stabilization of colloid suspensions alkanediylα,ω-bis (N-dodecyl-N, N'-dimethylammonium bromides), alkylammonium Geminis, were used as cationic surfactants, with dodecyl groups linked to both ends of α, ω- N,N'-dimethylamine chains separated by two or six methylene units, which act as spacers between the polar head groups. 12-6-12 Gemini surfactant was used with CMCs below the milimolar range, 5.0×10^{-4} mol/L in 5.0×10^{-3} mol/L KBr at 25°C.

2.3. The Process for Attaching Composite TiO_2-SiO_2 Nanoparticles

Titanium dioxide (TiO_2) crystals form a molecular machine powered by light. TiO_2 is a semiconductor charged by the ultraviolet photons. When these nanoparticles are charged, powerful oxidizing agents, namely hydroxyl radicals, are produced. These free radicals destroy the airborne germs and pollutants that circulate over the treated surface. Since nano TiO_2 has a high oxidative power, the composite layer of the TiO_2–SiO_2 should prevent possible attack on the cellulose by the h_{vb}^{+} generated by the TiO_2 under irradiation. The preparation of composite TiO_2–SiO_2 coatings on various surfaces including on cellulose surfaces was already investigated by several researchers [59-63]. Composite TiO_2-SiO_2 nanoparticles were used, where amorphous form of silica gel acted as a binding agent and at the same time prevented fibre against photocatalytic influence of photocatalyst. Tetraethyl orthosilicate (TEOS) and 3-glycidoxypropyltrimethoxysilane (GLYMO) were used in the research to reduce the possible negative effect on cellulose fibres. Metal alkoxides are popular precursors, because they react readily with water. The reaction is called hydrolysis, because a hydroxyl ion becomes attached to the metal atom [64]. Metal alkoxides are members of the family of metalorgnic compounds, which have an organic ligand attached to a metal or metalloid atom. The most thoroughly studied example is silicon tetraethoxide (or tetraethoxysilane, or tetraethyl ortosilicate, TEOS), $Si(OC_2H_5)_4$.A basic of the process is controlled hydrolysis of precursor and solvent where silica gels are formed, which is followed by condensation of disperssed phase, resulting in nanoclusters which grow and eventually link to form a solid gel. The particle size and shape are affected by numerous parameters [65-71]:

- ***Solvent;*** rection is the fastes when using methanol and the slowest when using n-buthanol. The smalles particles are obtained in the case of using methanol and the largest in the case of using n-buthanol [66].
- ***Si-alkoxides;*** reaction is the fastest in the case of using tetramethyl ortosilicate and he reaction is very slow in the case of using tetrapenthyl ortosilicate. Particles obtained in the case of using tetramethyl ortosilicate are very small (less than 0.2 μm) and they are bigger in the case of using tetrapenthyl ortosilicate (2 μm) [66].

- ***Reaction ratio (R);*** R is a ratio between concentrations of a water and TEOS. Any kind of increasing ratio R (H2O/TEOS) leads to increase in particle size. For preparing smaller particles value R (H2O/TEOS) has to be reduced [65-67, 69].
- ***Temperature***; the higher the temperature, the smaller the particles [65-67, 69].
- ***Catalyst;*** reaction of making particles can take place in alkalic or in acid pH range. In the case of alkalic catalysis ammonia can be used [71] and in the case of acid catalysis acetic acid can be used [68]. Acid or base catalysts can influence both the hydrolysis and condensation rates and the structure of the condensed product. Acid-catalysed condenstion is directed preferentially toward he ends rather than the middles of chains, resulting in more extended, less highly branched polymers. Base-catalysed condensation (as well as hydrolysis) should be direced toward he middles rather the ends of chains, leading to more compact, highly branched species [sol-gel science].

However, it can generally be said that sol-gel derived metal oxide networks, under acid-catalyzed conditions, yield primarily linear or randomly branched polymers which entangle and form additional branches resulting in gelation. On the other hand, metal oxide networks derived under base-catalyzed conditions yield more highly branched clusters which do not interpenetrate prior to gelation and thus behave as discrete clusters (Figure 4).

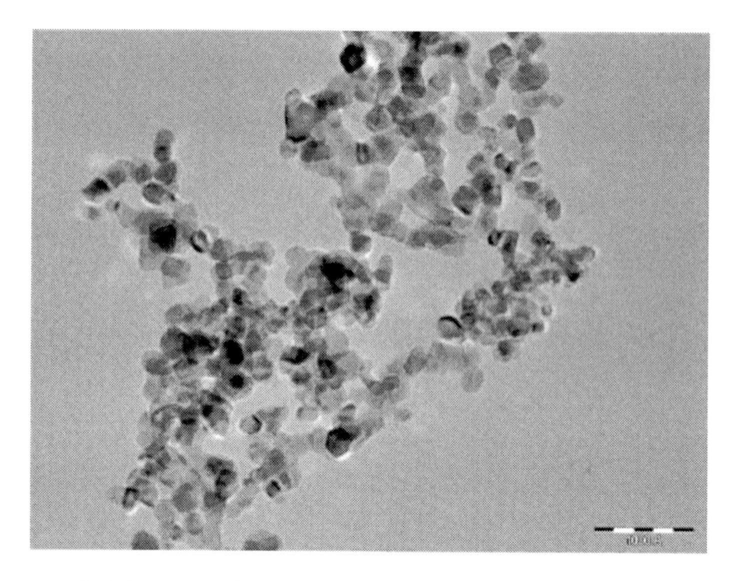

Figure 3. Agglomerated TiO_2 nanoparticles.

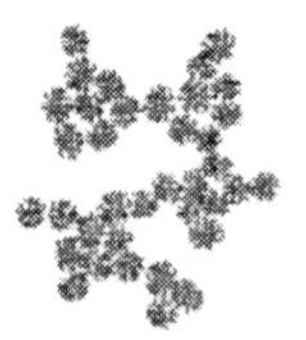

Figure 4. Acid- and alkaline-catalyzed [41].

The use of coupling agent

A coupling agent is usually added to provide bonding between the organic and the inorganic phases, thereby preventing macroscopic phase separation. The adhesion mechanism is due to two groups in the silane structure. The $Si(OR_3)$ portion reacts with the inorganic reinforcement, while the organofunctional (vinyl-, amino-, epoxy-, etc.) group reacts with the organic material.

TiO_2 application in the textile industry

The outstanding photo-catalytic properties do recommend TiO_2 as an ingredient in surface coatings with self cleaning properties [72]. Numerous literature reports on the fabrication of TiO_2 thin films by sol-gel dip coating technique using many types of titanium alkoxides as precursors, e.g. titanium tetraethoxide [73], titanium ethoxide [74, 75], titanium propoxide [76], tetraisopropyl-orthotitanate [77-80], titanium butoxide [81, 82], tetra-n-butoxide [83].

Much work has been carried out depositing TiO_2 on heat resistant surfaces like glass, ceramics and silica by sol–gel methods. Temperatures up to 500 °C produce anatase or anatase/rutile clusters [1, 29-33, 84].

Limited processing temperatures available for cellulose substrates limits the degree of anatase crystallisation resulting in lower activity compared to films on cerramics, which can be processed at higher temperatures (higher degree of crystallisation means higher photoactivity) [85]. By applying the procedure to the textile surfaces some hindrances appear, i.e. high temperature treatment conditions, fibre damage risks, problems connected with durability of the modification, etc. However, some reports on TiO_2 application in the textile industry can be found [86, 87]. In addition, anatase nanocrystalline Titania coatings were produced on cotton fabrics from alkoxide solutions using a low-temperature sol–gel process under ambient pressure [88, 89].

Bozzi *et al.* used radio frequency plasma (RF-plasma), microwave plasma (MW-plasma), and vacuum-UV light irradiation as a pretreatment of synthetic textile surfaces allowing the loading of TiO_2 by wet chemical techniques. These loaded textiles showed a significant photo-oxidative activity under visible light in air under mild conditions, which discolored and mineralized persistent pigment stains contained in wine and coffee [86].

In addition to that, a study was also performed for modified cotton textiles. Samples of bleached and mercerized cotton textiles were activated by RF-plasma, MW-plasma and UV-irradiation introducing negatively charged functional groups to anchor TiO_2 on the textile surface [87].

Daoud and Xin used low-temperature sol–gel process from alkoxide solutions under ambient pressure to produce anatase nanocrystalline TiO_2 coatings on cotton fabrics. TiO_2 coatings of the anatase form were obtained via a classical hydrolysis and condensation reactions of titanium isopropoxide, followed by a boiling water treatment [88].

Insights were provided into the attachment of TiO_2 on cotton surfaces using chemical spacers. The TiO_2 binded to the cotton by chemical means and the textiles presented self-cleaning properties [89].

In our research we succeed in formation of uniform TiO_2 and SiO_2 nanoparticles directly on the fibres surface.

3. ANALYSIS

3.1. Particle Size Distribution

To determine the particle size distribution, dynamic light scattering (DLS) measurements were carried out. The analysis was performed using a Malvern light scattering unit, Zetasizer Nano series HT (Malvern, UK). Individual peaks in particle size distributions were derived from multi-modal correlation functions. According to the results of particle size distribution determination, the samples contain different scattering populations. The third peak in the case of unstable 5.0 mg/mL P25 aqueous suspension after adjusting the pH of suspensions to pH ~ 10, for instance, proves the presence of large agglomerates. Such behaviour was confirmed by ζ–potential measurements [90, 91]. These disappeared after addition of sufficient amounts of surfactant and after adjusting the pH to pH ~ 3. In such conditions, only two populations were present, which are stable against sedimentation. In the latter conditions, only two overlapping populations occurred. As a result of Gemini addition we can observe a decrease in aggregation, which was the highest in the case of 0.5 and 2.5 mg/mL TiO_2 P25 aqueous dispersion when 250 x 10^{-6} mol/L Gemini was used, when the system was almost completely dispersed. Two scattering populations were determined; at 78 - 95 and at 279 – 281 nm. Large agglomerates of size ~ 4700 nm disappeared. Satisfying results were obtained for 5.0 mg/mL TiO_2 P25 aqueous dispersion in the presence of 250 x 10^{-6} mol/L Gemini and for 5.0 mg/mL TiO_2 P25 aqueous dispersion after adjusting the pH to pH ~ 3, likewise. Results indicate that the presence of 1.0 x 10^{-6} mol/L Gemini wasn't enough for good particles distribution in 0.5 mg/mL TiO_2 P25 aqueous dispersion. In alkaline suspension (pH ~ 10) significant sedimentation and flocculation occurred. Such systems are not applicable for practical use, e.g. for fibres coating.

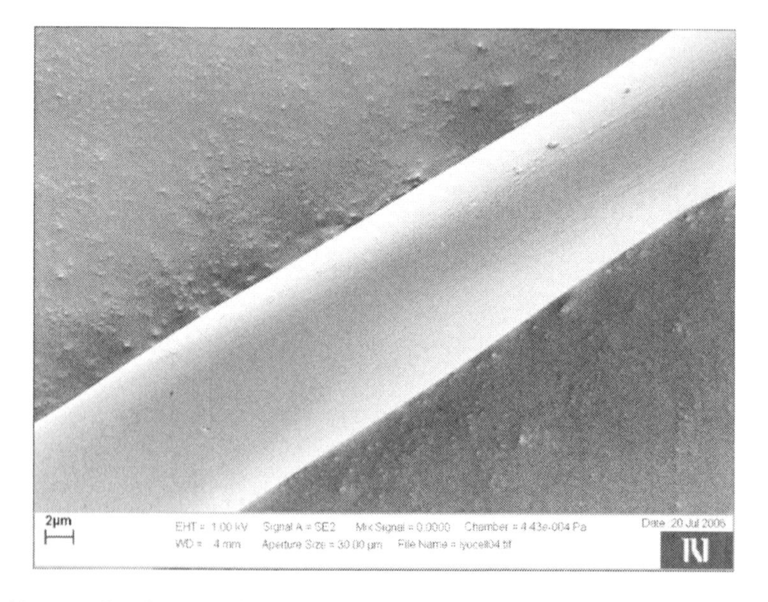

Figure 5. SEM image of surface morphology of untreated Lyocell fibre; taken at magnifications of 10×10^3.

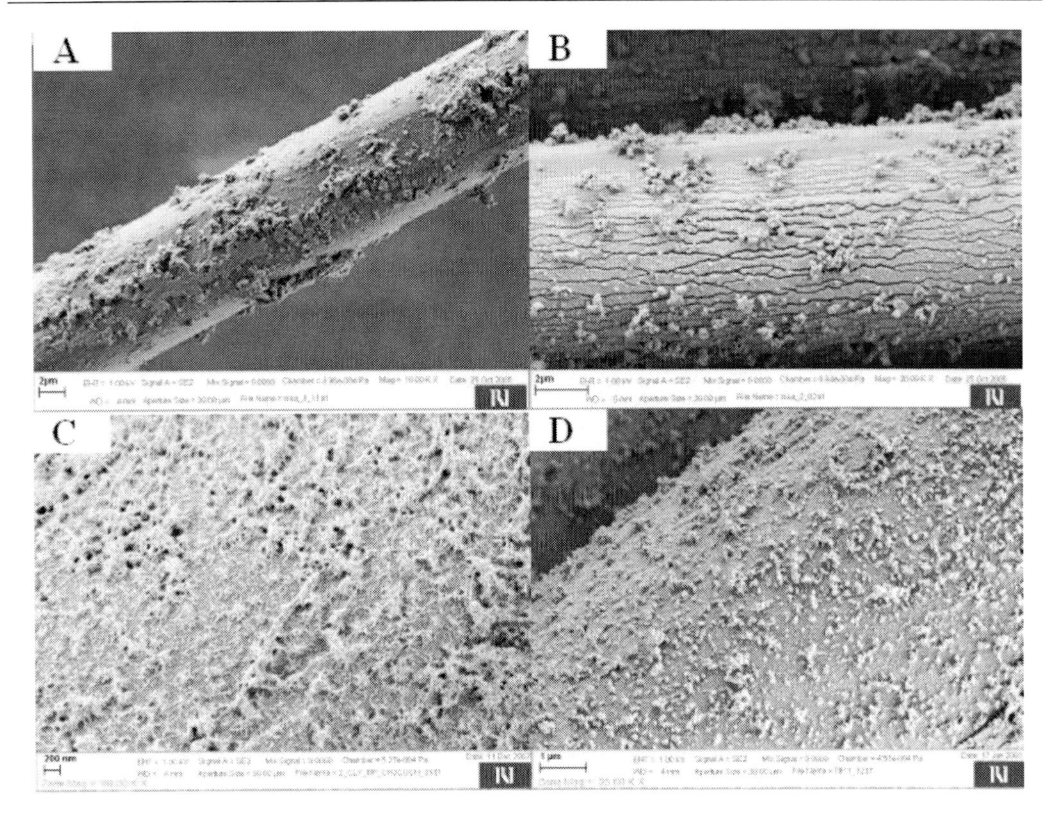

Figure 6. SEM images of surface morphologies of: (A) Fibres treated via acid catalyzed sol-gel process with addition of water; (B) Fibres treated via alkaline catalyzed sol-gel process with addition of water; (C) Fibres treated via alkaline catalyzed sol-gel process without added water and ; (D) Fibres treated via acid catalyzed sol-gel process without added water; taken at magnifications of 10, 20, 60 and 35×10^3, respectively.

3.2. Surface Morphology

The surface of untreated Lyocell fibres analyzed with scanning electron microscope (SEM) is smooth and without any particularity (Figure 5). TiO_2-modifications change fibers' surface. Figure 6A and 6B show the surface morphology of the fibres nanocoated via acid catalyzed sol-gel process with addition of water, and via alkaline catalyzed sol-gel process with addition of water. Acid catalyzed sol-gel process with addition of water resulted in a coating, which didn't coat the fibre surface entirely Figure 6A. On the Figure 6B cracks formed on the entire surface but we can already observe the formation of a defined structure. Incomplete formation of the coatings by the sol-gel processes with addition of water can be explained based on the fact that hydrolysis with addition of water was rapid and completed within seconds, due to the quick reaction between titanium isopropoxide and water upon contact. High reactivity was moderated by using sol-gel process without added water. Completely different surface morphologies were obtained using the same precursor and catalysts, only without added water, under conditions of slow hydrolysis (Figure 6C and 6D). For reaction of hydrolysis, moisture present in air was sufficient for it to proceed. A titanium network without agglomeration was formed via alkaline catalyzed sol-gel process without added water, which yielded nanocoatings with long, extended, thin structures (Figure 6C). That kind of coatings had large active surface, which plays an important role in the reactions of photocatalysis. Contrary, under alkaline conditions particles grow in size with decrease in number, since in the alkaline-catalyzed process a reaction of hydrolysis was slow and a reaction of condensation was rapid. Coatings with homogeneous distribution of spherical TiO_2 nanoparticles can be seen in Figure 6D. The obtained results are in agreement with previous reports about TiO_2 nanoparticles preparation [92, 93]; however we succeeded in preparation of TiO_2 nanoparticles with a narrow distribution of sizes directly on the surface of fibres.

Treatment with 5.0 mg/mL TiO_2 P25 aqueous suspension resulted in high agglomeration of TiO_2 P25 nanoparticles. Fibre surface in Figure 7A isn't coated entirely. This result is in accordance with the particle size distribution and ζ-potential analyses, where analysis demonstrated the presence of big agglomerates in the investigated suspension. The use of acid and alkaline 5.0 mg/mL TiO_2 P25 aqueous suspensions resulted in more homogeneous coatings at the fibre's surface (Figure 7B and 7C).

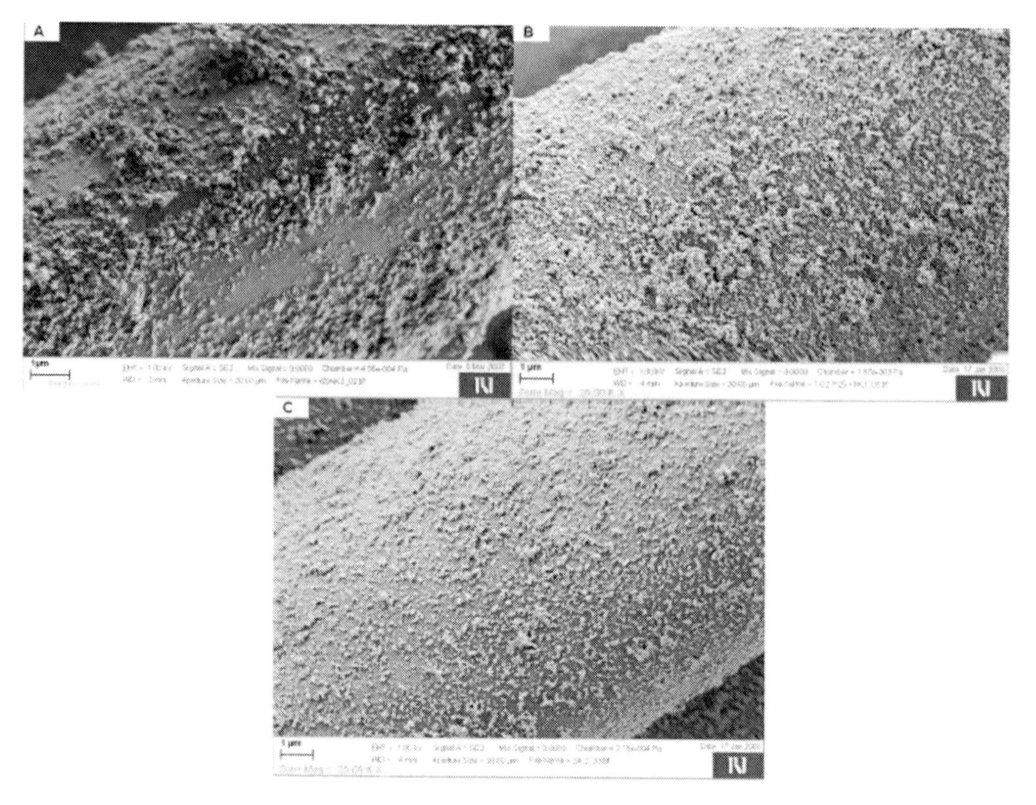

Figure 7. SEM images of surface morphologies of fibers treated in neutral (A), acid (B) and alkaline (C) aqueous suspensions of TiO_2 P25 nanoparticles.

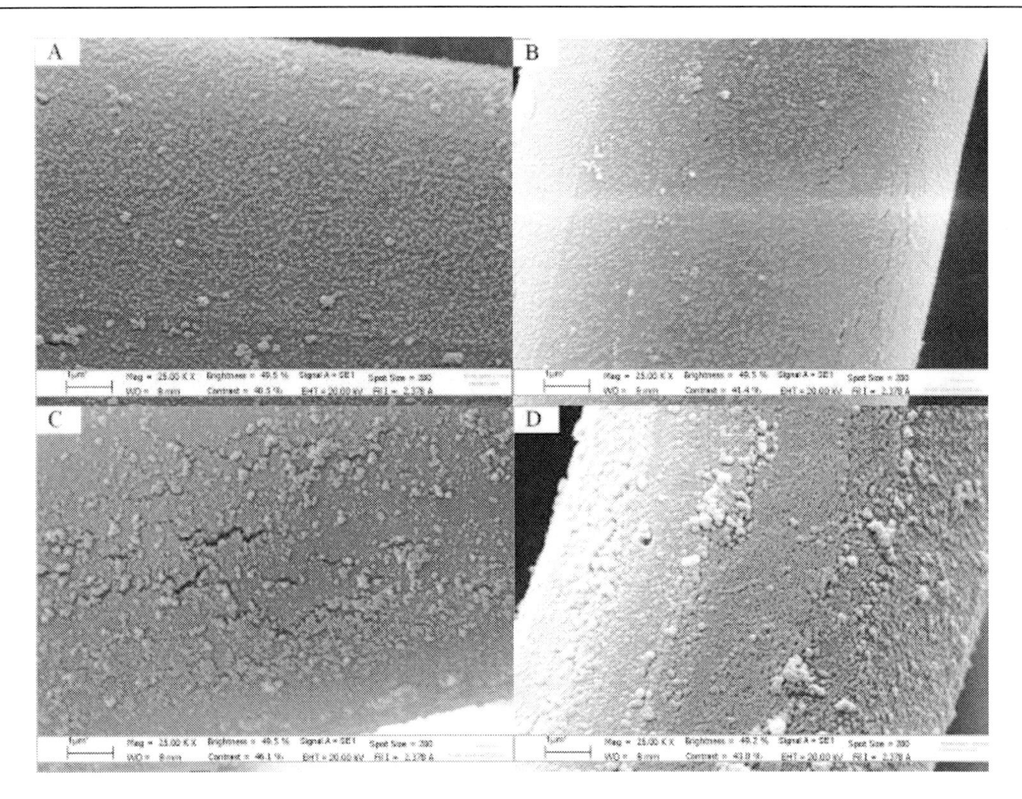

Figure 8. SEM images of surface morphologies of fibres treated with stable TiO_2 suspensions:
(A) 0.5 mg/mL TiO_2 P25 aq. suspension in addition of 1.0 x 10^{-6} mol/L Gemini
(B) 0.5 mg/mL TiO_2 P25 aq. suspension in addition of 250 x 10^{-6} mol/L Gemini
(C) 2.5 mg/mL TiO_2 P25 aq. suspension in addition of 250 x 10^{-6} mol/L Gemini
(D) 5.0 mg/mL TiO_2 P25 aq. suspension in addition of 250 x 10^{-6} mol/L Gemini

The use of TiO_2 P25-surfactant colloidal suspensions resulted in formation of more homogeneous coatings with more uniform particle distribution on the fibre's surface. 0.5 mg/mL TiO_2 P25 aqueous dispersion in addition of 1.0 x 10^{-6} mol/L Gemini yielded nanocoatings with poor density. At the surface of fibre some small agglomerates occurred, however the surface was completely covered with sufficiently dispersed nanoparticles (Figure 8 A). All the examined fibres were covered in the same way. Any agglomerates can't be seen at the surface of fibre, treated with 0.5 mg/mL TiO_2 P25 aqueous suspension in addition of 250 x 10^{-6} mol/L Gemini surfactant, occurred (Figure 8B). We can observe more mono-dispersed nanoparticles. Coatings were homogeneous. After the fibre surface treatment with 2.5 mg/mL TiO_2 P25 aqueous suspension in addition of 250 x 10^{-6} mol/L Gemini surfactant, several smaller clusters occurred. Coating wasn't regular, particle distribution wasn't uniform (Figure 8C). After the treatment in 5.0 mg/mL TiO_2 P25 aqueous suspension in addition of 250 x 10^{-6} mol/L Gemini, fibre surface was entirely covered, the level of density of yielded coating was high (Figure 8D).

Figure 9A and 9B show the surface morphology of fibres embedded in SiO_2 crust and fibres with TiO_2 nanoparticles bound by the coupling agent, respectively. On the surface of Lyocell fibres, which show regular cylindrical cross-sections with smooth surface and without any particularity when untreated, coating of spherical amorphous SiO_2 nanoparticles was formed. Particles were evenly distributed at the surface (Figure 9A). First layer of SiO_2

assures binding sites for TiO_2 nanoparticles. In the Figure 5B TiO_2 nanoparticles uniformly bounded on SiO_2 which is used for fibre protection against highly reactive TiO_2 nanoparticles can be seen.

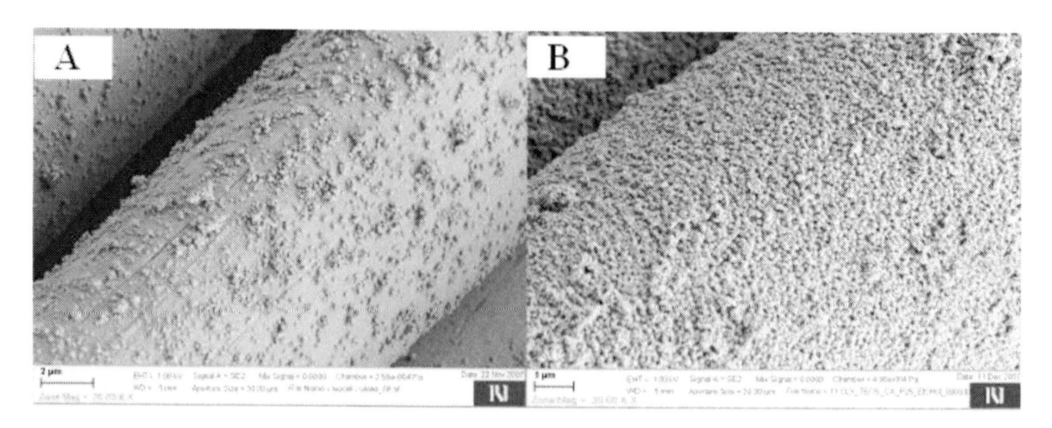

Figure 9. SEM images of the fibres surface morphology:
A Lyocell fibres embedded in SiO_2 crust; taken at magnification of 20×10^3
B TiO_2 nanoparticles bound on the layer of SiO_2; taken at magnification of 30×10^3

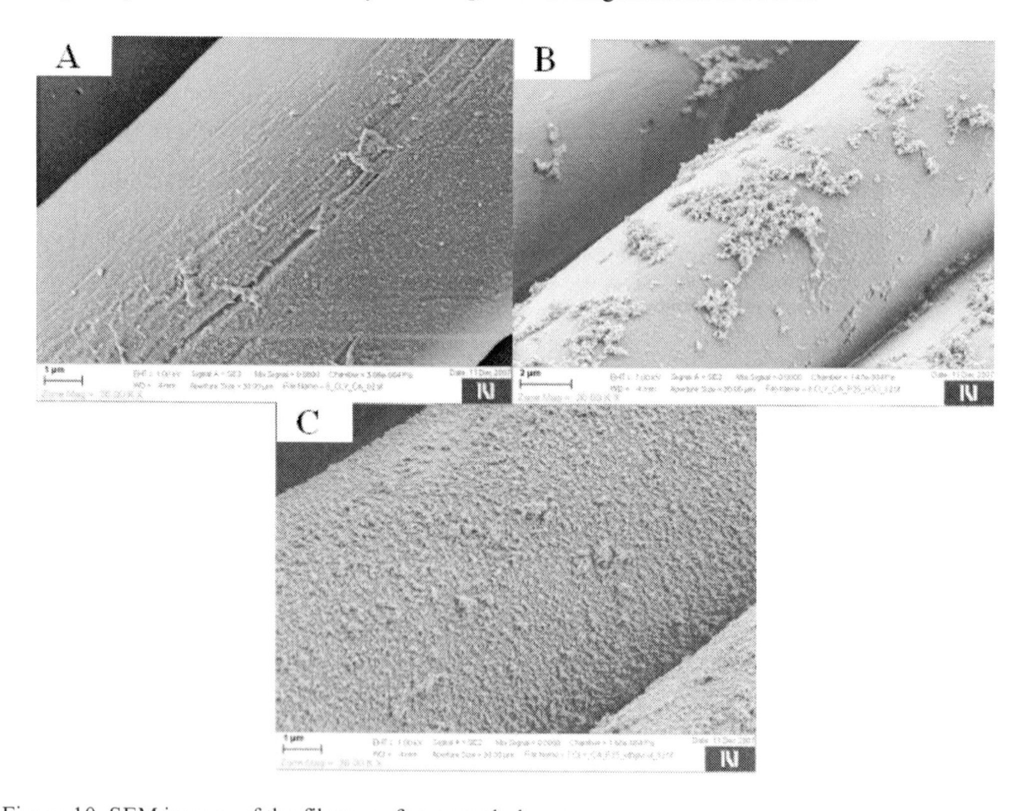

Figure 10. SEM images of the fibres surface morphology:
A coupling agent bound on the fibres surface; taken at magnification of 30×10^3
B TiO_2 nanoparticles bound by the coupling agent (treated by TiO_2 suspension in distilled water); taken at magnification of 20×10^3
C TiO_2 nanoparticles bound by the coupling agent (treated by TiO_2 suspension in ethanol); taken at magnification of 30×10^3.

The attachment of the coupling agent onto the surface of fibres was studied. In Figure 10A a layer of coupling agent, formed via sol-gel proces from precursor glymo, on the surface of fibre can be seen. During the functionalization of fibres with photocatalytically active TiO_2 nanoparticles, agglomerated clusters above 100 nm formed on the layer of coupling agent in the case of fibres treated with aqueous suspension of TiO_2 nanoparticles, prepared with distilled water (Figure 10B) due to the amphoteric character of TiO_2 nanoparticles, formed upon adsorption of water. Aqueous suspensions of TiO_2 nanoparticles are colloid suspensions where nanoparticles rapidly agglomerate, which results in reduction of the effective photocatalytical activity of surface. In the present research, the agglomeration of TiO_2 nanoparticles was controlled to increase the applicability of TiO_2 nanoparticles suspensions. To avoid agglomeration and to assure higher dispersability of TiO_2 nanoparticles, ethanol was used instead of distilled water. Agglomerates disappeared when ethanol was used. Coating with uniform distribution of the TiO_2 nanoparticles on the layer of the coupling agent has resulted after the treatment in suspension of the TiO_2 P25 nanoparticles in ethanol (Figure 10C). The fibre's surface was entirely coated with TiO_2 nanoparticles.

3.3. Self-Cleaning Efficiency

Breakdown of organic compounds in stain is not a single chemical transformation but a complex process. Various organic compounds undergo decomposition. Decomposition is a process that includes the physical breakdown and chemical transformation of complex organic molecules of material into simpler inorganic molecules [94]. In the decomposition process, different products are released: carbon dioxide (CO_2), energy, water, etc. As it slowly decomposes, it colours the soil darker. The decomposition will only take place if conditions are suitable. The rate of breakdown is greatly affected by the conditions.

Photocatalytic properties of TiO_2 coatings were investigated visually by determining stain discoloration, caused by the oxidation of the stain. TiO_2-treated Lyocell fabric was exposed to the sunlight for different periods of time. The results of the self-cleaning test, which is based on the photo-catalytic degradation of the organic dye-solution dropped on the surface of the untreated, and under different conditions nano-coated Lyocell fabrics are demonstrated in Figure 11. No decolouration of the test dyestuff was observed when the untreated sample (a) was used. However, TiO_2 nanocoated Lyocell fabrics, treated via sol-gel process from precursor TIP (b) and composite TiO_2-SiO_2 nanocoated Lyocell fabrics (c) displayed self-cleaning effect.

3.4. Anti-Microbial Efficiency

In the present work, bacterial adhesion on TiO_2 treated samples surfaces was studied. The gastrointestinal bacterium *Escherichia coli* (E. Colli) were chosen for the adhesion study. The bacterial strain E. Colli was obtained from DSMZ - the German Resource Centre for Biological Material. Bacteria was maintained on Nutrient agar (NA: 5.0 g of meat peptone, 3.0 g of meat extract, 0.5 g of NaCl, 5.0 g of D-glucose and 18.0 g of agar on 1.0 L of water).

Three to five colonies from a Nutrient agar (incubated at 37° C for 18 hrs) were touched with a loop and transferred to sterile phosphate buffered saline (PBS). The suspension was adjusted to give a turbidity equivalent to that of a 0.5 McFarland standard (1×10^8 CFU/mL). 0.1 mL of standardized organism suspension was transferred to a tube containing 9.9 mL (1:100 dilution) of nutrient broth and final suspension contained 1×10^6 CFU/mL for adhesion study.

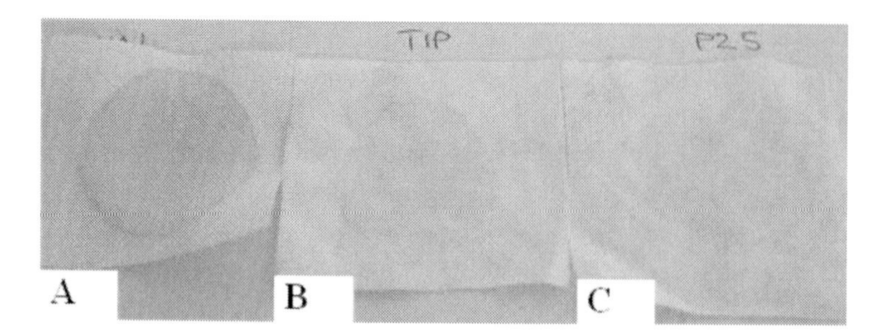

Figure 11. Unmodified (A), TIP- (B) and composite TiO_2 P25-SiO_2 (C) treated samples after 5 days of exposure to direct daylight

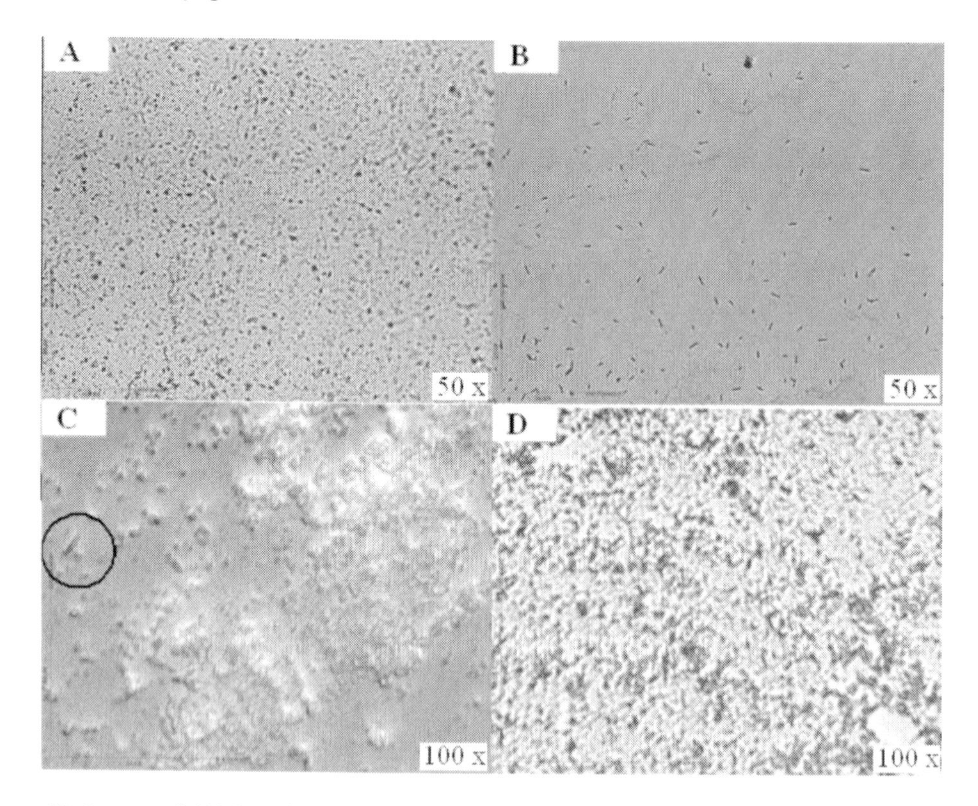

Figure 12. Images of: (A) Sample treated via alkaline catalyzed sol-gel process without added water, after treatment in nutrient broth without bacteria E. Colli; (B) Untreated sample after treatment in nutrient broth with bacteria E. Colli; (C) Sample treated via alkaline catalyzed sol-gel process with addition of water, after treatment in nutrient broth with bacteria E. Colli; (D) Sample treated via alkaline catalyzed sol-gel process without added water, after treatment in nutrient broth without bacteria E. Colli; taken at magnification of 50 and 100 x, respectively.

The samples were sterilized by using UV light at 420 nm for 2 hrs. In the diluted bacterial suspension the sterilized samples treated with TiO_2 were immersed and cultured under the same conditions as described above for 24 hrs. After incubation period the cultured samples were gently washed in PBS solution to remove non-adhered bacteria and colorized with methylene blue. Light microscope (Microscope Axiotech 25 HD at magnifications from 50 x to 1000 x with ZEISS high resolution camera AxioCam MRc; ZEISS program equipment KS 300 Rel. 3.0; "true color" analysis ZEISS) was used for observations of the adhesion of bacteria onto the samples treated with TiO_2.

The results of observations of the adhesion of bacteria onto the untreated and TiO_2 treated samples on light microscope are presented in Figure 12. Figure 12A shows TiO_2 treated sample after treatment in nutrient broth without bacteria E. Colli. TiO_2 nanoparticles have spherical shape and they entirely cover the surface of the sample. In Figure 12B E.Colli on the surface of untreated sample can be seen. E.Colli differs from TiO_2 nanoparticles in the rod-shape. In the Figure 12C one E. Colli attached to the surface treated via alkaline catalyzed sol-gel process with addition of water can be seen, which is due to insufficient covering of the surface with TiO_2. Any E. Colli can't be seen in the case of completely coated material surface treated via alkaline catalyzed sol-gel process without added water, after being exposed to the nutrient medium. The results indicate that for effective activity against the bacteria homogeneous coatings are required.

3.5. Mechanical Properties

In order to determine the influence of the photocatalytically active TiO_2 particles on the mechanical properties of Lyocell material, tenacity of single Lyocell fibres was measured.

The results of linear density (titre) and tenacity measurements of untreated, TiO_2 P25 treated and composite TiO_2 P25-SiO_2-treated single fibres are presented in Table 1. Linear density measured for raw Lyocell fibre was 1.17 dtex. These results are comparable with data in the literature [95]. The results indicate that any kind of treatment influences fibre properties. The highest linear density was observed when analyzing composite TiO_2 P25-SiO_2 coated fibre (1.39 dtex). TiO_2 P25- SiO_2 treatment increased linear density for 18.8%. Determined linear density for TiO_2 P25 treated Lyocell fibre was 1.22 dtex. Significant changes in mechanical properties occurred after self-cleaning test in the case of TiO_2 P25 coated fibres. The tenacity of TiO_2 P25 coated fibres was reduced for 21.9%. In the meantime, fibre mechanical properties have barely been changed in the case of TiO_2 P25-SiO_2-modified samples after 44 days exposure (4.5%) to daylight.

Table 1. Mechanical properties of untreated (1); TiO_2-treated fibres (2) and composite TiO_2-SiO_2 treated fibres (2) after 44 days of exposure to daylight.

Sample	Fineness		Tenacity	
	[dtex]	CV [%]	[cN/tex]	CV [%]
1	1.17	8.9	40.1	6.0
2	1.22	9.6	31.3	9.4
3	1.39	9.8	38.3	6.7

The results verified that the reactions which evolve from excitations in nano TiO_2 lead to the minor degradation of cellulose material, due to the high oxidative power of TiO_2. The results obtained by mechanical properties determination are in agreement with the findings of Winkler [64]. The composite TiO_2 - SiO_2 layer prevented the fibres against negative influence of photocatalytically active TiO_2.

Conclusion

In this research cellulose fibres were used as a starting material, since pre-treatment wasn't necessary which shortened the treatment process. Other fibres can also be coated, but have to be properly activated (if needed). In the case when substrate poses low number of surface functional groups it has to be activated by introducing some functional groups to anchor TiO_2 on the textile surface.

Different coating processes were used for nanocoatings formation. Nanocoatings obtained from pure aqueous suspensions of TiO_2 P25 nanoparticles were inhomogeneous with huge agglomerates; however by using stable suspensions of TiO_2 P25 nanoparticles, more homogeneous nanocoatings with uniform TiO_2 nanoparticles distribution were prepared. Significant differences between sol-gel from TIP precursor derived coatings were observed. Alkaline catalyzed sol-gel process yielded coatings, made of spherical TiO_2 nanoparticles, while nanocoatings with long, extended, thin structures resulted from acid catalyzed sol-gel process. The sol-gel processes with addition of water yielded coating, which didn't coat the fibre's surface entirely. High reactivity was moderated by the use of sol-gel process without added water, which resulted in more uniform coatings on the fibre's surface. An increase in available surface area, in particular, results in increased reactivity and functionality. Hence, particles agglomeration should be avoided, or somehow controlled.

The formation of mono-disperse nanoparticles on the fibre's surface will resulted in enhanced photocatalytic activity. With composite TiO_2-SiO_2 nanocoatings we succeed in prevention of fibres against high activity of TiO_2 nanoparticles.

References

[1] Wei, Q; Mather, R. *Technical Textiles International*, 2004, September, 21-23.

[2] Mahltig, B; Haufe, H; Böttcher, H. *J. of Mat. Chem.*, 2005, 15, 4385-4398.

[3] Sawhney, APS; Condon, B; Singh, KV; Pang, SS; Li, G; Hui, D. *Tex. Res. J.*, 2008, 78, 731-739.

[4] Xin, JH; Daoud, WA, Kong, YY. *Tex. Res. J.*, 2004, 74, 97-100.

[5] Onar, N; Ebeoglugil, MF; Kayatekin, I; Celik, E. *J. Of Appl. Poly. Sci.*, 2007, 106, 514-525.

[6] Abidi, N; Hequet, E; Tarimala, S; Dai, LL. *J. Of Appl. Poly. Sci.*, 2007, 104, 111-117.

[7] Gao, Y; Cranston, R. *Tex. Res. J.*, 2008, 78, 60-72.

[8] Peng, X; Xiaoyan, L; Wei, W; Shuilin, C. *J. Of Appl. Poly. Sci.*, 2006, 102, 1478-1482.

[9] Tarimala, S; Kothari, N; Abidi, N; Hequet, E; Fralick, J; Dai, LL. *J. Of Appl. Poly. Sci.*, 2006, 101, 2938-2943.

[10] Pezelj, E; Andrassy, M; Cunko, R. *Tekstil*, 2002, 51, 261-277.
[11] Fras, L; Kleinsschek, KS; Zabret, A; Stenius, P; Laine, J. 5th *World Textile Conference Autex*, 2005, 138-142.
[12] Hoffmann, MR; Martin, ST; Choi, W; Bahnemann, DW. *Chem. Rev.*, 1995, 95, 69-96.
[13] Kominami, H; Kato, J; Murakami, S; Ishii, Y; Kohno, M; Yabutani, K; Yamamoto, T; Kera, Y; Inoue, M; Inui, T; Ohtani, B; *Catalysis Today*, 2003, *84*, 181-189.
[14] Ohno, T; Sarukawa, K; Matsumura, M; *J. Phys. Chem., B*, 2001, *105*, 2417.
[15] Tanaka, K; Capule, MFV; Hisanaga, T. *Chem. Phys. Lett.*, 1991, *29*, 73.
[16] Sclafani, A; Herrmann, JM; *J. Phys. Chem.*, 1996, *100*, 13655.
[17] Rao, MV; Rajeshwar, K; Vernerker, VR; Dubow, J; *J. Phys. Chem.*, 1980, *84*, 1987.
[18] Nishimoto, S; Ohtani, B; Kajiwara, H; Kagiya, T. *J. Chem. Soc., Faraday Trans. I*, 1985, 81, 61.
[19] Guillard, C., Disdier, J., Herrmann, JM., Lechaut, C; Chopin, T; Malato, S; Blanco, J. *Catal. Today*, 1999, 54, 217.
[20] Heintz, O; Robert, D; Weber, JV. *J. Photochem. Photobiol., A Chem.*, 2000, 135, 77.
[21] Wang, R; Hashimoto, K; Fujishima, A. *Nature*, 1997, 388, 431.
[22] Nakamura, M; Sirghi, L; Aoki, T; Hatanaka, Y. *Surf.*, 2002, 507–510, 778.
[23] Chiang, K; Amal, R; Tran, T. *Adv. Environ. Res.*, 2002, 6, 471-485.
[24] Chen, D; Ray, AK. *Water Res.*, 1998, 32, 3223.
[25] Konstantinou, IK; Albanis, TA., *Appl. Catal. B Environ.*, 2004, 49, 1-14.
[26] Zielinska, B; Grzechulska, J; Grzmil, B; Morawski, AW. *Appl. Catal. B Environ.*, 2001, 35, 1–7.
[27] Stenehjem, E; Farika, M; Hovakeemian, B. *Synthesis and Photocatalytic Properties of Nanocrystalline Titania*, December, 2003, ENGR.
[28] Andrews, J. *Photocatalytically-active, self-cleaning aqueous coating compositions and methods*, United States Patent, New York, 31.12. 2002.
[29] Xie, Y; Yuan, C. *Rare Metals* (Beijing), 2004, 23(1), 20-26.
[30] Paz, Y; Luo, Z; Rabenberg, L; Heller, A. *J. Mater. Res.*, 1995, 10, 2842-2848.
[31] Pichat, P. Heterogenous Photocatalysis, In *Handbook of Heterogenous Catalysis,* Vol. IV, 2111-2122, Wiley-VCH, Weinheim/Germany (1997).
[32] Bahnemann, D. In *The Handbook of Environmental Chemistry*, 2, Part L, Springer-Verlag Berlin, 1999.
[33] Pichat, P; Disdier, J; Hoang-Van, C; Mas, D; Goutailler, G; Gaysse, C. *Catal. Today*, 2000, 63, 363-369.
[34] Ando, A. et.al: 82, (2001), 43-51 (Chem. Abstr. 136:348143).
[35] Bockelmann, D. Solare Reinigung verschmutzter Wässer mittels Photokatalyse, Cuvillier Verlag, 1994, Göttingen.
[36] Abrahams, J; Davidson, RS; Morrison, CL. *J. of Photochem.*, 1984, 26, 353-361.
[37] Shi, N; Fam, Z; Dong, Y; Shi, J; Hu, J. *Ind. Eng. Chem. Res.*, 1999, 38, 373-379.
[38] Attia, SM; Wang, J; Wu, G; Shen, J; Ma, J. *J. of Mat. Sci. and Tech.*, 2002, 18, 211-218.
[39] Caruso, AR; Antonietti, M. *Chem. of Materials*, 2001, 13, 3272-3282.
[40] *Colloid Chemistry of Silica*, Edited by H. E. Bergna. Washington: American Chemical Society, 1994.
[41] Brinker, CJ; Scherer, GW. *Sol-Gel Science: The Physics and Chemistry of Sol-Gel Processing.* San Diego: Academic Press, Inc., 1990.

[42] Lenza, RFS; Vasconcelos, WL. *J. of Non-Crystalline Solids*, 2000, 263 (1-3), 164.

[43] Schraml-Marth, M; Walther, KL; Wokan, A; Handy, BE; Baiker, A. *J. Non-Cryst. Solid*, 1992, 143, 93-111.

[44] Brunet, F; Cabane, B. *J. Non-Crystal. Solid*, 1993, 163, 211-225.

[45] Delange, RSA; Hehhink JHA; Kelzer, K; Burggraaf, AJ. *J. of Non-Crystalline Solids*, 1995, 191, 1.

[46] Avnir, D; Klein, LC; Levy, D; Shubert, V; Wojcik, AB. *Organo-Silica Sol-Gel Materials, The Chemistry of Organosilicon Compounds*, Part 2, Y. Apeloig and Z. Rapport, E's, Wiley & Sons, Chichester, 1997.

[47] Iler, RK. *The Chemistry of Silica*, Wiley: New York, 1979.

[48] Imae, T; Muto, K; And Ikeda, S. *Colloid Polym. Sci.,* 1991, 269, 43.

[49] Tehrani-Bagha, AR; Bahrami, H; Movassagh, B; Arami, M; Amirshahi, SH; Mengerc, FM. *Colloids and Surfaces A: Physicochem. Eng. Asp.*, 2007, 30, 121.

[50] Bacsa, RR; Kiwi, J. *Appl. Catal. B: Environm.*,1998, 16, 19.

[51] Bickley, RI; Gonzalez-Carreno, T; Lees, JS; Palmisano, L; Tilley, RJD., *J. Solid State Chem.*, 1991, 92, 178.

[52] Datye, AK., Riegel, G; Bolton, JR; Huang, M; Prairie, MR. *J. Solid State Chem.*, 1995, 115, 236.

[53] Vormberg, R., *Elements, Degussa Science Newsletter*, 2004, 9, 21-23.

[54] Shaw, DJ. *Introduction to Colloid and Surface Chemistry*, IInd Ed., Butterworths, London/Boston, p 167, 1970.

[55] Jiang, D. In: *Studies of Photocatalytic processes at nanoporous TiO_2 film electrodes by photo electrochemical techniques and development of a novel methodology for rapid determination of chemical oxygen demand*, Griffith University, Australia, 2004.

[56] Bajd, F. In seminar: *Interakcije med nanodelci* (Interactions between nanoparticles), 2007, University of Ljubljana, Faculty of Mathematics and Physics, Ljubljana, 18.

[57] Allen, NS; Edge, M; Ortega, A; Sandoval, G; Liauw, CM; Verran, J; Stratton, J; Mcintyre, RB. *Polym. Degrad. Stabil.*, 2004, 85, 927.

[58] Winkler, J. In: *Titanium Dioxide,* European Coatings Literature, 1st Ed., M. Vincentz, Hannover, Germany, 2003.

[59] Walther, KL; Wokaun, A. *J. of Non-Crystalline solids*, 1991, 134, 47.

[60] Lenza, RFS; Vasconcelos, WL., *Mat. Res.*, 2002, 5, 497.

[61] Zhongkuan, L; Cai, H; Ren, X; Liu, J; Hong, W; Zhang, P. *Mat. Sci. and Eng.: B*, 2007, 138, 151.

[62] Guan, K. *Surface and Coatings Tech.,* 2005, 191, 155.

[63] Yuranova, T; Mosteo, R; Bandara, J; Laub, D; Kiwi, J. *J. of Mol. Cat. A: Chem.*, 2006, 244, 160.

[64] Hu, MZC; Zielke, JT; Byers, CH; Lin, JS; Harris, MT. *J. of Mat. Sci.*, 2000, 35, 1957-1971.

[65] *Ullmann's encyclopedia of industrial chemistry*, 5th Ed., Wiley-VCH, Weinheim, A3 1993, 584-653.

[66] Stöber, W; Fink, A; Bohn, E. *J. Colloid Interf. Sci.*, 1967, 26, 62-69.

[67] Park, SK; Kim, KD; Kim, TH. *Colloid Surface A,* 2002, 197, 7-17.

[68] Karmakar, B; De, G; Ganguli, D. *J. Non-Cryst. Solids*, 2000, 272, 119-126.

[69] Kurumada, K; Nakabayashi, H; Murataki, T; Tanigaki, M. *Colloid Surface A*, 1998, 139, 163-170.

[70] Schubert, U; Husing, N. *Synthesis of Inorganic Materials,* Wiley-Vch, Germany, 2000, 200-208.
[71] Matsukas, T; Gulari, E. *J. Colloid Interface Sci.*, 1987, 124, 252-261.
[72] Hocken, J; Proft, B; Clean, B. surfaces by utilization of the photocatalytic effect, Sachtleben Chemie GmbH, Sachtleben Publications, Germany.
[73] Chrysicopoulou, P; Davazoglou, D; Trapalis, C; Kordas, G. *Thin Solid Films*, 1998, 323, 188-193.
[74] Harizanov, O; Harizanova, A. *Solar Energy Materials & Solar Cells,* 2000, 63, 185-195.
[75] Ozer, N; De Souza, S; Lampert, CM. *SPIE-Proce*eding, 1995, 2531,143-151.
[76] Bell, JM; Barczynska, J; Evans, LA; macdonald, KA; Wang, J; Green, DC; Smith GB. *SPIE-The International Society for Optical Engineering*, 1994, 2255, 324-331.
[77] Su, L; Lu, Z. *J. Phys. Chem. Solids*, 1998, 59 (8), 1175-1180.
[78] Kajihara, K; Nakanishi, K; Tanaka, K; Hirao, K; Soga, N. *J. Am. Ceramic Society*, 1998, 81, 2670-2676.
[79] Zaharescu, M; Crisan, M. *J. of Sol-gel Sci. and Tech.*, 1998, 13, 769-773.
[80] Avellaneda, CO; Pawlicka, A. *Thin Solid Films*, 1998, 335, 245-248.
[81] Phani, AR; Passacantando, M; Santucci, S. *J. of Physics and Chem. of Solids*, 2002, 63, 383-392.
[82] Wang, Z; Helmerson, U; Kall, PO. *Thin Solid Films*, 2002, 405, 50-54.
[83] Djaoued, Y; Badilescu, S; Ashirt, PV; Robichaud, J. Vibrational properties of the sol-gel prepared nanocrystalline TiO2 thin film, *Int.J.Vibr.Spec.*, www.ijvs.com .
[84] Holtzen, DA; Reid, AH. Titanium Dioxide Pigments, In *Coloring of Plastics, Fundamentals*, John Wiley & Sons, 2004.
[85] Kemmitt, T; Al-Sali, NI; Wateerland, M; Kennedy, VJ; Markwitz, A. *Current App. Phys.*, 2004, 4, 189-192.
[86] Bozzi, A; Yuranova, T; Kiwi, J. *J. of Photochem. and Photobio. A-Chem.* 2005, 172 (1), 27-34.
[87] Bozzia, A; Yuranova, T; Guasaquilloa, I; Laubb, D; Kiwi, J. *J. of Photochem. and Photobio. A: Chem.* 2005, 174, 156-164.
[88] Daoud, WA; Xin, JH. *J. of the American Ceramic Soc.*, 2004, 87, 5, 953.
[89] Meilert, KT; Laub, D; Kiwi, J. *J. Of Molec. Cat. A-Chem.*, 2005, 237 (1-2), 101-108.
[90] Veronovski, N; Andreozzi, P; La Mesa, C; Sfiligoj-Smole, M; Ribitsch, V. *Colloid Polym. Sci.*, 2010, 288, 387–394.
[91] Veronovski, N; Andreozzi, P; La Mesa, C; Sfiligoj-Smole, M. *Surface and Coatings Tech.*, 204, 1445-1451.
[92] Schubert, U; Husing, N; Lorenz, A. *Chem. Mate*r., 1995, 7, 2010.
[93] Sakka, S. *J. Sol-gel Sci. Tech.*, 1994, 3/2, 69.
[94] Juma, NG. In: *The pedosphere and its dynamics: a systems approach to soil science,* Vol 1, Edmonton, Canada, Quality Color Press Inc, 315, 2008.
[95] Čunko, R; Andrassy, M. Vlakna (Fibers), *Založba Zrinski Zagreb*, 2005.

In: The Sol-Gel Process
Editor: Rachel E. Morris
ISBN 978-1-61761-321-0

Chapter 16

SYNTHESIS AND CHARACTERIZATION OF SILICATE POLYMERS

Morten E. Simonsen, Camilla Sønderby and Erik G. Søgaard
Department of Chemistry, Biotechnology, and Environmental Engineering,
Aalborg University, Esbjerg, Denmark

ABSTRACT

The development of inorganic polymers is a new promising technology that may be used for many applications including inorganic coatings with protective or self-cleaning properties. In this chapter the development of an inorganic polymer based on amorphous silica and potassium hydroxide is described. The inorganic polymers are synthesized by a sol–gel process in which the silicate precursors are formed in-situ by dissolution of amorphous silica particles (microsilica, Elkem 983 U) by an alkaline solution. In this work potassium hydroxide with a molar concentration between 0.75 M and 4 M was used. In standard experiments equal amounts (w/w %) of silica and potassium hydroxide solution was used.

Using experimental data obtained from ESI-MS, FT-IR, XPS, NMR, SEM, and XRD a model for the gelation is suggested based on hydrolysis and condensation reactions occurring during synthesis. In addition the optimal composition of the binder system was determined from compressive strength tests and solubility experiments.

In the developed physico-chemical model the silica species polymerize through a condensation reaction resulting in formation of oligomers. The size and number of oligomers in the system increase until they extent throughout the solution and results in the formation of a gel. It was found that the dissolution proceeds for a longer time period when higher hydroxide concentrations are used resulting in higher dissolution and higher concentration of monomers. The gelation of the inorganic polymers synthesized from different concentrations of KOH was studied by viscosity measurements. It was found that two mechanisms contributed to the time required for gelation. At high hydroxide concentration the increased concentration of monomers in the system increase the rate of polymerization and therefore reduce the gelation time. At low hydroxide concentrations the surface charge of the silica particles is limited and the particles coagulate due to destabilization by the presence of potassium ions. A turn over point for these two mechanisms seems to be found at hydroxide concentrations between 0.75 and 1.5 M in the investigated systems.

INTRODUCTION

The development of inorganic polymers is a new promising technology that may be used for many applications including inorganic coatings with protective or self-cleaning properties. The inorganic polymers described here are synthesized from amorphous silica and an alkaline solution. In previous work amorphous silica (fumed silica, microsilica) has been used as one of the reactive components in the preparation of Geopolymers usually in combination with an alumina source in the form of fly ash or metakaoline (Rowles and O'Conner, 2003; Zivica, 2004; Fletcher et al., 2005; Davidovits, 2005; Brew and MacKenzie, 2007). The addition of amorphous silica in Geopolymer synthesis resulting in an increase in the Si/Al ratio has been found to increase the strength of the binder (Rowles and O'Conner, 2003; Fletcher et al., 2005). Moreover, microsilica is routinely used as an additive in high performance cement (Hjort et al., 1988).

In this chapter the development of an inorganic polymer based on amorphous silica and potassium hydroxide is described. The inorganic polymers are synthesized by a sol–gel process in which the silicate precursors are formed in-situ by dissolution of amorphous silica particles (microsilica, Elkem 983 U) by an alkaline solution. In this work potassium hydroxide with a molar concentration between 0.75 M and 4 M was used. In standard experiments equal amounts (w/w %) of silica and potassium hydroxide solution was used.

Using experimental data obtained from FT-IR, XPS, NMR, SEM, XRD a model for the gelation is suggested based on hydrolysis and condensation reactions occurring during synthesis. In addition the optimal composition of the binder system was determined from compressive strength tests and solubility experiments.

DISSOLUTION AND GEL FORMATION

Dissolution of Silica Particles

The synthesis of the inorganic polymers is considered to be a result of dissolution of the surface of the amorphous silica particles (Microsilica) by hydroxide resulting in formation of soluble silica species in the solution. The solubility of amorphous silica is highly dependent on the pH of the solution and a dramatic increase in the solubility is observed above pH 10 (Iler, 1979). Increasing temperatures will also increase the solubility of the amorphous silica. Dependent on pH different monomers will be present in the solution after dissolution (H_4SiO_4, $H_3SiO_4^-$, and $H_2SiO_4^{2-}$) (Iler, 1979). It is the polymerization of these monomers that eventually lead to the formation of inorganic polymers. The dissolution of the amorphous silica under alkaline conditions can be expressed by the following reactions.

$$SiO_2 + 2H_2O \rightarrow H_4SiO_4 \tag{1}$$

$$SiO_2 + H_2O + OH^- \rightarrow H_3SiO_4^- \tag{2}$$

$$H_3SiO_4^- + OH^- \longrightarrow H_2SiO_4^{2-} + H_2O \tag{3}$$

$$SiO_2 + 2OH^- \rightarrow H_2SiO_4^{2-} \tag{4}$$

These reactions suggest that H_2O and OH^- are consumed during the dissolution process resulting in a decrease in pH, which also was observed experimentally in Figure 1. In Figure 1 the pH evolution of the inorganic polymer synthesized from 0.75 – 4 M KOH are shown. From Figure 1 it is seen that the pH of the reaction solution changes from around 14 (dependent on the KOH concentration used in the synthesis of the inorganic polymer) to around 11.5. In Figure 2 the hydroxide concentration calculated from the pH measurements is shown as a function of the reaction time. The observed change in kinetics of the reaction is suggested to be due to interference from dissolution products (monomers, dimers and oligomers) and the changes in the particle size of the microsilica due to dissolution. The accessibility of the microsilica surface is expected to be a limiting factor for the dissolution due to the inhibition effects arising from the silicate species surrounding the particles. Moreover, it is seen that the dissolution proceeds for a longer time period when higher hydroxide concentrations are used resulting in higher dissolution. The increase in hydroxide concentration observed during the first 15 min can be ascribed to difficulties in pH measurements at high viscosity.

Investigations have also shown that the processing temperature affects the rate of dissolution (Simonsen et al., 2009b). An increase in the processing temperature from 25 to 50 °C resulted in a decrease in the reaction time from 10 hours to about 30 min. Although, the processing temperature has a great influence on the rate of polymerization the properties of the final inorganic polymer was not found to change significantly.

Another parameter which is of importance in the synthesis of inorganic polymers is the ratio of the mass content (m_{H2O}/m_{SiO2}). An increase in the water content results in a decrease in the hydroxide concentration due to volume changes. Thus, the dissolution of silica decreases thereby affecting the gel formation. Investigations have shown that inorganic polymers synthesized using m_{H2O}/m_{SiO2} ratio larger than 2 do not lead to gel formation (Simonsen et al., 2009b).

Mass Spectrometry

The dissolution of the amorphous silica particles is the first step in the synthesis of inorganic polymers. The dissolution or hydrolysis process is followed by condensation and polymerization reactions resulting in the formation of oligomers. This process releases the water that was initially consumed during dissolution. Thus, water only plays the role as reaction medium. The size and number of oligomers in the system will increase until they extent throughout the solution resulting in formation of a gel.

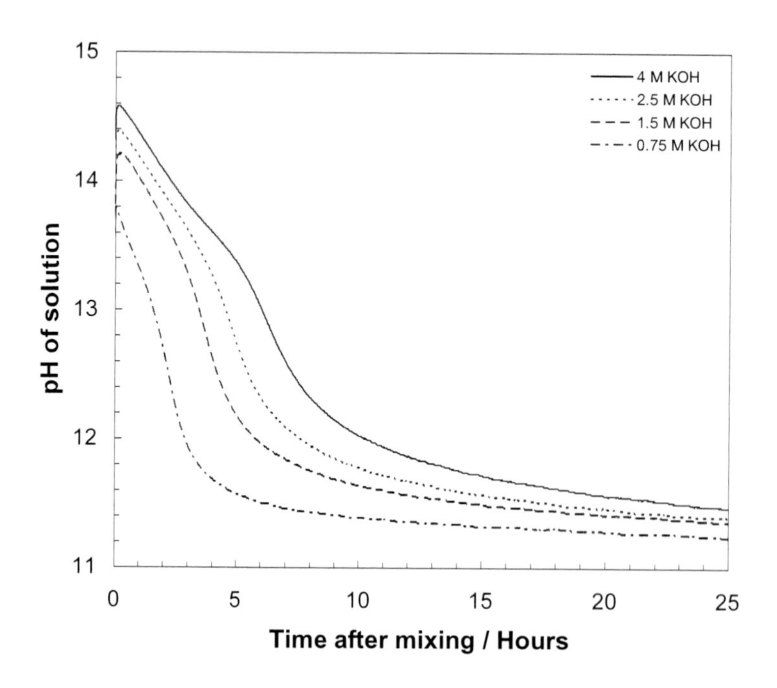

Figure 1. pH evolution during the dissolution process of inorganic polymers synthesized using 0.75 - 4 M KOH. Reprinted with kind permissions from Springer Science (Original printed in Simonsen et al., J Sol-Gel Sci Techno (2009) 50, 372-382). Copyright 2009 Springer.

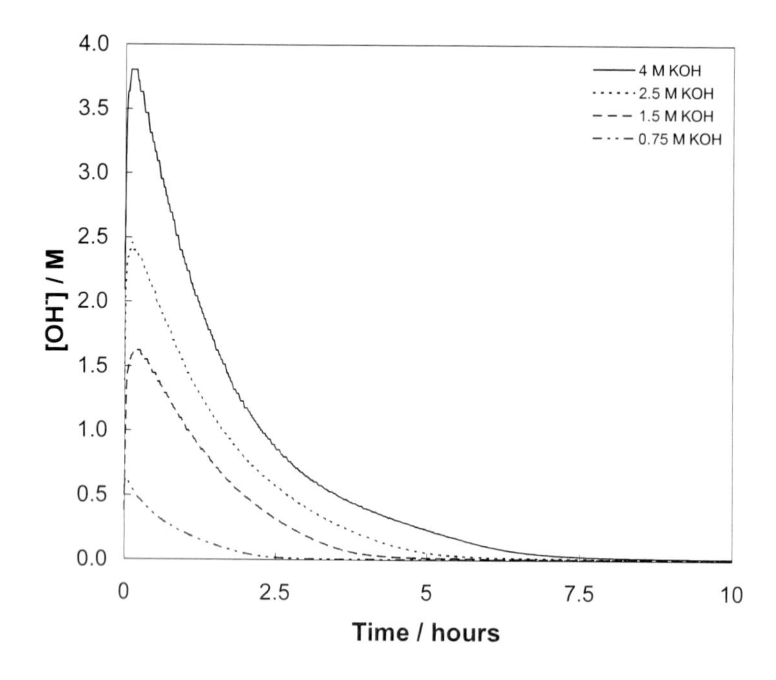

Figure 2. Hydroxide concentration as a function of the reaction time for different compositions of the inorganic polymers. The increase in hydroxide concentration observed during the first hour is due to difficulties in pH measurements at high viscosity. Reprinted with kind permissions from Springer Science (Original printed in Simonsen et al., J Sol-Gel Sci Techno (2009) 50, 372-382). Copyright 2009 Springer.

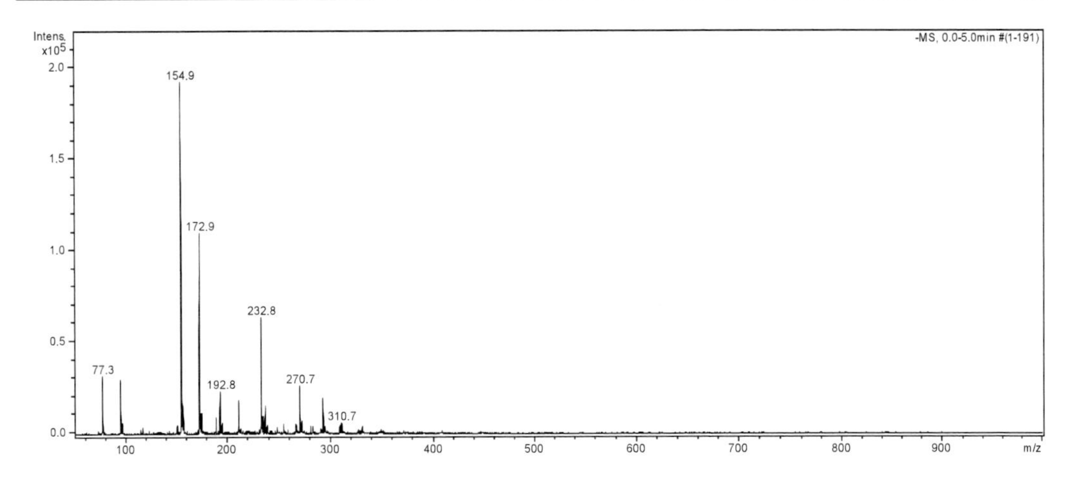

Figure 3. ESI-MS spectra of the inorganic polymer reaction solution obtained 8 hours after mixing. Reprinted with kind permissions from Springer Science (Original printed in Simonsen et al., J Sol-Gel Sci Techno (2009) 50, 372-382). Copyright 2009 Springer.

Electro spray ionization mass spectrometry (ESI-MS) has been used to identify the silicate species in the reaction solution during polymerization (Simonsen et al., 2009b, Simonsen and Søgaard, 2009). Investigations of inorganic polymers synthesized from microsilica and 4 M KOH have shown that all the observed molecule ions were found as single negatively charged species, which were obtained by the removal of a proton (H^+) from the molecule, thus leaving an anion that is subsequently detected.

From the ESI-MS spectrum of an inorganic polymer reaction solution (Figure 3) it is seen that the most dominant peak is located at 155 m/z, which can be assigned to a dehydroxylated dimer ($Si_2O_3(OH)_3^-$). The second most intensive peak is assigned to a fully hydroxylated dimer ($Si_2O_2(OH)_5^-$). Two monomeric species has been identified in the mass spectrum at 77 and 95 m/z corresponding to $SiO_2(OH)^-$ and $SiO(OH)_3^-$, respectively. The $SiO_2(OH)_2^{2-}$ ion (47 m/z) can not be observed in the mass spectra because the m/z value is below the detection range (50 - 2200 m/z). In addition a series of trimers has been identified at 233, 271 and 309 m/z. The most prevailing trimer is found at 233 m/z corresponding to a linear trimer with one oxo group or a cyclic trimer which usually is found in ^{29}Si NMR investigations of silicate systems. It is not possible to distinguish between these to molecules using ESI-MS because this technique does not give information on the interconnectivity of the atoms in the molecule (Eggers et al., 2005). However, from these results it is observed that no high intensity peaks could be assigned to fully hydroxylated linear trimers ($Si_3O_3(OR)_7^-$), suggesting that the trimers identified in this system under these conditions are present in the form of cyclic trimers. It should however be noted that Bussaian et al. (2000) have reported that the tendency of silica species not to appear as the fully hydroxylated ions increase with increasing silicate chain length. A series of tetramers can be identified at 293, 311, 331, 349, and 387 m/z, however the intensity of these peaks is below 10 % of the most intensive peak. In addition pentamers and hexamers have also been observed but again the intensity of these molecule ions was below 5 %. The identified silicate species and the suggested possible structures are listed in Table 1.

Table 1. Identified silicate species present in the reaction solution of the inorganic polymers synthesized from microsilica and 4 M KOH. The assigned molecule ions were confirmed by the isotopic pattern of the elemental composition of the ion. – H_2O denotes dehydroxylation resulting in the formation of an oxo group. Reprinted with permissions from Elsevier (Original printed in Simonsen et al., Inter J Mass Spec (2009) 1-2, 78-85). Copyright 2009 Elsevier.

Nr.	m/z	Compound	Intensity (%)	Possible structures
1	77	$SiO_2(OH)^-$	16.4	• - 1 H_2O
2	95	$SiO(OH)_3^-$	15.5	•
3	155	$Si_2O_3(OH)_3^-$	100.0	— - 1 H_2O
4	173	$Si_2O_2(OH)_5^-$	57.4	—
5	193	$Si_2O_3(OK)(OH)_2^-$	12.2	— - 1 H_2O
6	211	$Si_2O_2(OK)(OH)_4^-$	< 10	—
7	233	$Si_3O_4(OH)_5^-$	33.2	 △ ⌒ - 1 H_2O
8	249	$Si_2O_2(OK)_2(OH)_3^-$	< 5	—
9	271	$Si_3O_4(OK)(OH)_4^-$	13.8	 △ ⌒ - 1 H_2O
10	275	$Si_4O_7(OH)_3^-$	< 5	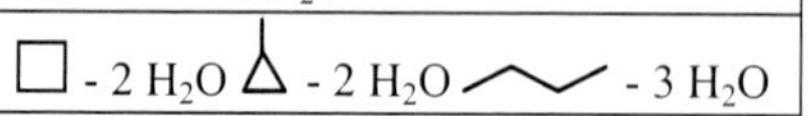 □ - 2 H_2O △ - 2 H_2O 〜 - 3 H_2O
11	293	$Si_4O_6(OH)_5^-$	10.4	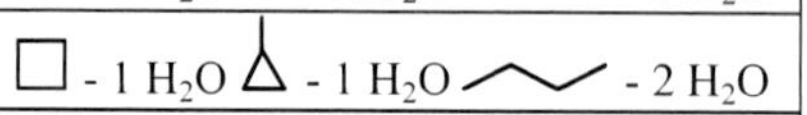 □ - 1 H_2O △ - 1 H_2O 〜 - 2 H_2O
12	309	$Si_3O_4(OK)_2(OH)_3^-$	< 5	 △ ⌒ - 1 H_2O
13	311	$Si_4O_5(OH)_7^-$	< 5	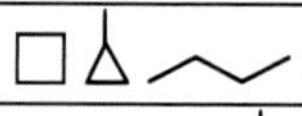 □ △ 〜 - 1 H_2O
14	331	$Si_4O_6(OK)(OH)_4^-$	< 5	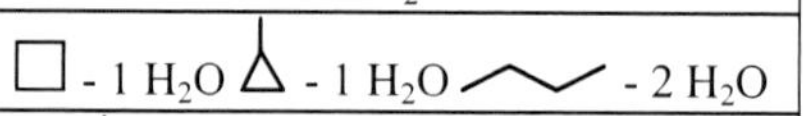 □ - 1 H_2O △ - 1 H_2O 〜 - 2 H_2O
15	349	$Si_4O_5(OK)(OH)_6^-$	< 5	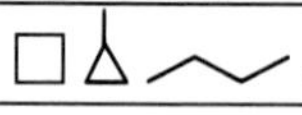 □ △ 〜 - 1 H_2O
16	371	$Si_5O_7(OH)_7^-$	< 5	
17	387	$Si_4O_5(OK)_2(OH)_5^-$	< 5	 □ △ 〜 - 1 H_2O
18	447	$Si_5O_7(OK)_2(OH)_5^-$	< 5	

The evolution in the intensity of the most dominant silicate species (155 m/z) in the reaction solution as a function of reaction time is shown in Figure 4. From Figure 4 it is seen that the intensity of the most dominant molecule ions increases up to 8 hours after mixing. The increase in the intensity is due to dissolution of microsilica leading to a greater amount of silicate species in the reaction solution. The pH change observed during the dissolution process suggests that the dissolution process proceeds for approx. 10 hours correlating with the observed increase in silicate species. The rapid drop in the intensity observed after 12 hours (Figure 4) is suggested to be due to polymerization of the silica species.

After 12 hours of reaction a broad band of low intensity signals were observed in the mass spectra in the spectral region from 550 up to 1200 m/z. These signals were assigned to silicate species containing 6 – 12 Si atoms. If the polymerization occur by addition of monomers it would be expected to see a change in the mass spectra leading to peaks of relatively high intensity at higher m/z values. However, no such peaks were observed in the mass spectra. The change in the intensity of the most dominant silicate species in the mass spectra suggests that these species may serve as building blocks during the polymerization. It is believed that the polymerization lead to silicate species with m/z values greater than 2200 making detection no longer possible using the ESI-MS method. Another possibility is that the silicate species polymerize on the surface of the undissolved microsilica particles encapsulating the microsilica particles. During gelation the polymer encapsulated sphere of the microsilica particles may increase in size resulting in the formation of a network between the microsilica particles leading to gelation.

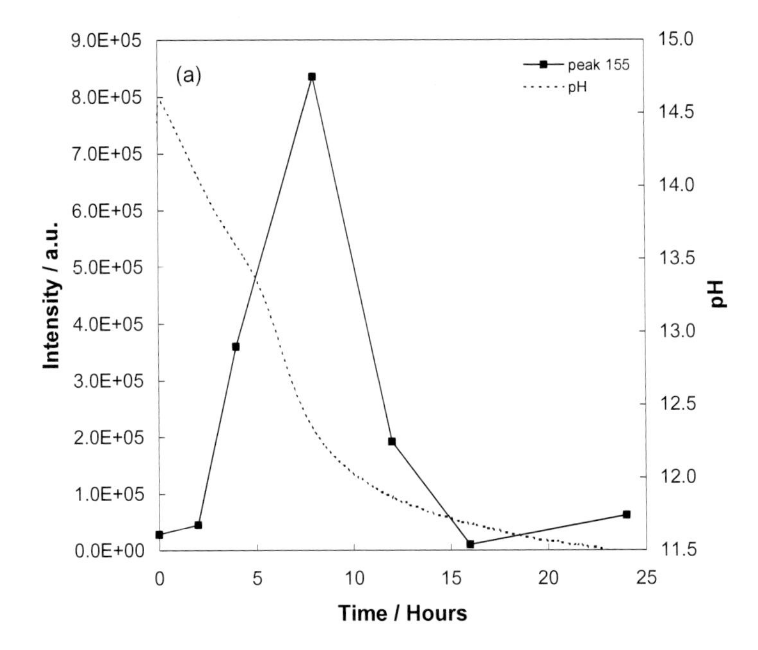

Figure 4. Evolution of the intensity of the most dominant molecule ion in the reaction solution of the inorganic polymer synthesized from microsilica and 4 M KOH. Reprinted with permissions from Elsevier (Original printed in Simonsen et al., Inter J Mass Spec (2009) 1-2, 78-85). Copyright 2009 Elsevier.

Viscosity Investigations

The viscosity of the reaction solution changes considerably during the dissolution and polymerization process in the synthesis of the inorganic polymers (Simonsen et al., 2009b). The viscosity increased rapidly during the initial mixing of silica and the hydroxide solution resulting in the development of a thick paste. After mixing the viscosity of the paste decreases as shown in Figure 5. The decrease in viscosity is thought to be due to two processes occurring during the initial dissolution step (Simonsen et al., 2009b).

- Adsorption of OH^- ions on the surface of the silica particles resulting in negatively charged particles. As the solution consists of only negatively charged particles surrounded by positively charged cations the particles will be repelled by one another and hence the viscosity decrease.
- Dissolution of the silica particles will cause the particles to decrease in size and number resulting in an additional decrease in viscosity.

The decrease in the viscosity is followed by period in which virtually no change in the viscosity was observed. The duration of the constant viscosity is dependent on the concentration of hydroxide used in the synthesis. Comparison of the viscosity development with the measured evolution in hydroxide concentration during synthesis suggests that the dissolution process continues in the stable viscosity period. The stable viscosity period is followed by a significant increase in viscosity due to polymerization of the silicate species leading to gelation.

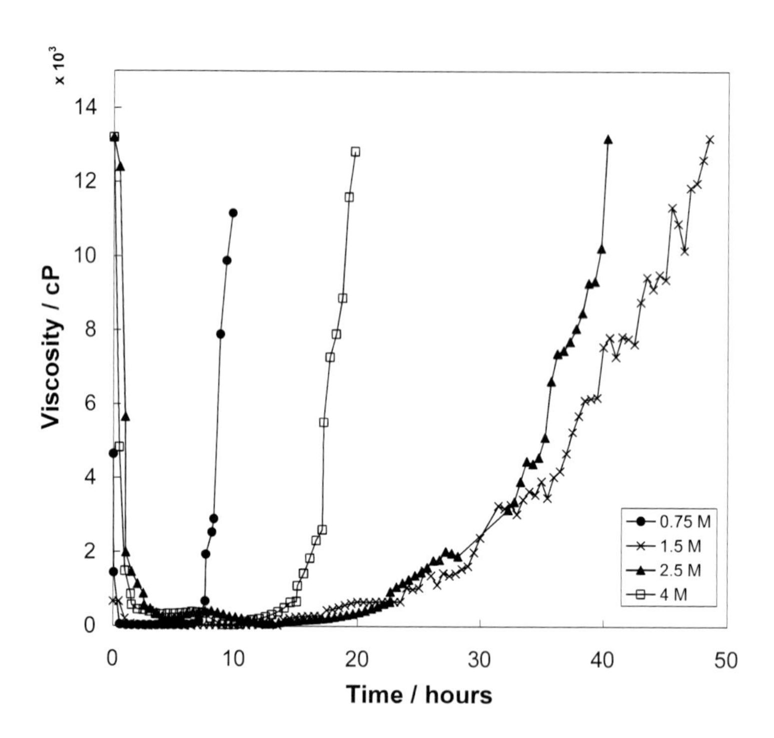

Figure 5. Viscosity development of different inorganic polymers during synthesis. Reprinted with kind permissions from Springer Science (Original printed in Simonsen et al., J Sol-Gel Sci Techno (2009) 50, 372-382). Copyright 2009 Springer.

Table 2. Particle properties and chemical composition of different Elkem microsilica.

Chemical composition	Elkem 940 U	Elkem 983 U	Elkem 995 U
Loss by combustion	< 3 %	0.53 %	Not tested
Silicium oxide (SiO_2)	> 90 %	98.6 %	99.5 %
Carbon	Not tested	0.24 %	Not tested
BET	17.4 m^2/g	13.5 m^2/g	42.6 m^2/g
Particle radius (DLS)	159.2 ± 0.6 nm	198.5 ± 1.5 nm	127.7 ± 0.5 nm

The stable viscosity period with low viscosity is elongated if the hydroxide concentration is decreased from 4 M to 1.5 M (Figure 5). This tendency was however not observed for the lowest hydroxide concentration, which has the shortest stable viscosity period. These results can be explained by two mechanisms occurring simultaneously (Simonsen et al, 2009b). At high hydroxide concentration the surface of the silica particles will stay negatively charged during the dissolution process. As the concentration of monomers in solution increase due to dissolution, polymerization and later gel formation will occur. Higher concentrations of monomers result in faster gel-formation. Hence, high hydroxide concentration results in faster gel-formation. That the 0.75 M hydroxide solution does not follow the general trend of faster gel formation with increasing concentration of hydroxide may be due to the fact that the amount of monomer produced is limited and that the hydroxide covered silica particles coagulate due to a lower surface charge destabilized by the presence of potassium ions.

In the synthesis of inorganic polymers many parameters influence the final properties. The time required for the inorganic polymer solution to form a continuous gel depends not only on the processing conditions but also the raw material (micro silica). The surface area of the particles which are dissolved has a great influence on the rate of dissolution and therefore on the gelation time of the inorganic polymer. The BET surface area provides an indication of how much surface area that participates in the heterogeneous reaction. The particle properties of different microsilica produced by Elkem are shown in Table 2. Comparison between the gelation times for inorganic polymers synthesized using Elkem 995 (42.6 m^2/g), Elkem 940 (17.4 m^2/g), and Elkem 983 (13.5 m^2/g) microsilica under similar conditions shows that the reaction of Elkem 995 microsilica (gelation time approx. 1.5 hours) was significantly faster followed by Elkem 940 (gelation time approx. 15 hours) and Elkem 983 (gelation time 20 hours). The average particle size is higher for Elkem microsilica 983 resulting in the lowest BET surface area of the three Elkem microsilica investigated and thus the lowest dissolution rate. The content of impurities which are present in the raw materials should also be seriously taken into account when modeling the gelation process.

Chemical Structure of Inorganic Polymers

FT-IR Investigation

Infrared spectroscopy is a useful analytical technique used for both qualitative and quantitative analysis of organic and inorganic materials. FT-IR spectroscopy allows differentiation of various types of bonds in a material on a molecular level. The inorganic polymers are composed from Si-O tetrahedrons which are connected via corner sharing bridging oxygen's. The connectivity of the tetrahedrons is specified by the number of bridging oxygen's. Tetrahedrons with *n* bridging oxygen's are denoted Q^n (n = 0, 1, 2, 3, or 4).

Amorphous SiO_2 is normally assumed to consist of Q^4 units forming a continuous random network. The Si-O stretching IR band of Q^4 units is located at 1100 cm^{-1}. FT-IR spectra of the inorganic polymers synthesized using different KOH concentrations are shown in Figure 6. The band at 1050 cm^{-1} is assigned to Q^3 units with one non-bridging oxygen (Si-O-NBO) per SiO_4 tetrahedron (Zholobenko et al., 1997). An increase in the hydroxide concentration shifts the position of the maximum absorbance of Si-O bands towards lower wave numbers, indicating the transformation of Q^4 units to Q^3 units. In Figure 6, the emerging of a new band centered around 900 cm^{-1} is observed, which is assigned to Si-O stretching with two non-bridging oxygen per SiO_4 tetrahedron (Q^2) (Serra et al., 2002). The observed increase in the signal at 2380 cm^{-1} can be assigned to CO_2 which have reacted with increasing amounts of KOH leading to the formation of HCO_3^-.

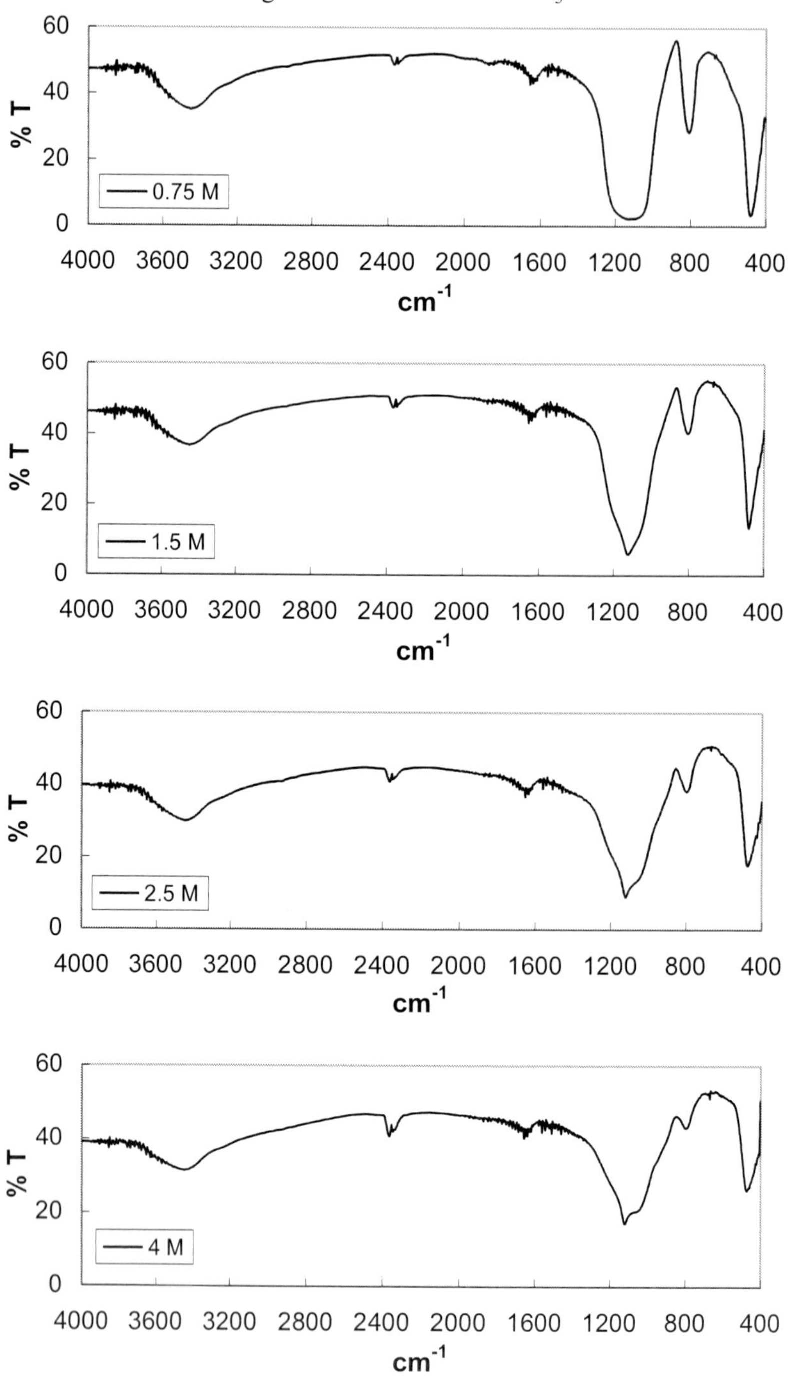

Figure 6. FT-IR spectra of the inorganic polymers synthesized using different KOH concentrations. Reprinted with kind permissions from Springer Science (Original printed in Simonsen et al., J Mater Sci (2009) 44, 8, 2079-2088). Copyright 2009 Springer.

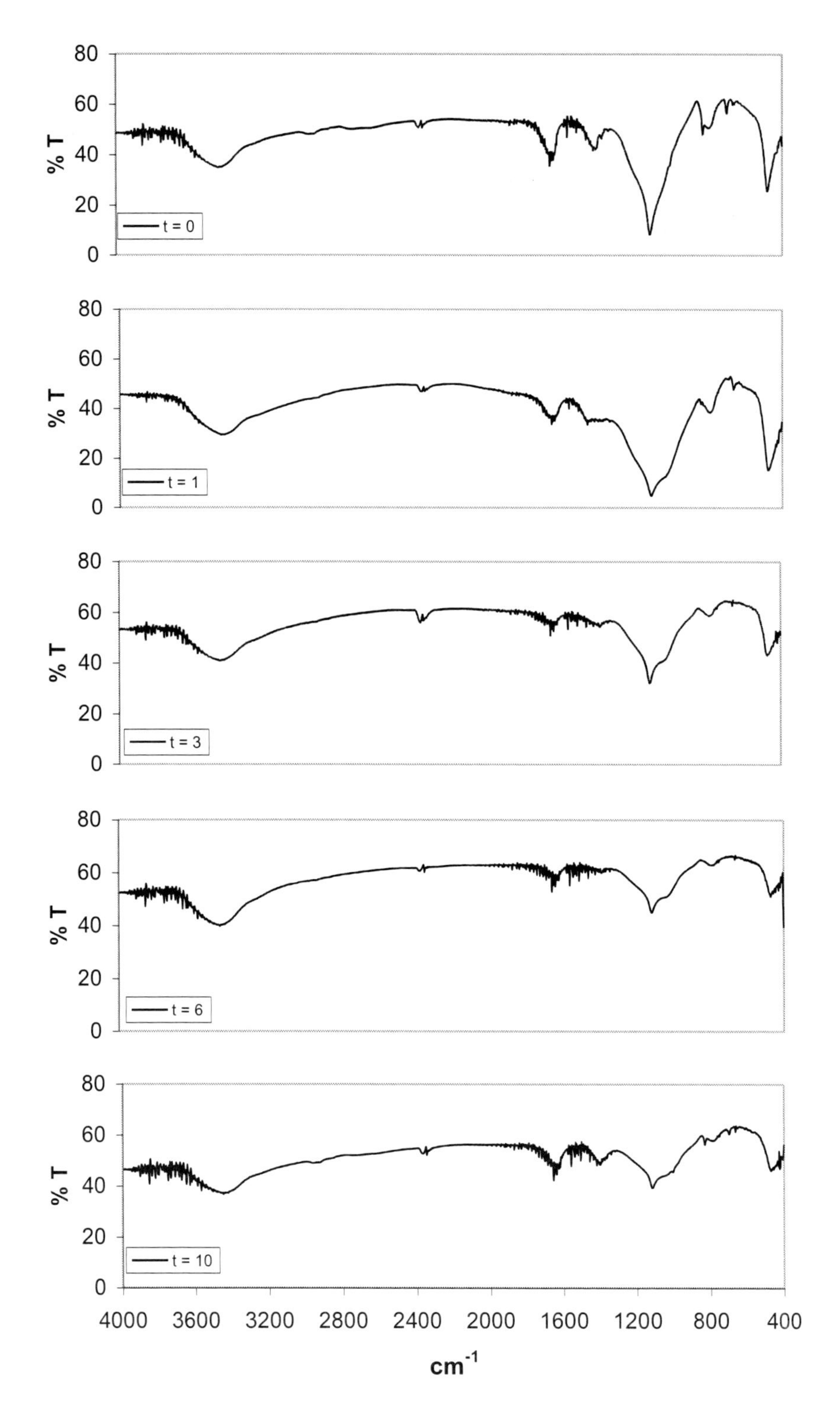

Figure 7. FT-IR spectra's of an inorganic polymer synthesized using 4 M KOH recorded at 0.5, 1, 3, 6, and 10 hours after mixing. Reprinted with kind permissions from Springer Science (Original printed in Simonsen et al., J Mater Sci (2009) 44, 8, 2079-2088). Copyright 2009 Springer.

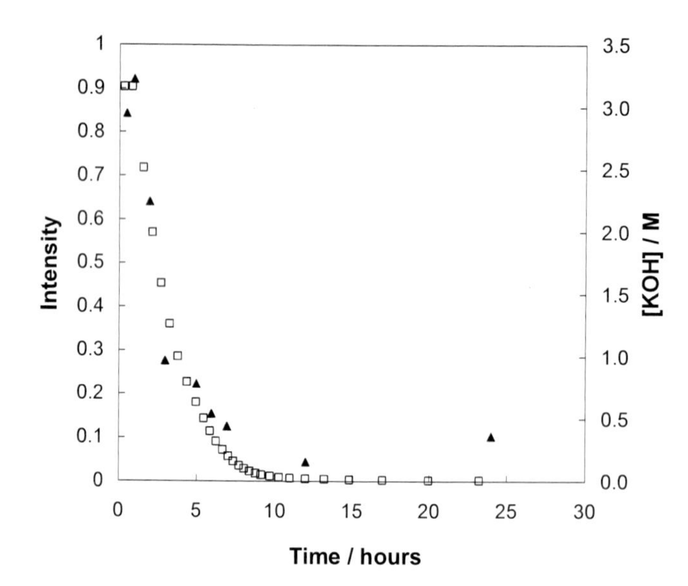

Figure 8. Intensity of the peak located at 1120 cm^{-1} as a function of the reaction time compared to the evolution in hydroxide concentration. Reprinted with kind permissions from Springer Science (Original printed in Simonsen et al., J Mater Sci (2009) 44, 8, 2079-2088). Copyright 2009 Springer.

The observed change in the FT-IR spectra from mainly consisting of Q^4 units to Q^3 and Q^2 units with increasing concentration of KOH has been used to investigate the changes in composition of the inorganic polymer during synthesis (Simonsen et al., 2009a). In Figure 7 the obtained FT-IR spectra's of an inorganic polymer synthesized using 4 M KOH recorded at 0.5, 1, 3, 6, and 10 hours after mixing. The intensity of the Q^4 peak located at 1120 cm^{-1} decrease significantly during the reaction, which is consistent with the dissolution model. In Figure 8 the intensity of the Q^4 peak located at 1120 cm^{-1} is plotted as a function of reaction time together with the evolution in hydroxide concentration. The two curves nearly overlap indicating that the change in intensity of the peak assigned to Q^4 units can be used to describe the dissolution process occurring during synthesis of the inorganic polymer.

The small increase in the intensity of the Q^4 peak after 24 hours may be due to formation of Q^4 units during the polymerization process leading to gelation. FT-IR investigation of inorganic polymer 3 weeks after preparation has shown that the peak increase further to approx. 0.4. The FT-IR spectrum of the inorganic polymers was found not to change further after the 3 weeks.

NMR Investigation

^{29}Si NMR has been used to study inorganic polymers synthesized from microsilica and KOH. However because of the low natural abundance for ^{29}Si (4.7 %) and the relatively long T1 relaxation times (120 s) the analysis is quite time consuming and long analysis time is required to obtain well resolved spectra. In Figure 9 the ^{29}Si NMR spectra of the inorganic polymers synthesized from microsilica and 0.75 – 4 M KOH are shown. In the ^{29}Si MAS NMR spectra the high intensity resonance located at -95 – -125 ppm can be assigned to Q^4

units originating from the microsilica framework (Hjorth et al., 1988). From the spectra it is seen that the Q^4 units is transformed into Q^3 units (-99 ppm) and Q^2 units (-89 ppm) as the concentration of KOH is increased from 0.75 to 4 M. The observed changes in the NMR spectra as a result of increasing KOH concentration correlated with the suggested dissolution process. Moreover in the NMR spectra of the inorganic polymer synthesized from 4 M KOH a resonance at -84 ppm may exist corresponding to Q^1 units (End groups of chains or dimers of SiO_4 tetrahedra). These findings correlate with the results obtained from the FT-IR analysis.

In Figure 10 the ^{29}Si MAS NMR spectra' of inorganic polymers synthesized using 4 M KOH obtained at different times after mixing are shown. As the NMR spectra are obtained under identically conditions the intensities can be compared directly (small deviations can however occur due to the packing of the rotor and different densities of the binder). Qualitative analysis of the inorganic polymer shows that Q^1, Q^2 and Q^3 units are formed during reaction with microsilica. The most significant changes are observed during the first 10 hours of reaction.

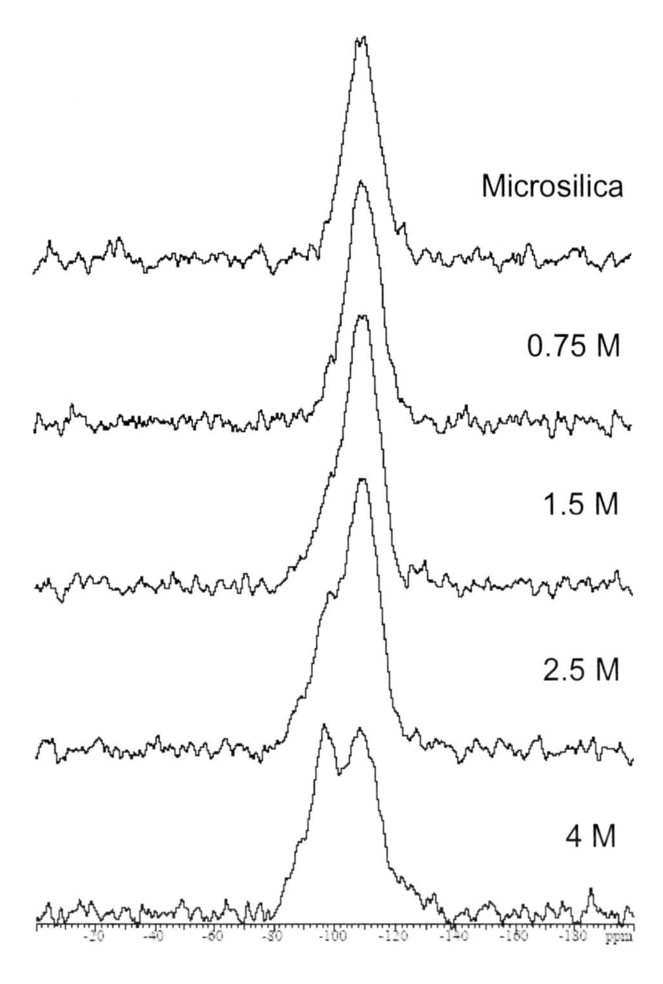

Figure 9. ^{29}Si MAS NMR spectra of the inorganic polymers synthesized using different KOH concentrations. The ^{29}Si MAS NMR spectra were recorded using a one pulse experiment on a Varian-INOVA 200 spectrometer (200 MHz, 4.7 T).

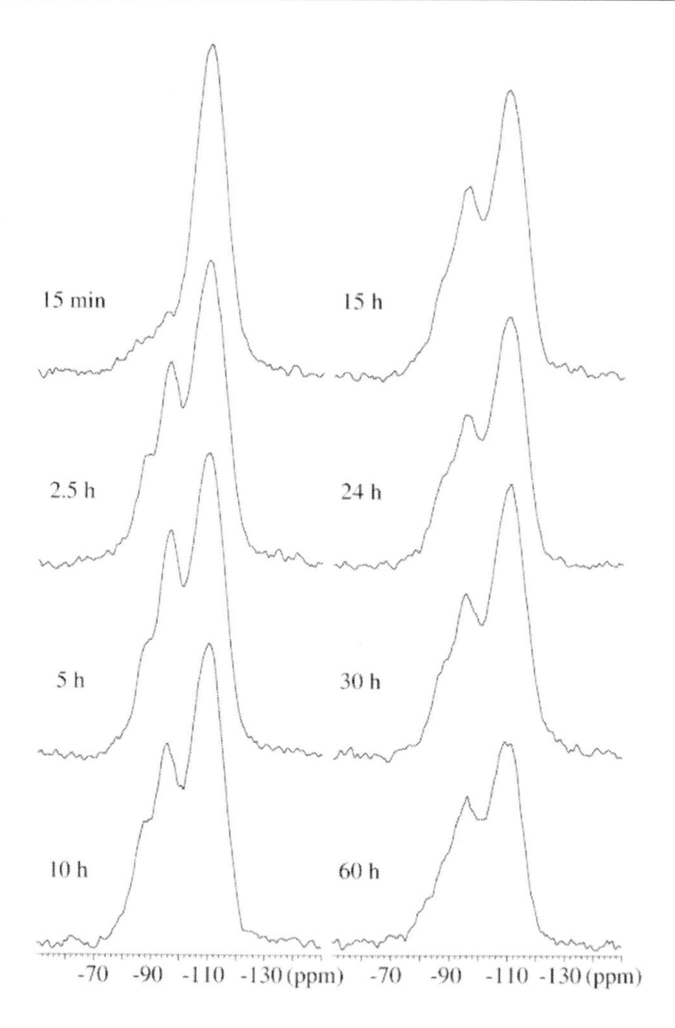

Figure 10. ^{29}Si MAS NMR spectra' of an inorganic polymer synthesized using 4 M KOH obtained at different times after mixing.

XPS Investigation

X-ray photoelectron spectroscopy (XPS) is a highly sensitive technique well suited to examine the composition and chemical states of materials. When a material is bombarded with X-rays, photoelectrons may be emitted from the topmost surfaces (typically 1–10 nm). All element yields photoelectrons of a specific binding energy, enabling elemental analysis. Furthermore, small changes in the chemical bonding environments result in small shifts in the photoelectron energy, thus allowing chemical information to be obtained. Among the chemical properties which may be evaluated from the binding energies are oxidation state, nearest neighbor atoms and type of bonding (Black et al., 2003a).

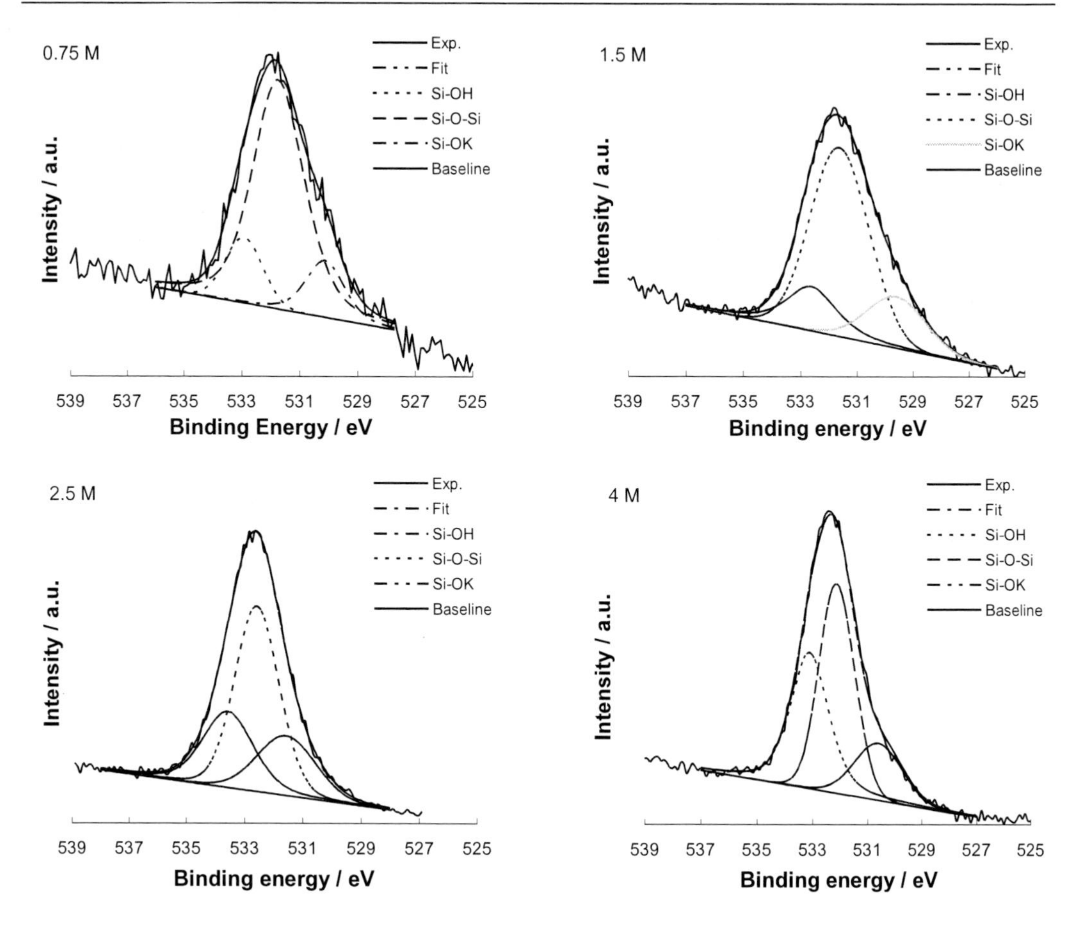

Figure 11. O1s XPS spectra of a freshly cleaved inorganic polymer synthesized using 0.75, 1.5, 2.5 and 4 M KOH respectively. Reprinted with kind permissions from Springer Science (Original printed in Simonsen et al., J Mater Sci (2009) 44, 8, 2079-2088). Copyright 2009 Springer.

XPS analyses of inorganic polymers have shown that the O1s spectra can be useful to obtain information on the degree of dissolution and polymerization of inorganic polymers (Simonsen et al., 2009a). In Figure 11 the O1s XPS spectra of freshly cleaved inorganic polymers synthesized from 0.75, 1.5, 2.5 and 4 M KOH are shown. The O1s peak has been proposed to be made up of oxygen in siloxane bonds (Si-O-Si), silanol bonds (Si-OH), and non-bridging oxygen Si-O-X (X = Ca, Na) (Miyaji et al., 1998; Black et al., 2003a). Normally, the peak ascribed to siloxane (Si-O-Si) bonds corresponding to the silicon skeleton in microsilica and the peak ascribed to silanol (Si-OH) bonds are reported to be located at 532 eV and 533 eV respectively (Paparazzo et al., 1992; Paparazzo, 1996; Miyaji et al., 1998; Black et al., 2003b). The peak corresponding to non-bridging oxygen is reported to be located at lower binding energies. The fitted peak information for the three types of bonds (Si-O-H, Si-O-Si, and Si-O-K) is shown in Table 3. From Table 3 it is seen that the percentage amount of Si-OH and Si-OK bonds increase with increasing hydroxide concentration used in the synthesis of the inorganic polymer.

Table 3. Fitted peak information for the three types of bonds (Si-O-H, Si-O-Si, and Si-O-K).

[KOH] M	$Area_{Si\text{-}OH}$ %	$Area_{Si\text{-}O\text{-}Si}$ %	$Area_{Si\text{-}OK}$ %
0.75	13.5	73.6	12.9
1.5	16.5	64.8	18.7
2.5	26.4	52.1	21.5
4	37.8	45.3	16.9

The percentage atomic content of the elements in the different inorganic polymers is shown in Table 4. The relative Si content of the inorganic polymer decrease as the KOH concentration used in the synthesis increase resulting in a higher O/Si ratio. Additionally, the K content also increase with the increase in KOH, with the exception of the inorganic polymer synthesized using 4 M KOH.

The O/Si ratios from the total fitted area of the O1s peak and the $Si2p_{3/2}$ peak using the atomic sensitivity factors for the instrument settings used in this investigation and microsilica as a reference are shown in Table 4. The O/Si ratio of the inorganic polymers increases significantly as the concentration of KOH used in the synthesis is increased. Microsilica is reported to consist of Q^4 units when examined by ^{29}Si NMR spectroscopy corresponding to a network of siloxane bonds resulting in an average formula of SiO_2 equivalent to a O/Si ratio of 2. The reaction mechanism for the preparation of the inorganic polymer is suggested to be a result of dissolution of the surface of the microsilica particles by hydroxide resulting in formation of soluble silica species in the solution. The O/Si ratio of the inorganic polymers prepared in this work changes from 2 – 2.6 as the KOH concentration is increased from 0.75 – 4 M. The increase in the O/Si ratio can be explained by the greater dissolution of SiO_2 particles leading to the formation of branched polymers. Depending on the nature of these polymers the O/Si ratio will be between 2 and 4 (O/Si ratio of monomers is equal to 4). SEM images of the inorganic polymers show that undissolved particles remain bonded in the matrix, so the obtained overall O/Si ratio of the inorganic polymer will be an average of the O/Si ratio for the remaining microsilica particles (O/Si ratio = 2) and the polymer binding the system together (O/Si ratio = 2 - 4). Hence, higher amount of polymer will lead to an increase in the overall O/Si ratio.

Table 4. Calculated percentage atomic content and O/Si ratio from the total fitted area of the O1s peak and the $Si2p_{3/2}$ peak using the atomic sensitivity factors for the instrument and microsilica as a standard.

[KOH] M	Si content %	O content%	K content %	Atomic ratio of O_{total}/Si	Atomic ratio of $O_{Si\text{-}O\text{-}Si}$ / Si ratio	Atomic ratio of $(O_{Si\text{-}OH} + O_{Si\text{-}OK})$ / Si
0.75	31.5	62.7	5.9	1.99	1.48	0.51
1.5	27.1	59.6	13.3	2.20	1.43	0.77
2.5	24.7	61.4	13.8	2.48	1.29	1.19
4	24.5	63.9	11.7	2.61	1.18	1.43

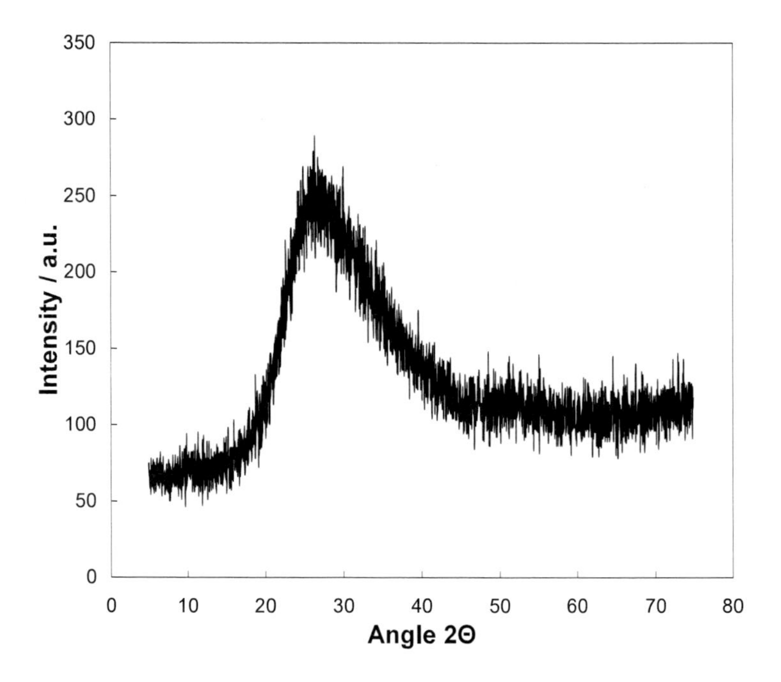

Figure 12. Typical XRD spectrum of the inorganic polymer.

XRD Analysis

Inorganic polymers are often described as X-ray amorphous (Duxson et al., 2007), since the main characteristic of the XRD spectra is a featureless bump centered at approx. 20–30° 2θ. A typical XRD spectrum of the inorganic polymers prepared in this work is shown in Figure 12. The XRD spectra of many amorphous materials appear almost identical. The similarities of these XRD spectra's are related to the characteristic bond length of inorganic oxide frameworks (Duxson et al., 2007). The Si-O-K bonding of potassium to the monomers and oligomers during the polymerization may lead to steric hindrance resulting in no long-range ordering (Simonsen et al., 2009a). Variation in the preparation temperature from 25 to 70 °C did not reveal any changes in the XRD spectrum of the inorganic polymer suggesting that higher processing temperature will lead to formation of crystalline material.

SEM Analysis

SEM imaging of the inorganic polymers support the described dissolution / gelation model as a significant change in the particle size of the silica particles is observed when different concentrations of hydroxide are used (Figure 13). SEM analysis also showed that the amount of polymer/binder in between the amorphous silica particles increased as the hydroxide concentration used in the synthesis increased resulting in a higher density of the material (Simonsen et al., 2009b). The undissolved particles remain bonded in the matrix, so that the strength and hardness of the microsilica particles correlates with the final strength.

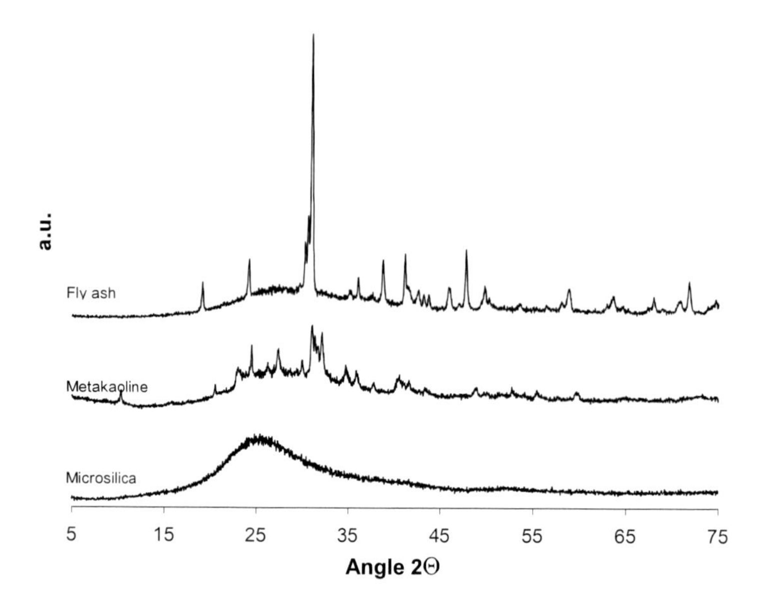

Figure 16. XRD pattern of the microsilica, metakaoline, and fly ash used in the synthesis of the inorganic polymers.

Compressive strength test of inorganic polymers synthesized using microsilica (Elkem 983 U) and KOH showed that the strength of the inorganic polymer was strongly related to the concentration of KOH used in the synthesis as shown in Figure 14. The optimal KOH concentration used in the synthesis of the inorganic polymer was found to be around 3.5 M, which resulted in a compressive strength in the region of 50 MPa (Simonsen et al., 2009b). Investigations showed that inorganic polymers synthesized at 25 °C reach its final strength after 7 days in comparison inorganic polymers synthesized at 50 °C obtain the final strength after only 4 days. However, it was found that the inorganic polymers obtained similar compressive strength. Thus, even though the temperature has a great influence on the curing rate the properties of the final inorganic polymer do not change significantly (Simonsen et al., 2009b).

Moreover, inorganic polymers synthesized from aluminosilicates (metakaoline (Metastar 501) and fly ash) were prepared. In Figure 15 a comparison of the compressive strength of inorganic polymers synthesized from metakaoline, fly ash, and microsilica is shown. From Figure 17 it is seen that the compressive strength of inorganic polymers synthesized from metakaoline and fly ash and a combination hereof obtain very low compressive strength compared to the inorganic polymers synthesized from microsilica. In Figure 16 the XRD patterns of microsilica, metakaoline, and fly ash are shown. From the XRD pattern it is seen that the microsilica is X-ray amorphous in comparison metakaoline and fly ash contain several crystalline peaks, which can be identified as quartz and mullite (Duxson et al., 2007). The presence of crystalline material will lead to lower dissolution and thus a lower amount of polymer resulting in lower compressive strength. Further investigations have shown that higher compressive strengths can be obtained by increasing the concentration of KOH and adding potassium silicate (waterglass) to the reaction solution. Moreover, investigations conducted by other groups have shown that variations in the metakaoline and fly ash used in the synthesis of inorganic polymers (geopolymers) highly influences the properties

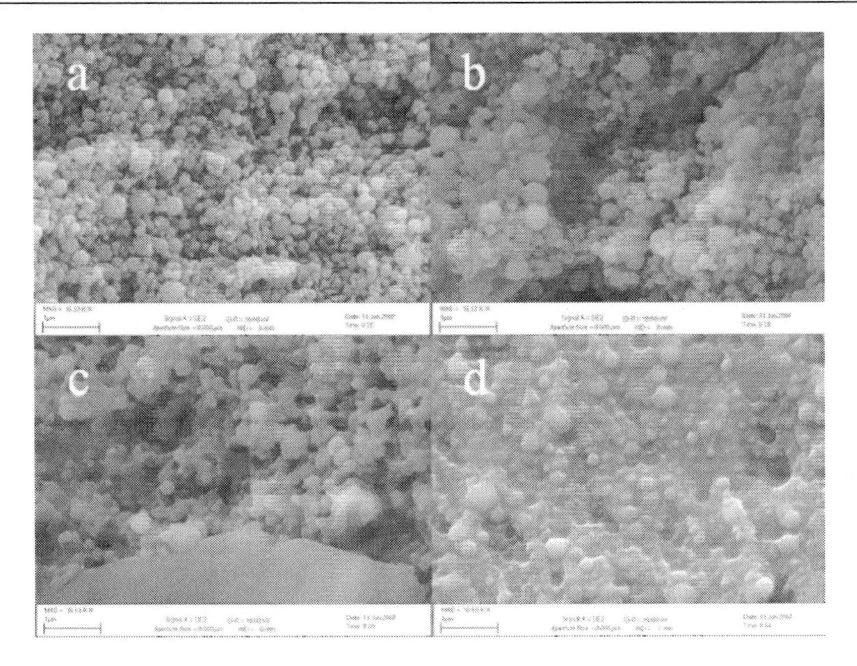

Figure 13. SEM images of inorganic binder at different potassium hydroxide concentrations. a) 0.75 M, b) 1.5 M, c) 2.5 M and d) 4 M. Reprinted with kind permissions from Springer Science (Original printed in Simonsen et al., J Sol-Gel Sci Techno (2009) 50, 372-382). Copyright 2009 Springer.

Physical and Mechanical Properties

The reaction of silicate or aluminosilicate with a highly concentrated alkali solution or silicate solution can produce inorganic polymers with properties similar to Portland cement depending upon the composition and curing technique (Komnitas and Zahataki, 2007). Variation in the preparation conditions of the inorganic polymers can result in a wide variety of properties, including high compressive strength, fire resistance and low thermal conductivity (Komnitas and Zahataki, 2007; Duxson et al., 2007). These advantages make inorganic polymers a promising technology for new construction materials even though the cost for manufacturing Portland cement is relatively low (0.05–0.08 USD/kg, year 2006) (Komnitas and Zahataki, 2007). Another application of inorganic polymer is inorganic coatings which may be used as a protective or self-cleaning coating.

Compressive Strength

Investigation of different compositions of the inorganic polymer has shown that drying of samples consisting only of inorganic polymers may lead to the formation of cracks in the material. The addition of sand or other materials such as glass fibers, iron, kaolin, or TiO_2 particles, increase the durability and strength of the material (Simonsen et al, 2009b). Addition of these materials reduces the strain in the material, and thus preventing the formation of cracks. In the reported data for the compressive strength quartz sand was added in order to prevent the formation of cracks.

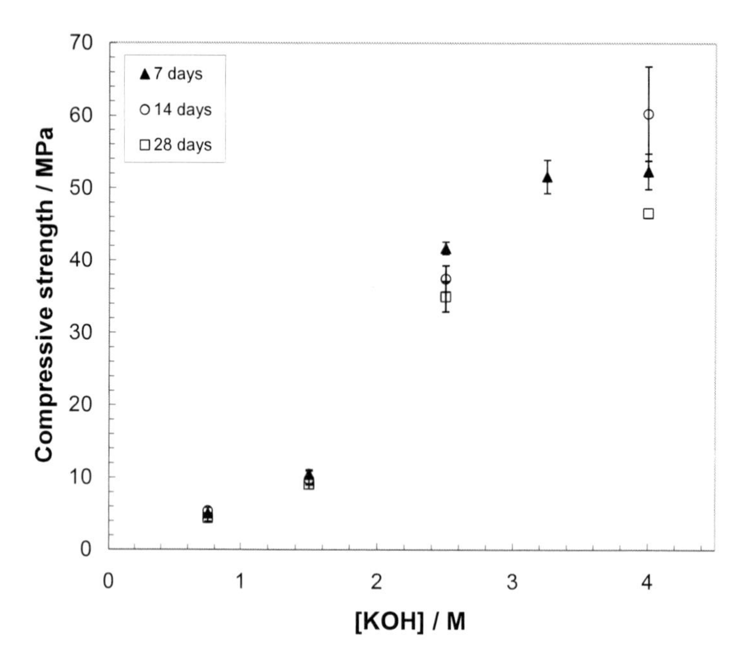

Figure 14. Compressive strength of the inorganic polymers synthesized using different KOH concentrations. The compressive strength test was performed with a piston speed of 10 mm/min. Reprinted with kind permissions from Springer Science (Original printed in Simonsen et al., J Sol-Gel Sci Techno (2009) 50, 372-382). Copyright 2009 Springer.

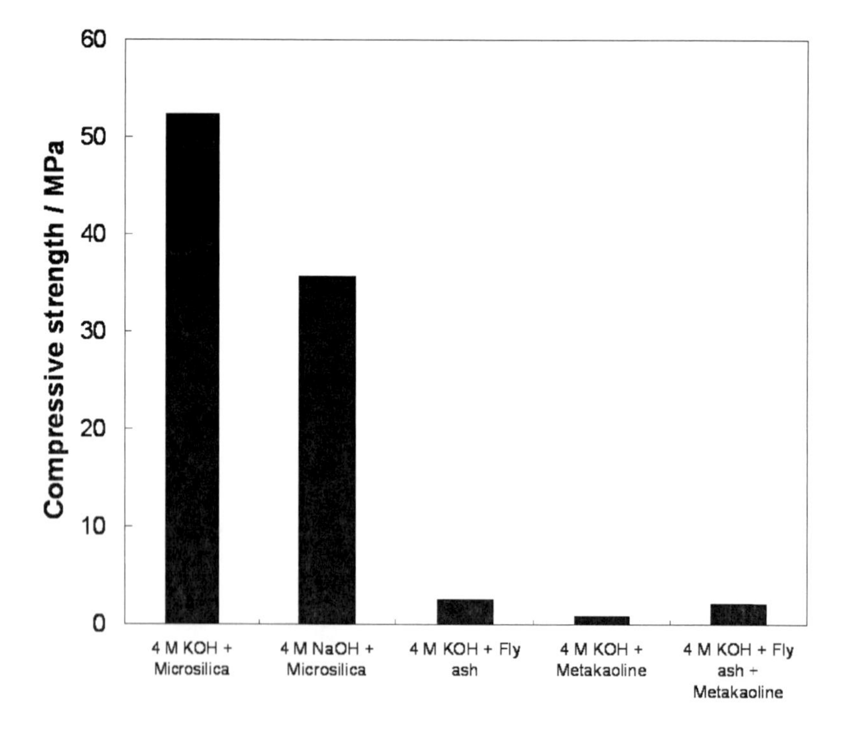

Figure 15. Compressive strength of inorganic polymers synthesized from different materials (Microsilica, metakaoline, and fly ash) after a curing period of 4 weeks. The compressive strength test was performed with a piston speed of 10 mm/min.

(Davidovits, 2005). In addition variation in the alkali solution used in the synthesis also influenced the strength of the resulting inorganic polymer. In general it was found in the present work that the compressive strengths of the inorganic polymers synthesized using KOH was higher than polymers synthesized from NaOH.

Thermal Conductivity and Temperature Resistance

Depending on the preparation conditions of the inorganic polymers different properties can be obtained. For instance an expanded inorganic polymer with low thermal conductivity can be obtained by microwave processing or rapid heat treatment of a slurry consisting of microsilica and ≥ 3 M KOH. In the microwave assisted preparation of the expanded inorganic polymer a frequency of 2.45 GHz was used in order to cause rapid evaporation of water from the system. The evaporation of water from the system freezes the system in a bobble structure as shown in Figure 17. Depending on intensity of the microwave irradiation and irradiation time different structures can be obtained. In Table 5 the thermal conductivity of the expanded inorganic polymer is compared to insulation materials such as Rockwool and expanded polystyrene. From the table it is seen that the expanded inorganic polymer have a thermal conductivity constant in the same order of magnitude as Rockwool and polystyrene. High temperature tests have shown that the expanded inorganic polymer is stable up to temperatures of at least 1100 °C. Moreover, the expanded inorganic polymer have a considerable strength of up to 2 MPa compared to the density which vary between 0.2 – 0.5 g/cm^3.

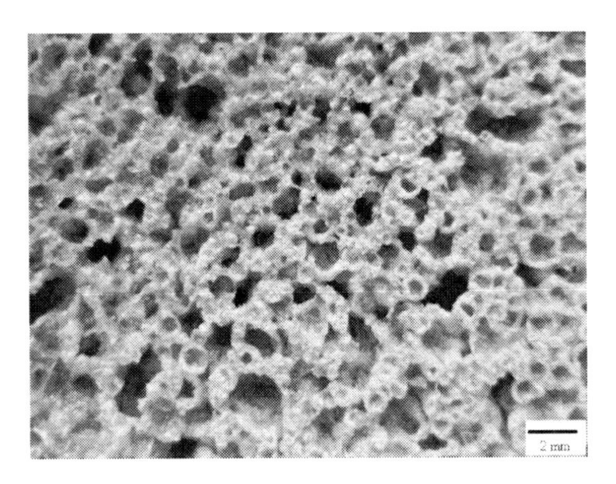

Figure 17. Characteristic bobble structure of an expanded inorganic polymer.

Table 5. Thermal conductivity of the expanded inorganic polymer in comparison to Rockwool and expanded polystyrene.

	Expanded inorganic polymer	Expanded inorganic polymer + perlit	Rockwool	Expanded Polystyrene
λ (W/m*°C)	0.064	0.072	0.050	0.0415

Table 6. Sabtility experiments of the inorganic polymers. In the table the stability of the inorganic polymers is classified in three categories; Not dissolved (ND), Partly dissolved (PD), and Dissolved (D).

Test solution	0.75 M	1.5 M	2.5 M	4 M
Dest. Water	ND	ND	ND	D
10^{-5} M NaCl	ND	ND	ND	D
10^{-3} M NaCl	ND	ND	ND	D
0.1 M NaCl	ND	ND	ND	D
10^{-5} M HCl	ND	ND	ND	D
10^{-3} M HCl	ND	ND	ND	PD
0.1 M HCl	ND	ND	ND	PD (gel formation)
10^{-5} M NaOH	ND	ND	ND	D
10^{-3} M NaOH	ND	ND	ND	D
0.1 M NaOH	D	D	D	D

Chemical Stability

The chemical stability of the inorganic polymers has been investigated using different test solutions. From Table 6 it is seen that the inorganic polymer synthesized from $\leq$ 2.5 M KOH does not dissolve unless the inorganic polymers are suspended in 0.1 M NaOH. In addition, it is seen that the inorganic polymers synthesized from the highest KOH concentration (4 M) do not form a stable binder. However, it was found that the inorganic polymers suspended in acid show higher stability. It is believed that the acid neutralizes excess hydroxide trapped in matrix of the inorganic polymer and thus stabilizes the inorganic polymer. In similar investigation inorganic polymers synthesized from metakaoline and fly ash using 4 M KOH was found to partly dissolve in water.

CONCLUSION

The synthesis of the inorganic polymer is a result of dissolution of the surface of the amorphous silica particles (Microsilica) by hydroxide resulting in formation of soluble silica species in the solution. ESI-MS investigation of the reaction solution shows that monomers, dimers, and trimers are the most dominant species in the solution before polymerization.

In the developed physico-chemical model the silica species polymerize through a condensation reaction resulting in formation of oligomers. The size and number of oligomers in the system increase until they extent throughout the solution and results in the formation of a gel. It was found that the dissolution proceeds for a longer time period when higher hydroxide concentrations are used resulting in higher dissolution and higher concentration of monomers. The gelation of the inorganic polymers synthesized from different concentrations of KOH was studied by viscosity measurements. It was found that two mechanisms contributed to the time required for gelation. At high hydroxide concentration the increased concentration of monomers in the system increase the rate of polymerization and therefore reduce the gelation time. At low hydroxide concentrations the surface charge of the silica particles is limited and the particles coagulate due to destabilization by the presence of potassium ions. A turn over point for these two mechanisms seems to be found at hydroxide concentrations between 0.75 and 1.5 M in the investigated systems.

FT-IR investigation of the inorganic polymers synthesized using different KOH concentration have shown that an increase in the hydroxide concentration shifts the position of the maximum absorbance of Si-O bands towards lower wave numbers, indicating the transformation of Q^4 units to Q^3 and Q^2 units. Furthermore, comparison of the evolution of the KOH concentration during synthesis and the change in intensity of the peak assigned to Q^4 units show that the change in the Q^4 peak can be used to describe the dissolution process occurring during synthesis of the inorganic polymer. ^{29}Si NMR investigation of the inorganic polymer showed similar results. XPS investigation of the different inorganic polymers showed that the total amount of oxygen and potassium present in the sample increased when higher concentrations of hydroxide were used in the synthesis. The O/Si ratio of the inorganic polymers changed from 2 – 2.6 when the KOH concentration was increased from 0.75 – 4 M. The increase in the O/Si ratio can be explained by the greater dissolution of SiO_2 particles leading to the formation of branched polymers.

The optimal potassium hydroxide concentration used in the synthesis of the inorganic polymer was found to be about 3.5 M, which resulted in a compressive strength in the region of 50 MPa. In addition it was found that the inorganic polymers synthesized from $\leq$ 2.5 M hydroxide have good chemical stability towards acid attacks and solutions with high ion strength.

REFERENCES

[1] Black, L; Garbev, K; Stemmermann, P; Hallam, KR; Allen, GC. *Cem. Concr. Res.,* 2003, 33, 899-911.

[2] Black, L; Stumm, A; Garbev, K; Stemmermann, P; Hallam, KR; Allen, GC. *Cem. Concr. Res.,* 2003, 33, 1561-1565.

[3] Brew, DRM; MacKenzie, KJD. *J. Mater. Sci.,* 2007, 42, 3990-3993.

[4] Bussian, P; Sobott, F; Brutschy, B; Schrader, W; Schüth, F. *Angew. Chem. Int. Ed.,* 2000, 39, 21, 3901- 3905.

[5] Davidovits, J. Geopolymer Chemistry and Sustainable Development, Proceedings of the World Congress Geopolymer, Saint Quentin, France, 28 June–1 July, 2005, Institut Géopolymére, Saint Quentin.

[6] Duxson, P; Fernández-Jiménez, A; Provis, J; Lukey, G; Palomo, A; Van Deventer, J. *J. Mater. Sci.,* 2007, 42, 2917-2933.

[7] Eggers, K; Eichner, T; Woenckhaus, J. *J. Mass Spectrom.,* 2005, 244, 72-75.

[8] Fletcher, RA; MacKenzie, KJD; Nicholson, CL; Shimada, S. *J. Euro Ceram. Soc.,* 2005, 25, 1471-1477.

[9] Hjorth, J; Skibsted, J; Jacobsen, H. *J. Cem. Concr. Res.,* 1988, 18, 5, 789-798.

[10] Iler, RK. The Chemistry of Silica, John Wiley & Sons, New York, 1979.

[11] Komnitas, K; Zahataki, D. *Miner. Eng,* 2007, 20, 1261-1277.

[12] Miyaji, F; Iwai, M; Kokubo, T; Nakamura, T. *J. Mater. Sci.: Mater. Med.,* 1998, 9, 61-65.

[13] Paparazzo, E. *Surf. Interface anal.,* 1996, 24, 729-730.

[14] Paparazzo, E; Fanfoni, M; Severini, E; Priori, S. *J. Vacuum Sci. Tech. A,* 1992, 10, 4, 2892-2896.

[15] Rowles, M; O'Conner, B. *J. Mater. Chem.,* 2003, 13, 1161-1165.

[16] Serra, J; González, P; Liste, S; Chiussi, S; León, B; Pérez-Amor, M; Ylänen, HO; Hupa, M. *J. Mater. Sci.: Mater Med.,* 2002, 13, 1221-1225.

[17] Simonsen, ME; Søgaard, EG. *Inter. J. Mass. Spec.,* 2009, 1-2, 78-85.

[18] Simonsen, ME; Sønderby, C; Li, Z; Søgaard, EG. *J. Mater. Sci.,* 2009, 44, 8, 2079-2088.

[19] Simonsen, ME; Sønderby, C; Søgaard, EG. *J. Sol-Gel Sci. Techno.,* 2009, 50, 3, 372-382.

[20] Zholobenko, VL; Holmes, SM; Cundy, CS; Dwyer, *J. Microporous Mater.,* 1997, 11, 83-86.

[21] Zivica, V. *Bull. Mater. Sci.,* 2004, 27, 2, 179-182.

In: The Sol-Gel Process
Editor: Rachel E. Morris

ISBN 978-1-61761-321-0

Chapter 17

MULTIENZYMATIC SYSTEM IMMOBILIZATION IN SOL-GEL SLIDES: DEVELOPMENT OF FLUORESCENT BIOSENSORS

I. Pastor*[*] *and A. Salinas-Castillo
Department of Chemistry Physics, University of Barcelona, Barcelona, Spain

ABSTRACT

One of the potential areas of research in the development of biosensors is the production of analytical devices based on the use of immobilized multienzymatic systems. Recently, the recovery of sol-gel chemistry has provided a new versatile method for immobilizing and stabilizing a wide range of enzymes and other biological molecules in transparent inorganic matrices. Compared to other immobilization methods, such as the adsorption to solid supports, the covalent attachment and the polymer entrapment, the sol-gel glasses show numerous advantages, including entrapment of a large amount of enzymes, thermal and chemical stability of the matrix, enhanced stability of the encapsulated biomolecules, excellent optical transparency and flexibility in controlling the pore size and the geometry. Furthermore, thanks to the porous nature of the matrix, the immobilized proteins remain accessible to interact with external specific analytes with negligible protein leaching, making possible the conduction of multienzymatic reactions. Indeed, the application of multienzymatic systems is an area of large interest in the development of biosensors, due to the fact that the product of one reaction can become the substrate of another one; regarding to this, the sol-gel process offers an excellent alternative for multienzymatic immobilization. In addition, thanks to the exceptional optical transparency that sol-gel matrices present, the use of fluorescent indicators coupled to enzymatic reactions is possible, which represents a successful advantage to solve some problems related with the specificity and sensitivity of this type of biosensors. In the present work, we revised the development of fluorescent multienzymatic biosensors based on sol-gel technology.

[*] Corresponding author: Isabel Pastor, E-mail: i.pastor@ub.edu, Fax: (+34) 934021231, Phone: (+34) 934020138

INTRODUCTION

Nowadays, the development of biosensors is the subject of an extensive research in different areas such as clinical diagnostic, food technology, bioprocesses monitoring and biomedical and environmental analysis (Borisov and Wolfbeis, 2008; Rich and Myszka, 2007; Xu et al., 2005; Velasco- García and Mottram, 2003; Wolfbeis et al., 2000). These devices are a specific type of chemical sensors that combine a biological component with a signal transducer. The transducer or the detector element can work in a physicochemical manner like optical, piezoelectric, electrochemical, etc., transforming the signal, resulting from the interaction of the analyte with the biological element, into another signal that can be more easily measured and quantified. The biological component can be an enzyme, an antibody, a polynucleic acid, or even the whole cell or the entire tissue in slices. Among these components, the enzyme-based biosensors have increased considerably their importance in the last years thanks to their exceptional properties, which include high specificity, rapid response and low cost manufacture (Choi et al., 2004). The configurations of enzymatic biosensors range from simple one-enzyme devices to multienzymatic systems. Generally, additional enzymes are necessary because the products obtained in the first enzymatic reaction are not detected directly by a transducer and/or because the sensitivity and selectivity of the recognition of the reaction product by the transducer is improved. Consequently, the use of more than one enzyme specie considerably extends the range of potential applications of biosensors for the detection of numerous biological substrates, but complicates the manufacture of these devices as well as their basic design (Wollenberger et al., 1993).

One of the most important steps in the development of multienzymatic biosensors is the immobilization of the multienzymatic system in an appropriate solid support which integrates the biomolecules maintaining their functionality, which allows performing the reactions in a sequential order while it provides accessibility towards the target analyte and the subsequent substrates (Xu et al., 2006). Diverse immobilization techniques have been applied, including adsorption to solid supports, covalent attachment and entrapment in polymers. In general, adsorption techniques are easy to perform, but the bonding of the enzyme is often weak causing leaching. Then, such biocatalysts lack the degree of stabilization, which is possible by covalent attachment and entrapment. On the other hand, covalent techniques are tedious and often require several chemical steps. Thus, one advantage of the immobilization is that prevents the leaching. However, it often leads to lack of activity and stability over time. That is the main reason why the development of biosensors is limited somewhat by the lack of a simple and generic immobilization protocol.

Recently, the use of sol-gel chemistry has become an excellent method for immobilizing and stabilizing enzymes and different biological compounds in inorganic matrices. Compared to other immobilization methods (Sakai-Kato and Ishikura, 2009), the sol-gel glasses show many advantages, including entrapment of a large amount of enzymes, thermal and chemical stability of the matrix, enhanced stability of the encapsulated biomolecules, excellent optical transparency and flexibility in controlling pore size and geometry. Furthermore, thanks to the porous nature of the matrix, the immobilized proteins remain accessible to interact with external specific analytes with negligible protein leaching, making possible the conduction of multienzymatic reactions (Pierre, 2004; Kandimalla et al., 2006; Jerónimo et al., 2007; Gupta and Chaudhury, 2007). Multienzymatic entrapment in the same sol-gel matrix allows sequential reactions to occur in a confined space, leading to high conversion efficiencies due to the proximity of the analytes and intermediates to the active sites of the enzymes (Jin and Brennan, 2002). Several types of enzymatic biosensors have been developed based on two enzymes sol-gel immobilization (Martinez-Pérez et al., 2003; Pastor et al., 2004; Cui et al., 2007; Singh et al., 2007). However, co-immobilization of three or more enzyme species in a single sol-gel glass has only been scarcely reported (Gallarta et al., 2007; Salinas-Castillo et al., 2008), due to the difficulty to get the optimal operational conditions for different enzymes in the same matrix. Besides, in spite of the sensitivity and reliability, which is higher for these biosensors than for the mono-enzymatic, bi-enzymatic or tri-enzymatic devices, the response time should be also higher due to the increased number of reactions that take place within the sol-gel matrix.

Other important issue in the development of biosensors is the transducer. There are several kinds of biosensors depending on the nature of the analytical signal transducer (Göpel et al., 1991; Thevenot et al., 2001). The use of optical techniques, where the information is gathered by the measurement of photons (rather than electrons as in the case of electrodes), represents a successful advantage to solve some problems related with the specificity and sensitivity, which usually occur in other class of biosensors. Specifically, this fact is relevant in those based on the measurement of absorbance, reflectance, or fluorescence emissions that occur in the ultraviolet (UV), visible, or near-infrared (NIR). In fact, fluorescence is so far the most often applied method, mainly by two reasons: on one hand, because it is possible to measure different kind of parameters (e.g. intensity, decay time, anisotropy, quenching efficiency, luminescence energy transfer) and, on the other hand, because it is one of the most specific and sensitive techniques (Göpel et al., 1991; Thevenot et al., 2001; Borisov and Wolfbeis, 2008). In consequence, a biosensor combining the properties of using sequential enzymatic reactions coupled to a fluorescent transducer with those of sol-gel immobilization, may offer the advantages of a sensitive, selective, low cost, and ready-to-use method for quantitative analytes determination.

We have developed several fluorescent biosensors based on the co-immobilization of different enzymes (two or three) in the same sol-gel matrix or by combining two sol-gel matrices. In the present review, the fluorescent detection is based on the use of molecular probes. We performed the detection of a reaction product obtained thanks to the oxidases like the Hydrogen peroxide (H_2O_2), using in all the cases the N-acetyl-dihydroxyphenoxazine (Amplex red). The non-fluorescent Amplex red reacts stoichiometrically with H_2O_2 in the presence of Horseradish peroxidase (HRP) to generate the red-fluorescent oxidation product Resorufine (Figure 1). Then, an enzyme is present in this type of systems. The other two enzymes are variable depending on the analyte we want to determine. The presence of H_2O_2 in the samples is one of the most important steps in our work. H_2O_2 can be generated, for example, by Uric acid in presence of Uricase or by Superoxide anion ($O_2^{\cdot-}$) in presence of Superoxide dismutase (SOD). Thus, we are developing three different fluorescent multienzymatic biosensors: for Uric acid using Uricase, HRP and the Amplex red probe (Martinez-Pérez et al., 2003); for Superoxide anion radical using SOD, HRP and Amplex red (Pastor et al., 2004); and a biosensor for Xanthine whose enzymatic system includes Xanthine oxidase (XO) which catalyzes the oxidation of Xanthine to Uric acid and Superoxide radical, SOD, HRP and Amplex red (Salinas-Castillo et al., 2008).

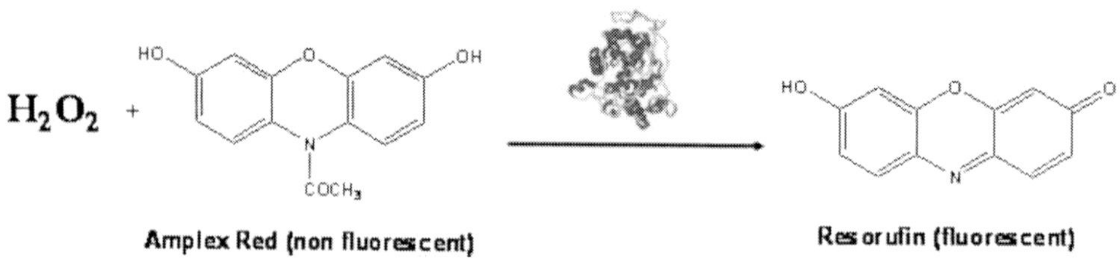

Figure 1. Scheme of the Resorufine formation from Amplex red and hydrogen peroxide catalysed by HRP.

Obviously, in the literature, it is possible to find others biosensors, which, in principle, look like multienzymatic fluorescent biosensors in sol-gel matrices. However, these biosensors work like unienzymatic biosensors. For example, Gallarta et al. in 2007 (Gallarta et al., 2007) developed a L-malic acid in wine biosensor, which was based on the co-immobilization of L-malate deshydrogenase (L-MDH), the NAD^+ cofactor and the Diaphorase enzyme (DI) in TMOS sol-gel matrices. Briefly, L-MDH reversibly catalyzes the reaction between L-malate and NAD^+, producing NADH, whose fluorescence (λ_{exc} =340 nm, λ_{em}=430 nm) could be directly related to the amount of L-malate. Afterwards, NADH is converted to NAD^+ by applying hexacyanoferrate (III) as an oxidant in the presence of DI. Another example is the multianalyte biosensors developed by Tsai and Doong (Tsai and Doong, 2004; Tsai and Doong, 2005). Both works describe the fabrication of array-based biosensors with an individual enzyme, which allows the simultaneous analysis of multiple samples in the presence of numerous unrelated analytes. These biosensors can immobilize a variety of enzymes into different spots to determine multi-analytes.

This review tries to describe the particularities, utilities and advantages and disadvantages of this kind of biosensors and re-examine the main elements to take in account in the fabrication of new devices as well as their most recent applications. Finally, we summarize the potential of the sol-gel technology in the development of novel fluorescent multienzymatic biosensors.

How to Develop Fluorescent Multienzymatic Biosensors Based on Sol-Gel Technology

Immobilization of Enzymes in Sol-Gel Slides

One of the key factors in the development of fluorescent multienzymatic biosensors is the immobilization of enzymes in the sol-gel matrix. Enzymes are highly sensitive molecules since they must work in appropriate conditions like at physiological pH and temperature. Inappropriate conditions or the presence of molecules like alcohol (e.g. methanol or ethanol) even in trace concentrations could produce the enzymatic denaturalization. However, the sol-gel technology implies generally the use of tetraalcohol ortosilicates like tetramethyl orthosilicate (TMOS) or tetraethyl orthosilicate (TEOS) which, at a first step, before gelation, are hydrolyzed, usually at acid pH, generating methanol or ethanol as secondary products. This means, that on one hand it is necessary to remove the alcohol generated during this process and, in the other hand, the pH must be changed to keep the enzyme in a more physiological state. There are several methods to carry out, e.g. by intensive stirred or sonication of the mixture during the hydrolysis reaction. Among them, the alcohol free sol-gel route described in Ferrer et al. in 2002 could be the most appropriate alternative. Following this route, it has been possible to entrap enzymes and bacteria in sol-gel matrices preserving their function and stability. Briefly, the protocol of immobilization consists in the mixture of sol-gel precursors (for example 4.46 mL of TEOS) with H_2O (1.44 mL) and 0.04 mL HCl (0.62 M) that were mixed under vigorous stirring in a closed vessel. After 1 h, 1 mL of the resulting solution is mixed with 1 mL of deionised water, and submitted to rotaevaporation for a weight loss of 0.62 g (i.e. 0.62 g are approximately the alcohol mass resulting from alkoxyde hydrolysis). Then, a volume of aqueous sol is mixed with an equal volume of buffered containing enzymes solution in an adequate mold at room temperature. Gelation occurs readily after mixing. Following gelation, the resulting matrices are wet aged in the phosphate buffer solution at 4°C during 48 h.

In spite of the fact that the described protocol above is well optimized for the immobilization of most of the biomolecules, in certain cases, it is necessary the presence of additives or to take additional precautions during the immobilization process because there are limitations on the sol-gel procedure which could interfere in the biosensor application. For example, in order to prolong the cracking time, it could be of interest to use cooler molds or to dope the sol-gel with a little percentage of PVA (polyvinylalcohol) (Tsai and Doong, 2004; Tsai and Doong, 2005).

Once the enzymes are entrapped in the sol-gel matrix, a preliminary analysis should be carried out to demonstrate that the immobilization process does not affect the enzymatic properties. Using appropriate conditions of transparency and the fluorescent spectras that the sol-gel matrices offer in the visible-UV, the spectra can be directly recorded in the sol-gel confirming the successful encapsulation of the enzymes. In recent studies developed by our group (Martinez-Pérez et al., 2003; Pastor et al., 2004; Salinas-Castillo et al., 2009; Pastor et al., 2010) the absorption spectra for different co-immobilized enzymes was showed (Figure 2). For example, in the simultaneous immobilization of enzymatic systems Uricase-peroxidase (UR-HRP) (Martinez-Pérez et al., 2003) (Figure 2A), superoxide dismutase-peroxidase (SOD-HRP) (Pastor et al., 2004) (Figure 2B) or xanthine oxidase-superoxide dismutase-peroxidase (XO-SOD-HRP) (Salinas-Castillo et al., 2009) (Figure 2C), the similarity between the absorbance values obtained for the free and encapsulated systems indicates that the yield of the protein encapsulation process is close to 100%. However, these results confirm the success of the enzyme immobilization but do not guarantee that the enzymes remain active upon encapsulation.

Fluorescence Detection

To verify the activity of the immobilized enzymes, we used the fluorescence of a molecular probe. The main advantage of using molecular probes instead to labels is that they are inert materials although they are able to react to a specific micro-environment or different chemical species such as ions, pH or oxygen. In that case, they are commonly called indicators. Fluorescence spectroscopy has been highly contributed by the design of novel fluorescent probes which has provided enormous advances in the bioscience research.

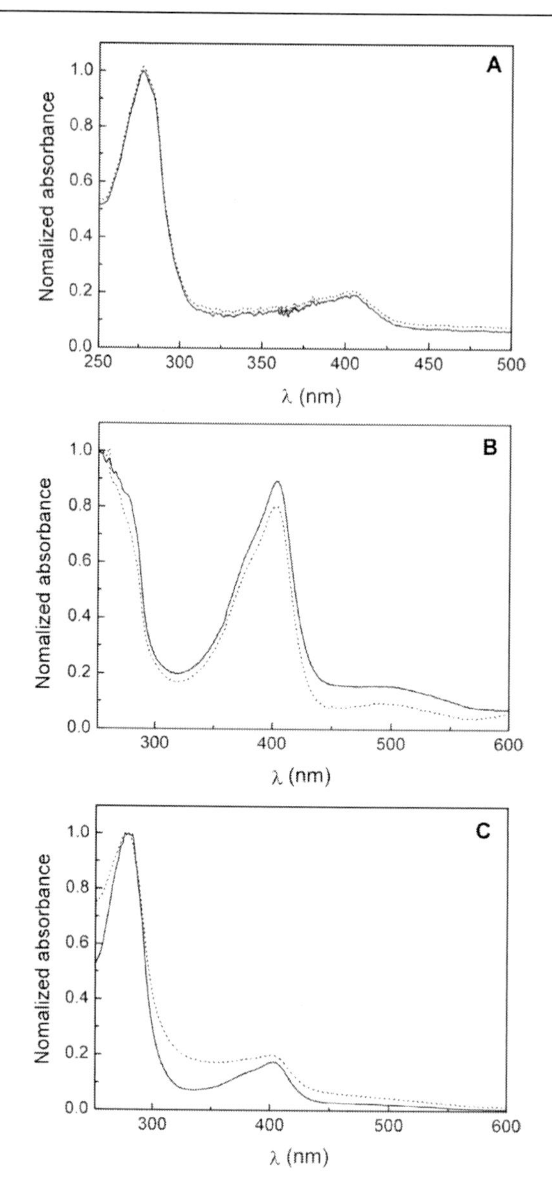

Figure 2. Normalized absorption spectra for different situations: (A) UR and HRP, (B) SOD and HRP and (C) XO, SOD and HRP. In each Figure the solid line represents the absorption spectra for the proteins in solution and the spotted line represents the proteins co-immobilized in sol-gel matrices.

At the moment, the monitoring of certain species of clinical importance such as oxygen molecules, protons ("pH"), carbon dioxide, H_2O_2, glucose, etc. is getting more and more interest within the clinical research since they are key metabolites which cannot be measured throughout luminescence spectroscopy. Consequently, these indicators allow us to visualize several species in a more precise way.

Molecular probes are also very useful to do time-course measurements of enzymatic reactions (i.e. oxidation of glucose; Glucose + O_2 → Gluconic acid + H_2O_2). The kinetics of such a reaction can be easily monitored by measuring the consumption of oxygen using a fluorescent oxygen probe by the formation of the acid using a pH probe or the formation of H_2O_2 using a specific probe.

In this review the probe selected was N-acetyl-dihydroxyphenoxazine (Amplex red), which could be entrapped in the sol-gel matrix together with the enzymatic system or staying in the solution. The nonfluorescent Amplex red reacts stoichiometrically with H_2O_2 in the presence of HRP to generate the red-fluorescent oxidation product Resorufine (see Figure 1).

Enzyme Activity and Characterization of the Sensor Performance in Solution

Another important point in the development of this sort of biosensors, that include several enzymes and, at least, involve one fluorescent probe, is to check how the analytical system operates in solution. This means, to check the concentration of the enzyme and fluorescent probe that is necessary, the time of reaction, the range of the sensibility, the specificity, the reproducibility, etc. when the system is in solution in order to optimize the biosensor in sol-gel matrix. These parameters are described along this report.

Enzyme Activity and Characterization of the Sensor Performance in Sol-Gel

Once the enzymatic systems were successfully encapsulated we must compare their capacity to quantify the analyte and the properties of the developed biosensor.

Firstly, whether two or more reactions are implicated in the analyte determination, it is very important to know the kinetics of the enzymatic system both in solution and when the enzymes are immobilized into a sol-gel matrix. For example, in the biosensors described in this work, the oxidation times of Amplex red were monitored at a fixed concentration of the analyte (either Uric acid, $O_2^{\cdot -}$ or Xanthine) and then compared with those obtained when the enzymatic system was in solution. As it is showed in the inset of Figure 3A the time of formation of Resorufine was equal in solution (~ 45 min) in the case of the Uric acid biosensor. That time corresponds with the oxidation time of Amplex red. However, the time of formation of Resorufine is slightly superior to the one in solution in the case of Xanthine and Superoxide biosensor (around ~60 min in sol-gel and 45 min in solution) (insets in Figure 3B and 3C, respectively). This is probably due to the slower diffusion of the Superoxide radical through the sol-gel matrix (Pastor et al., 2004). It is known that the surface of the sol-gel porous has around 15 % of negative charges (Badjić y Kostić, 1999; Eggers y Valentine, 2000), which could reduce the diffusion of negative analytes through the sol-gel matrix. This behaviour is showed for the determination of the Superoxide radical anion or Xanthine, where there is involved a superoxide radical too (Pastor et al., 2004; Salinas-Castillo et al., 2009; Pastor et al., 2010).

Secondly, standard curves for analytical systems in sol-gel, which correlate the signal of the fluorescence corresponding to the fluorescence signal with the analyte concentration, must be constructed (Figure 3). These standard curves give us relevant information about the analyte range of concentration where the developed biosensor could work, which corresponds with the detection limit that is possible to measure with the specific biosensor. In the examples described in this work, the methodology was similar for each analyte. Briefly, the enzymatic immobilized system was tested by means of immersion of the sol-gel slide in a solution containing the final oxidizable substrate, Amplex red, and the analyte in a specific interval of concentration (between 10 and 1000 nM) for Acid uric and Xanthine or the required elements to generate $O_2^{\cdot-}$, XO ($2 \cdot 10^{-4}$M), and Xanthine (between 10 and 1000 nM). The increase in the fluorescence signal, recorded at 590 nm, was proportional to the concentration of the analyte in the sample. The calibration curves were quite similar in all the cases with the corresponding system in solution. Their linear ranges were extended until very low concentrations of analyte up to 2 μM, presenting excellent correlation coefficients ($r^2 > 0.99$) and similar slopes (STD < 5%). From these curves, we can assert that the behaviour of the corresponding enzymatic systems in sol-gel and in solution is similar. The detection limit reported in these biosensors was 20 nM, mostly due to the spectroscopic characteristics of the Amplex red probe.

Thirdly, in our opinion, in the developing of any kind of biosensor is critical to study the reusability and the storage period of the sensor. The biosensors mentioned previously have a larger stability (up to 3 months) when they are stored at 4ºC and isolated from oxygen and light. In addition, this biosensor can be reusable. We propose that the reusability is one of the major advantages of the sol-gel immobilized enzymes. In the cases already reported, the same sol-gel matrix can be used at least, four times with no lack of enzymatic activity.

Finally, the possibility of co-immobilize the probe together with the enzymatic system in the same sol-gel matrix must be studied, because this simultaneous encapsulation prevents the biosensor reusability but facilitates its manipulation and offers a complete ready-to-use biosensor. This possibility have been successfully carry on in all the biosensors reported in this study, even in the trienzymatic immobilization reported in the developed of Xanthine biosensor (Salinas-Castillo et al., 2008). In all the cases, the encapsulation of the enzymatic system and the Amplex red probe had the same analytic range and the same limit of detection that when the Amplex red was not into the sol-gel.

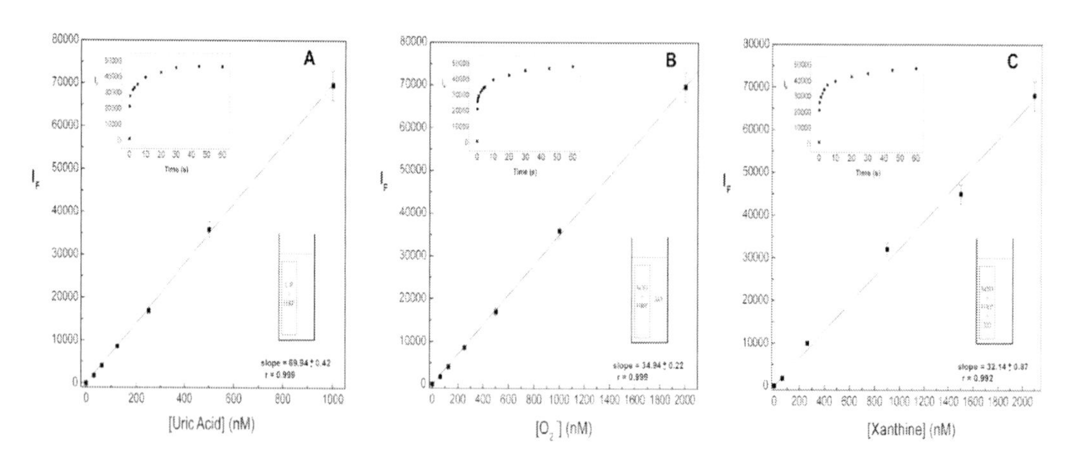

Figure 3. Standard curves obtained by the Amplex red assay for the three systems developed (three sets of triplicate samples were pooled to create the plots). The relative standard deviations (RSD) in the three curves were between 5 and 8%. Insets: time Resorufine formation at λ_{exc} = 530nm and λ_{em} = 590 nm, for sol-gel encapsulated systems 1–3 (T=37°C), respectively.

BIOSENSORS DEVELOPED USING THIS METHODOLOGY

A Reagent Less Fluorescent Sol-Gel Biosensor for Uric Acid Detection in Biological Fluids

Uric acid is the final oxidation product of Purine metabolism in humans and higher primates, which is a substance found in many foods. The detection of Uric acid in body fluids is a clinical indicator to diagnostic heart or chronic kidney diseases. There are an important number of biosensors developed to detect this analyte in biological samples (Akyilmaz et al., 2003; Wu et al., 2005; Perello et al., 2005; Wang et al., 2008; Ndamanisha and Guo, 2008; Chen et al., 2010). The Uric acid biosensor developed by Martinez-Pérez et al. in 2003 is based on the coupled UR-HRP reaction and the Amplex red probe. They reported the co-immobilization of both enzymes (UR-HRP) in the same sol-gel matrix, which is made from TEOS. The co-immobilized enzymatic system serves to prepare a reagent-less and ready-to-use fluorescent sol-gel biosensor for Uric acid determination with a high sensitivity (at nanomolar range). This sensibility allows to accurate the determination of Uric acid in a diluted solution of several biological fluids (serum, blood and urine), even in presence of interfering species (e.g. Ascorbic acid). The Table 1 compares the analytical range, detection limit and useful time of several uric acid biosensors.

Table 1. Analytical characteristics of different biosensors.

Reference	Technique	Limit detection (μM)	Analytical range (μM)
Uric acid Biosensor (Bi-enzymatic: UR-HRP)			
Chen et al.	Amperiometric	0.3	0-1.125
Wang et al.	Quartz crystal microbalance	2	5-1000
Ndamanisha and Gou	Electrochemical	1.8	60-390
Wu et al.	Chemiluminiscence	40	1-3000
Perello et al.	Chromatography	70	-
Martinez et al.	Spectrofluorometry	0.20	0-1
Akylmaz et al.	Amperiometric	-	0.1-0.5
Superoxide Biosensor (Bi-enzymatic: SOD-HRP)			
Chem et al.	Amperometric	0.073	0-0.6
Guo et al.	Amperometric	0.5	0.86-5.93
Campanella et al.	Amperometric	-	20-2000
Di et al.	Amperometric	0.1	0.2-1.6
Pastor et al.	Spectrofluorometry	0.02	0-1
Barbacanne et al.	Electron spin resonance	~0.001	-
Xanthine Biosensor (Tri-enzymatic: XO-SOD-HRP)			
Hlavay et al.	Quimioluminescence	0.55	3-216
Rehak et al.	Amperometric	20	Up to 1000
Zhao et al.	Amperometric	0.2	Up to 30
Dimcheva et al.	Amperometric	4.5	Up to 40
Kirgöz et al.	Amperometric	0.1	0.5- 40
Arslam et al.	Amperometric	1	10- 400
Rahman et al.	Amperometric	0.09	0.5-100
Villalonga et al.	Amperometric	150	310-6800
Salinas-Castillo et al.	Fluorescence	0.02	0-1

A Ready-to-Use Fluorimetric Biosensor for Superoxide Radical Using Superoxide Dismutase and Peroxidase Immobilized in Sol-Gel Glasses

Superoxide anion radical is produced in living systems as a reduced intermediate of molecular oxygen either through autooxidation processes or by enzymes involved during the normal metabolism. However, it is also involved in a number of pathological processes including aging, atherosclerosis, hypertension, diabetes, ischemic-reperfusion injury, neurodegenerative disorders, and cancer due to superoxide could appear under uncontrolled conditions causing reversible or irreversible damage to several biomolecules. This radical is quite reactive and is easily attacked by other active biomolecules (e.g., Nitric oxide) and scavenged by enzymes (e.g., Superoxide dismutase) and antioxidant agents (e.g., Ascorbate). Due to these reactions, its half-life is rather short and it is difficult to obtain precise measurements for superoxide radical concentration through a method capable to compete with these intra- or extracellular superoxide scavenging systems. Although numerous methods have been developed during last decades (Barbacanne et al., 2000; Di et al., 2004; Campanella et al., 2005; Guo et al., 2005; Chem et al., 2008), the fact is that, the measurement of the superoxide concentration in biological systems continues to be a challenging analytical problem. The biosensor developed by Pastor el al. in 2004 is based on the coupled SOD-HRP reaction, together with the use of the Amplex red probe. They reported the co-immobilization of both enzymes (SOD-HRP) in the same TEOS sol-gel matrix. Usually, physiological concentrations of superoxide are in the range of 10^{-10} M therefore, the high fluorescence quantum yield of the Resorufine allowed the determination of superoxide radical at submicromolar concentrations in biological fluids without sample pre-treatment. The Table 1 compares the analytical range, detection limit and useful time of this biosensor versus others superoxide radical sensors.

Immobilization of a Trienzymatic System in a Sol-Gel Matrix: A New Fluorescent Biosensor for Xanthine

Xanthine is a product generated during chemical degradation of Purine nucleotides that is found in DNA and RNA. It penetrates the cell membrane and accumulates in extracellular fluids. Quantitative determination of Xanthine in biological samples is of great interest for the diagnostic of several pathologies such as preinital asphyxia, adult respiratory distress syndrome, cerebral ischemia, tumour hyperthermia and pre-eclampsia. There are several methodologies to measure this analyte in biological samples (Hlavay et al., 1994; Rehak et al., 1994; Zhao et al., 1994; Dimcheva et al., 2002; Kirgöz et al., 2004; Rahman et al., 2004; Arslam et al., 2006; Villalonga et al., 2007), whose usual drawback is the laborious procedure for sample preparation. Salinas-Castillo et al. developed a fluorescent trienzymatic biosensor for Xanthine in 2009, which is based on the immobilization of three enzymes: Xanthine oxidase, SOD and Peroxidase (XO-SOD-HRP) in a single sol-gel matrix coupled to the used of Amplex Red probe. Due to the high sensitivity of this the biosensor, it is possible to determine Xanthine in biological fluids by dilution of the sample, reducing the interfering species to negligible levels. In Table 1 several xanthine sensors are compared with the Salinas-Castillo biosensor.

Multienzymatic System Immobilization in Sol-Gel Slides: Fluorescent Superoxide Biosensors Development

The last work developed in our group is an application of previous biosensors mentioned before, because both, the biosensor for superoxide radical (called system 1) and the biosensor for Xanthine (system 2) are based on the same scheme; therefore, both could be able to quantify superoxide radical. The results were compared with those obtained through a third system (system 3) which is constituted by two sol-gel slides. In one of them the enzymatic system SOD-HRP was immobilized simultaneously while in the other sol-gel matrix, the XO was immobilized alone. We investigated the ability of three sol-gel immobilized multienzymatic systems (systems 1, 2 and 3) coupled with the Amplex red indicator to determine superoxide radical and evaluate the scavenging capacity of several antioxidant compounds. This work showed that due to the characteristics displayed by system 1 and 3, both could be applied to study the scavenging properties and their specificity versus $O_2^{\cdot -}$ of four antioxidant species: (+)-cathechin and lauryl galate (LG) from natural extracts of green tea, and butylated hydroxytoluene (BHT) and 6-hydroxy-2,5,7,8-tetramethylchroman-2-carboxylic acid (Trolox), both typical synthetic antioxidants of reference. Among all of the antioxidants studied, (+)-cathechin is the most selective compound for superoxide radical.

Additional Remarks

In this review we have showed the potential use of the sol-gel technology in the development of multienzymatic biosensors. However, we want to highlight that the range of the experimental conditions covered by the studies showed here is only the first step that must be taken into account in the developing of this kind of biosensors. From our point of view it is important to emphasize the possibility of immobilizing two or more enzymes in the same sol-gel matrix, due to the high versatility of the sol-gel methodology. The high porosity of the sol-gel matrix allows the analytes to diffuse from the solution to the immobilized enzymes and to react. This simultaneous encapsulation allows the biosensor to be reused and facilitates its manipulation.

On the other hand, the excellent optical transparency of sol-gel matrices make possible the use of fluorescent indicators coupled to enzymatic reactions. The used of fluorescence together with the enzymatic system into sol-gel matrix can reduce the analytical range of the biosensors, due to the usual fluorescence quantum yield of probes. Thanks to the use of fluorescence together with the sol-gel methodology to immobilize enzymes we are able to develop highly sensible biosensors. Finally, the possibility to co-immobilize the probe together with the enzymatic system in the same sol-gel matrix offers a complete ready-to-use biosensor.

ACKNOWLEDGMENTS

We thank to Dr. C. Reyes Mateo for her interesting commentaries about this work, which have contributed in a great extent to the compression of our problems. IP thanks to Juan de la Cierva Program of Spanish Ministry of Science.

REFERENCES

[1] Akyilmaz E., Sezgintürk M. K. & Dinçkaya E. (2003). A novel biosensor based on urate oxidase-peroxidase coupled enzyme system for uric acid determination in urine. *Talanta, 61*, 73-79.

[2] Arslam, F., Yasar, A. & Kilic, E. (2006). An amperometric biosensor for xanthine determination prepared from xanthine oxidase immobilized in polypyrrole film. *Artif. Cell Blood Sub. Immobil. Biotechnol., 34*, 113-128.

[3] Badjic, J. & Kostic, N. M. (1999). Effects of encapsulation in sol-gel silica glass on esterase activity, conformational stability, and unfolding of bovine carbonic anhydrase II. *Chem. Mater., 11*, 3671–3679.

[4] Barbacanne, M. A., Souchard, J. P., Darblade, B., Iliou, J. P., Nepveu, F., Pipy, B., Bayard, F. & Arnal, J. F. (2000). Detection of superoxide anion released extracellularly by endothelial cells using cytochrome *c* reduction, ESR, fluorescence and lucigenin-enhanced chemiluminescence techniques. *Free Radic Biol Med, 29*, 388-396.

[5] Borisov, S. M. & Wolfbeis, O. S. (2008). Optical Biosensors. *Chemical Reviews, 108(2)*, 423-461.

[6] Campanella, L., Martini, E. & Tomassetti, M. (2005). Antioxidant capacity of the algae using a biosensor method *Talanta, 66*, 902-9011.

[7] Choi, M. M. F. (2004). Progress in Enzyme-Based Biosensors Using Optical Transducers. *Microchim. Acta, 148*, 107-132.

[8] Chen, X. J., West, A. C., Cropek, D. M. & Banta, S. (2008). Detection of the superoxide radical anion using various alkanethiol monolayers and immobilized cytochrome c *Anal Chem, 80*, 9622-9629.

[9] Chen, P. Y., Vittal, R., Nien, P. C., Liou, G. S. & Ho, K. G. (2010). A novel molecularly imprinted polymer thin film as biosensor for uric acid. *Talanta, 8*, 1145-1151.

[10] Cui, Y., Barford, J. P. & Renneberg, R. (2007). Development of a glucose-6-phosphate biosensor based on coimmobilized p-hydroxybenzoate hydroxylase and glucose-6-phosphate dehydrogenase. *Biosens Bioelectron, 22*, 2754-2758.

[11] Di, J., Bi S. & Zhang, M. (2004). Third-generation superoxide anion sensor based on superoxide dismutase directly immobilized by sol–gel thin film on gold electrode. *Biosens Bioelectron, 19*, 1479–1486.

[12] Dimcheva, N., Horazova, E. & Jordanova, Z. (2002). An amperometric xanthine oxidase enzyme electrode based on hydrogen peroxide electroreduction. *Z. Naturforsch, 57*, 883-889.

[13] Eggers, D. K. & Valentine, J. L. (2001). Molecular confinement influences protein structure and enhances thermal protein stability. *Protein Science, 10*, 250-261.

[14] Ferrer, M. L., del Monte F. Y. & Levy, D. (2002). A novel and simple alcohol-free sol-gel route for encapsulation of labile proteins. *Chem. Matter., 14*, 3619.

[15] Gallarta, F., Sáinz, F. J. & Sáenz, C. (2007). Fluorescent sensing layer for the determination of L-malic acid in wine. *Anal. Bioanal. Chem., 387*, 2297–2305.

[16] Göpel, W., Hesse, J. & Zemel, J. N. (1991). Eds. *Sensors: A Comprehensive Survey,* VCH Publishers: *Weinheim, Germany*, *Vol. 2*, Part 1, 1-27.

[17] Guo, Z., Chen, J., Liu, H. & Zhang, N. (2005). Electrochemical determination of superoxide based on cytochrome c immobulized on DDAB-modified power microelectrode. *Anal Lett, 38*, 2033-2043.

[18] Gupta, R. & Chaudhury, N. K. (2007). Entrapment of biomolecules in sol-gel matrix for applications in biosensors: Problems and future prospects. *Biosens. Bioelectron., 22*, 2387–2399.

[19] Hlavay, J., Haemmerli, S. P. & Guilbault, G. G. (1994). Fiberoptic biosensor for hypoxanthine and xanthine based on a chemiluminescence reaction. *Biosens. Bioelectron., 9*, 189–195.

[20] Jerónimo, P. C. A., Araújo, A. N., Conceicao, M. & Montenegro, B. S. M. (2007). Optical sensors and biosensors based on sol-gel films. *Talanta, 72*, 13-27.

[21] Jin, W. & Brennan, J. D. (2002). Properties and application of proteins encapsulated within sol-gel derived materials. *Anal. Chim. Acta, 461*, 1-36.

[22] Kandimalla, V. B., Tripathi, V. S. & Ju, H. (2006). Immobilization of biomolecules in sol-gels: biological and analytical applications. *Crit. Rev. Anal. Chem., 36*, 73–106.

[23] Kirgöz, U. A., Timur, S., Wang, J. & Teleforcu, A. (2004). Xanthine oxidase modified glassy carbon paste electrode. *Electrochem. Commun., 6*, 913–916

[24] Martinez-Perez, D., Ferrer M. L. & Mateo C. R. (2003). A reagent less fluorescent sol-gel biosensor for uric acid detection in biological fluids. *Anal. Biochem., 322*, 238–242.

[25] Ndamanisha, J. C. & Guo, L. P. (2008). Electrochemical determination of uric acid at ordered mesoporous carbon functionalized with ferrocenecarboxylic acid-modified electrode. *Biosens. Biolectron., 23*, 1680-1685.

[26] Pastor, I., Esquembre, R., Micol, V., Mallavia, R. & Mateo, C. R. (2004). A reagent less fluorescent sol-gel biosensor for uric acid detection in biological fluids. *Anal. Biochem., 334*, 335–343.

[27] Pastor, I., Salinas-Castillo, A., Esquembre, R., Mallavia, R. & Mateo, C. R. (2010). Multienzymatic system immobilization in sol-gel slides: Fluorescent superoxide biosensors development. *Biosens. Biolectron., 25*, 1526–1529.

[28] Perello, J., Sanchis, P. & Grasses, F. (2005). Determination of uric acid in urine, saliva and calcium oxalate renal calculi by high-performance liquid chromatrography/mass spectrometry. *J. Chromatogr. B, 824*, 175-180.

[29] Pierre, A. C. (2004). The sol-gel encapsulation of enzymes. *Biocatal. Biotransform., 22*, 145-170.

[30] Rahman, A., Won, M. S. & Shim, Y. B. (2007). Xanthine sensors based on anodic and cathodic detection of enzymatically generated hydrogen peroxide. *Electroanalysis, 19*, 631-637.

[31] Rehak, H., Snejdarkova, M. & Otto, M. (1994). Application of biotin-streptavidin technology in developing a xanthine biosensor based on a self-assembled phospholipid membrane. *Biosens. Bioelectron., 9*, 337–341.

[32] Rich, R. L. & Myszka, D. G. (2007). Higher-throughput, label-free, real-time molecular interaction analysis. *Anal. Biochem.*, *361*, 1-6.

[33] Sakai-Kato K. & Ishikura, K. (2009). Integration of biomolecules into analytical sustems by means of silica sol-gel technology. *Analytical Sciences*, *25*, 969-978.

[34] Salinas-Castillo, A., Pastor, I., Mallavia, R. & Mateo, C. R. (2008). Immobilization of a trienzymatic system in a sol-gel matrix: A new fluorescent biosensor for xanthine *Biosens. Biolectron.*, *24*, 1059–1062.

[35] Singh, S., Singhal, R. & Malhotra, B. D. (2007). Immobilization of cholesterol esterase and cholesterol oxidase onto sol-gel films for application to cholesterol biosensor. *Anal. Chim. Acta*, *582*, 335-343.

[36] Thevenot, D. R., Toth, K., Durst, R. A. & Wilson, G. S. (2001). Electrochemical biosensors: recommended definitions and classification. *Biosens. Bioelectron.*, *16*, 121-131.

[37] Tsai, H. & Doog, R. (2004). Simultaneous determination of renal clinical analytes in serum using hydrolase- and oxidase-encapsulated optical array biosensors. *Anal. Biochem*, *334*, 183-192.

[38] Tsai, H. & Doog, R. (2005). Simultaneous determination of pH, urea, acetylcholine and heavy metals using array-based enzymatic optical biosensor. *Biosens. Bioelectron.*, *20*, 1796-1804.

[39] Velasco Garcia, M. N. & Mottram, T. T. (2003). Biosensor technology addressing agricultural problems. *Biosystems Engineering*, *84*, 1-12.

[40] Villalonga, R., Camacho, C., Cao, R., Hernández, J. & Matías, J. C. (2007). Amperometric biosensor for xanthine with supramolecular architecture. *Chem. Commun.*, 942-944.

[41] Wang X., Yao Y., Zhang. J, Zhu Z. & Zhu J. (2008). Uric Acid Detection Using Quartz Crystal Microbalance Coated with Uricase Immobilized on ZnO Nanotetrapods. *Sensor. Mater.*, *20*, 111-121.

[42] Wolfbeis, O. S., Oehme, I., Papkovskaya, N. & Klimant, I. (2000). Sol-gel based glucose biosensors employing optical oxygen transducers, and a method for compensating for variable oxygen background. *Biosens. Bioelectron.*, *15*, 69-76.

[43] Wollenberger, U., Schubert, F., Pfeiffer, D. & Scheller, F. W. (1993). Enhancing biosensor performance using multienzyme systems. *Tibtech*, *11*, 255-262.

[44] Wu, F., Huang, Y. & Li, Q. (2005). Animal tissue-based chemiluminescence sensing of uric acid. *Anal. Chim. Acta*, *536*, 107-113.

[45] Xu, Z., Guo, Z. & Dong, S. (2005). Electrogenerated chemiluminescence biosensor with alcohol dehydrogenase and tris(2,20-bipyridyl)ruthenium(II) immobilized in sol-gel hybrid material. *Biosens. Bioelectron.*, *21*, 455–461.

[46] Xu, Z., Chen, X. & Dong, S. (2006). Electrochemical biosensors based on advanced bioimmobilization matrices. *Trac-Trend Anal. Chem.*, *25*, 899-908.

[47] Zhao, J., O'Daly, J. P., Henkens, R. W., Stonehuerner, J. and Crumbliss, A. L., 1996. A xanthine oxidase colloidal gold enzyme electrode for amperometric biosensor applications. *Biosens. Bioelectron.*, 11, 493–502.

In: The Sol-Gel Process
Editor: Rachel E. Morris

ISBN 978-1-61761-321-0

Chapter 18

DYE-DOPED SOL-GELS FOR THE OPTICAL SENSING TOWARDS ALIPHATIC AMINES

***Wenqun Wang*[1], *Kwok-Fan Chow*[1], *Alvin Persad*[1], *Ann Okafor*[1], *Andrew Bocarsly*[2], *Neil D. Jespersen*[1] and *Enju Wang*[1*]**
[1]Department of Chemistry, St. John's University, Jamaica, NY 11439, USA
[2]Department of Chemistry, Princeton University, Princeton, NJ 08540, USA

ABSTRACT

Bromocresol purple (CPR) and chlorophenol red (BCP) encapsulated in the sol-gel matrix were successfully applied to optical response towards aliphatic amines. The sol-gel matrix obtained by acidic hydrolysis of tetraethoxysilane and phenyltriethoxysilane in the presence of selected dyes was further spin-coated onto glass slides. The coating process was repeated as needed. The optical sensors with a single coating and double coatings showed similar long-term stability, but they differed in response times and sensitivities. These sensors' selectivity sequence towards amines is: methylamine > ethylamine > propylamine > butylamine > triethylamine. The response time for methylamine ($t_{95\%}$) is 75 s, and its detection limit is ~5 $\times 10^{-5}$ M in aqueous solution. The CPR and BCP doped sol-gel sensors showed good reproducibility, with a useful lifetime of four weeks.

Keywords: Aliphatic amines, methylamine, sol-gel sensors, bromocresol purple, chloro-phenol red

1. INTRODUCTION

Aliphatic amines are widely used in the crude oil, natural gas, food, agricultural, and pharmaceutical industries. Therefore, they can be found in their production waste stream. The

* Corresponding author: Tel. 718-990-5225, Fax: 718-990-1876, E-mail: wange@stjohns.edu

smell of old fish, urine, rotting flesh, and semen is mainly due to gaseous amines. Aliphatic amines are very toxic, and can cause environmental pollution and some serious diseases [1, 2].

The detection of aliphatic amine pollutants by sensitive and reversible sensors has become a field of great interest during the past several years. The application of sensors in analysis eliminates the need to transport specimens, reduces the effort required to identify the analyte, and provides immediate results. Electrochemical sensors, enzyme sensors and piezocrystal detectors have been reported in the literature [3-8]. The electrochemical sensors require reference and electrical signals, and they are not easy to manufacture and miniaturize. Several optical sensors have been studied by using conjugated macrocycles such as calixarene, porphyrins, and pH-sensitive dyes as the sensing reactants [9-17]. The interaction between the analyte and the reactive host compound incorporated into a polymer film involves a reversible chemical reaction. Most of these sensors showed higher response to long-chain amines, e.g. butylamine and hexylamines, and the responses to methylamine or ethylamine are poor. A recent paper by Oberg *et. al.* [18] reported an amine vapor sensor based on bromocresol green adsorbed on silica micro-spheres, and this sensor showed its best response to diethylamine and diethylamide, while having only a moderate response to *tert*-butylamine. No data was reported for methylamine and ethylamine.

Immobilization of a reagent into a sol-gel matrix has been widely been to develop optical sensors [19-25]. Compared to organic polymers, sol-gels as host matrices provide better stability, optical transparency, flexibility, and permeability to small ions and molecules. The matrix is porous enough to allow external analyte species to diffuse into the network and react with the entrapped reagent. In our laboratory we have demonstrated that doping appropriate pH indicators into sol-gel matrix yielded reproducible and fast response sensors to pH, hydrogen chloride gas, and ammonia gas [26-28]. Furthermore, we found that the addition of a small amount of phenyltriethoxysilane (Ph-triOES) can greatly enhance the entrapment of the phenol red dye in the TEOS-based sol-gel matrix, and therefore, the resulting sol-gel sensors have longer and more stable performance. This report will document our findings that other weak acidic dyes doped in TEOS and Ph-triOES sol-gel matrix show high response to small amine molecules, i.e. methylamine and ethylamine. The porous structure and polar nature of silica gel greatly enhance selectivity towards these amines.

2. Experimental

2.1. Chemicals and Reagents

Bromocresol purple, phenol red, bromothymol blue, cresol red and chlorophenol red, tetraethoxysilane (TEOS, > 99.9%), ethanol (99.5%), concentrated amine solutions (ranges 65% to 70% in aqueous solution or 99% pure) and amine hydrochloride salts (> 99%) (methylamine, ethylamine, diethylamine, triethylamine, propylamine) were obtained from Sigma-Aldrich. Phenyltriethoxysilane (Ph-triOES) (>97%) was obtained from Gelest, Inc. Nitrogen gas (99.9%) was obtained from Prest-o-Sales (Long Island City, NY). Hydrochloric acid and sodium hydroxide were purchased from Fisher Scientific. The dyes used are of the highest purity reagent grade with purity range from ~ 90 to 95%. All chemicals were used as obtained without further purification.

2.2. Preparations

2.2.1. Amine aqueous solutions

For aqueous amine solution response: amine solutions of 1.00 M were prepared by diluting the appropriate volume of each pure amine liquid or amine aqueous solution with de-ionized water. Other low concentrations in the range of 10^{-1} to 10^{-5} M of the amine solutions were prepared by serial dilution.

2.2.2. Amine gases samples

For amine gas response: first, the appropriate amount of the corresponding amine hydrochloride salt was dissolved into distilled water to make a 1.00 M solution. Lower concentrations of these salts in the range of 0.100 to 10^{-6} M were prepared by serial dilution. Then the amine salts were mixed with an equal number of moles of NaOH to produce the corresponding amine at the time of measurement.

2.2.3. pH 5 universal buffer

A universal buffer was prepared by dissolving 4.946 g boric acid, 28.400 g sodium sulfate, 6.560 g sodium acetate, 21.599 g sodium phosphate dibasic and 8.560 g sodium carbonate in 2.00 L volumetric flask with de-ionized water [26]. The solution was adjusted to pH 5 with 6 M sulfuric acid using an Orion pH-meter (Acurmet 20) and an Orion pH combination electrode, and then diluted to 2.00 L. Standard buffers of pH 4 (acetic acid-acetate), pH 7 (phosphates) and pH 10 (borate and carbonate) were used for the standardization of the pH electrode.

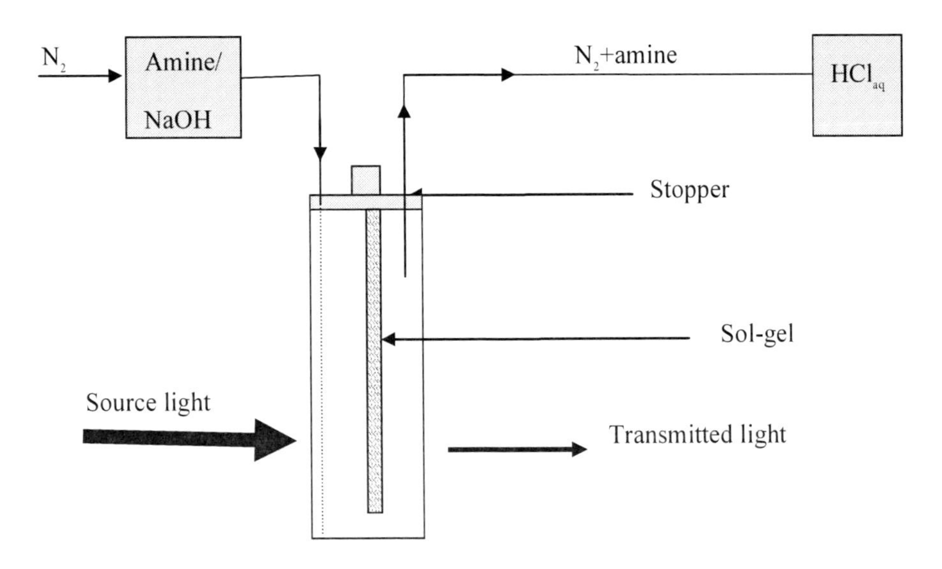

Figure 1. Schematic of the device for amine gas detection.

Table 1. Absorbance of various dye-doped sol-gel slides (double layered) in 0.1 M methylamine solution.

Dye	Wavelength (nm)	Absorbance change
Phenol red	560	0.063
Cresol Red	570	0.073
Bromothymol Blue	610	0.053
Bromocresol Purple	590	0.221
Chlorophenol	578	0.202

2.2.4. Preparation of dye solutions

Bromocresol purple or chlorophenol red stock solution was prepared by dissolving 1.00 mg of bromocresol purple or chlorophenol red in 200 mL of de-ionized water. These solutions were further diluted 1:4 with de-ionized water. A 3.00 mL aliquot of these diluted dye solutions were mixed with 50 μL of 0.100 M aqueous amine solution in a standard glass cuvette and the absorbance of the mixture was measured at the peak absorbance wavelength of the respective dye. De-ionized water was used as a reference.

2.2.5. Preparation of sol-gel sensors

The particular sol-gel composition is based on previous work using TEOS and a small amount of Ph-triOES as precursors [26-27]. Sol-gel solutions were prepared by mixing 20.0 mg of dye, 4.00 mL of tetraethoxysliane, 5.00 mL of ethanol, 1.60 mL of 0.100 M hydrochloric acid, and 200 μL of phenyltriethoxysilane. The mixture was stirred at room temperature for approximately 2 hours. 200 μL of the resulting sol-gel solution was placed on a pre-cleaned microscopic slide (7.5 cm × 2.5 cm) and then spin cast by using a homemade spinning machine at ~1000 RPM for 10 seconds for single layered slides. For double layered slides, the single layered slide was dried for two days under ambient conditions and a secondary coating was performed on top of it. For triple layered coatings, the process was repeated one more time. The coated slides were allowed to age for more than three days under ambient conditions before use. The resulting sol-gel slides were clear and transparent. The sol-gel slides were cut into small pieces to fit into a 1 × 1 × 4.4 cm^3 cuvette for the tests. Scanning Electron Microscopy (Polaron SEM system) and atomic force microscopy were used to study the morphology of the sol-gel surface. A typical double layer film had a thickness of approximately 2 μm. The AFM scan showed the surface of the sol-gel film is fairly rough.

2.3. Measurements

Absorbance measurements were recorded on a UV-Vis double beam (SHIMAZU DU-2000) or Cary 14 spectrophotometer. The sol-gel coated glass slides were placed in a commercial quartz cell (1cm × 1cm × 4.4 cm). In most cases, sol-gel coated glass slides were first soaked for at least 5 minutes in a pH 5.0 universal buffer to ensure a stable absorbance value before the first measurement was made. Absorbance spectra of sensor slides were taken from 350 nm to 700 nm. The UV/vis absorbance response of bromocresol purple doped sol-gel slide was recorded at 590 nm and that of the chlorophenol red doped sol-gel slides was recorded at 578 nm. The spectrophotometric base line was taken by setting zero absorbance at 700 nm with the double beam spectrophotometers. All measurements were done under ambient conditions that were approximately 23°C and 50-70% relative humidity.

For the detection of amines in solutions, tests were performed by immersing the sol-gel coated glass slides in 4.00 mL of an amine solution in a standard cuvette. For the methylamine gas response, the sol-gel coated glass slide was placed in the center of a sealed cuvette, the experiments were performed using a device depicted in Figure 1. The amine gas was generated by mixing 5.00 mL of various concentrations of methylamine hydrochloride solutions with 5.00 mL of the sodium hydroxide solution. Compressed nitrogen gas (90 mL/min) was used to purge the amine gases out of the mixture. The gas was then passed through the sample cuvette containing the sol-gel slide using an inlet tube extending to the bottom of the cuvette. An outlet tube directed the amine gases to the trap containing 0.1 M HCl solution. The absorbance at the absorption maxima was recorded for 10 minutes after the onset of the gas flow to ensure steady state response. The absorbance was measured against air as the reference.

3. Results and Discussion

3.1. Dye Selection

Dyes with pK_a values lower than 8.5 in aqueous or ethanol solutions and having distinctive color changes upon deprotonation were selected for incorporation into the silica sol-gel solution. After the sol-gel solution was cast on the glass and dried, these dye-doped sol-gel slides were tested with 0.1 M methylamine solution and the performance of each sensor was observed. When the amine in solution interacts with the dye, deprotonation of the dye occurs, which results in a color change of the sol-gel and a corresponding shift in the absorption maxima. UV/Vis absorbance spectra in the range of 350 nm to 700 nm were obtained for the freshly coated sol-gel slides and after they were treated with aqueous amine solution.

Bromocresol purple, phenol red, bromothymol blue, cresol red and chlorophenol red, having pK_a values 6-8 [29] in aqueous or ethanol solution, were successfully doped into sol-gels. These dye-doped sol-gels had significant responses towards methylamine. They all have clear color changes when exposed to methylamine. However, as shown in Table 1, the absorbance changes of phenol red, bromothymol blue and cresol red were much smaller than bromocresol purple and chlorophenol red. Therefore, bromocresol purple (BCP) and chlorophenol red (PCR) were chosen for further studies.

3.2. Response Principle

Bromocresol purple and chlorophenol red are pH indicators. They belong to sulfonphthalein dye family. They are weak organic acids, and their conjugate bases have quite different spectra from their acid form. When these dyes come in contact with amines, deprotonation occurs causing the changes of color. Both BCP- and PCR-doped sol-gel slides showed significant decrease of the peak at around 450 nm and a peak appeared in the range of 500 to 650 nm. The structures of the two dyes are shown below:

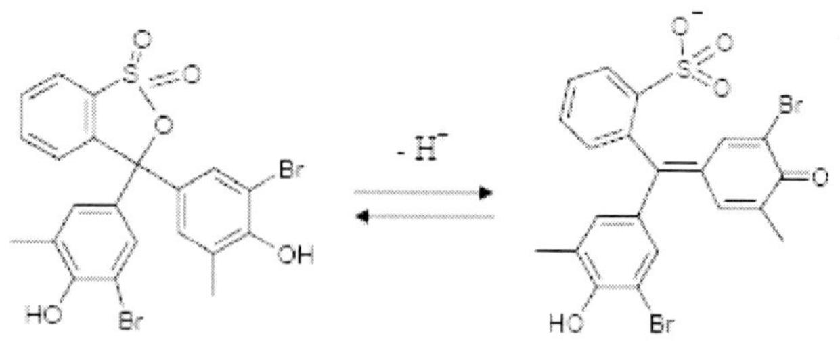

Bromocresol purple (BCP).

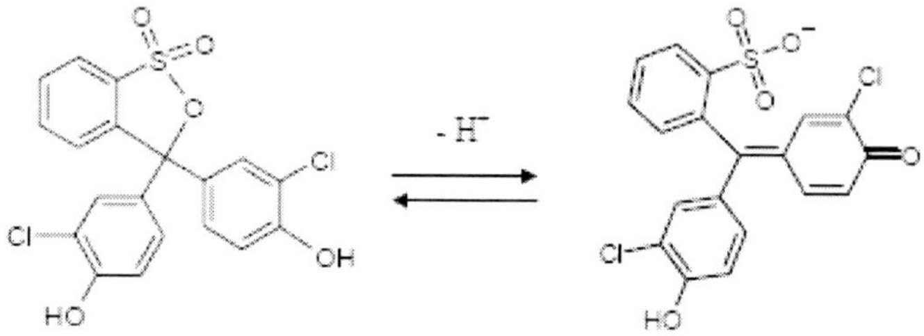

Chlorophenol red (CPR).

The sol-gel membrane response to methylamine can be described using the equation below:

$$CH_3NH_{2(aq)} + HIn_{(gel)} \leftrightarrows In^-CH_3NH_3^+{}_{(gel)} \tag{1}$$

Here, HIn is the protonated form of the dye, and In^- is its deprotonated form. The reaction equilibrium constant depends on the K_a^{HIn} values of the dye; a higher K_a^{HIn} (lower pK_a^{HIn}) value will result in a larger degree of deprotonation of the dye when exposed to methylamine of the same concentration:

$$K_{eq} = \frac{[In^-CH_3NH_3^+]}{[CH_3NH_2][HIn]} = \frac{K_a^{HIn}}{K_a^{CH_3NH_3^+}} \tag{2}$$

The sol-gel absorbance (A) is dependent on the dye concentration and the position of the deprotonation equilibrium. Higher amine concentrations result in a higher degree of deprotonation of the dye and a greater change in the absorbance. By using relative absorbance α [30] to represent the fraction of total dye concentration (C_T) in the deprotonated form ([C^-]), α can be related to the absorbance values at a given wavelength as follows:

$$\alpha = \frac{[In^-]}{[In_T]} = \frac{(A - A_0)}{(A_1 - A_0)} \quad (3)$$

A_0 and A_1 correspond to the absorbance of the completely deprotonated and protonated dye, respectively. The α value will be dependent only on the CH_3NH_2 concentration in the samples, but not on the sol-gel thickness or dye concentration in the sol-gel.

3.3. Influence of Gel Thickness on the Response

When making the sol-gels, single layer to triple layers of sol-gel coatings were prepared to achieve optimal sensor performance, such as film thickness, transparency, and long lifetime without cracking during measurements or storage. While the triple layer coated sol-gel films cracked within a week, single layer and double layer coated films can be stored over three months without cracking. Therefore, their response properties towards amines were evaluated in further detail. As shown in Figure 2, the response signal change of double layer coated sol-gels is almost twice as big as that of single layer coated sol-gels. On the other hand, the double layer coated sol-gels showed an increased response time. Since BCP-doped sol-gels (BCP-gel) showed identical trends, only the responses of CPR-doped sol-gels (CPR-gel) are shown in the graphs in this paper for clarity. The response times ($t_{95\%}$) of both BCP- and CPR-doped sol-gel slides towards the 0.100 M methylamine solutions were less than 50 seconds for single layer coated slides and less than 90 seconds for double layer coated slides. This indicates that the amine molecules need a longer time to diffuse through and to reach equilibrium in the thicker gels. Since the double layer coated slides showed acceptable response time and doubled sensitivity towards methylamine, these slides were further investigated for potential practical applications. Single layer coated slides were studied for the selectivity and response trends.

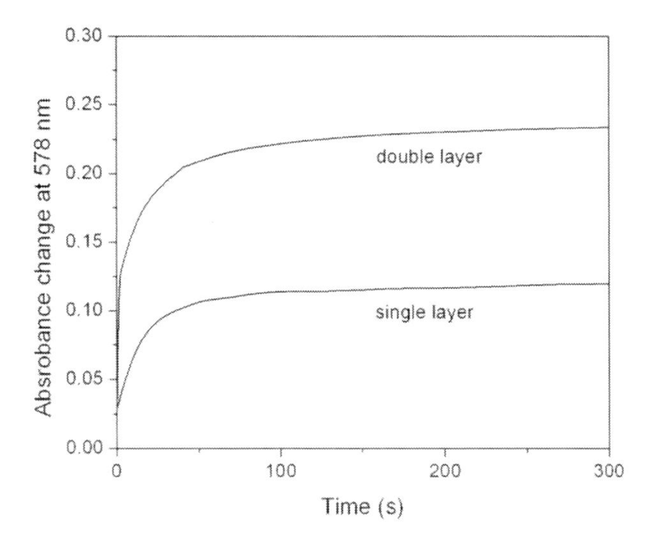

Figure 2. Time response of the single and double layer coated CPR-gel response to 0.100 M methylamine solution.

3.4. Response towards Different Aliphatic Amines

The response of sol-gel coated slides towards amines of various sizes was evaluated. Regardless of weather single coated or doubly coated, doped with BCP or CPR, all of these slides showed similar tendencies when exposed to 0.100 M aqueous amine solutions, i.e., the smaller the molecular size of the amine, the stronger the response of the sensors. The observed selectivity sequence is methylamine > ethylamine > propylamine > butylamine ≥ triethylamine. A typical response is illustrated in Figure 3 where a double layer coated CPR-gel was exposed to various 0.100 M amine solutions. Single layer coated sol-gels showed the same trend, but with half of the sensitivity.

The response times showed a similar trend, i.e., the response time became longer, as the molecular size of amine increased (Table 2). Furthermore, the response data (selectivity) expressed in the relative absorbance to various amines in aqueous solutions is shown in Table 3. Larger amines not only responded very weakly, they also took much longer to reach steady state.

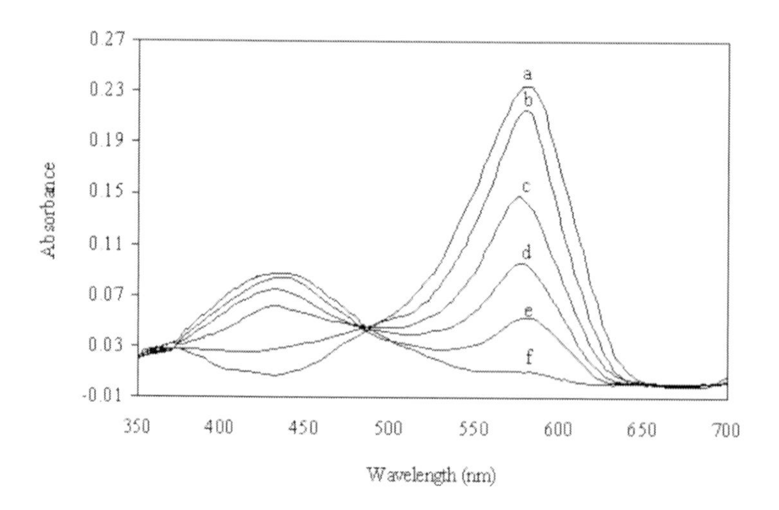

Figure 3. Spectra of a double layer coated CPR-gel slide when exposed to different 0.100 M amines in solutions: (a) methylamine; (b) ethylamine; (c) propylamine; (d) diethylamine; (e) triethylamine; (f) pH 5 buffer (blank).

Table 2. Response time ($t_{95\%}$) of double layer coated sol-gels towards 0.1 M amine solutions.

Amine in solutions	Response time (seconds)	
	BCP-gel	CPR-gel
Methylamine	78	89
Ethylamine	113	116
Propylamine	118	120
Butylamine	126	134
Triethylamine	196	206

Table 3. Comparison of absorbance data for dye solution and sol-gel response to 0.100 M various aqueous amines.

Amine in solution	α at 590 nm		α at 578 nm	
	BCP solution	BCP-gel	CPR solution	CPR-gel
None/Blank	0	0	0	0
Methylamine	1.00	0.85	1.00	0.90
Ethylamine	1.00	0.54	1.00	0.69
Propylamine	1.00	0.22	1.00	0.57
Butylamine	1.00	0.13	1.00	0.36
Triethylamine	1.00	0.10	1.00	0.21

Table 4. Absorbance data for BCP-gel and CPR-gel response to amine gases produced from 0.100 M amine hydrochloride salts.

Base in solution	α at 590 nm (BCP-gel)	α at 578 nm (CPR-gel)
Ammonium	1.00	1.00
Methylamine	0.97	0.94
Ethylamine	0.95	0.86
Propylamine	0.78	0.71
Diethylamine	0.25	0.30
Triethylamine	0.13	0.17

The above results are quite different from the behavior of these dyes in solution. No difference in absorbance was found when the diluted dye solution was treated with 0.100 M solutions of various amines (See Table 3). Even at 1 mM concentration, all the amines can deprotonate the dyes completely.

The response of sol-gel sensor to amines in the gaseous phase was also studied. Here, the gaseous amines were generated by mixing 0.100 M amine hydrochloride and 0.10 M of sodium hydroxide, and then purging the mixture with N_2 gas. The response sequence to amine gases was found to be: methylamine $\geq$ ethylamine > propylamine > butylamine > diethylamine > triethylamine as shown in Table 4.

All of the tested amines have about the same basicity (K_b values around 10.5) [29] and their interaction with the dye in aqueous solution is also very similar (Table 3). The differences in the response of amines at the sol-gels are attributed to the film porous solid structure, where diffusion played a very important role. From methylamine to butylamine, these linear amines increase in lipophilicity. And from ethylamine to triethylamine, the molecular size increases greatly along with the molecular mass. As the molecular size increased, the response of the dye-doped sol-gel films decreased as shown in a smaller absorbance increase at peak 590 nm for BCP-gels and 578 nm for CPR-gels (see Figure 3). It suggests that the sol-gel film plays a role as a molecular size selection barrier. Another factor that affects the response is the increased hydrophobicity of larger amines. The sol-gel matrix has been demonstrated to be a hydrophilic environment. In the sol, the pore solvents are aqueous alcohol, and subsequent drying leads to enrichment of water in the pores. When the residual solvent evaporates, some pores collapse around the dopant molecules and hydrated porous silica forms around the dopants. Amine molecules must pass through the hydrated pore to reach the doped sensing element [31-33]. The less hydrophilic amines should permeate the polar hydrophilic sol-gel matrix less easily, resulting in the decrease of the absorbance response and a longer response time.

3.5. Response towards Methylamine in Aqueous Solution and in Gaseous Phase

Since the sensor had the highest response to methylamine, further quantitative measurements were performed on BCP- and CPR-gels towards methylamine. The response curves expressed in relative absorbance versus log methylamine concentration are s-shaped curves. Figure 4 shows the response curves of CPR-gel. When CPR-gel slides were dipped in aqueous methylamine solution, there is a rapid absorbance change in the concentration range from 5 x 10^{-5} to 5 $\times 10^{-4}$ M, and the response is almost linear to log methylamine concentration. When BCP-gel slides were dipped in aqueous amine, rapid responses were found in the range of 5.0×10^{-5} M to 1.0×10^{-3} M and a slow absorbance change in the range of 1.0×10^{-3} M to 1.0 M. The lowest detectable concentration of aqueous methylamine is 5.0x10^{-5} M by the BCP-gel sensor, and 3.0 $\times10^{-5}$ M by the CPR-gel.

For the gas phase response (Figure 4), gaseous methylamine was generated by mixing methylamine hydrochloride with NaOH, and purged out of this solution using N_2 gas. The concentrations shown in the curves were the initial amine concentration. The response to initial methylamine in basic solution shifted to higher concentration ranges: 1 $\times10^{-3}$ M ~ 4 $\times10^{-3}$ M. The actual methylamine gas concentration in the N_2 gas stream was estimated to be 100 times less than that in solution-based on methylamine collected using HCl solution trap. Thus, the actual detection limit should be less than 10 ppm of methylamine. BCP-gel slide showed a similar response pattern and the response range was identical to the PCR-gels.

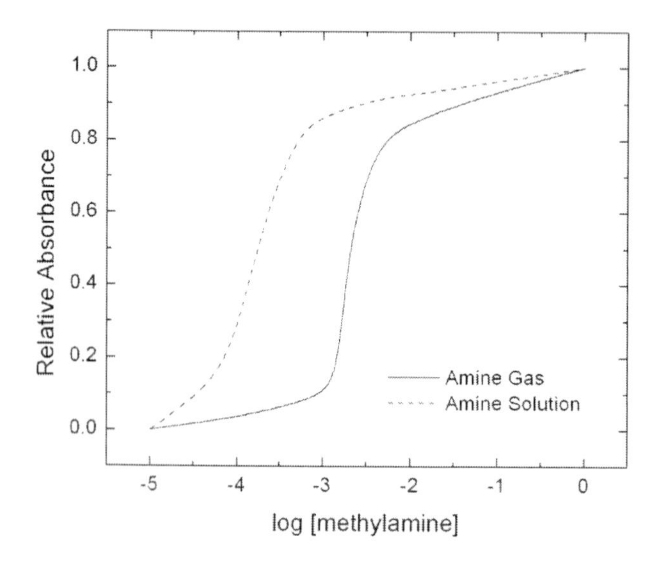

Figure 4. Calibration curves of the double layer coated CPR-gel response to methylamine in gas phase produced from the methylamine hydrogen chloride mixed with NaOH and in solution phase as dissolved amine in water.

The response time of the sensors to methylamine in solutions and in the gas phase were also evaluated. Response times ($t_{95\%}$) were measured on the double layer coated sensor slides. It was found that the response to gaseous methylamine is faster than in aqueous solution: $t_{95\%}$ = 89 s in solution phase vs. 51 s in the gas phase for CPR-gels, and $t_{95\%}$ =78 s vs 45 s for BCP-gels. This phenomenon may be attributed to faster diffusion of methylamine through the pores of the sol-gel matrix. In the solution phase, all of the amine molecules are available to diffuse into the sol-gel, however, the sol-gel matrix pores are saturated with water molecules. The larger and less polar amine molecules have to replace the small and polar water molecules in the polar sol-gel pores to reach the dye. Gas phase amines encounter a moisturized film and replacement of water molecules is not needed. In solution, the unfavorable water replacement process results in a slower penetrating of amines through the sol-gel matrix pores to reach the doped dye. Similar phenomenon has been observed in oxygen sensing using sol-gel matrix [34].

3.6. Reversibility of the Sensor

The sensors can be reversed with a low pH buffer (pH below 7). After contacting with a buffer of pH 5.0, the absorbance spectra were taken in the range from 350 nm to 700 nm on UV-Vis spectrometer. The maximum absorbance wavelength went back to 440 nm and the absorbance at 590 nm for BCP-gels and 578 nm for CPR-gels decreased to almost zero. The dyes converted back to their acid form by taking H^+ from the acidic buffer.

3.7. Reproducibility of the Sensor

3.7.1. Intra-day reproducibility

BCP-gel and CPR-gel slides were immersed in 0.100 M methylamine aqueous solution and absorbance of the slides were measured; then the slides were reversed with pH 5.0 universal buffer solution. The cycle was repeated every hour to check the intra-day reproducibility. The relative standard deviation (RSD% n = 8) was calculated and a value of 1.7 % for CPR-gel, 0.9 % for BCP-gel was obtained. This suggests that the sensors have excellent intra-day reproducibility.

3.7.2. Long-term reproducibility (lifetime of the sensor)

To investigate the lifetime of the sensors, the sol-gel coated slides were examined in 0.100 M methylamine solution for a period of 12 weeks. The absorbance of the slides was measured every week, and the slides were reversed with pH 5.0 universal buffer solutions after each measurement. Between each measurement, the slides were stored in ambient air, dried and shielded from the light. The test results showed that the slides were stable for the first four weeks. After four weeks, the absorbance responses started to decrease slightly. Therefore, a new calibration curve should be prepared before each quantitative testing.

CONCLUSION

Dyes with pK_a of 6-8 in aqueous or ethanol solution have been successfully encapsulated in the sol-gel matrix with TEOS and Ph-TriOES as precursors under acid hydrolysis conditions. Sol-gels doped with dyes of lower pK_a values (pK_a = ~6) give the most sensitive response toward amines. Double layer coated sensors showed enhanced response signal and slightly increased response time. Unlike many of the polymer membrane-based sensors, the sol-gel matrix described here functions as a molecular-size restricted sieve and the smallest methylamine gave the highest response. The CPR- and BCP-gel sensors showed good intra-day reproducibility, with a useful lifetime of four weeks.

ACKNOWLEDGMENTS

The authors thank Dr. Joseph Seraphin for taking the AFM images and Dr. Guozhi Wang for his technical advisement and proof reading of this paper.

REFERENCES

[48] Greim, H; Bury, D; Klimisch, HJ; Oeben-Negele M; Zeigler-Skylakakis, Z. Toxicity of aliphatic amines: structure-activity relationship, *Chemosphere*, 1998, 36, 271-295.

[49] Gong, WL; Sears, KJ; Alleman, JE; Blatchley, ER. Toxicity of model aliphatic amines and their chlorinated forms, *Environ. Toxicol. Chem.*, 2004, 23, 239-244.

[50] Opdycke, WN; Park, SJ; Meyerhoff, ME. Polymer-membrane pH electrodes as internal elements for potentiometric gas-sensing systems, *Anal. Chim. Acta,* 1983, 155, 11-20.

[51] Surmann, P; Peter, B. Carbon electrodes in amine analysis: effect of chemically-modified carbon surfaces on signal quality and reproducibility, *Electroanalysis*, 1996, 8, 685–691.

[52] Casella, IG; Rosa, S; Desimoni, E. Electro-oxidation of aliphatic amines and their amperometric detection in flow injection and liquid chromatography at a nickel-based glassy carbon electrode, *Electroanalysis*, 1998, 10, 1005–1009.

[53] Niculescu, M; Nistor, C; Frebort, I; Pec, P; Mattiasson, B; Csoregi, E. Redox hydrogel-based amperometric bienzyme electrodes for fish freshness monitoring, *Anal. Chem.,* 2000, 72, 1591–1597.

[54] Mirmohseni, A; Oladegaragoze, A. Construction of a sensor for determination of ammonia and aliphatic amines using polyvinylpyrrolidone coated quartz crystal microbalance, *Sens. Actuators B Chem*., 2003, 89, 164-172.

[55] Charlesworth, JM; McDonald, CA. A fiber optic fluorescing sensor for amine vapors, Sens. *Actuators B Chem*., 1992, 8, 137–152.

[56] McCarrick, M; Harris, SJ; Diamond, DJ. Assessment of a chromogenic calyx[4]arene for the rapid colorimetric detection of trimethylamine, *Mater. Chem*., 1994, 4, 217–221.

[57] Papkovsky, DB; Pomonarev, GV; Wolfbeis, OS. Longwave luminescent porphyrin probes. Spectrochim. *Acta*, Part A, 1996, 52, 1629-1638.

[58] Charlesworth, JM; McDonald, CA. A fibre-optic fluorescing sensor for amine vapors, *Sens. & Actuators,* B., 1992, 8, 137-142.

[59] Mohr, GJ; Demuth, C; Spichiger-Keller, UE. Application of chromogenic and fluorogenic reactands in the optical sensing of dissolved aliphatic amines, *Anal. Chem.*, 1998, 70, 3868-3873.

[60] Mohr, GJ; Tirelli, N; Spichiger-Keller, UE. Plasticizer-free optode membranes for dissolved amines based on co-polymers from alkyl methacrylates and the fluoro reactant ETHT 4014, *Anal. Chem.*, 1999, 71, 1534-1539 .

[61] Qin, W; Parzuchowski, P; Zhang, W; and ME; Meyerhoff, ME. Optical sensor for amine vapors based on dimer-monomer equilibrium of indium(III) octaethylporphyrin in a polymeric film, *Anal. Chem.*, 2003, 75, 332 - 340.

[62] Liu, CJ; Lin, JT; Wang, SH; Jiang JC; Lin, LG. Chromogenic calixarene sensors for amine detection, *Sens. and Actuators B*, 2005, 108, 521-527.

[63] Kang, Y; Meyerhoff, ME. Rapid response optical ion/gas sensors using dimer–monomer metalloporphyrin equilibrium in ultrathin polymeric films coated on waveguides, *Analytica Chimica Acta*, 2006, 565, 1-9

[64] Gräfe, A; Haupt, K; Mohr, GJ. Optical sensor materials for the detection of amines in organic solvents, *Analytica Chimica Acta*, 2006, 565, 42-47.

[65] Oberg, KI; Hodyss, R; Beauchamp, JL. Simple optical sensor for amine vapors based on dyed silica microspheres, *Sens. and Actuators B,* 2006, 115, 79-85.

[66] Collinson, MM. Sol-gel strategies for the preparation of selective materials for chemical analysis, *Crit. Rev. Anal. Chem.*, 1999, 29, 289–311.

[67] von Bultzingslowen, C; McEvoy, AK; McDonagh, C; MacCraith, BD; Klimant, I; Krause, C; Wolfbeis, OS. Sol–gel based optical carbon dioxide sensor employing dual luminophore referencing for application in food packaging technology, *Analyst*, 2002, 127, 1478-1483.

[68] Butler, TM; MacCraith, BD; McDonagh, CM. Development of an extended range fiber optic pH sensor using evanescent wave absorption of sol-gel entrapped pH indicators. *Proc. of SPIE,* 1995, 2058, 168-178.

[69] Makote, R; Collinson, MM. Organically-modified silicate films for stable pH sensors, *Anal. Chim. Acta,* 1999, 394, 195-200.

[70] Noire, MH; Bouzon, C; Couston, L; Gontier, J; Marty P; Pouyat, D. Optical sensing of high acidity using a sol–gel entrapped indicator *Sens. & Actuators B*, 1988, 51, 214-219.

[71] Lin, J; Brown, CW. Sol-gel glass as a matrix for chemical and biochemical sensing, *Trends Anal. Chem.*, 1997, 16, 200-211.

[72] Shahriari, MR; Ding, JY; in: Klein; LC. (Ed.), *Sol-gel optics: processing and applications*, Kluwer Academic Publishers, Boston, 1994, Ch. 13, 279.

[73] Wang, E; Chow, KF; Kwan, V; Chin, T; Wong, C; Bocarsly, A. Fast and long-term optical sensors for pH-based on sol-gels, *Anal. Chim. Acta,* 2003, 495, 45-50.

[74] Wang, E; Chow, KF; Wang, W; Wong C; Yee, C; Persad, A; Mann, J; Bocarsly, A. Optical sensing of HCl with phenol red doped sol-gels, *Anal. Chim. Acta,* 2005, 534, 301-306.

[75] Persad, A; Chow, KF; Wang, W; Wang, E; Okafor, A; Jespersen, N; Mann' J; Bocarsly, A. Investigation of dye-doped sol–gels for ammonia gas sensing, *Sens. Actuators B,* 2008, 129, 359-363.

[76] CRC *Handbook of Chemistry and Physics*, 84th ed. 2003, P8-48.

[77] Morf, WE; Seiler, K; Rusterholz, B; Simon, W. Design of a calcium-selective optode membrane based on neutral ionophores, *Anal. Chem.*, 1990, 62, 738-742.

[78] Hench LL; West, JK. The sol-gel process, *Chem. Rev.*, 1990, 90, 33-72.

[79] Samuel, J; Polevaya, Y; Ottolenghi, M; Avnir, D. Determination of activation energy of entrance into micropores: quenching of the fluorescence of pyrene-doped SiO_2 sol-gel matrixes by oxygen, *Chem. Mater.*, 1994, 6, 1457-1461.

[80] Dunn, B; Zink. JI. Probes of pore environment and molecule-matrix interactions in sol-gel materials, *Chem. Mater.*, 1997, 9, 2280-2291.

[81] McDonagh, C; MacCraith, BD; McEvoy. AK. Tailoring of sol-gel films for optical sensing of oxygen in gas and aqueous phase, *Anal. Chem.*, 1998, 70, 45-50.

In: The Sol-Gel Process
Editor: Rachel E. Morris
ISBN 978-1-61761-321-0

Chapter 19

SOL-GEL BASED SOLID STATE DYE LASER— PAST, PRESENT AND FUTURE

Uday Kumar
Indian Institute of Science Education and Research-Kolkata,
West Bengal, India

ABSTRACT

The first part of the chapter is associated with the necessary and sufficient back ground which is quite often needed to understand the rest of the matter. It starts from solution based dye laser system with detail discussion and ends on the concept of solid state dye laser.

The second part has been devoted only on the development of sol-gel based solid state dye laser. It begins with an important introductory section on sol-gel and then dye molecules impregnation process; spectroscopic properties, lasing performance, and photostability are discussed into detail in different sections. Ultimately, the chapter ends with open questions and future research direction.

PART-I: INTRODUCTION TO LASER/DYE LASER

1. Laser Fundamental

The phenomenon of Light Amplification by Stimulated Emission of Radiation has been abbreviated as LASER in the field of coherent optics. The high degree of coherence, directionality, monochromaticity and brightness are the most outstanding properties of a laser beam. These properties of laser beam have made it able for its wide applications. Basically, a laser involves use of fundamental processes of interaction between radiation and matter (atom, molecule, ion) [1]. These processes are described below:

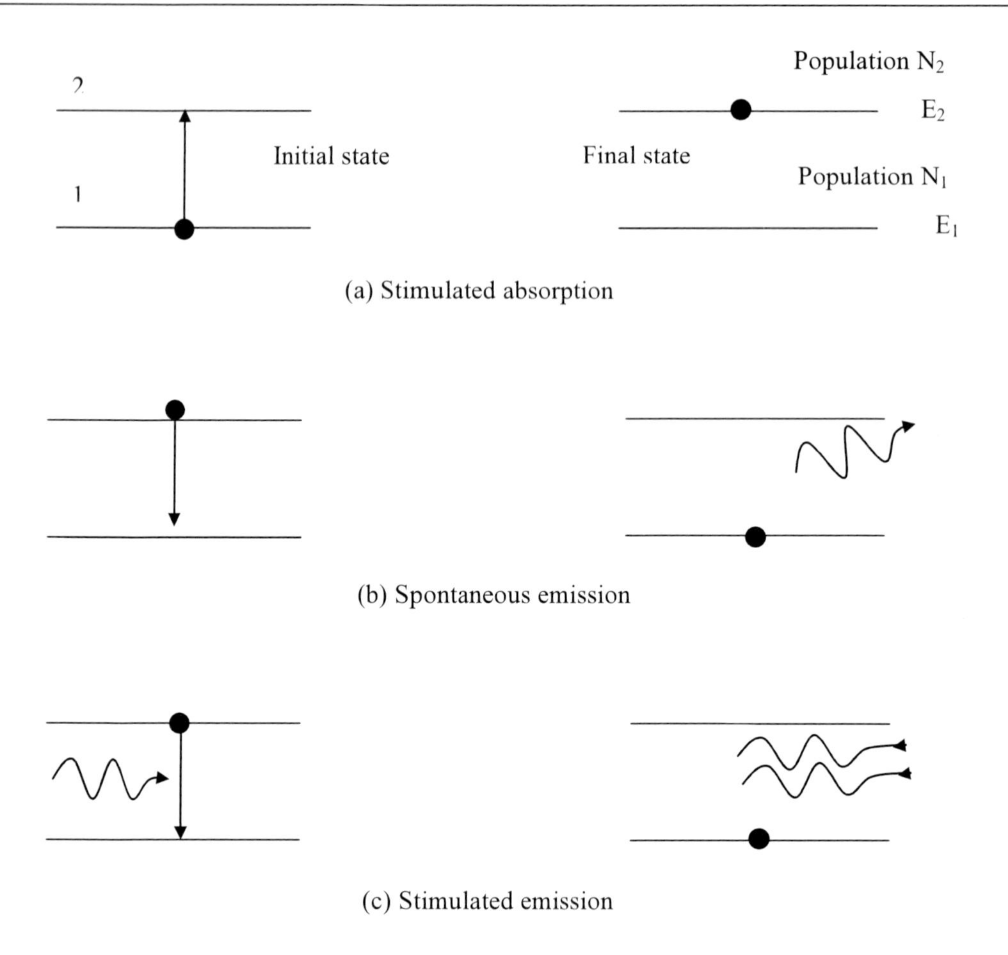

Figure 1. Fundamental processes in laser.

1.1. Fundamental Processes in Laser

a) Absorption

Consider two energy levels 1 and 2 of an atom with energies E_1 and E_2 respectively ($E_1 < E_2$). Initially, atom is in the energy level 1. When a photon of energy $h\nu = E_2 - E_1$ is incident on the atom, it gets absorbed and atom is elevated to the upper energy level 2 from energy level 1. The process is called as absorption.

b) Spontaneous/stimulated emission:

The atom in the upper energy level 2 is very unstable and it comes down to the lower energy level 1 by emitting radiation in the form of electromagnetic wave. The emission process may occur in two distinct ways.

(1) If the excited atom drops to the lower energy level 1 from higher energy level 2 by emitting a photon in an entirely random way then it is referred to as spontaneous emission. (2) On the other hand, if the excited atom is triggered to undergo $2 \rightarrow 1$ transition by the photons of energy $h\nu = E_2 - E_1$ then emission process results in coherent radiation consisting of one stimulated and one stimulating photons. These processes are shown in the **Fig. 1** (**a**, **b**, **c**).

The first successful operation of laser was demonstrated by Maiman in 1960 [2, 3] using ruby crystal. Since then a number of lasers are available covering the electromagnetic spectrum from sub millimeter to soft X-rays with varying output power between a few microwatt to 100 Terra Watt (Nova laser). Presently, lasers are available which can be used either in cw or in pulse mode with extremely short duration like few femto second.

1.2. Elements of a Laser

(i). An active medium, (ii). a pumping source and (iii). an optical resonator.

i. Active medium:

A liquid, solid or gas containing an active species (atoms, molecules or ions) can be used as active medium. These active species are origin of laser emission in laser devices.

ii. Pumping source:

This is required for the excitation of active species from lower to higher energy level. These species can decay through radiative and/or nonradiative processes. Out of these two processes, only radiative process, which can occur either by spontaneous or stimulated emission, is useful for lasing purpose. A number of pumping sources depending on the characteristics of active species are available.

iii. Optical resonator:

When an active medium containing active species is pumped, the gain per unit length of most active media is so small that very little amplification of a beam of light results from a single pass through the medium. The main function of an optical resonator is to provide substantial amplification of beam of light originating from the active medium. A most commonly used optical resonator consists of a pair of highly reflecting mirrors facing each other.

1.3. Types of Lasers

Lasers are classified into five groups depending on the active medium. These are as follow:

i. **Gas Laser:** The active species are in gaseous state in this class of lasers. Ions lasers (argon laser, krypton laser), atomic lasers (He-Ne laser), molecular lasers (CO_2 laser, N_2 laser) and excimer lasers are examples of this class of laser.
ii. **Solid state lasers:** Majority of solid-state lasers involves solid host materials in crystalline or amorphous state incorporated by active species. Ruby laser, Nd-YAG laser, Nd-Glass laser are the examples of this class of lasers.
iii. **Semiconductor lasers:** This class of lasers involves semiconductor materials (n-type, p-type) as an active material. The lasing action is obtained by electrical pumping.

iv. **Free Electron laser:** It is possible to design a laser in which energy levels of atom or molecules are not involved. Stimulated scattering from free electrons passing through spatially varying magnetic field has made such a laser possible. Free electron lasers are high power tunable lasers in which the kinetic energy of a beam of electrons is converted into light.

v. **Dye laser:** Organic laser dye dissolved in appropriate solvent is an active gain medium in dye-laser system. Because of broad spectral features of laser dyes that give rise to tunability and wide frequency span from UV to IR region, dye lasers have very important position in the family of lasers. Dye lasers output can be cw or pulsed depending on the pumping sources. Recently, the shortest and the longest dye laser wavelength limit are reported 196 and 1850 nm respectively [4-5].

2. Solution Based Dye Lasers

2.1. History and Important Features

The first time study of dye laser started with observation of stimulated emission from alcoholic solution of 3, 3-Diethyltricarbocyanine dye excited by intense pulse of giant ruby laser [6]. A number of cyanine dyes were used to obtain lasing action in the near IR region by Schafer, Schmidt and Volze [7]. The first dye laser operation in the visible region was reported by Schafer, Schmidt and Marth [8]. Soon a flash lamp pumped dye laser and a laser pumped dye laser were reported [9-10]. Since then the advance discovery of a number of new and better laser dyes have contributed to the development of newer and newer dye lasers. Presently, a number of efficient laser dyes belonging to different series such as xanthene, coumarin, oxazine etc. are available [11].

An attractive application of dye lasers is to produce ultra short pulses using modelocking or distributed feedback techniques [12-15]. The femtosecond light pulses have become an important tool to excite measure and control ultra fast processes in fundamental science and technology. These light pulses can be used to understand many of the photochemical processes. Due to capability of producing ultra short pulses as the main feature of dye lasers, the light pulse down to 6 fs and energy 5 kJ can be produced [16]. As we know, the lasing characteristics of the materials highly depend on the photophysical and photochemical properties of the active species and hence, a brief attention should be paid.

2.2. Photophysical and Photochemical Properties of Laser Dyes

Organic dyes are characterised by a strong absorption band in the visible region of the electromagnetic spectrum. They contain an extended system of conjugated bonds in their molecular structure. The absorption band of dyes depends on the number of conjugated bonds in their structure. The longest absorption band of dyes is attributed to the transition from electronic ground state S_0 to the first excited singlet state S_1. The transition moment for this process is typically very large. The reverse process $S_1 \rightarrow S_0$ is responsible for the spontaneous emission known as fluorescence and for the stimulated emission in dye lasers. But electronically excited molecule can return to the ground state by a number of pathways, either intermolecular or intramolecular, as shown schematically in the **Fig. 2** [17]. Intermolecular interactions can result in photochemical reactions, deactivation processes, and quenching processes. The two major photochemical reactions, oxidation and dimerization, cause destruction of the original dye molecule and permanent drop in the fluorescence output [17]. An intramolecular interaction includes photophysical processes such as radiative, nonradiative transitions and photochemical processes as isomerization, dissociation. The main important factor influencing the lasing action of dyes is the contribution of the radiative processes compared to nonradiative processes.

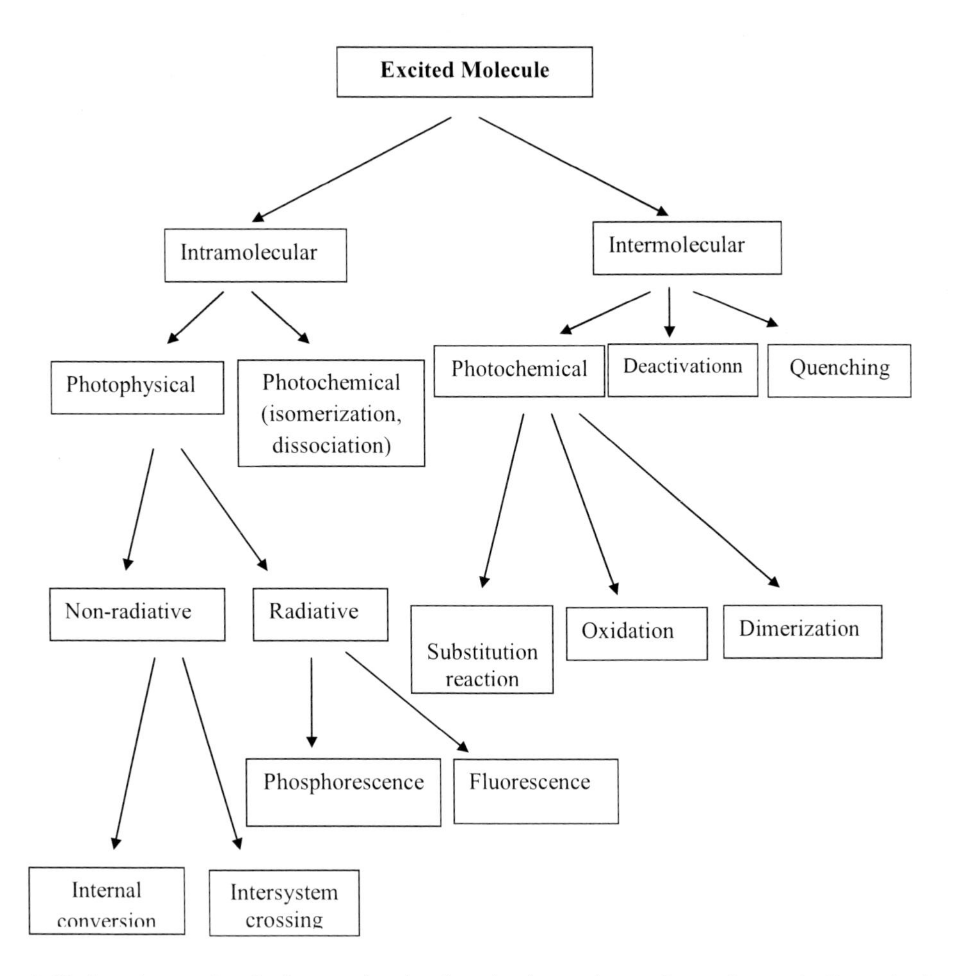

Figure 2. Various intermolecular/intramolecular deactivation pathways in an electronically excited molecule.

2.2.1. Radiative and Non-Radiative Processes

In order to understand these two photophysical processes it is suitable to consider the energy level diagram shown in the **Fig. 3**

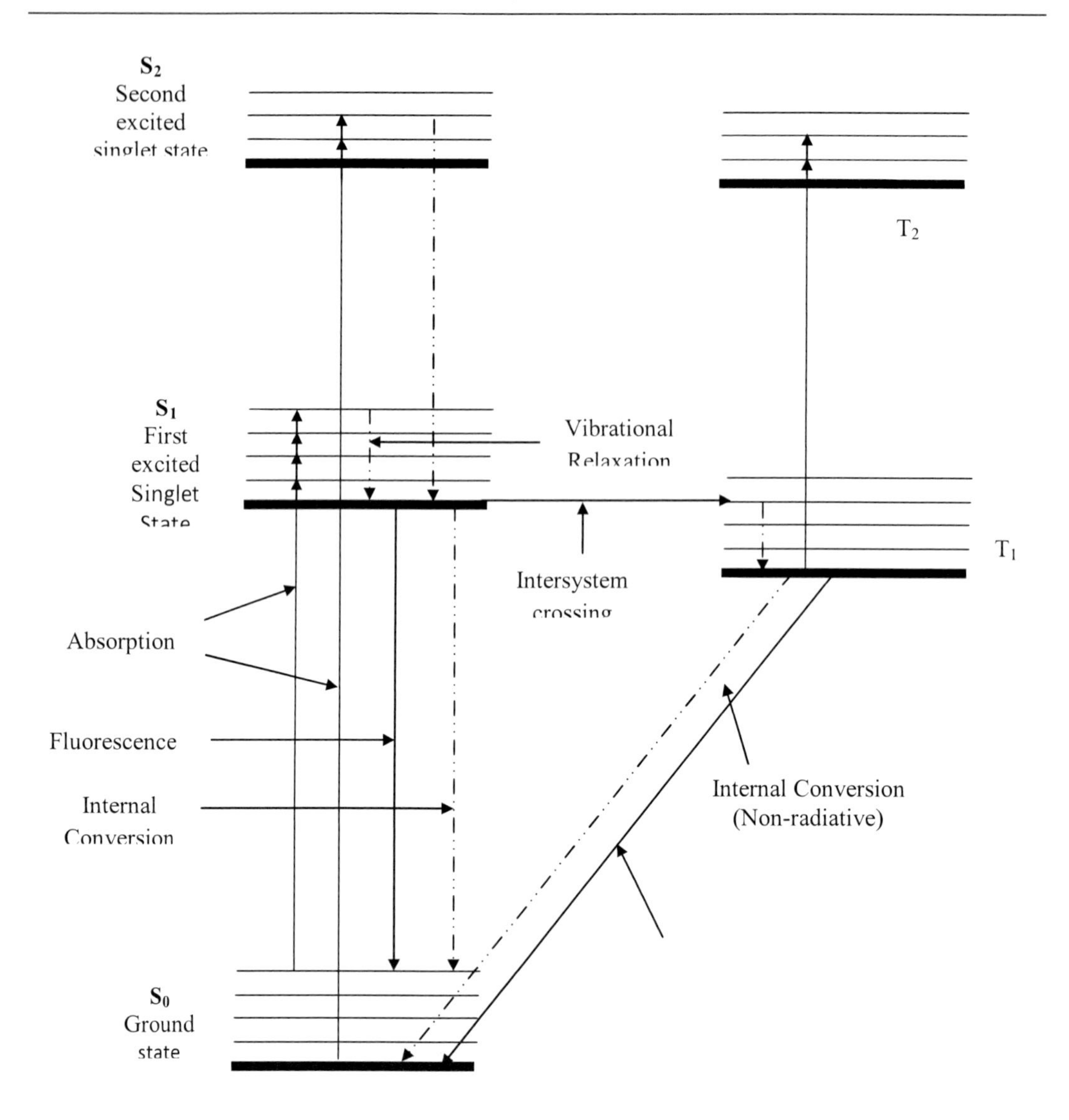

Figure 3. Photophysical processes in organic dye molecule.

When active medium of dye laser is excited by a suitable pump source, the molecules are excited typically to some higher level in the singlet state S_1; from which they relax within picosecond to the lowest vibronic level of S_1. Now the excited molecule can return to S_0 by emitting a photon whose energy is less than that of absorbed light. Such type of radiative $S_1 \rightarrow S_0$ transition is called fluorescence and is shifted to higher wavelength side compared to that of absorption. The process of fluorescence is spin allowed with fluorescence lifetime of the order of a few nanosecond. The ratio of the emitted and absorbed photons is termed as fluorescence quantum yield and its value is in between 0 and 1.

In addition, the molecule in first excited singlet state may undergo a nonradiative transition from S_1 to S_0 (state of the same multiplicity) or S_1 to T_1 (state of the different multiplicity) and finally to ground state S_0 via $S_1 \rightarrow T_1$ transition. These non-radiative transitions $S_1 \rightarrow S_0$ and $S_1 \rightarrow T_1$ are called as internal conversion and intersystem crossing respectively. These two processes reduce the population of S_1 state, which is available for radiative transition, and consequently, reduce the fluorescence quantum yield. Also the excited molecule in T_1 state (after intersystem crossing) may come down to the ground state by emitting a photon and this is known as phosphorescence. It is always at longer wavelength than fluorescence.

Some times, the molecule in the T_1 state may be again prompted, with the help of thermal energy from surrounding, to the isoenergetic point and cross back to the S_1 state and subsequently returning to S_0 by radiative process $S_1 \rightarrow S_0$. Such radiative transition is known as delayed fluorescence [18]. In some cases, encounter between two triplet molecules gives rise to formation of excited dimers, which is generally known as excimer having different emission characteristic than that of monomer species [18].

2.2.1. (A) Factors Influencing Radiative Transition

There are various intramolecular and environmental factors which affect the fluorescence properties of organic laser dyes and hence the lasing action.

a) Intramolecular factors:
   1) Structural mobility: This is one of the important factors influencing the fluorescence quantum yield of a dye molecule. A planer and rigid structure of a dye molecule favours high fluorescence quantum yield. While a flexible and non-rigid structure favours the enhancement in intramolecular rotation and vibration resulting dissipation of excitation energy by nonradiative process.
   2) Substitution Effect: The position of some specific functional groups [11] on an aromatic ring of the dye molecule has been found to be exerting predictable effects on the fluorescence properties of dye molecules. Particularly, halogen substitutions (Cl, Br, I, F) on the aromatic ring reduce the fluorescence efficiency. This effect is called as intramolecular heavy atom effect and the effect is due to the enhancement of intersystem crossing process [11].
   3) Other Intramolecular Processes: There are several intramolecular processes responsible for quenching of fluorescence. If a part of a dye molecule is strongly electron donating or withdrawing, reversible charge transfer may occur between this group and the excited chromophore resulting in the loss of electronic excitation [11]. Likewise, a substituent with a low-lying singlet or triplet state may quench the fluorescence via energy transfer [11].
   4) π electron distribution: The π electron distribution of the chromospheres also affects the rate of intersystem crossing. A dye with π electrons of the chromophore forming a loop when oscillating between the end groups shows a higher triplet yield than in related compound where this loop is blocked [11].

b) Environmental Effects:
   1) Effect of Viscosity: It is known that nonrigid structure and loss in planarity of dye molecule reduce the fluorescence yield. The planarity and rigidity of dye molecule can be maintained in ground state as well as in excited state by increasing the viscosity of the solvent or trapping dye molecules in solid matrix [11] which results in an increase in fluorescence yield.
   2) Effect of Polarity: The process of internal conversion in dye molecule has been found to be strongly dependent on solvent polarity [19-21]. A dye molecule

having positively charged chromophore must have an accompanying anion e.g. ClO_4^- or I^-. Because of quenching ability of these anions, the fluorescence efficiency of dye decreases. This quenching ability is highly dependent on the concentration of the dye and polarity of the solvent. In a polar solvent the dye molecules are fully dissociated and hence, the quenching anions have not sufficient time to reach the excited dye molecule during their lifetime, while the dye molecule is undissociated in a non-polar solvent and the anion can quench the excited state reducing the fluorescence efficiency. It has also been found that the photophysical properties of coumarin derivatives are affected by the polarity of the solvent [22, 23].

3) External Heavy Atom Effect: The solvent containing heavy atom (Cl, Br, I, F) enhances the spin orbit coupling and hence quench the fluorescence of the dye.
4) Hydrogen Bond Effect: Hydrogen bonding has been found to play an important role in the photophysics of the fluorescein derivatives. This type of bonding causes blue shift in fluorescence and reduces the fluorescence quantum yield.
5) Charge Transfer Effect: It is known for a long time that the fluorescence of dyes is quenched by certain anions [24]. The quenching efficiency depends strongly on the chemical nature of anion. The quenching ability decreases in the order: I^-, SCN^-, Br^-, Cl^-, ClO_4^- [11]. Because of quenching ability of these anions, the fluorescence efficiency of dye having positively charged chromophore decreases.
6) Energy transfer Effect: The excited state singlet or triplet, of a dye molecule can be quenched externally by energy transfer to the surrounding molecule, if the surrounding molecule has a level of energy equal to or lower than that of the state to be quenched [11]. Under favourable conditions such energy transfer can occur over distances up to about 10 nm [25-26]. This effect can be used to quench undesirable triplet states.
7) Excited state reactions: Some dyes become more basic or acidic on optical excitation (e.g. coumarins). They may pick up a proton from the surrounding solution or lose one to the solution [11]. These new molecular forms show different fluorescence properties [11].
8) Aggregation of dye molecules: Organic molecules in aqueous solutions tend to form dimers and higher aggregates. They are generally weakly fluorescent or non-fluorescent. This is attributed to the high dielectric constant of water, which reduces the Coulombic repulsion between identically charged molecules. Thus the equilibrium between monomers and dimers is shifted to the side of the latter with increasing dye concentration. The aggregation of dyes in aqueous solution can be reduced by addition of organic compounds.

2.3. Stimulated Emission and Laser Gain in Organic Dyes

The lasing action in organic laser dye takes place due to transition between first excited singlet and ground states. Under influence of intense pumping radiation, some dye molecules (say N_1) are excited to first singlet state (S_1). By fast relaxation, they come down to the lowest vibronic level of the same electronic state. The gain equation at the wavelength λ is then given by

$$g(\lambda) = N_1\,\sigma_{em}(\lambda) - N_0\,\sigma_{ab}(\lambda) - N_T\,\sigma_T(\lambda) \tag{1}$$

Where σ_{em}, σ_{ab} and σ_T are singlet emission, singlet absorption and triplet absorption cross-section respectively. N_0, N_1, and N_T are population of dye molecules in ground, singlet and triplet state respectively.

If $N_T = 0$ i.e. population of triplet state is zero then

$$g(\lambda) = N_1\,\sigma_{em}(\lambda) - N_0\,\sigma_{ab}(\lambda) \tag{2}$$

This is the simplest form of gain equation. From above equations it is evident that the population of first singlet state should be highest while those of ground and triplet states should be lowest in order to achieve high gain. In actual practice, dye solution filled in a cuvette is kept in the laser cavity. To sustain laser oscillation, the gain should be either greater or equal to losses taking place in the laser cavity. In such condition the laser starts oscillating and lasing action is obtained.

2.4. Pumping Schemes for Dye Laser

Various pumping schemes are shown in the **Fig. 4**. A simplified practical arrangement of a dye laser system consists of a square fluorimeter cuvette filled with dye solution, which is excited by the beam from a suitable laser. The resonator is formed by the two glass-air interfaces of the polished sides of the cuvette. The exciting laser and the dye laser beams are at right angles to each other. Similar type of arrangement using flow cell instead of fluorimeter cell is also possible.

A longitudinal pumping arrangement can also be used in which the exciting laser and the dye laser beams are in the same direction and along the same straight line.

To achieve better lasing performance, the dye solutions should satisfy the following conditions:

a) The dye solution should have sufficient absorption at pumping wavelength.
b) The dye solution must have reasonably high fluorescence quantum yield.
c) There should be a minimum overlap area between absorption and fluorescence spectra of dye because more overlapping area enhances the re-absorption of emitted intensity resulting losses in stimulated emission.
d) The dye solution should have high photochemical and thermal stability.

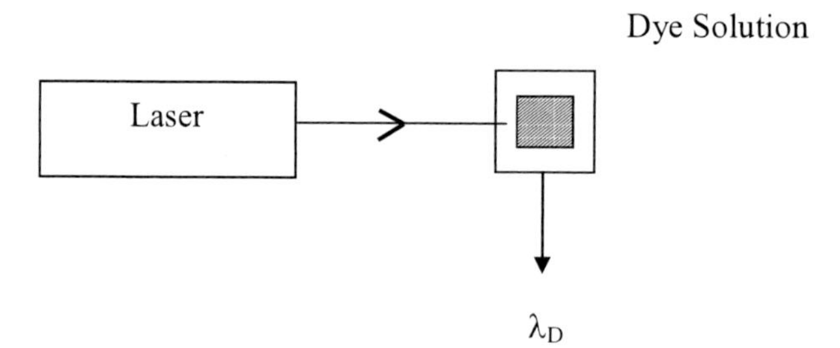

(a) Transverse pumping using a square fluorimeter cuvette

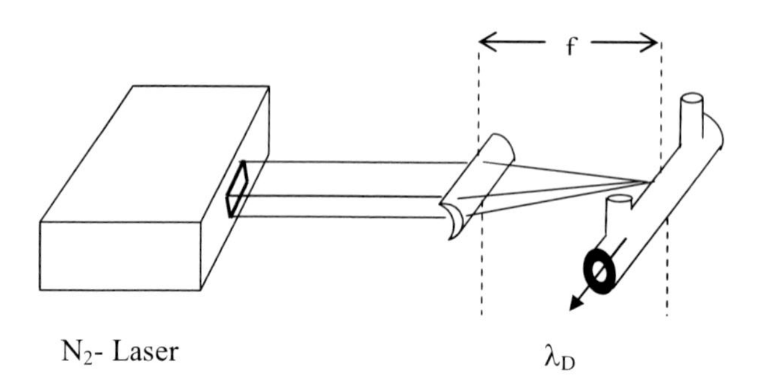

(b) Transverse pumping using a flow cell

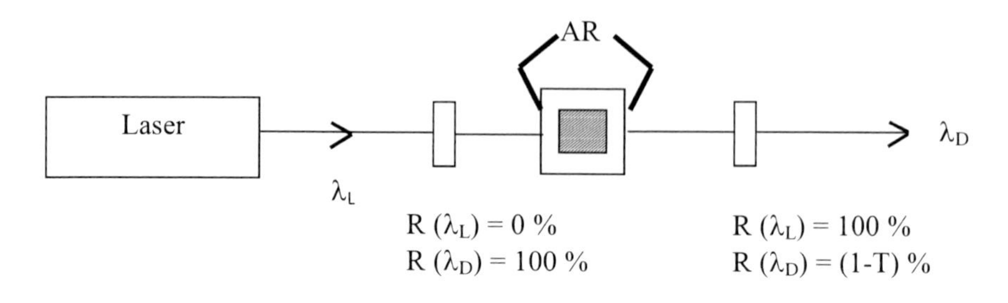

(c) Longitudinal pumping

Figure 4. Pumping arrangements for dye laser.

2.5. Concept of Dye Impregnation in Transparent Solid Matrix

In a conventional dye laser, organic dye dissolved in an appropriate solvent is used, as amplifying medium and it requires a flow system-containing reservoir, pump, tube etc. to maintain the circulation of dye solution through the cell/jet stream placed in the laser cavity. However, cumbersome system design, inflammable nature of solvent used, toxicity of dye and solvent limit the applications of these lasers. Entrapping laser dye molecules in a transparent solid material would eliminate many of these problems leading to the development of solid-state dye laser. A solid-state dye laser has various advantages over a liquid state dye laser. These are as follows:

1) The solid-state dye doped materials can be prepared in the form of films (free standing, coated on an appropriate substrate), cylindrical rod, square rod, optical fibre etc. These are helpful in the development of compact laser and optical devices.
2) The solid-state dye laser material is capable to provide high laser output power because of high gain compared to a liquid state laser.
3) The rotational relaxation process is one of the main reasons for the radiative energy loss in flexible molecules. Trapping such dye molecules in solid cage reduces the radiative loss resulting enhanced performance of dye doped solid-state active material.
4) Due to absence of translational motion of dye molecules in a solid cage, the intermolecular collisional loss is avoided and hence a solid-state dye laser material having high concentration can be obtained without noticeable quenching.
5) The decomposition products less interact with dye molecules and hence cannot facilitate further decomposition in solid-state dye laser resulting high phot-ostability.

3. Transparent Solids as an Alternative

The initiative work on solid-state dye laser was done by trapping Rhodamine dyes in PMMA material in 1967-1969 [27-29]. Since then a number of materials like various types of polymers, co-polymers, polycomposite glasses have been used as host matrices for laser dyes [30-37, 39-55]. Glasses being a hard and optically transparent material can be used as host matrix for laser dyes. But conventional glass preparation technique is not suitable for the introduction of organic laser dyes because of high processing temperature leading to the permanent destruction of dye molecules. The solution to this problem is the novel sol-gel technique, a low temperature technique for preparation of glasses [56-57].

3.1. Important Characteristics of a Laser Host Material

1. Low processing temperature - protecting dopant molecules from thermal degradation.
2. High transparency in the wavelength region of absorption and emission of the dopant.
3. Hard and rigid - easier in handling.
4. Long shelf life i.e. material degradation time should be as long as possible.
5. High laser damage threshold.

3.2. Choice of Sol-Gel Materials as Host Medium

The low processing temperature for the production of glasses (uncommon to conventional technique) by sol-gel technique makes the task of dye impregnation easier. The highly porous nature of these materials allows trapping of high dopant concentration with reduced possibility of aggregation. These materials have good mechanical properties, high transparency in the entire visible region and excellent optical quality. The inert nature of matrix and highest laser damage threshold are advantages of sol-gel glasses over polymers, co-polymers and polycomposite glasses [39]. On the basis of above mentioned reasons, the sol-gel materials can be chosen as one of the best host matrix for doping laser dyes.

Part-II: Sol-Gel Based Solid State Dye Laser

1. Introduction to Sol-Gel

1.1. General Introduction

Apart from conventional technique of glass preparation, sol-gel technique has proven to be much more useful. Through this technique, homogeneous inorganic oxide materials with desirable properties of hardness, optical transparency, chemical durability, tailored porosity, and thermal resistance, can be produced at room temperature, as opposed to the much higher melting temperatures required in the conventional glasses [58-60]. It is remarkable to note that the low processing temperature and porous characteristic of sol-gel materials produced by this technique allow incorporation of various organic and inorganic additives up to high dopant concentration during the process of formation of the glassy network.

The pioneering work by Avnir and coworkers in 1984 on introduction of laser dye molecules into glasses developed at low temperatures by sol-gel technique has opened gates to a new class of materials [56]. Since then continuous research works worldwide on development of newer and newer sol-gel materials [61-70] and dye doped sol-gel glass materials have been carried out world wide [71-88]. The properties of pure sol-gel glasses [58-60, 63, 69, 89-94], photophysical and lasing properties of dye impregnated sol-gel materials prepared by various routes have been investigated in detail [71, 73, 77-80, 82-85]. The effect of matrix parameters such as its composition, morphology, pore size on the spectral characteristics of the dye has also evaluated [69, 78, 79, 95, 96]. Sol-gel composites known as ORMOSILs (organically modified silicates) have been developed by using organic modifier for better compatibility of dye/host and to enhance the mechanical properties [97-100]. In many cases they have shown improved performance than pure sol-gel silica. The sol-gel technology, by which inorganic or composite organic-inorganic materials are made at relatively low temperature, consists of hydrolysis of the constituent molecular precursors and subsequent polycondensation to glass-like form.

1.2. General Chemistr

In terms of chemical mechanism, the sol-gel processing can be divided into two categories, namely aqueous route and alkoxide route. The former route employs the principles of colloid chemistry to generate colloid sized particles from ionic species in an aqueous medium, whereas the latter route generally requires the use of an alcohol as a mutual solvent for the organometallic precursor and water as hydrolysing agent. The alkoxide route has attracted more attention recently and this route has been adopted for the preparation of sol-gel materials in the present study. Metal alkoxides are most popular because they react readily with water. The most widely used metal alkoxides are the alkoxysilanes such as tetramethoxysilane (TMOS) and tetraethoxysilane (TEOS).

The sol-gel process involves two chemical reactions namely hydrolysis and condensation. Generally speaking, the hydrolysis reaction, through the addition of water, replaces alkoxide groups (OR) with hydroxyl groups (OH). Subsequent, condensation reactions involving the silanol groups (Si-OH) produces siloxane bonds (Si-O-Si) plus the by-product water.

$$\text{Hydrolysis: } Si(OC_2H_5)_4 + 4H_2O \rightarrow Si(OH)_4 + 4C_2H_5OH$$

$$\text{Condensation: } (OH)_3Si\text{-}OH + OH\text{–}Si(OH)_3 \rightarrow (OH)_3Si\text{–}O\text{–}Si(OH)_3 + H_2O$$

As the number of siloxane bonds increases, the individual molecules are bridged as a result of poly-condensation and jointly aggregate in the sol. These sol particles aggregate, or inter-knit into a network, a gel is formed. Upon drying, trapped volatiles (water, alcohols) are driven off and the network shrinks as further condensation can occur. However the characteristics and properties of a particular sol-gel inorganic network are related to a number of factors that affect the rate of hydrolysis and condensation reactions, such as, pH, temperature and time of reaction, reagent concentrations, nature and concentration of catalyst (acid/base), H_2O/alkoxide molar ratio [63, 68-69, 101-102]. Thus, by controlling these factors, it is possible to vary the structure and properties of the sol-gel materials over wide ranges [102].

1.3. Sol-Gel Processin

The preparation procedure of materials from alkoxides using sol-gel process can be divided into following steps as described below:

(1) Precursor preparation: In order to achieve the advantages of the improved homogeneity, the reacting species must be dispersed uniformly into a single-phase solution. This can be achieved by using a common solvent like methanol or ethanol for alkoxides, water and catalyst; otherwise these precursor components are immiscible.

It is important to realize that the choice of starting materials can have a marked effect on the resulting gel and ultimately on the glass. To achieve and maintain good homogeneity in multicomponent systems care must be taken to minimize self-condensation. This may be controlled by the processing conditions like the order of mixing, temperature and solvent type, as well as the choice of reactants/by using alkoxides of a similar reactivity [103].

To summarize, the careful choice of starting materials and mixing conditions enables a wide range of compositions to be prepared where the mixing of components should be uniform down to the molecular level giving processing advantages over conventional techniques.

(2) Sol to gel transition: With the presence of homogenizing agent (alcohol), hydrolysis is facilitated due to the miscibility of the alkoxide and water resulting silanol molecules. These silanol molecules take part in condensation reaction giving rise to siloxane unit. As the number of siloxane bonds increases, the individual molecules are bridged and jointly aggregate in the sol and after inter-knit into a network, a gel is formed.

(3) Forming product: The mobile characteristic of sol fascilitates to form potentially useful shapes like disc, square block, thick and thin films. Casting and coating are the most widely used techniques.

In casting technique, the sol is poured into a mould made of glass or polymer and then after appropriate drying samples are obtained in the form depending on the geometry of the mould used. For the preparation of freestanding film by this technique, a suitable shaped wire immersed into sol is withdrawn [89]. The composition of solution, drawing time of the substrate, and viscosity of the sol are very crucial parameters.

On the other hand, coating technique is well known for thin film coating on a substrate. Dipping process, lowering process and spinning process can be used to get uniform coating on an appropriate substrate [73, 78-79, 104].

(4) Drying: The drying of gel is the most critical stage for producing sol-gel glasses of any significant size. During this process, the formation of cracks in the wet gel is the main technological hurdle. A number of methods like strengthening the gel by reinforcement, increasing the strength of interparticle bonds, reducing the surface tension forces by enlarging the pore diameter, hypercritical drying have been proposed for drying of wet gel [59, 105].

During the initial stage of drying, the evaporation of liquid from the micropores in the gel causes large capillary stresses to develop, which can initiate cracking. These capillary forces depend on the rate of evaporation, which is related to the solvent vapour pressure, and also depend inversely on the pore size. Therefore, large pore sizes and a stronger gel network tend to reduce the degree of cracking. In the final stage of drying, cracking is a result of nonuniform contraction of the gel [106]. The dried gel produced by conventional drying is called xerogel.

By adding organic additives to the starting sol, the cracking tendency of gel on drying can be reduced. These organic additives are termed as drying control chemical additive (DCCA) e.g. formamide and glycerol [107-112]. These are capable of controlling the vapour pressure of the solvents in the pores. They can also control pore size and pore distribution in the gel. In hypercritical drying technique, the gel immersed in an appropriate solvent inside an autoclave is heated above the critical temperature and pressure [105]. The dried gel obtained by this method is called aerogel.

(5) Gel to glass transition: A dried gel may have some residual alkoxy and silanol groups. To get a densified glass, the gel is heated to high temperature leading to collapse of pores. This process is called as sintering. However, this temperature is much lower than that of the temperature involved in the conventional glass preparation.

1.4. Applications

The excellent properties such as high transparency, large porosity, high optical quality, good mechanical and thermal stability, inert nature etc. of sol-gel materials and advantages of processing technique over conventional method of glass preparation have led to a number of applications of these materials in various fields of science and technology. Practically, these are unlimited and it is difficult to list all of them. The specific uses of these sol-gel produced glasses are derived from the various material shapes generated in the gel state, i.e., monoliths, films, fibers etc. [60]. Many other special applications includes optics, protective and porous films, optical coatings, window insulators, dielectric and electronic coatings, reinforcement fibers, fillers, catalysts and host matrix for various types of dopants [60]. Some potential applications of doped sol-gel derived glasses are shown below [113-128].

1) Electrochromic, photochromic and gasochromic plates for smart windows
2) Sensors for environmental and biological impurities
3) Luminescent solar concentrators
4) Active waveguides
5) Quantum semiconductor dots
6) Materials for linear and nonlinear optics
7) Complexes of rare earth ions that can be used for diagnostics and biological markers
8) Solid-state tunable lasers

2. Synthesis of Laser Dye Impregnated Sol-Gel Material

2.1. Dye Impregnation in Sol-Gel Matrix

Preparation of sol-gel samples starts from sol state (sol precursor) and ends with solid state (solid xerogel). The final sol-gel product is highly transparent and porous in nature. Due to solution phase nature of starting precursor and porous nature of gel material, following two methods can be used for the dye incorporation. Because of low processing temperature of sol-gel materials, dye can be mixed with the starting sol. The dye mixed homogeneous coloured glass solution can be used for the sample preparation. This method of dye impregnation is called pre-doping method [38, 56-57]. On the other hand, due to porous characteristics of the materials, dye can also be trapped into the pores of the gel matrix by diffusion process and this is called post doping method [38-39]. Both these methods are potential and have been successfully utilized for dye impregnation [38-39]. A large number of literatures are available on impregnation of laser dyes in sol-gel glasses by pre-doping method [56-57, 64-80, 82-100]. However, less attention has been paid on post-doping. In this regard, efforts made by Rahn et.al and Deshpande et. al are really admirable [38-39]. It is also important to note that most of the synthesis of laser dye doped sol-gel materials from solid state laser application point of view are acid catalysed [56-57, 64-80, 82-100] and a scanty of reports are available on the synthesis of dye impregnated base catalysed sol-gel materials [39].

2.1.1. Pre Doping Method

The alkoxide ruote has been widely used for the preparation of dye impregnated sol-gel materials in the contest of sol-gel based solid state dye laser. Metal alkoxides are the most popular because they react readily with water. Alkoxysilanes such as tetramethoxysilane (TMOS) and tetraethoxysilane (TEOS) are the most widely used metal alkoxides. The sol is prepared by acid /base catalysed hydrolysis of alkoxysilane in the presence of distilled water and/or methanol as common solvent followed by condensation and polycondensation. Here the main function of catalyst is to promote hydrolysis reaction during sol-gel processing. An appropriate alkoxysilane, distilled water, alcohol and catalyst are used in certain molar ratio maintaining a certain order of mixing for sol the preparation. The mixture is then kept under stirring. After sometime, the mixture becomes transparent. Now an appropriate amount of alcoholic solution of dye can be mixed with the glass sol and the resultant homogeneous dye doped glass sol is used for the sample preparation. The concentration of dye solution or volumetric ratio of dye solution and glass sol can be adjusted to obtain the desired dopant concentration. This homogeneous coloured precursor sol can be poured into an appropriate mould to get desired shape of dye impregnated sol-gel sample. The mercury float method can be used for the synthesis of dye doped film. However, HCl catalysed sol does result in translucent film prepared by mercury float method. The thickness of the film can be controlled by controlling the viscosity of the dye doped glass sol. Film of thickness of few mm can be prepared by using suitable mould also. Dip coating or spin coating of dye mixed glass solution on a suitable substrate can also result in quite good quality film. The refractive indices of the substrate and the dye doped sol play an important role in the preparation of the film by either of the two methods (dip/spin).

In addition to acid catalysed sol-gel glass preparation, Deshpande et. al have used base with acid catalysed synthesis of dye doped sol-gel materials for the use in solid state dye laser as active material [38].

2.1.2. Post Doping Method

In this method, dye molecules are introduced into solid sol-gel matrices by dipping the solid sol-gel (undoped) into dye solution, where the dye is allowed to diffuse into the pores of the glass structure after synthesis [38-39]. An appropriate alkoxysilane, distilled water, alcohol and catalyst are used in certain molar ratio maintaining a certain order of mixing for sol preparation. The mixture then should be kept under stirring. After sometime, the mixture becomes transparent and then drying control chemical additive (formamide, glycerol etc.) should be added to the solution in certain molar ratio. Moreover, this resultant solution can be utilized for solid sample preparation as mentioned in section 2.1.1. Depending on your choice and requirement, the synthesized sample may be in the form of parralelopiped, free standing film or thin film coated on a suitable substrate. Now from here, the synthesized sample can be used either after high temperature densification to improve mechanical strength or as prepared for impregnating dye molecule in the pores of the sample [38-39]. Then either of the two synthesised samples can be used to immerse in dye solution of known concentration for different intervals of time. The no. density of dye molecules in the solid host can be calculated by difference method from the knowledge of absorption spectrum of the dye solution before and after the dipping of glass sample. However, an error in the concentration measurement in the range of 5 to 10% is expected. The concentration of dye solution and or time interval of dipping can be adjusted to obtain desired no. density in solid host.

2.2. Drying and Ageing of Samples

Since dye doped sol is converted into gel and this gel is very sensitive to drying and ageing. Therefore, it is advisable to seal the opening of the mould with Teflon tape. Few perforations should be made on the top of the Teflon tape to carry out controlled drying and ageing. In all cases, drying of samples should be carried out very carefully; otherwise cracks may appear on surface of the sample rendering these samples useless for the optical study.

3. Photophysical Properties, Lasing and Photostability of Laser Dyes in Sol-Gel

The lasing action of dye depends on its photophysical properties [38, 129]. Therefore, it is necessary to pay more attention to investigate the photophysical properties of dye/sol–gel elements before studying lasing action. Here more specifically I would like to describe absorption, fluorescence and time resolved characteristics of dye doped sol-gel host as photophysical properties. These properties are quite basic properties of laser dyes based on which lasing performance can be estimated even before doing real experiment. The materials with high extinction coefficient and high quantum yield are preferable candidates as good lasing performer.

3.1. Instrumental Techniques

The study of absorption spectrum can yield the information of extinction coefficient using Lambert-Beer's law. The scanning of absorption spectrum is done by commercial spectrophotometer available in the market. The Lambert law states that "the fraction of incident radiation absorbed by a transparent medium is independent of the intensity of incident radiation and that each successive layer of the medium absorbs an equal fraction of incident radiation." The Beer's law states that the amount of radiation absorbed is proportional to the number of molecules absorbing the radiation i.e. concentration of the absorbing species. The two are combined to yield Lambert-Beer's law [18] and can be stated mathematically as follow---

$$\log(I_0/I) = \varepsilon_v bc \quad (3)$$

where I_0 ,I, ε_v, b and c are incident intensity, transmitted intensity, molar extinction coefficient, optical path length and concentration of the absorbing species respectively. The left hand term of the above equation is defined as optical density of the absorbing species and its expansion consists many higher terms. For lower concentration of absorbing species, the higher terms can be neglected and the relation between optical density and concentration becomes linear. This is utilized for the calculation of molar extinction coefficient of the absorbing species. For more details readers are advisable to go through any available standard text book of spectroscopy.

The study of fluorescence spectrum basically can yield the knowledge about emission characteristics of the species. If we talk about the steady state fluorescence then spectral gain band width and fluorescence quantum yield are the important chatracteristics of the fluorescing species. The fluorescence spectrum can be scanned using commercial fluorimeter available in the market. The experimental setup of a fluorimeter is shown in the **Fig. 5**. It consists of a xenon lamp as an excitation source, an excitation monochromator, a sample compartment, an emission monochromator, a photomultiplier tube to detect signal, a photon counter and a recorder to get fluorescence spectrum or fluorescence data from photon counter can be stored in computer. The fluorescence quantum yield of an emissive species is defined as the ratio of the number of emitted photons to the number of absorbed photons by the species. The two well-known methods [130] for the measurement of fluorescence quantum yield value are (1) Transverse excitation emission geometry and (2) Front surface excitation emission geometry.

Transverse excitation emission geometry method is applicable to only low concentrated sample whose optical density is less in appropriate amount so that re-absorption can be avoided. The exciting source and detector are perpendicular to each other i.e. the sample is excited along one direction and fluorescence is collected along its perpendicular direction. Fluorescence spectra of sample and reference are recorded by exciting them with the same wavelength normally absorption maximum of sample. The fluorescence spectra are corrected for monochromator and detector sensitivity. A formula can be deduced from the definition of fluorescence quantum yield and readers are advised to see Demas' paper for more detail [130]. The area under curve of fluorescence spectra of sample and reference are used for fluorescence quantum yield calculation.

The formula for quantum yield calculation is given below

$$Q_s = (\Delta_{es} / \Delta_{er}) \times (\Delta_{ar} / \Delta_{as}) \times (n_s / n_r)^2 \times Q_r \quad (4)$$

Where $(\Delta_{es} / \Delta_{er})$ is the ratio of corrected area under emission curve of sample and reference and, $(\Delta_{ar} / \Delta_{as})$ is the ratio of corrected area under absorption curve of sample and reference. n_s and n_r are refractive indices of sample and reference respectively and Q_r is the fluorescence quantum yield of reference. However, this method is not useful for the measurement of quantum yield of high dopant concentration.

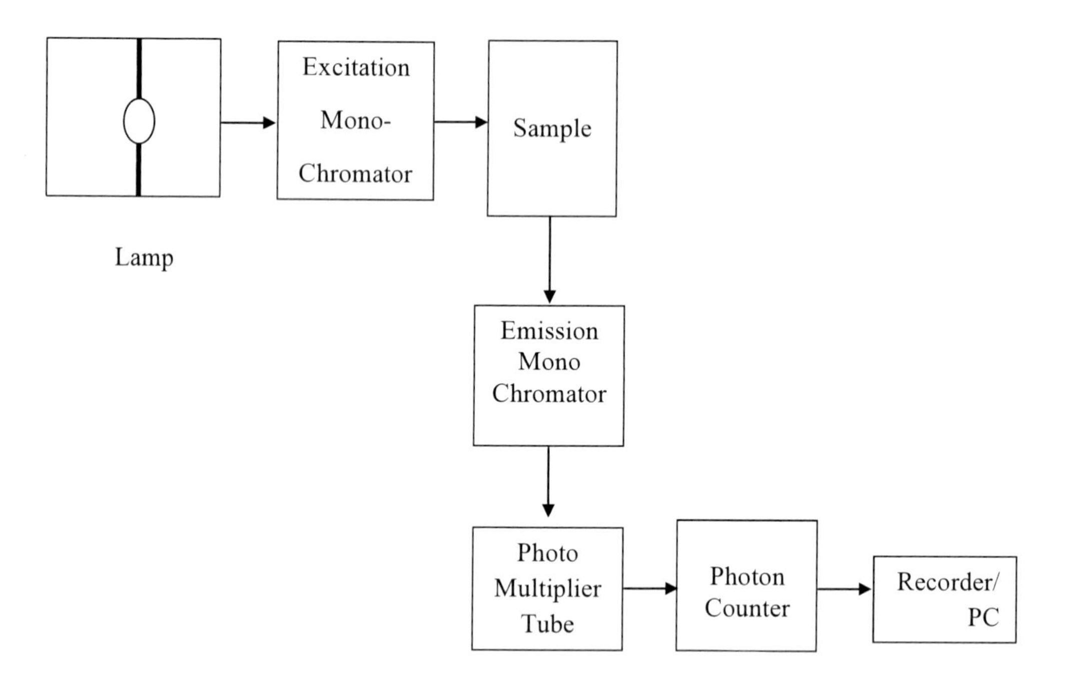

Figure 5. Fluorimeter Setup.

For high dopant concentration, front surface method is useful. The concentration of dopant is normally selected so that 99% or more exciting light is absorbed in a first few layers of the sample. In an experimental setup, the exciting light is allowed to incident on the surface of the sample at an angle of 45° to the normal to the surface drawn at the point of incidence and emitted intensity is collected along the normal direction by the detector. Similarly, the fluorescence of reference material with known quantum yield is recorded. These recorded fluorescence spectra of sample and reference are corrected for monochromator and detector sensitivity. The fluorescence quantum yield of sample is then calculated using following formula

$$Q_s = (\Delta_{es} / \Delta_{er}) \times (n_s / n_r)^2 \times Q_r \quad (5)$$

where symbols are having there usual meaning as earlier mentioned. While applying this method in actual practice, there is a need to take into account the relative absorbance of sample and reference at the exciting wavelength.

Another important parameter of fluorescing species is the fluorescence lifetime. The fluorescence lifetimes of dyes (monomer) are normally of the order of a few nanosecond. However, some fast components like dimmers and others may also exist with lifetime in the order of few to few hundred of picoseconds. There are two main dominant methods for the decay time measurement: time domain and frequency domain methods. In time domain method, the sample is excited with a pulse of light. The width of the pulse is made as short as possible and is preferably much shorter than the decay time of the sample. Time dependent intensity is measured followed by excitation pulse and the decay time is calculated from the slope of a plot of logI(t) vs t or from the time at which the intensity decreases to 1/e of initial intensity at $t = 0$. Another alternative method for measuring lifetime is the frequency domain method, also well known as phase modulation method. In this case, the samples are excited with intensity modulated light, normally sine wave modulation and intensity of incident light is varied at high frequency, in the order of Mega Hertz, such that its reciprocal frequency is comparable to the reciprocal of decay time. When the sample is excited in this way, the emission is forced to respond at the same frequency and consequently the time of emission from the sample is delayed with respect to excitation. This delay is measured in terms of phase shift and is used to calculate the lifetime of the fluorescent species involved in the measurement. More detail description is beyond the scope of the chapter and reader are suggested to read any text abailable or they can go through "Principle of fluorescence spectroscopy" written by J. R. Lakowicz [131].

3.2. Rhodamines Dyes

Rhodamine dyes constitute an important family of laser dyes belonging to xanthene class. These dyes have been widely studied in solution phase. The spectroscopic properties and laser emission maximum of many of these dyes have been found to be surprisingly dependent on solvent characteristics and acid/base environment. The change in absorption peak position as well as laser emission maximum is more pronounced with those dyes whose amino groups are not fully alkylated [11]. In addition, the quantum efficiency, which should be high for a good laser dye, depend largely on temperature and viscosity particularly in those xanthene dyes that contain mobile dialkyl substituents [11]. High quantum efficiency with such dyes can be achieved by dispersing these dye molecules in highly viscous host media. Rhodamine dyes that carry a free (nonesterified) COOH group can exist in several forms [11] such as Rhodamine-B.

As we know, high extinction coefficient and high quantum yield characteristics of a laser dye are favourable for good lasing performance. Normally, the spectral characteristics of monomer species in solution phase are considered as standard. These characteristics should be retained when laser dye molecules are doped into solid sol-gel materials from the point of view of good lasing active material. Therefore, it is necessary to investigate the photophysical properties of laser dye impregnated sol-gel materials into detail before studying the lasing action. As far as laser damage threshold is concerned which should be high for a good host; sol-gel material is considered to have the highest value among polymer, polycomposite glass and ormosil (organically modified silicate) [39]. But deviations in characteristic spectroscopic properties of dye in sol-gel element compared to that in solution are one of the important problems in the development of sol-gel dependent solid-state dye lasers. When fluorescent dyes are incorporated within the pores of silica gels, they usually show deviations from the characteristic fluorescent behaviour observed in the solution phase, which is typically considered as optimum. There are numerous experimental reports on the difficulties involved in retaining the optimum spectroscopic properties of organic dyes when incorporated in sol-gel silica [132-135]. The change in molecular environment (polarity, acid/base level) from sol-gel-xerogel transitions and matrix parameters (composition, morphology, porosity) have been found to be responsible factors for such deviations in spectroscopic properties of dye molecules in sol-gel glasses [71, 95, 136]. These changes affect the photophysical properties of dye in sol-gel solids and ultimately the lasing performance.

Rhodamine family composed of a number of laser dyes. However, Rhodamine-B (Rh-B) and Rhodamine-6G are most and well studied laser dyes. Rh-B is controversial laser dye due to its deactivation processes which exits in various molecular forms known as cation, zwiterion, lactone and dimer depending on molecular environment and dye concentration.

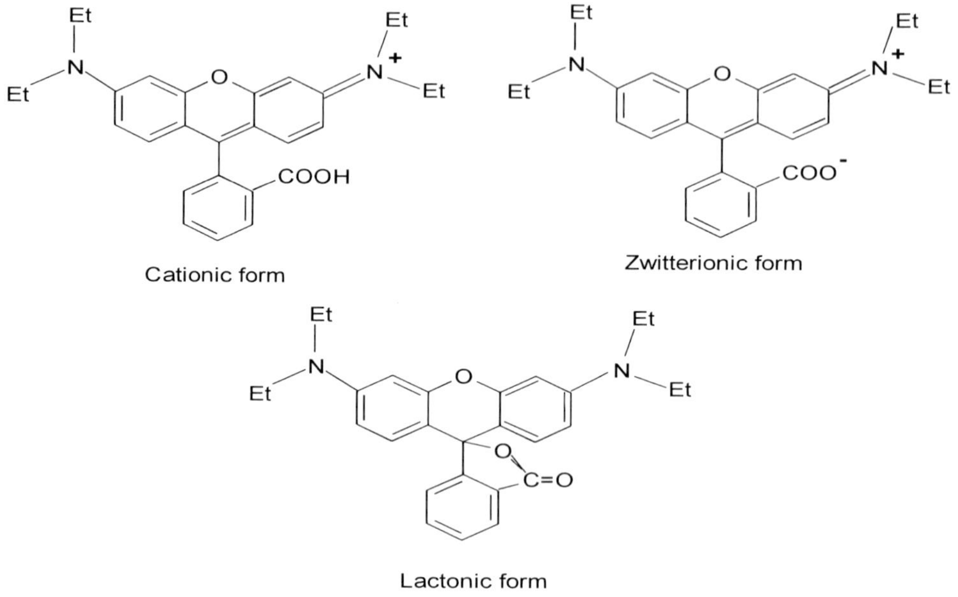

Cationic form

Zwitterionic form

Lactonic form

Figure 6. Molecular forms of Rhodamine-B.

Many reports on photophysical properties of acid catalyzed Rh-B contained sol–gel glass are available [79, 136, 95, 138]. The published reports related to photophysical properties of Rh-B in SiO_2 sol–gel matrices show the existence of dimers [79, 139] or lactone [136-137]. The reason for Rh-B dimer formation has been associated to water content in SiO_2 sol–gel matrix [79]. On the other hand,decrease in polarity from sol to xerogel [136] and increase in acid concentration during Rh-B doped SiO_2 sol–gel glass preparation [137] have been reported for Rh-B lactone formation. Takahashi et al. have shown that the fluorescence spectrum of Rh-B excited at 310 nm shows the emission band from ionic forms (cation, zwiterion) and a new band around 450 nm which similar to the Rh-B lactone form in solution [137]. However, it is well known that Lactone form has no contribution to the absorbance or fluorescence in the visible region and is colorless. In this contest, the new emission band at 450 nm of Rh-B in sol-gel (probably lactone emission band) with 310 nm excitation is not well understood and more detail study is required. The same authors have reported the dimer formation of Rh-6G in sol-gel glass prepared by high HCl concentration [137]. The smaller pore size in a doped glass prepared by high acid hydrolysis is the reason for the aggregate formation of Rh-6G molecules in the glass. Although these reports mention the poor photophysical behavior of Rh-B in SiO_2 sol–gel glass and consequently, such materials cannot be used as amplifying media in solid state dye laser.

Negishi et al. have studied Rh-B in SiO_2 and Si-Ti binary systems and photochemical stability has been reported to be superior in Si-Ti binary oxide system [136]. This enhanced photochemical stability of Rh-B in Si-Ti system is attributed to the existence of specific interactions between Rh-B molecules and Si-Ti binary oxide matrices [136]. As a result, Si-Ti oxide system seems to be better and suitable host matrix for encapsulation of Rh-B molecule. In the field of solid state laser, thin film having special advantages of waveguiding and low losses [80, 120]. A planar waveguide lasers can be developed based dye doped glass films prepared by sol-gel technique [80]. The observation of lasing effects on Rh-B, Rh-6G and SulfoRhodamine-640 doped waveguides prepared by sol-gel method has been reported [80] with gain coefficients in the range of 120-200 dB/cm. However, further research has been suggested in order to optimize the efficiency and photostability of the doped films by selecting proper laser dye and composition [80].

D. Lo et al. have studied fluorescence properties, lasing and narrow linewidth operation of sol-gel based solid state dye laser doped with many laser dyes [83, 140]. They prepared the dye doped samples by pre-doping method using formamide as drying control chemical additive. They reported fluorescencc and superradiance at 650 and 655 nm for Rhodamine-640 and 638 and 625 nm for Rhodamine-640 respectively. In case of Rh-6G in sol-gel, a narrow linewidth of 3.3GHz and efficiency 0.2% have also been reported [83] when the sample was pumped by the second harmonic of Nd:YAG laser and spectral response was detected and analysed by either a Burleigh WA 4500 wavemeter or a home made etalon/CCD system. As the narrow linewidth operation is one of the important performances of a tunable laser and therefore, this kind of study provide the information about tuning capability of the lasing material. Spectroscopic properties of Rh-6G in sol-gel prepared by different approach have also been studied [141].

Apart from this, the spectroscopic properties, lasing action and photostability of Rh-6G doped aluminosilicate xerogel prepared by pre-doping sol-gel technique have been reported by Mckiernan and his co-workers [72]. The Rh-6G/sol-gel sample (10^{-3}M) was excited by 351 nm laser beam and fluorescence spectra were recorded with a Spex 1702 single grating monochromator and detected by a RCA C31034 photomultiplier tube and photon counting. The fluorescence peak in front surface and transverse excitation emission geometry was found at 563 and 592 nm respectively. This red shift in fluorescence peak in later case could be associated with re-absorption effect (the emitted light must travel through about 3 mm inside the sample). In former case, a low energy shoulder at 599 nm was reported. However, no such shoulder type characteristic was found in later case. Authors did not point out about the nature of shoulder. In my opinion, this shoulder could be due to formation of fluorescent aggregate at high dye concentration. The free running laser spectra were collected by pumping transversely the samples using 511 or 540 nm laser output of Lambda physk dye laser. The band center varied from 565 to 572 nm and bandwidth varied from 7 to 10 nm. The photostability curve i.e. laser ouput intensity (along Y-axis) vs number of pump pulses (along X-axis) could neither be fitted single exponential nor double exponential in all experiments [72]. In my view, the reason for the photodegradation of dye molecules is not simple and further study is required to understand this complex degradation mechanism. Even the degradation mechanism of organic dye molecules is not well understood in solid state dye laser system till time. The author pointed out, a large amount of Rh-6G molecules remain non-encapsulated (might be adsorbed on the walls of the pores or channels in cross linked glass structure) and emission from these non-encapsulated Rh-6G molecules could be fading quickly while the encapsulated dye would continue active [72]. Here important to note, the authors have not identified the interaction between excited dye molecules and residual in the host matrix as one of the reason for degradation of dye molecules during pumping. This might be playing a crucial role.

Rahn et al. initiated a new method for impregnating organic laser dye molecules in sol-gel material mentioned earlier as post-doping method [39]. A base catalyst and formamide as drying control chemical additive (DCCA) were used for the synthesis of sol-gel glass. After drying and ageing at room temperature, these xerogel materials were stabilized and densified by heating at 800^0C to produce a porous material with better mechanical properties. Due to densification process, the pores of the material become smaller with thicker pores wall resulting enhanced mechanical strength. These densified sol-gel samples were used to immerse into ethanolic solution of Rhodamine-590 (Rh-6G) dye. They studied the performance of Rh-6G in PMMA (slope efficiency 17.6±2%, half life 4000±500 pulses), polycompotie glass (slope efficiency 19±2%, half life 1800±300 pulses) and sol-gel (slope efficiency 24±2%, half life 4500±900 pulses) and the best performance of Rh-6G was found in sol-gel. Pulse length (microsecond and nanosecond) dependent performance was also studied and Rh-6G was found comfortable with short pulse length. Why does it happen so? Triplet state lifetime of organic dyes is in the order of microsecond. Consequently, longer pulse lengths permit the buildup of long-lived triplet states and produce increased thermal lensing which results in inferior performance. Further, the long pulse length operation requires more pump pulse energy to reach threshold which causes more amount of pump pulse energy accumulation on the sample surface resulting enhanced amount of photodegradation per pulse (as photodegradation is dependent on accumulated pulse energy). Ahmad et al. [142] studied Rh-6G and Rh-B in methanol, starting sol precursor of sol-gel, wet gel (sol-gel) and in dried sol-gel materials and reported an improved slope efficiency (40%) and photostability (60 $GJmol^{-1}$) of Rh-6G in dried sol-gel compared to reported by Rahn et al. (efficiency-24±2%, photostability-48±12 $GJmol\text{-}^{1}$) [39]. Dyes were impregnated into the host by pre-doping method using two step method of synthesis of xeroge. The reduction in the preparation time of fabrication of the sol-gel monoliths is the beauty of this two step method. The authors have reported the highest value, among all studied dye doped hosts, of normalized photostability of 325 $GJmol^{-1}$ for Rh-6G and 410 $GJmol^{-1}$ for Rh-B in wet gel (semi solid) state of sol-gel. The reason for these large value of photostability determined in their work for the wet sol-gel material compared to the dried sol-gel material has been expected to be due primarily to reduced oxygen interactions in the water environment. However, contribution due to increased molecular movement inside the wet gel compared to dry sol-gel by diffusion process is obvious. May there be any other possible reason for increased photostability of dye molecule in wet gel phase? The better chemical compatibility of dye molecule in wet gel sample compared to dried one seems to be additional reason for the enhanced photostability of the dye molecules in the wet gel sample.

Normally, dye doped sol-gel material show lasing actions based on the amplified spontaneous emission under optical pumping. But the lasing action from dye/sol-gel element occurs at a fixed wavelength where the self-absorption of the spontaneous emission is minimized. Now the question, is it possible to modulate the emission wavelengths of dye doped thin film laser. The answer is yes, it is possible by constructing a tunable resonator into the device. Fukushima et al. constructed a resonator into the SiO_2/TiO_2 thin films using Au nanoparticles and studied the tunable amplification of the emission from Rh-B doped thin film prepared by pre-doping method [143]. A two beam interference of a laser was performed to produce nanoparticles in spatial periodical grating structure by photoreduction of chloroauric ion doped in the film. This periodical grating structure functions as a distributed feedback resonator and the modulation of the grating intervals leads to tunable amplification of the laser light emission. In this way, a feedback gained narrowing of laser emission with FWHM~1 nm compared to 20 nm FWHM of due to amplified spontaneous emission from Rh-B/sol-gel thin film was achieved [143] together with lasing peak tenability in 588 to 620 nm wavelength region. This is really good piece of work and further attention is needed.

The basic requirements for a solid matrix to be used as host for organic laser dyes molecules are transparency to pump and lasing wavelengths, high optical quality with low level of scattering, high laser damage threshold, and good thermal and photochemical stability. Inorganic sol-gel material retains wide range of transparency, high optical quality, controlled pore structure and high laser damage threshold as earlier discussed. However, polymers on the other hand have better compatibility with organic dyes and superior optical homogeneity [44, 144] apart from easy modification in the structure and composition which does allow their properties to be controlled and as a result can be optimized for specific applications [145]. But the thermal degradation under laser irradiation limits the photostability of dye doped polymeric materials. In addition, these polymers have relatively high value of thermal coefficient of refractive index which causes thermal lensing effects. These drawbacks of polymers limit their use as host matrix in the field of solid state dye laser. Is there any possibility to have a host matrix that can have benefits of sol-gel and polymer simultaneously? It will be very nice and fantastic to have materials like this. In this regard, development of inorganic-organic composite materials is a good kind of effort. Costela et al. have done good amount of work in this direction [146-148]. Recently, they have reported the lasing and photostability of Rh-6G dye impregnated sol-gel sample prepared by post-doping method [149] as reported by Rahn et al. [39] and Deshpande et al. [38]. First glass samples were prepared and then immersed into the solution of Rh-6G in mixture of polymerizable organic monomers, allowing Rh-6G and monomer molecules to get enter into the pores of the sol-gel matrix by diffusion process. In this way, the microenvironment in the pores is modified which may modify the performance of dye/sol-gel samples. A 35% of lasing efficiency and no change in output lasing intensity at peak even after 10^5 pump pulses with second harmonic of Nd:YAG laser 5mJ/pulse energy and 10Hz repetition rate have been reported. More importantly, the density of the xerogel is crucial which is related to pores distribution and pore size. Smaller pore size may not be suitable for the entry of monomers resulting less protection for dye molecule and photobleaching may enhance reducing the laser performance. We should always remember, the host material prepared in this process is biphasic. More research work is required along this direction.

Evidently, one can understand that modification in method of preparation may result into good performance for sol-gel based solid state dye laser in terms of lasing efficiency and photostability of active species. In this regard, Deshpande et al. have done enlarged amount of work using various Rhodamines dyes [38, 141, 150-152]. For dye molecule trapping into the pores of sol-gel materials, they used pre-doping and pot-doping methods. Acid catalysed hydrolysis of tetraethoxysilane in presence of primary alcohol as common solvent and water was carried out at room temperature for glass sol preparation. In pre-doping method, dye have been doped into two types of materials-(1) prepared using only acid catalysed and (2) prepared using a KOH /water solution in acid catalysed prepared glass sol [38, 150-152] and the performances (quantum yield, laser efficiency, photostability) of various Rhodamine dyes have been found better in sol-gel sample prepared by second method than first one. In parallel, they also used post-doping method for dye impregnation into the pores of the sol-gel materials. Formamide was used as DCCA in starting material precursor to control pores distribution and pores size. Among all the three types of sol-gel materials prepared by pre-doping and post-doping, the post-doping method have been identified as the best method for the preparation of Rhodamines dyes impregnated sol-gel materials for sol-gel based solid state dye laser [38, 150-152]. Material bulk density has been found to be playing important role in the lasing performances of these materials. The lowest bulk density of 1.41g/cm^3 for post-doped sample compared to 1.6g/cm^3 for pure acid catalysed (pre-doped) and 1.65g/cm^3 for KOH solution used along with acid catalysis (pre-doped) glass samples have been reported [150]. The photophysical properties and lasing performances of Rhodamine-B, Rhodamine-19 and Rhodamine-110 have shown the best performance in post doped samples among all the three types of samples as described above [38, 150-152]. Very recently, Deshpande et al. have reported the effect of glycerol, used as drying control chemical additive, on optical and mechanical properties of sol-gel materials and found highly dependent on the amount of glycerol used [153]. Just recently, Sharma et al. have studied the spectroscopic and lasing properties of many xanthenes dye encapsulated in sol-gel silica and polymer host and concluded better photostability of dyes in sol-gel compared to polymer [168].

3.3. Coumarin and Other Dyes in Blue-Green Region

Much research work has been done on solid state dye lasers based on dyes of the Rhodamine and Pyrromethene families, with emission in yellow and red spectral regions. Less attention has been paid to the development of solid-state dye lasers with emission in the technologically important blue-green spectral regions. Some studies on the lasing properties of a number of blue-green emitting dyes incorporated into sol-gel materials have appeared in last few years [75-76, 82-83, 85-87, 97, 140, 155-158].

D. Lo and his co-workers have done really a good amount of work on coumarin dyes doped sol-gel solid state dye lasers [75, 83, 140]. They incorporated coumarin-481 dye molecules into sol-gel sample and observed superradiant emission peaking around 535 nm with peak power 7MW, conversion efficiency of 36.6% and 89mJ per pulse enegy [75]. The FWHM of fluorescence was nearly 60 nm and the same of superradiance nearly 8 nm. In time domain, the fluorescence signal was about 20 ns FWHM. However, with focused beam excitation, the same duration was reduced to 12 ns and the peak intensity was more than one hundred thousand times stronger than that of fluorescence pulse. These have been achieved without the aid of a resonator cavity. They also reported the fluorescence and lasing properties of several coumarin dye doped in sol-gel materials prepared by pre-doping method using formamide as DCCA [140]. A narrow line-width operation from coumarin-460 doped into sol-gel prepared by pre-doping method was studied and lasing peak of 474 nm, efficiency

0.87% and laser line-width 4.5 GHz were observed [83] under excitation of 308 nm using double grating resonator cavity. Exalite-377 has also been impregnated into sol-gel by pre-doping method and spectroscopic properties and lasing have been observed [83, 155, 158]. The absorption/fluorescence peaks were 328 nm and a double peak structure 360 nm, 375 nm (with excitation 337 nm) respectively. Lasing action from E-377 have been seen under short pulse (1ns) nitrogen laser pumping at 337 nm [155]. With a grating as the wavelength selection element, the tenability from 367 to 387 nm was achieved. Even after reporting a slope efficiency of 34.7% using a resonator cavity consisting of two flat mirrors, the half-lives were 2000 and 700 pulses at 1 and 3.5 Hz respectively. Here it is important to note, the emission spectrum is dye concentration sensitive and in addition to normal lasing wavelength of ~360 nm, a superradiance peak wavelength of 378 nm could be seen sometimes depending on dye concentration in sol-gel [158]. Narrow line width studies of E-377 in sol-gel have also been reported with line width of 4.2 GHz [83]. Stilbene-420 is another important laser dye which is well known for spectral response in 400 to 550 nm region [140] with superradiance peak ~440 nm. Regarding detail lasing characteristics of this dye, very less information is available.

Recently few reports have shown improved performance of coumarin dyes in organically modified silicate (ORMOSIL) prepared by sol-gel method [159-160] with limited transmission in 200 to 400 nm region compared to sol-gel. The laser output energy of coumarin-440 doped in ORMOSIL was measured and no noticeable decrease in the output was observed even after several thousand of pump pulses (~5mJ/pulse, 1Hz) [159]. Further optimization in the preparation and the measurement process and technique may yield far better results.

Deshpande et al. have studied the spectroscopic properties of few important coumarin dyes in sol-gel glasses [135, 161-162] and the performances have been found to deviating compared to their solution counterpart which is normally considered as standard. These deviations in standard spectroscopic properties were only seen in case of sample prepared by pre-dope method and no such deviations could be seen in case of samples prepared by post dope method [161]. Here one important point, the dye molecules have been seen to coming out of the solid cage with drying time and accumulating on the surface of the samples. This might be due to the lack of chemical and mechanical compatibility. Thus the method of preparation, basically related to modification in micro molecular environment inside the host matrix, has been seen to have playing an important role in the performances of dye impregnated sol-gel solids in terms of spectroscopic properties and lasing action [135, 161-162]. In my knowledge and opinian, coumarin dyes normally show abnormal characteristics (leaching, bleaching, protonation in ground and excited state, aggregate formation etc) when incorporated in pure sol-gel glasses. These are due to lack of compatibility between dye molecule and host matrix. In this regards, ORMOSIL may be used instead of pure sol-gel as host matrix or dye impregnation by post dope method using dye polymer solution may enhance the performance. In the glass preparation for the use in post dope method, a careful and controlled use of DCCA is required to control the pore size and density.

3.4. Pyromethane and Few Other Dyes

Pyromethance constitutes an important class in the family of laser dyes. A number of reports are available on spectroscopic and lasing properties of these laser dyes in sol-gel [39, 142, 149, 160, 163-166]. In addition to Pyromethene-567 (P-567), Rahn et al. have studied

Perylene Orange (KF-241) and Perylene Red (KF-856) in various types of host matrices including sol-gel glass [39] and reported the best performance of P-567 in sol-gel among various kinds of hosts. All these dyes have been incorporated in the sol-gel cage by post-doping method. In case of P-567 in sol-gel, a slope efficiency of (16±2) %, lasing peak wavelength 556 nm and half-life of 3500±1000 pulses have been reported. However, KF-241 has the best performance in polycomposite glass with slope efficiency 29±2%, lasing wavelength 578 nm and half-life of 2300±200 pulses. KF-856 has shown almost comparable performance in polycomposite glass and ormosil and better than sol-gel. Pulse length dependent lasing study of all above mentioned dyes have been carried out and the best long pulse length laser performance was from Perylene Orange in polycomposite glass. Latter Ahmad et al. have observed enhanced slope efficiency 25% and photostability 40 $GJmol^{-1}$ of P-567 post-doped sol-gel sample impregnated with PMMA [142] compared to results published by Rhan et al. [39]. Costela et al. synthesized new kind of hybrid solid state dye laser materials based on highly porous silica aerogel and impregnated P-567 and P-597 by post-doped method [149]. The open porous network of the aerogel was saturated with laser dyes dissolved in aappropriate organic monomers and polymerization took place inside the silica structure. Under the demanding condition of tightly focused transversal pumping with 532 nm, 5 mJ pulses at 10 Hz repetition rate, the dye P-567 exhibited laser action with only 10% drop in the lasing output after 10^6 pump pulses in the same position of the sample. Normally, the output intensity of the laser with number of pump pulses either should decrease or at least remain constant (in ideal condition means no degradation). But here it is important to note, the laser output of P-567 has been increasing (with pump pulses) initially up to a maximum and then attains the initial value after nearly 10^6 pulses with fluctuating nature in intensity. A similar type of irregular behavior has been previously observed with other pyromethene dyes incorporated into silica gel matrices earlier [167]. No non-speculative answer is available and further study is required to clarify this point. One of the possible speculative explanations may be associated with higher dye concentration doped into sol-gel sample than optimum. However, others needed to study carefully. At this point, I am not in hurry to make any speculative statement. In my opinion much work along this direction with more photostable dyes including pyromethene is needed and a particular attention should be paid to enhance chemical and mechanical compatibility of dye doped host for the best development of sol-gel based solid state dye laser. Stilbene-420, LDS-759, LD-800, Perilymide dye LFR and sulfoRhodamine-640 have also been studied in sol-gel [80, 140] and very a few reports are available on these dyes in sol-gel. Therefore, further attention needed to pay.

3.5. Energy Transfer Based Solid State Lasers

Energy transfer between donor and acceptor dye molecules is an important mechanism for extending the lasing wavelength [168], enhancement in tunability [169], and also to increase the power of dye laser [170-173]. Many reports on energy transfer dye lasers are available with a number of donor-acceptor pairs using coumarins, xanthenes and other laser dyes [174-176]. Dye lasers based on energy transfer mechanism have also been reported using mixture of more than two dyes resulting wide tunability [177-178].

The transfer of electronic excitation energy from a donor to an acceptor molecule generally takes place either by a direct coupling (collisional transfer or resonance transfer of the excitation due to long range dipole-dipole interaction) or by a fluorescence- absorption process (radiative transfer), if the two spectra i.e. fluorescence of donor and absorption of acceptor, are sufficiently overlapped [25, 179]. The mechanism of energy transfer is dependent on method of excitation and fluorescence characteristics of donor [25, 131]. The main mechanisms of energy transfer that have been proposed are:

(I). Radiavle energy transfer: A radiative energy transfer may occur if the fluorescence of donor is absorbed by acceptor molecule. This is only possible if fluorescence spectrum of donor and absorption spectrum of acceptor sufficiently overlap. This process can be represented as

$$D^* \rightarrow D + h\nu \tag{6a}$$

$$A + h\nu \rightarrow A^* \tag{6b}$$

where the asterisk indicates an electronically excited state. In a radiative energy transfer process, the fluorescence lifetime of both the donor and the acceptor remain unchanged.

(II). Collisional energy transfer: This process is nonradiative and viscosity dependent. A very small value of critical distance (R_0) between a donor and acceptor molecules may favor this process. This process is mostly favourable to occur in gas and liquid phase. In case of solids, the diffusion controlled collision is restricted and hence the process too.

(III). Resonance energy transfer: A donor molecule in its excited state may transfer the excitation energy to an acceptor molecule in the ground state lying in the proximity through dipole-dipole type coupling if the proximity is defined as the separation between the two molecules typically of the order of less than 10 nm. The mechanism of energy transfer involved here is known as Forster resonance energy transfer. When both the two types of molecules are fluorescent, the term "fluorescence resonance energy transfer (FRET)" is often used instead, although the energy is not actually transferred by fluorescence mechanism. In FRET the radius of interaction is much smaller than wavelength of emitted light. In the proximity region as defined above, the excited donor molecule emits virtual photon those are absorbed by acceptor molecule in the ground state. These virtual photons are not detectable since their existence violets the laws of energy and momentum conservation and therefore, FRET is radiationless process. This process contributes the donor fluorescence quenching and enhances the acceptor fluorescence yield. The process of fluorescence quenching is defined as that competes with the spontaneous emission process and thereby reduces the lifetime of the excited state of the donor molecule. In this regard, the fluorescence quenching of donor because of interaction with acceptor can be expressed as

$$D^* + A \xrightarrow{K_{ET}} A^* + D \tag{7}$$

Where asterix indicates excited state and K_{ET} is the energy transfer rate constant. Energy transfer reduces the emission intensity from D^* and sensitizes emission from A^*.

From the knowledge of critical distance (R_0) between donor and acceptor dye molecules, and fluorescence lifetimes of donor and acceptor in the mixture of the two, the nature of excitation energy transfer can be understood. Forster proposed a theory for such a nonradiative energy transfer and has shown that critical transfer distance (R_0) between donor and acceptor molecules can be expressed by the relation

$$R_0^6 = \frac{9\times 10^3 \ln 10 K^2 \phi_D}{128\pi^5 n^4 N_0} \int \frac{F_D(\bar{\upsilon})\varepsilon_A(\bar{\upsilon})d\upsilon}{\upsilon_0^4} \tag{8}$$

where ν_0 is the wave number taken as average of absorption maximum of A and emission maximum of D, $F_D(\nu)$ is the emission spectra of the donor such that $\int F_D(\nu)d\nu = 1$, $\varepsilon_A(\nu)$ is the absorption spectrum of the acceptor, N_0 is Avogadro's number, n is the refractive index of the host matrix, k^2 is the square of the molecular orientation factor taken here as 2/3 and ϕ_D is the fluorescence quantum yield of pure donor. The critical distance R_0 is also known as Forster distance and is defined as the distance between donor and acceptor at which the probability of energy transfer is 50%. Further, it has also been shown by Forster that energy transfer phenomenon can be regarded as bimolecular process and therefore, Stern-Volmer relation for bimolecular quenching, given by

$$\frac{I_0}{I} - 1 = K_{SV}[A] \tag{9}$$

can be utilized for the experimental determination of energy transfer rate constant (K_{ET}) and critical transfer distance (R_0). Where I_0 and I are fluorescence intensity of donor species in absence and in the presence of acceptor species respectively, K_{sv} and [A] are rate constant for bimolecular quenching (determined by slope of Stern-Volmer plot) and acceptor concentration respectively. Now, K_{ET} and R_0 can be calculated using following formulae:

$$K_{ET} = K_{SV}/\tau_D \tag{10}$$

where τ_D is the lifetime of pure donor and

$$R_0 = 7.35/([A]_{1/2})^{1/3} \tag{11}$$

where $[A]_{1/2}$ is the concentration of the acceptor to decrease the fluorescence of donor by one half.

The transfer efficiency is an important parameter in the energy transfer process. Normally, it is measured using relative fluorescence intensity of the donor in the presence (I_{DA}) and absence (I_D) of the acceptor. The efficiency of energy transfer between donor and acceptor is given by equation [131]

$$E = \left(1 - \frac{I_{DA}}{I_D}\right) \tag{12a}$$

This is based on the fact that the loss of the relative intensity of the donor in the presence of acceptor is due entirely to energy transfer from the donor to the acceptor.

However, the transfer efficiency can also be calculated from the knowledge of lifetimes in the respective conditions i. e. the fluorescence lifetime of donor in the presence and absence of the acceptor using the expression given below.

$$E = \left(1 - \frac{\tau_{DA}}{\tau_D}\right) \tag{12b}$$

It is important to note, equations (12a) and (12b) are applicable to the donor-acceptor pairs if those are separated by a fixed distance. Clearly, the situation is found true in case of donor-acceptor in solution and dispersed in membrane. More over, the decay of the donor in the presence (τ_{DA}) and absence (τ_D) of acceptor should be single exponential for the calculation of transfer efficiency using equation (12b). The decay rate in the presence of acceptor will remain single exponential if and only if the distance between donor and acceptor is unique i.e. single distance. However, based on the quantum electrodynamical calculations, the radiationless (FRET) and radiative energy s are the short and long-range asymptotes of a single unified mechanism [180-181]. The study of energy transfer between donor and acceptor dye molecules in condensed phase has become a subject of interest after the pioneering work by Forster [182-183] and later by Dexter [184].

Yang and his worker have studied a number of dyes pair doped in ORMOSIL from solid state dye laser point of view [160, 163-166]. Initially, they studied the performance of pair of dyes as donor and acceptor doped in ORMOSIL by pre-doping method [165-166]. Laser dyes perylene orange (p-orange) or perylenne-red (p-red) or pyromethene-567 (p-567) as acceptor were co-doped with coumarin-440 (C-440) or coumarin-500 (C-500) as donor into ORMOSIL by pre-doped sol-gel process and relative concentration dependent laser performances have been studied [165]. Among all studied range of concentrations, a maximum transfer efficiency, transfer probability and energy transfer rate constant of 69.7%, 17.38×10^{-8} s^{-1} and 12.1×10^{-12} $M^{-1}s^{-1}$ respectively have been observed with C-500:p-orange/5:1 molar ratio of dye pair. For the same dye pair and concentration, the slope efficiency and laser threshold have been found equal to 9.9% and 115.1μJ respectively with 6 mm sample thickness pumped by 355 nm Nd:YAG laser in longitudinal configuration and cavity length 5cm (the output coupler with 60-70% transmission at the laser wavelength). But keeping the donor: acceptor molar ratio 5:1 constant for both the dyes pairs C-440:p-567 and C-500:p-567, the slope efficiency and laser threshold values have been reported to be 40.9%, 121.1 μJ and 37.3%, 156.9 μJ respectively. Obviously, low value of the efficiency is of no practical use. The slope efficiencies of p-red in the presence of C-440 and C-500 dyes improved at least 1-fold compared to that of p-red in the absence of coumarin dyes. However, only marginal improvements in slope efficiencies of p-567 could be seen in the presence of coumarin dyes. The photostability of p-567 in the absence/presence of C-440 was studied in the exposure of 254 nm UV lamp light with respect to time. The p-567 was found to be degrading within few hours, however, it starts degrading after 600h in the presence of C-400 dye. But in case of p-orange, no improvement in the photostability in the presence of coumarin dyes was observed, suggesting an investigation in the role of coumarin dyes. Is it due to interactions between triplet state of p-567 and ground state of coumarin dye molecules? Authors expressed, the energy compatibility and dye concentration may play an important role in enhancement of efficiency and photostability [165]. Several combination of two (p-567: p-red) and three (C-440: p-567: p-red) dyes have also been studied under the same conditions as reported earlier [165] except a difference in excitation wavelength (532 nm) and cavity length (3.5 cm) [166] and the best performance of C-440:p-567:p-red/5:2:1 was observed. The slope efficiency, laser threshold, tunable range, laser lifetime have been reported 20.8%, 111.9μJ, ~60 nm, >45000 pulses and 17.7%, 59.2μJ, ~24 nm, nil (not reported) for C-440: p-567: p-red/5:2:1 (molar ratio) and p-567: p-red/ 1:1 (molar ratio) respectively [166]. Several coumarin dyes as donor with p-567 (combination of two dyes) and with p-567: p-red (combination of three dyes) have been investigated with change in donor concentration while keeping other one (p-567) or two (p-567: p-red) dyes concentration constant [163-164]. The slope efficiency, laser threshold, tunable range and conversion efficiency can be seen in the **table 1**. In case of two dyes pair, the best performance can be noticed with C-440: p-567 in MTES (methyltriethoxysilane) and C-503: p-red/C-540A: p-red in VTES (vyniltriethoxysilane), while C-440: p-567: p-red in VTES could be seen in case of three dyes (**Table 1**). The mechanism responsible for the improvement in the laser efficiencies of p-red and p-567 in the presence of coumarin dyes was the quenching of the triplet-state of p-red and or p-567 by ground state coumarin dyes by a multiphoton process [163-164]. Thus excited state coumsrin dyes could transfer energy to the ground state p-red and or p-567, which reduced the triplet state absorption loss and increased the absorption cross section of p-red and or p-567, and as a result, improved the laser efficiency of p-red and or p-567. It was found that the effect of various coumarin dyes on the laser performances of p-red and p-567 varied sharply, depending on the dye pair combination, i.e. higher laser efficiencies of p-red in the presence of coumarin dyes were observed in nearly all cases while the slope efficiency of p-567 could only be increased by introducing C-440, C-500 andC-540A. It should also be noted that the laser output wavelength of p-red blue-shifted 6 ~ 11 nm in the presence of coumarin dyes as compared with that of p-red in the absence of coumarin dyes, while less than 1nm blue-shifts of the laser output wavelength of p-567 were only observed in the cases in which the laser efficiencies of p-567 were enhanced.

The effect of various coumarin dyes on the laser performances of p-red and or p-567 can be broadly understand based on the following two aspects:

The energy transfer efficiency between dye pairs and the improvement in the laser performances of p-red and p-567 are dependent on the energy level compatibility between the dyes, i.e. the energy level compatibility between the triplet-state p-red/p-567 and the ground state coumarin dyes together with the energy level compatibility between excited state coumarin dyes and ground-state p-red/p-567. For example, it had been observed previously that a small absorption peak of C-500 at about 920 nm was found which coincided with the triplet-state energy of p-567, thus there can be a possibility of triplet-state quenching of p-567 by C-500 [165]. It was suggested that the triplet-state quenching of p-red/p-567 by coumarin dyes might be varied, depending on the absorption of coumarin dyes in the longer wavelength and the triplet-state energy of p-red/p-567, which resulted in the energy level compatibility between the triplet-state p-red/p-567 and ground state coumarin dyes. It should also be noted that the energy level compatibility between excited state coumarin dyes and ground-state p-red/p-567 depended on the overlap between the emission spectra of coumarin dyes and the absorption spectra of p-red/p-567, extensively studied by other authors. It should be mentioned that the triplet-state yield p-567 was much less that of p-red, thereby resulting in less excited-state coumarin dyes via the multi-photon process, and the relative absorption of p-red in 400~ 500 nm was larger than that of p-567 in the same region, thereby resulting in a larger overlap between the emission spectra of the coumarin dyes [165].

Table 1A. Laser performance of P-red doped in VTES-derived ORMOSILs in the absence and presence of P-567 and coumarin dyes [164].

Dye pair (molar ratio)	Slope efficiency (%)	Laser Threshold (μJ)	Tunable range (nm)	Conversion efficiency (%)
C440:p567:p-red (2.5:2:1)	21.0	95.3	570.7-632	2.4
C440:p567:p-red (5:2:1)	15.4	115.7	573.8-642.8	1.5
C440:p567:p-red (7.5:2:1)	11.8	147.6	-	-
C460:p567:p-red (2.5:2:1)	17.6	114.2	582.0-639.2	1.4
C460:p567:p-red (5:2:1)	12.5	146.8	579.8-634.6	0.8
C503:p567:p-red (2.5:2:1)	17.6	105.4	575.8-637.2	2.4
C503:p567:p-red (5:2:1)	18.8	126.4	574.3-640.4	1.9
C503:p567:p-red (7.5:2:1)	16.4	121.3	-	-
C540A:p567:p-red (2.5:2:1)	21.0	100.6	562.3-642.4	2.4
C540A:p567:p-red (2.5:2:1)	16.0	110.3	577.6-647.6	2.1
C540A:p567:p-red (2.5:2:1)	16.4	124.7	573.1-638.4	1.4
p-567:p-red (2:1)	14.4	149.6	578.4-615.5	0.7
p-red	10.4	103.5	599.5-613.3	<0.2

Table 1B. Laser performance of P-red doped in VTES-derived ORMOSILs in the absence and presence of coumarin dyes [163].

Dye pair (molar ratio)	Slope efficiency (%)	Laser Threshold (μJ)	Tunable range (nm)	Conversion efficiency (%)
C440:p-red (10:1)	11.8	74.8	-	-
C440:p-red (20:1)	11.7	70.4	-	-
C440:p-red (40:1)	5.1	131.9	-	-
C460:p-red (1:1)	7.8	91.3	583.82-613.43	-
C460:p-red (2.5:1)	12.6	59.6	585.10-613.12	-
C460:p-red (5:1)	11.8	53.3	587.65-611.34	-
C460:p-red (7.5:1)	11.0	53.0	586.43-613.08	-
C503:p-red (1:1)	14.0	49.6	585.30-614.13	0.52
C503:p-red (2.5:1)	12.2	49.6	589.66-613.90	-
C503:p-red (5:1)	11.4	61.8	588.79-613.78	-
C503:p-red (7.5:1)	10.6	67.0	591.66-611.95	-
C540A:p-red (5:1)	15.5	46.8	-	-
C540A:p-red (10:1)	13.7	54.5	-	-
C540A:p-red (15:1)	10.1	65.2	-	-
p-red	7.3	118.9	599.49-613.25	<0.2

Table 1C. Laser performance of P-red doped in MTES-derived ORMOSILs in the absence and presence of coumarin dyes [163].

Dye pair (molar ratio)	Slope efficiency (%)	Laser Threshold (μJ)	Tunable range (nm)	Conversion efficiency (%)
C440:p-567 (2.5:1)	72.4	53.7	542.35-567.10	2.32
C440:p-567 (5:1)	74.7	47.0	540.95-572.16	2.2
C440:p-567 (7.5:1)	65.6	66.0	540.34-578.25	2.824
C460:p-567 (2.5:1)	62.4	52.0	-	-
C460:p-567 (5:1)	63.2	56.7	-	-
C460:p-567 (7.5:1)	58.4	66.3	-	-
C500:p-567 (5:1)	-	-	541.73-566.40	-
C503:p-567 (2.5:1)	63.7	49.0	-	-
C503:p-567 (5:1)	60.1	52.7	-	-
C503:p-567 (7.5:1)	55.7	55.2	-	-
C540A:p-567 (2.5:1)	76.3	50.0	541.47-571.63	-
C540A:p-567 (5:1)	52.3	51.4	541.03-572.77	-
C540A:p-567 (7.5:1)	48.4	54.0	540.77-574.07	-
p-567	64.2	57.8	542.34-566.93	1.427

Further more, the dye concentration is one of the very important parameters which can decide the energy transfer efficiency between dye pairs and the improvement in the laser performances of p-red and p567. This can be explained by the intermolecular distance between dye molecules and the long range dipole-dipole energy transfer mechanism. In other words, increasing the donor concentration beyond a particular value may cause the reduction in the effective distance between the coumarin molecules which may cause an energy transfer to take place within the donor system itself. It is also to be noted that too high coumarin dye concentration may influence the hydrolysis and poly-condensation processes of the precursors and can lead to more light scattering increasing optical losses [163-164]. As far as photostability of p-567 is concerned, a three fold improvement can be seen due to the presence of coumarin dyes. However, the presence of coumarin dyes would not largely improve the photostability of p-red (except for a marginal enhancement of 25%). This may be related to the photodegradation of p-red molecules. Is it due to the photochemical reaction related to the triplet state of p-red or some thing else? The photochemical reaction related to the triplet state of p-red does not play a major role in the photodegradation mechanism of p-red in the presence of p-567 and C-440 [163]. The photoreduction may have contributed a major role in photodegradation mechanism of p-red as recently suggested for perylene di-imide [185]. Three laser dyes p-567, p-red and C-540A pre-doped ORMOSIL based sol-gel planner waveguide has been constructed on glass substrate and distributed-feedback (DFB) laser action was achieved after pumping by Nd:YAG laser at 532 nm [160]. A wide narrow line-width output tuning in a range of more than 70 nm from 564.5 to 635.1 nm have been achieved by varying the period of the gain modulation generated by nanosecond Nd:YAG laser at excitation wavelength of 532 nm. Authors have claimed, this is the widest tunable range of dye solid state DFB laser materials ever reported in this region. Energy transfer based lasing action from Rh-6G/Rh-B and Rh-B/p-red has also been reported [186-187]. Rh-B/p-red/xerogel samples were prepared by pre-dope sol-gel method with varying relative concentration of dyes and the samples were excited with 6 ns 532 nm output wavelength of frequency doubled Nd:YAG laser using a 10 cm long stable plano-concave linear cavity [186]. A wide tunable range of lasing wavelength from 604 to 638 nm with slope efficiency in the range 8 to 23% has been observed from entire concentration range under study. Both radiative and non-radiave processes have been reported as responsible mechanism for the energy transfer. A energy transfer based lasing action from Rh-6G/Rh-B pre-doped sol-gel samples have been observed with efficiency 10.3 to 23.4%, pump threshold energy 2.47 to 3.5 mJ, lasing peak 614 to 630 nm and tunable range 2 to 8 nm due to variation in relative dyes concentration [187].

Overall it can be concluded, the hardest work is remaining to do is the development of highly photostable solid state dye laser with a workable efficiency and good conversion efficiency.

3.6. Nanoparticle of Organic Dye?

Organic nanoparticle is more fascinating and interesting topic in spectroscopy since early last decade. In organic molecules, electrons are localized in chemical bonds or just over one molecule and consequently, one cannot expect the effect of electron confinement in organic crystal like in metal or inorganic semiconductor nanoparticles. Nevertheless, organic nanoparticles show size effect on absorption and fluorescence spectra, fluorescence enhancement and nonlinear optical response [188-196]. Size dependent optical properties of organic nanoparticles and domain size have been reported by many authors [197-203]. The domain size of organic nanoparticle is quite different from that of the inorganic materials (from few nm to few hundreds nm) [197-203]. The luminescence properties of organic nanocrystals looks to be complicated itself compared to bulk counterpart due to flexible conformations, existence and modification in various inter and intramolecular overlapping structures depending on shape and size. Dual emission has been observed from perylene, pyrene, corone and anthracene nanocrystals recently [204-205] and the existence of two emissive centers has been proposed to explain this dual nature; one in the core and other one in the periphery of the nanoparticle [205]. This is an indication of the crucial and important role of heterogeneous microstructure of organic nanoparticles in complex luminescence characteristics. However, the role of surface effect cannot be ignored and a more detail study is required in this direction. An increase in absorption bandwidth with size of aromatic naocrystals trapped in thin polymer film has been observed with a sudden fall to zero over 12 nm [198] and this new phenomenon has been explained by taking into account the effect of surface charge (indicating importance of surface effect). In this regard, the advanced spectroscopic tools could be used to study these nanoparticles system. With single nanoparticle spectroscopy, two distinct types of nanoparticles were found with two different emission wavelengths (particles with blue emission and particles with red emission) and this difference in spectral characteristics was attributed to the presence of two morphological types of particles which was further verified by separating the two types of particles by centrifugation [191]. On the basis of available spectroscopic data of various organic compounds, it is concluded that the size effect on organic nanoparticles is related to some molecular conformation, packing, and elastic properties of nanoparticles [192]. This has yielded the idea to call this size effect as structural confinement, an analogue of electron confinement for metals and semiconductor [192]. Can these dye nanoparticles be used as active centre (source) in dye laser? In this regard, the heterogeneous structure of nanoparticle does seem to be the biggest hurdle and more effort is needed in this direction. In my knowledge, there is no report available on the lasing action from dye nanoparticle.

4. Open Question and Future Direction

It is remarkable to note that even after doing a considerable amount of work on the development of sol-gel based solid state dye lasers, a number of problems still remain to be solved.

1) Why do most of the organic laser dyes is less photostable? This first and major problem is associated to the photostability of dye laser. Normally, the photostability of dye laser is expressed in term of half life which is defined as the number of pump pulses required to decrease the laser output intensity to half of its original peak value. The half life of a dye laser should be as higher as possible and this is the biggest challenge in the field of solid state dye laser.
2) What is the theoretical understanding about the photodegradation of laser dye? Is there any model that can explain the photodegradation data of laser dye? In this regards (photodegradation of active species in sol-gel host), a large number of experimental reports are available. However, theorectical development is really needed. In my knowledge, no unified model is available which can explain the experimental data well.
3) What is the status of the solid state dye laser in the blue-green region? The blue-green region of the electromagnetic spectrum is of great technological importance. However, very a little attention has been paid to develop this region in the entire work done in the field of sol-gel based solid state dye laser. Keeping into mind the compatibility between laser dye and host matrix, the future work should be carried out.
4) The work on the use of more than two laser dyes in the energy transfer based sol-gel solid state dye laser should be carried out to get wide tunable range, highly efficient and highly photostable solid state dye laser. Particularly, pyromethene should be taken into consideration as being fairly highly photostable dye. Quantum dots are highly photostable. Using a combination of quantum dot and photostable dye with suitable spectroscopic properties, a highly photostable and efficient energy transfer sol-gel based solid state dye laser is expected.
5) Dye laser in infrared region? Technologically important infrared (IR) region has been ignored completely. Work with IR dyes should be carried out to cover the wavelength region from 800 to 1000 nm at least.
6) Is ormosil the best and suitable host for laser dye? ORMOSILs have shown good chemical compatibility and mechanical strength with dye molecules which have resulted in the enhancement of laser performance. And hence much attention in this direction is needed to pay i.e. ORMOSIL based development of solid state dye lasers.
7) Lasing action from nanoparticles of organic laser dyes? Laser emission from nanoparticle of organic dye is a quite fascinating idea and work along this direction may open a new pathway toward highly fluorescent hybrid sol-gel nanocomposite.

ACKNOWLEDGMENT

The completion of the work might not have reached up to the successful end without the help of Professor Aparna V. Deshpande (former head of the department of physics, Institute of chemical technology, University of Mumbai and my doctoral supervisor). I cannot express my feelings in words. Even I express my sincere thanks with regards to her.

REFERENCES

[1] Wilson J.; Hawkes J. F. B. *Optoelectronics-An Introduction*; Prentice-Hall of India Pvt. Ltd: New Delhi, India, 1996, 155-169
[2] Maiman T. H. *Nature*, 1960, 187, 493-494
[3] Maiman T. H. *Phys. Rev. Lett.* 1960, 4, 564-566
[4] Heitmann U.; Kotteritzsch M.; Heitz S.; Hese A. *Appl. Phys. B* 1992, 55, 419-423
[5] Stuke M. Dye lasers 25 *Years; Topics in Applied Physics;* Springer Verlag: Berlin, Germany, 1992; 70, 95-109
[6] Sorokin P. P.; Lankard J. R. *IBM J. Res. Develop.* 1966, 10, 162-163
[7] Schafer F. P.; Schmidt W.; Volze *J. Appl. Phys. Lett.* 1966, 9, 306-309
[8] Schafer F. P.; Schmidt W.; Marth K. *Phys. Lett. A* 1967, 24, 280-281
[9] Sorokin P. P.; Lankard J. R. *IBM J. Res. Develop.* 1967, 11, 148-148
[10] Pererson O. G.; Tuccio S. A.; Snavely B. B. *Appl. Phys. Lett.* 1970, 17, 245-247
[11] Schafer F. P. *Dye lasers*; Springer Verlag: Berlin, Germany, 1990; 1, 11-390
[12] Schmidt W.; Schafer F. P. *Phys. Lett.* A 1968, 26, 558-559
[13] Bradley D. J.; Durrant J. F.; O' Neill F.; Sutherland B. *Phys. Lett.* 1969, 30A, 535-536
[14] Arthurs E. G.; Bradley D. J.; Roddie A. G. *Appl. Phys. Lett.* 1972, 20, 125-127
[15] Bor Z. IEEE *J. Quantum Electronics* 1980, 16, 517-524
[16] Duarte, *High Power Dye Laser*; Springer Verlag: Berlin, Germany, 1991, 93-182
[17] Suratwala T.; Gardlund Z.; Davidson K.; Uhlmann D. R.; Watson J.; Bonilla S.; Peyghambarian N. *Chem. Mater.* 1998, 10, 199- 209
[18] Rohatgi-Mukherjee K. K. *Fundamentals of Photochemistry; New Age International* (P) Limited: New Delhi, India, 2002, 156-177
[19] Cesey K. G.; Quitevis E. L. *J. Phys. Chem.* 1988, 92, 6590-6594
[20] Onganer Y.; Quitevis E. L. *J. Phys. Chem.* 1992, 96, 7996-8001
[21] Chang T. L.; Cheung H. C. *J. Phys. Chem.* 1992, 96, 4874-4878
[22] Jones II G.; Jackson W. R.; Choi C.-yoo; Bergmark W. R. *J. Phys. Chem.* 1985, 89, 294-300
[23] Jones II G.; Jackson W. R.; Halpern A. M. *Chem. Phys. Lett.* 1980, 72, 391-395
[24] Pringsheim P. Fluorescence and Phosphorescence; *Interscience*: New York, USA, 1949, 322-323
[25] Forster Th. *Discuuss Faraday Soc.* 1959, 27, 7-29
[26] Kellogg R. E. *J. Lumin.* 1970, 1, 2, 435-447
[27] Soffer B. H.; McFarland B. B. *Appl. Phys. Lett.* 1967, 10, 266-267
[28] Peterson O. E.; Snavely B. B. *Appl. Phys. Lett.* 1968, 12, 238-240

[29] Noboikin Y. V.; Ogurtsova L. A.; Podgornyi A. P.; Porovskaya F. S. *Opt. Spectrosk.* 1969, 27, 307-311

[30] Ulrich R.; Weber H. P. *Appl. Opt.* 1972, 11, 428-434

[31] Itoh U.; Takakusa M.; Moriya T.; Saito S. Jap *J. Appl. Phys.* 1977, 16, 1059-1060

[32] Gromov D. A.; Dyumaev K. M.; Manenkov A. A.; Maslyukov A. V.; Matyushin G. A.; Nechitailo V. S.; Prokhorov A. M. *J. Opt. Soc. Am. B* 1985, 2, 1028-1031

[33] O'Connel R. M.; Saito T. T. *Opt. Eng.* 1983, 22, 292-297

[34] Acuna A. V.; Amat-Guerri F.; Costela A.; Douhal A.; Figuera J. M.; Florido F.; Sastre R. *Chem. Phys. Lett.* 1991, 187, 98-102

[35] Amat-Guerri F.; Costela A.; Figuera J. M.; Florido F.; Sastre R. *Chem. Phys.* Lett 1993, 209, 352-356

[36] Costela A.; Florido F.; Garcia-Moreno I.; Duchowicz R.; Amat-Guerri F.; Figuera J. M.; Sastre R. *Appl. Phys. B* 1995, 60, 383-389

[37] Bonder M. V.; Przhonskaya O. V.; Tikhonov E. A. Sov. *J. Quantum Electron* 1989, 19, 1415-1417

[38] Deshpande A. V.; Kumar U. *J. Non-Cryst. Solids* 2002, 306, 149-159

[39] Rahn M. D.; King T. A. *Appl. Opt.* 1995, 34, 8260-8271

[40] Costela A.; Garcia-Moreno I.; Barroso J.; Sastre R. *Appl. Phys. B* 1998, 67, 167-173

[41] Rahn M. D.; King T. A.; Gorman A. A.; Hamblett I. *Appl. Opt.* 1997, 36, 5862-5871

[42] Popov S. Appl. Opt. 1998, 37, 5449-6455

[43] Giffin S. M.; Mckinnie I. T.; Wadsworth W. J.; Woolhouse A. D.; Smitha G. J.; Haskell T. G. *Opt. Commun.* 1999, 161 (1, 2, 3), 163-170

[44] Duarte F. *J. Appl. Opt.* 1994, 33, 3857-3860

[45] Duarte F. J.; Costela A.; Garcia-Moreno I.; Sastre R.; Ehrlich J. J.; Taylor T. S. *Opt. Quantum Electron* 1997, 29, 461-472

[46] Cazeca M. J.; Jiang X.; Kumar J.; Tripathy S. K. *Appl. Opt.* 1997, 36, 4965-4968

[47] Reisfeld R.; Brusilovsky D.; Eyal M.; Miron E.; Burstein Z.; Irvi *J. Chem. Phys. Lett.* 1989, 160, 43-44

[48] Salin F.; le Saux G.; Georges P.; Brun A.; Bagnall C.; Zarzycki J. *Opt. Lett.* 1989, 14, 785-787

[49] Sarkisov S.; Curley M.; Wilkosz A.; Grymalsky V. Opt. Commun. 1999, 161, 132-140

[50] Gu Zu-Han; Peng G. D. *Opt. Lett.* 2000, 25, 375-377

[51] Lin L. T.; Bescher E.; Mackenzie J. D.; dai H.; Stafsudd O. M. *J. Materials Science* 1992, 27, 5523-5528

[52] Lettinga M. P.; Zuilhof H.; Zandvoort Marc A. M. *J. van Phys. Chem. Chem. Phys.* 2000, 2, 3697-3707

[53] Vijila C.; Ramalingam A. *J. Mater. Chem.* 2001, 11, 749-755

[54] Somasundaram G.; Ramalingam A. *J. Photochem. Photobiol. A* 1999, 125, 93-

[55] Somasundaram G.; Ramalingam A. *Chem. Phys. Lett.* 2000, 324, 25-30

[56] Avnir D.; Levy D.; Reisfeld R. *J. Phys. Chem.* 1984, 88, 5956-5959

[57] Avnir D.; Kaufman V. R.; Reisfeld R. *J. Non-Cryst. Solids* 1985, 74, 395-406

[58] Brinker C. J.; Scherer G. W. *Sol-Gel Science: The Physics and Chemistry of Sol-Gel* Processing; Academic Press, Inc.: New York, USA, 1990,25-241

[59] Brinker C. J.; Scherer G. W. *J. Non-Cryst. Solids* 1985, 70, 301-322

[60] Keefer K.D. in Silcon Based Polymaer Science: A Comprehensive Resource; J. M. Zeigler, F. W. G. Fearon; *ACS Advances in Chemistry* Ser. No.224; Americon Chemical Society: Washington DC, USA, 1990, 227-241
[61] Biazzotto J. C.; Sacco H. C.; Ciuffi K. J.; Ferreira A. G.; Serra O. A.; Iamamoto Y. *J. Non-Cryst. Solids* 2000, 273, 186-192
[62] Kobayashi Y.; Muto S.; Matsuzaki A.; Kurokawa Y. *Thin Solid Films* 1992, 213, 126-129
[63] Chaudhuri S. R.; Sarkar A. in *Sol-Gel Optics: Processing and Applications*; L. C. Klein, 1994, 83-101
[64] James P. F. *J. Non-Cryst. Solids* 1988, 100, 93-114
[65] Mackenzie J. D. *J. Non-Cryst. Solids* 1988, 100, 162-168
[66] Toki M.; Miyashita S. M.; Takeuchi T.; Kambe S.; Kochi A. *J. Non-Cryst. Solids* 1988, 100, 479-482
[67] NG L. V.; MeCormick A. V. *J. Phys. Chem.* 1996, 100, 12517-12531
[68] Tailler H. J.; Gobel R.; Hartung U. *J. Non-Cryst. Solids* 1988, 105, 162-164
[69] Kawaguchi T.; Hishikura H.; Iura J.; Kokubu Y. *J. Non-Cryst. Solids* 1984, 63, 61-69
[70] Mizuno T.; Nugget H.; Manage S. *J. Non-Cryst. Solids* 1988, 100, 236-240
[71] Levy D.; Anvil D. *J. Phys. Chem.* 1988, 92, 4734-4738
[72] Mackierran J. M.; Yamanoka S. A.; Dunn B.; Zink Z. I. *J. Phys. Chem.* 1990, 94, 5152-5654
[73] Reisfeld R. *J. Non-Cryst. Solids* 1990, 121, 254-266
[74] Canva M.; Georges P.; Brun A. *J. Non-Cryst. Solids* 1992, 147&148, 636-640
[75] Lo D.; Parris J. E.; Lawless J. L. *Appl. Phys. B* 1992, 55, 365-367
[76] Lam K.S.; Lo D.; Wang K. H. *Opt. Commun.* 1995, 121, 121-124
[77] Ammer F.; Penzkofer A.; Weidner P. *Chem. Phys.* 1995, 192, 325- 331
[78] Fujii T.; Nishikiori H.; Tamura T. *Chem. Phys. Lett.* 1995, 223, 424-429
[79] Nishikiori H.; Fujii T. *J. Phys. Chem. B* 1997, 101, 3680-3687
[80] Finkelstein I.; Ruschin S.; Soreck Y.; Reisfeld R. *Opt. Materials* 1997, 7, 9-13
[81] Garcia M J.; Ramirez J E.; Mondragon M. A. ; Ortega R.; Loza P.; Campero A. *J. Sol-Gel Sci. Technol.* 1998, 13, 657-661
[82] Lam K.S.; Lo D. *Appl. Phys. B* 1998, 66, 427-430
[83] Lo D.; Lam K.S.; Lam S.K.; Ye C. *Opt. Commun.* 1998, 156, 316-320
[84] Hungerford G.; Suhling K.; Ferreira J. A. *J. Photochem. Photobiol. A* 1999, 129, 71-80
[85] Lam S. K.; Zhu X.L.; Lo D. *Appl. Phys. B* 1999, 68, 1151-1153
[86] Pal S. K.; Sukul D.; Mandal D.; Sen S.; Bhattacharyya K. *J. Phys. Chem.* B 2000, 104, 2613- 2616
[87] Zhu X.L.; Lam S. K.; Lo D. *Appl. Opt.* 2000, 39, 3104-3107
[88] Qian G.; Yang Z.; Yang C.; Wang M. *J. Appl. Phys.* 2000, 88, 2503-2508
[89] Sakka S.; Kamiya K.; Makita K.; Yamamoto Y. *J. Non-Cryst. Solids* 1984, 63, 223-235
[90] Sakka S., Kamiya K. *J. Non-Cryst. Solids* 1982, 48, 31-46
[91] Gonzales-Oliver C. J. R.; James P. F.; Rawson H. *J. Non-Cryst. Solids* 1982, 48, 129-152
[92] Nogami N.; Moriya Y. *J. Non-Cryst. Solids* 1980, 37, 191-201
[93] James P. F. *J. Non-Cryst. Solids* 1988, 100, 93-114
[94] Gratz H.; Penzkofer A.; Weidner P. *J. Non-Cryst. Solids* 1995, 189, 50-54

[95] Soreck Y.; Reisfeld R.; Weiss A. M. *Chem. Phys. Lett* 1995, 244, 371-378
[96] Suratwala T.; Gardlund Z.; Davidson K.; Uhlmann D. R.; Watson J.; Bonilla S.; Peyghambarian N. *Chem. Mater.* 1998, 10, 190-198
[97] Knobbe E. T.; Dunn B.; Fuqua P. D.; Nishida F. *Appl. Opts.* 1990, 29, 2729-2733
[98] Altman J. C.; Stone R. E.; Dunn B.; Nishida F. *IEEE Photonics Technol. Lett.* 1991, 3, 189-190
[99] Larrue D.; Zarzycki J.; Canva M.; Georges P.; Bentivegna F.; Brun A. *Opt. Commun.* 1994, 110, 125-130
[100] Hu L.; Jiang Z. *Opt. Commun.* 1998, 148, 275- 280
[101] Prassas M.; Hench L. L. in *Ultrastructure Processing of Ceramics, Glasses, and Composites;* L. L. Hench and D. R. Ulrich; John Wiley and Sons: New York, USA, 1984, 100-137
[102] Brinker C. J. *J. Non-Cryst. Solids* 1988, 100, 31-50
[103] Yamane M.; Inoue S.; Nakazawa K. *J. Non-Cryst. Solids* 1982, 48, 153-159
[104] Schroeder H. in *Physics of Thin Films*; G. Hass and R. E. Thun; Academic press: New York, USA, 1969; 5, 87-140
[105] Iler R. K. *The Chemistry of Silica*; John Wiley & Sons: New York, USA, 1979, 462-599
[106] Zarzycki J. in *Ultrastructure Processing of Ceramics, Glasses, and Composites*; L. L. Hench and D. R. Ulrich; John Wiley and Sons: New York, USA, 1984, 27-49
[107] Wallace S.; Hench L. L. in *Better ceramics Through Chemistry*; C. J. Brinker, D. E. Clark and D. R. Ulrich; North Holland: New York, USA, 1984, 47-49
[108] Hench L. L. in *Science of Ceramic Chemical Processing*; L. L. Hench and D. R. Ulrich; John Wiley & Sons: New York, USA, 1986, 52-57
[109] Wang S. H., Hench L. L. in *Better ceramics Through Chemistry*; C. J. Brinker, D. E. Clark and D. R. Ulrich; North Holland: New York, USA, 1984, 71-75
[110] Hench L. L.; Orcel G. *J. Non-Cryst. Solids* 1986, 82, 1-10
[111] Orcel G.; Hench L. L. *J. Non-Cryst. Solids* 1986, 79, 177-194
[112] Mackenzie J. D. in *Science of Ceramic Chemical Processing*; L. L. Hench and D. R. Ulrich; John Wiley & Sons: New York, USA, 1986, 113-121
[113] Zayat M.; Reisfeld R.; Minti H.; Zastrow A. *Solar Energy Materials and Solar Cells* 1998, 54, 109-120
[114] Krasovec U. O.; Orel B.; Reisfeld R. *Electrochemical and Solid State Lett.* 1998, 1, 104-106
[115] Orel B.; Krasovec U. O.; Groselj N.; Kosec M.; Drazic D.; Reisfeld R. *Sol-Gel Sci. Technol.* 1999, 14, 291-308
[116] Orel B.; Groselj N.; Krasovec U. O.; Bukovec P.; Reisfeld R. *Sensors and Actuators B* 1998, 50, 234-245
[117] Reisfeld R.; Shamrakov D.; Jorgensen C. K. *Solar Energy Materials and Solar Cells* 1994, 33, 417-427
[118] Wolfbeis O.; Reisfeld R.; Oehme I. in *Sol-Gel Sensors. Optical and Electronic Phenomena in Sol-Gel Glasses and Modern Applications*; R. Reisfeld and C. K. Jorgensen, Structure and Bonding, Springer Verlag: 1996, 85, 51-55
[119] Reisfeld R.; Shamrakov D. *Sensors and Materials* 1996, 8, 439-443
[120] Sorek Y.; Reisfeld R.; Finkelstein I.; Ruschin R. *Appl. Phys. Lett.* 1995, 66, 1169-1171
[121] Reisfeld R.; Minti H.; Eyal M.; Chernyak V. *J. Nonlinear Optics* 1993, 5 339-342

[122] Reisfeld R. in *New Materials for Nonlinear Optics, Optical and Electronic Phenomena in Sol-Gel Glasses and Modern Applications*; R. Reisfeld and C. K. Jorgensen; Structure and Bonding, Springer Verlag: 1996, 85, 99-107
[123] Strek W.; Sokolnicki J.; Legendziewicz J.; Maruszewski K.; Nissen B.; Reisfeld R.; Pavich J. *Optical Materials* 1999, 13, 41-48
[124] Saraidarov T.; Reisfeld R.; Pietraszkiewicz M. *Chem. Phys. Lett.* 2000, 330, 515-520
[125] Reisfeld R. in *Optical and Electronic Phenomena in Sol-Gel Glasses and modern Applications*; R. Reisfeld and C. K. Jorgensen; Structure and Bonding, Springer Verlag: 1996, 85, 515-529
[126] Reisfeld R.; Seybold G. J. *Lumin.* 1991, 48&49, 898-900
[127] Reisfeld R.; Yariv E.; Minti H. *Optical materials* 1997, 8, 31-36
[128] Reisfeld R.; Yariv E. *Optica Applicata* 2000, 30, 481-490
[129] López Arbeloa F.; López Arbeloa T.; López Arbeloa I.; Costela A.; García-Moreno I.; Figuera J.M.; Amat-Guerri F.; Sastre R. *Appl. Phys. B* 1997, 64, 651-657
[130] Demas J. N.; Grosby G. A. *J. Phys. Chem.* 1971, 75, 991-1024
[131] Lakowicz J. R. *Principles of Fluorescence Spectroscopy*; Springer: Baltimore, Maryland, USA, 2006; 3rd edition, 98-472
[132] Monte F. del; Levy D. *J. Phys. Chem. B* 1998, 102, 8036-8041
[133] Monte F. del ; Levy D. *J. Phys. Chem. B* 1999, 103, 8080-8086
[134] Monte F. del; Mackenzie J. D.; Levy D. *Langmuir* 2000, 16, 7377-7382
[135] Deshpande A. V.; Panhalkar R. R. *J. Lumin.* 2002, 96, 185-193
[136] Negishi N.; Fujino M.; Yamashita H.; Fox M. A.; Anpo M. *Langmuir* 1994, 10, 1772-1776
[137] Takahashi Y.; Kitamura T.; Nogami M.; Uchida K.; Yamanaka T. *J. Lumin.* 1994, 60&61, 451-453
[138] Fujii T.; *Trends Photochem. Photobiol.* 1994, 3, 243-247
[139] Hinckley D.A.; Saybold P.G.; Borris D.P. *Spectrochim Acta* 1986, 42A, 747-754
[140] Lo D.; Parris J. E.; Lawless J. L. *Appl. Phys. B* 1993, 56, 385-390
[141] Deshpande A.V.; Panhalkar R.R. *Materials Letters* 2002, 55, 104-110
[142] Ahmad M.; King T.A.; Ko D.K.; Cha B.H.; Lee J. *J. Phys. D: Appl. Phys.* 2002, 35, 1473-1476
[143] Fukushima M.; Yanagi H.; Hayashi S., Suganuma N.; Taniguchi Y. *J. Appl. Phys.* 2005, 97, 106104-106107
[144] Rahan M. D.; King T. A. *J. Mod. Opt.* 1998, 45, 1259-1267
[145] Sastre R.; Costela A. *Adv. Mater.* 1995, 7, 198-202
[146] Costela A.; Garcia-Moreno I.; Gomez C.; Garcia O.; Sastre R. *Chem. Phys. Lett.* 2003, 369, 656-661
[147] Costela A.; Garcia-Moreno I.; Gomez C.; Garcia O.; Sastre R. *Appl. Phys. B* 2004, 78, 629-634
[148] Costela A.; Garcia-Moreno I.; Gomez C.; Garcia O.; Garrido L.; Sastre R. *Chem. Phys. Lett.* 2004, 387, 496-501
[149] Costela A.; Garcia-Moreno I.; Gomez C.; Garcia O.; Sastre R.; Roig A.; Molins E. *J. Phys. Chem. B* 2005, 109, 4475-4480
[150] Deshpande A. V.; Kumar U. *J. Lumin.* 2008, 128, 1121-1131
[151] Deshpande A. V.; Kumar U. *J. Non-Cryst. Solids* 2009, 355, 501-506
[152] Deshpande A. V.; Kumar U. *J. Lumin.* 2010, 130, 839-844

[153] Deshpande A. V.; Rane J. R.; Jathar L. V. *J. Sol-Gel Sci. Technol.* 2009, 49, 268-276
[154] Sharma S.; Mohan D.; Singh N.; Sharma N.; Sharma A. K. *Optik* 2010, 121, 11-18
[155] Ye C.; Lam K. S.; Chik K. P.; Lo D.; Wang K. H. *Appl. Phys. Lett.* 1996, 69, 3800-3802
[156] Lam K. S.; Lo D.; Wang K. H. *Appl. Opt.* 1995, 34, 3380-3383
[157] Weissbeck A.; Langhoff H.; Beck A. *Appl. Phys. B* 1995, 61, 253-255
[158] Ye C.; Lam K. S.; Lam S. K.; Lo D. *Appl. Phys. B* 1997, 65, 109-111
[159] Yang Y.; Qian G.; Su D.; Wang Z.; Wang M. *Mater. Sci. Engg. B* 2005, 119, 192-195
[160] Yang Y.; Lin G.; Xu H.; Wang M.; Qian G. *Opt. Commun.* 2008, 281, 5218-5221
[161] Deshpande A. V.; Kumar U. *J. Fluorsc.* 2006, 16, 679-687
[162] Deshpande A. V.; Jathar L. V.; Rane J. R. *J. Non-Cryst. Solids* 2010, 356, 1-7
[163] Yang Y.; Zou J.; Rong H.; Qian G. D.; Wang Z. Y.; Wang M. Q. *Appl. Phys. B* 2007, 86, 309-313
[164] Yang Y., Lin G.; Zou J.; Wang Z.; Wang M.; Qian G. *Opt. Commun.* 2007, 277, 138-142
[165] Yang Y.; Qian G.; Su D.; Wang Z.; Wang M. *Chem. Phys. Lett.* 2005, 402, 389-394
[166] Su D.; Yang Y.; Qian G.; Wang Z.; Wang M. *Chem. Phys. Lett.* 2004, 397, 397-401
[167] Nhung T. H.; Canva M.; dao T. T. A.; Chaput F.; Brun A.; Hung N. D.; Boilot J. P. *Appl. Opt.* 2003, 42, 2213-2218
[168] Sebastian P. J.; Sathinandan V. *Opt. Commun.* 1980, 32, 422-424
[169] Dunning F. B.; Stokes E. D. *Opt. Commun.* 1972, 6, 60-162
[170] Sasaki H.; Kobayashi Y.; Muto S.; Kurokawa Y. *J. Am. Ceram. Soc.* 1990, 73, 453-456
[171] Ali M.; Samir A.; Mitwally K. *Appl. Opt.* 1989, 28, 3708-3712
[172] Saito Y.; Nomura A.; Kano T. *Appl. Phys. Lett.* 1988, 53, 1903-1904
[173] Schmidt W.; Appt W.; Witterkiwndt N. Z. *Naturforsch* 1972, 27A, 37- 41
[174] Raju B. B.; Varadarajan T. S. *J. Lumin.* 1993, 55, 49-54
[175] Lang J. M.; Drickamer H. G. *J. Phys. Chem.* 1993, 97, 5058- 5064
[176] Ramalingam A.; Somasundaram G. *Part of the SPIE conference on Photopolymer Device Physics Chemistry and application*s IV, Quebec (Canada) 3417 (1998) 228-230
[177] Muto S.; Ushida K.; Kotaka A.; Ito C.; Inoha F. *Trans Inst Electron Commun. Engg.* Japan 1981, 64C, 509- 511
[178] Ahmed S. A.; Gregarly J. S.; Infante D. *J. Chem. Phys.* 1974, 61, 1584-1585
[179] Rivano V.; Mazzinghi P.; Burlamacchi P. *Appl. Phys. B* 1984, 35, 71-75
[180] Andrews D. L. *Chem. Phys.* 1989, 135, 195-201
[181] Andrews D. L.; Bradshaw D. S. Eur. *J. Phys.* 2004, 25, 845-858
[182] Forster T. *Ann. Phys.* (Leipzig) 1948, 6, 55-75
[183] Forster T. Z. *Naturforsch*, 1949, A4, 321-327
[184] Dexter D. L. *J. Chem. Phys.* 1953, 21, 836-850
[185] Tanaka N.; Barashkov N.; Heath J.; Sisk W. N. *Appl. Opt.* 2006, 45, 3846-3851
[186] Nhung T. H.; Canva M.; Chaput F.; Goudket H.; Roger G.; Brun A.; Manh D. D.; Hung N. D.; Boilot J. P. *Opt. Commun.* 2004, 232, 343-351
[187] Khader M. A. *Opt. & laser Technol.* 2008, 40, 445-452
[188] Li S.; He L.; Xiong F.; Li Y.; Yang G. *J. Phys. Chem. B* 2005, 108, 10887-10892
[189] An B. K.; Kwon S. K.; Jung S. D.; Park S. Y. *J. Am. Chem. Soc.* 2002, 124, 14410-14415

[190] Treussart F.; Botzung-Appert E.; Ha-Duong N.-T.; Ibanez A.; Roch J.; Pansu R. *ChemPhysChem* 2003, 4, 757-760

[191] Tian Z.; Huang W,; Xiao D.; Wang S.; Wu Y.; Gong Q.; Yang W.; Yao J. *Chem. Phys. Lett.* 2004, 391, 283-287

[192] Gutierrez M. C.; Hortiguela M. J.; Ferrer M. L.; del Monte F. *Langmuir* 2007, 23, 2175-2179

[193] Ferrer M. L.; del Monte F. *J. Phys. Chem. B* 2005, 109, 80-86

[194] Oaki, Y.; Imai, H. *Adv. Funct. Mater.* 2005, 15, 1407-1414

[195] J. Gesquiere; T. Uwada; T. Asahi; H. Masuhara; P. F. Barbara *Nano Lett.* 2005, 5, 1321-1325

[196] T. Asahi; T. Sugiyama; H. *Masuhara Acc. Chem. Res.* 2008, 41, 1790-1798

[197] Kasai H.; Oikawa H.; Okada S.; Nakanishi H. *Bull. Chem. Soc. Jpn.* 1998, 71, 2597-2601

[198] Takeshima M.; Matsui A. H. *J. Lumin.* 1999, 82, 195-204

[199] Horn D.; Rieger J. Angew. *Chem., Int. Ed.* 2001, 40, 4330-4361

[200] Xiao D,; Xi. L.; Yang W.; Fu H.; Shuai Z.; Fang Y.; Yao J. *J. Am. Chem. Soc.* 2003, 125, 6740-6745

[201] Kurokawa N.; Yoshikawa H.; Hirota N.; Hyodo K.; Masuhara H. *chemPhysChem* 2004, 5, 1609-1615

[202] Lagoudakis P. G.; de Souza M. M.; Schindler F.; Lupton J. M.; Feldmann J. Wenus J.; Lidzey D. G. *Phys. Rev. Lett.* 2004, 93, 257401-257404

[203] Patra A.; Hebalkar N.; Sreedhar B.; sarkar M.; samanta A.; Radhakrishnan T. *P. Small* 2006, 2, 650-659

[204] Katagi H.; Kasai H.; Okada S.; Oikawa H.; Komatsu K.; Matsuda H.; Liu Z.; Nakanishi H. *Jpn. J. Appl. Phys. Phys.* 1996, 35, L1364-L1366

[205] Seko T.; Ogura K.; Kawakami Y.; Sugino H.; Toyotama H.; Tanaka *J. Chem. Phys. Lett.* 1998. 291, 438-444

In: The Sol-Gel Process
Editor: Rachel E. Morris
ISBN 978-1-61761-321-0

Chapter 20

Preparation and Properties of ZnO Nanoparticles and Thin Films Deposited by the Sol-Gel Method

Wei-jie Song[1], Jia-Li[1], Yu-long Zhang[1], Rui-qin Tan[2] and Ye-Yang[1]
[1]Ningbo Institute of Material Technology and Engineering, Chinese Academy of Sciences
[2]Faculty of Information Science and Engineering, Ningbo University

Abstract

The use of sol-gel deposition for the fabrication of ZnO nanoparticles and thin films is reviewed. Low-agglomeration ZnO nanoparticles are obtained, and the effects of citric acid concentration, pH value, and various surfactants on grain size and distribution are discussed. ZnO thin films are then deposited and annealed in various environments, and the relationships between the annealing environments and the properties of the ZnO thin films are investigated. The roles that kinetic factors play in the film transformation process are also reviewed. Finally, the composition-related structural, microstructure, electrical, and optical properties of $Mg_xZn_{1-x}O$ thin films are discussed in detail.

1. Introduction

The first discussions of sol-gel material deposition can be traced to research done by Ebelman in 1846[1]. Over the years, the literature in this field grew to more than 50,000 papers published worldwide by the 1990s. Early work by these investigators and others led to a rapid expansion of research in this area. Because of their higher cost efficiency, excellent compositional control, and homogeneity on the molecular level, sol-gel processes have become an attractive method for the fabrication of nano-powders and thin films[2-5].

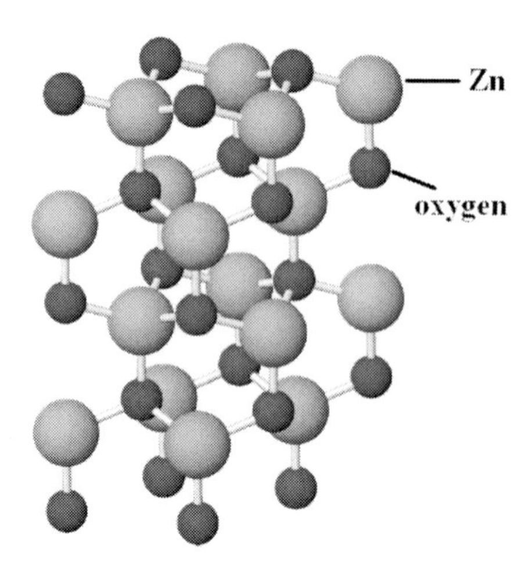

Figure 1. The crystal structure of ZnO .

ZnO is a direct-band-gap, II-VI semiconductor. Its bandgap is 3.3 eV at room temperature and 3.44 eV at 4 K. The bandgap is tunable by doping. For example, Cd doping can reduce the gap to approximately 3 eV, and addition of Mg can increase Eg to approximately 4 eV. ZnO has a wurtzite hexagonal structure with lattice constants a = b = 3.24 Å and c = 5.20 Å. The Zn atoms are tetrahedrally coordinated with four oxygen atoms. The crystal structure is shown in Fig. 1. Zn d-electrons hybridize with O p-electrons, forming predominantly covalent bonds with significant contribution from ionic bonding. Room-temperature UV luminescence of photons with nearly bandgap energy is attributable largely to an excitation state with a binding energy of 60 meV. The carrier effective mass of ZnO is equal to 0.24 m_0 and 0.59 m_0 for electrons and holes respectively. The electron mobility of a single crystal of ZnO is between 100 and 200 cm^2/V, and the hole mobility is close to 180 cm^2/V [6-8].

Zinc oxide exhibits a wide range of properties that make it attractive for a variety of applications, including UV light emitters, solar cells, piezoelectric transducers, gas sensors, transparent electronics, waveguides, and surface acoustic wave (SAW) devices[9-11]. Stoichiometry and homogeneity of composition are key to the potential applications of these materials. The size and shape of the ZnO particles and the microstructure of ZnO thin films are important parameters that need to be controlled.

In this research, ZnO nanoparticles and thin films are fabricated using the sol-gel method. The underlying chemical and physical aspects of the solution deposition of ZnO material will be discussed, with a focus on understanding the solution preparation, film deposition, and phase transformation processes that occur during film synthesis. In the first part of this study, ZnO nanoparticles were prepared using the sol-gel method and their properties were studied. In the second part, the effects on microstructural evolution and densification behavior of ZnO thin films were studied. Moreover, the electrical and optical properties of ZnO and $Mg_xZn_{1-x}O$ thin films were investigated in detail.

2. PREPARATION OF ZNO NANOPARTICLES USING THE SOL-GEL METHOD

Zinc oxide (ZnO) nanoparticles can potentially be used in photo-catalysis, composite materials, sensors, and dye-sensitized solar cells because of their excellent optical, electrical, mechanical, and chemical properties [12-15]. The physical and chemical properties of ZnO nanoparticles can be manipulated by controlling their microstructures to meet the requirements of specific applications [16-18]. An ideal preparation method would enable the particle size, size distribution, crystallinity, and morphology to be easily controlled [19-22]. In the present work, a surfactant-assisted complex sol-gel process using (Zn(NO3)2·6H2O) and citric acid (C6H8O7) was investigated for synthesizing uniformly distributed ZnO nanoparticles. Polyethylene glycol (PEG) 2000, ethylene glycol, and polyvinyl alcohol (PVA) 1750±50 were used as surfactants.

First, the effects of citric acid concentration on the grain size of the ZnO nanoparticles were investigated. The grain sizes obtained at each citric acid concentration were all approximately 20–30 nm. The ZnO nanoparticles obtained for different concentration ratios of Zn^{2+}/CA are shown in Fig. 2. Low-aggregation powders were obtained at $C_{Zn}{}^{2+}/C_{cit}=1:1.5$, as shown in Fig. 2b. The particles obtained at $C_{Zn}{}^{2+}/C_{cit}=1.5:1$ and $C_{Zn}{}^{2+}/C_{cit}=1:3$ are shown in Figs. 2a and Fig. 2c. Aggregation existed at both concentrations. According to the results reported by Monderlaers [23], citric acid has turned out to be an excellent complexing agent in preventing the formation of precipitates. Two types of carboxylate groups coexisted in the zinc-CA complex and were reported as asymmetric and symmetric structures [24]. In the two carboxylate group structures of the zinc-CA chelates, two carboxyls formed a coordination bond with one Zn^{2+} ion and prevented it from aggregating in the nucleation process. The addition of a smaller amount of citric acid (molar ratio of 1:1 to Zn^{2+}) led to complete dissolution. Conglomerations appeared when the amount of citric acid was inappropriate, as shown in Figs. 2a and 2c. In the present experiments, the ZnO nanoparticles obtained revealed low conglomeration at a Zn^{2+}/CA ratio of 1:1.5. The presence of an appropriate amount of citric acid also led to a higher viscosity and a faster increase in viscosity during evaporative gelation of the solution, which in turn enabled a faster solution-gel transition and prevented aggregation [25].

The effects of pH on the grain size and surface area of the ZnO nanoparticles are shown in Fig. 3. No deposits were observed during pH adjustment. It was observed that the grain sizes of the ZnO particles were minimal and the surface area was maximal at a pH of 1. This was because the citric acid was incompletely ionized without $NH_3{\cdot}H_2O$ and acted as a bridge between two Zn^{2+} ions to form a three-dimensional complex [25]. The grain sizes increased as ammonia was added drop by drop to the solution to adjust the pH to less than 7. A lower NH3·H2O/Zn(NO3)2 molar ratio resulted in increased supersaturation of the solution. Consequently, a large amount of ZnO precipitated from the solution by homogeneous nucleation in a short time, which inhibited the growth of ZnO nanoparticles and led to a decrease in ZnO grain size at a pH of 7. It was theorized that Zn2+ was stably complexed by ammonia for a high NH3·H2O/Zn(NO3)2 molar ratio. As a result, heterogeneous nucleation of ZnO seldom occurred and the particles grew slowly [26].

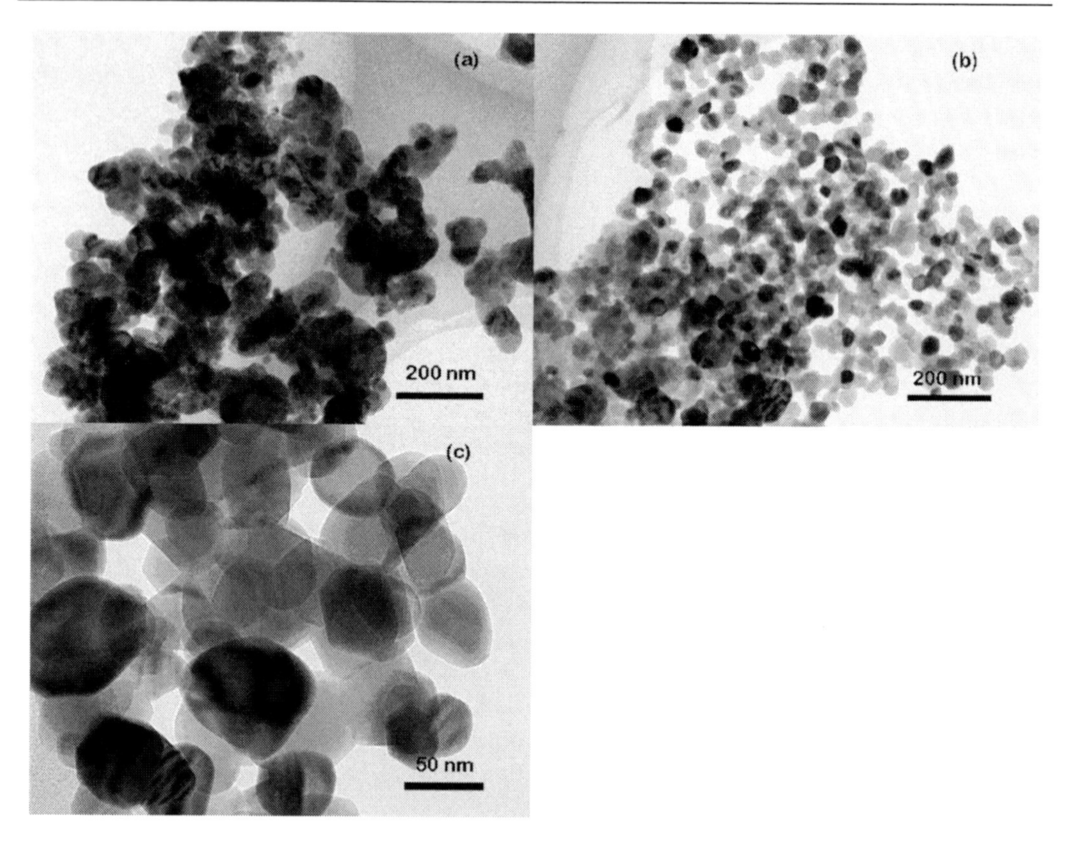

Figure 2. ZnO nanoparticles obtained via different concentration ratios of Zn/citric (a) C_{Zn}^{2+}/C_{cit}=1.5:1, (b) C_{Zn}^{2+}/C_{cit}=1:1.5, and (c) C_{Zn}^{2+}/C_{cit}=1:3.

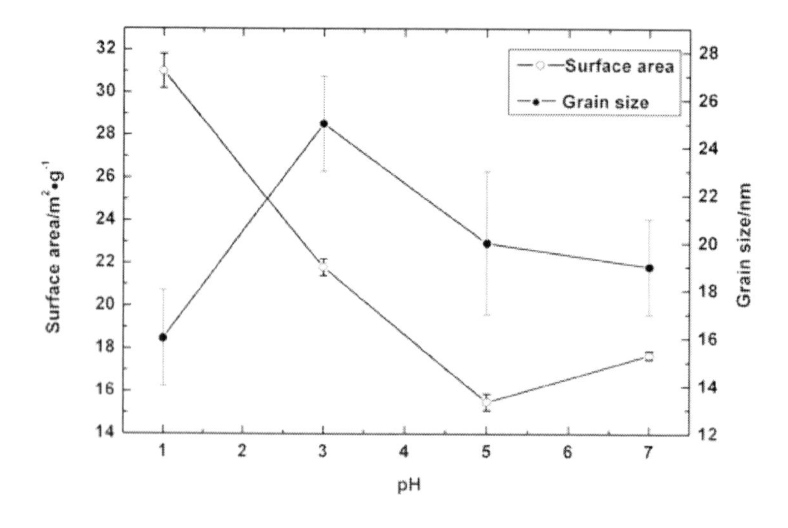

Figure 3. Surface area and grain size of ZnO nanoparticles prepared at a pH of 1, at a pH of 3, at a pH of 5, and at a pH of 7.

Various surfactants were used to prepare the ZnO nanoparticles to narrow the size distribution further. Fig. 4a shows SEM images of the ZnO nanoparticles prepared without surfactants. The ZnO nanoparticles prepared without surfactants grew asymmetrically, and the particle size distribution revealed substantial agglomeration. Figs. 4b, 4c, and 4d show SEM images of the ZnO nanoparticles prepared using ethylene glycol, PEG 2000, and PVA 1750 respectively as surfactants. It can be observed from Fig. 4b that the particles prepared with ethylene glycol revealed low conglomeration tendencies, but a wide size distribution from 20 nm to 200 nm. This result indicated that EG has an inhibiting effect on particle growth[27]. The particles were ball-shaped and showed low agglomeration when PEG 2000 was used as a surfactant. Substantial conglomeration of ZnO nanoparticles can be observed in Fig. 4d with polyvinyl alcohol 1750 used as a surfactant. These results can be explained by the fact that PVA were fully pyrolyzed at that temperature and the seeds were already touching each other [28]. It can be observed from XRD data that each ball was composed of a large quantity of nanoparticles approximately 5 nm in size. When PEG 2000 was used as a dispersant, the low aggregation could be explained by the steric hindrance mechanism. When the dispersant was added to the nitrate solution in the process of sol formation, the dispersant kept the sol particles separated in the solution because of polymeric dispersal of long-chain molecules [29]. After PEG 2000 was absorbed by the sol particles, the growth rate of the sol particles decreased correspondingly because their surface tension was reduced significantly[29-31]. Without PEG 2000, the primary sol particles from this solution were unstable against coagulation because their surface tension was sufficiently high. Steric hindrance was not observed among the sol particles, which means that they tended to agglomerate together. Therefore, the final ZnO particles prepared without PEG 2000 appeared to have poor dispersibility. It can therefore be asserted that PEG 2000 is the best surfactant for preparing uniform ZnO nanoparticles.

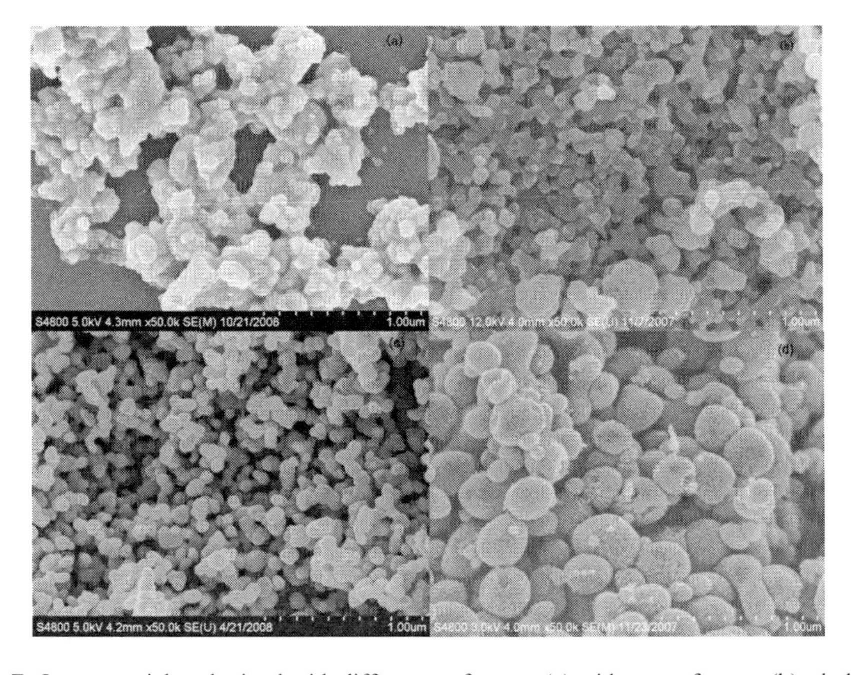

Figure 4. ZnO nanoparticles obtained with different surfactants (a) without surfactant, (b) ethylene glycol as surfactant, (c) polyethylene glycol (PEG) 2000, and (d) polyvinyl alcohol (PVA) 1750±50.

In summary, the surfactant-assisted complex sol-gel method for ZnO nanoparticle preparation has been successfully optimized. Low-agglomeration ZnO nanoparticles were obtained at C_{Zn}^{2+}/C_{cit}=1:1.5 at a pH value of one. ZnO nanoparticles with a narrow size distribution were obtained using PEG 2000 as a surfactant.

3. FILM DEPOSITION

Solution Preparation

The name "sol-gel process" comes from the transition of a system from liquid or "colloidal sol" to solid "gel" form[32]. The fabrication of solution-derived thin films usually begins with the synthesis of the precursor solution. A common precursor for producing zinc oxide is zinc acetate-2-methoxyethanol-monoethanolamine (MEA). This is followed by film deposition using spin coating or dip coating. Then, the film as deposited is subjected to a mild heat treatment at typical temperatures ranging from 200°C to 400°C. Finally, the amorphous film is further heat-treated at high temperature (600°C–1100°C), which converts it into a dense crystalline film. Three processes take place during the drying and subsequent baking of the film after the final coat of sol is applied: vaporization of the solvents, decomposition of the zinc acetate, and crystallization of the zinc oxide. Typically, the solution undergoes a series of hydrolysis and condensation reactions that causes the formation of a colloidal suspension [33]:

Hydrolysis:

$$(CH_3\text{-}\overset{\overset{\displaystyle O}{\|}}{C}\text{-}OZn)^+ + H^+ + 2(OH)^+ = CH_3COOH + Zn(OH)_2$$

Condensation:

$$Zn(OH)_2 \xrightarrow{O_2} ZnO + H_2O$$

Theory of Nucleation and Growth

Because pyrolyzed films are typically amorphous, film crystallization in solution-derived thin films occurs by a nucleation-and-growth process. The model for the nucleation and growth behavior of amorphous films was derived from the standard theory of crystallization in glasses.

In general, two events occur during the production process of chemical-solution-deposited thin films: nucleation and grain growth. According to classical nucleation theory, the free energy barrier for nucleation can be expressed as [34]:

$$G^* = \frac{16\pi\gamma^3}{3(\Delta G_v)^2} \cdot f(\theta), \quad (1)$$

where γ is the interfacial energy of the newly formed interface, ΔG_v is the driving force for crystallization, i.e., the free energy difference per unit volume for the amorphous film-crystalline film transformation, and $f(\theta)$ is a function related to the contact angle θ according to Eq. (2):

$$f(\theta)=\frac{2-3\cos\theta+\cos^3\theta}{4}. \tag{2}$$

For bulk nucleation (homogeneous nucleation) as occurs in most films, $\theta=\pi$ and $f(\theta)=1$, while for interface nucleation (heterogeneous nucleation) as occurs at the interface between films and substrates, $0<\theta<\pi$, and therefore $0<f(\theta)<1$. Obviously, the barrier height for bulk nucleation is higher than that for interface nucleation, and the difference can be expressed as:

$$\Delta G^*=\frac{16\pi\gamma^3}{3(\Delta G_v)^2}\cdot\left[1-f(\theta)\right] \tag{3}$$

The characteristics of the nucleation event are defined by the difference in barrier heights between bulk nucleation and interface nucleation, which is associated with ΔG_v, according to Eq. (3). Moreover, based on the standard approach to nucleation in glasses, Schwartz et al. [34] further suggested that ΔG_v is large and ΔG^* is small when crystallization occurs at a low temperature, and that therefore the barrier heights for both interface nucleation and bulk nucleation can be overcome, yielding a film which displays a fine-grained microstructure. In the opposite case, ΔG_v is small and ΔG^* is large, so that lower-energy interface-nucleation events become more important. When interface nucleation is dominant, the film prefers to grow in the c direction because the (002) plane has the minimum interface energy.

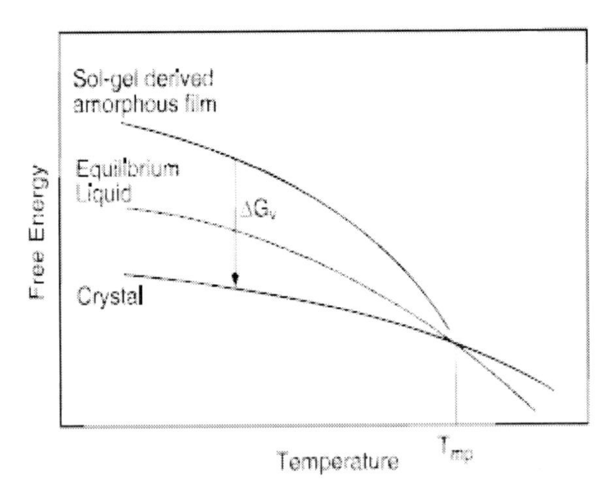

Figure. 5. Schematic of free energies of a solution derived amorphous film and crystalline film.

Impact of the Annealing Environment

It has been reported that the physical properties of ZnO films strongly depend, not only on growth techniques, but also on subsequent heat treatments such as annealing [35-44]. In this work, ZnO thin films were deposited and annealed in various environments using rapid thermal annealing (RTA). The aim of this research is to investigate in detail the relationships between the annealing environments and the properties of the ZnO thin films.

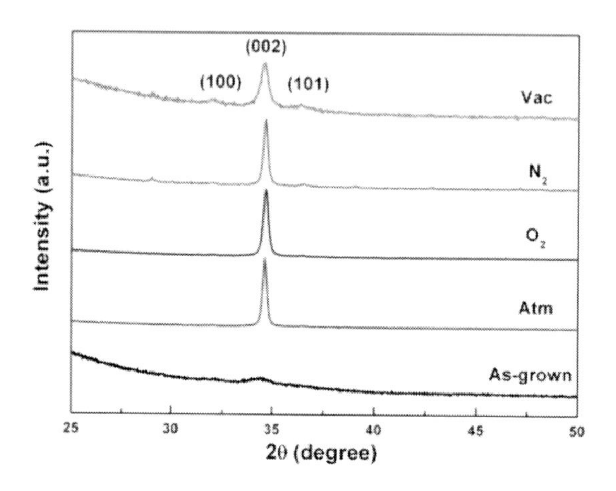

Figure 6. XRD patterns of ZnO thin films annealed in different ambient.

Fig. 6. shows XRD spectra of as-grown ZnO and ZnO thin films annealed at 750°C in vacuum, nitrogen, oxygen gas, and atmospheric environments. The as-grown ZnO films had only a small peak, which indicated weak crystallization. After subsequent thermal annealing in RTA, the intensity of the ZnO diffraction peak (002) strongly increased, and its peak width was narrowed. The observation of dominant (002) peaks at approximately 34.4° indicated that the films had been grown with a preferred c-axis orientation normal to the substrates.

To determine the degree of preferred orientation, the volume fraction $\alpha_{(002)}$ of c-axis-oriented grains in the ZnO films was defined as

$$\alpha_{(002)}\ (\%) = I_{(002)} / [I_{(100)} + I_{(002)} + I_{(101)}],$$

where $I_{(002)}$, $I_{(100)}$, and $I_{(101)}$ were the measured intensities of the (002), (100), and (101) diffraction peaks for the ZnO thin films [45]. It is shown that ZnO thin films annealed under atmospheric and O2 conditions were highly c-axis-oriented, with α(002) values of 97.9% and 97.8% respectively. After annealing in N2, the degree of c-axis orientation decreased to 93.9%, and the minimum value was obtained for the films annealed in vacuum. This implied that the crystal quality was better for ZnO thin films annealed in an abundant-oxygen environment.

Fig. 7. (a)–7(d) show SEM images of ZnO thin films annealed in air, N_2, O_2, and vacuum respectively. Different surface features can be observed in these images. From the figure, it can be seen that the ZnO film annealed in vacuum had the minimum grain size, with some micropores appearing on the surface. This implied that growth pressure has a dominant effect on the structure and surface morphology of ZnO thin films. According to Lin et al.[46], the concentration of Zn interstitial and O vacancy is proportional to $(P_{O_2})^{-\frac{1}{2}}$. This means that more defects ought to be formed at a lower growth pressure when annealing in vacuum. These defects could act as centers of nucleation and lead to a larger number of small grains.

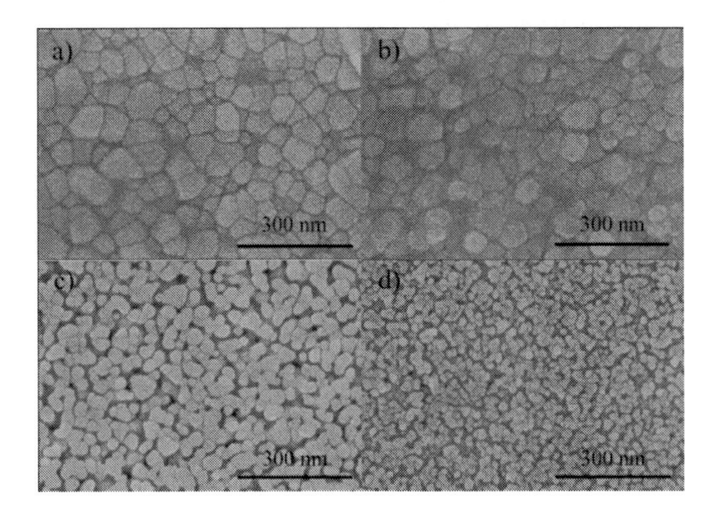

Figure 7. FE-SEM images of ZnO thin films annealed in different ambients (a) atm; (b) N_2; (c) O_2; (d) vacuum.

The resistivity of ZnO thin films annealed in different environments is shown in Fig. 8. The resistivity of the ZnO thin film annealed in vacuum was 0.095 Ω·cm, which was the minimum for the four environments. This decrease in resistivity might be attributable to an increase in severe oxygen vacancies which act as a pathway for the current. However, the resistivity increased with a further increase in oxygen concentration. The resistivity of ZnO thin films annealed in O_2 had the highest value, approximately 129.7 Ω·cm. This increase could be explained by the decrease in oxygen vacancies because more oxygen was absorbed during annealing in an abundant-oxygen environment.

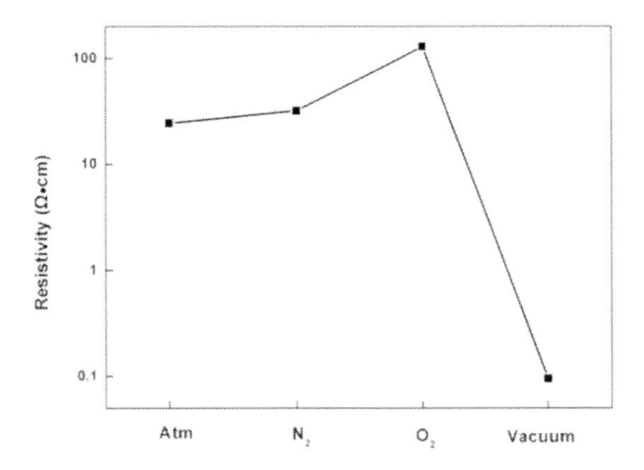

Figure 8. Resistivity of ZnO thin films annealed in different ambient.

To understand the differences among the electrical properties of the ZnO films annealed in various environments, X-ray photoelectron spectroscopy was performed on the ZnO surfaces annealed in O_2 and in vacuum. Zn 2p spectra of ZnO annealed in O_2 and in vacuum are shown in Figs. 9(a) and 9(c) respectively. The binding energies were calibrated by taking the carbon C 1s peak (284.6 eV) as reference. This approach indicated that the Zn in the films was in the 2^+ chemical state because the binding energy position of the spectra matched closely the standard data for zinc oxide.

XPS spectra for O 1s of ZnO annealed in O_2 and in vacuum are shown in Figs. 9(b) and 9(d). As reported in the literature, the O 1s peak could be fitted into three peaks. The lower binding energy (~530.0 ± 0.2 eV) corresponded to the O^{2-} ions in the wurtzite structure of ZnO. The higher-binding-energy component (~532.0 ± 0.2 eV) has usually been attributed to the presence of loosely bound oxygen species on the surface of the films which are associated with a specific species, e.g., -CO3, adsorbed H2O, or adsorbed O2. The moderate-binding-energy (~531.0 ± 0.2 eV) component was associated with O2- ions in oxygen-deficient regions within the ZnO matrix [47-48]. The presence or absence as well as changes in the intensity of this component may be related in part to variations in the concentration of oxygen vacancies. The XPS measurements reveal that the relative intensity ratio of the moderate component to total O 1s of ZnO annealed in vacuum was much larger than for ZnO annealed in O_2, which indicated a greater deficiency in the ZnO surface annealed in vacuum.

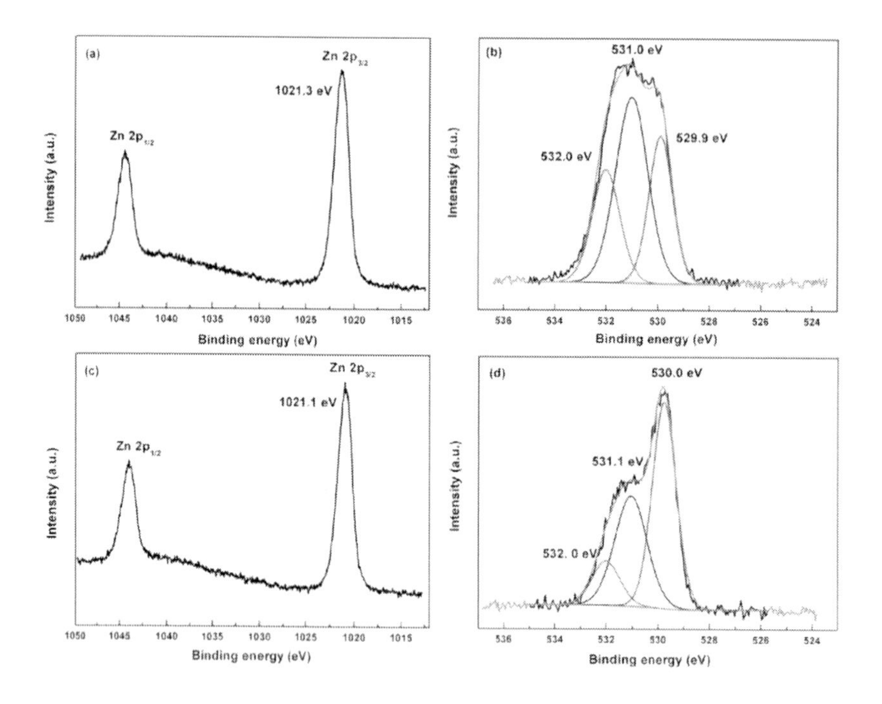

Figure 9. XPS spectra of the (a) Zn 2p region; (b) O 1s region of ZnO annealed in vacuum; (c) Zn 2p region; (d) O 1s region of ZnO annealed in O2.

The atomic concentration ratios of Zn to O were computed from the measured peak areas using the following sensitivity factors: Zn: 3.726; O: 0.78. The ratio of Zn to O for the ZnO film annealed in O_2 was 1.12, but this value increased to 1.58 after annealing in vacuum. This implied that oxygen atoms were pumped out easily and a zinc-rich thin film was formed by annealing in vacuum, which led to the existence of more zinc interstitial atoms within the crystal structure of the ZnO films. Therefore, the deficiencies of interstitial zinc and oxygen vacancies helped to increase the rate of electron transition and resulted in high carrier mobility in the ZnO films annealed in vacuum.

Fig. 10 shows the optical transmittance spectra in the 200–1000 nm wavelength range for ZnO thin films annealed in air, nitrogen, O_2, and vacuum. The transmittance value of the blank quartz substrate was eliminated for each sample. The average optical transmittance values of ZnO thin films annealed in air, N_2, O_2, and vacuum were approximately 83.5%, 80%, 82.6%, and 86.2% respectively in the visible range (400–800 nm). Sharp absorption edges were observed in the ultraviolet region for all samples. The sharp absorption edges occurred at a wavelength of approximately 379 nm, which corresponded to a band gap of approximately 3.27 eV. This indicated that the annealing environment did not change the band-gap value of ZnO thin films.

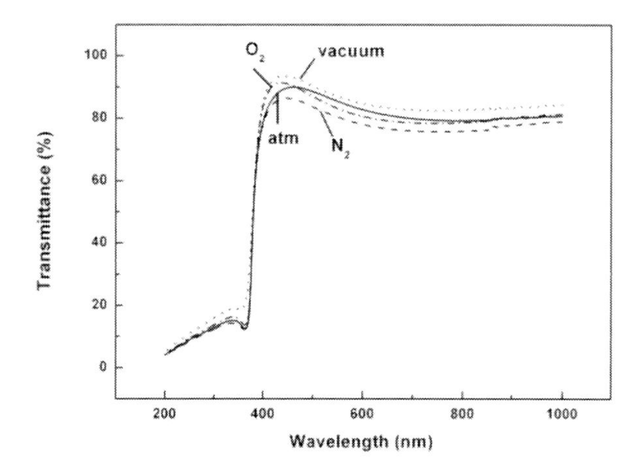

Figure 10. Optical transmittance spectra of ZnO thin films annealed in different ambient.

The effects of pH on the grain size and surface area of the ZnO nanoparticles are shown in Fig. 3. No deposits were observed during pH adjustment. It was observed that the grain sizes of the ZnO particles were minimal and the surface area was maximal at a pH of 1. This was because the citric acid was incompletely ionized without $NH_3 \cdot H_2O$ and acted as a bridge between two Zn^{2+} ions to form a three-dimensional complex [25]. The grain sizes increased as ammonia was added drop by drop to the solution to adjust the pH to less than 7. A lower $NH_3 \cdot H_2O/Zn(NO_3)_2$ molar ratio resulted in increased supersaturation of the solution. Consequently, a large amount of ZnO precipitated from the solution by homogeneous nucleation in a short time, which inhibited the growth of ZnO nanoparticles and led to a decrease in ZnO grain size at a pH of 7. It was theorized that Zn^{2+} was stably complexed by ammonia for a high $NH_3 \cdot H_2O/Zn(NO_3)_2$ molar ratio. As a result, heterogeneous nucleation of ZnO seldom occurred and the particles grew slowly [26].

In summary, different annealing environments had a substantial impact on the structural and electrical properties of ZnO thin films. All the annealed films were highly c-axis-oriented except for the film annealed in vacuum. ZnO thin films annealed in vacuum had the minimum resistivity value among all the films, 0.095 Ω·cm. XPS results indicated that more defects existed in ZnO surfaces annealed in vacuum than in those annealed in O_2.

$Mg_xZn_{1-x}O$ Alloy Film

Recently, the ability to widen the bandgap by forming a ternary alloy of MgxZn1-xO has generated interest in this material [49-56]. The lattice constants of MgxZn1-xO can be expected not to induce significant change compared to those of ZnO because of the similarity in ion radius between Mg2+ (0.57 Å) and Zn2+ (0.60 Å) [57]. The wide optical band gap of MgxZn1-xO alloy films has been successfully exploited in heterostructured devices such as mid-UV and deep-UV light detectors, UV-light-emitting diodes, and laser diodes. Moreover, $Mg_xZn_{1-x}O$ alloys possess attractive applications in MgxZn1-xO/ZnO heterostructures for optoelectronic and display devices [58-64].

In this study, the sol-gel technique is used to deposit $Mg_xZn_{1-x}O$ thin films. The composition-related structural, microstructural, electrical, and optical properties of $Mg_xZn_{1-x}O$ thin films have been investigated in detail.

Fig. 11 shows the XRD patterns of $Mg_xZn_{1-x}O$ thin films with various Mg contents. When $x \geq 0.2$, the intensity of the (002) peak is found to decrease with increasing Mg content, which indicates a decrease in crystallinity with high Mg content. Because the phase diagram of an MgO-ZnO binary system shows a thermodynamic solid solubility of approximately 4 mol% MgO in ZnO, a large dissimilarity in the structures between wurtzite ZnO and cubic MgO may result in a nonequilibrium phase for $Mg_xZn_{1-x}O$ for higher values of x beyond the solubility limit. Once Mg content exceeds 0.3, separation of MgO can be observed, as evidenced by the appearance of the MgO (200) peak.

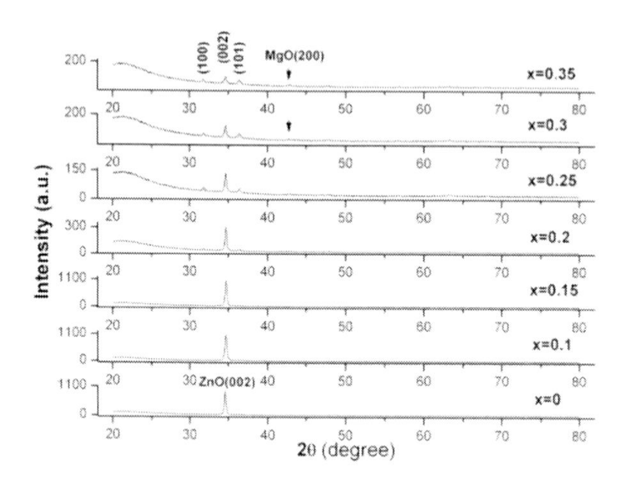

Figure 11. XRD patterns of $Mg_xZn_{1-x}O$ with different Mg content.

Fig. 12 displays the surface morphologies of $Mg_xZn_{1-x}O$ thin films. It is evident that for the films with x = 0 and x = 0.1, the crystallites are highly compact. When x = 0.15, some voids and micropores appear on the surfaces of the films. With increasing Mg content, these defects increased. By combining this result with the XRD result, it is possible to deduce that these defects may be caused by disequilibrium and phase separation due to excessive Mg content.

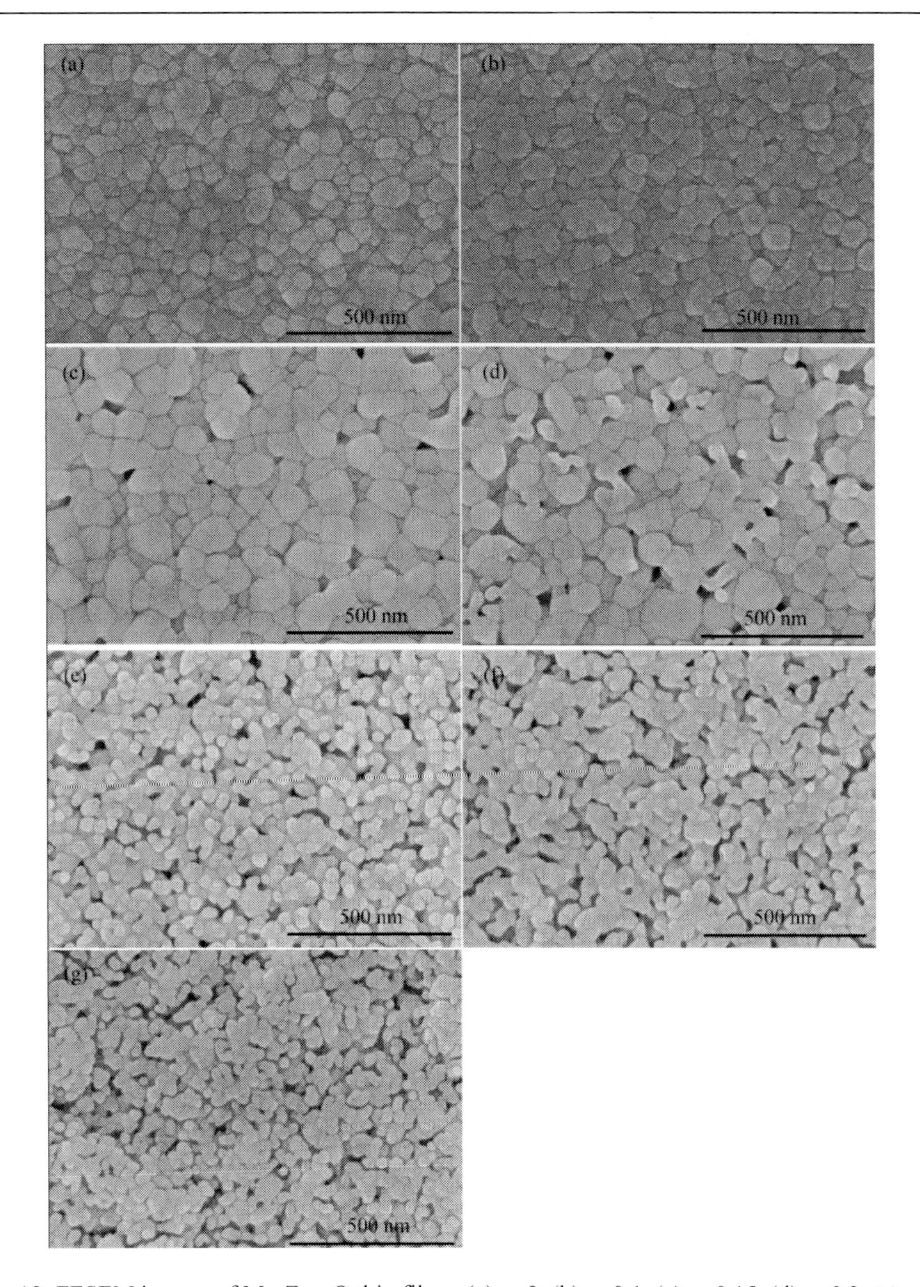

Figure 12. FESEM images of $Mg_xZn_{1-x}O$ thin films: (a) x=0, (b) x=0.1, (c) x=0.15, (d) x=0.2, (e) x=0.25, (f) x=0.3, (g) x=0.35.

Fig. 13 shows a plot of the optical transmittance spectra in the 200–800 nm wavelength range for $Mg_xZn_{1-x}O$ thin films. The films are more than 80% transparent in the visible region and have sharp band-edge absorption in the UV region. To show the result more clearly, the derivative of transmittance (T) against wavelength (λ) was computed. Fig. 14 shows plots of $dT/d\lambda$ versus λ for the $Mg_xZn_{1-x}O$ thin films. The sharp absorption edge in the transmittance spectra may result in a sharp peak in the plot of $dT/d\lambda$ versus λ. Fig. 14 shows that a pure ZnO thin film (x = 0) has an absorption peak around 379 nm (3.27eV). The peak value for x = 0.1 is 357 nm, which corresponds to a band gap of 3.47 eV. For the sample with Mg content x

= 0.15, the sharp absorption edge occurred at 347 nm, corresponding to an increased band gap value of 3.56 eV. A further increase in Mg content changed the band gap very little, which provides an additional evidence for Mg saturation in the $Mg_xZn_{1-x}O$ thin films.

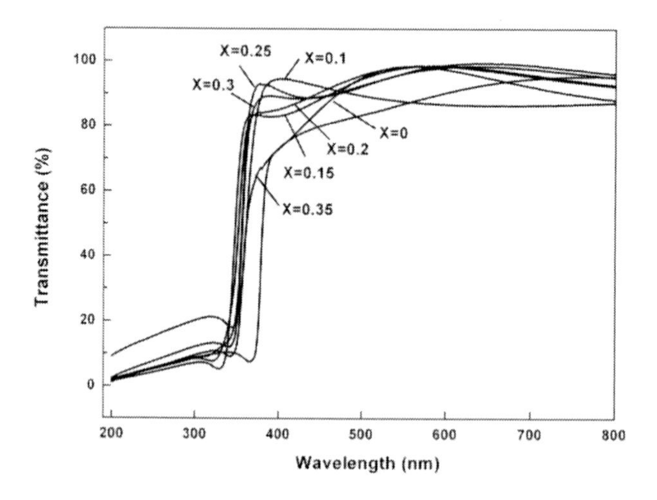

Figure 13. Optical transmittance spectra of $Mg_xZn_{1-x}O$ thin films

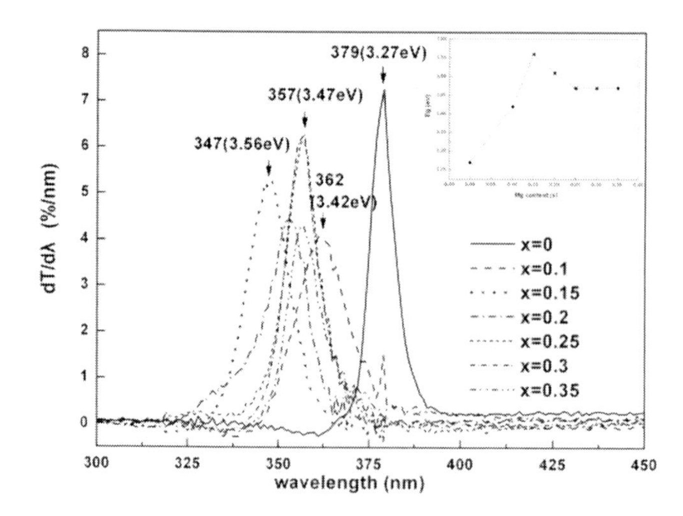

Figure 14. dT/dλ-λ vs. λ for MgxZn1-xO thin films with various Mg contents. (insert: variation of band gap value with Mg content in MgxZn1-xO thin films).

In summary, $Mg_xZn_{1-x}O$ thin films with various Mg contents have been fabricated successfully using a sol-gel technique. Separation of MgO is observed when $x \geq 0.3$. A decrease in crystallinity can be observed and voids and micropores appear on the surface of the films when $x \geq 0.2$. The maximum band gap value is 3.56eV, corresponding to an Mg content of x = 0.15. Further increases in Mg content change the band gap very little.

SUMMARY

Over the past decades, the development of ZnO thin films for electronic devices has progressed remarkably. The development of sol-gel processing routes has progressed, and high-quality ZnO materials can now be prepared. The improvement in film quality is due to the efforts made to understand the solution chemistry and other processes which affect thin-film microstructure. Further research in this field should involve using sol-gel deposition to prepare device-quality thin films that can be tested under application conditions.

ACKNOWLEDGMENT

This work was supported by the "Hundred Talents Program," the Chinese Academy of Sciences, the Zhejiang Natural Science Foundation (Y407364), and the Ningbo Natural Science Foundation (No. 2008A610047, No. 2009A610018). The authors would give thanks to the all measurers of Ningbo institute of material technology & engineering Chinese academy of Science.

REFERENCES

[1] Ebelman, JJ. *Ann.*, 1846, 57, 331
[2] Schwartz, RW. *Chem. Mater.*, 1997, 9, 2325
[3] Zhang, YL; Yang, Y; Zhao, JH; Tan, RQ. *J. Sol-Gel Sci. Technol.*, 2009, 51, 198.
[4] Kim, KT; Kim, CI. *J. Eur. Ceram. Soc.*, 2004, 24, 2613-2617.
[5] Madeswaran, S; Giridharan, NV; Jayavel, R. *Mater. Chem. Phys.*, 2003, 80, 23.
[6] Gray, TJ. *J. Am. Ceram. Soc.*, 1954, 37, 534.
[7] Supab, C. Ph. D. thesis, *University of Maryland*, 2001.
[8] Pan, HC. Ph. D. thesis, Northwestern University, 1990.
[9] Lima, SAM; Davolos, MR; Quirino, WG. *Appl. Phys. Lett.,* 2007, 90, 023503.
[10] Ghost, R; Mridha, S; Basak, D. *J. Mater. Sci.: Mater. Electron.*, 2009, 20, S371.
[11] Oh, BY; Jeong, MC; Moon, TH. *J. Appl. Phys.*, 2006, 99, 124505.
[12] Polarz, S; Neues, F. *J. Amer. Chem. Soc.*, 2005, 127, 12028.
[13] Narendar, Y; Messing, GL. *Catal. Today*, 1997, 35, 247.
[14] Xu, HY; Liu, XL; Cui, DL; Li, M. *Sens. Actuators B*, 2006, 114, 301.
[15] Sheng, X; Zhao, Y; Zhai, J; Jiang, L; Zhu, D. *Appl. Phys. A*, 2007, 87, 715.
[16] Lu, GQ; Lieberwith, I; Wegner, G. *J. Am. Chem. Soc.*, 2006, 128, 15445.
[17] Liang, JB; Liu, JW; Xie, Q; Bai, S; Yu, WC. *J. Phys. Chem. B*, 2005, 109, 9463.
[18] Zhang, J; Sun, LD; Yin, JL; Su, HL. *Chem. Mater.*, 2002, 14, 4172.
[19] Chen , S; Kumar, RV; Gedanken, A; Zaban, A. *Israel J. Chem.*, 2001, 41, 51.
[20] Wu, ZY; Bao, ZX; Zou, XP; Tang Mater, DS. *Sci. Technol.*, 2003, 19, 981.
[21] Hou, Y; Yang, M; Pang, GS; Feng, SH. *J. Mater. Sci.*, 2008, 43, 2149.
[22] Ohyama, M; *J. Am. Ceram. Soc.*, 1998, 81, 1622
[23] Werde, K; Mondelaers, D; Vanholand, G; Nelis, D. *J. Mater. Sci.*, 2002, 37, 81.
[24] Anedda, R; Cannas, C; Musinu, A. *J. Nanopart. Res.*, 2008, 10, 107.

[25] Hardy, A; Haen, JD; Van Bael, MK. *J. Sol-Gel Sci. Technol.*, 2007, 44, 65.
[26] Yu, K; Jin, ZG, X. Liu, X. *Mater. Lett.*, 2007, 61, 2775.
[27] Huang, ZB; Tang, FQ; Zhang, L. *Thin Solid Films*, 2005, 471, 105.
[28] Sang, WB; Fang, YY; Fan, JR. *J. Cryst. Growth*, 2007, 299, 272.
[29] Ferri, JK; Stebe, K. *J. Adv. Colloid. Interface Sci.*, 2000, 85, 61
[30] Zhang, YF; Zhang, JX; Lu, QM; Zhang, QY. *Mater. Lett.*, 2006, 60, 2443.
[31] Yang, J; Lian, JS; Dong, QZ. *Mater. Lett.*, 2003, 57, 2792.
[32] Piotr Kursa, Mater thesis, University of Alberta, 2007.
[33] Özgür, Ü; Alivov, YI; Liu, C; Teke, A. *J. Appl. Phys.*, 2005, 98, 041301.
[34] Shinobu, F; Chikako, S; Toshio. K. *Appl. Surf. Sci.*, 2001, 180, 341.
[35] Sagar, P; Shishodia, PK; Mehra, RM. *Appl. Surf. Sci.*, 2007, 253, 5419.
[36] Liu, DS; Sheu, CS; Lee, CT. *J. Appl. Phys.*, 2007, 102, 033516.
[37] Shinde, VR; Gujar, TP; Lokhande, CD, *Sol. Energy Mater. Sol. Cells.*, 2007, 91, 1055.
[38] Chen, CC; Yu, BH; Liu, JF. *Mater. Lett.*, 2007, 61, 2961.
[39] Tsai, HY. *J. Mater. Process. Technol.*, 2007, 192, 55.
[40] Ghosh, R; Paul, GK; Basak, Mater, D. *Res. Bull.*, 2005, 40, 1905.
[41] Lin, KM; Chen, YY. *J. Sol-Gel Sci. Technol.*, 2009, 51, 215.
[42] Hu, SY; Lee, YC; Lee, JW. *Appl. Surf. Sci.*, 2008, 254, 1578.
[43] Lee, YC; Hu, SY; Water, W. *J. Lumin.*, 2009, 129, 148.
[44] Wei, XQ; Man, BY. Liu, M. *Physica B.*, 2007, 388, 145.
[45] Lin, KM; Tsai, P. *Thin Solid Film*, 2007, 515, 8601.
[46] Lin, BX; Fu, ZX; Jia, YB. *Appl. Phys. Lett.*, 2001, 79, 943.
[47] Chen, M; Wang, X; Yu, YH. *Appl. Surf. Sci.*, 2000, 158, 134.
[48] Yang, XJ; Miao, XY; Xu, XL. *Opt. Mater.*, 2005, 27, 1602.
[49] Yang, H; Li, Y; Norton, DP. *Appl. Phys. Lett.*, 2005, 86, 172103.
[50] Tampo, H; Shibata, H; Matsubara, K. *Appl. Phys. Lett.*, 2006, 89, 132113.
[51] Guo, QX; Tanaka, T; Nishio, M. *Japan. J. Appl. Phys.*, 2007, 46, 560.
[52] Shigehiko, S; Takeo, H; Motoki, K. *IEEE Electron Device Letters*, 2007, 28, 543.
[53] Kazuto, K; Daisuke, T; Motoki, K. *Japan. J. Appl. Phys.*, 2007, 46, 865.
[54] Sharma, AK; Narayan, J; Muth, JF. *Appl. Phys. Lett.*, 1999, 75, 3327.
[55] He, YN; Zhang, JW. *Microelectronics Journal*, 2005, 36, 125.
[56] Vashaei, Z; Minegishi, T; Yao, T. *Journal of crystal growth*, 2007, 306, 269.
[57] Hwang, DK; Jeong, MC; *J. M. Myoung. Appl. Surf. Sci.*, 2004, 225, 217.
[58] Ghost, R; Basak, D. *J. Appl. Phys.*, 2007, 101, 023507.
[59] Ghost, R; Basak, D. *J. Mater. Sci: Mater Electron*, 2007, 18, 141.
[60] Qiu, MX; Ye, ZZ; He, HP. *Appl. Phys. Lett.*, 2007, 90, 182116.
[61] Sarver, JF; Katnack, FL; Hummel, FA. *J. Electrochem. Soc.*, 1959, 106, 960.
[62] Ohtomo, A; Kawasaki, M;.Koida, T *Appl. Phys. Lett.*, 1998, 72, 2466.
[63] Park, WI; Yi, GC; Jang, HM. *Appl. Phys. Lett.*, 2001, 79, 2022.
[64] Wang, MS; Kim, EJ; Kim, S. *Thin Solid Films*, 2008, 516, 1124.

In: The Sol-Gel Process
Editor: Rachel E. Morris

ISBN 978-1-61761-321-0

Chapter 21

WET SOL-GEL SILICA MICROSPHERES FOR THE SUSTAINED RELEASE OF HUMAN GROWTH HORMONE

Margherita Morpurgo**, *Andrea Mozzo, Camilla Ferracini, Mauro Pignatto and Nicola Realdon
Department of Pharmaceutical Sciences,
University of Padova, Padova, Italy

ABSTRACT

Sol-gel silica in the wet form formulated in the shape of microsphere was investigated for the controlled release of protein drugs and in particular human growth hormone (r-hGH). The influence of the gel SiO_2 content on protein conformation, load and release rate was investigated. Protein fold upon gel embedment was measured by circular dichroism analysis and was found to be unaffected up to a concentration of SiO_2 equal to 12% w/v, while minor loss of r-hGH α-helix occurred at higher silica concentration. Several r-hGH loaded microspheres containing 5, 8, 10, 12 and 15% SiO_2 (w/v) were synthesized using a surfactant-free W/O pseudo-emulsion method, purposely selected to minimize the risk of protein unfolding. The amount of protein that can be incorporated in the polymer and is released in a controlled way was found to depend on the matrix silica content, varying form 0.6 mg to more than 5.3 mg/ml of wet gel for the 5% and 12-15% SiO_2 formulations respectively. At low silica concentration, total release of r-hGH from microspheres occurs within 12 hours and it is mostly driven by diffusion through the gel pores. At higher SiO_2 content, release is significantly slower and it is mostly dominated by polymer erosion with a time scale that varies between 100 and 150 hours depending on the formulation under investigation.

* Department of Pharmaceutical Sciences University of Padova, Via Marzolo, 5; 35131 Padova, Italy, e. mail: margherita.morpurgo@unipd.it, Phone:(+39)-049-827-5339;Fax:(+39)-049-827-5366.

INTRODUCTION

Sol-gel silica is an inorganic polymer with high potentials in biomedicine. It is easy to produce from low cost precursors and it is highly versatile so that materials useful for multiple applications can be envisaged. Depending on the polymerising conditions, dry or wet silica based polymers with nano- meso- o micro-porosity are obtained [1]. Among the biomedical applications currently investigated are the controlled release of low molecular weight, polymer and protein drugs [2-5]. Depending on the polymer porosity, internal organization and composition, sol-gel silica based materials capable of improving the dissolution of poorly soluble drugs [6, 7] or matrices for the sustained release of highly soluble and unstable drugs have been described [4, 8-16]. Further advantage is that silica and its degradation products are well tolerated by the human body: parenteral administration of different types of synthetic silica polymers has been described in the literature and high tolerability was demonstrated [17-20]. Indeed the polymer is eroded into silicic acid (a compound that is abundant in mineral water) and it is then eliminated through the kidneys. Only minor and reversible toxicity was shown upon large dose administration.

In the context of protein drug formulation, the first evidence of the potential applicability of sol-gel silica as a matrix for controlled release came from the 2001 work of Santos and Ducheyne, who demonstrated that sol-gel silica xerogels were capable of releasing an embedded protein (Tgf-β) in its bioactive form [21]. However, the xerogels described in that work were not yet optimal as only a small fraction of the embedded biological activity was recovered upon release, indicating the need to increase pore size to allow a more efficient protein diffusion.

It was later observed that protein release from silica xerogels occurs secondary to both their diffusion from the matrix pores and polymer erosion. Therefore, the release can be improved by introducing synthesis or processing procedures that increase either polymer erodibility and/or porosity. To this end, several methods have been suggested. For example, faster releasing gels have been obtained by introducing additives in the matrix, by carrying out the polymerisation using pH and water to ratio (R) values that yield low degree of Si-OH condensation [22-24]. In another approach, the drying step, that is normally carried out to remove the alcohol liberated by condensation reactions has been substituted by a buffer-exchange one. In this case hydrogel-type wet formulations were obtained where the final silica network was surrounded by physiological and/or protein friendly buffers preserving its original porosity [20]. It was shown that when proteins were introduced in this kind of wet gel during its formation, their original fold remained unaffected and their stability towards heat and protease degradation was found to be enhanced. While the kinetics of protein release from these matrices varied with both protein type and gel properties, it was found that matrix erosion was relatively fast, so that in some cases it could dominate the overall process. These wet gels were also crushed into 100-300 micron microparticles and were injected sub-cutaneously into balb-C mice at a dose of about 160-540 mg SiO_2/kg: no apparent toxicity was observed and the gels were completely reabsorbed within 4-15 days, depending on the formulation type [20].

These preliminary data showed that the wet polymers had promising properties as protein drug slow releasing systems. However, these matrices are soft and easy to break, and are therefore difficult to implant or inject unless cast as films and introduced as a part of a composite implant [25]. If one were to think of wet sol-gel silica as a stand-alone matrix for the sustained release of proteins in substitution therapy, a microparticle injectable form must be obtained. Even if injectable microparticles can be obtained simply by crushing a wet monolith (as was done by Teoli for the toxicity and *in vivo* degradation studies), this procedure cannot be standardized and therefore is not practically applicable. A possible alternative is to obtain the wet polymer directly in the shape of microsphere by carrying out the polymerisation in a emulsified state [26-29]. This strategy was successfully adopted for the preparation silica microspheres embedded with low molecular weight compounds, and low poly-disperse microsphere formation was aided by the use of surfactants.

In this work we evaluated a novel strategy to obtain sol-gel silica microspheres embedded with proteins for their controlled release. To this end, the microsphere polymerisation method described in the literature was adapted to the special requirements of protein compounds, which generally suffer from instability when in non-physiological conditions. Wet silica formulations similar to those described by Teoli were therefore obtained in the shape of microsphere by carrying out the polymerisation a W/O pseudo-emulsion [29] environment, namely an unstable emulsion obtained by mechanical stirring only and without the use of surfactants. In fact, surfactant molecules can themselves induce protein denaturation because they can promote solvent exposure of otherwise hidden protein epitopes, thus initiating the unfolding process.

Microspheres obtained through this method were investigated for the entrapment and controlled release of recombinant human growth hormone (r-hGH), a protein that was chosen both as a model and because its administration in therapy could take advantage of a sustained release formulation.

Growth hormone (GH) is an important protein (MW 22130, pI 5.3) involved in metabolism and body growth, whose impaired body levels cause several dysfunctions in both children and adults. Administration of exogenous growth hormone has been traditionally used in substitution therapy of GH deficiency but recently it has also been expanded for the treatment of syndromes not directly related to lowered GH body levels that may require alternative administration routes than the classic parenteral one [30]. One example is its the promotion of tissue repair: in this context both systemic and local administration were shown to improve diabetic wound healing [31, 32]. When injected in the blood stream, GH has a very short half life, so that it must be administered on a daily base. Similarly, when administered locally on a wounded tissue, the activity of local proteases induce fast degradation and loss of activity. On the other hand, GH treatments are commonly long term ones, as they are aimed at treating chronic disorders. The combination of all the above elements justifies the search for controlled release strategies aimed at reducing the number of administrations and improve patient compliance.

Several approaches to obtain sustained release GH formulations are described in the literature [33] but only a small number of formulations reached clinical trials [34] and one product only (Nutropin depot® developed by Genentech and Alkermes) consisting of r-hGH formulated in poly-lacticglycolic acid (PLGA) microspheres [35, 36] reached the market (now removed) for parenteral administration despite injection-site reactions represented a drawback for its use. Therefore, the present lack of a sustained release formulation with optimal properties justifies the search for alternative approaches to be used in parenteral or local administration.

This work represents the first attempt to use wet sol-gel silica microspheres as an alternative strategy for protein drug formulation and in particular for the sustained release of growth hormone. The results here obtained with GH may have a general relevance as many of the findings here described may be translated to the entrapment of other bioactive proteins.

Materials and Methods

Materials and Instrumentation

Tetramethylorthosilicate (TMOS) was obtained by Acros Organics (New Jersey, USA). Recombinant human Growth Hormone (r-hGH) was kindly provided by Genentech (South S. Francisco, CA, USA). Fluorescein-5-isothiocyanate was from Molecular Probes (Eugene, Oregon, USA). All other chemicals were of analytical or reagent grade and were purchased from Sigma-Aldrich (S.Louis Missouri, USA). Fluorescence was measured on a Jasco FP-6200 fluorometer. The UV-VIS measurements were carried out on a Varian Cary50 Win UV spectrometer. HPLC analyses were performed on a Agilent 1200 series apparatus using a Phenomenex C18 column. Far-UV CD spectra were recorded on a Jasco J-810 spectropolarimeter. Microparticle images were recorded with a Leika Leitz DM IRB fluorescent microscope integrated with a Nikon CD camera.

Fluorescent Labeling of r-hGH

The growth hormone used for the *in vitro* release experiments was fluorescent labeled with the amine-group specific fluorescent reagent, fluoresceine-isothiocyanate (FITC) according to standard procedures. 50 µl of FITC solution (2.5 mg/ml DMSO) were added while gently stirring to 1 ml of r-hGH solution (3 mg/ml) in 10 mM phosphate, 150 mM NaCl, pH 7.4 buffer (PBS) and the reaction was incubated overnight at 4°C in the dark. NH_4Cl was later added to a final concentration of 50 mM and the solution was incubated for 2 hours at 4°C. The product was purified by gel filtration (NAP-5 column, Amersham Bioscience, Uppsala, Sweden) and concentrated by ultrafiltration using 10KDa cut-off Amicon Ultra® regenerated cellulose filters (Millipore, MA, USA). Protein concentration and fluorescein/protein ratio were estimated by measuring the absorbance at 280 and 495 nm according to standard protocols. The final product was sterile-filtered through 0.22 µm filters. These conditions lead to the binding of about 1 molecule of fluorescein each protein. Circular dichroism analysis showed that no conformational changes occurred upon FITC labeling (not shown).

Sol-Gel Silica Matrix Polymerization and Protein Entrapment

Gels were obtained from tetramethoxysilane (TMOS) according to a procedure already described [20, 37]. All synthetic steps were performed in a sterile environment to prevent microbial contamination. Partial hydrolysis of the alkoxysilane precursors was initially promoted by an acidic treatment at room temperature for 30 minutes using HCl (Si:HCl:H_2O molar ratios = 1: 6x10^{-6}: 1.25). This solution was then mixed under gentle vortex with PBS alone or containing the protein to obtain the final '*sol*', which was allowed to gel into microspheres according to the protocols described below. Several gels were obtained at different final SiO_2 content which varied between 5-15% w/v. The amounts of reagents used are summarized in **table 1**. Protein concentration in the reaction medium varied between 0.5 to 5.3 mg/ml gel. Protein-free gels were also obtained by replacing the protein solution with PBS buffer only. Depending on the test to be later carried out, we synthesized gels containing either unmodified or FITC-labeled GH or unmodified BSA. Gels containing the unmodified proteins were used for circular dichroism experiments.Protein release and matrix erosion tests were performed on gels embedded with FITC-labeled GH.

Table 1. Amounts of pre-activated TMOS "*Sol*" and of buffer solution used to obtain 1 ml of various SiO_2 content wet gels.

% SiO_2 (w/v)	5	8	10	12	15
μl of pre-hydrolysis Sol	142	171	284	346	425
μl of buffer (+/- r-hGH)	858	772	716	654	575
R (H_2O/SiO_2)	58.69	32.2	25.07	19.15	14.04

Gelation into Microspheres

In 50 ml polypropylene tubes, the '*sol*' (1 ml) was added dropwise into a 10x volume of paraffin oil which was agitated until gelation using a palette system (500 rpm). After polymerization the particles were separated by centrifugation. Residuals of paraffin oil were removed by diethyl-ether washes (2x with 20 ml), each one followed by a centrifugation step. After the final centrifugation, diethylether was removed and the particles were rinsed once with 5 ml of PBS or 10 mM TRIS, 150 mM NaCl, pH 7.4 buffer (TBS) and then re-suspended in the minimal amount of buffer to prevent drying; the remaining traces of organic solvent were removed from the aqueous solution by 30 minute evaporation *under vacuo*. All preparations were analysed by light/fluorescence microscope to assess microparticle shape and average diameter using the ImageJ program. Fluorescence analysis of the washing solutions was carried out to quantify the amount of protein not incorporated during the preparation process.

Circular Dichroism (CD)

The effect of gel entrapment on the conformation of h-GH and serum bovine albumin (BSA) as a further model protein was verified by CD. Protein (0.15-0.3 mg/ml of gel) loaded gels at SiO_2 content between 5 and 15% w/v were prepared directly in the analysis quartz cuvette. Far-UV CD spectra were recorded at room temperature (25°C) within 2 hours from gel formation and were compared to those of the same protein in buffer solution. Data were collected from 200 nm to 260 nm, at 0.2 nm intervals with 20 nm/min scan speed, 2 nm bandwidth, and 16 s response: CD values are expressed as molar residue ellipticity in units of deg cm^2 $dmol^{-1}$.

Protein Release and Matrix Erosion

GH Release experiments were performed in sterile environment [20, 37] by immersing the microspheres at 37°C, into 1000x volume of sterile TBST buffer (TRIS 10mM, NaCl 150 mM, Tween 80 0.4%, pH 7.4). The amount of release buffer and the sampling schedule were selected in order to allow free dissolution of the gel SiO_2 in sink conditions without reaching saturation (120 mM) for silicic acid. Samples were maintained under constant agitation in a shaking thermostatic bath. At scheduled times, aliquots (1 ml) of the release medium were removed and replaced with fresh buffer. FITC-protein concentration in the release samples was quantified by fluorescent measurement (λex. 495 nm, λem. 525 nm) according to a calibration curve measured in parallel. Fluorescent data were confirmed by RP-HPLC (phenomenex C-18, eluted with a H_2O/CH_3CN (+0.05%TFA) gradient).

All release samples were analyzed for their silica content using the molibden-blue colorimetric test [38]. Briefly, to 1 ml of sample, 40 μl of the molybdate reagent (12.3g of $(NH_4)_6MoO_{24}.4H_2O$ in 100 ml H_2O, added of 100 ml of 4.5M H_2SO_4) are added, mixed and allowed to stand for 15 min. 40 μl of oxalic acid solution (10 g/100 ml H_2O) are then added, followed immediately by 40 μl of ascorbic acid solution (2.8 g/100 ml H_2O). The blue silicomolibdic complex is formed within 40 min and quantified spectrophotometrically at 810 nm. A calibration curve is obtained in the same way using standard solutions (0 to 30 μM).

Results and Discussion

1. Gels Macroscopic Features

Gel time (**table 2**) varied between 1.5 and 5 minutes and was found to be inversely correlated to SiO_2 content. This is expected as the reaction rate accelerates with the reaction components concentration. A linear correlation between SiO_2 content and *1/gel-time* was observed (not shown). Gel consistency is also strongly affected by SiO_2 content: low silica containing gels are soft and require special care during handling to avoid them to break. This is a problem when preparing the microspheres as the method involves vigorous mixing and several centrifugation- re-suspension steps. In our hands 5% is the minimal SiO_2 concentration at which we managed to obtain good quality microspheres. At higher SiO_2 concentration, sample manipulation becomes easier and good quality microspheres were obtained in all preparation batches.

Table 2. SiO_2 content and gel polymerisation time.

% SiO_2	5	8	10	12	15
gel time (sec)	315	180	150	110	95

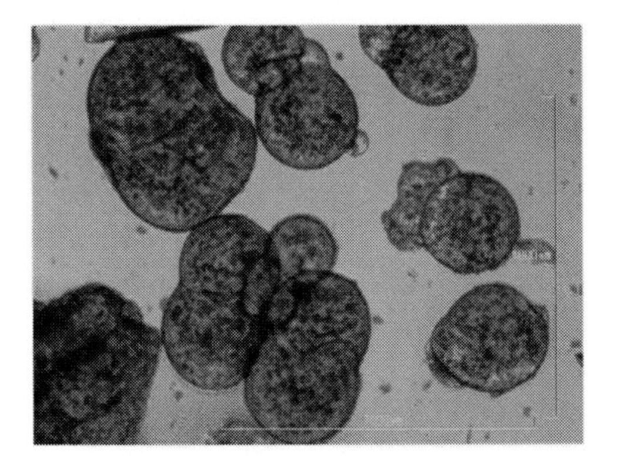

Figure 1. Light and fluorescent images of 15% SiO_2 microspheres loaded with fluorescently labelled r-hGH. Size bar: 1000 μm.

Images of sample microspheres loaded with fluorescent r-hGH are shown **in Figure 1**.

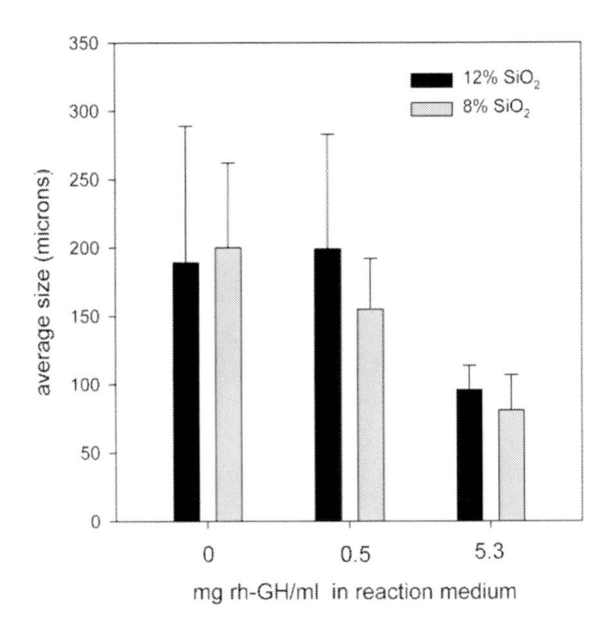

Figure 2. Effect of reaction medium GH concentration on the size of 8% and 12% SiO_2 microspheres. Particle size was measured upon microscopy image analysis using ImageJ software.

The effect of both silica and protein content on microsphere size and shape was investigated. Microsphere size was found to be unaffected by silica content. The radius of protein-free microspheres was 400 +/- 180, 200+/- 100 and 320 +/- 150 μm for 5 12 and 15% SiO_2 respectively. It appears that protein-free microspheres are rather large and the differences registered at different SiO_2 content are not statistically relevant, this also because of their large polydispersivity. Both large size and polydispersivity are probably due to the lack of surfactant in the bi-phase polymerization medium. As mentioned above, this experimental detail together with the choice of using the totally non polar paraffin oil as the external phase was specifically adopted to avoid potential protein denaturation (which could occur in the presence of surfactants) and to maximize its localization in the hydrophilic milieu of the polymerising sol. In the absence of a surfactant the system is maintained in a pseudo-emulsified state only by the action of the rotating palette system. Microsphere size is therefore highly sensitive to the system variables including palette shape, its rotation speed and the oil to buffer volume ratio. For example, smaller microspheres are obtained at higher palette rotation speed. Nevertheless, once the conditions have been optimised, batches with reproducible properties are obtained.

When the protein-free buffer solution in the polymerising medium was substituted a with r-hGH-containing one, microspheres with significantly smaller size were obtained. Increasing protein concentration in the reaction medium not only reduced microsphere size but also their polydispersivity (**figure 2**). This phenomenon is likely due to a surfactant-like effect induced by the protein. Despite the protein surface activity emerging from the microsphere surface analysis, no evident protein surface compartmentalization occurred during polymerization. In fact, the fluorescent signal of the FITC-labelled GH appears homogeneously distributed within the gel both in the bulk and microspheres formulations of figure 1.

2. Microsphere Protein Loading Capacity

Microsphere loading capacity was found to be highly affected by the matrix silica content: the amount of protein that is retained after polymerisation increaes with the mount of silica (Table 3). In the case of the 12 and 15% SiO_2 preparations, quantitative protein incorporation was achieved at all of the GH concentrations tested. At 5% SiO_2, the loading capability was dramatically lower and only 0.6 mg GH/ml gel could be incorporated, the excess added in the polymerising mix being extracted during the washing procedures. We conclude that each matrix has a maximal loading capacity which is of 0.6 and 3.2 mg/ml gel in the case of the 5 and 8% SiO_2 formulations respectively, while in the case of the 12 and 15% ones it is higher than 5.3 mg/ml, the actual limit not been reached in the experimental conditions here adopted. It must be pointed out that in order to achieve a GH concentration of 5.3 mg/ml of final gel, we had to use a rather highly concentrated (9 mg/ml) fluoresceine-r-hGH PBS stock solution (see table 1), which was the most concentrated solution that we obtained in our laboratory. More concentrated stock solutions should be used in order to achieve higher loadings and to measure the maximal loading capability of the two high retaining gels (12 and 15% SiO_2) here investigated. From the above considerations it must be noted that the maximal protein load may not only be dictated by the gel properties but also by the protein solubility in aqueous buffers.

Table 3. Amounts of protein retained in the microspheres

% SiO_2	mg GH/g wet gel (5.3 mg/ml loading)
15	> 5.3
12	> 5.3
8	3.2
5	0.63

3. Protein Conformation in Solution and after Gel Embedment

One of the advantages of silica is that it is UV transparent: this allows the spectrophotometric evaluation of proteins also when they are entrapped in the sol-gel polymer. In order to evaluate how the embedment in gels differently formulated affects protein structure, we synthesized four different polymers - that varied for their final SiO_2 content - which were loaded with two different proteins, GH and BSA, the latter used as a control. Testing two different proteins was aimed at evaluating if the effects observed with one could or could not be considered of general value.

The circular dichroic spectra of r-hGH and BSA in PBS solution and after entrapment in 5, 8, 12 and 15% SiO_2 (w/v) wet gels are shown in **figure 3**. All gel-embedded BSA spectra are superimposable to the one of the PBS solution, indicating that this protein maintains its full conformation in all of the formulations tested. On the contrary, embedded r-hGH spectra superimposable to the solution one were obtained up to 12% SiO_2 only, while a deviation from the native conformation was observed in the 15% SiO_2 formulation, with about 15 % loss of alpha helix content (as calculated from the intensity of the 208 and 222 nm bands).

The differences registered between the two proteins demonstrate that in the sol-gel polymerizing conditions, GH is more sensitive to denaturation than the stable BSA. From a more general point of view, this result indicates that what is obtained with one protein cannot be directly translated to another one so that each protein needs to be tested individually.

In any case, the results indicate that the polymerization conditions are rather protein-friendly up to 12% SiO_2. It must also be pointed out that the spectra were recorded from gels obtained directly in the CD cuvette. These samples differ slightly from those used in the release assays as they were not rinsed with buffer immediately after polymerization and before the analysis. Therefore the methanol liberated during polymerization remained in the medium surrounding the protein. At 15% SiO_2 w/v, methanol concentration reaches 40% v/v and it is likely to affect protein structure. Therefore methanol, rather than physical confinement could be responsible for the slight GH denaturation observed in the 15% SiO_2 gel [39, 40]. To overcome this inconvenience, silica precursors with protein friendly leaving groups can be used instead of TMOS [40, 41]. CD spectra of the microparticle formulations could not be recorded because of excessive light scatter which disturbs the analysis. However, it is likely that equal if not minor protein conformation loss would occur in these formulations. In fact in this case it is expected the methanol to be extracted at least partially by the organic outer phase, therefore reducing its negative effect on protein conformation.

4. Matrix Erosion

Erosion experiments were carried out on microparticles at 5, 8, 12 and 15% SiO_2 w/v. As shown in figure 3, the rate of the process decreases with the increase in silica content between 5 to 12% SiO_2 and it is linear with the square root of time, as expected from a diffusion related phenomenon [20]. No difference was observed between 12 and 15 % SiO_2 formulations, suggesting that the two matrices have similar degree of cross-linking. The time necessary to achieve full matrix erosion (t-ex) was calculated from the square-root of time-linearized data and it is displayed in table 4. No difference was observed between protein-loaded and protein-free formulations, indicating that the effect induced by the protein on microsphere size (figure 2) does not alter the area exposed to the solvent.

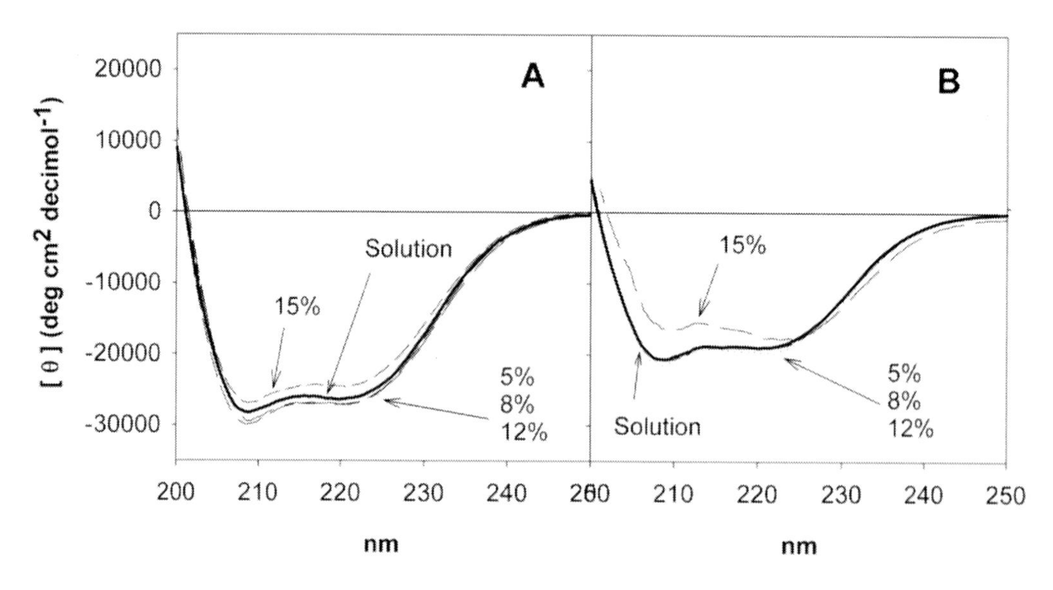

Figure 3. Circular dichroism spectra of BSA (A) and r-hGH (B) in solution in PBS buffer (heavy line) and embedded in wet sol-gel silicas at 5, 8, 12 and 15% SiO_2 concentration.

These results are in line with what previously obtained with wet gel monoliths at 4% and 12% SiO_2, which displayed 70 and 250 hours t-ex for erosion, respectively [20]. This finding could indicate that neither the gel shape nor the polymerisation method influence significantly gel erodibility. Indeed, it was shown that erosion of wet SiO_2 gels occurs not only from the outer gel surface but also from within the internal polymer network [20] because the matrix pores are large enough to allow free diffusion of the solvent. This could explain why the matrix shape and size does not influence the erosion speed. On the other hand, literature data showed that the emulsion polymerization method does affect sol-gel silica structure producing more retentive matrices than the ones obtained as monoliths. For example, Radin and Ducheyne compared the release of the low molecular weight drugs vancomycin and bupivacain from silica xerogels (dried silica) obtained in the shape of microspheres (through a similar method as the one described here) or microgranules obtained upon crushing and sieving dry monoliths [28]. Even if the time-scale of the release experiments was longer than what registered here because their particles were in the dry form, the microspheres were significantly more retentive than the crushed granules. Similarly, preliminary data obtained in our laboratory showed that wet microspheres obtained through a similar method as the one described here were more retentive and resistant to hydrolysis than corresponding microparticles (with the same amount of SiO_2 and size) obtained upon crushing manually larger monoliths [42]. It was then hypothesized that the organic solvent used for the microsphere washing steps, while extracting the methanol liberated upon the condensation reactions also promotes to some degree the matrix compaction giving rise to gels with smaller pores.

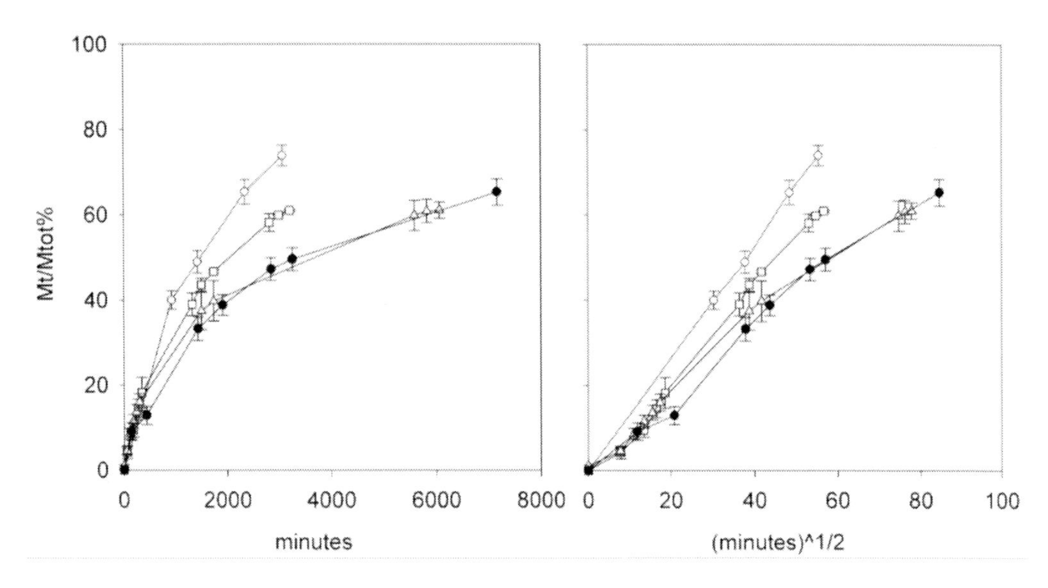

Figure 4. Erosion profile of microspheres polymerised at different SiO_2 concentrations (5% (○), 8% (□), 12% (Δ) and 15% (●) w/v). Data have been plotted against time (A) and time-square-root (B). Each curve represents the average of multiple experiments in which protein-free and r-hGH-loaded microspheres were tested. Bars represent the standard deviation.

The fact that the microspheres here obtained have a similar crodibility as their corresponding monoliths can therefore be explained by the combination of two phenomena acting in two different directions: on one hand, the emulsion polymerisation method, including the organic washes step, generates more compact and resistant gels; on the other hand the reduction in size associated with the change from monolith to microspheres with the subsequent increase in the area available to the solvent, accelerates all diffusion-related phenomena, including erosion.

Table 4. Time necessary for matrix erosion at different SiO_2 content. Values were obtained upon extrapolation of the erosion curves plotted against the square root of time.

% SiO_2	*t-ex* erosion (h)
15	263±17
12	246±14
8	124±6
5	87±1

5. Protein Release

5.1. Method Validation

Along the release assays, the concentration of r-hGH in the medium was measured indirectly by quantifying the amount of the fluoresceine covalently bound to the protein. This method is faster than others that rely on direct protein quantification (e.g. HPLC, or immunoenzymatic methods) and it is more convenient when a large number of experiments is carried out simultaneously. The method is also sensitive: a linear response is observed down to as low as 0.2 μg protein/ml. However before carrying out the release assays, the fluorescence method had to be validated in order to ensure that the entire fluorescence signal liberated along the release assay was due to the fluoresceine labelled-r-hGH molecule and not to its degradation products (fluoresceine-labelled low MW peptides or fluoresceine alone). Validation was carried out by comparing the data of a 24 hours release experiment analysed through HPLC and fluorescence. The results (figure 4) clearly show that the two methods are equivalent.

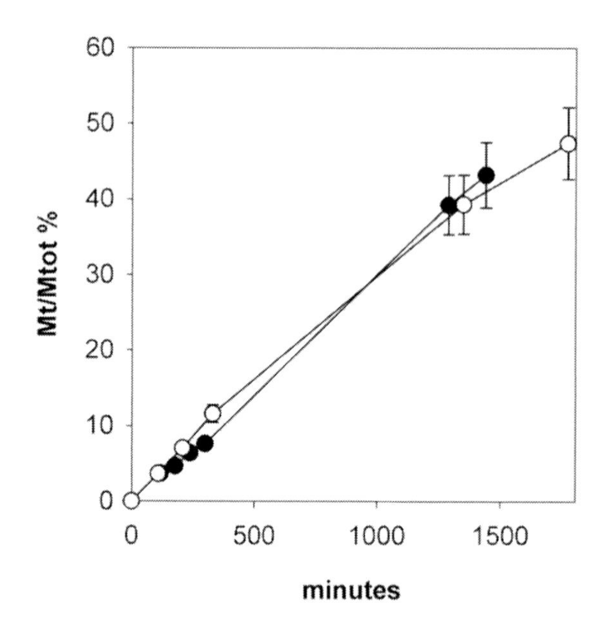

Figure 5. Comparison between the early release of r-hGH (15% SiO_2 microspheres, first 24h) as measured by HPLC (•) and FITC-fluorescence (o).

5.2. r-hGH Release from Different Formulations

Figure 5 shows the results of the r-hGH release experiments depicted in parallel with the erosion ones. In all cases, r-hGH release was faster than matrix erosion. This is more evident for the 5% gel: protein release ended within 10 hours whereas the time required for total erosion was 3.5 days. This indicates that the pores of the 5% SiO_2 matrix are larger than the protein radius and, at the same time, that there is no chemical interaction between polymer and the protein. The difference in GH release and matrix erosion for the 8, 12 and 15% gels is of minor entity than that observed for the 5% gel. In this case, the pore size is probably of the same order of the protein and erosion starts to be dominant over diffusion. These data explain also the differences observed among the four gels in terms of their loading capability (table 3).

When comparing these results with what reported by Teoli for the monolith wet gels, it appears that r-hGH is retained less than the model proteins avidin, Ribonuclease A (Rnase) and bovine serum albumin (BSA) confirming that each protein behaves differently and no general rule can be given when it comes to predicting the interaction with sol-gel silica polymers [20, 43]. In our experience, however, acidic molecules (as is the case for r-hGH) are retained less than basic ones by sol-gel silica polymers [15] and this likely because no ionic interaction can occur.

Table 5. Time necessary for total GH release from gels at different SiO_2 content. Values were obtained upon extrapolation of the erosion curves plotted against the square root of time.

% SiO_2	*t-ex* GH release (h)
15	142,9±13,8
12	137,2±6,7
8	98,8±7,8
5	10,7±2,2

It is also interesting to note that the release curve from the 15% SiO_2 microspheres flattens below the 100% protein release. This suggests that partial denaturation of the protein has occurred either in the gel or during the release experiment. The first hypothesis would support what observed by CD analysis. In this respect, it is also noteworthy that 12 and 15% SiO_2 gels do not differ dramatically in release or erosion rates and both display a very high loading capability (**table 3**). Therefore, even if from the loading and erosion point of view the two formulations are equivalent, both CD and release data indicate the 12% one is superior because it does not induce protein denaturation.

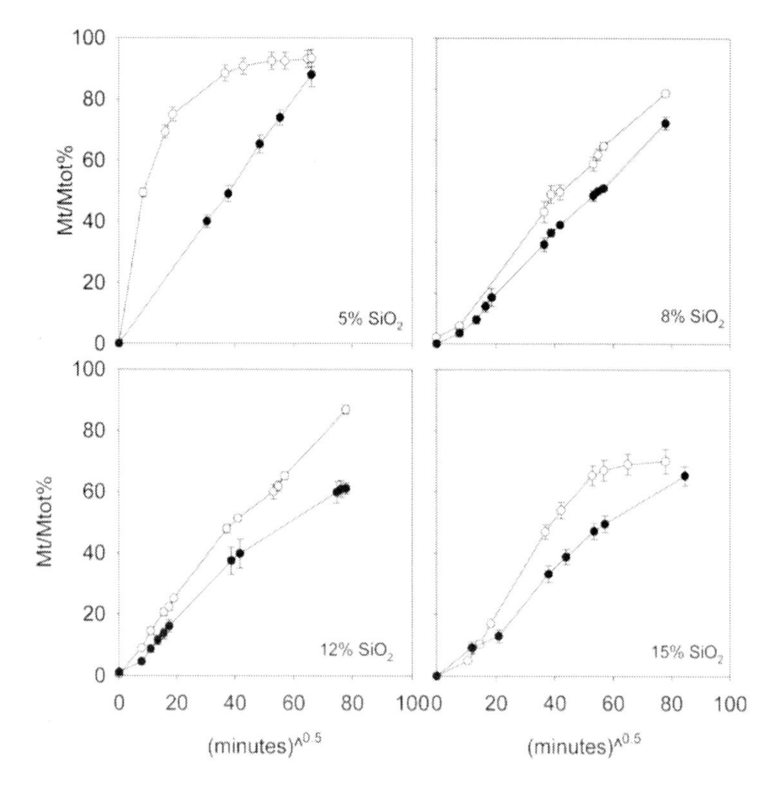

Figure 6. r-hGH release (o) and Silica erosion (•) from 5%, 8%, 12% and 15% SiO2 microspheres. Release experiments were carried out in duplicate. Bars represent the standard deviation.

Conclusion

Wet sol-gel silica microspheres are promising formulations capable of entrapping relatively large amounts of r-hGH and releasing it in a controlled manner. The microspheres are easy to obtain from low cost precursors. Protein loading capacity can be increased by increasing the SiO_2 content and the release rate can be modulated by varying simple parameters during microparticle preparation. Simultaneous evaluation of the formulation through different analytical procedures allowed assessing their potentials from different points of view. Formulations have been compared for their loading and releasing properties, whereas CD analysis allowed determining the protein folded status, a property that might affect not only the total biological activity releasable into the outside medium but also the immunological response upon patient administration.

While the 12 and 15% SiO_2 gels showed to have the best properties in terms of loading capacity and sustained release, the circular dichroism data showed that in the 15% one only, partial r-hGH denaturation occurred. This is likely due to the large amount of methanol liberated during condensation, even if further investigation is needed to clarify this point. However, if this were the case, 12% w/v is the maximum silica concentration usable when embedding GH as the protein drug. Alternatively, silica precursors others than TMOS that release protein-friendly leaving groups instead of methanol could be used [25].

While keeping in mind that each protein system needs to be investigated individually, some of the data here reported have general value and suggest that microspheres of silica gel in it's the wet form are promising matrices for the sustained release of protein drugs. The protein-loaded microspheres here described are completely biodegradable and have a diameter that is small enough (less than 150 μm) to pass through 24-30 gauge needles and in principle are injectable. More toxicity studies must be carried out before this material could be approved for parenteral administration. In the meanwhile, other administration routes might be conceivable, among which the topic one. In fact, wet sol-gel silica embedded proteins already proved to have higher stability and to be totally preserved from protease degradation, so that inserting protein-loaded wet-silica microspheres within a topic gel would permit a sustained release of intact drug, without the need of repeated administrations. Indeed, evaluation of such a formulation for the treatment of diabedic ulcers is currently underway in our laboratory.

Acknoledgments

This research was funded by the Italian Minister of Research and education (PRIN and ex 60%). We thank Genentech for kindly providing r-hGH.

References

[1] Brinker, C. J. & Scherer, G. W. (1989). *Sol-Gel Science. The Physics and Chemistry of Sol-gel Processing*. San Diego, CA: Academic Press. 1-11.

[2] Rigby, S. P., Fairhead, M. & van der Walle, C. F. (2008) Engineering silica particles as oral drug delivery vehicles. *Current Pharmaceutical Design. 14,* 1821-31.

[3] Unger, K., Rupprecht, H., Valentin, B., & Kircher, W. (1983) the use of porous and surface modified silicas as drug delivery and stabilizing agents. *Drug Dev Ind Pharm. 9,* 69-91.

[4] Bottcher, H., Slowik, P., and Sub, W. (1998) Sol-gel carrier system for controlled drug delivery. *Journal of Sol-Gel Science and Technology. 13,* 277-81.

[5] Kortesuo, P., Ahola, M., Karlsson, S., Kangasniemi, I., Kiesvaara, J., and Yli-Urpo, A. (1999) Sol-gel-processed sintered silica xerogel as a carrier in controlled drug delivery. *J Biomed Mater Res. 44,* 162-7.

[6] Smirnova, I., Suttiruengwong, S., Seiler, M., and Arlt, W. (2004) Dissolution rate enhancement by adsorption of poorly soluble drugs on hydrophilic silica aerogels. *Pharm Dev Technol. 9,* 443-52.

[7] Mellaerts, R., Aerts, C. A., Van Humbeeck, J., Augustijns, P., Van den Mooter, G., and Martens, J. A. (2007) Enhanced release of itraconazole from ordered mesoporous SBA-15 silica materials. *Chem Commun (Camb)*, 1375-7.

[8] Kortesuo, P., Ahola, M., Karlsson, S., Kangasniemi, I., Kiesvaara, J., and Yli-Urpo, A. (1999) Sol-gel-processed sintered silica xerogel as a carrier in controlled drug delivery. *J. Biomed. Mater. Res. 44,* 162-67.

[9] Ahola, M. S., Sailynoja, E. S., Raitavuo, M. H., Vaahtio, M. M., Salonen, J. I., and Yli-Urpo, A. U. (2001) In vitro release of heparin from silica xerogels. *Biomaterials. 22,* 2163-70.

[10] Barbè, C., Bartlett, J., Kong, L., Finnie, K., Lin, H., Larkin, M., Calleja, S., Bush, A., and Calleja, A. (2004) Silica Particles: A novel Drug Delivery System. *Advanced Materials. 16,* 1959-66.

[11] Calleja, S., Bush, A., Flannigan, S., and Barbé, C. Micro-Encapsulation and Controlled Release from Silica Particles. in 31st Annual meeting of the Controlled Release Society. 2004. Honolulu, Hawaii.

[12] Doadrio, A. L., Sousa, E. M., Doadrio, J. C., Perez Pariente, J., Izquierdo-Barba, I., and Vallet-Regi, M. (2004) Mesoporous SBA-15 HPLC evaluation for controlled gentamicin drug delivery. *J Control Release. 97,* 125-32.

[13] Czarnobaj, K. and Lukasiak, J. (2004) In vitro release of cisplatin from sol-gel processed porous silica xerogels. *Drug Delivery. 11,* 341-44.

[14] Morpurgo, M., Teoli, D., Palazzo, B., Bergamin, E., Realdon, N., and Guglielmi, M. (2005) Influence of synthesis and processing conditions on the release behavior and stability of sol-gel derived silica xerogels embedded with bioactive compounds. *Farmaco. 60,* 675-83.

[15] Roveri, N., Morpurgo, M., Palazzo, B., Parma, B., and Vivi, L. (2005) Silica xerogels as a delivery system for the controlled release of different molecular weight heparins. *Anal Bioanal Chem. 381,* 601-6.

[16] Maver, U., Godec, A., Bele, M., Planinsek, O., Gaberscek, M., Srcic, S., and Jamnik, J. (2007) Novel hybrid silica xerogels for stabilization and controlled release of drug. *Int J Pharm. 330,* 164-74.

[17] Kortesuo, P., Ahola, M., Karlsson, S., Kangasniemi, I., Yli-Urpo, A., and Kiesvaara, J. (2000) Silica xerogel as an implantable carrier for controlled drug delivery--evaluation of drug distribution and tissue effects after implantation. *Biomaterials. 21,* 193-98.

[18] Lai, W., Garino, J., Flaitz, C. M., and Ducheyne, P. (2000) Physiological removal of silicon from bioactive glass. *Bioceramics. 192-1,* 581-84.

[19] Lai, W., Garino, J., and Ducheyne, P. (2002) Silicon excretion from bioactive glass implanted in rabbit bone. *Biomaterials. 23,* 213-17.

[20] Teoli, D., Parisi, L., Realdon, N., Guglielmi, M., Rosato, A., and Morpurgo, M. (2006) Wet sol-gel derived silica for controlled release of proteins. *J Control Release. 116,* 295-303.

[21] Santos, E. M., Radin, S., and Ducheyne, P. (1999) Sol-gel derived carrier for the controlled release of proteins. *Biomaterials. 20,* 1695-700.

[22] Viitala, R., Jokinen, M., Tuusa, S., Rosenholm, J. B., and Jalonen, H. (2005) Adjustably bioresorbable sol-gel derived SiO2 matrices for release of large biologically active molecules. *Journal of Sol-Gel Science and Technology. 36,* 147-56.

[23] Viitala, R., Jokinen, M., Maunu, S. L., Jalonen, H., and Rosenholm, J. B. (2005) Chemical characterization of bioresorbable sol-gel derived SiO2 matrices prepared at protein-compatible pH. *Journal of Non-Crystalline Solids. 351,* 3225-34.

[24] Viitala, R., Jokinen, M., and Rosenholm, J. B. (2007) Mechanistic studies on release of large and small molecules from biodegradable SiO2. *Int J Pharm. 336,* 382-90.

[25] Conconi, M. T., Bellini, S., Teoli, D., de Coppi, P., Ribatti, D., Nico, B., Simonato, E., Gamba, P. G., Nussdorfer, G. G., Morpurgo, M., and Parnigotto, P. P. (2009) In vitro and in vivo evaluation of acellular diaphragmatic matrices seeded with muscle precursors cells and coated with VEGF silica gels to repair muscle defect of the diaphragm. *J Biomed Mater Res A. 89,* 304-16.

[26] Barbè, C., Kong, L., Finnie, K. S., Calleja, S., Hanna, J. V., Drabarek, E., Cassidy, D. T., and Blackford, M. G. (2008) Sol-gel matrices for controlled release: from macro to nano using emulsion polymerisation. *Journal of Sol-Gel Science and Technology. 46,* 393-409.

[27] Barbè, C. and Bartlett, J., Controlled release ceramic particles, compositions thereof, processes of preparation and methods of use. 2001-2007: US, patent # US 7258874

[28] Radin, S., Chen, T., and Ducheyne, P. (2009) The controlled release of drugs from emulsified, sol gel processed silica microspheres. *Biomaterials. 30,* 850-8.

[29] Teoli, D., Parisi, L., Bellini, S., Realdon, N., Conconi, M., Parnigotto, P. P., and Morpurgo, M. Sol-Gel Based microparticles and films for sustained Delivery of Proteins and Growth Factors. in 1st conference on Innovation in Drug Delivery: from biomaterials to devices. 2007. Naples, Italy.

[30] Wit, J. M. (2000) The use of GH as pharmacological agent (minireview). *Endocr Regul. 34,* 28-32.

[31] Gimeno, M. J., Garcia-Esteo, F., Garcia-Honduvilla, N., San Roman, J., Bellon, J. M., and Bujan, J. (2003) A novel controlled drug-delivery system for growth hormone applied to healing skin wounds in diabetic rats. *J Biomater Sci Polym Ed. 14,* 821-35.

[32] Garcia-Esteo, F., Pascual, G., Garcia-Honduvilla, N., Gallardo, A., San Roman, J., Bellon, J. M., and Bujan, J. (2005) Histological evaluation of scar tissue inflammatory response: the role of hGH in diabetic rats. *Histol Histopathol. 20,* 53-7.

[33] Brodbeck, K. J., Pushpala, S., and McHugh, A. J. (1999) Sustained release of human growth hormone from PLGA solution depots. *Pharm Res. 16,* 1825-9.

[34] Clemmons, D. R. (2007) Long-acting forms of growth hormone-releasing hormone and growth hormone: effects in normal volunteers and adults with growth hormone deficiency. *Horm Res. 68 Suppl 5,* 178-81.

[35] Johnson, O. L., Cleland, J. L., Lee, H. J., Charnis, M., Duenas, E., Jaworowicz, W., Shepard, D., Shahzamani, A., Jones, A. J., and Putney, S. D. (1996) A month-long effect from a single injection of microencapsulated human growth hormone. *Nat Med. 2,* 795-9.

[36] Cleland, J. L., Mac, A., Boyd, B., Yang, J., Duenas, E. T., Yeung, D., Brooks, D., Hsu, C., Chu, H., Mukku, V., and Jones, A. J. (1997) The stability of recombinant human growth hormone in poly(lactic-co-glycolic acid) (PLGA) microspheres. *Pharm Res. 14,* 420-5.

[37] Conconi, M. T., Bellini, S., Teoli, D., de Coppi, P., Ribatti, D., Nico, B., Simonato, E., Gamba, P. G., Nussdorfer, G. G., Morpurgo, M., and Parnigotto, P. P. (2008) In vitro and in vivo evaluation of acellular diaphragmatic matrices seeded with muscle precursors cells and coated with VEGF silica gels to repair muscle defect of the diaphragm. *J Biomed Mater Res A.*

[38] Koroleff, F., in *Methods of Seawater Analysis,*, K.K. Grasshoff K, and Ehrhardt M,, Editor. 1983, Wiley-VCH: Weinheim. p. 174-83.

[39] Ellerby, L. M., Nishida, C. R., Nishida, F., Yamanaka, S. A., Dunn, B., Valentine, J. S., and Zink, J. I. (1992) Encapsulation of proteins in transparent porous silicate glasses prepared by the sol-gel method. *Science. 255,* 1113-5.

[40] Besanger, T. R., Chen, Y., Deisingh, A. K., Hodgson, R., Jin, W., Mayer, S., Brook, M. A., and Brennan, J. D. (2003) Screening of inhibitors using enzymes entrapped in sol-gel-derived materials. *Anal Chem. 75,* 2382-91.

[41] Cruz-Aguado, J. A., Chen, Y., Zhang, Z., Brook, M. A., and Brennan, J. D. (2004) Entrapment of Src protein tyrosine kinase in sugar-modified silica. *Anal Chem. 76,* 4182-8.

[42] Morpurgo, M., Pignatto, M., Braggion, E., Mozzo, A., Franceschinis, E., and Realdon, N. Erosion and Release of rhGH from Wet-Sol-Gel Silica in Injectable Form. in 36th Annual Meeting & Exposition of the Controlled Release Society (CRS). 2009. Copenhagen, Denmark.

[43] Teoli, D., Realdon, N., Conconi, M. & Morpurgo, M. (2006). VEGF Embedded Sol-gel Derived Silica Polymers as Supports for the Neovascularization of Bioengineered Implants. in 33rd annual meeting of the Controlled Release Society. Vienna, Austria.

In: The Sol-Gel Process
Editor: Rachel E. Morris

ISBN 978-1-61761-321-0

Chapter 22

SOL-GEL SYNTHESIS AND STUDY OF SILICA NANOPARTICLES AS POTENTIAL CARRIERS FOR DRUG DELIVERY

***Elena V. Parfenyuk**[*1]* **and Bouzid Menaa**[†2]*
[1]Institute of Solution Chemistry of the Russian Academy of Sciences, Russian Federation
[2]Fluorotronics, Inc., Vista, CA, USA

ABSTRACT

In the present time most researchers believe that progress in medicine is connected with the development of drug delivery systems using nanoparticles. Very important drugs are immune-modulators which are natural or synthetic compounds. They often have protein nature. The efficiency of immune-modulator delivery into cells using nanocarriers will depend on both physical-chemical and biological properties of the drug, and the properties of nanocarriers themselves. Silica materials with different surface functionalities have been synthesized by sol-gel method. Physical properties of the silica materials in general have been studied to date to enlighten the molecular interactions and biocompatibility between host matrices and biomolecules (enzymes, proteins). These properties will be reported in this Chapter. We will also mainly report the work that is a part of original study concerning with the potential application of silica particles as nanocarriers of immune-modulator. For instance, for drug delivery applications, the adsorption ability of the silica materials has been studied and will be also described specifically using human serum albumin as a model compound of the drug. The studies of physical and adsorption properties of the silica materials as well as their interactions with immune cells *in vitro* is in the forefront of the research in the field as it will allow

* Institute of Solution Chemistry of the Russian Academy of Sciences, 1 Akademicheskaya Str. Ivanovo 153045, Russian Federation, Fax : + 7 0932 237 8509; E-mail: evp@isc-ras.ru
† Fluorotronics, Inc., 2453 Cades Way, Building C, Vista, CA 92081, USA, Ph: +32 4 759 28 957; Email: bouzid.menaa@gmail.com

designing and selecting the most promising, and efficient silica carrier of immune-modulator proteins for drug delivery.

Keywords: porous silica nanoparticles, adsorption properties, proteins, silica/protein interactions, biocompatibility, drug nanocareers, drug delivery applications

1. INTRODUCTION

The sol-gel process has been described in detail by Brinker and Scherrer [1]. Typically, alkoxide monomers (tetramethoxysilane, TMOS or tetraethylethoxysilane, TEOS) undergo hydrolysis in the absence or presence of an acid or base catalyst to form silanols. The silanols link together through polycondensation to form a siloxane network characterizing the inorganic silica matrix. Residual alcohols (e.g., methanol in the case of TMOS) are then released upon polycondensation to form methanol or ethanol depending on the silica precursors used (TMOS or TEOS, respectively). The preparation of inorganic silica gels has been extended to novel hybrid nanomaterials in which organic and inorganic species are mixed at the molecular level [2]. The pioneering works on those materials were leaded by Schmidt [3] and Wilkes [4]. Organic-inorganic hybrids or ORMOSILs (ORganically MOdified SILicates) concern the modification of TMOS or TEOS by adding organically modified silanes $R_nSi(OCH_3)_{4-n}$, to undergo hydrolysis and polycondensation using the same protocol as described earlier for the unmodified TMOS glass and to obtain sol-gel matrices such as [(100-x) TMOS: (x) $R_nSi(OCH_3)_{4-n}$, n = 1, 2, 3, R = functional organic group, x = mol% composition] where -Si-O-Si- is the siloxane network in which some of the bridging oxygens of a SiO_4 tetrahedron are substituted by one, or several organic groups.

The special physical and chemical properties of silica are quite attractive from the viewpoint of the compatibility with biomolecules and living cells. The thermodynamic stability of the Si–O bond is of 452 kJ/mol indicates a strong inertness that exclude interference with enzymes and functions typical of differentiated cells. The approach to sol-gel silica loaded with biological systems, particularly enzymes, has produced important and valuable implementations in biotechnology [5-14] and open tremendous possibilities to nanomedicine with the synthesis of functionalized silica-based nanoparticles as drug nanocareers for instance.

To achieve the entrapment of an active protein in silica nanoparticles, it is necessary to maintain the active conformation of the protein within the nanoporous materials. The sol-gel method is compatible with the preservation of biomolecules or cells upon encapsulation or adsorption. The mild conditions associated with the synthesis at ambient temperature make the sol-gel route to offer new possibilities in biotechnology.

The biomolecules entrapped in ormosil nanoparticles can be efficiently employed in drug delivery and other pharmaceutical and medical applications [15]. The organically modified and anticancer drug doped nanoparticles (diameter ca. 30 nm) obtained through aqueous dispersion have been used in photodynamic cancer treatment [16]. Water-insoluble photosensitizing anticancer drug, 2-devinyl-2-(1-hexyloxyethyl) pyropheophorbide, was entrapped in non-polar core of micelles by hydrolysis of triethoxyvinylsilane. The resulted nanoparticles were spherical, highly monodispersed, and stable in aqueous system.

In a recent review, Vallet-Regi et al. [17] showed how in the last few years there has been an exceptional growth in research focused on drug delivery systems. This explosion has been induced by the features and possibilities that these systems offer to biomedicine, such as the possibility of several drugs to be administrated using new therapies improving efficacy and safety. Silica-based ordered mesoporous materials have experienced an impressive consideration in drug delivery applications since they were first employed as drug delivery systems back in 2001. The possibility of tuning the physicochemical properties of the matrices together with the development of stimuli-responsive technologies are two of the main reasons of the great scientific attention on these bioceramics as conventional drug delivery systems.

In general, the size and chemical nature of the pores silica-based sol-gel matrix can be tailored to conform the biomolecule size and to avoid the protein or enzyme leaching. The modification of TMOS or TEOS surface with organoalkoxysilanes offers the possibilities of changing the properties of the materials (hydrophobicity, polarity and surface charges, pores size, steric effects). Changing the ratio of tetraalkoxysilane to organotrialkoxysilane can control the cation exchange capacity, polarity of porous surface, the hydrophobicity/ hydrophilicity [18] and consequently the local environment of the protein. We can then associate the physical properties and structure of the host materials to adsorption process, drug or biomolecule viability, biocompatibility.

One of the big challenge that we are reporting here is the physical and adsorption properties of the silica materials as well as their interactions with immune cells *in vitro* that is in the forefront of the research in the field as it will allow designing and selecting the most promising, and efficient silica carrier of immune-modulator proteins for drug delivery [19]. For instance, for drug delivery applications, the adsorption ability of the silica materials has been studied and will be also described specifically using human serum albumin as a model compound of the drug.

2. Characteristics of Silica Materials

Silica materials with different surface chemical functionalities were synthesized by sol-gel method using tetraethoxysilane (TEOS) as precoursor. Polyethyleneimine (PEI), 3-(aminoporopyl) triethoxysilane (APTES), methyltriethoxysilane (MTEOS) serve as modifiers:

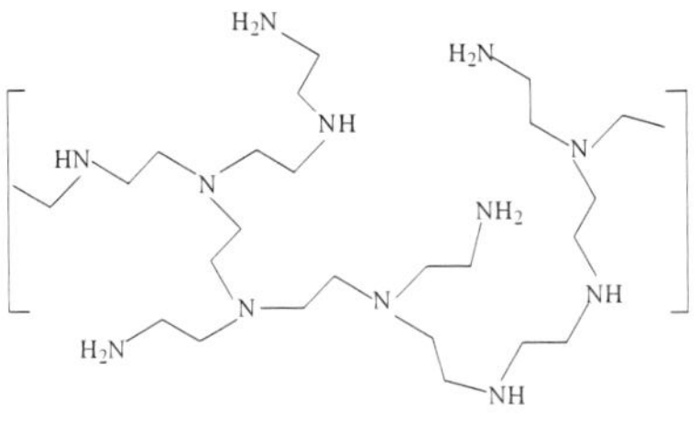

PEI 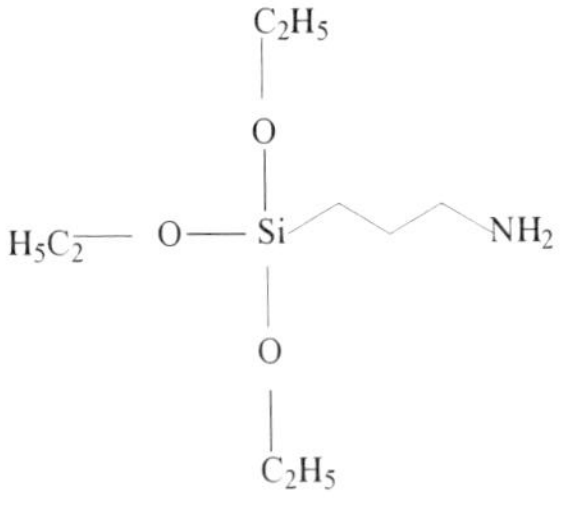 APTES 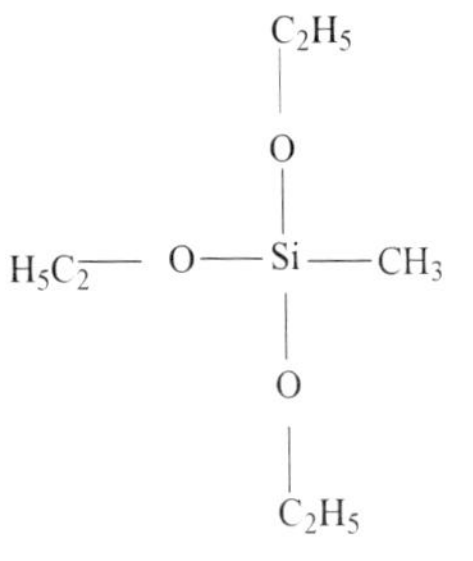 MTEOS

Unmodified silica containing surface hydroxyl groups was prepared by hydrolysis and followed by condensation of the hydrolysis products of TEOS under basic conditions. Amino groups were introduced on silica surface by co-hydrolysis of TEOS and APTES or by adding of PEI during the TEOS hydrolysis. Hydrophobic methyl groups were introduced by co-hydrolysis of TEOS and MTEOS.

2.1. Characterization of Silica-Based Nanoparticles via Fourier Transform Infrared Spectroscopy and Elemental Analysis the Synthesized Silica Materials

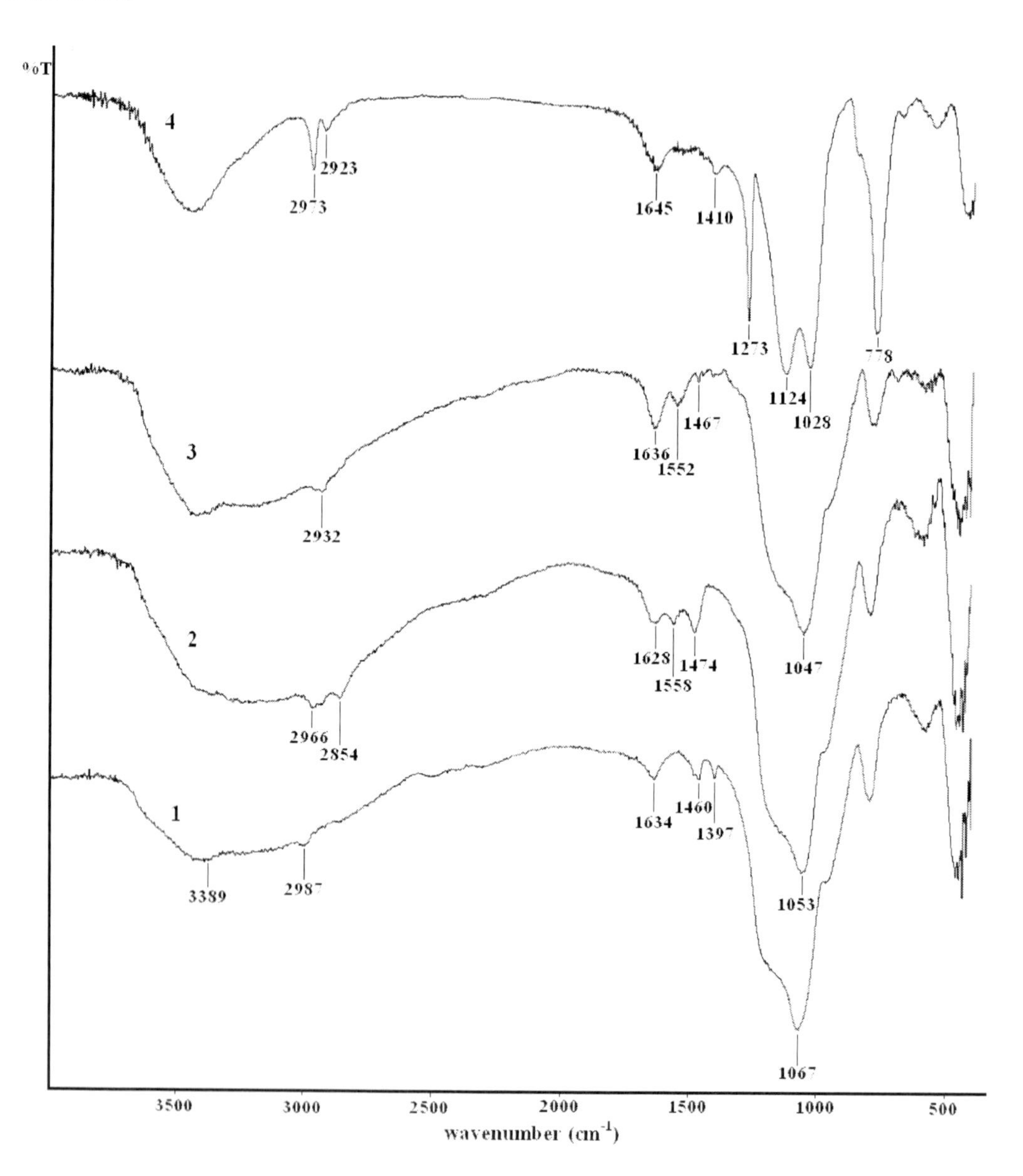

Figure 1. IR spectra of silica materials: 1- unmodified silica; 2- PEI-modified silica; 3- APTES-modified silica; 4- methyl modified silica.

The presence of the functional groups was confirmed by infrared absorption spectroscopy (IR). IR spectra were recorded using an Avatar 360 FT-IR ESP spectrometer at room temperature. The spectra were recorded in the range of 4000 to 400 cm^{-1}. The samples were examined as KBr disks. Figure 1 shows IR spectra of the synthesized powders. As it can be seen from Figure 1, new peaks at 2966, 2854, 1558 and 1474 cm^{-1} appear in the spectrum of the PEI-modified silica in comparison with the unmodified silica. In the spectrum of the APTES-modified silica new peaks at 2932, 1553 and 1467 cm^{-1} are observed. According to literature data [20,21], the peaks in 3000-2800 cm^{-1} region can be assigned to symmetrical and asymmetrical C-H stretching modes (ν_{s}, ν_{as}) of CH_2 and CH_3 groups. The peaks at 1474 cm^{-1} and 1467 cm^{-1} are associated with asymmetrical C-H deformation vibrations (δ_{as}) of the alkyl groups. New peaks at 1558 cm^{-1} and 1553 cm^{-1} are assigned to deformation vibrations of amino groups [20,22]. These results testify qualitatively about the modification of the silica materials. The IR spectrum of the methyl-modified silica powder indicates the introduction of CH_3 groups in the silica network. Pronounced peaks at 2972 cm^{-1} and 1410 cm^{-1} are associated with asymmetrical and symmetrical C-H stretching vibration of CH_2 and CH_3 groups. A strong peak at 1275 cm^{-1}can be attributed to symmetric C-H bending vibrations of methyl groups [22,23]. The appearance of additional peaks at 1124 cm^{-1} in the Si-O-Si stretching region (1200-1000 cm^{-1}) indicates that the organic groups have been introduced into the silica network via non-hydrolyzable Si-C covalent bonds [24].

The elemental analysis data (Table 1) confirm indirectly the introduction of the functionalities in silica matrix.

Table 1. Elemental analysis data of the synthesized silica materials.

Sample	(N)%	(C)%	(O)%	(H)%
Unmodified silica	0.002	0.075	0.362	0.178
APTES-modified silica	2.976	9.474	5.772	2.684
PEI-modified silica	8.942	17.486	14.825	4.471
Methyl Modified silica	0.000	3.393	5.139	0.983

2.1. Characterization of Silica-Based Nanoparticles via Small Angle X-Ray Scattering (SAXS)

The synthesized silica powders were investigated by SAXS. The experiments were performed with a diffractometer DRON-2 (Russia) operating at 40 V (Cu-Kα radiation, λ = 0.154 nm). The spectra were recorded in the 2θ range of 8 to 57° with 2θ step size of 1°. All the spectra show a broad halo due to amorphous silica. Thus, the obtained results confirm the amorphous properties of the synthesized powders. As an example, SAXS spectrum of unmodified silica is presented in Figure 2. The obtained result is very important because according to literature data [25, 26], only amorphous silica particles are not classifiable as carcinogen to humans.

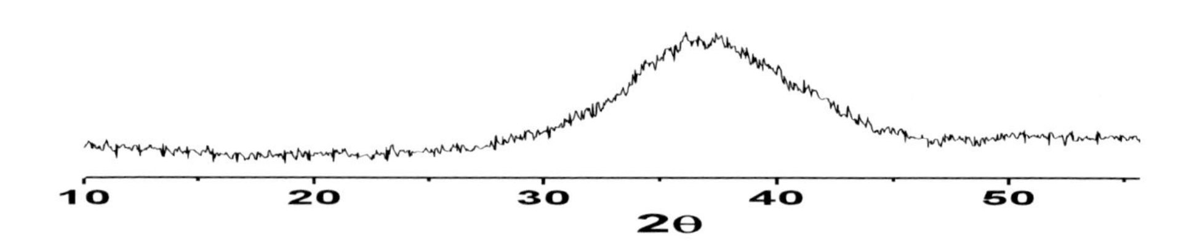

Figure 2. SAXS spectrum of unmodified silica.

2.3. Surface Properties and Porosity of Silica-Based Nanoparticles

The introduction of different chemical functionalities should have an influence on the surface physical properties of the studied silica materials (specific surface area, pore volume and size as well as pore size distribution). These properties were determined with a Micromeritics ASAP 2010 analyzer. The amount of N_2 gas adsorbed at various partial pressures ($0<P/P_0<1.0$) served to determine the BET surface area. The pore size distributions were plotted according to Barrett-Joyner-Halenda (BJH) method using desorption isotherms. The obtained data are presented in Table 2.

Table 2. Surface physical properties of silica materials.

Samples	Specific Surface area ($m^{2/}g$)	Pore volume (cm^3/g)	Average pore diameter (nm)
Unmodified silica	108	1.071	33.3
APTES-modified silica	85	0.813	33.1
PEI-modified silica	12	0.043	-
Methyl modified silica	811	0.949	4.3

All the synthesized materials except PEI-modified silica are mesoporous. According to the obtained data, PEI-modified silica material can be considered as non-porous.

The modification of the silica surface by methyl groups results in significant decrease of pore size (from 33.3 to 4.3 nm) and increase in the specific surface area (from 108 to 811 m^2/g) of the modified silica relative to the unmodified one. This tendency is similar to that reported previously [27]. In addition the hysteresis profile for methyl modified silica (Figure 3) corresponds to type H2 by IUPAC [28]. Thus methyl modified silica material has bottle-neck pore shape with narrow size distribution (Figure 4).

The introduction of amino groups into the silica network (APTES-modified silica) results in a decrease of specific surface areas and pore volume in comparison with unmodified silica. As can be seen from Figure 3, these materials show typical H3 shape associated with adsorption into slit-shaped pores or into the space between parallel platelets [28] and wide pore size distributions (Figure 4).

Thus, the reported results showed that a nature of modifier has strongly influence on structure peculiarities and physical chemical parameters of the modified silica materials.

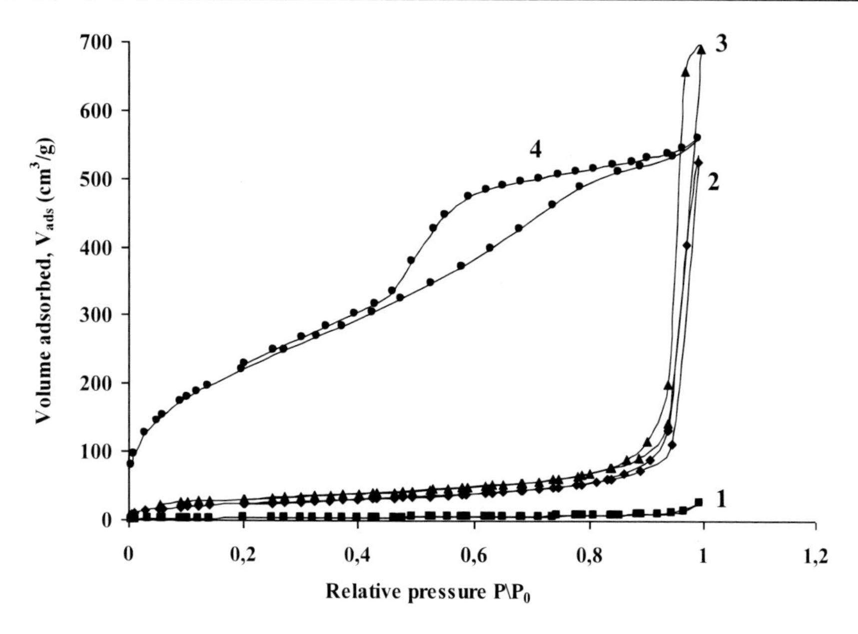

Figure 3. N2 adsorption-desorption isotherms for the PEI-modified (1), APTES-modified (2), unmodified (3) and methyl-modified (4) silica powders.

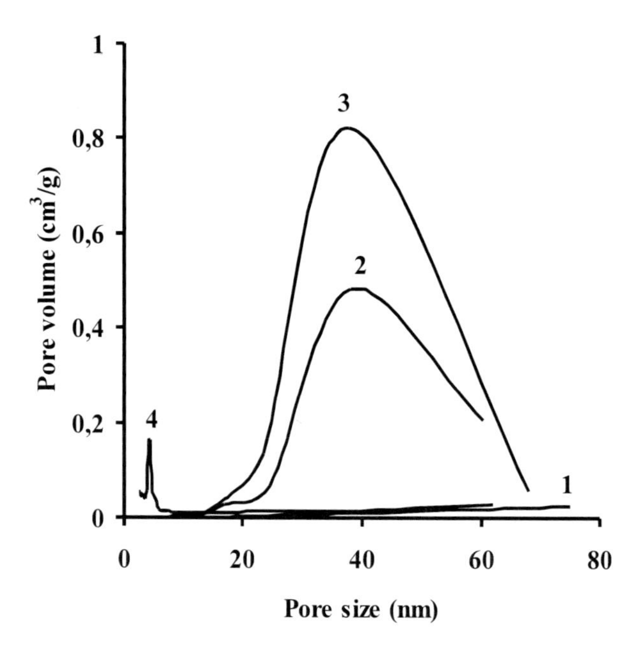

Figure 4. Pore size distribution for the PEI-modified (1), APTES-modified (2), unmodified (3) and methyl-modified (4) silica powders.

2.4. Biocompatibility of Silica-Based Nanoparticles

The special physical and chemical properties of silica are quite attractive from the viewpoint of the compatibility with biomolecules and living cells. The thermodynamic stability of the Si–O bond is of 452 kJ/mol, which indicates a strong inertness that excludes interference with enzymes and functions typical of differentiated cells. The approach to sol-gel silica loaded with biological systems, particularly enzymes, has produced important and valuable implementations in biotechnology [5–14] and open tremendous possibilities to understand the protein folding process. Recently, we have shown that apomyoglobin (ApoMb) and cytochrome *c* (Figure 5) was able to adsorb in organically-modified nanoporous sol-gel glasses [29].

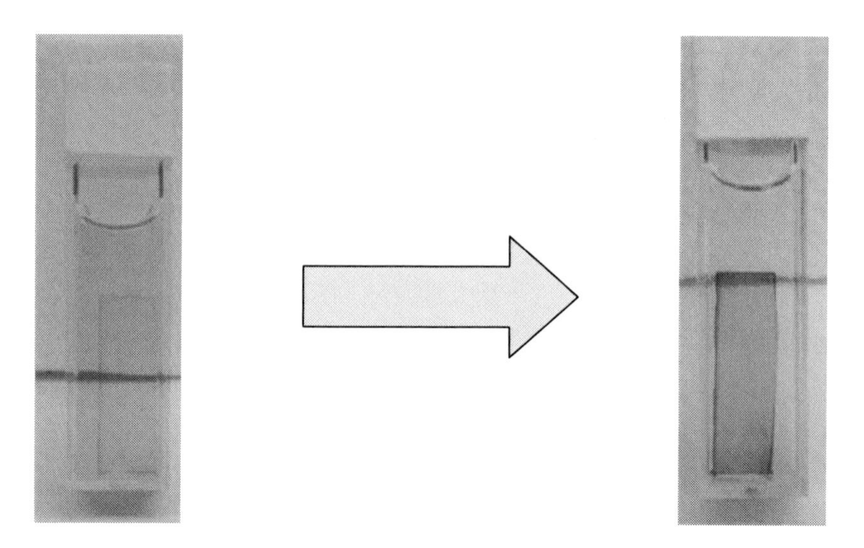

Figure 5. Adsorption of bioactive cytochrome C onto organically-modified nanoporous silica sol-gel glass.

Apomyoglobin (a model protein used to probe the protein folding process) was also encapsulated during in organically-modified nanoporous sol-gel glasses, keeping its folded state and suggesting the enhancement of the biological activity of the proteins in amorphous silica-based support as function of functionalized silica surface with hydrophobic groups. The protein helical structure was determined via Circular Dichroism Spectroscopy with signal in the far-UV region at 222 nm (Figure 6) [30, 31]. It is worth noting that nanoporous sol-gel glasses are constituted of silica nanoparticles that are consequently biocompatible as regards of biomolecules, proteins and enzymes and silica constitute therefore a viable surface and ideal support for protein bioencapsulation. Silica nanoparticles (~20 nm) characterized by Atomic Force Microscopy in nanoporous sol-gel glasses as shown Figure 7.

The biomolecules entrapped in ormosil nanoparticles can be efficiently employed in drug delivery and other pharmaceutical and medical applications [15]. The organically modified and anticancer drug doped nanoparticles (diameter ca. 30 nm) obtained through aqueous dispersion have been used in photodynamic cancer treatment [16]. Water-insoluble photosensitizing anticancer drug, 2-devinyl-2-(1-hexyloxyethyl) pyropheophorbide, was entrapped in non-polar core of micelles by hydrolysis of triethoxyvinylsilane. The resulted nanoparticles were spherical, highly monodispersed, and stable in aqueous system.

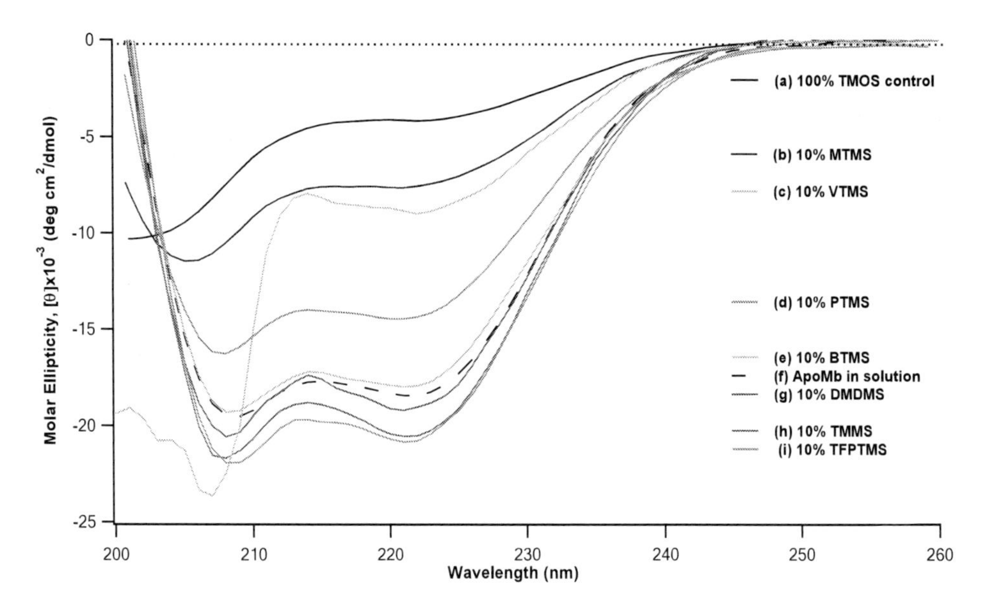

Figure 6. CD spectra in the far-UV region for apoMb in unmodified TMOS glass and modified glasses TMOS:$R_n(OCH_3)_{4-n}$ (solid curves) and in dilute solution (dashed curve) (*f*). All spectra were measured in the presence of potassium phosphate (10 mM, pH 7). The helicity of encapsulated apoMb increases with the hydrophobicity characterized by the increase of alkyl chain length (*b; d; e*) of the organic modifier R, its fluorine functionalization with trifluoropropyltrimethoxysilane (TFPTMS) (*i*) or the decrease of siloxane network n>1 with dimethyldimethoxysilane (DMDMS) and trimethylmethoxysilane (TMMS) (*b; g; h*); the spectra for apoMb encapsulated in the vinyl-functionalized silica glass (VTMS) (*c*) showed also that the protein folds [27,29-31].

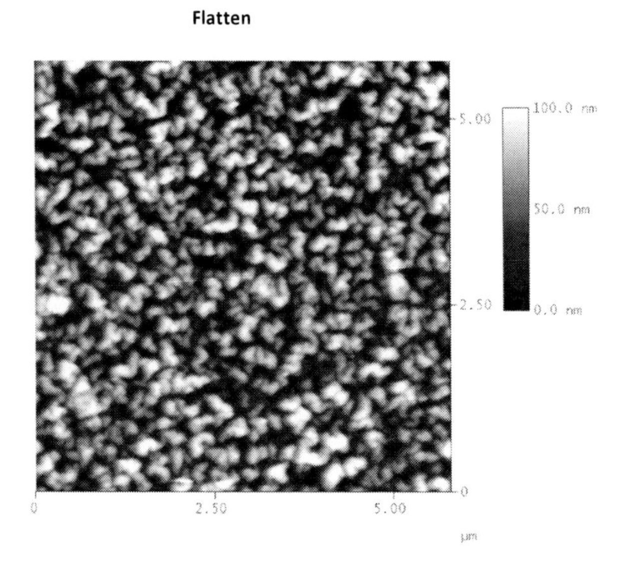

Figure 7. Atomic Force Microscopy image showing 20 nm silica particles in organically modified glass TMOS:$RSi(OCH_3)$ [R=Ethyl] with encapsulated apoMb [27].

3. Adsorption and Desorption of Human Serum Albumin onto the Silica Materials

Because drug is high expensive, the adsorption ability of the silica materials has been studied using human serum albumin as a model compound of the drug. The adsorption experiments were performed in 50 mM phosphate buffer (pH=7.4).

The adsorption isotherms were assessed by a depletion method by measuring the bulk protein concentrations before and after adsorption. An amount of silica material was suspended in respective albumin solution. The resulting mixture was stirred gently for 24 hours at room temperature until equilibrium was reached. Then the suspension was centrifuged in an Eppendorf Minispin Plus microcentrifuge at 10 000 rpm. The supernatant was carefully removed from the test tubes and the concentration of the protein the supernatant was measured spectrophotometrically (using a double-beam spectrophotometer Carry 100) at a wavelength of 278 nm and an emission wavelength range of 200-500 nm.

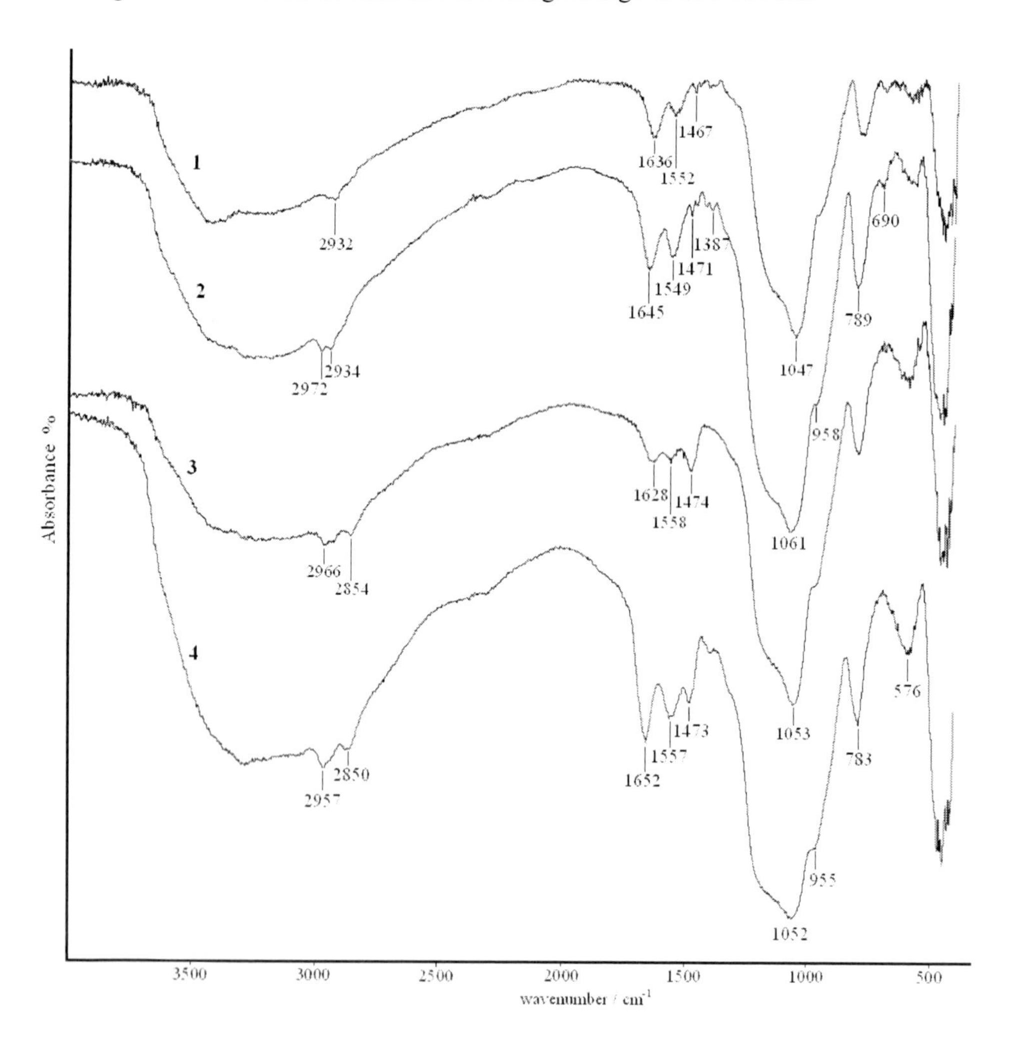

Figure 8. IR-spectra of APTES-modified silica (1 and 2) and PEI-modified silica (3 and 4) before and after adsorption of human serum albumin.

Desorption of HSA from the silica materials is tested with two methods. Amount of free HSA in the supernatant was measured spectrophotometrically after storage of the HSA covered particles suspension during 24 hours and after their washing with buffer.

In order to confirm the adsorption of the protein onto surfaces of the silica materials, the dispersion phases of the suspensions after adsorption were separated, washed with buffer and dried. Then the IR spectra of the silica materials after adsorption were recorded. As an example, the spectra of APTES- and PEI-modified silica materials before and after adsorption of HSA are presented in Figure 8. The spectra show protein characteristic amide I band (1700-1630 cm^{-1}) assigned to C=O, C-N and C-N-H stretch vibrations and amide II band (1570-1510 cm^{-1}) associated with NH/CN deformation vibrations [32, 33].

Figure 9 shows the adsorption isotherms of human serum albumin onto the synthesized silica materials. It should be noted that for unmodified, APTES-modified and methyl modified silica HSA adsorption isotherms show defined plateaus. However, desorption experiments did not result in detectable changes in the surface concentration of HSA. This suggests irreversible adsorption at these conditions, although a hysteresis effect upon desorption cannot be excluded (i.e., slow desorption at longer times). It should be noted that irreversibility of adsorption for different proteins on solid surfaces has been found in numeral experimental studies [34-39]. The irreversible adsorption is explained by multiple contact formation of protein molecule with adsorbent surface. Even if adsorbed protein interacts with adsorbent surface due to non-covalent interactions (electrostatic, hydrophobic, van der Waals) and the effective contribution from each of these contacts is relatively small, the total driving force for such adsorption will be very large (and negative). In addition, structural alterations of protein molecule induced by adsorption may give substantial contribution to total energy of the protein adsorption [35]. Thus, the energy barrier to desorption is seems prohibitively high.

Maximum capacity of adsorbed layer on the synthesized particles is very important from the point of view of the aims of this study. These values for the synthesized material were estimated by means of Scatchard analysis of the following equation

$$q_{ads} = \frac{q_{\max} \cdot C \cdot K}{(1 + C \cdot K)},$$

where C is the protein concentration after adsorption, K is the empirical constant that has no physical meaning in our case. The estimated $q_{\max}$ values are presented in Table 3.

Table 3. Maximal capacity of adsorbed HSA layer on the synthesized materials.

Sample	Unmodified silica	APTES-modified silica	Methyl modified silica
q_{max}, $mg \cdot m^{-2}$	0.44±0.04	0.28±0.04	0.18±0.02

As can be seen from Table 3 and Figure 9, the order of affinity of HSA to the synthesized materials is the following:

methyl modified < APTES-modified < unmodified < PEI-modified

According to the molecular dimensions of the native HSA molecule, a closed packed monolayer of the protein is calculated to be in the range of 1.9–7.6 mg/m^2 [40]. Thus, the plateau values of the adsorption isotherms of HSA indicate that completion of an adsorbed monolayer does not occur. It may be caused by electrostatic repulsion of neighboring negative charged protein molecules in the monolayer. Another reason may be adsorption-induced structural changes of the protein molecules. Therefore, it might be possible that due to more extensive unfolding, the protein is more smeared out over the silica surfaces, thereby reducing the remaining area on which the adsorption can take place [41].

The isotherm of HSA adsorption on PEI-modified silica (Figure 9, b) is S-type. According to literature data such type of the adsorption curve may be associated with reorientation of the adsorbed protein molecules with increasing surface coverage from "side –on" to "end-on" [35]. Another reason may be that the attractive interactions between HSA and PEI-modified silica surface are not limited by monolayer. PEI is practically neutral at pH ~10.5 whereas it possesses considerable positive charge density in the neutral and acidic pH range due to protonation of its amino groups [42]. APTES-modified silica also contains the protonated surface amino groups at pH=7.4. However PEI is branched polymer possessing primary, secondary and tertiary amino groups.

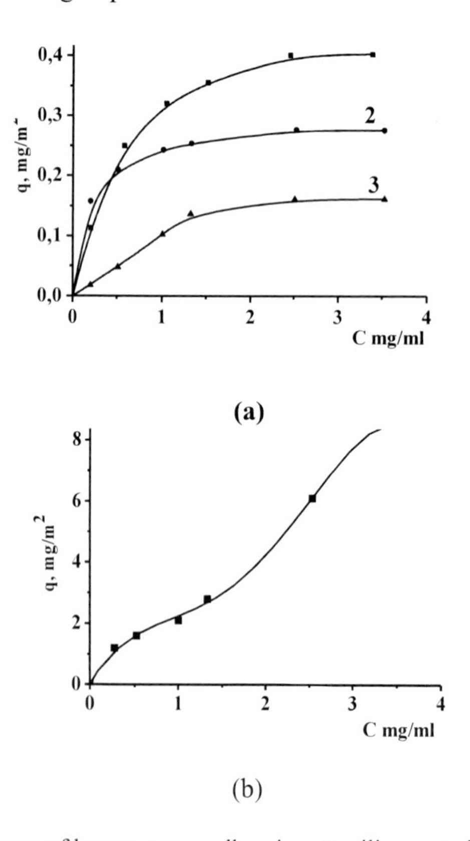

Figure 9. Adsorption isotherms of human serum albumin onto silica materials: 1 – unmodified silica; 2 –APTES-modified silica; 3-methyl modified silica (a). Adsorption isotherm of HSA onto PEI-modified silica (4, b).

Therefore, PEI-modified silica has a greater quantity of amino groups per m^2 of surface area in comparison with APTES-modified silica. This fact is confirmed by calculation on the basis of the data obtained by elemental analysis method (Table 1). The quantity of functional groups per gram and m^2 of surfaces of the synthesized silica materials was calculated as:

$$Q = \frac{Content of element\%}{Ar(element)} \times \frac{1}{100\%}$$

The obtained data are presented in Table 4.

Table 4. The quantity of functional groups per gram and m^2 of silica materials.

Samples	Quantity of functional groups (mmol/g)	Quantity of functional groups (mmol/m^2)
Unmodified silica	1,78	0,016
APTES-modified silica	2,13	0,025
PEI-modified silica	6,39	0,532
Methyl modified silica	2,83	0,003

The larger quantity of binding centers onto PEI-modified silica surface may lead to the higher affinity of HSA in comparison with APTES-modified silica material. It is expected that the adsorption of HSA onto APTES- and PEI-modified silica materials is mainly governed by electrostatic forces because at pH=7.4 the protein molecules are negatively charged (pI_{HSA} =4.7-4.9 [43,44]). However the unmodified silica surface is negatively charged under these conditions [45,46]. From the adsorption isotherms, it is clear that electrostatic repulsions are not sufficient to preclude the adsorption of HSA. This primarily indicates that the interaction is not electrostatically driven. According to literature data [40,47], the other important factor governing the adsorption of proteins onto solid surfaces is structural changes in the protein molecules [27]. These structural changes promote adsorption because a loss in secondary structure results in a gain in conformational entropy of the protein especially in the case of the proteins with low structural stability as HSA [47,48]. They are able to compensate the electrostatic repulsion by structural changes. A higher flexibility in the protein structure may lead to an optimization of the interactions between the proteins and sorbent surface even under unfavorable electrostatic conditions. On the hydrophobic surfaces it is expected that adsorption is promoted due to hydrophobic interactions. These hydrophobic interactions emanate from the gain in entropy of water molecules when they are replaced by proteins at the sorbent surface. As it can be seen from Figure 9 (a), the least amount of HSA is adsorbed onto methyl modified silica. It is generally believed that the structural changes are stronger at hydrophobic surfaces [41, 49-51]. Therefore, it might be possible that the methyl modified surface the protein is after a certain time more smeared out over the sorbent surface, thereby reducing the remaining area on which the adsorption can take place. Another reason may be the least amount of functional groups on the surface of the synthesized hydrophobic material (Table 4).

The average size of the modified silica particles after adsorption of HSA as well as their size distribution was determined by proton correlation spectroscopy (PCS). As an example, the data on the APTES-modified silica with adsorbed HSA are presented in Figure 10.

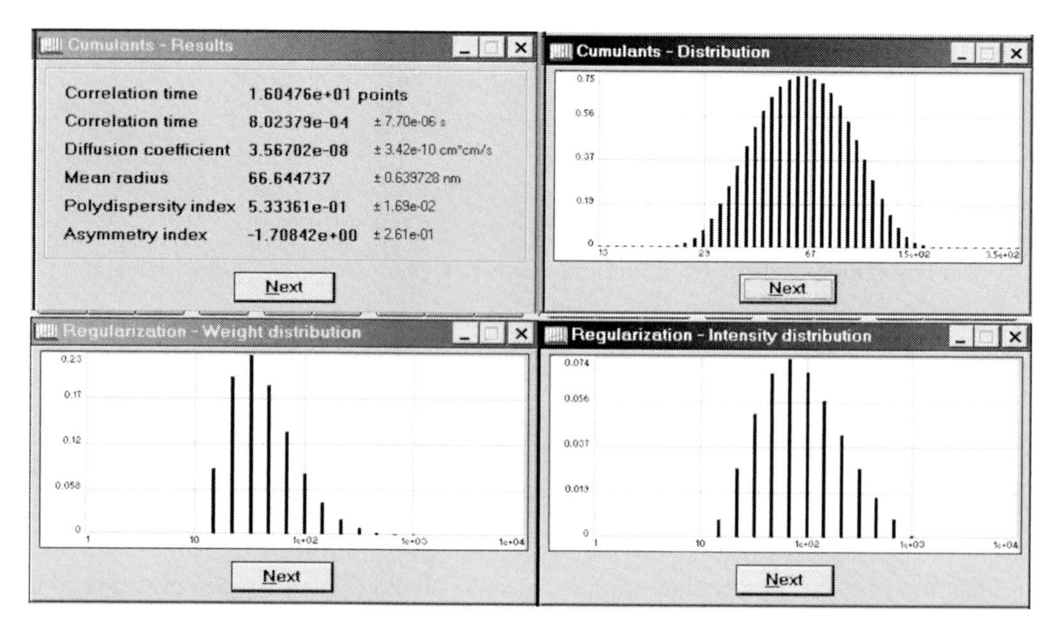

Figure 10. PCS results of APTES-modified silica after adsorption of HSA.

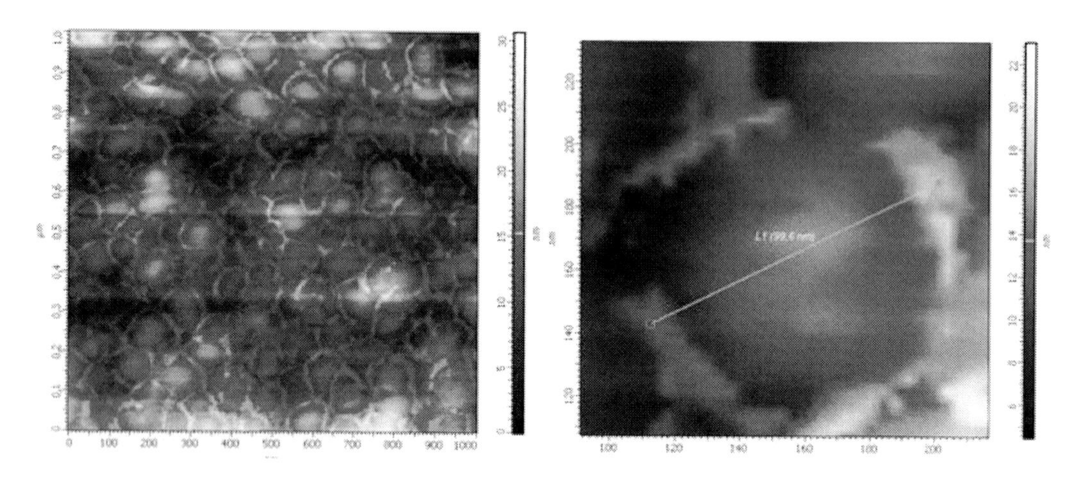

Figure 11. AFM images of PEI-modified silica particle after adsorption of HSA.

The formed particles are polydisperse with an average radius of 66.6±0.6 nm. For the HSA adsorbed onto PEI-modified silica and methyl modified silica, the particle average radius is 47.2±0.5 nm and 50.0±0.5 nm, respectively. These data are in good agreement with the results obtained by atom force microscopy (AFM) method. Figure 11, an example, shows AFM images of PEI-modified silica particle after adsorption of HSA.

As it has been mentioned above, upon adsorption onto solid surfaces, proteins may undergo significant structural changes. It is not good for biomedical applications because the protein molecules may loose their biological functionality. It is known that the adsorption-induced changes in protein structure depend on both physical and chemical properties of the protein and sorbent surface.

Various studies have shown that larger structural changes are induced upon adsorption onto hydrophobic sorbent surfaces than to hydrophilic surfaces [41,49-51]. Studies concerning protein adsorption onto hydrophilic surfaces emphases on crucial role of electrostatic interactions between the protein and sorbent surface. Strong electrostatic attraction between opposite charged protein molecule and solid surface often leads to more extensive structural changes in the protein structure [45,52]. Molecular dynamics simulations also predict that the shape of cytochrome c molecules is less spherical on the hydrophilic surface due to polar interactions between the protein and the surface [53]. However, as has been mentioned above, conformational changes also depend on the intrinsic properties of the protein. With respect to interactions with hydrophilic surfaces, proteins are determined as structurally stable, "hard", and labile, "soft" [48]. HSA is a "soft" protein [47,48] which undergoes conformational changes more easy upon adsorption due to larger flexibility of its molecules in comparison with "hard" proteins.

A powerful tool for performing thermodynamic investigations of adsorption-induced changes in protein structure is differential scanning calorimetry (DSC) [38, 40, 52, 54-56]. As it has been indicated in work [40], the basic idea of the DSC studies is that thermal denaturation will not be observed if the protein has been denatured upon adsorption/ desorption. Therefore, the adsorbed protein is considered totally or partly denatured if the expected melting transition of the protein is not observed or reduced. It is assumed that thermal denaturation of the adsorbed molecules leads to similar denatured state as in solution. One of the parameters characterizing adsorption-induced conformational changes of protein molecule is temperature of denaturation transition (T_d). It is defined as the temperature at which a local maximum occurs in the DSC curve. Interactions between protein molecule and sorbent surface may result in a shift of T_d. That is the reason why monitoring the denaturation temperature of a protein in the adsorbed state gives information on protein-surface interactions and the influence of adsorption on the protein's conformational stability. According to the literature, the adsorption-induced structural changes lead to both a decrease [52,55] and an increase [51,56] in the denaturation temperature of protein as compared to that of the protein in solution. A higher T_d for adsorbed protein implies that the adsorption-induced perturbed state has a higher thermal stability than the native state.

Figure 12 shows DSC curves for the synthesized silica powders after adsorption of HSA. The DSC curves testify about dehydration process of the samples upon heating which attains their maxima in the range of 75-96°C.

Against the dehydration process, the endothermic peak at 55.8°C is observed on the DSC curve of the sample with the APTES-modified silica (Figure 12, Inset). Our DSC measurements indicate that this transition is reversible. The estimated enthalpy value of the transition is 11 J/g. The peak at 55.8°C can not be assigned to the APTES- modified silica nanoparticles themselves because no endothermic or exothermic transitions over given temperature range are observed for the modified silica powders (the data are not presented here). DSC experiment showed that the transition at 55.8°C is irreversible. According to literature data, the denaturation temperature of HSA in solution at pH=7.4 is 63.2±0.4°C [57].

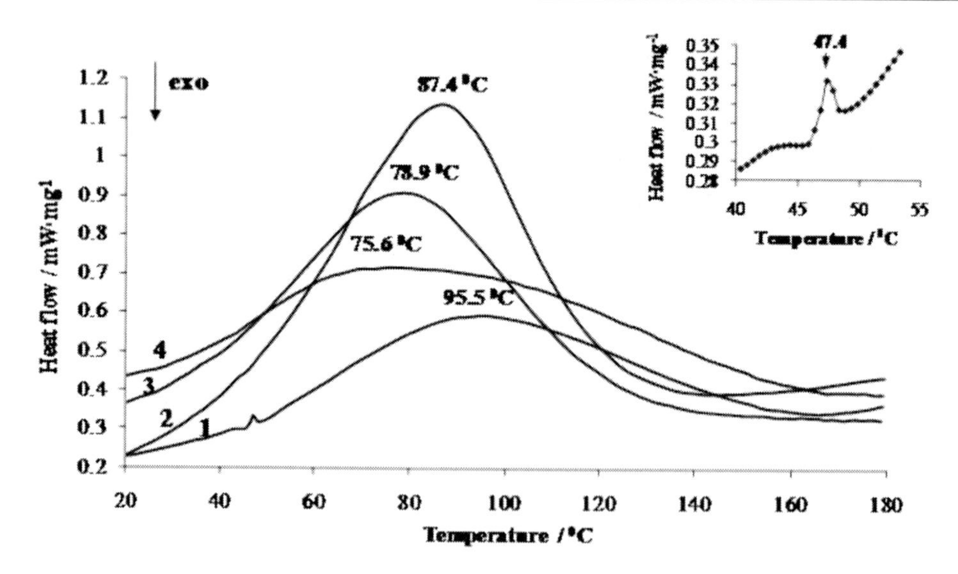

Figure 12. DSC curves of APTES-modified (1), PEI-modified (2), methyl modified (3) and unmodified (4) silica powders after adsorption of HSA.

Thus, the adsorption of HSA on APTES-modified silica nanoparticles is accompanied by a 7.4°C decrease in $T_{d,}$ indicating that the adsorbed state of the proteins is destabilized. It means that the native structure of the protein undergoes definite changes at adsorption.

It should be noted that no analogous transition peak was detected for HSA absorbed onto other studied silica materials. It is likely the mentioned above structural alterations has already realized upon adsorption.

Thus, the obtained results showed that adsorption of human serum albumin onto the synthesized silica materials depends on their chemical surface functionalities. It is different both quantitatively and qualitatively. The largest amount of the protein is adsorbed onto PEI-modified silica. However the protein undergoes the least structural changes upon adsorption onto APTES-modified silica. These materials were chosen as potential drug nanocarriers.

4. Silica Materials as Drug Nanocarriers. Interactions between Different Types of Immune Cells and Organically-Modified Silica Nanoparticles

In the present time most researchers believe that progress in medicine is connected with the development of drug delivery systems using nanoparticles. Very important drugs are immune-modulators which are natural or synthetic compounds. They often have protein nature. Efficiency of drug delivery system depends on both pharmacology activity of the drug and properties of its nanocarrier. It is necessary that the nanocarrier does not influence on viability and functionality of cells. Therefore interactions between different types of immune cells and organically-modified silica nanoparticles were studied.

APTES-modified and PEI-modified silica particles were chosen as potential nanocarriers. Average size of the nanoparticles in isotonic solution was determined by spectroturbidimetric measurements. Turbidity spectra of suspensions of the modified silica powders in isotonic solution (0.3 g/30 ml) were recorded with spectrophotometer SF-26. As silica particles form aggregates upon drying, the suspensions were subjected to self-sedimentation during 2-4 days. The turbidity spectra of the supernatants were recorded using 10 mm quartz cuvettes in the range of 200-400 nm. The turbidity was calculated as

$$\tau_M = 2.3\, D_\lambda \,/\, l,$$

where D_λ is the optical density at wavelength λ; l is the layer thickness.

Particle average radius was calculated as

$$r = \alpha\, \lambda_0 \,/\, 2\pi\, \mu_0$$

where λ_0 is the average wavelength in the range of measurements; μ_0 is the medium refractive index [58,59]. The value of refractive index of disperse phase which is necessary for this calculation is reported in [59]. The uncertainty in the radius value is 10 %.

To determine the level of immunocompetent cells interacting with the silica nanoparticles, sheep immunoglobulin (IG) labeled with fluoresceine isothiocyanate (FITC) was immobilized onto the nanoparticles. Immobilization of labeled immunoglobulin onto the modified silica nanoparticles (100 and 200 nm) was carried out from isotonic solution. To suspension of the nanopatricles of definite average size a solution of the protein was added (2:3 v/v). The amount of FITC-positive cells was estimated by flow cytometry method.

We studied two different types of immune cells: the peripheral blood lymphocytes and the peritoneal macrophages. These cells populations represent the two main types of immune cells participating in reaction of adoptive (the lymphocytes) and innate (the macrophages) immunity. The heparinized peripheral blood and peritoneal fluid of women who underwent laparoscopic investigation were used as the material. The standard procedure of isolation of the peripheral lymphocytes and the peritoneal macrophages by centrifugation in density gradient of Ficoll-Urografin (d-1,078) was performed. Enriched fractions of peripheral lymphocytes and peritoneal macrophages at concentration of $2x10^6$ cells/ml in RPMI 1640 culture media were studied.

To estimate the level of cells reacting with the silica nanoparticles, 100 μl lymphocytes or macrophages were incubated with 3 and 5 μl of the immunoglobulin-immobilized silica particles for 1 and 24 h at 37°C and 5% CO_2. The level of immunocompetent cells reacted with the silica nanoparticles was estimated by flow cytometry as the amount of fluorescence-bright cells. To estimate the rate of unspecific staining due to spontaneous reaction of lymphocytes and macrophages with the sheep immunoglobulins, cells were incubated in the presence FITC-conjugated sheep IgG antibodies only and the percentage of fluorescence-bright cells was assessed by flow cytometry (negative control). The flow cytometry analyses were performed on FACScan (Becton Dickinson, San Hose, USA) using CellQuest Pro software (Becton Dickinson). Data from forward versus side scatter was obtained to analyze the CD45+CD14- lymphocyte and CD45+CD14+ macrophages populations. The data were presented as the percentage of the FITC-stained cells in lymphocyte population.

The immunocompetent cells were cultured with the silica nanoprticles in vitro in RPMI 1640 medium in presence of the silica nanoparticles of different size (100 nm or 200 nm in diameter) for 1 and 24 h at 37°C and 5% CO_2. Concentration of the nanoparticles in the culture media was 400 μg/10^6 cells/ml. This concentration was chosen according to the results reported by Lucarelli M. et all [60] and was the highest dose of the silica nanoparticles in series of the functional experiments with human macrophages. As the control experiment, cells were incubated in RPMI 1640 medium in the absence of the nanoparticles at the same conditions. After incubation the viability of cells was estimated using the trypan blue dye exclusion assay. The expression of the surface molecules HLA-DR, CD11b, CD95, SR-A and SR-B by macrophages after incubation with the silica nanoparticles was defined by flow cytometry. The level of bactericidal activity of the macrophages after incubation with the silica nanoparticle was estimated by the test of spontaneous (NBTsp) and zymosan stimulated (NBTst) reduction of nitrotetrazolium blue.

The following conclusions were made on the basis of the obtained results:

Firstly, according to trypan blue dye exclusion assay data, the viability of the lymphocytes and phagocytes after incubation with the APTES-modified silica nanoparticles did not change significantly. The viability percentage was found to be 85-95% after incubation of the cells with the silica nanoparticles both of 200 nm and 100 nm size. On the contrary the viability of the lymphocytes and macrophages after their incubation with PEI-modified silica nanoparticles was very low (less that 25%). Literature data indicate that definite amounts of PEI are toxic to some cells and might be a result of the strong adhesion of PEI to the outer cellular membranes [61,62]. So we excluded this type of nanoparticles from our further investigations.

Secondly, the level of the lymphocytes reacting with the nanoparticles is very small (from 3% to 10%) and it is independent on the time of incubation. So, we can conclude that the lymphocytes do not react intensively with the silica nanoparticles (Figure 13).

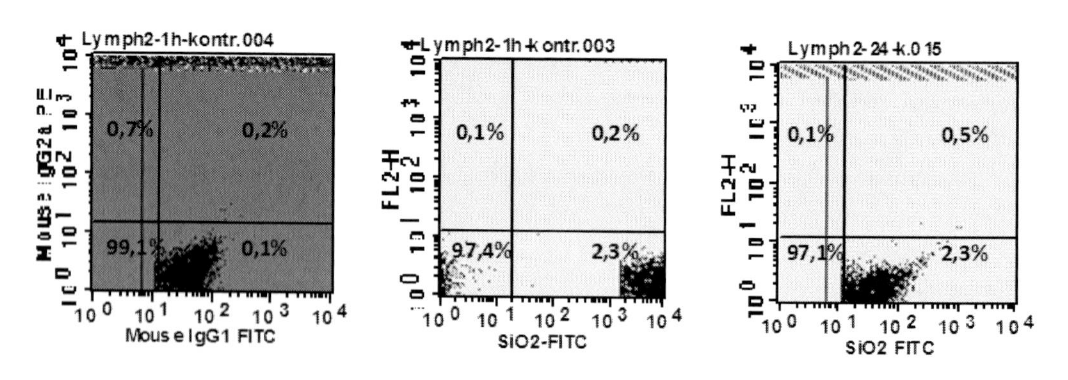

Figure 13. Flow cytometry analysis data representing the intensity of interaction of the peripheral blood lymphocytes with the silica nanoparticles (A – negative control, the cells were incubated in the presence of sheep FITC-labeled immunoglobulin, B – distribution of the lymphocytes according to FITC-negative (low left corner) and FITC-positive cells (low right corner) after 1-hour incubation of the lymphocytes with the modified silica nanoparticles, C - distribution of the lymphocytes according to FITC-negative (low left corner) and FITC-positive cells (low right corner) after 24-hour incubation of the lymphocytes with the modified silica nanoparticles).

On the contrary the peritoneal macrophages intensively react with the silica nanoparticles. After incubation for 1 hour, the presence of the silica nanoparticles was found on the surface membrane and/or in intracellular space of approximately 60-70% of the macrophages. After incubation for 24 hours, the amount of the macrophages reacted with the silica nanoparticles was more than 80%. Thus, according to our results cellular uptake of the silica nanoparticles by the macrophages is time-depending (Figure 14).

Thirdly, the obtained results indicated that the functional activity of the macrophages depends on the size of modified silica nanoparticles and incubation time. The elevation of HLA-DR and SR-A molecules expression as well as the increase of spontaneous bactericidal activity of the macrophages after their incubation with the modified silica nanoparticles with a diameter of 200 nm for 1 hour was observed (Figure 15). It is likely that the particles with a diameter of 200 nm are recognized by the macrophages as foreign substance and led to activation of the cells.

The silica nanoparticles with a diameter of 100 nm do not influence the macrophages viability and functional activity. After 24-hour of cultivation of the peritoneal macrophages with the particles of both sizes we did not find significant changes on the functional receptors expression on the surface of the macrophages (the data are not presented here).

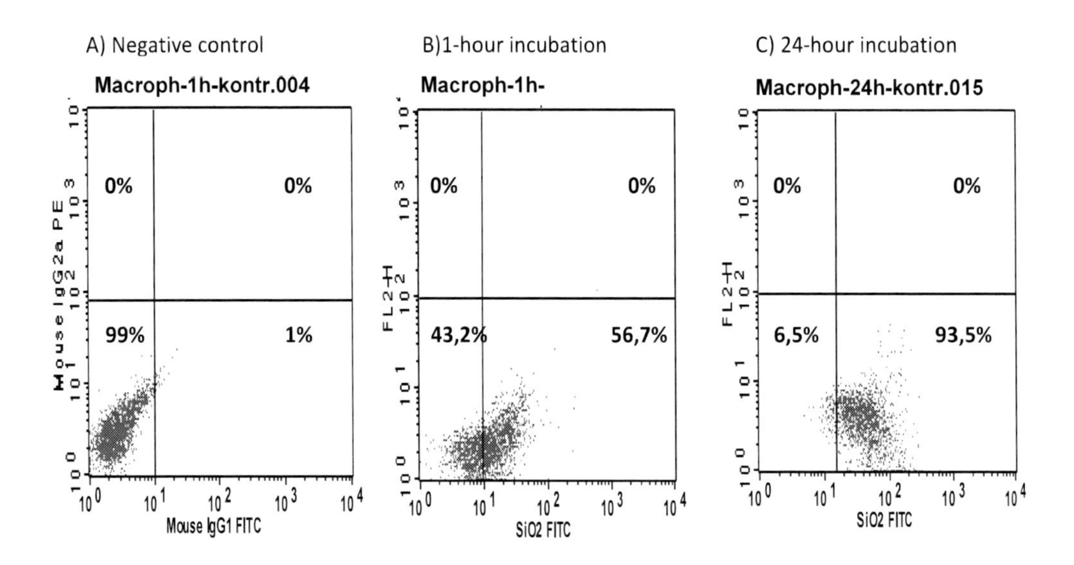

Figure 14. Flow cytometry analysis data representing the intensity of interaction of peritoneal macrophages with silica nanoparticles (A – negative control, the cells were incubated in the presence of sheep FITC-labeled immunoglobulin, B – distribution of the macrophages according to FITC-negative (low left corner) and FITC-positive cells (low right corner) after 1-hour incubation of the macrophages with the modified silica nanoparticles, C - distribution of the macrophages according to FITC-negative (low left corner) and FITC-positive cells (low right corner) after 24-hour incubation of the macrophages with the modified silica nanoparticles).

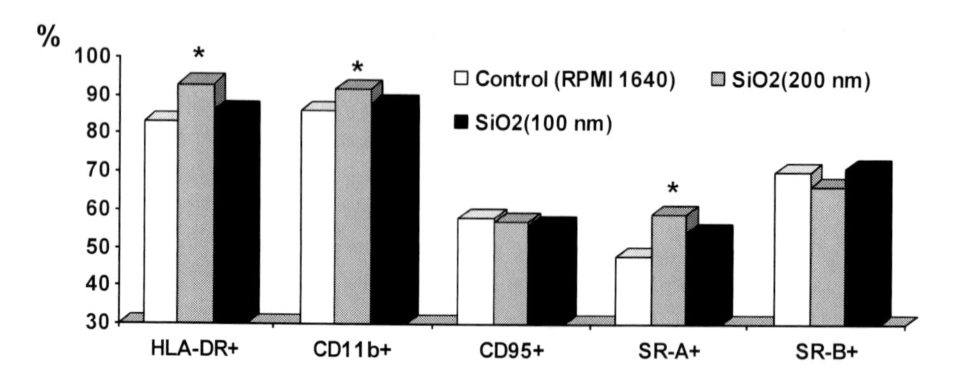

Figure 15. Characteristic of surface expression of function molecules by the peritoneal macrophages after its 1-hour incubation in the presence of the modified silica nanoparticles of different size (100 and 200 nm), (* - differences between control and experimental data are statistically significant (*- $p<0.05$)).

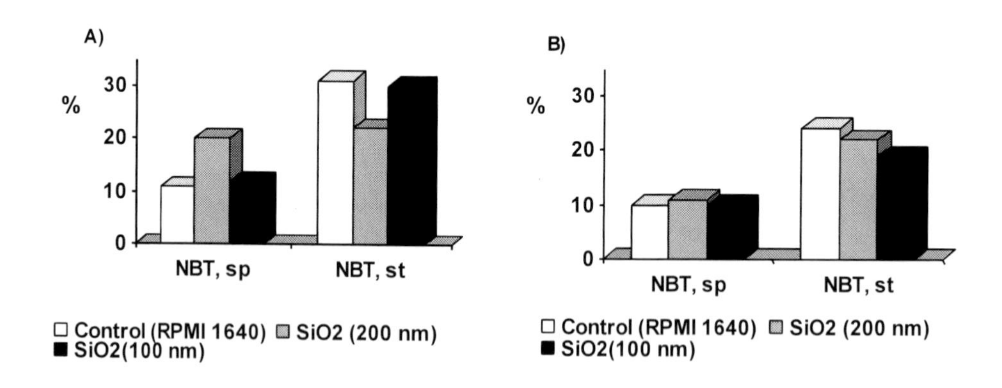

Figure 16. Characteristic of spontaneous and stimulated bactericidal activity of the peritoneal macrophages after its incubation in the presence of the modified silica nanoparticles of different size (100 and 200 nm): A) 1 hour incubation, B) 24-hour incubation, (* - differences between control and experimental data are statistically significant (*- $p<0.05$)).

Fourthly, according to standard NBT-test, the level of spontaneous and stimulated bactericidal activity of the peritoneal macrophages depends on the size of modified silica nanoparticles and incubation time. As can be seen from Figure 16, after 1 hour incubation of the macrophages with the 200 nm silica nanoparticles the significant increase of spontaneous bactericidal activity of the macrophages is observed.

However, no changes were found after 1-hour incubation of the macrophages with the silica nanoparticles with 100 nm in diameter. after 24-hour of cultivation of the peritoneal macrophages with the particles of both sizes, no significant changes in the level of spontaneous and stimulated bactericidal activity of the macrophages were observed.

CONCLUSION

Nanoparticles technology has experienced rapid growth in recent years with its broad application in drug delivery, imaging and diagnosis [63-68]. Nanoparticles have significant adsorption capacities due to their relatively large surface area, therefore they are able to bind or carry other molecules such as chemical compounds, drugs, probes and proteins attached to the surface by covalent bonds or by adsorption. Hence, the physicochemical properties of nanoparticles, such as charge and hydrophobicity, can be altered by attaching specific biomolecules (proteins, enzymes...). The functionality of nanoparticles is thus enhanced or changed. The efficacy of nanoparticles for any application depends on the physicochemical characteristics of both biomolecules and surface modifiers.

The studies of physical and adsorption properties of the silica materials as well as their interactions with immune cells *in vitro* are at the forefront of the research in the field as it will allow designing and selecting the most promising, and efficient silica carrier of immune-modulator proteins for drug delivery. The adsorption ability of the silica materials has been studied using human serum albumin as a model compound of the drug. The adsorption experiments were performed in 50 mM phosphate buffer (pH=7.4). The obtained results showed that adsorption of human serum albumin onto the synthesized silica materials depends on their chemical surface functionalities. It is different both quantitatively and qualitatively. The largest amount of the protein is adsorbed onto PEI-modified silica. However, the stability of the protein is necessary, therefore APTES-modified silica in which the protein undergoes the least structural changes were chosen as potential drug nanocarriers.

We studied two different types of immune cells: the peripheral blood lymphocytes and the peritoneal macrophages. These cells populations represent the two main types of immune cells participating in reaction of adoptive (the lymphocytes) and innate (the macrophages) immunity. We estimated the level of cells reacting with the silica nanoparticles. We concluded that the viability of the lymphocytes and phagocytes after incubation with the APTES-modified silica nanoparticles did not change significantly. The level of the lymphocytes reacting with the nanoparticles is very small and it is independent on the time of incubation. So, we can conclude that the lymphocytes do not react intensively with the silica nanoparticles. The functional activity of the macrophages depends on the size of modified silica nanoparticles and incubation time. Indeed, we found that the particles with a diameter of 200 nm are recognized by the macrophages as foreign substance and led to activation of the cells. In fact, the level of spontaneous and stimulated bactericidal activity of the peritoneal macrophages depends on the size of modified silica nanoparticles and incubation time.

More importantly, the application of nanoparticles in drug delivery is challenged by thepotential destruction of protein function when nanoparticles enter the human body. Therefore, it is essential to understand in detail the effects of protein-nanoparticle adsorption, identifying which properties of nanoparticles determine their tendency to perturb protein conformation. In general, structural changes occur due to the intrinsic properties of the protein combined with features of the nanoparticles, such as surface chemistry.

REFERENCES

[1] Brinker, CJ.; Scherer G. W. Sol–gel science: the physics and chemistry of sol–gel processing; Academic Press: San Diego, CA, USA, 1990.

[2] Sanchez, C.; Ribot, F. Design of hybrid organic-inorganic materials synthesized via sol-gel chemistry. *New J. Chem.* 1994, 18, 1007-1047.

[3] Schmidt, H. New type of non-crystalline solids between inorganic and organic materials. *J. Non-Cryst. Solids* 1985, 73, 681– 691.

[4] Huang, H.; Orler, B.; Wilkes, GL. Ceramers: hybrid materials incorporating polymeric/oligomeric species with inorganic glasses by a sol gel process. 2. Effect of acid content on the final properties. *Polym. Bull.* 1985, 14, 557–564.

[5] Avnir, D.; Ottolenghi, M.; Braun, S.; Zusman, R. Doped sol–gel glasses for obtaining chemical interactions, US Pat. 1994, 5, 292,801.

[6] Avnir, D.; Braun, S.; Lev, O.; Ottololenghi, M. Enzymes and other proteins entrapped in sol-gel materials. *Chem. Mater.* 1994, 6, 1605-1514.

[7] Livage, J. Bioactivity in sol-gel glasses. *C. R. Acad. Sci.* Paris, 1996, 322, 417-427.

[8] Braun, S.; Rappoport, S.; Zusman, R.; Avnir, D.; Ottololenghi, M. Biochemically active sol-gel glasses: the trapping of enzymes. *Mater. Lett.* 1990, 10, 1-5.

[9] Rao, MS.; Dave, BC. Thermally-regulated molecular selectivity of organosilica sol-gels. *J. Am. Chem. Soc.* 2003, 125, 11826-11827.

[10] Livage, J.; Roux, C.; Da Costa, JM; Desportes, I.; Quinson JF. Immunoassays in sol-gel matrixes. *J. Sol–Gel Sci. Technol.* 1996, 7, 45-51.

[11] Dunn, B.; Zink, JI. Probes of pore environment and molecular matrix interactions in sol-gel materials. *Chem. Mater.* 1997, 9, 2280-2291.

[12] Dave, BC.; Dunn, B.; Valentine, JS.; Zink. Sol-gel encapsulation methods for biosensors. *J. Anal. Chem.* 1994, 66, 1120A-1127A.

[13] Gill, I.; Ballesteros, A. Encapsulation of biologicals within silicate, siloxane, and hybrid sol-gel polymers: an efficient and generic approach. *J. Am. Chem. Soc.* 1998, 120, 8587-8598.

[14] Avnir, D.; Braun, S. Biochemical Aspects of Sol–Gel Science and Technology; Kluwer Academic Publishers: Amsterdam, Netherlands,1996; 143pp.

[15] Jain, TK.; Roy, I.; De, TK.; Maitra, A. Nanometer silica particles encapsulating active compounds: a novel ceramic drug carrier. *J. Am. Chem. Soc.* 1998, 120, 11092–11095.

[16] Roy, I.; Ohulchanskyy, T.; Pudavar, HE.; Bergey, EJ., Oseroff, AR., Morgan, J.; Dougherty, TJ.; Prasad, PN. Ceramic-based nanoparticles entrapping water-insoluble photosensitizing anticancer drugs: a novel drug-carrier system for photodynamic therapy. *J. Am. Chem. Soc.* 2003, 125, 7860–7865.

[17] Manzano, M.; Vallet-Regi, M. New developments in ordered mesoporous materials for drug delivery. *J. Mater. Chem.* 2010, 20, 5593-5604 [and references therein].

[18] Schmidt, H. New type of non-crystalline solids between inorganic and organic materials. *J. Non-Cryst. Solids* 1985, 73, 681– 691.

[19] Kulikova, GA.; Parfenyuk, EV.; Ryabinina, IV.; Antsiferova, YS.; Sotnikova, NY.; Posiseeva, LV.; Eliseeva, M.. *In-vitro* studies of interaction of modified silica nanoparticles with different types of immunocompetent cells. *J. Biomed. Mater. Res., Part A*, 2010, 95A, 435-439.

[20] Kazitsina, LA.; Kupletskaya, NB. Application of UV-, IR- and NMR spectroscopy in organic chemistry. Moscow: High School; 1971. (in Russian).

[21] Brunner, H.; Vallant, T.; Mayer, H.; Hoffmann, H. Stepwise Growth of Ultrathin SiOx Films on Si(100) Surfaces through Sequential Adsorption/Oxidation Cycles of Alkylsiloxane Monolayers. *Langmuir* 1996, 12, 4614-4617.

[22] Kitadai, N.; Yokoyama, T. ; Nakashima, S. ATR-IR spectroscopic study of L-lysine adsorption on amorphous silica. *J. Colloid and Interface Sci.* 2009, 329, 31–37.

[23] Gao, Y.; Heinemann, A.; Knott, R.; Bartlett, J. Encapsulation of Protein in Silica Matrices: Structural Evolution on the Molecular and Nanoscales. *Langmuir* 2010, 26, 1239-1246.

[24] Dong, H. Organic-Inorganic Hybrid mesoporous Silica Materials and Their Application as Host Matrix for Protein Molecules. PhD Thesis, Drexel University., 2002.

[25] Warheit, DB. Inhaled amorphous silica particles: what do we know about their toxicological profiles? *J. Environ. Pathol. Toxicol. Oncol.* 2001, 20, 133-141

[26] Elias, Z.; Poirot, O.; Danigere, MC.; Terzetti, F.; Marande, AM.; Dzwigaj, S.; Pezerat, H.; Fenoglio, I.; Fubini, B. Cytotoxic and transforming effects of silica particles with different surface properties in Syrian hamster embrio (SHE)cells. *Toxicol in Vitro* 2000, 14, 409-422.

[27] Menaa, B.; Herrero, M.; Rives, V.; Lavrenko, M.; Eggers, D. K. Favourable influence of hydrophobic surfaces on protein structure in porous organically-modified silica glasses. *Biomaterials* 2008, 29, 2710–2718.

[28] Sing, K. S. W.; Everett, D. H.; Haul, R. A. W.; Moscou, L.; Pierotti, RA.; Rouquerol, J.; Siemieniewska, T. Reporting physisorption data for gas/solid systems with special reference to the determination of surface area and porosity (Recommendations 1984). *Pure Appl. Chem.* 1985, 57, 603–619.

[29] Menaa, B.; Torres, C.; Herrero, M.; Rives, V.; Gilbert, ARW; Eggers, DK. Protein adsorption onto organically modified silica glass leads to a different structure than sol-gel encapsulation. *Biophys. J.* 2008, 95, L51-L53.

[30] Menaa, B.; Menaa, F.; Aiolfi-Guimaraes, C.; Sharts, O. Silica-based nanoporous sol-gel glasses: from bioencapsulation to protein folding studies. *Int. J. Nanotech.* 2010, 7, 1-45.

[31] Menaa, B; Miyagawa, Y.; Takahashi, M.; Herrero, M.; Rives, V. Bioencapsulation of apomyoglobin in nanoporous organosilica sol-gel glasses: influence of the siloxane network on the conformation and stability of a model protein. *Biopolymers* 2009, 91, 895-906.

[32] Beauchemin, R.; N'soukpoe -Kossi, C. N.; Thomas, T. J.; Thomas, T.; Carpentier, R.; Tajmir-Riahi, H. A. Polyamine Analogues Bind Human Serum Albumin. *Biomacromolecules* 2007, 8, 3177-3183.

[33] Sukhishvili, SA.; Granick, S. Adsorption of human serum albumin: Dependence on molecular architecture of the oppositely charged surface. *J. Chem. Phys.* 1999, 110, 10153-10161.

[34] Soderquist, ME.;Walton AG. Structural changes in proteins adsorbed on polymer surfaces. *J. Colloid Interface Sci.* 1980, 75, 386-395.

[35] Haynes, CA.; Norde, W. Globular proteins at solid/liquid interfaces. *Colloid. Surf. B: Biointerfaces* 1994, 2, 517-566

[36] Retzinger, GS. Adsorption and coagulability of fibrinogen on atheromatous lipid surfaces. *Atherosclerosis, Thrombosis and Vascular Biology* 1995, 15, 786-792.
[37] Engel, MFM; van Milero, CPM; Visser, AJWG. Kinetic and structural characterization of adsorption-induced unfolding of bovine α-lacalbumin. *J. Biol. Chem.* 2002, 277, 10922-10930
[38] Koutsopoulos, S.; van der Oost, J.; Norde, W. Structural features of a hyperthermostable endo-b-1,3-glucanase in solution and adsorbed on 'iInvisible'' particles. *Biophys.J.* 2005, 88, 467-474.
[39] Vertegel, AA; Siegel, RW; Dordick, JS. Silica nanoparticle size influences the structure and enzymatic activity of adsorbed lysozyme. *Langmuir* 2004, 20, 6800-6807.
[40] Welzel, PB. Investigation of adsorption-induced structural changes of proteins at solid/liquid interfaces by differential scanning calorimetry. *Thermochim. Acta* 2002, 382, 175–188.
[41] Buijs, J; Norde, W; Lichtenbelt, J. Changes in the secondary structure of adsorbed IgG and F(ab') studied by FTIR spectroscopy. *Langmuir* 1996, 12, 1605-1613.
[42] Mészáros, R; Thompson, L; Bos, M; de Groot. P. Adsorption and Electrokinetic Properties of Polyethylenimine on Silica Surfaces. *Langmuir* 2002, 18, 6164-6169.
[43] Putnam, FW. (Ed.), The Plasma Proteins, Vol. I, 2nd Edition, Academic Press, New York, 1975, pp. 133–181.
[44] Houska, M; Brynda, E. Interactions of proteins with polyelectrolytes at solid/liquid interfaces: sequential adsorption of albumin and heparin. *J. Colloid. Interf. Sci.* 1997, 188, 243-250.
[45] Rezwana, K; Meier, LP; Gauckler, LJ. Lysozyme and bovine serum albumin adsorption on uncoated silica and AlOOH-coated silica particles: the influence of positively and negatively charged oxide surface coatings. *Biomaterials* 2005, 26, 4351–4357.
[46] Iler, RK. The chemistry of silica. John Wiley & Sons, Inc. N.Y., 1979. Vol. 1.
[47] Norde, W. Adsorption of proteins from solution at the solid-liquid interface. *Adv Colloid Interface Sci.* 1986, 25, 267-340.
[48] Jin, W; Brennan, JD. Properties and applications of proteins encapsulated within sol–gel derived materials. Anal. Chim. Acta 2002, 461, 1-36.
[49] Tunc, S; Maitz, M; Salzer, R. Conformational changes during protein adsorption. FT-IR spectroscopic imaging of adsorbed fibrinogen layers gerald steiner. *Anal Chem.* 2007, 79, 1311-1316.
[50] Engel, MFM; van Mierlo, CPM; Visser, AJWG. Kinetic and structural characterization of adsorption induced unfolding of bovine α-lactalbumin. *J. Biol. Chem.* 2002, 277, 10922-10930.
[51] Koutsopoulos, S.; vander Ost, J., Norde, W. Adsorption of an endoglucanase from the hyperthermophilic *Pyrococcus furiosus* on hydrophobic (polystyrene) and hydrophilic (silica) surfaces increases protein heat stability. *Langmuir* 2004, 20, 6401-6405.
[52] Larsericsdotter, H.; Oscarsson, S.; Buijs, J. Thermodynamic analysis of proteins adsorbed on silica particles: electrostatic effects. *J. Colloid Interface Sci.* 2001, 237, 98–103
[53] Tobias, DJ; Mar, W; Blasic, JK; Klein, ML. Molecular dynamics simulations of a protein on hydrophobic and hydrophilic surfaces. *Biophys. J.* 1994, 71, 2933-2941.

[54] Brandes, N.; Welzel, PB.; Werner, C; Kroh, LW. Adsorption-induced conformational changes of proteins onto ceramic particles: Differential scanning calorimetry and FTIR analysis. *J. Colloid and Interface Sci.* 2006, 29, 956–969.

[55] Billsten, P; Carlsson, U; Jonsson, BH; Olofsson, G; Höök, F; Elwing, H. Conformation of Human Carbonic Anhydrase II Variants Adsorbed to Silica Nanoparticles. *Langmuir* 1999, 15, 6395-6399.

[56] Vermonden, T.; Giacomelli, C. E.; Norde, W. Reversibility of Structural Rearrangements in Bovine Serum Albumin during Homomolecular Exchange from AgI Particles. *Langmuir* 2001, 17, 3734-3740.

[57] Picó, GA. Thermodynamic features of the thermal unfolding of human serum albumin. *Int. J. Biol. Macromol*, 1997, 20, 63-73.

[58] Wallach, ML.; Heller, W. Theoretical investigations on the light scattering of colloidal spheres. XII. The determination of size distribution curves from turbidity spectra. *J. Chem. Phys.* 1961, 34, 1796-1802.

[59] Khlebtsov, BN.; Khanadeev, VA.; Khlebtsov, NG. Determination of the size, concentration, and refractive index of silica nanoparticles from turbidity spectra. *Langmuir* 2008, 24, 8964-8970.

[60] Lucarelli M, Gatti AM, Savarino G, Quattroni P, Martinelli L, Monari E, Borasch D. Innate defence functions of macrophages can be biased by nano-sized ceramic and metallic particles. Eur Cytokine Network 2004; 15: 339-346.

[61] Chollet, P; Favrot, M; Hurbin, A; Coll, J.-L. Side-effects of a systematic injection of linear polyethylenimine-DNA complexes. *J. Gene Med.* 2001, 4, 84-91.

[62] Fischer, D; Bieber, T; Li, Y; Elsasser, HP; Kissel, T. A novel nonviral vector for DNA delivery based on low molecular weight, branched polyethylenimine: effect of molecular weight on transfection efficiency and cytotoxicity. *Pharm. Res.* 1999, 16, 1273–1279.

[63] Nel, A; Xia, T; Madler, L; Li, N. Toxic potential of materials at the nanolevel. *Science* 2006, 311, 622-627.

[64] Fischer, HC; Chan, WC. Nanotoxicity: The growing need for in vivo study. *Curr. Opin. Biotechnol.* 2007, 18, 565-571.

[65] Nie, S; Xing, Y; Kim, GJ; Simons, J. W. Nanotechnology applications in cancer. *Annu. Rev. Biomed. Eng.* 2007, 9, 257-288.

[66] Marcato, PD; Duran, N. New aspects of nanopharmaceutical delivery systems. *J. Nanosci. Nanotechnol.* 2008, 8, 2216-2229.

[67] McBain SC; Yiu HH; Dobson J. Magnetic nanoparticles for gene and drug delivery. *Int. J. Nanomedicine* 2008, 3, 169-180.

[68] De Jong, WH; Borm, PJ. Drug delivery and nanoparticles: applications and hazards. *Int. J. Nanomedicine* 2008, 3, 133-149.

In: The Sol-Gel Process
Editor: Rachel E. Morris
ISBN 978-1-61761-321-0

Chapter 23

BORDER BETWEEN SOL-GEL AND WET PRECIPITATION IN CALCIUM PHOSPHATES SYNTHESIS

Alexandra Bucur and Raul Bucur
National R-D Institute for Electrochemistry and Condensed Matter (NIRDECM) Timisoara, Condensed Matter Dept, Timisoara, Romania

ABSTRACT

Calcium phosphate ceramics have been intensively studied during the last decades because of their great potential use for human dental and bone implants. The most used synthesis methods in the field are the wet chemistry methods, namely sol-gel and wet precipitation, due to a series of advantages comparing with other methods. The sol-gel synthesis method involves the formation of a colloidal sol which will then turn into a gel, this method requiring no special energy conditions for the formation of the desired compound; meanwhile, the chemical precipitation method means the system needs to be offered special conditions in order for the precipitation to take place, like certain values for the pH or for temperature. Based on previous working experience and observations, we would like to try a different, less-conventional approach, which consists in the study of the line between the two chemical synthesis routes. We have started the synthesis like a classic wet precipitation, with calcium chloride and phosphoric acid as calcium and phosphorus precursors, but we did not precipitate the desired calcium phosphate phase by pH control, but allowed the reaction to take place in time, in aqueous media at room temperature and pressure, approach which is specific to the sol-gel way. This new approach is based on the notice that though at the first sight they seem very similar, the experience proves us that actually a very fine border exists between them, and that is what we wanted to see and prove. The question that arises is: what is the actual border between sol-gel and wet precipitation? The idea of this question actually came from the study of the literature, which presented inconsistencies and differences in the opinions of different researchers in the field and inappropriate use of terms when describing synthesis work, probably due to insufficient knowledge and differentiation between one method or another.

INTRODUCTION

Synthesis of ceramic materials of calcium phosphate-type is an area of great interest for present and mostly for future applications in the biomedical field, this type of substances (hydroxyapatite, tricalcium phosphates and other members of the family) being naturally present in large amounts in human hard tissues, namely bone tissue and dental tissue [1]. Being the most important inorganic constituents of the hard tissues, calcium phosphates have long been considered very attractive in the field of hard tissue repairs, due to qualities like: very good biocompatibility and osteoconductivity, the ability to bond directly to the bony tissue without any connective tissue, non-toxicity or good dissolution behavior in the physiological environment [2]. Among the large variety of methods involved in the achievement of calcium phosphate bioactive materials, the sol-gel and the chemical precipitation methods are two of the most widely used, due to a series of advantages which will be briefly presented. The challenge for scientists and engineers in the biomedical field is to understand the natural processes that take place in our bodies, which give birth to bones and teeths without high temperature treatment or any other kinds of special treatments, in long periods of time - months or years. At present, because of different production processes, calcium phosphate ceramics have different physicochemical, mechanical and biological properties [3], which are not identical to the properties of natural materials. Achievement of the "perfect" synthetic material, which would possess all the characteristics of the natural tissues, is the ultimate challenge for researchers in the field of human hard tissue repair materials.

The both sol-gel and wet precipitation methods are part of the wet-chemistry techniques family. The sol-gel method is one of the oldest synthesis routes (more than 100 years old), starting with observations of the hydrolysis of some orthosilicates [4], and among the most promising future routes for hybrid materials manufacture. Through hydrolysis and condensation reactions, the precursors, which are initially mixed in an aqueous/alcoholic/ other media, will form a colloidal solution containing solid particles suspended in a liquid phase, named the *sol*. This sol will further transform into a solid network (macromolecules) immersed in the liquid phase – *the gel*. By drying the gel in constant conditions (temperature, pressure), the liquid phase is removed and a porous material will be formed. Further thermal treatment of the material leads to further polycondensation and improvement of the mechanical properties. Comparing with other methods, the sol-gel synthesis provides advantages like: an extremely intimate contact between reactants (at molecular level) and milder conditions comparing with other synthesis routes [5]. Materials obtained by the sol-gel route have a very high level of purity and homogenous composition, which makes them suitable for a very broad range of applications [6].

One fundamental application of the sol-gel route is the synthesis of ceramic materials for bone replacements, where materials could be obtained at room temperature, similar to natural processes occurring inside the human body. So far, calcium phosphate ceramic bioactive materials have been synthesized by a number of methods, of which sol-gel has been given the most attention due to its specific simple nature and best results [7]. The calcium precursor used in the synthesis can be either an inorganic one (oxide, hydroxide, carbonate, nitrate, etc) [8, 9], or an organic one (like calcium diethoxide) [10, 11]. The regular phosphorus precursor is phosphoric acid [12], but also monoacid diammonium phosphate or triethyl phosphate [2], monoacid dipotassium phosphate [9], and other precursors can be used.

The precipitation from solution is the second widely used and also very promising synthesis method, because of a series of reasons like the intimate contact between reactants and the opportunity to control the product morphology and phase composition, or the distinct feature of obtaining the hydrated form of the material, as a result of synthesis, with important effects on the properties of the material [13].

Chemical wet precipitation technique involved in the synthesis of calcium phosphates includes the pH control, very important because it dictates the nature of the species present in the system (degree of neutralization of the phosphoric acid is pH-dependant) [6]. Depending on the experimental parameters (like pH or temperature) and how the synthesis is conducted, monophasic materials like β-tricalcium phosphate or hydroxyapatite can be obtained, or biphasic (sometimes even more than 2-phased) mixtures of the two can be formed [14]. This method currently involves the addition of the phosphorus precursor over the calcium precursor, and pH adjustment using ammonia derivatives (hydroxide, nitrate, etc) – if necessary (if the calcium precursor consists in calcium hydroxide – for instance, there is no need for further pH adjustment). The next step after precipitate filtration, washing and drying, is represented by the thermal treatment of the precipitate up to the temperature where the desired phase(s) is(are) present and stable. The firing process has an important contribution in achieving the desired morphology and degree of crystallinity or porosity, due to the influence on the state of water inside the structure of the material. The wet precipitation has been successfully used and has a long history in the synthesis of materials for implants, due to the above-mentioned qualities of controlling the phase morphology and composition, reliability and fastness being also attributes to be taken into consideration.

EXPERIMENTAL

The purpose of this work was to achieve tricalcium phosphate from solution at room temperature without using the traditional pH precipitation condition, with minimum of energy requirements and despite some opinions that tricalcium phosphate cannot be precipitated under these circumstances [15]. Calcium chloride ($CaCl_2$ anhydrous, Reactivul Bucuresti) and phosphoric acid (H_3PO_4 85%, Merck) were used as calcium and phosphorus precursors. The calcium chloride was dissolved in distilled water and used as aqueous solution, and the phosphorus precursor was used as received. The synthesis procedure presented here started with the regular steps of adding the phosphorus precursor dropwise onto the calcium precursor, in a molar ratio of 1.5 (Ca:P), under magnetic stirring. In order to fulfill the purpose of this study, the usual control by pH adjustment step was excluded, and also temperature control; the initial mixture of precursors, with a pH of 0.5, was only magnetically stirred for 30 minutes at room temperature. The mixture was investigated at room temperature by x-ray diffraction using a PANalytical Xpert PRO MPD diffractometer with Cu Kα incident radiation and PIXcel detector, working conditions 45 kV and 30 mA, with a step size of 0,016°/step and counting time of 50 sec/step. The scanning range used was 10-100° 2theta, but some results are presented only for the range 10-70° 2theta (not necessary to use results from the range 70-100° 2θ). The mixture was applied on the x-ray diffraction zero-background silicon holder as liquid phase, then dried in place and investigated by XRD method.

The software used for data analysis and phase identification was X'Pert High Score Plus and ICDD database, version 2008.

After a period of 2 weeks at room temperature allowed for ageing, the mixture was again subjected to XRD investigation. Before using x-ray diffraction, we have used the simplest way of investigation, the visual inspection. At the beginning, eye observation of the studied sample presented the formation of a white sol, which during the 2 weeks ageing time transformed into a white gel, and then into paste. The paste continued to dry slowly and was again investigated by XRD after 2 months at room temperature. After all this time, the paste had transformed into a solid material, agglomerated in clusters, but easy to crush.

Results and Discussion

After the 30 minutes magnetic stirring of the calcium chloride and phosphoric acid mixture, a white milky gel started to form, meaning the chemical interaction (hydrolysis) of the precursors has started. According to [16], "the chemical interaction between basic powdered solids and phosphoric acid solutions is a central step in the phosphate hardening mechanism, which results in the generation of a new gel-like phase".

The x-ray diffraction pattern presented in figure 1, recorded after the 30 minutes stirring, is characteristic of an amorphous compound, which was expected, the peaks that are present being characteristic to the Pt substrate (due to the equipment used). Yet, there is a clear tendency of formation of a peak, with the maximum around the 2θ value of 31.2°, which is the typical reflection of the $Ca_3(PO_4)_2$ low temperature anhydrous β phase. In our case this calcium phosphate phase is not yet present as a distinct phase, but only we can see the start of formation of this compound, due to the fact that at room temperature, with no pH adjustment or thermal treatment, the reaction will need more energy for completion, which is counted in terms of longer time.

One characteristic feature of phosphoric compounds is their ability to polymerize and form different structures that are disordered and mixed, conventionally assigned to a class of non-crystalline solids [16], and we think that this has been happening in our case, too. Phosphorus is capable to form a hybridized sp^3 bond, involving an oxygen atom with a coordination number of 4. This will result in the buildup of a PO_4^{3-} tetrahedral group, which further will become the basic structural unit for phosphoric acid compounds, both in liquid and solid phase. An important role in the chemistry of the disordered state of dissolved phosphoric acids and their ceramic-forming ability is the hydrogen bonds, which form in an aqueous environment and lead to networks of H-bridged chains. The PO_4^{3-} tetrahedra contained in phosphoric acid compounds are linked by hydrogen bonds and in time combine into polymeric structures of different types. This is what we can also see in the present case.

In our previous experience [17, 18] we noticed that time needed for the maturation of compounds and proper accomplishment of the chemical reaction in the field of sol-gel synthesis of calcium phosphates family compounds is longer, meaning tens or even hundreds of hours.

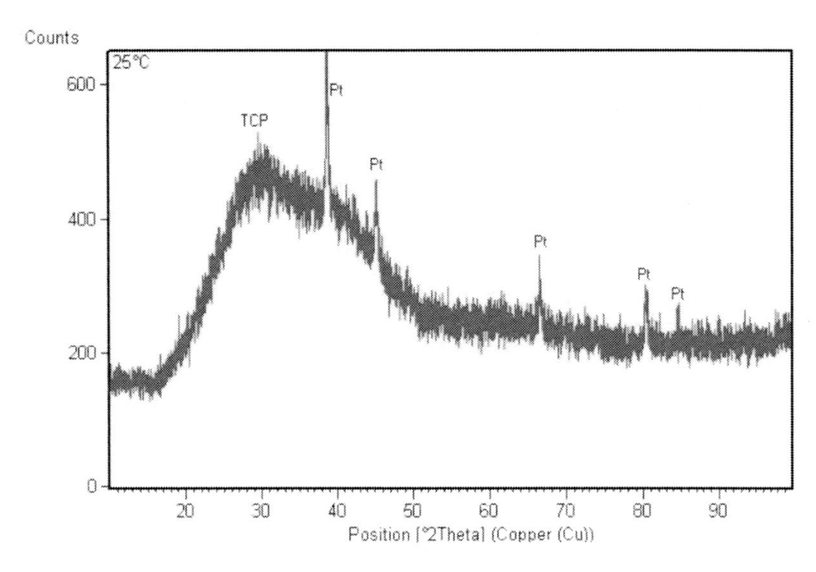

Figure 1. X-ray diffraction pattern of precursors mixture after 30 minutes stirring.

Considering the previous statement, we allowed the reaction in the gel phase to continue at room temperature and pressure, in a closed container. The hydrolyzed phosphorus sol interacted with the Ca sol (probably as Ca^{2+} in water), forming derivatives containing Ca-O-P bonds. Further drying of the material causes the condensation of these units and formation of more bonds of type Ca-O-P in the dry gel. Non-crystalline solids like our sample generally display a multiple-linked three-dimensional lattice, formed by condensation from aqueous solution at temperatures equal or close to room temperature [16].

During a 2 weeks time period, the white paste kept on drying, the obvious changes in the sample appearance being monitored again by x-ray diffraction (Figure 2 and 3). We presume that after this time, our sample is a mixture of gel, glassy phase and some small microcrystalline formations, based on PO_4^{3-} tetrahedral group and Ca-O-P bonds disordered formation. Indeed, the morphology of the sample has changed during this period of time, the crystallinity increased considerably and a new clearly-observable phase appeared, $CaClH_2PO_4$, presuming the equation:

$$CaCl_2 + H_3PO_4 \rightarrow CaClH_2PO_4 + HCl \tag{1}$$

Figure 3 presents a closer look at the pattern of the sample after 2 weeks, the y-axis values being considered only for a smaller domain, for visual purposes. The pattern revealed the existence of $CaHPO_4$ in small amounts (from decomposition of $CaClH_2PO_4$), $Ca_2P_2O_7$ (polycondensation reaction) and unreacted phosphoric acid and very small tracks of calcium chloride (not represented on the pattern due to very small intensity). We also found the calcium phosphate, this time as clear peak (even if small intensity). The surprising thing (contradiction with basic chemistry principles) is that the XRD patterns reveal the presence of this compound, meaning the sol-gel characteristic approach of using long maturation times can suppress the pH adjustment stage. Another surprising thing is the presence of calcium pyrophosphate, because this compound, theoretically, should be formed from reactions taking place at high temperatures.

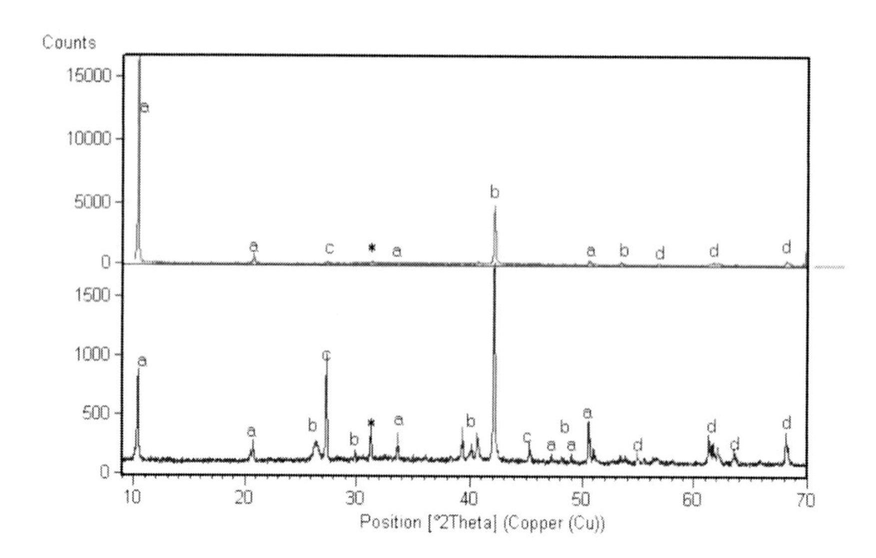

Figure 2. X-ray diffraction pattern of precursors mixture after 2 weeks (up) and after 2 months (bottom) at room temperature a) $CaClH_2PO_4$, b) $CaHPO_4$ c) $Ca_2P_2O_7$ d) H_3PO_4 *) $Ca_3(PO_4)_2$.

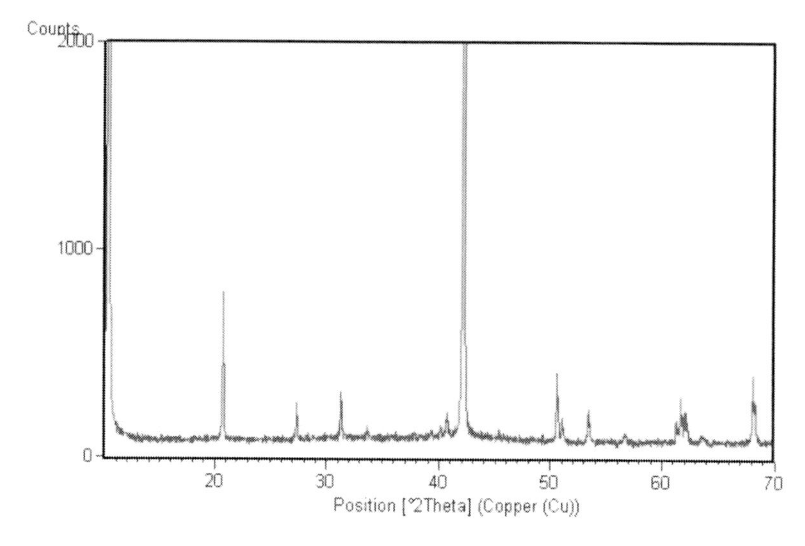

Figure 3. Pattern of the sample after 2 weeks maturation time, with y axis domain in the field 1-2000 counts for better visualization.

The presence of small amounts of unreacted precursors in the analyzed sample is not surprising, considering basic chemistry principles regarding pH and species existence domains in solution. According to the order used for precursors addition, it would be a natural thing to find at least phosphoric acid in the XRD pattern of the sample, if not also calcium chloride, because the reaction should be (theoretically) / is (practically) very slow with no "help" from outside. A plausible explanation for the presence of unreacted calcium chloride and phosphoric acid would be the gradual accumulation of reaction products, which inhibits the chemical interaction of the reactants due to an enhanced diffusion barrier [16].

The XRD pattern will give us an indication of how slowly the reaction takes place, by the fact that small peaks belonging to calcium chloride and phosphoric acid are present in the sample even after 2 months maturation time (phosphoric acid peaks are represented in figure 2). The crystallinity degree increased considerably, this time we can see clear and well defined peaks.

We expect changes in our sample due to the fact that stability of H-bonded compounds deposited from aqueous media may vary widely in time. What can we see in the pattern is the decrease of the intensity of the most intense peak of 10.35° 2θ belonging to $CaClH_2PO_4$, meaning this phase is "consuming" itself in time, and giving birth to $CaHPO_4$ by decomposition (as previously stated), according to the equation:

$$CaClH_2PO_4 \rightarrow CaHPO_4 + HCl \tag{2}$$

The previous statement is supported by the presence in the pattern of peaks ascribed to $CaHPO_4$. In the same time, some small peaks belonging to $CaClH_2PO_4$ grow in intensity in the pattern of the sample after 2 months comparing with 2 weeks time sample, because reaction given by eq. (1) is not completed by 2 weeks time, but continuing due to the presence of the precursors in the sample. At the same time, we could notice on the superposed patterns representation (not presented here) that the intensity of the peaks ascribed to the unreacted precursors is lower for the sample maturated for 2 months comparing with the sample after 2 weeks. The same thing happened to the intense peak at the value of 42.13° belonging to $CaHPO_4$, whose intensity decreases around three times in the sample after 2 months comparing with the sample after 2 weeks. This is happening because other crystalline phases are formed based on the slow decomposition of the monocalcium phosphate.

The presence of the calcium pyrophosphate phase in the sample and XRD pattern may be the result of pyrophosphoric acid (formed in time by slow dehydration of phosphoric acid) reaction with calcium chloride, according to the equation:

$$H_4P_2O_7 + 2CaCl_2 \rightarrow Ca_2P_2O_7 + 4HCl \tag{3}$$

A small number of peaks could also be ascribed to chloroapatite (ICDD no. 1011), defined as $CaCl_2.3Ca_3(PO_4)_2$, chemical formula that would be in good agreement with the use of calcium chloride as precursor and the expected phenomenon of tricalcium phosphate crystallization.

The number of peaks present in the pattern of the 2 weeks sample is considerably smaller than the number of peaks in the pattern after 2 months. A noticeable effect is the shift in the value of 2θ for almost all the peaks after 2 months, some of them being shifted towards right and some towards left (shifts in the range 0.1-0.6). What that means is local order in the sample, some crystalline fractions suffering contractions and some – relaxations. These shifts are given by the changes in the crystallization system suffered by crystalline microdomains, because at the beginning the predominant crystallization system is the monoclinic one, then new phases appear that crystallize in hexagonal and orthorhombic systems.

We presumed that due to the intimate nature of the reactions and phenomena taking place, and also looking at the shape of the peaks in the x-ray diffraction patterns, that the crystalline particles that grow in our sample are in the nanometric dimensions domain. In order to prove our assumption, we have calculated the crystallites size for the $CaClH_2PO_4$ phase that appeared in our system after 2 weeks, for the most intense peak of that particular phase (10.3181° 2θ), using the Scherrer formula [19] (this formula is only applicable for nano-scale particles):

$$d = \frac{0.9 \cdot \lambda}{B \cdot \cos\theta} \tag{4}$$

where 0.9 is the value ascribed to the shape factor K (typical value), d is the crystallite size of the phase of interest, λ is the wavelength of the incident radiation Cu K_{α}, B is the full width at half maximum intensity (FWHM) and θ is the Bragg's angle. The calculated crystallite size is 18.14 nm, which proves our assumption. Calculating the crystallites dimensions for the same phase after 2 months time, we found the exact same value of 18.14 nm. The same tendency was observed in the case of the peak at 42.1332° 2θ belonging to the $CaHPO_4$ phase, the crystallites dimensions remain at the same value of 15.88 nm (calculated with Scherrer) during the ageing and drying period of 2 months.

Formation of the calcium phosphates from solution under the specified conditions may be regarded as an oversaturation of the solution of orthophosphoric acid salts that crystallize to form an intergrown crystalline solid, the phenomena that constitute the base for further development of crystalline phases being nucleation (given by oversaturation) and growth.

Now to come back to the question that was the subject of this study: what is the border between sol-gel and wet precipitation? We notice the differences, for instance in the period of time required by the 2 methods in order to achieve the desired phase and the pH adjustment stage presence or absence, and the similarities, for instance in the fact that both ways are wet chemistry and both presume a very intimate contact between reactants. The differences mentioned above cannot be considered as a border between the two methods, because in some cases pH adjustment is used in sol-gel, and sometimes similar periods of time are used for sol-gel as for the wet precipitation [2, 20]. Unfortunately, in our opinion there are cases when the two synthesis routes do not have clearly defined identities and sometimes they are a reason of confusion for some researchers, that is the reason why we decided to ask this question to the scientific community and ourselves. The border is actually dependent on these factors, but not given by them. The sol-gel method is defined by the formation of the sol and then gel; for that reason, the solution has to be allowed to maturate in constant conditions, which will not happen in wet precipitation, who takes place fast. The chemical reactions that happen in order for the sol and gel to form are slow reactions; the equilibrium is constantly changing and if we have a look at the insights, things are a lot more complicated than they look at the first view (and we could see this in our study). The intermediate species are formed continuously by hydrolysis and condensation, and disappear after a while along with the formation of a new compound, like it was proved in our experimental investigations. All these transformations are strongly influenced by the water state in the intimate structure of the materials, coordinated by the central cation [13].

The border can then be defined, in our opinion, by the formation of this particular state of the matter which is described by the *sol* and *gel* terms. We believe literature is sometimes confusing regarding the 2 mentioned synthesis methods, in our opinion the pH adjustment excludes the need for long-time maturation or aging of the solution, respectively long-time aging does not require pH adjustment (if pH adjustment is performed, the reaction is fast and very little changes will be expected in time; also, the pH adjustment will prevent the formation of the sol, because it will quickly lead to the precipitation of the desired compound/phase).

CONCLUSION

The work described here is synthesis of calcium phosphate using a method that is not exactly sol-gel, nor wet precipitation. After the precursors were mixed, the pH adjustment stage was excluded and the mixture of reactants was allowed to react at room temperature and pressure in a closed container for long time, without further implication of any thermal treatment scheme. Investigation by x-ray diffraction after 30 minutes magnetic stirring, 2 weeks and 2 months maturation times revealed the very slow formation and crystallization of calcium phosphate β phase. The multi-phased crystalline material contains crystallites of nanometric size dimensions.

This synthesis approach could be promising and might bring interesting results and very useful in the medical field, but unfortunately the time necessary for chemical reactions completion is too long to be used in the current synthesis of materials. Future investigation ideas were drawn from this study, consisting in combinations of features of the 2 synthesis routes compared here. Design of the appropriate conditions in respect to the desired properties has to take into account the control of the new phase formation rate (chemical interaction between the powdered phase and liquid component) and the rate of structure formation. The chemical reaction rate is strongly influenced by the choice of reactants, their properties and reactivity, and by the relationship between chemical properties of the central coordinating Ca cation and bonding properties of the phosphate system.

The approach we have used for this study allowed us to understand more of the intimacy of the calcium phosphate formation reaction in aqueous media and helped us find a plausible answer to our question.

The answer to the question in the title of this study is not an easy one, but our opinion is that the border between sol-gel and wet precipitation lies around the itself name of the former, being given by that special state called sol (and then gel) and the conditions that allow its formation. In the present study we could actually see how complex, and at the same time slow, the reactions in the sol-gel method can be, and we could understand the differences comparing with simple precipitation. The most important attribute of the sol-gel route is the fact that it is the closest to the nature among the all known synthesis routes for biocompatible materials, because nature creates human bones in a very similar manner, in times of months and years, under moderate conditions of supersaturation [21].

REFERENCES

[1] Putlyaev, V. I. & Safronova, T. V. (2006). *Biomaterials,* 3-4, 99-102.

[2] Guzmán Vásquez, C., Piña Barba, C. & Munguia, N. (2005). *Revista mexicana de fisica, 51(3)*, 284-293

[3] Jinlong, N., Zhenxi, Z. & Dazong, J. (2001). *J Mat Synth Proc, 9(5)*, 235-240.

[4] Hench, L. L. & West, J. K. (1990). *Chem Rev,* 90(1), 33-72.

[5] Brinker, C. J. & Scherer, G. W. (1990). Sol-Gel Science - *The Physics and Chemistry of Sol-Gel Processing*, New York, Academic Press.
[6] Beganskienė, A., Dudko, O., Sirutkaitis, R. & Giraitis, R. (2003). *Mat Sci (Medžiagotyra), 9(4)*, 383-386.
[7] Roach, P., Eglin, D., Rohde, K. & Perrez, C. C. (2007). K *J Mater Sci: Mater Med, 18*, 1263-1277.
[8] Gan, L. & Pilliar, R. (2004). *Biomaterials, 25*, 5303-5312.
[9] Feng, W., Mu-Sen, L., Yu-Peng, L. & Sheng-Song, G. (2005). *J Mat Sci Letters, 40*, 2073-2076
[10] Fellah, B. H. & Layrolle, P. (2009). *Acta Biomaterialia, 5*, 735-742.
[11] Hesaraki, S., Safari, M. & Shokrgozar, M. A. (2009). *J Mater Sci: Mater Med, 20(10)*, 2011-2017.
[12] Caroline Victoria, E. & Gnanam, F. D. (2002). *Trends Biomater Artif Organs* 16(1), 12-14.
[13] Malysheva, A. Y. & Beletskii, B. I. (2001). *Glass and Ceramics, 58(3-4)* , 147-149.
[14] Borodajenko, N., Salma, K. & Berzina-Cimdina, L. (2008)., *The 11'th European Powder Diffraction Conference, Poster session*, Book of Abstracts pg. 126.
[15] Stoia, M., Ionescu, M., Stefanescu, O., Murgan, R. & Stefanescu, M. (2008). *Chem Bull POLITEHNICA Univ (Timisoara), 53(67)*, 1-2, 204-207.
[16] Karpukhin, A., Vladimirov, V. S. & Moizis, S. E. (2005). *Refractories and Industrial Ceramics, 46(3)*, 180-186.
[17] Morar, I. R., Ioitescu, A. & Doca, N. (2009). *J Optoelectron Adv Mat Symposia, 1(1)*, 17-19
[18] Ioitescu, A., Vlase, G., Vlase, T., Ilia, G. & Doca, N. (2009). *J Therm Anal Cal, 96(3)*, 937-942.
[19] Shih, W. J., Wang, J. W., Wang, M. C. & Hon, M. H. (2006). *Mat Sci Eng, C26*, 1434-1438.
[20] Sanosh, K. P., Chu, M. C., Balakrishnan, A., Lee, Y. J., Nim, T. N. & Cho, S. J. (2009). *Current Appl Phys, 9,* 1459-1462.
[21] Leonor, I. B., Azevedo, H. S., Pashkuleva, I., Oliveira, A. L., Alves, C. M. & Reis, R. L. (2004). *Learning from nature how to design new implantable biomaterials,* Kluwer Academic Press Publishers, 123-150, printed in Netherlands.

In: The Sol-Gel Process
Editor: Rachel E. Morris

ISBN 978-1-61761-321-0

Chapter 24

MAGNESIUM-DOPED BIPHASIC CALCIUM PHOSPHATE NANOPOWDERS VIA SOL-GEL METHOD

Iis Sopyan,[1,*] Toibah Abdul Rahim[1] and Zainal Arifin Ahmad[2]
[1]Department of Manufacturing and Materials Engineering, Faculty of Engineering, International Islamic University Malaysia (IIUM),
PO Box 10, 50728 Kuala Lumpur, Malaysia
[2]School of Materials and Mineral Resources Engineering, University Sains of Malaysia, Nibong Tebal, Penang, Malaysia

ABSTRACT

Biphasic calcium phosphate (BCP) is a mixture of non-resorbable hydroxyapatite (HA) and the resorbable tricalcium phosphate (TCP) is an interesting material for bone implant as it shows biocompatibility and bioactivity to tissue bone. More efficient bone repair was widely been known in BCP than HA alone. Good implant materials should be biodegradable as it can degrade inside the bone and defect simultaneously with the formation of a new bone.

In this study BCP has been doped with magnesium through sol-gel method. Doping of magnesium ions into BCP will results in biological improvement as the ion will cause the acceleration of nucleation kinetics of bone minerals. Magnesium depletion adversely affects all stages of skeletal metabolism, leading to decrease in osteoblastic activities and bone fragility.

Magnesium–doped biphasic calcium phosphate (Mg-BCP) powders were successfully prepared using $Ca(NO_3)_2.4H_2O$ and $(NH_4)_2HPO_4$ as the precursors and $Mg(NO_3)_2.6H_2O$ as the source of the dopant. Morphological evaluation by FESEM measurement showed that the particles of Mg-BCP were tightly agglomerated, with primary particulates of 50-150 nm diameters. FESEM result also showed that doping of magnesium into BCP particles caused fusion of particles leading to more progressive densification of particles as shown by higher concentration of magnesium doping. Successful incorporation of Mg into BCP lattice structure was confirmed by higher

---

* Corresponding author : Email: sopyan@iiu.edu.my

crystallinity of Mg-BCP and by shifting of tricalcium phosphate (TCP) peaks in XRD patterns to higher 2θ angles as the Mg content increased. XRD and FTIR measurement showed that the increment of crystallinity was directly proportional to the amount of the dopant. Both analyses also revealed that TCP appeared only after calcination of 700°C and above. All the powder exhibited highly crystalline BCP characteristics after calcination at 900 °C. XRD analysis revealed that β-TCP peak increased in intensity with the increased level of doped Mg, meanwhile HA peak was almost no change in intensity. With the increasing Mg concentration into the BCP, the solubility limit of the Mg in the β-TCP decreases and Mg starts segregated as free MgO or incorporated into the HA. A significant contraction has been observed in the calculated lattice parameter which may reflect the addition of Mg into the β-TCP phase and the differences of the lattice parameters and *c/a* ratio of β-TCP were much bigger than that of HA. FT-IR analysis confirms the formation of biphasic mixtures of HA and Mg stabilized β-TCP when calcined at high temperatures as bands of HPO_4^{-2} and $P_2O_7^{-4}$ decreased. Thermal analysis showed that the particles crystallize faster with more magnesium added. This study showed that magnesium doping into BCP through sol-gel method has improved crystal growth and fusion of BCP particulates.

INTRODUCTION

Biphasic calcium phosphate (BCP) (Kivrak and Tas, 1998) is a mixture of non-resorbable hydroxyapatite (HA) and the resorbable tricalcium phosphate (TCP) is an interesting material for bone implant as it shows biocompatibility and bioactivity to tissue bone. More efficient bone repair was widely been known in BCP than HA alone. Good implant materials should be biodegradable as it can degrade inside the bone and defect simultaneously with the formation of a new bone. Thus, the β-TCP is widely used as a biodegradable bone substitutes as it gives rise to extensive bone remodeling around the implant (Tas, et al., 1997). On the other hand, when used as biomaterial for alveolar ridge augmentation, the rate of biodegradation of β-TCP has been shown to be too fast (Kivrak, et al., 1998). Thus, in order to balance the non-biodegradability of HA which is more stable phase and at the same time to slow the rate of biodegradation of β-TCP, the interest of biphasic calcium phosphate (BCP) concept have been studied by many research groups.

Pure BCP and BCP doped by magnesium were prepared via a novel sol-gel method as we utilized non-alkoxide metals as the raw materials. Several advantages gained by producing ceramic powder through the sol-gel method are good homogeneity of powder (Bezzi, et al., 2003; Gibson, et al., 2001), nanosize dimensional of the primary particles (Bezzi, et al., 2003), and high reactivity (Bezzi, et al., 2003) compared to conventional methods such as solid-state reaction, hydrothermal, and wet chemical precipitation (Suchanek, et al., 1998).

Various research groups have attempted to dope calcium phosphate materials with magnesium (Zyman, et al., 2008; Kalita, et al., 2007; Landi, et al., 2006; Kannan, et al., 2005; Gibson, et al., 2002; Fadeev, et al., 2003) for bone implant materials. Doping of magnesium ions into BCP will results in biological improvement as the ion will cause the acceleration of nucleation kinetics of bone minerals (Landi, et al., 2006). Magnesium depletion adversely affects all stages of skeletal metabolism, leading to decrease in osteoblastic activities and bone fragility. Moreover, the incorporation of Mg ions into the TCP phase will produce more stable phase composition after heat treatment which compose HA and Mg-TCP.

In order to obtain BCP materials, the thermal analysis on Mg-doped BCP has been studied. Here we produced Mg-doped biphasic calcium phosphate (Mg-BCP) at different Mg concentration (varied from 0-15%) through a sol-gel method. The calcination temperature was varied from 500 to 1000°C. Their physico-chemical characterizations are reported as well.

MATERIALS AND METHODS

Materials

Mg-doped BCP powders were prepared by sol-gel method using calcium nitrate tetrahydrate (Merck KGaA, Germany) and di-ammonium hydrogen phosphate (Merck KGaA, Germany) as the precursors for calcium and phosphorus respectively. Magnesium nitrate tetrahydrate (R & M Chemicals, UK) was used as the source of the dopant. EDTA (Merck KGaA, Germany) was used the chelating agent to avoid immediate precipitation of calcium ions to occur during the experiment. Urea (R & M Chemicals, UK) was used as the gelling agent during the reaction.

Experimental Procedure

In order to produce 25 gram of Mg-doped BCP powders, a 250 mL of ammonium solution (11%, R & M Chemicals, UK) was heated at 60 °C, and 90.5 g of EDTA was added while stirring until it dissolved and gave clear solution. Into this, 100 mL of aqueous solution of 65 g of $Ca(NO_3)_2.4H_2O$ and 50 mL of $Mg(NO_3)_2.6H_2O$ were poured, and then 20 g of $(NH_4)_2HPO_4$ and 22.6 g of urea were subsequently added. The addition of $Mg(NO_3)_2.6H_2O$ was based on its molar percentage (mol %) that will dope into the BCP. The mixture was then refluxed at ~100°C for 3-5 hours while stirring until a white gel of Mg-doped BCP mixture were obtained. The obtained gel was then dried at 340°C under ambient air and subsequently subjected to heat treatment under flowing air. The synthesis of Mg-doped BCP was carried out at 11 different molar concentration of Mg which varied from 0.01 mol% to 15 mol%. Figures 1 and 2 show the flow chart of preparation of Mg-doped BCP via sol-gel method and pictures of white gel, black gel and Mg-doped BCP powder.

Powder Characterization

The degree of crystallinity of the calcined powders and the presence of β-TCP and HA phases was identified by X-ray diffractometer (XRD) (Shimadzu, XRD 6000). For the measurements, X-ray diffraction spectra of the pure and Mg doped BCP powders were collected over the 2θ range of 20-50°C using copper Kα radiation with a step scan rate of 2° in 2θ min-1. The crystallite sizes of the synthesized powders were measured by using Scherer's equation and Nanoparticle Sizer (Malvern Instruments, Zen 1600). For IR data, the powders were directly placed onto the attenuated total refractance (ATR) holder of the FT-IR spectrometer (Perkin Elmer FT-IR, Spectrum 100 Series) in order to follow the chemical evolution from the gel to the BCP and to determine the presence of anions partially substituting PO_4^{3-} and/ or OH^- groups. Thermogravimetric and differential analysis (TG/ DTA) was performed on the synthesized Mg-doped BCP powders, white gel and black gel in ambient air by using TG/ DTA instruments (Perkin Elmer, Pyris Diamond TG/DTA) with 5°C/min heating rate. The specific surface area of the powders was measured via a nitrogen adsorption method using the Brunauer-Emmet-Teller (BET) Surface Area Analyzer (Quantachrome Instruments, Autosorb-1). Samples outgassing were performed at 350°C for 3 hours before the analysis.

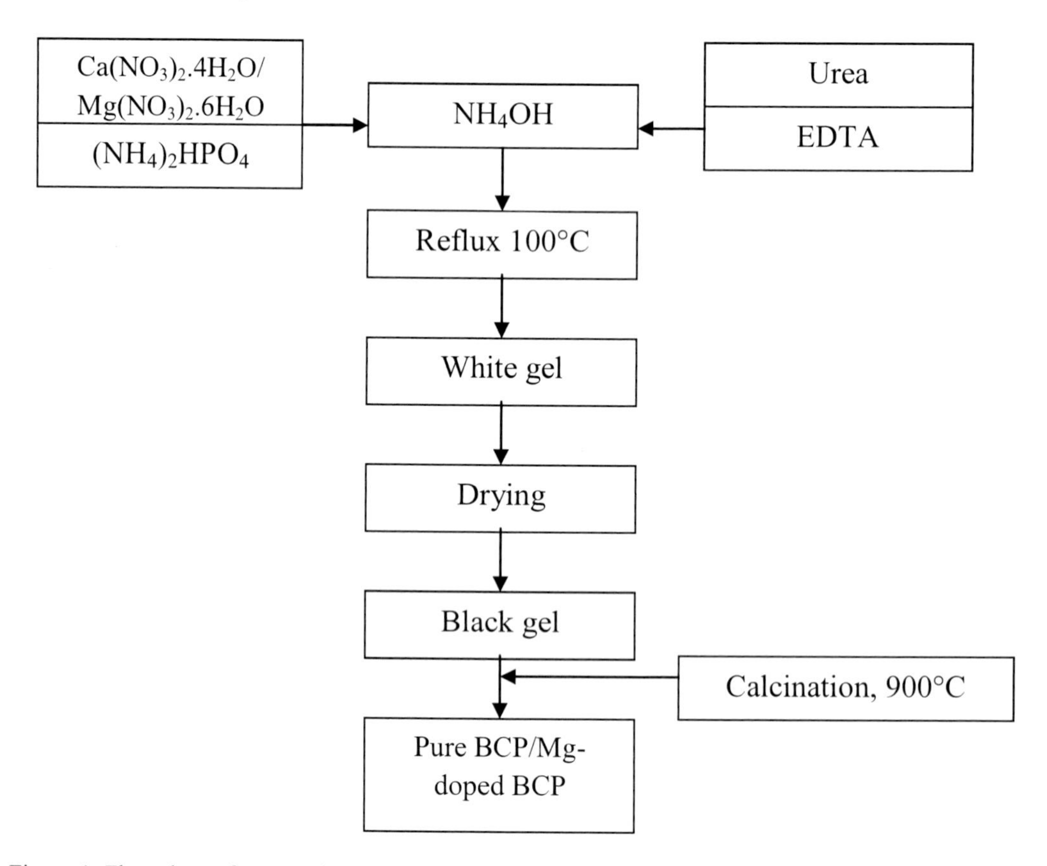

Figure 1. Flow chart of preparation of Mg-doped BCP via sol-gel method.

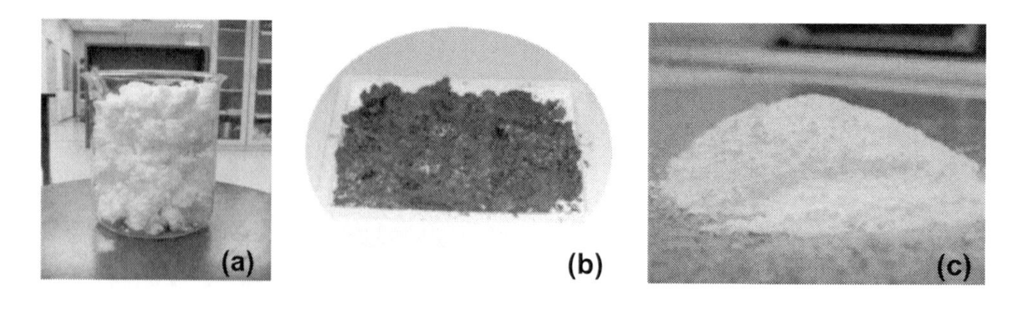

Figure 2. Figure of white gel (a), black gel (b) and Mg-doped BCP powder (c).

RESULTS AND DISCUSSIONS

Characterization of White Gel by Using TG/ DTA

$$5Ca(NO_3)_2 + Mg(NO_3)_2 + 3\,(NH_4)_2HPO_4 + 4NH_4OH \rightarrow Ca_{(5-x)}Mg_x(PO_4)_3OH + 10NH_4NO_3 + 3H_2O \quad \text{(Eq. 1)}$$

After refluxing the reaction mixture at 100°C, white gel was formed. The reaction of hydroxyapatite formation can be expressed in Equation 1. TG/DTA was performed on the white gel, black gel and the powders to study the thermal transformation process by using a heating rate of 5°C/min. A total multi step weight loss of ~100% for TG/ DTA of white gel is presented in Figure 3. The figure shows that the first drop occur at temperature ~100°C was due to evaporation of water. Then it has experienced ca. 10% of weight loss which contributed to removal of water from the gel. The second drop starting at temperature ~100°C and experienced ca. 50% of weight loss. It can be attributed to organic molecules which are nitrate and urea. But then at higher temperature, T~300°C, the decomposition of acetates and hydroxide occurred. At this temperature, the white gel experienced exothermic process as the calcination take place. Calcination will change the amorphous phase of the gel to crystalline phase of BCP.

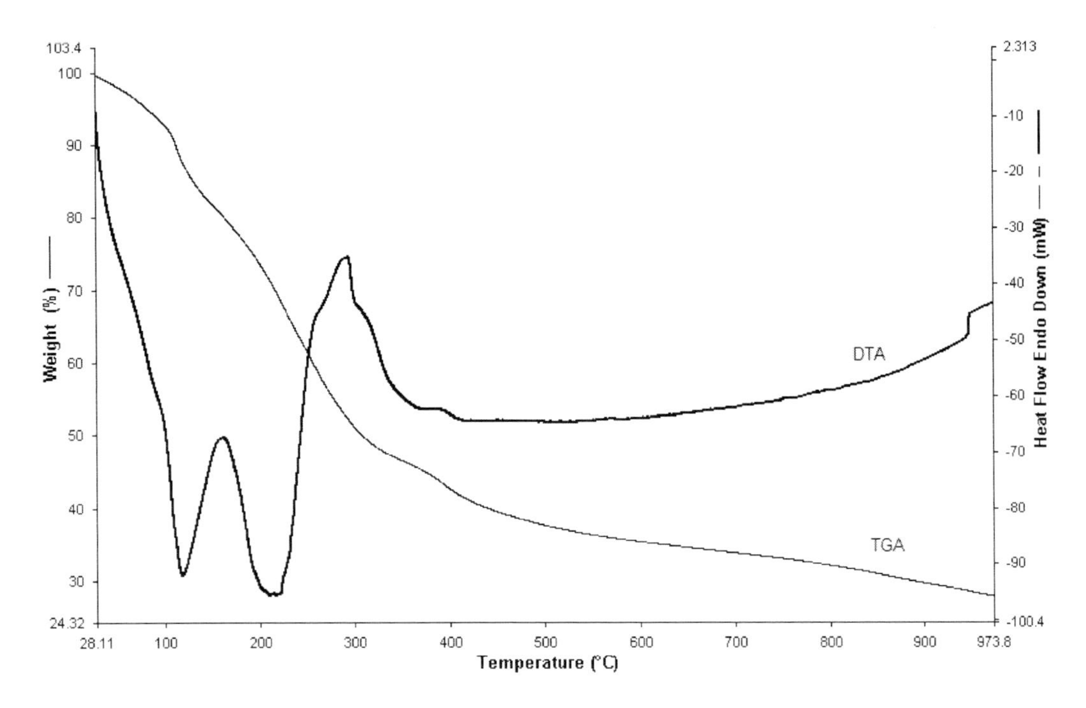

Figure 3. TG/DTA of white gel of BCP.

Characterization of Mg-Doped Biphasic Calcium Phosphate Powder

The XRD patterns for the as synthesized powders at different Mg percent calcined at 900°C are showed in Figure 4. Calcination of the powders at this temperature has indicated the formation of β-TCP phase together with HA phase as the main phases. Identification of phases in the calcined powders was achieved by comparing the obtained XRD patterns with PDF Card No. 09-432 for HA and 09-169 for β-TCP. The presence of β-TCP phase has confirmed the formation of biphasic mixtures in the synthesized powder (Kannan, et al., 2005). The presence of Mg ions has stabilized the β-TCP structure (Landi, et al., 2006) as the β-TCP peak increased with the increasing percent of Mg doped (Figure 4). The increase in β-TCP peak can be correlated with the incorporation of Mg into the β-TCP phase. Moreover, due to the existence of Mg, the diffraction peaks of β-TCP planes of Mg-doped BCP powders shifted to larger angles in the pattern, compared with that of pure BCP powder. The favorable incorporation of Mg into β-TCP phase is attributed to substitution of smaller Mg^{2+} (0.86Å) into Ca^{2+} (1.14Å). Ryu et al. (2006) stated that β-TCP contains a cation site of octahedral coordination which is smaller than cation site of HA, which is more suitable for the smaller Mg^{2+} ion than the Ca^{2+} ion. Though, when the Mg concentration was increased up to 10% and above, the solubility limit of the Mg in the β-TCP phase decreased and Mg started to segregate as free MgO or will be incorporated into the HA phase (Ryu, et al., 2006). Thus, at higher concentration of Mg (10% Mg-BCP and 15% Mg-BCP), MgO phase at around 43° in 2θ was observed.

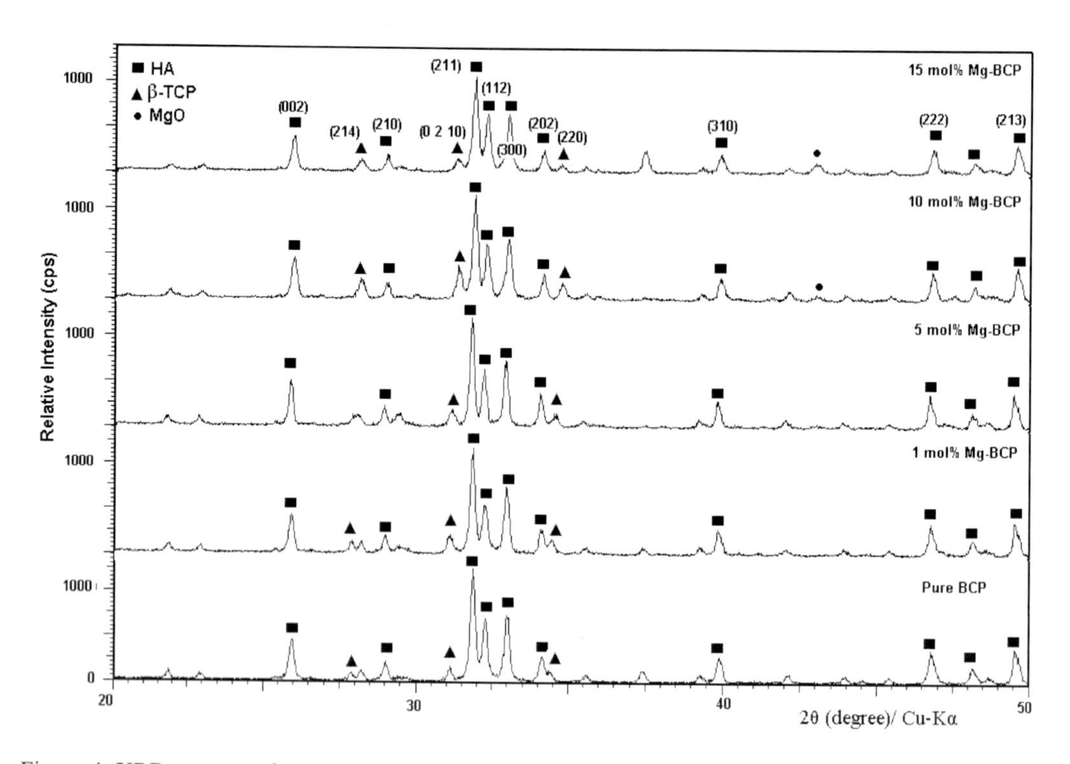

Figure 4. XRD patterns of pure BCP and Mg-doped BCP at various concentrations.

XRD patterns for selected sample pure BCP and 10 mol% Mg-doped BCP calcined at different temperature condition are presented in Figures 5 and 6. XRD analysis for both samples showed that powder calcined at 500ºC has presented a diffraction pattern of a poorly crystalline apatite as it is still in amorphous state. The transformation of amorphous phase with broaden peak to crystalline phase occurred in between 500 to 700ºC. The appearance of β-TCP has started to be detected at 700°C and 800°C for sample with 10 mol% of Mg and pure BCP, respectively. This result also showed that β-TCP peak was started to be visible clearly at 800°C for 10 mol% Mg-doped BCP compared to pure BCP. This finding is in accordance to thermogravimetric analysis (TGA) which inferred that an increase in Mg content resulted in lower crystallization temperature of BCP. This study has also been done for other Mg- doped BCP and they showed the same result. At 900ºC, all the synthesized powders have showed the existence of β-TCP phase at highest resolution of peaks that confirmed the formation of biphasic mixtures at the optimum temperature. However, when the calcination temperature was increased up to 1000°C, the XRD pattern of the powder has showed improvement in peak intensity of HA phases and MgO phase only while β-TCP remained unchanged. This result suggests that at higher calcination temperatures, reversible transformation of β-TCP phase into HA phase has occurred.

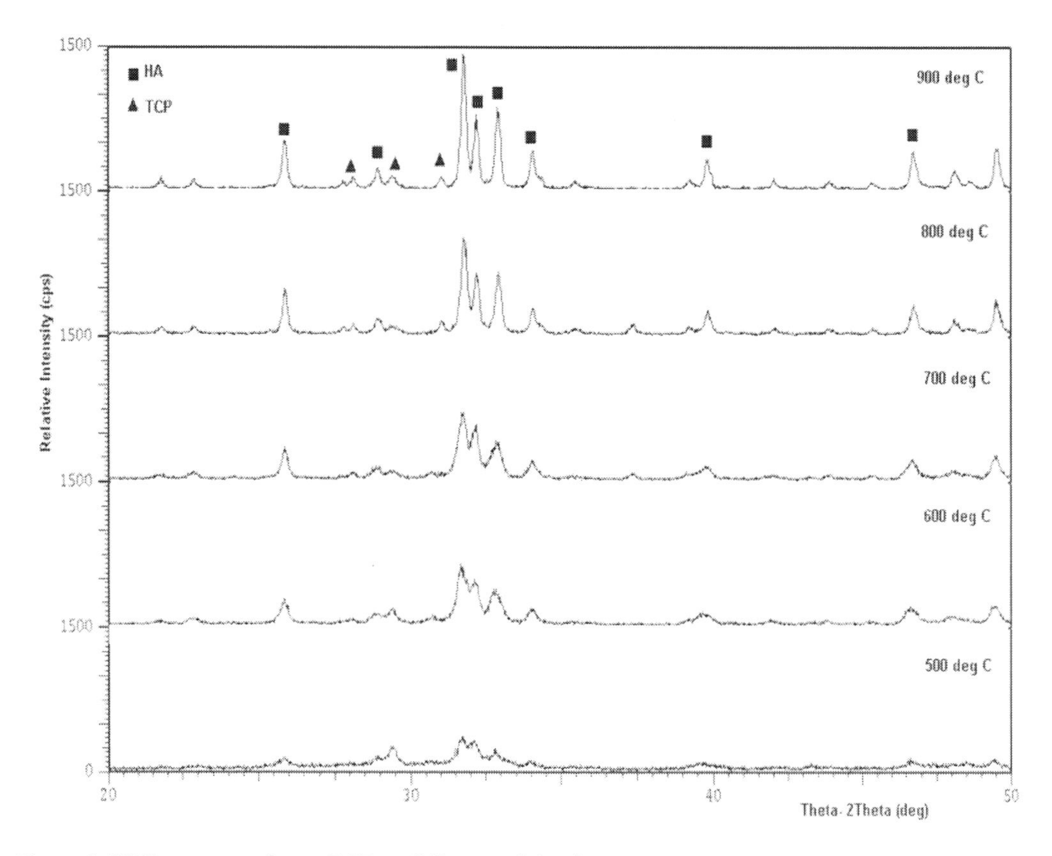

Figure 5. XRD patterns of pure BCP at different calcination temperatures.

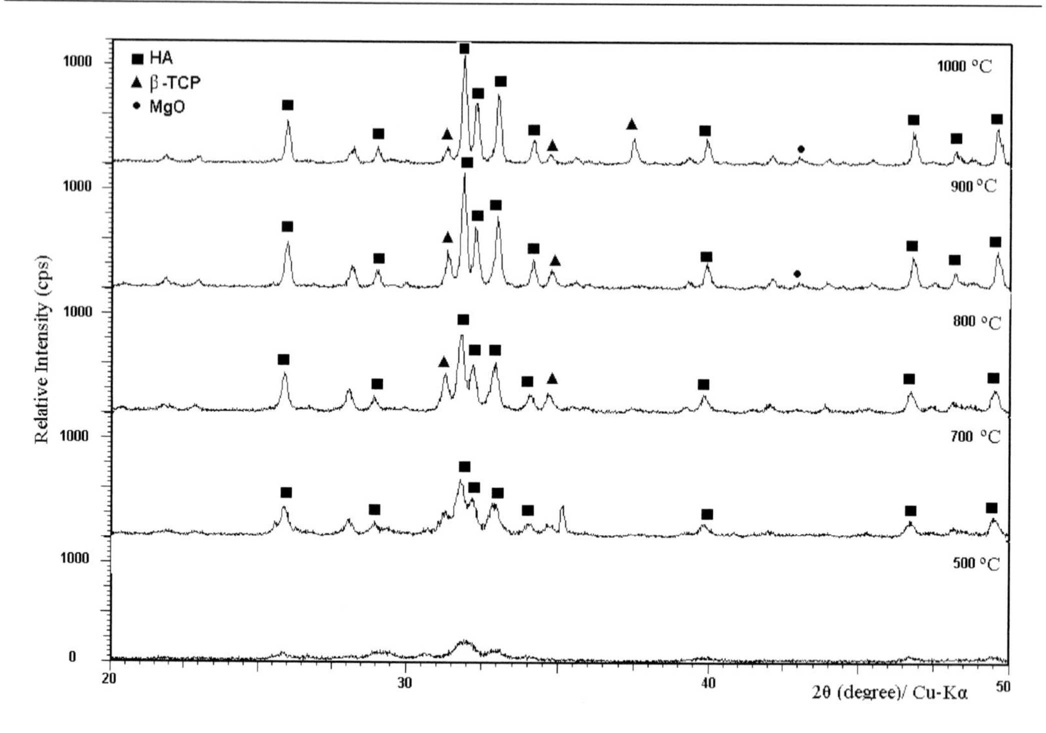

Figure 6. XRD patterns of 10 % Mg-doped BCP at different calcination temperatures.

Figure 7. FESEM images of pure BCP (a) and 15 mol% Mg-doped BCP (b) powder.

FESEM images of pure BCP and Mg-doped BCP powders showed nanometric primary particles of about 75-150 nm with globular shaped (Figure 7). The nanosized primary particles were agglomerated into micrometric aggregates. Furthermore, from the FESEM observations, powder that content 15 mol% Mg-doped tend to form flat like structure in shape. FESEM revealed that doping Mg into BCP particles caused fusion of particles leading to more progressive interparticulate fusion of particles as shown by 15% Mg doping. This showed that Mg has shown its function as a sintering additive which improved sintering behavior of BCP, thus possibly leading to improvement in mechanical strength.

The change in the microstructures of the powders could be observed clearly as the calcination temperature was increased from 500 to 1000°C for 10 mol% Mg-doped BCP powder. For the lowest calcination temperature; 500°C, the microstructure of the powders appeared to be composed of very small individual crystallites in nano size (~21.52 nm, as measured by Scherrer'equation), spherical shapes and tightly agglomerated. However, when the calcination temperature was increased up to 600°C, the spherical-shaped crystals began to exhibit an early stage of sintering with only small regions of densification of particles. At 700°C calcination temperature, the individual particles tend to melt together as more progressive densification of particles occurred at higher temperature. The fusion of particles can be seen clearly when the 10 mol% Mg-doped BCP powder was calcined at 900°C. Figure 8 shows the microstructure of powder that has been calcined at 500°C, 700°C and 900°C. From the figure, we can conclude that an increase in calcination temperature has increased the coarsening of the microstructure of the powder particles as the spherical-shapes crystals growth well as they melt and diffuse together. This is in good agreement with the XRD analysis which indicates that at higher calcination temperatures, the crystallinity of the 10 mol% Mg-BCP was increased with the increasing crystallite size.

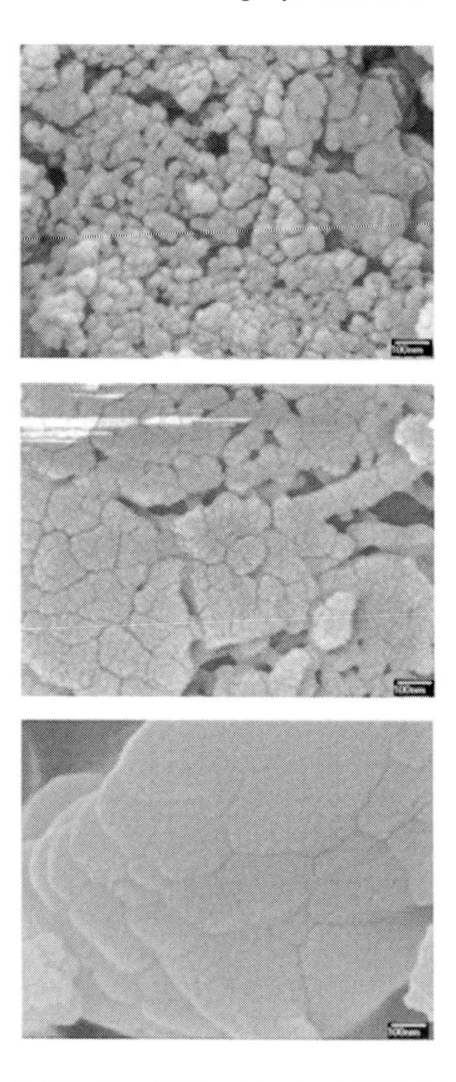

Figure 8. FESEM pictures of 10% Mg-doped BCP calcined at (a) 500°C, (b) 700°C and (c) 900°C.

The average crystallite size of the samples was calculated via the Scherrer equation (Li, et al., 2008; Sasikumar, et al., 2006). The results are shown in Table 1. The analysis showed that the particle size increased with increased in Mg content. The result was reliable with the crystallinity of the synthesized powders. The surface area analysis determined using the BET method has been done on pure BCP, 5 mol% and 10 mol% Mg-doped BCP and the results obtained are 8.98, 5.82 and 6.56 m_2/g, respectively. Moreover, the particle size of these powders also measured by using Nanosizer and the results are not consistent with the surface area analysis and XRD analysis. The measured particle size by Nanosizer was respectively about 3-4 times larger than the particle size measured by XRD (Scherrer's equation). This might be due to the measurement of agglomerates particle instead of single particle by the Nanosizer. Figures 9 and 10 show the particle size distribution of 5 mol% Mg-doped BCP and pure BCP measured by Nanosizer.

Table 1. The average crystallite size of the samples was calculated via the Scherer's equation.

Mg content (mol %)	Average crystallite size (nm) (Scherer's equation)
0	46.58
1	46.60
5	46.86
10	48.67
15	47.18

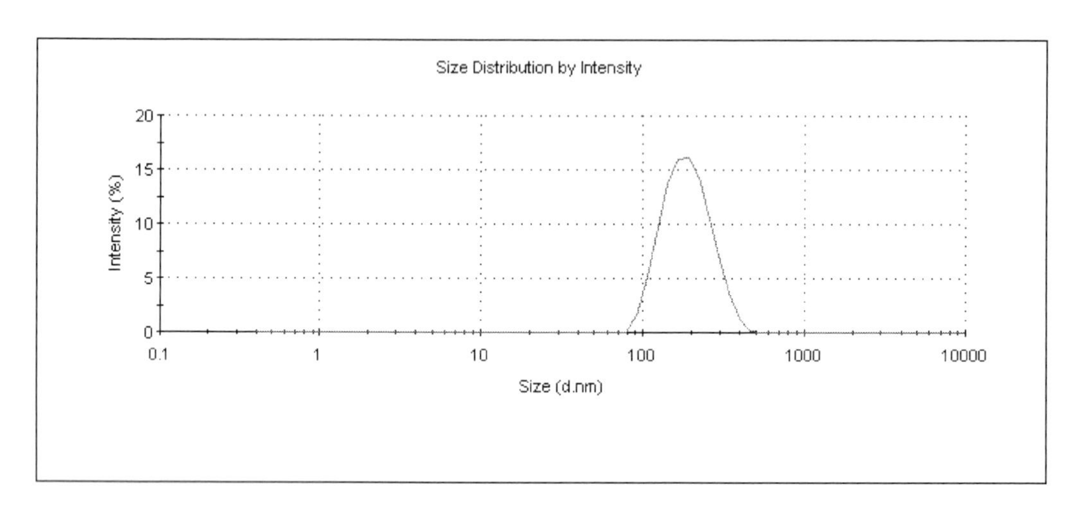

Figure 9. Particle size analysis of 5 mol% Mg-doped BCP powder measured by Nanosizer.

The calculated lattice parameters for the HA phase and β-TCP phase for pure BCP and Mg-doped BCP (1, 10 and 15 mol%) powders calcined at 900°C are presented in Table 2. The calculated values of lattice parameters have indicated the presence of biphasic phase in the pure BCP and Mg-doped BCP. For the hexagonal apatite phase that represent the HA phase, the lattice parameter ***a*** decreased with increasing Mg percent up to 15 mol%. The similar trend goes to ***c***-axis lattice parameter of the HA phase. For the rhombohedral phase of β-TCP with respect to PDF Card No. 09-169, it is noted that both lattice parameters, ***a*** and ***c*** decreased with increasing % Mg doping for all samples. Moreover, a significant contraction has been observed in the calculated lattice parameter which may reflect the substitution of smaller ionic radius of Mg^{2+} than in Ca^{2+} ion in the β-TCP phase, which coincided with other previous research work (Ryu, et al., 2006). The differences of the lattice parameters and ***c/a*** ratio of β-TCP were much bigger than that of HA. This shows that Mg has substituted Ca in the β-TCP phase.

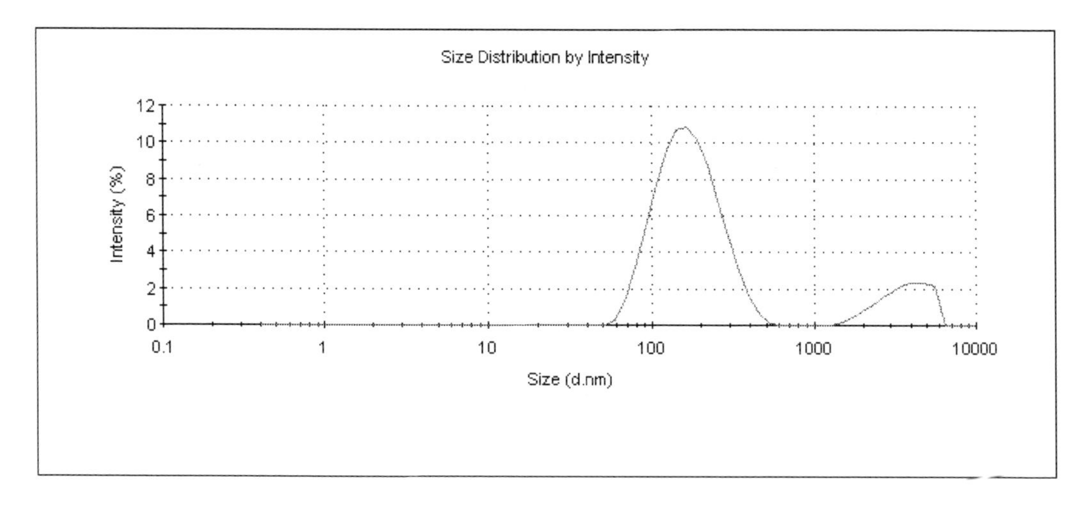

Figure 10. Particle size analysis of pure BCP powder measured by Nanosizer.

Table 2. Calculated lattice parameters for different Mg-doped BCP powders calcined at 900°C.

Lattice parameters		1 mol% Mg-BCP	10 mol% Mg-BCP	15 mol% Mg-BCP
HA	*a*	9.4115	9.4018	9.4073
	c	6.8748	6.8725	6.8737
	c/a	0.7305	0.731	0.7307
β-TCP	*a*	10.3224	10.3100	10.3181
	c	37.6490	37.6285	37.5814
	c/a	3.6473	3.6497	3.6423

FT-IR study showed the presence of hydroxyl (OH) stretch mode bands at ~3570 cm^{-1} (Figure 11). The band at ~627 cm^{-1} is derived for librational modes of OH groups in BCP. Phosphate bands are observed at 563 cm^{-1}, 599 cm^{-1}, 961 cm^{-1}, 1024 cm^{-1} and 1085 cm^{-1}. The spectra also showed that the peak resolution of OH and PO_4 bands are viewed less intensity with the increased Mg percent concentration. Besides, HPO_4^{-2} band at 875 cm^{-1} tend to decrease as the Mg concentration increased. This showed that the condensation of HPO_4^{-2} complete faster once we doped with high Mg content. FT-IR spectra for pure and 10 mol% Mg-doped BCP calcined at different temperatures are presented in Figures 12 and 13. The spectra in Figure 13 for 10 mol% Mg-doped BCP show that C-O band at 1300-1600 cm^{-1} was decreased in its resolution as the calcination temperature increased. Besides, the decrease in carbonate band has increased the OH and PO_4 bands' intensity. This shows that the as synthesized BCP powder involved a substitution of carbonate groups for hydroxyl (A-type) and the phosphate (B-type) groups. Moreover, the spectrum revealed the presence of hydrogen phosphate groups (HPO_4^{-2}) and pyrophosphate ($P_2O_7^{-4}$) bands are evident at 500°C for the peak at 875 cm^{-1} and 715 cm^{-1}, respectively. However, these bands tend to decrease as the calcination temperatures were increased. This confirms the formation of biphasic mixtures of HA and Mg stabilized β-TCP in the synthesized powder when calcined at high temperature.

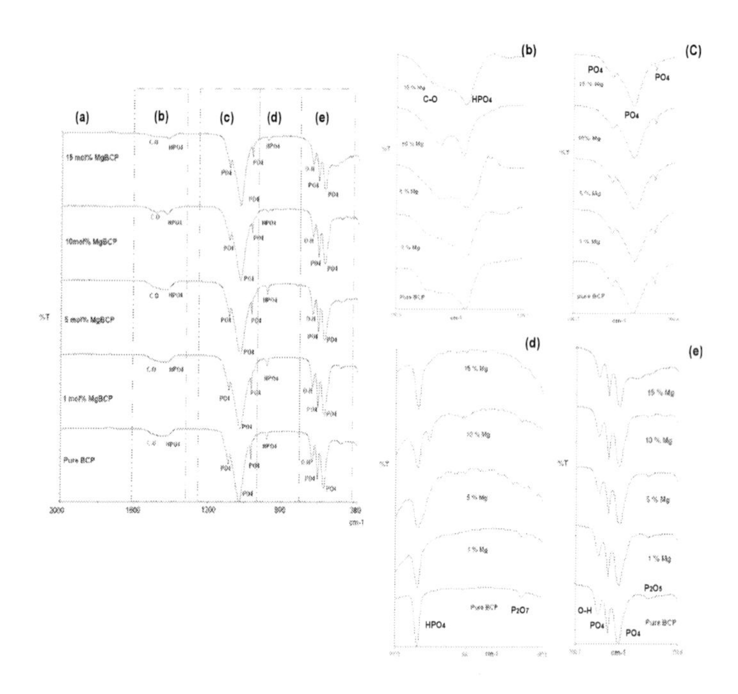

Figure 11. FT-IR spectra of pure BCP and Mg-BCP powders at various concentrations.

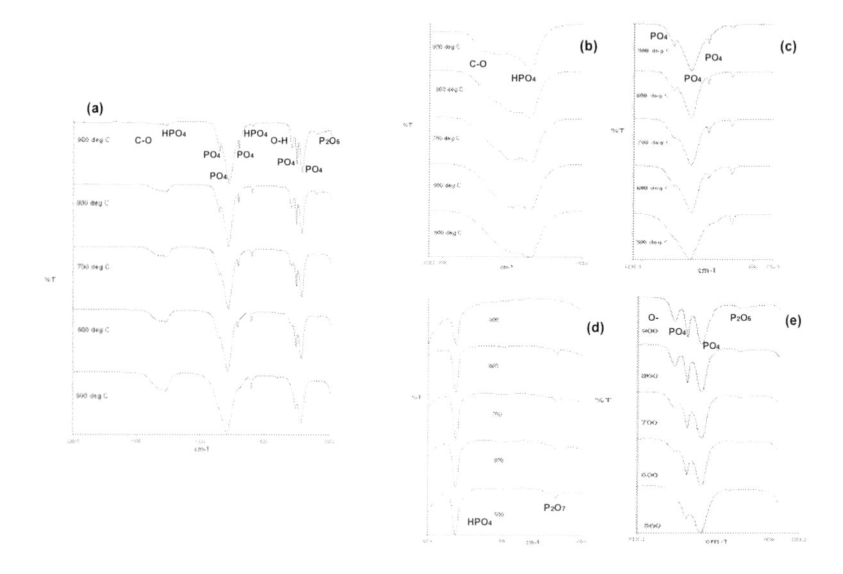

Figure 12. FT-IR spectra of pure BCP at various calcinations temperature.

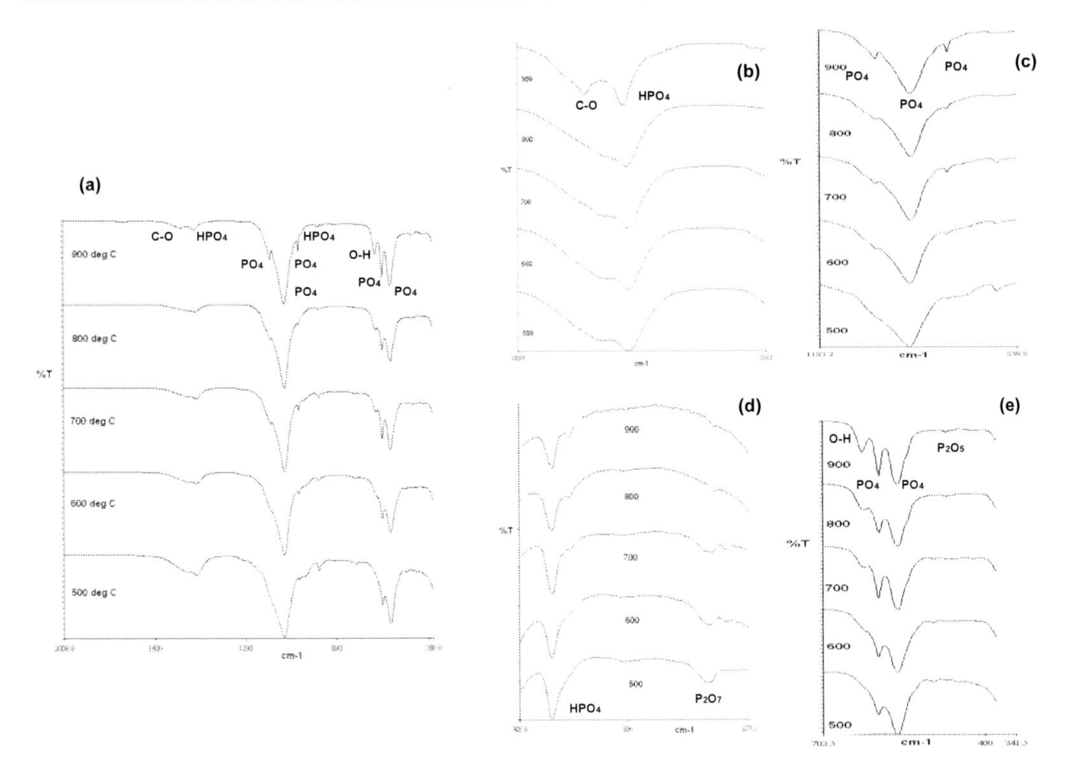

Figure 13. FT-IR spectra of 10 mol% Mg-doped BCP at various calcinations temperature.

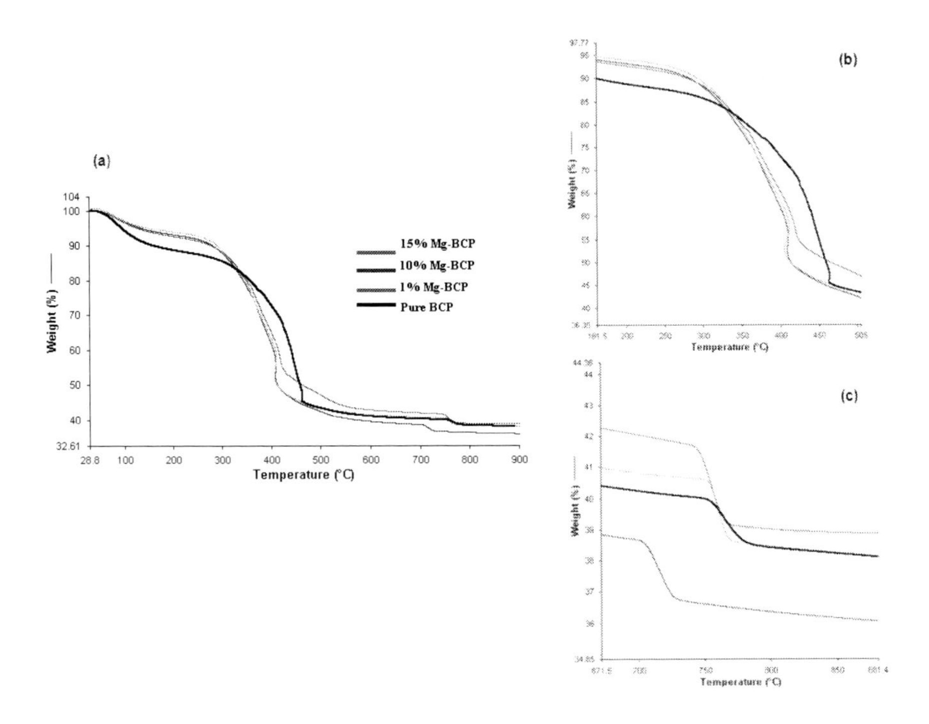

Figure 14. TG/ DTA of (a) black gel of pure BCP and Mg-doped BCP, (b) crystallization of HA and (c) formation of BCP/ Mg-doped BCP.

The TGA plots report the weight loss for the synthesized powders of pure BCP and Mg-doped BCP is illustrated in Figure 14. The graph shows three main weight losses in the samples while the temperature ranged from 30°C to 1000°C. The first drop occurs at temperature ~30 to ~250ºC which consistent with the removal of interstitial water. The plots revealed that the calcinations have changed the amorphous phase of the dried gel to crystalline phase of BCP within the temperature range between 250-500°C in which all samples have showed a similar weight loss of about ~50%. The weight loss was due to evaporation of the residual solvents and vigorous decomposition of the organic and inorganic precursors (Milev, et al., 2003) at 440°C. In a study by Landi, et al. (2008), it was stated that at ~400°C, the condensation reaction of hydrogen phosphate ions has took place. This reaction will increase the formation of TCP upon heating with the increasing Mg level doped. This statement is in good agreement with our study. The decomposition of HPO_4^{-2} and $P_2O_7^{-4}$ bands to biphasic mixtures (Kannan, 2005; Lazic, et al., 2001) is apparent by the observed similar weight losses of about ~2% for all samples within the temperature range ~740 to ~800°C. This result was in good agreement with FT-IR results as it shows the decrease in intensity of HPO_4^{-2} and $P_2O_7^{-4}$ bands as the calcination temperature increased. The difference in weight loss (250-800°) with varied Mg substitution witnessed in the TG curves is also a supporting evidence for the stabilization role of Mg in the condensation of HPO_4^{-2} (Kannan, 2005). The decomposition of HPO_4^{-2} to $P_2O_7^{-4}$ and $P_2O_7^{-4}$ to $Ca_3(PO_4)_2$ can be represented by following equations (Lazic, et al., 2001):

$$2\,HPO_4^{-2} \rightarrow P_2O_7^{-4} + H_2O \qquad \text{(Eq. 2)}$$

$$P_2O_7^{4-} + 2OH^- \rightarrow 2PO_4^{3-} + H_2O \qquad \text{(Eq. 3)}$$

In this study, we are suggesting that the decomposition of $P_2O_7^{4-}$ does not only produce TCP alone but also P_2O_5. The reaction takes place according to the reaction in equation (4):

$$3Ca_2P_2O_7 \rightarrow 2Ca_3(PO_4)_2 + P_2O_5 \qquad \text{(Eq.4)}$$

This result is in accordance to FT-IR result which revealed the presence of P_2O_5 band at 400-450 cm^{-1}. This stable phase of P_2O_5 has promotes crystal growth of the synthesized powder.

From the TGA analysis, it also shows that the decomposition temperature of HPO_4^{-2} and $P_2O_7^{-4}$ bands to biphasic mixtures occur at lower temperature for 10 mol% Mg-doped BCP. This confirmed the formation of β-TCP phase is higher at this Mg concentration as Mg act as sintering additive (Kalita, et al., 2006). This result was supported by XRD result as it shows highest intensity of β-TCP phase at 10 mol% Mg- doped BCP compared to other samples. Table 3 presents the starting temperature for crystallization of HA and the decomposition temperature of pyrophosphate to biphasic mixtures of pure BCP and Mg-doped BCP powders. The result revealed that doping of Mg up to 10 mol% has reduced the crystallization temperature of HA and formation temperature of biphasic mixtures. However, when Mg level was increased up to 15 mol%, the crystallization and formation of biphasic mixtures temperature decreased again. This is presumably due to the solubility limit of the Mg in the β-TCP phase was decreased. The expelled Mg from β-TCP started to segregate as free MgO or will be incorporated into the HA phase. This result is in good agreement with XRD and FT-IR findings.Conclusion

Mg–doped BCP powders were successfully syntheized via sol-gel method using $Ca(NO_3)_2.4H_2O$ and $(NH_4)_2HPO_4$ as the precursors and $Mg(NO_3)_2.6H_2O$ as the source of the dopant. All the powder exhibited highly crystalline BCP characteristic after calcination at 900 °C. XRD analysis revealed that β-TCP peak increased in intensity as Mg was doped, and that the Mg doping was into the β-TCP phase of BCP rather than to HA phase. With the increasing Mg concentration into the BCP, the solubility limit of the Mg in the β-TCP decreases and Mg started to segregate as free MgO or incorporated into the HA. A significant contraction has been observed in the calculated lattice parameter which may reflect the incorporation of Mg into the β-TCP phase. The differences of the lattice parameters and ***c/a*** ratio of β-TCP were much bigger than that of HA. FT-IR analysis confirmed the formation of biphasic mixtures of HA and Mg stabilized β-TCP in the synthesized powder when calcined at high temperatures as bands of HPO_4^{-2} and $P_2O_7^{-4}$ decreased.

Table 3. Formation of BCP/ Mg-BCP temperatures at various Mg content,.

Mg content (mol %)	Starting temperature for Crystallization of HA	Formation of BCP/Mg-BCP (°C)
0	466	751
1	411	745
10	410	701
15	423	746

REFERENCES

[1] Aoki, S., Yamaghuci, S., Nakahira, A. & Suganuma, K. (2004). Preparation of porous calcium phosphate using a ceramic foaming technique combined with a hydrothermal treatment and the cell response with incorporation of osteoblast like cells. *J Cer Soc Jpn*, *112*, 193-199.

[2] Bezzi, G., Celotti, G. Landi, E, La Toretta, T. M. G., Sopyan, I. & Tampieri, A. (2003). A novel sol–gel technique for hydroxyapatite preparation. *Materials Chemistry and Physics*, *78*, 816-824.

[3] Fadeev, V., Shvorneva, L. I. & Barinov, S. M. et al. (2003). Synthesis and structure of Magnesium-substituted hydroxyapatite. *Inorganic Materials*, *39(9)*, 947-950.

[4] Gibson, I. R., Ke, S. & Best, M. et al. (2001). Effect of powder characteristics on the sinterability of hydroxyapatite powders. *Journal of Materials Science: Materials in Medicine, 12*, 163-171.

[5] Gibson, I. R. & Bondfield, W. (2002). Preparation and characterization of Mg/ carbonate co- substituted hydroxyapatites. *Journal of Materials Science: Materials in Medicine, 13*, 685-693.

[6] Kalita, S. J. & Bhatt, H. A. (2007). Nanocrystalline hydroxyapatite doped with magnesium and zinc: Synthesis and characterization. *Materials Science and Engineering C, 27*, 837-848.

[7] Kannan, S., LemoS, I. A. F., Rocha, J. H. G. & Ferreira, J. M. F. (2005). Synthesis and characterization of magnesium substituted biphasic mixtures of controlled hydroxyapatite/β-tricalcium phosphate ratios. *Journal of Solid State Chemistry*, *178*, 3190-3196.

[8] Kivrak, N. & Tas, A. C. (1998). Synthesis of calcium hydroxyapatite-tricalcium phosphate (HA-TCP) composite bioceramic powders and their sintering behavior. *J Am Ceram Soc, 81(9)*, 2245-2252.

[9] Landi, E., Logroscino, G., Proietti, L., Tampieri, A., Sandri, M. & Sprio, S. (2008). Biomimetic Mg-Substituted hydroxyapatite: from synthesis to in vivo behavior. *J Mater Sci: Mater Med*, *19*, 239-247.

[10] Lazic, S., Zec, S., Miljevic, N. & Milonjic, S. (2001). The effect of temperature on the properties of hydroxyapatite precipitated from calcium hydroxide and phosphoric acid. *Thermochimica Acta*, *374*, 13-22.

[11] Li, R., Seeherman, H. & Tofighi, A. (2008). US20080096797 A1.

[12] Milev, A., Kannangara, G. S. K. & Ben-Nissan, B. (2003). Morphological stability of hydroxyapatite precursor. *Materials Letters*, *57*, 1960-1965.

[13] Ryu, H. S, Hong, K. S., Lee, J. K. & Kim, D. J. (2006). Variations of structure and composition in magnesium incorporated hydroxyapatite/ β-tricalcium phosphate. *J. Mater. Res., 21(2)*, 428-436.

[14] Sasikumar, S. & Vijayaraghavan, R. (2006). Low temperature synthesis of nanocrystalline hydroxyapatite from egg shells by combustion method. *Trends Biomater. Artif. Organs, 19(2)*, 70-73.

[15] Suchanek, W. & Yoshimura, M. (1998). Processing and properties of hydroxyapatite-based biomaterials for use as hard tissue replacement implants. *J. Mat. Research*, 13(No. 1), 94-117.

[16] Tas, A. C., Korkusuz, F., Timucin, M. & Akkas, N. (1997). An investigation of the chemical synthesis and high temperature sintering behavior of calcium hydroxyapatite (HA) and tricalcium phosphate (TCP) bioceramics. *J Mater Sci: Mater Med*, *8*, 91-96.

[17] Xiu, Z., Lu, M. & Liu, S., et al. (2005). Barium hydroxyapatite nanoparticles synthesized by citric acid sol-gel combustion method. *Materials Research Bulletin, 40*, 1617-1622.

[18] Zyman, Z. Z., Tkachenko, M. V. & Polevodin, D. V. (2008). Preparation and characterization of biphasic calcium phosphate ceramics of desired composition. *J Mater Sci: Mater Med,* 19: 2819-2825

In: The Sol-Gel Process
Editor: Rachel E. Morris

ISBN 978-1-61761-321-0

Chapter 25

HIGHLY STABLE TiO_2 SOL WITH HIGH PHOTOCATALYTIC PROPERTIES

***S. Rahim*[1], *M. Sasani Ghamsari*[1,2], *S. Radiman*[1] *and A. Hamzah*[3]**

[1]School of Applied Physics, Faculty of Science & Technology, National University of Malaysia (UKM),43600, Bangi, Selangor, Malaysia

[2]Solid State Lasers Research Group, Laser & Optics Research School, NSTRI, 11365-8496, Tehran, Iran

[3]School of Bioscience & Biotechnology, Faculty of Science & Technology, National University of Malaysia (UKM),43600, Bangi, Selangor, Malaysia

ABSTRACT

Improved sol-gel method has been applied to prepare highly crystalline TiO_2 colloidal nanoparticles with high photocatalytic reactivity. The precursor solution contained titanium (IV) isopropoxide (TTIP), 2-propanol. The sol was obtained through the hydrolysis of TTIP under the optimized conditions. FTIR, TEM and XRD were used to study the morphology, size, shape and crystallinity of prepared TiO_2 sol. Experimental results have shown that the prepared sample has an anatase structure. It has a narrow size distribution between 2–5 nm which has been confirmed by X-ray diffraction pattern. To demonstrate the photocatalytic properties of TiO_2 colloidal nanoparticles, the Methylene Blue (MB) photodecomposition test has been used. In this approach, the obtained photodecomposition reaction rate of Methylene blue under UV light is high and shows the prepared colloidal TiO_2 sample has enough potential for photocatalytic applications. The photocatalytic property of the dried sol has been evaluated by inactivation test of Escherichia coli.

Keywords: Nanomaterials, TiO_2 nanocrystals; Sol-gel preparation; Photocatalytic reactivity.

1. INTRODUCTION

Titanium dioxide is one of the most important semiconductor materials and has gained much attention due to its potential application in nanotechnology [1]. In recent years, titanium dioxide (TiO_2) nanoparticles have been widely used in different applications such as photocatalysis, nanomedicine, solar cells and catalyst support [2-4]. The colloidal form of this material has an important role in the preparation of self-cleaning glass, heat mirror, optical waveguide, and has been used as a biosensor for human serum albumin [5-7]. TiO_2 nanoparticles are usually synthesized by various preparation methods [8-10]. Among them, the sol-gel method is a low cost and reliable technique and has been widely used [11]. In the sol-gel method, several chemical reactions take place, and it is a challenge for researchers to control these reactions and obtain the controlled size, crystallinity, morphology and physical properties of the prepared nanomaterials by this method. The complication of the sol-gel method increases about TiO_2 which is due to chemical activity of Ti^{+4} ions source. Normally, titanium alkoxides are used as ion sources of Ti^{+4} in the sol-gel preparation of TiO_2 nanomaterial. These alkoxides are vigorously hydrolyzed with water or alcohols. Therefore, the preparation of highly stable TiO_2 sol by the sol-gel technique is not simple and usually, the resultant material (TiO_2) is amorphous [12]. On the other hand, the synthesis of highly stable TiO_2 sol with high crystallinity is very important from a technology point of view [13-15]. Despite several approaches for synthesis of highly crystalline TiO_2 sol [16-21], more research works must be done to establish a semi-standard preparation method of highly crystalline TiO_2 sol with potential application in different cases. In this approach, we have successfully prepared narrow size distributed colloidal TiO_2 nanoparticles with high crystallinity and the photocatalytic activity of obtained material has been tested to confirm its properties.

2. MATERIALS AND METHODS

In this work, hydrolysis and precursor solution were prepared separately in order to control hydrolysis and condensation process as well as to control colloidal size. A mixture of 2-propanol and titanium isopropoxide (TTIP) was prepared and used as precursor solution. This precursor solution was added dropwise to the hydrolysis solution (mixed solution of 2-propanol and water). Before that, the hydrolysis solution was heated at temperature 60–80°C under vigorous stirring. The resultant of these two mixtures was kept standing for 2 hours and a transparent (bluish) TiO_2 sol was obtained. Some details can be found in our last paper [22]. Fourier Transform Infrared (FTIR) spectroscopy was used for the detection of the presence of functional group in the compound. Particle size of this colloidal sample was characterized by using transmission electron microscopy (CM12 (Philips)) operating at a 100 kV accelerating voltage. The crystallinity of colloidal TiO_2 was followed by X-ray diffraction (XRD: Bruker AXS-D8). The XRD powder which was used for X-ray characterization was obtained by drying of sol in a vacuum oven at 100°C. Photocatalytic experiment was performed by degradation of methylene blue under ultraviolet (UV) irradiation. 5.0 ml of TiO_2 colloidal sample was mixed with 95 ml of a 15μM methylene blue aqueous solution for the photocatalytic activity tests. The mixture was placed on a magnetic stir plate with continuous stirring and illuminated with UV light (wavelength 365 nm) 10 cm from above. The methylene blue has an adsorption peak at 665nm in the visible spectrum. After a period of UV irradiation, the decreasing of methylene blue concentration was measured from the changes of the absorption intensity by UV-vis spectrometer (Perkin Elmer) which is reflecting to the photocatalytic activity.

3. Results and Discussion

The infrared spectrum of the synthesized colloidal TiO_2 sol in the range 4000–400 wavenumber has been shown in Figure 1. This spectrum is to identify the chemical bonds as well as functional groups in the compound. The large broad band at 3500 – 3000cm^{-1} can be assigned the OH stretching frequencies of alcohol. These bands correspond to O-H vibration of the Ti-OH group and H_2O molecules. The sharp bands at 1600cm^{-1} and 1384 cm^{-1} can be assigned to the bending mode of nitrate group as a result of nitric acid addition. The low energy region (below 1000cm^{-1}) indicated the bands due to stretching mode of Ti-O and Ti-O-Ti bond of a titanium dioxide network. The UV-Vis absorption of colloidal sample was shown in Figure 2. Inset in Fig 2 shows a photograph of colloidal TiO_2 nanoparticles, which were prepared by sol gel method. It is transparent and bluish in color showing a stable colloid system [23]. It can be seen that the absorption edge of anatase TiO_2 NCs is at about 325 nm. The corresponding energy of the absorption edge of anatase TiO_2 NCs has been calculated by the equation $E = h\upsilon$. It is about 3.81 eV which is 0.71 eV higher than that of bulk anatase TiO_2 (~3.1eV). This blue shift of the absorption edge is attributed to the quantum confinement effect [16]. TEM image of the sample showed that the colloidal TiO_2 nanoparticles have narrow size distribution and no agglomeration occurred as seen in Fig 3. The size distribution of the sample was in the range 2–5 nm as shown in Figure 3. To deal with the photocatalytic properties of colloidal TiO_2, we have used decolorization (mineralization) mechanism of methylene blue by using UV-vis absorption spectra. The maximum absorption wavelength of methylene blue at 665 nm has been recorded in terms of time of UV light irradiation. It could be noted that the absorption of methylene blue is decreasing (mineralized) with increasing of irradiation time by UV light and highly depends on experimental conditions (pH, MB concentration, temperature) [24,25]. In order to have a baseline for determination of MB concentration after irradiation and in the present of colloidal TiO_2, we have recorded the absorbance of MB solutions on stabilized experimental conditions without any photocatalytic material. The relationship between absorbance and MB concentration was shown in Figure 4. Beer-Lambert law is significant to determine the changes in MB concentrations because the absorbance (A) that we get from UV-vis spectrometer is directly proportional to MB concentration as shown below:

$$A = \varepsilon bc \tag{1}$$

where ε is molar absorbtivity with units of L.mol^{-1}.cm^{-1}, b is the path length of the sample and c is the concentration of the compound in solution, expressed in mol.L^{-1}. According to equation 1, the average obtained absorbtivity for MB was $2.3\times10^{+6}$ L.mol^{-1}.cm^{-1}. The photo-degradation reaction rate constant k in this approach was gained after fitting by power law equation. Jing et. al. [26] has used zero order reaction which is most suitable to describe the change of MB concentration:

$$-dc/dt = k, \quad (2)$$

where k is the reaction rate constant. By integrating Eq (2), it becomes

$$\mathrm{Ln}\,(C_o/C) = kt \quad (3)$$

where C_o is the initial concentration of the MB solution and C is the MB solution concentration at t time. From figure 5, we have found that the kinetic coefficient of degradation (k) was 0.0044/min. This rate is higher than reported rate by other [8]. Photocatalysis can be typically divided into three stages. First step is associated with generation of electron/hole ($\mathbf{e}^-_{\mathbf{CB}} - \mathbf{h}^+_{\mathbf{VB}}$) pairs by irradiation with light having photonic energy greater or equal to the existing band gap (~3.1 eV for titania). In the second step, the generated carries are migrating from the crystalline interior to the surface of photocatalytic semiconductor material. The third step will be become by redox interactions at the particle surface between the contaminant and the free electron/hole pairs that survive the migration [27]. The generation reaction of electron-hole due to light irradiation has been shown in Eg. (4). It is broadly accepted that photogenerated holes ($\mathbf{h}^+_{\mathbf{VB}}$) can contribute in the direct oxidation of H_2O and OH^- ions reaction and leads to production of hydroxyl radicals ($OH^\bullet$) [28]. This hydroxyl radicals ($OH^\bullet$) are produced from (bulk solution) or terminal hydroxyl groups (catalyst surface) as shown in Eqs. (5) and (6) [28]. Superoxide radicals often result from the interplay between the photogenerated electrons ($\mathbf{e}^-_{\mathbf{CB}}$) and molecular oxygen. The addition of peroxides increases the occurrence of reaction (7) and the presence of hydroxyl radicals, thereby increasing the degradation kinetics [29,30].

$$TiO_2 + h\nu \rightarrow TiO_2 + e^-_{CB} + h^+_{VB} \quad (4)$$

$$h^+_{VB} + OH^- \rightarrow OH^\bullet \quad (5)$$

$$h^+_{VB} + H_2O \rightarrow OH^\bullet + H^+ \quad (6)$$

$$e^-_{CB} + O_2 \rightarrow O^{-2} \quad (7)$$

As it has been explained by Coutinho [30], the degradation of the MB can be described in terms of the elementary mechanisms as shown in Eqs. (8)–(10). The direct reaction of the dye with photogenerated holes can be followed on the base of a process which is similar to the photo-Kolbe reaction or oxidation through successive attacks by hydroxyl radicals or superoxide species [31]. The hydroxy radical in particular is an extremely strong non-selective oxidant that has shown to lead to the partial or complete oxidation of many organic chemicals [32].

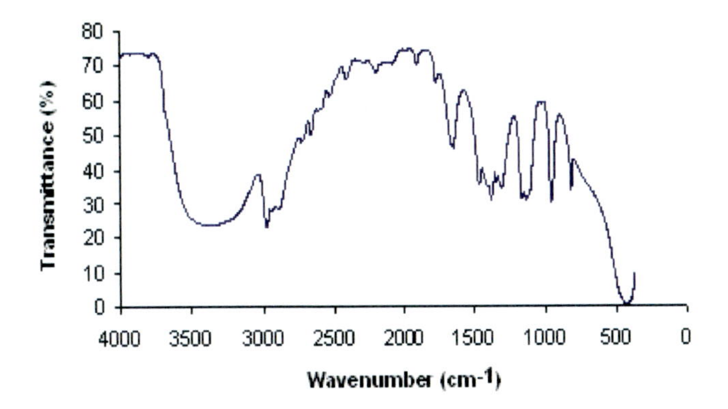

Figure 1. FT-IR spectrum of colloidal TiO_2 nanocrystals.

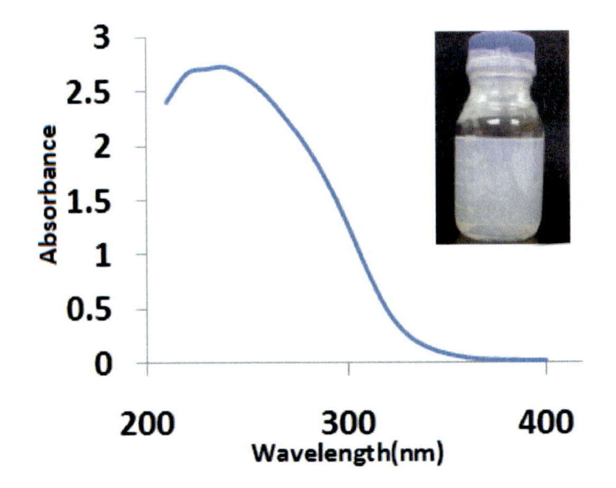

Figure 2. Absorption spectrum of colloidal TiO_2 nanocrystals.

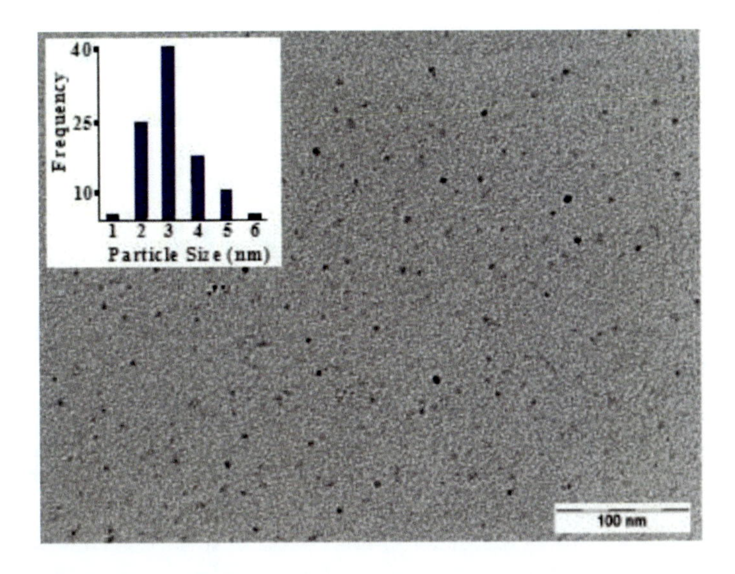

Figure 3. TEM image for colloidal TiO_2 sample.

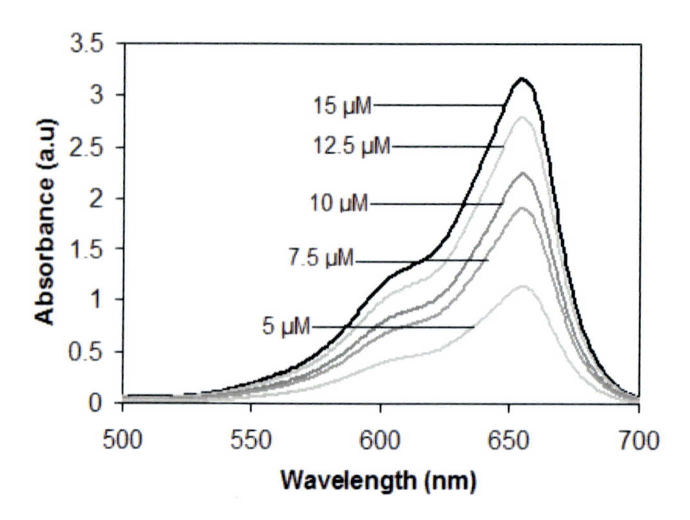

Figure 4. Absorption spectra of MB at different concentrations.

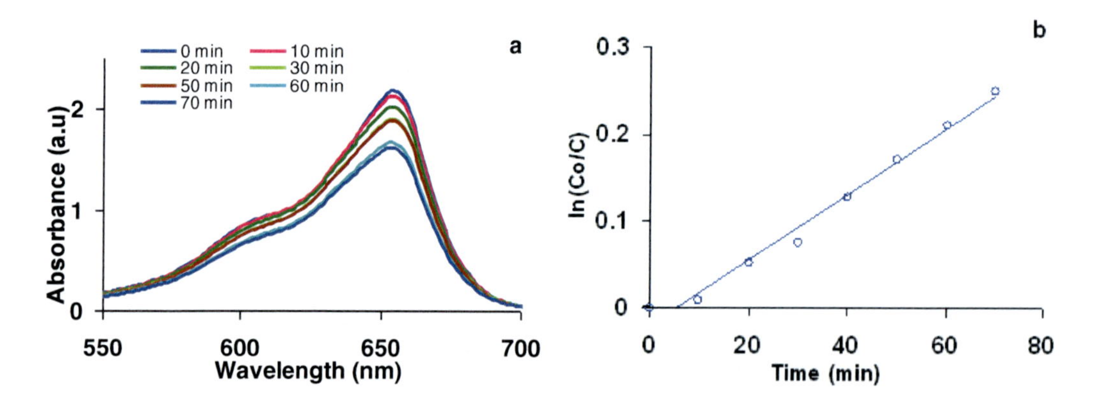

Figure 5. (a) Absorption spectra of MB degradation reaction by colloidal TiO_2 at different UV-exposure times. (b) Determination of the rate constant of the MB test for the colloidal TiO_2 nanoparticles.

$$OH^{\bullet} + MB \rightarrow \text{colorless product} \tag{8}$$

$$h_{VB}^{+} + MB \rightarrow \text{colorless product} \tag{9}$$

$$O^{-2} + MB \rightarrow \text{colorless product} \tag{10}$$

These reactions indicated that the MB molecules were attacked in the presence of colloidal TiO_2. In this process, the organic day (methylene blue) have to be adsorbed on the catalyst surface where it is either attached by reactive oxygen species (ROS) or/and it is directly oxidized by donating electrons to trapped holes in the photocatalyst surface. As we know, photocatalytic activity by TiO_2 initiated by an absorption of a photon (excitation) from UV light energy where the band gap energy is higher or equal to the TiO_2 semiconductor (3.1eV for anatase TiO_2), producing electron-hole pairs. Instead, just few nanoseconds are needed for these electron and hole pairs to recombine on the surface. They can also be trapped in the surface state where they can react with organic molecules especially oxygen. Therefore, the electron and hole pairs can initiate redox reaction leading to decomposition of methylene blue (organic compound). Ohtani et al. [33], shown that the amorphous of TiO_2 nanoparticles can not present photocatalytic reactivity. These result confirms that our sample consist of crystalline TiO_2. There is another reason that can be considered as a criterion to confirm the crystallinity of our sample. This criterion is pH. Generally the observed pH dependence of the photocatalytic reaction can be understood in terms of the adsorption of both the molecule of medium material and hydroxyl ions onto the charged TiO_2 surface [34]. When TiO_2 nanoparticles have amorphous structure, the density of surface charge is very low and photocatalytic properties are negligible. Therefore, the observation of photocatalytic behavior in very low or high pH can not confirm high crystallinity of photocatalytic material. Coutinho [30] has explained that in the presence of water the surfaces of the metal or semiconductor oxides are hydroxylated. Depending on the pH, these surfaces groups may add or attract protons. Therefore, the rate of photocatalytic reaction increases if we use acidic or basic solutions. Our photocatalytic experiment has been down in alcohol solution. The pH of solution was 8. It means that our sample has high crystallinity to show the photocatalytic reactivity in non basic or acidic solution.

In order to understand the scientific routes of these results we have to present a brief description of sol-gel process. As it has been explained in last our paper [35], the titanium has two different coordination number in the oxide crystal (6-fold Ti-coordination), and in the initial alkoxide precursor (4-fold Ti-coordination). However, when the alkoxide reacts with water, the metal ion increases its coordination by using its vacant d-orbitals to accept oxygen lone pairs from nucleophilic ligands (such as OH groups) by coordination expansion. Consequently, titanium ions in solution exist as 6-fold coordinated structures with $Ti(O)_a(OH)_b(OH_2)_{6-a-b}$ composition where a and b depend on the processing conditions. If the reaction is performed in an acidic medium, only -OH_2 and -H groups will be present. Thus, the composition reduces to $[Ti(OH)_x(OH_2)_{6-x}]^{(4-x)+}$. These six-fold structural units undergo condensation reaction to become the octahedra in the final crystalline structure. In the present study, the hydrolysis and condensation reactions occur at a low pH and a small amount of water has been added to control the hydrolysis reaction. Therefore, all produced water by condensation reaction has been used by hydrolysis reaction. As a result, the overall chemical reaction can be given as;

$$Ti(OR)_4 + 2H_2O \rightarrow TiO_2 + 4ROH \tag{11}$$

Consequently, the real structure has been Ti-O-Ti [36]. This structure is responsibility for the catalytic reactions and properties. In order to evaluate the crystallinity of dispersed TiO_2 nanomaterial inside the sol, the sample has been dried and the X-ray diffraction pattern of extracted TiO_2 powder from the sol has been recorded and illustrated in Figure 6. From that figure one can find that the prepared TiO_2 nanopowder has broad peak which is indicating that TiO_2 nanoparticles have nanosize. The peaks intensity was not too high just because the sol was dried only at low temperature (100°C) for 2 hours. The pattern also confirmed that the extracted TiO_2 powder contains the single phase which is anatase. The formation of anatase phase is significant for better catalysis activity compare with brookite or rutile phase [37]. An average size of 4 nm was calculated by using Scherrer equation [38] which has good agreement with TEM image of highly stable TiO_2 sol. It is well distributed as no agglomeration occurred. In order to evaluate the photocatalytic properties of extracted TiO_2 powder from colloidal sample, *Escherichia coli (E. coli)* were used for photocatalytic degradation studies. Single colonies were isolated from nutrient agar plate (5g/L Yeast Extract, 10g/L Tryptone, 15g/L agar) to inoculate 10 mL of nutrient agar liquid media (10g/L Tryptone, 5 g/L Yeast Extract) in a 250 mL Erlenmeyer flask. The flask was incubated overnight at 37^0C on an incubator shaker at 150 rpm around 18-24 hours. Cells were harvested by centrifuging at 4000 rpm at 4°C for 15 minutes. Then, the pellets of *E.coli* were washed twice using normal saline and suspended in normal saline. Bacterial cell density (OD≈ 0.5) in liquid media was estimated using a UV-vis spectrophotometer (at $\lambda = 550$ nm). In photocatalytic experiments, 0.1, 1.0 and 2.5 g/L of aqueous TiO_2 suspension in normal saline was prepared immediately prior to photocatalytic reaction and kept in the dark room. 10% of fresh standard inoculums (approximately 10^6 cfu/mL) were inoculated to 200 mL glass beakers containing 80 mL of sterilized normal saline and 10 mL of TiO_2 suspension. The slurry of TiO_2-cells was intensively mixed with magnetic stirrer and illuminated with fluorescent light (240V,50Hz) at position 20 cm from above as shown in Figure 7. An *E coli* suspension without TiO_2 was illuminated as a control. Samples were taken at 1 hour intervals for 6 hours. Viable concentration of *E coli* was measured with spreading plate method after serial dilution of the sample in normal saline. All plates were incubated at 37°C around 18-24 hours and the colonies were counted using colony counter. An inactivation rate was determined as survival rate of bacteria that was defined by a simple formula shown as follows:

$$\text{Survival rate} = \frac{x}{y} \times 100 \tag{12}$$

where x and y are the cfu/mL before and after exposure, respectively. Figure 8 shows the graph of photocatalytic inactivation of *E.coli* by three titanium dioxide concentrations which are 0.1, 1.0 and 2.5 g/L. Experiments were also done in the dark and without titanium dioxide as control. The survival percentage of *E.coli* was decreasing with illumination time for all TiO_2 concentrations. However, the optimum inactivation of *E.coli* (10^5 cfu/mL) was achieved in the presence of 1.0 g/L titanium dioxide where is 60 % of *E.coli* was inactivated. This value does not correspond to a total absorption of the lamp-emitted photons and also does not correspond to the optimum titanium dioxide concentration [39]. At high concentration of titanium dioxide, the light scattering and shadowing effect is dominant, reducing the extent to which the fluorescence light can reach all the particles in suspension [40]. Furthermore, Rafael et al [41] suggested that this phenomenon is dependent on the experimental setup. Meanwhile, Sunada et al [42] found that the mechanism of photocatalyst through *E.coli* illumination with TiO_2 surface could be divided into three stages: (1) the outer membrane of *E.coli* was attack and decomposed partially by reactive species such as OH^-, O_2 and H_2O_2; (2) disordering of the inner membrane leading to the peroxidation of the membrane lipid thus killing the cell and; (3) decomposition of the dead cell. If the fluorescence illumination continued for a sufficiently long time, the *E.coli* was found to mineralize completely into CO_2, H_2O and other mineral compound.

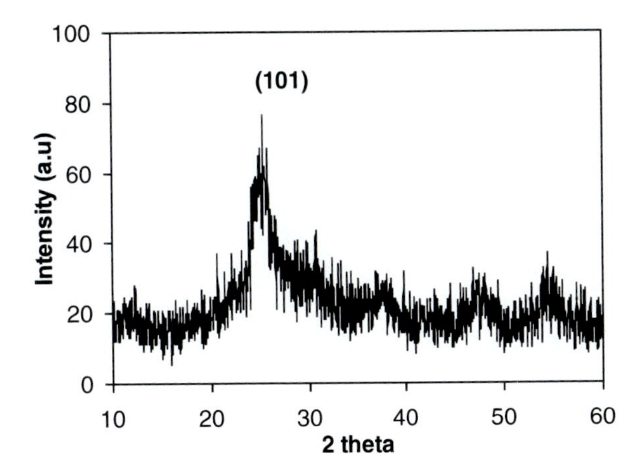

Figure 6. XRD pattern of the prepared colloidal TiO_2.

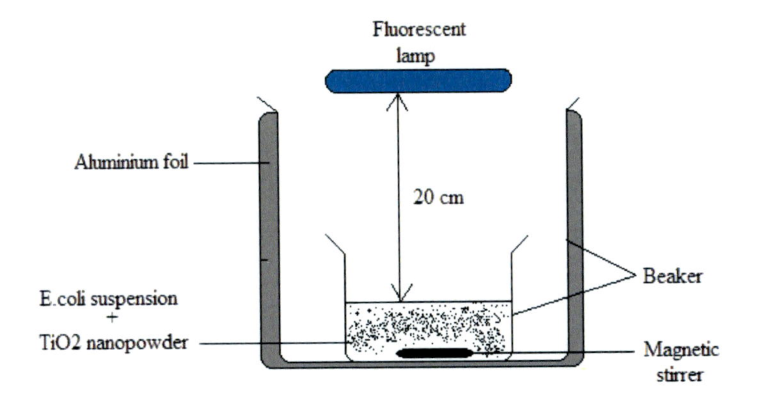

Figure7. Experimental setup for inactivation test of *Escherichia coli* by TiO_2 nanocrystals reaction under fluorescent lamp.

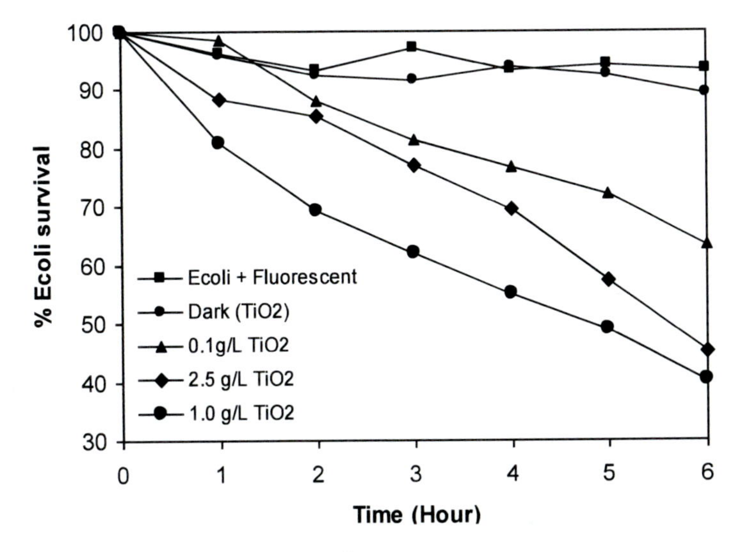

Figure 8. Photocatalytic inactivation of *E.coli*, 10^5 cfu/mL, control (dark and without TiO_2) and reaction in the absence of TiO_2.

CONCLUSION

In this study, the highly stable TiO_2 sol was successfully prepared by the sol gel method. Experimental results have shown that the prepared sample has an anatase structure, which has been confirmed by X-ray diffraction pattern. The size distribution of prepared material is in the range of 2–5 nm and no agglomeration occurred. The photodecomposition test of methylene blue has proved that the prepared colloidal TiO_2 nanoparticles by our method have high crystallinity and good photocatalytic properties. In conclusion, the results reported above suggest that titanium dioxide sol that synthesized in our sol gel process may be a viable material for inactivation of E.coli by photocatalysis process. The optimum inactivation of E.coli (10^5 CFU/mL) was achieved in the presence of 1.0 g/L titanium dioxide.

ACKNOWLEDGMENTS

This work was supported by University Kebangsaan Malaysia (Grant No; UKM-OUP-NBT-27-138/2008).

REFERENCES

[1] Ch.O., Robichaud, A. E., Uyar, M. R., Darby, L. G. & Zucker, M. R. (2009). Wiesner, *Env. Sci. Tech., 43*, 4227–4233.

[2] Thevenot, P., Cho, J., Wavhal, D., Timmons, R. B. & Tang, L. (2008). Nanomedicine: Nanotechnology, *Biology and Medicine*, *4*, 226-236.

[3] Wang, H., Liu, Y., Xu, H., Dong, X., Shen, H., Wang, Y. & Yang, H. (2009). *Renewable Energy*, *34*, 1635-1640.

[4] Panagiotou, G. D., Petsi, T., Bourikas, K., Kordulis, C. & Lycourghiotis, A. (2009). *J Catalysis, 262*, 266-279.

[5] Liu, Z., Zhang, X., Murakami, T. & Fujishima, A. (2008). *Solar Energy Materials and Solar Cells*, *92*, 1434-1438.

[6] Karasiński, P. & Rogoziński, R. (2008). *Optics Communications*, *281*, 2472-2480.

[7] Kathiravan, A. & Renganathan, R. (2008). Colloids and Surfaces A: *Physicochem. Eng. Aspects*, *324*, 176-180.

[8] Thongsuwan, W., Kumpika, T. & Singjai, P. (2008). *Current Applied Physics*, *8*, 563-568.

[9] Kim, II. D., Rothschild, A., Yang, D. J. & Tuller, H. L. (2008). *Sensors and Actuators B 130*, 9-13.

[10] Euvananont, C., Junin, C., Inpor, K., Limthongkul, P. & Thanachayanont, C. (2008). *Ceram. Inte.*, *34*, 1067-1071.

[11] Aegerter, M. A., Almeida, R., Souta, A., Tadanaga, K., Yang, H. & Watanabe, T. (2008). *J Sol-Gel Sci Technol, 47*, 203-236.

[12] Livage, J., Henry, M., Sanchez, C. (1988). *Progress Solid State Chemistry*, *18*, 259-341.

[13] Wu, J., Lü, X., Zhang, L., Huang, F. & Xu, F. (2009). *Eur. J. Inorg. Chemi.*, *19*, 2789-2795.

[14] Liau, L. C. K. & Chou, W. W. (2009). *Microelectronic Engineering*, *86*, 361-366.

[15] Luo, W., Li, R., Liu, G., Antonio, M. R. & Chen, X. (2008). *J. Phys. Chem.* C 112, 10370–10377.

[16] Wang, P., Wang, D., Li, H., Xie, T., Wang, H., Du, Z. (2007). *J. Colloid and Inter. Sci.* 314, 337–340.

[17] Lu , Y., Hoffmann, M., Yelamanchili, R. S., Terrenoire, A., Schrinner, M., Drechsler , M., Möller, M. W., Breu, J. & Ballauff, M. (2008). *Macromol. Chemi. and Phys.*, *210*, 377-386.

[18] Kawasaki, S. I., Xiuyi, Y., Sue, K., Hakuta, Y., Suzuki, A. & Arai, K. (2009). *J. Supercritical Fluids, 50*, 276-282.

[19] Colomer, M. T., Guzman, J., Moreno, R. (2010). *J. Am. Ceram. Soc.*, *93*, 59-64.

[20] Li, S., Li, Y., Wang, H., Fan, W. & Zhang, Q. (2009). *Eur. J. Inorg. Chem.*, 4078-4084.

[21] Chen, Z., Zhao, G., Li, H., Han, G. & Song, B. (2009). *J. Am. Ceram. Soc.*, *92*, 1024-1029.

[22] Mahshid, M., Askari, M., Sasani Ghamsari, M., Afshar, N., Lahuti, S. (2009). *J. Alloys & Comp.*, *478*, 586-589.

[23] Sasani Ghamsari, M. & Bahramin, A. R. (2008). *Materials Letters*, *62*, 361-364.

[24] Hoffmann, M. R., Martin, S. T., Choi, W. & Bahnemann, D. W. (1995). *Chem. Rev.* 95, 69-96.

[25] Xu, N. , Shi, Z., Fan, Y., Dong, J. Shi, J. & Michael, Z. C. (1999). *Hu, Ind. Eng. Chem. Res.*, *38*, 373-379.

[26] Jiang, X. & Wang, T. (2009). *Environment Sci. & Tech.*, *41*, 4441-4446.

[27] Yang, H., Zhang, K., Shi, R., Li, X., Dong, X. & Yu, Y. (2006). *J. Alloys Compounds, 413*, 302-306.

[28] Ishibashi, K. I., Fujishima, A., Watanabe, T. & Hashimoto, K. (2000). *Electrochem. Commun.*, *2*, 207–210.

[29] Parida, K. M., Sahu, N., Biswal, N. R., Naik, B. & Pradhan, A. C. (2008). *J. Colloid Interface Sci.*, *318*, 231-237.

[30] Coutinho, C. A. & Gupta, U. K. (2009). *J. Colloid Interface Sci.*, *333*, 457-464.

[31] Yang, D., Ni, X., Chen, W. & Weng, Z. (2008). *J. Photochem. Photobiol., A: Chem. 195*, 323–329.

[32] Sarathy, S. R. & Mohseni, M. (2007). *Environ. Sci. Technol.*, *41*, 8315-8320.

[33] Ohtani, B., Ogawa, Y. & Nishimoto, S. (1997). *J. Phys. Chem. B*, *101*, 3746-3752.

[34] Wang, Y. (2000). *Water Res.*, *34*, 990-994.

[35] Mahshid, S., Askari, M., Sasani Ghamsari, M. (2007). *J Mate. Proc. Tech.* 18, 296-300.

[36] Sugimoto, T., Okada, K. & Itoh, H. (1997). *J Colloid Interface Sci.*, *193*, 140-143.

[37] Berry, R. J. & Mueller. M. R. (1994). *Microchemi. J.*, *50*, 28.

[38] Xiong, G., Zhi, Z., Yang, X., Lu, L. & Wang, X. (1997). *J. Mater. Science Lett.*, *16*, 1064.

[39] Benabbou, A. K., Derriche, Z., Felix, C., Lejeune, P. & Guillard, C. (2007). *Appli. Catal. B*: *Env.*, *76*, 257-263.

[40] Coleman, H. M., Marquis, C. P., Scott, J. A., Chin., S. S. & Amal, R. (2005). *Chem. Eng. J. 113*, 55-63.

[41] Rafael, V. G., Javier, M., Carlos, S. & Cristina, P. (2009). *Catalysis Today*, in press.

[42] Sunada, K., Watanabe, T. & Hashimoto, K. (2003). *J. Photochemi. and Photobio*. A: Chem., *156*, 227-233.

In: The Sol-Gel Process
Editor: Rachel E. Morris
ISBN 978-1-61761-321-0

Chapter 26

SOL STABILIZATION IN SOL-GEL PROCESS

***M. Sasani Ghamsari*[1,3]*, H. Mehranpour*[2] *and S. Radiman*[3]**
[1]Solid State Lasers Research Group, Laser & Optics Research School, NSTRI, Tehran, Iran
[2] Faculty of Materials Science & Engineering, Sharif University of Technology, Tehran, Iran
[3]Applied Physics School, Faculty of Science & Technology, National University of Malaysia (UKM), Bangi, Malaysia

1. INTRODUCTION

The sol-gel process as a chemical route is widely used in the fields of materials and engineering. Historically, the sol-gel techniques have been developed during the past 40 years as an alternative process for glasses and ceramics production [1]. It is reported that the sol-gel process had been introduced by Ebelman [2], who had prepared a transparent SiO_2 thin film from slow hydrolysis of an ester of silicic acid in 1845. In comparison with other techniques such as CVD or sintering, the sol-gel technique is a very simple process with considerable advantages. This process is low cost, has high controllability and potential application for preparation of different materials and devices. These specifications of the sol-gel process caused it to be employed as a unique method for preparation of metallic, ceramic, hybrid, composite, fiber, and glass substances with different size and morphology with a wide range of applications. During several decades, different forms of the sol-gel process have been introduced by researchers. These methods are different in some aspects of sol-gel principles. But, the preparation of sol or gel as an intermediate stage of the process is common in all of them. In figure (1), the different kind of sol-gel processes are shown. The hydrolysis and condensation of metal alkoxide can be completely considered as a fully sol-gel process. In such a method, which is used primarily for the fabrication of materials (typically a metal oxide), the process is started by preparation of a chemical solution which acts as the precursor for an integrated network (or *gel*) of either discrete particles or network polymers. Typical precursors are metal alkoxides and metal chlorides, which undergo various forms of hydrolysis and polycondensation reactions.

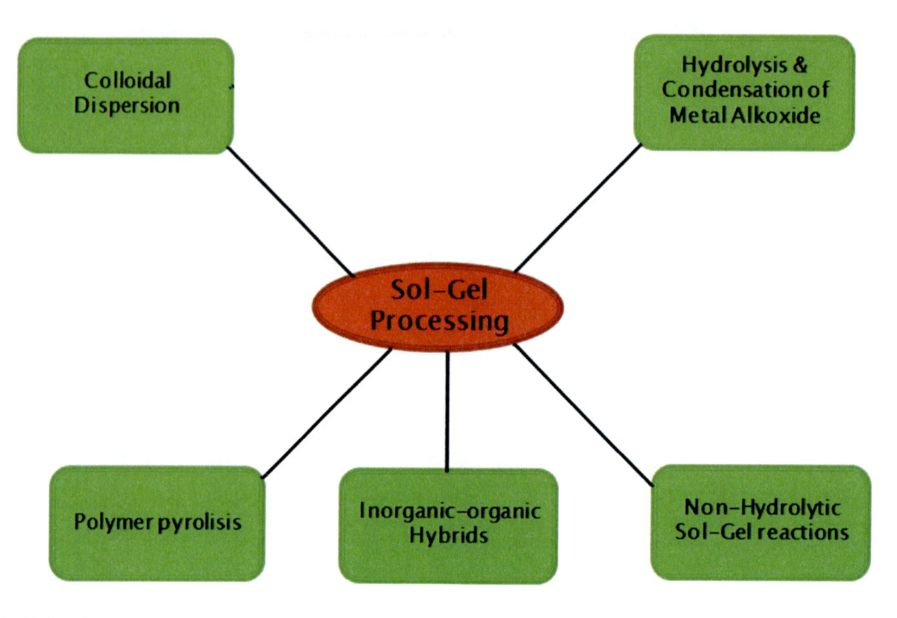

Figure 1. Sol-gel processes.

On the base of sol-gel chemistry, the formation of a metal oxide involves connecting the metal centers with oxo (M-O-M) or hydroxo (M-OH-M) bridges [3]. This connection will cause the generation of a group of metal-oxo or metal-hydroxo polymers in solution. Thus, the sol evolves towards the formation of a gel-like diphasic system containing both a liquid phase and solid phase whose morphologies range from discrete particles to continuous polymer networks. This process is illustrated in figure (2).

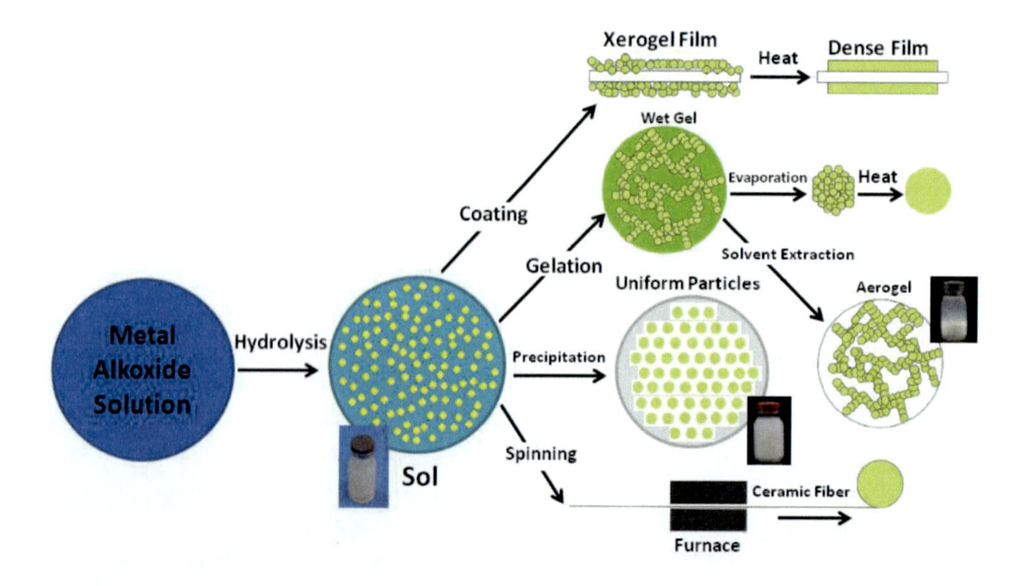

Figure 2. Metal alkoxoide sol-gel process for preparation of different materials and devices.

The prepared sol by this method can be employed for production of different materials such as powders, fibers and aerogels or devices such as thin and thick films. From a technological point of view, the ability to use one process in many high-tech applications is very important, and this is the main reason for global approaches to the sol-gel technique [4]. In figure (3) the number of publications per year on sol-gel is shown [5].

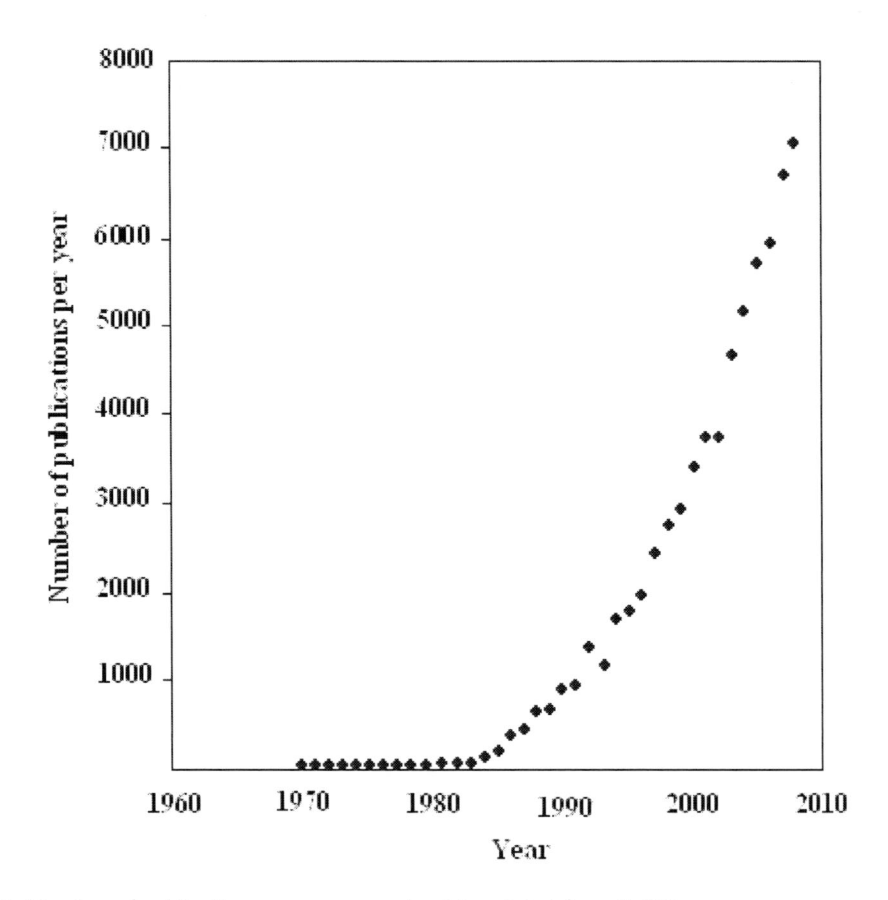

Figure 3. Number of publications per year on sol-gel [reprinted from Ref.5].

It can be found from this figure that the scientific efforts for using the sol-gel technique has exponentially increased with time. In order to better understand the sol-gel process, we present an example. The single crystalline composition of $3Y_2O_3$: $5Al_2O_3$ which is commonly called yttrium aluminum garnet (YAG) with the cubic garnet structure and doped with a transition metal or lanthanide elements, plays an important role in laser technology and is widely used as an important solid-state laser material and window material for a variety of lamps. The first step for the synthesis of YAG single crystal begins with the preparation of YAG powder from Al_2O_3 and Y_2O_3 powders by a solid-state reaction, which is usually done by extensive mechanical mixing and lengthy heat treatments above 1600°C [6]. The prepared powder is changed to YAG single crystal by the Czochralski method, which is a very expensive method. Recently, the sol–gel process had been developed for preparation of nanocrystalline YAG powder using mixtures of inorganic salts of the respective elements, and YAG single crystal is produced by a simple heat treatment at 1500°C [7]. The cost of the sol-gel method is very low in comparison with this thermal method. Therefore, the sol-gel process helps us to save time and energy, which are very important from a mass production point of view. There are two other important sol-gel processes that are used for sol preparation. The first is the non-aqueous sol-gel method and the second one is the colloidal technique [8]. In the nonaqueous sol-gel technique, an organic solvent which has exclusion of water is used for the transformation of the precursor to take place. Therefore, instead of inorganic metal salts and metal alkoxides which are used in aqueous sol-gel chemistry, the number of different precursors such as metal acetates and metal acetylacetonates are available [8]. In addition the organometallic compounds can also be put on the list of non-aqueous sol-gel precursors. As it has been defined, the organometallic compounds have a direct metal-carbon bond and this is different with metal organic compounds that have a metal-hetero atom (usually oxygen) bond. Then, it has been accepted that the process is rather based on thermal decomposition than sol-gel [8]. In non-aqueous sol-gel technique, two main research directions were followed. In the first one, the preparation of metal oxide gels is the main goal. In the second one, the synthesis of metal oxide powders was the target. Synthesis of titanium oxide and ZnO nanoparticles can probably be considered as earlier examples of the second approach [9,10]. These efforts for producing nanomaterials by non-aqueous sol-gel made a foundation for the intensified search of new synthesis routes to metal oxide nanoparticles under nonaqueous and/or nonhydrolytic reaction conditions. Nowadays, the non-aqueous/non-hydrolytic sol–gel is applied to prepare a family of metal oxide nanoparticles. The application of this method for preparation of simple binary metal oxides, more complex ternary or multi-metal and doped systems has immensely grown [11]. In the case of the colloid, the volume fraction of particles (or particle density) may be so low that a significant amount of fluid may be needed to remove initially for the gel-like properties to appear. This can be accomplished in any number of ways. The simplest method is to allow time for sedimentation to occur, and then decanting the remaining liquid. Centrifugation can also be used to accelerate the process of phase separation. In next section we will present some important applications of sol-gel method for preparation of thin films which are used in optical and optoelectronic devices.

2. Sol-Gel Applications

Due to simplicity, cost and reproducibility and reliability of this technique, it is used for preparation and fabrication of many materials and devices. As has been discussed by Belleville [12], polymeric or colloidal sol-gel solution are two the type that can be chosen for the preparation of sol-gel coatings. In both of these coating solutions the hydrolysis and the condensation of metallic precursors (salts or alkoxides) are used to prepare the colloidal or polymerized particles. The size of colloidal particles is normally between 1 nm to a few hundreds of nm and the dispersed polymeric species in a liquid medium have different sizes which depend on the catalyst used in sol-gel reactions. The colloidal route is convenient for preparing unstressed thick films or multilayer coatings, thus preventing crazing and peeling phenomena. In figure (4) the schematic applications of sol-gel process has been shown. It can be seen from this figure that the use of sol-gel technique in optics, photonics and electronics is very popular and thick or thin film of optical or electronic materials are synthesized by this method.

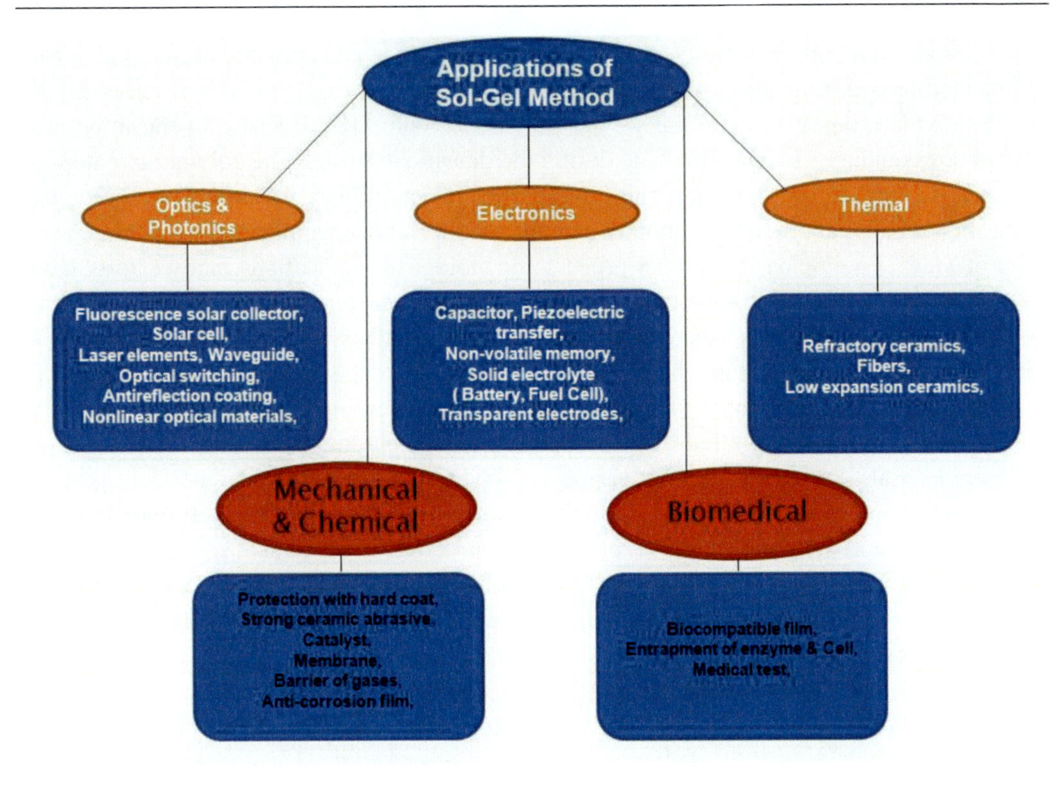

Figure 4. Applications of sol gel method [13].

In optics, the employment of sol-gel process to fabricate waveguides, thin film coating, optical switching and nonlinear optical materials is very important because, it is very simple method with high controllability for preparation of material with designed properties and characteristics. In the emergence of nanotechnology the stability of properties in nanoscale and synthesis of nanomaterials with high lifetime and mass production of devices with high yield are current challenges which can be surmounted by high controllable and low cost process such as sol-gel method. Anti-microbial, hygiene and anti-fingerprinting nanocoatings, self-cleaning, superhydrophilic nanocoating for metals, glasses and polymers, anti-fog nanocoating, multifunctional nanocoating for scratch and abrasion resistance, corrosion protection nanocoating and many other kinds of coating are several examples of nanothin film production by sol-gel technique. These examples are very good illustrations about potential applications of stabilized sols. In all these nanocoatings, we need to have a very stable sol with reasonable physical and chemical properties such as viscosity, purity, particle size distribution, stability, lifetime, and so on. Optics is a one of the fields that the sol-gel is intensely employed to fabricate optical devices. The waveguides, antireflection and high-reflection optical coatings, optical switching and nonlinear optical thin film materials are popular sol-gel derived thin films. The properties of prepared sol for these applications have a critical role on the optical properties of obtained materials by sol-gel process [14-16]. For other instance, the perovskite thin film materials are the most interesting materials widely used for optical dielectric layers. The employment of sol-gel technique for preparation of thin film of materials highly depends to sol-gel chemistry and stabilization of prepared sol [17]. In this way, the highly laser-damage resistant optical coatings is very important and can be generally produced by an elevated purity control, but their abrasion resistance after room deposition is often poor. Generally, a good abrasion resistance and a high refractive index can be obtained by a densification step (UV irradiation or thermal curing) which is necessary to finally provide a dense layer in the sol-gel polymeric route [18]. Because of the important internal stress induced by the shrinkage during the densification step, the polymeric materials are usually not utilized for preparing thick films or multilayer stacks [12]. Biological application of sol–gel products is another example of sol stablized application from biological point of view. To employ the sol-gel process for biological applications many efforts have extensively been done during the past decade [19-21]. Bioreactors, biosensors and bio-chips are devices that can be easily made via the sol-gel process. It has been demonstrated that sol–gel can be served as a applicable technique for preparation of the bioactive materials such as glass, ceramic, cements or coatings and host matrices which are used for biological materials, from proteins to whole cells [22-31]. In addition, the employment of sol-gel technique to prepare materials for production of various biochemicals, for growing and detecting of viruses and for production of antiviral vaccines, are other kinds of offere useful applications of sol–gel materials. The surface properties of sol-gel materials are important to biocompatibility of cell. There are several important parameters such as type and the density of surface charge, balance between the hydrophilicity and the hydrophobicity on surface, the chemical structure and functional groups, surface topography and roughness that affect the surface properties of sol-gel prepared thin film for the biological applications. It means that the preparation of biomaterials by sol-gel needs to satisfy biological requirements and new items must be considered [32]. Functionalization of sol-gel derived materials is one of the important strategies in sol-gel process. The processes of functionalization are very different. For example, the mesostructured materials are one family of sol-gel derived materials that can be prepared by combination of sol-gel chemistry and self-assembly routes using templates. The tuning in the pore size, shape and material architecture of the mesostructured materials can be obtained by using of larger inorganic templates, organic templates or biotemplates [12]. In addition, several methods are served to prepare the mesostructured materials with designed properties, for instance, the evaporation-induced self-assembly (EISA) procedure is a traditionally technique for production of a series of the mesostructured materials family. This kind of mesostructured materials are prepared by mixing an inorganic or/and hybrid precursors in the presence of surfactant, water, solvent and catalysts. Sol stability and uniformity are two important factors that affect the final properties of the mesostructured materials by this method. Transparent hierarchical hybrid functionalized membranes are considered as new kinds of this group of sol-gel derived materials and synthesized by situ generation of mesostructured hybrid phases inside a non-porogenic polymeric host-matrix. EISA in a non-porogenic hydrophobic polymeric medium is a novel concept to allow synthesizing of new hybrid materials characterized by hierarchical structures with macro- and mesoscopic domains. Following this newly proposed route, a large variety of macro-organizations can be obtained under one-pot synthesis conditions, with specific mesostructures, depending on the surfactant/polymer/solvent affinities [12]. It has been demonstrated that the control of the multiple affinities existing between organic and inorganic-components renders it possible to manage the nanobuilding and the length scale partitioning of hybrid nanomaterials with tuned functionalities and desirable size organization from angstrom to centimeters. Thus, the introduction of hydrophobic and solvophobic organic polymers as additional compounds in the one-pot surfactant templated synthesis of mesoorganized organo-silicas induces extra hierarchical levels [12]. During last decade, different functionalization processes have been developed and currently are used in several industries. One of the common applications of sol-gel derived functionalized materials is shown in figure (5).

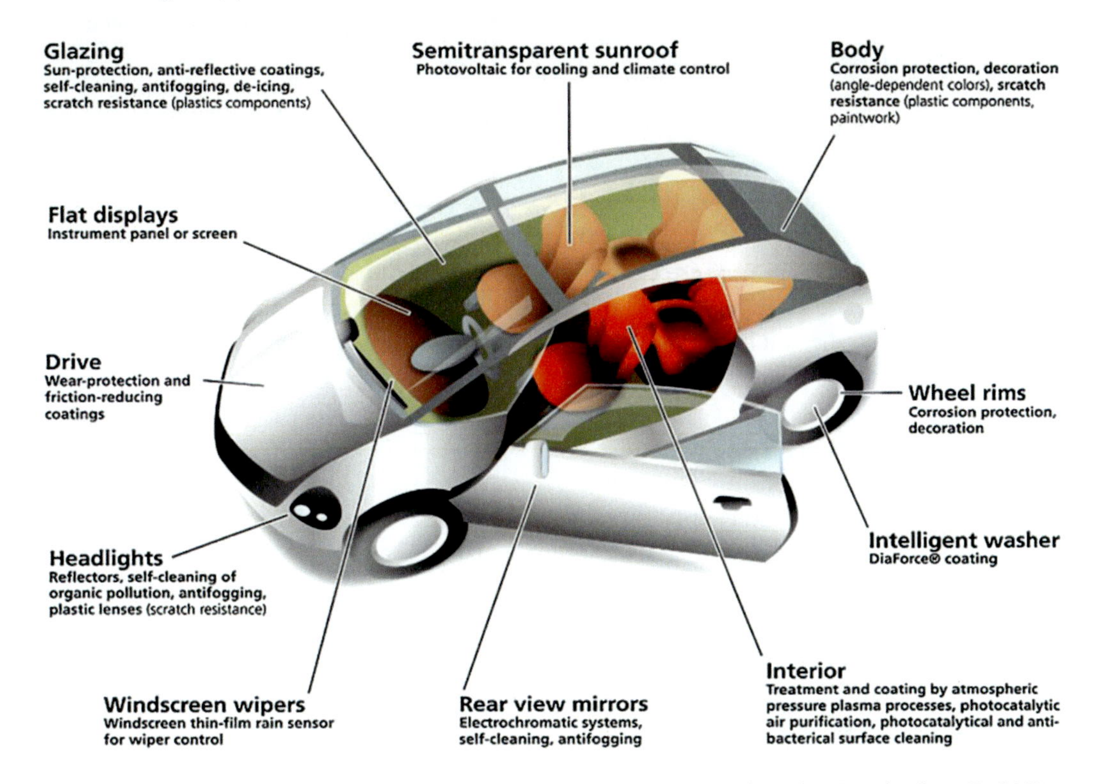

Figure 5. Typical applications for functional coatings in automotive engineering [reprint from Ref.33].

Semiconducting organic-inorganic hybrid materials are very important materials that were prepared by modified sol-gel process. This modification is due to weakly or strong chemical bonds between semiconductor organic and inorganic building blocks in sol-gel reactions. The development of such materials, which have already found numerous applications as organic diodes, solar cells and transistors, is one of the big achievements of sol-gel science. The notion is to create materials with new combinations of properties by combining inorganic and organic building blocks on a molecular level. In some cases, the sol stabilization has direct effect on the properties of obtained hybrid material. As mentioned above, the sol stabilization has critical role on the sol-gel process and we need to give deep understanding about stabilization process and factors that are affecting it. In the following, we will highlight the important parameters. This is done by first introducing sol-gel chemistry.

3. Sol-Gel Chemistry

Sol is a colloidal suspension of solid particles in a liquid phase. When the dispersed particles are very small, the Brownian motion causes them to be suspended solid in a liquid matrix. Turner [34] has explained that a gel can be considered as a solid material network which is containing a liquid component. To prepare a sol by sol-gel technique, two different methods which are termed as aqueous and non-aqueous procedures can be employed. In aqueous or alkoxide method metal alkoxides with a general formula as $M(OR)_x$ are used as the most typical sol-gel precursors. In aqueous procedure, the hydrolysis of a metal alkoxide, and further condensation reactions in the present of water or will occurred. The general reaction, for this kind of sol-gel process is given by:

$$M(OR)_x+H_2O=HO\text{-}M(OR)_{x-1}+HOR \quad (1)$$

Normally, the equilibrium condition is lost during a completely reaction between water and metal alkoxide. This case can be achieved by adjusting the concentration ratio of reactants. This parameter is very effective for controlling the properties of obtained material by sol-gel process. There are several parameters that must be considered for using alkoxides precursors. Namely, the solubility, the volatility and the oligomerization capability are these parameters. Oligomerization of the metal alkoxide affects the homogeneity of the final products and in the absence of steric or inductive effects; oligomerization is a dominant feature of metal alkoxide. Steric and inductive effects are two parameters that represent the physical prosperities of metal alkoxide. Bradly [35] has stated that the physical properties of metal alkoxide are highly affected by factors such as size and shape of the alkly group (R), atomic radius and the coordination number of the metal. Due to the high electronegativity of oxygen, the metal-oxygen bonds ($M^{\delta+}$-$O^{\delta-}$-C), are expected to posses significant ionic character. This leads to some new important factors such as inductive effect of alkyl group. The presence of p-orbital oxygen in bonding with d-orbital in metal for early transition metals and finally the formation of oligomeric species through alkoxo bridges can be an effective controlling parameters of decay of the polarity in meal-oxygen bond. One of the characteristic features of metal alkoxides is their ability to exchange alkoxo groups with alcohols, and this has been widely exploited for the synthesis of new homo- and hetero-leptic alkoxide derivatives of various *s*- , *d*- , *f*- , and *p*-block elements [35]. The exchange reaction that can be considered as follows:

$$M(OR)_x + n\grave{R}OH \leftrightarrow M(OR)_{x-n}\,(O\grave{R})_n + nROH\uparrow \quad (2)$$

If the alcohol $\grave{R}OH$ has a higher boiling point than ROH , then the desired product can be easily obtained by shifting the equilibrium of Eq. (2) by removing ROH by fractional distillation. As has been explained by Bradly [35], many of the final products (particularly those containing sterically congested and chelating alkoxo ligands) prepared by this route assume special significance because of their reduced molecularity, enhanced solubility (in organic solvents), and volatility as well as novelty in structural features. Three important factors such as the steric demands of the alkoxo groups (OR and $O\grave{R}$), the relative O-H bond energies of the reactant and product alcohols and the relative bond strengths of the metal–alkoxo bonds of the reactant and product alkoxides are affecting the extent of substitution in alcoholysis reactions. In general, the interchange tendency of alkoxy groups increases from tertiary to secondary to primary groups [35].The interchange tendency of alkoxy group in the case of titanium alkoxides, $Ti(OR)_4$ has been evaluated and found the interchangeability of alkoxo groups in alcoholysis reactions can be followed as $MeO^- > EtO^- > PriO^- > ButO^-$. The presence of vacant d-orbitals in most of the metals offers a facile initial step in a nucleophilic attack (S_N2) of an alcohol molecule on the metal alkoxide (Eq. 2), and thus low activation energy for alcohol interchange involving a four-membered cyclic transition state seems reasonable.

The exchange ability of alkoxo groups leads to a new kind of sol-gel process which is termed as non-aqueous sol-gel. This kind of sol-gel technique has been extensively studied by researchers [36-42]. Nonaqueous (or non-hydrolytic) sol-gel processes in organic solvents, generally under exclusion of water, are able to overcome some of the major limitations of aqueous systems, and thus represent a powerful and versatile alternative [38]. Obviously, in nonaqueous processes the formal oxygen that is required for the formation of M–O–M bonds is not provided by any added water. Therefore, it must either stem from the precursor used or be provided in some form from the organic medium [42]. In fact, often these cases cannot be clearly distinguished, and the mechanisms for this oxygen transfer are manifold. Since, the non-aqueous sol–gel route involves the cleavage of a carbon–oxygen bond; the chemistry of these processes is based on reactions at the frontier of organic and inorganic chemistries. A classification of the different possible mechanisms involved can be made according to the following general trends: (i) The metal precursors react with the solvent by a ligand exchange, followed by one of the condensation reactions. (ii) The solvolysis of the metal complex leads to the formation of a hydroxyl group (non-hydrolytic hydroxylation reactions) which further leads to a reaction with a metal precursor or another hydroxyl group forming M–O–M bonds. (iii) Finally, more complex mechanisms such as Guerbet-like reactions or aldol condensations can take place [36]. These reactions imply several steps and/or require a concerted mechanism of many molecular species. They have recently been evidenced in the formation of several metal oxides nanoparticles.

3.1. Steric Effects

The steric effect as a one of the common concepts in chemistry is widely used. This effect describes the effect on a chemical or physical property (structure, rate, or equilibrium constant) are occurring due to introduction of substituents that have different steric specifications [43]. The steric effect is basically associated with the fact that atoms and larger parts of a molecule occupy a certain region of space. When atoms or groups are brought together, hindrance will be induced, resulting in changes in shape, energy, reactivity, etc. There are several parameters that are affected on steric effects. These parameters such as the sum of nonbonded repulsions, bond angle strain, and bond stretches or compressions are the main effective steric parameters. For the purpose of correlation analysis or linear free-energy relation, various scales of steric parameters have been proposed in the literatures. The construction of a scale for the steric effect of different substituents has been introduced by Taft [44]. This construction has been based on the rate constants of acid-catalyzed hydrolysis of esters in aqueous acetone. It was shown that $\log(k/k_0)$ was insensitive to polar effects, and thus, in the absence of resonance interactions, this value can be considered as being proportional to steric effects (and any others that are not field or resonance effects) [43]:

$$\log \frac{k}{k_0} = E_S^{Taft} \tag{3}$$

where (k/k_0) refers to the ratio of acid-catalyzed hydrolysis rate constant and $\log_{10}k = \rho\sigma+\delta E_s$. In this equation σ is the polar substituent constant; ρ is the sensitivity factor for the reaction to polar effects and δ sensitivity factor for the reaction to steric effects. However, E_s is reproducing the steric effect of substituent, but not perfectly, well. To find a better describing of the steric effects, many attempts have been done and different E_s have been proposed. Some of them are the modifications of E_s. The most important of modified E_s are E_s' and E_s^c [45]. E_s' is similar to E_s but is defined on the basis of more unified reactions and over wider range of substituents.

The corrected steric constant Es^c includes an additional term in order to correct the hyper-conjugation effect of α-hydrogen atoms $Es^c = Es–0.306(3–n\text{H})$ where the nH, is the number of α-H atoms. A very similar steric constant taking into consideration the hyper-conjugation effect of both α-H and α-C was proposed by Palm and coworkers [45]. In this scale, hydrogen is taken to have a reference value of E_s=0. Steric factors and solvent effects have important roles in sol-gel process.

3.2. Inductive Effect

Inductive or electronic effect is another important parameter that can be used to control the sol-gel reactions. The rate of hydrolysis reaction is directly proportional with inductive effect. In order to show the inductive effect several electronic substituent constant (σ) were defined that have been represented the both global and particular electronic effects. The σ values obtained unambiguously from experimentally accessible data or from the many possible reaction series are called primary values and the corresponding set primary standard. The σ values derived from the primary values, by rescaling with modified ρ constants or correlation equations, are called secondary values and the corresponding set secondary standard [46]. Taft [47] has proposed the experimental methods to determine the steric effect. These methods based on the principle of the linearity of free energies (LFE). This has been carried out for an enormous number of chemical processes and among them it is possible to pick out those in which the influence of one of the effects predominates [48]. It has been proposed that in aliphatic compounds there are powerful steric effects of substituents which must be separated from the electronic. He proposed assessing steric effects using the hydrolysis of carboxylic acid esters. The inductive constants of aliphatic substituents were assessed at the same time [46];

$$\sigma^* = \frac{1}{2.48}\left[\log\left(\frac{k_X}{k_{Me}}\right)_B - E_S\right] \tag{4}$$

where σ* is the constant for a substituent depending exclusively on the overall polar effect of the substituent (corresponding rate constant k), E_s is the constant of the substituent reflecting its steric effect, k_X and k_{Me} denote the rate constants of acid- and base-catalyzed hydrolysis or esterification of substituted and unsubstituted esters, respectively. The index B refers to the reaction of alkaline hydrolysis of carboxylic acid esters identical in all other respects. The factor 2.48 is a constant introduced with the aim of making the polar scale of σ* within the limits of the same scale as applied by Hammett [49]. Since base-catalyzed hydrolysis involves both inductive and steric effects and acid-catalyzed hydrolysis involves only the steric effect, removing the steric effect leaves only the inductive effect, assuming that in both reactions the steric effects are the same. When defining the σ* values as a measure of inductive effects, the choice in reactivity types is such that specific steric, resonance, and other effects are apparently constant and a linear free energy relationship of the Hammett type holds [46]. Recently, the fundamental characteristics of the constituent atoms are used as base for estimation of the σ* inductive constant of substituents in the frame of an additive model [50]:

$$\sigma^* = \sum_{i=1}^{n} \frac{\sigma_i^A}{r_i^2} = 7.84 \times \sum_{i=1}^{n} \frac{\Delta\chi_i \times R_i^2}{r_i^2} \quad (5),$$

where the sum runs over all the atoms of the substituent, r_i is the distance of the ith atom of the substituent to the reaction center, and R_i is the covalent radius of the atom. σ^A is an empirical atomic parameter reflecting the ability of an atom to attract (or donate) electrons and its values were estimated by multivariate regression analysis on Taft σ* constants for several substituents. The values were found to have good correlation with the difference in electronegativity Dc between a given atom and the reaction center, reflecting the driving force of electron density displacement and, with the square of the covalent radius of the atom, reflecting the ability to delocalize the charge. The Taft–Lewis inductive constant σ_I was proposed [51] to measure the inductive effect in aliphatic series on a scale for direct comparison with aromatic s values, derived from σ* constant as [46]:

$$\sigma_I = 0.45 \times \sigma^* = \frac{1}{5.51} \times \left[\log\left(\frac{k_X}{k_H}\right)_B - \log\left(\frac{k_X}{k_H}\right)_A \right] \quad (6).$$

The derived ρ value of 5.51 was later modified by Taft into ρ=6.23 [52]. In this way, σ_I values can also be considered as a measure of the inductive effect of substituents bonded to aromatic carbons. These σ_I values of Taft and Lewis were used as a basis set to obtain a large number of inductive constants [53,54]. The useful inductive constant must be selected for sol-gel reactions.

3.3. Gelation

Gelation occurs when the links are formed between sol particles which were produced by hydrolysis and condensation. During the gelling time an extensive giant spanning cluster reaches across the containing vessel. At this point the mixture has a high viscosity so that it does not pour when the vessel is tipped [55]. Many sol particles are still present as such, entrapped and entangled in the spanning cluster. The viscosity of this initial gel is high and its elasticity is low. Normally, the exothermic or endothermic transition cannot be observed at the gel point. There is not also any discrete chemical change. Only the viscosity change rapidly. Spanning of clusters are continuing and further cross-link and chemical inclusion of isolated sol particles will be happened by the following of gelation. This leads to an increase in the elasticity of the sample. Precise definition of terms such as gel-point and gelation time is elusive. The classical theory of gelation has been addressed by Flory and Stockmayer [56]. This pioneer work has answered why the fraction of all the possible bands that can form in a polymerizing system is necessary for gelation. This theory shows that the polymerization actually needs a minimum density of bands to be formed before an infinitely large molecule appear. In simple expression, the growth of particle leads to increases in the density of growing chains at the periphery of a growing polymer particle which can be predicted only by the considering of chain extension and branching. This will eventually lead to overcrowding and the prediction of a density which increases indefinitely with particle size and is thus unsatisfactory [55]. On the base of random filling of sites and formation of bond between sites on lattices several percolation models have been introduced to analyze the mechanism of gelling phenomenon. In these percolation models, the gelation is defined as a formation of systems composed of sol particles and this is associated by a lack of any correlation between successive bond formations in such models. Therefore it is difficult to account them for using as gelling models [55]. However, the formation of gels can be described on the base of sol particles linking. The latter process is also described by a kinetic model in which the rate of change in the number of clusters ns of size s is described by the Smoluchowski Equation in terms of their formation by aggregation of clusters. There are several parameters that are considered to control the gelling process. Kind of alkoxide, pH and concentration ratio is the main parameters that have important role on gelation. In figure (6), the influence of pH has been illustrated. It can be found that in one critical pH the gelling time is very long and we can prepare a sol with long time stability. To obtain this critical point, we need to adjust other parameters. This subject will be discussed in the next part of the chapter.

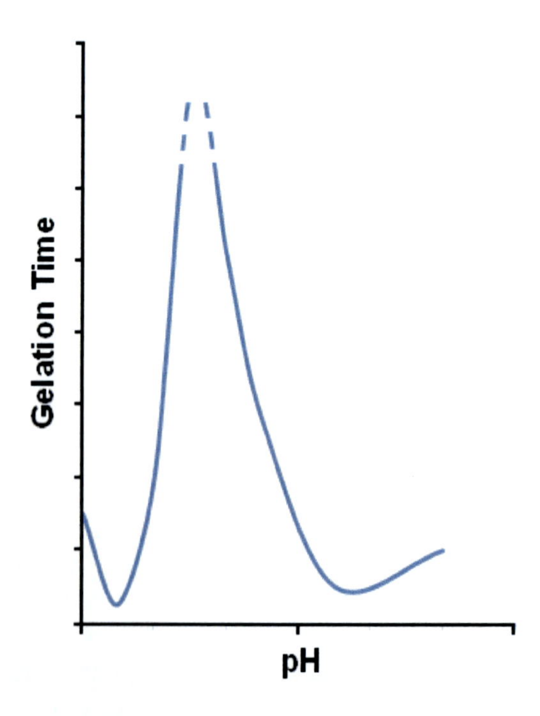

Figure 6. Schematic illustration of pH effect on gelling time.

4. Sol-Gel Transition Kinetic

In the last part, it has been explained that Smoluchowski model was used to describe the kinetics of gelling process. After near one century, the Smoluchowski model is subject of many scientific researches [58-62]. The kinematic analysis of gelling is very important for controlling the sol-gel process and consequently, is much important for stabilization of sol. Similar to many various phenomena with quite different nature, the gelling process can be classified in the aggregation–fragmentation processes. Due to important role of gelation on the characteristics of sol-gel products, many research attempts have been done to study the gelling phenomena on the base of Smoluchowski model and the aggregation–fragmentation process [63-83]. The evaluation of gelling process in the frame of the aggregation–fragmentation has been extendedly carried out by Lushnikov [85-90], who has tried to analyzed the formation of a gel in a disperse system wherein binary coagulation alone governs the temporal changes of particle mass spectra is studied under the assumption that the coagulation kernel is proportional to the product of masses of coalescing particles. According to Lushnikov approach the kinetics of aggregation-fragmentation processes can be formulated on the base of several parameters such as time dependent average number of g-mers in the system ($n_g(t)$), or time dependent the concentrations $C_g(t)= n_g(t)/V$, where V is the volume of the system proportional to M, the total number of monomeric units. If the sol-gel process occurs in a closed system, the total number of monomeric unit (M) is constant, because, there is not any input (sources) or output (sinks) of mass to be conserved during the process. It can be thermodynamically assumed that the average numbers of g-mers are proportional to M, and if not, the respective concentrations are simply equal to zero as $M \to \infty$. The famous Smoluchowski equation is one example of the respective kinetic equations which are consequences of that thermodynamically hypothesis. In describing the kinetics of the coagulation process, the simplest example of which is the evolution of a system of monomeric units that are able to form g-mers resulting from a chain of binary irreversible coalescence processes given by $(n_1)+(n_2) \to (n_1+n_2)$. In the simplest case of spatially uniform systems Smoluchowski equation is written [87]:

$$\frac{dc_g}{dt} = \frac{1}{2}\sum_{l=1}^{g-1} k(g-l,l)\times c_{g-l} \times c_l - c_g \sum_{l=1}^{\infty} k(g,l)\times c_l \qquad (7).$$

Here the coagulation kernel $K(g,l)$ is the transition rate for the process $(g)+(l) \to (g+l)$. The right-hand side of equation (7) has formed by two different terms. The first term describes the gain in the g-mer concentration due to coalescence of $(g-l)$- and l-mers while the second one is responsible for the losses of g-mers due to their sticking to all other particles. Lushnikov [87], explained that the Smoluchowski equation leads to some unpleasant consequences. In particular, if the coagulation kernel $K(g,l)$ is a homogeneous function of the masses g and l, i.e., $K(ag,al)=a^{\lambda} K(g,l)$, and $\lambda>1$, the moments of particle mass spectrum behave reasonably only during a finite interval of time. Moreover, there are not physical principles that would forbid the exponent λ to exceed unity and quite realistic coagulation kernels with $\lambda>1$are not rarities.

Unusual features of Smoluchowski equation led to propose two different strategies about sol-gel transition kinetics. As has been explained by Lushnikov [87]; the first one sand the most widespread assumes that after the critical time the coagulation process instantly transfers large particles to a gel state, the latter being defined as an infinite cluster. This gel can be either passive (it does not interact with the coagulating particles) or active (coagulating particles can stick to the geld). In the latter case the gel should be taken into account in the mass balance and no paradox with the loss of the total mass comes up. Still neither this definition nor the post gel solutions to the Smoluchowski equation give a clear answer to the question, what is this, the gel?

In the second scenario, it has been considered that the coagulation process is occurring in a system of a finite number M of monomers enclosed in a finite volume V. In this case any losses of mass are excluded "by definition." The gel appears as a *single* giant particle of mass g comparable to the total mass M of the whole system. The consideration of finite systems calls for an alternative approach. Although the solution to the evolution equation for $K \propto gl$ had been found, it contained a recurrence procedure that did not open a gate for a straightforward asymptotic analysis of the result. The main goal of Lushnikov [87], was to find the exact analytical expression for the particle mass spectrum in the finite coagulating system with the kernel $K(g,l)=2gl$. From the first sight, the coagulation process cannot lead to something wrong. Indeed, let us consider a *finite* system of M monomers in the volume V. If the monomers move, collide, and coalesce on colliding, the coagulation process, after all, forms one giant particle of the mass M. The concentration of this M-mer is small, $c_M \propto 1/M$. Better to say, it is zero in the thermodynamic limit V, $M \to \infty$, $M/V=m<\infty$. In other words, no particles exist in coagulating systems after a sufficiently long time. The description of the coagulation process in terms of occupation numbers (numbers of g-mers considered as random variable are used to extract the total particle number ($n_g(\tau)$) as follows:

$$\bar{n}_g(t) = \frac{M!}{2\pi i} a_g(t) \oint \frac{dz}{z^{M-g+t}} e^{G(z,t)} \quad (8),$$

where $\tau = t/V$. In order to solve the Eq. (8), the formal series has been introduced by Lushnikov [87];

$$D(z,t) = \sum_{g=0}^{\infty} \frac{z^g}{g!} e^{g^2 \cdot \tau} \quad (9).$$

Here, z is a formal variable whose power g just defines the coefficient before z^g. By solving these equations, it can be finally found that:

$$\bar{n}_g(\tau) = C_M^g \times e^{(g^2 - 2Mg + g)\tau} \times (e^{2\tau} - 1)^{g-1} \times F_{g-1}(e^{2\tau}) \quad (10).$$

This equation change to new one under thermodynamic limits:

$$\bar{n}_g(t) \approx \bar{n}_g^s(t) = M \frac{g^{g-2}}{g!} e^{-2gt} (2t)^{g-1} \quad (11).$$

The notation $\bar{n}_g^s(t)$ stands for the exact solution to the Smoluchowski equation. It is seen that the concentration $C_g(t) = \bar{n}_g^s(t)/\bar{V}$ can be introduced as an independent function of *M*. The total mass concentration can be given as:

$$m(t) = 1 - \mu_c(t) \quad (12)$$

Where the mass concentration is described by the function $\mu_c(t)$ at the critical time (t_c). The total particle number concentration can also be obtained as:

$$n(t) = m(t) - t(m^2(t)) \quad (13).$$

Finally, at the critical point ($t=t_c$ or $\varepsilon=0$), the total particle number can be expressed as:

$$\bar{n}_g(t) = \frac{M}{\sqrt{2\pi g^5}} e^{\frac{-g^3}{8M^2}} \quad (14).$$

The total mass concentration analysis has shown that the transition kinetics depends to several parameters such as concentration of monomers *M,* volume of system (*V*), thermodynamically conditions of system and so on. Therefore, the control of size and density of monomers inside the sol-gel system are very important to keep sol-gel transition under control. Lushnikov [87], showed that the second moment of the particle mass spectrum, for $K(g,l) \propto gl$ leads to a strange result:

$$\phi_2(t) \propto \frac{1}{t_c - t} \quad (15),$$

where the critical time t_c depends on the initial mass spectrum, and this is not yet all. After the critical moment $t=t_c$ the mass concentration exponentially drops down with time. At $t.>t_c$ the particle number concentration $n(t) = \sum_g C_g(t)$ crosses the *t* axis and becomes negative.

5. Sol-Gel Uniformity

As has been explained in previous part, the sol-gel kinetic transition depends highly on the density of primary monomer in the system. On the other hand, the size of monomer is another parameter that affects the kinetic of sol-gel transition. By definition, a monodispersed monomer system is a number of particles with uniform size, shapes, structure. Due to their characteristics, they are considered as a special material in the sol-gel systems. Since, the preparation of uniform sol and its stabilization is the main purpose of our approach, the production of monodispersed with size and concentration controlled monomers is very important. For this purpose it is necessary to focus on the monodispersed monomers formations. It means that the mean diameter and structure of the monodispersed monomers must be perfectly controlled. Two of the most important factors which severely affect the final particles size are nucleation and growth of particles in the solution. Knowing these factors would cause to better control and better stability. In this chapter the nucleation and the growth of particles in the sol-gel system are briefly explained.

5.1. Nucleation

Particles are formed and dispersed during the nucleation and the growth process. The nucleation of particles is the process which depends on the initial characteristic of monomers which determines the initial crystal structure. Usually, the final shape and structure of particles are determined by the growth of particle. Nucleation of particles happens due to clustering of agglomeration of molecules in a supersaturation melt solution or vapor [91,92]. At the initial stage of nucleation, there is no distinct structure and particles do not have the crystalline structure. This issue make the nucleation complicate and a good general agreement on nucleation is really hard to gain. It means that in the first stage of nucleation the particles without any crystalline character are formed. At this stage the particle is unstable and can solve easily in the solution. At the next stage, stable and crystalline particles are made. This mechanism of particles formation can be referred as "primary" and "secondary" nucleation and could be considered as follows (See figure 7) [91-93]. In the next parts of this chapter these various kinds of nucleation will be explained.

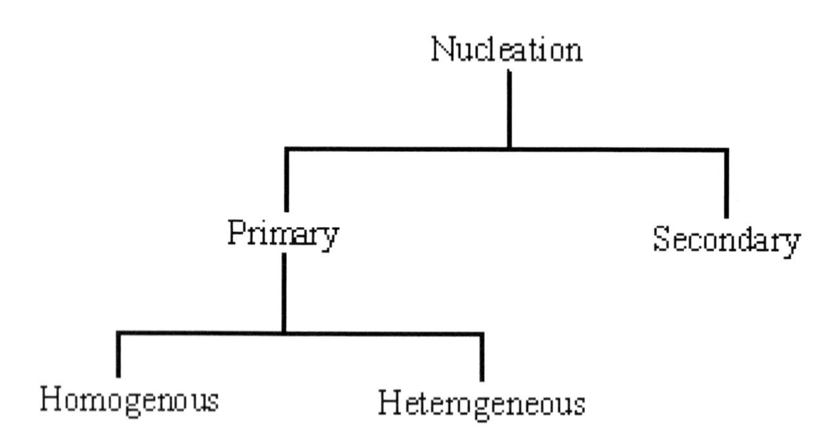

Figure 7. Schematic diagram of nucleation process in sol-gel systems.

5.1.1. Primary Nucleation

As has been shown in figure 8, the primary nucleation has two different kinds that can be categorized as heterogeneous and homogenous nucleation processes. These two kinds of nucleation play an important role in the stabilization of sol and application of obtained material by this technique. Therefore, it is necessary to consider them as important stages in sol-gel process.

5.1.2. Homogenous Nucleation

Classical nucleation theory has been explained by Gibbs, Volmer, Becker and Doring, which can be correctly used for description of the nucleation of the particles in the homogenous system [94]. The foundations of this theory are based on the continuous reaction and formation of enough monomers to produce the primary particles. It has been assumed that the particle has spherical shape. The transformation free energy of this process is ΔG which can be described as follows [95-98]:

$$\Delta G= 4\pi r^2\gamma+4/3\pi r^3\Delta G_V \tag{16}$$

In this equation ΔG_V is the bulk free energy per volume which is difference between the old and new phase and γ is the surface free energy per unit area. The Gibbs equation (21) can be also obtained by considering the free energy of number of atoms in the monomer from supersaturation solution at temperature T [95-98]:

$$\Delta G= -n\Phi + \gamma A \tag{17}$$

Where Φ is the chemical potential difference for a monomer between the supersaturated solution and the solution in equilibrium with the monomer and A is the surface area of the monomer:

$$\Phi = \mu - \mu_\infty = kT \ln\left(\frac{x}{x_\infty}\right) \tag{18}.$$

If we combine the equation of (16) with (17) and (18), the following equation is obtained [95-98]:

$$\Delta G = -nkT \ln\left(\frac{x}{x_\infty}\right) + 4\pi r^2 \gamma \tag{19},$$

where x/x_r is called the supersaturation ratio (S), x is the mole fraction of the monomer in the solution and x_r is the mole fraction of the monomer at the saturated stage in the solution. Alternatively, the supersaturation ration can be explain as term of concentration and become C/C_∞ where C_∞ in the solubility of the monomers in the solution. So the supersaturation ratio S is:

$$S = \frac{c}{c_\infty} \tag{20}.$$

From the Eq(16) the critical value of cluster is given by[95-98]:

$$\frac{d\Delta G}{dr} = 0 \tag{21},$$

and so:

$$r_C = \frac{-2\gamma}{\Delta G_V} \qquad (22).$$

From equation (19):

$$r_c = \frac{2\gamma V_m}{kT \ln S_c} \qquad (23),$$

and

$$\Delta G_c = \frac{16\pi\gamma^3 V_m^2}{3(kT \ln S_c)^2} \qquad (24),$$

which k is Boltzmann constant per molecule ($1.3805\times10^{-23}JK^{-1}$), V_m is the molecular volume and S_c is the supersaturation ratio at the supersaturation stage [95-98]. The behavior of the created particle severely depends on the size. The critical size means the minimum size of stable particle. The particle can be formed in any radius but the particle which has $r>r_c$ can grow to the final size. Below the critical radius (r_c) the particle dissolve or evaporate from the solution. From the free energy diagram, we can easily see that the particle size larger than r_c can grow and Eq.(24) can show the amount of the critical free energy for the formation of particle with r_c radius. For the formation of the particle with r_c the free energy of the system must change [95-98]. The energy of the system at the constant temperature and pressure is constant but this does not mean that the energy level in all part of the system is the same. This energy does not show that all parts of system have the same energy, maybe there will be some unaccommodating in the energy in some part of system because of statistical distribution of energy, molecular velocity or in those supersaturated regions where the energy level rise temporarily to the high value nucleation will be happened. This ΔG_{crit} which was obtained in Eq.(24) is necessary to form a stable nucleus. The rate of the nucleation J as number of particles formed is explained by the use of Arrhenius equation [99, 100]:

$$J = A \exp\left(-\frac{\Delta G}{kT}\right) \qquad (25),$$

From the Eq (24) and (25) the rate of the nucleation can be obtained [99,100]:

$$J = A \exp\left[\frac{-16\pi\gamma^3 V_m^2}{3k^3 T^3 (\ln S)^2}\right] \qquad (26).$$

This equation shows that there are three main factors affecting the nucleation process: temperature (T); supersaturation ration (S); and interfacial tension (γ) [99,100]. It can be theoretically found that the speed of nucleation increases extremely with increasing of supersaturation. But, in experiment, after critical supersaturation the nucleation rate decrease exponentially. In this case, a `critical` supersaturation could be exist in the region of S~4, but it is also clear that nucleation could happened at any moment of $S > 1$. Temperature has the important role on the nucleation process. According to Eq.(23), the critical size of particle depends on temperature and the volume free energy (ΔG_V) is a function of supercooling (ΔT) [101,102]. The volume free energy is given by:

$$\Delta G_V = \left(-\frac{\Delta H_f \Delta T}{T^*} \right) \quad (27),$$

where T^* is solid-liquid equilibrium temperature. Also temperature can increase the supersaturation ration S so that the solution reach the supersaturation faster and nucleation occurs earlier. Generally, nucleation rate changed rapidly with a slight change of $\Delta G_c/kT$ near the maximum supersaturation. In other word, $\Delta G_c/kT$ is almost constant for a given material. From equation (24) $\Delta G_c/kT \approx T^3(\ln S_c)^{-2}$ so it can be said that $T^3 (\ln S_c)^2 \approx constant$ in a given system and if the temperature goes up in the system the supersaturation ratio reduce) [101,102]. This issue could be easily seen in the Sugimoto results [102]. He made the monodisperse AgBr particles by gradual release of silver ions from Ag^+ gelatin complex. His experimental results have been obtained at four temperatures of 25, 30, 35 and 40°C the supersaturation ratio has been reduced from 3.61, 3.35, 3.03 and 2.89 respectively. Also the nucleation period speed increase with the temperature from 7.74 to 0.767s [102].

5.1.3. Heterogeneous Nucleation

Impurities in the system can effect on the nucleation of monomer in the solution. However, in some cases they reduce the nucleation process rate, but generally they act as an accelerator. The foreign impurities can induce the nucleation energy at degree of supercooling which is lower than those required for spontaneous nucleation. This means that the overall free energy for the critical nucleus formation under heterogeneous condition ΔG_{crit}, must be less than free energy under homogeneous condition [103-105];

$$\Delta G_{crit}(Het) = F(\Phi)\, \Delta G_{crit}(Hom) \quad (28)$$

which the factor $F(\Phi) < 1$ is accounting the decreased energy barrier to nucleation due to a foreign solid phase. No general rule applies and each case must be considered separately.

5.1.4. Secondary Nucleation

Secondary nucleation is used to describe the nucleation in supersaturated conditions. When the particles at the supersaturation point are formed, they are not crystalline and can be dissolved by the Ostwald ripening again. At this time they are in their metastable condition but after a time they growth and become crystalline and also stable at the point of the secondary nucleation happened [105-107]. There are several reviews that were deal with the secondary nucleation. Strickland-Constable [108] and McCabe [109], showed that the secondary nucleation was dominant at low supersaturation. Also they have shown that there is a connection between the secondary nucleation and the growth of particles. It was shown that the secondary nucleation "contact nucleation" is one of common process on the producing of monomers and play the most effective role. In a colloidal suspension the particles collide with each other or on the vessel walls. So the secondary particles may be formed. It was proved that two potential source cause to make the contact nuclei. Firs one is the crystal lattice structure. Small pieces of the crystal may be dislodged from the surface by contact. In this case the high contact energy is needed. The second potential source of the contact nuclei is the hypothetical growth layer that separates the solution phase from the lattice structure. Ottens and de Jong [110] developed a model of contact nucleation, with the expression for the nucleation rate of crystal between size L and L+dL as;

$$dB = k(C - C^*)^q \omega_L E_L n \times dL \qquad (29),$$

where ω_L is the crystal-impeller contact frequency and E_L the impact energy associated with the contact.

5.2. Growth

As the particles reach on the stable level, they begin to grow to the final size. The final size and shape of particles depend on the surface of primary particles. The growth at first, begin from the surface which has the minimum energy. The growth of monodispersed particles was first generally described by Lamer [111]. In the one monodispersed system, the monomers species were created from some involved reactions in the system. No nucleation could be happened in this stage, even if the concentration of monomers goes above the solubility level of bulk solid [112-114]. The concentration of monomers increases until it reaches a critical level for nucleation, when the nucleation visually starts to be observed. In this stage, the nucleation speed is low so the concentration of monomers still increases. Finally the concentration of monomers reaches at the point (maximum supersaturation). At this level, the generation of particles and the consumption of monomers for nucleation is balanced. After that, the concentration of monomers decreases due to the increasing consumption of monomers by growth. When the concentration of monomers goes down to the critical supersaturation level again, the nucleation is stopped and the stable particles can grow. When the concentration of the monomers reaches to the solubility level, the particles growth ended. In the first zone of Lamer diagram (I), the supersaturation of monomers increases until the critical concentration. In the second zone (II), the nucleation of particles starts and in the third zone (III), the growth of the particles continue until the concentration of monomer ions reduce to the equilibrium solubility[112-114]. The growth of particles is controlled by the diffusion flux of molecules to the particles, or by the rate of condensation reaction between the particles and the solute [112-115]. Kim [116] prepared TiO_2 nanoparticle from the hydrolysis of TEOA with using a semi-batch method. By controlling the reactor type, flow rate and reaction temperature mono disperse TiO_2 were produced and they result fit the theoretical curve for growth under diffusion control. The obtained results by Kim [116], showed that the flow rate and temperature the produced the smallest size (12.7nm) of particles were 1.1 cc/min. Also, Jean and Ring [117] examined the kinetics of growth of titania (made by hydrolysis of tetraethoxytitanate) particles and found that the diffusion coefficient was 10^{-9} cm^2/s which was obtained by Brownian diffusion coefficient for 4μm particles. Keshmiri and Kesler [118] made a thorough study on the monodispersed submicrometer YSZ (yttria-stabilized zirconia) spheres through sol-gel processing via controlled colloidal formation precipitation. They suggested that the predominant mechanism for uniform growth of the monodispersed spheres is followed by attachment of ultrafine particles onto the larger sphere. Figure (8), shows the different formation stage of YSZ using LaMer theory and proved by SEM images.

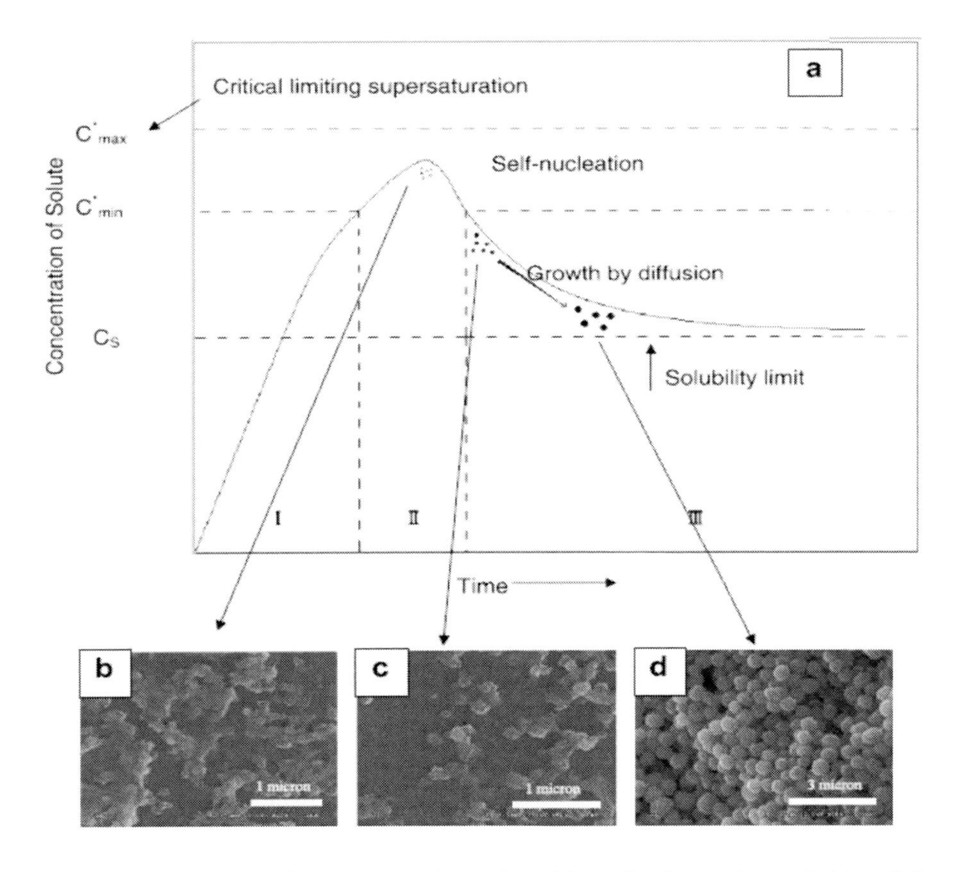

Figure 8. Correlation between solute concentration and particle nucleation, and growth (a), and the corresponding SEM images of the obtained YSZ particles for each region [118].

In our recent report [119], the Sugimoto model has been used to describe the nucleation and the growth of TiO_2 nanoparticles. The basis of this experiment is nucleation and growth of particles which are controlled by diffusion and their growth follow the LaMer diagram (see figure 9).

Our result are S_m=3.19, r^*=4.01nm and $\Delta G^*=1.16\times10^{-16} J$, r_0=8.2nm, $Q_0=7.03\times10^{-8}$ mol.dm^{-3}s^{-1}, $\dot{\upsilon}= 2.93\times10^2$ nm^{3}s^{-1} and D = 6.18×10^{-5} cm^{2}s^{-1}. The value of D is in a reasonable range for diffusion of ions or complexes in an aqueous solution. Moreover, this coefficient is approximately equal with the diffusion coefficient of Ti ions and was obtained by Shingyouchi [120]. Thus, it is clear that the predominant growth mechanism of nanoparticles is that the small particles nucleate and aggregate. The growth of the primary particles may obey the classical mode. But, the aggregation into large particles describes by the models which are introduced by other. Also with these models we can predict that the final size of particles and study on how the particles can growth for the best properties.

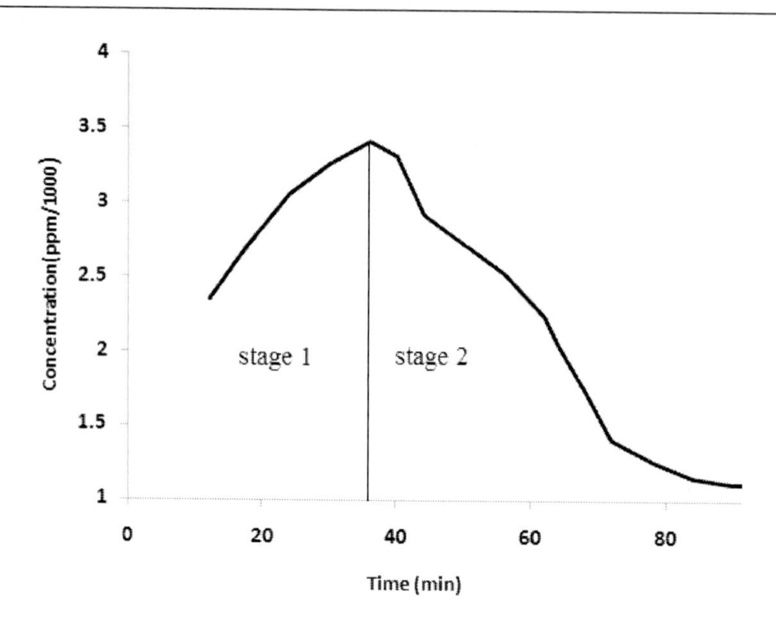

Figure 9: Changing in concentration of [Ti] using to obtain the nucleation and growth parameters with Sugimoto Model.

6. Sol Stability

It has been mentioned that the stabilization of sol is very important for many applications of sol-gel process. There are two different classes of parameters that have affected the stability of a sol. First class contains a number of factors that depend to physical and chemical properties of raw materials which are used for sol-gel technique. For example, the steric and inductive effects are two important chemical and physical parameters that belong to the first class. It has been discussed that the physical properties of the metal alkoxides $[M(OR)x\ ]n$ are largely influenced by the size and shape of the alkyl (R) group as well as by the valency (x), atomic radius, stereochemistry, and coordination number of the metal. Owing to the high electronegativity of oxygen (3.5 on Pauling scale), the M–OR bonds (in alkoxides of metallic elements) would be expected to possess significant ionic character [35]. When the rate of condensation or other reactions is affected indirectly by the functional groups attached to the metal center, polymerization sol-gel process occurred. For example, the polymerization of sol with a butoxy group on precursor is slower than other. In addition, the polarity of the metal–oxygen bond is another factor to control the polymerization of sol. This parameter is a postulated factor to explain the formation of oligomeric species through alkoxo bridges. This latter tendency, which leads to coordination polymerization, is a dominant feature of metal alkoxides unless inhibited by steric and electronic factors.

Generally, it can be considered that the extent of aggregation (n) for an appropriate stoichiometry $[M(OR)x\ ]n$ is dependent on the following considerations: (i) aggregation increases as the metal atom becomes more electron deficient; (ii) aggregation increases when the larger size of the metal atom is used (the greater tendency is introduced by increasing the degree of association (n) which is due to forming alkoxo bridged systems) (iii) the steric effects of the alkyl substituents, which with increasing steric demand inhibit aggregation and have been found to be of greater importance than the electronic nature of the substituents in determining the ultimate extent of aggregation [35]. The second class of polymerization factors contains a series of parameters which depends on the process conditions. In the following, we will present some of these parameters that have very important roles in stabilization process.

6.1. Surface Charge Density

In the most points of a solid surface which contact to the electrolyte solution, the positive or negative electrical charges which are affected by the physical and chemical conditions in the interfacial region can be considered. Surface electrical charges exist due to one or more of the following mechanisms [96,121]:

- favorable adsorption of ions
- Dissociation of surface charged species
- Isomorphic substitution of ions
- Accumulation or depletion of electrons at the surface
- Physical adsorption of charged species onto the surface.

The dominance of each mechanism depends to physical and chemical conditions that were established on the interface of particles and solution. For hydrous oxide, experimental studies have shown that hydrogen and hydroxide ions have important role in determining both the surface charge and potential of all inorganic oxides. It was seen that if the concentration of acid or base change, a little variation can be observed in surface or potential charge. Therefore, the electrical double layer at the powder/liquid solution can be considered by the unequal adsorption of H^+ and OH^-, which establish the charge on the particles by the MOH bonds on the surfaces of the particles. This is schematically represented as [121-123]:

$$M{-}OH + H^+ \leftrightarrow M{-}OH_2^+ \tag{30}$$

$$M{-}OH \leftrightarrow M{-}O^- + H^+ \tag{31}$$

where $M{-}OH_2^+$ and $M{-}O^-$ represent the positive and the negative surface sites [96,121]. Depending on the metal atom, the protons are added or removed from the surface of oxide (the activity of the MOH group). The difference between these two mechanisms depends on mechanism of the hydroxo complex formation. The surface charge σ_0 on oxide sites given by the difference in adsorption density of hydrogen (Γ_{H+}) and hydroxide ions (Γ_{OH-}) respectively, and it is defined as [96,121]:

$$\sigma_0 = F(\Gamma_{H^+} - \Gamma_{OH^-}) = F(\Gamma_{MOH_2^+} - \Gamma_{MO^-}) \tag{32}$$

where F is Faraday constant, σ_0 is commonly expressed in $\mu C/cm^2$ and Γ_{H+} and Γ_{OH-} are in mol/cm^2 . Since both of the surface electrical charge and the ψ_0 are determined by the adsorption of hydrogen and hydroxide ions they called the Potential- determining ions (PDI).

When the surface parameters are zero, the solid is said to be at its point of zero charge (PZC). It designates the equilibrium suspension pH when the surface charge density σ_0 at the particles/aqueous solution interface become zero. At pH>PZC, equation 32 predominates, and the particle is negatively charged, whereas at pH<PZC the particle is positively charged. For several oxides, the pH of PZC is given in Table 1. The magnitude of the surface potential depends on the departure of the pH from the PZC and that potential attracts appositely ions that may be present in solution. Thus, a suspension is stable only if the pH in kept sufficiently far from the PZC.

Table 1. Point of Zero Charge (PZC) of various aqueous ceramic suspension.

Point of Zero Charge (PZC) of Various Aqueous Ceramic Suspension	
Materials	**pH of PZC**
Silica, quartz	2-3
Tin oxide (cassiterite)	4.5
Zirconia	6
Chromium oxide	6.5
Geothite (FeOOH)	6-7
Boehmit(AlOOH)	6.8-7.6
Hematite (synthetic)	7.8-8.8
Alumnia (α and γ)	8-9
Calcit (CaCO3)	9.5
Magnesia	12
Mulite ($3Al_2O_3$. 2 SiO_2)	7.5
Cordierite	2-3

To maintain electro neutrality at the particle/water interface the charge is balanced by an equal and opposite charge in the mobile aqueous phase. This compensation solution charge is created by the adsorption of electrolyte ions at the solution site of the interface. There is very important item that can be termed as electrical bouble layer. In a charge density of a solid surface, an electrostatic force between the solid surface and the charged species near the surface charge exist and lead to segregate positive and negatively charged species. However, there also exist Brownian motion and entropic force, which homogenize the distribution of various species in the solution. Both surface charge determining ions and counter-ions always exist in the solution, which have opposite charge of the determining ions. Although charge neutrality is maintained in a system, distributions of charge determining ions and counter-ions in the near of the solid surface are inhomogeneous and very different. Combination of three forces mainly controlled the distributions of both ions [96,121,122]:

1. Coulombic force or electrostatic force
2. Entropic force or dispersion
3. Brownian motion

The combined result is that the concentration of counter ions is the highest near the solid surface and decreases as the distance from the surface increases, whereas the concentration of determining ions changes in the opposite manner. Assuming surface charge is positive. Such inhomogeneous distributions of ions in the proximity of the solid surface lead to the formation of so-called double layer structure. According to the divination of double layer, the double layer include two layers, Stern layer and Gouy layer, Helmholtz plane_ separated these two layer . Standard theory explain that the potential drops linearly through the tightly bond layer of water and counter-ions. Between the solid surface and the Helmholtz plane is the Stern layer, where the electric potential drops linearly through the tightly bound layer of solvent and counter-ions. Beyond the Helmholtz plane (at h=H) until the Gouy layer or diffuse double layer at this point counter ions are at the average concentration level in the solvent. In the Gouy layer, the counter ions mobile freely and the electric potential does not decrease linearly [96,121,122]. By considering the diffuse double layer in the interface of particle, the electric potential drops approximately as following:

$$V_R \propto e^{-k(h-H)} \tag{33}$$

where H is the thickness of the Stern layer, k is called as the Debye-Hückel screening length and is also used to describe the thickness of double layer, and k is given by:

$$k = \sqrt{\frac{F^2 \Sigma_i c_i z_i^2}{\varepsilon \varepsilon_0 R_g T}} \tag{34}$$

The distance between any two particles are large enough so that the charge distribution on particle surface is not influenced by other particles. In this equation F is Faraday's constant, ε_o is the permittivity of vacuum, ε is the dielectric constant of the solvent, and C_i and Z_i are the concentration and valence of the counter-ions of type i. This equation indicates that the electric potential at the near of solid surface reduce with increasing of concentration, and increases with an increased dielectric constant of the solvent exponentially. Higher concentration and valence state of counter-ions would result in a reduced thickness of both Stern layer and Gouy layer. In theory, the Gouy diffusion layer would end at a point where the electric potential reaches zero, which would be the case only when the distance from the solid surface is infinite. However, in practice, double layer thickness is typically of approximately 10 nm or larger [96,121,122].

One of the interactions between particles is directly connect with the surface charge and the electric potential adjacent to the interface. The electrostatic repulsion between two particles gets out from the electric surface charges, which are weaker to a varied extent by the double layers. When two particles are far apart, there will be no overlap of two double layers and electrostatic repulsion between two particles is zero. However, when two particles approach one another, double layer overlaps and a repulsive force develops.

6.2. Van Der Waals Attraction Potential

Beside the electrical forces between particles and solvent, there is another important force that must be considered for sol stabilization. When particles are dispersed in a solvent and the size of the particles are small, the force between the particles mainly includes van der Waals attraction force and Brownian motion and the influence of gravity becomes negligible. Furthermore, if the particles have spherical Van der Waals force is a week force and becomes significant only at a very short distance. Brownian motion ensures the nanoparticles colliding with each other all the time. The combination of van der Waals attraction force and Brownian motion would result in the formation of agglomeration of the nanoparticles [96,121,122]. Sum of the molecular interaction for the pairs of molecules in the solution compose the Van der Waals interaction and we can assume that between two nanoparticles of one molecule in each particle, as well as to all pairs of molecules with one molecule in a particle and one in the surrounding medium such as solvent. Addition of all the van der Waals interactions between two molecules which have the spherical shape with radius, r, and they would separated by a distance, S, gives the total interaction energy or attraction potential;

$$F_A = - A/6 \{2r^2/ (S^2+4rS) + 2r^2/(S^2+4rS+4r^2) + \ln [(S^2+4rS)/(S^2+4rS+4r^2)]\} \qquad (35)$$

The negative sign shows the attraction nature of between two particles, and A is a positive constant termed the Hamaker constant of a given colloidal dispersion and is a material property (which has a magnitude on the order of 10^{-19} to 10^{-20} J). If the geometrical approximation is used A can express by solvent-solvent interaction and particle-particle interaction: A= ($A_S^{1/2} - A_p^{1/2}$) where A_S and A_P are the Hamaker constant for the solvent-solvent interaction and particle-particle interaction respectively. Table 2 listed some Hamaker constants for a few common materials. For example, when the separation distance between two equal sized spherical particles are significantly smaller than the particle radius, i.e., S/r << 1, the simplest expression of the van der Waals attraction could be obtained:

$$F_A = - A\, r/12\, S \qquad (36)$$

Table 2. Hamaker constants for some common materials.

Material	$A_i(10^{-20})$J
Metals	6.2-45.5
Gold	45.3
Oxides	10.5-15.5
Al_2O_3	15.4
MgO	10.5
SiO_2 (fused)	6.5
SiO_2 (quartz)	8.8
Ionic crystals	6.3-15.3
CaF_2	7.2
Calcite	10.1
Polymers	6.15-6.6
Polyvinyl chloride	10.82
Polyethylene oxide	7.51
Water	4.35
Acetone	4.20
Carbon tetrachloride	4.78
Chlorobenzene	5.89

Furthermore, it should be noted that the interaction between two molecules are obviously different from that between two particles. Van der Waals interaction energy between two molecules can be simply represented by:

$$F_A \propto - S^{-6} \tag{37}$$

Although the nature of the attraction energy between two particles is the same as that between two molecules, integration of all the interaction between molecules from two particles and from medium results in a totally different dependence of force on distance. The attraction force between two particles decays much slowly and extends over distances of nanometers. As a result, a barrier potential must be developed to prevent agglomeration. Two methods are widely applied to prevent agglomeration of particles: electrostatic repulsion and satiric exclusion [96,121,122]. The total interaction between two particles, which are electrostatic, stabilized, is the combination of van der Waals attraction and electrostatic repulsion:

$$V= V_A + V_R \tag{39}$$

The electrostatic stabilization of particles in a solution is successfully explained by the DLVO theory which is the abbreviation of the creators Derjaguin, Landau, Verwey, and Overbeek. The forces between two particles in a solution are assumed to the combination of van der Waals attraction potential and the electric repulsion potential. In this theory there should be some important assumptions as follows:

1. The surface of solid must be infinite
2. Surface charge density is uniform
3. No re-distribution of surface charge, i.e., the surface electric potential remains constant.
4. No change of concentration profiles of both counter-ions and surface charge determining ions, i.e., the electric potential remains unchanged, and
5. Solvent exerts influences via dielectric constant only, i.e., no chemical reactions between the particles and solvent.

As we see, some of the considerations do not adapt with the real situation and far from the real view of two disperse particle in a suspension. For example, charge density changed when two charged particles are move toward together and get close. However, the DLVO theory works with these assumptions and helps very well to explain the interactions between two approaching particles, which are electrically charged, and thus is widely accepted in the research community of colloidal science [96,121,122]. Figure (14), shows the effect of adding Vander Waals forces and the electric double layer repulsion, and the combination of the two opposite potentials as a function of distance from the surface of a spherical particle˙ Vander Waals forces and the electric double layer repulsion do not exist and become zero at the distance far from the solid surface. Near the surface is a deep minimum in the potential energy produced by the Van der Waals attraction. A maximum is produced a little farther away from the surface, while the electric repulsion potential is dominant. The maximum is also called repulsive barrier. If the barrier is greater than ~10 kT, where k is the Boltzmann constant, the collisions of two particles produced by Brownian motion will not overcome the barrier and agglomeration will not occur. Also, the overall potential is affected severely by the concentration and counter-ions because the electric potential is dependent on the concentration and valence state of counter-ions as given in equations (33) and (34), and the Van der Waals attraction potential is almost independent of the concentration and valence state of counter-ions. When the distance of two particles or the surfaces are larger than the combined thickness of two electric double layers of two particles, there would be no overlap of diffusion double layers, and thus there would not be any interaction between two particles. However, when two particles move toward each other and get closer and the two electric double layers overlap, a repulsion force is developed. As the distance decreases, the repulsion increases and in the one point at the $S_0=r+r_0$ reaches the maximum.

Such a repulsion force can be explained in two ways. One is that the repulsion force is obtained from the overlap of electric potentials of two particles and two is the osmotic flow. When one particle reaches the other one the concentrations of ions between two particles (their double layers) overlap increases significantly, since each double layer would retain its original concentration profile. As a result, the original equilibrium concentration profiles of counter ions and surface charge determining ions are ruined. To retain the original equilibrium concentration, there must be solvent to flow into the region where the two double layers overlap. Such an osmotic flow of solvent effectively repels two particles apart, and the osmotic force disappears only when the distance between the two particles becomes equal to or becomes larger than the sum of the thickness of the two double layers [96,121,122]. Although many important assumptions of the DLVO theory are not satisfied in the really colloidal systems, in which small particles dispersed in a diffusive medium, the DLVO theory is still valid and has been widely applied in practice, as far as the following conditions are met [96,121,122]:

1. Dispersion is very dilute, so that the charge density and distribution on each particle surface and the electric potential in the proximity next to each particle surface are not interfered by other particles.
2. No other force is present besides van der Waals force and electrostatic potential, i.e., the gravity is negligible or the particle is significantly small, and there exist no other forces, such as magnetic field.
3. Geometry of particles is relatively simple, so that the surface properties are the same over the entire particle surface, and, thus surface charge density and distribution as well as the electric potential in the surrounding medium are the same.
4. The double layer is purely diffusive, so that the distributions of counter-ions and charge determining ions are determined by all three forces: electrostatic force, entropic dispersion, and Brownian motion.

REFERENCES

[1] Y. Dimitriev, Y. Ivanova, R. Iordanova, History of sol-gel science and technology, *J. Uni. Chemi. Tech. and Metal.* 43 (2008) 181-192.

[2] J.J. Ebelman, *Ann. Chim. Phys.* 16 (1846) 129.

[3] J. Livage, M. Henry, C. Sanchez, sol-gel chemistry of trasition metal oxides, *Progr. Solid State Chem.* 18 (1988) 259-341.

A. Walcarius,; M.M. Collinson, Analytical chemistry with silica sol-gels: Traditional routes to new materials for chemical analysis, *Annu. Rev. Anal. Chem.* 2 (2009) 121–143.

B. Lind, S. D. Gates, N. M. Pedoussaut, T. I. Baiz, Novel Materials through Non-Hydrolytic Sol-Gel Processing: Negative Thermal Expansion Oxides and Beyond, *Materials* 3 (2010) 2567-2587. (Open access)

[4] F. Ivanauskas, A. Kareiva, B. Lapcun, Computational modeling of the YAG synthesis, *J. Math. Chem.* 46 (2009) 427–442.

[5] J. G. Hou, R.V. Kumar, Y. F. Qu , D. Krsmanovic, Crystallization kinetics and densification of YAG nanoparticles from various chelating agents, *Materials Research Bulletin* 44 (2009) 1786–1791.

[6] G. Clavel, E. Rauwel, M.-G. Willinger, N. Pinna, Non-aqueous sol–gel routes applied to atomic layer deposition of oxides, *J. Mater. Chem.* 19 (2009) 454–462.

[7] Mahshid, M. Askari, M. Sasani Ghamsari, Synthesis of TiO_2 nanoparticles by hydrolysis and peptization of titanium isopropoxide solution, *J. Mate. Proc. Tech.* 18 (2007)296-300.

[8] M. Vafaee, M. Sasani Ghamsari, Preparation and characterization of ZnO nanoparticles by a novel sol–gel route, *Materials Letters* 61 (2007) 3265–3268.

[9] T. K. Tseng, Y. S. Lin, Y. J. Chen, H. Chu, A Review of Photocatalysts Prepared by Sol-Gel Method for VOCs Removal, *Int. J. Mol. Sci.,* 11(2010) 2336-2361.

[10] Ph. Belleville, Functional coatings: The sol-gel approach, *C. R. Chimie* 13 (2010) 97–105.

[11] S. Sakka (ed.), *Handbook of sol-gel science and technology, Processing, Characterization and Application*, Kluwer Academic Publisher, Boston/London, 2005.

[12] M. Sasani Ghamsari, M. Vafaee, Sol-gel, *Mater. Lett.* 62 (2008) 1754-1756.

[13] M. Dutta, S. Mridha, D. Basak, Effect of sol concentration on the properties of ZnO thin films prepared by sol–gel technique, *Applied Surface Science* 254(2008) 2743-2747.

[14] P. P. Kiran, B. N. Sh. Bhaktha, D. N. Rao," Nonlinear optical properties and surface-plasmon enhanced optical limiting in Ag–Cu nanoclusters co-doped in SiO_2 Sol-Gel films, *J. Appl. Phys.* 96 (2004) 6517-6723.

[15] R. Ashiri, A. Nemati, M. Sasani Ghamsari, H. Adelkhani, Characterization of optical properties of amorphous $BaTiO_3$ nanothin films, *Journal of Non-Crystalline Solids* 355 (2009) 2480–2484.

[16] Y.J. Guo, X.T. Zu, X.D. Jiang, X.D. Yuan, W.G. Zheng, S.Z. Xu, B.Y. Wang ,Laser-induced damage mechanism of the sol , Nucl. Inst. Met. Phys. Res. Sec. B: *Beam Interactions with Materials and Atoms* 266 (2008) 3190-3194.

[17] P. C. A. Jeronimo, A. N. Araujo, M. C. B. S. M. Montenegro, Optical sensors and biosensors based on sol-gel films: Review, *Talanta* 72 (2007) 13–27.

Lukowiak, W. Strek, Sensing abilities of materials prepared by sol-gel technology, *J. Sol-Gel Sci. Technol.* 50 (2009) 201–215.

[18] D. Anir, T. Coradin, O. Lev, J. Livage,“ Resent bio-applications of sol-gel materials, *J. Mater. Chem.* 16 (2006) 1013-1030.

[19] M.F. Desimone, G.S. Alvarez, M.L. Foglia, L.E. Diaz, Development of sol-gel hybrid materials for whole cell immobilization, Recent Pat Biotechnol. 3 (2009) 55-60.

[20] C. Zolkov, D. Avnir, R. Armon, Tissue-derived cell growth on hybrid sol–gel films, *J. Mater. Chem.* 14 (2004) 2200-2205.

[21] T. Cohen, J. Starosvetsky, U. Cheruti, R. Armon, Whole cell Imprinting in sol-gel thin films for bacterial recognition in liquids: Macromolecular Fingerprinting, *Int. J. Mol. Sci.* 11(2010) 1236-1252.

[22] C. Zolkov, D. Avnir, R. Armon, Sol-gel based poliovirus-1 detector,. *J. Virol. Methods,* 155 (2009)132–135.
C. Pierre, The sol-gel encapsulation of enzymes, *Biocatalysis and Biotransformation* 22 (2004) 145-170.

[23] T.R. Besanger, J. D. Brennan, Entrapment of membrane proteins in sol-gel derived silica, *J. Sol-Gel Sci. Technol.* 40 (2006) 209-225.

[24] M. L. Ferrer, D. Levy, B. Gomez-Lor, M. Iglesias, High operational stability in peroxidase-catalyzed non-aqueous sulfoxidations by encapsulation within sol–gel glasses, *J. Mol.Catal. B: Enzym.* 27 (2004) 107-111.

[25] Bogdanoviciene, A. Beganskiene, K. Tonsuaadu, J. Glaser, H-J. Meyer, A. Kareiva, Calcium hydroxyapatite, $Ca_{10}(PO_4)_6(OH)_2$ ceramics prepared by aqueous sol–gel processing,*Mater. Res. Bull.* 41 (2006) 1754-1762.

[26] D. M. Carta, D. Pickup, J. C. Knowles, E. Smith, R. J. Newport, Sol–gel synthesis of the P_2O_5–CaO–Na_2O–SiO_2 system as a novel; bioresorbable glass, *J. Mater. Chem.* 15 (2005) 2134-2140.

[27] T. Coradin, J. Livage, Aqueous silicates in biological sol-gel applications new perspectives for old precursors, *Acc. Chem. Res.*, 40 (2007) 819–826.

[28] K. Bewilogua, G. Brauer, A. Dietz, J. Gabler, G. Goch, B. Karpuschewski, B. Szyszk, Surface technology for automotive engineering , *CIRP Annals - Manufacturing Technology* 58 (2009) 608–627.

[29] C.W. Turner, sol-gel process principles and applications, *Ceramic Bulletin* 70 (1991) 1487-1490.

[30] D. C. Bradly, R. C. Mehrotra, I.P. Rothwell, A. Singh, *Alkox and Aryloxo derivative of metals*, Academic Press Inc. (2001).

[31] G. Clavel, E. Rauwel, M.-G. Willinger, N. Pinna, Non-aqueous sol–gel routes applied to atomic layer deposition of oxides, *J. Mater. Chem.* 19 (2009) 454-462.
Vioux, Nonhydrolytic Sol-Gel Routes to Oxides, *Chem. Mater.* 9 (1997) 2292-2299.

[32] M. Niederberger, Non-aqueous sol–gel routes to metal oxide nanoparticles, *Acc. Chem. Res.* 40 (2007)793-800.

[33] N. Pinna and M. Niederberger, Surfactant-free nonaqueous synthesis of metal oxide nanostructures, *Angew. Chem., Int. Ed.* 47(2008)5292-5304.

[34] M. Niederberger, G. Garnweitner, N. Pinna and G. Neri, Non-aqueous routes to crystalline metal oxide nanoparticles- Formation mechanisms and applications, *Prog. Solid State Chem.* 33(2005)59-70.

[35] N. Pinna, G. Garnweitner, M. Antonietti, M. Niederberger, Non-aqueous synthesis of high-purity metal oxide nanopowders using an ether elimination process, *Adv. Mater.* 16 (2004) 2196–2200.

[36] M. Niederberger, N. Pinna (Eds.*), Metal oxide nanoparticles, formation, assembly and application,* Springer-Verlag London Limited, 2009.

[37] M. Torrent-Sucarrat, S. Liu, F. De Proft, Steric Effect: partitioning in atomic and functional group contributions, *J. Phys. Chem. A*, 113(2009) 3698-3702.

[38] R. W. Taft, *Steric effects in organic chemistry*, Ed. M. S. Newman, Wiley, New York, 1956.

[39] M. Hirota, K. Sakakibara, T. Yuzuri, S. Kuroda, Re-examination of steric substituent constants by molecular mechanics, *Int. J. Mol. Sci.* 6 (2005) 18-29.

[40] R. Todeschini, V. Consonni, *Molecular Descriptors for Chemoinformatics: I & II,* Wiley-VCH, Verlag, 2009.

[41] V. I. Galkin, R. D. Sayakhov, R. A. Cherkasov, Steric effects: the problem of their quantitative assessment and manifestation in the reactivity of hetero-organic compounds, *Russian Chemical Reviews* 60 (1991) 815-829.

[42] R W Taft, Polar and steric substituent constants for aliphatic and o-Benzoate groups from rates of esterification and hydrolysis of esters, *J. Am. Chem. Soc*, 74(1952) 3120–3128.

[43] K. Johnson (Ed.), *The Hammett Equation*, Cambridge University Press, New York, 1983.

[44] A.R. Cherkasov, Inductive. descriptors. 10. Successful years in QSAR, *Curr. Comput. – Aided Drug Des.* 1 (2005) 21-42.

[45] R.W. Taft, I.C. Lewis, The general applicability of a fixed scale of inductive effects. II. Inductive effects of dipolar substituents in the reactivities of m- and p-substituted derivatives of benzene, *J. Am. Chem. Soc.*, 80 (1958) 2436-2443.

[46] R.W. Taft, Sigma values from reactivities, *J. Phys. Chem.* 64 (1960) 1805-1815.

[47] M. Charton, The estimation of Hammett substituent constants, *J. Org. Chem.* 28 (1963) 3121-3124.

[48] M. Charton, Definition of .inductive: substituent constants, *J. Org. Chem.* 29 (1964) 1222-1227.

[49] J. D. Wright, N. A. J. M. Sommerdijk, *Sol-gel materials: chemistry and applications*, CRC Press, 2001.

[50] P.J. Flory, *Principle of polymer chemistry*, Cornell University Press, 1953.

[51] M.V. Smoluchowski, *Z. Phys. Chem.* 92 (1917)129.

[52] Boris L. Granovsky · Michael M. Erlihson, On Time Dynamics of Coagulation-Fragmentation Processes, *J. Stat. Phys.* 134 (2009) 567–588.

[53] J. Bertoin, Two solvable systems of coagulation equations with limited aggregations, *Ann. I. H. Poincaré – AN* 26 (2009) 2073-2089.

[54] M. R. Yaghouti, F. Rezakhanlou, A. Hammond, Coagulation, diffusion and the continuous Smoluchowski equation, *Stochastic Processes and their Applications* 119 (2009) 3042-3080.

[55] V. N. Kolokoltsov, The central limit theorem for the Smoluchovski coagulation model, *Probab. Theory Relat. Fields* 146 (2010) 87-153.

[56] A.C. McBride, A.L. Smith, W. Lamb, Strongly differentiable solutions of the discrete coagulation_fragmentation equation, *Physica D* (2010) in press.

[57] J. Laurenzi1, S. L. Diamond, Kinetics of random aggregation-fragmentation processes with multiple components, *Phys. Rev. E* 67 (2003)051103-051117.

Boudaoud, J. Bico, B. Roman Elastocapillary coalescence: Aggregation and fragmentation with a maximal size, *Phys. Rev. E* 76 (2007) 060102-060107.

[58] H. J. herrmann, D. P. Landau, New universality class for kinetic gelation, *Phys. Rev. Lett.* 49 (1982) 412-415.

[59] E. M. Hendriks, M. H. Ernst, R. M. Ziff, Coagulation equations with gelation, *J. Statis. Phys.* 31(1983)519-563.

[60] J. G. Crump, J. H. Seinfeld, On existence of steady-state solutions to the coagulation equations, *J. Colloid and Inter. Sci.* 90 (1984) 469-476.

[61] M. H. Ernst, R. M. Ziff, E. M. Hendriks, Coagulation processes with a phase transition, *J. Colloid and Inter. Sci.* 97 (1984) 266-277.

[62] R. M. Ziff, M. H. Ernst, E. M. Hendriks, A transformation linking two models of coagulation, *J. Colloid and Inter. Sci.* 100 (1984) 220-223.

[63] P. G. J. Van Dongen, M. H. Ernst, On the occurrence of a gelation transition in smoluchowski's coagulation equation, *J. Statis. Phys.* 44 (1986) 785-792.

[64] B. Lu, The evolution of the cluster size distribution in a coagulation system, *J. Statis. Phys.* 49 (1987) 669-684.

[65] P. G. J. Van Dongen, M. H. Ernst, Tail distribution of large clusters from the coagulation equation, *J. Colloid and Inter. Sci.* 115 (1987) 27-35.

[66] J. E. Martine, D. Adolf, J. P. Wilcoxon, Viscoelasticity near the sol-gel transition, *Phys. Rev. A* 39 (1989)1325-1332.

[67] B. J. McCoy, G. Madras, Evolution to similarity solutions for fragmentation and aggregation, *J. Colloid and Inter.* Sci. 201 (1998) 200-209.

[68] S. C. Davies, J. R. King, J. A. D. Wattis, Self-similar behaviour in the coagulation equations, *J. Eng. Math.* 36 (1999) 57-88.

[69] M.A. Herrero, J.J.L. Velázquez, D. Wrzosek, Sol–gel transition in a coagulation–diffusion model, *Physica D* 141 (2000) 221–247.

[70] R. Botet, M. Płoszajczak, Universal features of the order-parameter fluctuations: Reversible and irreversible aggregation, *Phys. Rev. E* 62 (2000) 1825-1841.

[71] B. J. McCoy, G. Madras, Discrete and continuous models for polymerization and depolymerization, *Chemi. Eng. Sci.* 56 (2001) 2831-2836.

[72] G. Madras, B. J. McCoy, Numerical and Similarity Solutions for Reversible Population Balance Equations with Size-Dependent Rates, *J. Colloid and Inter. Sci.* 246 (2002) 356-365.

[73] [B. J. McCoy, A population balance framework for nucleation, growth, and aggregation, *Chemi. Eng. Sci.* 57 (2002) 2279-2285.

[74] G. Madras, B. J. McCoy, Distribution kinetics of Ostwald ripening at large volume fraction and with coalescence, *J. Colloid and Inter. Sci.* 261 (2003) 423-433.

[75] F. Leyvraz, Scaling theory and exactly solved models in the kinetics of irreversible aggregation, *Physics Reports* 383 (2003) 95-212.

[76] R. Li, B.J. McCoy, R. B. Diemer, Cluster aggregation and fragmentation kinetics model for gelation, *J. Colloid and Inter. Sci.* 291 (2005) 375-387.

[77] F. Leyvraz, Scaling theory for gelling systems: Work in progress, *Physica D* 222 (2006) 21-28.

[78] G. Madras, B. J. McCoy, Kinetics and dynamics of gelation reactions, *Chemi. Eng. Sci.* 62 (2007) 5257-5263.

A. Lushnikov, From Sol to Gel Exactly, *Phys. Rev. Lett.* 93 (2004) 198302-198305.

A. Lushnikov, Exact kinetics of sol-gel transition, *Phys. Rev. E* 71 (2005) 046129-046139.

A. Lushnikov, Exact kinetics of sol-gel transition in a coagulating mixture, *Phys. Rev. E* 73 (2006) 036111-036122.

[79] A. Lushnikov, Gelation in coagulating systems, *Physica D* 222 (2006) 37-53.

[80] A. Lushnikov, Evolution of coagulating systems, *J. Colloid Interface Sci.* 45 (1973) 549-59.

[81] H. Vehkamäki, *Classical nucleation theory in multicomponent systems*, Springer-Velrag, (2006) p. 119.

[82] M. N. Rahaman, Ceramic Processing and Sintering, Marcel Deker, (2003) p. 530.

[83] J. P. van der Eerden, O. S. L. Bruinsma, " *Science and technology of crystal growth*, Kluwer Academic Publisher, (1995) p. 53.

[84] J. Schmelzer, J.W. P. Schmelzer, *Nucleation theory and applications*, Willy-VCH, (2005) p. 74.

[85] D. Kashciev, *Nucleation basic theory with application*, B.H., Bulgaria, (2000) p. 113.

[86] G. Cao, *Nanostructures & nanomaterials: synthesis, properties & applications*, Imperial College Press, (2004) p. 58.

[87] U. Gasser, E. R. Weeks, A. Schofeld, P. N. Pusey, D. A. Weitz, Real- pace imaging of nucleation and growth in colloidal crystallization ",*Science*, 292 (2001) 258-262.

[88] Humphrey J. Maris, Introduction to the physics of nucleation, *C. R. Physique,* (2006) 946–958.

[89] S. Auer, D. Frenkel, Prediction of absolute crystal nucleation rate in hard-sphere colloids, *Nature* 409 (2001) 1020-1023.

[90] J. W. Mullinw, *Crystallization*, Butterworth-Heninemann, Forth Edition, (2001) p. 135.

[91] T. Sugimoto, F. Shiba, Spontaneous nucleation of monodisperse silver halide particles from homogeneous gelatin solution II: silver bromide , *Physicochemical and Engineering Aspects* 164 (2000) 205–215.

[92] R. W. Cahn, P. Haasen, *Physical metallurgy,* Elsevier Science B. V., 3 (1996) p. 679.

A. G. Jones, *Crystallization Process System,* Butterworth-Heinemann, (2002) p. 125.

[93]] M. Kulmala, K. E. J. Lehtinen, and A. Laaksonen, "Cluster activation theory as an explanation of the linear dependence between formation rate of 3nm particles and sulphuric acid concentration, *Atmos. Chem. Phys*. 6 (2006) 787–793.

[94] Y. Tai, C.-Y. Shih, A new model relating secondary nucleation rate and supersaturation, *J. Crys. Growth* (1996) 186-189.

Gadomski, On the kinetics of polymer crystallization: a possible mechanism, *J. Mole. Liqui*. 86 (2000) 237-247.

[95] . Palberg, Crystallization kinetics of repulsive colloidal spheres, *J. Phys. Cond. Mat.* (1999) 323-360.

[96] H. Garabedian, R.F. Strickland-Constable, Collision breeding of crystal nuclei: Sodium chlorate. II, *J. of Crystal Growth* 12 (1972) 53-56.

[97] [109] W.L. McCabe, Crystal growth in aqueous solutions:1-Theory, *Ind. Eng. Chem.* 21 (1929) 30-33.

[98] Erroll P. K. Ottens, Esso J. de Jong, A model for secondary nucleation in a stirred vessel cooling crystallizer, *I*nd. Eng. Chem. 12 (1973) 179-184.

[99] V. LaMer, R. Dinegar, Theory, Production and mechanism of formation of monodispersed hydrosols, *J. Am. Chem. Soc*. 72 (1950) 4847-4854.

[100] J. K. Beattie, Monodisperse colloids of transition metal and lanthanide compounds, *Pure & Appl. Chem*. 61(1989) 937-941.

[101] Destree, J. B. Nagy, Mechanism of formation of inorganic and organic nanoparticles from microemulsions, *Colloid and Interface Science*, (2006) 353-367.

[102] E. Finney, R. G. Finke, Nanocluster nucleation and growth kinetic and mechanistic studies: A review emphasizing transition-metal nanoclusters, *J. Colloid and Inter. Sci.* 317(2008) 351-374.

[103] J. Brinker, G. W. Scherer, *Sol-gel science: the physics and chemistry of sol-gel processing*, Academic Press Inc., (1990) p. 279.

[104] K. Do Kim, H. T. Kim, Formation of monodisperse TiO_2 nanoparticles generated without HPC dispersant by controlling reaction method and temperature, *J. Indu. and Eng. Chemi.* 6(2000) 281-286.

[105] J. H. Jean, T. A. Ring, *Langmuir*, 2 (1984) 221-225.

[106] M. Keshmiri, O. Kesler, Colloidal formation of monodisperse YSZ spheres: Kinetics of nucleation and growth, *Acta Materials*, 54 (2006) 4149–4157.

[107] H. Mehranpour, M. Askari, M.S. Ghamsari, H. Farzalibeik, Application of Sugimoto model on particle size prediction of colloidal TiO_2 nanoparticles, *Nanotech* 2010, vol. 1, Chapter 3, 436 – 439.

[108] K. Shingyouchi, A. Makishima, M. Tutumi, S. Takenouchi, S. Konishi, Determination of diffusion, *Journal of Non-Crystalline Solids* 100(1988) 383-387.

[109] Burtrand I. Lee, Edward J. A. Pope, *Chemical processing of ceramics,* Marcel Decker, 1994, p. 160.

G. Jones, *Crystallization process Systems*, Butterworth-Heinemann, 2002, p. 155.

In: The Sol-Gel Process
Editor: Rachel E. Morris
ISBN 978-1-61761-321-0

Chapter 27

SOL GEL TITANIA: LEADER HETEROGENEOUS PHOTOCATALYST FOR WASTE REMEDIATION

***N. N. Binitha*[*1,2], *Z. Yaakob*[1] *and M.R. Resmi*[2]**

[1]Department of Chemical and Process Engineering, Faculty of Engineering and Built Environment, National University of Malaysia, 43600 UKM Bangi, Selangor, Malaysia

[2]Department of Chemistry, Sree Neelakanta Government Sanskrit College Pattambi, Palakkad-679306, Kerala, India

ABSTRACT

Chemical industry is facing major challenges to dispose environmentally hazardous waste materials. Heterogeneous catalysis is a fascinating field which offers eco-friendly treatment steps and minimum waste; heterogeneous photocatalytic degradation of pollutants being the most desired method in this direction. Titania is well known as a cheap, stable, nontoxic, and efficient photocatalyst without secondary pollution. However, pure TiO_2 materials usually have very low quantum efficiency and poor performance in the visible region of light, thus restricting its extended applications in photocatalysis. Use of nonmetals or other metals as dopant systems is the most practicable approach for shifting the absorbance of TiO_2 to the visible region. These dopant materials bring forth new properties and enhanced activity due to structural and electronic modifications on TiO_2. Sol-gel synthesis of metal oxides is a widely accepted method for the preparation of such materials, particularly to incorporate ions in the gels. This short communication discusses the sol - gel preparation of TiO_2, its properties, and photocatalytic applications in waste remediation. A brief review on the modifications adopted for developing TiO_2 photocatalysts of improved performance is presented with examples.

[*] Corresponding author: Ph: +91 466-2212223. Fax: +91 466-2212223 Email:binithann@yahoo.co.in

1. INTRODUCTION

"Waste" describes material that was not used for its intended purpose or unwanted material produced as a consequence of another process. In the chemical industry, disposal issue of the waste products is a challenge. Some of these materials are not biodegradable and often leads to waste disposal crisis and environmental pollution. If improperly managed, this waste can pose dangerous health and environmental consequences. A variety of physical, chemical and biological methods are presently available for treatment of industrial waste. Biological treatment is a proven technology and is cost effective. However, it is found that some of the industrial waste pollutants such as dyes are intentionally designed in such a way to resist aerobic microbial oxidation and are converted to toxic or carcinogenic compounds. Physical methods such as ion exchange, adsorption, air stripping, etc., are also ineffective on pollutants which are not readily adsorbable or volatile, and have the further disadvantage that they simply transfer the pollutants to another phase rather than destroying them causing secondary pollution. This leads to search for highly effective method to degrade the industrial effluents into environmentally compatible products.

Heterogeneous photocatalytic reaction is one potential technology that can be used to destroy the pollutant waste materials using semiconductor catalyst under light irradiation. The use of semiconductor particles as photocatalysts for the degradation of toxic organic chemicals continues to be an active area of investigation [1–3]. Present communication tries to highlight the improvements of modified titania (TiO_2) for its applications in pollutant abatement.

2. TiO_2 PHOTOCATALYST

Among the semiconductor photocatalysts (oxides, sulfides, *etc.*), TiO_2 has been most extensively investigated due to its high photocatalytic activity, chemical and biological stability, insolubility in water, acidic and basic media, non-toxicity, high quantum yield and availability. The discovery of photoelectrochemical water splitting on reduced titania surfaces during 1970s was the instrumental in establishing a new research field centered round the generation of hydrogen from water by using photo-energy from sunlight absorbed on to the photoelectric or photocatalytic surfaces of oxide materials. With proper manipulation of the microstructure, crystalline phase and/or addition of proper impurities or surface functionalization, TiO_2 can be used for the degradation of wide range of pollutants. Among the different properties, crystal structure is one of the most important parameters that influence catalyst activity.

TiO_2 exists in three forms: anatase, rutile and brookite. Crystals of anatase and rutile TiO_2 are tetragonal in shape whereas brookite has a rhombic shape [4]. Among these forms, anatase and rutile TiO_2 are used for photocatalytic applications. Researchers proposed that anatase TiO_2 shows higher activity than rutile TiO_2 [5-8]. Possible reasons for its higher activity include the fact that anatase TiO_2 is less dense than rutile TiO_2. Hence anatase TiO_2 is more porous and has higher surface area than rutile TiO_2 and so it is more suitable for catalytic applications. Another possible reason for higher photocatalytic activity of anatase TiO_2 may be due to its higher Fermi energy level. Fermi energy of anatase TiO_2 is about

0.1eV higher than rutile TiO_2 [9]. This possibly increases the photocatalytic activity of anatase TiO_2.

3. DOPING

A serious disadvantage of TiO_2 photocatalysis would be that only UV light can be used for photocatalytic reactions because of the relatively high intrinsic band gap of anatase TiO_2 (3.2 eV). Thus only 5% of the incoming solar energy on the earth's surface can be utilized for photocatalysis over unmodified TiO_2. Therefore, it is of great interest to find ways to extend the absorption wavelength range of TiO_2 to visible region without the decrease of photocatalytic activity [10]. Doping is the process of intentionally introducing impurities into an extremely pure semiconductor to change its properties. Via the inclusion of specific dopants it should be possible to improve the efficiency of the photocatalytic behavior by creating new band structures or by suppressing the recombination of photogenerated electron–hole pairs to improve quantum efficiency.

There are two types of doping, metal and non metal doping. Transition-metal element doping has proved to be partially successful because it may induce thermal instability of TiO_2, and the dopant sites, also serving as carrier-recombination centers, can reduce the photocatalytic efficiency [11]. Transition metal doping can also have the adverse effect on the photocatalytic activity through the formation of localized d-states in the band gap of TiO_2 [12]. Shockley–Read–Hall model describes that the localized d-states act as a trapping site that capture electron from conduction band or hole from valence band [13]. Therefore transition metal doping shortens the lifetime of mobile carriers and ultimately reduces the activity of a photocatalyst. Non-metal doping is the alternate being pursued for improving the visible light response of TiO_2. To date, the desired band gap narrowing of TiO_2 can be better achieved by using anionic dopant species, such as carbon, nitrogen, fluorine, sulfur etc, rather than metals [14–19].

3.1. Nitrogen Doped TiO_2

Among the different nonmetals, nitrogen doped TiO_2 is the widely studied photocatalyst because N dopant is found to be the most successful anion dopant for decreasing the band gap and increasing the photocatalytic activity. Sato [12] reported for the first time that a TiO_2-based material from the mixtures of a commercial titanium hydroxide and ammonium chloride calcined at about 400˚C showed higher photocatalytic activity in the visible-light region. Asahi et al. [14] reported that nitrogen-doped TiO_2 could induce the visible-light activity in which nitrogen atoms substituted small quantity of oxygen atoms (0.75%), and the doped nitrogen was responsible for the visible-light sensitivity due to the narrowing of the band gap by mixing the N2p and O2p states. It has initiated a new research area to extend the photo absorbance into visible-light region using nitrogen-doped TiO_2. Nitrogen-doped TiO_2 have been produced through different processes, such as hydrolytic process [20-21], mechanochemical technique [22-24], reactive DC magnetron sputtering [25], high temperature treatment of titania under NH_3 flow [14, 15], sol-gel method [26], solvothermal process [27], and calcination of a complex of Ti & with a nitrogen- containing ligand [28]

4. Sol Gel Method

The synthesis of metallic oxides by the sol–gel process is presently widely accepted method for the preparation of such materials, in particular when it is desired to incorporate other ions in the gels [29,30]. Several important features characterize these solids, which make them differ considerable from those prepared by more conventional methods. Sol–gel processing enables materials to be mixed on an atomic level and thus crystallization and densification to be accomplished at a much low temperature. The advantages of the sol–gel process in general are high purity, homogeneity and low temperature. For a lower temperature process, there is reduced loss of volatile components and thus the process is more environmental friendly. The final properties shown by these materials depend on the hydrolysis conditions, the pH of the gelling solution, the nature of the salt added before gelling [31]. When TiO_2 is prepared by sol–gel method, it can be obtained with a crystallite size in the range of nanometers. There are successful reports on the preparation of visible light active nitrogen doped TiO_2 via sol - gel method using, NH_3, Urea, Hexamethylenetetramine etc as dopants.

5. Mechanism of Photocatalysis over TiO_2

When photocatalyst TiO_2/ doped TiO_2 absorbs Ultraviolet (UV)/visible radiation from sunlight or illuminated light source, it will produce pairs of electrons and holes. The electron of the valence band of TiO_2 becomes excited when illuminated by light. The excess energy of this excited electron promoted the electron to the conduction band of TiO_2 therefore creating the negative-electron (e^-) and positive-hole (h^+) pair. The energy difference between the valence band and the conduction ie the band gap is 3.2 eV.

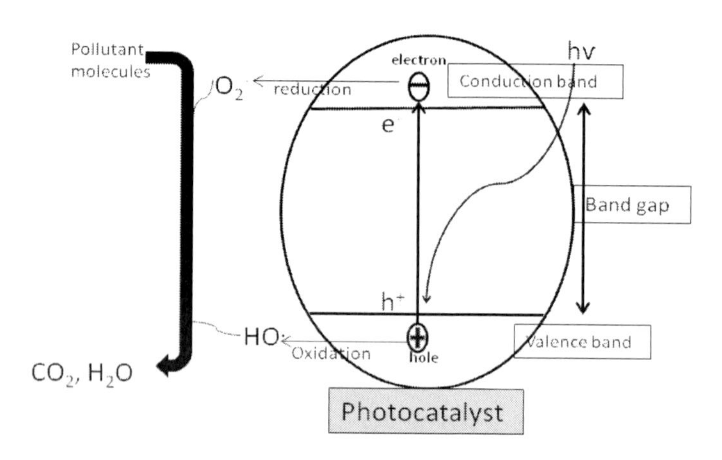

Figure 1. Mechanism of Photocatalysis.

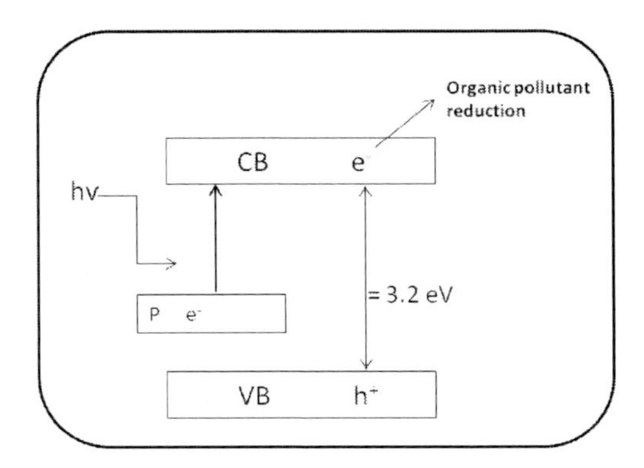

Figure 2. Reduced band gap of N-doped TiO_2 by the formation of a new N2p band above the O2p valence band.

The positive-hole of TiO_2 breaks apart the water molecule to form hydrogen gas and hydroxyl radical. The negative-electron reacts with oxygen molecule to form super oxide anion. This cycle continues when light is available. The resulting •OH radical, being a very strong oxidizing agent (standard redox potential +2.8 V) [32], can oxidize all organic compounds to the mineral final products, *i.e.* CO_2, H_2O and corresponding inorganic ions.

Asahi et al. [14] reported the narrowing of the band gap by mixing the N2p and O2p states (similar is the case for other non metals also). However, some studies show two absorption edges in the UV visible absorbance/reflectance spectra, one around 400 due to the band structure of original TiO_2 and second one around 530 nm which is attributed to the newly formed N2p band which is located above the O2p valence band

The nature of N dopant revealed from X-ray photoelectron spectroscopic (XPS) studies shows that N exists in substitutional as well as interstitial positions.

6. Photoactivity Studies

Since the discovery of the photosensitization effect of the TiO_2 electrode on the electrolysis of water into H_2 and O_2 by Honda and Fujishima in 1972 [33], photocatalysis by TiO_2 semiconductors has been widely studied with the aim of efficiently converting light energy into reliable and effective chemical energy [34–37]. Table 1 shows the photoactivity results on the degradation of pollutants over some selected modified TiO_2 catalysts [38-47].

System	Photoactivity studies
Ir and Co ionized TiO_2 nanotube	Production of stoichiometric hydrogen and oxygen by the splitting of water under the visible light irradiation.
phosphor-doped TiO_2 nanoparticles	photodegradation of methylene blue and 4-chlorophenol under visible light
S^{6+}-doped TiO_2	Shows strong visible-light induced antibacterial effect.
N-Ni co doped TiO_2 catalyst	Exhibits an enhancement of 200% in visible light photoactivity for the decomposition of formaldehyde Compared with P25.
Fe/TiO_2 binary catalyst	Effectively degrades o-cresol

(Continued)

System	Photoactivity studies
phosphate-modified nano TiO_2 films	3 times active for the degradation of ethanol when compared to P25
Fe^{3+} and Zn^{2+} doped TiO_2 nanotube (TNT)	Modification of TNT with Fe^{3+} ions induced improvement of photocatalytic activity for acetaldehyde under visible-light irradiation as well as ultraviolet (UV) irradiation. TNT modified with Zn^{2+} ions showed the largest enhancement of photocatalytic activity under UV irradiation
TiO_2-modifield mesoporous MCM-41	Exhibits relatively high photocatalytic activity in oxidative decomposition of phenol
Gd-doped nano TiO_2	Photoactivity for the degradation of X-3B and Rhodamine B was found to be higher than that of P25.
Nano TiO_2 doped with boron and nitrogen	Reusable photocatalysts for MO degradation under visible light

Conclusion

In summary, we briefly reviewed the recent developments on environmental pollutant degradation giving emphasis to heterogeneous photocatalysis over TiO_2. The need of modifications on TiO_2 for visible activity is understood from the mechanism of photocatalysis. Among the modifications adopted, non metal doping, especially nitrogen doping is found to fit for this application. The excellent activity after nitrogen doping is a result of the decreased band gap attributed to the N 2p band. The photoactivity is found to be very good after doping, especially during co doping of metals and nonmetals.

References

[1] Hoffmann, MR; Martin, ST; Choi, W; Bahnemann, DW. *Chem. Rev.,* 1995, 95, 69-96.

[2] Herrmann, JM. *Catal. Today,* 1999, 53,115-129.

[3] Mathews, RW. *J. Catal.*, 1988, 111, 264-272.

[4] Yamamoto, S; Sumita, T; Sugiharuto; Miyashita, A; Naramoto, H. *Thin Solid Films,* 2001, 401, 88-93.

[5] Bickley, RI; G onzalez-Carreno, T; Lees, JS; Palmisano, L; Tilley, RJ. *J. Solid State Chem.*, 1991, 92, 178-190.

[6] Augustynski, J. *Electrochem. Acta.*, 1993, 38, 43–46.

[7] Karakitsou, KE; Verykios, XE. *J. Phys. Chem.*, 1993, 97, 1184–1189.

[8] Ding, Z; Lu,G; Greenfield, P. J. Phys. Chem. B., 2000, 104, 4815–4820.

[9] Porkodi, K; Daisy Arokiamary, S. *Materials Characterization.*, 2007, 58, 495–503.

[10] Vinodgopal, K; Kamat, PV. *Environ. Sci. Technol.* 1995, 29, 841-845.

[11] Choi, W; Termin, A; Hoffmann, MR. *J. Phys. Chem.*, 1994, 98, 13669-13679.

[12] Sato, S; Chem. *Phys. Lett.*, 1986, 123, 126-128.

[13] Zhu, YF; Zhang, L; Yao, WQ; Cao, LL. *Appl. Surf. Sci.*, 2000, 158, 32-37.
[14] Asahi, R; Morikawa, T; Ohwaki, T; Aoki, K; Taga, Y. *Sci.*, 2001, 293, 269-271.
[15] Irie, H; Watanabe, Y; Hashimoto, K. *J. Phys. Chem. B.*, 2003, 107, 5483-5486.
[16] Yu, JC; Yu, JG; Ho, WK; Jiang, ZT; Zhang, LZ. *Chem. Mater.*, 2002, 14, 3808-3816.
[17] Kim, H; Choi, W. *Appl. Catal. B: Environ.*, 2007, 69, 127-132.
[18] Ohno, T; Akiyoshi, M; Umebayashi, T; Asai, K; Mitsui, T; Matsumura, M. *Appl. Catal. A: Gen.*, 2004, 265, 115-121.
[19] Demeestere, K; Dewulf, J; Ohno, T; Salgado, PH; Van Langenhove, H. *Appl. Catal. B*: Environ., 2005, 61, 140-149.
[20] Noda, H; Oikawa, K; Ogata, T; Matsuki, K; Kamada, H. *Chem. Soc. Jpn.*, 1986, 8, 1084-1090.
[21] Salthivel, S; Kisch, H. *Chem. Phys. Chem,* 2003, 4, 487-490.
[22] Yin, S; Yamaki, H; Komatsu, M; Zhang, QW; Wang, JS; Tang, Q; Saito, F; Sato, T. *J. Mater. Chem.*, 2003, 13, 2996-3001.
[23] Yin, S; Yamaki, H; Zhang, QW; Komatsu, M; Wang, JS; Tang, Q; Saito, F; Sato, T. *Solid State Ionics.*, 172, 2004, 205-209.
[24] Wang, JS; Yin, S; Komatsu, M; Zhang, QW; Saito, F; Sato, T. *Appl. Catal. B. Environ.*, 52, 2004, 11-21.
[25] Lindgren, T; Mwabora, JM; Avendano, E; Jonsson, J; Hoel, ACG. Granqvist, SE. *Lindquist, J. Phys. Chem.*, B 107, 2003, 5709-5716.
[26] Burda, C; Lou, YB; Chen, XB; Samia, ACS. Stout, J; Gole, JL. *Nano Lett.*, 3, 2003,1049-1051.
[27] Aita, Y; Komatsu, M; Yin, S; Sato, T. *J. Solid State Chem.*, 177, 2004, 3235-3238.
[28] Sano, T; Negishi, N; Koike, K; Takeuchi, K; Matsuzawa, S. *J. Mater. Chem.*, 14, 2004, 380-384.
[29] Lopez, T; Sanchez, E; Bosch, P; Meas, Y; Romez, R. *Mater. Chem. Phys.*, 1992, 32, 141
[30] Pecchi, G; Reyes, P; Orellana, F; Lopez, T; Gomez, R; Fierro, JLG. *J. Tech. Biotechnol.*, 1999,74, 1.
[31] Kumar Krishnankutty Nair, P. *Appl. Catal. A.*, 1994, 119, 163-174
[32] Janes, DR; Nigel, JB. *Rev. Water Res.*, 2001, 35, 2101-2111.
[33] Fujishima, A; Honda, K. *Nature.*, 1972, 23, 37-38.
[34] Kawai, T; Sakata, T. *Chem Phys Lett.*, 1980, 72, 87-89.
[35] Kraeutler, B; Bard, AJ. *J Am Chem Soc.*, 1978, 100, 2239-2240.
[36] Fox, MA. *Acc Chem Res.*, 1983, 16, 314-321.
[37] Fox, MA; Dulay, MT. *Chem Rev.*, 1993, 93, 341-357.
[38] Khan, MA; Yang. OB. *Cat. Today.*, 2009, 146, 177-182.
[39] Lin, L; Lin, W; Xie, JL; Zhu, YX; Zhao, BY; Xie, YC. *Appl. Catal. B.*, 2007, 75, 52-58
[40] Yu, JC; Ho, W; Yu, J; Yip, H; Wong, PK; Zhao, J. *Environ. Sci. Technol.*, 2005, 39, 1175-1179.
[41] Zhang, X; Liu, Q. *Applied Surface Science.* 2008, 254, 4780–4785.
[42] Pal, B; Hata, T; Goto, K; Nogami, G. *J. Molecular Catalysis A: Chemical.*, 2001, 169,147–1550
[43] Korosi , L; Oszko, A; Galbacs G; Richardt, A; Zollmer, V; Dekany, I. Applied Catalysis B: *Environmental.*, 2007, 77, 175–183.

[44] Murakami, N; Fujisawa, Y; Tsubota, T; Ohno, T. Applied Catalysis B. *Environmental.* 2009, 92, 56–60.

[45] Zheng, S; Gao, L; Zhang, Q; Guo, J. *J. Mater. Chem.*, 2000, 10, 723-727.

[46] Xu, J; Ao, Y; Fu, D; Yuana, C. Colloids and Surfaces A: *Physicochem. Eng. Aspects.*, 2009, 334, 107–111.

[47] Gombac, V; De Rogatis, L; Gasparotto, A; Vicario, G; Montini, T; Barreca, D; Balducci, G; Fornasiero, P; Tondello, E; Graziani, M. *Chemical Physics.*, 2007, 339, 111-123.

In: The Sol-Gel Process
Editor: Rachel E. Morris

ISBN 978-1-61761-321-0

Chapter 28

CONTROLLING SILICA NETWORKS USING BRIDGED ALKOXIDES FOR HYDROGEN SEPARATION MEMBRANES

Toshinori Tsuru*, Hye Ryeon Lee and Masakoto Kanezashi
Department of Chemical Engineering, Hiroshima University,
Kagamiyama Higashihiroshima, Japan

ABSTRACT

Tetraethoxysilane (TEOS) is a commonly used precursor for preparation of sol-gel derived silica membranes. In this chapter, we propose the use of a new type of alkoxide, a bridged alkoxide which contains siloxane bonding or organic group between 2 silicon atoms, such as bis (triethoxysilyl) ethane (BTESE) and 1,1,3,3-tetraethoxy-1,3-dimethyldisiloxane (TEDMDS), for the development of a highly permeable hydrogen separation membrane. The concept for improvement of hydrogen permeability of a silica membrane is to design a loose silica network using bridged alkoxides, i.e., to shift the silica networks to a larger pore size for an increase in H_2 permeability. BTESE silica membranes showed approximately one order magnitude high H_2 permeance (0.2-1.0x10^{-5} mol/(m^2·s·Pa)) compared with previously reported TEOS-derived silica membranes, and high permeance ratios of H_2 to SF_6 ($\alpha(H_2/SF_6)$=1,350-36,300) with low permeance ratio of H_2 to N_2 ($\alpha(He/N_2)$ ~10). TEDMDS membranes also showed loose silica networks. The present result confirms the new concept of controlling silica networks for designing silica membranes.

INTRODUCTION

Amorphous silica network structures allow permeation of the smallest molecules such as helium (kinetic diameter: 0.26 nm) and hydrogen (0.289 nm), but not larger molecules such

* Corresponding author: E-mail: tsuru@hiroshima-u.ac.jp

as nitrogen (0.364 nm). [1] Silica membranes, which can maintain their thermally stable amorphous structures in a wide temperature range up to 1000°C, have drawn a great deal of attention for hydrogen separation membranes. Since silica membranes for hydrogen separation, which were prepared on porous substrates such as α-alumina and vycor glass, were first reported in 1989 by thermal CVD [2] and in 1990 by sol-gel processing [3]. In general, thermal CVD membranes show high selectivity, while sol-gel silica membranes show high permeability. Oyama and co-workers [4, 5] reported on high-temperature (HT, 600°C) CVD derived ultrathin silica membranes on graded mesoporous supports. The HT-CVD membrane showed a high hydrogen permeance in the order of 10^{-7} $mol{\cdot}m^{-2}{\cdot}s^{-1}{\cdot}Pa^{-1}$ with H_2/CH_4 selectivity above 1,000 at 600°C. Compared to CVD derived silica membranes, sol-gel derived silica membranes show approximately one order magnitude high hydrogen permeance with moderate hydrogen selectivity (<1,000) [6-9]. Tetraethoxysilane (TEOS), $Si(OC_2H_5)_4$ consisting of mono silicon, is a commonly used precursor for preparation of sol-gel derived silica membranes. In the process of the hydrolysis and polymerization reaction of TEOS, the (≡Si-O-) unit can be a minimum for amorphous silica networks. Base on molecular dynamics (MD) simulation, the amorphous silica structure had a large number of pores with a pore size of less than 0.25 nm, through which He (kinetic diameter: 0.26 nm) and H_2 (kinetic diameter: 0.289 nm) molecules cannot permeate [10].

In this chapter, a novel strategy of designing silica networks using bridged alkoxides which contains siloxane bonding or organic group between 2 silicon atoms, such as bis (triethoxysilyl) ethane (BTESE, $(C_2H_5O)_3{\equiv}Si{-}C_2H_4{-}Si{\equiv}(OH_5C_2)_3$) and 1,1,3,3-teraethoxy-1,3-dimethyldisiloxane (TEDMES, $(C_2H_5O)_2(CH_3)Si{-}O{-}Si(CH_3)(OH_5C_2)_2$), is introduced for the development of a highly permeable hydrogen separation membrane. [11-13] As schematically shown in Figure 1, bridged alkoxides such as BTESE and TEDMES will form more loose silica networks than TEOS, since ≡Si-C-C-Si≡ and ≡Si-O-Si≡ are minimum units and form silica networks as a spacing unit for BTESE and TEDMES-derived silica networks, respectively. The concept for improvement of hydrogen permeability of a silica membrane is to design a loose silica network using bridged alkoxides, i.e., to shift the silica networks to a larger pore size for an increase in H_2 permeability.

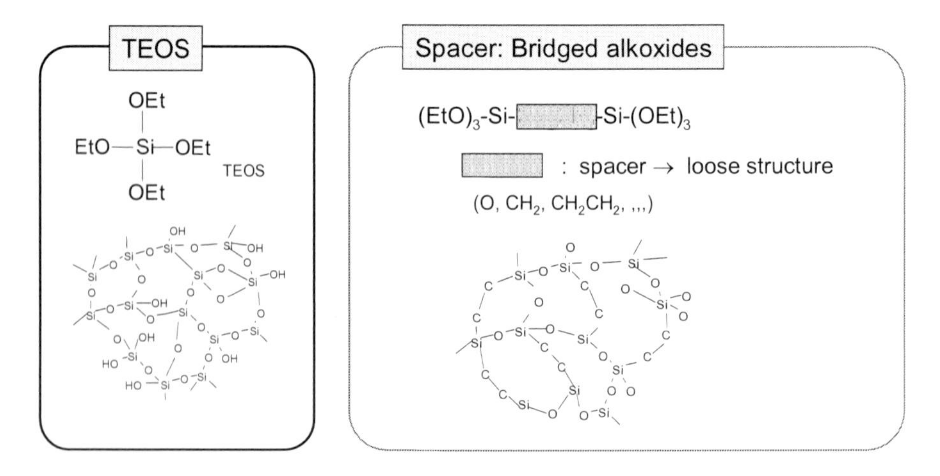

Figure 1. Strategy of bridged alkoxides for controlling silica networks.

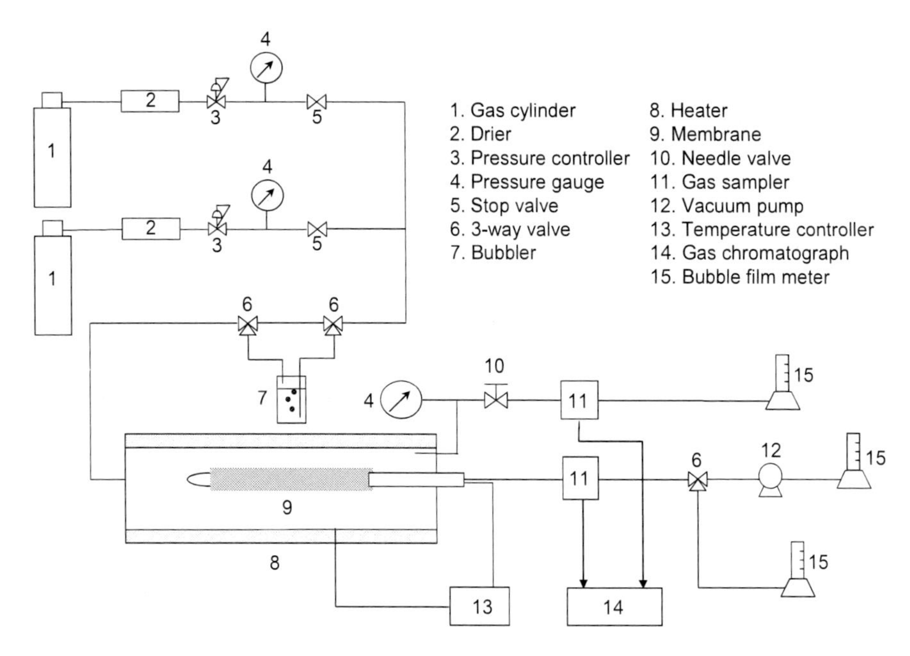

Figure 2. Schematic experimental apparatus for gas permeation.

EXPERIMENTAL

Silica polymeric sols were prepared from BTESE and TEDMES in ethanol solutions. BTESE polymer was prepared by hydrolysis and polymerization reaction of BTESE in ethanol with water and HCl as a catalyst in the molar ratio of BTESE/EtOH/H_2O/HCl= 1/28/6/0.1 at 25°C. A specified amount of TEDMDS and water with acetic acid as acid catalyst in ethanol, were stirred at 60°C for 24h with the composition of the solution adjusted at TEDMDS: H_2O: CH_3COOH = 1: 20: 0.2 in a molar ratio. Two types of porous α-alumina supports were used for coating silica sols: outer diameter of 3 mm (pore size: 150 nm, NOK Corp., Japan) and outer diameter of 10 mm (pore size: 1μm). The BTESE and TEDMDS silica polymeric sols were coated on SiO_2-ZrO_2 intermediate layers fabricated on the outer surface of cylindrical α-alumina supports, and fired from 300-500°C.

Figure 2 shows schematic diagram of experimental apparatus for gas permeation/separation measurement and hydrothermal stability test. Single gas of industrial grade (He, H_2, Ar, N_2, C_3H_8, SF_6) was fed on the outside (upstream) of a cylindrical membrane at 200-500 kPa, keeping the downstream at atmospheric pressure.

RESULT AND DISUCSSION

Characterization of TEDMES and BTESE-Derived Membranes

Figure 3 shows FE-SEM image of the cross section of TEDMDS-derived silica membrane, showing crack-free continuous silica layer for selective hydrogen separation could be successfully formed on silica-zirconia intermediate layers. The two layers are the α-alumina layer and intermediate layer with separation layer. There was no obvious TEDMDS-derived silica layer on the surface of intermediate layer. The intermediate layer with separation layer was thinner than 200 nm on the top of the porous α-alumina layer.

Figure 4 shows the depth profile of a BTESE-derived silica membrane by XPS. The concentration of Zr that is attributed to SiO_2-ZrO_2 layer was detected from the membrane surface and decreased with an increase of depth because of approaching to α-alumina support. The concentration of Si that is attributed to both BTESE-derived silica and SiO_2-ZrO_2 layer showed high near surface and also showed a similar trend to the detection curve of Zr. The concentration of C atoms, which is attributed to C in Si-C-C-Si bonds of BTESE-derived silica, showed high near surface and decreased to 0 at 40 nm from the membrane surface. On the contrary, the concentration of Al atoms, which is obtained from Al particle layer and alumina support, was 0 at the depth from 0 to 30 nm and increased at 30 nm from the membrane surface. These results suggest that BTESE-derived silica layer with thickness of approximately less than 40 nm was formed inside the SiO_2-ZrO_2 intermediate layer, probably due to the penetration of BTESE sol inside the intermediate layer. The thickness of less than 40 nm is so thin that it can be considered as "molecular net".

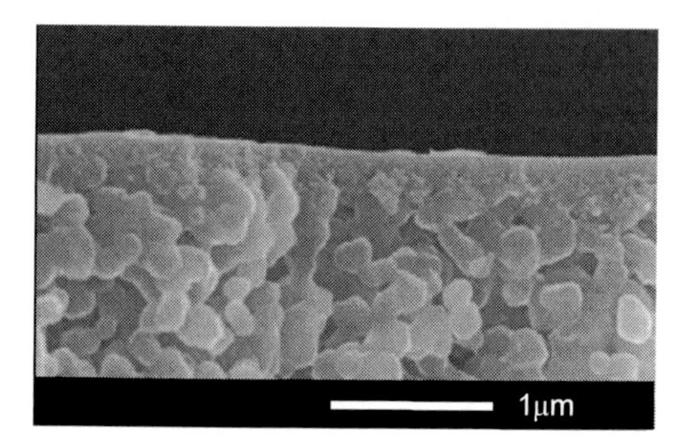

Figure 3. FE-SEM image of cross section of TEDMDS-derived silica membrane.

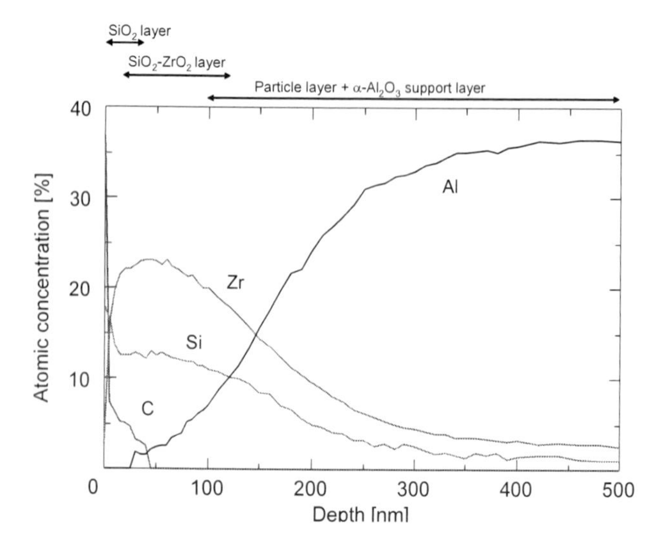

Figure 4. XPS depth profile of BTESE-derived membranes.

Permeation Properties of BTESE and TEDMES-Derived Membranes

Figure 5 shows gas permeance for TEDMDS-derived and TEOS-derived silica membrane [14] at 200°C as a function of kinetic diameter. It should be noted that TEOS-derived silica membranes were fired at 550°C in air atmosphere. TEDMDS derived silica membranes fired at 300°C showed H_2/N_2 permeance ratio of 11.6 with H_2/SF_6 permeance ratio of 260, while that fired at 450°C showed H_2/N_2 permeance ratio of 8.3 with H_2/SF_6 of 335. All gas permeances for TEDMDS-derived silica membrane fired at 450°C were higher than those fired at 300°C, however, similar pore size distributions were observed despite the different calcination temperatures. This is because thermo-gravimetric change was less than 6% between 300°C and 450°C. Both TEDMDS-derived silica membranes showed slightly high H_2 permeance than He, which is governed by Knudsen permeance through the loose amorphous networks. On the other hand, TEOS-derived membrane showed high selectivity of H_2/N_2 of 340 with H_2/SF_6 permeance ratio above 1,000. TEOS-derived silica membrane showed high He permeance than H_2 due to molecular sieving mechanism, which is different permeation characteristic from TEDMDS-derived membranes. Judging from the H_2/N_2 and H_2/SF_6 permeance ratio for TEDMDS-derived and TEOS-derived membranes, the average pore size can be estimated to be approximately in the range of 0.5-0.6 nm for TEDMDS-derived membrane and approximately 0.3 nm for TEOS-derived membrane. The difference in these pore size distributions is probably because amorphous silica structures created by TEDMDS, which consists of siloxane unit (–Si–O–Si–) are expected to be much looser than those created by TEOS with mono silicon.

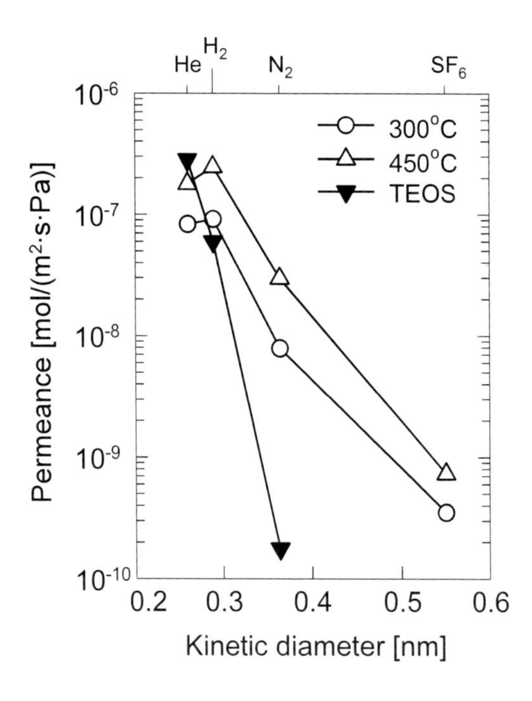

Figure 5. Gas permeances for TEDMDS-derived [13] and TEOS-derived [1] silica membrane at 200°C as a function of kinetic diameter. a function of kinetic diameter. as a function of kinetic diameter.

Figure 6 shows gas permeances for BTESE-derived silica (BTESE-1), TEOS-derived silica [14], MFI-type [15] and DDR-type [16] zeolite membranes at 200°C as a function of kinetic diameter. BTESE-1 membrane shows high hydrogen permeance of 1 x 10^{-5} mol·m^{-2}·s^{-1}·Pa^{-1} with high H_2/SF_6 permselectivity of 1,000 and low H_2/N_2 permselectivity (~10) [11]. On the other hand, a silica membrane (TEOS) [14] shows one order low H_2 permeance compared to BTESE-1 membrane with high H_2/N_2 permselectivity of 1,000. It is quite difficult to obtain the exact average pore size of sol-gel derived silica membranes due to the pore size distributions comprising of silica networks and inter-particle pores (grain boundaries) [1]. Gas permeances for zeolite membranes, which have defined zeolite channels such as MFI-type (~0.56 nm) [15] and DDR-type zeolite (~0.45 nm) [16], as a function of kinetic diameter are also shown in the same figure. Intercrystalline gaps for both zeolite membranes were plugged by counter diffusion CVD method, which is an effective tool to modify defects and intercrystalline gaps in a membrane while avoiding blocking the zeolite pores. Both zeolite membranes (MFI-type, DDR-type) show low H_2 permeance due to the thick zeolite layer (~10 μm). A MFI-type zeolite membrane shows H_2/SF_6 permselectivity of 300 with H_2/N_2 permselectivity of 5. On the other hand, DDR-type zeolite membrane shows high H_2/SF_6 permselectivity of above 10,000 with H_2/N_2 permselectivity of 20. It should be noted the permeance of SF_6 for the DDR-type zeolite membrane was less than detection limit (<10^{-11} mol·m^{-2}·s^{-1}·Pa^{-1}) [16]. Judging from the H_2 permselectivity (H_2/N_2, H_2/SF_6) for silica and zeolite membranes, the order of average pore size can be estimated as follows; TEOS-derived silica membrane (~0.3 nm)<DDR-type zeolite membrane (~0.45 nm)<BTESE-derived silica membrane (~0.5 nm)<MFI-type zeolite membrane (~0.56 nm).

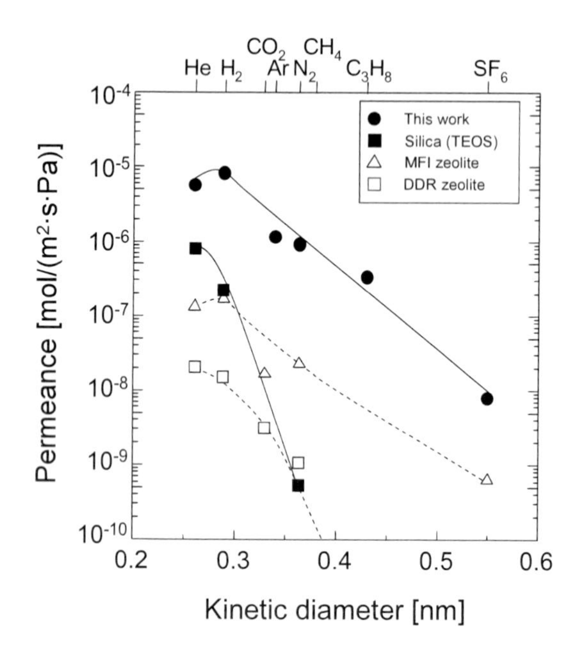

Figure 6. Gas permeances for BTESE-derived silica (M-1), TEOS-derived silica [14], MFI-type [15] and DDR-type [16] zeolite membranes at 200°C .as a function of kinetic diameter.

CONCLUSION

A novel strategy of designing silica networks using bridged alkoxides was introduced for the development of a highly permeable hydrogen separation membrane. Gas permeation experiments confirmed that bis (triethoxysilyl) ethane (BTESE) and 1,1,3,3-tetraethoxy-1,3-dimethyldisiloxane (TEDMDS) created more looser networks that TEOS with mono silicon because of bridged structures (≡Si-C-C-Si≡, ≡Si-O-Si≡).

REFERENCES

[1] Tsuru, T. (2008). Nano/subnano-tuning of porous ceramic membranes for molecular separation, *J. Sol-Gel. Technol.*, *46*, 349.

[2] Gavalas, G., Megiris, C. E. & Nam, S. W. (1989). Deposition of H_2 Permselective SiO_2 Films, *Chem. Eng. Sci.*, *44*, 1829.

[3] Kitao, S., Kameda, H. & Asaeda, M. (1990). Gas separation by thin porous silica membrane of ultra fine pores at high temperature, *MAKU (Membrane)*, *15*, 222.

[4] Lee, D., Zhang, L., Oyama, S. T., Niu, S. & Saraf, R. F. (2004). Synthesis, characterization, and gas permeation properties of a hydrogen permeable silica membrane supported on porous alumina, *J. Membr. Sci.*, *231*, 117.

[5] Gu, Y. & Oyama, S. T. (2007). High molecular permeance in a poreless ceramic membrane, *Adv. Mater.*, *19*, 1636.

[6] Uhlhorn, R. J. R., Keizer, K. & Burggraaf, A. J. (1992). Gas transport and separation with ceramic membranes. Part II. Synthesis and separation properties of microporous membranes, *J. Membr. Sci.*, *66*, 271.

[7] de Lange, R. S. A., Keizer, K. & Burggraaf, A. J. (1995). Analysis and theory of gas transport in microporous sol-gel derived ceramic membranes, *J. Membr. Sci.*, *104*, 81.

[8] Asaeda, M. & Yamasaki, S. (2001). Separation of inorganic/organic gas mixtures by porous silica membranes, *Sep. Purif. Technol.*, *25*, 151.

[9] Yoshioka, T., Asaeda, M. & Tsuru, T. (2007). A molecular dynamics simulation of pressure-driven gas permeation in a microporous potential field on silica membranes, *J. Membr. Sci.*, *293*, 81.

[10] Castricum, H. L., Sah, A., Kreiter, R., Blank, D. H. A., Vente, J. F., ten Elshof, J. E. (2008). Hybrid ceramic nanosieves: stabilizing nanopores with organic links, *Chem. Commun.*, 1103.

[11] Kanezashi, M., Yada, K., Yoshioka, T. & Tsuru, T. (2009). Design of silica networks for development of highly permeable hydrogen separation membranes with hydrothermal stability, *J. Am. Chem. Soc.*, *131*, 414.

[12] Kanezashi, M., Yada, K., Yoshioka, T. & Tsuru, T. (2010). Design of silica networks for development of highly permeable hydrogen separation membranes with hydrothermal stability, *J. Membr. Sci.*, *348*, 310.

[13] Lee, H. R., Kanezashi, M., Yoshioka, T. & Tsuru, T. Preparation of hydrogen separation membranes using disiloxane compounds, *Desalination and Water Treatment* in press.

[14] Tsuru, T., Shintani, H., Yoshioka, T. & Asaeda, M. (2006). A bimodal catalytic membrane having a hydrogen-permselective silica layer on bimodal catalytic support: Preparation and application to the steam reforming of methane, *App. Cataly. A: General 302*, 78.

[15] Kanezashi, M. & Lin, Y. S. (2009). Gas permeation and diffusion characteristics of MFI-type zeolite membranes at high temperatures, *J. Phys. Chem. C, 113*, 3767.

[16] Kanezashi, M., O'Brien-Abraham, J., Lin, Y. S. & Suzuki, K. (2008). Gas permeation through DDR-type zeolite membranes at high temperatures, *AIChE J., 54*, 1478.

In: The Sol-Gel Process
Editor: Rachel E. Morris
ISBN 978-1-61761-321-0

Chapter 29

An Introduction to the Statistical Theory of Polymer Network Formation

Pierre Gilormini*
Laboratoire Proc´ed´es et Ing´enierie en M´ecanique
et Mat´eriaux, CNRS,
Arts et M´etiers ParisTech, Paris (France)

Abstract

A short but detailed introduction to the statistical theory of polymer network formation is given, including gel formation, gel structure, and sol fraction. Focus is put on the use of probability generating functions, and results that are of interest for polymer network elasticity are emphasized. Detailed derivations are supplied, and a simple 6-step procedure is provided, so that the reader is able to adapt and apply the theory to his own chemical systems, even if examples are given on polyurethanes essentially.

Keywords: gelation, network, polymers, polyurethane.

1. Introduction

The statistical theory of polymer network formation may sometimes appear as too complicated, which limits its extensive use. The aim of this chapter is to give a short but detailed introduction to this theory, including gel formation, gel structure, and sol fraction, that is both straightforward and ready for use. An exhaustive account would be beyond an introductory chapter and extensive details can be found elsewhere, but it seemed useful to favor step-by-step applications of the theory to simple examples. The exposition uses slightly simplified notations, even if generality is lost temporarily, and differs on some points from original papers, but results are unchanged. Indications are given on various extensions of the

---

* # E-mail address: pierre.gilormini@paris.ensam.fr

theory, but emphasis is put on basic ideas, with examples given on stepwise polymerization, and on polyurethanes especially. Moreover, stress is put on the results that are useful to the understanding of polymer network elasticity. The theory presented here uses probability generating functions, but some correspondences with the recursive approach of Macosko and Miller [1, 2] are given, which is summarized in a review by Queslel and Mark [3].

This Chapter has been inspired by the long review paper that K. Du–sek published in 1989 [4], but his more recent papers have also been useful, especially one [5] where distributions of functionalities are considered. The paper that Ilavsk'y and Du–sek [6] published in 1983, where the entanglement factor is detailed, has also been very helpful. Some results of these articles are recovered below. The reader will also find complementary material in a book [7] published in 2002, and of course the seminal paper by Gordon [8] remains essential to the mathematics involved in the theory, but beyond the elements that are necessary below. These are only very few examples among numerous papers that have been published on the subject.

In what follows, the basic mathematical concepts that are necessary are presented first. Then, they are applied to the simplest case of a single type of monomer unit, before a simple polyurethane case is considered, without functionality distribution. This limitation is then removed before a summary of the method is given with a practical 6-step procedure. Finally, this procedure is applied to cases of various complexities in order to illustrate its possibilities. Even if telechelic polymers only are considered in this Chapter, without cyclization, to keep things simple, quite exhaustive derivations are given in this restrictive context, and it is hoped that they provide the reader with the elements that allow a further exploration of more elaborate theories of polymer network formation.

2. Mathematical Background

A very elementary part of the theory of stochastic branching processes is detailed here, with limitation to the only notions that are applied below extensively, and an exhaustive account can be found in the reference monograph by T.E. Harris [9], for instance. This theory is applied here to the growth of a random graph, and the simplest image that may be used in such an exposition is that of a family tree. Such trees are examples of stochastic graphs when one assumes that distribution of males and females at each generation obeys a random process. An essential phenomenon in this context is the possible ultimate extinction of a family.

Slightly special family trees are considered here, since the children of females only are of interest[1]. In this matriarchal society, a family tree begins with a primitive mother as an ancestor, and the offspring of females only are recorded at each subsequent generation.

Therefore, ultimate extinction of a family is defined by all branches of the tree leading to males. Consider, for instance, that any primitive mother has 3 children and that any daughter (or granddaughter, great-granddaughter, etc.) has 2 children. If p denotes the probability that a child is a female, then 1 - p is the probability that the child is a male. The children of a

[1] This contrasts with the very beginning of the theory, at the end of the 19th century, when F. Galton and H.W. Watson studied the extinction of family names. One reason for the present choice is that m'ere, in French, means both mother and mer, thus making a connection between genealogy and polymer science...

primitive mother may be either 3 sons (and the family extincts immediately), or one daughter and two sons, or two daughters and one son, or three daughters. The probability of having 3 daughters is p^3 and the one for three sons is $(1 - p)^3$. When daughters have brothers, several possibilities yield from birth order. If a daughter has two brothers, for instance, she may be the oldest child, the youngest, or intermediate, with a probability of $(1-p)^2 p$ in each case. Similarly, three possibilities come up when there is one son and two daughters, with probabilities of $(1 - p)p^2$. The whole set of possible combinations can be summarized as the coefficients of the following polynomial of a dummy variable z:

$$F_0(z) = (1 - p)^3 + 3(1 - p)^2 pz + 3(1 - p) p^2 z^2 + p^3 z^3 \# \quad (1)$$

Figure 1. Various sets of children that a primitive mother (gray) may have, with either sons (white) or daughters (black). Birth order is not considered, that would triple each of the two central cases.

In polynomial $F_0(z)$, the coefficient of z^k equals the probability for a primitive mother (hence subscript 0) to have k daughters, according to the above discussion. Not only does it gather the set of probabilities in a single expression (hence its name: probability generating function), with $F_0(1) = 1$ consequently (the sum of all probabilities equals one), but this polynomial also has a very concise form:

$$F_0(z) = (1 - p + p z)^3 \quad (2)$$

as can be checked easily by developing the right-hand side. More generally, if the number of children of a primitive mother, and consequently her maximum number of daughters, is f, then the probability to have k daughters is the coefficient of z^k in $(1 - p + p z)^f$. In the $f = 3$ case detailed above, the various possibilities can be represented graphically as in Figure 1.

The same procedure can be followed to count the possibilities that a daughter has when she becomes a mother (with two children): two boys (probability $(1 - p)^2$, with immediate extinction of the descendants), two girls (probability p^2), or one boy and one girl (with two cases depending on the birth order, and therefore a total probability of $2(1 - p)p$).

Figure 2 shows these three possibilities, with probabilities that are gathered in the following polynomial:

$$F(z) = (1 - p)^2 + 2(1 - p)p z + p^2 z^2 = (1 - p + p z)^2 \quad (3)$$

where the coefficient of z^k is the probability for a non-primitive mother to have k daughters. The average number of daughters per mother can be deduced from the three plots shown in Figure 2 by summing the product of the number of daughters by the probability in each case:

$$0\times(1-p)^2 + 1\times 2(1-p)p + 2\times p^2, \text{ that is } 2p, \text{ which is also } F\#(1),$$

where F # denotes the derivative of F. In the more general case where a primitive mother may have f children, the possibilities that are offered to the females in the following generations are given by the coefficients of (1-p+p z) f-1 and the average number of daughters per mother is (f -1)p.

Even if a primitive mother has daughters, extinction of the family is nonetheless possible, since all branches of the family tree may lead to males. The probability e that the descendants of a non-primitive mother extinct can be obtained by considering the three situations shown in Figure 2: it is equal to the probability for a mother to have sons only, plus the probability that she has one son and one daughter whose descendants extinct (equal to e times the probability p to have a daughter), plus the probability that she has two daughters whose descendants extinct (equal to e 2 times the probability p 2 to have two daughters).

Figure 2. Various sets of children that a non-primitive mother may have: either two sons, or one son and one daughter, or two daughters.

Therefore, the probability of extinction obeys the following equation

e = (1 - p) 2 + 2(1 - p)p e + p 2 e 2 or, equivalently, e = F (e) (4)

with two roots: e = 1, that does not depend on p and is consequently excluded since it predicts extinction even when each generation has daughters only, and

e = # 1 p - 1 # 2 # (5)

It can be observed first that extinction is less probable when daughters get more probable, as expected. If it is impossible to have sons (daughters only, p = 1) extinction is impossible (e = 0) but if sons and daughters are equally probable (p = 1#2) extinction is unavoidable (e = 1) because the average number of daughters per mother (2p) is 1, reaching the limit of a stable population growth. If p is lower than 1#2, extinction is even more a certitude, and the values of e given by (5) that are larger than 1 must be interpreted as being equal to 1 (extinction is sure: e = 1). It is now possible to compute the probability s that the whole family of a primitive mother extincts, by weighting the probability of occurrence of each plot in Figure 1 by e raised to a power equal to the corresponding number of daughters, since the latter are as many non-primitive mothers whose descendants may extinct:

s = (1 - p) 3 + 3(1 - p) 2 p e + 3(1 - p)p 2 e 2 + p 3 e 3 or, equivalently, s = F 0 (e) (6)

with s = 0 if p = 1 and s = 1 if p # 1#2, using the corresponding e values, which extends the above conclusions to the whole tree.

In the more general case with f children per primitive mother, the probability that the descendants of a non-primitive mother extinct is obtained the same way as above and is the root (non equal to 1) of (1 - p + p e) f-1 = e or, equivalently, e

1 f-1 - p e + p - 1 = 0 (7)

which can be recast into

p = e 1 f-1 - 1 e - 1 that gives p = 1 # f-1 k=1 e k-1 f-1 (8)

after simplification by e-1, by taking u = e

1 f-1 and i = f-1 in u i
-1 = (u-1)# i
k=1 u k-1 ,

which is easy to obtain. The final one-to-one relation between e and p shows that the solution is unique, here again e decreases when p increases (since p decreases when e increases), and one has e = 0 when p = 1 and e = 1 when

p = 1 f – 1 (9)

since the denominator is then the sum of f - 1 terms that are all equal to 1. Now that the probability of extinction e for a non-primitive mother is obtained, the probability s that the number of descendants of a primitive mother is finite is obtained as above, by considering the various possibilities at the first generation and weighting by e raised to the corresponding number of daughters, which gives

s = (1 - p + p e) f = F 0 (e) = e f f-1 (10)

where the last expression has been obtained by noting that F 0 (z) = F (z) f f-1 . Therefore, the extreme probability values for the whole tree (descendants of a primitive mother), s = 0 and s = 1, are obtained in the same conditions as for any subtree (descendants of a non-primitive mother), e = 0 and e = 1, as already noted in the f = 3 case above.

Figure 3. Monomer (left) and trimer (right) in the case of a single, trifunctional, kind of reactive molecule. Terminal reactive groups are shown as corners, links are shown as white squares.

Consequently, s = 0 (the family is immortal certainly) if p = 1, and s = 1 (the family tree is inevitably finite) if p # 1#(f - 1).

The above results are sufficient to pass now from preliminary genealogical considerations to polymer applications.

3. A First Application

Consider the condensation of a very large number of monomer units with 3 equally reactive functional groups. Between the initial stage where no functional group has reacted yet and the final stage where reaction is complete, a mixture of molecules with various numbers of connected monomer units develops. Figure 3 shows a star-shaped monomer, with three arms of equal length, for instance, and a trimer. If cyclization is excluded, the various molecules formed can be represented as graphs similar to the family trees discussed in the previous Section, as illustrated in Figure 4. In such trees, the number of mothers is equal to the number of monomer units involved, and the numbers of daughters and sons are equal to the numbers of reacted and unreacted functional groups, respectively. A polymer molecule can be represented by several graphs, depending on the monomer unit that is taken as the primitive mother, and this affects the number of generations, as illustrated also in Figure 4.

Figure 4. Graphs representing the two molecules of Figure 3: primitive mother with sons only (monomer), and two possibilities for three-mother families (trimer), with different choices for the primitive mother and either two or three subsequent generations.

Since all functional groups are assumed to have the same reactivity and the number of monomer units considered is very large, the probability p to have daughters corresponds here to the probability that a group has reacted, and is therefore equal to the fraction x of groups that have undergone reaction at the stage considered. Consequently, x, the extent of reaction, is henceforth used instead of p. It should be noted that the purpose here is a statistical description of the distribution of polymerous molecules in the system, taking advantage of an initial number of monomers so large that it can be considered as infinite; the purpose is not to follow the evolution of given molecules when x increases from 0 to 1.

The extinction of a family that has been discussed in the previous Section corresponds now to a molecule being comprised of a finite number of monomer units, i.e., belonging to the sol phase. In opposition, a mother with an infinite number of descendants corresponds to a monomer unit in the gel phase. The condition for gelation to occur can thus be deduced directly from the analysis in the above Section, where it has been found that an infinite family is possible for p > 1#2: gelation occurs when the fraction of reacted groups reaches the value x g = 1#2. More generally, relation (9) leads to the Flory condition [10] for f-functional monomer units:

$$x\ g = 1\ f - 1\ \# \qquad (11)$$

Once the gel has formed, the fraction of sol (soluble) phase, which is the fraction of monomer units that belong to molecules of finite size, can be obtained easily: it is merely equal to s defined by (10), which can be rewritten in the present context as

$$s = (1 - x)\ 3 + 3(1 - x)\ 2\ x\ e + 3(1 - x)\ x\ 2\ e\ 2 + x\ 3\ e\ 3 = \#\ 1\ x - 1\ \#\ 3 \qquad (12)$$

with the last expression, also given by Flory [11], obtained by using (5) where x replaces p. Actually, the expanded expression of s in (12) adds up the respective probabilities for a monomer unit to have from 0 to 3 reacted groups with, in each case, these groups giving rise to finite subtrees, which is precisely the probability for a monomer unit selected at random from the system to belong to the sol phase. Thus, not only does (12) give the sol fraction, which decreases from 1 (up to the gel point, 0 # x # x g , e must be taken equal to 1) to 0 (full conversion, x = 1), but its various terms also give some information about the structure of the sol phase. The simplest is the first term, equal to F 0 (0), which is merely the fraction of unreacted monomers. The second term counts the number of terminations of the finite trees, where a single group has reacted and two remain unreacted, the third term gives the length of the branches, since twice reacted monomer units form chains, and the last term counts nodes, with three reacted groups from which three chains stem. Of course, this information is statistical only, and relates to averages. These actually are number averages here, because the latter yield directly from the statistical approach and are given by simple formulae, whereas molar or mass fractions are considered in the following Sections. They can be deduced from number averages, and the three quantities coincide in the present simple case of a single type of monomer unit.

By definition, a monomer unit that belongs to the gel phase has at least one reacted group with infinite continuation, i.e., a series of linked monomer units connects it to the boundaries of the reaction vessel, in practice. If it has only one such reacted group, it belongs to a dangling chain, since the other possible reacted groups have finite continuations.

If the monomer unit has two reacted groups with infinite continuations, it belongs to an elastically active chain. It gives rise to a dangling chain if the third group has reacted (with finite continuation). Finally, three reacted groups giving rise to as many infinite chains define a node of the elastically active network. The fractions of these various types of monomer units in the gel phase can be evaluated easily by considering Figure 1, where the leftmost scheme is excluded since belonging to the gel phase requires at least one reacted group, of course. Using the probabilities e and 1-e for finite and infinite continuations, respectively,

d = 3(1 -x) 2 x(1 - e) + 2 × 3(1 -x) x 2 (1 - e)e + 3 ×x 3 (1 - e)e 2 = (1 - e)F # 0 (e) (13)

gives the fraction of monomer units that belong to dangling chains. Coefficients 2 and 3 in front of the second and third terms come from the possible choices for the reacted group (among 2 or 3) with infinite continuation. The first term corresponds to monomer units that terminate dangling chains, the second one to intermediates along branches, and the last term to (elastically inactive) nodes belonging to dangling chains. The replacement of e with (5) gives

d = 3 # 1 x - 1 # 2 # 2 - 1 x # (14)

for x > 1#2 only, since d = 0 when there is no gel. This fraction reaches a maximum for x = 3#5 before decreasing back to 0 when x = 1 (all chains are elastically active when reaction is complete). Similarly, the fraction a of monomer units that belong to elastically active chains but are not nodes of the active network is deduced from the two rightmost schemes in Figure 1, since at least two reacted groups are required, and is given by

a = 3(1 - x) x 2 (1 - e) 2 + 3 × x 3 (1 - e) 2 e = 3 # 1 x - 1 ## 2 - 1 x # 2 (15)

for there are 3 possible choices for the reacted group with finite continuation in the third scheme in Figure 1. The two terms in (15) again have immediate interpretations: the first one corresponds to monomer units that are intermediates between triple nodes and from which an unreacted arm stems, whereas the second term is for monomer units from which dangling chains originate. Finally, the fraction of monomer units that form elastically active

0 0.2 0.4 0.6 0.8 1 0.5 0.55 0.6 0.65 0.7 0.75 0.8 0.85 0.9 0.95 1 s, d, a, n x s d a n

Figure 5. Variations of the various fractions of trifunctional monomer units when polymerization proceeds beyond the gel point: those belonging to the sol (s), the dangling chains (d), the elastically active chains (a), the active nodes (n).

nodes is obtained by considering the rightmost scheme in Figure 1 only, since 3 reacted nodes are necessary, each with infinite continuation:

n = x 3 (1 - e) 3 hence n = # 2 - 1 x # 3 # (16)

Additional terms would appear for functionalities larger than 3, because more than 3 infinite chains may stem from an active node; this will be considered in a later Section. All these fractions, which are defined for x > x g , which fulfill the balance equation for the total number of monomer units s + d + a + n = 1, and which are illustrated in Figure 5, can be gathered in the following polynomial

F 0 (z) = s + dz + az 2 + nz 3 = [1 - x + xe + x(1 - e)z] 3 = F 0 (e + (1 - e)z) (17)

where the coefficient of z k (with a (1 - e) k term) gives the probability for a monomer unit to have k links with infinite continuations.

Let now these results be applied to the theory of rubber elasticity. First, the number # e of moles of elastically active chains is expressed from the molar mass M of the monomer and the mass density # of the mixture, for a given extent of reaction x. The number of elastically active chains is equal to 3n#2 times the total number of monomer units in the system, since each active chain has its two ends connected to a trifunctional active node, which gives, using (16),

e = 3 2 # M # 2 - 1 x # 3 (18)

with (3#)#(2M) moles of elastically active chains per unit volume when reaction is complete and each chain is made of two arms of a monomer unit connected by one link, that is the shortest possible chain length. The expression given by Dossin and Graessley [12], and used by [3] and [6] for instance, for the elastic shear modulus of the network is

G = # f n 2 - h # n RT V + # T e RT (19)

with our notations, where V is the gel volume, T denotes absolute temperature, R the gas constant, f n the average functionality of active nodes, (with f n = f = 3 in the present case). Parameter h, introduced in [12], varies from 0 to 1 and allows a continuous variation from the affine theory of rubber elasticity to the phantom network theory of Flory [13].

Parameter #, introduced by Langley [14], accounts for an entanglement effect between elastically active chains in the case of a perfect network, if they are long enough. Factor T e , between 0 and 1, weights this effect when the network is imperfect; it is proportional to the probability that two elastically active chains cross, and is therefore equal to the square of the ratio between the lengths of the elastically active chains in the imperfect and perfect networks, respectively.

In the simple case considered in this Section, the entanglement factor is obtained easily by taking a monomer arm as unit length (the unit can be chosen arbitrarily, since a length ratio is computed), assuming that all arms (3 per monomer unit) have the same lengths. Each monomer unit belonging to an elastically active chain without being an active node contributes with 2 unit lengths, and each node contributes with 3, which gives 2a + 3n moles of unit lengths per initial mole of monomers. In a perfect network, all nodes are active, leading to 3 moles of unit lengths per initial mole of monomers. The ratio of these two quantities raised to power 2 gives the T e coefficient:

T e = # 2 3 a + n # 2 = 1 x 2 # 2 - 1 x # 4 (20)

which evolves from 0 at the gel point (x = 1#2) to 1 at full conversion (x = 1). Figure 6 compares the variations of T e and n during gel growth: the two quantities are close and consequently the values of the h and # parameters will have a moderate influence on the shape of the variations of the shear modulus G in the present simple case, and will affect its amplitude essentially.

The above entanglement factor can readily be extended to the more general case of f - functional star-shaped monomer units where the f # 2 arms have the same length. The molar fraction of monomer units with k branches having infinite continuations is the coefficient of z k in

F 0 (z) = [(1 - x + ex) + x(1 - e)z] f = # f k=0 C k f (1 - x + xe) f-k x k (1 - e) k z k (21)

where C k f denotes a binomial coefficient, and they belong to elastically active chains if k # 2, otherwise they are connected to the network on one side only and consequently belong to dangling chains. Each of these monomer units contributes with k arms to the elastically active chains and, using an arm as unit length, the total length of the elastically active chains is

#(r# x) = # f k=2 C k f k (1 - x + xe) f-k x k (1 - e) k # (22)

0 0.2 0.4 0.6 0.8 1 0.5 0.55 0.6 0.65 0.7 0.75 0.8 0.85 0.9 0.95 1 n and T e x n Te

Figure 6. Variations of the fraction n of monomer units that are elastically active nodes of the network and of the entanglement factor T e during gel growth, in the simple case of one kind of trifunctional monomer unit.

Taking the derivative of (21) with respect to z leads to two equivalent expressions

F # 0 (z) = f[(1 - x + ex) + x(1 - e)z] f-1 x(1 - e) = = # f k=1 k C k f (1 - x + xe)

f-k x k (1 - e) k z k-1 (23)

which, when setting z = 1, give

fx(1-e) = # f k=1 k C k f (1-x+xe) f-k x k (1-e) k = f(1-x+ex) f-1 x(1-e)+# (24)

hence

$$\#(r\# x) = fx(1 - e)[1 - (1 - x + ex) f\text{-}1] = fx(1 - e)[1 - F (e)] = fx(1 - e) 2 \qquad (25)$$

by using (7a). Consequently, #(1# 1) = f and finally

$$T e = x 2 (1 - e) 4 \# \qquad (26)$$

It may be noted that functionality does affect T e through e, and the previous result (20) is recovered when f = 3.

4. A Simple Polyurethane Case

Consider a polyurethane produced by reaction between bifunctional diols (one OH group at each end of the molecule) and triisocyanates (three NCO groups that are likely to create links by reacting with with diol OH groups). Let nD and n I denote the molar fractions of diols and triisocyanates in the mixture, with nD + n I = 1. The ratio r = [NCO]#[OH] = (3n I)#(2n D) identifies the proportions of functional groups in the mixture and r # 1 will be assumed here, which means that there will never be an excess of NCO groups 2 . Let x denote the fraction of reacted NCO groups; the fraction of reacted OH groups is, therefore, equal to rx, with a fraction of 1 - r OH groups left unreacted when all NCO groups have reacted (x = 1), since each reaction consumes one group of each type. If, here again, equal reactivities are assumed, which excludes that the last NCO group is less reactive after two groups of an isocyanate molecule have reacted, for instance, the results given heretofore can be adapted readily.

The polynomials that gather the probabilities that a triisocyanate unit selected at random from the mixture has 0, 1, 2, or 3 reacted groups and for a randomly picked diol to have 0, 1, or 2 reacted groups are

$$F 0I (z) = (1 - x + x z) 3 \text{ and } F 0D (z) = (1 - rx + rx z) 2 \qquad (27)$$

by adapting (2) to probabilities p = x and p = rx, respectively. The polynomials that describe the probabilities to have more than one reacted group in a triisocyanate or diol unit are

$$F I (z) = (1 - x + x z) 2 \text{ and } FD (z) = 1 - rx + rx z \qquad (28)$$

respectively, by adapting (3). The condition for extinction is obtained by writing that the possible additional links of a triisocyanate that has already reacted once and the possible second link of an already reacted diol have finite continuations:

$$e I = (1 - x) 2 + 2(1 - x)x e D + x 2 e 2 D \text{ and } e D = 1 - rx + rx e I \qquad (29)$$

where the occurrence of e D in the first expression and of e I in the second owes to the fact that the descendants of a triisocyanate reacted group are the descendants of the linked diol, and vice versa. This renders the alternation between isocyanate and diol units in polyurethane. These two equations can also be written more concisely as

e I = F I (e D) and e D = FD (e I) (30)

and their solutions between 0 and 1 (excluding the trivial e I = e D = 1 solution) are

e I = # 1 - 1 r x 2 # 2 and e D = 1 - 2 x + 1 r x 3 # (31)

The gel condition e I (x) = e D (x) = 1 is therefore obtained when reaction has proceeded up to

x g = 1 # 2 r # (32)

As a consequence, a gel is obtained for complete reaction (x = 1) if the mixture is such that 1#2 < r # 1.

2 In the study of a mixture with an excess of NCO groups, r can be defined as [OH]#[NCO] in order to keep x, still defined as the fraction of reacted majority groups, and rx in the same ranges.

More generally, if f-functional molecules are reacted with g-functional molecules, with f # 2 and g # 2, the mixture is described by the r = (f n f)#(g n g) ratio and the extinction probabilities are given by

e f = (1 - x + x e g) f-1 and e g = (1 - rx + rx e f) g-1 (33)

which can also be rewritten, by developing the terms in parentheses,

e f = 1 + f-1 # k=1 C k f-1 (e g - 1) k x k and e g = 1 + g-1 # l=1 C l g-1 (e f - 1) l (rx) l(34)

which are also equivalent to

e f - 1 e g - 1 = f-1 # k=1 C k f-1 (e g - 1) k-1 x k and e g - 1 e f - 1

= g-1 # l=1 C l g-1 (e f - 1) l-1 (rx) l (35)

leading to

(f - 1)x + f-1 # k=2 C k f-1 (e g - 1) k-1 x k # × × # (g - 1)rx + g-1

l=2 C l g-1 (e f - 1) l-1 (rx) l # = 1 # (36)

The gel condition e f = e g = 1 is thus obtained when x takes the value

x g = 1 # r(f - 1)(g - 1) # (37)

which is a classical result (see [4] or [7], for instance), and (32) is recovered if f = 3 and g = 2. Therefore, a gel is obtained when reaction is complete if

r > 1 (f - 1)(g - 1) # (38)

If a triisocyanate molecule has no link with infinite continuation, it belongs to the sol phase and this occurs with a probability of F 0I (e D), by adapting (10) to the triisocyanatediol alternation, and similarly for a diol molecule with a probability of F 0D (e I). These two number fractions can be used to compute the mass fraction m s of sol phase, by weighting with the mass fractions of isocyanate and diol in the system

m I = M I n I

M I n I + MDnD

and mD = MDnD

M I n I +MDnD (39)

where M I and MD denote the molar masses of the two components, hence

m s = F 0I (e D) m I + F 0D (e I) mD (40)

and, therefore,

m s = # -1 + 1 r x 2 # 3 m I + # 1 - 2 x + 1 r x 3 # 2 mD (41)

which does give m s = 1 at the gel point (x = 1# # 2r). At the end of reaction (x = 1), m s is not equal to the initial excess mass fraction of diol (1 - r)mD if r < 1: for non stoichiometric conditions, the gel at the end of reaction coexists with a sol phase that contains molecules where isocyanates and diols are linked. Once the gel is formed, the various number fractions of interest are given by the coefficients of

F 0I (z) = [1 - x + xeD + x(1 - e D)z] 3 and # F 0D (z) = [1 - rx + rxe I + rx(1 - e I)z] 2 (42)

by adapting (17) to triisocyanate and diols, respectively, still keeping in mind that an isocyanate extends toward a diol, and vice versa. Of course, the above sol fractions are obtained as the constant terms in (42),

F 0I (0) = F 0I (e D) and

F 0D (0) = F 0D (e I), and the

number of moles of triple nodes in the gel phase, which are tri-reacted isocyanates, is readily obtained from the coefficient of z 3 in

F 0I (z): n = x 3 (1 - e D) 3 n I = # 2 - 1 r x 2 # 3 n I # (43)

The 3n#2 moles of elastically active chains in the mixture, which has a mass density #, lead to the following number of moles of elastically active chains per unit volume:

e = 3 2 # n M I n I +MDnD = 3# r 2rM I + 3MD # 2 - 1 r x 2 # 3 # (44)

The molar fraction of the elastically active network, including both chains and nodes, is deduced readily from the coefficients of z 2 and z 3 in (42), respectively:

m a = [3(1 - x + xeD)x 2 (1 - e D) 2 + x 3 (1 - e D) 3]m I + r 2 x 2 (1 - e I) 2 mD (45)

or, equivalently

m a = # 2 - 1 rx 2 # 2 ## -1 + 2 rx 2 # m I + 1 x 2 mD ## (46)

Finally, the mass fraction of dangling chains, excluding the isocyanate units they stem from, which belong to active chains, is given by the z terms in (42) :

m d = 3(1 - x + xeD) 2 x(1 - e D)m I + 2(1 - rx + rxe I)rx(1 - e I)mD (47)

which, from (31), gives

m d = # 2 - 1 rx 2 ## 3 # -1 + 1 rx 2 # 2 m I + 2 x # 1 - 2 x + 1 rx 3 # mD # (48)

and it can be checked that m s + m a + m d = 1 does apply.

The entanglement factor can also be obtained, since the total length of elastically active chains per mole of mixture adds up the contributions of the three arms of each elastically active triple node, the two arms of intermediate triisocyanates along elastically active chains, and the diols in the latter chains. Therefore, the same constitutive elements as in (45) are obtained and give

#(r# x) = 3#x 3 (1 - e D) 3 n I + 6#(1 - x + xeD)x 2 (1 - e D) 2 n I + r 2 x 2 (1 - e I) 2 nD(49)

by taking a diol as unit length, assuming that all diols have the same length, and denoting # the length of a triisocyanate arm (assumed all equal) with this unit. Replacing e I and e D in the above expressions, and using n I = 2r#(2r + 3) et nD = 3#(2r + 3), which are readily obtained, (49) is recast as

#(r# x) = 3 x 2 # 2 - 1 rx 2 # 2 2# + 1 2r + 3 (50)

hence the entanglement factor:

T e = # #(r# x) #(1# 1) # 2 = 1 x 4 # 2 - 1 rx 2 # 4 # 5 2r + 3 # 2 (51) which does not depend on the relative length # of triisocyanate arms compared to diols.

This is not surprising: be it in the imperfect or perfect network, an elastically active chain is a repeated sequence of one isocyanate arm followed by one diol followed by one isocyanate arm, and this defines another unit length (equal to 2# + 1 times the previous one) that vanishes when the ratio of the total lengths in the imperfect and perfect networks is performed. Consequently, the lengths of isocyanate arms and diols are not relevant in the T e factor. In the examples considered below, # is used nevertheless as a temporary and convenient variable in the intermediate # computation. In stoichiometric conditions (r = 1), T e and n have similar magnitudes during gel growth, which is similar to the result obtained in the simple homopolymer case of the previous Section, since (51) and (43) are equivalent to (20) and (16) where x is replaced by x 2 , which modifies the shapes of the curves in Figure 6 marginally, with now x between # 2 and 1. Moreover, off stoichiometry but at full conversion (x = 1), Figure 7 shows similar trends for n and T e . Consequently, the same conclusions as in the previous Section are obtained as far as the influences of parameters h and # are concerned.

The entanglement factor can also be calculated in the general case of star-shaped f - functional units reacting with g-functional units, with f and g larger than 2. For the total length of elastically active chains, one gets

f (r# x) = fx(1 - e g)[1 - F f (e g)]n f = fx(1 - e g)(1 - e f)n f

And # g (r# x) = #grx(1 - e f)[1 - F g (e f)]n g = #grx(1 - e f)(1 - e g)n g (52)

as the contributions of f-functional and g-functional units, respectively (note that rx replaces x for the latter), using an arm of f-functional molecule as unit length, with # denoting the length of an arm of g-functional molecule. Therefore,

#(r# x) = # f (r# x) + # g (r# x) = fg rx(1 - e g)(1 - e f) # + 1 gr + f (53)

0 0.2 0.4 0.6 0.8 1 0.5 0.55 0.6 0.65 0.7 0.75 0.8 0.85 0.9 0.95 1 n/ni]

and T e r Te n/n i Te

Figure 7. Fraction n#n I of triisocyanates that are nodes of the active network (unbroken line) and entanglement factor T e (broken line), for full conversion with an excess of OH, when composition of the triisocyanate-diol mixture evolves. The entanglement factor T e is also shown (dotted line) for the diisocyanate-triol mixture that is considered later.

using n f = gr#(gr + f) and n g = f#(gr + f), which yield from n f + n g = 1 and r = (fn f)#(gn g). Consequently, #(1# 1) = fg(# + 1)#(g + f) and

T e = r 2 x 2 (1 - e g) 2 (1 - e f) 2 # g + f gr + f # 2

with f # 3# g # 3# r = fn f

gn g # 1 (54)

which does not depend on # , as expected. This expression simplifies slightly when one type of units is bifunctional. Assume for instance that g = 2; the contribution of bifunctional molecules is directly given by the coefficient of z 2 in

#

F 0g (z), that is

$$ \# g = \# r 2 x 2 (1 - e f) 2 n g = \#(1 - e g) 2 n g \tag{55} $$

using e g = 1 - rx + rxe f , as given by (29). Therefore,

$$ \#(r\# x) = \# f (r\# x) + \# g (r\# x) = 2f(1 - e g) 2 \#f + 2 2r + f \tag{56} $$

using n f = 2r#(2r+f) and n g = f#(2r+f). Consequently, #(1# 1) = 2f(#f +2)#(2+f) and

$$ T e = (1 - e g) 4 \# 2 + f 2r + f \# 2 $$

with f # 3 and r = fn f

$$ 2n g \# 1 \tag{57} $$

which recovers (51) for f = 3. Finally, (57) can also be obtained directly from (54) by changing g into 2 and using (1 - e g) = rx(1 - e f) that yields from (34). It should be noted, however, that these operations do not change (53) into (56) because a bifunctional molecule is not star-shaped, actually. Consider now that f = 2, the bifunctional molecules are minority, which leads now to

$$ T e = r 2 (1 - e f) 4 \# g + 2 gr + 2 \# 2 $$

with g # 3 and r = 2n f

$$ gn g \# 1 \tag{58} $$

which can also be deduced from (54) by changing f into 2 and using (1 - e f) = x(1 - e g).

Thus, having either majority or minority bifunctional units leads to different entanglement factors and, consequently, network properties.

5. A More General Polyurethane Case

In order to illustrate the possibilities of the theory further, consider now a more complex system with a distribution of functionalities: tri-, bi-, and monofunctional isocyanates, as well as bi- and monofunctional diols. Let # I3 , # I2 , and # I1 denote the fractions of each isocyanate type, #D2 and #D1 the fractions of diols, with # I3 +# I2 + # I1 = 1 and #D2 + #D1 = 1. Let again n I and nD denote the molar fractions of isocyanates and diols, whatever their types, with n I +nD = 1, and assume equal reactivities. The ratio between the numbers of NCO and OH groups in the mixture, that is again supposed not to exceed 1, and that is also the ratio of the number of NCO and OH groups that have reacted, is given by

$$r = f I f D n I nD \tag{59}$$

using the average functionalities f I = 3 # I3 + 2 # I2 + # I1 and f D = 2 #D2 + #D1. The probability that an isocyanate unit selected at random from the mixture has k reacted groups combines the probabilities to get each type of isocyanates and therefore is the coefficient of z k in the following polynomial:

$$F\ 0I\ (z) = \#\ I3\ (1 - x + xz)\ 3 + \#\ I2\ (1 - x - xz)\ 2 + \#\ I1\ (1 - x + xz) \tag{60}$$

and similarly for a diol selected at random:

$$F\ 0D\ (z) = \#D2\ (1 - rx + rxz)\ 2 + \#D1\ (1 - rx + rxz)\ \# \tag{61}$$

Computing the probabilities for an isocyanate or a diol to have reacted more than once must owe to the fact that, for instance, a randomly picked NCO group has a probability 3# I3 #f I to belong to a trifunctional isocyanate. Similarly, it has a probability 2# I2 #f I to belong to a bifunctional isocyanate, and a probability # I1 #f I to belong to a monofunctional isocyanate.

Therefore, the probabilities for a unit to have established more than one link are given by the following polynomials:

$$F\ I\ (z) = 1\ f\ I\ [3\#\ I3\ (1 - x + xz)\ 2 + 2\#\ I2\ (1 - x + xz) + \#\ I1\ ] \tag{62}$$

and

$$FD\ (z) = 1\ f\ D\ [2\#\ D2\ (1 - rx + rxz) + \#D1\ ] \tag{63}$$

for isocyanates and diols, respectively. More generally, F (z) deduces from F 0 (z) as

$$F\ (z) = F\ \#\ 0\ (z)\ F\ \#\ 0\ (1) \tag{64}$$

where F #

0 (z) is the derivative of F 0 (z). This concise relation is a further interest of using probability distribution functions to describe network formation, and it can be checked that (62) and (63) are recovered, as well as (3) and (28).

The probabilities of extinction e I and e D are given by the coupled equations e I = F I (e D) and e D = FD (e I), since it suffices to express that the additional links of a unit that has already reacted once have finite continuations, with an alternation of isocyanate and diol. Therefore,

$$e\ I = 1\ f\ I\ [3\#\ I3\ (1 - x + xeD\ )\ 2 + 2\#\ I2\ (1 - x + xeD\ ) + \#\ I1\ ] \tag{65}$$

and

e D = 1 f D [2# D2 (1 - rx + rxe I) + #D1] # (66)

The special role played by monofunctional molecules may be noted: they contribute to the extinction probability with their share in the average functionality, since they cannot establish new links after they have reacted once. These two equations give immediately

e I - 1 e D - 1 = 3# I3 x 2 (e D - 1) + 6# I3 x + 2# I2 x f I

and e D - 1

e I - 1 = 2#D2 rx f D (67)

hence

e D = 1 - 2 3 3# I3 + # I2 # I3 x + f I f D 6# I3 #D2 1 r x 3

and e I = 1 -

f D 2#D2 1 - e D r x # (68)

If there is no monofunctional unit, x = r = 1 leads to e D = e I = 0: a perfect network is obtained, since no link has finite continuation. In contrast, if monofunctional units are present, they are dangling chains in the network, which is not perfect although the reaction is complete in stoichiometric conditions, and the extinction probability is not zero.

The fraction of reacted NCO groups at the gel point, obtained for e D = e I = 1 to have infinite chains, is

x g = # 1 4r f I f D (3# I3 + # I2) #D2 (69)

and therefore a gel is obtained when reaction is complete if

r > f I f D 4(3# I3 + # I2) #D2 # (70)

For instance, if # I3 = # I2 = 1#2 and #D2 = 1 (then # I1 = #D1 = 0), i.e., for an equal mixture of bi- and trifunctional isocyanates plus bifunctional diols only, a gel is obtained at the end of reaction if r > 5#8 = 0#625. Thus, distributed functionalities shift the gel point, which is 0#5 without bifunctional isocyanates as shown in the previous Section, to higher values. This is due to rapidly saturated units, because of their small functionality, coming into play and slowing down the network formation.

Since probabilities e D and 1-eD merely have to be given to links with finite and infinite continuations, respectively, the fractions of the various types of isocyanate units when a gel has formed (active network nodes, elastically active chains, dangling chains, sol fraction) are given by the coefficients of the following polynomial:

F 0I (z) = # I3 [1 - x + xeD + x(1 - e D)z] 3 + # I2 [1 - x + xeD + x(1 - e D)z] 2 +

+ # I1 [1 - x + xeD + x(1 - e D)z] (71)

and for diols, similarly:

F 0D (z) = #D2 [1 -rx+rxe I +rx(1-e I)z] 2 +#D1 [1 -rx+rxe I +rx(1-e I)z] # (72)

For instance, the total mass fraction of sol is

m s = 1

M # #D1 (1 - rx + rxe I)n DMD1 + #D2 (1 - rx + rxe I) 2 nDMD2+

+ # I1 (1 - x + xeD)n I M I1 + # I2 (1 - x + xeD) 2 n I M I2 +

+# I3 (1 - x + xeD) 3 n I M I3 # (73)

using the molar mass of each component and the average molar mass of the mixture

M = nD (#D1MD1 + #D2MD2) + n I (# I1 M I1 + # I2 M I2 + # I3 M I3) # (74)

The number of moles of trifunctional isocyanate units that have formed triple nodes in the gel is

n = # I3 x 3 (1 - e D) 3 n I = r f D # I3 r f D + f I x 3 (1 - e D) 3 (75)

since n I = rf D #(rf D +f I) and nD = f I #(rf D +f I), and therefore the number of moles of elastically active chains per unit volume is

e = # M r f D r f D + f I # 3# I3 + # I2 - f I f D 4#D2 1 rx 2 # # (76)

The total length of elastically active chains, which generalizes (49), is obtained by adding up the contributions of the three arms stemming from the network nodes, the two arms of tri- and bifunctional isocyanates along elastically active chains, and the diols in the latter chains:

#(r# x) = 3#x 3 (1 - e D) 3 # I3 n I + 6#(1 - x + xeD)x 2 (1 - e D) 2 # I3 n I +

+ 2#x 2 (1 - e D) 2 # I2 n I + r 2 x 2 (1 - e I) 2 #D2 nD (77)

using the same unit length as in the previous Section. This expression can be simplified with (68):

#(r# x) = f I f 2 D 4 #D2 (1 - e D) 2 2# + 1 r f D + f I # (78)

The reference length is then, using (68) again:

#(1# 1) = f I f 2 D 4 #D2 # 2 3 3# I3 + # I2 # I3 - f I f D 6# I3 #D2 #2 2# + 1 f D + f I (79)

and T e is obtained by squaring the ratio of these two quantities:

T e = (1 - e D) 4 # f D + f I r f D + f I # 2 # 6 # I3 #D2

4#D2 (3# I3 + # I2) - f D f I # 4 (80)

which does not depend on # . The last term disappears and f D = 2 when there is no mono-functional component: this generalizes (57) where the average functionality of isocyanates f I replaces the integer functionality g.

Thus, the expressions are more involved in the case of distributed functionalities, but of course the results of the previous Section are recovered when # I3 = #D2 = 1 and therefore

I2 = # I1 = #D1 = 0.

6. Concise Procedure

As could be observed, the statistical theory of network formation cannot be reduced to a small set of general formulae, for they differ in each case considered. Empirical adaptation of known results to new conditions, even close apparently, is risky and should be avoided.

Nevertheless, the procedure to follow can be adapted in a very systematic manner, since it proceeds in 6 steps:

1. Write, for each chemical species, the polynomial that uses as coefficients the probabilities to have the various numbers of reacted groups.
2. Deduce, by taking and normalizing the derivative, the polynomial that uses as coefficients the probabilities, for each chemical species, to have additional reacted groups after a first group has reacted.
3. Use this polynomial to compute the probability that a reacted group leads to a finite branch, for each chemical species.
4. Deduce the gel condition by writing that these probabilities reach the critical value of 1, below which infinite chains are allowed.
5. Rewrite the initial polynomial for each chemical species, replacing the dummy variable z by the expression e + (1 - e)z that uses the probability of extinction of the other chemical species.
6. Deduce the fractions of interest in the mixture while reaction proceeds: sol fraction, active nodes, elastically active chains, dangling chains, trapped entanglements, etc.

In the present exposition, notations differ slightly from those employed in the literature, where several dummy variables z are introduced simultaneously, or where the coefficient of z k is defined from the k-th derivative of the polynomial, for instance, which did not seem necessary here. Moreover, the gel condition has been introduced more simply than by using additional polynomials or computing a determinant, as is frequently done since [8]. Despite these changes, the results are the same as in the literature, for instance (14) corresponds to (3.33) in [7] and (38) to (3.86), using the reciprocal definition of r. Similarly, (68) corresponds to (67) of [5] in the #D1 = 0 case, using the reciprocal definition of r too, (69) corresponds to (64), (73) to (68) and (76) to (74). These results are also recovered in page 314 of [4] but by reversing the roles of isocyanates and diols, just as in the appendix of [6].

7. EXAMPLES OF STRAIGHTFORWARD APPLICATION OF THE PROCEDURE

The 6-step procedure of the previous Section is applied below to three cases with various complexities, in a straightforward manner which assumes that the above theory has been studied.

7.1. Polyurethane Considered in [6]

The mixture is composed of triols with either 3 or 2 functional groups, and bifunctional diisocyanates 3 . Therefore, with r = 2n I #((# + 2)n T), using # for # 3 for brevity and consequently # 2 = 1 - #, step 1 gives

F 0I (z) = (1 -x+xz) 2 F 0T (z) = #(1- rx+ rxz) 3 + (1 -#)(1- rx+ rxz) 2 (81)

hence (step2)

F I (z) = F # 0I (z) F #

0I (1) = 1 - x + xz F T (z) = F # 0T (z) F #

0T (1) = 3#(1 - rx + rxz) 2 + 2(1 - #)(1 - rx + rxz) # + 2 (82)

leading to e I = F I (e T) and e T = F T (e I), which give (step 3)

e I = 1 - 2 3 2# + 1 #rx + # + 2 3#r 2 x 3 and e T = 1 - 1 - e I x (83)

with gelation for (step 4)

x g = # 1 2r # + 2 2# + 1 # (84)

Moreover (step 5),

F 0I (z) = F 0I (e T + (1 - e T)z) = [1 - x + xe T + x(1 - e T)z] 2 and

F 0T (z) = F 0T (e I + (1 - e I)z) = #[1 - rx + rxe I + rx(1 - e I)z] 3 +

$$+ (1 - \#)[1 - rx + rxe\ I + rx(1 - e\ I\ )z]\ 2 \qquad (85)$$

3 Fractions of additional monofunctional components are also considered in [6], but are not supplied; they are ignored here.

give, for instance (step 6)

$$n = \#\ r\ 3\ x\ 3\ (1 - e\ I\ )\ 3\ n\ T\ (86)\ \text{and}$$

$$m\ s = 1\ M\ [(1 - \#)(1 - rx + rxe\ I\ )\ 2\ n\ T\ M\ T2 +$$

$$+ \#(1 - rx + rxe\ I\ )\ 3\ n\ T\ M\ T3 + (1 - x + xe\ T\ )\ 2\ n\ I\ M\ I\ ] \qquad (87)$$

with

$$M = n\ T\ [(1 - \#)M\ T2 + \#M\ T3\ ] + n\ I\ M\ I\ \# \qquad (88)$$

One also has $T\ e = [\#(r\#\ x)\#\#(1\#\ 1)]\ 2$ with

$$\#(r\#\ x) = 3r\ 3\ x\ 3\ (1 - e\ I\ )\ 3\ \#n\ T + 6(1 - rx + rxe\ I\ )r\ 2\ x\ 2\ (1 - e\ I\ )\ 2\ \#n\ T +$$

$$+ 2r\ 2\ x\ 2\ (1 - e\ I\ )\ 2\ (1 - \#)n\ T + \#x\ 2\ (1 - e\ T\ )\ 2\ n\ I \qquad (89)$$

by counting the constitutive elements of elastically active chains. The arms of triols are assumed to all have the same length, taken as unit, and # denotes the length of a diisocyanate. Using $n\ T = 2\#[(\#+2)r+2]$ and $n\ I = (\#+2)r\#[(\#+2)r+2]$, deduced from the definition of r and from $nD + n\ I = 1$, this expression simplifies into

$$\#(r\#\ x) = (1 - e\ I\ )\ 2\ \# + 2\ (\# + 2)r + 2\ r\ (\# + 2) \qquad (90)$$

giving, finally

$$T\ e = r\ 2\ (1 - e\ I\ )\ 4\ \#\ \# + 4\ (\# + 2)r + 2\ \#\ 2 \qquad (91)$$

which does not depend on # . It may be noted that this result can also be obtained by replacing the integer functionality g in (58) by the average functionality # + 2.

These results do recover those given in the appendix of [6] when monofunctional units are ignored, with the reciprocal definition for r. Similarities may also be noted with the results given above in the last polyurethane case studied above, when monofunctional molecules are ignored: the gel condition, for instance, or the number of active nodes. For the latter, rx replaces x, in addition to the roles of alcohols and isocyanates being reversed, what could be missed in an empirical adaptation to the present chemical system.

In the special case where all triols are trifunctional (# = 1), there is also a similitude with the first, simple, polyurethane case considered above, by reversing the roles of the two components since the majority component is trifunctional now. For instance, the entanglement factor (91) becomes

T e = 1 r 2 x 4 # 2 - 1 rx 2 # 4 # 5 3r + 2 # 2 (92)

instead of (51). As shown in Figure 7, T e is larger than with reversed functionalities, which allows more significant changes of the G(r) function by tuning parameters h and # in (19).

7.2. Tetrafunctional Monomer

Up to this point, functionalities lower than 4 were considered, but it is interesting to observe the qualitative changes that are induced by a functionality of 4. In this Section, the simple case of the condensation of a 4-functional monomer is considered. For convenience, a star-shape with all arms of the same length is assumed. The procedure starts from (step 1)

F 0 (z) = (1 - x + xz) 4 hence (step 2) F (z) = F #

0 (z) F # 0 (1) = (1 - x + xz) 3 (93)

and (step 3) condition e = F (e) leads, after simplification by 1 - e, to a second degree equation with respect to 1 - e (note that it is often simpler to compute 1 - e than e):

x 3 (1 - e) 2 - 3x 2 (1 - e) + 3x - 1 = 0 (94)

where a single root ensures that e = 0 when x = 1 (for a perfect network, all chains are infinite and extinction probability is zero):

1 - e = 3x - # x(4 - 3x) 2x 2 (95)

which does lead to the expected gel condition (11) when 1 - e = 0, i.e., x g = 1#3 (step 4).

Finally (step 5),

F 0 (z) = F 0 (e + (1 - e)z) = [1 - (1 - e)x + (1 - e)xz] 4 (96)

leads immediately (step 6) to the sol fraction s, the fractions of monomers involved in dangling chains d and of elastically active chains a (uncounting nodes):

s = [1 -(1-e)x] 4 d = 4[1 -(1-e)x] 3 (1-e)x a = 6[1 -(1-e)x] 2 (1-e) 2 x 2 (97)

and, eventually, the fractions of monomers that are active nodes of the network, with either 3 or 4 reacted groups, are given by:

n 3 = 4[1 - (1 - e)x](1 - e) 3 x 3 and n 4 = (1 - e) 4 x 4 # (98)

Therefore, the total fraction of monomers that are active nodes, and their average functionality are given by n = n 3 + n 4 = [4 - 3(1 - e)x](1 - e) 3 x 3

and f n = 3n 3 + 4n 4

n = 4 3 - 2(1 - e)x 4 - 3(1 - e)x # (99)

Figure 8 shows that this average functionality, which of course is 4 for a perfect network (x = 1), takes its lowest possible value of 3 at the gel point (x = 1#3) and varies nonlinearly when reaction proceeds.

The respective weights of the two terms in the expression (19) of the shear modulus are interesting to discuss. When h varies from 0 to 1, the first term evolves from f n n#2, the number of elastically active chains (since each one connects two nodes, and f n n chains stem from the latter) to (f n #2 - 1)n, the value given by the phantom network theory [13]. The entanglement factor is obtained easily from the total length of elastically active chains, which can be written as follows by using the length of a monomer arm as unit:

3 3.2 3.4 3.6 3.8 4 0.3 0.4 0.5 0.6 0.7 0.8 0.9 1 fn x

Figure 8. Variation of the average functionality of active nodes during condensation of tetrafunctional monomers.

0 0.5 1 1.5 2 0.3 0.4 0.5 0.6 0.7 0.8 0.9 1 n, f n n/2, (fn/2-1)n, Te x n Te fn n/2 (fn/2-1)n

Figure 9. Variations of the fraction of monomers that are active nodes (unbroken line), of the two limit values of the first term in the expression of the shear modulus (broken lines), and of the entanglement factor (dotted line), while condensation of tetrafunctional monomers proceeds.

#(x) = 4n 4 + 3n 3 + 2a = 4 (1 - e) 2 x (100)

thus #(1) = 4 and

T e = x 2 (1 - e) 4 (101)

in concordance with (26). Figure 9 shows the variations of the two parts of the shear modulus, including the two limit cases for the first one. For comparison, the variations of the fraction of monomers that are active nodes is also shown. It can be observed that the term from the affine theory retains part of the downward concavity of the variations of the number of active nodes, which was absent in the cases studied up to this point with functionalities below 4. In contrast, the term from the phantom network theory exhibits an upward concavity. Its values are close to half the values for the affine theory, being exactly one half for perfect networks only (which is expected for tetrafunctional nodes). Its values are found close to the entanglement factor.

7.3. Polyurethane Considered in [15]

The system is composed of bifunctional diols and a mixture of tri- and quadriisocyanates. With r = (4 - #)n I #(2n D), where # = # I3 for conciseness and therefore # I4 = 1 - #, step 1 leads to:

F 0I (z) = (1 - #)(1- x + xz) 4 +#(1- x + xz) 3 F 0D (z) = (1 - rx + rxz) 2 (102)

thus (step 2)

F I (z) = F # 0I (z) F #

0I (1) = 4(1 - #)(1 - x + xz) 3 + 3#(1 - x + xz) 2 4 - # FD (z) = F #

0D (z) F # 0D (1) = 1 - rx + rxz # (103)

Hence e I = F I (e D) and e D = FD (e I), which give readily (step 3)

e D = 1 - 3(4 - 3#)rx - # # 8(1 - #)rx 2

and e I = 1 - 1 - e D rx (104)

since a single root of the second degree equation to solve ensures e D = e I = 0 if x = r = 1 (no finite chain in the perfect network), with

= (4 - #) r [16(1 - #) - 3(4 - 5#)rx 2] # (105)

The gel point yields, when e D = e I = 1 (step 4):

x g = # 1 6r 4 - # 2 - # # (106)

Next (step 5),

F 0I (z) = F 0I (e D + (1 - e D)z) =

= (1 - #)[1 - x + xeD + x(1 - e D)z] 4 + #[1 - x + xeD + x(1 - e D)z] 3

and # F 0D (z) = F 0D (e I + (1 - e I)z) = [1 - rx + rxe I + rx(1 - e I)z] 2 (107)

give, for instance (step 6),

n = [4(1 -#)(1-x+xeD)x 3 (1-eD) 3 +#x 3 (1-eD) 3 + (1-#)x 4 (1-eD) 4]n I (108)

since active nodes are formed by triisocyanates and quadriisocyanates, and the latter may have either 3 (first z 3 term in

#

F 0I (z)) or 4 (single z 4 term in
#
F 0I (z)) reacted NCO groups.
This expression can also be recast as

n = (1 - e D) 3 x 3 [4 - 3# - 3(1 - #)(1 - e D)x]n I # (109)

The sol fraction is given by

m s = 1 M # (1 - rx + rxe I) 2 nDMD+

+(1 - #)(1 - x + xeD) 4 n I M I4 + #(1 - x + xeD) 3 n I M I3 # (110)

with

M = nDMD + (1 - #)n I M I4 + #n I M I3 # (111)

All these results agree with appendix 2 of [5] in the # = 1 case.
The entanglement factor T e = [#(r# x)##(1# 1)] 2 can be deduced from the terms of
#
F 0I (z) and
#
F 0D (z) where z is raised to at least 2, which gives

#(r# x) = 4#(1 - #)×

× [x 2 (1 - e D) 2 + 3(1 - x + xeD)x(1 - e D) + 3(1 - x + xeD) 2]x 2 (1 - e D) 2 n I +

+ 3# #[x(1 - e D) + 2(1 - x + xeD)]x 2 (1 - e D) 2 n I + r 2 x 2 (1 - e I) 2 nD (112)

assuming that all arms of isocyanate units have the same length, which is a fraction # of the length of a diol. Using n I = 2r#(2r + 4 - #) and nD = (4 - #)#(2r + 4 - #), which yield from the definition of r, the previous equation reduces to

#(r# x) = 2# + 1 2r + 4 - # (1 - e D) 2 (4 - #) (113)

with, consequently, #(1# 1) = (2# + 1)(4 - #)#(6 - #) and

T e = # 6 - # 2r + 4 - # # 2 (1 - e D) 4 (114)

which does not depend on # . Here again, T e is obtained from (57) if the average functionality (f = 4 - #) replaces the integer functionality of the component that is not bifunctional.

When x = 1 (reaction is complete) and # = 0, i.e., there is no triisocyanate, therefore all nodes are tetrafunctional and r = 2n I #nD , the gel point is obtained for r = 1#3 and one has merely e D = - 1 2 + # 1 r - 3 4 and n = (1 - e D) 3 (1 + 3e D)n I (115)

with an entanglement factor given by

T e = 9 (1 - e D) 4 (r + 2) 2 (116)

which is a special case of (57) when f = 4. Figure 10 illustrates the variations of the number fraction of active nodes, with downward concavity for large r values, and of their average functionality

f n = n I n [4(1 - e D) 4 + 12e D (1 - e D) 3] = 4 1 + 2e D

1 + 3e D # (117)

The above relations lead to the curves in Figure 11, where the two parts of the expression of the shear modulus are shown, with two limit cases for the first one, as well as the entanglement factor. Trends can be compared with those already found in the previous Section, where 4-functional units were also present, but now when r varies. The approximate 1#2 ratio between the two variants for the first part of the shear modulus is observed, but the downward concavity found for n#n I is significantly weakened by the product with n I = 2r#(2r+1) and f n . Moreover, the entanglement factor now differs significantly from the other curves.

These results may also be compared with those given by the recursive approach of [2] as summarized in appendix 2 of [3], for instance. The same definition is used for r, and P (F out B) corresponds to e D . Actually, relations (70) and (78) of [3] do coincide with (104) when # = 0 and # = 1, respectively. Similarly, the numbers of active nodes obtained in [2] agree with (109). In contrast, the entanglement factor in [3] misses the (2 + f)#(2r + f) squared term that was obtained in (57) from different molar fractions of constituents being involved in the perfect and imperfect networks off stoichiometry. This may be due to different definitions of the reference network, and the two approaches nevertheless lead to the same results in stoichiometric conditions (r = 1).

0 0.2 0.4 0.6 0.8 1 0.3 0.4 0.5 0.6 0.7 0.8 0.9 1 n/ni
and f n-3 r n/ni fn-3

Figure 10. Fraction n#n I of isocyanates that are active nodes of the network (unbroken line) and average functionality of these nodes (broken line), when reaction is complete, as functions of the composition of a quadriisocyanate-diol mixture.

0 0.2 0.4 0.6 0.8 1 1.2 1.4 0.3 0.4 0.5 0.6 0.7 0.8 0.9 1 fn n/2, (fn/2-1)n, Te r fn n/2 (fn/2-1)n Te

Figure 11. Limits of the first term in the expression of the shear modulus (unbroken lines), for h = 0 (upper curve) and h = 1 (lower curve), for complete reaction, as functions of the composition of a quadriisocyanate-diol mixture. The entanglement factor T e is also shown (broken line).

8. CONCLUSION

An introduction has been given to the statistical theory of polymer network formation that uses probability generating functions, starting from the few mathematical notions that are required and ending with a simple procedure that can be followed in many cases of practical interest. It has been shown on several detailed examples of various complexities that, even if the theory may seem complicated at first sight, its use is quite simple in many circumstances. Emphasis has been put here on connections with the theory of rubber elasticity, with special attention devoted to the trapped entanglement factor, but other applications can be considered, of course, by taking advantage of the various statistical features of the gel and sol phases that are obtained. The variation of the molecular weight distribution in the latter with reaction extent, for instance, is of fundamental importance. The present exposition is far from being exhaustive and has been limited to a mere introduction. Examples have been given on stepwise polymerization of homopolymers of various uniform functionalities, and on different polyurethane systems, including distributed functionalities, but random crosslinking of linear polymer molecules has not been considered, for instance. This important process, of which vulcanization of rubber is an example, has a close connection with complete condensation of (minority) tetrafunctional and (majority) bifunctional units, as demonstrated by P.J. Flory [16]. Equal reactivities have also been assumed in this Chapter, whereas this may not apply to important systems; the examples considered in [17] and [7], where two reactivities are considered, are interesting in this respect. More importantly, cyclization has been neglected, although intramolecular reactions may play an important role in many cases. This phenomenon leads to somewhat more elaborate developments than what could be covered in this Chapter, and the reader may refer to the work of R. Stepto ([18], for instance) and to the review by K. Du–sek [4] for more details.

REFERENCES

[1] Macosko, C.W.; Miller D.R. Macromolecules 1976, 9, 199-206.
[2] Miller D.R.; Macosko C.W. Macromolecules 1976, 9, 206-211.
[3] Queslel J.P.; Mark. J.E. Adv. Polym. Sci. 1984, 65, 135-176.
[4] Du–sek K. In Telechelic Polymers: Synthesis and Applications ; Goethals, E.; Ed.; CRC Press: Boca raton, FL, 1989; pp 289-360.
[5] Du–sek K.; Du–skov'a M.; Fedderly J.J.; Lee G.F.; Hartmann B. Macromol. Chem. Phys. 2002, 203, 1936-1948.
[6] Ilavsk'y M.; Du–sek K. Polymer 1983, 24, 981-990.
[7] Pascault J.-P.; Sautereau H.; Verdu J.; Williams R.J.J. Thermosetting Polymers ; Marcel Dekker Inc.: New York, NY, 2002.
[8] Gordon M. Proc. Roy. Soc. London 1962, A268, 240-259.
[9] Harris T.E. The Theory of Branching Processes , Springer: Berlin, 1963.
[10] Flory P.J. J. Amer. Chem. Soc. 1941, 63, 3083-3090.
[11] Flory P.J. J. Amer. Chem. Soc. 1941, 63, 3091-3096.
[12] Dossin L.M.; Graessley W.W. Macromolecules 1979, 12, 123-130.
[13] Flory P.J. Proc. Roy. Soc. London 1976, A351, 351-380.

[14] Langley N.R. Macromolecules 1968, 1, 348-352.
[15] Fayolle B.; Gilormini P.; Diani J. Colloid Polym. Sci. 2010, 288, 97-103.
[16] Flory P.J. Principles of Polymer Chemistry , Cornell University Press: Ithaca, NY (1953).
[17] Ilavsk'y M.; – Somv'arsky J.; Bouchal K.; Du–sek K. Polym. Gels Networks 1993, 1, 159- 184.
[18] Cail J.L.; Stepto R.F.T. Polym. Bull. 2007, 58, 15-25.

INDEX

A

B

C

D

E

H

I

J

K

L

M

N

O

P

Q

R

S

T

Y

Z

oxide in Fe_2O_3/SiO_2 catalyst in a silica gel matrix: Optical and magneto- optical propertiestransfer science derived zinc oxide buffer layer for use in random laser media–gel single-layer SiO_2 acid and base thin films *in organic solvents: synthesis* coefficient of titanium ion in TiO_2---SiO_2 wet gel prepared from metal alkoxides during leaching